SPACE SHUTTLE

THE HISTORY OF THE
NATIONAL SPACE TRANSPORTATION SYSTEM

THE FIRST 100 MISSIONS

The Space Shuttle Orbiter Columbia on Pad 39A just prior to the first flight in April 1981. (NASA photo KSC-81PC-0136)

DENNIS R. JENKINS

"... in technical pursuits, the first idea put into practice can
never be perfect – and can never mean the end of the work."

— *Dr. Walter R. Dornberger, 1964*

" ... wingless would be only a stunt ..." (NASA photo KSC-81PC-0136)

© 1992, 1996, 2001 by Dennis R. Jenkins

Researched, written, and published in the
United States of America by
Dennis R. Jenkins
8700 Ridgewood Avenue, Unit B-406
Cape Canaveral, Florida 32920-2017 USA

Cover imageset by
Broadfield Imaging Corporation
824 Palmetto Avenue
Melbourne, Florida 32901 USA

US Book Trade Distribution by
Voyageur Press
123 N. Second Street
Stillwater, MN 55082 USA
Toll Free: 800-888-9653
books@voyageurpress.com

ANNIVERSARY
1981-2001

Printed through
World Print Ltd.
Kwong Sang Hong Centre
151-153 Hoi Bun Road
Kwun Tong, Kowloon
Hong Kong

Printed in China

European edition published by
Midland Publishing
4 Watling Drive
Hinckley LE10 3EY, England
Tel: 01455 233 747 Fax: 01455 233 737
E-mail: midlandbooks@compuserve.com
(ISBN 1 85780 116 4)

Midland Publishing is an imprint of
Ian Allan Publishing Ltd.

UK and European distribution by
Midland Counties Publications
4 Watling Drive
Hinckley LE10 3EY, England
Tel: 01455 233 747 Fax: 01455 233 737
E-mail: midlandbooks@compuserve.com

Library of Congress Catalog Card Number: 00-091732

Library of Congress Cataloging-in-Publication Data:
    Jenkins, Dennis R.
        Space Shuttle
        The History of the National Space Transportation System

    ISBN: 0-9633974-5-1   Hardcover, Third Edition
                           First Printing, April 2001
                           Second Printing, June 2001

Prepared using an Apple Power Macintosh G4™
    Dual 500MHz 7400 CPU upgrade courtesy of Newer Technology
    Pages composed in QuarkExpress® 4.11
    Artwork composed in Adobe Photoshop® 5.5
    Set in Linotype Centennial from Adobe
    Halftones and line art scanned using an Epson Expression 1600
    Transparencies scanned using a Nikon Super CoolScan LS-1000

TABLE OF CONTENTS

OV-102, Columbia, is backed out of the assembly facility in Palmdale for her overland move to Edwards AFB on 8 March 1979. (Boeing North American photo B02832-0006)

OV-099, Challenger, is rolled-out from the North American assembly facility at Plant 42 in Palmdale, California, on 30 June 1982. (Boeing North American photo B02841-0002)

*"It is better to attempt a gigantic endeavor but fall slightly short
than to attempt very little and be highly successful."*

PREFACE

INTRODUCTION TO THE THIRD EDITION

When I wrote the first mini-edition of what became this book back in 1988, the response was overwhelmingly positive, leading me to write the first of the hard-cover editions. But at the time, publishers said that there would never be a market for more than 500 books on Space Shuttle, and all refused to publish it. Believing that there was a story to be told, I published the book myself. By late 1999, two editions and more than 50,000 copies later, I was being asked when the next edition would come out. The answer is – now.

This is more than a minor update to the previous edition – it is nearly a complete rewrite. More data on early concepts has come my way, and I felt it needed to be included. I also decided to delve further into the politics and funding issues surrounding the decision to build Space Shuttle, so the 1970–71 period is covered in far greater detail this time around. The vehicle has undergone some significant modifications during the late 1990s, and I have tried to cover all of them in at least some detail. And of course I have added the next 25 missions, now up through the 100th flight.

As with any work that claims to be 'complete,' I am sure there are still many holes that I have missed – or in some cases, chosen not to cover. I have tried to answer much of the constructive criticism offered by various historians and in the sci.space newsgroups, but I have still elected to narrowly limit the scope of the book to the development of manned lifting-reentry vehicles, and have completely ignored the space capsules that proved so successful on Mercury, Gemini, and Apollo. Space limitations led to deleting my previous brief coverage of European and Soviet designs, although Mark Wade had graciously written an excellent section on Buran. I have also not covered designs subsequent to Space Shuttle unless they are somehow directly related – I will leave the current crop of hopefuls to the next author or historian. Still, I hope this work proves a useful history and reference to those interested in the subject.

PREFACE

This is a brief history of the technology that began the true 'space age.' That it does not dwell on the adventures and accomplishments of Mercury, Gemini, and Apollo, reflects the transitory nature of those projects. Although all were exciting, and contributed greatly to man's – and the United States' – self-esteem and technological base, they actually had little bearing on the long-term exploration of space. The Space Shuttle is giving man his first 'routine' access to space, in much the same way that the DC-3 gave the first routine access to air travel. But Shuttle is very much a compromise vehicle, the outcome of both technological and economic limitations, as well as the shortsightedness of men lacking a vision of the inevitable. It has cost well over $100,000 million – and the lives of seven astronauts – to achieve the first 100 flights as we strive for the future.

For the most part this publication intentionally ignores the immense work that went into the design, development, testing, and operation of the vast infrastructure required to support the Space Shuttle program. This is not meant to minimize the efforts of the thousands of individuals – including myself – who contributed to the ground support equipment, facilities, and simulators. It is simply that the intent of this publication is to document the development of the flight vehicle itself. The vehicle, in retrospect, may have been the simpler part of the effort.

Although an attempt is made to tell the complete story of the vehicle's development, several facts must be stated. First, this history spans a period of almost eighty years, and the development of the actual vehicle we call Space Shuttle started over three decades ago. Much of the early documentation has been destroyed as a part of the normal 'house-keeping' performed by the companies and agencies involved. An overwhelming percentage of the remaining documentation concerns the Rockwell studies and proposals since they were the eventual winner, although other companies (particularly Grumman) maintain exceptional history offices to assist researchers in their quests. Secondly, some of the more interesting research projects, particularly during the late-1950s and early-1960s, were classified by the Department of Defense, and the documentation was destroyed before it was ever made public. The companies and government agencies involved in the projects frequently no longer exist, making research impossible. Nevertheless, with a few exceptions, it is believed the story contained herein is mostly complete and accurate.

On 22 February 1990, Robert L. Crippen, then NASA Space Shuttle Director, issued a memo stating that due to the new 'mixed fleet' strategy of using expendable boosters to supplement the Shuttle, the nomenclature 'National Space Transportation System' would no longer be used, and the current nomenclature is simply 'Space Shuttle Program.'

The future beckons. The Dream is Alive ...

Dennis R. Jenkins
Cape Canaveral, Florida
December 2000

ACKNOWLEDGMENTS

A project like this could not possibly be accomplished without help from a great many people. As always, a few contributed more than their fair share, going well beyond what could reasonably be expected: Tony Landis, Grant Cates, Jorge Frank, Kim Keller, Julie Kramer, and Chris Hansen – amongst others – definitely fall into this category. The entire effort would not have been possible without the assistance of the KSC Technical Library Documents Section staff (Donna Atkins, William Cooper, Jane Page, and Dorothy Price), chasing through the stacks to find twenty year old proposals. The staffs at the various NASA History Offices provided a tremendous amount of guidance, counseling, encouragement, and valuable data. Special thanks to my mother, Mrs. Mary E. Jenkins, for putting up with me the entire time I was concentrating on this effort. I would also like to thank Jim Cumbre, Winston R. Davis, Cheryl Gumm, and a host of others, for finally settling once and for all if Mike Adams was awarded Astronaut Wings for X-15 flight #191 (he was, posthumously, on 15 November 1967). My sincere apologies to anybody I have forgotten to list here. My thanks to you all. After the overwhelming support I received, any mistakes that remain are completely my own.

James C. Adamson (USA), Donna Atkins (KSC), Cheryl Agin-Heathcock (DFRC), James F. Aldridge (ASC/HO), Louise Alstork (NASA/HO Editor), Nadine Andreassen (NASA/HO Program Assistant), Dave Arnold, Liem Bahneman , Kevin J. Barré (LMMSS/MAF), Jack Bassick (David Clark Company), Brian Bateman (KSC), Judith Bauer (Aerojet), John V. Becker, Bob Biggs (Rocketdyne), Larry Biscayart (DFRC), Thomas M. Boyle, Walter J. Boyne, Robert Bradley (SDAM), Roy Bridges (KSC/CD), Alan Brown (DFRC/PAO), Lee Browndorf (VLS), Colin Burgess, Daniel K. Carpenter (JSC/PA), Grant Cates (KSC), Mark Cleary (45SW/HO), Michelle Cokley, Sue Cometa (Rockwell International), William Cooper (KSC), Diana G. Cornelisse (ASC/HO), Keith L. Cowing, Amos Crisp (MSFC, for his efforts to save Pathfinder), Robert L. Crippen, Virginia Dawson, Dwayne A. Day, Melodie de Guibert (Thiokol Propulsion), Bill Dearing (KSC), Peter Deidle (DFRC), Hugo Delgado (KSC), Todd A. Downey (JSC), Alfred C. Draper III, Bernadette R. Draper, Bruce P. Dunn, Monica Dwyer-Abress, Jerry C. Elliott (JSC), Ralph Esposito (KSC), Michael Finneran (LaRC), Cynthia A. Fontenot (JSC), Jorge R. Frank (JSC), Colin Fries (NASA/HO Contract Archivist), Stephen J. Garber (NASA/HO), David A. Gerlach (JSC), Sandy Gettings, Patti Gibbons, Marta G. Giles (JSC), Richard David Glueck (my co-conspirator in setting up sci.space.history), Phil Green (Rockwell International), Kay Grinter (KSC), Roger Guillemette (Florida Today), Don Haley (DFRC), Richard P. Hallion (USAF/HO), Chris Hansen (JSC), Sharon Hansen (LMSSC/Michoud), Roselle Hanson (KSC), Cheryl Heathcock (DFRC), Wesley B. Henry (USAF Museum), Tom A. Heppenheimer, Bruce W. Hess (ASC/HO), Waymon Higgins (Surfside), Sara B. Hill (Boeing), Dill Hunley (DFRC/HO), Frank Izquierdo (KSC), Ken Jenks, Frederick A. Johnsen (AFFTC and then DFRC), Knut Jorgensen, Fred H. Jue (Rocketdyne), Mark Kahn (NASA/HO Contract Archivist), Kim Keller (KSC), Peter E. Kirkup (Grumman), William J. 'Pete' Knight, Trevor Kott (JSC), Julie Kramer (JSC), Marion A. Lanasa, Tony Landis (DFRC), Roger D. Launius (NASA Chief Historian), Neil Lewis, Elaine E. Liston (KSC/HO), Denny Lombard (Lockheed Martin), Mike Lombardi (Boeing Historian), Lois Lovisolo (Grumman), Scott Lowther, Stella Luna (JSC), June Malone (MSFC), T.K. Mattingly, Patricia M. McGinnis (Boeing), Charlien McGlothin (MSFC), Elric N. McHenry (JSC), Larry Mead (Grumman), Jay Miller, Jim Mistrot (JSC), Robert Mitchell (WSTF), Mike Moore (LMTAS), JoAnn Morgan (KSC), Claude S. Morse (AEDC), Thomas L. Moser, Robert D. Mulcahy (SMC/HO), Valerie Neal (NASM), Dennis Newkirk, Brian Nicklas (LaRC, courtesy of the Smithsonian Institution), Jane Odom (NASA/HO Archivist), Jane Page (KSC), Terry Panopalis, Margaret Persinger (KSC), Robert Pierce (LMTO), Dorothy Price (KSC), Rachelle Raphael, Frederick Raymes,

Anne B. Ribble (IBM), A.L. Salandra (Bell Aerospace Textron), Kristen A. Schario (AFRL/PROP), Larry E. Scheikart (University of Dayton), L. Pat Scott (LMTO), Mary Shafer (DFRC), Pat Sheeney (LTV), Erik Simonsen (Boeing), Jean Simpson (DoD/JSC), Henry Spencer, MSgt. Terry W. Steffey (Holloman AFB), Rick Sturdevant (AFSPC/HO), Glen E. Swanson (JSC/HO), Kevin C. Templin (JSC), Brian C. Thomson, Richard H. Truly, Emma M. Underwood (AEDC/ACS), Lisa Vasquez-Morrison (JSC), Mark Wade, Robert Walz (Boeing), Douglas K. Ward (JSC), Cathy Watson (LaRC), James Wentworth, Mary N. Wilkerson (JSC), Rodney Wilks (Thiokol), Leslie Williams (DFRC), Rodger Williams (AIAA), and all of the participants in the sci.space.* newsgroups. A special note of thanks to Jack Putnam and Roger Kasten, Jr. at Newer Technology for providing the fast processor upgrades for my Macintosh G4 that powered me through putting it all together.

Discovery (OV-103) and the first Shuttle Carrier Aircraft (N905NA) depart Edwards AFB on the way to the Kennedy Space Center on 2 November 2000 after the 100th Space Shuttle mission (STS-92). It had been an interesting 30 years. (NASA photo EC00-0311-13 by Tony Landis)

This book is dedicated to the memories of the eleven
United States Astronauts who have perished in space flight,
and flight-preparation accidents.

Godspeed

Roger B. Chaffee

Virgil I. Grissom

Edward H. White II

Apollo 1 (AS-204), 27 January 1967

Major Michael J. Adams

X-15 Flight 3-65-97, 15 November 1967

Gregory Bruce Jarvis

Sharon Christa McAuliffe

Ronald Erwin McNair

Ellison S. Onizuka

Judith Arlene Resnik

Francis R. 'Dick' Scobee

Michael John Smith

STS-33/51-L (*Challenger*), 28 January 1986

Dr. Eugen Sänger. (Messerschmitt-Bölkow-Blohm, Gmbh)

BACKGROUND

Reusable space vehicles have been discussed for almost eighty years; Robert Goddard, Konstantin Eduardovich Tsiolkovskiy, and Hermann Oberth[*] wrote of them during the 1920s. In Germany, others put form as well as thought to the theoretical concept of winged rockets. A Viennese scientist, Max Valier,[†] believed that rocket engines would eventually replace the airplane's internal combustion engine and lead by natural evolution to winged spacecraft that would travel back and forth to Earth orbit and the planets.[1]

Working in the mountains of western Germany, Valier produced several manned rocket-powered gliders eventually working up to a 100-pound, seven-foot long vehicle he called *Stork*. This glider used a 165-lbf engine to get airborne, but flight testing was short and inconclusive. So, during 1929, Fritz von Opel, who had funded many of Valier's experiments, produced a much larger glide-rocket that flew for about 10 minutes at 100 mph. The 600 pound vehicle crashed while attempting to land, and the ensuing fire almost cost von Opel his life in what was most likely the first manned rocket flight.[2]

During 1925, Walter Hohmann, a German civil engineer, published a book entitled *The Attainability of Celestial Bodies*.[‡] This work examined one aspect of space flight in particular – the derivation of optimum transfer trajectories for flights from Earth to other planets. However, Hohmann also examined problems involved in returning to Earth, and was especially concerned about the effects of atmospheric heating during reentry. Although he failed to come up with any specific solutions to the problem, Hohmann theorized several techniques that returning spacecraft might use, including variable-geometry wings and external insulation. Unfortunately, none of the technologies available at the time would have permitted the building of a reentry vehicle, even if a means of launching it had existed.[3]

SÄNGER AND BREDT

The greatest contributions to these early studies were undoubtedly made by one man – Eugen Sänger – who was inspired by the work of Oberth, Valier, and Hohmann. While Sänger was a doctoral candidate at the Viennese Polytechnic Institute during 1929, he conceptualized the development of a spacecraft capable of flying into low-earth[§] orbit and returning to normal aircraft-type landings. From these studies emerged a 1931 concept dubbed Silverbird, a winged vehicle propelled by a rocket engine burning liquid oxygen and kerosene, capable of reaching Mach 10 at altitudes in excess of 100 miles. This first Silverbird had a fairly simple 'spindle-shaped' (Sänger's description) fuselage and straight-wings of low-aspect ratio with sharp leading edges and wedge airfoil sections.[4]

A detailed explanation of this concept was contained in the *Techniques of Rocket Flight*, published[#] privately by Sänger in late-1933. Sänger further elaborated in the following year that a modified version of the vehicle with a hypersonic lift-to-drag (L/D) ratio of 5.0:1 could obtain Mach 13 velocities at the moment of fuel exhaustion followed by a deceleration to a prolonged Mach 3.3 cruise. An operating altitude of 160,000 feet and a range of 3,100 miles were projected. In collaboration with mathematician Irene Bredt, whom he later married, Sänger continued to refine variations of this design for the next 30 years.[5]

---

[*] Oberth was one of the founding members of the German Society for Space Travel (*VfR – Verein für Raumschifart*), more often referred to as the German Rocket Society.

[†] During 1924 Valier privately published a book entitled *The Advance into Space* where he detailed some space flight concepts. Tragically, Valier died in a laboratory accident in 1930 when an experimental rocket engine exploded on its test stand.

[‡] For simplicity's sake, the accepted English translations of all titles will be used.

[§] Low-earth (known at the time as near-earth) orbits (LEO) are generally considered to be at altitudes between 100 and 400 miles, while mid-earth (MEO) orbits reach out to 1,000 miles or higher. Geosynchronous Earth orbits (GEO) are at 22,300 miles.

[#] Interestingly, Sänger used the same publisher as Valier – it took four years to pay off the resulting printing charges.

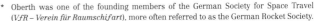

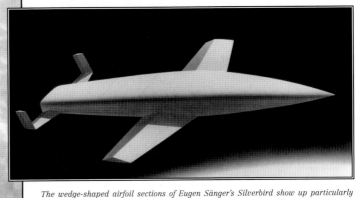

The wedge-shaped airfoil sections of Eugen Sänger's Silverbird show up particularly well in the illustration above. The artist's concept at left depicts the Amerikä Bomber proposed for use by the Luftwaffe to bomb New York City during World War II. Although the state-of-the-art did not allow Sänger to actually build his creation, the general concept would shape many of the early reusable lifting-reentry spacecraft concepts developed in America, Europe, and the Soviet Union. (Messerschmitt-Bölkow-Blohm, Gmbh)

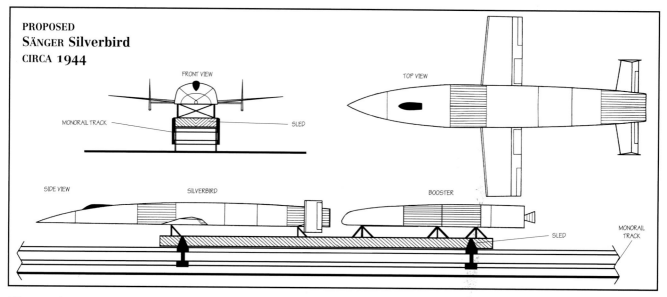

FRONT VIEW

TOP VIEW

MONORAIL TRACK   SLED

SIDE VIEW   SILVERBIRD   BOOSTER

SLED

MONORAIL TRACK

This version of Eugen Sänger's Silverbird was dubbed the Amerikä Bomber and was intended to bomb New York City during World War II. Fortunately for the United States, the Luftwaffe expressed no serious interest in the concept, although it is unlikely the 660-pound payload could have caused any serious damage except, perhaps, to morale. (Dennis R. Jenkins)

Towards the end of the 1930s, the pair had developed the Silverbird concept into a flat-bottom half-ogive fuselage using low-aspect ratio wings with wedge airfoils, and a similar horizontal stabilizer with twin endplate vertical stabilizers. Bredt estimated that this vehicle would have a supersonic L/D of 6.4:1, and later testing revealed surprisingly good low-speed stability and decent landing characteristics, with an actual subsonic L/D of 7.5:1.

This Sänger-Bredt Silverbird was theoretically capable of carrying an 8,300-pound payload into low-earth orbit using a 200,000 lbf liquid-fueled rocket engine which was ignited in flight after launch by a Mach 1.5 rocket sled. Following deployment of the payload, the Silverbird would return to Earth in a long series of semi-ballistic skips, each skip being shallower and shorter than the previous, until the vehicle glided to a conventional landing. Alternately, the Silverbird could follow a sub-orbital trajectory and deliver a 16,000-pound payload to a point halfway* around the world from its launch site using the skip-glide (dynamic soaring) technique. Skip-gliding would lose favor during the 1950s when studies by NACA Ames showed the technique would expose the airframe to prolonged and extreme reentry heating, a point apparently overlooked (or severely underestimated) by Sänger and Bredt.

During World War II, Sänger and Bredt concentrated the majority of their work on the design and development of more conventional high-speed aircraft and propulsion systems. Nevertheless, they found time to continue study of the Silverbird by advocating it as an *Amerikä Bomber* to the German Luftwaffe. The vehicle they proposed was 91.8 feet long with a wingspan of 49.2 feet and a launch weight of 200,000 pounds. A maximum speed of 13,500 mph and a range of 14,500 miles were predicted. The vehicle was to be boosted in a westerly direction to Mach 1.5 by a rocket sled riding a monorail track approximately 1.8 miles in length. Following lift-off, the aircraft would be powered by an internal liquid-fuel rocket engine at a 30-degree angle until fuel exhaustion occurred at an altitude of about 100 miles. Then the vehicle would commence a series of aerodynamic skip-glides across the upper atmosphere, delivering its 660 pounds of bombs on New York City, finally slow-

ly decelerating to settle into the lower atmosphere. At the end of its flight, the vehicle would enter a continuously descending glide, and eventually land at a prepared base 12,000 miles from its launch site. There it would be launched in the opposite direction for its return mission, ending up at its original launch site.[6]

While little official support was forthcoming, the pair published a report in 1944 under the title *Concerning Rocket Propulsion for Long-Range Bombers*, and several copies of this report were captured by American, British, and Soviet technical intelligence teams. Reportedly, Soviet Premier Josef Stalin was so impressed by the report that he sent a team into western Europe to locate and kidnap Sänger and Bredt. The plan was foiled by the French intelligence service, and the Sängers continued their work, first in France and subsequently in West Germany.[7]

In 1964, shortly before his death, Sänger was busily at work as a consultant to MBB-Jünkers, developing a delta-wing vehicle (the RT-8-01) that bore a resemblance to his

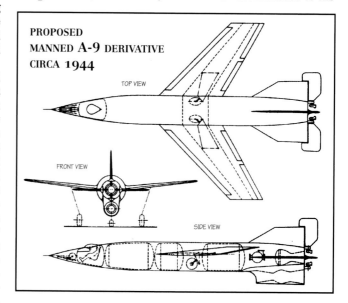

PROPOSED
MANNED A-9 DERIVATIVE
CIRCA **1944**

TOP VIEW

FRONT VIEW

SIDE VIEW

The manned A-9 appeared to be an early attempt to build a hypersonic research vehicle. Using its own rocket motor, the A-9 would launch vertically, then transition to horizontal flight using a ramjet located in the ventral stabilizer. The vehicle never got past the concept stage during World War II. (U.S. Army)

---

* Hence the term 'antipodal glider' which is frequently applied to the Sänger designs. Webster's defines antipodal as "…a point on the opposite side of the Earth or moon…"

work of the previous three decades. This design exercise continued for several years as the Sänger-I before being dropped due to a lack of funding, although an advanced outgrowth known as Sänger-II would briefly appear on the scene during the 1980s.

EARLY EXPERIMENTS

Although much of the early work on reusable spacecraft followed the inspiration of the Sänger-Bredt Silverbird, several German rocket specialists from Pëenemunde brought to the United States under the auspices of Project Paperclip had already participated in attempts to develop and fly lifting-reentry vehicles.

Reportedly, one such effort was the A-9/A-10 combination of 1944, which was envisioned by its developers as a large booster (the A-10) topped by a winged second stage (A-9) capable of delivering a one ton warhead to a target 3,000 miles away. The A-10 would boost the A-9 into the upper atmosphere, where the A-9 would fire its engine, continue downrange in a ballistic arc, then transition to a Mach 3.5 glide towards its target.

Major General Walter R. Dornberger, the commander of Pëenemunde, and Dr. Krafft Ehricke, both of whom who will play other roles in this history after the war as employees of Bell Aircraft, contributed to the design of a manned version of the winged A-9. The aircraft would be equipped with a retractable tricycle landing gear, larger swept wings, pressurized cockpit, longer body (fuselage) to hold a tank of liquid acetylene, and a ramjet integrated into the ventral stabilizer.

The manned A-9 vehicle would be launched vertically like the A-4 using the main rocket motor. At an altitude of 65,000 feet and a velocity of 2,250 mph the vehicle would transition to horizontal flight using the ramjet. The 3,500 pounds of acetylene would allow the aircraft to sustain its Mach 3.4 cruising speed for almost 30 minutes, allowing a range of over 1,100 miles. The estimated landing speed was a very conservative 100 mph. Studies were also undertaken on substituting tetranitromethane and visol propellants for the normal liquid oxygen and alcohol used in the main rocket engine, boosting the range of the vehicle to almost 2,000 miles.

The exact purpose of the concept was not recorded in the U.S. Army reports generated after the war that gave the description of the aircraft along with a three-view drawing, but the vehicle had no obvious payload leading to speculation that it was strictly a research aircraft. There are various references to mating the manned A-9 with the proposed A-10 booster stage, but there is no concrete evidence that this was ever seriously considered. An orbital version of the manned A-9, launched by a large two-stage A-11/A-12 booster capable of lifting 60,000 pounds to low-Earth orbit, was also reportedly under study at the war's end, although neither project progressed pass the casual study stage.[8]

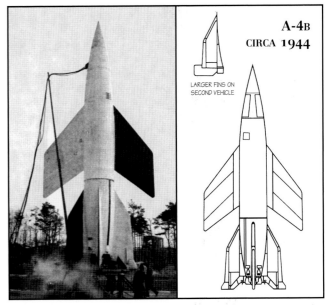

Ludwig Roth's A-4b swept-wing V2 – two were launched in 1945. (U.S. Air Force)

In a related, but completely separate effort, Ludwig Roth supervised the design and manufacture of two swept-wing derivatives of the A-4 (V2). The first of these winged A-4b ('bastard') vehicles exploded shortly after launch on 8 January 1945. The second vehicle, launched 24 January, successfully began a Mach 4 glide before one of the wings separated from the vehicle, probably from unexpectedly high flight loads, and the A-4b broke up.[9]

The German winged reentry research efforts were terminated as the war in Europe came to an end in 1945, and neither the United States, nor the Soviet Union, seemed particularly interested in continuing it.

During late-1949, Hsue Shen Tsein, a professor at the California Institute of Technology (CalTech),[*] concluded that a sufficient technological base existed to develop a Mach 12 transcontinental passenger aircraft that used rocket engines burning liquid fluorine and liquid hydrogen. The vehicle would have had a range of approximately 3,000 miles, over 1,800 of which were during a glide from its maximum altitude of 140,000 feet. A maximum velocity of 9,000 mph would be achieved, but a landing speed of only 150 mph was predicted, along with surprisingly good low-speed stability characteristics.[10]

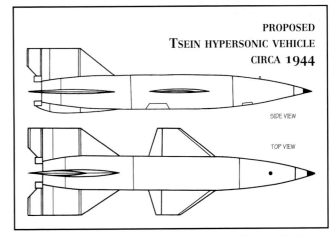

The hypersonic research vehicle proposed by Tsein would have been 78.9 feet long and 16.5 feet high with small trapezoid-shaped wings spanning 18.9 feet. The vehicle would weigh 96,500 pounds, including 72,400 pounds of propellants. A single pilot would have been accommodated in a cramped pressurized cockpit and would have used a periscope-like device for visibility. (U.S. Air Force)

---

* Tsein had been a student of Dr. Theodore von Kàrmàn during the 1930s at the Guggenheim Aeronautical Laboratory, which was part of CalTech. Kàrmàn and his students formed GALCIT, which received contracts from the Army Air Corps to design and construct liquid- and solid-fuel rocket engines for possible military applications. Tsein would later return to the People's Republic of China and lead the PRC's effort to develop a series of ballistic missiles and space launch systems. Kàrmàn and another student, Frank J. Malina, went on to form Aerojet Engineering Corporation (later Aerojet General), which specialized in rocket engine development and manufacture. GALCIT, which was an acronym formed from Guggenheim Aeronautical Laboratory (GAL) and the California Institute of Technology (CIT), was reorganized as the Jet Propulsion Laboratory (JPL) in 1944, and is currently operated by CalTech under contract to NASA.

VON BRAUN'S DREAM

In the United States during the years immediately following World War II, both scientists and military leaders recognized that the ability to launch payloads into orbit would have important implications. Several lines of thinking emerged. First, the development of relatively small satellites to perform practical, Earth-oriented tasks in orbit such as communication, scientific research, and reconnaissance was foreseen by such visionaries as Arthur C. Clarke. Second, further consideration led some to conclude that permanent manned bases in orbit were "... not necessary for most activities envisioned there: rendezvous of the rockets and satellites themselves is sufficient for most purposes." But this perspective and its appearance in literature did nothing to deter yet a third line of development – the need for a space station, and the 'shuttle' vehicles necessary to build and maintain it. Among the supporters of this concept were Dr. Wernher von Braun and the Germans from Pëenemunde, anxious to continue the work they had begun before World War II.

The First Symposium on Space Flight was held on 12 October 1951 at the Hayden Planetarium in New York City. Several papers from the symposium were subsequently published by Collier's magazine under the title *Man Will Conquer Space Soon* – contributors included von Braun, Joseph Kaplan, Heinz Haber, Willy Ley, Oscar Schachter, and Fred Whipple. Topics ranged from manned orbital space stations and orbiting astronomical observatories, to problems of human survival in space, lunar ventures, and questions regarding international law and sovereignty in space.[11]

The symposium concluded that the United States could launch an artificial Earth-orbiting satellite by 1963, and mount a 50-man expedition to the Moon in late 1964, all at a cost of approximately $4,000 million. A large space station would be needed to support the lunar mission, and this aspect of the plan would be the subject of much research and debate over the next ten years.

Dr. von Braun was completely serious. In retrospect many of the designs and concepts presented at the symposium were 'brute-force' approaches to space travel, but the point of the symposium was to show that the technology existed to attempt it. And the authors, all of whom had impeccable credentials, believed it did – although all admitted it was in its infancy. Their goal was to exploit the available technology, then continue to refine it.

The first step proposed by von Braun and Ley was the launching of a 10-foot cone-shaped vehicle carrying three rhesus monkeys. It was to orbit for 60 days at an altitude of 200 miles – unfortunately for the monkeys, no recovery capability was included. Following a successful demonstration of the monkey's ability to live in space for a prolonged period, approval would be given for a full-scale manned space program. It would take ten years, and culminate with a manned lunar base.

The reusable 'ferry rocket' proposed by von Braun was 265 feet tall and 65 feet in diameter at its base. The primary structure was steel alloy, and its gross lift-off weight (GLOW) was over 14,000,000 pounds. This compared to the later Saturn V, also largely a von Braun design, that stood 363 feet high and had a GLOW of 6,423,000 pounds, benefiting from ten years of technological advances.

The first stage of the ferry rocket was powered by 51 rocket engines, each developing 550,000 lbf. They were to burn 10,500,000 pounds of nitric acid and hydrazine in a 50:50 mixture. Although equipped with wings, the first stage was expendable, at least during the early portions of the program. The second stage, also expendable, had 34 engines producing 100,000 lbf each. The winged third stage was just over 77 feet long with swept wings that spanned 156 feet. It used five engines with a total of 440,000 lbf. A crew of ten and 72,000 pounds of payload could be carried in this stage, which was designed to return to a conventional horizontal landing.

This vehicle was a refinement of the rocket originally proposed by von Braun in his book *The Mars Project*, and was designed expressly to ferry the materials and personnel necessary to construct the 250-foot diameter space station also proposed at the symposium. This space station popularized the 'wheel' shape so dramatically portrayed in the Arthur C. Clarke and Stanley Kubrick movie *2001: A Space Odyssey*.

Because of its size, a large, isolated, site was deemed necessary to prepare and launch the ferry rocket. For safety considerations, only sites where the entire trajectory were over open ocean were considered by von Braun. Two locations were investigated – Johnston Island in the Pacific Ocean, and the Air Force Proving Grounds at Cape Canaveral, Florida. The latter was recommended based largely on its proximity to the Redstone Arsenal where von Braun was employed and most rocket research was undertaken during the 1950s.[12]

While none of these ambitious concepts resulted in specific development programs, they did offer an opportunity for engineers to evaluate the current state-of-the-art in reentry and propulsion technologies, as well as fueling the imagination of the American public and a generation of science-fiction writers.

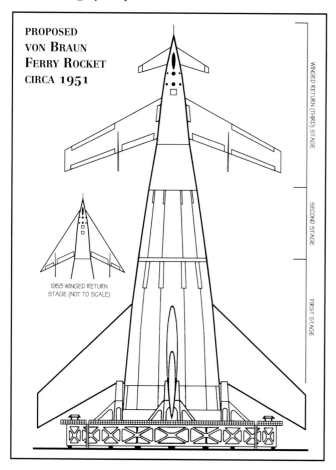

PROPOSED
VON BRAUN
FERRY ROCKET
CIRCA 1951

WINGED RETURN (THIRD) STAGE

SECOND STAGE

FIRST STAGE

1953 WINGED RETURN STAGE (NOT TO SCALE)

Wernher von Braun's 'ferry rocket' was the size of a Saturn V. Its approach to construction was crude, but von Braun and others believed it would have worked. (NASA)

THE FIRST X-PLANES

For the thirty years beginning in the mid-1940s, the United States sponsored a series of research aircraft commonly known as the 'X-planes.' The early X-planes, such as the Bell X-1 and Douglas D-558-I, gave aviation its first experience with controlled supersonic flight. Air Force Captain Charles E. Yeager became the first human to purposely break the sound barrier on 14 October 1947 when the Bell Aircraft Company XS-1 achieved Mach 1.06 at 43,000 feet. It took six additional years before NACA' research pilot A. Scott Crossfield exceeded Mach 2 in the U.S. Navy-sponsored Douglas D-558-II Skyrocket.[13]

By 1956, similar research aircraft had achieved velocities in excess of Mach 3, and had flown to altitudes over 126,000 feet.[†] The X-planes were among the first vehicles to encounter the kinds of control difficulties that would demand the development of thruster reaction controls and require complex alloy structures to withstand the temperatures of high-speed flight. Other benefits from the early X-plane projects included insight into the problems of inertial coupling, exhausts impinging upon control surfaces, and a better appreciation for the complex physiological protection necessary for crewmen.

But none of these early X-planes, later called 'Round One,' were able to significantly exceed Mach 3, somewhat limiting their continued usefulness. On 4 October 1951 Robert J. Woods, one of the founders of Bell Aircraft and co-designer of the X-1, presented a paper to the NACA Committee on Aerodynamics that highlighted the need for a hypersonic research aircraft. Woods also proposed establishing a study group to evaluate the problems expected to be encountered during prolonged hypersonic flight.[14]

Ten days later, Woods presented a concept developed by Dornberger, now working for Bell, for a rocket-powered hypersonic research vehicle with novel variable-geometry wings capable of velocities in excess of 4,000 mph at altitudes of between 50 and 75 miles. Actually, this was not the first such proposal; during an informal meeting of the NACA Subcommittee on Stability and Control in June 1951, Maxwell W. Hunter II, an engineer from Douglas Aircraft Company, had suggested that hypersonic research be initiated to aid the designers of the ICBMs.[15]

Hypersonic research was indeed initiated during the early-1950s. For the most part this research did not directly influence the development of lifting-reentry vehicles, but some of the data nevertheless proved useful to the cause. Primarily this involved information on reentry heating accumulated during flights of highly instrumented missiles (such as the Nike-Nike and the Lockheed X-17) that achieved speeds in excess of Mach 14 and generated thermal data that engineers subsequently applied to various lifting-reentry shape designs. The ICBM programs also yielded valuable data on large rocket engine design, which would greatly contribute to Space Shuttle.

Independently, two engineers at the High-Speed Flight Research Station[‡] (HSFRS), Hubert M. 'Jake' Drake and L. Robert Carman, released an ambitious proposal for a small hypersonic research vehicle on 21 May 1952. Entitled *A Suggestion of Means for Flight Research at Hypersonic Velocities and High Altitudes*, they predicted that a vehicle with a 100,000-pound gross weight could obtain Mach 6.4 at 660,000 feet and sustain Mach 5.3 for over one minute. The vehicle would have been powered by a rocket engine burning liquid oxygen and water-alcohol propellants. Using this as a carrier aircraft, a vehicle the same general size and weight as the Bell X-2 could be launched at Mach 3 and 50,000 feet, attaining velocities of Mach 10 at altitudes approaching 1,000,000 feet. It was felt that the second stage could maintain Mach 8 for over one minute.[16]

Although not known at the time, this general concept of a two-stage vehicle, with both the orbiter and booster being reusable, would shape most early Space Shuttle studies.

On 20 May 1952, David Stone, an engineer with the Piloted Aircraft Research Division of NACA Langley, released a somewhat more conservative proposal. Stone felt that the Bell X-2 itself could be utilized for research flights approaching Mach 4.5 at altitudes of 300,000 feet. In order to achieve this significant increase in performance[§] Stone proposed fitting two JPL-4 Sergeant solid fuel rockets under the vehicle's fuselage, and also installing a reaction control system in the nose, tail, and wingtips.[17]

---

* The National Advisory Committee for Aeronautics (NACA) became the National Aeronautics and Space Administration (NASA) on 1 October 1958.

† Although many X-planes explored high-speed/altitude flight environments, others were concerned with far less glamorous areas of aerodynamics and have received substantially less publicity, although their contributions were in several cases greater.

‡ The High-Speed Flight Research Station (HSFRS) became the High-Speed Flight Station (HSFS) in 1954, the Flight Research Center (FRC) in 1959, and the Hugh L. Dryden Flight Research Center (DFRC) in 1976. In 1981 it was administratively absorbed into the Ames Research Center and changed its name to the Ames-Dryden Flight Research Facility (usually DFRF). In 1994, it again became simply DFRC.

§ The unmodified X-2 eventually recorded an unofficial altitude record of 125,907 feet on 7 September 1956 with Captain Iven Kincheloe at the controls. On 27 September 1956 it achieved an unofficial speed record of Mach 3.196 before pilot Captain Milburn G. Apt lost control of the vehicle and was killed, effectively ending the X-2 project.

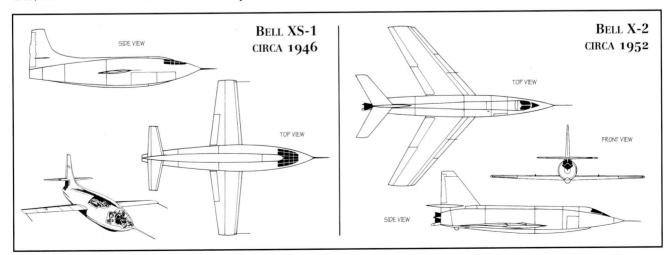

BELL XS-1
CIRCA 1946

BELL X-2
CIRCA 1952

The Bell XS-1 (later redesignated X-1) was the first aircraft to break the 'sound barrier' in level flight on 14 October 1947. Nine years later, Captain Milburn G. Apt became the first person to break Mach 3 – unfortunately Apt was killed when he lost control of the X-2 during the record-breaking flight. (Jay Miller Collection)

On 8 September 1952, NACA formed the hypersonic research study group suggested by Woods a year earlier. Chaired by Clinton E. Brown, this group made a number of profound recommendations for future NACA high-speed research projects. Although they realized that significant problems remained to be solved, the group was generally optimistic that hypersonic research vehicles could be developed in the near future. On 23 June 1953, after reviewing both the Drake-Carman and Dornberger concepts, the group recommended adopting David Stone's modified X-2 proposal, although this decision would be overcome by events before any hardware was completed.

Meanwhile, on 24 June 1953, the NACA Committee on Aerodynamics endorsed Woods' earlier proposal for the development of a hypersonic research aircraft that could study the problems of flight at high-altitude (12–50 miles) and very high-speed (Mach 4–10). After the resolution was ratified by the NACA Executive Committee during July, concepts were solicited from the various NACA laboratories' and contractors. In October 1953, the Air Force Scientific Advisory Board Aircraft Panel, chaired by Dr. Clark B. Millikan, also recommended the development of a hypersonic research vehicle capable of at least Mach 7 performance. The proposal was discussed during a routine meeting of the NACA Inter-Laboratory Research Airplane Panel held in Washington D.C. on 4 February 1954. Hartley A. Soulé, who had directed the cooperative USAF/NACA research aircraft program since 1946, indicated the panel found that a completely new manned research aircraft was needed, as opposed to the modified X-2 endorsed by the Brown group in 1952. The NACA Headquarters referred the entire matter to the four research laboratories (Langley, Ames, Lewis, and JPL) for review and comment.[18]

Researchers, especially at NACA Langley, became concerned about how to stabilize a vehicle at hypersonic speeds. Stability difficulties had already been encountered with the X-1 and X-2 at Mach numbers substantially lower than those expected of the new aircraft. The problem appeared to be solved by Langley researcher Charles H. McLellan with a scheme to replace the normally thin

supersonic-airfoil section of the tail surfaces with a 10-degree wedge shape, similar to that proposed by Eugen Sänger two decades earlier.[†] McLellan's calculations indicated that, in addition to allowing considerably smaller control surfaces, the basic shape should prove substantially more effective in eliminating the disastrous directional stability decay encountered by the X-1 and X-2. Besides the wedge airfoil shape, it was also discovered that the ability to vary the wedge's angle-of-attack was highly desirable to permit a variation of the stability derivatives, providing greatly increased research flexibility. Perhaps even more importantly, it would also constitute a means of quick recovery from a divergent maneuver, and potentially permit a wider envelope of reentry profiles. It was also noted that since aerodynamic control disappeared above 200,000 feet, some form of reaction control thrusters would need to be provided, in addition to the normal aerodynamic control surfaces.[19]

Researchers at Langley proposed that the new vehicle should have a secondary mission to explore space-related technologies, particularly concerning reentry techniques. Interestingly, many other researchers dismissed this role since they believed manned space flight was many decades in the future. Nevertheless, Langley spent considerable time during 1954 on studies into reentry stability and thermodynamics.[20]

Two structural design approaches to overcome reentry heating concerns were debated from the beginning of the studies. One was a largely conventional low-temperature design of aluminum alloy or stainless steel protected from the high-temperature environment by a layer of external insulation. The other approach used an exposed 'hot-structure' in which no attempt was made to provide thermal protection, but which used materials and a design approach that permitted high structural temperatures to be tolerated. This latter concept had originally been conceptualized by a group of Bell engineers led by Wilfred Dukes during the early 1950s.[‡]

Other concepts investigated included the use of wing structures protected by semi-flexible non-load-bearing metallic alloy heat shields known as 'shingles,' and a 'water-wall' technique, also developed at Bell, that used a liquid flowing between two superalloy skins as the basic insulator and was in an early state of development. It was anticipated that this technique could be used in areas around the cockpit, and that a variation could also be used in the wing leading edges. The water-wall technique would later find an application during studies for the still-born X-20 Dyna-Soar, while the exotic superalloy shingles would evolve indirectly into the thermal protection system tiles used on the Space Shuttle.

The peak-heating environment for Mach 7 research flights involved temperatures in excess of 3,500 degF. With appropriate allowances for heat radiation, this would result in skin temperatures as high as 2,000 degF in thin metal wings, far in excess of the 1,200 degF limit of the best available superalloy – Inconel-X. Researchers at NACA Langley discovered, however, that the use of

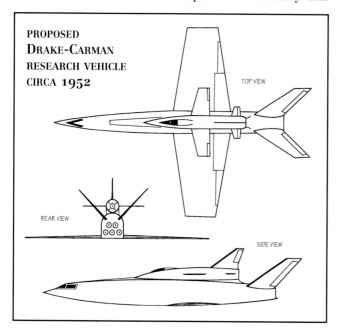

PROPOSED
Drake-Carman
Research Vehicle
Circa 1952

TOP VIEW

REAR VIEW

SIDE VIEW

This vehicle's booster was approximately 100 feet long with a 66-foot wing span and used five large liquid-fuel rocket engines to provide thrust. The single-engine second-stage high-Mach research vehicle was approximately 46 feet long with a 20-foot wing span. (U.S. Air Force)

---

* NACA/NASA installations are known by a distinct set of titles. 'Laboratories' used to be the largest permanent installations, and are now known as 'Centers,' with the exception of the Jet Propulsion Laboratory (JPL). Smaller installations were initially known as 'Stations,' but are now referred to as 'Facilities.'

† Sänger probably had chosen the wedge sections in order to simplify calculations, not necessarily because he had understood the associated aerodynamic benefits.

‡ In addition to the technical challenges, a bureaucratic challenge arose when the Air Force decided to classify all hot-structure research as 'secret,' severely limiting its availability to NACA and university researchers.

thick skins (≈0.10-inch) would provide a sufficient heat-absorption capacity to keep the skin temperatures within acceptable limits. This convenient solution was possible only if the period of peak heating was kept short – either by reentering at a very high angle-of-attack, or by the use of large speed brakes. Otherwise the reentry heat loads would exceed the hot-structure's allowable limits.[21]

During subsequent studies, it was discovered that hot-structure designs presented their own problems. The unequal heating of the upper and lower surfaces of the wings during some flight profiles resulted in intolerable thermal stresses in most designs. To solve this, designers devised new wing shear members which permitted the entire wing to deform both spanwise and chordwise with asymmetrical heating. Although this technique eliminated most gross thermal stresses, local thermal stress problems still existed in the vicinity of the stringer attachments. Further study indicated that the proper selection of stringer proportions and spacing would produce a design free from thermal buckling.[22]

Similar difficulties were also encountered during the design of the wing's corrugated-web beams and leading edge. Differential heating of the latter was discovered to produce changes in the natural torsional frequency of the wing unless some form of flexible expansion joint was incorporated into its design. The hot leading edge was discovered to expand faster than the remaining structure so that compression was introduced, which destabilized the section as a whole and reduced its torsional stiffness. To negate this phenomenon, the entire leading edge of the wing was segmented and attached to flexible mounts. The Langley Research Airplane Study Group, led by John V. Becker, incorporated all of these findings into a preliminary configuration which had a pronounced resemblance to the ultimate X-15 – an Inconel-X superalloy aircraft having a cruciform tail with a wedge profile and a short-span mid-mounted trapezoid-shaped wing.

A meeting sponsored by NACA on 9 July 1954 was attended by representatives from the Navy and various Air Force organizations, including the Wright Air Development Center (WADC), Air Research and Development Command (ARDC), and the Air Force Scientific Advisory Board (SAB). The Director of the NACA, Hugh L. Dryden, confirmed that studies, although incomplete, indicated that the desired research vehicle was technically feasible. At this meeting, the Office of Naval Research (ONR) announced that it had already contracted to the Douglas Aircraft Company for conceptual studies of a Mach 7+ research vehicle known as the D-558-III. Douglas had reported that the desired speeds, along with altitudes of 700,000 feet, were possible, but that it had not completely analyzed the thermo-structural problems associated with atmospheric reentry.[23]

Round Two

The meeting concluded with an agreement that NACA should release the results of the various studies to the aerospace industry. On 5 October 1954, the Committee on Aerodynamics met to further consider the question of a hypersonic research aircraft, and during this meeting various historical and technical reports were evaluated by Committee members including Walter C. Williams, and NACA research pilot Scott Crossfield. Although there was some initial opposition to the idea, the Committee finally endorsed the need for the project. Accordingly, Hugh Dryden initiated a joint effort with the Air Force and Navy to produce a suitable technical specification upon which

The experimental aircraft fleet at the High-Speed Flight Station circa 1957. Clockwise from left front: X-1A, D-558-I, XF-92A, X-5, D-558-II, X-4, and the X-3 in the center. These aircraft explored a variety of aeronautical technologies including variable-geometry wings (X-5) and the delta-wing planform (XF-92A). The D558-II was the first aircraft to exceed Mach 2 in level flight; despite its rakish appearance, the X-3 was a disappointing performer The XF-92 was not a true X-plane since it was nominally a prototype fighter that eventually evolved into the F-102 and F-106. (NASA photo E-2889)

manufacturers could base proposals. As finally adopted, the research specification called for a design speed of 4,500 mph and an altitude capability in excess of 250,000 feet.[24]

A formal presentation was given by the NACA to the Department of Defense (DoD) Air Technical Advisory Panel on 14 December 1954. The specification was approved with the stipulation that NACA should have technical control of the project, but that DoD would have all administrative and contractual responsibility. The Air Force's Air Materiel Command (AMC) issued formal 'invitations to bid' to a dozen prospective airframe contractors on 30 December 1954. Seventeen days into the new year, NACA was informed that the research aircraft would be identified as Air Force Project 1226, and would carry the official designation of X-15. The X-15 would be known as 'Round Two' in the X-plane evolution.

A top-level summary of the various NACA studies was presented at a bidder's conference that was held at Wright-Patterson AFB on 18 January 1955. Although many companies had initially expressed an interest, only Bell, Douglas, North American, and Republic actually submitted proposals. Of these, North American's design appeared most likely to succeed, largely because it used the most conservative design approach. Towards the end of September 1955, North American was notified that it had won the X-15 contract.

The North American design team, led by Harrison 'Stormy' Storms, Jr. and Charles Feltz, had accepted an extraordinarily difficult task when they agreed to design, build, and flight test the new hypersonic research aircraft. Although giving the initial appearance of a rather simple configuration, the X-15 was actually the most complex single-seat aircraft of its time. Much of the basic research into structures and materials had been accomplished during the NACA studies that led up to the X-15 contract, but a great deal of applied engineering remained to be done. Additionally, significant effort was required in the area of human factors, as well as into fabrication and assembly techniques. Eventually, some 2,000,000 engineering man-hours and over 4,000 wind-tunnel hours were devoted to finalizing the X-15 vehicle configuration. On 11 June 1956,

Like many of the early X-planes, the X-15 was air-launched – in this case by one of two early Boeing NB-52s operated by NASA. (SDAM Collection)

The X-15 used a novel landing gear that included skids in the back and normal tires at the nose – limiting the aircraft to landing on dry lakebeds. (NASA photo E-7411)

North American received approval to start construction on the first of three X-15s, and the total manufacturing and checkout process consumed just over two years.

The first of the three X-15s eventually arrived at Edwards AFB, California, during the summer of 1959. The X-15 was a striking aircraft, with a polished black Inconel superalloy skin spanning 22.3 feet across the wings and stretching 50.25 feet. The majority of the airframe was constructed of various titanium alloys, although some aluminum and stainless steel was used in low-heat areas (around the cockpit, etc.). The short-span trapezoid wings had extremely thin (5 percent) sections and an aspect ratio of 2.5:1. The leading edge was thin, but not particularly sharp, and the trailing edge was blunt, with a thickness of approximately 2.0 inches at the root and 0.375 inch at the tip. The two slab-stabilators had the same airfoil section as the wings, and could be deflected symmetrically for pitch control or differentially for roll control.[*]

The upper and lower vertical stabilizers had a fixed section next to the fuselage, and an all-moving section outboard (above or below) that served as rudders. The movable section of the lower stabilizer was jettisonable, since the rear skid-type landing gear was not long enough to clear it when in place. The front landing gear was of a conventional dual-wheel type.

The X-15's cockpit was unconventional in that it was equipped with three control sticks. One was the normal

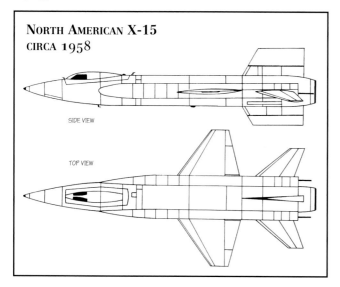

NORTH AMERICAN X-15

CIRCA 1958

SIDE VIEW

TOP VIEW

The external appearance presented by the X-15 was of a rather conventional mid-wing monoplane, but it was its internal arrangement, and the extensive use of exotic materials in its construction, that differed substantially from contemporary production aircraft. (U.S. Air Force)

center-mounted stick that was used with the aerodynamic surfaces during low-speed atmospheric flight. The second was a side-stick mounted on the right side of the cockpit that was mechanically linked to the center stick. This was among the first applications of a side-stick controller, and after the pilots got used to it, the center control stick was frequently removed as redundant. Another side-stick controller, mounted on the left side of the cockpit, was used during high-speed or high-altitude portions of flight to operate hydrogen-peroxide reaction control thrusters mounted in the nose, tail, and each wingtip.

A single Reaction Motors Incorporated XLR99-RM-1 rocket engine with 57,000 lbf furnished 87 seconds of powered flight. The XLR99 was the first high-thrust rocket engine capable of being throttled, with a range from 30 percent (in later versions) to 100 percent of rated power. It was also the first large rocket engine capable of being restarted in flight, although this capability was seldom used by the X-15. The combustion chamber was regeneratively cooled by passing fuel through 196 tubes built into the walls of the nozzle. The engine used liquid oxygen and anhydrous ammonia for propellants in a 1.25:1 ratio by weight, and 1,445 gallons of oxidizer and 1,034 gallons of fuel were carried internally. The entire XLR99 installation weighed 915 pounds. This was by far the most advanced rocket engine built until the development of the Space Shuttle Main Engines (SSME), surpassing the F-1 engines used on the Saturn V in complexity, if not sheer power.

This was reflected in the continuing difficulties encountered during its development, although once delivered the engine proved to be remarkably trouble-free. The total cost, including development, for ten XLR99-RM-1 engines was $68,323,030, plus a fee of $6,014,000 to Reaction Motors. Ironically, the engine cost was over five times the original estimated cost for the entire X-15 program, and the fee paid to Reaction Motors was greater than the original estimated cost of the entire engine research, development, and production program.

On 10 March 1959, following five months of extensive systems and vibration testing on the ground at Edwards AFB, North American test pilot Scott Crossfield[†] took the first X-15 (56-6670) on its initial captive-carry flight

---

[*] Although commonplace on today's high-performance aircraft, this system was considered innovative when the X-15 was designed, being amongst the first application of differentially moving slab-stabilators on a high-performance vehicle. A significant amount of data had been accumulated by North American on this concept during the development of the three YF-107A fighter prototypes and the design of the still-born XF-108 escort fighter.

[†] Crossfield had left NACA to join North American in order to be more closely associated with the X-15 program. This decision later haunted him when NASA and the Air Force decided that all record flights would be flown by government research pilots, effectively leaving Crossfield out of the record books.

under the wing of an NB-52. Several additional captive flights followed, culminating in the X-15's first unpowered flight on 8 June 1959. The first powered flight was made by Crossfield on 17 September 1959, but since the new XLR99 was not yet ready, the flight was conducted with two of the much-used XLR11 engines.* The XLR99 made its flight debut on 15 November 1960 and attained a speed of Mach 2.97 at an altitude of 81,200 feet, using somewhat less than 50 percent of available power.

By the end of 1961, the X-15 had already attained its Mach 6 design goal and reached altitudes in excess of 200,000 feet. During August 1962 an inertial navigation system being designed for Dyna-Soar was installed on the first X-15. On 22 August 1963, NASA test pilot Joseph A. Walker took the third X-15 (56-6672) to an altitude record of 354,200 feet (67+ miles).

John B. McKay in the second X-15 (56-6671) survived the collapse of the left main landing skid during a landing attempt on 9 November 1962. This accident essentially destroyed the aircraft, but a decision was made to rebuild it into an advanced configuration known as the X-15A-2 under a $5,000,000 contract to North American. Major Robert A. Rushworth conducted the modified aircraft's first powered flight on 25 June 1964, achieving Mach 4.59 and 83,300 feet. A flight on 18 November 1966 with Captain William J. 'Pete' Knight resulted in an unofficial speed record of Mach 6.33.

This record was accomplished, in part, due to the use of an experimental ablative coating developed by the Martin Marietta Corporation under the name MA-25S. This material, consisting of a resin base, a catalyst, and a glass bead powder, was designed to be sprayed onto the exterior of a vehicle to protect it from the extreme heat generated by high-speed flight within the atmosphere or during reentry. All ablative coatings, MA-25S included, are designed to be expendable, sacrificing themselves in order to minimize heating effects on the structure they protect. MA-25S was intended to be removed by solvents following a flight, then replaced prior to the next mission. Unfortunately, this proved to be extremely labor intensive and was a nearly impossible task.

Various other ablative coatings had been evaluated for use on the X-15A-2 including Emerson Electric Thermo-Lag T500, Dow Corning DC-325, Armstrong Cork #2755, NASA Purple Blend, Molded Refrasil Phenolics, and the General Electric Century Series materials. On several occasions these ablators had been applied to various surfaces of the X-15 and tested for their heat protection qualities. And, although Thermo-Lag T500 had proven the most effective, it was also the most difficult to work with and very seldom achieved a uniform quality, explaining why it was superseded by the Martin-developed product.

The success achieved by Knight's unofficial flight gave NASA the confidence to explore the maximum speed potential of the research vehicle. The X-15A-2 was carefully prepared for the flight, consisting mostly of the 700-plus man-hours needed to refurbish the MA-25S ablative coating, and attaching a non-functional model of a proposed future scramjet.

On 3 October 1967, the X-15A-2 set a world absolute

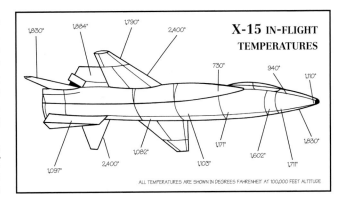

Temperatures experienced during X-15 high-speed flight ranged up to 2,400 degF at a velocity of 4,500 mph. The X-15 carried no external insulation (except for the X-15A-2) and absorbed the entire heat load with its innovative Inconel-X hot-structure airframe construction. (NASA)

speed record of 4,520 mph (Mach 6.7) with the recently-promoted Major Pete Knight at the controls. This record would stand until the return of the Space Shuttle Orbiter *Columbia* from its first mission in 1981. But a subsequent post-flight examination revealed a number of serious problems – the high temperature and extremely high speed airflow had combined to char and pit the MA-25S material so badly that it was deemed impossible to restore it for use on any further flights. Additionally, the airframe had suffered serious damage, particularly along the forward fuselage chines and around the ventral stabilizer, where the dummy hypersonic scramjet had been attached for flow tests. Nevertheless, these couple of flights using the MA-25S ablator provided a great deal of much needed data for designers of future spacecraft, and resulted in a reluctance by NASA to rely on ablators for Space Shuttle. The damaged X-15A-2 was subsequently repaired and returned to Edwards AFB, but did not fly again before the X-15 program was terminated. It is currently on display at the Air Force Museum.

Major Michael J. Adams was killed on 15 November 1967 when the third X-15 (56-6672) crashed during a high altitude research flight. It was the only fatality during what is arguably the most productive flight research program ever undertaken. At the time of the accident, the third X-15 was fitted with the Honeywell MH-96 self-adaptive flight control system originally designed for the X-20 Dyna-Soar. This system was intended to make the vehicle easier to control with either the aerodynamic surfaces or the reaction thrusters, or a combination of both. On the flight that killed Adams, a series of events caused Adams to become disoriented and lose control, eventually leading to a saturation of the MH-96 system computer, resulting in loss of effective control. Notwithstanding the tragic end of the MH-96 flight test program, the lessons learned would benefit future winged spacecraft designers.

Nine more flights were completed before the X-15 flight program ended on 24 October 1968 with William Dana at the controls of flight number 199. A 200th flight was attempted and aborted before launch on several occasions, once due to a snow storm at Edwards! The X-15 constituted a major step on the path to a lifting-reentry spacecraft, examining many of the problems associated with returning from orbit, including structural heating, aerodynamics, structural loading, and guidance and control. The X-15 flight data revealed hypersonic heating rates some 30 percent lower than predicted, leading to a reexamination and correction of the predictive methods. The X-15 pilots had repeatedly demonstrated the ability to

---

* The Reaction Motors Incorporated (later absorbed by Thiokol) XLR11 seems to have been used by almost every rocket-powered research aircraft flight tested at Edwards. This was a four chamber engine, and the chambers could be ignited individually, allowing total thrust to be tailored for a given flight profile. Propellants were liquid oxygen and diluted ethyl alcohol. Early versions produced 1,500 lbf per chamber, while later versions increased this to 2,400 lbf per chamber. The entire engine, including its associated liquid oxygen turbopump, weighed approximately 350 pounds.

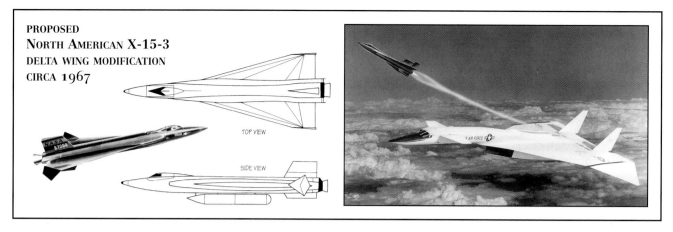

PROPOSED
NORTH AMERICAN X-15-3
DELTA WING MODIFICATION
CIRCA 1967

TOP VIEW

SIDE VIEW

The proposed modified X-15-3 was a significant departure from the original X-15 concept. The original short-span trapezoidal wing was replaced by a slender delta-wing with end-plates, and a modified North American XB-70A carrier aircraft was to be used instead of the venerable Boeing NB-52. It was anticipated that this would provide a significantly improved speed increment since launch was to be at fairly high supersonic speeds (approaching Mach 2), instead of ≈450 mph from the subsonic NB-52. (North American Rockwell)

make precision unpowered reentries and landings. Although not often emphasized, the biomedical aspects of the program also proved extraordinarily productive – pilot biological signs were closely monitored via telemetry during all flights, and the X-15 was the first aircraft to expose a pilot to weightlessness for any length of time.

During December 1968, the *Deutsche Gesellschaft für Raketentechnik und Raumfahrt* awarded John V. Becker and the X-15 team the Eugen Sänger Medal, created to honor significant contributions in the field of reusable or winged spacecraft.

Twelve men flew the X-15 – Michael J. Adams, Neil A. Armstrong, A. Scott Crossfield, William H. Dana, Joe H. Engle, William J. 'Pete' Knight, John B. McKay, Forrest S. Peterson, Robert A. Rushworth, Milton O. Thompson, Joseph A. Walker, and Robert M. White. Six of them were awarded Astronaut Wings* for their flights. During 199 flights, the three X-15s spent eighteen hours above Mach 1; more than twelve hours above Mach 2; almost nine hours above Mach 3; nearly six hours above Mach 4; slightly over one hour in excess of Mach 5; and just a few minutes over Mach 6. The entire X-15 program, inclusive of engine and support expenses, cost over $150,000,000. In all, it was well worth the expense as the X-15 program contributed significantly to the U.S. manned space program in general, and was the only existing data base on winged manned reentry vehicles available when the development of the Space Shuttle was begun during the

1970s. The other surviving X-15 (56-6670) was presented to the National Air and Space Museum in 1968, where it is currently on display in the main gallery.

As with many research projects, the X-15 spawned plans to utilize the vehicle for purposes other than those for which it was designed. By the mid-1960s, fully 65 percent of X-15 flights involved using the aircraft as a carrier for special experiment packages. Several interesting proposals were also forthcoming for possible adaptations of the X-15 vehicle itself.

One of these involved rewinging the third X-15 with a slender delta-wing, but the idea was abandoned following the loss of the X-15-3 in 1967. The delta wing was considered the next logical step in the X-15 program, and to this end, North American had spent four years and 300 hours of wind tunnel time defining the configuration. The wind tunnel tests included both thermodynamic and aerodynamic loading, and it was concluded that the wing could use leading edges constructed of columbium superalloy, with Renè-41 on the upper and lower surfaces. Various illustrations have shown the delta-wing X-15A-3 being launched from the back of an XB-70A† Valkyrie carrier aircraft, presumably at high supersonic speeds. As later proven with the Lockheed D-21 drone project, the prospects for a successful supersonic launch were dubious.

A more radical proposal involved a proposal to build a two-seat X-15B that would have been launched into orbit on the back of two modified Navaho boosters. Other variations to this theme involved using a cluster of four Titan I boosters to achieve an early manned presence in space. This general concept was also advanced as part of the Air Force's 'Man-in-Space-Soonest' studies. The X-15B proposals were circulating well before the *Sputnik* launch in 1957, and although the Soviet launch stimulated activity on it, no hardware was constructed, and the proposal never made it further than paper and wind-tunnel studies.

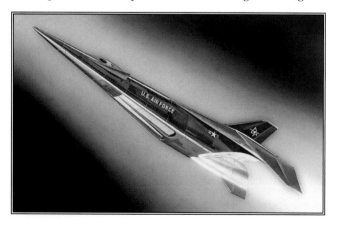

One of the most radical advanced X-15 proposals was this delta-wing configuration that was to have been powered by a pair of ventral scramjets. Noteworthy are the canards mounted just under the windscreen, which were used for trim during transonic and low-supersonic flight, and retracted into the fuselage during the high-speed portions of flight. (North American Rockwell)

---

*   The first 'Astronaut Wings' flight was by White on 17 July 1962, to an altitude of 314,750 feet. Subsequently Rushworth, Engle, Knight, and Adams also completed flights above 264,000 feet (50 miles), which entitled them to wear Astronaut Wings according to the definitions laid down by the Air Force and Navy. Interestingly, NASA required flights to be above 327,360 feet (62 miles), the international standard, so only Walker (354,200 feet) qualified, since McKay and Dana exceeded 50 but not 62 miles.

†   Actually, during the early-1960s, various organizations within the Air Force envisioned using derivatives of the B-70 for launching space payloads of up to 15,000 pounds. It was estimated that this would save $2,630 million over 15 years compared to more conventional expendable launch vehicles. Another variation of this concept, dubbed B-70 RBSS (Recoverable Booster Space System) would have launched Dyna-Soar type manned vehicles. This proposal went far enough that it was briefed to a Senate subcommittee, but was shelved shortly thereafter as the XB-70 program waned and was eventually cancelled.

However, even before the X-15 flew for the first time, various organizations were evaluating vehicles capable of far greater performance. By January 1957, the NACA Ames Aeronautical Laboratory had conceived a piloted Mach 10 demonstrator powered by the same XLR99 used in the X-15. The Ames proposal stated that the "... vehicle should be capable of developing the highest possible lift-drag ratios consistent with design requirements imposed by considerations of aerodynamic heating, stability and control, and structural strength." The proposal seemed aimed primarily at acquiring data for the Air Force, probably for the HYWARDS demonstrator: "Militarily, the information obtained from flights of such an airplane would be of great and immediate value. ... Commercial applications ... are more tenuous."[25]

The configuration chosen by Ames was a flat-top wing-body with drooped wing tips. A combined elevon-split-flap control system was incorporated into the trailing edge of the wingtips, and preliminary analysis showed the configuration was stable in flight up to Mach 10. Aerodynamic heating was to be controlled by a combination of radiation from the surface and an internal cooling system. Ames suggested that the "... vehicle appears to pose difficult but solvable structural problems."

The vehicle had an overall length of 70 feet, with a wingspan of 25 feet. The fuselage was to be 50 feet long with a maximum diameter of 6.7 feet. The nominal weight of the design in gliding flight was 15,000 pounds, with a wing loading of 20 pounds per square foot. It was expected that landing speeds would be on the order of 200 mph, and the landing gear was similar to that on the X-15 (rear skids and nose tires).[26]

Engineers at Ames proposed two alternate methods of launching the vehicle. Drop tests to prove the low-speed aerodynamics would be conducted from a Convair B-36 bomber; a large booster powered by the 150,000 lbf Rocketdyne XLR89-NA-1 engine from the Atlas ICBM would be used for higher speed flights. These latter flights would have originated from Cape Canaveral, Florida, with booster separation occurring at 100,000 feet and Mach 6. The vehicle would then fire its XLR99 and eventually land at Edwards AFB after a sub-orbital flight.[27]

By 3 September 1957 Ames had considerably refined its approach – and sales pitch. The X-15 was eating up all the available funding for research aircraft, and Ames was feeling largely left out since most of the X-15 activity was at Langley and the High-Speed Flight Station. Part of the introduction to the September study report indicated that the research airplane program "... will continue to be effective only so long as foresight and judicious planning provide research airplanes sufficiently advanced over existing service airplanes that research may maintain a substantial lead over practical need. ... It is not sufficient to think of a new research airplane simply in terms of one that will fly faster, by a reasonable margin, than the Mach number of 7 that is expected of the X-15, but rather of one that will fly enough faster to permit exploration of important problems sufficiently different from those of the X-15 and preceding research aircraft to warrant the current high cost of a new research-airplane project."[28]

Although the report, which was classified 'secret' at the time, never directly stated so, it is obvious that the motivation of the Ames researchers was to build a flight research vehicle as part of HYWARDS, which would soon be incorporated into Dyna-Soar. There were numerous mentions of "... military airplanes capable of hypersonic, long-range flight ..." that effectively skirted the security veil surrounding the BoMi, RoBo, and HYWARDS projects.

In the September 1957 report, Ames discussed two configurations, known as 'A' and 'B,' – each could carry a single pilot, his support equipment, and 1,200 pounds of research instrumentation. Both configurations had relatively low wing loadings (≈20 psf) and used a single XLR99 engine from the X-15 program. Both configurations also required large external boosters to act as first stages in order to attain the 18,000 feet-per-second design velocity.[29]

The structural design was similar for both configurations, with a structure that was thermally insulated from the outer skin and actively cooled where necessary. This was a significant departure from the hot-structure being used on the X-15, which Ames believed (correctly, as it turned out) was approaching its maximum allowable temperatures and speeds. Research at Ames confirmed, again, that the region of highest heating is generally the high-pressure side of the vehicle.

In the 'A' configuration, Ames removed the fuselage from this region by locating it totally above the wing, which acted as a heat shield. The lower surface of the wing was essentially flat to facilitate the prediction of pressure distributions and heating rates. By using the shielding effect of the wing to keep the fuselage relatively cool, the problems of insulating the pilot and instrumentation was greatly simplified – it was expected that simple external insulation would be sufficient for this design, and

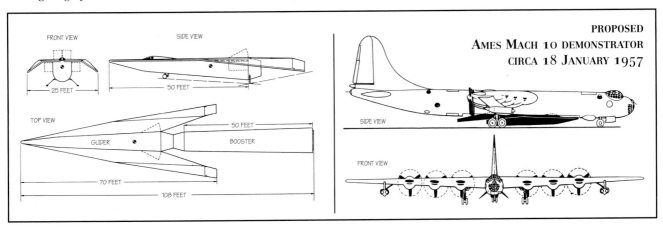

In the January 1957 proposal, the NACA Ames Mach 10 demonstrator used a first stage powered by a Rocketdyne XLR89, while the research vehicle itself was powered by a modified version of the Reaction Motors XLR99. Alternately, the second stage could be air-launched from a Convair B-36, although this resulted in a significant decrease in terminal velocity (3,400 mph versus 7,500 mph). Note the control surfaces on the top and side of the research vehicle's fuselage. (NASA)

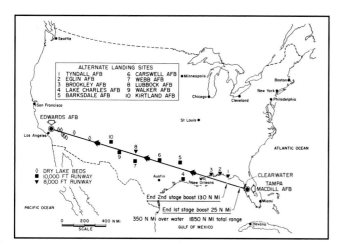

The high-speed flight path as envisioned in January 1957 used a launch point on the west coast of Florida with a landing at Edwards AFB. Preliminary low-speed (<Mach 6) flights would use the High Range that had been developed for the X-15. (NASA)

that an active cooling system was not required. Tip cones were to be used to provide static stability and control, both directional and longitudinal. The trailing-edge flaps deflected with the cones and were mainly used for transonic and low-speed longitudinal control.

Configuration 'B' was essentially the same design proposed in the original January 1957 report. This design combined the flow field of a slender expanding fuselage with the lower-surface flow field of the wing to achieve a higher coefficient of lift at a lower angle-of-attack. This resulted in a higher hypersonic L/D ratio, which designers expected would minimize the effects of heating due to generally slower speeds during reentry. Locating the fuselage below the wing also cambered the configuration, making it essentially self-trimming at the maximum L/D, significantly reducing drag due to trimming effects. The drooped, toed-in wingtips were used to achieve static stability and control at hypersonic speeds. These were supplemented by a retractable ventral stabilizer to improve stability below Mach 6. Trailing-edge elevons were provided at the wing tips, and flaps were located on the afterbody. The flaps could be operated together as speed brakes or individually to provide directional control.

The Ames study was notable in that it delved deeply into the various possibilities to protect an airframe from high heating. Three basic approaches were investigated – an unprotected hot-structure composed of high-temperature metallic and ceramic materials; an internally cooled structure made up of various high-strength materials; and a structure protected from the heat by external insulation. Although the last two approaches were expected to involve greatly increased structural complexity, the potential weight savings intrigued researchers.

Since Inconel-X was well understood at this point, and was being used in the construction of the X-15 vehicles, Ames concentrated on more advanced alloys. The commercially available L-605 cobalt-based alloy demonstrated adequate mechanical properties (did not become brittle) and oxidation resistance (did not ablate) up to about 1,800 degF. Some molybdenum alloys retained good strength characteristics even above this temperature, but oxidized rapidly without protective coatings, none of which existed at the time. Ceramic materials were not seriously considered for structural elements since at the time they tended to be brittle and had minimal load-carrying abilities. They were, however, considered essential for small, hot areas such as the leading edge of the wings.[30]

An investigation of active cooling systems yielded promising results to keep temperatures within the ability of existing hot-structure technology, but left little hope that a workable vehicle could be designed around them. The problem was simple – coolant is heavy. For the small research aircraft being contemplated, it was estimated that over 2,000 pounds of coolant – water or liquid helium – would be needed. The study eventually concluded that "... unless high heat-capacity coolants can also be used efficiently for propulsion, the weight of a design in which surfaces are directly cooled appears prohibitive."

Preliminary investigations into insulating the airframe from heat effects looked promising. Researchers realized that in addition to its heat-insulation value, any protective layer must also be capable of withstanding the effects of the high-Mach airstream while operating at surface equilibrium temperatures. Because of high temperatures and anticipated noise levels of over 170 dB, any external insulation would require a high degree of structural integrity. The most promising design for temperatures in excess of 1,800 degF appeared to be a metallic superalloy outer skin with closely spaced 'stand-offs' that allowed a thick layer of Thermoflex insulation between the outer skin and the titanium airframe structure underneath. It was estimated that this combination would add about two pounds per square foot to the total weight of the aircraft.

Because of differences in thermal expansion between the insulating and primary structures, expansion joints would be required at appropriate intervals. These, in turn, could lead to surface irregularities that could significantly increase local heating rates at high speed, and would have to be designed and maintained carefully. The study noted that the "... successful development of a smooth and lightweight insulating structure with adequate life expectancy is believed to represent a major development effort." Truer words were never written.[31]

By this time the original B-36 carrier aircraft had given way to the NB-52s that had been modified for the X-15 program. When air launched from the NB-52, the vehicle was expected to be capable of velocities approaching Mach 8, offering a means for preliminary checkout and handling quality tests. Preliminary feasibility studies were also conducted of mounting two 125,000-lbf solid rockets on the vehicle and air launching the combination from the NB-52. This would have permitted some additional speed increment while waiting for the booster to be designed and built.

Air launch operations were expected to be based at Edwards AFB to take advantage of the infrastructure developed for the X-15. Longer range flights would be launched over the Pacific Ocean where radar and telemetry systems at Pt. Mugu and Vandenberg AFB could be used, with landings again taking place at Edwards. For the highest speed flights, one possibility investigated was launching the vehicle from a ship off the coast of Cape Canaveral. Initial landings would take place at one of the Air Force bases in Florida, and the radar and telemetry installed to support ICBM testing on the Atlantic Test Range would be used. As flight distances grew longer, the ship would simply sail further out to sea. Eventually, flights would be launched from very downrange in the Atlantic, and tracked until they were over Cape Canaveral. Tracking would then be intermittent until the vehicle landed at Edwards after a suborbital flight.[32]

The later study had added the 'A' configuration largely in response to wide-spread criticisms regarding the initial high-wing configuration proposed in January. Any high wing configuration ran contrary to popular reasoning that

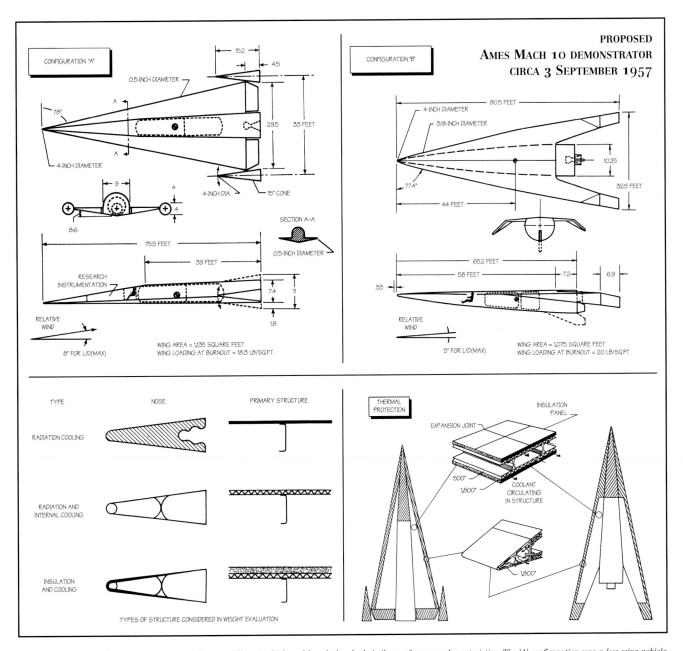

By 3 September 1957 the Ames study was carrying two different vehicles, although they had similar performance characteristics. The 'A' configuration was a low-wing vehicle that was largely an answer to criticisms of other researchers to the original high-wing vehicle, which was now dubbed configuration 'B.' The B-36 carrier aircraft had given way to a Boeing NB-52 since that is what the X-15 program was actually using, and a variety of booster stages were considered for the higher-speed flights. (NASA)

it exposed too much of the structure to the severe heat environment. Ames continued to champion the high-wing design, arguing that "... it is apparent that the flat-top arrangement is the more efficient," claiming tests showed the design offered a 35 percent increase in performance.[33]

This design brought to a head the internal debate within the NACA on the relative merits of high-wing versus flat-bottom-low-wing designs, a battle that the low-wing advocates eventually won, as evidenced by Space Shuttle. The detailed investigations into the concept of providing external insulation for high-speed aircraft continued at Ames, which remains NASA's primary center for thermal protection systems. Although the Ames design was not seriously pursued, it did serve several important purposes, including some much needed laboratory research that would be used on later programs. All of this led to the first serious attempt to build a lifting-reentry vehicle as part of 'Round Three,' better known as Dyna-Soar.

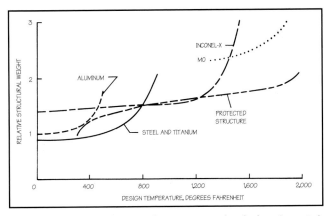

This was the chart used in the 3 September 1957 report to describe the various materials considered for the airframe and their maximum allowable temperature for any given relative structural weight. Notice that Inconel-X has almost twice the heat resistance of steel for the same relative weight – explaining why it was used in the X-15 hot-structure. A protected structure of some description was necessary when the heat limits of Inconel-X were surpassed at roughly 1,600 degF. (NASA)

John V. Becker. (NASA photo L-70-5770)

CHAPTER II

DYNAMIC SOARING

SKIP-GLIDING

One of the more serious unknowns regarding Eugen Sänger's Silverbird designs was the effect of aerodynamic heating on the structure during its prolonged skips. The X-15 would – at least in part – provide answers, but other projects were being pursued in parallel.

In the immediate post-war era, three basic vehicle types were considered feasible for high-speed global flight – ballistic, boost-glide, and skip-glide. The ballistic vehicle leaves the atmosphere at an angle relative to the Earth's surface that requires the least energy input for a given flight path – best exemplified by the ICBM. A boost-glider is accelerated to a speed and altitude such that the vehicle is at its maximum hypersonic lift-to-drag (L/D) ratio, and the unpowered portion of the flight is continuously maintained at the maximum L/D, with the altitude decreasing as the vehicle slows down due to aerodynamic drag. On the other hand, a skip-glide vehicle (such as the Silverbird) is boosted to its maximum speed and altitude along a ballistic trajectory – as the vehicle reaches its maximum L/D during its return to Earth at the end of its first ballistic flight phase, it 'skips' upward into its next ballistic flight phase. Each upward skip results in a lower altitude than the last since the vehicle is unpowered and is using a natural phenomenon to maintain flight. This sequence continues until the vehicle no longer has sufficient power to leave the sensible atmosphere, at which time it lands.[1]

In 1955, and again in 1956, H. 'Harvey' Julian Allen and his colleagues at NACA Ames published classified analyses of the two glide techniques – the researchers dismissed the ballistic rocket as unsuitable for manned operation within the foreseeable future. The conclusion was that the skip-glide vehicle had a longer range for any given velocity increment than an equivalent boost-glide concept. It was noted, however, that the skip-glide vehicle would experience significantly higher heating rates than the boost-glide concept, and would also experience greater magnitude and longer duration acceleration (g) forces. This was because the transfer of kinetic energy to heat occurred in

abrupt pulses during the skipping phases of flight – the first skip, of course, was the most severe, but all of the early skips had to endure extreme heating. The high heating rate of the skip-glide vehicle would necessitate more thermal protection, creating a heavier vehicle, probably negating the small advantage in predicted range. The boost-glider, on the other hand, gradually converted kinetic energy to heat over the entire flight trajectory, resulting in a relatively low level of heating. The 1955 NACA report concluded that the boost-glide concept appeared to hold the most promise for an operational military system. The report noted, however, that even the boost-glide vehicle would be of limited usefulness since it could not "... be a very maneuverable vehicle in the usual sense and thus is, perhaps, limited to use for bombing and reconnaissance."[2]

The reports went on to describe analytical and wind tunnel research that had been conducted on possible vehicle configurations. The conclusion reached at NACA Ames was that any boost-glide vehicle should possess a hypersonic L/D of 5 or 6 with an extremely large wing area. The heating of the wing leading edge was found to be greatly reduced by a highly-swept planform, but nevertheless, temperatures of over 3,100 degF would be experienced at roughly Mach 12. The researchers determined that "... wing leading edge heating constitutes one of the serious problems in the design of this type of aircraft ..." but concluded that the average heating of the rest of the airframe was at least an order of magnitude less than the leading edge.[3]

These reports were the beginning of the end for the skip-glide technique first proposed by Sänger, and within a few months of the report's release, almost all work on skip-gliding ceased in favor of boost-glide concepts. It should be noted that although the vehicles described in most of these concepts are called 'gliders,' this description was only appropriate because they were not continually powered. With some exceptions all of these vehicles carried turbojet engines that would extend the terminal portion (landing) of their flight, and frequently included rocket engines to act as upper stages so that they could achieve higher orbits.

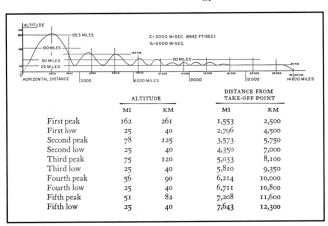

| | ALTITUDE | | DISTANCE FROM TAKE-OFF POINT | |
	MI	KM	MI	KM
First peak	162	261	1,553	2,500
First low	25	40	2,796	4,500
Second peak	78	125	3,573	5,750
Second low	25	40	4,350	7,000
Third peak	75	120	5,033	8,100
Third low	25	40	5,810	9,350
Fourth peak	56	90	6,214	10,000
Fourth low	25	40	6,711	10,800
Fifth peak	51	82	7,208	11,600
Fifth low	25	40	7,643	12,300

This composite chart shows how skip-gliding worked. Each peak was lower than the peak before it, although the lows remained constant until very late in the flight. This chart was created by Willy Ley based on calculations done by Eugen Sänger and Irene Bredt during World War II. (Willy Ley / The Viking Press)

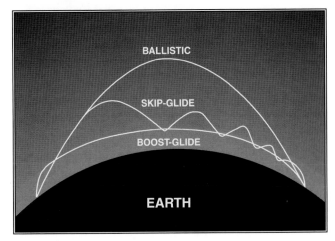

The chart above (reconstructed from a poor microfiche copy) shows the difference between the three concepts for high-speed global flight studied by NACA Ames during 1955 and 1956. (NACA/RM-A55L15)

BoMi, RoBo, and Brass Bell

After World War II, Dr. Walter R. Dornberger, former director of the Pëenemunde rocket test center, and Dr. Krafft A. Ehricke, a close associate, emigrated to the United States and joined the staff of Bell Aircraft Company. While at Bell, they advocated the development of a Silverbird-like flat-bottom lifting-reentry vehicle, and during early-1952, even journeyed to France in a vain attempt to persuade Eugen Sänger and Irene Bredt to come to the United States to join Bell.

Early Bell skip-glide designs were simple variations of the Sänger-Bredt Silverbird, but the wedge-profile straight wings soon gave way to the delta planform more typical of later concepts. During these studies the company devoted a great deal of attention to the problems of atmospheric reentry, particularly the field of thermal protection, where Bell researchers explored both active (e.g., liquid circulation) and passive (i.e.; radiative, heat-sink, and ablative coatings) schemes in considerable detail.

On 17 April 1952, Bell proposed building a piloted Bomber-Missile (dubbed BoMi) to the Air Force. This original BoMi was a two-stage vehicle featuring a large five-engine, delta-wing booster, and a smaller three-engine double-delta glide-rocket. The mated pair would use the booster's engines for two minutes, then the glide-rocket would separate and ignite its engines – the booster would fly back to its base for a runway landing and subsequent reuse. During the period of maximum aerodynamic pressure (max-q), the glide-rocket would throttle down two of its engines to ease air-loads, and the basic flight profile resembled the skip-glide technique described by Sänger and Bredt for the Silverbird.

The booster stage was nearly 120 feet in length with a wing span of 60 feet, and carried a crew of two. Most of the airframe was to be constructed of various aluminum alloys, but used titanium alloy hot-structure wing leading edges. The upper stage was 60 feet in length with a 35-foot wing span, and carried a single pilot and a payload consisting of 4,000 pounds of nuclear weapons. It was to be built entirely of titanium alloys and use a Bell-designed active cooling system. The gross lift-off weight (GLOW) of the mated vehicle, including its weapons load, was projected to be 800,000 pounds and both stages used nitrogen tetroxide (N2O4) and undimethylhydrazine (UDMH) propellants. During May 1952, Bell requested $398,459 from the Air Force to initiate a year-long feasibility study into the BoMi concept.[4]

There appeared to be no military usefulness to an orbital vehicle, so this BoMi was a sub-orbital design having a 3,300 mile range, and capable of Mach 4 at 100,000 feet. Two nuclear weapons would be carried in an A-5 Vigilante-style rear ejecting bomb bays. But Bell engineers were also looking towards space – an orbital version, with a 144-foot long lower stage built entirely of titanium, was also proposed. The new upper stage would be 75 feet long, carry a payload of 14,000 pounds, and be covered with a graphite-epoxy ablative heat shield on a honeycomb backing. It was hoped that the ablative coating could be sprayed-on after each mission to renew the heat shield, but this was one of the more serious unknowns in the vehicle's development – later experience with the X-15A-2 would prove disappointing. The propellants for the orbital version were changed to liquid oxygen (LO2) and liquid hydrogen (LH2), which provide a somewhat higher specific impulse at the expense of greatly increased complexity and weight – especially considering that no large LO2/LH2 engines had yet been designed or built.[5]

An initial review of Bell's BoMi concept by the Air Research and Development Command (ARDC) was completed on 10 April 1953, and discovered several serious flaws. The most important of these was that the cooling difficulties had not been adequately addressed, and that Bell's lift-to-drag ratio figures were hopelessly optimistic. Engineers at the ARDC also seriously questioned how such a vehicle would be controlled in flight. The ARDC pointed out that Bell's proposal duplicated several parts of the Atlas ICBM and FEEDBACK orbital reconnaissance satellite projects.[*] The 3,300-mile range was also considered inadequate for intercontinental operations and the aircraft appeared unable to turn around in flight. Still, the proposal seemed to offer a reconnaissance capability far in advance of the FEEDBACK program, and the aircraft could also provide an excellent test vehicle for future hypersonic programs.[6]

Notwithstanding the technical misgivings of the ARDC engineers, on 1 April 1954 the Air Force awarded Bell a $220,000 one-year contract to study

This was the final version of Bell's proposed BoMi/RoBo with a circular cross-section fuselage, mid-mounted wing, and canted wingtips. Note the lack of any viewing ports or periscopes for the pilot. The small stage attached to the aft of the glider is a 'trans-stage' that was added late in the program for a proposed orbital version launched by an expendable booster. The glider was 60 feet long, and the entire stack was 134 feet long. (Bell Aerospace Textron)

* Since both of these projects were highly classified at the time, it is unlikely that Bell was aware they were duplicating on-going research. In any case, Bell's proposed application of this research was certainly unique.

Weapons System MX-2276 based on the skip-glide concept. MX-2276 was to be capable of performing both reconnaissance and bombardment roles, and was to have a maximum velocity of 15,000 mph at 259,000 feet with a projected range of around 12,000 miles. These requirements meant a much larger vehicle that the one Bell originally proposed. The increase in size caused Bell to abandon the large fly-back first stage in favor of expendable rocket-powered booster stages based largely on the technology being developed for the ICBM programs.[7]

The official contract expired in May 1955, but Bell continued with company funds and by 1 December 1955, a grand total of $420,000 had been expended on the various BoMi and MX-2276 studies.* During this period Bell engineers studied the reports generated by NACA Ames and came to much the same conclusion. The skip-glide BoMi was replaced by a generally similar configuration that used a boost-glide flight profile.[8] Although the Bell engineers had no way of knowing it, BoMi would indirectly have a significant influence on future Space Shuttle concepts.

The Air Force also requested that Boeing include the boost-glide concept in its studies for Weapons System MX-2145 which had been initiated in May 1953 to investigate possible follow-ons for the Convair B-58 Hustler. Boeing's limited investigation of the concept indicated that it would be far easier to design a vehicle that orbited the Earth, as opposed to turning around and returning after reaching its target. Boeing also stressed the difficulties in designing a structure that could withstand the extreme heat and aerodynamic stresses expected on a boost-glide flight, but recommended continued studies because of the system's great military potential.[9]

On 4 January 1955 the Air Force issued the system requirements (SR-12) for a reconnaissance vehicle with a range of 3,000 miles at altitudes over 100,000 feet. The Wright Air Development Center (WADC) took SR-12 and used it to establish the operational requirements for System 118P. Several contractors quickly expressed initial interest, including Bell which received a $125,000 contract on 21 September 1955 to investigate applying the technologies being developed for BoMi/MX-2276 to System 118P.+ Bell's design included a two-stage launch vehicle that boosted a glider to a velocity of Mach 15 and an altitude of 165,000 feet. It was suggested by Bell that the System 118P effort be broken into a three-phase project – the first phase being a 5,000-mile range vehicle, the second having a range of 10,000 miles, and the final phase being a vehicle with global (orbital) range. A detailed proposal relating to the phasing was presented by Bell on 1 December 1955, and the suggestion was subsequently accepted by the Air Force.[10]

The MX-2276 and System 118P efforts were combined on 20 March 1956 when the Air Force awarded Bell a $746,500 (later raised to $1.2 million) contract for Reconnaissance System 459L, also known as Brass Bell. This was in response to general operations requirement (GOR-12), which was issued on 12 May 1955 as an outgrowth of SR-12. This requirement defined a manned high-altitude reconnaissance platform that was to be available to operational Air Force units in the third-quarter of 1959.[11]

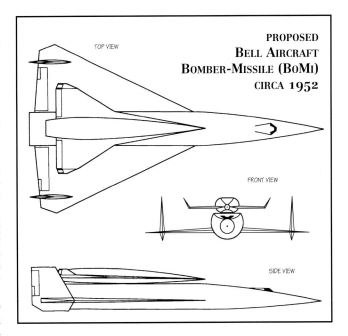

PROPOSED
Bell Aircraft
Bomber-Missile (BoMi)
CIRCA **1952**

Many variations to the BoMi theme were explored during the project's three year definition phase. Early studies used some sort of recoverable booster topped by a delta-winged glide-rocket, although later derivatives substituted expendable boosters for the recoverable lower stage. This was similar to the evolutionary process that the Space Shuttle would go through 15 years later. (redrawn from Bell documents by Dennis R. Jenkins)

During November 1956 the Air Force asked NACA to review the on-going boost-glide studies at Bell and Boeing, and Hugh Dryden formed a steering committee within NACA to evaluate the efforts and recommend an approach to hypersonic and orbital flight research. By December 1956, the operational Brass Bell vehicle was envisioned using a launch vehicle powered by the propulsion system being designed for the Atlas ICBM. A speed of 12,000 mph, an altitude of 170,000 feet, and a maximum range of 6,300 miles were expected. Bell engineers reasoned that with the addition of two more boosters (resulting in a trimese configuration similar to the much-later Titan III), the range of Brass Bell could be extended to 11,500 miles with a maximum speed of 15,000 mph, although little actual investigation into this possibility had been performed.[12]

While the Air Force was busy channeling most of Bell's work towards the development of a boost-glide reconnaissance system, it had not abandoned the application of the concept towards a bombardment platform. On 19 December 1955, the Air Force had asked the aerospace industry to undertake analysis and preliminary design tasks for a manned hypersonic bomber. Six companies – Boeing, Convair, Douglas, McDonnell,‡ North American, and Republic – responded to the request and undertook company funded studies. On 12 June 1956, SR-126 was released for the formal study of a Rocket-Bomber (RoBo) and three contractors, Convair, Douglas, and North American, were each awarded study contracts totaling $860,000 that ran through December 1956. The purpose of the study was to determine the feasibility of a manned hypersonic bombardment system, and one of the technologies to be investigated was a boost-glide vehicle similar to BoMi and Brass Bell. The anticipated payload of such a bomber varied from 1,500 to 25,000 pounds, and the vehicle was also supposed to have a secondary strategic reconnaissance capability. The contractors continued to study the concept with company funds, and by the end of 1957, a total of $3.2 million had been expended by all parties involved.[13]

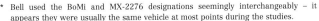

---

*   Bell used the BoMi and MX-2276 designations seemingly interchangeably – it appears they were usually the same vehicle at most points during the studies.
+   This was actually funded as an extension to the MX-2276 study contract, eliminating most of the procurement red tape.
‡   The McDonnell Aircraft Company and the Douglas Aircraft Company did not merge to form McDonnell Douglas Corporation until 28 April 1967. McDonnell Douglas, in turn, was purchased by The Boeing Company in 1996.

HYWARDS

In support of both RoBo and Brass Bell, the Air Force initiated the Hypersonic Weapon And Research and Development Supporting System (HYWARDS) program. This was formalized by SR-131, released by the ARDC on 6 November 1956, and HYWARDS itself was designated System 455L. The intent was to provide research data on aerodynamic, structural, human factor, and component problems associated with very high-speed (Mach ≈15) atmospheric flight and reentry. HYWARDS was also to serve as a test vehicle for the development of subsystems to be employed in future boost-glide weapons systems.[14]

There were four likely engine choices for the HYWARDS vehicle – the first was an exotic fluorine-ammonia 35,000 lbf engine under development at Bell; the second was the 55,500 lbf sustainer engine from the Atlas ICBM; next was the 60,000 lbf engine from the Titan ICBM; while the fourth choice was the 57,000 lbf XLR99 used by the X-15. It was expected that one of these engines would propel HYWARDS to a velocity of 8,200 mph and an altitude of 360,000 feet. Initial flights would be air-launched similar to the X-15, while later flights would be vertically launched using a suitably modified ICBM. The piloted vehicle could theoretically be modified to achieve orbital velocities later in the test program if desired.

Two different NACA groups, one from Langley and the other from Ames, conducted investigations into possible configurations for the HYWARDS vehicle. The Langley group was led by John V. Becker, who had been largely responsible for the shape of the X-15 research vehicle. The final Langley report was issued on 17 January 1957, and contained the surprising revelation that the design speed of HYWARDS should be increased to Mach 18, at a somewhat lower altitude.

Analysis had indicated that this was the speed at which boost-glide vehicles approached their maximum heating environment since the rapidly increasing flight altitudes necessary for speeds above Mach 18 caused a reduction in heating rates, and the heating rate in space itself was negligible. The heating analysis conducted by Peter F. Korycinski and Becker confirmed major advantages for a

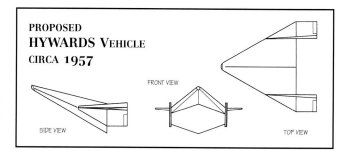

John Becker attempted to design a vehicle that could withstand the thermodynamics of reentry by using a hot-structure combining radiative cooling with internal heat absorption, similar to the X-15, but employing higher temperatures and using water as a heat sink. (U.S. Air Force)

configuration having a flat bottom surface for the delta-wing, with the fuselage located in the relative cool shielded area on the leeside (top) of the wing. This flat-bottomed design had the smallest possible critical heating area for a given wing loading, thus reducing the amount of actual heat-shield needed, whether this be ablative coatings, a hot-structure, or other techniques. In this respect, the configuration differed considerably from the earlier Bell designs which had used a mid-mounted wing. This was one of the first clear indications that aerodynamic design decisions could significantly alleviate some of the heating and structural concerns associated with hypersonic flight.

Interestingly, the Langley study reached exactly the opposite conclusions expressed in the two 1957 reports issued by NACA Ames that advocated a high-wing design. The research team led by Alfred Eggers and H. 'Harvey' Julian Allen at Ames also designed a HYWARDS vehicle, essentially an outgrowth of the earlier Ames Mach 10 demonstrator vehicle. Their proposal called for a maximum speed of approximately Mach 10, but the initial Ames vehicle had a range of only 2,000 miles, compared to 3,200 miles of the Langley design. To produce the highest possible L/D ratio, the Ames design made use of the favorable interference lift that occurred when the pressure field of the underslung fuselage impinged on a high-mounted wing. Unfortunately, in this concept the entire fuselage with its special cooling requirements was located in the hottest region of the flow field. The additional weight required to keep the airframe cool quickly outweighed any advantage in having a higher L/D.

A mid-term review of the RoBo program was conducted on 20 June 1957, and the various contractors presented their proposals. Both Bell and Douglas favored three-stage boost-glide vehicles, while Convair proposed a basically similar vehicle with a third stage equipped with an on-orbit rocket engine and turbojets for use during landing. North American proposed a fairly conventional two-stage vehicle, Boeing offered an unmanned boost-glider called a 'glide-missile,' and Republic wanted to build a small unmanned vehicle that greatly resembled the XF-103 powered by a hypersonic ramjet engine after being launched by an unidentified booster. After a review of these proposals, the Air Force decided that the boost-glide concept was feasible and could be developed into an operational weapons system by 1970. Nevertheless, there was a great deal of concern about propulsion, since there were no large man-rated boosters* yet, and also about developing a suitable physiological environment for the pilot. Further study concluded that an experimental vehicle could probably be flown in 1965, an intermediate-range vehicle in 1968, and a full RoBo-type system by 1974 – a very optimistic schedule for such an advanced project.[15]

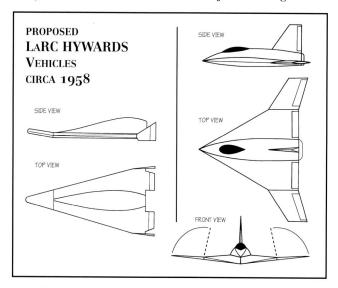

Designs by Langley's Becker for the proposed Minimum Manned Orbital Mission (later Project Mercury). The vehicle at right makes a ballistic reentry at 90 degrees angle-of-attack, followed by transition to normal flight attitudes and landing. The glider at left embodies all of the design features developed by the Becker team – high-lift, high-drag, L/D ≈1.5, maneuverable reentry, radiation-cooled metal skin, etc. Neither vehicle was built, but conceptual features of both were used in both Dyna-Soar and Shuttle. (U.S. Air Force)

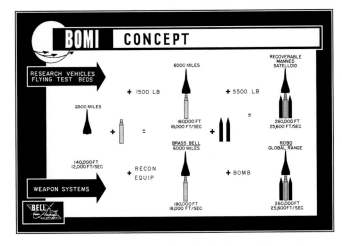

The three-phase BoMi concept lived until the end. The glider could be mated to one or three boosters, plus a trans-stage, greatly increasing its speed and range potential. Note that both research and military versions of each were proposed. (Bell Aircraft)

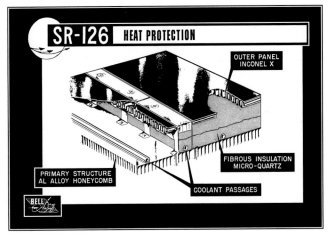

A cross-section of the 'water-wall' thermal protection system shows the coolant passages that kept the inner skin cool. Note the thick layer of Micro-Quartz insulation between the Inconel-X outer skin and the aluminum inner skin. (Bell Aircraft)

The Final RoBo Configurations

Little remains of the RoBo studies since they were all extremely classified at the time and most were subsequently destroyed prior to declassification. Nevertheless, portions of various Bell and Convair reports survived.

On 29–30 July 1957, Bell's Walter Dornberger presented *An Approach to Manned Orbital Flight* to the Air Force Scientific Advisory Board which relied heavily on the BoMi/RoBo designs that had been investigated for the previous five years. In fact, many of the charts were identical to the presentations made the previous month at the final RoBo briefing to the ARDC.[16]

Bell continued to call the vehicle BoMi, and still followed the three-phase development plan outlined during 1955. The first phase of this plan would develop the basic glider – still using a severely swept delta planform and a hypersonic L/D of 4 or 5. The first use of the glider would be as an extension of the existing X-planes program (Round Three), using an NB-52 carrier aircraft as a first stage. The glider would be equipped with a fluorine-ammonia rocket engine and could attain a maximum speed of just over 8,000 mph at 140,000 feet, with a range of 2,500 miles. The second phase added a single liquid-fueled booster for a vertical launch up to 180,000 feet – the glider would then nose-over into level flight and attain a maximum speed of 12,000 mph (18,000 fps) and a range of over 6,000 miles. This vehicle could carry 1,500 pounds of research instrumentation, or alternately fulfill the reconnaissance requirements of Brass Bell. The final vehicle added two additional boosters to create a three-stage[+] system that could carry 5,500 pounds up to 260,000 feet and orbital velocities. This version could be used as the "first recoverable manned satellite," and could also fulfill the requirements for the RoBo bomber with global range.[17]

The orbital design had a GLOW of 740,800 pounds and used fluorine-ammonia propellants in all three boosters. The manned hypersonic glider weighed 24,000 pounds including the 5,500-pound nuclear weapon. The glider was 60 feet long, and the overall length of the system was 134 feet. Bell believed that the 'parallel-tandem' arrangement of the boosters minimized the overall size of the

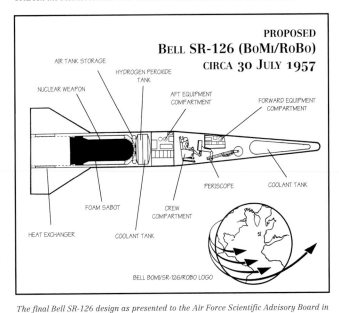

The final Bell SR-126 design as presented to the Air Force Scientific Advisory Board in July 1957. This design had a crew of one and carried a single 4,000-pound nuclear weapon in an aft-ejecting bomb bay. The vehicle was truly a glider, and was not equipped with any type of engines other than a small reaction control system. Note the two large tanks for the liquid-metal cooling system. The vehicle was expected to have a single-orbit capability. (Bell Aircraft)

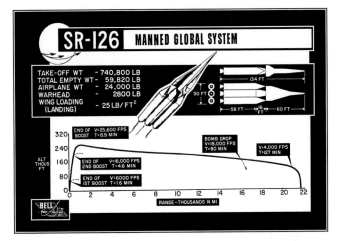

The final Bell RoBo configuration featured global range using three rocket boosters ignited in parallel. This chart shows the flight profile flown by the boost-glide vehicle, with 22,000 nautical miles (one orbit) being completed 127 minutes after launch. It is interesting to note that the projected weapons release was almost three-quarters of the way through the flight (about 17,000 nm). Even Eugen Sänger had realized that some targets might be 'too close' for this type of bombardment. (Bell Aircraft)

---

* Both the Atlas and Titan ICBMs would later have man-rated versions in support of the NASA Mercury and Gemini programs, and these, along with the later Saturn I and Saturn V boosters, were proposed for most orbital projects during this time period.

† Not truly three stage, as defined today – the three boosters were arranged much like a Titan III, and all three were ignited on the ground.

stack, gave the entire system greater stiffness, and reduced the interaction effects at booster separation. The maximum acceleration forces during the ascent would be just over 4-g for short periods, followed by less than 0.10-g during the gliding deceleration.[18]

Bell had learned a great deal in the five years its engineers had been investigating boost-glide concepts and recognized aerodynamic heating as the major challenge to building such a vehicle. Analysis showed that potential interaction between the shock wave and boundary layer air on top of the wing could raise temperatures in this region significantly, but they were still below those expected for the bottom surface of the wing. These same analyses showed that the 2,200 degF maximum temperature on the bottom of the wing would occur approximately 45 minutes after lift-off. Temperatures increased closer to the leading edge, with a maximum of 4,500 degF expected near the leading edge.[19]

Three different solutions to leading edge heating were investigated by Bell. The furthest developed consisted of cooling the leading edge with either liquid lithium or liquid sodium. Alternately, Bell considered a hot-structure similar to the one being used on the X-15, or a 'sweat cooling' system that was considered to require too much coolant to be practical. A double-wall hot-structure had been developed for the remaining airframe that consisted of a light heat-resistant alloy outer wall that could withstand high temperature but carried no load, a layer of insulation such as Dyna-Flex, and a water-cooled aluminum inner wall that carried the structural load. Both the liquid-metal and water-cooled structures expended vapor overboard as heat was absorbed by the coolant. Surprisingly, the weight of this structure, including the necessary coolant supply, was comparable to conventional aluminum aircraft structures. Both of the actively-cooled systems had been extensively tested by Bell, which believed they were technically feasible for large-scale applications. For instance, the leading edge liquid sodium cooling loop had already successfully operated over 600 hours at 1,200 degF, and was being tested at 1,600 degF. Silicon carbide leading edge samples had been successfully tested to 4,000 degF. A two-foot square sample of the water-wall skin structure had been successfully tested for 90 minutes at full intensity heating.[20]

Bell stated that "[T]he technical status of the program is that there are no major problem areas to which engineering solutions have not been found. In most cases these engineering solutions have already been reduced to practical design." In many cases this was backed up by solid evidence. The basic nose shape had been tested at the Air Force Arnold Engineering Development Center at temperatures as high as 10,000 degF and velocities of Mach 16. The glider configuration had been tested in the NACA Ames hypersonic wind tunnel at up to Mach 10, and various pieces had been free-flight tested on sounding rockets above Mach 12. Panel flutter had been investigated in the NACA Langley wind-tunnel at up to Mach 4 and temperatures of 850 degF.[21]

The final glider was capable of carrying a 5,500-pound payload in addition to its pilot. In the RoBo version this payload was a nuclear weapon that included its own navigation system for final course corrections during free fall. But Bell was looking forward and believed the vehicle could be used to send men into orbit. By replacing the weapon with a small rocket engine, propellants, and instrumentation, Bell devised the Space-BoMi (S-BoMi).

Other changes included the addition of a reaction control system to provide control at altitudes over 270,000 feet,* generally similar to the monopropellant system

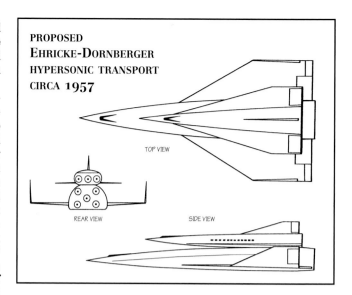

Krafft Ehricke and Walter Dornberger collaborated on this passenger version of Bell's BoMi in 1957. The vehicle was to be complete a 3,000 mile flight in 75 minutes, in an era when the jet-powered Boeing 707s were just beginning to enter service. Note the upswept nose on each stage, a characteristic of most early hypersonic designs. (U.S. Air Force)

developed for the X-15. A liquid-fuel auxiliary power unit was also added. Surprisingly, Bell considered sealing the crew cabin to be a major technical challenge, and expected a leak rate of 10 cubic feet per minute at a 5 psi pressure differential. It was interesting to note that the expected leak rate was higher than the crew consumption rate. A stored gas supply was provided to replenish the cabin.[22]

The weight of the orbital subsystems varied directly as a function of time on orbit. For a two-hour mission (one orbit) the extra weight was only 2,000 pounds – for a 36 hour mission (24 orbits) the additional weight climbed to 30,000 pounds. Most of this was for reaction control propellant (14,000 pounds) and APU fuel (12,000 pounds), although the pilot would require additional supplies also.[23]

Analysis indicated that the optimum configuration was to replace the weapon with the appropriate subsystems and still leave 1,500 pounds for research instruments. This would allow the vehicle to attain an altitude of 480,000 feet (90 miles) at a velocity of 25,640 fps. At the maximum lift coefficient (at this altitude, S-BoMi would still be 'flying' instead of 'orbiting') the vehicle could remain aloft of 10 hours – roughly 7 orbits. Very little control of this time aloft was given to the pilot – he could alter the lift coefficient to reduce the time to 5 hours, but otherwise could not come home whenever he wanted. Alternately, additional propellants could be carried that would allow the S-BoMi to reach an elliptical orbit of 50 by 300 miles for two orbits.[24]

Bell's vision of what the S-BoMi could be used for was truly enlightening, and very accurate. The primary uses were, of course, to evaluate the physiological aspects of space flight on the pilot, followed by measuring the effects of space on various components of the vehicle and its subsystems. But other uses foreseen by Bell included the development of long-range communications equipment, gathering environmental data, and investigating space navigation methods. Even at this early time, Bell acknowledged that much of what S-BoMi could accomplish could also be done by unmanned scientific satellites, but argued that the presence of man would provide a hedge against unanticipated problems or situations.[25]

---

* The 260,000-foot operational altitude of the RoBo design had been chosen partially to avoid needing reaction controls.

Bell felt that their knowledge of hypersonic gasdynamics was sufficient to allow them to complete the design of the suborbital BoMi within two years. But Bell noted several areas that required further study for the orbital S-BoMi. The primary one, unsurprisingly, was continued development of structural insulation and cooling techniques, followed by the development of accurate orbital navigation equipment. It was also expected that additional aerodynamic problems would result from the interaction of the vehicle skin and the atmosphere in its 'free molecular' state – the composition of the atmosphere at the 50-mile altitude proposed for BoMi was relatively well understood, but at 300 miles was essentially unknown.[26]

The final Convair RoBo design built upon the boost-glide concepts that had been under study since 1952, and resulted in an unusual vehicle configuration. The glider used a severely swept (75-degree) delta wing with a swept trailing edge and wing tips that could fold down to become vertical stabilizers during the relatively-slow portion of terminal flight. The glider carried a crew of one, and was equipped with a small rocket engine for use during high speed flight to change course and assist in deorbit (i.e., landing sooner or later than would naturally occur). Two turbojet engines for use during slow-speed flight and landing were also carried. The weapon had its own set of conventionally-shaped delta wings and was carried externally attached to the back of the glider with a large fairing that extended the glider's wings to meet the leading edge of the weapon's wings. A similar faring extended the fuselage contours so that the glider-weapon appeared to be a single 105-foot long vehicle at launch. The missile was equipped with a rocket engine that acted as an additional propulsion stage after the boosters had been jettisoned.[27]

The glider weighed 22,900 pounds and carried 11,150 pounds of fuel and coolant, plus a single pilot and his equipment. The missile weighed 25,517 pounds, and the fairings added another 5,533 pounds for an all-up weight of 65,100 pounds. The glider had 1,470 square feet of effective wing area when the tips were in the up position – the missile added 880 square feet while it was attached, and the fairing provided another 486 square feet.[28]

Several different boosters were studied early in the program including two- and three-stage boosters using LO2/RP-1, fluorine (F2) and hydrazine (N2H4), and nuclear rockets. The difference in efficiency of these boosters was telling – all were sized to carry a 65,000 pound payload, but the LO2/RP-1 booster was 190 feet long and had a gross lift-off weight (GLOW) of 4,085,500 pounds; the 2-stage all-nuclear system was 195 feet long but had a GLOW of only 442,750 pounds; while the three-stage N2/N2H4 booster was only 80 feet long and weighed 888,880 pounds.[29]

The booster finally selected used a unique arrangement. The first stage was a cluster of four solid rockets with a total weight of 1,027,800 pounds, with a liquid-fuel second stage powered by F2/N2H4 engines that weighed 356,600 pounds. The glider/missile combination sat atop this stage and used the missile's F2/N2H4 engine as a third stage.[30]

PASSENGERS IN SPACE

During late-1957, Bell's Dornberger and Ehricke began to collaborate on a two-stage passenger-carrying version of the BoMi concept. The stages were to be mounted in piggyback fashion, with the lower stage having five rocket engines, and the passenger stage having three. Initially, each stage was envisioned as a straight-winged vehicle, but both stages eventually emerged with delta-planforms. It

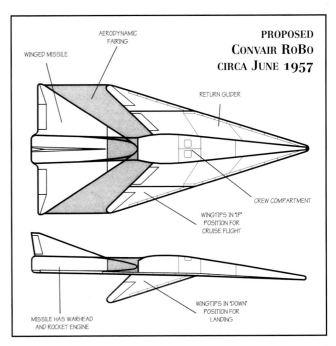

Convair's RoBo effort was accomplished at the Fort Worth Division, not San Diego, and the approach was certainly unusual. A large winged missile was mounted behind the manned boost-glider and integrated into the vehicle with a large aerodynamic fairing. Both the missile and fairing were dropped at the target, leaving the boost-glider to streak for home. As could be expected, a very similar-appearing vehicle would be proposed as the Dyna-Soar III two years later. (Convair via the AFHRA Collection)

was anticipated that the vehicle would take-off vertically with both stages firing until 130 seconds after launch when the lower stage would separate and glide back to land. The passenger stage would continue, completing a 3,000 mile flight in about 75 minutes at an altitude of 150,000 feet. The passenger stage seated 20 at roughly double the regular airfare, and the vehicle would maintain at least 0.25-g during the entire flight to ensure passenger comfort. Although there was some initial interest from various airlines, none were willing to commit large amounts of development funds on such a radical and untried concept, especially when jet travel itself was not well proven.[31]

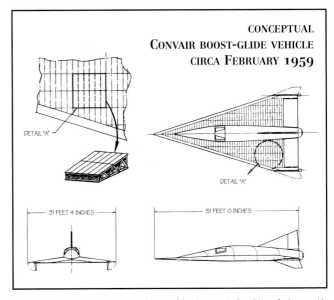

This was a 'typical boost-glide vehicle' used by Convair in San Diego during an Air Force contract for the design, fabrication, and testing of refractory alloy airframe components intended for boost-glide vehicles. Under this contract, Convair tested structures made of several molbdenum and niobium superalloys at temperatures up to 2,500 degF. This data was in direct support of the RoBo and Brass Bell programs, and apparently the tests were successful. (Convair via Bob Bradley)

Nevertheless, in March 1960 Bell Aircraft announced plans for a hypersonic passenger transport that could be operational in the mid-1980s. The first stage was a conventional-looking delta-winged aircraft powered by six large air-breathing engines that would operate in three different propulsion cycles – normal turbojet up to 50,000 feet; a transitional jet/ramjet phase; then full ramjet propulsion to 120,000 feet and Mach 5.2. A sub-orbital 'aerospacecraft,' derived from the S-BoMi designs, would then ignite its engines, launching down rails embedded in the back of the first stage vehicle. The aerospacecraft would have a maximum altitude of 210,000 feet, and reach velocities of 15,000 mph. Auxiliary turbojet engines allowed maneuvering during the landing phase, which would take place at any conventional airport. The concept died before any serious design work took place.

THE SLEEK AND WINGED SPACECRAFT CALLED DYNA-SOAR

On 4 October 1957, the Soviet Union orbited the first Earth artificial satellite – *Sputnik*. The shock to the western governments was profound, and less than a week later, the Air Force consolidated Brass Bell, RoBo, and HYWARDS into a single three-step development program called Dyna-Soar (for Dynamic Soaring – as Eugen Sänger had referred to his skipping reentry technique).[*]

Separately, the NACA Hypersonic Research Steering Committee met on 15 October 1957 at NACA Ames to determine the direction of the 'Round Three' research airplane, and three different approaches to manned space flight were proposed. A minority, led by Maxime A. Faget from NACA Langley, argued for a purely ballistic Allen-type blunt reentry shape – essentially what later emerged as the capsule for Mercury and Gemini. Another minority favored the 'lifting-body' approach of tailoring the design of a blunt reentry shape to provide a modest lift-to-drag ratio[†] which permitted limited maneuvering during the reentry profile. The remainder of the conference endorsed the concept of a flat-bottom hypersonic glider as described by John Becker from NACA Langley.

On 21 December 1957, the ARDC issued System Development Directive 464L for 'Step I' of Dyna-Soar, a small single-seat hypersonic boost-glide demonstrator intended to support future weapons system development. The directive set July 1962 as the target date for the first atmospheric flight test of the Dyna-Soar I (DS-I) vehicle.

The goals of the first step in Dyna-Soar development was to build a 'conceptual test vehicle' to obtain data in a flight regime significantly beyond that of X-15, while also providing a means to evaluate various military subsystems. A velocity of 12,250 mph and an altitude of 170,000 feet using the booster selected for HYWARDS were anticipated. This phase of Dyna-Soar also incorporated the Round Three NASA hypersonic research program goals, established during the October meeting at Ames. The second step of Dyna-Soar would involve the same high-altitude reconnaissance objectives as the earlier Brass Bell program. A two-stage booster would propel the vehicle to 15,000 mph at an altitude of 350,000 feet, enabling it to glide 5,750 miles. The system would be capable of providing high quality photographic and radar intelligence information, and would also be capable of performing limited bombardment missions. The final 'Step III' vehicle incorporated most of the capabilities previously envisioned for RoBo, and encompassed a more sophisticated vehicle that would be capable of orbital flight while also performing strategic reconnaissance and bombardment missions.[32]

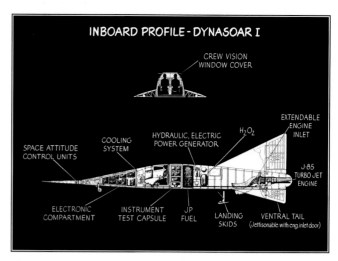

This was the General Dynamics proposal for Dyna-Soar as of 26 March 1958. The proposal was heavily based on the final RoBo (SR-126) 'semi-satellite rocket bomber.' Convair estimated the cost of Phase I at $22,118,000, Phase II at $406,457,000, and Phase III at $174,325,000. Both DS-II and DS-III had military capabilities, including the ability to carry nuclear bombs and various reconnaissance systems. A new 100-foot resolution sighting radar was also proposed. (Convair via the Bob Bradley Collection)[56]

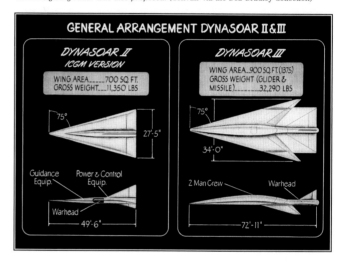

Because of the amount of technical advances necessary to develop Dyna-Soar, the ARDC reasoned that it was not possible to proceed directly into building the Step I vehicle. Therefore, a two-phase program was initiated where the first phase would involve validating various assumptions, theories, and gathering data from previous boost-glide programs such as BoMi and RoBo. The primary results expected from this phase included a better definition of the exact flight profile a boost-glide vehicle should fly. The second phase would refine a preliminary vehicle design, establish more precise performance requirements, and define subsystems and research instrumentation. While this two-phase approach would consume the better part of 12–18 months, it was expected that preliminary work on the Step II and III requirements could be undertaken in parallel. The initial schedule showed that testing the Step I vehicle at near orbital velocities would begin in 1966, followed by Dyna-Soar II (Brass Bell) in 1969, and Dyna-Soar III (RoBo) in 1974.[33]

By 25 January 1958, the Air Force had screened a list of 111 potential bidders for the conceptual demonstrator,

---

* Other sources have identified this as short for 'Dynamic Ascent and Soaring Flight.'
† Interestingly, later NASA capsule designs had substantial increases in their lift-to-drag ratios. In fact, Apollo, with an L/D of 0.8:1 rivaled some of the early lifting-bodies in cross-range capability, although it was never used operationally.

and ten aerospace companies were selected to receive RFPs, including Bell, Boeing, Chance Vought, Convair, Douglas, General Electric, Lockheed, Martin, North American, and Western Electric. Later, three additional large aerospace contractors, McDonnell, Northrop, and Republic, were also added to the list, although several teaming agreements resulted in only nine actual contenders for the research and development contract.[34]

The various proposals were submitted to the source selection board in March 1958, and represented two basic approaches to early orbital flight. The first, dubbed the 'satelloid' concept, used a glider that would be boosted to a velocity of 17,400 mph and an altitude of 400,000 feet, thereby achieving global range as a satellite. The second approach used a glider with a slightly higher L/D that could circumnavigate the Earth using a boost-glide profile after being boosted to only 300,000 feet.

Three of the contractors offered the first approach, with Lockheed proposing a 5,000-pound delta-wing glider, but the proposed Atlas booster lacked sufficient throw-weight to achieve global range. North American dusted off the two-seat delta-wing X-15B with a unique liquid-fueled stage-and-a-half booster using an expendable fuel tank, an indication of things to come for Space Shuttle. Republic proposed a 16,000 pound delta-winged glider powered by a three-stage solid propellant booster and capable of carrying a single large space-to-ground missile. Interestingly, the X-15-derived vehicle was the only one offered by any competitor that did not use a delta planform.[35]

The other six contractors opted for the high L/D boost-glide concept. Boeing and Vought joined forces to offer a small 6,500 pound delta-winged glider utilizing a cluster of Minuteman motors to form a booster stage. The Boeing-Vought glider was only capable of carrying a 500 pound payload, which included the pilot and his pressure suit, leaving little capacity for anything else. An 11,300 pound delta-wing glider incorporating air-breathing landing engines was proposed by Convair, although somewhat interestingly,[*] no booster system was investigated. The Douglas design used a 13,000 pound arrow-wing glider boosted by three modified Minuteman stages burning in parallel. The addition of another stage would allow the design to achieve orbit, but the initial glider did not contain the proper life-support systems for prolonged orbital flight. The Martin-Bell team designed a 13,300 pound glider with a two man crew that used a modified Titan ICBM as a launch vehicle. McDonnell also offered an arrow-wing design, but opted for a modified Atlas booster. Northrop offered a 14,200 pound glider launched by a hybrid solid rocket booster that used a liquid oxidizer.[36]

While the initial competition among the contractors was ongoing, the Air Force and NACA consummated an agreement outlining the civilian agency's participation in the Dyna-Soar program. The agreement was signed on 20 May 1958 and stated that NACA would provide "... technical advice and assistance ..." while the Air Force would handle funding, management, and program direction. The support agreement was reaffirmed by the new NASA on 14 November 1958.

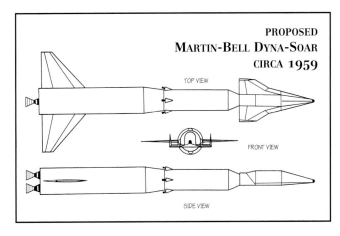

PROPOSED
MARTIN-BELL DYNA-SOAR
CIRCA 1959

The Martin-Bell Dyna-Soar proposal used a modified Titan ICBM as a booster. Forty-foot wings were added to assist in counteracting the lift generated by the glider's wing-body. The Dyna-Soar glider itself bore a strong resemblance to the vehicle that would evolve from the winning Boeing-Vought team. (Martin Marietta Corporation)

This technical advice had actually begun before the source selection board met in March 1958. John Becker from NACA Langley had recommended the rejection of the high L/D ratio, structurally complex, water-cooled glider proposed earlier for HYWARDS. Instead, he favored a small, relatively simple, radiative-cooled shape with a hypersonic L/D of 2:1, increasing to 5:1 at subsonic speeds. Such a vehicle, in Becker's opinion, could be developed more quickly and with less technical risk, and also greatly ease the problem of finding a booster with adequate throw capability. The basic vehicle could serve equally as well as an advanced prototype for the boost-glide weapons system, and also for a future maneuverable, landable, space reentry system. This was at odds with Allen at NASA Ames who believed a hypersonic L/D of at least 5.0 was required for his high-wing concept.

Of the competitors, only the Boeing-Vought[†] and Martin-Bell[‡] teams actually attempted the design of a true orbital spacecraft, with the others envisioning some form of hypersonic research vehicle that could eventually spawn an orbital spacecraft. Not content to pursue this intermediate approach, on 16 June 1958 the Air Force funded the Boeing-Vought and Martin-Bell teams to perform more detailed studies. As it happened, only Boeing-Vought offered a small vehicle along the lines proposed by Becker – however, the design contained several flaws, mainly a use of aerodynamic design features that aggravated the

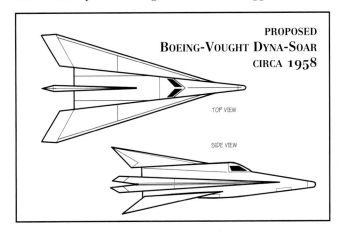

PROPOSED
BOEING-VOUGHT DYNA-SOAR
CIRCA 1958

The original Boeing-Vought Dyna-Soar proposal had a single large centrally-located dorsal stabilizer and two canted ventral fins located mid-way on each side of the wing-body. This conceptual design was substantially modified during the development phase, and the final vehicle bore no resemblance to this proposal. (U.S. Air Force)

---

* Interesting because Convair was the prime contractor for the Atlas ICBM proposed as a booster by several other competitors. The Atlas was rather quickly phased out of service as an ICBM, but was used for the later Project Mercury flights, and continues to be an important space-launch vehicle.

† Boeing, Chance-Vought, Aerojet, General Electric, Ramo-Woodridge, and North American Aviation.

‡ Martin, Bell, American Machine and Foundry, Bendix, Goodyear, and Minneapolis-Honeywell.

EVOLUTION OF THE
BOEING DYNA-SOAR
CIRCA 1959–62

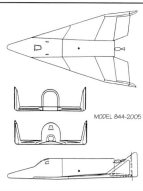

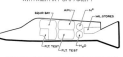

MODEL 814-1047-1
EARLY SUBORBITAL GLIDER
WITH MILITARY CAPABILITY

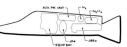

MODEL 814-1047-2
LOW ALTITUDE TEST AND
LANDING DEVELOPMENT GLIDER

- MANNED ORBITAL GLIDER FOR RESEARCH
  & SYSTEMS DEVELOPMENT
- MAXIMUM 684-POUND WARHEAD
  CAPABILITY IN MILITARY STORES BAY
- SEVEN VERNIER ROCKETS WITH
  10-150 FPS ΔV CAPABILITY
- 1,100-POUND FLIGHT TEST EQUIPMENT
  CAPABILITY

- TWO 2,600 LBS. EACH J-85
  AIR-BREATHING TURBOJET ENGINES
- 470-POUNDS OF JP-4 FUEL PROVIDING
  6 MINUTES AT FULL MIL. POWER
- APPROXIMATELY 16-CUBIC-FEET OF
  CONDITIONED VOLUME PROVIDED IN
  EQUIPMENT BAY FOR UP TO 400-POUNDS
  OF FLIGHT TEST EQUIPMENT

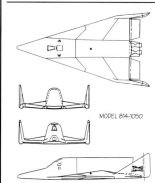

MODEL 814-1050

MODEL 844-2005

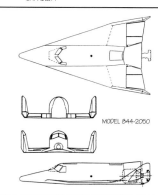

MODEL 844-2050

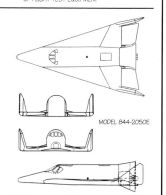

MODEL 844-2050E

The humor of the program name was not totally lost on the Air Force, evidenced by the dinosaur drawing found in the archives at the Air Force Historical Research Agency – note the green alien in the saddle. The Model 814-1047 shown at top was one of the earliest 'final' designs, and both the suborbital (-1) and low-speed (-2) variants are shown. Note the capability for a 684-pound nuclear weapon in the suborbital version, a clear indication of its potential military use – it is ironic that this payload was almost identical to the 660-pound load on Eugen Sänger's original Amerikä Bomber. The lower series consists of the last of the Model 814 variants (-1050) from March 1959, followed by the 844-2005 variant proposed during Phase Alpha in July 1959. This model had a thinner wing to eliminate pitching moment concerns, and larger aerodynamic control surfaces. The 844-2050 was the model shown at the full-scale mockup review in September 1961 and featured a revised windscreen and rear fuselage. The 844-2050E was the model that was entering production as the program was cancelled. The ramp on the aft fuselage was enlarged to cure a stability problem uncovered during wind-tunnel tests. (U.S. Air Force via the AFHRA Collection)

reentry heating problem. The Martin-Bell team initially used a higher-L/D mid-wing design that also would have serious heating problems, but the source selection board considered Bell's experience with hot-structure design from the earlier BoMi and System 118P efforts sufficient to overcome the potential risks. Both teams were subsequently briefed by Becker and other NASA researchers on the results of the various Langley and Ames studies.

In consideration of anticipated funding constraints, the Air Force completed a preliminary development plan that supplanted the three step approach that had been approved in October 1957. The program now consisted of two phases, the first of which was to evaluate aerodynamic characteristics, pilot performance, and subsystem operation of a military test vehicle. Interestingly the term 'experimental prototype' was directed to be used in place of the earlier 'conceptual test vehicle.'

To accomplish these goals, both Boeing-Vought and Martin-Bell envisioned Dyna-Soar as a manned flat-bottom glider with a highly-swept delta-wing weighing between 7,000 and 13,000 pounds capable of a velocity of 17,000 mph at 300,000 feet. Assuming a March 1959 approval of the preliminary development plan, the Air Force believed that unpowered air-drop tests of the DS-I vehicle could begin in January 1962, followed by manned suborbital test flights in July 1962 with full orbital operations in October 1963. Weapons systems studies would be conducted concurrently with the vehicle development, with the initial operational capability of an armed Dyna-Soar II set for late-1967. The DS-II vehicle was expected to be used for reconnaissance, air defense, space defense, and bombardment missions. Eventual weapons envisioned for DS-II included space-to-space, space-to-air, and space-to-ground guided missiles, as well as conventional bombs.[37]

On 23 April 1959, the DoD Director of Research and

Engineering, Dr. Herbert F. York, attempted to establish new goals for Dyna-Soar I – providing non-orbital research into hypersonic flight at velocities up to 15,000 mph. No new booster derivatives would be developed; instead the vehicle was to be launched on an existing Air Force or NASA booster. Secondary objectives, given time and funding, were the testing of various military subsystems and the attainment of orbital velocities. Not surprisingly, the Air Force did not endorse these objectives. In order to formalize the conditions established by GOR-92, on 7 May 1959 the ARDC issued System Requirement 201 to replace the SR-126 that had been issued for RoBo in 1956. The SR-201 firmly established the purpose of Dyna-Soar I to "determine the military potential of a boost-glide weapon system and provide research data on flight characteristics up to and including global [orbital] flight." This controversy over what Dyna-Soar was supposed to represent would continue to plague the program, and eventually lead to its demise.[38]

On 29 October 1959, yet another review of Dyna-Soar requirements was begun, and again Air Force officials expressed the need for an orbital vehicle with real military capabilities. By 1 November, a revised development plan had been completed, returning to the three step approach. Step I was a manned glider ranging in weight from 6,570 to 9,410 pounds to be launched to suborbital velocities by a modified Titan I booster. The second step involved using a Titan II booster to attain orbital velocities and some sort of limited military capability. Step III was a full orbital weapons system using a Titan III booster. It was now anticipated that the first of 19 air-drop tests would occur in April 1962, with July 1963 marking the first unmanned suborbital flight. Eight manned suborbital flights were scheduled to begin in May 1964. Total cost was expected to be $493.6 million, including the development costs of the Dyna-Soar vehicle. The first manned orbital flight as part of Step II would be launched from Complex 40 at Cape Canaveral Air Force Station during August 1965. This plan was approved by the Air Force Weapons Board on 2 November 1959.[39]

The Boeing-Vought team was declared the winner of the Dyna-Soar competition on 9 November 1959, with contract AF33(600)-39831 signed on 11 December 1959.[*] At the same time, Martin Marietta received a contract to develop a man-rated version of the Titan II booster. Dyna-Soar was officially designated System 620A on 17 November 1959. Subsequently, on 27 April 1960, the Air Force ordered ten 'production' Dyna-Soar vehicles, and assigned them serial numbers 61-2374/83. The procurement schedule called for two vehicles to be delivered during 1965, four to be delivered during 1966, and two during 1967. The other two airframes were to be used for static tests and unmanned drops, and most probably would have been completed in 1965. This was followed on 6 December 1960 by two more contracts – one to Honeywell for the primary flight system, and to RCA ten days later for the communication and data links.[40]

Phase Alpha

The subsequent effort to develop the Dyna-Soar vehicle consumed the better part of two years. The first phase of this work was, logically enough, known as Phase Alpha. A myriad of designs had been explored by the proposal contractors, and an ad hoc committee known as the 'Alpha Group' was established with the responsibility of comparing engineering data and designs relating to the definition of the orbital Dyna-Soar vehicle. The vehicle that eventually emerged[+] was a slender delta planform with a rounded and slightly upward tilting nose, and twin endplate vertical stabilizers. The airframe consisted primarily of an exotic Renè-41 superalloy hot-structure with a molybdenum superalloy heat shield on the lower surface. It was felt that this would provide protection for the 10,000 pound vehicle to approximately 2,700 degF. The wing leading edges were coated molybdenum segments that could withstand temperatures up to 3,000 degF, while the severest reentry heating, up to 4,300 degF, would be experienced by the reinforced graphite and zirconia rod nose cap.[41]

During early-1960, the Air Force announced a series of reentry tests using a recoverable RVX-2[‡] nose-cone. The RVX-2 vehicle was scheduled to be boosted by an Atlas ICBM to speeds of about Mach 22, and would provide critical heating and aerodynamic tests of some Dyna-Soar subsystems. LaRC's Becker and McLellan came up with a rather unlikely constellation of tests that hung off the otherwise conical RVX-2 vehicle. About half the proposed tests related directly to the Dyna-Soar I vehicle, while the rest were more general in nature, including a number of

---

* Vought's involvement with the Dyna-Soar program eventually dwindled to the design and fabrication of the high-temperature nose-cap. This work would later serve Vought (later LTV) well when asked to develop the carbon-carbon nose cap and wing leading edges for Space Shuttle.

+ Somewhat ironically, the vehicle bore a greater resemblance to the Martin-Bell proposal than it did to the winning Boeing-Vought design.

‡ The original RVX (or RVX-1) was the first ablative-protected ICBM nose-cone recovered after a long-range flight.

§ The term 'air-launched' was substituted for 'air-drop' when a rocket engine was added to allow evaluation of the vehicle at higher velocities than possible in unpowered flight. Initially this engine would be an existing powerplant, most probably the XLR11 used in the X-1, but when the retro-package was added to the Dyna-Soar design it was felt that a modified version of it could be used for the air-launched flight – thereby providing some operational data on the retro-package itself.

Boeing model showing Dyna-Soar under the wing of a B-52C. The support equipment shown was generally similar to that used by the X-15 project. (Jay Miller Collection)

lifting-reentry shapes and inlet configurations, as well as several afterbody designs. Unfortunately for the Dyna-Soar team, the Air Force later cancelled the RVX-2 flights due to funding constraints, although Becker tried unsuccessfully to continue them under NASA auspices.

Yet another development plan released on 1 April 1960 further elaborated on the three-step program presented in November 1959. The strictly suborbital Step I was now directed towards achieving four primary objectives – exploring maximum heating regions during reentry flight profiles, investigating maneuverability during reentry, demonstrating conventional landing techniques, and an evaluation of the ability of man to function usefully during prolonged hypersonic flight. The 20 air-launched[§] test flights were now scheduled to begin in July 1963 with the much-used XLR11 rocket engine providing velocities of Mach 2 at altitudes up to 80,000 feet. Beginning in November 1963, a series of five unmanned sub-orbital flights using Titan I boosters would be launched from Cape Canaveral towards Mayaguana in the Bahamas and Fortaleza, Brazil. These would be followed by 11 manned flights, progressively increasing velocity to 19,000 fps.

The second step had been subdivided into Step IIA, intended to gather data on orbital maneuvering velocities and military subsystems – and Step IIB, which was an interim operational system capable of performing orbital

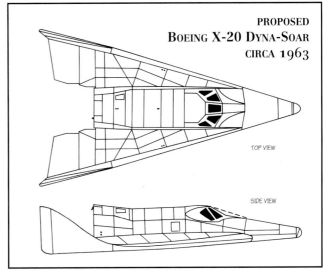

PROPOSED BOEING X-20 DYNA-SOAR CIRCA 1963

TOP VIEW

SIDE VIEW

The final Boeing Model 844-2050E X-20 Dyna-Soar configuration just before the project was cancelled in December 1963. (Boeing Military Airplane Company)

---

strategic reconnaissance and satellite inspection missions. A fully operational weapons system was the goal of Step III. At this point there was not a great deal of detail concerning these flights, except that they would most probably use a Titan III launch vehicle.[42]

On 26 April 1961, the Dyna-Soar office elaborated on the previous flight test plan – following 20 air-launched tests starting in January 1964, two unmanned launch tests would occur in August 1964, with the first of 12 manned suborbital flights in April 1965 on top of a modified Titan II. The first manned flights were expected to achieve speeds of between 11,000 and 15,000 mph. The plan also detailed follow-on tests, including one-orbit flights from Cape Canaveral to Edwards starting in April 1966. An interim operational vehicle, capable of fulfilling reconnaissance, satellite inspection, space logistics, and bombardment missions was anticipated by October 1967.

The complete weapons system, including new space-to-ground and space-to-space missiles, could be available by late 1971. Very shortly after this plan was released, funding constraints forced the combining of the goals for Steps I and IIA, although the actual flight plan did not change.[43]

However, the Air Force Space Systems Division (SSD) announced its own manned SAtellite INspector program, called SAINT II, on 19 May 1961. The proposed SAINT II demonstrator was a two-man lifting-body derived from the unmanned SAINT I satellite inspector vehicle* then under development by SSD. SAINT I involved an orbital system capable of identifying and destroying satellites in low-earth orbit, and the SAINT II proposal involved a

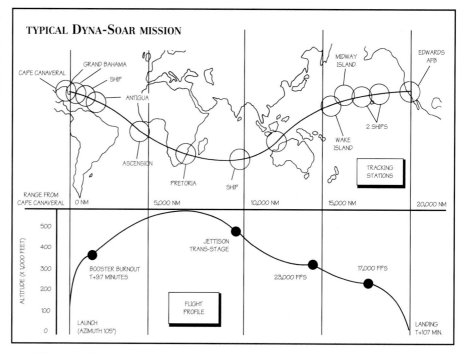

TYPICAL DYNA-SOAR MISSION

A typical once-around mission was envisioned as follows: the Dyna-Soar would be launched from Complex 40 at Cape Canaveral aboard its Titan IIIC launch vehicle, achieving orbital insertion 9.7 minutes into flight at an altitude of 320,000 feet and a speed of 16,670 mph (24,450 fps). The Dyna-Soar would remain at orbital altitude, performing its mission for approximately 12,000 miles, beginning its return to Earth approximately 13,000 miles downrange. Reentry would occur roughly 2,000 miles later at a velocity of 16,000 mph. The vehicle would land at Edwards AFB 107 minutes after launch, approaching the lakebed at 250 mph, touching down at 175 mph, with a run-out distance of 2,750 feet. (U.S. Air Force)

manned vehicle capable of performing precise orbital rendezvous and also of fulfilling limited space logistics missions. The lifting-body vehicle would be able to maneuver during reentry and accomplish conventional landings at predetermined sites. SAINT II was to be launched by a Titan II with a new fluorine and hydrazine powered upper stage called Chariot. Twelve manned orbital missions were scheduled, with the first unmanned flight occurring early in 1964, and the initial manned launches set for later that same year. SAINT II was to have both a low-earth and a mid-earth capability.†

SSD officials listed several reasons why the initial Dyna-Soar I configuration could not carry out the intended SAINT II missions, mainly its limited payload potential and its inability to obtain mid-earth orbits. They also felt that the reentry velocity of Dyna-Soar could not be significantly increased because of material limitations of the airframe, and the inability to adapt the configuration to ablative heat-shield protection. The estimated cost for SAINT II was $413,800,000 in FY62 through FY65, and the program was considered a serious contender for funds by the Dyna-Soar project office.[44]

Dyna-Soar Description

By the summer of 1961, Boeing had made significant progress on the basic design of the initial Dyna-Soar I glider. Wind tunnel research was continuing on configuration studies, and Boeing's various subcontractors were assisting in numerous material and subsystem tests. A full-scale mock-up at Boeing's Seattle facility was reviewed by the Air Force and NASA on 11 September 1961. During the mock-up inspection, the Air Force

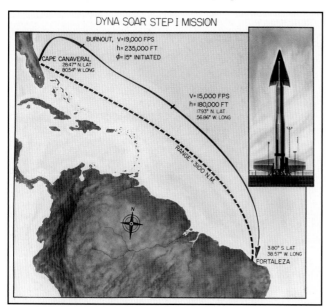

DYNA SOAR STEP I MISSION

This map shows the planned short-range test flights from Cape Canaveral to Fortaleza, Brazil that were originally planned for Dyna-Soar Step I. The vehicle would be traveling 13,000 mph at an altitude of 235,000 feet at engine burn out. (Tony Landis Collection)

* The unmanned SAINT I was cancelled in mid-1961, prior to its first flight.
† At the time, these were called near-earth (100–400 miles) and far-earth (400–1,000 miles) orbits; the modern terms are used for the convenience of the reader.

directed Boeing to equip the vehicle for multi-orbit, as opposed to single-orbit, operations. This required the installation of a more sophisticated guidance system, as well as the installation of a retro-package for de-orbit. Previously, a single-orbit Dyna-Soar mission did not require a de-orbiting system, largely because the flight profile was simply a ballistic around-the-world trajectory. Two different de-orbiting systems were examined, one with a small rocket engine in the glider's transition stage, the other with a new 16,000-lbf fourth stage added to the Titan III booster (creating a Titan IIIC). This fourth stage (called a 'trans-stage') could be used for precise orbital insertion during ascent, and then remain attached to the glider and restarted to provide a de-orbit burn. Sizing the stage correctly could also allow easier access to mid-earth orbits in the future. This latter method was subsequently chosen for the production Dyna-Soar vehicles.[45]

The final Dyna-Soar design was the result of over 13,000 hours of wind tunnel tests that included approximately 1,800 hours of subsonic, 2,700 hours of supersonic, and 8,500 hours of hypersonic time in almost every high-speed wind- and shock-tunnel in the United States. The wing planform ended up as a pure delta with a leading-edge sweep of 72.8 degrees. The glider had a hypersonic L/D of 1.5:1 and a hypersonic lift coefficient of 0.6 with an expected cross-range capability of 1,750 miles. The radiation cooled structure was designed to survive four flights. A maximum of 1,000 pounds could be carried in a 75-cubic-foot payload compartment.[46]

At this point, the Dyna-Soar glider was 35.3 feet long, spanned 20.4 feet, and had approximately 345 square feet of wing area. Aerodynamic control of the vehicle was provided by conventional trailing-edge elevons that were hydraulically actuated and could provide roll as well as pitch control. The Honeywell MH-96 self-adaptive flight control system, identical to the one that later contributed to the crash of the third X-15, was to be used. The glider weighed 10,830 pounds empty and 11,390 pounds all-up.

The Dyna-Soar was designed to be statically stable in the normal range of reentry and subsonic glide conditions. The wing section on the early Dyna-Soar configuration used a double-wedge upper surface and flat under surface that provided good hypersonic flight characteristics and simple manufacturing. But this design required the addition of retractable stabilizers for low speed flight, so the upper wing surface was modified to improve low-speed handling without compromising hypersonic stability. This modification, however, resulted in some transonic instabilities that were eliminated by the addition of an aft fuselage ramp that gave the final Dyna-Soar its distinctive appearance.

The internal structure of the glider differed greatly from conventional aircraft of the period, and consisted of a truss framework with fixed and pinned joints in square and triangular elements that looked similar to bridge construction. This truss framework was made of Renè-41 superalloy that could resist temperatures of up to 1,800 degF. A program was developed to expand the information base on Renè-41, and resulted in new techniques for the manufacture, welding, and extruding of this high-strength material.

The internal truss structure was covered by a series of Renè-41 panels to form the load-bearing airframe. Each Renè-41 panel was corrugated to add stiffness to the structure and allow expansion during reentry heating.

These panels also formed the inside layer of the Dyna-Soar heat shield. A layer of Dyna-Flex or Micro-Quartz insulation[*] covered the Renè-41 panels to protect them against heat transfer from the outer heat shield. Special attention was given to heat leakage at the expansion joints, access panels, landing gear doors, and hinge area at the elevon control surfaces.

The outer heat shield would have been made up of sections of D-36 columbium that were attached to the underlying Renè-41 panels using a stand-off clip design. Although the D-36 columbium would have less strength at high temperatures than the molybdenum chosen for the leading edges, it could be machined and welded more easily during manufacturing.

A major problem facing Boeing was potential oxidation of the refractory metals used in the heat shield. These special alloys would begin to oxidize and break down after exposure to high heat loads which could have led to structural failure. The answer was the development of an oxidation resistant silicide coating. A fluidized bed technique was developed to meet the production needs of Dyna-Soar, and was used to coat both D-36 columbium and TZM molybdenum panels. A final coating of Synar-silicon carbide applied over the silicide coating would give Dyna-Soar its distinctive black color. These coatings would have had to be replaced after each flight. but tests conducted on a four panel heat shield showed that this could be relatively easily accomplished.

A Dyna-Soar lifts-off from Cape Canaveral's Launch Complex 40 on top of a modified Titan II in this illustration. As the weight of the Dyna-Soar glider grew, the Titan II launch vehicle would give way to a Titan III, which included two large strap-on solid rocket motors, and finally to a Titan IIIC. Thought was also given to using a Saturn IB later in the program. (Boeing Military Airplane Company)

---

[*] Dyna-Flex is also known by the name Cerrachrome, while Micro-Quartz goes by the name Q-Fiber Felt. Both are a fibrous batt insulating material that continues to be used in a variety of high-temperature applications.

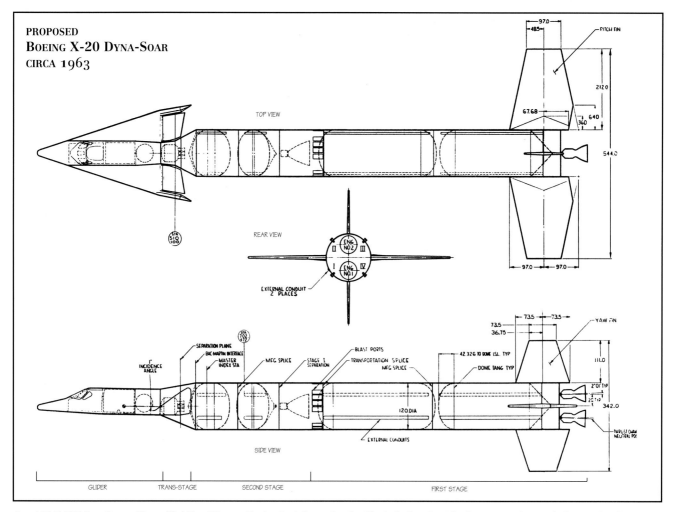

PROPOSED
BOEING X-20 DYNA-SOAR
CIRCA 1963

TOP VIEW

REAR VIEW

SIDE VIEW

GLIDER | TRANS-STAGE | SECOND STAGE | FIRST STAGE

A model 844-2050 Dyna-Soar and its modified Titan II booster. The drawing indicates that the glider had a launch weight of 10,000 pounds. Note the large pitch and yaw stabilizers that have been added to the Titan II, a vehicle that normally does not have any external stabilizers. (Boeing Military Airplane Company)

The two parts of the heat shield that would receive the highest heat levels would be the wing leading edge and the nose-cap. The leading edge components were constructed from TZM molybdenum – a half-titanium, half-molybdenum alloy with small amounts of zirconium added. Two different designs were undertaken for the nosecap, with the early Boeing-designed baseline becoming a contingency design after Ling-Tempo-Vought (LTV) proved to have a better concept. The LTV design that was chosen for production consisted of a siliconized graphite structure overlaid with zirconia tiles that were restrained by zirconia pins. In case of cracks in the structure, the tiles and pins were held in place by platinumrhodium wire. The Boeing contingency design used a single-piece zirconia structure reinforced with platinum-rhodium wire. During the molding process, shaped tiles were cast in the outside surface to allow thermal expansion and to control possible cracks from spreading.

The cockpit windows were the largest designed for a manned spacecraft at the time and presented a certain amount of concern within Boeing and the Air Force. With temperatures expected to reach 2,000 degF in the cockpit area, a D-36 columbium heat shield was designed to cover the forward three windows. The single window on each side would remain uncovered for the entire flight since they were not subjected to high heating rates. After reentry the heat shield covering the front windows would be jettisoned to allow the pilot increased forward visibility for landing. In the event that the heat shield did not jettison as

planned, NASA test pilot Neil Armstrong conducted tests with a modified Douglas F5D Skyray that demonstrated a pilot could still land with side vision only.

The crew compartment would have been a welded aluminum structure pressurized with a mixed oxygen-nitrogen atmosphere at 7.5 psi. The single pilot was accommodated on a Weber rocket-powered ejection seat that provided meaningful escape only at subsonic speeds. Interestingly, the seat was successfully tested on the rocket-sleds at Holloman AFB during 1964, after the X-20 had already been cancelled.[47] The glider was controlled by conventional rudder pedals and a side stick controller, and the seat was adjustable to different positions for boost, on-orbit, and reentry conditions.

The pilot would have faced an instrument panel similar to other contemporary research aircraft with one notable exception – the Energy Management Display Indicator (EMDI). Developed by General Precision Instruments, the display allowed the pilot to maneuver within the thermal and structural limit of the vehicle. The display was a 4-inch monochrome cathode ray tube with transparent overlays that moved along with the forward flight of the glider. The EMDI also displayed information that would have allowed the pilot to select contingency landing sites along the reentry profile. Six pilots had already been selected to fly Dyna-Soar – Captain Albert Crews, Major Henry Gordon, Captain William J. 'Pete' Knight, Major Russell Rogers, Major James Wood, and NASA research pilot Milton O. Thompson.[48]

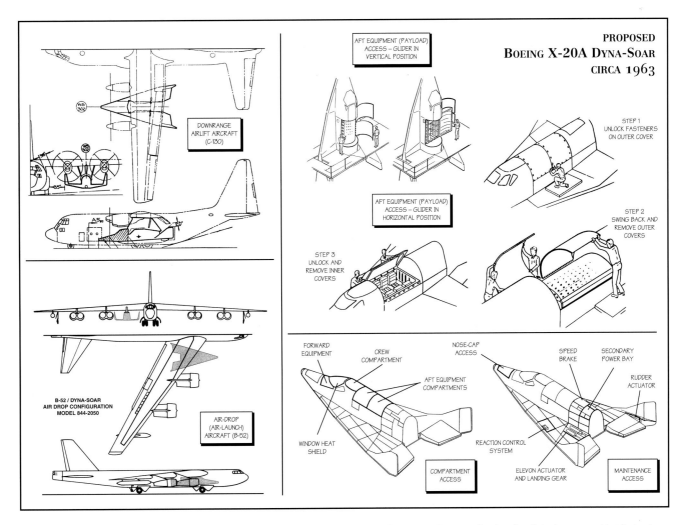

Various aspects of the Dyna-Soar program are shown. Much like the X-15, the X-20 would have used a Boeing B-52 as a carrier aircraft early in the program. Most Boeing drawings show either a G- or H-model B-52, but in actuality a B-52C was selected for modification, although it was never completed. A C-130 was also going to be used in case the X-20 landed at one of the downrange island contingency airfields. The aft equipment compartments (also called payload bays, although they could not be opened on-orbit, at least initially) could be accessed when the vehicle was either horizontal or vertical. (Boeing Military Airplane Company)

The Dyna-Soar glider would have used a Minneapolis-Honeywell inertial guidance unit that was essentially a modified version of the system used on the Atlas-Centaur launch vehicle. Twenty-four test flights were successfully completed aboard a McDonnell NF-101B at the Gulf Test Range located at Eglin AFB, and several production units were ready when the program was cancelled – these units were later flown aboard the X-15 aircraft with generally satisfactory results.

The onboard power for Dyna-Soar would have been provided by two Sundstrand auxiliary power units (APU) that were part of an integral power generation and cooling system that used liquid oxygen and liquid hydrogen to power the APUs and help cool onboard instruments. A Garrett cooling system used liquid hydrogen to extract heat from the cockpit and equipment bay. Redundant cooling loops transferred heat from the electrical generators and APUs to a hydrogen and glycol-water heat exchanger. Another development in the effort to keep the cockpit and equipment bay cool was the invention of the water-gel cold plate. This heatsink used a gel mixture of 95 percent water and 5 percent cyanogum-41 jelling agent routed through a series of wicks that provided proper distribution of the water-gel during ascent, on-orbit, and reentry phases.

Dyna-Soar used a three-point landing skid arrangement since conventional rubber tires on aluminum or steel wheels could not withstand the expected tempera-ture extremes in the landing gear bays. The nose and two main gear struts would have been constructed of Inconel-X, and the Goodyear-developed main landing skids resembled stiff wire brushes constructed of Renè-41 wire bristles wound around a series of longitudinal rods. The Bendix nose skid was a single-piece Renè-41 forging. Tests were conducted on concrete and asphalt runways with surprisingly good results, although all of the initial landings would be on the dry lakebeds at Edwards AFB.

The X-20A glider would have been boosted into orbit atop a Titan IIIC that was purpose-built for the Dyna-Soar program. The Titan IIIC consisted of a strengthened Titan II core with the addition of two 5-segment, 120-inch solid propellant boosters – and would go on to be a workhorse heavy-lift launch vehicle. The attached trans-stage could be used for precise orbital insertion during ascent, and then remain attached to the glider and restarted to provide a de-orbit burn. On-orbit control would have been provided by a Bell Aerosystems reaction control system similar to the ones used on the X-15 and Mercury spacecraft.

The X-20 would also carry an emergency escape motor located in the transition section which could provide emergency escape during most of the Titan IIIC boost phase. The Thiokol XM92 solid propellant four-nozzle engine produced 40,000 lbf for 13.4 seconds. This escape motor would also have been used to propel the X-20A to supersonic speeds during the later stages of the air-launch program.

But Why ?

On 7 October 1961 yet another plan to restructure Dyna-Soar was unveiled, this time including the development of a mid-earth orbit demonstration vehicle in addition to the low-earth vehicle originally envisioned. This plan eliminated the entire suborbital test phase and reduced the air-launched program to 15 flights. The Air Force anticipated the first unmanned orbital flight in November 1964, and the initial piloted orbital flight in May 1965 on top of a Titan IIIC. The next five flights would be piloted multi-orbital missions. The ninth test flight, scheduled for June 1966, would be an unmanned exploration of the velocities needed for mid-earth missions, and nine subsequent piloted flights would demonstrate potential military missions such as satellite inspection and orbital reconnaissance. The flight test program was to terminate in December 1967 with a total program cost of $921 million, but no specific production schedules or operational scenarios were included in the revised plan.[49]

Also in October 1961, an Air Force management group severely criticized the SAINT II program by insisting that the projected number of flight tests and the proposed funding limit were too unrealistic. This review killed the SAINT II effort, and as a result of the controversy, the Air Force prohibited further use of the SAINT moniker.

Secretary of Defense Robert S. McNamara endorsed the latest Dyna-Soar restructuring on 23 February 1962. Dyna-Soar was now officially a research and development program to explore and demonstrate maneuverable reentry of a piloted orbital glider which could execute precision runway landings at preselected sites on Earth. After considering various designations (including XJN-1, and XMS-1 for 'Experimental Manned Spacecraft – One'), the Dyna-Soar officially became the X-20 on 19 June 1962. Subsequently, the name Dyna-Soar was also made official. By this time it had become apparent that the availability of adequate funding for the Titan IIIC launch vehicle, which was appropriated separately, was rapidly becoming the major constraint to Dyna-Soar development. The first unmanned X-20 test was scheduled to occur on the fourth Titan IIIC development flight, but launch dates could not be determined until booster availability was ascertained. Funding for the Titan IIIC was formally approved by Congress on 15 October 1962, and shortly thereafter a revised launch schedule for the X-20 was released.[50]

But Dyna-Soar suddenly found itself with a new competitor – Blue Gemini. On 18 January 1963, Secretary McNamara directed a comparison study between the X-20 and NASA's Gemini to determine which represented the more feasible approach to a military capability. A major study item was to be Gemini's two man crew, versus the single-seat X-20. The Air Force had just signed an agreement with NASA to allow Air Force crews to fly on selected Gemini flights, but the Air Force Chief of Staff, General Curtis LeMay, insisted the purpose of Air Force participation in NASA Gemini was limited to obtaining experience and information concerning manned space flight. McNamara, however, was not convinced that the Air Force could demonstrate a pressing requirement for any manned orbital capability.[51]

Nevertheless, on 26 March 1963 Boeing was awarded a $358,076,923 supplemental contract for the continued design, manufacture, and testing of the X-20, although by this time there were persistent rumors of program cancellation. The contract covered the conversion of a B-52C (53-0399) for the air-launched test flights, and the modification of Launch Complex 40 at Cape Canaveral AFS to support the Titan IIIC with the Dyna-Soar glider attached. Neither was ever completed.

Meanwhile, Dyna-Soar was again having to restructure to meet funding reductions. While final budget figures were still pending for FY64, the impact of the October 1961 redirection was becoming more apparent. The first development plan had included definitive military objectives leading to the development of orbital reconnaissance and bombardment vehicles. These goals had been altered, and the major emphasis placed on the development of sub-orbital and orbital research vehicles. By mid-1963, the Department of Defense was seriously questioning the need for Dyna-Soar, and it appeared that the alternatives for the X-20 had been severely narrowed – direct the program towards achieving military goals, or terminate it.

A military test program was subsequently identified for Dyna-Soar, and would have consisted of six X-20A flights – four for testing the reconnaissance and satellite inspection equipment, followed by two operational satellite inspection demonstration missions. The cost to develop and test a militarized X-20A would total $228 million, in addition to the basic Dyna-Soar costs.[52]

The Air Force also completed a study concerning the use of a proposed X-20B for anti-satellite missions. Two flights would be added to the basic X-20 program to demonstrate the operational capability at an additional cost of $227 million. To conduct a 50-flight operational Dyna-Soar program would cost $1,200 million during fiscal years 1965–72. A mid-earth satellite inspection version, designated X-20X, had a 14-day endurance with a two-man crew and could inspect targets as high as 1,000 miles. The initial flight of the X-20X was projected for September 1967 and would need $350 million in additional funding.[53]

Although military missions for the Dyna-Soar were finally identified, convincing Washington that they were valid proved more difficult. A military presence in space could be more rapidly, and much more economically, achieved by participating in NASA Gemini. For instance, minor changes in equipment and flight profiles, costing only $16.1 million, could allow testing of military subsystems aboard Gemini during long-duration (≈14 days) missions. The main advantage to Gemini was that it was lighter than the X-20 and consequently could carry more fuel for orbital maneuvering, or a larger payload, while using the smaller, and substantially less expensive Titan II

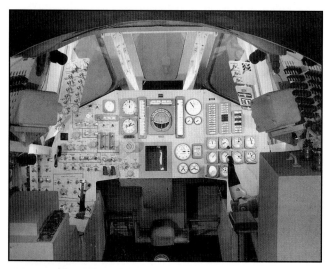

The final version of the X-20 cockpit as displayed at the Air Force Association convention in Las Vegas during 1962. (Jay Miller Collection)

booster. The inherent advantage of the Dyna-Soar vehicle was its maneuverability during reentry which meant that it could return to base quicker, and also change landing sites during reentry if needed. Dyna-Soar also could potentially perform true military missions, while Gemini would always be limited to subsystems testing.[54]

The Air Force continued to argue that both programs should be allowed to continue. However, when Deputy Defense Secretary Harold Brown recommended a permanently manned military space station, serviced by modified Gemini capsules, the X-20 had been dealt a death blow, and in the end, neither Dyna-Soar or Blue Gemini came to be.

Cancelled Now, Forever More

On 10 December 1963, Secretary of Defense McNamara cancelled Dyna-Soar in favor of model testing via the ASSET program, and reappropriated the X-20 funding to the Manned Orbiting Laboratory (MOL).* Thus ended the first serious attempt to build a reusable manned orbital spacecraft. At the time of its cancellation, Dyna-Soar was three years from its first flight – $410 million had already been spent on its research and development, and another $373 million was expected to be spent prior to first flight. As an aside, although NASA had supported the Air Force's participation in the Gemini project, it in no way concurred with the decision to cancel Dyna-Soar. NASA Associate Administrator for Advanced Research and Technology, Dr. Robert L. Bisplinghoff, pointed out that advanced studies had repeatedly shown the importance of developing the technology for maneuverable hypersonic vehicles with high-temperature, radiative-cooled metal structures. Ground-based test facilities were unable to accurately simulate this lifting reentry environment, and consequently, the Dyna-Soar flights were necessary to provide such data. Even if Dyna-Soar had never flown an operational mission, it would have provided extremely valuable information on reentry flight control and heating problems, something that was seriously lacking during the development of the Space Shuttle ten years later.[55]

---

* The MOL program itself would also be cancelled before its first flight. It is ironic that the launch complex (SLC-6) constructed at Vandenberg AFB to support the MOL program was later modified to serve as the west coast Space Shuttle launch site. The STS-33/51-L *Challenger* accident caused the Air Force to mothball and later cancel plans to use the launch complex. During 1990 it was proposed to rebuild SLC-6 to support Titan IV, a variation of the booster originally intended for Dyna-Soar and MOL. This was also cancelled prior to completion. To date, over $7,000 million has been spent on the facility to support various projects, yet the only launches to take place from it have been several small Lockheed Launch Vehicle (LLV) boosters. The complex is currently being modified by Boeing to support the Delta IV EELV program.

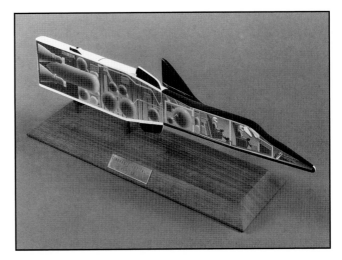

An 'Advanced Orbital Vehicle' version of Dyna-Soar is shown in the cutaway model. Note the two man crew and the configuration of the transition section. A Genie-derived abort solid rocket motor was located above and below the front of the transition section, replacing the original Thiokol XM92 motors inside the transition section on the X-20 – freeing up room for other systems. (Boeing History Office photo 2A130260)

It was unusual to see the full-scale mockup without the front windscreen heat shield – this gives a good opportunity to see the shape of the front glass. (Jay Miller Collection)

This mockup of the final X-20 glider was displayed at various trade shows and other events. Although the skip-glide concept had long been abandoned, the display stand still shows the characteristic 'skips' described by Eugen Sänger along with the description 'piloted research aerospacecraft.' (Tony Landis and Jay Miller Collections)

R. Dale Reed holding a model of the M2-F1. (NASA photo E-16475)

OF BLUNT SHAPES AND LIFTING-BODIES

During 1951 H. 'Harvey' Julian Allen developed the concept of a blunt-body reentry shape while working at NACA Ames. At the time, the popular conception of a spacecraft was along the lines of the German V2 – a slender cylinder with a pointed nose and sweptback fins. Experience showed that this shape would develop a strong 'attached' shockwave streaming from the nose as it reentered the atmosphere, with the associated high heating eventually melting the structure. Allen, together with colleague Alfred Eggers, demonstrated that a blunt object would develop a 'detached' shockwave that would carry most of the heat load away from the body with it, and that residual heating would be well within the capacity of existing materials technology to deal with.[1]

Allen's work was highly classified at the time and was first put to use on the nuclear warheads for the Atlas and Titan ICBMs. By mid-1958 Maxime A. Faget and other researchers at NACA Langley concluded that "… the state of the art is sufficiently advanced so that it is possible to proceed confidently with a manned satellite project based upon the ballistic reentry type vehicle." Faget estimated that the blunt-body could successfully withstand reentry, but would subject the astronaut to over 8.5-g. While not considered ideal, slightly modified versions would form the basis for the capsules of Project Mercury.[2]

With this discovery one of the major stumbling blocks to developing a useful reentry shape disappeared. But Allen's symmetrical blunt bodies had one very significant military drawback – they had extremely low L/D ratios and followed generally ballistic flight paths that did not permit much variation in landing site. Fortunately, other researchers at Ames, Clarence Syvertson, George Edwards, and George Kenyon, realized that the blunt body shape could be modified to generate a modest lift-to-drag ratio (on the order of 1.5:1) that enabled potential landing sites to be several hundred miles cross-range. This was particularly important to polar operations, where the natural rotation of the Earth moves the launch site considerably cross-range during even a short orbital flight – high cross-range allows a vehicle to land at its launch site after a single orbit.

This concept was first presented in a paper presented by Thomas J. Wong at the same conference when Max Faget detailed his blunt-body capsule design. A 30-degree half-cone wingless reentry vehicle was shown to have sufficiently high lift and drag coefficients that the maximum deceleration during reentry would be limited to approximately 2-g. This configuration also allowed a lateral reentry path deviation of ±230 miles, and a longitudinal variation of 700 miles. Unfortunately, this vehicle would weigh substantially more than the pure blunt-body shape, and the throw-weight of the Redstone and Atlas boosters being used for Project Mercury did not allow such luxuries.[3]

This greatly interested the Air Force, which subsequently funded several small research programs, including one with The Aerospace Corporation.* The Air Force had several reasons for its interest. First, a maneuverable ICBM warhead would greatly reduce the effectiveness of any anti-ballistic missile (ABM) system. Secondly, the ability to select a landing site away from the ground track of a photo-reconnaissance satellite would allow more opportunities for a film-return vehicle to be recovered.

There were other perceived benefits from the lifting-body design. For any given planform size (wing area), the lifting-body offered a large internal volume when compared to a capsule or wing-body shape. But this would prove misleading, as later events would show – although the volume of the vehicle tends to be large, it is composed of many areas with complex curves that do not necessarily lend themselves to being efficiently filled with normally square (cube) equipment. Even propellant tanks presented a challenge since they needed to be odd shapes.

In short order, four very different lifting-body shapes were proposed, including the Ames M1, Langley HL-10, Langley 'lenticular,' and The Aerospace Corporation A3. Each one of these proposals represented a unique and distinctive design approach.

The Ames M1 began as a modified 13-degree half-cone, flat on the top, with a rounded nose to reduce heating and an L/D of 0.5:1 at hypersonic speeds (although this later increased to 1.4:1). The design did have one major failing, however – at subsonic speeds it demonstrated a pronounced tendency to tumble end-over-end. In fact, the shape had virtually no subsonic L/D and could not land horizontally. Eventually the Ames' researchers discovered that most of the stability problems were cured by modifying the aft end with body flaps that looked much like a badminton shuttlecock, creating the M1-L.[4]

This design was further modified in late 1958 when the half-cone shape was gradually expanded in an effort to improve stability – creating the M2 shape. Soon a protruding canopy, twin vertical stabilizers, and various control

---

* The Aerospace Corporation, headquartered in El Segundo, California, is a not-for-profit company chartered to provide technical consulting to the Department of Defense, in particular, the Air Force. The contract referenced here was in addition to on-going evaluations conducted by Aerospace on all DoD and NASA shuttle studies, as well as other space-related efforts. The Aerospace Corporation is still deeply involved in most aspects of the DoD space program.

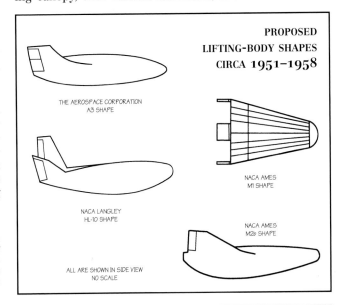

PROPOSED
LIFTING-BODY SHAPES
CIRCA 1951–1958

THE AEROSPACE CORPORATION
A3 SHAPE

NACA LANGLEY
HL-10 SHAPE

NACA AMES
M1 SHAPE

NACA AMES
M2b SHAPE

ALL ARE SHOWN IN SIDE VIEW
NO SCALE

surfaces were added, leading to it being called the M2b 'Cadillac.' This is the M2 shape most are familiar with.[5]

Langley's HL-10 (the 'HL' stood for horizontal lander), developed by a team headed by Eugene S. Love, was a more traditional shape with a flattened and rounded delta-body and sharply upswept tips with a single centrally mounted vertical stabilizer.

Another shape proposed by Langley was dubbed the 'lenticular.' This shape transitioned to horizontal flight by extending control surfaces after following a reentry profile similar to a symmetrical capsule. Donna Reed, the wife of aerospace engineer R. Dale Reed at NASA FRC, called the shape a 'powder puff' – others called it a flying saucer.[6]

The A3 (later called the SV-1 by Martin) was developed by a team led by Frederick Raymes at The Aerospace Corporation in El Segundo, California. This shape had a severe delta planform with pronounced rounding and twin vertical stabilizers, and was specifically tailored to meet anticipated Air Force cross-range requirements for film-return vehicles.[7]

As early as November 1960, Martin Marietta had used a variation of the Ames M1 in studies for the SAMOS film-return surveillance satellite. Engineers under the direction of Hans Multhopp eventually rejected the Ames M1 shape, opting instead for the A3 configuration developed by The Aerospace Corporation, but modifying it to an even more streamlined variation known as the A3-4, or SV-5 ('space vehicle five').[8]

From hypersonic wind-tunnel and hypervelocity test facility experiments, researchers predicted each of these lifting-bodies had acceptable hypersonic characteristics. However, serious questions remained about their subsonic and transonic performance. In an attempt to answer these questions, actual prototypes would need to be tested.[9]

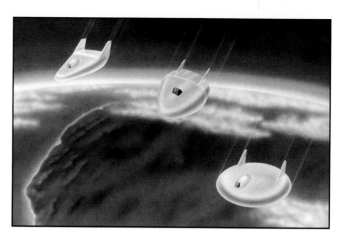

This 12 October 1962 drawing shows the Ames M2, Ames M1-L, and Langley lenticular lifting-body shapes. Dale Reed and Dick Eldredge at the FRC proposed flight testing each of them using a common internal structure with unique aeroshells. Unfortunately, only the M2-F1 was actually built and tested. (NASA photo EC62-175)

PRECISION RECOVERY

The ASSET program of 1961–65 (see page 49) had furnished a great deal of information on the aerothermodynamics and aeroelastic characteristics on a generic wing-body shape reentering Earth's atmosphere from near-orbital velocities. During 1966–67 the Air Force tested a totally different kind of reentry vehicle – an ablative-cooled lifting-body. Unlike ASSET, which would sacrifice an optimum aerodynamic configuration in favor of a large internal volume (to contain instrumentation), the PRIME (precision recovery including maneuvering entry) shape emphasized aerodynamics with a carefully derived external shape. Again, unlike the structures and heating research oriented ASSET, Project PRIME explored the problems of maneuvering reentry, including pronounced cross-range maneuvers up to 710 miles off the ballistic track. A secondary objective was to acquire design data and to develop configuration characteristics pertinent to possible future manned lifting reentry vehicles.[10]

ASSET, PRIME, and PILOT, a low-speed piloted demonstrator of the same shape, were all part of Project START (spacecraft technology and advanced reentry tests), and were officially designated Program 680A.

During late 1964 the Air Force selected the SV-5D, a subtle variation of the basic Martin SV-5 shape, for the PRIME reentry vehicle. Martin Marietta had been working on the SV-5 since 1962, using a combination of Air Force contracts and corporate funds. Most of this effort was in support of Air Force contract AF04(695)-103, also known as Project M-103, to evolve a practical lifting-body reentry vehicle configuration. Project PRIME eventually consumed over 2,000,000 engineering man-hours, including configuration and material studies, wind- and shock-tunnel work, and conducting 50 low-speed glide flights using a recoverable model suspended from a ballute. Besides being the PRIME vehicle contractor, Martin was also responsible for the associated aerospace ground equipment, guidance system, tracking and command systems, and the recovery and instrumentation systems.[11]

Each of the PRIME vehicles was 6.66 feet long, 2.1 feet high, spanned 3.8 feet, and had a 34-inch cross-section. A hypersonic L/D of 1.3:1 was expected, and since the vehicle would be recovered in flight, the subsonic L/D was immaterial. During each flight, approximately 240 aerodynamic, thermal, and guidance system measurements would be transmitted to the ground via VHF telemetry.

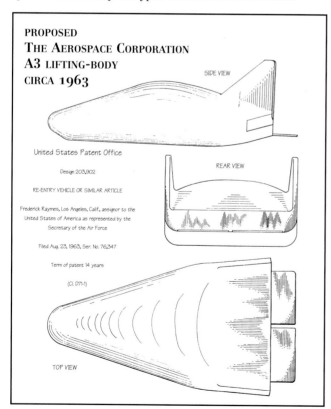

PROPOSED
THE AEROSPACE CORPORATION
A3 LIFTING-BODY
CIRCA 1963

SIDE VIEW

United States Patent Office

Design 203,902

RE-ENTRY VEHICLE OR SIMILAR ARTICLE

Frederick Raymes, Los Angeles, Calif., assignor to the United States of America as represented by the Secretary of the Air Force

Filed Aug. 23, 1963, Ser. No. 76,347

Term of patent 14 years

(Cl. D71-1)

REAR VIEW

TOP VIEW

Frederick Raymes of The Aerospace Corporation developed the A3 lifting-body in response to evaluating the Ames M1 shape. This design further tailored the 'blunt' lifting-body concept to achieve even more cross-range potential, an essential Air Force requirement. (Courtesy of Frederick Raymes)

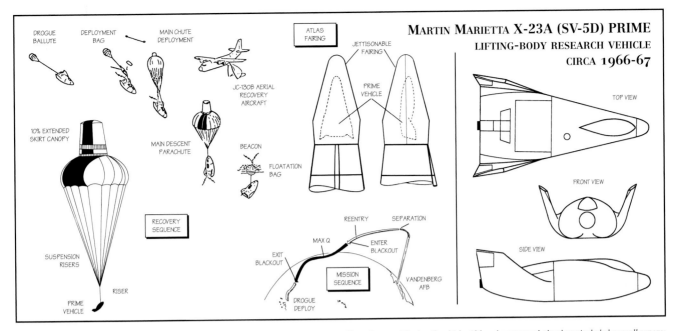

MARTIN MARIETTA X-23A (SV-5D) PRIME
LIFTING-BODY RESEARCH VEHICLE
CIRCA 1966-67

Four PRIME vehicles were built, but only three were launched since all objectives were met without the use of the fourth vehicle. Although unmanned, the shape included a small canopy area on the forward fuselage. The three PRIME vehicles were protected by a jettisonable fairing during ascent atop Atlas boosters from Vandenberg AFB, California. (U.S. Air Force)

Guidance for each mission was provided by an on-board inertial navigation system – however during the final portion of the flight course corrections were uplinked from the ground to ensure the vehicle arrived at the correct location for airborne recovery.[12]

The SV-5D (later designated X-23A) was an 890 pound lifting-body constructed primarily of 2014-T6 titanium alloy, with some beryllium, stainless steel, and aluminum used where appropriate. The airframe was manufactured in two major sections – the aft main structure and a removable forward 'glove section.' The stabilizers were a bonded steel skin over honeycomb panels, except at the antenna window where a beryllium hot-structure was used.[13]

The structure was completely covered with a Martin Marietta-developed ablative heat shield consisting of ESA-3560HF for the larger flat surface areas and a more robust ESA-5500M3 for the leading edges. This ablator used a silica-nylon fiber interlaced through a silicone-based honeycomb to hold the char. The nose cap was constructed of carbon-phenolic. The thickness of the ablators varied between 0.8 and 2.75 inches, depending on anticipated local heating conditions.[14]

Since the PRIME test flights would all terminate at approximately Mach 2 with the deployment of a drogue ballute, adequate stability and control could be maintained with two hydraulically actuated 12-inch square lower flaps, and fixed upper flaps and rudders. A nitrogen gas reaction control system was provided for maneuvering outside the atmosphere. The lower flaps were beryllium structures 0.5 inch thick with a 1.65 inch thick carbon-phenolic composite skin covered by 0.3 inch of ESA-3560HF. The fixed, deflected upper flaps provided aerodynamic trim at high angles of attack, and the fixed, deflected rudder maintained directional stability during hypersonic and supersonic flight. As the drogue ballute deployed, its cable would slice through the upper structure of the main equipment compartment where a 47-foot diameter recovery parachute was stored. Once the recovery chute was deployed, the SV-5D hung in a tail-down attitude awaiting aerial-retrieval by a JC-130B Hercules.[15]

On 21 December 1966, the first PRIME vehicle (FV-1) was launched from Space Launch Complex Three East (SLC-3E) at Vandenberg AFB on a trajectory simulating a reentry from low-earth orbit with a zero cross-range maneuver. This flight was used to demonstrate pitch-only maneuvers, which were initiated during reentry using the two body flaps on the lower surface of the afterbody. The performance of the Atlas and SV-5D was nominal through the boost and reentry phases resulting in a ballute deployment 4,300 miles downrange and within 900 feet of the preselected recovery location near Kwajalein Island (actually, Roi Namur). The ballute deployed successfully at 99,850 feet, and after the SV-5D had descended to a little less that 45,000 feet, the main recovery parachute began to deploy but did not complete its extraction sequence. The vehicle fell into the Pacific and was lost. However, all flight objectives (except recovery) had been accomplished, and over 90 percent of all possible telemetry had been received.[16]

The second vehicle was launched from SLC-3E on 5 March 1967, and successfully completed a 654 mile cross-range maneuver – the first for a returning spacecraft. This flight used differential movement of the lower body flaps to bank the vehicle at hypersonic speeds. Due to a failure in the parachute separation process (several stringers had failed to be cut, resulting in the vehicle being suspended in a manner the recovery JC-130B could not snag) the vehicle was also lost in the Pacific.[17]

On 19 April the third SV-5D was launched in a trajectory that simulated reentry from low-earth orbit with a maximum

The recovered third PRIME X-23A displays the effects of reentry heating on its ablative coating. (Martin Marietta Corporation)

(710 mile) cross-range maneuver. A hypersonic L/D of 1.0:1 was actually achieved at velocities in excess of Mach 25. The performance of the PRIME vehicle and all of its subsystems was perfect, and this time everything worked well for the recovery. The waiting JC-130B successfully snagged the SV-5D at 12,000 feet less than five miles from its preselected recovery site. A complete inspection by a Martin-USAF team revealed that the vehicle was in satisfactory shape to be launched again if needed. This was an important, if somewhat unheralded, milestone in demonstrating the potential reusability* of lifting-reentry spacecraft.[18]

Satisfied with the results of the first three flights, and facing a cutoff of funding, the Air Force cancelled the last PRIME flight. The recovered third vehicle is on exhibit at the Air Force Museum at Wright-Patterson AFB, Ohio. Total costs of the PRIME flight tests, including the Atlas boosters, was $70.5 million.[19]

Manned Lifting-Bodies[†]

R. Dale Reed, an aerospace engineer at NASA's Flight Research Center (FRC) had been following the development of the lifting-bodies with considerable interest, noting that while the hypersonic flying qualities of the design were no longer in question, there was still considerable doubt concerning their low-speed stability. In February 1962 Reed, an avid model builder, built a 24-inch model of the M2 shape which he launched from a 60-inch wingspan radio-controlled mothership. During several of the drop-tests, Donna Reed used an 8-mm home movie camera to record flights which were later shown to Alfred Eggers and the FRC director, Paul Bikle. The results were encouraging enough for Bikle to authorize a six month feasibility study of a lightweight manned M2 glider, construction of which was subsequently authorized in September 1962.[20]

The glider was conceived as a two-part vehicle. The first was the internal structure that also provided mounting points for the landing gear and pilot seat. Over this would go the aerodynamic shell. In this manner various lifting-body shapes could be tested simply by constructing a new external shell. The M2 shape was the first shape chosen to be tested, but Reed anticipated testing the M1-L and lenticular shapes as well. This plan eventually was dropped and the program went straight into the 'mission weight' phase having only tested the M2-F1. Victor Horton, Dick Eldredge and Dick Klein supervised the construction of the tubular steel frame, while Gus Briegleb of the nearby Sailplane Corporation of America built the plywood outer shell.[21]

Technicians set aside floor space in the FRC hangar, walled it off with canvas, and put up a sign reading 'Wright's bicycle shop.' The project rapidly became a center-wide effort, as many of the engineers and technicians were members of the Experimental Aircraft Association and only too happy to help, mostly on their own time.[22]

The project was financed locally out of discretionary funds, largely because Paul Bikle feared that NASA Headquarters would not support the effort. Part of his concern was the fact that a major contractor had indicated it might cost over $150,000 to construct the glider, a fair amount of money in the early 1960s. By using in-house personnel, the M2-F1 was completed in early 1963 for a little less than $30,000.

The glider measured 20 feet long, 10 feet high, and 14 feet wide. It had two vertical stabilizers, each fitted with a small elevon outboard that were quickly nicknamed 'elephant ears.' The main body had trailing edge flaps for trimming purposes and the undercarriage was borrowed from a Cessna 150. Complete with its pilot, the M2-F1 weighed only 1,138 pounds, although this was significantly more than the 600 pounds originally estimated. At first the pilot had no means of escape, other than bailing over the side, but the M2-F1 was subsequently fitted with a lightweight rocket-propelled Weber zero-zero ejection seat.[23]

By March 1963, wind tunnel tests at Ames were completed with encouraging results. Back at Edwards, ground tows behind various FRC trucks were accomplished, with the first occurring on 5 April 1963. These were designed to lead to captive flights using a Paresev-like canopy, but none of the FRC vehicles proved fast enough to get the vehicle airborne.

In typical FRC fashion, engineers determined that a high-performance tow car was the answer and quickly consulted with various automobile racing aficionados who lived around the area. The FRC then proceeded to buy a stripped-down Pontiac convertible with a 455 cubic-inch engine, four-barrel carburetor, and a four-speed stick shift. The team then turned the car over to Bill Straup's hot-rod shop[‡] in Long Beach where the engine was 'tuned,' rollbars and radio equipment installed, and the right-hand bucket seat turned around to face aft. The car's hood and trunk were then painted high-visibility yellow and NASA emblems applied to the doors. The first Paresev flight occurred in June 1963, and a little over four hours of flight time was accumulated in 100 tows.[24]

The first piloted lifting-body free flight occurred on

The radio-controlled mothership was used for a fairly long time, and is shown here in 1968 with the M2-F2 and various Hyper III models. From left to right; Richard C. Eldredge, R. Dale Reed, James O. Newman, Bob McDonald. Reed can be seen holding the original M2-F1 model on page 32. (NASA/DFRC photo ECN-2059)

---

* It should be noted that other reentry vehicles have been successfully reused, including a Gemini capsule (GT-2) that was reflown (unmanned) to test heat shield modifications for the proposed MOL logistics vehicle (which was heavily based on Gemini).

† For a truly excellent account of the manned lifting-body program at Edwards, see R. Dale Reed with Darlene Lister, *Wingless Flight: The Lifting Body Story*, SP-4220, (NASA, Washington DC, 1997).

‡ Mickey Thompson's funny-car shop has always been credited with these modifications. But Dale Reed in *Wingless Flight* sets the record straight – Thompson only provided the wheels and tires when the program changed to racing 'slicks' instead of street tires.

16 August 1963, when NASA research pilot Milton O. Thompson cast off from a Douglas R4D (Navy version of the C-47/DC-3) and dropped 10,000 feet to land on Rogers Dry Lake. Like most subsequent flights, the M2-F1 was trailing the R4D on a 1,000-foot tow cable at about 120 mph. The glider had approximately two minutes of free flight before touching down at between 85 and 90 mph. On 3 September 1963, FRC unveiled the glider to the news media, where it immediately became a hot item in the popular press.[25]

In addition to Thompson, six other test pilots flew the little M2-F1 – William Dana, Captain Jerauld Gentry, Donald Mallick, Bruce Peterson, Donald Sorlie, and Colonel Chuck Yeager. Over the course of its test program, the engineers at the FRC made numerous modifications to the glider, including a 24-lbf solid-fuel rocket that was added to assist in the pre-landing flare maneuver, and Cessna 180 landing gear that ultimately replaced the original Cessna 150 units. Eventually the M2-F1 completed over 100 flights and 400 ground tows before being retired to the Smithsonian's National Air and Space Museum.[26]

The M2-F1 proved that a lifting-body shape could fly safely at subsonic speeds. Encouraged by the results of this very limited flight test series, Dale Reed and the engineers at the FRC pursued the building of rocket-powered lifting-bodies to evaluate performance up to Mach 1.5 and 60,000 feet. By early 1963 preliminary studies were underway on an air-launched, rocket-powered, lifting-body built largely with off-the-shelf systems and equipment. This vehicle was intended to expand the envelope of lifting-body research into the low-supersonic and transonic speed regions, and also to evaluate the landing behavior of a 'mission-weight' lifting-body.[27]

The original M2-F1 had a wing loading (the term was still used, even if the vehicle did not have wings) of only 20 percent of that expected of a space-rated vehicle, and NASA had recognized from the outset that it would be considerably more difficult to land than a mission-weight version. Although both vehicles would have about the same L/D ratio, the lightweight vehicle had an inherently shorter time between beginning the landing flare and touchdown. The longer interval of the mission-weight vehicle was desirable from a pilot's point of view, but on the negative side was the fact that a heavier vehicle lands much faster.[28]

Plans were made to procure two identical rocket-powered M2-F2 vehicles, but when NASA Headquarters approved the program it decided the best approach was to evaluate two competing shapes. The original intent was to build an almost exact replica of the M2-F1, but this soon proved to be impossible. Researchers at NASA Ames wanted to delete the 'elephant ear' elevons since they would be impractical on an actual reentry vehicle (they could not be adequately insulated and would burn off). Also, the center-of-gravity would move considerably if a rocket engine was installed in the aft fuselage, so the cockpit had to be moved much further forward.[29]

In February 1964 the FRC released an RFP for the construction of two mission-weight, transonic lifting-body gliders – one based on the M2-F2 shape, the other on the HL-10. The competitors had five weeks to submit detailed

The M2-F1 used simple Cessna landing gear and had large 'elephant ear' elevons on the outside of the vertical stabilizers. Although difficult to see, lettering above the small rocket engine on the back reads "instant L/D." (NASA/DFRC)

technical and cost proposals. Only five companies actually submitted responses and, on 2 June 1964, the FRC awarded a fixed-price contract to the NorAir Division of Northrop to build the vehicles. Northrop was to deliver the M2-F2 in late-spring 1965, with the HL-10 following in the late-fall. In August 1964 it was decided to incorporate provisions for the installation of an XLR11* rocket engine.[30]

One of the factors that had most influenced the decision to select Northrop was the close personal relationship enjoyed between NASA and Richard Horner, Executive Vice President of Northrop – Horner had worked with Bikle at Edwards, then transferred to NASA Headquarters before joining Northrop, and he assured the FRC that Northrop could build the two gliders cheaply. The other four contractors had estimated costs of up to $15 million each to build the two gliders, but Northrop ultimately completed both for slightly under $2.5 million.[31]

While the mission-weight program was beginning, an opportunity almost presented itself. Four surplus Little Joe boosters were available at the NASA White Sands Test Facility. The solid-fuel Little Joe had been developed to launch boiler-plates of the Apollo command module escape system. Dale Reed began looking into the possibility of mounting an M2-F2 on top of a Little Joe – this would allow Mach 6 performance for short periods of time. As it happened, some surplus Gemini parachutes were also available, providing a way to slow down after reentry. By scrounging this equipment – all essentially free for the taking – the researchers at the FRC began to think they had a viable program. But as often happens, there were unknown complications. The Little Joe boosters had a shelf life – and it had expired. Before they could be used the boosters would need to be inspected – at an estimated cost of $1,000,000 each. The idea quickly faded.[32]

The M2-F2 was rolled out of the Northrop plant in Hawthorne on 15 June 1965, and trucked to Edwards the next day. Of conventional aluminum construction, the M2-F2 was 22 feet long, spanned 9.6 feet, and weighed 4,630 pounds without its single XLR11. A full-span ventral flap controlled pitch, while split dorsal flaps controlled roll (lateral) motion through differential operation and pitch trim through symmetrical operation. Twin ventral flaps provided directional (yaw) control and also acted as speedbrakes. The M2-F2 had a stability augmentation system to assist the boosted control system damp out unwanted vehicle motions. Four throttleable

---

*   The Reaction Motors XLR11 (the Navy had a very similar XLR8) had first been used on the XS-1, and went on to power most every rocket research aircraft tested at Edwards. After the X-15 switched to the more powerful XLR99, many of the XLR8/11 engines were donated to various air museums. When the lifting-body program decided to use the engines, NASA removed them from the museums, refurbished them, and returned them to flight status. When the lifting-body project ended in 1975 the engines were returned to the museums they had been borrowed from.

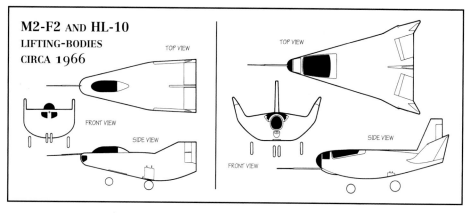

M2-F2 AND HL-10
LIFTING-BODIES
CIRCA 1966

TOP VIEW

TOP VIEW

FRONT VIEW

SIDE VIEW

SIDE VIEW

FRONT VIEW

The relative shapes of the two 'mission-weight' lifting-bodies is shown here. By the end of the flight test program the HL-10 was considered the more practical of the two designs and had flown the fastest and highest. (NASA)

hydrogen peroxide rockets, rated at 400 lbf each, could provide 'instant thrust' during the pre-landing flare. A modified Weber zero-zero ejection seat from a Convair F-106 Delta Dart was provided in case the pilot needed to get out in a hurry. When the M2-F2 arrived at Edwards, it was placed next to the M2-F1 for a family photograph – although the vehicles were generally similar in size, they also differed greatly. The M2-F2 lacked the 'elephant ears,' had an extended boat-tail, a canopy located much further forward, and was 10 times heavier.[33]

On 23 March 1966, the M2-F2 completed its first captive-flight under the wing of the NB-52 carrier aircraft, but there were still some nagging concerns that the lifting-body might fly upward and impact the NB-52 after it was released. In Dale Reed's book, Jerry Gentry is quoted as saying: "There was no question which way you were going when the B-52 dropped you … One guy used to say that if they dropped a brick out of the B-52 at the same time [he was] released, [he'd] beat the brick to the ground."[34]

The first gliding free-flight, with Milt Thompson at the controls, occurred on 12 July 1966. The M2-F2 dropped away from the NB-52 at 45,000 feet while flying at 450 mph, and Thompson made several 90 degree turns and practice landing flares before landing the vehicle at 200 mph on Rogers Dry Lake. During the next four months, 13 additional glide flights would be flown by Thompson, Gentry, Bruce Peterson, and Donald Sorlie. The glide tests confirmed the predictions that the vehicle would suffer

The HL-10 under the wing of the NB-52B. Note the mission marks on the fuselage of the NB-52 and how the canopy on the lifting-body opens. The flush canopy helped maintain excellent drag characteristics. (NASA/DFRC)

from poor lateral-directional stability, particularly at low angles-of-attack and high speeds.[35]

The M2-F2 was grounded on 21 November 1966 to allow the installation of the XLR11 engine. On 2 May 1967 the M2-F2 made its first flight carrying, but not using, the XLR11 with Jerry Gentry at the controls. On the morning of 10 May 1967, during the vehicle's first powered flight, Bruce Peterson started to exit the second of two planned S-turns only to find the M2-F2 rolling and banking wildly. Peterson managed to regain control of the vehicle, but to compound his problems, a rescue helicopter positioned itself directly in front of the M2-F2, and Peterson was forced to try and avoid the helicopter in addition to controlling the M2-F2. Peterson attempted to flare onto the lakebed, but bounced back into the air, then tumbled end-over-end' at more than 250 mph, seriously injuring himself, and almost destroying the M2-F2.[36]

This was the unfortunate end of the M2-F2's flight test program. The airframe was subsequently shipped back to Northrop where a two-month inspection showed it was rebuildable. On 28 January 1969 it was announced that the vehicle would be rebuilt as the M2-F3 incorporating a large central vertical stabilizer in addition to the upswept wingtip stabilizers. This center stabilizer proved quite effective as a large 'flow fence' to improve lateral stability. The task of rebuilding the airframe was split between Northrop and workers at the FRC in an effort to keep the expenses down – nevertheless, it took almost three years and cost nearly $700,000.[37]

Bill Dana took the rebuilt M2-F3 on its first glide flight on 2 June 1970 and found the handling qualities of this version were much improved. On 13 December 1972 the M2-F3 attained Mach 1.6, its fastest flight, and a week later, on 21 December reached 71,493 feet, its highest. During the latter part of the test program, the M2-F3 was used to check out a reaction roll-control system and a rate command augmentation system – systems that might have potential use on the Space Shuttle then being designed. The vehicle was retired after completing 43 flights – 16 as the M2-F2 and 27 as the M2-F3, and subsequently joined the collection of the National Air and Space Museum.

At one point Northrop submitted an unsolicited $200 million proposal to build a version of the M2 shape that would be launched by a Titan booster. Although interesting, no action was taken by the government mainly because there were no funds, and the proposal faded from sight.

Concurrently with the M2 flight tests, the FRC was also flying the HL-10. This aluminum vehicle was 22 feet long and spanned almost 15 feet across its aft fuselage. The control system consisted of upper body surface and outer stabilizer flaps for transonic and supersonic trim, blunt trailing edge elevons, and a split rudder on the central vertical stabilizer. Like the M2-F2 it featured a landing rocket and control augmentation system.[38]

Internally, the M2-F2 and HL-10 were very similar, with nearly identical subsystems and structural details. Both

---

* Video footage of this accident was seen weekly at the beginning of the <u>Six Million Dollar Man</u> television series. Bruce Peterson mostly recovered, becoming the Director of Safety at the FRC, and continued on limited flight status in the Marine Reserves.

vehicles used a riveted aluminum alloy semi-monocoque forebody, the aft structure was basically an aluminum box with side fairings, and two full-depth keels extended from the crew compartment to the extreme rear. The box at the rear was the attachment point for the vertical stabilizers and rocket engine, and also supported the non-load bearing outer skin panels.[39]

Like the M2-F2, provisions were incorporated in the HL-10 for an XLR11 rocket engine, but this was not fitted when the vehicle was rolled-out of the Hawthorne plant on 18 January 1966. The HL-10 was shipped to NASA Ames for testing in its 40 by 80-foot full-scale wind tunnel – revealing some flow separation over the outer vertical stabilizers. Engineers at Ames and Northrop did not consider this particularly serious, and cleared the HL-10 for flight.[40]

Bruce Peterson took the vehicle on its first glide flight on 22 December 1966. During the three minute descent Peterson discovered that he had minimal lateral control over the HL-10, but nevertheless succeeded to land the lifting-body. The flow separation problem had turned out to be much more serious than initially thought, and NASA Ames embarked on yet another round of wind tunnel tests. After these tests, the leading edge of the outer vertical stabilizers was modified to direct more airflow over the control surfaces, pointing out the significance of seemingly minor shape changes. The HL-10 would be grounded for 15 months while engineers corrected the problem.[41]

During its next flight, on 15 March 1968, Jerry Gentry found the modifications worked well and the HL-10 handled nicely. On 13 November 1968, John Manke made the vehicle's first successful powered flight (an earlier attempt had failed when the XLR11 did not ignite properly), reaching Mach 0.84 using only two of the four thrust chambers. The HL-10 went supersonic for the first time on 9 May 1969, marking the first supersonic flight of any of the manned lifting-bodies. On 18 February 1970, Peter C. Hoag reached Mach 1.86, and nine days later Bill Dana topped out at 90,303 feet. The HL-10 thus became the fastest and highest flying of the lifting-bodies. A total of 35 flights were conducted during the basic research program.[42]

Later NASA undertook two additional flights using the HL-10 in an attempt to determine whether the complexity and weight of landing engines could be justified on the upcoming Space Shuttle Orbiter. The XLR11 was replaced by three Bell Aerosystems 500-lbf hydrogen peroxide thrusters that would be fired as the vehicle passed through 6,500 feet on the way to landing. The engines would reduce the approach angle from the normal 18 degrees to 6 degrees, and also boost the airspeed to over 350 mph. At 200 feet above the lakebed the pilot would shut down the engines, extend the landing gear, and make a routine landing. Peter Hoag completed the two flights on 11 June and 17 July 1970, the latter marking the last flight of the HL-10.[43]

The general consensus was that the benefit of landing engines, namely the ability to make a missed approach, was outweighed by the complexity of the engine installation and the increased pilot workload due to shallower descent angles and higher approach speeds. The HL-10 program was terminated and its basic shape was deemed the best performer of the lifting-bodies, becoming the basis for several early space shuttle concepts. The HL-10 was later loaned to the California Museum of Science and Industry, where it was seriously damaged while being lifted into position for display. The vehicle was returned to Edwards for repair, and on 3 April 1990 the freshly restored HL-10 was put on display at Dryden.

The M2-F2 under the NB-52A (52-003) The High and the Mighty One during 1966. The B-52 wing pylon was the same one used by the X-15 program, and a special adapter was built that allowed the lifting-bodies to be carried. (NASA photo ECN-1436)

One other lifting-body would be built and tested at Dryden. By 1969-70 Dale Reed had become interested in the 'race-horse' lifting-bodies, such as the FLD-7 (which became the X-24B) and the Hyper III configuration developed by NASA Langley. These shapes had hypersonic L/Ds as high as 3.0:1 allowing cross-ranges of over 1,500 miles to either side of the reentry path. It can be argued that these designs were not true lifting-bodies since each used some sort of deployable wing for low-speed flight and landing (often called, incorrectly, variable-geometry wings).[44]

True to form, Reed and Dick Eldredge began by flying radio-controlled models of the Hyper III shape, but using a Rogallo Limp Wing parachute for recovery. They also designed and built a special twin-engine 14-foot span model mothership to carry the lifting-bodies to altitude. What they quickly confirmed – no big surprise – was that the Hyper III needed either a gliding parachute or deployable wings in order to land safely. Three different wing

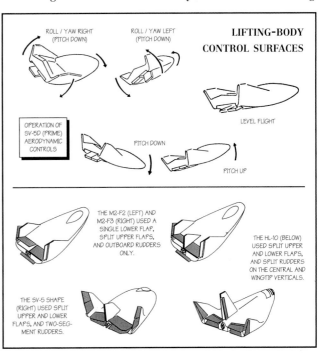

Each of the four lifting-body shapes used a unique control surface arrangement, although all four were essentially similar in concept (top). The M2 was the only shape not to have split lower flaps for pitch control, using a single centrally-located body-flap in their place. (U.S. Air Force)

The M2-F3 drops away from the NB-52B (52-008) during 1971. Note the position of the lower body flap and the pylon adapter on the NB-52. (NASA photo EC71-2774)

configurations were soon experimented with. The first was a pair of switchblade wings that pivoted out of slots in the lower part of the body, resulting in a swept wing configuration. The second was a single-piece wing that pivoted in the center and was stowed along the top of the fuselage – the right wing exited towards the rear of the vehicle, while the left wing swung forward. This resulted in a straight-wing configuration after deployment.[45]

The third type of wing was a Princeton Sailwing that had been tested at NASA Langley with promising results. The Sailwing was two D-shaped spars stowed in slots in the body and deployed like a switchblade wing, with trailing edge cables pulling taut from a tip rib and stretching upper and lower fabric membranes from the spar to the cable. The fabric surfaces curved upward, like a hand glider, forming a cambered airfoil and producing positive-lift airflow over the wing.[46]

NASA Langley conducted wind-tunnel tests on the Hyper III without wings up to Mach 4.6, then subsonic tests using each of the three wing types. The one-piece pivoted wing proved to be the most effective, and Dale

The Hyper III is shown on 28 July 1969 with Daniel C. Garrabrant standing alongside. The single-piece pivot wing is installed, and the Princeton Sailwing is on the ground. (NASA photo E-20464)

Reed proceeded with plans for a full-scale light-weight vehicle similar in concept to the M2-F1. There was one significant difference, however. This time Reed proposed an unpiloted vehicle – it would be flown via remote control from a stationary cockpit on the ground.[47]

The Hyper III was manufactured in the shops at the FRC in order to keep costs down. The finished vehicle was 35 feet long and 20 feet wide at the tail surfaces. The fuselage was a Dacron-covered steel-tube frame, the nose was made from molded fiberglass, and the four tail surfaces were constructed from aluminum sheet metal. Parts were scrounged from many sources. For instance, the elevon actuators and the battery-driven hydraulic pump were surplus units from the PRIME program. A recovery parachute was installed that could safely lower the vehicle in the event of an emergency.[48]

Bruce Peterson was at the controls of a borrowed Navy SH-3 helicopter when the Hyper III was dropped for its first flight on 12 December 1969. While attached to the end of its 400-foot cable, the Hyper III refused to track straight, and Peterson struggled to maintain the proper orientation. When the helicopter reached 10,000 feet the Hyper III was released. For this flight the retractable wing had been fixed in its deployed position. Milt Thompson controlled the Hyper III from the ground cockpit, gliding it three miles north, then through 180-degree turn back to the landing site. As the vehicle approached the landing site, control transferred to Dick Fischer, and the Hyper III made a perfect landing.[49]

Post-flight analysis indicated that the subsonic L/D of the Hyper III was much lower than expected – closer to 4.0:1 instead of the 5.0:1 that had been calculated. The concept of using a ground-based cockpit had proven acceptable, although there were some anxious moments as procedures were worked out. Although only a single flight had been made, the initial phase of the program was considered a success.[50]

Dale Reed began making plans for piloted flights of the Hyper III using a Grumman SA-16B Albatross as the carrier aircraft (the NB-52 was much too fast for the fabric covered glider). But NASA Headquarters rejected the plans, and the Hyper III never flew again.[51]

AIR FORCE EFFORTS

Although heavily involved in supporting the M2-F2 and HL-10 lifting-body programs, the Air Force also conducted independent research into the lifting-body concept. In fact, as early as the mid-1950s, the Air Force had begun investigating lifting-body designs (SAMOS, SAINT, etc.) for a variety of purposes, including use as a film-return* vehicle. The majority of these configurations ranged from 'low' L/D ratios of approximately 1.0:1, to 'higher' L/D ratios[†] of around 3.0:1, although some that used variable-geometry planforms ventured into higher L/Ds.

The Air Force Flight Dynamics Laboratory[‡] (AFFDL) examined a wide variety of shapes beginning in 1959 and continued until the X-24B demonstrator was built. These shapes tended to reflect the relatively high L/D ratios

---

* This was long before reconnaissance satellites had electronic surveillance equipment of sufficient resolution to satisfy intelligence needs. Today almost all satellites digitally encode their data and return it to Earth via telemetry, eliminating the need for film-return vehicles.

† All things being relative. These are extremely low numbers when compared to conventional aircraft, but much higher than Allen-type blunt reentry vehicles.

‡ The Air Force FDL was established on 8 March 1963 at Wright-Patterson AFB from the reorganization of the earlier Directorate of Aeromechanics.

preferred by Alfred C. Draper, Jr., a respected researcher at the AFFDL, and many of the early vehicles represented interesting approaches to the problem of lifting-reentry and orbital flight. The WADD-II* was a relatively low L/D configuration called the 'lead sled' by Draper that had originally appeared as an early suggestion to the Alpha Group evaluating Dyna-Soar. This was followed by the WADD-III, a delta planform with extremely low wing loading and was thus known as the 'lite kite.' One of the more interesting ideas was the MDF-1,+ a novel lifting-body designed around the subsonic 'Clark-Y' airfoil developed by Virginius E. Clark during the early-1920s. This shape had surprisingly good hypersonic and low-speed stability characteristics, and was shown to the SSD during the design configuration phase of the earlier PRIME program, although it was rejected in favor of the A3-4/SV-5D.[52]

As early as 1962, Draper proposed the flight testing of manned prototypes of several of the designs being developed. The Air Force, much to the dismay of both Draper and NASA, declined. Nevertheless, Draper and his associates went forward with their laboratory studies, generating a large series of tailored body shapes. Some of the proposals reentered the atmosphere as lifting-bodies but deployed variable-geometry wings during the transonic phase of flight to significantly increase their L/D ratios.[53]

In addition to fixed and variable-geometry designs, the group also explored 'interference' configurations that used complex undersurface designs to position shock flows for favorable increases‡ in lift. Similarly, the group investigated Nonweiler or 'caret' wave-rider configurations which made use of favorable flow interference to increase their hypersonic L/D ratios. This approach continued to be pursued during 1991 in support of the X-30 NASP, prior to its cancellation.[54]

Interestingly, the Draper group received many of its ideas for interference configurations from the propulsion community. Specifically, these came from the inlet designers where interference is an accepted fact of life. Draper and his team soon discovered, however, that these interference configurations usually demanded a very complicated lower surface configuration that aggravated aerothermal

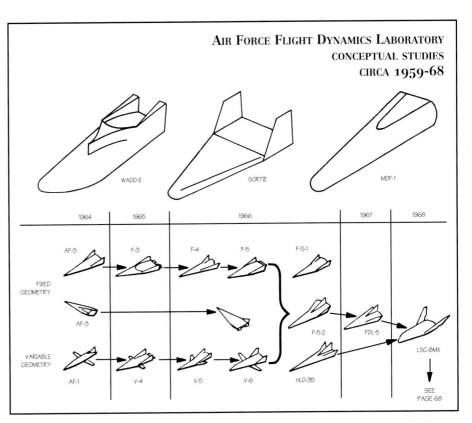

An Air Force Flight Dynamics Laboratory team led by Alfred C. Draper designed a variety of lifting-body and wing-body shapes in an attempt to find the optimum planform for a maneuvering reentry vehicle to meet Air Force requirements. (U.S. Air Force)

problems by creating intersections and junctures, and research into them was apparently discontinued.[55]

The early AFFDL designs were completed by 1964 and fell into three general categories. AF-1 was a variable-geometry wing design with a single vertical stabilizer. As with all the variable-geometry wing designs, AF-1 kept the wings in a retracted position inside the fuselage during the critical reentry period, deploying them during atmospheric maneuvers to increase the supersonic and subsonic L/D ratios. Subsequent members of this category during 1966 included the V-4, V-5, and V-6, all having twin vertical stabilizers and low-mounted wings. The second category contained wave-rider configurations starting with AF-3. This category had the least impact on later designs due to the complications of protecting the vehicle from reentry heating, as discovered during evaluations of the earlier Ames Mach 10 demonstrator concept. The last general category was a series of shapes that served as the foundation for the later X-24B/FDL-7 design, and was characterized by AF-5. This design evolved through the F-3, F-4, and F-5 shapes during 1966, finally contributing to the FDL-5 design of 1967. Seven unique designs were part of a series labeled FDL-1 through FDL-7.§ These configurations were all suited to hypersonic aircraft capable of flight from Mach 4 to orbital velocities, but were tailored primarily for aircraft operating in the Mach 8–12 regime.[56]

The Air Force hoped that these shapes would prove useful for sustained hypersonic-cruise aircraft using some form of air-breathing propulsion (probably scramjet), as well as for unpowered boost-glide vehicles capable of landing at virtually any convenient airfield. Unfortunately, at the time there were no suitable engines for such a high speed vehicle, although research apparently continued at a low priority. It is possible that the much-rumored Aurora hypersonic reconnaissance vehicle (i.e.; SR-71 replacement) is an outgrowth of this research.

---

* WADD was the Wright Air Development Division, the predecessor to the current Aeronautical Systems Division.

+ MDF was Draper's organizational symbol.

‡ The North American XB-70A bomber used a somewhat similar technique, folding its wingtips down at high speeds to capture the shock wave generated by its nose. This increased the vehicle's overall L/D, and also improved its directional stability characteristics. Although extremely successful on the XB-70A, it would prove a great deal more difficult to use this technique as speeds increased into the hypersonic regime.

§ *An observation by John V. Becker:*
"The Air Force FDL X-24 series of so-called 'lifting-bodies' were much less drastic departures from conventional delta-winged aircraft than the NASA M2b or HL-10. They were really flat-bottomed delta's with a fuselage blended into a thickened upper surface. Hence their generally better handling qualities and higher L/Ds."

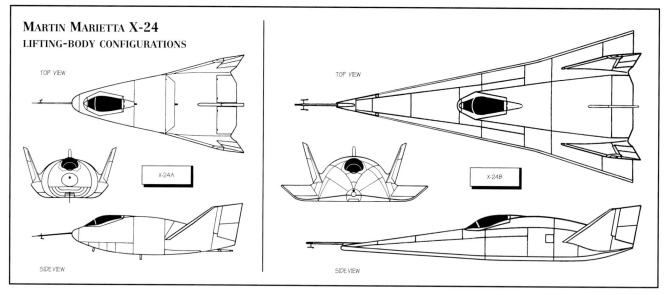

MARTIN MARIETTA X-24
LIFTING-BODY CONFIGURATIONS

TOP VIEW

X-24A

SIDE VIEW

TOP VIEW

X-24B

SIDE VIEW

The sleek skin of the X-24B was 'gloved' onto the X-24A to create an entirely new shape – the only 'race-horse' lifting-body to be extensively flight tested. In many ways the X-24B bridged the gap between wing-bodies and lifting-bodies since its flat-bottom fuselage spread into small wing-like surfaces at the tips. (U.S. Air Force)

PILOT

While NASA was testing the M2 and HL-10 vehicles, the Air Force was building its own piloted SV-5D lifting-body demonstrator. Project PILOT (for piloted low-speed tests) was part of the ongoing START program that included PRIME and ASSET. Those tests demonstrated the shape's hypersonic stability, but the Air Force also wished to evaluate its low-speed handling and landing qualities. This would provide the last part of a knowledge base that would contain data for the A3-4/SV-5 shape from the Mach 25 achieved by the SV-5D (X-23A), all the way to approach and landing speeds with the SV-5P (X-24A).

The objectives of PILOT were to investigate the static and dynamic stability characteristics and to verify control techniques for the vehicle during the low-supersonic, transonic, and landing approach speed regimes. Additional data was to be collected on various aerodynamic forces and pressures, hinge moments, man-machine interrelationships, and the energy management requirements for lifting-body designs. Post-flight analysis was to look at the correlation of flight data with wind tunnel predictions.[57]

The program got off to a slow start and by December 1964 still did not have official approval. At the time PILOT was facing a number of serious questions involving its procurement and management structure, the number of test

The X-24A also was air-launched from the NB-52s, but used a different adapter to attach it to the X-15 pylon. (NASA photo E-20215)

vehicles to be procured, and last but not least, which actual airframe contractor would be selected. By this time the procurement method followed by the FRC for the mission-weight M2 and HL-10 vehicles appeared quite attractive since it promised to cost substantially less than a typical Air Force contract. The Aerospace Corporation was against <u>any</u> procurement, but Major General Benjamin I. Funk, commander of the SSD, assured the START program office that the SV-5 configuration would be flight tested in the low-speed regime regardless of Aerospace's opinions.[58]

The next year was spent on funding and political issues revolving around the number of vehicles, the anticipated cost of the program, and the level of AFFTC and FRC support required. Additional problems were encountered during early wind tunnel tests of the piloted version of the SV-5 shape when it became evident that the larger canopy required for manned use resulted in interference that created instability in some flight regimes. The cockpit's canopy enclosure was redesigned and the problem was minimized, but not completely eliminated. Other problems included difficulties with working out control laws for the six additional control surfaces on the aft fuselage and flow differences caused by the increased scale of the vehicle.[59]

The Air Force hoped to capitalize on the 'cheap' manufacturing that Northrop had used for the NASA lifting-bodies, and released requests for proposals to both Northrop and Martin Marietta for the fabrication of a single rocket-powered SV-5P vehicle. The proposals were received in February 1966 and quickly evaluated – as a result, Martin Marietta was informed that it had been selected. The SV-5P was ordered on 2 March 1966. On 29 April 1966, the project office conducted an initial design review with Martin engineers and recommended several minor modifications that were easily incorporated into the vehicle. Developed under the direction of Martin engineers 'Buz' Hello and Lyman Josephs, construction was well under way by May and the vehicle took a little over a year to complete. On 11 July 1967 the vehicle was officially designated X-24A.[60]

The X-24A (66-13551) was rolled-out at the Martin Marietta plant in Baltimore (Middle River) on 3 August 1967. The vehicle spanned 13 feet with an overall length of 24 feet, and was basically a four-times larger version of the PRIME SV-5D shape. Power was supplied by the seemingly irreplaceable XLR11. The vehicle had an empty weight of less than 6,000 pounds, increasing to 11,000 pounds with fuel and a pilot. Since testing was to be limited to less than Mach 2, the X-24A was of conventional aluminum construction, with no special attention paid to heat protection. The basic aerodynamic control system consisted of eight movable surfaces on the aft end of the vehicle. Pitch control was derived from the symmetrical deflection of the lower and/or upper flaps depending upon flight conditions. Differential deflection of the flaps provided the primary roll control. Pitch or roll commands, which caused either lower flap to fully close, resulted in control being transferred to the corresponding upper flap through a simple clapper mechanism. The two pairs of rudders were deflected symmetrically as a bias feature with directional control being provided by the deflection of the surfaces in unison.[61]

The X-24A would not fly immediately . Instead, it was delivered to NASA Ames for wind tunnel testing to establish a baseline reference to compare against the flight test data. Wind tunnel testing commenced on 27 February 1968 and lasted for about two weeks. While the X-24A was in the wind tunnel, researchers took the opportunity to measure the aerodynamic effects of an ablative heat shield. Surface measurements from the recovered PRIME X-23A were used to generate a rough texture that was glued to the exterior of the X-24A. Testing revealed that this caused a significant reduction in L/D that would make landing following reentry much more difficult. Although this was an interesting data point, and used during the development of space shuttle, no further work on it was done using the X-24A. It did, however, raise yet more doubts about the usefulness of an ablative heat shield on any operational lifting-reentry vehicle.[62]

On 15 March 1968 the X-24A was delivered to Edwards where it joined the HL-10 and M2-F2. Miscellaneous static ground tests needed to be accomplished, and it wasn't until almost a year later that the X-24A was cleared for flight. Taxi tests were made during the first week of March 1969 using two small hydrogen peroxide thrusters (not the XLR11). Low- and high-speed taxi tests of the X-24A mated to the NB-52 carrier aircraft were conducted on 2 April, and a captive-carry flight was made on 4 April.[63]

Finally, on 17 April 1969 Captain Jerauld Gentry made the maiden glide flight, the first of nine unpowered flights. Gentry also flew the first powered flight on 19 March 1970, reaching Mach 0.87. On 14 October 1970, 23 years to the day after Chuck Yeager first broke the sound barrier, John Manke piloted the X-24A on its initial supersonic flight. A typical flight consisted of 2.5 minutes of powered operation followed by a 5 minute glide to landing – during the course of its test program, the X-24A accumulated a total of 2 hours, 54 minutes and 28 seconds of flight time during 28 missions. A maximum speed of Mach 1.60 and an altitude of 71,400 feet were attained. The last portion of the X-24A test program was dedicated to simulating space shuttle approaches and landings, and the vehicle completed its last flight on 4 June 1971.[64]

While the X-24A was being built, Martin Marietta completed two jet-powered low-speed lifting-bodies based on the SV-5 shape as a company-funded venture. The SV-5Js, each powered by a single 3,000-lbf Pratt & Whitney J60 turbojet, were to be used as astronaut trainers at the

The X-24B prior to its first captive carry flight on 19 July 1973. Note the helium tanks on the back of the NB-52 pylon, and the exposed XLR11 installation on the X-24B itself. (AFFTC photo 2143-73 by Sgt David A. Greso via the Terry Panopalis Collection)

Air Force Test Pilot School. These vehicles were considered to be significantly underpowered and never flew, remaining at the Martin plant at Middle River. During 1968, the Air Force Systems Command solicited suggestions for possible uses for the two unused SV-5J airframes and the AFFDL responded on 23 January 1969 with a proposal to modify one of them into the FDL-7 configuration for flight testing. In order to minimize modification costs, it was proposed that the basic SV-5J aft body structure and vertical stabilizers be retained, with the FDL-7 shape being 'gloved' over the existing forebody. This configuration was generally referred to as the FDL-8.[65]

Wind tunnel tests were conducted to validate the resulting aerodynamic configuration with satisfactory results. The proposed FDL-8 was to be jet powered and air-launched from one of the NB-52s. As the studies matured, however, the advantages of rocket propulsion became apparent and led to the selection of the XLR11. Once that decision was made, the use of the X-24A airframe instead of one of the SV-5Js became the next logical step. As the X-24A was already nearing the completion of its flight test program, it was felt that the modification could make use of many already proven systems, and reduce the overall modification effort and subsystem build-up costs.[66]

Following the last X-24A flight, the vehicle was stored at Edwards for six months while contract negotiations and final design details were completed. On 1 January 1972, the Air Force awarded* Martin Marietta a $1.1 mil-

Like the other lifting-bodies the X-24B was constructed primarily of aluminum. This was made possible since these were low-speed demonstrators and were not intended to withstand reentry. (NASA photo E-25214 via the Terry Panopalis Collection)

---

* $550,000 of this had actually come from the NASA FRC budget on 11 March 1971, since NASA was also very interested in obtaining flight test data on the AFFDL shape in support of the ongoing Space Shuttle program.

lion contract to modify the X-24A into the X-24B (also known as the SV-5P-2). The vehicle was transported by C-141 to the Martin facility in Denver on 15 December 1971, and entered a special manufacturing fixture on 7 April 1972. When the modified aircraft was rolled out on 11 October 1972, it had grown over 14 feet in length and ten feet in span – weight was up to 13,700 pounds. The sleek new shape had a double-delta planform swept 78 degrees, and a nose that was tilted up three degrees.[67]

The vehicle was returned to Edwards on 24 October, had the XLR11 installed, and began several months of ground testing. John Manke completed the first X-24B glide flight on 1 August 1973 and its first powered flight on 15 November. On 25 October 1974, Lieutenant Colonel Michael V. Love accelerated to Mach 1.76, the fastest X-24B flight. Manke also took the X-24B to 74,130 feet on 22 May 1975, marking its highest flight. The basic X-24B research flight program had included 30 flights accomplished over 24 months by three pilots.[68]

By the summer of 1975 the Space Shuttle was well into its operational definition phase, and its designers were again debating whether to provide air-breathing landing engines. The primary concern was if low L/D reentry shapes could successfully complete unpowered landings on confined hard runways. John Manke and Mike Love were convinced they could, and on 5 August 1975 Manke guided the X-24B to a perfect landing on Edwards runway 04. Two weeks later, Love duplicated the performance, both landings occurring within a few feet of the anticipated touchdown point. On 23 September 1975, Bill Dana completed the last powered X-24B flight, also marking the end* of rocket-powered research flights at Edwards AFB.[69]

Six checkout flights to familiarize three possible future Space Shuttle pilots with precision unpowered landings were subsequently undertaken. Each of the pilots, Einar Enevoldson, Thomas C. McMurtry, and Captain Francis Scobee, performed two glide flights from 45,000 feet before the X-24B was retired on 26 November 1975.[70]

The complete X-24B flight test program consisted of 36 air launched flights, 24 of them powered, that accumulated 3 hours 46 minutes and 43.6 seconds of flight time. The X-24B, along with the HL-10, proved significant in resolving the final questions of how to fly and land a space shuttle. The X-24B, and one of the stillborn SV-5Js rebuilt to resemble the X-24A, are on exhibit in the Air Force Museum at Wright-Patterson AFB. The other SV-5J is on display at the Air Force Academy in Colorado Springs.[71]

HYFAC, NHFRF, AND OTHER DREAMS

Several X-24B-derived follow-ons were proposed to continue research in support of the Space Shuttle, much as the X-15 had supported Project Apollo. Many (perhaps most) of these were referred to as the X-24C.

The FRC proposed a relatively austere air-launched vehicle with a Mach 8 potential and 40 seconds of sustained Mach 6+ cruise. NASA Langley proposed the Mach 12 Hypersonic Facilities Aircraft (HYFAC) and the somewhat less ambitious Mach 8 High-Speed Research Aircraft (HSRA). Both of these designs had provisions for evaluating prototype air-breathing scramjets. Draper and his colleagues at the AFFDL estimated that their entire X-24C program could be accomplished for $60–70 million using surplus XLR99 engines.[72]

During 1972, the Air Force proposed two other test vehicles, one air-breathing and one rocket powered. The air-breather looked a lot like a delta-winged F-15 fighter

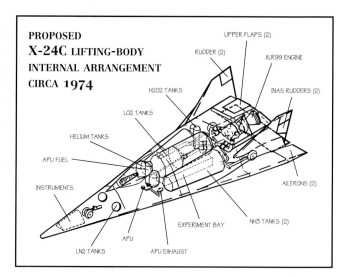

PROPOSED X-24C LIFTING-BODY INTERNAL ARRANGEMENT CIRCA 1974

This was an early proposal for an X-24C powered by a modified XLR99 engine instead of scramjets or larger rocket engine that powered most later proposals. (U.S. Air Force)

with end-plate vertical stabilizers, and was optimized for studying the Mach 3–5 environment. The rocket-powered proposal from the AFFDL, called the Incremental Growth Vehicle (IGV), would initially be optimized for Mach 4.5 missions, but later could be modified for Mach 6, and still later for Mach 9 research. McDonnell Douglas received a six-month contract to assist the AFFDL in defining the IGV in early 1973. The IGV was optimized for aerodynamic research and had a common structural core and subsystems, but much of the exterior could be changed to test various shapes and materials. The aircraft was expected to have a low-mounted delta planform, with later versions using a set of external propellant tanks that were carried as a 'vee' mounted ahead of the nose and stretching over the top of the wings. The IGV would have used versions of the XLR99 engine, and flights were expected to use the High Range developed for the X-15 program.[74]

Generally the Air Force studies tended to emphasize configurations suitable for military missions such as reconnaissance, while the NASA designs were more applicable to long-range hypersonic transports.

Among the latter, the Hypersonic Research Airplane (HRA) was a study conducted by NASA Langley with support from the Vought Corporation. A fundamental objective of the proposed vehicle was to provide a versatile research capability with the least expenditure of funds. The HRA was intended to study hypersonic stability, scramjet integration, and thermal protection systems. A 10-foot long payload bay was provided, and the outer wing panels were designed to be replaced to allow a variety of configurations and structures to be tested. The HRA would be launched from the same NB-52s that had supported the X-15 program, and would have a maximum speed of Mach 9 with a 60-second cruise capability at Mach 7. The vehicle was 50 feet long with a wing span of 24.2 feet, and weighed about 60,000 pounds.[74]

The HRA made extensive use of Lockalloy, a beryllium-aluminum alloy (Be-38Al) developed by Lockheed. This material had a high specific heat and stiffness at a low density – much higher than the Inconel-X used in the X-15. Lockheed had already successfully constructed a ventral stabilizer for the Mach 3 YF-12A interceptor from the material. The majority of the airframe was to be construct-

---

* Rocket-powered research is making a comeback with the X-33 and X-43 programs early in the 21st century.

ed as a Lockalloy heat sink with a shell attached to ring frames on 20-inch centers. The payload bay was to be constructed of aluminum, and was recessed from the outer mold line 6 inches on the top surface and 4 inches on the bottom. This allowed the thermal protection system materials to be mounted on stand-offs in the recessed areas.[75]

The HRA continued in parallel with other X-24C studies through mid-1978. A great deal of analytical work was accomplished, and some preliminary wind- and shock-tunnel work was apparently completed, but no actual hardware was ever constructed.

As late as July 1974, NASA and the Air Force jointly conducted a series of conceptual studies exploring various options for an air-breathing hypersonic aircraft. Surfacing from these studies was the realization that the FDL-8 (X-24B) shape appeared ideal. Two versions of this configuration, one with dual cheek-type air intakes for supersonic jet engines, and the other powered by a modified version of the XLR99, were released for comment in late-1974. NASA and the Air Force subsequently established the 'X-24C Joint Steering Committee,' which promptly rejected the relatively conservative vehicles proposed by NASA and the AFFDL, and started forming its own conclusions about the future of hypersonic research.[76]

Out of this committee came the National Hypersonic Flight Research Facility (NHFRF – pronounced 'nerf') concept in July 1975. NASA forecast a $200 million program involving construction of two aircraft with 200 flights over a ten-year period – NASA and the Air Force would start funding the program in 1980 with the first flight in 1983. Initially the vehicle built upon the proven X-24B shape, but as the program became steadily more ambitious, NHFRF slowly began to change configuration.[77]

The vehicle would have a top speed of Mach 8, with an extended Mach 6 cruise. One of the primary areas of research envisioned for the vehicle was advanced propulsion systems such as scramjets – during preliminary studies, several of the designs proposed by various manufacturers bore more than a passing resemblance to the final Aerospaceplane configurations from the early 1960s. After reviewing the preliminary studies, NASA Langley selected (contract NAS1-14222) the Lockheed Skunk Works to conduct a detailed analysis and preliminary design (and probably to eventually build the vehicles).[78]

The study was exhaustive, and ran from November 1975 until at least January 1977, and possibly longer. Initially Harry Combs and his Lockheed team focused on defining a feasible propulsion system for the hypersonic vehicle. It was obvious that rocket engines still provided the only reasonable means of achieving the desired

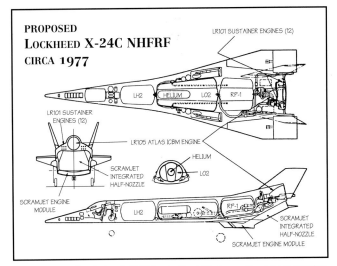

It is interesting that during the X-24C study Skunk Works considered the LI-900 thermal protection system as the highest risk, noting that it had the "highest man-hours per square foot refurbishment" and that it "jeopardizes the two-week turnaround time schedule." Nevertheless, it was eventually selected for Space Shuttle. (Lockheed Martin Skunk Works via the Tony Landis Collection)

speeds, but the 40-second steady-state Mach 6 cruise required much lower thrust than the initial acceleration, leading to the concept of using separate cruise engines.[79]

Engines considered during the study included extended-nozzle XLR99s, XLR11s, Rocketdyne LR105 sustainer engines from the Atlas ICBM, and LR101 vernier engines from the same missile. The problem was illustrated by the fact that either the XLR99 or LR105 was necessary to provide the initial speed increment, but to sustain Mach 6 cruise and a dynamic pressure of 1,000 psf at 90,000 feet required only 16,000 lbf. Under those conditions the minimum thrust from the XLR99 was 29,500 lbf – the LR105 was even worse at 46,000 lbf. Without separate cruise engines, this would have required the use of speed brakes during the entire cruise portion of flight, leading to higher propellant consumption and some interesting structural heating problems.[80]

The choice of cruise engines was based on many items besides available thrust. The weight of the engine installation, and the mix of propellants required (which might drive separate tanks, etc.). In the end it appeared that a LR105/LR101 combination was the best. Both engines had been man-rated for the Mercury program, although by 1977 the engines being produced were decidedly different than the 1961 engines had been.[81]

As with all proposals for air-launched vehicles using the NB-52 aircraft, the X-24C/NHFRF was limited to

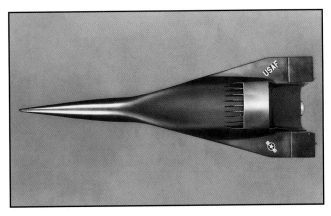

This was what the Lockheed Skunk Works thought the X-24C might look like. Note the integrated scramjet module under the aft fuselage and the large trailing-edge elevons. A single rocket engine was in the fuselage to provide the initial boost to high velocity. (Lockheed Martin Skunk Works via the Tony Landis Collection)

57,000 pounds. To achieve this, Skunk Works proposed Lockalloy as the major structural material. The entire skin area, including leading edges, was Lockalloy of sufficient thickness to act as a heat sink and not require the use of a thermal protection system. However, some mention was made in the study of the possible use of LI-900 tiles (a variation of what was eventually used on the Space Shuttle Orbiter) in very high heat areas.[82]

During 1977 the estimated costs of NHFRF escalated past $500 million, leading NASA Headquarters to cancel the effort in September 1977. NASA acting Associate Administrator for Aeronautics and Space Technology, James J. Kramer, stated that "… the combination of a tight budget and the inability to identify a pressing near-term need for the flight facility had led to a decision by NASA not to proceed with a flight test vehicle at this time…". It was exactly the same rationale that had caused the demise of Dyna-Soar ten years earlier.[83]

This was not the end, however. Hypersonic advocates within NASA and the Air Force still wanted a new research vehicle, and a much more austere Hypersonic Technology Integration Demonstrator effort began. In many regards this looked like an evolution of the delta-wing X-15 proposals, but was in reality a new design.[84]

The stated purpose of HYTID was "… to accelerate the development and demonstration of technology for future military systems designed to operate within the atmosphere at speeds between Mach 4 and Mach 8."[85]

The HYTID design contained sufficient flexibility to accommodate aerodynamic drag, thermal interference heating, and flight control perturbations that would probably accompany the various experiments it was meant to test. In addition, the forebody of the vehicle was designed to provide the appropriate inlet precompression properties required to test advanced scramjet engines, while the afterbody was shaped for the expected engine exhaust expansion requirements.[86]

Over 20 vehicle concepts were studied before selecting a baseline configuration. Computations were performed to derive vehicle size and volumetric requirements. Several preliminary weight and balance iterations were then performed to optimize the size and arrangement of the lifting and control surfaces, fuselage, landing gear, propellant tanks, and subsystems. Propellant and ancillary fluid usage sequences were defined for trajectory analysis of the rocket cruise, maximum Mach, and scramjet cruise missions.[87]

Then reality set in. The team had originally assumed they would be able to modify a B-52G as a carrier aircraft, thereby gaining about 15,000 pounds in available capability. The cost of this modification was prohibitive (not to mention SAC was not keen on giving up a relatively new B-52G). This resulted in falling back to the same NB-52B used by the X-15 and lifting-body programs, and its associated 57,000 pound weight limit. As noted by the designers, this "… made the HYTID design somewhat more difficult."[88]

Regardless, a concept definition was developed that was generally similar to an earlier NASA X-24C configuration except for the addition of wing-tip vertical end plates to improve directional stability throughout the entire Mach range. This outgrowth of the original X-24C lifting-body configurations had evolved from increasing the vehicle fineness ratio in order to decrease the airbreathing propulsion system size required for hypersonic cruise, and also to improve the low-speed landing characteristics.[89]

The initial boost to the desired speed was provided by a single XLR99, while two XLR11 engines provided cruise thrust to allow more time for experiments. This was one of the same configurations investigated by the Skunk Works during its X-24C studies. Allowances were made for uprating the XLR99 at some point in the future to provide better performance. A four-scramjet module, almost two feet high, could be installed under the aft fuselage. A pair of large jettisonable propellant tanks could be carried under the fuselage, very much like the X-15A-2.

The design criteria established for the HYTID was that the vehicle be capable of 100 flights at Mach 6 for a duration of up to 40 seconds. The primary heat-sink structure consisted primarily of aluminum and titanium alloys assembled using TIG or plasma welding to eliminate mechanical fasteners. This essentially cool structure would be covered with thermal protection systems, including the same variety of Martin Marietta-developed ablators used on several X-15A-2 flights, and nomex felt blankets very similar to those eventually used on the Space Shuttle Orbiter.[90]

In the end, nothing came of HYTID since there was no funding available to support new research aircraft. The design lingered on through most of the mid- and late-1970s before finally fading from the scene.

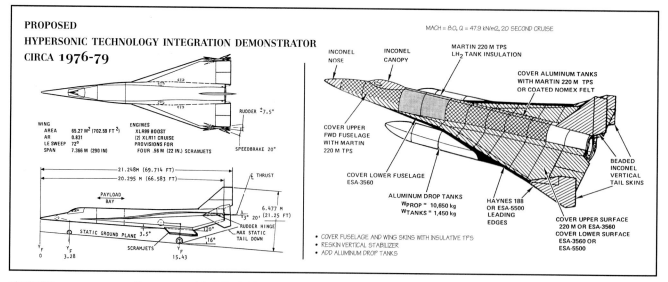

The HYTID bore an uncanny resemblance to some of the X-15A-3 delta-wing proposals, although in reality it shared little with them. The mix of rocket engines is noteworthy, combining the tremendous thrust of the XLR99 for the boost phase with the tried-and-true XLR11 for sustained cruise at Mach 6. (U.S. Air Force via the Scott Lowther Collection)

MINI-SHUTTLE

During August 1972 Milt Thompson and Joe Weil of the NASA FRC advocated construction of a 36-foot long, manned 'mini-shuttle' to study the Space Shuttle Orbiter's handling performance in the Mach 5-to-landing flight regime. The proposed single-seat vehicle had a 23-foot wingspan and weighed 30,320 pounds at launch. Air-launched from the NB-52, this vehicle could fly in direct support of Space Shuttle development, validating wind-tunnel and performance predictions at hypersonic to subsonic velocities, including the approach and landing phase.[91]

Thompson and Weil had estimated that an in-house designed, XLR99-powered, mini-shuttle could be built for $19.7 million, but critics, ignoring the FRC record of low-cost projects, argued this figure was closer to $150 million. Various other configurations, some using the XLR11, were also investigated, but none were approved for construction.[92]

The Dryden-built M2-F1 lifting-body poses with the Space Shuttle Orbiter Enterprise on the ramp at the Dryden Flight Research Center. (NASA photo EC81-16287)

Despite strong backing from Northrop, Martin Marietta, and North American Rockwell, the proposed mini-shuttle succumbed to the cost argument. By this time, the Space Shuttle had become a somewhat controversial project, and neither NASA, nor the contractors, were willing to push too hard. Interestingly, no one seems to have proposed a sub-scale unmanned shuttle reentry vehicle to be flown like the earlier ASSET and PRIME shapes. It only later came to light that the Soviets had launched several sub-scale models (BOR-5) of the *Buran* orbiter prior to committing the actual design to its first orbital flight. The Orbiter's actual hypersonic, supersonic, and transonic performance remained unproven until the first Mach 25 manned reentry from space. In any case, in 1981 *Columbia* would prove that the aerodynamic performance of the Space Shuttle Orbiter was completely satisfactory.[93]

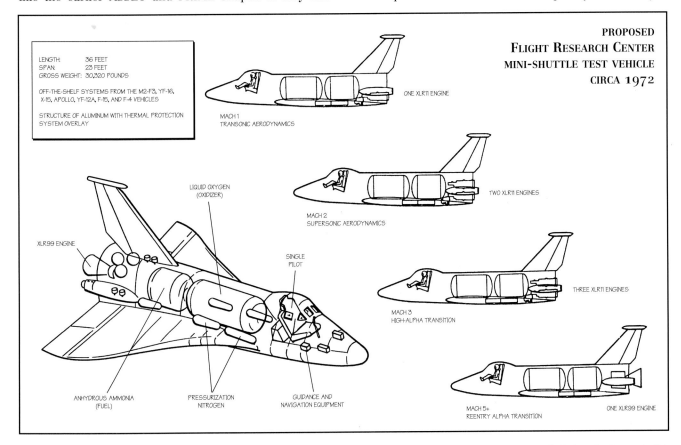

The NASA Flight Research Center proposed building a 'mini-shuttle' in order to test the basic aerodynamic stability of the orbiter from approximately Mach 5 to landing speeds. This was similar to what the Soviets did to test aspects of the Buran design. The proposal was rejected for cost and political reasons, and the actual performance characteristics of the Orbiter were not confirmed until Columbia returned from space in April 1981. (U.S. Air Force)

Alfred C. Draper. (Courtesy of Alfred C. Draper III and the Draper Family)

ASSET

The cancellation of Dyna-Soar was a significant blow to the advocates of lifting reentry – within both the government and industry. But even before Dyna-Soar's cancellation, it had been decided that, given McNamara's reluctance to pursue a manned orbital capability, the best that could probably be accomplished in the short term was a series of unmanned tests to evaluate some of the concepts that had been expected to be proved by Dyna-Soar. Hopefully, the results could be used to build a larger system sometime in the future.

The Air Force Flight Dynamics Laboratory (AFFDL) at Wright-Patterson AFB initiated two such projects during August 1959. Projects 1366 (Aerodynamics and Flight Mechanics) and 1368 (Structural Configuration Concepts for Aerospace Vehicles) were a small part of the on going Mechanics of Flight (Program 750A) applied research program. It appeared that the advent of new high-temperature materials and the development of a sophisticated guidance package for the Scout booster would permit the development of large-scale hypersonic test models representative of possible reentry shapes. These models could be launched using Scout or Blue Scout boosters to obtain detailed data during actual reentry flights.[1]

The AFFDL originally envisioned the gliders as simple wing-body vehicles with a pronounced keel on the ventral surface. But engineer Alfred Draper recognized that the keel design would introduce a dihedral effect problem and also introduce serious heating problems during reentry. Draper successfully argued that the gliders should be built to the WLB-1 design that had recently been developed by the AFFDL. The final configuration bore a slight resemblance to the Dyna-Soar glider with a flat-bottom delta planform, rounded nose cap, rounded wing leading edges, and a tilted ramp-nose for hypersonic trim. The similarity to Dyna-Soar allowed designers to take advantage of the large body of wind- and shock-tunnel data that was being accumulated by the Dyna-Soar project, although much of the detailed data was being closely held by Boeing.[2]

Initial attempts to begin a program to build and fly seven gliders were thwarted by a lack of available boosters, but by April 1960 it appeared that a somewhat reduced effort could proceed, and contracting efforts were begun. On 31 January 1961 the effort was designated Project 1466 within Program 750A and named Aerothermodynamic/Elastic Structural Systems Environmental Tests (ASSET). During April 1961, McDonnell Aircraft signed a cost-plus-fixed-fee contract (AF33(616)-8106) with the Research and Technology Division of the AFFDL for the development of six experimental gliders that would be the major thrust of the project. The stated purpose of ASSET was to "... assess, through free-flight tests, the applicability and accuracy of theories, analytical prediction methods, and experimental techniques available for the solution of reentry problems in structures, aerothermodynamics and aerothermoelastics."[3]

There were two slightly different glider configurations – aerothermodynamic structural vehicles (ASV), and aerothermoelastic vehicles (AEV). McDonnell built four of the former and two of the latter. The gliders were 5.7 feet long and weighed 1,130 pounds (ASV) or 1,225 pounds (AEV) fully equipped, dependent upon the instrumentation installed. All had a sharply swept (70 degrees) low aspect-ratio delta wing that spanned 4.6 feet and provided 14 square feet of area. The flight control system used hydrogen peroxide reaction thrusters maintained flight attitude after launch. Three of the ASV gliders would be launched using a DSV-2F Thor* first-stage with an Aerojet AJ10-118 (Delta) upper stage (creating a DSV-2G); the first ASV and both AEV gliders used standard DSV-2F Thor boosters without an upper stage.[4]

Although superficially similar, and sharing common subsystems, the two types of gliders differed completely in mission and research capabilities, and this was reflected in their totally different flight profiles. The ASVs had a mission to determine temperature, heat flux, pressure distribution, and to evaluate materials and structural concepts during hypersonic gliding reentry. These were boosted to altitudes of 190,000 to 225,000 feet, and velocities from 11,000 to 13,250 mph giving them a range varying from 1,000 to 2,300 miles. On the other hand, the AEVs were boosted to 168,000 and 187,000 feet at a velocity of 8,900 mph, obtaining ranges of 620 and 830 miles. Their mission was to study various panel flutter phenomenon, and the effect of hypersonic flight on aerodynamic control surfaces.[5]

The X-15 had pioneered the use of Inconel-X and the A-12/SR-71 program was demonstrating the large-scale use of titanium as a structural material. ASSET would give

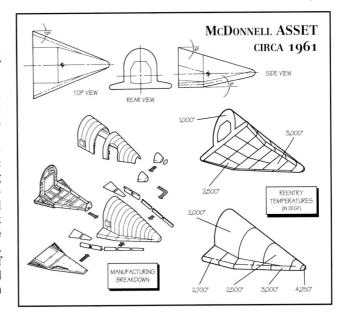

The four ASVs encountered temperatures up to 4,250 degF on their zirconium nose caps. Various other exotic materials were also used in the vehicle's construction, including molybdenum (up to 3,000 degF), graphite (also 3,000 degF), columbium (2,500 degF), L-605 cobalt (2,000 degF), and titanium (1,000 degF). (U.S. Air Force)

---

* These Thor Intermediate-Range Ballistic Missiles (IRBM) were part of a group returned from the United Kingdom after having stood nuclear alert for several years and were available at extremely reasonable costs. Contracting problems with NASA (who managed the Delta upper stage) resulted in the first ASV being launched without it.

the first real-world exposure to a new generation of refractory materials such as zirconium and carbon, and superalloys such as molybdenum and columbium. Advanced thermal insulating materials were also tested. Contractors other than Martin also provided materials to be tested – Boeing, Bell, and Solar all provided panels or nose caps that were used on the ASSET gliders. Through these tests ASSET provided a wealth of data that contributed to the development of more advanced materials, primarily composites and carbon-carbon that would be available when Space Shuttle development began later in the decade.[6]

The major problem facing the ASSET team was the development of reliable test equipment (primarily transducers and telemetry) and determining realistic test conditions. Defining the ideal test environment was more difficult than it might seem – when it appeared the projected temperatures were just about right, the structural loads or oxidation levels might be totally wrong. It was a careful balancing act intended to obtain as much useful data as possible in a short flight test program. And some of the tests were complicated. For instance, testing the zirconium nose cap involved achieving a maximum temperature of 4,250 degF using a heating rate of 70 degF per second at an oxidizing atmosphere of 220,000 feet, then holding approximately 4,000 degF for 17 minutes during reentry. But the idea was to obtain useful and realistic test data that could be compared with similar data obtained in hypersonic shock- and wind-tunnel at NASA Langley, the Air Force's Arnold Engineering Development Center, and the Naval Ordnance Laboratory.[7]

It had been expected that the first glider would be flown during January 1963, but the change from Scout to Thor boosters somewhat delayed the program. The change was welcome, however, since the Thor allowed higher velocities and altitudes than would have been possible using Scouts. The first launch (ASV-1) finally occurred on 18 September 1963, just a few months prior to Dyna-Soar's cancellation. Interestingly, no formal exchange of ideas or data had transpired between the two

programs. Four more of the vehicles were launched during 1964, and the last glider (ASV-4) was launched on 23 February 1965. All were launched from Cape Canaveral and flew downrange towards the south-south-east.[8]

Although the second vehicle (an ASV) was destroyed when its booster failed, and several vehicles were not recovered from the Atlantic around Antigua, the program was considered highly successful, and demonstrated that winged reentry vehicles could successfully traverse the upper atmosphere. The total cost of ASSET, including the Thor boosters, was $41 million ($21,236,000 for the gliders and operations, and $19,861,000 for the boosters).[9]

AMERICAN INDIAN

Not directly connected in any obvious way to the development of the space shuttle, the Air Force's Navaho program of 1946–58 contains many parallels to Shuttle and set in place many of the development activities that were later used by shuttle. North American Aviation was one of seventeen potential contractors asked to submit designs in response to Air Force requirements for a long-range missile capable of delivering nuclear warheads. On 29 March 1946, North American was issued a $500,000 letter contract (W33-038-ac-14191) for a one year study of a supersonic guided missile with a 500 mile range. Designated MX-770, the missile was initially to carry a 2,000 pound nuclear warhead, although this was increased to 3,000 pounds on 26 July 1946. North American was also provided with several German A-4 and V1 weapons for evaluation, and was tasked with the development of an 'Americanized' sub-scale version of the A-4. Called the Nativ, this missile would to be used for testing various subsystems for the Navaho. The Nativ eventually flew on 26 May 1948, and a total of six launches were attempted – three got off the launch pad, but only one was considered a success. An inauspicious beginning to the American missile program, but a great deal was learned nevertheless.[10]

During July 1947, the Air Force ordered North American to redirect its on-going work towards a three-phase development process that would yield short-range (175–500 mile), intermediate-range (500–1,500 mile), and long-range (1,500–5,000 mile) missiles, although official interest in the short-range version quickly faded. By March 1948 this effort had been directed exclusively

Several of the ASSET vehicles were recovered after their flights and survive today. The ASV-3 vehicle (left) is in the Air Force Museum at Wright-Patterson AFB, Ohio. Note the Dyna-Soar model in the background. An unidentified ASSET is on display (above) at the Air Force Space Museum at Cape Canaveral, AFS. Seeing these vehicles in person relates just how small they were. (Dennis R. Jenkins)

towards a 5,000-mile range supersonic missile called Navaho I, capable of carrying a 10,000 pound nuclear warhead. It was expected that the vehicle would be propelled by a liquid-fueled rocket after being launched by a ground-based rocket-sled or from a larger aircraft.[11]

By October 1949, work on the Navaho had progressed to the point that a final decision on the flight profile and propulsion system had to be made. Two distinct possibilities existed – a two-stage ballistic rocket or a rocket-boosted ramjet-powered cruise missile. The science of pure ballistic flight was in its infancy, and there was a great deal of concern over whether it could be perfected within a reasonable time period. Not wanting to direct all of its efforts towards a single technology, the Air Force decided to pursue both concepts. Convair was directed to concentrate its efforts on the ballistic concept that eventually resulted in the Atlas ICBM – North American was to pursue the winged cruise missile. The decision was subsequently made to use a large unmanned rocket-powered aircraft as a booster for a ramjet-powered cruise missile.[12]

Throughout this development process, the entire Air Force missile program was undergoing constant reorganization and Navaho was no exception. Work was repeatedly redirected between the definitive 5,000 mile-range weapon, and two interim 1,500 mile versions as priorities changed at the Pentagon. Work was divided between three projects: the original MX-770; a follow-on MX-771A which competed the supersonic Navaho II with the subsonic Martin Matador; and MX-775B which competed the definitive Navaho III with the supersonic Northrop Boojum. There was also a constant shifting between the Air Force's desire for air-launched version of Navaho and ground-based derivatives.[13]

At least the last problem was resolved in August 1950 when the Air Force directed that all work on the air-launched version be terminated, and efforts concentrated exclusively on the ground-based derivative. But this in itself presented a problem – the ground-based version required the use of a rocket-powered first stage to get the cruise missile to a velocity sufficient to ignite the ramjets. Initial estimates suggested that the booster would need two 120,000 lbf rocket engines, although this was subsequently raised to two 135,000 lbf units. No rocket engines of such power were available in the United States at the time.

This caused North American to enter the large rocket engine business, and the Rocketdyne Division was born. The new company quickly established a location in Santa Susana, California, that would become the home to the most extensive rocket engine test facilities in the country until the advent of the National Space Transportation Laboratory* in the 1960s. The engine developed for Navaho itself would continue to be refined, and variants of it were eventually adapted for use in Thor, Jupiter, Atlas, Delta, Saturn I, and Saturn IB space launch vehicles. Rocketdyne and its facilities would later play a key role in the development of the Space Shuttle Main Engine.[14]

The guidance system for Navaho also presented unique challenges. The original contract called for a circular-error-probability (CEP) of less than 1,500 feet at a range of 5,000 miles, although this was later relaxed to three nautical miles as part of the Tea-Pot Committee recommendation. To develop such a sophisticated guidance system, North American created the Autonetics Division and adapted the inertial guidance system developed by MIT's Charles Draper (who would later play a key role in the

North American built the Westinghouse J40 turbojet-powered X-10 as an aerodynamic and systems testbed for the cruise component ('upper stage') of the SM-64 Navaho missile. At a later date the X-10 itself was considered a cruise missile candidate, armed with a nuclear warhead and capable of taking off and flying to its target under its own power. The successful development of the Atlas and Titan ICBMs precluded any significant work on this concept. The X-10 successfully contributed to the development of the much larger Navaho missile, although that program would subsequently be cancelled in favor of the rocket-powered ballistic missiles. The X-10 verified the aerodynamics of the cruise component, as well as its complex navigation system. Thirteen X-10s were manufactured, and ten were actually flown – only a single example still survives, displayed at the Air Force Museum. An SM-65 Navaho missile is on display outside the gate at the Cape Canaveral AFS, Florida. (U.S. Air Force via the AFHRA Collection)

development of space shuttle guidance systems) for use in Navaho. The initial results were less than spectacular since the early gyroscopes (indeed all gyroscopes until the advent of laser-ring gyros) tended to drift unacceptably. To deal with this anomaly, Autonetics adopted an optical star-tracker to update the inertial position periodically by locating known stars. This same technique would later be employed on the Space Shuttle Orbiter.[15]

A Navaho G-26 test vehicle on its booster at Cape Canaveral before launch. The Navaho used two ramjet engines, while the booster used two liquid-propellant rocket engines. This 'piggy-back' concept would be featured in many of the subsequent two-stage-to-orbit space shuttle studies. (North American Aviation via the Terry Panopalis Collection)

---

* Now the Stennis Space Center, Mississippi.

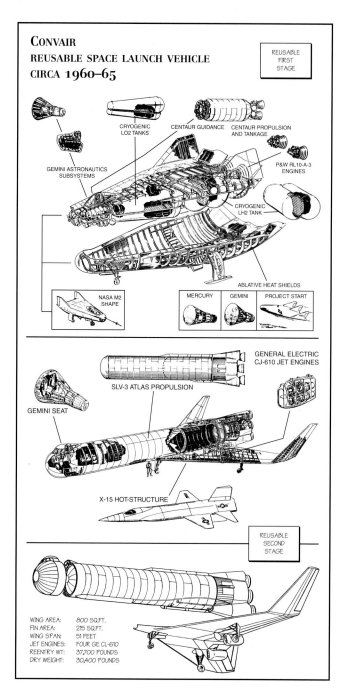

CONVAIR
REUSABLE SPACE LAUNCH VEHICLE
CIRCA 1960–65

REUSABLE
FIRST
STAGE

CRYOGENIC LO2 TANKS
CENTAUR GUIDANCE
CENTAUR PROPULSION AND TANKAGE

GEMINI ASTRONAUTICS
SUBSYSTEMS

P&W RL10-A-3
ENGINES

CRYOGENIC
LH2 TANK

ABLATIVE HEAT SHIELDS

NASA M2
SHAPE

MERCURY | GEMINI | PROJECT START

GENERAL ELECTRIC
CJ-610 JET ENGINES

SLV-3 ATLAS PROPULSION

GEMINI SEAT

X-15 HOT-STRUCTURE

REUSABLE
SECOND
STAGE

WING AREA: 800 SQ.FT.
FIN AREA: 215 SQ.FT.
WING SPAN: 51 FEET
JET ENGINES: FOUR GE CL-610
REENTRY WT: 37,700 POUNDS
DRY WEIGHT: 30,400 POUNDS

Towards the end of the study, the Convair concept for a reusable Atlas became quite sophisticated and included an orbital version of the NASA M2 lifting-body – the M2 sat on top of the winged atlas and the pair was launched vertically. The first-stage booster was equipped with jet engines and would fly back to its launch site for a horizontal landing, while the M2 would glide to its landing site after reentry. Some thought was given to replacing the Atlas sustainer engine with a Rocketdyne H-1 from the Saturn program (Convair via the AFHRA Collection)

Thus, the Navaho gave birth to several key technologies[*] and companies that would play major roles in the development of the Space Shuttle. But Navaho itself never got past the flight test phase for several reasons. One was unforeseen development problems with both the ramjet propulsion and the inertial guidance systems. But more importantly were breakthroughs that allowed lightweight hydrogen bombs that proved small enough to be launched by the ballistic missiles of the era. The ICBM subsequently became the strategic weapon of choice for the Air Force. Navaho was cancelled in August 1957 after the construction of 13 sub-scale X-10 prototypes and seven full-scale XSM-64 (XB-64A) development vehicles.

AEROSPACEPLANE

As it happened, Dyna-Soar was not the only orbital system being considered by the Air Force during the late 1950s and early 1960s. Another project, conducted quietly, was the Air Force Aerospaceplane (not to be confused with the 1980's X-30 National Aero-Space Plane – NASP). The history of Aerospaceplane (abbreviated ASP) is one of the more confusing paths on the road to space shuttle. Partially because it was shrouded in secrecy, but perhaps more because it was generally unorganized and fragmented, it is difficult to trace the true progress of the ASP studies. Seemingly, a great deal of money was expended on the project, and at least seven major aerospace contractors were involved in it at some time or another.

The story begins with a 1957 DoD Study Requirement (SR-89774) that investigated the concept of reusable space boosters. Contracts were issued to several contractors to determine the feasibility of a reusable space launch vehicle (RSLV) – each focusing on a different existing or proposed booster. In all cases, the contractor was to examine the minimal modifications needed to permit winged, fly-back recovery with a horizontal landing.[16]

Boeing concentrated on the Saturn family of boosters being developed by NASA. Interestingly, similar work was also accomplished by Boeing for NASA under contract NAS8-5036 through the end of 1962. Martin and North American also investigated reusable Saturns as part of the NASA studies. The Boeing study concluded that adding wings to the first stage of the Saturn C-5 was feasible, and could result in a fully-reusable fly-back first stage by 1970. The concept was considered economically feasible given a launch rate of 24 per year, but resulted in a 22 percent reduction in payload capacity to low-earth orbit. One interesting item from the later NASA-sponsored studies is that Boeing situated the proposed KSC runway for the booster in roughly the same location as the shuttle landing facility would be built a decade later.[17]

The Atlas booster was the subject of studies at Convair. Although initiated as part of SR-89774 in 1957 these efforts continued through at least the end of 1965 as part of on-going NASA space station resupply and Air Force recoverable booster studies. Martin Marietta was asked to investigate modifications to the Titan II booster, and over 20,000 manhours were invested in a configuration that ultimately included a large set of aft-mounted wings, forward canards, and a pair of turbojet engines in nacelles mounted about mid-body.[18]

These studies all concluded that reusable boosters could substantially reduce the cost of putting payloads into orbit. This was encouraging since some factions within the Air Force wanted a permanent manned presence in space, and others wanted to use space as a rapid way to get supplies from one point on the Earth to another (antipodal delivery). The budding national intelligence community was also interested in reducing the cost of launching the future spy satellites.

But others had grander plans. Begun by Weldon Worth while he was Director of the Aero Propulsion Laboratory at Wright-Patterson AFB, Aerospaceplane was probably the first major attempt to actually develop a large com-

---

* Some published reports (such as James N. Gibson, *The Navaho Missile Project: The Story of the "Know-How" Missile of American Rocketry*, Schiffer Publishing, 1996), have credited the Navaho as being the forbearer of almost everything that has subsequently happened in worldwide aviation. Although this is highly overstating the case, Navaho nevertheless played an important role in many other projects, even if the SM-64 itself was cancelled.

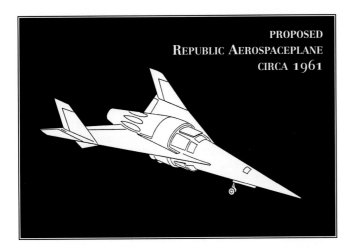

This was an early single-stage-to-orbit Aerospaceplane concept from Republic's Alexander Kartveli. This design soon gave way to the tee-tail design shown on page 57, but would make a brief comeback in 1965 when the Air Force requested proposals for scramjet test vehicles. The test vehicle would have taken-off under its own power using a turbojet accelerator. At approximately Mach 3, the scramjet engine would ignite, allowing speeds of Mach 12. The scramjet combustor encircles the fuselage and used a self-regulating exhaust plug nozzle in the scramjet mode. Flaps around the fuselage regulated the airflow into the engine at high speeds. (U.S. Air Force)

pletely-reusable logistical vehicle capable of flying into space and returning using lifting reentry techniques.[19]

Prior to mid-1960 the project was generally called Spaceplane (no Aero) and focused on feasibility studies of potential single-stage-to-orbit (SSTO) concepts. Envisioned as a large winged vehicle that would take-off and land horizontally from conventional runways, most of the Spaceplane concepts used a complex multi-phase propulsion system that would have extracted gaseous oxygen from the atmosphere, liquefied it, and combined it with stored fuels to become an air-breathing hybrid rocket.[20]

Concurrently with (or perhaps as part of) Spaceplane, various studies were initiated into the development of liquid-air cycle engines (LACE) and air-collection engine systems (ACES). These systems seemed to provide a way to lower the gross lift-off weight of an SSTO vehicle by allowing it to launch without most of the oxidizer necessary for the ascent to orbit. Instead, the ACES would liquefy oxidizer from the atmosphere while the vehicle gathered speed.

ACES used a series of heat exchangers where liquid hydrogen in thin tubes counterflowed against a stream of incoming air. The low temperature of the liquid hydrogen caused the air to turn into a liquid, which was stored in low pressure tanks. This 'liquid air' was then pumped under high pressure into the combustion chambers like any other oxidizer. Not as efficient as liquid oxygen, but then, the vehicle did not have to carry the oxidizer (which is much heavier than fuel) from the ground either.

During late 1960 and early 1961 Marquardt engineers successfully demonstrated the basic engine concept using prototype Garrett AiResearch heat exchangers at company test facilities in Saugus, California. Engines up to 275 lbf were successfully operated for more than five minutes at a time using oxidizer liquefied from ambient sea-level air. The presence of nitrogen in the air caused a slight degradation of performance in these early tests, but since there is a significant difference in the boiling point of nitrogen and oxygen, the engineers soon discovered how to control the process to yield a higher oxygen content. Research on this system continued under Air Force contract until at least 1967 when it was known as LACES (liquid air collection engine system), although little documentation seems to have survived.[21]

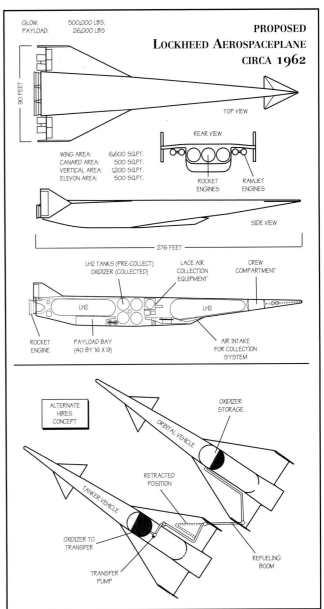

Most of the Lockheed effort was placed on the vehicle shown at top, while a lesser amount was expended on the HIRES concept that attempted to transfer oxidizer from one vehicle to another at speeds between Mach 4 and Mach 6. The baseline vehicle used a turbo-LACE system from take-off to Mach 3.5, supplemented by a ramjet from Mach 2 to Mach 9 and a nitrogen expander rocket from Mach 3.5 to Mach 5.5, and finally a high-expansion-ratio rocket from Mach 5.5 to orbital velocity. Air was collected by LACE during ascent for use by the final rocket engine. (Lockheed-California Company)[21]

Convair Spaceplane

One of the contractors that studied the Spaceplane was the San Diego Division of Convair. This contract was awarded in late 1957, but was terminated in July 1959 for unknown reasons. Nevertheless, the company continued to evaluate the concept using a combination of internal funding and small Air Force contracts intended to study specific subsystems. Eventually Convair received Aerospaceplane contracts that built on the technology base established during these studies.

Several variations of the Convair concept were proposed, including a 'one-stage-to-orbit' vehicle and a 'versatile' Spaceplane that combined an SSTO capability with the additional ability to act as a booster for second stage vehicles. Both unmanned expendable and manned recoverable second stages were investigated. Interestingly, in most configurations the second stage was carried, completely enclosed, in the large payload bay of the SSTO vehicle. Convair confirmed that gross lift-off weight and total propellant consumption were the two primary factors affecting orbital payload performance – single-stage-to-orbit configurations were shown to be even more sensitive to these factors, leading Convair to adapt a version of the ACES engine system. The study concluded that the SSTO Spaceplane could place about 18,000 pounds into a 500-mile orbit with a GLOW of 480,000 pounds. At the same GLOW, a two-stage version using ACES could place almost 90,000 pounds into orbit.[22]

One of the problems encountered by Convair, and indeed all of the Spaceplane contractors, was what engines to use. At the time there were only two reasonable choices: liquid-fuel rocket engines and ramjets. Ramjets are air-breathers and seemed ideal for accelerating in the atmosphere while ACES collected oxidizer to feed the rockets. But ramjets had a major drawback – they could only operate below Mach 4, and Spaceplane needed to accelerate to much greater speeds before using its rocket engines. Some configurations used hybrid engines that operated as ramjets below Mach 4 and turned into inefficient* rockets above Mach 4. This would be a major stumbling block for the first year or two of the studies.

The aerodynamic configuration seemed easier to define. Spaceplane used a notched delta planform that allowed the ramjets to be located with the inlets in the wing pressure field and the exhausts near the trailing-edge of the wing. The three-inch leading edge radius and 70-degree leading-edge sweep was a compromise between leading edge stagnation temperature and drag. The basically poor structural planform was offset by the large thickness ratio – 14 percent at the root with maximum thickness at the trailing edge and good fuel distribution. About a quarter of the air collected by ACES was

stored near each wing tip to provide inertia relief at maximum weight. It is interesting to note that at launch, all available tankage was filled with liquid hydrogen, and as the hydrogen was burned the tanks were purged and filled by liquid air generated by ACES.[23]

Convair estimated, extremely optimistically as it turned out, that between 2 and 20 million pounds would need delivered to orbit per year beginning in 1970, and a figure of 10 million pounds was used for comparison purposes during the study. Each payload was expected to weigh between 10,000 and 100,000 pounds, with a majority falling towards the lighter end of the range. Most military payloads were expected to be launched to orbits below 1,000 miles, while commercial payloads were expected to be geosynchronous. As with many early space shuttle studies (which this was not, but it was certainly a precursor), Convair justified the $2,900 million in development costs for the vehicle against a predicted savings in launch operations costs. In this case, Convair estimated that based on a system life of ten years the Spaceplane would cost only $6,000 million to operate, versus $23,000 million for an equivalent number of expendable boosters.[24]

Enter the Scramjet

But a new propulsion concept was also being developed, one that was again resurrected for use on NASP during the 1980s. The scramjet had first been discussed during early 1958 in a technical paper from the NACA Lewis Research Laboratory that attempted to address the fact that conventional ramjets could not efficiently operate much above Mach 4 due to pressure and temperature problems in the combustion process. The scramjet solved

---

* Inefficient because the combustor and nozzle were not optimized for rocket performance since they also had to operate as ramjets.

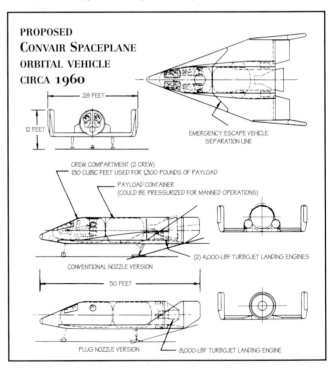

PROPOSED
CONVAIR SPACEPLANE
CIRCA 1960

FORWARD FUEL BULKHEAD

ENGINES 129 FEET

FORWARD PROPULSION BAY

WING AREA 632 SQ.FT.

TURBOFAN ENGINES

LIQUID AIR

LH2

PAYLOAD

ROCKET PROPELLANT EXPENDABLE TANKS

ROCKET ENGINES

GROSS TAKE-OFF WEIGHT WITH PAYLOAD = 450,000 POUNDS

270 FEET

FOUR-MAN CREW COMPARTMENT

55.8 FEET

S P A C E P L A N E
RECOVERABLE BOOSTER STUDY

PROPOSED
CONVAIR SPACEPLANE
ORBITAL VEHICLE
CIRCA 1960

28 FEET

12 FEET

EMERGENCY ESCAPE VEHICLE SEPARATION LINE

CREW COMPARTMENT (2 CREW) 130 CUBIC FEET USED FOR 1,300 POUNDS OF PAYLOAD

PAYLOAD CONTAINER (COULD BE PRESSURIZED FOR MANNED OPERATIONS)

CONVENTIONAL NOZZLE VERSION

(2) 4,000-LBF TURBOJET LANDING ENGINES

50 FEET

PLUG NOZZLE VERSION

8,000-LBF TURBOJET LANDING ENGINE

The Convair Spaceplane had a gross take-off weight of 450,000 pounds using conventional air-breathing propulsion. The LACES system generated liquid air from the atmosphere during ascent. At approximately 66,000 feet and a velocity of 5,300 fps the engine switched cycles to burn the stored fuel and oxidizer. At this point the aircraft weighted over 1,000,000 pounds – the LACES had collected approximately 550,000 pounds of liquid air during ascent. (Convair via the Robert Bradley Collection)

The Spaceplane had a large payload compartment that could alternately carry this small orbital vehicle. Much like later space shuttle studies, this orbital component carried all of its propellants in external tanks alongside the fuselage on top of the wings. (See insert in the drawing to the left.) These were jettisoned as the vehicle began its reentry. Note the emergency crew escape capsule integrated into the forward section of the vehicle. (Convair via the Robert Bradley Collection)

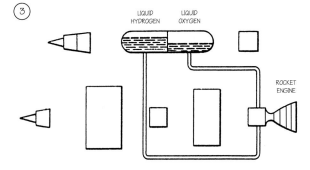

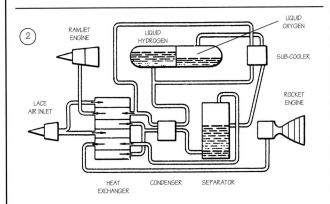

From take-off to 2,300 fps and 30,000 feet the primary propulsion was from the rocket engine. Air came in through the LACE air inlet and entered the heat exchanger where it was cooled to a very low temperature and then condensed to liquid form in the condenser. This liquid air was used as an oxidizer in the rocket engine. LH2 was used to cool the incoming air, and also as a fuel in the rocket engine. In addition, some LH2 was bypassed from the heat exchanger to the ramjets, mainly to reduce their drag by allowing them to operate at partial thrust.

From 2,300 fps to 5,300 fps and 60,000 feet, power came from the ramjets, and the LACE was used to collect LO2 by passing the liquid air through a separator that produced fairly pure LO2 and liquid nitrogen as a by-product. The nitrogen was dumped overboard or used to purge various systems if required, and the LO2 was saved.

When the Spaceplane reached 5,300 fps and 60,000 feet the LACE system was shut down and the aircraft continued to accelerate using the ramjets. After the Spaceplane passed through approximately 70,000 feet, all power was provided by the rocket engine operating in the conventional mode where liquid oxidizer and fuel were taken from onboard tanks. By the time the engines switched to their rocket mode the aircraft weighed over 1,000,000 pounds – 550,000 pounds more than it weighed at take-off. The Spaceplane eventually reached 24,500 fps and 1,500,000 feet – orbital velocity and altitude. At this point the vehicle weighed approximately 200,000 pounds, and sufficient oxidizer and fuel remained to perform on-orbit maneuvers, and de-orbit. The vehicle landed as a glider.

On 23 February 1962, Archibald Gay, an engineer at Convair in San Diego, applied for a patent on a 'method and apparatus for coordinating propulsion in a single stage space flight.' It would take over a decade, but on 4 September 1973 the U.S. Patent Office granted General Dynamics Corporation patent #3,756,024 covering the concept. The engine described here was generally similar to the other LACES/ACES concepts being investigated by Marquardt and Rocketdyne, so this (very simplified) description is applicable to those designs as well. For the purposes of the patent, the Spaceplane shown here was 235 feet long, a wing span of 133 feet, and had a gross lift-off weight of 450,000 pounds. This was with the liquid hydrogen tanks full of fuel, but essentially no oxygen in the oxidizer tanks.
(General Dynamics/Convair via Bob Bradley and the AFHRA Collections)

these by achieving supersonic combustion. Testing was initiated in long-duration facilities where proof-of-principle experiments confirmed the feasibility of supersonic burning – prolonged full-scale demonstrations of supersonic combustion, however, would prove more elusive. Most of the work accomplished on the early scramjets was done by Antonio Ferri, initially at NACA Langley, then continued after moving to the General Applied Science Laboratory in Ronkobkoma, New York. Researchers Fred Billing and Gordon Dugger at the Johns Hopkins Applied Physics Laboratory also continued to refine and expand the scramjet concept, with help from several airframe and engine manufacturers, particularly Marquardt.

During this research Ferri established a close working relationship with one of aviation's most prolific innovators, Alexander Kartveli, the chief engineer at Republic Aviation. One of the more exotic combinations proposed by these two men was the stillborn hybrid-ramjet-powered XF-103 interceptor that was expected to fly at over 2,500 mph in a day when the North American F-100 Super Sabre was just barely capable of breaking the sound barrier.

By early 1960, the scramjet appeared to be a viable propulsion system, and one that would solve some of the technical difficulties being encountered during the Spaceplane studies, which were subsequently cancelled. In their place, towards the end of 1960,[*] the Wright Air

Development Center initiated the Aerospaceplane program which was intended to develop a single-stage-to-orbit aircraft using scramjet propulsion as part of Project 651A. Requirements for Aerospaceplane included a crew of three and a 40 by 25-foot payload bay that was 10 feet high.[25]

The scramjet and Aerospaceplane seemed made for each other, and Republic quickly entered the picture with a large bullet-shaped delta-winged vehicle with a series of intakes and exhausts for the scramjet propulsion system around a circular center fuselage and a single rocket engine in the extreme tail. This eventually gave way to a more conventional design with swept-wings and a tee-tail, but with the now-classic scramjet external-combustion shape on the fuselage bottom directly under the wing. Boeing, Convair, Douglas, Goodyear, and Martin also undertook preliminary studies of an Aerospaceplane under contracts from the Air Force totaling $20 million. The studies seemed to indicate that an operational system could be developed by 1966–68 if a sufficiently high priority was assigned to the project.[26]

The propulsion system for the Aerospaceplane involved several propulsion cycles, including turbojets that operated up to Mach 3, a modified ramjet where fuel was burned subsonically in the combuster and expanded supersonically to produce thrust up to Mach 6, and finally, a hybrid scramjet-rocket up to orbital velocity. Scramjets maintain a supersonic flow in the combustor – the fuel is then injected and burned at supersonic velocities, usually outside the combustion chamber. The idea of extracting oxidizer from the atmosphere via ACES was retained however, and most

---

* In November 1960 the Air Force Scientific Advisory Board recommended that a project be initiated to investigate "… the concept of an orbital aircraft" and this formed the genesis of the Aerospaceplane studies.

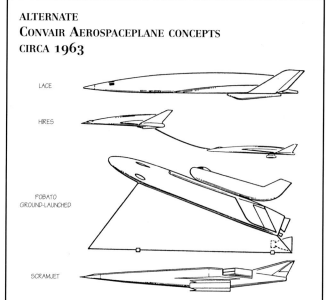

ALTERNATE
CONVAIR AEROSPACEPLANE CONCEPTS
CIRCA 1963

LACE

HIRES

POBATO
GROUND-LAUNCHED

SCRAMJET

This transparent model of an early Convair Aerospaceplane concept shows the arrangement of the major internal components. Like most of the original scramjet-powered vehicles, this design has a series of inlets and exhausts around the outside of a basically circular fuselage. Unfortunately, no specific data could be found regarding the design, although it most probably used a LACES/ACES propulsion system since Convair was heavily involved in that technology. (Convair via the SDAM Collection)

concepts continued to liquefy air at lower altitudes and store it for use as the vehicle began to leave the sensible atmosphere. Testing showed that hydrocarbon-fueled scramjets would be capable of sustained flight to Mach 8, while fuels such as liquid hydrogen would be required to achieve orbital velocities. ACES concepts continued to be developed by Convair, Linde (a division of Union Carbide), Marquardt, and Rocketdyne.[27]

ACES was not the only concept investigated during the Aerospaceplane studies. Most contractors did parametric evaluations of conventional concepts that carried all of the propellants from the ground – termed POBATO (propellants on-board at take-off). A much more bizarre concept was called HIRES (hypersonic in-flight refueling system), and designers at Convair, Douglas, and North American each considered trying to refuel the Aerospaceplane in flight at Mach 6. This actually advanced far enough that test flights were discussed using two of the X-15s flying formation to validate the idea. Fortunately for the X-15 program, the refueling demonstration was never attempted.[28]

Various contracts were also issued to develop specific advanced subsystems for the Aerospaceplane. For instance, the Air Force awarded a contract to General Electric for the development of an advanced integrated flight control system called GENAC (General Electric Navigation and Control). This system also included the ability to conduct orbital rendezvous maneuvers, but included the assumption that "… a man is aboard to perform critical logic decisions and to implement the complex docking

and coupling maneuvers associated with an arbitrary rendezvous." Different versions of this system were developed for (at least) the Douglas and Convair Aerospaceplane concepts.[29] Additionally, Martin Marietta received a contract from the Air Force Flight Dynamics Laboratory Structures Division to build a full-scale wing-fuselage structure representative of those being considered for Aerospaceplane.[30]

As might be expected, serious development difficulties quickly arose, effectively stalling the program. In late 1962 the project baseline was changed from the single-stage concept to a less-challenging two-stage design. Seven contractors, Boeing, Convair, Douglas, Goodyear, Lockheed, North American, and Republic undertook extensive studies of the revised vehicle. On 21 June 1963 three of the contractors, Convair, Douglas, and North American, received $500,000 contracts from the Air Force Aeronautical Systems Division for detailed development planning. From the results of these studies, the Air Force then believed that an operational system could be developed by 1970 at a cost of some $3,000 million.[31]

It was not to be however. As early as December 1960, the Air Force Scientific Advisory Board had warned that too much emphasis was being placed on the operational aspects of Aerospaceplane and not enough on the development effort. By October 1963, the SAB had concluded that the state-of-the-art was insufficient for the development of a reusable spacecraft, and that the Air Force had not adequately established any requirements for a fully reusable space launcher. Congress subsequently deleted all FY64 funding for the project.[32]

There had been significant engine development testing in long-duration facilities at velocities up to Mach 8, but it had become clear that many problems still needed to be overcome. Further, adequate testing facilities were largely non-existent – national attention was firmly fixed on the race to the moon, and few funds were available to build them. As a result, the first 'golden age' of scramjet development ended with the cancellation of Aerospaceplane.

The SR-651 studies were allowed to continue until the end of 1964, but were redirected towards developing various technologies for sustained hypersonic flight within the atmosphere including thermal protection systems and the scramjet propulsion system (and the associated LACES concept). The results of the study are still classified.[33]

Although the exact progress of Aerospaceplane is still classified, it is unlikely that any serious hardware development (except for ACES prototypes and the Martin structure) actually took place. Nevertheless, there was considerable enthusiasm for the Aerospaceplane inside the Air Force since a reusable launch system would offer great flexibility for a variety of military missions, including the orbital supply of space stations, astronaut rescue, and servicing satellites. The system also potentially offered a rapid method of getting critical military payloads from one location on Earth to another.

Despite the official cancellation of the program, during FY64 three more Aerospaceplane studies were ordered from Convair (actually, General Dynamics Astronautics), Douglas, and North American. These studies, not intended to lead to any hardware development, were initiated by the Air Force Aeronautical Systems Division. But in July 1964 all responsibility for Aerospaceplane-related work was transferred to the new Space Systems Division (SSD). This meant that any potential single-stage-to-orbit (or any other reusable concept) vehicle had to compete with ongoing SSD studies, which were generally aimed at improving ballistic launch systems.[34]

In 1965 the Air Force sought $100–150 million for a sub-scale scramjet demonstrator that could be launched from the NB-52 or a modified XB-70A. Using the NB-52 was expected to be less expensive, but required the addition of a booster rocket to obtain the desired speeds. This program was seen as complementing the NASA hypersonic research engine (HRE) that was going to be tested by the X-15 before that flight program was terminated. The HRE was designed to operate at Mach 8, while the Air Force was setting its sights on Mach 12 operation. Both Marquardt and Republic expressed considerable interest in this program, but funds were never found and the Air Force quietly shelved the concept. Interesting, a similar vehicle (the X-43A Hyper-X) is finally scheduled to be flight tested in 2001.[35]

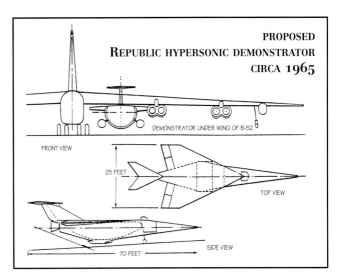

PROPOSED
REPUBLIC HYPERSONIC DEMONSTRATOR
CIRCA 1965

Republic resurrected its final Aerospaceplane concept in 1965 when the Air Force expressed an interest in developing a Mach 12 hypersonic scramjet demonstrator. This design is very similar to the Aerospaceplane concept shown below, but was a great deal smaller to allow it to be air-launched from a B-52 carrier aircraft. The Air Force could not find the $100 million necessary to build this vehicle and its Marquardt scramjet engine, so realistic flight testing of a scramjet would have to wait until 2001. (Republic Aviation)

RECOVERABLE ORBITAL LAUNCH SYSTEM

As Aerospaceplane was being cancelled, a new project was gearing up that looked remarkably similar. Dr. Ernst A. Steinhoff had begun working on Aerospaceplane-type projects at the RAND Corporation in 1961. As time passed and potential development problems with the single-stage-to-orbit concepts began to be recognized, Steinhoff became convinced that the vehicle would never materialize. One of the points Steinhoff made was that such projects depended to a large extent on the availability of full-scale (or at least, large-scale) test facilities, which were non-existent. For instance, it would take years to build a large-scale facility capable of testing aerodynamic config-

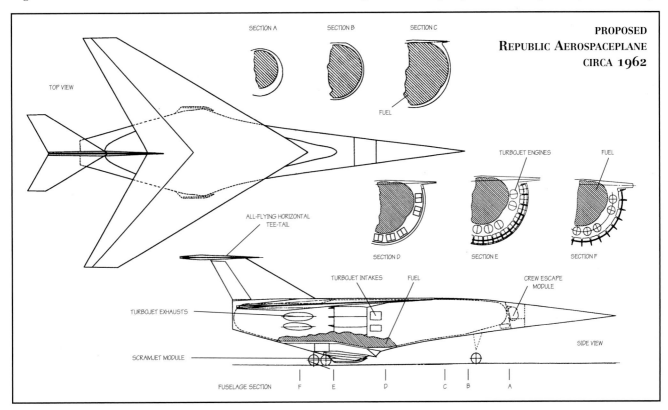

PROPOSED
REPUBLIC AEROSPACEPLANE
CIRCA 1962

By 1963 the Aerospaceplane had emerged from secrecy, and Marquardt even used this illustration in an advertisement in *Aviation Week* looking for advanced propulsion engineers, The ad began "Propulsion for the Aerospaceplane concept is one of a number of major contributions to advanced technology by engineers and scientists at ..." (Marquardt Corporation)

Convair's final (1964) Aerospaceplane concept (above and below) was 264.5 feet long with a 94.7-foot wingspan (to the tips of the vertical stabilizers). The vehicle used an ACES powerplant coupled to six scramjet engines. Although the vehicle was capable of single-stage-to-orbit performance, the baseline vehicle used a lifting-body second stage carried recessed into the top of the Aerospaceplane. Both stages carried three man crews. The mated vehicle took-off horizontally using a 4,400-foot take-off roll. The first stage was capable of ferry flights of 3,500 miles, but normally launched the second stage approximately 350 miles from the launch site. (Convair via the SDAM Collection)[90]

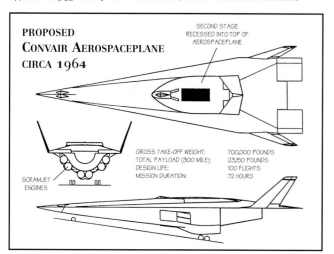

PROPOSED
CONVAIR AEROSPACEPLANE
CIRCA 1964

SECOND STAGE
RECESSED INTO TOP OF
AEROSPACEPLANE

SCRAMJET
ENGINES

GROSS TAKE-OFF WEIGHT: 700,000 POUNDS
TOTAL PAYLOAD (300 MILE):   23,150 POUNDS
DESIGN LIFE: 100 FLIGHTS
MISSION DURATION: 72 HOURS

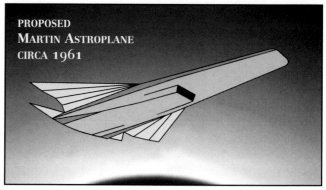

PROPOSED
MARTIN ASTROPLANE
CIRCA 1961

In 1961 Martin suggested their proposed nuclear-powered Astroplane as a possible Aerospaceplane. This aircraft was approximately the same size as the XB-70A experimental bomber, and was capable of using existing Air Force runways. Extendable flexible wings were housed in an enlongated fuselage, and a 'magneto-hydrodynamic' (MHD) engine used nitrogen liquefied from the atmosphere by a LACES-type system. Even given the high-tech nature of Aerospaceplane, this concept seemed a little too advanced and was not seriously pursued (U.S. Air Force)[44]

urations beyond Mach 4.5.* Steinhoff felt that a better case could be made for a large Mach 3 aircraft launching a smaller second-stage booster since only modest technological advances would be required.[36]

In early 1963 Steinhoff decided to move to Holloman AFB, New Mexico, as the chief scientist of the Air Force Missile Development Center (AFMDC), although it would be late December 1963 before he actually arrived because of various bureaucratic hurdles.† Nevertheless, during February and March 1963 Steinhoff and a small team from Holloman worked out the details of a Recoverable Orbital Launch System (ROLS) study. This adds to the confusion regarding Aerospaceplane since an identical title had been used for some previous studies and the term was used generically to discuss many concepts being proposed during the late 1950s and early 1960s.[37]

The concept pursued by the Missile Development Center was centered around Steinhoff's convictions that a two-stage system with a staging velocity of Mach 3 was the most sensible approach. Interestingly, by this time the Aerospaceplane project had also settled on a two-stage approach. The ROLS vehicle envisioned at Holloman had an atmospheric offset range of approximately 2,850 miles during first stage flight – in other words, the first stage could cruise that far from its take-off location before launching the second stage.‡ This potentially allowed the second stage to be placed into any orbit to "... intercept, inspect, and/or 'negate' unknown orbiting vehicles ..." Steinhoff also flatly rejected the HIRES and POBATO approaches, convinced that the ACES/LACE concept being developed for Aerospaceplane was the best method.[38]

Thus ROLS took form as a two-stage vehicle with an air-breathing first stage using ACES, and a conventional rocket-powered second stage. The pair would take-off horizontally and use aerodynamic lift until reaching the Mach 3 staging velocity. The orbital injection plane could be up to 2,850 miles from the launch site, and the second

---

* Interestingly, Steinhoff pointed this out in April 1963 – in August 1963 the Air Force cancelled a program to build a large-scale Mach 14 wind tunnel at the Arnold Engineering Development Center, a major blow to the Aerospaceplane project.

† It should be noted that the Aerospaceplane was being developed by the Wright Air Development Center (shortly to become the Aeronautical Systems Division) – a separate entity within the Air Force from the Missile Development Center (shortly to become the Space Systems Division).

‡ At least one individual pointed out that if the cruise feature was removed there was very little revolutionary in the approach. Of course, as another pointed out about aerospaceplanes in general – and not without some levity – take away 'space' from Aerospaceplane and all you have is 'aeroplane.'

stage would be at least partially (more preferably, fully) reusable with a short turn-around time between launches. The concept was to make maximum use of available technology and existing facilities – the ACES concept and large Mach 3 booster notwithstanding.[39]

True to the idea, however, the AFMDC tried to leverage off other ongoing work. For instance, in addition to ACES there was considerable interest in some of the engines being developed for the supersonic transport. These included the General Electric ultra-high-temperature stoichiometric turbojet, an advanced Curtiss-Wright turbojet, and finally the Pratt & Whitney STF-219L turbofanjet. The basic design for the first-stage vehicle was taken largely from the final round of Aerospaceplane studies – the Convair Sears Haack-type body was selected rather than essentially circular body first proposed by Republic.[40]

Considerable doubt still existed over the ability of ACES to actually become an operational system, and the AFMDC awarded contracts to Boeing, Linde, and Air Products & Chemicals for further definition of the system. The final reports indicated that it was theoretically possible to manufacture workable systems, and Linde even produced a laboratory model that appeared to offer promise. Other doubts existed over the projected structural weight of the large first stage, and a contract was awarded to North American to perform a detailed analysis based on their experience with the XB-70 bomber.[41]

By the middle of 1964, however, funding was becoming a critical issue. It became apparent that the interest in a reusable booster was small, and that additional support would need to be found from other sources. One possibility was using the basic first-stage design for something besides space launches. By deleting the ACES equipment and installing a payload bay it was estimated that the Mach 3 vehicle could carry 200,000 pounds of cargo for 2,850 miles. This would allow some of the vehicle development costs to be absorbed by others, although it would probably have compromised the vehicle's design to a great extent.[42]

Work continued, and by the end of 1964 ROLS had evolved into a vehicle with an 825,000-pound gross take-off weight with a range of 2,800 miles. This range allowed the vehicle to reach more than 50 percent of all satellites in any orbit from a single launch site, and 100 percent of all satellites in low- or moderate-inclination orbits from two launch sites located at approximately 41 degrees north and separated by 90 degrees longitudinally. The two-stage system was capable of placing 21,701 pounds into low-earth orbit. The expected launch rate would require 3 or 4 vehicles, each capable of 3,000 flights over its lifetime.[43]

While interesting, the Holloman studies did not continue for long after the end of 1964 as the group was absorbed into the mainline Space Systems Division work flow. No actual development work was accomplished on ROLS, and instead SSD efforts appear to have concentrated on recoverable versions of existing launch vehicles.

MANNED RECONNAISSANCE VEHICLE

While the Air Force pursued its various boost-glide programs and Aerospaceplane, other studies were undertaken not specifically related to either one. For instance, on 4 February 1958, the San Diego Division of Convair was asked to study a Manned Orbiting Reconnaissance System to be launched into a 400-mile orbit using a modified Atlas ICBM. This manned reconnaissance vehicle (MRV) was to be capable of remaining in orbit for 7 days and carry a one- or two-man crew plus reconnaissance equipment

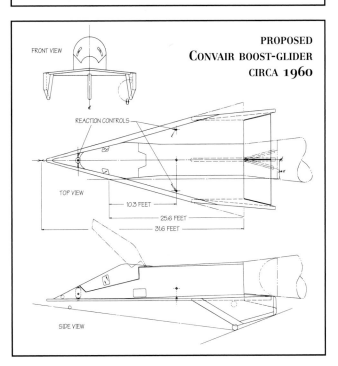

The diminutive Manned Reconnaissance Vehicle (MRV) proposed by Convair came in both single- and two-seat varieties, although both were identical externally. The landing gear housed in the ventral stabilizers was unusual. (Convair via the SDAM Collection)

PROPOSED CONVAIR MRV CIRCA 1958

PROPOSED CONVAIR BOOST-GLIDER CIRCA 1960

The MRV was used as the basis for another study a year later. What documentation could be found indicated this study was used to determine pressure thresholds on a generic boost-glide vehicle. Although they never intended to build the vehicle, Convair nevertheless updated it significantly from the 1958 baseline. (Convair via the SDAM Collection)

consisting of a single camera. This was neither a boost-glide or skip-glide concept, and was operationally more similar to space shuttle – using a ballistic trajectory into orbit, then reentering and gliding to a landing.[44]

The study concluded that a one-man vehicle would weigh 22,755 pounds, while a two-person version would add 2,335 pounds. Landing weights were 6,150 and 6,800 pounds, respectively. The Atlas launch vehicle would need to be equipped with an upper stage that weighed about 15,000 pounds. Both versions of the vehicle were externally identical – the glider was 34.75 feet long and used a 75-degree delta planform with a span of 20 feet and a total area of 375 square feet. Two ventral stabilizers housed the main landing gear and provided stability during reentry, although their potential heating problems were seemingly ignored. Interestingly, reaction controls for use on-orbit

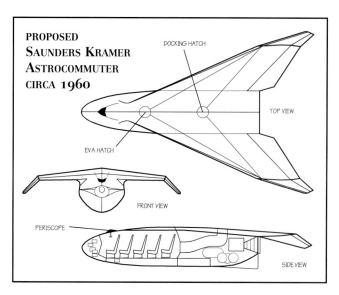

PROPOSED
SAUNDERS KRAMER
ASTROCOMMUTER
CIRCA 1960

DOCKING HATCH

TOP VIEW

EVA HATCH

FRONT VIEW

PERISCOPE

SIDE VIEW

Saunders Kramer's Astrocommuter *was to ferry personnel to his proposed space station. Lockheed Missiles and Space Company, Kramer's employer, never embraced the concept.* (Lockheed Missiles and Space Company)

were rejected as being too heavy and complex, and a gryoscopic momentum system was selected instead.[45]

The crew capsule was located in the middle of the vehicle, at the approximate center of gravity, and was a round cylinder with spherical ends that measured 6 feet in diameter and 7.2 feet long. This was sufficient for the seat, and also to provide a small area where the pilot could "stretch, relieve some of the boredom, and even perform maintenance task such as replacing an instrument light bulb…" The pilot was provided with a periscope that could be used to aim the camera, and also to provide forward visibility during landing. The capsule was suspended within the aeroshell by a series of nylon webs in an attempt to minimize the g-loading on the crew, and the crew seats could be reclined for launch. One of the major concerns expressed in the report was the inability to completely seal the capsule and the resultant air leakage during the week on orbit – additional oxygen and nitrogen was carried to replenish the contents lost through leakage.

Guidance was to be through a onboard inertial-type system, although data signals were also expected to be received from ground stations to provide "precise corrections." A monopropellant APU provided power while on-orbit, although a battery was also provided for emergencies. Convair anticipated skin temperatures of over 2,900 degF at the nose and 2,500 degF on the leading edges, falling to only 1,170 degF some 20-feet from the leading edge. The materials expected to be used included a ceramic nosecap, and ceramic, cermetic, or molybdenum leading edges over an Inconel-X or Hastelloy R235 structure.[46]

This three-month study did not result in any particular interest from the unknown client (probably an internal company customer). A subsequent 1960 study refined the concept to study pressure threshold predictions under an Air Force contract. The 1960 version featured a much more conventional cockpit for the two-man crew, and was described as a boost-glider.

OTHER EARLY EFFORTS

The ten years ending in 1965 saw a significant amount of progress towards a lifting-reentry spacecraft. Although the two major development efforts – Dyna-Soar and Aerospaceplane – were cancelled prior to any significant

hardware being completed, a great deal of basic and applied research had been accomplished. Most of this progress had been made towards military systems mostly because the Air Force was struggling to find a reason for a manned presence in space. Plus, NASA was largely preoccupied with its presidential mandate to land on the Moon.

Nevertheless, as the various NASA centers began to complete their Apollo assignments, greater emphasis began to be placed on future reusable spacecraft. These were often seen as logistics vehicles for the large orbital space station NASA hoped to build, but in many ways it did not matter. The basic technologies were the same to build any kind of reusable spacecraft, and the hard choice about payload capacity and size could wait a while longer.

In anticipation, almost every aerospace company and government research laboratory began investigating applicable concepts and technologies.

Boeing, for example, developed an entire series, the F-6 family, of small winged lifting-reentry vehicles having more sharply swept wings than the abortive Dyna-Soar vehicle. Boeing engineers called these 'swallow' configurations, mainly because the planform resembled the tail of a Swallow in flight. LaRC also designed several lifting-reentry concepts in addition to its HL-10 lifting-body shape.[47]

During 1960 Saunders Kramer, an engineer at the Lockheed Missiles and Space Company, designed his Astrocommuter to ferry personnel to and from a proposed space station. The delta-winged reentry vehicle was 46 feet long, with a wingspan of 28.3 feet, and was equipped with two turbojet engines that allowed 48 minutes (≈356 miles) of flight during landing. The Astrocommuter used an early version of the Saturn I booster. Kramer expected that development would cost approximately $135 million, and the first flight was anticipated in early 1966. Neither NASA nor Lockheed ever officially embraced the concept, which quickly faded from the scene.[48]

Around the same time, Martin Marietta promoted a fully-reusable TSTO concept called Astrorocket. This company-funded program began in 1961 as part of the Astroplane* development effort, and ran through 1965. During the development of Astroplane, Martin began to recognize the limitations on gross take-off weight imposed by the horizontal take-off (HTO) mode of operations. Martin felt this limitation (based primarily on runway strength and length) was driving Aerospaceplane to exceedingly sophisticated solutions such as the ACES engine system. Studies were therefore initiated to explore the potential of using vertical take-off techniques that would use more conventional rocket technology. The objective was to define a vehicle that required only a modest advancement of the state-of-the-art while maintaining the desirable economic characteristics of Aerospaceplane.[49]

Over 100,000 engineering manhours were spent evaluating single-stage-to-orbit horizontal take-off and horizontal landing (i.e., Aerospaceplane) concepts, two-stage vertical take-off and horizontal landing systems using air-augmented rocket engines in the first stage, and a variety of two-stage vehicles using conventional rocket propulsion. The study included the development of vehicle sizing and cost estimating computer programs, evaluating storable and cryogenic propellants, in-flight abort analysis, escape system designs, and potential military uses.[50]

Some of the groundrules for the study were noteworthy. The nominal mission was to launch 50,000 pounds of

---

* The nuclear-powered Astroplane was presented to the Air Force as a possible Aerospaceplane concept. See page 58 for an illustration.

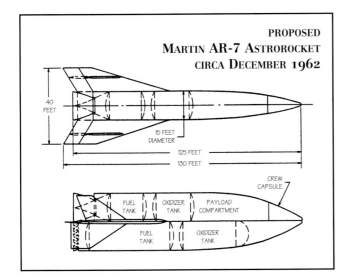

PROPOSED
MARTIN AR-7 ASTROROCKET
CIRCA DECEMBER 1962

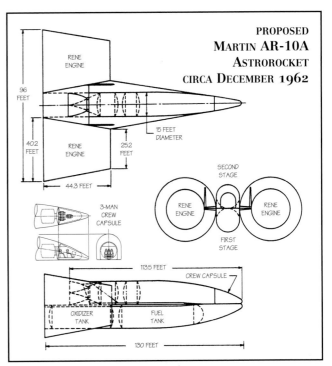

PROPOSED
MARTIN AR-10A
ASTROROCKET
CIRCA DECEMBER 1962

By December 1962, Martin had evaluated over 20 vehicle configurations designed to lift 50,000 pounds into a 345-mile orbit. The two chosen for further study were the AR-7 (above) and the AR-10A (left).The AR-7 used two stages that had 4:1 tangent ogive nose shapes drooped together at the stage separating plane, and were attached belly-to-belly. The first stage used a 'five ring thrust-plate' engine producing 2,284,000-lbf with an exit diameter of 15 feet. The thrust-plate engine was selected to obtain a very short nozzle length to assist in managing the center-of-gravity. The second stage used a 570,000-lbf plug-nozzle engine. The AR-10A used two innovative RENE air-augmented rocket engines for first stage propulsion. The rocket engine was buried in the large circular shrouds that doubled as wings on the first stage. The first stage produced 2,045,000-lbf at lift-off, but this increased rapidly to over 5,504,040 lbf at 23,500 feet altitude. Thrust vector control was provided by vanes located within the shrouds. The second stage used a 456,266-lbf plug-nozzle engine. In this configuration the stages were mounted belly-to-back, with the first stage having a high wing, and the second stage using a low wing. Both stages used similar crew compartments that sat three and provided a rest area behind the flight deck. Underneath the aerodynamic skin, the crew compartment was actually an emergency escape capsule. (Martin Marietta)

payload into a 345 mile circular orbit with the vehicle capable of supporting a crew of three for a week. It was highly desirable, but not mandatory, to be able to deliver a reduced payload into higher orbits. There were no restrictions on gross take-off weight, and the only restriction on landing weight was that a standard Strategic Air Command runway (12,000 feet long and 150 feet wide) be able to handle the vehicle. All liquid-propellant stages were manned and recoverable; solid-propellant stages were unmanned and expendable "… since the economic value of recovering solid-propellant motors appears questionable."[51]

One of the more interesting concepts evaluated during the study was the rocket engine nozzle ejector (RENE). Martin had been developing air-augmented rocket technology since the late 1950s, although others had been investigating the concept since the mid-1940s. Essentially RENE used a conventional rocket engine that was located ahead of a large mixing chamber. Atmospheric air entered the mixing chamber via a large inlet and mixed with the exhaust stream of the rocket engine. During the mixing process, part of both the kinetic and chemical energy of the fuel-rich rocket exhaust transferred to the incoming air resulting, under most flight conditions, in an increase in total thrust. Martin actually built a prototype of the RENE engine using a 2,000-lbf rocket that was successfully tested at the Arnold Engineering Development Center during 1962 under conditions simulating flight at Mach 2 and 40,000 feet. Under some conditions up to a 50-percent increase in thrust was observed, but the system proved remarkably sensitive to inlet configuration and airflow through the inlet (similar to any supersonic or hypersonic air-breathing engine). Although Martin continued to work on the concept, its complications were thought to outweigh its benefits.[52]

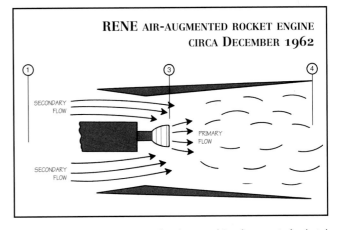

RENE AIR-AUGMENTED ROCKET ENGINE
CIRCA DECEMBER 1962

During the late 1940s various researchers began studying the concept of a ducted rocket – adding atmospheric air to a rocket exhaust in an attempt to increase thrust. Martin began developing a version of this technology, dubbed RENE, during the late 1950s and applied the concept to the AR-10 version of the Astrorocket in 1962. In contrast to normal ducted rockets, RENE was characterized by a low ratio of secondary-to-primary mass flow (approximately 3:4), a supersonic mixed stream, and a divergent shroud. The two latter characteristics eliminated the need for a complex variable-geometry shroud. The figure above illustrates the RENE mode of operation. Atmospheric air enters a large inlet at station 1, diffuses to station 3 where it mixes with the fuel-rich exhaust stream of the rocket engine, and finally exits the mixing chamber at station 4. (Martin Marietta)

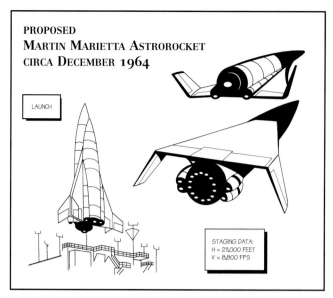

PROPOSED
MARTIN MARIETTA ASTROROCKET
CIRCA DECEMBER 1964

LAUNCH

STAGING DATA:
H = 211,000 FEET
V = 8,800 FPS

By 1964 the Astrorocket look a great deal more conventional, although at the time it was still a fairly radical design. Both stages used delta wings and were fitted with air-breathing engines for use during approach and landing. The 2,500,000 pound vehicle was launched vertically with both stages firing, and separation occurred at approximately 211,000 feet. Both stages would return to the launch site runway under their own power. (Martin Marietta)

In the end Martin decided that a relatively conventional two-stage vehicle with a gross lift-off weight of 2,500,000 pounds was the best configuration. A large delta-winged booster with a high mounted wing carried a small orbiter with a low mounted delta-wing to an 8,800 fps staging velocity. The three-man orbiter could remain in orbit for up to two weeks, while the fly-back booster used large turbojet engines to return directly to its landing site. The integrated vehicle was to be launched vertically from Cape Canaveral in Florida.[53]

The general configuration of Astrorocket was similar in many respects to the multitude of designs that would follow it during Air Force and NASA studies on the way towards developing space shuttle. But although it gave a fuzzy look into the future, Astrorocket was never seriously considered for further development.

The NASA Manned Spacecraft Center (MSC)* and the Marshall Space Flight Center (MSFC) joined together during the early 1960s for several joint study efforts on vehicles that greatly resembled space shuttles. Seven large aerospace companies participated in these studies including Boeing, Convair, Douglas, Lockheed, Martin, McDonnell, and North American. The studies did not directly design an operational vehicle, but focused on evaluating whether the state-of-the-art existed to <u>attempt</u> development of a fully reusable spacecraft. The outcome was encouraging enough that the NASA *ad hoc* Committee on Hypersonic Lifting Vehicles endorsed the development of a two-stage-to-orbit (TSTO) 'shuttle-craft' in June 1964.[54]

REUSABLE AEROSPACE PASSENGER TRANSPORT

During June 1962 MSFC had awarded the Spacecraft Organization of the Lockheed-California Company a $428,000 contract (NAS8-2687) to study a re-usable ten-ton orbital carrier vehicle. The study was scheduled to run for six months – under various guises it eventually ran for three years. The first part of the study evaluated several different vehicle concepts, including a fully-reusable three-stage vehicle that stood 274 feet high and weighed 2,250,000 pounds. Each of the winged stages

returned to its launch site for a conventional horizontal landing. This vehicle could deliver 40,000 pounds to a 200-mile orbit, although most designs in the study could only deliver 20,000 pounds. Other concepts included two- and three-stage horizontal take-off vehicles, all equipped with winged stages that returned to their launch site. Alternate concepts studied for comparison included vehicles that did not use wings, but returned via parasails or other recovery methods. Many of the orbital vehicles – unsurprisingly – bore a resemblance to the Boeing Dyna-Soar glider, although all were considerably larger.[55]

MSFC subsequently expanded the effort to include four $342,000 Reusable Aerospace Passenger Transport (RAPT) study contracts. Convair (NAS8-11463), Lockheed (NAS8-11319), Martin Marietta (NAS8-20277), and North American (NAS8-5037) were awarded contracts to study various aspects of transporting passengers and light cargo to and from Earth orbit in support of large-scale space operations. All of the concepts had flight elements that were capable of lifting-reentry and aircraft-type horizontal landing at their launch site. The baseline mission to was place two crew members, ten passengers, and 6,615 pounds (three metric tons) of cargo into a 300 mile orbit. None of these studies were intended to result in the development of an actual vehicle, but were meant to evaluate various aspects of future space transportation systems.

The Convair and Lockheed contracts were interrelated, with Convair investigating second-stage orbital vehicles and Lockheed concentrating on first-stage boosters. All three of the RAPT contractors used these baseline vehicles to perform their economic and operational analysis which were released during mid-1965. Several of these concepts included rail-launched vehicles that were also evaluated during the Air Force ROLS efforts.[56]

The Lockheed economic analysis was interesting. Four concepts were evaluated – the existing Saturn IB, a reusable first stage with a new expendable second stage, a reusable first stage that used an existing expendable launch vehicle as a second stage, and the baseline fully reusable system. All of the concepts had roughly the same payload capacity, and rates of up to 16 launches per month were used for the analysis. No research and development costs were assessed against the existing systems such as the Saturn IB, but all R&D costs ($2,800 million) were amortized over a ten year period for the new vehicles. The Saturn IB had the highest recurring cost – $34.9 million per launch. Lockheed estimated that this would be reduced as the production rate was increased, but at no time did it drop below $18 million per launch. Both of the reusable first stage and expendable second stage options ran between $18 million and $4 million per launch, although Lockheed believed the $4 million figure was unrealistically low. Lockheed estimated the cost of the fully reusable system could be as low as $1 million per launch, almost regardless of the launch rate.[57]

Another factor Lockheed considered was the possible attrition rate and its impact on launch costs. The baseline included one launch failure per 1,000 attempts for the fully reusable system. The cost of driving this attrition rate lower (say, 1 in 10,000) did not significantly alter the economics of operating the vehicle, but drove development costs upward by an order of magnitude. Conversely, allowing the attrition rate to rise to 1 loss in 100 launches drove the operating costs up over 33 percent while only

---

* MSC, located outside Houston, Texas, was renamed the Lyndon B. Johnson Space Center (JSC) on 17 February 1973.

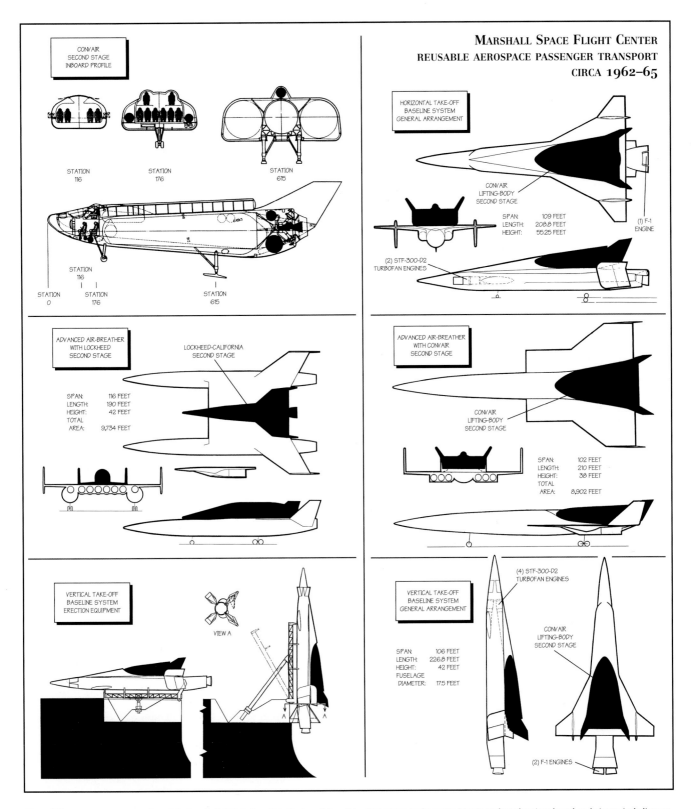

Many different concepts were evaluated during the RAPT studies. Note the two air-breathing first-stages in the center. Martin evaluated various launch techniques, including normal air-breathing horizontal take-off, rocket horizontal take-off, rocket vertical take-off (the two illustrations on the bottom), and a ground-based accelerator sled (next page). The results of these contracts were sent to the DoD/NASA Aeronautics and Astronautics Coordinating Board (AACB) Subpanel on Reusable Launch Vehicle Technology after the subpanel was established in August 1965, and formed the basis of several of their baseline concepts. (Lockheed and Convair)

slightly lowering development costs. Therefore, Lockheed determined that 1 loss per 1,000 flights seemed – from an economic perspective – to be a reasonable compromise.[58]

An evaluation of the economics of developing new engines revealed that using existing F-1 and J-2 engines from the Saturn program resulted in the highest operational costs despite the fact that no development costs were required. The trouble was primarily with the J-2, and replacing it with a new high-performance LO2/LH2 engine in the second stage significantly lowered recurring costs. Interestingly, the engine technology chosen was the innovative aerospike concept developed by Rocketdyne, much like was later proposed by Chrysler during the alternate space shuttle concept studies. Replacing the F-1

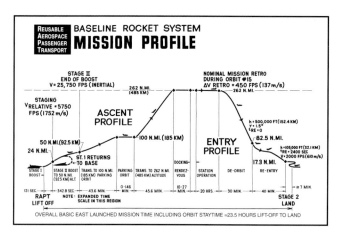

BASELINE ROCKET SYSTEM
MISSION PROFILE

The mission profile chart largely ignores the question of how the composite vehicle initially took-off, but shows the rest of the flight in considerable detail A normal space station resupply mission was expected to take only 15 orbits, or slightly less than one day. This was one contributor to the expected 16 mission per month launch rate. (Lockheed)

with a new LO2/LH2 engine lowered the operating costs but never recouped the development investment.[59]

The baseline horizontal take-off vehicle included a Lockheed-designed winged fly-back first stage that was 208.8 feet long with a 109-foot wingspan. The large cranked-arrow wing provided 4,977 square feet of surface area. A single F-1 and two H-2 engines produced ascent thrust, while four STF-200-D2 turbojets were used during the 300 mile cruise back to the launch site. At lift-off the first stage weighed 1,168,400 pounds. An alternate vertical take-off booster was generally similar but used two F-1 engines and weighed 1,338,000 pounds. A roll of 11,000 feet was required for the horizontal take-off variant – upon returning the first stage required a landing roll of 7,000 feet after touching down at 190 knots. Alternately, a rocket sled could provide the initial energy for a horizontal take-off, similar to a version also proposed during the Air Force ROLS studies.[60]

The Convair-designed lifting-body second stage used a new-development Rocketdyne aerospike engine. A single large LO2 tank was located long the centerline of the vehicle, flanked by LH2 tanks on either side. Contrary to many lifting-body designs, all of the propellant tanks were circular in cross-section with a constant taper in diameter along their length. The crew and passengers rode in a

This was how the ground accelerator sled for the reusable aerospace passenger transport would have looked. The rocket sled would release the vehicle at 650 fps (≈430 mph), then deploy a large scoop that would act as a water brake to slow the sled. The same water brake was capable of stopping the entire vehicle under abort conditions. This was the preferred launch method for the RAPT concept. (Martin Marietta)

pressurized compartment in the extreme nose, and a hatch was provided to allow docking with a space station. The 6,615 pounds of cargo was carried in a dorsal compartment that had a hatch at the forward end that opened into the crew compartment, and a hatch at the aft end that allowed access to bulkier payload packages.[61]

The second stage would be constructed from 718 nickel alloy and Ti-8-1-1 titanium alloy. An ablative 'slipper' covered the entire bottom of the vehicle to protect the structure during reentry. This limited temperatures to no greater than 740 degF – about the maximum the titanium shell could withstand. The 4-inch radius stabilizer leading edges were made from Hastelloy X, with René-41 used for the rest of the stabilizer surface.[62]

Martin was primarily interested in comparing possible launch modes for the RAPT vehicle. Surprisingly, given the amount of attention Lockheed and Convair paid to conventional horizontal take-off, this was not one of the modes evaluated by Martin. This is even more surprising since the early part of the studies had emphasized horizontal takeoff because of its inherent passenger safety and comfort, and its similarity to commercial airliners. Instead, Martin compared rocket sled-accelerators and a vertical takeoff systems using the same combination of existing and new vehicles defined by Lockheed and Convair. The purpose of the analysis was to compare the relative merits, particularly economic and safety, of the two launch methods. The differing launch modes drove Martin to optimize each vehicle configuration to account for differences in velocity requirements, wing design criteria, fly-back propulsion, abort modes, and escape systems. This data was used internally to update the configurations, then fed back to Lockheed and Convair, with the resulting data compared against each other to ensure consistency.[63]

Reliability and safety comparisons established abort methods, procedures, and system requirements for each of the launch modes. Two types of aborts were considered – composite and separated stages. Martin also evaluated operational considerations based on cabin arrangements, critical work loads, and launch site layouts for each mode. A two-week turn-around per vehicle was assumed, and analysis were conducted based on 2, 4 and 8 vehicle fleets at launch rates of 4, 8, and 16 per month.[64]

Martin expected a mission abort to occur approximately every 20 flights (0.95 mission reliability), usually due to propulsion system failure. The vehicles were designed to withstand an engine failure at lift-off and still be able to recover intact. Interestingly, Martin believed that the fly-back jet engines could be used to provide thrust in the event of an engine failure at take-off – although never stated this would have meant that the jet engines had to be running at lift-off since it would have been impossible to start them in time to provide much assistance. Martin also concluded that total loss of thrust with less than 1,000 feet of altitude would result in the loss of the vehicle since no viable abort scenarios could be established. The catastrophic loss of a vehicle was put at 1 in 1,000 missions.[65]

The sled launch technique involved accelerating the mated vehicle down a long track until it reached 650 fps. At that speed the vehicle would be released unless the pilot opted to abort. The sled would be slowed (either on its own, or with the vehicle in the event of an abort) by a large water brake at the far end. Acceleration during the take-off run and deceleration during any abort was limited to 2-g. The water brake used a 250 square-foot scoop on the sled itself that could be variably deployed to maintain the 2-g limit.[66]

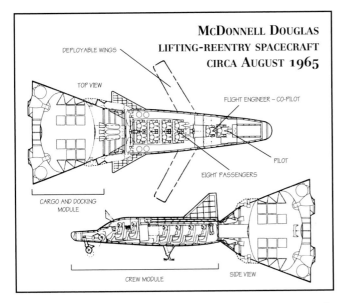

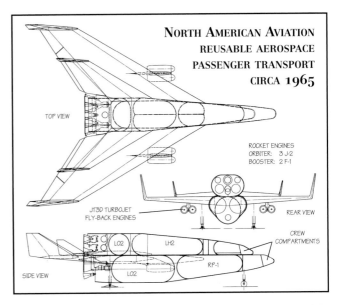

One of the vehicles studied by McDonnell in 1965 would be investigated again in the future – by McDonnell and North American during the NASA Phase A studies, and again by Chrysler during the NASA ASSC studies. (McDonnell Aircraft Corporation)

The North American first stage configuration was significant in that it provided the foundation of several North American ILRV Phase A space shuttle proposals. Also note the similarity to the design on page 74. (North American via the AFHRA Collection)

The North American study was generally similar to the others, but did not involve interfacing between contractors. North American was free to develop its own concepts within the guidelines provided by NASA, but ended up with generally similar configurations. The concept selected by North American was a two-stage horizontal take-off vehicle. The first stage used turbojet engines to return to the launch site where it landed on skid-type landing gear. The lifting-body second stage did not contain fly-back engines, and was a true glider during reentry and landing. North American purposely did not consider ACES or HIRES propulsion systems in an effort not to duplicate on-going Air Force studies (Aerospaceplane, ROLS).[67]

One of the parameters investigated by North American was the use of variable-geometry wings. Variable geometry could be used to improve the hypersonic and/or subsonic aerodynamic characteristics of a vehicle. This could result in three possible benefits. First, it could improve the L/D ratio and thereby increase the landing footprint (cross-range) and reduce the fly-back penalty somewhat. Second, it could be used to improve the stability and control of the vehicle by allowing the center-of-gravity to be varied. Finally, it could be used to reduce wing loading, and thereby lower reentry heating and loading. North American found, however, that all of these benefits could be attained using fixed-geometry designs, and that from an aerodynamic point of view "… these modifications do not provide any necessary improvement." Additionally, all of the variable-geometry concepts carried substantial weight penalties and were much more complex. The one possible exception was the use of flexible (or inflatable) deployable wings similar to those proposed by Martin for Astroplane and championed by Goodyear. However, North American did not believe that inflatable structures offered much advantage since they could not be used at hypersonic speeds where most of the difficulties were thought to lie.[68]

MSFC was not the only NASA center studying reusable spacecraft, although they were largely at the forefront because their primary development work on Apollo had ended, or soon would. Conversely, the Apollo work at the MSC was just picking up and left little time for worrying about the future. Nevertheless, MSC issued a contract (NAS9-3562) to McDonnell to define some preliminary mission requirements for a future lifting-reentry spacecraft. Again, this study was never intended to result in an actual vehicle being developed.

McDonnell studied four different vehicles. A ballistic capsule was essentially a scaled-up Gemini heat shield with a conical forebody (that looked very much like an Apollo), with a 0.6:1 maximum L/D. A modified HL-10 shape had a maximum hypersonic L/D of about 1.1:1, limited mainly by downward control deflection, and a 4.7:1 subsonic L/D. A variable-geometry design provided by NASA Langley provided a maximum hypersonic L/D of 1.8:1 and a 8:1 subsonic L/D. Another wing-body configuration designed by McDonnell provided an L/D of 2.7:1 at hypersonic velocities, and 4.5:1 subsonically.[69]

The study included an evaluation of possible thermal protection techniques for each vehicle. The ballistic vehicle, unsurprisingly, continued to use an ablative heat shield. But the other vehicles could use either of two methods. The radiative structural heat protection concept used a metallic 'shingle' over a thick layer of fibrous insulation that covered a water-cooled structure. A similar ablative structural heat protection concept added a layer of ablative protection on top of the metallic shingle. This allowed the use of less expensive structural materials for the primary airframe, and also attempted to minimize the oxidation of the shingles during reentry.[70]

An economic analysis yielded some surprising results. Total program costs (research, production, and operations) for a baseline space station resupply mission indicated that the ballistic capsule was significantly less expensive than the reusable designs ($6,738 per pound, versus $7,304 to $9,005).[71] But this analysis included only the costs of the reentry vehicle – and did not include the cost of the necessary boosters. If all costs had been considered, it is likely the results would have been significantly different.

These studies all served their purpose and provided a better understanding of some of the concepts being explored by the government and contractor researchers. Many of the configurations evaluated during these studies were subsequently used by the DoD/NASA Aeronautics and Astronautics Coordinating Board (AACB) Subpanel on Reusable Launch Vehicle Technology when they were established in August 1965.

From 1964 through 1968 Lockheed participated in a series of studies for a multipurpose reusable spacecraft (MRS) for the Air Force Flight Dynamics Laboratory at Wright-Patterson AFB. A carefully tailored body allowed Lockheed to delete the normal endplate stabilizers seen on the earlier lifting-body designs. This particular design was tested in various wind tunnels at speeds from 200 mph to Mach 20 and was found to have generally acceptable handling qualities. The vehicle was powered by a fluorine-hydrogen rocket engine, with a small jet engine mounted directly above it providing power during fly-back and landing. Note the air intake at the base of the vertical stabilizer, and the complex multi-lobe propellant tanks necessary to fit into the lifting-body shape. Although this concept did not result in any specific development program, the information gained during its investigation helped both the Air Force and Lockheed during early space shuttle studies. (Lockheed Martin via the Tony Landis Collection)

AIR FORCE REUSABLE LAUNCH VEHICLES

Meanwhile, the termination of Aerospaceplane led the Department of Defense to redirect most of its hypersonic research studies towards an examination of possible manned hypersonic launch platforms. Within the Air Force Flight Dynamics Laboratory, this effort spawned the multipurpose reusable spacecraft (MRS), reusable launch vehicle (RLV),* and reusable space launch vehicle (RSLV) studies (see page 52) between 1964 and 1968. These were subsequently put under the auspices of the joint DoD/NASA Aeronautics and Astronautics Coordinating Board's (AACB) Subpanel on Reusable Launch Vehicle Technology. The Subpanel was officially established on 24 August 1965, and was composed of joint-chairmen from the Air Force and NASA, along with eight DoD and ten NASA representatives. The Subpanel issued their final report in September 1966 after having examined various candidate reusable launch vehicles making use of hypersonic air-breathing and rocket-powered stages.[72]

Three classes of vehicles were defined by the AACB Subpanel, with Class I being rudimentary spacecraft that could perform simple orbital missions as early as 1974. Lifting-reentry vehicles (Class II) with more capabilities would become operational during 1978, with still more advanced vehicles (Class III) by 1982.[73]

It was envisioned that the Class I spacecraft would resemble the stillborn Dyna-Soar glider in general size and configuration, and would be launched by existing boosters such as the Titan IIIM or Saturn IB. The Titan version would have a gross lift-off weight of 1,820,000 pounds when equipped with a small lifting-body orbiter capable of carrying six astronauts. Total development costs for this version were estimated at $700 million in FY65 dollars, with another $19 million required to fulfill the baseline operational missions for ten years. The Saturn version would also need $700 million for development, but would cost $36 million to operate over the same ten year period. Surprisingly, the Saturn was slightly lighter, weighing 1,296,000 pounds at lift-off. New development of an advanced expendable booster was also considered for Class I, and it was estimated that this configuration would cost $2,000 million to develop, but only $14 million to operate. A radical, partially reusable booster concept was also

proposed involving a winged, fly-back lower stage topped by an additional expendable stage attached to the lifting-body orbiter. It was expected that this vehicle could be developed for $2,500 million, and would cost $15 million to operate. The GLOW for the advanced expendable booster was 873,000 pounds, while the mostly reusable concept weighed 1,200,000 pounds at lift-off.[74]

Class II would evolve into a fully reusable system using high-efficiency liquid hydrogen fueled rocket engines for both the orbiter and a fly-back booster. The orbiter would feature air-breathing engines to assist in atmospheric flight and landing operations, and would evolve to include some sort of dual-cycle turbofan-ramjet propulsion for extended range within the atmosphere. Two different concepts were investigated in detail, with the first having a single booster stage and a GLOW of 1,643,500 pounds. It was expected to cost $3,000 million to develop, but only $4 million to operate over ten years. The second concept utilized a horizontal take-off booster instead of being vertically launched, and would cost $4,000 million to develop. This design would have a profound influence on the early development of space shuttle concepts.[75]

The final Class III vehicle would have been larger and incorporated scramjet technology in both stages. The only concept examined for Class III involved horizontal take-off, and the projected lift-off weight was 965,000 pounds without a payload. Since the technology did not exist to actually develop this design, no cost data was prepared.[76]

The Subpanel concluded that numerous technical risks and cost uncertainties required resolution, but that other factors, notably an expected increase in manned Earth orbital activity, encouraged the development of reusable space vehicles. At the time, the AACB panelists could not identify any single concept capable of satisfying the anticipated future needs of both NASA and the DoD, and thus the Subpanel summarized a variety of proposed systems. In general, the subpanel recommended partially reusable systems, believing that they would cost less to develop than fully reusable vehicles. Their development costs could also be amortized in a shorter period of time

---

* The exact topics and results of these classified studies could not be determined, but it is known that Lockheed designed a stage-and-a-half shuttle-like vehicle during the RLV studies under contract AF33(615)-67-C-1885 and Convair studied the T-18 system.

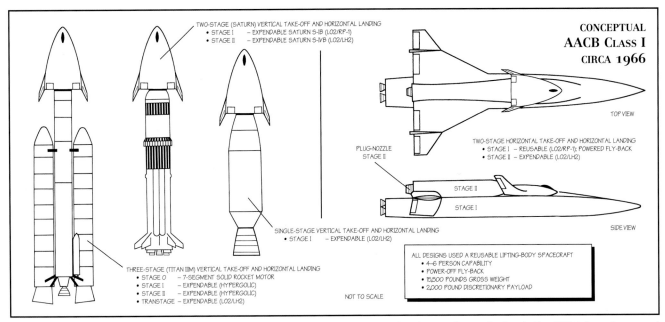

Three of the four Class I concepts examined by the AACB involved 'throw-away' boosters. The cheapest and quickest to make operational used a Titan IIIM similar to that proposed for MOL – although certification of this booster was never completed, it has always been felt it would be relatively easy to man-rate if required. Next was a modified Saturn IB, followed by an entirely new advanced expendable launch vehicle. The most advanced of the Class I designs was the mostly reusable horizontal take-off – horizontal landing vehicle shown at the extreme right, very similar to the ones used by the RAPT study at MSFC. In all these concepts, a horizontal-landing lifting-body was used, although other alternatives were also examined. (U.S. Air Force)

with fewer flights, thus making such vehicles competitive with expendable systems by late 1975. The Subpanel believed that fully reusable vehicles could be developed at a later time, after some operational experience had been gained with the partially reusable concept.[77]

While true competitiveness with expendable launch vehicles would ultimately prove unattainable for the space shuttle, at least on a purely economic basis, the overall argument made a great deal of sense, and would also prove to be remarkably clairvoyant. Nevertheless, fascination with fully reusable two-stage systems would continue to dominate both Air Force and NASA space shuttle R&D activities for another five years.

AFFDL AND STAGE-AND-A-HALF

In late-1968, the Air Force Flight Dynamics Laboratory contracted with Lockheed and McDonnell Douglas to investigate various space shuttle concepts. This study included the first applications of the 'stage-and-a-half' concept that was first advocated by Robert Salkeld and also during the earlier Air Force RLV study. This began the Integral Launch and Reentry Vehicle (ILRV) effort, a phrase that would be discontinued by the Air Force after NASA began the Phase A studies under the same terminology. The Air Force would also use the results of these ILRV studies as the DoD input to the Space Task Group in June 1969 (see page 79).[78]

Lockheed's proposed StarClipper orbiter was an elegant delta-wing vehicle measuring 186.5 feet long and spanning 106 feet from the upper tip of each up-turned wingtip. Small variable-geometry wings were included in the design to allow slower approach and landing speeds. The orbiter was nestled between two 23.67-foot diameter fuel tanks that formed an extreme 'vee' configuration around the orbiter's nose. Essentially, this proposed orbiter drew on two streams of AFFDL work. First, the development of the stage-and-a-half concept was a continuation of the RLV study that had culminated in a paper prepared by Alfred Draper and Charles Cosenza for the American Institute of Aeronautics and Astronautics in May 1968; sec-

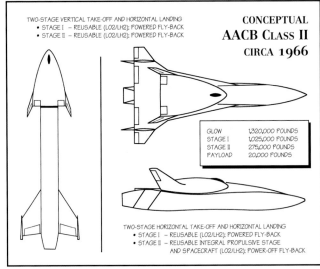

Both of the Subpanel's Class II concepts were fully reusable and based on a lifting-body orbiter with no alternate designs considered. It was estimated that each of these designs would cost $4,000 million to develop, and would cost less than the Class I concepts to operate. (U.S. Air Force)

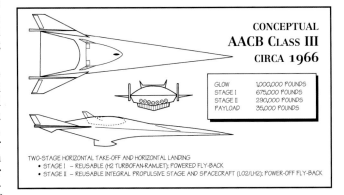

The Subpanel realized that technology had not advanced to the point of allowing the development of the Class III concepts, but nevertheless, proposed this fully-reusable horizontal take-off and landing design as representative of possibilities for the early-1980s. (U.S. Air Force)

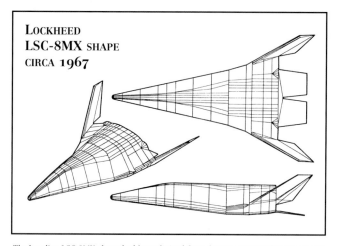

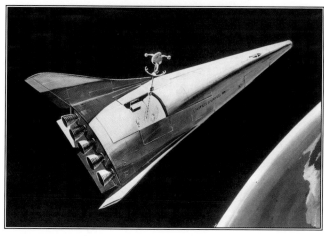

The baseline LSC-8MX shape had been derived from designs originated inside the Air Force Flight Dynamics Laboratory at Wright-Patterson AFB, Ohio. (Lockheed)

NASA released this conceptual space shuttle that was obviously based on the StarClipper orbiter design. (NASA photo S70-486-V via the SDAM Collection)

ondly, the general shape of the orbiter was derived from the FDL-5LD and FDL-8H shapes developed by the AFFDL in 1967, although by now it was known as the LSC-8MX.[79]

The vehicle was to have a launch weight of 3,500,000 pounds with a 50,000 pound payload carried in a 22 by 60 foot payload bay. Five engines burning liquid oxygen and liquid hydrogen provided 5,130,000 lbf in a vacuum. The external fuel tanks were to be jettisoned approximately 2,500 miles downrange at an altitude of 320,000 feet and a speed of 13,500 mph. Upon reentry, the shape was expected to have a 1.8:1 hypersonic L/D, increasing to 6.8:1 at subsonic speeds. Landing speeds of 180 mph were estimated, and air-breathing engines were provided to enable go-arounds if required. Even at this early date two different launch sites were envisioned – one at Vandenberg AFB, California, and one at Cape Canaveral AFS, Florida. The two launch sites were a direct result of having expendable hardware that had to be jettisoned during ascent. The fully reusable shuttle concepts being investigated by NASA were referred to as 'all azimuth' capable since they did not discard any hardware in flight and could safely (in theory) overfly anywhere.[80]

Noteworthy in the Lockheed concept were the number of similarities between the capabilities sought of this vehicle and those of the eventual Space Shuttle, namely reduced operating and developmental costs via a partially throw-away configuration, and the rather large payload bay size. Lockheed also submitted this design to NASA as part of the Phase A space shuttle competition.[81]

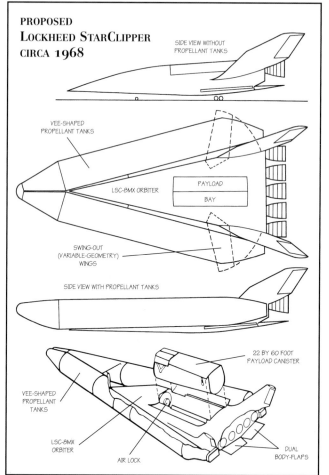

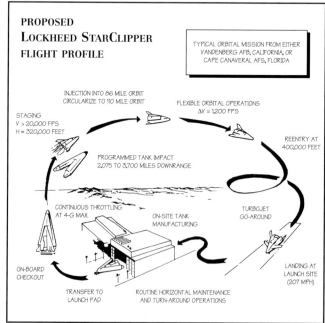

The Lockheed design was significant because it was the first major concept that moved part of the propellants externally into throw-away tanks. In the case of the StarClipper, only the liquid hydrogen fuel was moved outside – competitor McDonnell Douglas would take the idea one step further and move everything outside. The Lockheed proposal, along with the one from McDonnell Douglas, was ahead of its time, since nobody had yet admitted that a partially throw-away vehicle might ultimately be more economical to design, develop and operate than a completely reusable one. (Lockheed Missiles and Space Company)

The McDonnell Douglas Model 176 utilized a 'parallel tankage' concept, as opposed to the vee-shaped fuel tanks proposed by Lockheed, but was noteworthy since both propellants were moved outside the orbiter. The end result was a rather ungainly looking vehicle in its launch configuration, but one that had an excellent payload fraction. The orbiter was 130 feet long with small low-mounted delta-wings and 'vee' stabilators on the upper fuselage sides. A variable geometry wing was contained within the fuselage during ascent, on-orbit and reentry, being deployed when the orbiter descended below 100,000 feet to improve the low-speed L/D ratio.[82]

Fuel tanks, 150 feet long and 24 feet in diameter, were mounted on either side of the orbiter, while 73 feet long oxidizer tanks were attached on the orbiter's top and bottom. These were referred to as 'tip tanks' by McDonnell Douglas. The system would be launched vertically, with the external tanks being discarded during ascent when their propellants were expended.

The original baseline McDonnell Douglas vehicle carried a crew of three and a 25,000-pound 15 by 30 foot payload. A total of 7,400 square feet of wetted area was provided by the orbiter wing-body and small variable-geometry wing. Five 879,000 lbf high-chamber pressure (HiPC) engines provided power during ascent. These engines used liquid oxygen and liquid hydrogen, and were considered to be low-risk development items based heavily on Saturn technology. The vehicle had a gross lift-off weight of 3,683,000 pounds.[83]

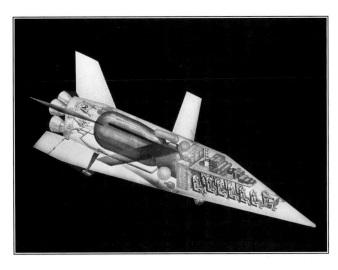

NASA released this artist concept of early variation of the McDonnell Douglas orbiter. Note the propellant tanks are in the main body and passengers are in the forward fuselage – this was a common configuration during the RAPT studies. Unlike later variations, the deployable wings were mounted high on the fuselage. (McDonnell Douglas)

Several different variations of the concept were also proposed, most varying the orbiter's fineness ratio and changing the size of the orbiter and propellant tanks in an attempt to find the optimum payload fraction. Vehicles as small as 95.3 feet, and as large as 164 feet long were investigated. Some of these included payload bays measuring 15 by 60 feet and capable of carrying 50,000 pounds.

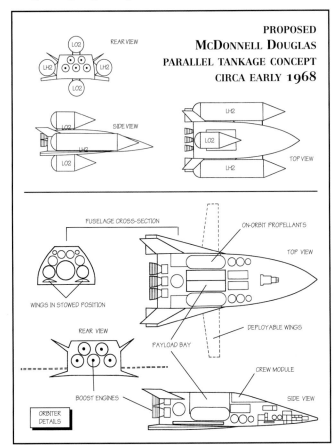

PROPOSED
McDonnell Douglas
PARALLEL TANKAGE CONCEPT
CIRCA EARLY **1968**

The McDonnell Douglas concept for the Air Force study was ungainly, but was noteworthy since it moved all of the vehicle's propellants outside the orbiter into expendable tanks. It would be several years before the mainstream shuttle design teams came up with the same idea to lower weight and costs. This design was also submitted to NASA, which showed little interest in pursuing it since it did not fit the role of a 'fully reusable' vehicle. Noteworthy is the large wing-body area, in addition to the deployable variable-geometry wing. (McDonnell Douglas Astronautics Company)

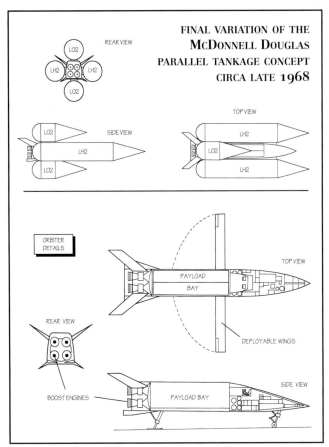

FINAL VARIATION OF THE
McDonnell Douglas
PARALLEL TANKAGE CONCEPT
CIRCA LATE **1968**

The final evolution of the McDonnell Douglas ILRV design differed considerably from the original. Notice that the later orbiter is somewhat shorter than the LH2 tanks flanking it, while the original orbiter is longer than its tanks. It was not the orbiter size that changed (both orbiters were 130 feet long), but the estimates of the amount of propellants required to obtain orbit, that grew larger. The relative shape of the LO2 tanks did not change substantially, although both sets of tanks increased in diameter, from 21.6 to 24 feet. (McDonnell Douglas Astronautics Company)

PROPOSED
McDonnell Douglas ILRV
CIRCA LATE 1969

The McDonnell Douglas design matured substantially in the year between the original ILRV study and late 1969. The vehicle had a GLOW of 4,587,540 pounds, including 50,000 pounds of payload. The baseline vehicle had a cross-range of 625 miles, but this could be extended to 2,300 miles by sacrificing 4,000 pounds of payload capacity. A 90 flight per year manifest could be sustained with five boosters and three orbiters. McDonnell Douglas estimated the entire system could be operational 71 months after approval. (McDonnell Douglas via the AFHRA Collection)

MODEL 176

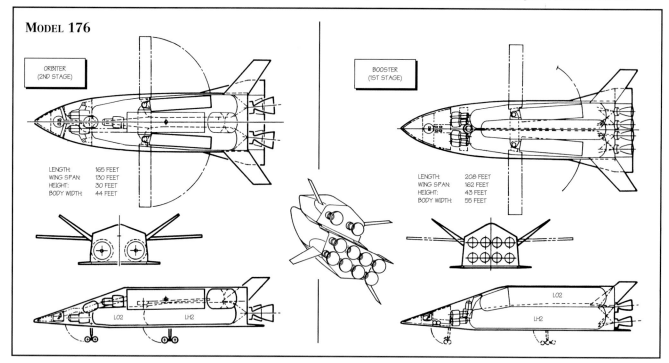

ORBITER (2ND STAGE)

LENGTH:	165 FEET
WING SPAN:	130 FEET
HEIGHT:	30 FEET
BODY WIDTH:	44 FEET

BOOSTER (1ST STAGE)

LENGTH:	208 FEET
WING SPAN:	162 FEET
HEIGHT:	43 FEET
BODY WIDTH:	55 FEET

LO2   LH2

LO2   LH2

STACKING AT VAFB SLC-6

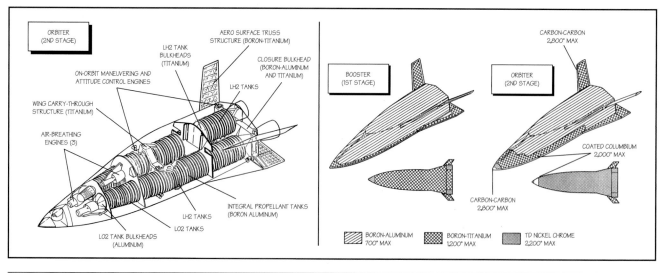

ORBITER (2ND STAGE)

AERO SURFACE TRUSS STRUCTURE (BORON-TITANIUM)

LH2 TANK BULKHEADS (TITANIUM)

CLOSURE BULKHEAD (BORON-ALUMINUM AND TITANIUM)

ON-ORBIT MANEUVERING AND ATTITUDE CONTROL ENGINES

LH2 TANKS

WING CARRY-THROUGH STRUCTURE (TITANIUM)

AIR-BREATHING ENGINES (3)

INTEGRAL PROPELLANT TANKS (BORON ALUMINUM)

LH2 TANKS

LO2 TANKS

LO2 TANK BULKHEADS (ALUMINUM)

CARBON-CARBON 2,800° MAX

BOOSTER (1ST STAGE)

ORBITER (2ND STAGE)

COATED COLUMBIUM 2,000° MAX

CARBON-CARBON 2,800° MAX

BORON-ALUMINUM 700° MAX	BORON-TITANIUM 1,200° MAX	TD NICKEL CHROME 2,200° MAX

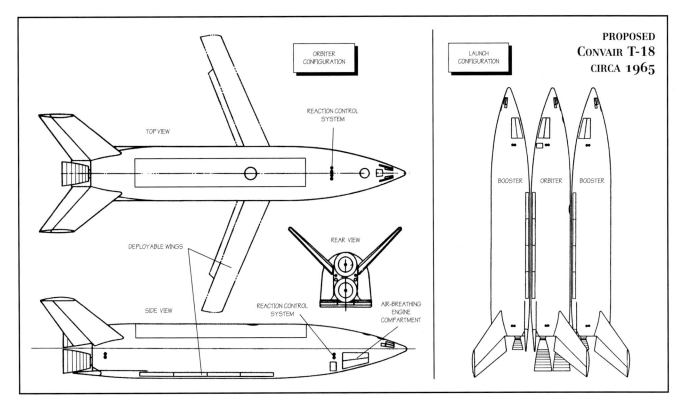

The T-18 shape had first been proposed by Convair during a 1965 study for the Air Force Space and Missile Systems Organization. Although similar in general appearance to the BAC-developed MUSTARD vehicle, Convair was seemingly unaware of the concurrent European studies. This triamese concept would be proposed during the Air Force ILRV studies, and again during Phase A of the NASA space shuttle studies. (Convair via the Bob Bradley Collection)

McDonnell Douglas also followed Lockheed's lead and submitted this design to NASA as part of the Phase A space shuttle competition.

The Convair Division of General Dynamics in San Diego developed one of the more unique concepts presented in the ILRV studies. The basic vehicle had been designed in 1965 as part of a classified study under contract AF33(615)-67-C-1885. The T-18 triamese configuration used three vehicles – two boosters and an orbiter – that were externally identical. The orbiter featured a 15 by 60-foot payload bay plus on-orbit propellant tanks. The boosters replaced the payload bay with an additional large propellant tank. The basic structure for both the booster and orbiter was identical, and the triamese concept was developed by Convair with the aim of reducing costs. The economy was achieved on the basis of commonality between the booster and orbiter elements such that only a single vehicle needed developed. Since, in general, the orbiter had the more severe requirements, the boosters were 'over engineered' as a result of the commonality.[84]

At launch the three elements were attached together and propellant crossfeed was employed so that all engines were operating using propellants from the booster tanks. Staging occurred at 8,000 fps with the boosters performing a gliding reentry, deploying their wings, and cruising back to the launch site using retractable jet engines. The orbiter continued firing its own engines using internal fuel until reaching the desired orbit. After completing its mission, the orbiter performed a lifting reentry, deployed its wings and jet engines, and returned to its launch site.[85]

The T-18 system (later called the FR-1 configuration, with T-18 referring to the actual vehicle design) was designed to boost 50,000 pounds into a 115 mile polar orbit from Vandenberg AFB. Each element used two high-pressure LO2/LH2 engines, with the orbiter having larger bell nozzles to provide altitude compensation. The orbiter

was designed with sufficient cross-range (1,500 miles) to allow a return to the launch site at least once every 24 hours while in polar orbit. Up to 80,000 pounds could be launched into due-east orbits from Cape Canaveral.[86]

Convair submitted a modified version of this basic design, by then evolved into the FR-3A and FR-4, to NASA as part of the Phase A space shuttle competition.

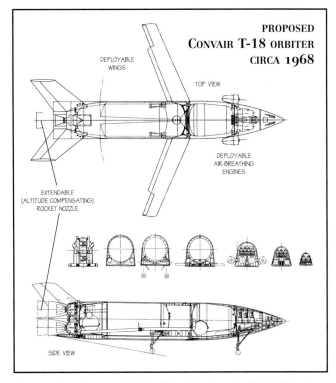

Convair performed a great deal of detailed engineering on the T-18 vehicle under a classified Air Force contract. (Convair via the Bob Bradley Collection)

PROPOSED
CONVAIR FR-1 SYSTEM
CIRCA LATE 1969

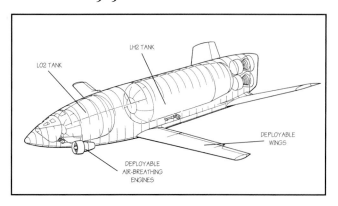

LO2 TANK

LH2 TANK

DEPLOYABLE WINGS

DEPLOYABLE AIR-BREATHING ENGINES

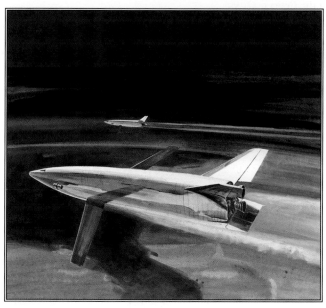

The T-18 booster was essentially a flying propellant tank with a small crew compartment in the nose and two large rocket engines in the tail. The use of only two rocket engines is the easiest way to distinguish a T-18 from the later FR-3 and FR-4 concepts offered to NASA. When the wing was in the stowed position it prevented the landing gear from deploying, but this was not considered an issue since the vehicle could not fly subsonically (or land) without the wing being deployed. If it failed the crew would be forced to eject. (Convair via the Bob Bradley Collection)

A T-18 booster deploys its air-breathing engines after separating from the orbiter. (Convair photo 1033288 via the SDAM Collection)

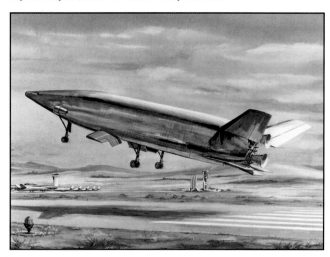

A nice series of artist concepts showing a T-18 booster coming in for a landing. Note the stacked vehicle in the background awaiting launch, and another booster taxing on the ramp. The T-18 boosters only used two turbofans instead of the four used by the larger boosters proposed to NASA. (Convair photos 102420B and 102557B via the SDAM Collection)

CONVAIR T-18 BOOSTER
STRUCTURAL DETAILS

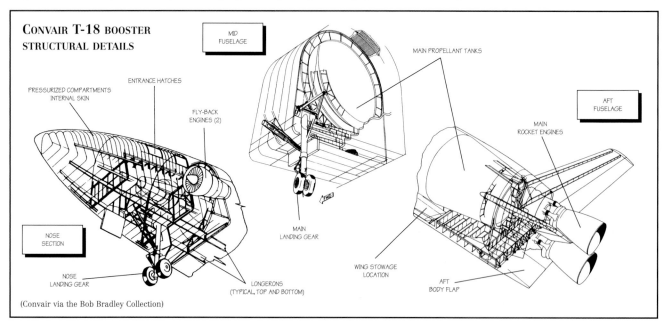

MID FUSELAGE

MAIN PROPELLANT TANKS

AFT FUSELAGE

PRESSURIZED COMPARTMENTS INTERNAL SKIN

ENTRANCE HATCHES

FLY-BACK ENGINES (2)

MAIN ROCKET ENGINES

MAIN LANDING GEAR

WING STOWAGE LOCATION

AFT BODY FLAP

NOSE SECTION

NOSE LANDING GEAR

LONGERONS (TYPICAL, TOP AND BOTTOM)

(Convair via the Bob Bradley Collection)

THE LITTLE GUYS

What is interesting about the vast majority of these efforts is the fact that most of the actual designs were originated inside government research laboratories. With a couple of notable exceptions, particularly Boeing and Martin, industry waited for the government, whether it was NASA or the Department of Defense, to furnish a conceptual design which industry then refined and enhanced. This was particularly true for the configurations studied by Lockheed and McDonnell Douglas. This relationship came about primarily due to the schedule and political situation surrounding the early U.S. manned space program (Mercury, Gemini, and Apollo), and would continue throughout the space shuttle development cycle. The few truly original designs submitted by contractors during the shuttle development effort were usually not well received.[87]

One inventive designer was Darrell Romick at Goodyear Aerospace. Beginning in 1950, Romick proposed a variety of space vehicles for a multitude of purposes, including several that looked very much like a space shuttles. As early as 1956 he had proposed vertical take-off two- and three-stage winged recoverable vehicles to support orbiting space stations. By August 1960 this had evolved into a three-stage horizontal take-off concept. Later in the year the design was considerably refined into a two-stage horizontal take-off vehicle 282 feet long with a 170-foot wing span. The orbiter had a payload bay 8 feet in diameter and 15 feet long. The mated vehicle weighed 1,600,000 pounds at launch and both stages were powered by LO2/LH2 engines. From the available documentation it does not appear that any of Romick's designs were taken very seriously by NASA or the Air Force, and they remain but interesting footnotes in history.[88]

Another exception to government invention was Robert Salkeld. In 1965 Salkeld proposed several different vehicles that all used conventional aircraft, usually deriv-

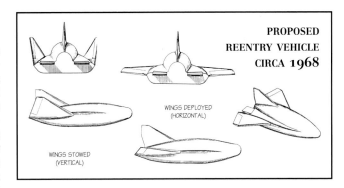

PROPOSED
REENTRY VEHICLE
CIRCA 1968

WINGS DEPLOYED (HORIZONTAL)

WINGS STOWED (VERTICAL)

This vehicle was patented by a group of North American engineers led by Frederick Raymes in August 1968. The wings folded up and out of the way during reentry, similar to that used by the Soviet Spiral prototype. (Courtesy of Frederick Raymes)

atives of the Lockheed C-5 Galaxy, as first stages. The concept that was the primary focus of the studies was a small stage-and-a-half orbiter that fit inside the cargo bay of a modified C-5. The ability to launch with little preparation, and to be based almost anywhere in the world, provided a quick response capability to orbital emergencies that should have appealed to the Air Force.[89]

The spacecraft carried a crew of two, in addition to three passengers and sufficient equipment to perform emergency on-orbit rescues or repair satellites in low-earth orbit. The spacecraft would be ejected through the rear doors of the C-5, aided by a drag-chute to stabilize the spacecraft until engine ignition. A single liquid-fuel rocket engine was fed propellants from two external drop tanks and two Minuteman solid fuel boosters could provide additional thrust if necessary.

Various growth versions of this concept were also proposed, most using larger derivatives of the C-5 along with larger orbiters. However, neither NASA nor the Air Force displayed any interest in the concept, and none of these designs progressed beyond paper studies

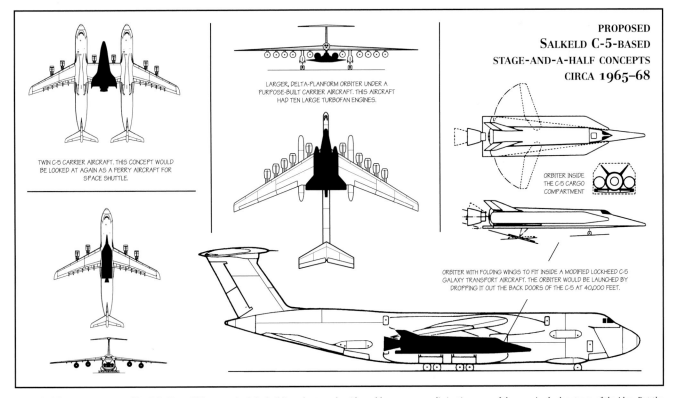

PROPOSED
SALKELD C-5-BASED
STAGE-AND-A-HALF CONCEPTS
CIRCA 1965–68

TWIN C-5 CARRIER AIRCRAFT. THIS CONCEPT WOULD BE LOOKED AT AGAIN AS A FERRY AIRCRAFT FOR SPACE SHUTTLE.

LARGER, DELTA-PLANFORM ORBITER UNDER A PURPOSE-BUILT CARRIER AIRCRAFT. THIS AIRCRAFT HAD TEN LARGE TURBOFAN ENGINES.

ORBITER INSIDE THE C-5 CARGO COMPARTMENT

ORBITER WITH FOLDING WINGS TO FIT INSIDE A MODIFIED LOCKHEED C-5 GALAXY TRANSPORT AIRCRAFT. THE ORBITER WOULD BE LAUNCHED BY DROPPING IT OUT THE BACK DOORS OF THE C-5 AT 40,000 FEET.

Several of the concepts proposed by Salkeld would have required the building of extremely wide and long runways, eliminating some of the perceived advantages of the idea. But the modified Lockheed C-5-based concepts would have provided significant flexibility in locations for basing and launching the small rescue spacecraft carried in its cargo bay. (NASA)

The early 1960s were a time of great debate over the possibility of a manned space station. Three NASA Centers, the Manned Spacecraft Center at Houston, the Langley Research Center in Virginia, and the Marshall Space Flight Center in Hunstville, all either designed, or let contracts for the design, of various space station concepts. Most of these concepts had some variety of 'resupply' vehicle associated with them. Many of the resupply vehicles were variations of either Gemini or Apollo capsules, but many others investigated lifting-reentry vehicles, and a sample of these concepts are presented on this and the next page.

McDonnell Aircraft Company
CIRCA 1963

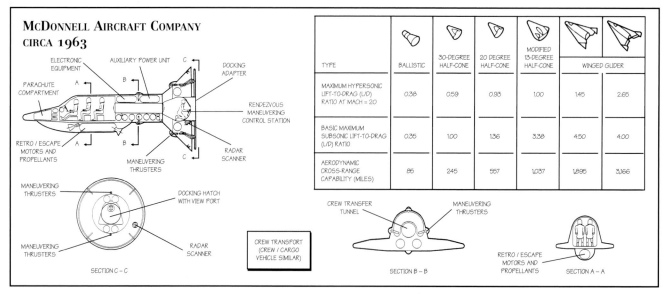

TYPE	BALLISTIC	30-DEGREE HALF-CONE	20 DEGREE HALF-CONE	MODIFIED 13-DEGREE HALF-CONE	WINGED GLIDER	
MAXIMUM HYPERSONIC LIFT-TO-DRAG (L/D) RATIO AT MACH = 20	0.38	0.59	0.93	1.00	1.45	2.65
BASIC MAXIMUM SUBSONIC LIFT-TO-DRAG (L/D) RATIO	0.35	1.00	1.36	3.38	4.50	4.00
AERODYNAMIC CROSS-RANGE CAPABILITY (MILES)	85	245	557	1,037	1,895	3,166

The McDonnell Aircraft Company released their Proposal for an Operations and Logistics Study on 20 March 1963. McDonnell proposed to study several concepts for a logistics spacecraft to find the design most compatible with the requirements for: docking and cargo transfer; crew and cargo transport; reentry touchdown control; and recovery. A small winged spacecraft was one of the favored designs (a modified Gemini capsule and a much-modified Gemini called Big-G were the others), and came in two varieties. The first variety (illustrated above) was a Crew Transport which could ferry four men plus a flight crew of two. The second derivative replaced the aft-mounted Docking Adapter with a large cargo module, and could ferry four men plus sufficient cargo for the expected duration of the crew on-orbit. The cargo module was left attached to the space station to serve as a storage assembly, while the winged vehicle returned carrying the space station's previous crew. Both vehicle derivatives were launched by Titan IIIM or Saturn IB boosters. Various docking techniques were to be investigated, including the use of closed-circuit television and fully automated concepts to replace the need to have an astronaut manually dock the vehicles. It must be remembered that docking on-orbit was still largely unproven at this time. Cross-range was considered an important factor in the logistics spacecraft design, and also in the selection of an orbit for the space station itself. It was felt that if a sufficient cross-range could not be achieved by the logistics vehicle (if a Gemini was selected due to cost considerations, for example), then the orbit for the space station should be adjusted to ensure sufficient crew return opportunities in the event of an on-orbit emergency (McDonnell Aircraft Company).

North American Aviation
CIRCA 1963

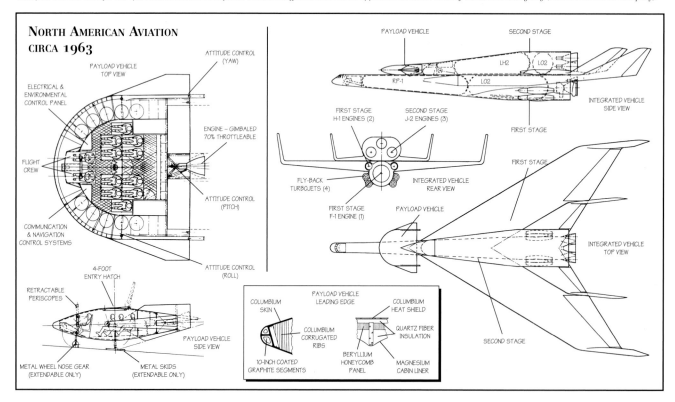

The Information Systems Division of North American Aviation released their Proposal for an Operations and Logistics Study of an Manned Orbiting Space Station on 18 March 1963. Two different logistics vehicles were presented, one based heavily on the Apollo spacecraft, and the other a derivative of an advanced reusable orbital carrier specifically designed for efficient logistics operations. This design was a horizontal-launched, two-stage system with a separate payload vehicle. The gross lift-off weight of the integrated vehicle was 1,210,000 pounds, and all three stages were manned and reusable. The two boosters, mounted top-to-bottom, were parallel-staged, with the winged payload vehicle mounted directly ahead of the second stage. The first stage was approximately 108 feet long, and the first and second stage vehicles could boost 25,000 pounds into near-earth orbit. The engines were derivatives of the F-1, H-1, and J-2 used in the Saturn program. The payload vehicle had a basic body diameter of 25 feet and carried a flight crew of two, along with ten passengers and a limited amount of cargo. All the stages used unconventional metal landing gear, loosely based on X-15 experience, to survive very high speed landing. (North American)

MSC CONCEPTS
CIRCA 1969–70

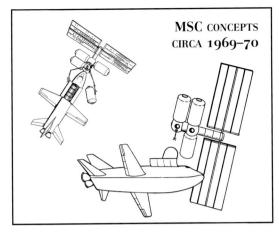

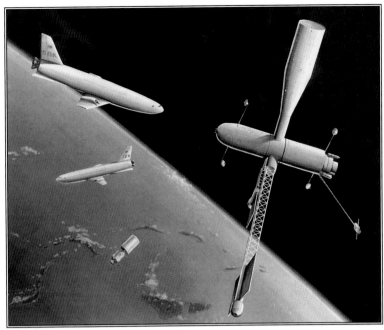

The 1969 illustration (right) is from an in-house study conducted by the Engineering and Development Directorate of the Manned Spacecraft Center (MSC) in Houston. Notice the early delta-shaped horizontal stabilizer on each vehicle. Clark Covington and his staff at MSC had spent the better part of a year investigating space stations and the vehicles necessary to resupply them. The two orbiters pictured here are derivatives of the space shuttle concepts being investigated by North American under the Phase A contract (page 87). The illustrations shown above are from 1970, as evidenced by the change in stabilizer shape. (NASA)

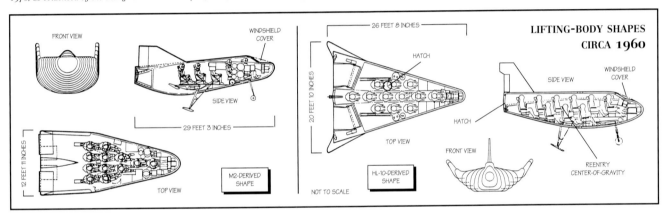

LIFTING-BODY SHAPES
CIRCA 1960

The Langley Research Center (LaRC) and its contractors investigated variants of the HL-10 and M2 lifting-body shapes for use as Space Station logistics vehicles. The shapes shown above are representative of the concepts proposed during the early-1960s. Although the Manned Spacecraft Center (MSC) rejected the lifting-body concept at the time, the Johnson Space Center (JSC) has since embraced it as a possible Assured Crew Return Vehicle (ACRV) for the year 2000 version of the International Space Station. (NASA)

LOCKHEED
CIRCA 1963

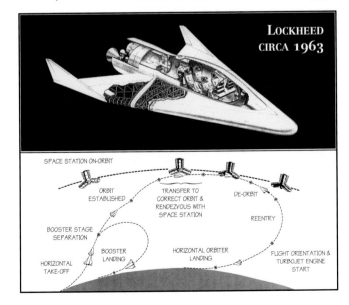

Lockheed's 1963 space station resupply study featured a two-staged winged vehicle that was an outgrowth of a 1958 effort undertaken with Hughes Aircraft under a classified Air Force contract. Both the orbiter and booster were fully reusable after horizontal launch and landings. (Lockheed-California Company)

SCIENCE FICTION
2001: A SPACE ODYSSEY
PAN AMERICAN WORLD AIRWAYS
ORION passenger SHUTTLE
CIRCA 1968

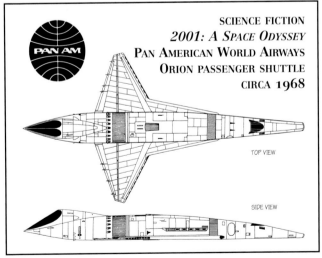

An optimistic glimpse into the future was provided by Arthur C. Clarke and Stanley Kubrick in the critically acclaimed 1968 movie 2001: A Space Odyssey. Space travel was shown as routine, with a large rotating space station being assembled in Earth orbit, serviced by commercial passenger shuttle craft such as the Pan American World Airways Orion. The movie was released in a new format, 70-mm Cinerama, which could only be fully enjoyed at theaters specially equipped to project it. Pan Am actually began taking non-binding reservations for flights on Orion after the movie's release. (Pan Am Archives)

Maxime A. Faget. (NASA photo S81-30585)

LET THE GAMES BEGIN

By mid-1965 the concept of a fully reusable lifting-reentry spacecraft had been seriously investigated for over a decade. Interestingly, the two primary development programs had both been initiated by the Air Force – BoMi, and Aerospaceplane. Each had taken very different approaches to the problem of lifting-reentry. BoMi had begun as a large fully reusable two-stage-to-orbit (TSTO) system but had evolved into the small Dyna-Soar glider launched by an expendable booster. Aerospaceplane was originally envisioned as a complex single-stage-to-orbit (SSTO) system, but soon evolved into a TSTO system when it became clear that SSTO was simply too difficult for existing technology.

Both of these programs were cancelled for various reasons. There is little doubt that Dyna-Soar would have succeeded in producing a workable system, albeit one that did not have a clear mission. Nevertheless, the little glider would have produced a wealth of information on lifting-reentry technology and operations. The ultimate success of Aerospaceplane is more open to debate – the technological challenges were immense, the concepts ill-defined, and the program management indecisive and unfocused. Failure was probably inevitable.

During all of this the NASA space centers had been largely preoccupied with Apollo and the presidential mandate to land on the Moon. What NASA support there had been for Dyna-Soar and Aerospaceplane had come from the aeronautical centers (Ames and Langley) and the Flight Research Center at Edwards where Apollo was much less an issue. As work on Apollo began to wind down the space centers began looking toward the future. This was especially true at the Marshall Space Flight Center (MSFC) simply because their work (designing the Saturn boosters), of necessity, began to end before that of the Manned Spacecraft Center (MSC) or the Launch Operations Center (LOC – later the Kennedy Space Center, KSC).

Based on the results from the various reusable aerospace passenger transport (RAPT) studies than ran from 1962 through 1965, the MSFC Future Projects Office determined that a fully-reusable TSTO concept was the most desirable. The Air Force was not so easily convinced based largely with the difficulties encountered by BoMi, RoBo, and Aerospaceplane. The Air Force Flight Dynamics Laboratory, which was sponsoring an independent set of integral launch and reentry vehicle (ILRV) studies, was not sure the state-of-the-art existed to reasonably design a fully reusable TSTO system. These disparate ideas were evident when the AACB ultimately released their report, which contained a variety of partially reusable and fully-reusable concepts for the future.

As is often the case, the conclusions from this first round of studies differed depending upon your point of view. If the criterion was to achieve the lowest possible cost per flight, then the fully reusable TSTO concept was the most reasonable. The downside was that it carried, by far, the highest development cost and the most potential risk. On the other hand, if the goal was to develop an early capability at minimal cost, then a small Dyna-Soar-type glider boosted by a Titan IIIM or Saturn IB appeared the

best. A wide variety of options existed between these two extremes, allowing one to choose a level of reusability that balanced available development funds with out-year operational costs. At Lockheed, Maxwell W. Hunter II – who had been heavily involved in the StarClipper stage-and-a-half concept – pointed out that TSTO concepts could only be expected to pay for themselves if they were used 100 times per year or more. If you expected to use the vehicle only once or twice per month, then stage-and-half or partially expendable concepts made a great deal more sense. It was an argument that would be buried in statistical analysis for the next few years. All of this assumed, of course, that one believed reusable systems brought any perceived benefit over expendable systems at all.

A few high-level NASA managers were becoming interested in reusable spacecraft – the most influential was George E. Mueller, the head of the Office of Manned Space Flight at NASA Headquarters. His domain included all of Apollo, but more important, he was a major proponent of a large space station as the next logical program for NASA. And he understood that low-cost reliable launch vehicles were one of the keys to making a space station a reality. During January 1968 Mueller hosted a one-day symposium at NASA Headquarters where industry was invited to discuss their current thinking about reusable launch vehicles with an audience made up of senior NASA and Air Force officials – other members of industry were excluded mainly because much of the data presented was proprietary.[1]

Martin Marietta was the first to present, and showed a very conservative variation of the AACB Class I concept based on a Dyna-Soar glider with a Titan IIIM booster. Lockheed presented Max Hunter's StarClipper stage-and-a-half design – perhaps the most interesting aspect of Lockheed's presentation was a claim that advanced avionics would allow StarClipper to launch within one hour of arriving at the launch site. Mueller was listening. McDonnell Douglas showed the same external-tank concept they were designing for the Air Force ILRV study. General Dynamics/Convair was the only company to discuss a true TSTO design, although it had three elements. Again, it was the same basic design that was being studied under the Air Force ILRV contract. The triamese concept was not presented directly, but was included during an in-depth discussion of the relative costs of developing and operating various types of reusable launch vehicles. What was most interesting about these designs was that all of them were based on some variation of a lifting-body, although the General Dynamics and McDonnell Douglas designs both used deployable wings at low speeds.[2]

Again, it came down to a matter of choice. Did NASA want low out-year operating costs, or low near-term development costs. Nobody knew the answer yet, but most within NASA were opting for the former. Despite the best efforts of NASA management, nobody could come up with a convincing new goal to replace landing on the Moon. None of the immediate projects envisioned by the NASA leadership – space stations, continued Moon landings, or trips to Mars – captured the imagination of Congress or the American people. Instead, the war in Vietnam was dividing the Nation and driving a robust economy to the brink of collapse.

In August 1968 George Mueller received an award from the British Interplanetary Society for his work on Apollo. In his acceptance speech he talked about space stations being the next logical step for the United States, and the ability to resupply them economically:[3]

"Essential to the continuous operation of the space station will be the capability to resupply expendables as well as to change and/or augment crews and laboratory equipment. ... Our studies show that using today's hardware, the resupply cost for a year equals the original cost of the space station ... Therefore, there is a real requirement for an efficient earth-to-orbit transportation system – an economical space shuttle. ... The shuttle ideally would be able to operate in a mode similar to that of large commercial air transports and be compatible with the environment at major airports. ... Interestingly enough, the basic design ... for an economical space shuttle ... could also be applied to terrestrial point-to-point transport."

Mueller also mentioned a concern: "One problem is, of course, the germination period of from 7 to 15 years for new designs. Jet power, available in 1946, came into commercial use on the Boeing 707 in 1958. Driving against traditional time lags, the Saturn V system has been developed and used within nine years of its conception. It is reasonable to conclude, then, that a space shuttle development program, initiated now [1968], could not be brought to fruition before the end of the 1970s ..."[4]

This was probably the first significant 'official' use of the term 'space shuttle,' although Max Hunter and a few others had already used 'shuttlecraft' and other variations for several years. Mueller again used the term while addressing the National Space Club in Washington during November. By 1969 'space shuttle' had become the standard terminology (finally replacing the awkward ILRV moniker), and on 5 January 1972 President Richard M. Nixon announced the development of a Space Shuttle, ensuring the term would continue to be used.[5]

Separate from, yet intimately related to, the efforts to define what sort of reusable launch vehicle was desirable were the efforts to determine the future of the entire American space program. The peak funding for the Apollo program occurred in FY65, when the NASA budget reached $5,250 million. Almost immediately the budget began declining, and by FY69 was down to $3,953 million – a 25 percent decline in only four years. In 1965 over 377,000 contractors were working on NASA contracts, in addition to 33,000 government employees. By 1969 the number of contractor personnel was down to 186,000, although the NASA jobs had held relatively steady. A new project would need to be found – quickly.[6]

CHOOSING A CONFIGURATION – PHASE A

Amid the drama of the approaching Moon landing, and the uncertainty in Washington concerning budgets and future directions, NASA continued to make plans. In September 1966, the AACB had concluded that "no single, most desirable vehicle concept could be identified ... for satisfying future DoD and NASA objectives." Shortly after the AACB had released their report, representatives from NASA Headquarters, MSC, and MSFC met to plan a joint study of reusable launch vehicles – the meeting was intended to create a common set of requirements and goals for a reusable launch vehicle for NASA. Daniel Schnyer from the Office of Manned Space Flight observed that "we have a

vast store of knowledge to draw on, and should now be able to get together and decide on an agency concept for the entire logistics system." He was, of course, referring to the fact that any reusable transport was still viewed primarily as a way to supply an orbiting space station.[7]

The meeting was sufficiently noteworthy that Max Akridge, a representative from MSFC, later described it as "the beginning of the space shuttle as such."[8] But in many ways it was still premature. Neither NASA nor industry truly understood what would be required of a space shuttle. The space station it was supposed to service had not been designed, so nobody knew how large it was, or how many flights per week (or month or year) would be required to maintain it. It was still undecided if the station itself would be launched by the Saturn Vs (production of which effectively ended in 1968), or if the space shuttle itself would need to be designed to carry components of it to orbit and assemble them.

Engineers and scientists from the various NASA centers reviewed the results of earlier Air Force and NASA studies (such as Aerospaceplane and the MSFC RAPT concepts) and began to refine a set of requirements for a space shuttle. The Air Force also participated in these discussions, but could never reach agreement with their NASA counterparts on exactly what a space shuttle should look like.

On 30 October 1968, the Manned Spacecraft Center and the Marshall Space Flight Center issued a joint request for proposals (RFP) for an eight-month study of an integral launch and reentry vehicle (ILRV), reusing the same terminology originally coined by the Air Force. The RFP gave each contractor a month to describe what topics they believed should be studied before the government selected four winning companies.[9]

Both Centers emphasized the fact that these studies would not necessarily result in an actual development contract being awarded, and also stated that the studies should concentrate on "... economy and safety rather than optimized payload performance ...". This meant that aircraft-like operations were desirable, and that any vehicle should be able to be launched with 24 hours notice. All of the studies were to be based on the anticipated requirement to place between 5,000 and 50,000 pound payloads into orbits between 115 and 345 miles, with 25,000 pounds into a 300 mile orbit being the baseline for comparisons. The payload bay was to provide a volume of at least 3,000 cubic feet, and the vehicles were to be able to return at least 2,500 pounds from orbit. A cross-range of 450 miles was required for safety considerations, although it was generally known that certain factions (specifically, the Air Force) desired a far greater capability.[10]

Five large aerospace companies, the Convair Division of General Dynamics, Lockheed, McDonnell Douglas, Martin Marietta, and North American Rockwell expressed an interest in participating in the study, which was 'Phase A' of the Space Shuttle design and development cycle.

In the anticipated four phase process, Phase A was labeled advanced studies, Phase B was project definition, Phase C was actual vehicle design, and Phase D was to be production and operations. As the program progressed this changed slightly, with Phase A later being termed preliminary analysis instead of advanced studies, and Phases C and D being combined. The rationale behind this Phased Project Planning (PPP) approach, which had been formally adopted by NASA in August 1968, was to foster competition throughout the development process, eventually reducing the number of competitors as the phases advanced. It was felt that this would leave only those contractors that could reasonably be expected to accomplish the task by the time Phase C (design) contracts were awarded, and that one ultimate winner would be selected for Phase D.

On 30 November 1968 the five companies submitted their study proposals to a joint MSC/MSFC source selection board. Convair proposed to focus on TSTO concepts, primarily variations of the triamese vehicle they had first proposed to the Air Force several years earlier. Lockheed wanted to refine the StarClipper design, including alternate boosters that would yield a sort-of two-and-a-half-stage concept. McDonnell Douglas proposed to study variations of their parallel tankage concept, as well as running independent analysis of the other current concepts and comparisons with the 'Big G' Gemini derivative being proposed as a space station logistics vehicle. Martin Marietta was a new player, leveraging off the Astrorocket and Astroplane designs they had been working on using company funds – their proposed study would concentrate on a truly unique TSTO design. North American wanted to study low-cost expendable boosters with reusable upper stages. NASA spent the next two months evaluating the proposals.[11]

But not everybody agreed that a space shuttle was necessary. After Richard Nixon was elected president in November 1968, he commissioned a Task Force on Space chaired by Charles H. Townes. The group reviewed many aspects of the on-going and planned future space program, and issued their report on 8 January 1969. When it came to any proposed new launch vehicle intended to reduce the cost of accessing space, the report concluded: "We do not recommend initiation of such a development, but study of the technical possibilities and rewards [should continue]." The warning fell largely on deaf ears.[12]

In February 1969, NASA awarded four ten-month $300,000 study contracts to General Dynamics/Convair, Lockheed, McDonnell Douglas, and North American Rockwell for the Phase A studies, although Martin Marietta

would also participate using company funds. The rationale for not accepting Martin's proposal is unclear, but was most probably based on concerns over the unusual vehicle configuration. In any case, NASA had always maintained that only four contractors would participate in Phase A, with two continuing to Phase B. NASA Langley monitored* the McDonnell Douglas contract, while MSFC monitored Lockheed and General Dynamics and MSC performed the same task for North American. Additionally, the Air Force and The Aerospace Corporation functioned in oversight roles, examining NASA's activities and coordinating these studies with their own in-house efforts and the ongoing stage-and-a-half studies being conducted by the AFFDL. The contractor reports were due by the end of the year.

All of this led directly to the Space and Missiles Systems Organization (SAMSO) issuing another set of Air Force ILRV study contracts. The four $100,000 contracts were issued to the same companies participating in the NASA Phase A studies – General Dynamics/Convair, Lockheed, McDonnell Douglas, and North American. The studies were for "an assessment of the utility of a broad range of spacecraft and launch vehicle concepts."[13]

The STG and SSTG

As a follow-on to the Townes report, on 13 February 1969 President Nixon set up the Space Task Group (STG)[†] within the National Aeronautics and Space Council to prepare a recommendation on future space activities and a proposed budget for them. The final report was due by 1 September 1969. The STG was chaired by Vice President Spiro T. Agnew, and included Secretary of the Air Force Robert C. Seamans, Jr.[‡] and NASA Administrator Thomas O. Paine. Robert Mayo, the Director of the Bureau of the Budget (BoB); Glenn Seaborg, the Chairman of the Atomic Energy Commission (AEC); and U. Alexis Johnson, the Undersecretary of State for Political Affairs, were observers. The STG held their first meeting on 7 March 1969, and during a second meeting on 22 March, Seamans announced that there appeared to be "considerable military interest" in a space shuttle and suggested another joint DoD/NASA panel to study it further.[14]

Less than three years earlier a similar joint study (the

---

* The various NASA Centers were responsible for technical monitoring only. All the contracts were issued and administered by the Manned Spacecraft Center (MSC).

† Use caution when reading this section since it references both the STG (Space Task Group) and the SSTG (Space Shuttle Task Group), two decidedly separate, but related, entities.

‡ Actually, Nixon had appointed the Secretary of Defense, Melvin Laird, to the STG, but Laird delegated the responsibility to Seamans.

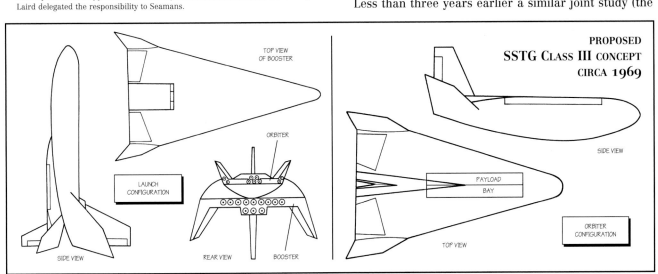

The SSTG devised a designation system to describe the configurations of future space shuttle vehicles. Class I consisted of reusable orbiters using expendable boosters, most probably derived from existing Titan IIIM or Saturn IB stages. This approach would return the most expensive portion of the vehicle (the orbiter), while minimizing overall development costs. Class II included stage-and-a-half vehicles with a recoverable orbiter that did not require a booster, but with throw-away propellant tanks (similar to what Lockheed and McDonnell Douglas had proposed to the Air Force during the 1968 ILRV study). Class III concepts were fully reusable TSTO vehicles, and this was the approach that SSTG thought all future space shuttle development efforts should take. To illustrate their point, the SSTG showed this NASA Langley concept based on the well known HL-10 lifting-body shape. This design had an orbiter 130 feet long with a payload bay that measured 15 by 60 feet. The recoverable booster measured 236 feet long and was powered by 12 boost engines. (NASA)

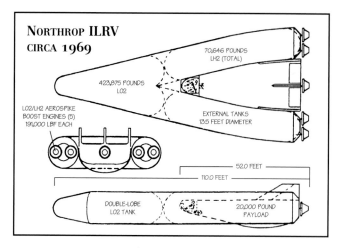

NORTHROP ILRV
CIRCA 1969

70,646 POUNDS
LH2 (TOTAL)

423,875 POUNDS
LO2

LO2/LH2 AEROSPIKE
BOOST ENGINES (5)
191,000 LBF EACH

EXTERNAL TANKS
13.5 FEET DIAMETER

52.0 FEET

110.0 FEET

DOUBLE-LOBE
LO2 TANK

20,000 POUND
PAYLOAD

Although Northrop did not directly participate in the Air Force or NASA ILRV studies, the company did design this stage-and-a-half design that used an M2 lifting body shape along with large external propellant tanks. (Northrop Corporation)[100]

AACB) had not been able to find a design concept that could satisfy both DoD and NASA requirements. This time the procedure would be different. The DoD and NASA would separately work to better define their requirements and determine the characteristics of a system that would best meet their needs. A joint study group would then convene in an attempt to meld these requirements into a single operational concept, and present their results to the STG no later than 15 June 1969. The objective of the joint study was "to assess the technical feasibility and economic sensibility of such a system."[15]

On 5 May 1969 George Mueller called representatives of General Dynamics, Lockheed, McDonnell Douglas, and North American together to change the rules for the Phase A studies. The modest 25,000 pound payload capability was being raised to 50,000 pounds. This reflected the fact that instead of being primarily a logistics vehicle for some future space station, the space shuttle was now envisioned as a satellite launcher – and many satellites would need large upper stages in order to reach higher orbits. It was also largely based on what Mueller knew the final joint DoD/NASA study was going to say since he was a co-chairman of the joint study group. The contractors were also asked to study increasing the payload bay diameter to 22 feet, enabling shuttle to carry the payloads designed to fit on top of the Saturn S-IVB stage.[16]

Mueller also presented several NASA-developed concepts for advanced on-board avionics systems. In particular, he believed that the use of computers would allow the rapid on-board checkout of the vehicle before a mission, eliminating many of the people that were required to prepare and launch Saturn, and therefore reducing processing costs. This was similar to what Lockheed had briefed in January 1968. Mueller also wanted the computers to digest and summarize in-flight data, presenting it in an easily assimilated format to the crew. The contractors took this briefing and went back to work on their studies – all would be different than originally envisioned.[17]

New directions issued on 20 May 1969 involved radical changes for some of the contractors. For instance, instead of expendable boosters, North American was asked to concentrate on a fully reusable two-stage system that had originally been conceived by Maxime Faget at MSC.* McDonnell Douglas also switched to two-stage systems, in this case based on the HL-10 lifting-body shape originally designed at NASA Langley. Lockheed and Convair continued with their original studies, although Convair turned to

a two-element version of its triamese concept. Another change occurred on 6 August when NASA decided to abandon looking at partially-reusable concepts such as StarClipper. Lockheed, therefore, began applying the basic LSC-8MX design to a fully-reusable two-stage system.[18]

In the meantime, a Space Shuttle Task Group (SSTG) headed by Leroy E. Day[†] had been set up to provide NASA's input to the STG. This represented the first truly comprehensive examination by NASA of the space shuttle concept, although at times they seemed slightly out of sync with recommendations being made by Mueller to the Phase A contractors. On 12 June 1969, the SSTG issued its final report concluding that any space shuttle should be capable of performing six different types of missions: (1) logistical support of a low-earth orbiting space station, including ferrying personnel, supplies, and propellants to the space station, and experiment data and manufactured products from the station back to Earth, (2) placement and retrieval of satellites into low-earth orbit, (3) propellant delivery to spacecraft in low-earth orbits, (4) low-earth orbit satellite servicing and maintenance, (5) delivery of other payloads to low-earth orbits, and (6) short (one to three days) duration manned orbital research missions.[19]

The report expressed a decided preference for fully or near-fully reusable vehicles, but realized that the real challenge was to design an orbital vehicle flexible enough to perform these diverse missions, and still economical enough to develop and operate. As such, the SSTG listed a number of important trade studies that it suggested should be conducted, including:[20]

- partially or fully reusable systems,
- piloted fly-back booster versus expendable boosters,
- winged (or wing-body) versus lifting-body configurations,
- off-the-shelf engines or new design propulsion systems,
- low (≈265 miles) versus high (1,726 miles) cross-range capability, based primarily on DoD requirements,
- small (≈15,000 pound) versus large (50,000+ pound) payload capability, again based on DoD requirements,
- vertical versus horizontal launch, and
- sequential staging versus parallel-burn.

Although the SSTG report was supposed to detail NASA's requirements, Day and others knew what the Air Force wanted and found the points hard to ignore – hence the recommended trade studies. At this point, the largest uncertainty concerned the planform for the proposed orbiter. Max Faget at MSC thought that a simple straight-winged orbiter would be satisfactory, but there was beginning to be a large body of opinion that disagreed. The lifting-body supporters were enthusiastic about the results of flight tests on the NASA Langley HL-10 shape, which was rapidly winning supporters at Edwards. But the most vocal opposition came from the Air Force. The AFFDL preferred sharply-swept delta planforms, and Alfred Draper and his colleagues published numerous papers and studies in the aerospace trade journals on the results of the unclassified portions of their research. The various contractors were studying the application of variable-geometry wings (similar to those used on the F-111), and even wings stowed within

---

* See the DC-3 section of this chapter for more information on Faget's concept.

† Day, who later became the Deputy Director for Space Shuttle development, was reluctantly reassigned from the historic Apollo project to head the SSTG.

‡ This was an attempt to capitalize on the low-speed capabilities of straight, or slightly-swept wings, but overcome their inherent problems at high speeds by hiding them within the fuselage until late in the descent.

SYSTEM DESIGNATION	PR-1	PR-2	FR-1	FR-2	FR-3
(1) THRUST AND WEIGHTS IN MILLIONS OF POUNDS (2) INCLUDING 50,000 POUND PAYLOAD (3) DEPENDS UPON AERODYNAMIC CONFIGURATION				**STG CLASSES CIRCA 1969**	
SYSTEM CONFIGURATION					
OVERALL DIMENSIONS (FT)					
LENGTH	250	130	189	207	205
WIDTH (MAX.)	85	64	77	58	58
LIFT-OFF PARAMETERS					
GROSS WEIGHT [2]	7.99	3.30	3.49	2.64	2.67
THRUST (S.L.)	13.50	4.00	4.78	3.92	3.95
ORBITER VEHICLE					
GROSS WEIGHT [2]	0.71	0.51	1.0	0.67	0.73
PROPELLANT WEIGHT	0.48	0.19	0.75	0.46	0.51
THRUST (VAC.)	0.50	4.7	1.4	1.4	1.2
HYPERSONIC L/D [3]	~2.1	~2.2	~1.8	~2	~2
LANDING WEIGHTS					
BOOSTER ELEMENT	N/A	N/A	0.20	0.25	0.28
ORBITER ELEMENT [2]	0.23	0.31	0.27	0.21	0.21

The joint report expected there would be between 30 and 70 missions per year in the ten years since 1975, and that the system would cost between $4,000 million and $6,000 million to build. This mission model included "only those flights required for existing, approved, or high-priority planned programs." Since this excluded a number of known uses for shuttle, the report surmised that including other missions could raise the flight rate to "about 140 per year." The cost of getting a pound of payload into low-earth orbit was expected to drop from $800 for a one-way trip on an expendable launcher, to between $50 and $100 for a round trip on the space shuttle. A similar reduction from $10,000 per pound into synchronous orbit to less than $500 per pound was also forecast.[24]

The joint study concluded that expendable stages used to launch a reusable orbiter were too expensive to operate and should not be examined further. Two types of partially reusable configurations (designated PR-1 and PR-2) were examined: PR-1 was a reusable orbiter and expendable boosters (Titan IIIM or Saturn IB), while PR-2 was based roughly on the Lockheed StarClipper and McDonnell Douglas parallel-tankage concepts. Three fully-reusable configurations (FR-1, FR-2, and FR-3) were also examined, based mostly on two- and three-element variations

the fuselage until they were needed. This latter concept included the obvious solution of swinging the wing horizontally inside the fuselage, and also a concept where the wings were folded upward flush with the sides of an ungainly slab-sided lifting-body orbiter.[‡]

Cross-range was also a problem, with Faget and MSC wanting as little cross-range as was required for safety considerations, while the DoD desired a large cross-range for their optimum mission profile. MSC also preferred a small (≈15,000 pound) payload, but this seemed difficult to justify on economic grounds, and almost all independent studies had indicated that approximately 25,000 pounds was the smallest payload that should be designed for, and DoD was known to want a minimum of 65,000 pounds.

A few days after LeRoy Day released the final SSTG report, the joint DoD/NASA report was issued: "The Joint DoD/NASA Space Transportation System (STS) Study concluded that the development of an STS is needed to provide a major reduction in operating costs and an increased capability for national space missions. ... Partially or fully reusable STS configurations meet the requirements with a 50,000-pound payload capability to and from polar or space station orbits." Note that the 50,000 pound requirement was to a 115-mile polar orbit from Vandenberg.[21]

The report continued that "Payload volumes of 10,000 cubic feet and required diameters of 15 feet and lengths of 60 feet are included." The payload sizes were based on various factors. The 15-foot diameter was a NASA requirement for "space station logistics support" and also an Air Force requirement for "satellites such as advanced versions of TACSAT or of surveillance systems." The 60-foot length was driven strictly by the Air Force "for ocean surveillance spacecraft ... or two medium-altitude surveillance satellites."[22]

Air Force requirements also drove the desired cross-range. "... 1,400 nautical miles for single-pass request surveillance mission with return to Washington, D.C. Cross-range capability of 1,900 nautical miles would allow assured return within 12 hours from sun-synchronous orbits and is therefore a desired capability; 1,500 nautical miles is the selected design value and 2,000 nautical miles or more are desirable."[23]

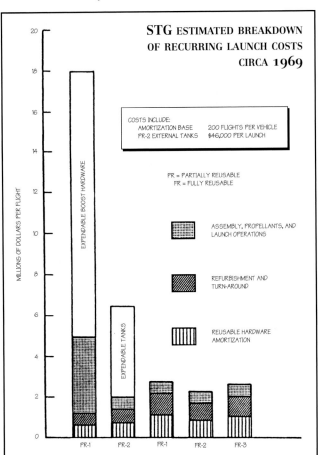

COSTS INCLUDE:
AMORTIZATION BASE   200 FLIGHTS PER VEHICLE
PR-2 EXTERNAL TANKS $46,000 PER LAUNCH

PR = PARTIALLY REUSABLE
FR = FULLY REUSABLE

ASSEMBLY, PROPELLANTS, AND LAUNCH OPERATIONS

REFURBISHMENT AND TURN-AROUND

REUSABLE HARDWARE AMORTIZATION

STG ESTIMATED BREAKDOWN OF RECURRING LAUNCH COSTS CIRCA 1969

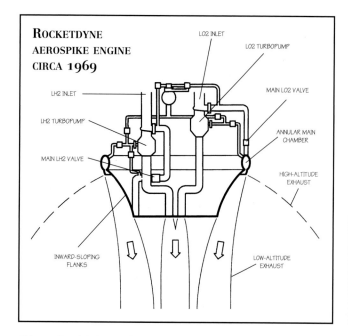

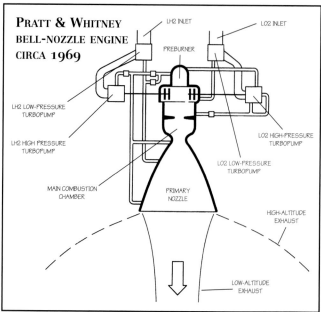

This was the original aerospike engine developed by Rocketdyne. The exhaust flows along the ramps on the inside, constrained only by atmospheric pressure on the outside. (Dan Gauthier via Tom Heppenheimer/NASA)

An early Pratt & Whitney high-pressure space shuttle main engine concept with a bell-nozzle. Note the use of a single preburner at the very top of the engine. (Dan Gauthier via Tom Heppenheimer/NASA)

of the Convair triamese concept. The report recommended the development of either PR-2 or any of the fully-reusable concepts.[25]

A NASA decision on 6 August to officially abandon looking at partially reusable concepts was based largely on the stated preference of the SSTG and Joint DoD/NASA study for fully reusable vehicles. The partially-reusable designs had been an attempt to minimize development costs with the realization that operational costs would be somewhat higher – although still far lower, at least at reasonably high yearly flight rates – than expendable boosters. This seemed to work reasonably well when the space shuttle was being considered primarily as a logistics vehicle for space stations. But as shuttle began to morph into a general-purpose launch vehicle, NASA began to envision extremely high flight rates and decided to accept the higher development cost of a fully reusable system for the promise of substantially lower operational costs. It was a choice that would come back to haunt the program a couple of years later.[26]

On 15 September 1969 the Space Task Group headed by Spiro Agnew delivered their final report. As part of an overarching National space program, the STG recommended[*] a reusable launch vehicle that would (1) provide a major improvement over the present way of doing business in terms of cost and operational capability, (2) carry passengers, supplies, rocket fuel, other spacecraft, equipment, or additional rocket stages to and from LEO on a routine, aircraft-like basis, and (3) be directed toward supporting a broad spectrum of both DoD and NASA missions.[27]

As conceptualized in the STG report, a reusable space transportation system would consist of (1) a reusable chemically-fueled space shuttle operating between Earth and LEO in an airline-type mode, (2) a chemically-fueled space tug or vehicle for moving people and equipment to different Earth orbits and as a transfer vehicle between the lunar-orbit base and the lunar surface, and (3) a reusable nuclear stage for transporting people, spacecraft, and supplies between Earth orbit and lunar orbit and between LEO and geosynchronous orbit and for other deep space missions.[28]

At the end of October 1969, the contractors began submitting their Phase A reports to NASA, while the Air Force contractors began doing the same to SAMSO. All of the studies included a proposed vehicle configuration, the results of the various trade studies, estimated development costs, proposed management structures, and an evaluation of future operational concepts.

Main Engine Phase A

The 1960s had seen important advances in the field of large rocket engines, particularly the use of liquid hydrogen (LH2) as a fuel. During early 1967 the Air Force initiated an advanced rocket engine development program that would evaluate even more advanced concepts that might be applicable towards future launch vehicles. Two companies, Pratt & Whitney and Rocketdyne, were awarded contracts to build experimental versions of two very different designs.

At Rocketdyne, the chosen technology was called the aerospike engine.[†] This was an attempt to solve the inescapable limitations of the bell-shaped nozzle normally used on rocket engines. The bell nozzle has to be tailored for the atmospheric pressure it operates in – short nozzles work best at high atmospheric pressures (i.e., low altitudes), while longer nozzles are more efficient in the near vacuum of space. Various attempts at 'extendible' nozzles had proven complex and heavy, although several small ones were in production. The aerospike engine[‡] eliminated the bell nozzle entirely and used atmospheric pressure as the outer wall, thereby creating a constantly variable nozzle area based solely on the atmospheric pressure around it. The engine used a ring-shaped combustion chamber surrounding a central body that resem-

---

[*] It should be noted that Robert Seamans was not totally convinced that the technology was available to build a reasonable space shuttle, and he expressed these concerns in a letter to the Vice President on 4 August 1969.

[†] The aerospike engines tested by Rocketdyne in the late 1960s differed considerably from the linear aerospike engines developed for the X-33 and VentureStar RLV program in the 1990s, although the basic theory remained the same.

[‡] This is also called a 'plug-nozzle.'

bled an upside-down cone with inward-sloping flanks and a central vent. The turbine exhaust flowed through the vent while the main engine exhaust expanded against the flanks on the inside and the atmosphere on the outside. The drawback was that the aerospike was slightly less efficient at any specific altitude than a well-optimized bell-nozzle engine, but was more efficient overall assuming an SSTO-type mission profile.[29]

Pratt & Whitney was assigned to investigate improvements in conventional high-pressure rocket engines using bell nozzles. The J-2 engine developed for the Saturn program had operated at a chamber pressure of 763 psi – the new engine was designed to go as high as 2,740 psi, a substantial increase. If this could be achieved it would allow a much more powerful engine to be produced in a very compact, lightweight package. The engine was designated XLR129-P-1 and was supposed to generate 250,000 lbf.[30]

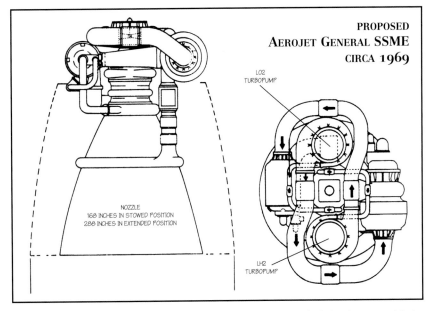

Aerojet General is the forgotten contractor in the SSME competition. This was the design they proposed during Phase A of the competition. (Aerojet General)

During the various phases of the Space Shuttle design competition, the airframe contractors were never allowed to design or propose their own main engines. Instead they were directed to use the current baseline engine being developed under a separate competition conducted by MSFC. In late 1968, Aerojet General, Pratt & Whitney, and Rocketdyne were selected to participate in the Phase A space shuttle main engine (SSME) studies.* All of the companies were considered qualified for the task. Pratt and Rocketdyne had each built large-scale rocket engines for past Air Force or NASA programs, while Aerojet had built the hydrogen-fueled NERVA nuclear engines in collaboration+ with Westinghouse, and also manufactured the main engines for the Titan family of expendable boosters.

But Rocketdyne suffered a blow when NASA decided that the aerospike technology was too immature for the upcoming space shuttle. Unlike normal rocket engines, but very much like scramjets, the aerospike needed to be carefully and totally integrated into the airframe. The only design feature that was clearly defined in the SSME specification was that the nozzle was required to be a bell-type – precluding any competitor from using the more advanced, but substantially higher risk, aerospike nozzle in their proposal.[31]

The engine specified for use during the vehicle Phase A competition was a high-pressure regeneratively cooled engine using LO2 and LH2 propellants. The engine baselined for the orbiter had a two-position extendible nozzle, with expansion ratios of 58:1 and 120:1, and provided a gimbal capability of seven degrees in both the pitch and yaw axes. This engine produced approximately 590,000 lbf in a vacuum and 510,000 lbf

at sea level. The engine was throttleable from 73 to 100 percent of rated power to provide a means to limit the maximum dynamic pressure (max-q) and g-loading (max-g) during ascent, and was also designed to operate at a 10 percent thrust level during on-orbit operations.[32]

The version specified for the booster had a fixed nozzle with a 5:1 expansion ratio, and was rated at 500,000 lbf (sea level). It is interesting to note that the concept of operating the main engines over their rated thrust to provide an abort margin in case of a main engine failure had not yet been proposed. An alternate engine configuration, based heavily on the XLR129 demonstrator, providing 415,000 lbf using a slightly smaller fixed nozzle was also available – as it would happen, the majority of the Phase A airframe contractors opted for this slightly smaller and lighter alternate engine.[33]

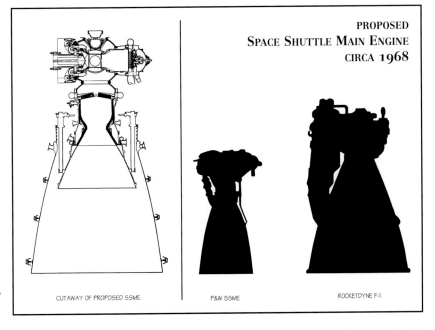

The size difference between the proposed Pratt & Whitney Space Shuttle Main Engine and the Rocketdyne F-1 that powered the first stage of the Saturn V is clearly illustrated here. Although significantly smaller and lighter, the SSME would also be much less powerful, producing only 500,000 pounds-thrust, compared to 1,522,000 lbf for the F-1. (Pratt & Whitney, A United Technologies Company)

---

* The engine competition also consisted of four phases, roughly paralleling the vehicle development process.

+ Aerojet built the rocket engine portion; Westinghouse was responsible for the nuclear portions.

McDonnell Douglas (NAS9-9204)

As with all of the Phase A studies, the McDonnell Douglas Astronautics Company effort was divided into essentially two parts. The first part contained the results of the original study that had begun prior to the redirection from George Mueller after the SSTG report. Early in the Phase A studies, McDonnell Douglas presented a concept dubbed MURP (manned upper reusable payload) to NASA for comment. The lifting-body shape was derived from a configuration originally developed at NASA Langley, but extensively modified by McDonnell during the MSFC RAPT studies. The design used a stowable variable-geometry wing to enhance its transonic and subsonic stability. Seats for ten passengers and a crew of two were provided atop a small forward payload area. Four pad-abort solid rockets were attached, two on either side, and provided a means to separate the entire personnel module that would then descend on a large parachute stored in the extreme rear of the vehicle. The main payload bay had 3,000 cubic feet of volume and occupied the majority of the mid-fuselage. A single 16,700-lbf turbojet engine was located on the lower aft centerline to provide a go-around option during landing operations. The landing gear was reminiscent of the X-15, with a two-wheel nose gear and retractable tail skids for the main gear. Thermal protection was also reminiscent of the X-15A-2, with a sprayed-on ablative-type coating over a hot-structure airframe. MURP was ultimately rejected from further consideration due to poor landing visibility, anticipated heating problems, and an overly complicated wing-pivot mechanism – nevertheless, a modified version of MURP would later reappear in the Chrysler alternate space shuttle concepts (ASSC) study (see page 123).[34]

When George Mueller redirected the Phase A studies, McDonnell Douglas switched to examining fully-reusable two-stage concepts based largely on the NASA Langley HL-10 lifting-body shape. The final report was completed in late November 1969 emphasizing a 'point' TSTO vehicle, with several other configurations based on parametric excursions from the point design. The overall goal to minimize the vehicle's operational costs was to be achieved through high system reliability, vehicle recoverability, and rapid ground turn-around made possible through modular component design and the use of an integrated on-board self-test and checkout system. A 25,000 pound payload could be delivered or returned in a 15 by 30 foot payload bay. The total system had a gross lift-off weight of nearly 3,400,000 pounds.

Thirteen vehicle configurations, all with a lifting-body orbiter, were considered by McDonnell Douglas during Phase A. The majority of these were intended to find the ultimate configuration of the recoverable booster, and the best mounting location for the orbiter (on top, on bottom, semi-submerged, etc.). Of the thirteen configurations, only five were deemed worthy of further study – a large lifting-body booster with the orbiter mounted semi-submerged in the bottom, a variable-geometry wing booster with the orbiter mounted recessed in the top, a large delta-winged booster with the orbiter mounted top-to-bottom, another variable-geometry wing booster with the orbiter mounted top-to-bottom, and a large clipped delta-winged booster with the orbiter mounted bottom-to-bottom. This last configuration was ultimately chosen for detailed study.

The orbiter was 107 feet long, and based heavily on the HL-10, although the aft-fuselage base area was modified slightly to accommodate a pair of 415,000 lbf engines, McDonnell Douglas opting for the less powerful, but small-

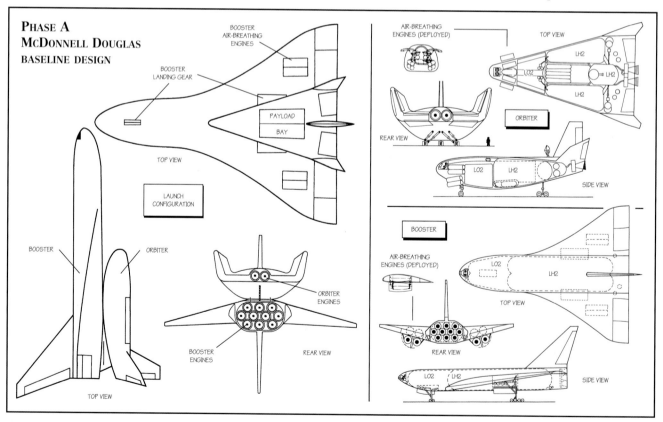

The McDonnell Douglas Phase A design bore a remarkable resemblance to the Class III concept (see illustration on page 79) defined by the Space Shuttle Task Group (SSTG) headed by Leroy E. Day. There were several major differences, however, since this design was much further developed, and the orbiter's size and propulsion systems differed significantly. Note the air-breathing engines on both stages. (McDonnell Douglas Astronautics Company)

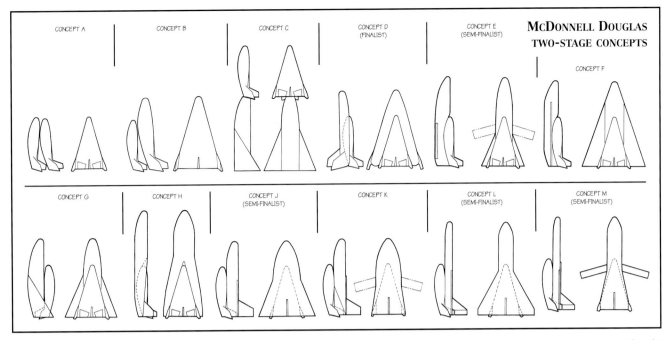

CONCEPT A | CONCEPT B | CONCEPT C | CONCEPT D (FINALIST) | CONCEPT E (SEMI-FINALIST) | **MᶜDONNELL DOUGLAS TWO-STAGE CONCEPTS**

CONCEPT F

CONCEPT G | CONCEPT H | CONCEPT J (SEMI-FINALIST) | CONCEPT K | CONCEPT L (SEMI-FINALIST) | CONCEPT M (SEMI-FINALIST)

The McDonnell Douglas engineering staff investigated thirteen basic configurations before selecting the one proposed. All of these configurations used an orbiter derived from the NASA Langley HL-10, although the boosters came in a variety of shapes. (McDonnell Douglas Astronautics Company)

er and lighter alternate engines. The geometric center of the payload bay was located on the vehicle center of gravity to permit the option of returning from orbit with or without the full payload. Alternate orbiter configurations, all still based on the HL-10 shape, were developed that allowed 50,000 pound payloads of either 15 by 60 feet or 22 by 30 feet. The boost propellant tanks were integral with the primary body structure to maximize volume, with the forward and aft tank bulkheads providing vehicle structure. The LO2 tank was forward of the payload bay with the tank skins conforming to the mold shape on the fuselage sides and bottom. The LH2 tankage was composed of three tanks, the two side tanks providing the mold line on the sides, top, and bottom. The inboard walls of these tanks also provided the thrust structure for the boost engines. The third LH2 tank was located aft of the payload bay and acted as a collector for the boost engines. A crew of two would be housed in a shirt-sleeve environment in the extreme nose of the orbiter.

The orbiter's thermal protection system consisted of titanium, Renè-41, nickel-chromium, and columbium-752 shingles over a layer of fibrous batt insulation.[*] An alternate configuration was considered, using hardened compacted fiber (HCF) plates in place of the nickel-chrome and columbium shingles, but was not recommended since the plates were heavier than the equivalent alloy shingles and not as durable. Columbium shingles were preferred over nickel-chrome because of the extensive experience acquired in their manufacture and use during the earlier ASSET program.

The large booster was a 195-foot long clipped-delta configuration with ten boost engines similar in type to those used on the orbiter. The 15-percent thick delta wing contained the landing gear, air-breathing engines, and JP-4 fuel. The LO2 tank was in the forward fuselage to minimize engine gimbal angles for initial thrust vector and center of

PHASE A MᶜDONNELL DOUGLAS BASELINE ORBITER

ACOUSTIC FATIGUE PANELS
THRUST STRUCTURE
CONTROL SURFACES AND SUPPORT STRUCTURE
LEADING EDGES AND THERMAL PROTECTION SYSTEM PANELS
PAYLOAD BAY
AIR-BREATHING ENGINES
FRAME AND MANUFACTURING SPLICES
ORBITER INTERFACE ATTACH FITTINGS AND BACKUP STRUCTURE
CREW COMPARTMENT
MAIN LANDING GEAR
TANK SECTION
TENSION COMPRESSION AND SHEAR PANELS
NOSE LANDING GEAR

After investigating the 13 configurations, these are the vehicles McDonnell Douglas concentrated on for the rest of the study. (McDonnell Douglas Astronautics Company)

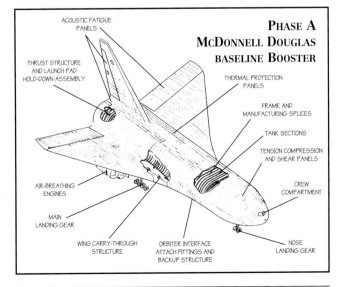

PHASE A MᶜDONNELL DOUGLAS BASELINE BOOSTER

ACOUSTIC FATIGUE PANELS
THRUST STRUCTURE AND LAUNCH PAD HOLD-DOWN ASSEMBLY
THERMAL PROTECTION PANELS
FRAME AND MANUFACTURING SPLICES
TANK SECTIONS
TENSION COMPRESSION AND SHEAR PANELS
CREW COMPARTMENT
AIR-BREATHING ENGINES
MAIN LANDING GEAR
WING CARRY-THROUGH STRUCTURE
ORBITER INTERFACE ATTACH FITTINGS AND BACKUP STRUCTURE
NOSE LANDING GEAR

---

[*] Although each company differed at times in their terminology, in general 'shingles' were metallic panels attached to stand-offs over a layer of fibrous batt insulation. This differs from 'plates' or 'tiles' that were a ceramic-like insulator usually applied directly to the vehicle structure, as used on the eventual Space Shuttle Orbiter.

gravity control during boost, and also to locate the empty vehicle center-of-gravity further forward. The propellant tank aft bulkheads were convex in order to minimize residual propellants. The vehicle basic body structure was provided by the propellant tank walls with integrally machined rings and stringers. A two man crew was seated on a flight deck in the forward fuselage ahead of the LO2 tank. The booster used a titanium structure with titanium and Renè-41 shingles for thermal protection. The leading edges of the wings and empennage were also to be of Renè-41.

The operational mission profile called for the mated vehicles to be launched vertically, powered only by the booster engines. The vehicles would then separate at an altitude of approximately 210,000 feet at a speed of 6,100 mph. The orbiter engines would be ignited for insertion into a 50 by 115 mile orbit, which would then be circularized using the engines at 10 percent thrust. After proper phasing, the orbiter would be raised to the 300 mile mission altitude. Reentry would be accomplished at a flight path angle of -1.5 degrees. A banked attitude and 50 degrees angle-of-attack would be maintained during pullout until the lower fuselage surface temperature reached ≈2,200 degF – this temperature would be maintained by varying the bank angle until load factors became prohibitive. A roll maneuver to wings level would take place at speeds between 4,600 and 2,800 mph in order to stretch the range and achieve approximately 450 mile crossrange. When the speed reached 2,650 mph, the orbiter was pitched down and allowed to descend using the profiles developed by the HL-10 during landing trials.

After separation, the booster would pitch to 50 degrees angle-of-attack while inverted and continue in this mode through ballistic apogee. A roll-out to upright flight would then be performed to allow a pullout above the thermodynamic boundary. Upon reaching subsonic horizontal flight, the booster would deploy its air-breathing engines and fly back to a landing somewhere near the launch complex.

It was estimated that the development program, including two flight-test orbiters and two boosters, would cost $5,946 million and take only 21 months. At a launch rate of 12 flights per year, using three orbiters and two boosters, the cost-per-pound to orbit would be $119 – at 100 flights per year, the cost dropped to $67 per pound.

Studies were also conducted on finding an all-azimuth launch site to replace both the Kennedy Space Center and Vandenberg AFB. Over 100 sites were examined, with McConnell AFB, located just outside Wichita, Kansas, being preferred. Its location was such that no serious geographic constraints would be encountered (mountains, lakes, etc.), and adequate abort landing sites were located along probable launch routes. The final report recommended that further studies be performed, particularly of climatic conditions, to assess the advisability of an entirely new launch facility.

ORBITER FERRY CONFIGURATION

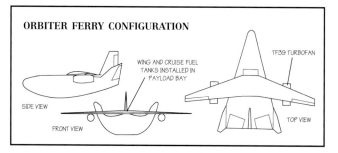

For ferry operations, a swept-wing and two 22,500 lbf GE TF39 turbofans was installed on a special pallet in the payload bay. (McDonnell Douglas Astronautics Company)

PHASE A
McDONNELL DOUGLAS

Various details of the McDonnell Douglas design were presented in their proposal. (McDonnell Douglas Astronautics Company)

ALTERNATE BOOSTERS

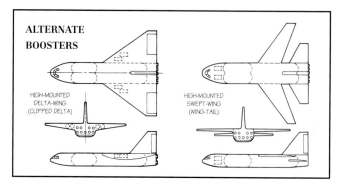

THERMAL PROTECTION SYSTEM

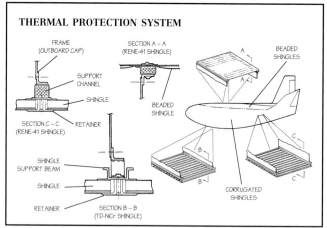

PAD ERECTION TECHNIQUE

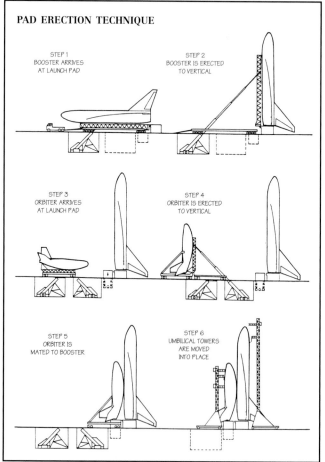

North American Rockwell (NAS9-9205)

The original Phase A investigation proposed by the Space Division of North American Rockwell was dedicated to economic comparisons of reusable logistic vehicles with expendable boosters or partially expendable booster-derived concepts. Perhaps more so than the other Phase A contractors, North American was significantly impacted by George Mueller's redirection. Instead of expendable boosters, the company was asked to examine fully-reusable two-stage concepts centered around a configuration that had been conceived by MSC under the direction of Maxime Faget.[35]

As envisioned by MSC, the straight-wing orbiter could deliver a 12,500 pound payload to low-earth orbit, but the basic aerodynamic shape and weight-to-planform area relationships could be parametrically enlarged to obtain data on vehicles capable of delivering larger payloads. These vehicles could be configured to deliver up to 50,000 pounds to a 300 mile circular orbit inclined at 55 degrees. All were designed for vertical take-off. The orbiter was mounted forward on the booster, providing a forward center-of-gravity during the ascent phase of the mission. A large planform and flat base area enabled the orbiter to enter the atmosphere at a high angle-of-attack – essentially emulating a blunt-body that minimized potential structural heating and lessened the need for any external thermal protection system. The integrated vehicle had a GLOW of 4,476,000 pounds.

The preferred orbiter spanned 146 feet with 2,830 square feet of effective wing area. A fuselage 202 feet long contained LO2 and LH2 tanks, a 15 by 60 foot payload bay, room for ten passengers, and a flight crew of two. The initial configuration included a unique low-mounted delta-shaped horizontal stabilizer, although later designs used a straight horizontal, similar in planform to the main wing. Two 510,000-lbf baseline boost engines were mounted side-by-side in the extreme aft fuselage. Four P&W JT3D-7 turbojet engines, each rated at 19,000 lbf, were provided for landing and ferry operations. The air-breathing engines were mounted in fixed titanium nacelles located on the upper surface of the wing at roughly mid-span. It was noted that the use of future high thrust-to-weight ratio turbofan engines, then under development by Pratt & Whitney and General Electric, could result in a weight reduction of almost 10,000 pounds, most of which could translate directly into payload capacity.

The two crew members were located forward in an aircraft-type cockpit using contemporary aircraft avionics and control systems. Ten passengers were located immediately behind the flight deck, and they and their equipment were considered to be part of the payload weight allowance. An access tunnel was provided between the payload bay and the passenger compartment. The payload bay was mounted at the orbiter center-of-gravity to provide a vehicle that could be trimmed on reentry and during cruise with, or without, a payload. The orbiter was designed to allow an airlock located in the payload bay to rotate in order to provide a docking interface with a space station. The location of the payload bay required the use of two LH2 tanks, one forward of the bay and one aft – the forward LH2 tank was integral with the LO2 tank, the propellants being separated by a common bulkhead.

The propellant tanks, constructed of fusion-welded

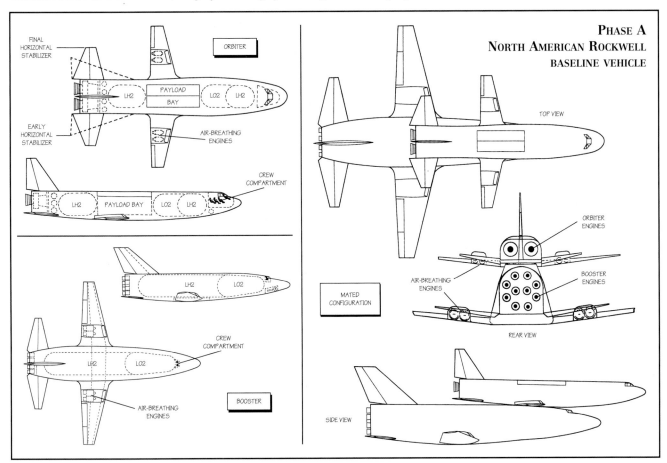

This concept was originally conceived by Maxime Faget at MSC, although North American significantly refined it during the Phase A studies. Note the difference in the horizontal stabilizer between the original and the final designs – some drawings show a different variation of the delta shape. (North American Rockwell)

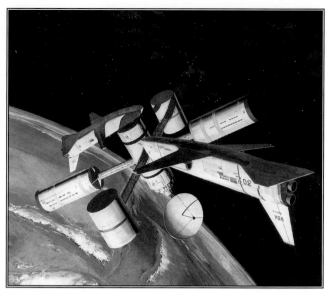

The future, circa 1969. A large space station in the background, and a straight-wing North American orbiter releasing a satellite. (NASA photo S69-55878 via the SDAM Collection)

A space stations with two straight-wing orbiters docked. These orbiters have wing-fuselage fillets, something that did not get baselined until Phase B. (North American Rockwell)

2219-A1 aluminum, were suspended and therefore not subjected to any aerodynamic or thermodynamic loads. The primary loads were carried by a skin-stiffened 6A1-4V titanium structure that, in order to withstand the expected reentry thermal environment, was protected on the lower sides and bottom of the fuselage by LI-1500, a silica-based* insulation material developed by the Lockheed Missiles and Space Company. Although Lockheed was competing with North American, LI-1500 was being developed under a separate contract to NASA, and therefore could be used by any of the competitors. A bare titanium hot-structure was considered satisfactory for the top of the orbiter because of the milder thermal environment. Fiberglass insulation was used extensively inside the orbiter to minimize heat transfer to the propellant tankage and crew compartment.

The 280-foot long booster used a large propellant tank constructed of 2219-A1 aluminum alloy as the primary load-carrying structure. The LO2 tank was located forward of the LH2 tank with a common bulkhead separating the propellants. An aluminum truss and reinforced phenolic-polyamide honeycomb-sandwich substructure was used to maintain the desired flat base area, and was also used to bond LI-1500 external insulation tiles to the bottom and sides of the booster. The 244-foot span wings were constructed of 6A1-4V titanium, with air-breathing engine nacelles located on their upper surface. Booster main propulsion was to be eleven baseline 500,000 lbf engines. Four P&W JT9D-15 turbojets would burn 57,800 pounds of JP-4 during a 310 mile cruise back to the launch site. The study noted the possible weight savings obtainable by using gaseous hydrogen instead of JP-4/5 for the air-breathing engines. Interestingly, the proposed booster was unmanned, although provisions for a two man flight crew were included for ferry flights.

Several minor variations on the basic shape for both the orbiter and booster were presented at various times during the study contract, the most notable difference being the size and shape of the horizontal stabilizer. On most early vehicles the horizontal stabilizer was a delta

shape, extending well forward, almost to the trailing edge of the main wing. A more conventional empennage was used on later designs.

The impacts of potential systems failures were also analyzed during the study, and indicated that failure of a single engine on either the booster or the orbiter would not prevent the successful completion of the primary mission. However, a system failure requiring premature separation would necessitate action to limit reentry loads on the orbiter. For low-altitude mission termination, the orbiter would remain at low altitude, returning to a contingency landing site at the earliest feasible opportunity. In the case of a high-altitude mission termination, the orbiter engines would be used to guide the vehicle through an optimum sub-orbital trajectory in order to minimize aerodynamic and thermal loads on the orbiter. It was noted that the failure of both orbiter main engines would undoubtedly expose the orbiter to excessive loads, and most probably result in a water landing. This analysis would be repeated many times during the course of the Phase A and Phase B studies, and always reached much the same conclusions. This was the main reason everybody was leaning towards a three- or four-engine orbiter.

This August 1969 photo shows the early horizontal stabilizer configuration, as well as the relatively exposed location of the air-breathing engines on top of the wings of each element. (NASA photo S69-39844)

It was expected that the design and production of the initial hardware would consume 4.5 years, after which six flight vehicles would be built at the rate of two per year. A launch rate of 50 flights per year could be supported using the six vehicles. Almost unrealistically, North American estimated that only 200 personnel would be required for maintenance and launch operations using airline-type schedules.

---

* This material would later evolve into the 'tiles' used by the production Space Shuttle Orbiters, and its development had been funded jointly by Lockheed and NASA. The designation includes the density of the material, thus LI-900 tiles weigh 9 pounds per cubic foot, LI-2200 weigh 22 pounds per cubic foot, etc.

PAD ERECTION OPTIONS

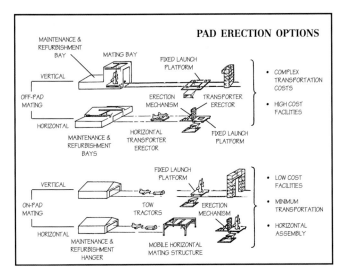

BOOSTER AIR-BREATHING ENGINE OPTIONS

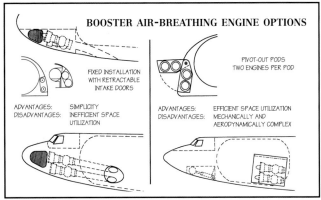

MISSION PROFILE

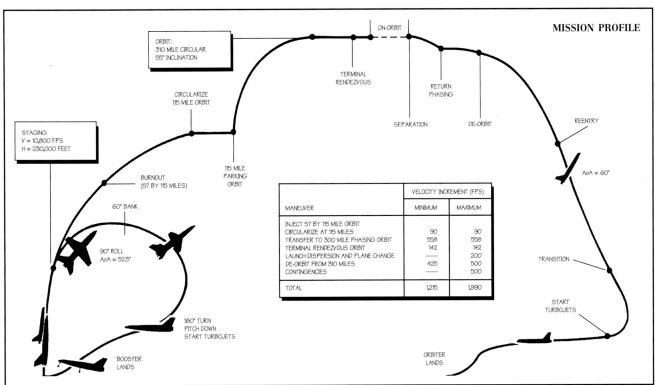

ORBIT:
310 MILE CIRCULAR
55° INCLINATION

ON-ORBIT

TERMINAL
RENDEZVOUS

RETURN
PHASING

REENTRY

CIRCULARIZE
115 MILE ORBIT

SEPARATION

DE-ORBIT

AoA = 60°

STAGING:
V = 10,800 FPS
H = 230,000 FEET

115 MILE
PARKING
ORBIT

BURNOUT
(57 BY 115 MILES)

60° BANK

TRANSITION

90° ROLL
AoA = 52.5°

START
TURBOJETS

180° TURN
PITCH DOWN
START TURBOJETS

BOOSTER
LANDS

ORBITER
LANDS

MANEUVER	VELOCITY INCREMENT (FPS)	
	MINIMUM	MAXIMUM
INJECT 57 BY 115 MILE ORBIT		
CIRCULARIZE AT 115 MILES	90	90
TRANSFER TO 300 MILE PHASING ORBIT	558	558
TERMINAL RENDEZVOUS ORBIT	142	142
LAUNCH DISPERSION AND PLANE CHANGE	—	200
DE-ORBIT FROM 310 MILES	425	500
CONTINGENCIES	—	500
TOTAL	1,215	1,990

ASCENT THERMAL ENVIRONMENT

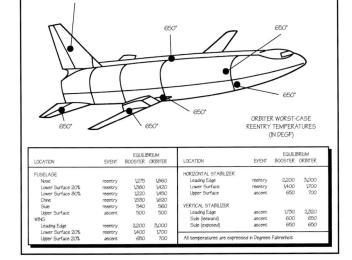

650°

ORBITER WORST-CASE
REENTRY TEMPERATURES
(IN DEG F)

LOCATION	EVENT	EQUILIBRIUM BOOSTER	ORBITER	LOCATION	EVENT	EQUILIBRIUM BOOSTER	ORBITER
FUSELAGE				HORIZONTAL STABILIZER			
Nose	reentry	1,275	1,360	Leading Edge	reentry	2,200	3,000
Lower Surface 20%	reentry	1,380	1,420	Lower Surface	reentry	1,400	1,700
Lower Surface 80%	reentry	1,220	1,450	Upper Surface	ascent	650	700
Chine	reentry	1,530	1,620				
Side	reentry	540	560	VERTICAL STABILIZER			
Upper Surface	ascent	500	500	Leading Edge	ascent	1,730	2,320
WING				Side (leeward)	ascent	600	650
Leading Edge	reentry	2,200	3,000	Side (exposed)	ascent	650	650
Lower Surface 20%	reentry	1,400	1,700				
Upper Surface 20%	ascent	650	700	All temperatures are expressed in Degrees Fahrenheit			

PAYLOAD CONFIGURATION OPTIONS

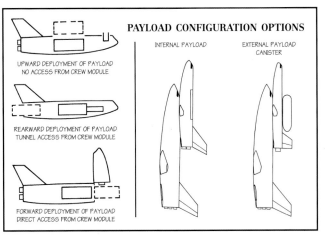

INTERNAL PAYLOAD

EXTERNAL PAYLOAD
CANISTER

UPWARD DEPLOYMENT OF PAYLOAD
NO ACCESS FROM CREW MODULE

REARWARD DEPLOYMENT OF PAYLOAD
TUNNEL ACCESS FROM CREW MODULE

FORWARD DEPLOYMENT OF PAYLOAD
DIRECT ACCESS FROM CREW MODULE

As part of the Phase A study, North American Rockwell investigated several options on where to carry, and how to deploy, the payloads carried by the orbiter. It was decided that the best way to off-load internal cargo was through upward opening payload bay doors. (North American Rockwell)

Lockheed (NAS9-9206)

The Space Systems Division of the Lockheed Missiles and Space Company (LMSC) released its Phase A final report on 22 December 1969. The study reflected over ten years of LMSC interest in reusable launch vehicles and space operations concepts, during which the company had expended significant resources studying various reusable lifting-entry spacecraft configurations and aerothermal transfer technologies.[36]

Within LMSC, Max Hunter was probably the strongest advocate of reusable spacecraft and had been greatly involved in shaping the earlier StarClipper concept. Several different concepts were investigated based on the changing directions provided by George Mueller. One was a stage-and-a-half vehicle using external drop-tanks that had been developed during the Air Force studies, and the other was a pure triamese concept featuring a high degree of commonality between the orbiter and booster primary structures. Midcourse in the study, the drop-tank concept was abandoned as the cost trends resulting from very high traffic rates became apparent and the SSTG recommended against such designs.

Although the study, led by Wilson B. Schramm, contained an in-depth analysis of two-stage and triamese 25,000 pound payload concepts, it concluded "... the best insurance against development risk would seem to be an approach based on large system size with growth potential ...". With this in mind, the final report centered around a 50,000 pound payload. The orbiter could accommodate payloads measuring either 15 by 60 or 22 by 30 feet, with Lockheed preferring the former. A potential existed for a maximum 1,725 mile cross-range, but Lockheed noted that by reducing it to 460 miles, between 8,000 and 10,000 pounds could be eliminated from the thermal protection system. The proposed two-stage configuration was preferred over the dissimilar tri-amese since it displayed significant advantages in reliability and safety, and would also cost some 20 percent less to operate over the life of the vehicle.

Lockheed made extensive use of a computer-based system synthesis model called MAGIC that allowed the inter-related effects of various parameters to determine launch system sizing. As an example, orbiter thermal protection system weight was strongly influenced by wing loading, which in turn was derived from the planform geometry in relation to the weight of primary structure and subsystems. MAGIC allowed engineers to run multiple 'what-if' scenarios in order to achieve the optimal system sizing. The program had been in use at LMSC on other programs for about two years, and had yielded generally good correlation between models, wind-tunnel testing, and flight tests of various aircraft.

The orbiter (model LS-112) was aerodynamically similar to the stage-and-a-half LSC-8MX vehicle proposed to the Air Force during 1968. The major difference was that this orbiter had three main engines instead of five, and carried all of its propellants internally, as opposed to the unique vee-shaped propellant tanks of the earlier Air Force design. The orbiter featured non-integral cryogenic tankage which was thought to have somewhat less manufacturing risk, and offer better inspection and maintenance characteristics. A graceful blended delta-wing again ended in wingtips up-swept to form vertical stabilizers. The relatively large stabilizers improved the subsonic L/D, as well as enhancing hypersonic stability. The extreme nose contained a crew module with seating for four, followed by a tank containing 46,508 pounds of LH2. Two kidney shaped tanks, each containing 145,404 pounds of LO2, were below, behind, and outboard of the LH2 tank, flanking the payload bay. Additional LH2 was carried in two tanks behind the payload bay.

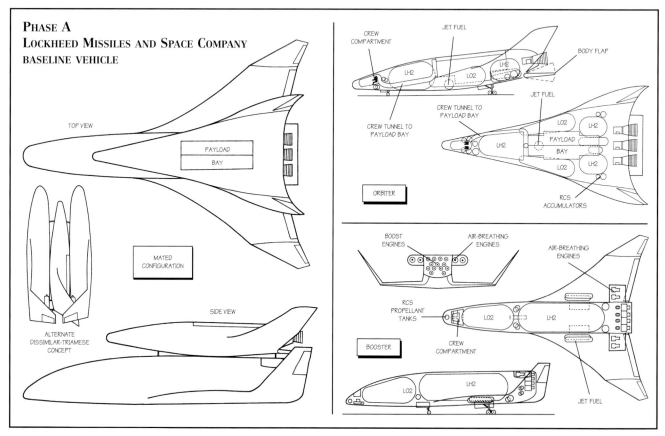

PHASE A
LOCKHEED MISSILES AND SPACE COMPANY
BASELINE VEHICLE

TOP VIEW

PAYLOAD BAY

ALTERNATE DISSIMILAR-TRIAMESE CONCEPT

MATED CONFIGURATION

SIDE VIEW

CREW COMPARTMENT

JET FUEL

LH2

LH2

LO2

BODY FLAP

JET FUEL

CREW TUNNEL TO PAYLOAD BAY

CREW TUNNEL TO PAYLOAD BAY

LO2

LH2

LH2

PAYLOAD BAY

LH2

LO2

LH2

ORBITER

RCS ACCUMULATORS

BOOST ENGINES

AIR-BREATHING ENGINES

AIR-BREATHING ENGINES

RCS PROPELLANT TANKS

LO2

LH2

BOOSTER

CREW COMPARTMENT

LO2

LH2

JET FUEL

Both the orbiter and the booster were to be of conventional aluminum construction. A tantalum-tungsten alloy (Ta-10W) capable of withstanding 2,750 degF was to be used for the orbiter nose cap. LI-1500 tiles were bonded directly to the orbiter and booster external structures, or alternately, a combination of corrugated Renè-41 and nickel-chromium/columbium-752 superalloy shingles could be supported on posts with Micro-Quartz or Dyna-Flex insulation packaged between the shingle and the structural panel.

The booster for the 50,000 pound payload orbiter was 237 feet long, had an empty weight of 322,550 pounds, and featured low-mounted delta wings that swept up at the tips to form vertical stabilizers. Liquid oxygen was carried in a conical tank forward, with LH2 in a cylindrical tank midships. A crew of two was carried in the extreme nose. Thirteen of the alternate 415,000-lbf engines were mounted high in the fuselage to minimize center-of-gravity offset effects with the orbiter attached. Four 38,600 lbf turbofan engines were on fixed mounts on the extreme rear fuselage, and JP-4 fuel was carried in four wing tanks.

By using a large number (13) of engines in the booster, it was felt that the probability of total mission success was greatly improved – mimicking an approach usually taken by the Russians. Simulations showed that two engines could be lost as early as ten seconds after lift-off and the system could still obtain orbit if the remaining 11 engines were throttled up to 115 percent, a capability not offered by any of the Phase A baseline engines. An abort-to-orbit could be achieved under the same conditions with as many as five engines out. The LMSC study noted that the probability of total mission success was approximately 0.995, with a single catastrophic engine failure being the predominate concern. Even so, the analysis concluded that there would be 190 catastrophic failures (i.e.; loss of vehicle and/or crew) per one million flights.

A total of five orbiters and five boosters would be required for the flight test program, with each stage (orbiter and booster) flying 175 horizontal test flights and 25 vertical flights. Two orbiters and two boosters would be refurbished for later use in the operational fleet. Total research, development, test, and evaluation (RDT&E) costs were estimated at $5,510 million. It was anticipated that vehicle turn-around could normally be accomplished in sixteen 8-hour shifts, with the possibility of expediting it to 10 shifts if required. Based on 1,000 flights, total recurring cost per flight was predicted to be $1,255,000 per flight, with the operating cost per pound of payload to be $15.10 – an additional $10 per pound was required to amortize RDT&E costs. To support this flight schedule, Lockheed proposed building five orbiters at $61 million each, and two boosters at $98 million each, to supplement the prototypes refurbished from the development testing.

The Lockheed study concluded that the basic aerodynamic technology existed to build a fully reusable space shuttle, but recommended additional study on orbiter landing modes and aborts, providing a payload bay larger than either of the 15 by 60 or 22 by 30 foot ones envisioned, critical systems redundancy, and the use of hydrogen as a fuel for the air-breathers instead of conventional JP-4.

One ironic feature of the LMSC study was to point out a discrepancy between the Air Force and NASA operational

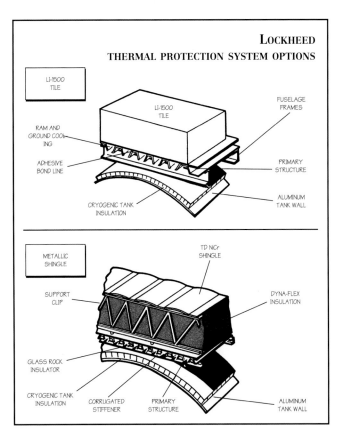

Lockheed was developing the silica-based tiles under a separate NASA contract, but nevertheless proposed an alternate TPS that used metallic shingles. Interestingly, Lockheed noted several drawbacks to the tiles, mainly their fragility. (LMSC)

cost models – using the NASA model, a work force of 1,500 was necessary for the operational era, instead of the 400+ force predicted by the Air Force model. Lockheed noted it was "… hard to envision such an expenditure …" for an operational system.[*]

As it ended up, Lockheed did not totally give up on the stage-and-a-half concept that it had proposed to the AFFDL and abandoned during Phase A. Lockheed would continue to develop and refine the concept in-house, and the design would show up again during the ASSC study later in the competition.

Unlike the eventual OPFs, that are separate from the VAB, most early concepts expected to construct extensions onto the building itself. Here is a Lockheed design being serviced, although it may not be the Phase A concept. (NASA photo KSC-70PC-251)

---

[*] In 2001, United Space Alliance (USA) – a joint venture between Boeing and Lockheed Martin – employs over 5,000 people at the Kennedy Space Center to process the Space Shuttle system. This does not include personnel at JSC, MSFC, and elsewhere that are also directly involved in the shuttle program.

General Dynamics / Convair (NAS9-9207)

General Dynamics / Convair released its final Phase A technical report on 31 October 1969. The studies had centered around two different vehicles, designated FR-3A and FR-4, that NASA had indicated an interest in during the mid-term review held earlier in the year. Both vehicles had a 15 by 60 foot payload bay and were capable of lifting a 50,000 pound payload to the required 300 mile orbit. Late in the study, NASA asked about the feasibility of carrying a 22 by 30 foot payload, and Convair concluded the FR-4 could be modified relatively easily for the task, while the FR-3A would require considerable work.[37]

The shape of the FR-3A orbiter, and of all the FR-4 components, was adopted directly from the FR-1 system that GD/Convair had developed during 1968 for the Air Force ILRV studies. The general shape had first been proposed by Convair as the T-18 during a 1965 study for the Air Force Space and Missile Systems Organization. The FR-1/T-18 was a system of three identically-sized elements (triamese) with all engines ignited prior to lift-off and propellant cross-feed between the elements during ascent. This concept was extremely similar to the MUSTARD vehicle conceived by BAC in England during the early 1960s, although it does not appear that either company was aware of the other's on-going work.

The final approach for NASA differed from previous designs in that the all of the earlier efforts had been marked by a conscious attempt to reduce development costs by utilizing three substantially similar elements. Engineers within NASA believed that this approach severely compromised the design of both the booster and orbiter – the new concept retained the notion of aerodynamic similarity between the stages, but rejected outright commonality. Maximum performance, minimum weight, and low recurring costs were now considered more important than minimal development costs, and identical sizing of the stages was incompatible with these goals.[38]

The body of the FR-3A booster consisted of a constant-section with a gently tapered forebody terminating in a hemispherical nose. The forebody shape was parabolic in planform with a partially straight-lined lower surface ramp in the profile view. The constant-section consisted of a flat-bottom with sides that tapered inward at a 12-degree angle with a full upper radius that was designed to allow maximum usage of state-of-the-art cylindrical or truncated conical cryogenic dewars. The flat-bottom improved the hypersonic L/D, and also allowed fairly easy storage of the switchblade-type wings. The sides sloped inward toward the top to improve the hypersonic L/D, and to reduce reentry heating on the side surfaces. The vee tail was attached high on the afterbody to provide hypersonic and transonic stability. Hypersonic roll control was to be achieved via differential deflection of the ruddervators on the vee tail.

The general arrangement of both the FR-3A and FR-4 orbiters consisted of a nose compartment 14 feet long that had provisions for a two-man flight crew in conventional side-by-side seating, and contained all essential avionics. Immediately aft of the fail-safe pressure bulkhead which terminated the crew compartment was a seven-foot long subsystems compartment housing the environmental, electrical, and hydraulic subsystems. The air-breathing engines were housed immediately aft of the subsystems compartment, but forward of the LO2 tank. Two Rolls-Royce RB211-22 turbofans were designed to be rotated to their flying position by a hydraulically-actuated double-acting mechanism similar in design to the wing-fold mechanism of carrier-based fighter aircraft. The main LO2 tank formed an integral part of the airframe, and was to use internal frames and external stringers with monocoque domes on the ends. The FR-4 was provided with 10,075 cubic feet of LO2 storage, while the smaller FR-3A had 7,618 cubic feet. The area below the LO2 tanks was used for small on-orbit propellant tanks. Behind the LO2 tank was the 15 by 60 foot payload bay.

The wing-pivot bulkhead was located approximately

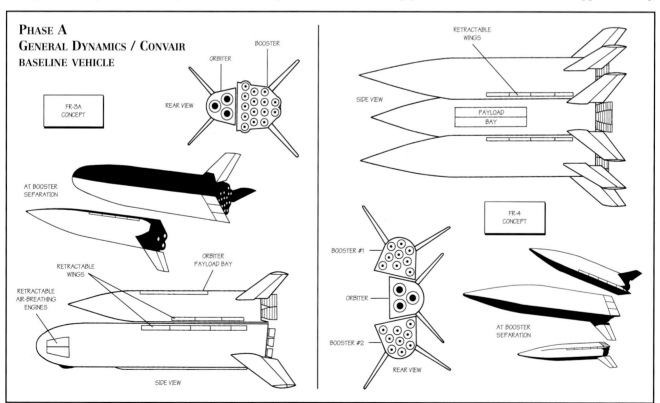

PHASE A
GENERAL DYNAMICS / CONVAIR
BASELINE VEHICLE

FR-3A CONCEPT

ORBITER

BOOSTER

REAR VIEW

AT BOOSTER SEPARATION

ORBITER PAYLOAD BAY

RETRACTABLE WINGS

RETRACTABLE AIR-BREATHING ENGINES

SIDE VIEW

RETRACTABLE WINGS

SIDE VIEW

PAYLOAD BAY

FR-4 CONCEPT

BOOSTER #1

ORBITER

BOOSTER #2

REAR VIEW

AT BOOSTER SEPARATION

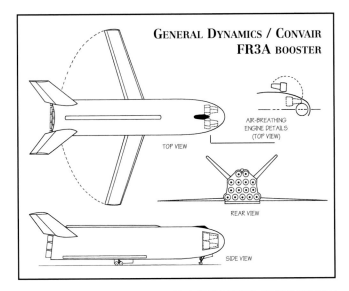

GENERAL DYNAMICS / CONVAIR
FR3A BOOSTER

TOP VIEW

AIR-BREATHING
ENGINE DETAILS
(TOP VIEW)

REAR VIEW

SIDE VIEW

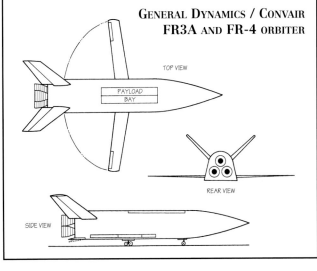

GENERAL DYNAMICS / CONVAIR
FR3A AND FR-4 ORBITER

TOP VIEW

PAYLOAD
BAY

REAR VIEW

SIDE VIEW

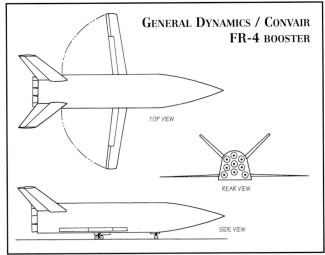

GENERAL DYNAMICS / CONVAIR
FR-4 BOOSTER

TOP VIEW

REAR VIEW

SIDE VIEW

diameter pressurized tube that provided access for the crew from the flight-deck to the payload bay. This tube was routed from the pressure bulkhead at the rear of the crew module, along the bottom centerline through the subsystem compartment, to the bulkhead at the aft end of the air-breathing engine compartment. At this point, it swung out around the lower forward LO2 tank dome, along the lower left side of the compartment between the lower structure and the LO2 tank then swung up around the aft LO2 tank dome and into the payload bay. The payload bay end terminated in a flexible 42-inch diameter tube that could be attached to a payload (such as Spacelab) to provide shirt-sleeve access.

The basic orbiter structural materials were aluminum alloy for the propellant tanks and main body structure, titanium alloy for the lower heat shield supports and thrust structure, and boron-aluminum composite in longitudinal high-stress areas such as beam caps and stiffeners. The projected reentry thermal protection system consisted of Micro-Quartz and Dyna-Flex insulation covered by post-supported shingles – cobalt superalloy on the lower surface and titanium-811 on the upper and side surfaces.

The FR-3A booster was considerably larger than either the FR-1/T-18 or FR-4 boosters since the entire boost function was contained in one vehicle instead of two as in the

mid-bay, and supported the deployable wings with large clevis fittings at the outboard end of the carry-through truss structure. At the forward and aft ends of the payload bay, heavy bulkheads were provided to support the two nose and two aft landing gear assemblies. These installations were to be completely outside the basically circular structure of the payload compartment. The nose and aft landing gear concept was very similar to the B-52 quadricycle arrangement, except the wide track (26 feet) negated the need for outriggers. The landing gear could not be extended until after the wings were deployed, but this was not considered a problem, since no attempt to make a gear-down landing with the wings stowed was ever anticipated. The wing was deployed using screw-jacks driven by redundant hydraulic motors in a fashion similar to the F-111 variable-sweep system.

The main LH2 tank was located directly aft of the payload bay, and was similar in construction to the LO2 tank. A total of 19,140 cubic feet of LH2 storage was provided for the FR-4, and 15,050 for the FR-3A. The compartment aft of the LH2 tank contained the thrust structure, main engine gimbal support, stabilizer attach structure, and propellant lines with a non-structural outer fairing surrounding the entire aft compartment. Three of the alternate baseline 415,000 lbf engines were supported at gimbal points on the thrust structure and were protected during reentry by an aft body-flap extension of the lower-fuselage.

Other design details on the orbiter included a 30-inch

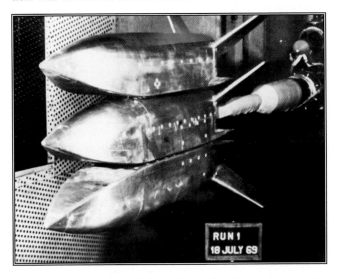

RUN 1
18 JULY 69

Even the Phase A proposals were backed up by a considerable amount of engineering. Here a model of the FR-4 concept is being tested in a wind tunnel. A FR-3A model underwent similar testing, as had the earlier T-18/FR-1. (Convair via the SDAM Collection)

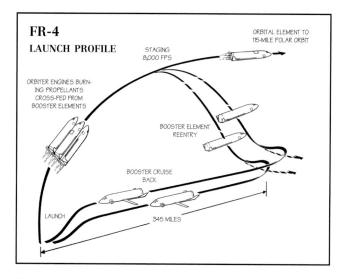

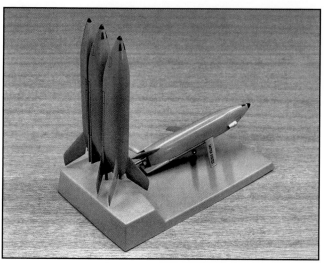

The triamese FR-4 launch profile included the ability for the boosters to recover at the launch site, roughly 345 miles back uprange. The FR-3A mission profile is detailed below. (General Dynamics / Convair)

Model showing the launch configuration of the triamese, as well as the landing configuration of one of the booster elements. This appears to be an earlier T-18 model, as proposed to the Air Force in 1965. (Convair via the SDAM Collection)

other designs (resulting in a 'biamese' configuration). The booster was generally derived from the FR-1 shape, but had a blunter nose and steeper side slopes. The cross-section was flat-bottomed with a full upper radius to accommodate the largest possible cylindrical tank. Sufficient space was reserved in the nose for the four Rolls-Royce RB211-56 turbofans and their JP-4 fuel, as well as various avionics subsystems. The two man flight crew was situated in a compartment elevated above the nose section to improve visibility during landing operations. Again, a vee-tail was used, and no elevons were fitted, roll control being achieved with the ruddervators. The LO2 and LH2 tanks were constructed of T2021 or T2022 aluminum alloy with a thermal protection system of titanium-752 alloy shingles suspended on long studs over Micro-Quartz insulation.

The dual FR-4 boosters were generally similar to the orbiter, except the payload bay was omitted, and the LO2 and LH2 tanks shared a common, center bulkhead to maximize propellant storage. The booster stages were also somewhat larger, and housed nine of the 415,000-lbf

main engines each. Three Rolls-Royce RB211-56 turbofans were needed on each stage to handle the 356,000 pound fly-back weight. In this system, one booster was mounted on each side of the orbiter, and all three stages ('triamese') were ignited upon lift-off, ensuring that all systems were working before committing to launch. Cross-feeding allowed the orbiter to use propellants from the boosters during ascent, saving its limited propellant load for the final boost into space and on-orbit operations.

Various studies of fixed-wing alternatives were made, with the conclusion that there was little overall weight difference between fixed and stowed wings. Because of the largely unknown transonic stability problems that might be encountered by a fixed-wing vehicle, all configurations proposed by Convair used switchblade-type stowed wings. The studies further indicated that delta-wings, or double-deltas, would reduce the aerodynamic center shift during transonic transition, but they were heavy and inefficient during subsonic cruise back to the launch site since their aspect-ratio was low.

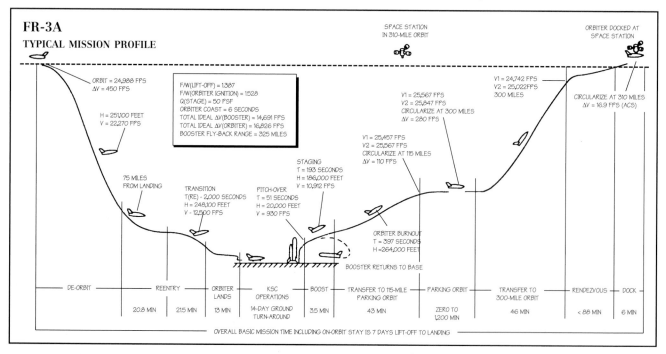

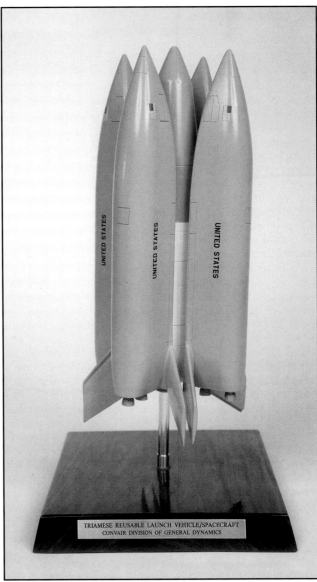

The basic triamese concept gave rise to several variations, most notably the biamese presented in the main text. Other derivatives included the five-element design at right, and the triamese at left both used booster stages derived from the baseline triamese concept coupled with an unmanned cargo element, most probably to launch large space stations segments, or the proposed vehicle for a piloted mission to Mars. (General Dynamics/Convair via the SDAM Collection)

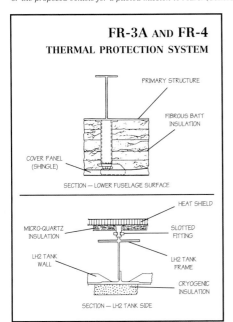

FR-3A AND FR-4
THERMAL PROTECTION SYSTEM

PRIMARY STRUCTURE

FIBROUS BATT
INSULATION

COVER PANEL
(SHINGLE)

SECTION — LOWER FUSELAGE SURFACE

HEAT SHIELD

MICRO-QUARTZ
INSULATION

SLOTTED
FITTING

LH2 TANK
WALL

LH2 TANK
FRAME

CRYOGENIC
INSULATION

SECTION — LH2 TANK SIDE

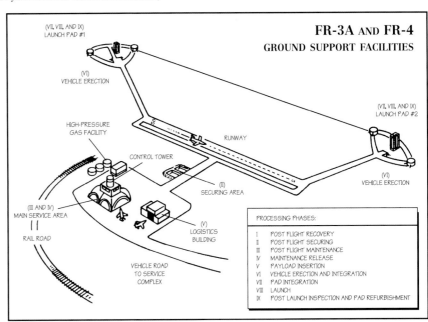

FR-3A AND FR-4
GROUND SUPPORT FACILITIES

(VII, VIII, AND IX)
LAUNCH PAD #1

(VI)
VEHICLE ERECTION

(VII, VIII, AND IX)
LAUNCH PAD #2

HIGH-PRESSURE
GAS FACILITY

RUNWAY

CONTROL TOWER

(VI)
VEHICLE ERECTION

(II)
SECURING AREA

(III AND IV)
MAIN SERVICE AREA

(V)
LOGISTICS
BUILDING

RAIL ROAD

VEHICLE ROAD
TO SERVICE
COMPLEX

PROCESSING PHASES:	
I	POST FLIGHT RECOVERY
II	POST FLIGHT SECURING
III	POST FLIGHT MAINTENANCE
IV	MAINTENANCE RELEASE
V	PAYLOAD INSERTION
VI	VEHICLE ERECTION AND INTEGRATION
VII	PAD INTEGRATION
VIII	LAUNCH
IX	POST LAUNCH INSPECTION AND PAD REFURBISHMENT

Martin Marietta (NO NASA CONTRACT ISSUED)

Certainly, the most unusual proposal to be generated out of the NASA Phase A studies was the Martin Marietta Spacemaster. This is perhaps fitting, since Martin was the only contractor to fund the study totally with company funds, NASA declining to issue a Phase A contract for unknown reasons. Martin proposed a fully reusable system based on an orbiter and fly-back booster but it was the nature of the booster that was unusual.[39]

The orbiter was a severely-swept blended double-delta planform spanning 107 feet, with the wingtips canted almost straight up to act as vertical stabilizers. The 181.2-foot long fuselage housed a large LH2 tank forward of a payload bay able to accommodate canisters measuring either 15 by 60 feet or 22 by 30 feet. Maximum payload capacity was 25,000 pounds. Twin LO2 tanks were in the blended section of the wing/fuselage on either side of the payload bay, on the vehicle center-of-gravity. Smaller LO2 and LH2 tanks for use on-orbit were carried aft of the main LO2 tanks. Two 415,000 lbf boost engines were housed in the aft fuselage and a crew of two was carried in the extreme nose. A landing speed of under 207 mph was projected, as was a cross-range capability of 1,700 miles. This version of the orbiter had no provisions for air-breathing engines.

The proposed baseline first stage booster consisted of twin fuselages, 28 feet in diameter and almost 197 feet long. These were connected by a forward wing that spanned 70 feet between the fuselages and contained eight gaseous hydrogen (GH2) powered Rolls-Royce RB211 turbofans. A rear wing had a span of 148 feet with 20 degrees of anhedral on the outer panels and used a NACA-2415 section. A total of fourteen (seven per fuselage) 415,000 lbf boost engines were provided. A crew of two was carried in the extreme front of the left fuselage. Each fuselage was

composed of an integral aluminum LO2 tank forward of the front wing, and a cylindrical integral LH2 tank between the front and aft wing. Reaction control systems were carried forward of the LO2 tank, the front landing gear was between the two tanks below the front wing, and the main engines and rear landing gear were carried behind the LH2 tank. Twin swept-back vertical stabilizers rose slightly over 24 feet above each fuselage. The orbiter was suspended from the forward and aft wings, and the system had a gross lift-off weight of 3,500,000 pounds.

A larger version of the same basic design, this one with a booster 204.2 feet long, spanning 160.8 feet, and powered by sixteen 415,000 lbf engines was also proposed. The orbiter for this version would have four GH2-burning Rolls-Royce RB162 landing engines under the forward section of the payload bay that could be lowered into the airstream when needed. This orbiter would have been slightly larger, growing to 186.8 feet long and spanning 110.3 feet, and using three boost engines instead of the baseline two. This version would have had a gross lift-off weight of 4,146,000 pounds with a payload capacity of 50,000 pounds.

Other booster designs were studied, all based on a single conventional fuselage, with a variety of wing configurations including straight, delta, double-delta, variable-geometry, and tapered aft wings with forward canards. The alternative booster chosen for study used a tapered aft wing and forward canard configuration with 13 boost engines. This vehicle would have been 258.5 feet long with a 156-foot wingspan and a single large vertical stabilizer that was centrally located above the aft wing. Several positioning and mating options were also studied with booster-top-to-orbiter-bottom ultimately being selected. This configuration, although the second choice during Phase A, would become the favored approach for the McDonnell Douglas/Martin Marietta team during the later Phase B competition.

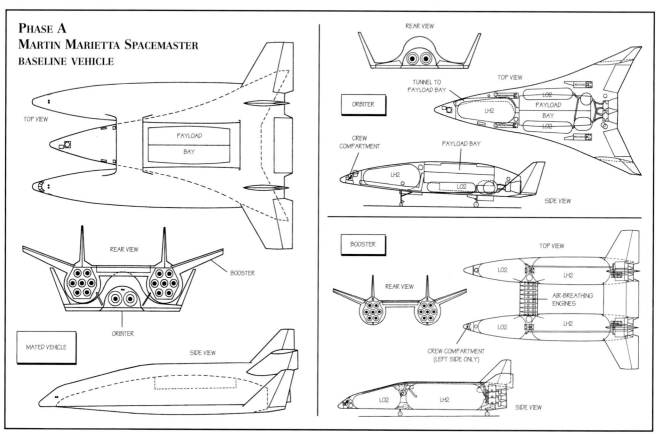

PHASE A
MARTIN MARIETTA SPACEMASTER
BASELINE VEHICLE

TOP VIEW

PAYLOAD BAY

REAR VIEW

BOOSTER

ORBITER

MATED VEHICLE

SIDE VIEW

REAR VIEW

TUNNEL TO PAYLOAD BAY

TOP VIEW

ORBITER

LO2
LH2
PAYLOAD BAY
LO2

CREW COMPARTMENT

PAYLOAD BAY

LH2

LO2

SIDE VIEW

BOOSTER

REAR VIEW

TOP VIEW

LO2   LH2

AIR-BREATHING ENGINES

LO2   LH2

CREW COMPARTMENT (LEFT SIDE ONLY)

LO2   LH2

SIDE VIEW

GROUND HANDLING

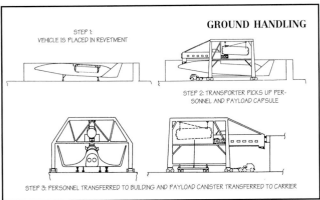

STEP 1: VEHICLE IS PLACED IN REVETMENT

STEP 2: TRANSPORTER PICKS UP PERSONNEL AND PAYLOAD CAPSULE

STEP 3: PERSONNEL TRANSFERRED TO BUILDING AND PAYLOAD CANISTER TRANSFERRED TO CARRIER

LAUNCH CONFIGURATION

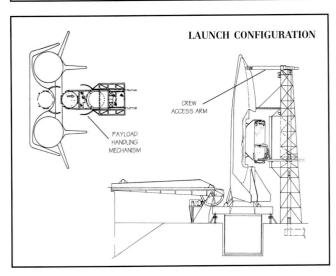

CREW ACCESS ARM

PAYLOAD HANDLING MECHANISM

BASELINE ORBITER STRUCTURAL ARRANGEMENT

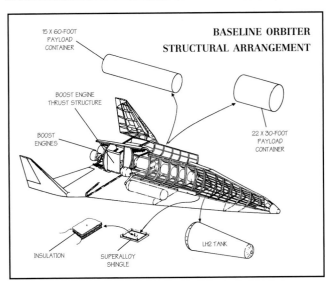

15 X 60-FOOT PAYLOAD CONTAINER

BOOST ENGINE THRUST STRUCTURE

BOOST ENGINES

22 X 30-FOOT PAYLOAD CONTAINER

INSULATION

SUPERALLOY SHINGLE

LH2 TANK

BASELINE BOOSTER STRUCTURAL ARRANGEMENT

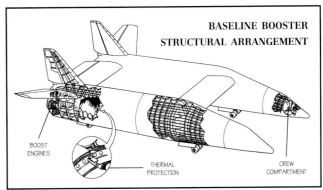

BOOST ENGINES

THERMAL PROTECTION

CREW COMPARTMENT

PHASE A
MARTIN MARIETTA *SPACEMASTER*

ALTERNATE MOUNTING LOCATIONS

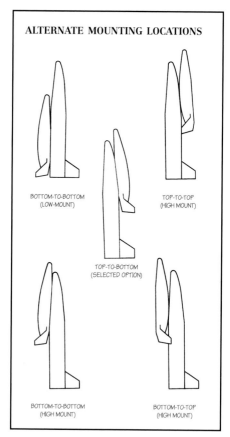

BOTTOM-TO-BOTTOM
(LOW-MOUNT)

TOP-TO-TOP
(HIGH MOUNT)

TOP-TO-BOTTOM
(SELECTED OPTION)

BOTTOM-TO-BOTTOM
(HIGH MOUNT)

BOTTOM-TO-TOP
(HIGH MOUNT)

ALTERNATE BOOSTER DESIGNS

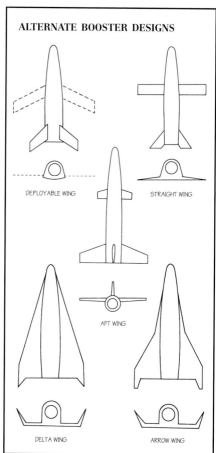

DEPLOYABLE WING

STRAIGHT WING

AFT WING

DELTA WING

ARROW WING

BASELINE VEHICLE

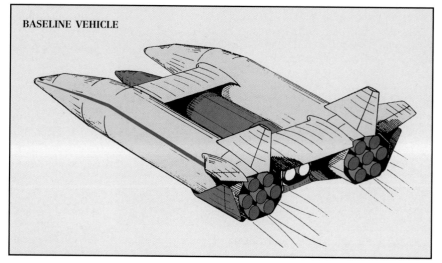

ALTERNATE BASELINE VEHICLE

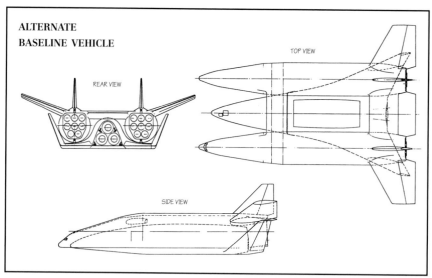

REAR VIEW

TOP VIEW

SIDE VIEW

The alternate baseline vehicle was visually very similar to the baseline, except that it was slightly larger and could lift 50,000 pounds of payload, instead of 25,000 pounds. Air-breathing engines were also provided for the orbiter. In both designs the payload bay was accessible when the vehicles were mated. (Martin Marietta)

ALTERNATE BASELINE VEHICLE

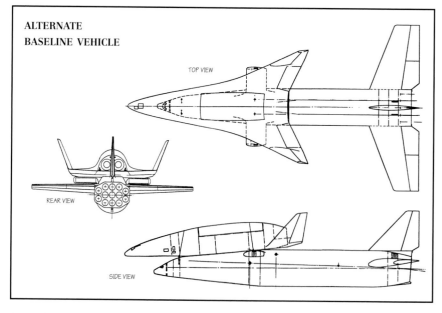

TOP VIEW

REAR VIEW

SIDE VIEW

Martin also investigated other booster designs, and provided extensive data for this more conventional concept. A slight variation of this booster design would again be proposed by the McDonnell Douglas/Martin Marietta team during Phase B. (Martin Marietta)

POLITICAL CONSIDERATIONS

The concept of a fully reusable space transportation system to make access to low-earth orbit more routine and less expensive finally began gaining momentum during 1969 as part of NASA's ambitious post-Apollo plans. These plans were fueled by the astounding success of the first lunar landing in July, and rapidly grew to proportions unthought-of, even during the peak of the 'space race.' When the Vice-President's Space Task Group (STG) report was released in September 1969, it had offered a choice of three long-range space plans:[40]

- an $8,000–10,000 million per year program involving a manned Mars expedition, a manned space station in lunar orbit and a 50-person Earth-orbiting space station serviced by a reusable space shuttle,
- an intermediate program, costing less than $8,000 million per year, that deleted the lunar-orbiting space station, but kept the other elements, or
- a relatively modest $5,000 million per year program that would embrace an Earth-orbiting space station and the space shuttle.

When President Nixon rejected all of these proposals, NASA was left with the task of identifying a project that could both gain enough political support for approval, and still be sizable enough to keep its development engineers and contractors usefully occupied during the 1970s. The space shuttle emerged as that project because it was the logical first step to implementing any future plans for a space station. It was also advertised as an 'economical' way to continue sending man into space, a feature that was very important during the budget conscious period that saw the Vietnam War and a major economic recession.

According to an analysis by the President's Science Advisory Committee (PSAC), 12 different launch systems could be replaced "with a STS used jointly by both DoD and NASA as a national transportation system capability." The space shuttle would also offer "provision for national security contingencies by the ready availability of transportation to orbit on short notice, with sufficient maneuverability and cross-range capability for a variety of missions."[41]

During 1970 the space shuttle concept was formally designated as the 'Space Transportation System,' and shortly thereafter Robert Seamans and Thomas Paine established the joint USAF-NASA Shuttle Coordination Board, better known as the 'STS Committee.' The intended goal of the Space Transportation System* was to provide "... an economical capability for delivering payloads of men, equipment, supplies, and other spacecraft to and from space by reducing operating costs an order of magnitude below those of present systems...". The committee would continue to review the shuttle program and recommend various changes in direction that would ensure the system would meet the operational needs of both DoD and NASA.[42]

Once it had decided that the Space Shuttle was its top priority program, NASA Headquarters unsuccessfully attempted to get the project approved by the White House during the fall of 1970. A new NASA Administrator, James C. Fletcher, was appointed in early 1971, and was determined to get the space shuttle initiative approved that year. NASA calculated that in order to get White House and Congressional approval to develop the new Space Transportation System they would have to significantly reduce the cost of access to space, and also get the DoD to commit to use the system for all of its launch needs once it became available. Thus, providing a vehicle that met all of the DoD requirements became a key element in the strategy for program approval. In addition, developing easy access to space was just the first step in NASA's ultimate goal, the development of a permanent manned space station.[43]

While meeting DoD requirements implied a large, high-performance space shuttle, NASA also had to meet its commitment to the Office of Management and Budget (OMB) to make access to space more economical. The Space Shuttle was the first NASA manned space program to be subjected to formal economic analysis and requirements. After an internal study by NASA failed to impress the OMB, 11-month contracts were issued in June 1970 to The Aerospace Corporation for estimating payload and launch vehicle costs, to the Lockheed Missiles and Space Company for analyzing the impacts of Shuttle capabilities on reducing payload delivery costs, and to Mathematica for a comparison of total costs of carrying out likely future NASA, DoD, and commercial space missions using the shuttle and other launch vehicles. NASA Deputy Administrator George M. Low later commented that the Mathematica analysis' impact on the Space Shuttle decision was "... influential and unfortunate ...". This was because although the analysis concluded the Space Shuttle could indeed significantly reduce the costs of launching payloads, the existence of an official study making such claims forced NASA to publicly maintain that Shuttle was a good investment on economic grounds.[44]

It became clear, even during NASA's early in-house analyses, that any economic justification depended on the Space Shuttle being the only U.S. launch vehicle during the 1980s and later. In particular, it was crucial for NASA to gain agreement from the national security community to use the Space Shuttle to launch all military and intelligence payloads, which were projected to be roughly one-third of all future space traffic. Thus DoD support was crucial on both political and economic grounds.[45]

But there were dissidents as well. Several 1970 studies by the think-tank RAND Corporation stated that Space Shuttle "... development is not easy to justify ...". Based primarily on traffic rates derived from conservative options in the STG and DoD space plans, it was projected that the Space Shuttle would show a net transportation cost savings of $2,800 million in the fifteen years ending in 1990. RAND noted, however, that the planned development cost of $9,000 million would require a peak civilian space budget in excess of $7,000 million in FY75, and that the funding required in any one year for the Shuttle alone would be in excess of the entire NASA budget for FY71. The RAND study investigated alternate space plans that might be adopted to alleviate budget peaks by slipping various elements in the national space plan, but noted that none of them resulted in a savings sufficient to compensate for space shuttle development costs through 1990. Also, while a savings of $2,800 million seemed large, total costs for all of the programs involved ranged from approximately $75,000 million to over $140,000 million (FY75–90), and any technical difficulties or overruns in the programs could negate all of the savings, or make it appear small by the time it was realized.[46]

The RAND study also discussed possible cost savings as a result of designing future satellites to utilize the excess

---

* The Space Shuttle subsequently added 'National' to its official title, becoming the National Space Transportation System (NSTS), although it continued to be known as the STS. In 1990 is would be renamed simply Space Shuttle Program.

capacity aboard partially loaded Space Shuttle flights, by employing the orbiter to recover and reuse satellites, and by servicing satellites on-orbit. The conclusion was that the savings of between $150 million and $200 million might strengthen the economic rationale for the space shuttle. While the study concentrated on a vehicle with a 50,000 pound* payload, it noted that there was little difference in total space transportation costs through 1990 for design payload payloads as low as 25,000 pounds, as long as the payload bay remained at 15 by 60 feet or larger.[47]

A mission cost analysis revealed that with a 50,000-pound payload shuttle, the cost of launching a pound into orbit was roughly $70 (each launch costing $3.52 million), not including amortizing RTD&E expenses of $8,735 million. For a vehicle capable of carrying only 25,000 pounds, the cost went up to $115 per pound ($2.87 million per launch), but development expenses were only $7,400 million. The unanswered question was how often a full payload bay (by weight) would ever be required. For comparisons, a Saturn V was estimated to cost $215 million for the vehicle and $40 million for each launch, while a Titan IIIM was estimated at $26 million for the vehicle and launch operations combined.[48]

The RAND study concluded that the development risks and costs for Space Shuttle were too high to justify its development during the 1970s. A low-earth orbiting space station was considered potentially more useful, and RAND was convinced that it was more economical to launch and service the space station using the existing fleet of Titan and Saturn boosters, or derivatives thereof. Modified Apollo capsules would be used to transport men and material to the station and return them to Earth. This report was in a minority, since the President's Space Task Group, DoD, President's Scientific Advisory Council, NASA, and many engineering and scientific organizations (i.e., the AIAA, etc.) all identified the space shuttle as an important element in any future national space effort and supported its immediate development.[49]

Although the Space Act of 1958 had mandated that NASA "... cooperate with other nations ...", the Mercury, Gemini, and Apollo programs were largely untouched by international cooperation. This was mainly because the 'race' for the Moon was conducted on a timetable dictat-

ed by President Kennedy, and left precious little time for the added complexity of an international partnership to put men on the Moon before 1970. Nor were there any obvious partners. When Project Apollo was begun during 1961, the United Kingdom, Canada, and Italy had begun a few modest space programs, but while scientific in nature, their contributions did not involve any significant technologies needed to get men into space.[50]

Outside of the United States, and its archrival Soviet Union, there were no national space agencies in place. Nowhere in western Europe, Japan, or Canada was there a competitor or partner to join the Moon race. Except for Swedish Hasselblad cameras (purchased commercially), and a small Swiss instrument package placed on the Moon to measure solar wind, international involvement in the American space program was limited to watching Neil Armstrong walk on the Moon on worldwide television.

By the end of the 1960s, international interest in space was decidedly different than it had been at the beginning of the decade. National space programs were in place in the major European countries, and many of their resources had been pooled into two multinational agencies – ESRO (European Space Research Organization) to conduct unmanned science programs, and ELDO (European Launcher Development Organization) to building the ill-fated Europa launch vehicle.

After the White House had rejected the Space Task Group's recommendations in 1969, NASA Administrator Thomas Paine had set out, seemingly with President Nixon's blessing, to explore the possibility of international cooperation in post-Apollo manned spaceflight participation. Reaction by prospective partners to the NASA invitation to cooperate in the Space Shuttle and Space Station projects was mixed. The Australians dropped out early, deciding that they could not afford the investment required. The Japanese, who in 1969 had established a national space agency, were unable to reach a domestic consensus on what their role should be in any international effort. Canada selected its niche early, proposing to develop and build a robotic manipulator system (later renamed Canadarm) to deploy and retrieve payloads from the Orbiter's payload bay.[51]

There was a great deal of international politics involved in the European negotiations. Part of the problem was that it became de facto U.S. policy to discourage Europe from developing their own launch vehicle, with the idea that European satellites could be launched on Shuttle, further reducing the cost of operating the system. But the Department of State had reservations about launching foreign satellites intended to serve only foreign markets on a U.S.-government controlled vehicle. This was finally overcome on 1 September 1971 when the U.S. set forth a policy on the availability of U.S. launchers for European satellites. At the same time it was made clear that any cooperation on shuttle or space station was not linked to the European nations developing their own launch capability.[52]

Initial discussions with the European nations included three possible development efforts: (1) a so-called 'sortie can,' also known as a research and development module (RAM), (2) a space tug that could be used to maneuver satellites in orbit after launch from a space shuttle, and (3) developing and manufacturing parts of the space shuttle itself. Initially the United States seemed open to any of these ideas, and the Europeans firmly embraced the idea of

UNITED STATES
LAUNCH VEHICLES
CIRCA 1970

SCOUT   DELTA   ATLAS   ATLAS CENTAUR   TITAN III   SATURN IB   SATURN V

---

* RAND Corporation was an Air Force project (begun at Douglas Aircraft Company), and utilized the figure most desirable to the DoD.

developing the Space Tug, even beginning preliminary analysis of the concept during late 1971. Tentative agreements were also reached where European companies would be encouraged to team with U.S. contractors to bid on portions of the space shuttle vehicle itself. It is estimated that European governments and companies spent about $20 million on these studies.[53]

But things rapidly changed. Most of the internal White House and NASA Headquarters documents use terms such as 'confusing' to describe official U.S. policy during late 1971 and early 1972. In April 1972 the Air Force advised NASA that it was worried about possible national security issues if the Space Tug was built in Europe, and would not guarantee purchasing a foreign-built tug if it did not meet its specifications. By June 1972 the United States was discouraging the Europeans from participating in any way with the development of the space shuttle or the space tug, and was trying to limit European participation to the development of the sortie can. In a meeting with a delegation from the European Space Conference (a semi-predecessor of ESA), Herman Pollack from the State Department said "I must report that the conditions the United States find necessary may diminish the attractiveness to Europe of participating in the Shuttle items."[54]

Europe went away very unsure of the United States motivations. Ultimately, the Europeans did agree to develop the sortie can, renamed Spacelab – a manned reusable laboratory to be carried in the Space Shuttle payload bay. But the decision was not an easy one, and resulted almost entirely from West Germany's strong desire to get involved in manned space flight, and its willingness to finance 52 percent of Spacelab's costs. There was also a strong feeling throughout Europe that it could not afford to be left further behind in space exploration. Although unmanned projects were generally more desirable economically, they did not afford Europe a place as a major space power. Even so, Spacelab could not be sold to the European governments strictly on its own merits, and it became part of a package deal that included a consolidation of ESRO and ELDO into a single European Space Agency (ESA), augmenting the planned French-led Ariane booster program and various science satellites.[55]

Still, none of the United States' allies were allowed to get heavily involved in the development of the vehicle itself. And although it was felt they were technically capable, detenté with the Soviet Union had not developed to the point of inviting them to participate in anything more involved than the politically-correct Apollo-Soyuz test mission planned for 1975.

DoD Requirements

Accommodating all potential Department of Defense missions required an orbiter that could handle payloads up to 60 feet long, could launch 40,000 pounds into polar orbit, and over 65,000 pounds into a due-east orbit. This was significantly more than the payload specified by the NASA Phase A RFP, and more than seemed possible while still keeping the vehicle size within the realm of possibility. Of the critical design parameters of the Space Shuttle, only the maximum payload width of 15 feet was based primarily on a potential NASA requirement, this being a projection that any future space station would be built using modules of that diameter. The NASA was seemingly even somewhat unsure of this requirement, and was willing to accept payload bay diameters as small as 12 feet, while at the same time asking various contractors to investigate ones as large as 22 feet.

But the single requirement that had the most impact on the Space Shuttle design was that for high cross-range, which is the ability to maneuver to either side of the vehicle's ground track during reentry. The Air Force wanted a 1,265 to 1,726 mile[*] cross-range capability to allow a quick return from orbit to secure military airfields. In particular, the Air Force wanted to be able to launch the Space Shuttle into polar orbit from Vandenberg AFB, have it rendezvous with an already orbiting reconnaissance satellite, service it, and return after a single revolution to Vandenberg. The landing strip would have moved some 1,265 miles to the east as the Earth rotated during the 90 minute flight. The capability involved considerable costs, dictating a vehicle with a much higher hypersonic L/D, as well as a significantly more robust[†] thermal protection system.[56]

The Aerospace Corporation conducted a classified study in mid-1970 to evaluate the results of the various Air Force and NASA ILRV studies. Considerable effort was directed toward understanding the induced aerothermodynamic environment that results from reentry, and the effects of the cross-range required for DoD missions. Of principle concern was the heating prediction methodology, the effects of turbulent heating estimates and transition criteria on the vehicle heating, and the effect of cross-range on the thermal protection system. A comparison of four methods of skin friction predictions was made with experimental data from various on-going programs to assess the accuracy of the methodologies. This initial comparison indicated that the Spalding-Chi correlation gave the best prediction, with errors within 20 percent of actual observed temperatures.[57]

The peak radiation equilibrium surface temperature on the vehicle centerline was determined to be primarily a function of the angle-of-attack, and predictions of the four different estimating techniques differed by as much as 300 degF. The effect of cross-range was more difficult to determine due to the differences in methodologies used by the various USAF and NASA contractors. It was determined, however, that thermal protection system weight increased approximately as the fourth-root of total cross-range, due primarily to longer heating and soak times. The controlling variable was determined to be flight time measured from initial reentry to pullout.[58]

A great deal of work was done on an engine-out analysis which verified and correlated the work done by several contractors in the USAF ILRV study. Many ascent profiles were investigated, along with a variety of different scenarios concerning how and when an engine could fail. Abort-to-orbit was considered feasible if the remaining engine of a two-engine orbiter could still provide a minimum of 55 percent[‡] of the thrust provided by both engines normally. Analyses also indicated that sequential[§] ignition resulted in a slightly superior payload capability in com-

---

[*] This requirement varied between 1,100 nm and 1,500 nm frequently during the Phase A and Phase B competition, finally standardizing on the 1,100 nm (1,265 mile) figure – sufficient for a once-around-abort – in late 1971.

[†] This is because much of the cross-range is achieved at extremely high speeds, exposing large portions of the airframe to reentry thermal effects for a longer period of time than short cross-range reentries.

[‡] This can be accomplished by running the remaining engine at greater than its rated thrust capability (110 percent), although this capability was not included in the baseline Phase A Space Shuttle Main Engine requirements.

[§] Ignition was generally not truly sequential since the orbiter engines were ignited prior to separation to verify that they were functioning correctly. In theory this would allow the booster to return the orbiter safely to an alternate landing site in the event of an engine failure during ignition. However, several contractors did propose truly sequential staging where the orbiter was separated prior to ignition. Initially NASA tended to prefer sequential staging based on prior experience with expendable boosters.

parison with simultaneous ignition of both the booster and orbiter. For a constant lift-off weight, a two-element configuration (such as Phase B) was thought to have a greater payload capacity than a three-element design (i.e.; Convair's triamese).[59]

The Aerospace Corporation also attempted to determine the effects of space radiation on various materials being considered for the construction of the orbiter. This was particularly important since no manned spacecraft had ever been reused (at least for a second manned flight), and few vehicles had been exposed to space then brought back into the oxidizing atmosphere of Earth repeatedly. Analytical techniques were used to predict the response of materials to the environment, and the results of underground nuclear testing when available, were used to supplement and verify these analytical results. The number of materials directly investigated was not extensive, but generally included at least one representative of each class of metallic skin material, i.e., metal matrix composites, light metal alloys, superalloys, dispersion strengthened alloys, and coated refractory alloys. With the exception of the composites, the analytical results indicated that no significant degradation of properties resulted from exposure to anticipated radiation levels. Very little work was actually accomplished on the cyclic effects of space-Earth exposure, mainly because adequate test facilities did not exist, and insufficient data was available for computer modeling.[60]

An extensive horizontal flight test program, simulating reentry conditions, was thought to be desirable by The Aerospace Corporation engineers, and a maximum speed of Mach ≈2.0 was judged to be feasible with the orbiter taking off under its own power using the air-breathing propulsion system. The study also found that constraining the Initial Operational Capability (IOC) date to December 1977 (as the Air Force desired) represented a high risk approach requiring "... rigid discipline in achieving key program milestones ...". The study concluded that a more realistic schedule would slip the IOC by 15 to 18 months.[61]

But, although DoD requirements drove several important aspects of the space shuttle design, it was not clear how strong military interest really was. Robert Seamans saw "... no pressing need ..." for the shuttle, but characterized it as "... a capability the Air Force would like to have ...". Thus, although it drove several significant design requirements, the Air Force was still not totally committed to the concept of a common space transportation system, and stated that it would continue to develop and purchase its own expendable (Titan and Atlas) boosters.[62]

The Department of Defense also did not want to contribute any substantial amount of money to the Space Shuttle development effort, other than funds to build a launch complex at Vandenberg AFB. Even then, the Air Force was not sure when the complex would be built, or how many flights a year would be launched from it. Interestingly, NASA did not seem concerned about this lack of Air Force money. Having just NASA responsible for funding simplified the budgetary approval process in Congress, and eliminated the danger of having a program dependent upon two sources of funding with double the danger of budget cuts.[63]

In 1971 the Air Force finally agreed not to develop any new expendable boosters, although they would continue to purchase existing designs. Leaders of the national security establishment (DoD, CIA, NSA, etc.) communicated their support of the Space Shuttle program both to the White House and to Congress later in 1971. But later, the

decision to replace all launch systems with the Space Shuttle, would have a significant impact to the DoD, and everyone else desiring access to space, after the Challenger accident in 1986 grounded the Space Shuttle fleet for almost three years.[64]

DISSENSION WITHIN NASA – THE MSC DC-3

But even within NASA there was no firm consensus of what a Space Shuttle should look like. At the Manned Spacecraft Center in Houston, engineers led by Mercury capsule designer Max Faget still wanted a straight-winged fully reusable vehicle. Other NASA centers supported partially or completely recoverable versions of the Saturn family of boosters, and there was still some minor support of a Aerospaceplane single-stage-to-orbit vehicle.[65]

On 23 January 1969 MSC started an in-house design study of Space Shuttle concepts. The study was initiated primarily due to growing concerns over the development of requirements for the upcoming Phase B RFP, with specific concerns centered around the lack of convergence of the design requirements, the high development and total program costs and risks, and the long development time. The basic philosophy of the MSC in-house study was to lower the development cost and risk, and to develop a system that could be operational by the end of 1975 – the year the last Apollo flight was scheduled. The study was variously known as the MSC-12.5k or MSC-15k (denoting payload capacity), and ultimately as the DC-3 because of its relative simplicity.

The basis of the study was a fully-reusable two-stage system where both elements used classic straight wings. Faget had been working on this general configuration since 1968, and North American was using it as the baseline for their revised Phase A studies (see page 87). The spacecraft engineers at MSC, led by Faget, designed an orbiter with an 8 by 30 foot payload bay capable of carrying* 10,000, 12,500, or 15,000 pounds of payload. Both the orbiter and fly-back booster would utilize as much existing spacecraft and aircraft hardware as possible in an attempt to lower development time, risks, and costs. The orbiter would have a low cross-range (≈230 miles), and the booster would be recovered down-range, then flown back to the launch site. Somewhat after the fact, this design would officially become the MSC-001 (evolving into the MSC-002), the first of many NASA design efforts on the road to the eventual Space Shuttle.

The vehicle emerged in a 27 April 1970 report as two vertically launched, fully reusable stages with the orbiter mounted to the top of the booster in a long overlap, or 'piggy-back,' arrangement. The orbiter contained two boost engines, two on-orbit maneuvering engines – all using LO2/LH2, and six air-breathing landing engines burning conventional JP-4/5. The boost engines for both the orbiter and booster were a modification of the XLR129 demonstrator producing 297,000 lbf in a vacuum. This engine was selected over the existing Saturn J-2 and F-1 engines based on various weight studies as well as vehicle and trajectory trade-off studies and analyses. The 15,000 lbf Pratt & Whitney RL10 from the Centaur upper stage was selected as the on-orbit maneuvering engine, although there were some concerns relative to life expectancy and maintenance.

---

* This was finalized by the designers at 15,000 pounds during the last two weeks of the study, although Maxime Faget still opposed any increase from the original 10,000 pounds, regardless of the economic considerations.

Twenty-seven air-breathing engine combinations for the orbiter and 28 combinations for the booster, representing four different engine manufacturers, were studied. The engine ultimately selected for use on the orbiter was the new Rolls-Royce RB162-86 rated at 5,250 lbf at sea level. The P&W ADVTF-B (a proposed engine) of 17,850 lbf was selected for the booster mostly because it permitted a cleaner wing installation than any competitor.

The orbiter LO2 and LH2 tanks were carried internally below the payload bay and the JP-4 was carried in an aircraft-style wet wing. The general configuration was conventional with a straight-wing spanning 90.83 feet with a 14 degree leading edge sweep. A conventional empennage arrangement was provided with the horizontal stabilizer spanning 43.66 feet with a ten degree leading edge sweep. The vertical stabilizer stood 41.66 feet high and was angled back 45 degrees. The six air-breathing engines were housed in titanium pods on top of the wings at roughly mid-span.

The orbiter fuselage was made up of the propellant tank structure with longerons adjacent to the payload bay. The integrally strengthened tank shell was formed by two overlapping cylindrical sections joined at a common keel web. A flat common bulkhead of honeycomb sandwich construction separated the propellants. The integral stiffening was external to the tank and provided a matrix for support of the thermal protection system. The orbiter forward fuselage was an aluminum alloy semi-monocoque. The TPS on the underside and lower half of the fuselage sides was a new silica-based external surface insulation on a fiberglass honeycomb sandwich substrate. The nose cap, underbody chine lines, and all leading edges were made of pyrolyzed carbon laminate. The upper portion of the fuselage was covered by titanium shingles. The aft section of the body was devoted to both the main and on-orbit propulsion installation with the engines supported on a titanium truss reinforced with fiber composite tape. The tankage for the on-orbit engines was located above the main tanks and behind the payload bay.

The fly-back booster was a straight-winged vehicle spanning 141 feet with a 14 degree sweep on the leading edge providing 2,840 square feet of area. The 203-foot long fuselage housed 11 boost engines of the same basic type as used by the orbiter. Four ADVTF-B turbofans were mounted on top of the wings in nacelles having closures to protect them during boost and reentry. Provisions were included for a flight crew of two. A pair cylindrical main propellant tanks provided the primary structure for carrying body bending and axial loads. The LO2 was located forward of the LH2 with a sandwich common bulkhead separating the propellants. The tank pressure shell was supplemented by longitudinal stiffeners and ring frames that also supported the thermal protection panels and insulation.

The leading edge of the booster wing, as well as other high heat areas such as the body chine line, nose cap, and leading edge of the stabilizer were constructed of pyrolyzed carbon laminate. Regions of the wing with less severe heating were covered with the same silica-based surface insulation used on the orbiter, while the entire upper surface of the wing was a bare titanium hot-structure.

The reference mission included a circular 310-mile orbit inclined 55 degrees. The on-orbit lifetime of the orbiter was seven days and a maximum launch rate of 30 flights per year was projected using a fleet of six orbiters and four boosters, with each orbiter having a 100 mission life. Both the orbiter and booster were to be equipped with automatic approach and landing systems. The DC-3 was to be capable of being launched on 48 hours notice for a rescue mission, if required.

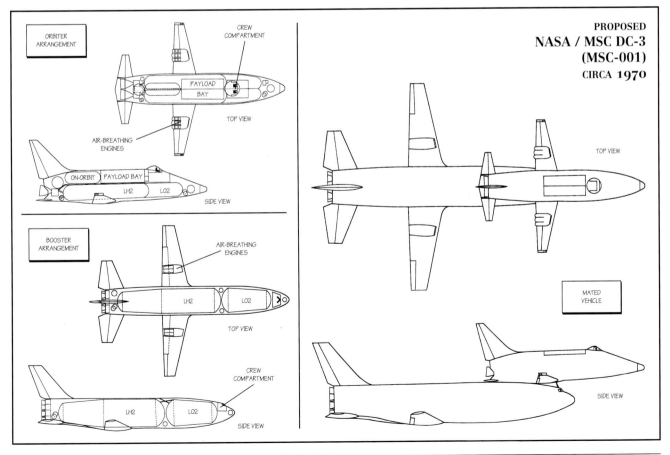

PROPOSED
NASA / MSC DC-3 (MSC-001)
CIRCA **1970**

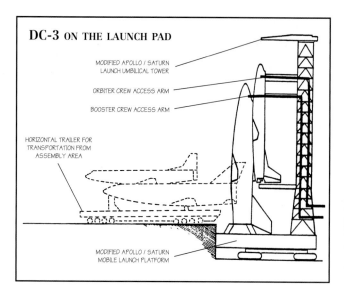

DC-3 ON THE LAUNCH PAD

MODIFIED APOLLO / SATURN
LAUNCH UMBILICAL TOWER

ORBITER CREW ACCESS ARM

BOOSTER CREW ACCESS ARM

HORIZONTAL TRAILER FOR
TRANSPORTATION FROM
ASSEMBLY AREA

MODIFIED APOLLO / SATURN
MOBILE LAUNCH PLATFORM

Maxime Faget's DC-3 was designed to make maximum use of Apollo launch facilities at the Kennedy Space Center including the Mobile Launch Platform (MLP), crawler-transporter, and Launch Umbilical Tower. (NASA)

One of the DC-3 (MSC-002) boosters returns to its launch site at the Kennedy Space Center in Florida. The mounting plate on top of the forward fuselage for attaching the orbiter is noteworthy. (NASA photo S70-H-122)

A one-tenth scale model of the DC-3 orbiter was dropped from an Army CH-54A helicopter beginning on 4 May 1970 at Fort Hood, Texas, and later at the White Sands Missile Range in New Mexico. The drop tests were designed to evaluate the ability to transition from a high angle-of-attack reentry to a level cruise attitude, the stability of the vehicle in stalled conditions, and to obtain a wide variety of free-flight data to assist in analytical aerodynamic transition simulations and prediction techniques. The model was approximately 13 feet long, had an eight-foot wingspan, and weighed 600 pounds. Constructed of aluminum, fiberglass, and styrofoam, the model was dropped from altitudes up to 12,000 feet, and used a parachute recovery system.[66]

Total RDT&E costs for the orbiter were estimated at $2,770 million, with the first orbiter costing $171.2 million. The booster would need $3,142 million in non-recurring development costs, and each article would cost $236 million. It was noted that these "... predictive estimates of the DC-3 shuttle costs contain many uncertainties ..."

In early 1969, the AFFDL had became aware of the Faget design and issued two analytical reports, one in

June and the other during November, that pointed out the difficulties of constructing a straight-wing design that could withstand the thermal and aerodynamic effects of reentry. Charles Draper followed this with two AIAA papers in October 1970 and January 1971 that completely rejected the straight-wing approach and showed how a delta-wing planform would furnish an orbiter with far better safety and cross-range capabilities. Advocates of Faget's design countered that the AFFDL did not appreciate NASA's expertise in low L/D designs that used exceptionally high angles-of-attack, a critical point in safely operating the MSC orbiter. But there were other critics of the MSC design approach, primarily from NASA's own Flight Research Center, which favored a lifting-body approach. The Assistant Secretary of the Air Force, Michael Yarymovych, also argued that a straight-winged orbiter could not meet the Air Force's need for high cross-range capabilities and large payload capacity.[67]

Nevertheless the MSC-001 design, and its derivatives, formed the core of subsequent MSC space shuttle studies until the designers were almost forcibly made to recognize that the vehicle possessed several fatal flaws, particularly

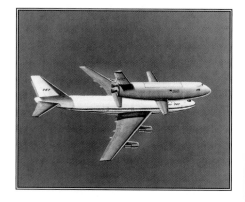

An indication of things to come. A variation of the DC-3 orbiter is carried atop a Boeing 747 in an illustration accomplished to demonstrate the relative size of the orbiter. Since the orbiter was equipped with air-breathing engines for ferry flights, there was no need for this combination to ever exist, and NASA explained the illustration as 'artistic license.' Note the repositioned air-breathing engines on the orbiter – exhaust ducts in the upper forward fuselage just under the cockpit. (NASA photo S70-204X)

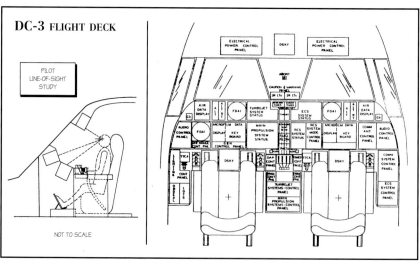

DC-3 FLIGHT DECK

PILOT
LINE-OF-SIGHT
STUDY

NOT TO SCALE

The DC-3 flight deck was conventional in most details but had some novel features, such as the 'microfilm data display and keyboard' in front of each pilot. Note the throttles on the center console, a feature lacking in the production orbiters. (NASA)

The straight-wings and generally common shapes of orbiter and booster show up well in these early NASA illustrations of the DC-3 concept. By this time the air-breathing engines had been moved from the top of the wings to a location in the forward fuselage, just ahead of the exhaust ducts that can be seen on the nose of both the orbiter and booster. Note the simple lifting frame (at left) that was used to mate the vehicles. (NASA photos S69-4056 and S69-4054)

A 1/10-scale model of the DC-3 orbiter was dropped by an Army CH-54A on several occasions both at Fort Hood, Texas, and the White Sands Missile Range in New Mexico. Although the model was lowered by parachute from a preset altitude, it often suffered fairly heavy damage, as shown in this photograph of an early drop test recovery. Fortunately, the styrofoam nose was designed to be easily replaced. Note the engine intakes in the extreme nose. (NASA photos S70-44262 and S70-15020 via Jason Wentworth)

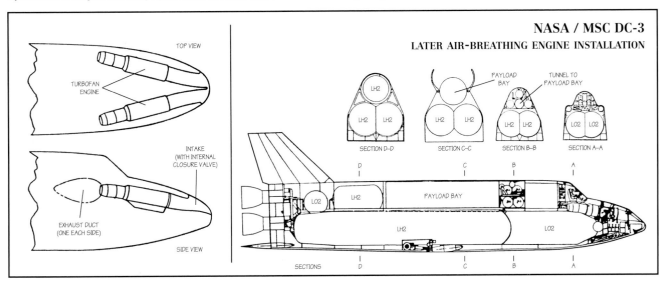

The were at least two configurations for the air-breathing engines on the DC-3 design. One had a pair turbojet engines located in pods on top of each wing at about mid-span. The other had a pair of large turbofans mounted in the forward fuselage with an air intake in the extreme nose and the exhaust along each side of the fuselage just behind the cockpit. (NASA)

A slightly different Blue Goose from those shown below. This variant has a large module under the fuselage holding four air-breathing engines, and rearranged propellant tanks in the fuselage. (NASA photo S70-6121 via the SDAM Collection)

a tendency to spin at hypersonic velocities and a suspected inability to withstand the thermal environment during reentry. A review of subsequent configurations shows that NASA designers varied such values as aspect ratio, wing sweep and tail location, using parametric manipulation, to derive a whole family of orbiters and launch vehicles. Even after January 1971, when NASA officially shifted its studies to delta planforms and began concentrating on partial-

ly reusable stage-and-a-half concepts, the basic Faget vehicle would not go away. Indeed, the last appearance of the Faget concept came at the very end of 1971 with the design of the MSC-043 configuration (see page 148).

There was also a competing MSC design, this one dubbed the Blue Goose by its designers. First appearing in the fall of 1970, this design initially proposed the radical idea of having a sliding mechanism built into the vehicle so that the entire wing could be translated 12 feet fore and aft to compensate for center of lift shifts. It was quickly realized that this was impractical, and various attempts to provide conventional solutions to the problem were applied to the same basic design.

Two different versions of the design seem to have been proposed, one with a tee-tail and the other with a more conventional stabilizer arrangement. Both designs featured variable-geometry canard surfaces mounted low on the nose. A deorbit engine, protected during ascent by a 'swing-down' nose-cap, was in the extreme nose, followed by a crew module, and then a large payload compartment. An LO2 tank was above and just forward of the wing, followed by a LH2 tank and finally by two rocket engines in the extreme aft fuselage.

It was determined that this configuration would have extreme heating and structural problems, and the design was abandoned. Besides, by this time the MSC-040 series of designs and the external tank concept had caught the imagination of the MSC design teams, and almost all efforts were concentrated on refining various excursions from the basic MSC-040A.

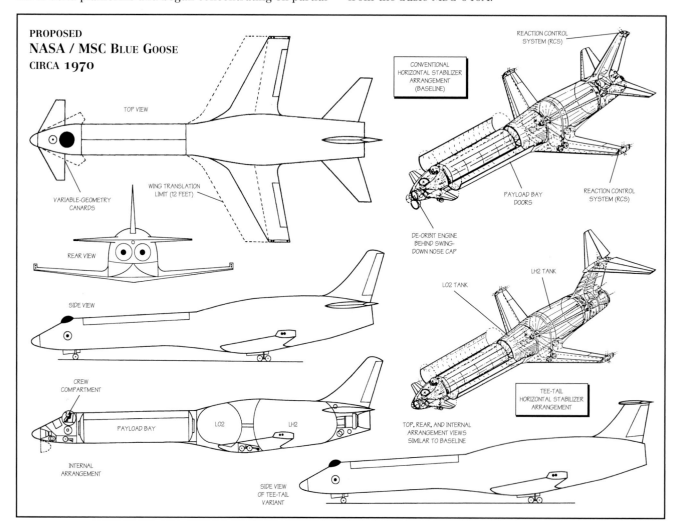

PROPOSED
NASA / MSC BLUE GOOSE
CIRCA 1970

The Phase A studies, supported by more than 200 man-years of engineering effort backed up by extensive wind tunnel testing, materials evaluations, and structural design, resulted in the evaluation of four basic baseline vehicles. These included straight-wing, delta-wing, and stowed-wing/variable-geometry wing vehicles as well as lifting-bodies. It was found that the powerful boost engines baselined by MSFC did not necessarily provide the propulsion capabilities needed for the vehicles being investigated since the majority of the contractors opted for the less powerful alternate engine design. This was partially because of their lighter weight, but also as a result of abort profile studies that indicated that a larger number of less powerful engines were preferable over a smaller number of more powerful engines during emergencies.

The Phase A studies, combined with the Air Force ILRV studies, confirmed that cross-range was the major sticking point when trying to consolidate the Air Force requirements with the needs of NASA. The Air Force wanted the space shuttle to be able to land at its launch site after only one orbit. During this time the Earth had rotated approximately 1,265 miles, meaning the returning spacecraft had to be able to fly at least that distance during reentry. On the other hand, NASA simply wanted an opportunity to land back at the launch site only once every 24 hours, requiring a relatively modest 265 miles in cross-range capability. However, many emergency scenarios required at least 450 miles cross-range to enable a return-to-launch site abort capability, leading to most designs providing at least that amount of cross-range.

The lifting-body concept was found to be the most poorly suited to space shuttle applications, primarily because the body shape did not lend itself to efficient packaging and installation of a large payload bay, propellant tanks, and major subsystems. The complex double curvature of the body resulted in a vehicle that would be difficult to fabricate, and further, the body could not easily be divided into subassemblies to simplify manufacture. Also, the lifting-body's large base-area yielded a relatively poor subsonic L/D, resulting in a less attractive cruise capability and poor low-speed' flight characteristics.

The variable-geometry designs were found to have many attractive features, including low inert/burnout weight and the high hypersonic L/D needed to meet the maximum cross-range requirements. In addition, the stowed wing approach permitted the wing to be optimized for the low-speed flight regime. Drawbacks included a high vehicle weight to projected planform area ratio, which would result in a higher average base temperature relative to either the straight or delta-wing designs. In addition, significantly increased design and manufacturing complexity would result from the mechanisms required to operate the wing and transmit the flight loads to the primary structure. The maintenance required between flights was expected to be high, and insufficient data existed to reliably determine potential failure modes which were thought to be numerous.

Faget was not a believer in lifting-reentry, despite his proposed design having wings – he still held to the idea of a high-drag blunt body. To operate his design as a blunt

body, Faget proposed to reenter the atmosphere at an extremely high angle-of-attack with the broad lower surface facing the direction of flight. This would create a large shock wave that would carry most of the heat around the vehicle instead of into it. The vehicle would maintain this attitude until it got below 40,000 feet and about 200 mph, when the nose would come down and it began diving to pick up sufficient speed for level flight. The vehicle would then fly to the landing site, touching down at about 140 mph. Since the only 'flying' was at very low speeds during the landing phase, the wing design could be selected solely on the basis of optimizing it for subsonic cruise and landing – hence the simple straight wing proposed by Faget, and used on the North American Phase A design and the MSC DC-3. The design did have one major failing, at least in the eyes of the Air Force – since it did not 'fly' during reentry, it had almost no cross-range capability.[68]

Faget had convinced many that his simple straight-wing concept would be more than adequate. But not everybody agreed. In particular, Charles Cosenza and Alfred Draper at the AFFDL did not accept the idea of building a shuttle that would come in nose high, then dive to pick up flying speed. With its nose so high the vehicle would be in a classic stall, and the Air Force disliked both stalls and dives, regarding them as preludes to an out-of-control crash. Draper preferred to have the shuttle enter its glide while still at hypersonic speeds, thus maintaining much better control while still continuing to avoid much of the severe aerodynamic heating.[69]

But if the shuttle was going to glide across a broad Mach range, from hypersonic to subsonic, it would encounter another aerodynamic problem: a shift in the wing's center of lift. At supersonic speeds the center-of-lift is located about midway down the wing's chord (the distance from the leading to the trailing edge) – at subsonic speeds it moves much closer to the leading edge. Keeping an aircraft in balance requires the application of an aerodynamic force that can compensate for this shift. The MSC Blue Goose accomplished this in an extreme manner by translating the entire wing fore-aft as the center-of-lift changed.

The Air Force had extensive experience with this phenomena, and had found that a delta planform readily mitigated most of the problem. Draper proposed that both stages of any reusable shuttle should use delta wings instead of straight ones. Faget disagreed, pointing out that his design did not 'fly' at any speed other than low subsonic – at other speeds it 'fell' and was not subject to center-of-lift changes since it was not using lift at all.[70]

To achieve landing speeds low enough to satisfy Faget, a delta wing would need to be very large since deltas tend to land 'hot' under the best of circumstances. A straight wing with a narrow chord, argued Faget, would be light and have relatively little area that needed a thermal protection system. To achieve the same landing speed, a delta would be large, add considerable weight, and greatly increase the area that required thermal protection.[71]

Draper argued that delta wings had other advantages. Since it was relatively thick where it joins the fuselage, delta wings offered more room for landing gear and other systems that could be moved out of the fuselage. Its sharply-swept leading edge produced less drag at supersonic speeds, and its center-of-lift changed slowly compared to a straight wing. But the delta offered one other advantage – one that became increasingly importantly as the military became more interested in using space shuttle. Compared to a straight wing, a delta produces consid-

---

* This is somewhat contrary to the results achieved during the HL-10 program at Edwards AFB, but it was expected that the shape of a full-scale shuttle would be compromised sufficiently to degrade its stability to the point of being unacceptable. Plus, any space shuttle would weigh significantly more than even the 'mission weight' lifting-bodies tested at Edwards, resulting in a higher wing loading.

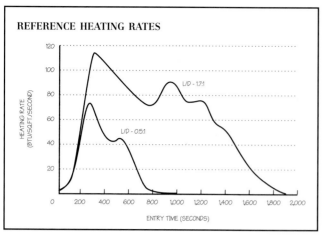

REFERENCE HEATING RATES

Comparison of heating rates for the Phase B baseline straight-wing low-L/D orbiter versus the delta-wing high-L/D concept. The straight-wing design had low lift (L/D=0.5:1) and little cross-range, but its reentry was brief and limited the heating rate. The delta-wing orbiter had much higher lift (1.7:1) with a substantial cross-range. Its reentry, however, was prolonged and imposed both a high heating rate and a high total heat. (NASA)

erably more lift at hypersonic speeds. This allowed a returning shuttle to achieve a substantial cross-range during reentry, a capability highly prized by the Air Force.[72]

Not accepting Faget's position that his vehicle did not 'fly' at high speeds, Draper argued that any straight wing concept would require additional aerodynamic surfaces (called canards) on the forward fuselage to compensate for the shifting center-of-lift. Canards produce lift of their own, and this would cause the main wings to be mounted further rearward, where they would probably disturb the center-of-gravity. It was a complicated problem, and one that would take at least another year to resolve.[73]

To help resolve the controversy, the two concepts baselined for study during Phase B included a Faget-style straight-wing low cross-range orbiter, and a Draper-supported delta-wing high cross-range orbiter. The straight-wing orbiter would be configured to provide design simplicity, low weight, decent handling, and good landing characteristics. The vehicle would be designed to enter at a high angle-of-attack to minimize heating and to facilitate the use of a heat shield constructed from materials available in the early 1970s. The delta-wing orbiter would be designed to provide the capability to trim over a wide angle-of-attack range allowing initial reentry at a high-alpha to minimize the severity of the reentry environment, then transitioning to a lower angle-of-attack to achieve a higher cross-range. Phase B would use these two government-provided concepts as departure points for continued evaluation and subsystem studies.

Refining the Configuration – Phase B

During February the STS Committee decided that any future space shuttle program would "be managed by NASA" and would "be generally unclassified." This fit better within the NASA scheme of doing business than a possible classified program would have. Nevertheless, many NASA officials, and most contractors, would maintain security clearances since classified items (such as the Air Force payload manifest, and some technology issues) would continue to fall under veils of security.[74]

On 20 February 1970 NASA issued a request for proposals for Phase B of the space shuttle studies. Competing companies had 30 days to respond, then the source selection board would choose two winners for 11 month contracts. Winnowing the field to a pair of contractors had

always been expected, and after the final Phase A reports were submitted, a 'teaming dance' began among the aerospace companies. Every company wanted to participate in the studies, and the way that most likely assured continued participating was to team with other companies.[75]

NASA issued two contracts on 6 July 1970 based on an evaluation of the Phase A results, and the companies' responses to a Statement of Work (SOW) issued in February 1970. The contractors selected to compete during Phase B included a team consisting of McDonnell Douglas and Martin Marietta; and North American Rockwell, who was joined by General Dynamics/Convair as a risk sharing subcontractor. This phase was to involve a more detailed analysis of mission profiles, and a preliminary look at the vehicles needed to complete them. Each contractor team was initially funded at $8 million, although this was subsequently raised to $10.8 million each.[76]

The program-level requirements established for Phase B by the Space Shuttle System Program Definition, dated 1 June 1970, included:[77]

- the shuttle shall be a fully reusable two-stage vehicle,
- the integrated vehicle shall be launched vertically, and landed horizontally,
- the initial operational capability (IOC) shall be in late calendar year 1977,
- the reference mission shall be a 310-mile 55-degree inclination circular orbit from a launch site located at 28.5 degrees north latitude (KSC),
- the payload bay shall have a clear volume 15 feet in diameter by 60 feet in length, with a reference mission capacity of 15,000 pounds,
- the space shuttle shall be capable of operating within the payload range from zero to maximum capability,
- two orbiters shall be developed, one capable of 230 miles and the other of 1,726 miles hypersonic cross-range,
- the mission duration (from lift-off to landing) shall be seven days,
- all vehicle stages shall be capable of ferry flights between airports,
- the booster and orbiter shall have a go-around capability during landing operations,
- launch rates will vary from 25 to 75 per year,
- total turnaround time from landing to launch shall be less than two weeks,
- both elements shall provide a 'shirt sleeve' environment with trajectory load factors of less than 3-g,
- a 43-hour turnaround capability shall be provided for rescue missions, and
- all subsystems, except primary structure and pressure vessels, shall be designed to fail-operational after failure of the most critical component, and to fail-safe for crew survival after failure of the two most critical components.

The high cross-range orbiter would need to achieve a hypersonic L/D in excess of 1.8:1 during major portions of its atmospheric reentry in order to meet the 1,726 mile (1,500 nm) cross-range requirement. This meant that the vehicle would expose substantial areas of its surface to high heating rates for a long period of time compared to the low cross-range orbiter design. Consequently, the high cross-range vehicle would have to be provided with substantially greater thermal protection. In fact, the total heat load experienced by the high cross-range orbiter was predicted to be five to seven times greater than the low cross-range design. Also, although the payload bay was sized to accommodate most anticipated NASA and DoD payloads, the reference capacity of 15,000 pounds was considerably less (and unacceptable) than the 65,000 pounds required by the Air Force.

There was quite a bit of interest by the various NASA Centers on the relative merits of the air-breathing engine installation proposed for the shuttle vehicle. The Flight Research Center at Edwards was readying an abbreviated flight test series using the HL-10 lifting-body to study the effects of landing engines on the pilot workload, while the Lewis Research Center* in Cleveland, Ohio, studied the relative advantages of turbofans versus turbojets, and jet-fuel (JP-4/5) versus gaseous hydrogen (GH2).

Since no actual vehicle existed in May 1970 to make realistic comparisons against, NASA Lewis picked some seemingly arbitrary numbers to illustrate the differences inherent to the various fuels and propulsion plants. The most impressive of the results showed that, based on needing 100,000 lbf for the booster stage, switching from JP-4/5 to GH2 could save up to 36,000 pounds in engine and fuel weight based on an anticipated 1972 state-of-the-art turbofan engine. A weight reduction of between 5,600 and 14,000 pounds was predicted for the orbiter. This savings was even more dramatic if it is remembered that a 50,000 pound reduction in combined vehicle weight actually translated into a savings of between 120,000 and 200,000 pounds in gross lift-off weight, due primarily to the reduction in propellants needed to accelerate the vehicle during ascent. For this reason, most contractors would opt for GH2 powered air-breathing engines during the early part of Phase B, even though none actually existed+ in 1970.[78]

Other requirements were levied by the various Centers involved in the oversight of the Space Shuttle design contracts, including:[79]

- each element (orbiter and booster) shall have a two-man flight crew, and be flyable under emergency conditions by a single crewman,
- the stages shall be capable of positive separation without the use of special separation rocket systems of the type employed on Saturn,
- in-flight refueling (subsonic or supersonic) shall not be used to meet design mission requirements,
- the booster and orbiter shall be capable of pilot-controlled landings under FAA Category-2 conditions,
- the vehicle shall incorporate on-board provisions to quickly and easily place the vehicle in a safe condition following landing to permit crew and passenger egress,
- there shall be no propellant cross-feed between elements (booster and orbiter), and
- the vehicle elements (booster and orbiter) shall be capable of landing on runways no longer than 10,000 feet.

Both contractors would start with the baseline vehicles that were selected as a result of the Phase A studies. These designs included a low cross-range orbiter that was based heavily on the MSC-002 (DC-3) that Maxime Faget had designed at MSC, although by this time it was becoming increasingly evident that the straight-wing design suffered some serious handicaps during reentry and hypersonic flight and could not meet the Air Force requirements. The other design was a high cross-range

orbiter that used a severely clipped delta-wing, and was thought to be superior during the reentry phase, but at this point had a very poor payload fraction (total payload versus total vehicle weight). No boosters were specified by NASA, the configuration and design of these being left to the contractor teams.

One of the first steps in the Phase B studies would be to 'size' the vehicles which involved determining the boost/payload performance, the optimum staging velocity, the thrust-to-weight ratio at lift-off, the number of engines, etc. Economics was continuing to play an major role in the design of space shuttle – the primary consideration was beginning to shift from providing the most economical long-term operational system, to reducing the near-term development costs.

Main Engine Phase B

The space shuttle main engine (SSME) that was to be investigated during Phase B of the engine competition was an evolutionary version of the Phase A alternate engine design that had been preferred by a majority of the airframe contractors. The high chamber-pressure, bell-nozzle engine used LH2 and LO2 in a 6:1 ratio by weight and produced 415,000 lbf at sea level – roughly 477,000 lbf in a vacuum. A combustion chamber pressure of 3,000 psi was specified – a significant advanced over any production engine, and also an improvement over the existing XLR129 demonstrator engine.

The power head assembly was identical for the orbiter and booster, although the nozzle design differed. A two-position nozzle with a maximum 120:1 expansion ratio was to be used in the orbiter, while the booster would use a fixed nozzle. Thrust vector control over a range of ±7 degrees was obtained with a self-contained gimbal actuation system mounted on the engine itself. The main engines could no longer be used at a ten percent thrust rating for on-orbit maneuvering, and therefore separate orbital maneuvering engines would need to be provided by the airframe contractors. The various analyses concerning abort modes that had been conducted by the Phase A contractors, as well as independent groups, had finally convinced MSFC that a capability to run the engines at more than their maximum thrust rating was required, at least for short periods of time.[80]

In February 1970, MSFC had issued a request for proposals that would lead to the award of three contracts for Phase B of the engine competition. Since there were only three major rocket engine manufacturers in the country, it was pretty well assured that the contracts would go to Aerojet General, Pratt & Whitney, and Rocketdyne, which they did – each company received $6 million to study engine concepts and produce some prototype hardware.

Aerojet brought significant experience from building the experimental 1,500,000-lbf M-1 engine for NASA in the mid-1960s. Rocketdyne had built the 1,522,000 lbf F-1 engine for the first stage of the Saturn V and the 230,000 lbf J-2 used in the second stage of the Saturn IB and the second and third stages of the Saturn V. Pratt & Whitney had been under a separate Air Force contract for several years to develop and test the XLR129 high-chamber pressure demonstrator engine. The challenge during this competition was not to build a larger, more powerful engine, but to build a small, compact, reusable engine that could be throttled during ascent to provide some measure of control over the maximum dynamic pressure and speed of the vehicle.

---

* Now the John H. Glenn Research Center at Lewis Field.

+ The engine manufacturers estimated that it would not be difficult to convert existing turbojet or turbofan engines from JP-4/5 to gaseous hydrogen, and in fact had already run limited tests on several prototypes.

Pratt & Whitney was on a roll, however. During 1970 they successfully demonstrated a hydrogen turbopump that produced over 100 horsepower per pound – more than five times the efficiency demonstrated by the J-2 turbopump that had been considered state-of-the-art only a few years before. Pratt then hooked the new turbopump up to the XLR129 combustion chamber and produced a 350,000 lbf engine. Although the XLR129 combustion chamber had only been designed for 2,740 psi, Pratt conducted over 200 test firings with the chamber operating at 3,000 psi.[81]

Rocketdyne was feeling the blow of the aerospike being eliminated from the competition. In order to catch up to Pratt & Whitney, Rocketdyne committed $3 million in company funds to build a full-scale test version of the SSME that could demonstrate a thrust of 415,000 lbf, stable combustion, a chamber pressure of 3,000 psi, and adequate cooling. The one thing that time did not allow was the building of realistic turbopumps, so the engine would be fed propellants from tanks under high pressure. Rocketdyne began with an injector based on the unit from the well-proven J-2, and began building their engine. A location called the Nevada Field Laboratory, some 20 miles from Reno, was the site of initial testing. Testing soon revealed that the basic design could operate at 505,700 lbf at 3,172 psi, with an exhaust velocity of 14,990 feet per second – all substantially better than the XLR129 was capable of doing at the time.[82]

But in late January 1971 NASA changed the requirements – substantially. Instead of 415,000 lbf, NASA now needed 550,000 lbf, due mainly to a decision to increase the shuttle payload capacity to 65,000 pounds to meet DoD requirements. This change suddenly catapulted Rocketdyne into the lead. Although they had not demonstrated this level of thrust, they were much closer than P&W. But in a sense it was all misleading – Rocketdyne still had not developed a turbopump, and this would become their Achilles' heel.[83]

About this time in the development of the main engines, a decision was reached to provide a digital engine controller. Digital control was selected over the more classical engine control concepts of valve sequencing or analog computer control because of a perceived life-cycle cost advantage, as well as more precise control during throttling. Since the controller could be relatively easily reprogrammed, operational changes could be rapidly and economically introduced into the engines as experience was gained with their operations. At this point the engine was still in an early state of development, so the design could still be optimized to take advantage of the additional flexibility offered by a digital controller. The digitally based system also demonstrated the most economical means of providing the sophisticated monitoring and redundancy management schemes required for the complex operation sequence of the engine. An added benefit to using a digital controller was that it would finally be possible to throttle over a thrust range of 50 to 115 percent of rated thrust, although the airframe contractors were requested not to use in excess of 105 percent during abort scenario planning, and subsequent study showed no need to use less than 65 percent.[84]

Revision

A revision to the Space Shuttle System Program Definition Document was issued on 1 September 1970, giving the two contractor teams just enough time to prepare new configurations for their 90-day reviews, each scheduled for mid-October 1970. The revision to the specification increased the desired payload from 15,000 pounds to 25,000 pounds for the same reference orbital mission, although this still did not satisfy the DoD requirement. Standard jet fuel (JP-4/5) was now required for the air-breathing engines, where previously it had been up the contractors if they wanted to use jet fuel or some other propellant – most had preferred GH2. This switch was driven by safety considerations, since NASA wanted to ensure that the main LH2 tanks were purged of residual hydrogen as quickly as possible. Apparently, little consideration was given to carrying small auxiliary hydrogen tanks specifically for the air-breathers. A capability was also to be provided to carry passengers in a self-contained pressurized payload bay module.[85]

Coincident with the 90-day reviews, NASA issued another revision to the Phase B specification, this one dated 13 November 1970. This change was in direct response to questions posed by the contractors during the reviews, and also to correct various ambiguities contained in the earlier releases. By this time both contractor teams, as well as NASA itself, had conducted numerous trade studies on variations of the baseline shuttle systems, with one of them being to determine the optimum number of engines needed by the orbiter to safely complete its mission.

Both contractor teams and NASA agreed that the optimum payload performance was achieved with the use of two main engines on the orbiter. The actual payload-to-weight ratio difference between a two and three engine orbiter was somewhat over 20 percent, a rather significant figure. The payload decrement resulted primarily from the additional engine and engine installation weight inherent in a three-engine arrangement. The two-engine orbiter also was considered better from a cost viewpoint, since it was less expensive to acquire and also less expensive to maintain. The major factor that drove the ultimate selection of a three-engine configuration was safety. In a two-engine configuration, the loss of an engine resulted in a 50 percent loss of power, and even if the remaining engine could be throttled to 105 percent, the orbiter had still lost 47 percent of its available thrust. This made aborting in some flight regimes very tricky. On the three-engine configuration, the loss of one engine, and the capability to throttle the other two to 105 percent, resulted in a loss of only 30 percent of available thrust, greatly simplifying abort planning.[86]

During late December 1970, the status of the space shuttle program within NASA changed considerably. Up until this time, the project had been managed by LeRoy Day and the Space Shuttle Task Group. Now the shuttle received a separate program office, headed by Charles J. Donlan, the Deputy Associate Administrator for Manned Space Flight at NASA Headquarters. Donlan reported directly to Dale D. Meyers, who was the Associate Administrator for Manned Space Flight. This elevated shuttle to the same status that had been enjoyed by such programs as Mercury, Gemini, and Skylab. But the Space Shuttle Program Office at NASA Headquarters would not last long.[87]

So in mid-1970 NASA entered Phase B with grand ambitions of developing a fully-reusable two-stage system to fulfill an unrealistically high flight rate that everybody had repeated so many time they were beginning to believe its validity. As 1970 drew to a close NASA began evaluating the initial results from the two Phase B contractors. Considering the only manned hypersonic aircraft yet flown was the diminutive X-15 from a decade earlier, the initial round of space shuttle proposals were truly fantastic.

McDonnell Douglas / Martin Marietta (NAS9-10959)

The Phase B team led by McDonnell Douglas and Martin Marietta also included the TRW Electronic Systems Group, Pan American World Airways, Raytheon, Sperry, Norden, Hamilton Standard, and LTV. McDonnell Douglas was responsible for the orbiter design and systems integration, while Martin Marietta concentrated on the design of the booster. In a December 1970 report to NASA this team presented their version of each orbiter configuration, along with two different derivatives of a booster to launch them. Both orbiters provided a 15 by 30 foot payload bay capable of carrying 15,000 pounds, which did not meet the Phase B specification for a 15 by 60 foot payload bay, but satisfied the original weight requirement. The team indicated that it would be possible to scale up the designs to meet the later weight requirement, but that they felt it presented an unreasonable risk, and that 15,000 pounds was a sufficient payload to justify development, with the additional capacity coming after sufficient operational experience was gained.[88]

The low cross-range orbiter was a Faget-style straight-wing vehicle spanning 113.8 feet and weighing 188,600 pounds empty. The fuselage was 147.6 feet long and 21 feet wide. Two 415,000 lbf boost engines were mounted vertically on the vehicle centerline similar to the BAC Lightning, and four 18,000 lbf air-breathing turbofans were mounted in twin pods on each side of the orbiter mid-fuselage for landing and ferry operations. LO2 and LH2 tanks sat end-to-end under the cockpit and payload bay, with the LO2 tank located forward for better center of gravity control. A secondary LH2 tank was located behind the payload bay on top of the main LH2 tank. The cross-range capability was estimated to be 230 miles (substantially less than the Apollo capsule was capable of), with a powered landing speed of 204 mph. The wings and fuselage consisted of a titanium alloy hot-structure, with the lower surfaces covered with columbium shingles. Cobalt superalloy was used for the control surfaces, and the vertical stabilizer was a hot-structure made of nickel superalloy. Carbon-carbon composite was used on the wing leading edges, and the semi-blended chine was an unshielded columbium superalloy hot-structure.

The high cross-range orbiter was a sleek blended-delta spanning 97.5 feet and weighing 203,500 pounds empty. The total fuselage length for this orbiter was 171 feet. Again, two 415,000 lbf boost engines were mounted vertically on the centerline. Four air-breathing turbofans were mounted in pairs in each wing, being deployed downward into the airstream when needed during landing. Two LO2 tanks were contained in the lower forward fuselage, with two LH2 tanks located side-by-side under the payload bay. A secondary LH2 tank was mounted above the LO2 tanks, but below the cockpit. It was projected this design would meet the 1,726 mile cross-range requirement, and the power-on landing speed was estimated to be 186 mph. The cryogenic propellant tanks were to be of fusion-welded aluminum construction, with the rest of the orbiter structure made up primarily of titanium alloys. Cobalt superalloy shingles covered the lower wing surfaces, and the wing leading edges were of columbium superalloy construction. Nickel superalloy shingles covered the sides of the forward fuselage. The vertical stabilizer had a nickel superalloy hot-structure with a carbon-carbon leading edge. The possible use of boron-epoxy and boron-aluminum was suggested if the technologies matured quickly enough.

The proposed booster was generally similar to one of the Martin Marietta Phase A alternate designs. The booster for the straight-wing orbiter was 220.2 feet long with a large aft mounted wing spanning 151 feet, and an empty weight of 452,803 pounds. Small canards on the forward fuselage contained ten 18,000-lbf turbofan engines for landing operations. Thirteen 415,000-lbf boost engines provided the power to carry the orbiter to staging velocity. The aluminum booster had a fuselage composed of integral LH2 and LO2 tanks with separate bulkheads, and the wings and canards used conventional spar and rib construction. The main body diameter of 33 feet was based on the fact that the tooling already procured for the Saturn V

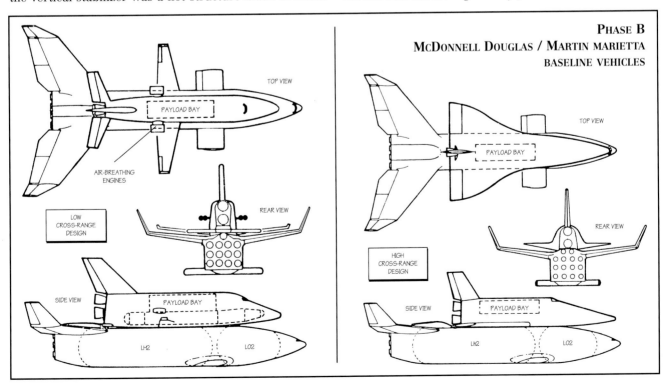

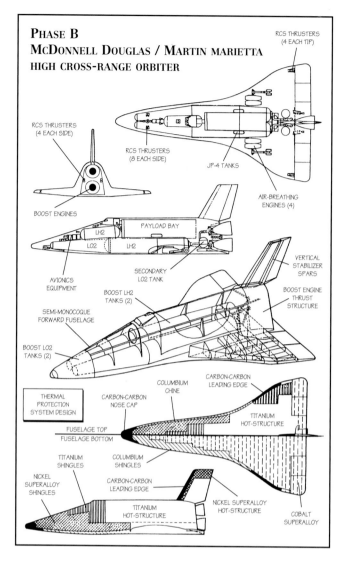

Phase B
McDonnell Douglas / Martin marietta
High cross-range orbiter

RCS THRUSTERS
(4 EACH TIP)

RCS THRUSTERS
(4 EACH SIDE)

RCS THRUSTERS
(8 EACH SIDE)

JP-4 TANKS

AIR-BREATHING
ENGINES (4)

BOOST ENGINES

PAYLOAD BAY

LH2

LO2

LH2

VERTICAL
STABILIZER
SPARS

AVIONICS
EQUIPMENT

SECONDARY
LO2 TANK

BOOST ENGINE
THRUST
STRUCTURE

BOOST LH2
TANKS (2)

SEMI-MONOCOQUE
FORWARD FUSELAGE

BOOST LO2
TANKS (2)

CARBON-CARBON
LEADING EDGE

COLUMBIUM
CHINE

CARBON-CARBON
NOSE CAP

TITANIUM
HOT-STRUCTURE

THERMAL
PROTECTION
SYSTEM DESIGN

FUSELAGE TOP
FUSELAGE BOTTOM

TITANIUM
SHINGLES

COLUMBIUM
SHINGLES

NICKEL
SUPERALLOY
SHINGLES

CARBON-CARBON
LEADING EDGE

TITANIUM
HOT-STRUCTURE

NICKEL SUPERALLOY
HOT-STRUCTURE

COBALT
SUPERALLOY

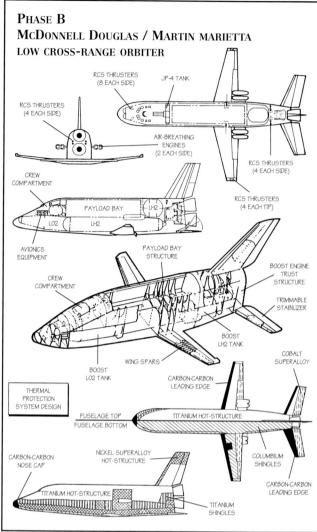

Phase B
McDonnell Douglas / Martin marietta
Low cross-range orbiter

RCS THRUSTERS
(8 EACH SIDE)

JP-4 TANK

RCS THRUSTERS
(4 EACH SIDE)

AIR-BREATHING
ENGINES
(2 EACH SIDE)

RCS THRUSTERS
(4 EACH SIDE)

CREW
COMPARTMENT

RCS THRUSTERS
(4 EACH TIP)

PAYLOAD BAY

LH2

LO2

LH2

AVIONICS
EQUIPMENT

CREW
COMPARTMENT

PAYLOAD BAY
STRUCTURE

BOOST ENGINE
TRUST
STRUCTURE

TRIMMABLE
STABILIZER

BOOST
LH2 TANK

WING SPARS

BOOST
LO2 TANK

COBALT
SUPERALLOY

CARBON-CARBON
LEADING EDGE

THERMAL
PROTECTION
SYSTEM DESIGN

FUSELAGE TOP
FUSELAGE BOTTOM

TITANIUM HOT-STRUCTURE

CARBON-CARBON
NOSE CAP

NICKEL SUPERALLOY
HOT-STRUCTURE

COLUMBIUM
SHINGLES

CARBON-CARBON
LEADING EDGE

TITANIUM HOT-STRUCTURE

TITANIUM
SHINGLES

was able to produce cryogenic tanks of that size. The booster for the high cross-range orbiter was generally similar in configuration, only slightly larger and heavier.

Several variations were studied for the booster, some with the aft wings mounted on top of the fuselage, others using low mounted aerosurfaces. Also evaluated was the best position of the vertical control surfaces, with some designs having a central vertical stabilizer (as in Phase A), others having wing-tip mounted verticals. The final configuration of a high mounted wing with the outboard tips bent upward to serve as vertical control surfaces was found to provide the least interference with the orbiter at launch, and still provided adequate stability for the booster.

Several unmanned second stages were also proposed to supplement the orbiter for heavy lift duties. These second stages included a modified Saturn S-IVB stage with solid rocket motors to augment thrust and capable of carrying 120,000 pounds into orbit, a modified Saturn S-II stage with a nuclear-powered upper stage capable of delivering 130,000 pounds, and an all new second stage built with components derived from the orbiter and booster capable of carrying 170,000 pounds. The total cost to develop the modified S-IVB stage was estimated at $82 million and a recurring cost per flight of $23.7 million yielded a cost-to-orbit of $198 per pound. The modified S-II/nuclear upper stage would need $106 million in development funds, but would boost a pound into orbit for $186. The shuttle-derived stage would cost $320 million to develop, but

would lower the per pound figure to $103. All were considered low risk technology, and the all-new stage was thought to provide more growth potential and longevity.

The operational baseline for the proposed vehicle included just ten days for turnaround between landing and launch. Three possible launch sites were evaluated – Kennedy Space Center, Vandenberg AFB, and the White Sands Missile Range. Since KSC was to use existing Apollo facilities, it was estimated it would only take five years and $86.75 million to be modified. Vandenberg and White Sands would require new facilities, taking six to seven years to build, and costing $285 million at Vandenberg and $317 million at White Sands. All three sites offered adequate abort options for the high cross-range orbiter, but only KSC had abort options for a first revolution return of the low cross-range version.

The McDonnell Douglas high cross-range orbiter riding atop its Martin Marietta booster launches from KSC in this NASA-supplied artist concept. Note the large forward canards on the booster – these housed the ten large air-breathing turbofan engines. Interestingly, the empty weight of this booster design was essentially the same as the fully-loaded gross weight of an early Boeing 747. The booster carried over 140,000 pounds of jet fuel for its cruise back to the launch site. (NASA photo S71-26563)

BOOSTER AIR-BREATHING PROPULSION SYSTEM

AFT JP-4 FUEL TANKS

FORWARD JP-4 FUEL TANKS

FUEL TRANSFER LINES

AIR-BREATHING ENGINES (10)

BOOSTER WING POSITION TRADE STUDY

The high cross-range orbiter approaches for a landing. Note the deployed air-breathing engines under the wing, and the long distance from the nose to the cockpit – this would have made manual landings challenging for the crew. (NASA photo S71-20741)

The Martin Marietta booster returns for a landing. Although difficult to discern in this drawings, the leading edge of the canard is open to allow it to function as an air intake for the turbofan engines. Note the landing gear configuration. (NASA photo S71-20743)

THERMAL PROTECTION SYSTEM
ORBITER LEADING EDGE CONCEPTS

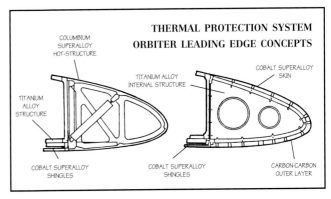

COLUMBIUM SUPERALLOY HOT-STRUCTURE

TITANIUM ALLOY INTERNAL STRUCTURE

COBALT SUPERALLOY SKIN

TITANIUM ALLOY STRUCTURE

COBALT SUPERALLOY SHINGLES

COBALT SUPERALLOY SHINGLES

CARBON-CARBON OUTER LAYER

ALTERNATE UPPER STAGES

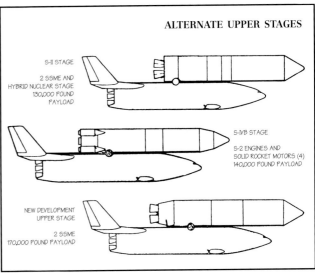

S-II STAGE
2 SSME AND HYBRID NUCLEAR STAGE
130,000 POUND PAYLOAD

S-IVB STAGE
S-2 ENGINES AND SOLID ROCKET MOTORS (4)
140,000 POUND PAYLOAD

NEW DEVELOPMENT UPPER STAGE
2 SSME
170,000 POUND PAYLOAD

LOW CROSS-RANGE ORBITER
CREW AND PASSENGER ACCOMMODATIONS

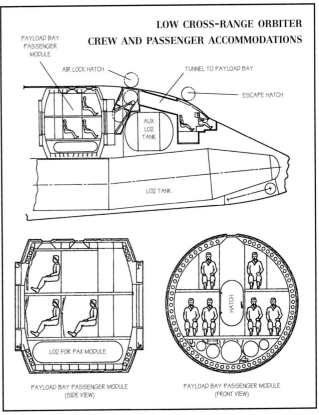

PAYLOAD BAY PASSENGER MODULE

AIR LOCK HATCH

TUNNEL TO PAYLOAD BAY

ESCAPE HATCH

AUX. LO2 TANK

LO2 TANK

LO2 FOR PAX MODULE

PAYLOAD BAY PASSENGER MODULE (SIDE VIEW)

HATCH

PAYLOAD BAY PASSENGER MODULE (FRONT VIEW)

PHASE B
McDonnell Douglas / Martin marietta

ALTERNATE DELTA-WING BOOSTER

LENGTH: 211.4 FEET
SPAN: 150.8 FEET

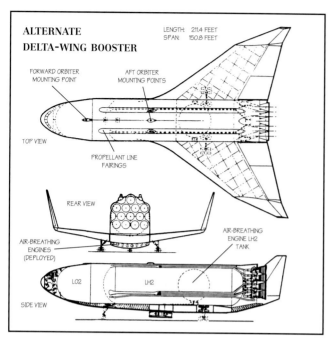

FORWARD ORBITER MOUNTING POINT
AFT ORBITER MOUNTING POINTS
TOP VIEW
PROPELLANT LINE FAIRINGS
REAR VIEW
AIR-BREATHING ENGINES (DEPLOYED)
AIR-BREATHING ENGINE LH2 TANK
SIDE VIEW
LO2
LH2

ALTERNATE FOLDING-WING ORBITER

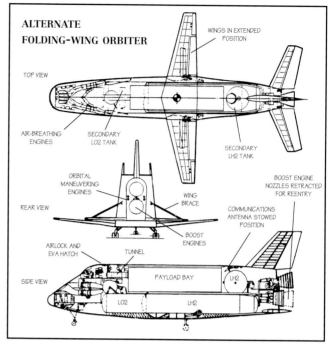

WINGS IN EXTENDED POSITION
TOP VIEW
AIR-BREATHING ENGINES
SECONDARY LO2 TANK
SECONDARY LH2 TANK
ORBITAL MANEUVERING ENGINES
REAR VIEW
WING BRACE
BOOST ENGINE NOZZLES RETRACTED FOR REENTRY
COMMUNICATIONS ANTENNA STOWED POSITION
AIRLOCK AND EVA HATCH
TUNNEL
BOOST ENGINES
SIDE VIEW
PAYLOAD BAY
LH2
LO2
LH2

The folding-wing orbiter was 170.2 feet long and spanned 113.5 feet when the wings were lowered. Total planform area was 5,432 square feet. (McDonnell Douglas)

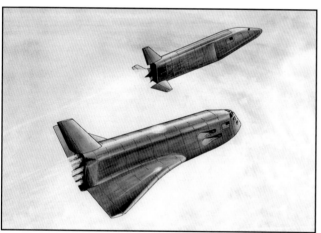

McDonnell Douglas did investigate other boosters during the early part of Phase B, before deciding upon the Martin marietta concept that was baselined. Here is a large delta-wing booster. (McDonnell Douglas via Jason Wentworth)

Another alternate booster, this time a delta-wing with upturned wing tips. Note that there are three launch pads – McDonnell Douglas wanted to build the LC-39C facility that had been part of early Apollo planning. (NASA via Jason Wentworth)

One of the early McDonnell Douglas orbiters being serviced at KSC. There is another shuttle stacked and ready for launch on the pad in the background, and a booster flying overhead as it returns from a mission. As was shown in most of the early illustrations, servicing was very aircraft-like. (NASA photo 70-H-1004)

The Martin-designed booster proposed during Phase B was a derivative of the alternate concept proposed by Martin Marietta during the Phase A studies. This same basic design returned (above) in late 1988 as part of the Martin Marietta proposal for the U.S. Air Force's Advanced Launch System (ALS). (Martin Marietta photo CN3257-88)

North American Rockwell (NAS9-10960)

The team headed by North American Rockwell included the Convair Division of General Dynamics, International Business Machines Corporation (IBM), American Airlines, and Honeywell. Rockwell was primarily concerned with orbiter design and the integration of all the system components. General Dynamics concentrated on the design of the booster, while IBM developed the flight control system for both. American Airlines provided an 'airline' scheduling view of things (after all, this was supposed to be an 'operational' system). On 10 November 1970 they released their baseline configuration for the Phase B competition.[89]

The low cross-range orbiter (model NAR-130-G, and later, NAR-130-J) was true to the MSC baseline Faget-style straight-wing vehicle. The design had a 230 mile cross-range and could carry 45,000 pounds into a near-earth orbit. The body shape and propellant tank arrangements were directed toward design simplicity, ease of manufacture, maximum packaging efficiency, and minimum total weight. A large leading edge fillet at the intersection of wing and fuselage was used to reduce the interference flow between the wing leading edge and body during reentry. The basic load-carrying structure was constructed of 6A1-4V titanium alloy, with a radiant heat shield constructed of various metallic superalloys shingles including columbium-129y (extreme nose), Haynes-188 (forward fuselage), and Inconel-718 (aft fuselage). The nose section had provisions for the crew of four and ten passengers, nose landing gear and vehicle equipment including power supplies and consumables. The passengers were seated on a mid-deck directly beneath the flight deck. The payload bay was 15 feet in diameter and 60 feet long, and was connected to the flight deck via an air lock.

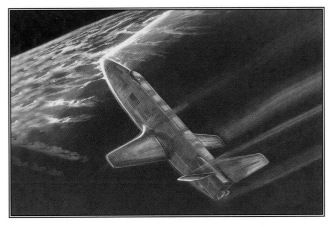

Artist concepts of the two original baseline North American orbiters. The major improvement, over the Phase A design, on the straight-wing orbiter was the slight blending of the main wing into the fuselage. This eliminated a interference that was causing significant heating rates at this location. Note the booster falling away from the delta-wing orbiter, below. (North American Rockwell)

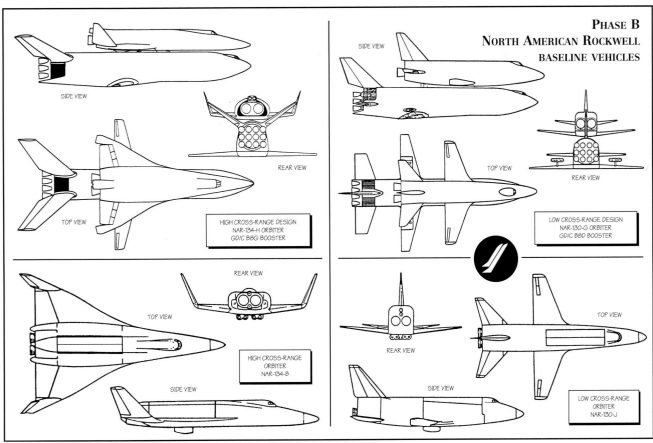

PHASE B
NORTH AMERICAN ROCKWELL
BASELINE VEHICLES

SIDE VIEW

SIDE VIEW

TOP VIEW

REAR VIEW

HIGH CROSS-RANGE DESIGN
NAR-134-H ORBITER
GDIC B8G BOOSTER

LOW CROSS-RANGE DESIGN
NAR-130-G ORBITER
GDIC B8D BOOSTER

REAR VIEW

TOP VIEW

HIGH CROSS-RANGE
ORBITER
NAR-134-B

SIDE VIEW

TOP VIEW

REAR VIEW

SIDE VIEW

LOW CROSS-RANGE
ORBITER
NAR-130-J

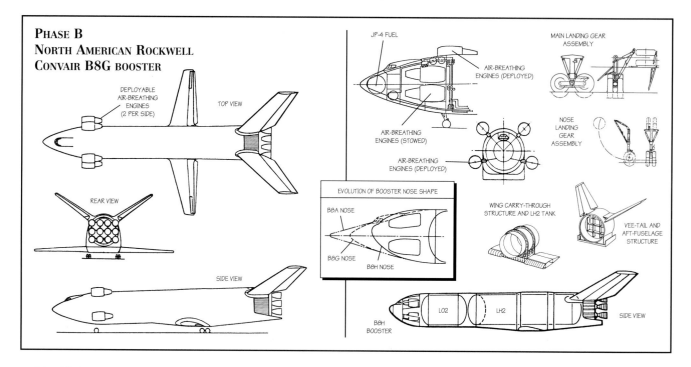

PHASE B
NORTH AMERICAN ROCKWELL
CONVAIR B8G BOOSTER

Liquid oxygen was carried in two floating 2219-T81 aluminum alloy tanks that were circular in cross-section and slightly tapered at the extreme front end. They were un-insulated and located just beneath the forward portion of the payload bay. The wing carry-though structure consisted of front and rear spars and provided sufficient volume for four P&W JTF22B-2* engines installed in rotating nacelles. The engines were to be modified to burn GH2, thus eliminating the need to carry jet fuel, although this violated the new rules. The aft fuselage contained a floating LH2 tank and was constructed mainly from 2219-T81 aluminum, except for the main thrust structure, which was titanium alloy. Structural provisions were incorporated in the fuselage for mounting independently-hinged horizontal stabilizers, the vertical stabilizer, on-orbit propellant tankage, and the main engine thrust structure.

This orbiter was expected to achieve a hypersonic L/D of approximately 0.56:1, and was designed to reenter the atmosphere at an extremely high angle-of-attack (≈60 degrees). At this attitude, a large majority of the aerodynamic heating was expected to be across the base of the vehicle, the upper surface being in the 'shadow' of the pri-

mary shock system. As a result, only limited thermal protection was required on the sides and upper surfaces of the fuselage, wing, and empennage. There was some concern over shock interactions and interferences on the forward portion of the wing, but tests conducted in NASA wind tunnels at Mach 10 found that shock interference effects were only slightly more severe than that observed for delta configurations. The vehicle had an outstanding L/D of 8.2:1 at subsonic speeds, and was expected to land at 155 mph. Of interest was a prediction that the payload capacity of the straight-wing orbiter would fall by 11 percent if the engines delivered a specific impulse one percent under that quoted by the engine interface control document.

The peak radiation equilibrium temperatures for the underside of the straight-wing orbiter entering at a 60-degree angle occurred during the pullout maneuver. It was confirmed during thermodynamic testing that leading edge blending was beneficial in reducing interference heating on the wing and fuselage at high angles of attack. The highest predicted surface temperatures, 3,000 degF, were observed on the wing and horizontal stabilizer leading edges. The fuselage lower surface encountered sustained temperatures slightly in excess of 1,400 degF, while the leading edge of the vertical stabilizer reached 875 degF because of interference flow. These temperatures were considered well within the capacity of existing materials technology to deal with.

The other orbiter was a delta shape (NAR-134-B, and later, NAR-134-H) with the wing tips turned up to act as vertical stabilizers. The orbiter was 192.3 feet long, spanned 126.6 feet, and was 49.3 feet high. The fuselage was divided into three major sections – the nose section, payload bay, and main engine bay. The nose section had sufficient volume for the LH2 tank, crew and passenger compartments, and the nose landing gear. The shape was based on providing the necessary aerodynamic characteristics, resulting in the crew compartment being located well aft of the nose. A floating 2219-T81 aluminum alloy

Convair spent a great deal of time in the wind tunnel trying to determine the optimum configuration for the booster. These models of the near-final delta-wing booster show the various locations for air-breathing engines that were evaluated. Also note the smoothing of the wing-fuselage juncture in the model at top left. (General Dynamics/Convair)

---

*   This was a non-augmented version of the F100/F401 engine then under development by Pratt & Whitney for the Air Force/McDonnell Douglas F-15A Eagle and Navy/Grumman F-14B Tomcat jet fighters.

LH2 tank was located directly aft of the nose landing gear and below the crew compartment. The crew and passengers were accommodated in a single pressurized compartment, with an air lock provided for access to the payload bay or for docking with a space station.

The main LO2 tanks were also constructed of 2219-T81 aluminum alloy, and were located on either side of the payload bay. This provided a relatively stable center-of-gravity location, with only a slight aft travel during burn of the main propellants. The area below the payload bay and LO2 tanks was used for the wing carry-through structure, the air-breathing engine installation, and landing gear. The P&W JTF22B-2 engines were again fueled with GH2 and installed in pairs in swing-down nacelles. The engine location near the vehicle center-of-gravity was planned so that if the requirement for air-breathing engines was eliminated, the engines themselves could be removed with little or no effect on the vehicle's balance. Two 415,000 lbf boost engines were provided side-by-side in the extreme rear fuselage. The empty orbiter weighed 217,732 pounds and a 20,000 pound payload could be accommodated. It was believed that this design could meet the 1,726 mile cross-range capability specified in the RFP. The fuselage, wings, and thrust structure were a hot-structure design constructed of 6A1-4V titanium, with Haynes-188 and Inconel-718 shingles used in areas of extremely high heating.

The high cross-range orbiter's aerodynamic configuration was designed to satisfy criteria based on the anticipated reentry flight profile, and also to provide low drag and aerodynamic stability during the entire flight regime. To achieve low peak heating, it was necessary to provide for low planform loading and the ability to trim over a wide angle-of-attack range. Exploiting this trim capability, the HCR orbiter's baseline entry mode was to enter at a high angle-of-attack (55 degrees) and to pitch down (35 degrees) after peak heating. The vehicle was then banked to achieve the desired cross-range. Transition to a lower angle-of-attack occurred at high supersonic speed, followed by a subsonic approach to the landing site. The HCR orbiter had a projected 0.7:1 hypersonic L/D at a 55 degree angle-of-attack, increasing to 2.2:1 at 20 degrees alpha. The maximum subsonic L/D was 6.9:1. With the vehicle trimmed at a 15 degree angle-of-attack, its landing speed was predicted to be 132 mph.

In its efforts to carry the orbiter to the desired staging velocity, the booster operated as a rocket powered launch

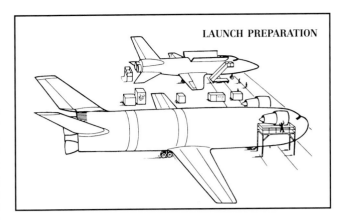

vehicle for approximately three minutes, as a hypersonic glider for about ten minutes, and as a subsonic aircraft for approximately 1.5 hours per mission. The straight-wing booster (Convair model B8G, and later, B8H) used with the high cross-range orbiter spanned 142 feet and was 257 feet long. A vee-tail was incorporated to minimize the aerodynamic effects of the orbiter upon the vertical control surfaces. It is interesting to note that the total empennage area was slightly larger than the total wing area (2,356 square feet versus 2,001), and that the total planform area of 10,760 square feet was provided mainly by the fuselage underbody. The straight wing was chosen primarily for simplicity and low weight, taking into

Artist concept of the final Convair delta-wing booster and North American high cross-range orbiter during ascent. (Convair photo 111124B)

Almost all of the early Phase A and Phase B designs used air-breathing engines during landing operations. This North American design had turbofan engines that swung out of the upper fuselage beside the payload bay. (NASA photo S71-17816)

Convair B8G booster being serviced. Note the air-breathing engines in the deployed position around the nose. The sheer size of the booster was deceiving, although the large trucks in this drawing give some sense of scale. (Convair photo 107824B)

The final North American Rockwell design for Phase B included a Convair B9U delta-wing booster that could be used to launch either the straight-wing low cross-range orbiter (NAR-130-J) or the delta-wing high cross-range orbiter (NAR-161-B). The delta-wing orbiter had given up its turned-up wing-tip vertical stabilizers for a single conventional tail by the final January 1971 variant. (North American Rockwell)

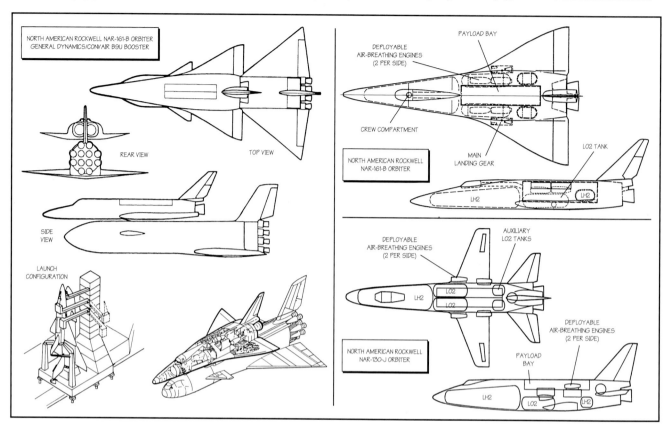

NAR-161-B
HIGH CROSS-RANGE ORBITER

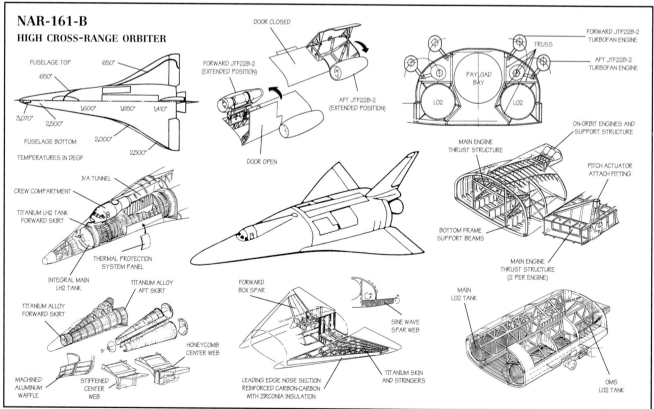

North American Rockwell continued to refine the two orbiter designs until March 1971. The basic layout of the vehicles did not radically change, except for the change from wing-tip vertical stabilizers to a single central dorsal stabilizer on the delta-wing orbiter (above). Even at this point Rockwell had not completely given up on the low cross-range orbiter (next page), and devoted a fair amount of its engineering effort towards the refinement of this design, along with its associated booster, although nobody seriously believed it would ever be built. Note the three on-orbit engines located under the vertical stabilizer on the delta-wing orbiter. (North American Rockwell)

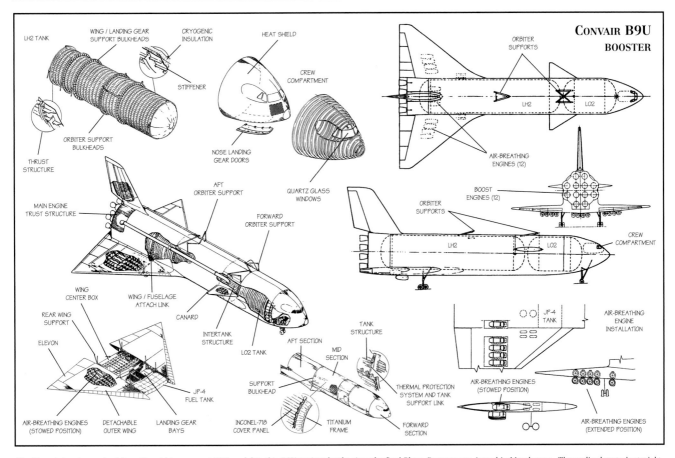

CONVAIR B9U

BOOSTER

LH2 TANK

WING / LANDING GEAR SUPPORT BULKHEADS

CRYOGENIC INSULATION

HEAT SHIELD

CREW COMPARTMENT

STIFFENER

ORBITER SUPPORT BULKHEADS

THRUST STRUCTURE

NOSE LANDING GEAR DOORS

ORBITER SUPPORTS

LH2

LO2

AIR-BREATHING ENGINES (12)

MAIN ENGINE TRUST STRUCTURE

AFT ORBITER SUPPORT

QUARTZ GLASS WINDOWS

FORWARD ORBITER SUPPORT

BOOST ENGINES (12)

ORBITER SUPPORTS

LH2

LO2

CREW COMPARTMENT

WING CENTER BOX

WING / FUSELAGE ATTACH LINK

CANARD

INTERTANK STRUCTURE

TANK STRUCTURE

REAR WING SUPPORT

ELEVON

LO2 TANK

AFT SECTION

MID SECTION

JP-4 TANK

AIR-BREATHING ENGINE INSTALLATION

AIR-BREATHING ENGINES (STOWED POSITION)

DETACHABLE OUTER WING

JP-4 FUEL TANK

LANDING GEAR BAYS

SUPPORT BULKHEAD

INCONEL-718 COVER PANEL

TITANIUM FRAME

THERMAL PROTECTION SYSTEM AND TANK SUPPORT LINK

FORWARD SECTION

AIR-BREATHING ENGINES (STOWED POSITION)

AIR-BREATHING ENGINES (EXTENDED POSITION)

The Convair booster evolved from the mid-term report B8G model to this B9U variant by the time the final Phase B report was issued in March 1971. The earlier booster's straight-wings and unique vee-tail had given way to a large delta-wing with small delta-shaped canards, and a more conventional vertical stabilizer. (Convair via the Bob Bradley Collection)

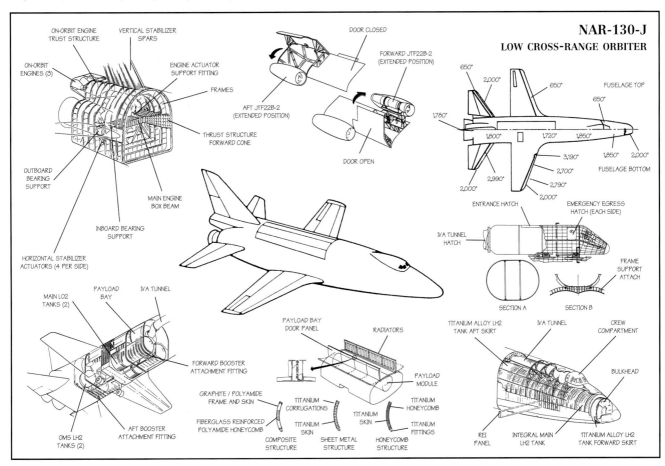

NAR-130-J

LOW CROSS-RANGE ORBITER

ON-ORBIT ENGINE TRUST STRUCTURE

VERTICAL STABILIZER SPARS

DOOR CLOSED

FORWARD JTF22B-2 (EXTENDED POSITION)

ON-ORBIT ENGINES (3)

ENGINE ACTUATOR SUPPORT FITTING

FRAMES

AFT JTF22B-2 (EXTENDED POSITION)

THRUST STRUCTURE FORWARD CONE

DOOR OPEN

OUTBOARD BEARING SUPPORT

MAIN ENGINE BOX BEAM

INBOARD BEARING SUPPORT

650°

2,000°

650°

FUSELAGE TOP

650°

1,780°

1,800°

1,720°

1,850°

1,850°

2,000°

3,190°

2,700°

FUSELAGE BOTTOM

2,990°

2,790°

2,000°

2,000°

HORIZONTAL STABILIZER ACTUATORS (4 PER SIDE)

ENTRANCE HATCH

EMERGENCY EGRESS HATCH (EACH SIDE)

IVA TUNNEL HATCH

FRAME SUPPORT ATTACH

SECTION A

SECTION B

MAIN LO2 TANKS (2)

PAYLOAD BAY

IVA TUNNEL

PAYLOAD BAY DOOR PANEL

RADIATORS

TITANIUM ALLOY LH2 TANK AFT SKIRT

IVA TUNNEL

CREW COMPARTMENT

FORWARD BOOSTER ATTACHMENT FITTING

PAYLOAD MODULE

BULKHEAD

GRAPHITE / POLYAMIDE FRAME AND SKIN

TITANIUM CORRUGATIONS

TITANIUM HONEYCOMB

FIBERGLASS REINFORCED POLYAMIDE HONEYCOMB

TITANIUM SKIN

TITANIUM SKIN

TITANIUM FITTINGS

REI PANEL

INTEGRAL MAIN LH2 TANK

TITANIUM ALLOY LH2 TANK FORWARD SKIRT

OMS LH2 TANKS (2)

AFT BOOSTER ATTACHMENT FITTING

COMPOSITE STRUCTURE

SHEET METAL STRUCTURE

HONEYCOMB STRUCTURE

consideration that the booster's hypersonic characteristics need not be optimized and that the low hypersonic L/D of 0.5:1 was satisfactory since cross-range was not a criterion. The vehicle was expected to achieve a subsonic L/D of 6.7:1, assuring a satisfactory cruise capability. The low cross-range orbiter used an earlier booster design (Convair model B8D) that featured a conventional vertical and horizontal stabilizer arrangement.

Twelve 415,000 lbf boost engines were clustered in three rows of four in the aft fuselage of the booster. Four large air-breathing turbofans were mounted behind swing-away doors around the extreme nose. This location was chosen primarily in an attempt to bring the center of gravity forward and help compensate for the large weight of the rocket engines in the aft fuselage. The propellant tanks were cylindrical 31-foot diameter load-carrying structures with no common bulkheads. The booster reentered the atmosphere at a 52.5 degree angle-of-attack after separation, reaching a maximum load factor of 4-g at 156,000 feet. The maximum dynamic pressure experienced by the booster was predicted to be 189 pounds per square foot. The booster glided until the air-breathing engines were started at an altitude of 22,000 feet, after which it returned 310 miles to its launch site. The booster had an estimated landing speed of 130 mph.

The proposed trajectory for either orbiter/booster combination was optimized to minimize gross lift-off weight. Staging occurred at 230,000 feet altitude and a velocity of 7,000 mph. Orbiter ignition occurred seven seconds after separation, and the orbiter continued to an initial orbit of 52 by 115 miles. The orbiter then coasted to apogee, and transferred itself into a 310 mile circular orbit using minimal fuel. The maximum total ascent time was 24 hours. Either orbiter was capable of remaining docked to a space station for a maximum of five days with its systems powered down. Only the high-cross-range orbiter had a sufficient cross-range capability (385 miles) to guarantee a safe return to the launch site within 24 hours of a decision to reenter, the low-cross-range version having to abort to an alternate landing site. The nominal reentry would require 219 miles of cross-range.

The North American Rockwell final report was issued on 26 March 1971, and although the two orbiters were retained fairly much intact from the November 1970 baseline, the proposed booster had changed radically.

The 269-foot long booster (Convair model B9U) now had a delta-wing that spanned 143.5 feet, a single vertical stabilizer mounted in the conventional location, and the entire booster stood 102 feet high while sitting on its tricycle landing gear. The low-mounted delta wing was situated well back on a basically circular cross-section fuselage, and small delta-shaped canards were set back about 50 feet from the nose. Twelve 550,000 lbf boost engines provided ascent thrust, although these engines were baselined for the ASSC studies and not for Phase B. Twelve P&W JTF22A-4 engines fueled by conventional JP-4 were housed in three swing-down nacelles under the aft fuselage and wings, and provided the capability for the booster to fly back to its launch site.

Changes to the low cross-range orbiter were limited primarily to subsystems and materials, probably due to the fact that it was only being carried as a reference and nobody seriously believed that it would ever be selected for production. The high cross-range orbiter (by now model NAR-161-B) had been refined substantially internally in terms of its subsystems, although the only external change was the deletion of the up-turned wingtips in favor of a more conventional vertical stabilizer mounted on the fuselage centerline.

The Phase B studies were detailed enough to allow Rockwell to build this full-scale crew cabin mock-up at its Downey, California, facility. Note the full-scale mural on the wall, and the scattering of piece-parts around the room. (NASA photo S71-32953)

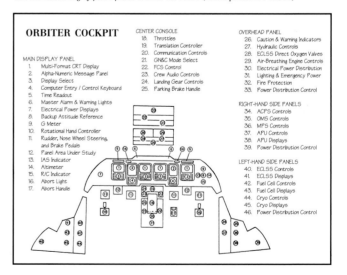

The orbiter and booster shared a common cockpit concept, although the details differed based on the quantity of boost and air-breathing propulsion instruments required. (North American Rockwell)

The relative sizes of the General Dynamics / Convair B8G and B9U boosters is shown in this comparison against a Boeing 747-100, a Boeing 707-320, and a Saturn V. (North American Rockwell)

An artist concept of the North American/Convair vehicle being towed from a hanger next to the VAB at KSC. This was an unusual configuration in that a vee-tail Convair B8G, instead of the more usual B8D, booster is shown with a NAR-130 low-cross range orbiter. Note the relative size of the pickup truck in the foreground – the Phase B designs were truly enormous, especially the booster. (NASA)

The artist was busy. Here a slightly earlier orbiter configuration (note the more pointed nose) is on top of a B8D with the conventional empennage. NASA and the contractors generated literally hundreds of these drawings for the public and to present to the various Congressional committees, etc. to support the yearly budget process. Often the drawings were only vaguely similar to actual configurations. (NASA photo S70-52784)

One interesting item was contained in the 'Facilities Utilization and Manufacturing Plan' that was a required deliverable with the proposal. Convair, who wrote the booster plan, noted that "No existing building in the United States is adequate for Booster final assembly and checkout. Any site will require construction of a new building or substantial modification of an existing building." Convair had narrowed its choices down to three sites – KSC, the Michoud Assembly Facility (where Saturn stages were built and, currently, the External Tank is manufactured), and the Tulsa airport. Tulsa was on the list because of a deal with teammate American Airlines where the airline would build the new building, then lease it to NASA for booster assembly. After the boosters had been built, American would add it to their large maintenance base in Tulsa.

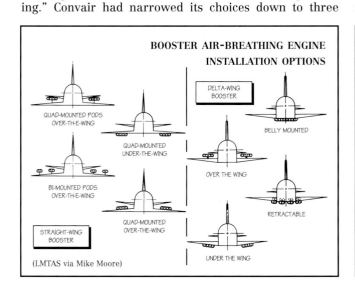

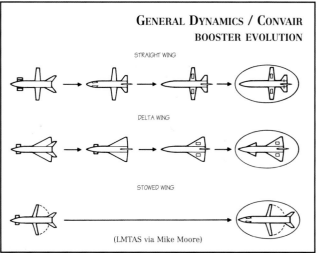

As North American did for the orbiter, Convair built a full-scale mockup of the cockpit of the booster. Note that there is barely a round dial-type instrument in sight – a truly advanced cockpit design for 1970 – and the side-stick controller integrated onto the outer armrest of each seat. (Convair photos 113407B and 113394B)

Reorganization

On 10 June 1971 NASA announced that the overall responsibility for space shuttle program management (including systems engineering and coordination of field center activities) would be assigned to the Manned Spacecraft Center (MSC) in Houston. This took many by surprise since Apollo had been managed by a strong program office based at NASA Headquarters in Washington DC. This decision reflected a conclusion on the part of NASA management that the responsibilities for integration and coordination of the Space Shuttle Program should be carried out directly by a field center that had sufficient technical competence to run the program. And a dynasty at MSC was born under the 'lead center' concept.[90]

Because of their technical capabilities and unique experience in the manned space program, the only two centers seriously considered for this assignment were MSC and MSFC. The decision to give it to MSC was made on the basis that MSC had the most experience with the element of space shuttle expected to be the most difficult to develop – the orbiter. NASA management also felt that by giving MSFC the responsibility for the development of the SSME, along with continuing Skylab responsibilities, the workload balance would be better if MSC received the overall program management assignment. The choice acknowledged that MSC was already responsible for the most visible part of the integrated vehicle – the orbiter. Other Centers' roles became increasingly minor, the exception being MSFC which retained responsibility for all propulsion related tasks.[91]

Concurrent with Phase B, NASA continued in-house space shuttle studies using engineering design support from the MSC, MSFC, and NASA Langley, although surprisingly NASA Ames does not seem to have been involved in any actual vehicle designs. The concept of paying contractors for Phase A and B studies at the same time NASA was conducting in-house studies might appear as unnecessary duplication. But part of NASA's rationale for continuing in-house work was that such an effort helped improve the technical competency and expertise of its own staff, placing them in a better position to review and monitor the activities of the contractors. As a general rule, NASA did not apply in-house results directly against the contractor's efforts, primarily because the agency considered them more of an internal learning curve instead of a true basis for comparison.[92]

By late 1969 the Air Force was expressing a decided preference to construct the orbiter using a conventional aluminum airframe with whatever thermal protection system was required to protect it. This was based as much as not on an anticipated titanium shortage, as well as a shortage of machine tools required to manufacture large pieces out of titanium alloys. Therefore, trade studies were conducted by NASA and several contractors to determine the impact to the space shuttle program if it was decided to construct an aluminum orbiter. It was estimated that the combined weight of the structure and thermal protection system would be approximately 15 percent less for titanium than for aluminum as the primary structure. However, the cost for the titanium structure was approximately 300 percent greater, and posed a potentially greater production development risk.[93]

Most of the proposed titanium vehicles had relied on 'hot-structure' designs similar to what John Becker at NASA Langley had designed for the X-15 research airplane. Where temperatures got excessive, this hot-struc-

ture was covered by metallic superalloy shingles over batt insulation materials. But hot-structures and metallic shingles had drawbacks – the structures were complex, and the high-refractory alloys such as columbium and molybdenum oxidized readily at high temperature and required coatings to resist this. Nevertheless, there appeared to be little alternative.

A solution began to emerge as NASA and Lockheed succeeded in developing a new thermal insulator made of interlaced fibers of silica that could be molded into 'tiles' that could be glued to the outside of a vehicle. Preliminary work showed that these tiles could withstand temperatures of 2,500 degF, making them suitable for everything but the extreme nose and wing leading edges. The outer surface of these tiles radiated heat way from the surface, much like the metal shingles, but the thickness of the tile kept the structure underneath cool enough that it could be made from aluminum. Since it was essentially a ceramic, the tiles did not oxidize and were also very light weight – as little as 15 pounds per cubic foot. Early versions of these tiles had already been proposed various times, but were always thought to carry a high risk since they were still little more than a laboratory experiment. This perception would rapidly change.[94]

The studies concluded that the thermal protection system of a titanium orbiter would weigh between 5,000 and 10,000 pounds less since it could endure heat loads* as high as 850 degF, as opposed to 300 degF for aluminum. However, an aluminum orbiter would be approximately 4,000 pounds lighter to begin with since its basic airframe was lighter than an equivalent titanium structure. Because there was less TPS on the titanium orbiter, it was expected to cost less to refurbish between missions, but preliminary cost data showed that the aluminum orbiter could be constructed for $80 million less. These various studies showed that there was virtually no meaningful difference between the orbiters, and this led most of the contractors to propose aluminum orbiters in the later Phase B extension and ASSC reports

As early as the end of 1969 when George M. Low, former director of the Apollo spacecraft design effort, had become Deputy Administrator of NASA. there were rumblings that the space shuttle was too expensive. Although fully behind the concept of a shuttle, Low cautioned that while fully reusable systems might be more economical to operate, their associated technical risks and multiple parallel development efforts all mitigated in favor of smaller, partially reusable systems. There were also growing external pressures, having to do more with economic realities than technical risks, that conspired against the fully reusable systems. Belatedly, NASA was finally being forced to listen – at least a little.

Alternate Concepts

One of the great drawbacks to fully reusable configurations, such as the abortive Aerospaceplane, had been the extremely high RDT&E costs inherent in their development and procurement. Because of this trend, maneuvering orbital reentry vehicle designs had always tended to be small, as evidenced by Dyna-Soar. Thus large concepts such as Aerospaceplane surprised nearly no one, particularly R&D pragmatists, when they fell by the way-

---

* This not only reduced the amount of area that needed protected by the TPS, but also reduced the required effectiveness of the remaining TPS since more heat could be absorbed by the orbiter's airframe.

side. In their 1968 AIAA research paper, AFFDL's Cosenza and Draper had argued convincingly that such partially expendable concepts as stage-and-a-half would make larger orbiters more attractive, freeing up internal volume for payload, and thus not requiring the development of vehicles large enough to carry their engines, fuel, and payload all internally within a fully reusable structure.

Almost concurrently with the issuance of the Phase B contracts, NASA decided to look into partially reusable concepts. Since this mainly involved replacing the booster element, MSFC took the lead. The alternate space shuttle concepts (ASSC) study was initiated on 6 July 1970 in parallel with the Phase B studies of fully reusable shuttle configurations. The intent of the ASSC study was to define and investigate various alternatives to the fully-reusable Phase B concepts so that meaningful comparisons of technical concerns and issues, operational approaches, and cost factors could be made. It also allowed NASA a fallback position if space shuttle funding was reduced, which was looking more and more likely.

MSFC issued $1.9 million study contracts to the Chrysler Space Division and the Lockheed Missiles and Space Company – subsequently, MSC issued a $4 million ASSC contract[*] to a combined Grumman/Boeing team. The initial matrix of alternate concepts to be studied consisted of 29 configurations in three general categories: 'fractional' stages (stage-and-a-half), semi-reusable (expendable booster/reusable orbiter), and two-stage fully reusable alternatives similar to the Phase B vehicles. MSFC also baselined a different LO2/LH2 engine for the ASSC studies, essentially an evolution of the original Phase A engine. Producing 550,000 lbf at sea level using a variable-position bell-nozzle, this engine was also used by North American in some of their Phase B concepts. Chrysler, however, did not feel bound by the engine choices specified by MSFC, and pursued a unique propulsion system (interestingly, with the help of Rocketdyne, which was one of the SSME competitors).[95]

Ultimately, ASSC concluded that the fully reusable TSTO vehicles being developed under Phase B were the 'best' since they gave the greatest assurance of achieving low cost per flight into orbit, and would have the lowest 'total' program[+] costs. However if peak yearly funding and/or development risk was to be minimized, consideration should be given to a phased development approach of a reusable orbiter with an expendable booster with the option of developing a recoverable booster later.

Chrysler (NAS8-26341)

Chrysler proposed what was surely the most unique alternate shuttle concept. Looking like an overgrown Apollo command module, SERV (single-stage earth-orbital reusable vehicle) had a diameter of nearly 90 feet and was 66.5 feet high. A payload bay measuring 23 feet in diameter and 60 feet in length could hold 116,439 pounds destined for a 115 mile orbit at a 28.5 degree inclination.[96]

Although it seems odd today, Chrysler was deeply involved in the space program during the late 1960s. The

Chrysler Corporation Space Division (CCSD) was formed in 1962 to manufacture the first stage for the Saturn I. Total production was two Saturn I boosters, and 12 improved Saturn IBs used during the Apollo program.

The CCSD study was headed by Charles E. Tharratt, with significant help from William R. Bladwin, John H. Wood, and Arthur P. Raymond, III. Feeling that they were lacking some key technical expertise, CCSD brought in several companies from outside Chrysler for assistance. Robert E. Schnurstein represented Rocketdyne, which was subcontracted to provide engineering support on the development of the unique propulsion system proposed for SERV. In addition, the Detroit Diesel Allison Division of General Motors provided support in the form of parametric engine data for advanced technology direct-lift gas turbine engines. The thermal protection system was devised by the Systems Division of AVCO.

The SERV concept had actually begun during early 1969 and placed "... special emphasis on operational flexibility and reduced dependence on uncertain technological advancement in a near-term development environment." Chrysler considered SERV to be "a logical extension of the Apollo program and draws on the accumulated knowledge and experience that currently exists ..."

Chrysler elected to propose a truly unique propulsive system – the 12-module LO2/LH2 integral aerospike engine was 87.4 feet in diameter, 8.2 feet long, developed 5,400,000 lbf, and a had specific impulse of 346.7 seconds. The engine was fully integrated into the SERV airframe, forming the base closure of the vehicle and using the entire base of the vehicle. Each of the 12 interconnected modules contained a set of two-stage turbine-driven pumps connected to a common manifold on the high-pressure side of the pumps. In this manner if one turbopump should fail during the boost phase, the remaining 11 turbopumps could be run at greater than 100 percent

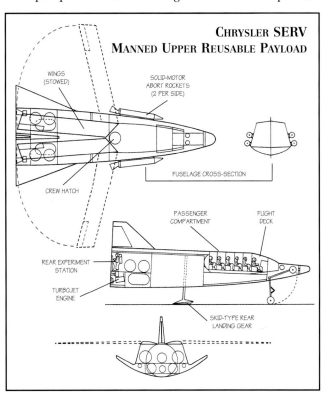

Chrysler elected to use a modified version of a NASA Langley-designed (and modified by both McDonnell Douglas and North American) orbiter as the manned upper stage for the SERV concept. The basic SERV vehicle was unmanned. (Chrysler Space Division)

---

* It is difficult to distinguish if this was actually an ASSC contract, or simply a late Phase B award. The documentation shows it was under the auspices of ASSC, but Larry Mead, the Grumman Proposal Manager, indicates it was actually closer in concept to a Phase B effort, although this could be explained by the fact that MSC was running it as opposed to MSFC. The significant funding attached to the contract, however, makes it appear to be a late Phase B award.

† Total program costs included research, development, test, production, and all recurring operational costs.

speed (they were designed for limited operations at 120 percent) and feed all twelve modules. The engine was capable of being throttled back to 18 percent of its nominal thrust, and in fact, due to its high specific impulse, it had to throttle back to 20 percent just prior to 'max-q' to avoid over-accelerating the SERV.

Protective doors covered the aerospike ramp during reentry operations. During initial tests these doors did not behave as predicted, and were deleted during subsequent ground tests. Since the exact cause of their initial failure could not be adequately determined in the short time available to the test team, it was felt that the doors might represent a development risk, although it was not thought to be great. Initial test data indicated the doors had a detrimental impact on nozzle efficiency, but further analysis concluded that a properly designed door had the potential for a significant improvement in installed nozzle efficiency.

The engine had been developed by Rocketdyne during their aerospike work prior to the technology being banned from space shuttle applications. A prototype of the aerospike engine integrated into a 27-inch diameter SERV model was tested at the Arnold Engineering Development Center during December 1970, and the test data indicated that the performance generally matched predictions, and no major development risks were expected by Rocketdyne. However, this was precisely the type of engine that MSFC was trying to avoid developing since the NASA thought that the inherent risks to the design were greater than available funding and time would allow. Nevertheless, Rocketdyne continued to refine this engine even after they began pursuing normal bell-nozzle concepts for the space shuttle main engine proposals. A renewed interest in the basic concept of the aerospike engine was expressed by several Advanced Launch System (ALS) contractors in the late 1980s, and a variation of the concept – the linear aerospike – was selected

to power the Lockheed Martin X-33 reusable launch vehicle technology demonstrator and the proposed VentureStar™ in the 1990s.

Chrysler continued to develop the SERV concept at a low level during the Phase A and Phase B space shuttle studies, although this was apparently done using only corporate funds since no record of a government contract could be found. When NASA decided to issue the ASSC studies, one was awarded to Chrysler since SERV was definitely an 'alternate' concept.

The gross lift-off weight of SERV was 4,500,000 pounds, including 116,439 pounds of payload, but not including the MURP or external payload containers. A total of 45,000 pounds of cargo could be returned from space. Twenty-eight JP-4 fueled turbojet lift-engines provided deceleration and landing propulsion. The SERV booster could be launched unmanned into orbit to be unloaded by an already orbiting 'space-tug,' or could carry its own winged piloted spacecraft. The spacecraft was a version of a shape developed at NASA Langley, then modified by both McDonnell Douglas and North American over the course of the space shuttle studies. The version proposed by Chrysler could be configured to carry cargo or up to ten passengers, or some combination of both, and looked much like the one proposed by McDonnell Douglas to MSFC in August 1965 (see page 65). Interestingly, however, the basic design came via North American, who assisted Chrysler during the ASSC studies.

SERV would have been launched from LC-39 facilities that were built to support Project Apollo (and currently used by Space Shuttle). Minor modifications to the VAB and mobile launch platform (MLP) were required, as was a totally new launch umbilical tower (LUT). The VAB would be used for servicing SERV and MURP, including the installation of cargo. SERV would not use the two existing launch complexes, being launched instead from

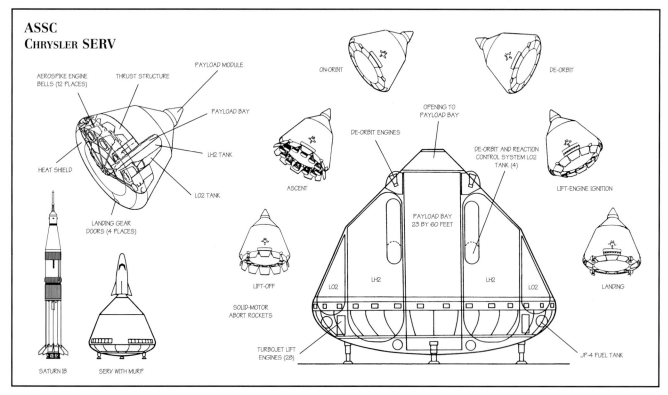

ASSC
CHRYSLER SERV

The basic SERV vehicle could accomplish many missions, depending upon its configuration. A large payload bay internal to SERV could be supplemented by either small or large cargo containers attached to the vehicle's nose. These containers were intended for odd-sized payloads, or payloads containing hazardous materials that were undesirable to carry internally. The aerospike engine was integrated into the outer perimeter of the vehicle. Note the size comparison with the Saturn IB. (Chrysler Corporation)

two large (and mostly empty) concrete pads located 1.75 miles from the VAB along the crawlerway.

Upon command to return to earth, SERV would reorient itself into the familiar Apollo command module reentry attitude, and enter the atmosphere semi-ballistically. With SERV, however, the targeted landing point was designed to be land instead of water. In fact, it was to be the same pads used for launch. Using existing Apollo-style reentry navigation and guidance technology, it was projected SERV would be within four miles of the landing pad when it reached 25,000 feet. At that altitude, air intakes and doors for exhaust efflux would open, and the four groups of seven air-breathing lift engines would ignite, providing SERV with a soft landing. MURP landings would occur at the 10,000-foot 'skid' strip on the Cape Canaveral Air Force Station adjacent to KSC.

Construction of SERV would have been at the Michoud Assembly Facility (MAF) used to construct the Saturn V first stage. The vehicle would then be transported by a Bay-class vessel owned by the West India Shipping Company that was 266 feet long with a 51-foot beam and a six-foot draft. 'Pin-on' type sponsons would increase the beam to 89 feet, enabling it to carry SERV and still pass through the locks and bridges between Michoud and KSC with no route modifications.

Plans called for a 12 month preliminary design effort, followed by ten months of detailed design. One structural test article would be constructed, and 20 months of static testing would be performed, along with a single propulsion test article that was being tested concurrently. Three flight test vehicles would then be procured. Total development time would be four years, and cost $3,565 million. Each production flight vehicle would cost $350 million in FY71 dollars, and have a 10 year or 100 flight life. This proposal received little publicity, and even less official support, perhaps straying a little too far from the established norm to be seriously considered by the government. Nevertheless, it provided a brief look at an interesting alternative.

Various details from the SERV concept from 1969 through the Alternate Space Shuttle Studies of 1970–71. (Chrysler Corporation)

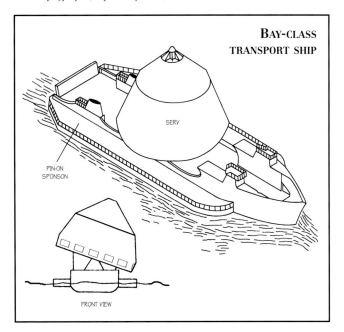

BAY-CLASS
TRANSPORT SHIP

SERV

PIN-ON
SPONSON

FRONT VIEW

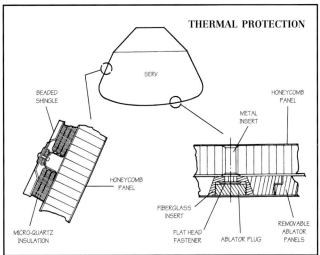

THERMAL PROTECTION

SERV

BEADED
SHINGLE

HONEYCOMB
PANEL

MICRO-QUARTZ
INSULATION

HONEYCOMB
PANEL

METAL
INSERT

FIBERGLASS
INSERT

FLAT HEAD
FASTENER

ABLATOR PLUG

REMOVABLE
ABLATOR
PANELS

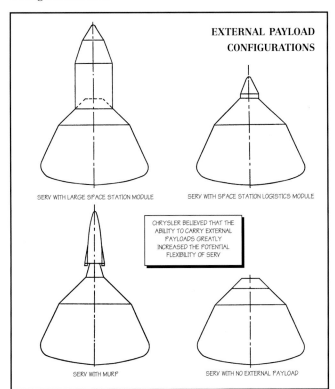

EXTERNAL PAYLOAD
CONFIGURATIONS

SERV WITH LARGE SPACE STATION MODULE

SERV WITH SPACE STATION LOGISTICS MODULE

CHRYSLER BELIEVED THAT THE ABILITY TO CARRY EXTERNAL PAYLOADS GREATLY INCREASED THE POTENTIAL FLEXIBILITY OF SERV

SERV WITH MURP

SERV WITH NO EXTERNAL PAYLOAD

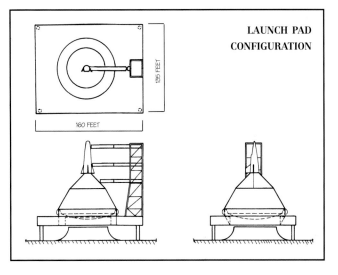

LAUNCH PAD
CONFIGURATION

135 FEET

160 FEET

There were several locations proposed to launch SERV. The favored was a set of new austere pads located 1.75 miles from the VAB. The one illustrated above was modified from the existing LC-39 complex, including a possible new LC-39C facility. (Chrysler)

Lockheed (NAS8-26362)

The Lockheed Missiles and Space Company ASSC study was released on 4 June 1971. The concept that Lockheed expended roughly 75 percent of its efforts on was a stage-and-a-half vehicle that could later be converted for use in a fully reusable two-stage system. In fact, four different configurations were evaluated – the LS-200-10 was a true stage-and-a-half vehicle, the LS-200-8 and LS-400-5 were stage-and-a-half vehicles that could be converted to two-stage, and the LS-400-7A was a true two-stage vehicle. The stage-and-a-half concept was based largely on the results of the 1968 ILRV study done for the Air Force Flight Dynamics Laboratory. In fact, the stage-and-a-half concept presented in the ASSC study had been under continuous development at Lockheed for more than five years and had accumulated more than 3,700 hours of wind tunnel testing, with almost 2,800 hours of that being directed towards understanding the aerodynamics of the vehicle, and the rest dedicated to thermodynamic testing.[97]

The stage-and-a-half vehicle was a direct descendant of the original LSC-8MX vehicle and consisted of a 156.5-foot long, delta-body orbiter with two large, expendable drop tanks. The orbiter contained all main propulsion systems, some ascent propellants, the payload, crew, and all supporting systems. The expendable drop tanks, 192 feet long, with a diameter of 27 feet, contained most of the ascent propellants (both LO2 and LH2) that were cross-fed into the orbiter main engines. There were no systems within the drop tanks except for propellant instrumentation, and external insulation. It should be noted that there was <u>no</u> booster, the orbiter being launched vertically under its own power.

The perceived advantages to stage-and-a-half were:

- only one complex vehicle (the orbiter) needed developing, resulting in lower development costs, lower total program costs, lower annual peak funding, and low sensitivity of program costs to vehicle changes,
- only one flight test program, crew training program, etc. would be required,
- safety would be enhanced by virtue of there being only one crew, and the fact that all engines would be ignited and checked out prior to lift-off,
- no advanced aerodynamic parallel staging technology development was required,
- the system could be converted to a true two-stage system, allowing a future reduction in recurring costs.

The only significant advantages the LS-400-7A two-stage vehicle had over the stage-and-a-half concept were lower recurring costs due to the use of all fully reusable elements, and no launch azimuth restraints as long as both stages had an intact abort capability. Since the technical feasibility of the two-stage concept had been established in earlier studies, and basic design issues were not central to the ASSC study, the approach taken by Lockheed for the two-stage design was to rely upon sizing and design layout data for use in cost and performance comparisons with the stage-and-a-half concept. Lockheed felt that developing the stage-and-a-half vehicle would lower all development* risks, and their associated costs,

---

* NASA did not necessarily share this point of view. It was felt that the tank-orbiter aerodynamic integration would be very critical and could consume considerable time and money to get right.

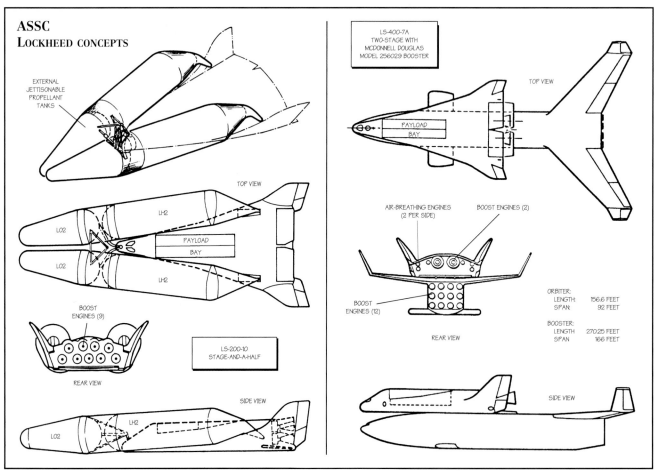

ASSC
LOCKHEED CONCEPTS

EXTERNAL JETTISONABLE PROPELLANT TANKS

TOP VIEW

LH2
LO2
PAYLOAD BAY
LO2
LH2

BOOST ENGINES (9)

REAR VIEW

LS-200-10
STAGE-AND-A-HALF

SIDE VIEW

LH2
LO2

LS-400-7A
TWO-STAGE WITH
MCDONNELL DOUGLAS
MODEL 256029 BOOSTER

TOP VIEW

PAYLOAD BAY

AIR-BREATHING ENGINES (2 PER SIDE)
BOOST ENGINES (2)

BOOST ENGINES (12)

REAR VIEW

ORBITER:
LENGTH   156.6 FEET
SPAN   92 FEET

BOOSTER:
LENGTH   270.25 FEET
SPAN   166 FEET

SIDE VIEW

---

thereby allowing NASA to build a usable space shuttle within the available budget, and at the same time reserve the option for a fully reusable vehicle when funding could be found to develop the booster.

This concept was to develop the drop tanks and operate the orbiter as a stage-and-a-half vehicle before developing a reusable booster and converting the orbiter to a two-stage vehicle. The concept could be accomplished with either a low or high degree of commonality, although Lockheed emphasized a design with maximum commonality and minimum conversion cost. The expected advantages to this approach were that peak annual funding would be reduced since only one stage would be in development at any one time, the ultimate achievement of a fully reusable two-stage system would be realized and with it the attendant low recurring program cost. At the same time, an early operational capability as a stage-and-a-half would be retained, and the delay of the final booster development permitted the incorporation of proven orbiter technology and flight data in an advanced booster design with improvement in the overall system.

The LS-200 stage-and-a-half orbiter was a delta-body having high volumetric efficiency, large leading edge radii, and a steeply sloped upper forward body to provide satisfactory pilot visibility. Comparison of the proposed delta-body orbiter to a wing-body vehicle showed that the LS-200 was 20,000 pounds lighter, and that it would save $500 million in total program costs. The delta-body represented the lowest aerothermodynamic development risk and could accomplish over 1,726 miles of cross-range. The sloped vertical stabilizers provided directional stability, longitudinal stability, and increased subsonic and hypersonic lift. Avionics equipment was to be located in the extreme nose, followed by the crew compartment which was joined to the 15 by 60 foot payload bay by a crew transfer tunnel. Payload erection and deployment equipment was located at the forward end of the payload bay.

Six P&W JTF22A-4 air-breathing engines were located in the bottom of the vehicle, and deployed downward into the airstream for subsonic flight, landing, and ferry missions. Nine of the ASSC baseline 550,000 lbf boost engines were installed at the extreme rear of the vehicle.

The orbiter airframe was a conventional aluminum stringer and bulkhead structure attached to the thrust structure. It could be disassembled for complete removal and replacement of the non-integral internal propellant tanks. The internal propellant tanks were arranged in two sets, with the LO2 tank aft and the LH2 tank forward. The titanium thrust structure transmitted the main engine thrust loads directly to the orbiter propellant tanks, the drop tanks, and the aft orbiter airframe shell and stabilizers. The upper surface was a titanium hot-structure capable of temperatures up to 1,000 degF. The lower surfaces, stabilizers, base heat shield, and leading edges would be protected with LI-1500 silica tiles. The nose cap was to be coated tantalum alloy and silicone capable of sustaining temperatures in excess of 3,000 degF. This was one of the few instances of proposing to use the Lockheed-developed tiles on leading edge surfaces – the material would ultimately prove much too fragile for this application. Total gross lift-off weight was estimated at 3,816,420 pounds.

Lockheed anticipated using a slightly modified version of the launch umbilical towers built during Apollo, to service the orbiter on the launch pad. All nine engines were ignited on the pad at lift-off. Staging occurred at 12,250 mph with only three engines burning, producing near zero dynamic pressure (zero-q). The external tanks were separated by releasing the forward attach pin, and letting aerodynamic forces carry the front of the tanks up

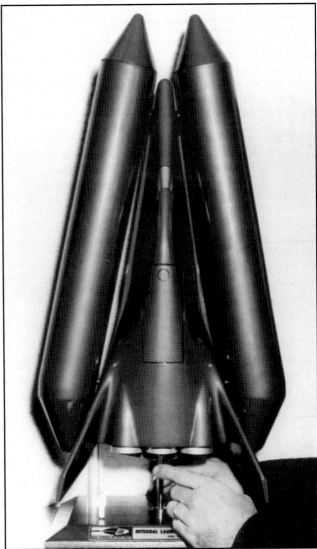

The LS-200-8 was a stage-and-a-half vehicle that could be converted to a two-stage system. (Lockheed Missiles and Space Company)

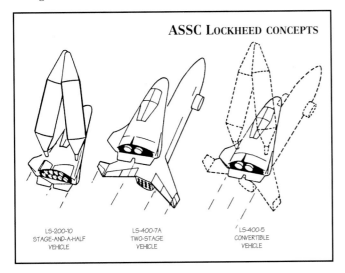

ASSC LOCKHEED CONCEPTS

LS-200-10
STAGE-AND-A-HALF
VEHICLE

LS-400-7A
TWO-STAGE
VEHICLE

LS-400-5
CONVERTIBLE
VEHICLE

The ASSC guidelines had asked the contractors not to conduct detailed subsystem design unless that subsystem was unique to the alternate concept. Therefore Lockheed did not design a booster, relying instead on the vehicle being proposed by the Phase B team of McDonnell Douglas and Martin Marietta. This shows the relative merits of the three LMSC concepts. (Lockheed Missiles and Space Company)

LOCKHEED
ALTERNATE ORBITER DESIGNS

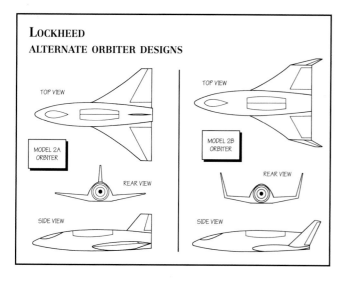

Various other designs were investigated by Lockheed, including the Model 2A and Model 2B orbiters. Alternate boosters were also considered such as a Saturn S-IC stage. (Lockheed Missiles and Space Company)

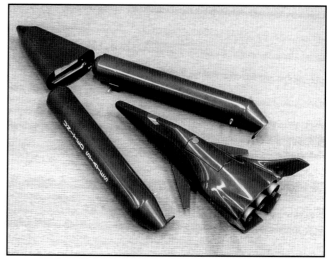

The Lockheed external tank concept lived a long life – this model was dated 1974, although exactly why it was built and for whom remains unknown. Since it does not have Air Force markings, it was probably a NASA study. (Lockheed via the SDAM Collection)

and over the orbiter, pivoting on the rear attach points. When the tanks were roughly vertical to the orbiter, the rear attach points (which also contained the propellant feed lines) would be released, and the orbiter pitched slightly downward, allowing the tanks to pass over the orbiter and burn up in the atmosphere. Total tank separation time was roughly 10 seconds. The orbiter continued using internal propellants until sufficient velocity was achieved to place it into the desired orbit.

The directions of the ASSC studies indicated the contractors were not to do detail subsystem design unless the subsystem was unique to the alternate concept. Therefore, Lockheed did not spend much effort to determine the booster configuration that would be used to launch the orbiter if it was upgraded to a true two-stage system. The booster used in the study was a scaled-up version of the Phase B configuration being proposed by McDonnell Douglas and Martin Marietta. The Martin booster appealed to Lockheed because of its wide-spread vertical stabilizers that would allow the orbiter external tanks to be jettisoned if required. Additionally, the general design of the booster was felt to offer more freedom in potential designs of the orbiter since it was intended to

launch a variety of vehicles, not just the orbiter being proposed by McDonnell Douglas. It was expected that the Lockheed orbiter would be attached in the same location as the Phase B orbiter.

For use in the two-stage system, the Lockheed orbiter would not carry its external tanks – seven of the nine rocket engines, along with two air-breathing engines would be removed. Internal propellant tankage would be increased to feed the remaining engines during second stage ascent, and the configuration of the internal tanks would be reversed to assist in controlling center-of-gravity shifts. The stage-and-a-half orbiter would carry an 11,500 pound weight penalty; the result of a thrust structure designed to handle nine main engines, instead of the two required for a two-stage system. This limitation could be overcome by redesigning the thrust structure in the event additional new orbiters were built.

The baseline included the production of two ground test orbiters and three flight test orbiters, which would be carried over into the operational phase. Three additional production orbiters would also be procured. Cumulative program costs for the stage-and-a-half system, including main engines and facilities, was expected to be $8,000 million with the first orbital mission occurring in April 1978. The expendable drop tank cost accounted for 24 percent of the total program cost. The average recurring cost for 445 flights was estimated at $7.1 million per flight, with the 446th flight costing $6.3 million. Peak annual funding was estimated at $1,160 million.

Lockheed considered the stage-and-a-half concept to be safer to develop and operate, and thought that it could be available for operations earlier than a two-stage system. It also avoided many of the technical and management problems associated with a two-stage system. As an example, Lockheed cited that there would be no need for commonality management or technical compromise. Additionally, there would be no complex interactive load path, aerothermodynamic, performance, or launch facility interfaces between stages.

LOCKHEED LS-400-7A

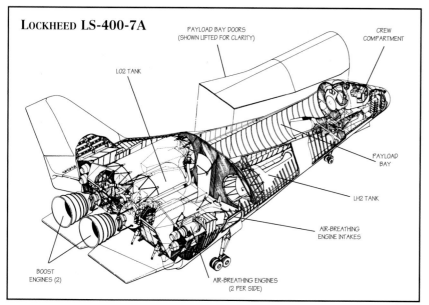

The inclusion of the Grumman/Boeing team in the ASSC studies has always been the subject of some confusion. Part of this stems from the fact that their contract was issued by MSC, not MSFC – the contract was also for more than twice the amount of the other ASSC contracts. In many respects, it appears that MSC decided to issue a third Phase B contract* and chose to use the ASSC study to accomplish it. The initial phase of the Grumman/Boeing study was directed toward definition and comparison of promising alternatives to the fully reusable concept including stage-and-a-half, expendable first stages, and variants of the two-stage reusable system. To accomplish this, Grumman conducted a limited design effort on two orbiters that roughly resembled the Phase B baseline vehicles, one a low cross-range, straight-wing orbiter, and the other a high cross-range delta-wing vehicle. Larry Mead was in charge.[98]

One notable item about the Grumman team was the inclusion of two European companies with significant roles. The Dassault company of France shared responsibility for the thermal protection system with AVCO, and Dornier of Germany would provide many of the orbiter subsystems. Other teammates, besides Grumman (orbiter and overall management) and Boeing (booster and operations), were Aerojet General (cryogenic tankage), AVCO (thermal protection), Eastern Airlines (operations and maintenance), General Electric (avionics), and Northrop ('orbiter technology' … probably advanced structures).

On 23 December 1970 the Grumman/Boeing ASSC team received MSC approval to conduct parallel studies of reusable two-stage configurations using externally mounted orbiter LH2 tanks. Design reviews during January and March 1971 indicated significant weight, development, and cost advantages for the external LH2

---

* Grumman would also participate in the Phase B prime and double prime studies, and even called them such.

---

tank orbiter and heat-sink booster concept. During a subsequent study effort, both two and three-engine orbiters embodying the external LH2 tank/heat-sink booster design were compared to a representative internal LH2 tank configuration. The results of these studies were reviewed by NASA during April 1971, and as a result, Grumman and Boeing were authorized to conduct a much more detailed study of a three-engine external LH2 tank orbiter/heat-sink booster vehicle. During this detailed study, the Grumman/Boeing team published 82 trade study reports, performed over 1,800 hours of wind tunnel testing on key configuration features, held 25 formal technical interchange meetings with NASA, and conducted 280 informal briefings.

The idea behind the external tank design was to shrink the size of the orbiter by moving the bulky LH2 tanks outside its airframe, thus reducing the physical size and weight of the orbiter as much as possible. This had the effect of altering the energy balance between the orbiter and booster from what was considered an acceptable staging velocity by the MSFC engineers. Instead of the 6,800 mph staging velocity that NASA desired, Grumman concluded the external tank orbiter should stage at, or slightly below, 4,750 mph. This was driven primarily by the fact that the orbiter had a significantly improved mass fraction and could produce more of the total ΔV, and that the booster could in turn be made smaller since it would have to carry less jet fuel for its flight back to the launch site.

Early in the study every attempt was made to use the baseline 550,000 lbf main engines specified by MSFC for the ASSC studies. Since slightly over 1,000,000 lbf was needed by the orbiter, this resulted in a two engine configuration dubbed H-32. It was quickly discovered during detailed analysis that abort constraints (abort-once-around in the event of a single engine failure) would force the staging velocity back up to 6,250 mph since lowering it resulted in the need for more OMS propellants, which increased the weight of the orbiter dramatically. When the weight of the orbiter increased, it increased the rela-

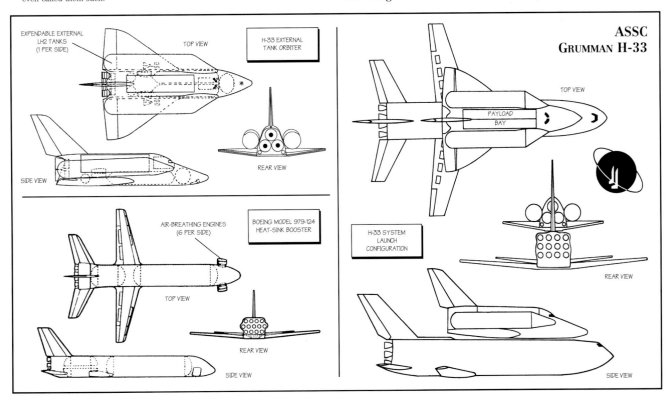

A very early Grumman high cross-range orbiter from October 1970. Compare with the orbiter at right – what a difference a month makes. This orbiter was a true wing-body instead of the pseudo-lifting-body Model 518. (NASA photo 70-H-1255)

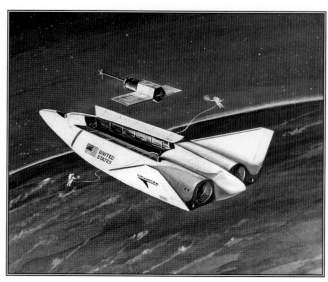

Noteworthy on this early (November 1970) high cross-range orbiter is the large 'boat-tail' between the boost engines, a feature that would probably have caused serious exhaust impingement problems. (NASA photo S70-51684)

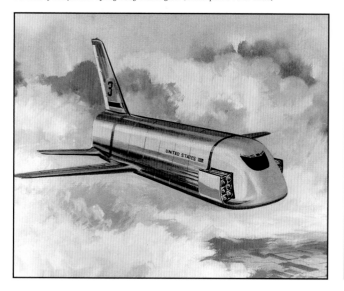

The Boeing H-33 booster flies back to its launch site after orbiter separation. Noteworthy are the six Pratt & Whitney JTF22A-4 air-breathing engines on either side of the nose in their extended position. (Grumman Aerospace Corporation)

Separation. However, this is a bit higher and calmer than separation would probably have been in the real world. Note that neither the booster nor orbiter engines are thrusting, and no reaction control jets are firing either. (NASA photo 70-H-1260)

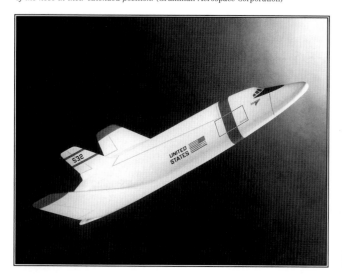

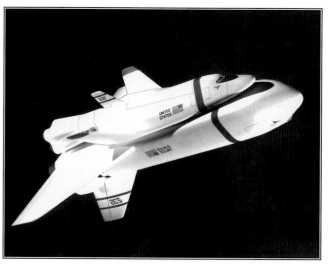

The Model 532 high cross-range orbiter was a minor evolution of the Model 518 vehicle. Noteworthy is the 'transtage' attached to the rear of the orbiter while it is still mounted to the booster. The booster appears to be almost identical to the Model 518 booster except for the bulge in the upper rear fuselage, indicating that the propulsion system may have changed. (Grumman Aerospace Corporation)

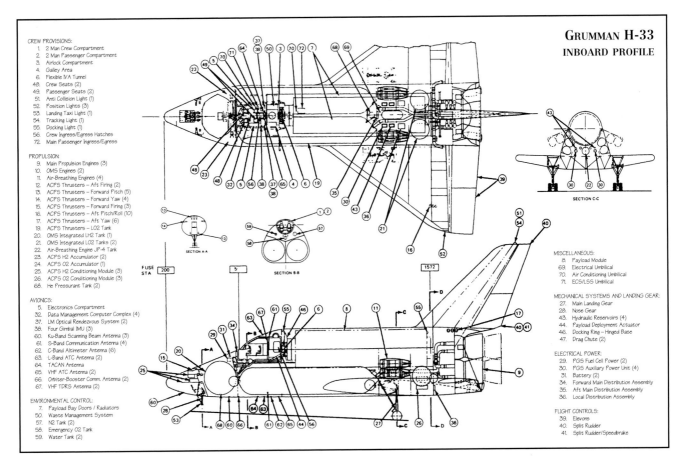

CREW PROVISIONS:
1. 2 Man Crew Compartment
2. 2 Man Passenger Compartment
3. Airlock Compartment
4. Galley Area
6. Flexible IVA Tunnel
48. Crew Seats (2)
49. Passenger Seats (2)
51. Anti Collision Light (1)
52. Position Lights (3)
53. Landing Taxi Light (1)
54. Tracking Light (1)
55. Docking Light (1)
56. Crew Ingress/Egress Hatches
72. Main Passenger Ingress/Egress

PROPULSION:
9. Main Propulsion Engines (3)
10. OMS Engines (2)
11. Air-Breathing Engines (4)
12. ACPS Thrusters – Aft Firing (2)
13. ACPS Thrusters – Forward Pitch (5)
14. ACPS Thrusters – Forward Yaw (4)
15. ACPS Thrusters – Forward Firing (3)
16. ACPS Thrusters – Aft Pitch/Roll (10)
17. ACPS Thrusters – Aft Yaw (6)
19. ACPS Thrusters – LO2 Tank
20. OMS Integrated LH2 Tank (1)
21. OMS Integrated LO2 Tanks (2)
22. Air-Breathing Engine JP-4 Tank
23. ACPS H2 Accumulator (2)
24. ACPS O2 Accumulator (1)
25. ACPS H2 Conditioning Module (3)
26. ACPS O2 Conditioning Module (3)
68. He Pressurant Tank (2)

AVIONICS:
5. Electronics Compartment
32. Data Management Computer Complex (4)
37. LM Optical Rendezvous System (2)
38. Four Gimbal IMU (3)
60. Ku-Band Scanning Beam Antenna (3)
61. S-Band Communication Antenna (4)
62. C-Band Altimeter Antenna (6)
63. L-Band ATC Antenna (2)
64. TACAN Antenna
65. VHF ATC Antenna (2)
66. Orbiter-Booster Comm. Antenna (2)
67. VHF TDRS Antenna (2)

ENVIRONMENTAL CONTROL:
7. Payload Bay Doors / Radiators
50. Waste Management System
57. N2 Tank (2)
58. Emergency O2 Tank
59. Water Tank (2)

SECTION A-A
SECTION B-B
FUSE STA 200
5
1572

SECTION C-C

MISCELLANEOUS:
8. Payload Module
69. Electrical Umbilical
70. Air Conditioning Umbilical
71. ECS/LSS Umbilical

MECHANICAL SYSTEMS AND LANDING GEAR:
27. Main Landing Gear
28. Nose Gear
43. Hydraulic Reservoirs (4)
44. Payload Deployment Actuator
46. Docking Ring – Hinged Base
47. Drag Chute (2)

ELECTRICAL POWER:
29. PGS Fuel Cell Power (2)
30. PGS Auxiliary Power Unit (4)
31. Battery (2)
34. Forward Main Distribution Assembly
35. Aft Main Distribution Assembly
36. Local Distribution Assembly

FLIGHT CONTROLS:
39. Elevons
40. Split Rudder
41. Split Rudder/Speedbrake

tive size and weight of the booster, negating most of the advantages of the external tank concept. So Grumman and Boeing decided to switch to the Phase B baseline 415,000 lbf engines – permitting the use of three engines in the orbiter and greatly simplifying the abort profiles. The selection of the smaller engine, twelve of which would also be used in the booster, allowed a reduction in the gross lift-off weight of the system by as much as 1.3 million pounds compared with the baseline fully reusable concept, while at the same time reducing the staging velocity back to 4,750 mph. The final three-engine orbiter configuration was designated H-33. A fully-reusable three-engine orbiter, and its associated booster, were also designed by Grumman and designated G-3. These designs were maintained primarily for comparison purposes, giving Grumman and Boeing easy access to data they felt was comparable to what the Phase B contractors were obtaining from their designs.

The 157-foot long H-33 orbiter was powered by three 415,000 lbf main engines and had a blended 55-degree delta-wing that spanned 97 feet. The swept-back vertical stabilizer was 61.25 feet high. Four P&W JTF22A-4 air-breathing engines were installed in two fuselage bays over the wing box beam and vehicle center-of-gravity, and deployed upward. The engines could be removed and the internal volume used for special tankage or other uses with no effect on vehicle balance. A cross-range capability of 1,265 miles was projected. The landing speed was estimated at 207 mph and a ferry range of 345 miles could be achieved using the air-breathers. The orbiter had an empty weight of 197,000 pounds, and two 30-foot drag chutes were provided to reduce the landing roll.

Two internal un-insulated tanks held 705,696 pounds of LO2, and two external tanks each carried 59,537 pounds* of LH2 each. The internal LO2 tanks could be removed through the nose section manufacturing break for repair and refurbishment. Each external tank was a 14.8-foot cylinder with simple conical forward and aft domes. The overall length of each tank was 102 feet, including the 17.7-foot aluminum alloy sheet-frame-stringer nose fairing. The tank shell was a welded monocoque structure of 2219-T87 aluminum alloy sheet, 0.057 inch thick, and each tank had a dry weight of 10,349 pounds. The external tanks were covered with a 0.75-inch coating of NOPCO BX250A foam to provide cryogenic insulation and a 0.1-inch layer of Pirex-250 ablator along the inboard surface to protect against interference heating.

The crew compartment was directly forward of the

The H-33 space shuttle lifts-off from its launch site at Kennedy Space Center, Florida. (Grumman photo 711665)

---

*   It needs to be remembered that hydrogen, with an atomic weight of one, is extremely light, therefore the amount of oxidizer and fuel seem disproportionate, but aren't.

payload bay and was provided with an airlock for access to payloads. The two cylindrical LO2 tanks ran beneath the crew compartment and the forward section of the payload bay. The basic body had a constant triangular cross-section with straight sides arranged around the LO2 tanks and payload bay. The orbiter main structure was to be a titanium and composite semi-monocoque using manufacturing techniques developed for the F-14 Tomcat fighter. The propellant tanks and crew compartment were to be constructed of 2219-T87 aluminum alloy. A thermal protection system was applied to all exterior areas except the Inconel-718 rudder and the replaceable titanium payload bay doors. Where surface temperatures were under 1,900 degF, metallic shingles backed with Micro-Quartz insulation were used. Elsewhere the TPS consisted of silica-based tiles, except for the nose cap and wing leading edges which were carbon-carbon. The underlying aluminum structure was maintained below 300 degF in the fuselage and 600 degF in the wing.

The Boeing-designed straight-wing H-33 booster was 245 feet long, spanned 177.5 feet, and had an empty weight of 220,135 pounds. The aero surfaces were based on large airplane technology (namely, the Boeing 747) and were of conventional two-spar design and fabricated primarily of aluminum and titanium. Twelve 415,000 lbf boost engines provided ascent thrust, and 12 GE F101 or P&W JTF22A-4 turbofan engines were located in 'pop-out' pods on the side of the nose, allowing the booster to fly back to its launch site. Sufficient jet fuel was carried so that a ferry range of 443 miles could be obtained. The 33-foot diameter fuselage featured separate LH2 (forward) and LO2 (aft) tanks incorporating proven Saturn S-IC design and manufacturing techniques. The LO2 tank was located aft to reduce fuselage structural weight, eliminate geysering, minimize pogo susceptibility, and allow a straight-through passage of the wing carry-through structure. Thermal protection during reentry was provided by the heat capacitance of the basic struc-

The H-33 orbiter could be launched using either the Boeing-designed straight-wing booster, or atop a modified Saturn V first stage. By the time Grumman had completed their ASSC studies, it was obvious to all involved that space shuttle would look vastly different than had been envisioned only 18 months earlier. (NASA photo S71-40393)

ture, hence the term' 'heat-sink'. This feature of a single shell structure reacting to both flight loads and heat loads significantly reduced the weight of the booster.

It was estimated that the RDT&E costs would be $2,674.3 million for the orbiter, $32.6 million for the external tanks, $2,180.6 million for the booster, and $892.9 million for flight test and management – for a total system RDT&E cost of $5,780.4 million. The government would furnish the main engines and other items worth an additional $1,016.5 million. Production orbiters would cost $615 million apiece, with the boosters costing $273.6 million each. Costing data indicated that the external tanks could be produced for $740,000 per pair (first units would cost $1.022 million). Total 'out of pocket' costs associated with each H-33 orbital flight were estimated to be $4.2 million, or $5.3 million if the cost of the fleet was included and amortized over a life of 500 flights per element (except external tanks, which were throw-away).

It was anticipated that with a development go-ahead in March 1972, the first orbiter would be completed in June 1973, with the first booster following in April 1976. The initial horizontal flight tests would be conducted during August 1976 for the booster and October 1976 for the orbiter. The first manned orbital test flight would be conducted in April 1978, with a fully operational system in place by July 1979. At peak traffic rates for a mature system, it was estimated that 650 people could support direct operations, with the total rising to 1,000 when engineering support, quality control, and other technical support personnel were included. If all other support personnel (fire, mail, janitorial, etc.) were included, the total rose to about 3,000 people to support 75 launches per year. This was an average cost (1970 dollars) per launch for manpower of $540,000.

---

* This was generally known in the aerospace industry as a 'hot-structure' (a term popularized by the Air Force/North American X-15 program), but Grumman preferred to call it a heat-sink.

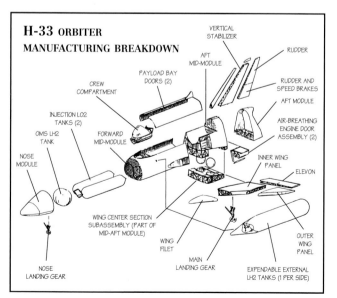

H-33 ORBITER
MANUFACTURING BREAKDOWN

VERTICAL STABILIZER
AFT MID-MODULE
RUDDER
PAYLOAD BAY DOORS (2)
CREW COMPARTMENT
RUDDER AND SPEED BRAKES
AFT MODULE
INJECTION LO2 TANKS (2)
OMS LH2 TANK
FORWARD MID-MODULE
AIR-BREATHING ENGINE DOOR ASSEMBLY (2)
NOSE MODULE
INNER WING PANEL
ELEVON
WING CENTER SECTION SUBASSEMBLY (PART OF MID-AFT MODULE)
WING FILET
MAIN LANDING GEAR
OUTER WING PANEL
NOSE LANDING GEAR
EXPENDABLE EXTERNAL LH2 TANKS (1 PER SIDE)

ANOTHER REVISION

As part of the Phase B study, the contractors had evaluated various ejection systems, and Rockwell* did a detailed trade study on three different designs – ejection seats similar to the seats installed on the triple-sonic Lockheed A-12/SR-71 series aircraft, encapsulated ejection seats of the type used on the North American XB-70 experimental bomber, and a separable crew compartment similar in concept to the one used on the General Dynamics F-111 and early Rockwell B-1A aircraft. The only system that could provide protection for more than the two man flight crew was the separable crew compartment, and that would raise the development costs of the orbiter by almost $300 million and add over 14,000 pounds[†] to its empty weight. All the systems examined had limitations in their ability to provide successful escape, and all would require a significant advance warning of an impending hazard from reliable data sources.

Another revision of the Space Shuttle System Program Definition document was issued in April 1971 stating (¶ 1.3.6.2.1): "… provisions shall be made for rapid emergency egress of the crew during development test flights …" The objective was to offer the crew some protection, though limited, from risks during the test flights. The philosophy was that after the test flights, all unknowns would be resolved, and the vehicle would be certified for 'operational' use like an airliner. Rockwell eventually selected modified SR-71 seats for installation in *Enterprise* and *Columbia* during the atmospheric and orbital flight test series. The ejection could be initiated by either crew member, and could be used in the event of uncontrolled flight, on-board fire, or impending landing on unprepared surfaces or water. The escape sequence required approximately 15 seconds for the crew to recognize the situation, initiate the ejection sequence, and get a safe distance away from the vehicle.[99]

Although the seats were originally intended for use during first stage ascent, or during gliding flight below 100,000 feet, subsequent analysis shows the crew would be exposed to the solid rocket booster and main engine exhaust plumes if they ejected during ascent. During descent, the seats provided meaningful protection from about 100,000 feet to landing. The end result was that a decision was made not to provide a crew escape system for operational shuttle flights. Although this would be highly criticized in the press after the STS-33/51-L (*Challenger*) accident in 1986, perhaps a more enlightened viewpoint was expressed by astronaut Robert L. Crippen: "… I don't know of an escape system that would have saved the crew from the particular incident that we just went through …"

Both the McDonnell Douglas and Rockwell teams put a great deal of study into the process of separating the stages at the appropriate time. Criteria included the need to exclude any expendable hardware which would preclude the attainment of an all azimuth launch capability, the exclusion of pyrotechnics and expendables such as staging motors, and the need for an inherent capability to provide positive separation, not only at normal staging, but also under abort conditions. All three contractors eventually settled on very similar concepts that involved using the orbiter main engines at about 50 percent power to provide separation force. Both the orbiter and booster auxiliary propulsion engines were used to provide attitude control, and mechanical linkages attached to the booster ensured the appropriate clearances were maintained during separation.

The separation sequence would be initiated when a low-level sensor in either the LO2 or LH2 tank in the booster was uncovered near the end of the boost phase. Uncovering the low-level sensor activated a run-out clock, that, in proper time sequencing, sent out commands to ignite the orbiter main engines and to shut down the booster engines. This sequencing also initiated the mechanical separation of the two vehicles. Alternately, the same sequence could be initiated by the flight crew, except that the booster would simply throttle down its main engines instead of shutting them off. This would allow the vehicles to separate, but also allow the booster to continue burning propellant in preparation for a normal reentry and landing. In this case the orbiter would continue to fire its engines until a sufficient amount of propellant was used to allow landing, either at a contingency field or back at the launch site.

There was also considerable concern about the lack of high-speed thermodynamic data available to the designers on both teams. The X-15 research airplane had explored the envelope up to Mach 6, but that was the extent of the manned aircraft experience, and both the orbiter and booster would achieve velocities greatly in excess of that speed. The ASSET and PRIME (X-23A) programs had provided some data on lifting-reentry, although it was not very applicable to the configurations being studied. Nevertheless, NASA and both contractor teams believed that the data accumulated by the X-15 and X-23A could be extrapolated for use, and that the expense of a new manned research aircraft could not be justified (actually, neither could the time), and one was never seriously considered for development.

The three major competitors line up for comparison at the end of Phase B. From left to right: Saturn V; North American high cross-range proposal, Grumman/Boeing H-33 ASSC vehicle; McDonnell Douglas/Martin Marietta high cross-range concept; and a Boeing 747-100 jumbo-jet. (Grumman Corporation)

---

*   The results of the McDonnell/Martin study could not be ascertained.

†   1989 estimates were that to retrofit an escape module on the existing orbiters would add over 30,000 pounds to the empty weight of each orbiter.

Dr. James C. Fletcher. (NASA photo 666081)

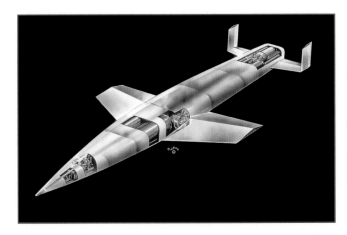

The illustration at right shows the Americä Bomber proposed by Eugen Sänger to the Luftwaffe during World War II, along with its 'skip-glide' trajectory. The chart shows the vehicle being launched at the extreme left, heading in a clockwise direction to its maximum altitude of almost 1,000,000 feet (285 kilometers). In this particular case, the trajectory takes the aircraft completely around the world instead of only half-way. A different view of essentially the same vehicle is shown above. A single pilot is in the extreme forward fuselage, followed by propellant tanks, lading gear, more propellant tanks, and finally the single rocket engine. The wedge-shape cross-section of the wings shows up well in both illustrations. (Sänger Archives via Terry Panopalis)

Clearly illustrating why the front windscreens needed a protective cover, this X-20A Dyna-Soar glows red-hot during the early part of its atmospheric reentry. (NASA/DFRC via Tony Landis)

The first X-15 leaves the NB-52A for flight 1-7-12 on 12 May 1960. The X-15 was the first hypersonic aircraft, and provided a great deal of data for the eventual development of Space Shuttle. (U. S. Air Force photo KE-15378 via the Terry Panopalis Collection)

An artist concept of the orbital M2 shape on top of a Titan II launch vehicle just after launch from Cape Canaveral AFS, Florida. Note how the adapter fairs the aft end of the lifting-body into the Titan. (Tony Landis Collection)

The modified X-15A-2 (56-6671) under the wing of the NB-52B. Note the large external propellant tanks unique to the X-15A-2. (NASA via the Terry Panopalis Collection)

The plywood M2-F1 (white) and the aluminum M2-F2 at the FRC in 1966. Note the relative location of each canopy. (NASA photo ECN-1107)

The HL-10 in 1990 showing off its recent restoration prior to being put on display at the Dryden Flight Research Center. (NASA via the Terry Panopalis Collection)

The X-24A on the ramp at the Flight Research Center – the yellow NASA tail band was standard at the time. (NASA photo ECN-1853 via the Terry Panopalis Collection)

The X-24B approaches Edwards AFB during 1975. Note the exposed XLR11 engine in the aft body. (NASA photo EC75-4643)

A rather unlikely concept showing version of the Lockheed StarClipper lifting-body shuttle carrying a fuselage full of passengers, although an attempt has been made to show propellant tanks. (NASA photo S70-487V via the SDAM Collection)

NASA drop tested this one-tenth scale model of the MSC-001 (DC-3) orbiter at White Sands Missile Range during May 1970. The 13-foot long model weighed 600 pounds and was lifted by an Army CH-54 Skycrane. (NASA photo S70-36663 via Jason Wentworth)

The Convair triamese concept (T-18 shown here) was proposed to the Air Force as early as 1965 and again in 1968 and 1969. It was also proposed to NASA during Phase A. (General Dynamics/Convair via the SDAM Collection)

An early North American / Convair Phase B low cross-range design as it would have looked during launch at KSC. Noteworthy is the complete lack of weather protection on the launch pad. (General Dynamics/Convair photo 108031B via the SDAM Collection)

February 1971 NASA drawing of a straight-wing orbiter servicing a space station. Note that the orbiter is not docked to the station, and the astronaut doing an EVA. By this time NASA had given up on a Saturn V-launched station. (NASA photo S71-557X)

The North American Phase B high cross-range design leaving the VAB for the trip to the launch pad at KSC. North American expected to use a service structure mounted on the Mobile Launch Platform. (Convair photo 112899B via the SDAM Collection)

General Dynamics/Convair model B8G booster from the early Phase B studies. Note the four large turbofan engines deployed around the nose. (NASA photo S70-46782)

A trio of Boeing Aerospaceplane designs. The top vehicle uses an expendable stage, and both top and bottom use a small orbital vehicle. (Boeing History Office photo 2A131290)

An early North American Phase C concept. Note the RCS pods on the wing tips and compare to the illustration on page 181. The hump on top of the fuselage covering the manipulator arm was only used by North American. (North American via the SDAM Collection)

McDonnell Douglas Phase B Extension – the missing segment from the lower payload bay door was not explained. Two remote manipulator arms are shown, a capability theoretically available on the production orbiters. (McDonnell Douglas photo D4C-90429)

Very-large VIRTUS aircraft proposed by NASA Langley to ferry the orbiter and conduct the early drop tests. The carrier would have been 294 feet long with a wingspan of 472 feet. Power came from four 40,000-lbf turbofans. (Tony Landis Collection)

Almost all the early artist concepts show orbiters being serviced on the ramp, or in very spartan hangers, much like transport aircraft from the 1960s. This was a late McDonnell Douglas Phase B Extension design. (McDonnell Douglas via the SDAM Collection)

THE RULES OF THE GAME CHANGE

The Phase B studies resulted in some truly ambitious two-stage vehicles designed to meet NASA's stated preference for a fully reusable space shuttle. All of the concepts presented by the contractors would have been expensive and contained potentially large development risks, even if the contractors were unwilling to fully admit it. In retrospect, it is questionable if any of the fully-reusable two-stage concepts *could* have been built given the technology of the era. In 1968, Pete Knight had just managed to take the X-15A-2 to 4,520 mph – a world speed record. The X-15 weighed about 30,000 pounds fully loaded. The Phase B boosters, which were supposed to fly much faster, would weigh over 400,000 pounds empty and 3,000,000 pounds fully loaded. On Knight's flight the ablative heat shield had partially failed, resulting in significant damage to the Inconel-X hot-structure. It did not bode well.

In fact, the entire issue of thermal protection was unsettled. It was not at all apparent that an adequate method of protecting either the booster or orbiter was at hand. The X-15 flights, and indeed the returning Apollo capsules, spent considerably less time in high thermal environments than anticipated for shuttle – and neither the X-15 or Apollo had a truly reusable thermal protection system. Charles Donlan, the NASA Space Shuttle program manager, later commented that the proposed reusable thermal protection systems all had significant problems. For one, imagine covering an aircraft larger than a 747 with metallic shingles made of exotic alloys that had previously only been used in small turbine blades.[1]

The size of the booster itself was also a potential problem, being larger than any aircraft yet built. If Boeing thought designing a 300,000-pound (empty) 600-mph jetliner – the 747 – was difficult, imagine attempting a 400,000-pound vehicle designed to go ten times as fast. To reduce the weight of the booster, and the orbiter in some cases, the contractors had proposed using integral tankage where the cryogenic propellant tanks carried part of the vehicle's weight and aerodynamic loads. The resulting stresses would tend to produce small leaks, possibly leading to a buildup of hydrogen under the skin – when heated by thermal transfer during reentry, this could have been disastrous.[2]

The development of the space shuttle main engines was also compromised since they were meant to power both the booster and orbiter – they could, therefore, not be optimized for either. Then there was the issue of actually separating two winged vehicles at 6,000 mph. The experience base consisted entirely of four launches during 1966 of a D-21 drone from a Lockheed M-21 Blackbird at speeds less than 2,000 mph. Three of the launches were successful – the fourth had resulted in the loss of both vehicles and the death of one crew member.[3] This brought up how to provide meaningful escape for the pilots of the booster during ascent, a problem that Charles Donlan did not believe was ever satisfactorily solved.[4]

Yes, hindsight indicates that the technical risks were great, even if the money could be found – and that was in serious doubt. Funding issues were beginning to force NASA to pay greater attention to criticism of the fully reusable concept, but in early-1971 the ultimate configuration of the space shuttle and any possible expendable components remained an open question.

Between 1970 and 1972 Donald Rice served as the assistant director of the Economics, Science, and Technology Programs Division at the Office of Management and Budget (OMB). Although admitting that the Mathematica economic analysis had been very thorough, Rice questioned some of the premises used during its development – in particular, the operating cost estimates provided by NASA. Rice had other worries as well, especially the possibility of cost overruns – Apollo had overrun its initial estimate by 75 percent between 1963 and 1969. If Shuttle did the same, its development cost could exceed $20,000 million. Even Mathematica admitted it was difficult to see how such an expenditure could be justified.[5]

In early-May 1971 NASA announced that it intended to issue the Phase C request for proposals as soon as August. The RFP would be based on a projected peak funding level of $2,300 million per year during the mid-1970s, requiring a total NASA budget of $4,500–5,000 million per year to maintain a balanced science and application program. Exactly why NASA believed this level would be approved is questionable – the fiscal year 1971 appropriation was only $3,270 million, and the FY72 request contained a similar number.

On 17 May 1971 OMB officials told NASA that its budget would remain essentially constant for the next five years – no large increases should be expected. An in-house OMB analysis had concluded that the fully reusable space shuttle was simply not cost competitive with the existing Titan III launch vehicle. On the same day, Donald Rice sent the NASA Administrator a letter that reiterated a $3,200 million budget ceiling. This was the third major budget blow for the agency within 18 months. In late 1969 the Bureau of the Budget had cut the NASA budget request by over $500 million, forcing then-administrator Tom Paine to abandon all hopes of a manned mission to Mars, and to concentrate on a space station and space shuttle. During the summer of 1970 the agency received more cuts, effectively cancelling the station. Now the shuttle was in doubt.[6]

The new budget numbers were a drastic blow because it meant the agency could not carry out the space shuttle development program it had been planning for the previous two years. If limited to the $3,200 million budget that was finally approved for FY72, the most NASA could hope to put into shuttle development, and still maintain a balanced science and application program, was roughly $1,000 million a year for five years. This brought a new sense of urgency to some of the partially-expendable ideas examined under the ASSC contracts.[7]

The funding limit left NASA with enough funds to develop an orbiter, but not the booster to go with it – therefore, most concepts for lowering costs involved replacing the fly-back booster with some sort of expendable stage. This did not appear technically feasible, however, given the large size of the Phase B orbiters. The problem was resolved in June 1971 with a decision by

This is the design that started NASA seriously listening to the external tank concept, although it was not the first proposal to use it. The Grumman H-33 orbiter moved its liquid hydrogen into liquid tanks flanking each side of the fuselage. This allowed a smaller and lighter orbiter, which in turn allowed a lower staging velocity, allowing a smaller booster. In the end all of the orbiter propellants would be moved into external tanks. (NASA photo S71-38215)

NASA to endorse some variation the 'external tank' concept publicly advocated by Draper and Cosenza at the AFFDL during 1968, and initially investigated in detail by both Lockheed and McDonnell Douglas during the Air Force RLV and ILRV studies as early as 1965. This concept moved the large LH2 tanks outside the orbiter airframe, and made them expendable – allowing a smaller and lighter orbiter. This resulted in a significant reduction in development costs, but with a corollary increase in costs each time the orbiter was launched. This concept had been revisited by Grumman during the ASSC studies, marking the first time that the MSC had officially sanctioned a partially reusable concept. Once it had gained NASA's ear, Grumman vigorously pursued the external tank concept, and the Grumman proposal manager, Larry Mead, continued to argue that it was the economically

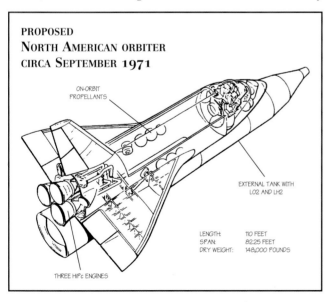

PROPOSED
NORTH AMERICAN ORBITER
CIRCA SEPTEMBER 1971

ON-ORBIT PROPELLANTS

EXTERNAL TANK WITH LO2 AND LH2

LENGTH: 110 FEET
SPAN: 82.25 FEET
DRY WEIGHT: 148,000 POUNDS

THREE HiPc ENGINES

Although it would not be decided for six more months, this is essentially the configuration that would ultimately be built – a three-engine delta-wing orbiter that carried all of its propellants in an external tank. (North American Rockwell)

preferred method. It would be economics, not technology, that eventually convinced NASA to listen more closely to what the AFFDL and Grumman was telling it.[8]

It should be noted that there were two very different external tank philosophies. The first was best exemplified by the Lockheed StarClipper and its derivatives. These were true stage-and-a-half vehicles and did not use a booster at all. Instead, a good percentage of the LH2 was moved into external tanks, and the vehicle took off under the power of its main engines burning the propellants in the external tanks – when these were expended, the tanks were jettisoned and the vehicle continued using internal propellants. Only a single vehicle and set of engines were used. Although moving most of the LH2 externally made the vehicle smaller than the fully reusable two-stage concepts, the StarClipper was still very large since it was, in essence, a single-stage-to-orbit design.

The other approach was best exemplified by the Grumman ASSC orbiter. In this case there was still a separate booster – it could be a fully-reusable fly-back vehicle, or some sort of expendable stage. After riding the booster out of most of the atmosphere, the orbiter would fire its own engines. The breakthrough here (four years after others had proposed it) was moving the liquid hydrogen outside the orbiter into expendable tanks, allowing a smaller and lighter orbiter.

Liquid hydrogen is very bulky, but very light. Although it made up only one-sixth of the propellant load by weight (the rest was LO2), it accounted for nearly 75 percent of the volume. Moving it outside allowed the basic orbiter to be much smaller, resulting in a lighter structure, less thermal protection, etc. Since the LH2 was so lightweight, the external tanks were also light, and they needed little thermal protection since they only had to withstand a single use.[9]

Grumman brought another new idea to the table – reducing the staging velocity. In designing the two-stage vehicles, the standard procedure was to choose a velocity that allowed the lowest weight for the mated pair. At a higher staging velocity, the first stage became truly enormous – at a lower velocity, the orbiter required more propellants and hence became larger and heavier. The groundrules developed by NASA during Phase B set the staging velocity at about 6,800 mph. With the LH2 carried in lightweight external tanks, Grumman calculated that it would be advantageous to move yet more propellant into these tanks and lower the staging velocity. This meant the orbiter would need to carry more LO2 internally but, since LO2 is fairly dense this did not drive the size of the orbiter significantly. The staging velocity suggested by Grumman was only 4,750 mph, greatly easing the task of designing the booster (although building a large hypersonic aircraft would still have been a challenge).[10]

What made the Grumman argument so convincing was how they presented it. Although other contractors had proposed similar ideas over the years, it was always very difficult to compare a design from one contractor with that from another. That is because each contractor used their own groundrules and criteria to predict weights, engine efficiencies, etc. Frequently these were different for each study (often imposed by the government, at least partially). During their ASSC study, Grumman had developed both a fully-reusable two-stage design and a partially-reusable external tank concept using the same rules – making it easy to compare them accurately. Lockheed had done much the same, but with decidedly less result.[11]

Grumman also presented an economic analysis for each design – again, developed with identical rules. The development cost for the external tank design was almost $1,300 million less ($6,497 versus $7,777 million), and peak funding was also reduced ($1,850 versus $2,200 million). Surprisingly, the cost per flight was actually slightly less than the fully-reusable vehicle ($4.22 versus $4.29 million). Although the external tanks, which would be thrown away on each flight, cost $740,000 a pair, Grumman estimated that other savings would more than offset this – primarily a reduction in the total amount of propellants needed, and eliminating much of the refurbishment required for the booster's thermal protection system. The cost of the tanks were expected to come down as the flight rate increased, primarily as a result of manufacturing efficiencies.[12]

GAINING GROUND

Although NASA planners still preferred the fully reusable approach, economic considerations, supported by outside analysis, clearly dictated the program would use some variation of the external tank concept. The major question now was the general configuration of the orbiter, and the degree to which the vehicle would carry its own propellants. The first NASA in-house design to address external tankage was the MSC-020 in May 1971. This was a straight-winged vehicle 106 feet long capable of carrying a 20,000 pound payload in a 15 by 30 foot payload bay, still reflecting Max Faget's desire for a simple configuration. The orbiter housed four main engines and an internal LO2 tank, and was mounted piggyback on a single large solid rocket booster (SRB). The LH2 tank was mounted in front of the SRB and remained attached to the orbiter after the booster was staged. In this design, the mated vehicle would lift-off with only the SRB thrusting, and the orbiter engines would be ignited either slightly before or immediately after SRB separation – a 'series-burn' concept. The SRB would deploy a parachute and be recovered, but the hydrogen tank was expendable.[13]

The Grumman ASSC orbiter had only moved the LH2 outside the orbiter into external tanks. Further analysis showed that it would be even more cost effective to place both the LO2 and LH2 externally in a single, expendable structure. It is ironic that this was largely the same conclusion that the AFFDL and McDonnell Douglas had reached at least three years earlier during the Air Force ILRV studies, and hinted at as early as the 1965 RLV study. This led directly to the MSC-021 design, also conceived in May 1971. The straight-winged orbiter was shortened to 90 feet in length and contained a single main engine and a 20,000 pound payload. Both the LO2 and LH2 tanks were mounted ahead of the single solid rocket booster, but this time they were attached to it by a truss structure reminiscent of several Soviet boosters. During 1971 and 1972, the engineers at MSC would design 53 orbiters in-house, although many were simply variations on a theme (larger payload, greater cross-range, etc.). Interestingly, a few additional designs would show up as late as 1978.[14]

While still evaluating the last of Max Faget's straight-wing orbiters, MSC began concentrating on delta-wing vehicles, beginning with the MSC-022B. The two most influential of the delta-wing designs were the MSC-036 series and the MSC-040 series. The MSC-036 had the same three-engine configuration used on the production orbiter, but had a 15 by 40 foot payload bay capable of

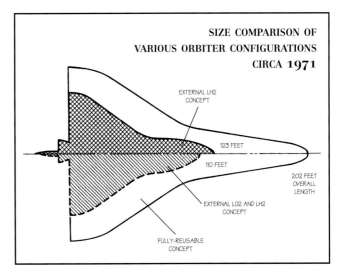

This drawing illustrates the large size reduction attained by moving the propellants into external tanks. The most dramatic reduction came from moving the liquid hydrogen externally – moving the oxygen provided a more modest reduction in planform area and overall volume. (North American Rockwell)

carrying 20,000 pounds and made use of improved J-2S engines from the Saturn program. After an interim look at the MSC-037 family, with three J-2S engines and a 15 by 60 foot payload bay, engineers concentrated on a family of orbiters with a 15 by 60 foot payload bay (but only a 25,000-pound capacity) and four J-2S engines. This was the MSC-040 series, and was very close to what would actually be built. The desire for an increased payload capacity (40,000 pounds) soon led to the MSC-040C powered by three new high chamber-pressure (HiPc)* engines. The MSC-040C design first appeared in August 1971 and would become the basis for the eventual Orbiter.[15]

The small payload bay in many of the concepts did not make the Air Force happy. In a 21 June 1971 letter to Dale D. Meyers, the NASA Associate Administrator for Manned Space Flight, Grant L. Hansen, the Assistant Secretary of the Air Force for Research and Development provided an assessment of the impacts of a 40-foot long payload bay instead of the desired 65-foot one. Hansen concluded that 71 of 149 planned Air Force payloads for the 1981–1990 timeframe could not be accommodated by the space shuttle. This would necessitate keeping the Titan III in production, and maintaining launch facilities on both coasts, negating much of the theoretical economic benefit of the shuttle. Hansen also noted that payloads that required a high-energy upper stage would require two shuttle missions (one for the upper stage, and one for the payload) and on-orbit assembly – something the Air Force believed was operationally unacceptable. Hansen was less concerned about slightly lowering the diameter of the payload bay, but concluded that a 15 by 60-foot configuration was still the preferred design.[16]

Throughout the first half of 1971, NASA engineers had met with their Air Force counterparts to work out the technical details for a production orbiter – the two main points of contention were cross-range and payload capability. The Air Force was still insistent upon being able to land at the launch site after a single orbit, and was committed to a 40,000 pound payload into polar orbit (equating to 65,000 pounds into a due east orbit). Fortunately, the Air Force did not seem to express any particular opinion on the external tank concept – not surprising since the idea had been initially developed during studies conducted for the military.[17]

---

* The HiPc engine would develop into the space shuttle main engine (SSME) as the program progressed.

NASA MANNED SPACECRAFT CENTER

SPACE SHUTTLE DESIGNS

CIRCA 1970–72

The following pages detail the major versions of the MSC shuttle designs from 1970 through 1972. The table below contains additional information. It should be remembered that although the designs may have been developed for launch on a particular booster, almost any combination was considered feasible, and many of the designs may well have been used on a variety of boosters.

Some of the designs could not be ascertained from existing documentation, hence the missing numbers in the table below. Other designs underwent such rapid change that it is possible that the description here does not match the design at all points in its life, although it is believed that these descriptions represent the final configuration of each designs.

No data could be ascertained for the vehicles not listed (e.g.; MSC-011 through MSC-019, etc.). (NASA)

VEHICLE	LANDING WEIGHT	WING	PAYLOAD SIZE	PAYLOAD WEIGHT	FEATURES
MSC-001	70,000	ST (AR7)	8 BY 30	15,000	INTERNAL LH2 AND LO2, AIR-BREATHERS, TWO ENGINES, REUSABLE BOOSTER
MSC-002	70,000	ST (AR7)	8 BY 30	15,000	INTERNAL LH2 AND LO2, AIR-BREATHERS, TWO ENGINES, REUSABLE BOOSTER
MSC-004-1	70,000	ST (AR7)	8 BY 30	15,000	INTERNAL LH2 AND LO2, TWO ENGINES, PAYLOAD BAY TUNNEL, REUSABLE BOOSTER
MSC-004-2	70,000	ST (AR7)	8 BY 30	15,000	INTERNAL LH2 AND LO2, TWO ENGINES, REUSABLE BOOSTER
MSC-007	75,000	ST (AR7)	12 BY 30	15,000	INTERNAL LH2 AND LO2, TWO ENGINES, REUSABLE BOOSTER
MSC-008-2	72,000	ST (AR7)	13 BY 17	12,000	INTERNAL LH2 AND LO2, TWO ENGINES, REUSABLE BOOSTER
MSC-009	78,000	ST (AR7)	10 BY 30	15,000	INTERNAL LH2 AND LO2, TWO ENGINES, REUSABLE BOOSTER WITH SRBs
MSC-010	90,000	ST (AR7)	10 BY 30	25,000	INTERNAL LH2 AND LO2, TWO ENGINES, REUSABLE BOOSTER WITH SRBs
MSC-010-1	92,000	ST (AR7)	15 BY 30	25,000	INTERNAL LH2 AND LO2, TWO ENGINES, REUSABLE BOOSTER WITH SRBs
MSC-020	130,000	ST (AR7)	15 BY 30	20,000	EXTERNAL LH2, INTERNAL LO2, FOUR ENGINES, SRB
MSC-021	85,000	ST (AR7)	15 BY 40	20,000	EXTERNAL LH2 AND LO2, ONE ENGINE, SRB
MSC-022	95,000	ST (AR5)	15 BY 40	20,000	EXTERNAL LH2 AND LO2, ONE ENGINE, SRB
MSC-022A	100,000	45° LE SW	15 BY 40	20,000	EXTERNAL LH2 AND LO2, ONE ENGINE
MSC-022B	105,000	DELTA	15 BY 40	20,000	EXTERNAL LH2 AND LO2, ONE ENGINE
MSC-023	135,000	DELTA	15 BY 60	40,000	EXTERNAL LH2 AND LO2, ONE ENGINE, REUSABLE BOOSTER
MSC-024	125,000	ST (AR5)	15 BY 60	40,000	EXTERNAL LH2 AND LO2, ONE ENGINE, STRETCHED MSC-022
MSC-025	130,000	45° LE SW	15 BY 60	40,000	EXTERNAL LH2 AND LO2, ONE ENGINE, STRETCHED MSC-022A
MSC-026	125,000	DELTA	12 BY 40	40,000	EXTERNAL LH2 AND LO2, ONE ENGINE, REUSABLE BOOSTER
MSC-027	95,000	DELTA	12 BY 40	40,000	GLIDER, EXTERNAL MAIN ENGINE, SRB
MSC-028	128,000	DELTA	15 BY 40	40,000	EXTERNAL LH2 AND LO2, ONE ENGINE, SHORTENED MSC-023
MSC-029	142,000	DELTA	12 BY 40	40,000	GLIDER, OMS TANKS AMIDSHIPS, EXTERNAL MAIN ENGINE
MSC-030	103,000	ST (AR5)	15 BY 40	20,000	EXTERNAL LH2 AND LO2, THREE J-2S ENGINES
MSC-031	133,000	ST (AR5)	15 BY 60	40,000	EXTERNAL LH2 AND LO2, THREE J-2S ENGINES
MSC-032	130,000	DELTA	15 BY 40	40,000	EXTERNAL LH2 AND LO2, THREE J-2S ENGINES, SRB
MSC-033	100,000	DELTA	12 BY 40	20,000	EXTERNAL LH2 AND LO2, ONE ENGINE, MODIFIED MSC-026, SRB
MSC-034	95,000	DELTA	15 BY 30	20,000	EXTERNAL LH2 AND LO2, ONE ENGINE, SHORTENED MSC-023, SRB
MSC-035	95,000	45° LE SW	12 BY 40	20,000	EXTERNAL LH2 AND LO2, ONE ENGINE, MODIFIED MSC-033
MSC-035A	135,000	45° LE SW	12 BY 60	40,000	EXTERNAL LH2 AND LO2, ONE ENGINE, STRETCHED MSC-035
MSC-036	110,000	DELTA	15 BY 40	20,000	EXTERNAL LH2 AND LO2, THREE J-2S ENGINES, SRBs
MSC-036A	110,000	DELTA	15 BY 40	20,000	EXTERNAL LH2 AND LO2, THREE J-2S ENGINES, SRBs
MSC-036B	110,000	DELTA	15 BY 40	20,000	EXTERNAL LH2 AND LO2, THREE J-2S ENGINES, SRBs
MSC-036C	110,000	DELTA	15 BY 40	20,000	EXTERNAL LH2 AND LO2, THREE J-2S ENGINES, PRESSURE FED BOOSTER
MSC-037	147,000	DELTA	15 BY 60	40,000	EXTERNAL LH2 AND LO2, THREE UPRATED J-2S, RECOVERABLE BOOSTER
MSC-037A	147,000	DELTA	15 BY 60	40,000	EXTERNAL LH2 AND LO2, THREE SUPER UPRATED J-2S, SRBs
MSC-038	100,000	DELTA	15 BY 40	20,000	EXTERNAL LH2 AND LO2, ONE HIPC ENGINE, SRBs
MSC-039	113,000	DELTA	15 BY 40	20,000	EXTERNAL LH2 AND LO2, THREE J-2S, PRESSURE FED BOOSTER
MSC-040	140,000	DELTA	15 BY 60	25,000	EXTERNAL LH2 AND LO2, FOUR J-2S, PRESSURE FED BOOSTER
MSC-040A	140,000	DELTA	15 BY 60	25,000	EXTERNAL LH2 AND LO2, FOUR J-2S, PRESSURE FED BOOSTER
MSC-040B	140,000	DELTA	15 BY 60	25,000	EXTERNAL LH2 AND LO2, FOUR 'SWING' J-2S, PRESSURE FED BOOSTER
MSC-040C	190,000	DELTA	15 BY 60	40,000	EXTERNAL LH2 AND LO2, THREE HIPC ENGINES, SRBs
MSC-040C-2	190,000	DBL DELTA	15 BY 60	40,000	EXTERNAL LH2 AND LO2, THREE HIPC ENGINES, WING GLOVE, TWIN TAILS, SRB
MSC-040C-3	190,000	50° DELTA	15 BY 60	40,000	EXTERNAL LH2 AND LO2, THREE HIPC ENGINES
MSC-040C-4	190,000	60° DELTA	15 BY 60	40,000	EXTERNAL LH2 AND LO2, THREE HIPC ENGINES
MSC-040C-5	190,000	50° DELTA	15 BY 60	40,000	EXTERNAL LH2 AND LO2, THREE HIPC ENGINES, CANARD, TWIN TAILS
MSC-040C-6	190,000	DBL DELTA	15 BY 60	40,000	EXTERNAL LH2 AND LO2, THREE HIPC ENGINES, CANARD, TWIN TAILS
MSC-041	114,000	30° LE SW	15 BY 60	15,000	EXTERNAL LH2 AND LO2, THREE J-2S ENGINES, PRESSURE FED BOOSTER
MSC-041A	114,000	30° LE SW	15 BY 60	15,000	EXTERNAL LH2 AND LO2, THREE J-2S ENGINES, PRESSURE FED BOOSTER
MSC-042A	110,000	DELTA	15 BY 60	25,000	GLIDER, TITAN IIIL6 BOOSTER
MSC-042B	110,000	30° LE SW	15 BY 60	25,000	GLIDER, TITAN IIIL6 BOOSTER
MSC-043	83,000	30° LE SW	10 BY 30	27,000	GLIDER, TWO 'SWING' HIPC ENGINES ON ET, PRESSURE FED BOOSTER
MSC-044	100,000	60° DELTA	10 BY 30	25,000	EXTERNAL LH2 AND LO2, TWO HIPC ENGINES, PRESSURE FED BOOSTER
MSC-047	185,000	DELTA	15 BY 60	40,000	EXTERNAL LH2 AND LO2, TWO HIPC ENGINES, TWIN TAILS, SRBs
MSC-048	205,000	DBL DELTA	15 BY 60	40,000	EXTERNAL LH2 AND LO2, TWO HIPC ENGINES, TWIN TAILS, SRBs
MSC-048A	195,000	DBL DELTA	15 BY 60	40,000	EXTERNAL LH2 AND LO2, FOUR J-2S ENGINES, TWIN TAILS, SRBs
MSC-049	205,000	DBL DELTA	15 BY 60	40,000	EXTERNAL LH2 AND LO2, THREE HIPC ENGINES, TWIN TAILS, 156-IN SRBs
MSC-049A	215,300	DBL DELTA	15 BY 60	40,000	EXTERNAL LH2 AND LO2, THREE HIPC ENGINES, TWIN TAILS, 178-IN SRBs
MSC-051	165,000	33° DELTA	15 BY 60	25,000	GLIDER, CANARD, 156-IN SRBs
MSC-052	175,000	35° DELTA	15 BY 60	25,000	GLIDER, THREE HIPC 'SWING' ENGINES ON ET, CANARD, 149-IN SRBs
MSC-053	185,000	35° DELTA	15 BY 75	25,000	GLIDER, FOUR HIPC 'SWING' ENGINES ON ET, CANARD, 120-IN SRBs
MSC-054	185,000	35° DELTA	15 BY 75	25,000	GLIDER, FOUR HIPC 'SWING' ENGINES ON ET, CANARD, 140-IN SRBs

ALL SIZES IN FEET, ALL WEIGHTS IN POUNDS

WING LEGEND:
ST	=	STRAIGHT-WING
LE SW	=	LEADING EDGE SWEEP
DELTA	=	DELTA-WING
DBL DELTA	=	DOUBLE DELTA-WING

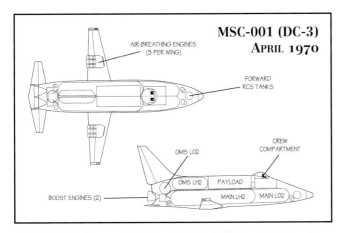

MSC-001 (DC-3)
APRIL 1970

AIR-BREATHING ENGINES (3 PER WING)

FORWARD RCS TANKS

CREW COMPARTMENT

OMS LO2

OMS LH2

PAYLOAD

BOOST ENGINES (2)

MAIN LH2

MAIN LO2

The MSC-001, also known as the DC-3, was the first serious in-house look MSC engineers took at developing an orbiter. This design is discussed in further detail on page 102.

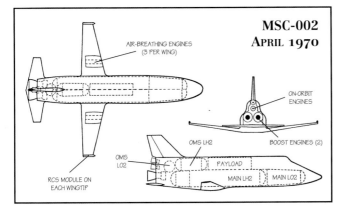

MSC-002
APRIL 1970

AIR-BREATHING ENGINES (3 PER WING)

ON-ORBIT ENGINES

OMS LH2

BOOST ENGINES (2)

OMS LO2

PAYLOAD

MAIN LH2

MAIN LO2

RCS MODULE ON EACH WINGTIP

The MSC-002 was a refinement of the earlier MSC-001, and is usually the design that is shown when referring to the DC-3. Primary improvement was in the area of the cockpit.

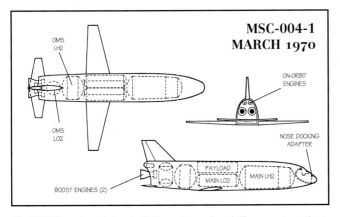

MSC-004-1
MARCH 1970

OMS LH2

ON-ORBIT ENGINES

OMS LO2

NOSE DOCKING ADAPTER

BOOST ENGINES (2)

PAYLOAD

MAIN LO2

MAIN LH2

The MSC-004 was begun before the DC-3 report was released. There were two variants, one with a tunnel connecting the cockpit and payload bay (below), and one without (above).

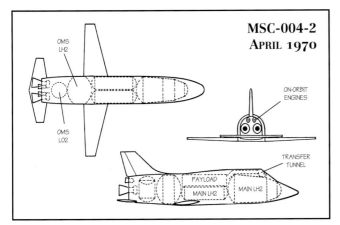

MSC-004-2
APRIL 1970

OMS LH2

ON-ORBIT ENGINES

OMS LO2

TRANSFER TUNNEL

PAYLOAD

MAIN LH2

MAIN LH2

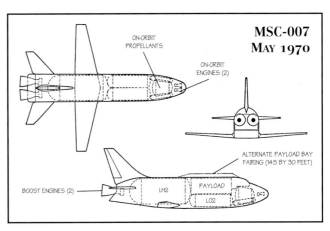

MSC-007
MAY 1970

ON-ORBIT PROPELLANTS

ON-ORBIT ENGINES (2)

ALTERNATE PAYLOAD BAY FAIRING (14.5 BY 30 FEET)

BOOST ENGINES (2)

LH2

PAYLOAD

LO2

The MSC-007 had two different payload bay door configurations that allowed it to carry over-sized payloads if required. The payload bay itself had grown in size to 12 by 30 feet.

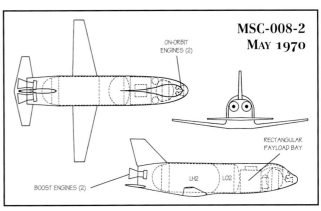

MSC-008-2
MAY 1970

ON-ORBIT ENGINES (2)

RECTANGULAR PAYLOAD BAY

BOOST ENGINES (2)

LH2

LO2

This design attempted to correct some center-of-gravity problems that had been experienced with the earlier designs, but provided an oddly shaped payload bay as a result.

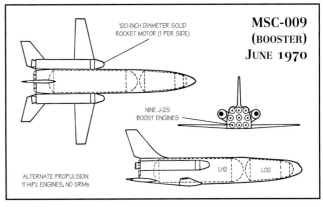

MSC-009
(BOOSTER)
JUNE 1970

120-INCH DIAMETER SOLID ROCKET MOTOR (1 PER SIDE)

NINE J-2S BOOST ENGINES

ALTERNATE PROPULSION: 11 HiPc ENGINES, NO SRMs

LH2

LO2

The MSC-009 was one of the first applications of solid rocket motors to the shuttle program. Several variations of where to attach the SRMs to the booster were investigated.

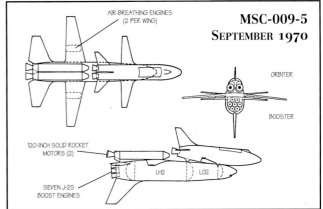

MSC-009-5
SEPTEMBER 1970

AIR-BREATHING ENGINES (2 PER WING)

ORBITER

BOOSTER

120-INCH SOLID ROCKET MOTORS (2)

LH2

LO2

SEVEN J-2S BOOST ENGINES

MSC-010
September 1970

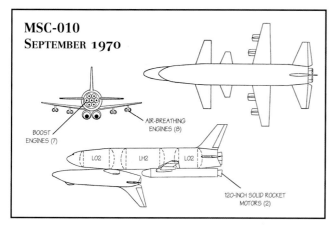

This was a refinement of the MSC-009-5 design, and continued to use two 120-inch SRBs. The payload was beginning to approach usable proportions, a total of 25,000 pounds.

MSC-021
May 1971

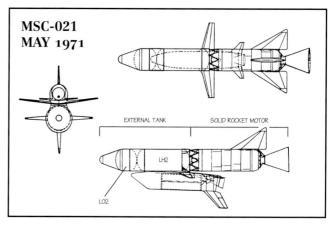

This was a refinement of the MSC-020, and was notable since it moved all propellants out of the orbiter into an external tank. Again, a single solid rocket motor was used.

MSC-020
May 1971

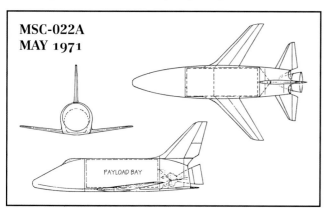

This was the first NASA concept to investigate an 'external tank', with the LH2 being carried in a large tank below the orbiter and in front of the single solid rocket motor.

MSC-023 / MSC-024
May 1971

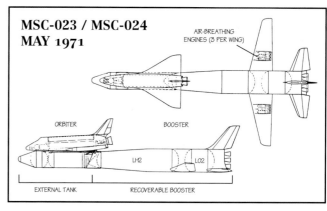

This was basically a MSC-022B with a payload bay measuring 15 by 60 feet instead of 15 by 40 feet. The MSC-024 was a 'stretched' version, with a payload bay 80 feet long.

MSC-022A
May 1971

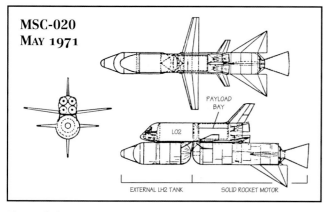

The basic MSC-022 fuselage found itself attached to a variety of different wings, such as the swept wings of the MSC-022A (above) and the delta-wing of the MSC-022B (below).

MSC-025
May 1971

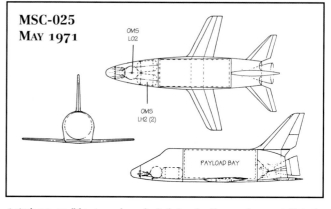

A single 550,000-lbf engine and a 15 by 60-foot payload bay was incorporated into the MSC-025 and MSC-026, which were incremental improvements on the MSC-022 series.

MSC-022B
May 1971

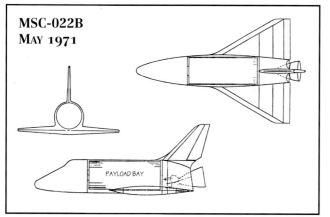

MSC-026
June 1971

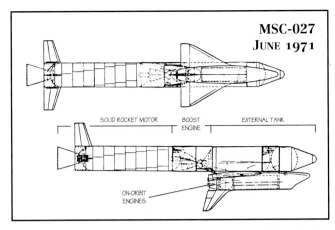

MSC-027
JUNE 1971

SOLID ROCKET MOTOR   BOOST
                      ENGINE   EXTERNAL TANK

ON-ORBIT
ENGINES

Taking the external tank concept one step further, the MSC-027 moved the main engine out of the orbiter. A 12 by 40 foot payload bay and two OMS engines occupy the fuselage.

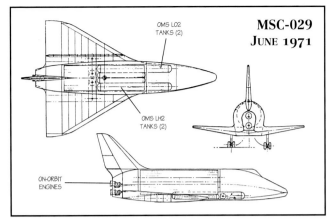

MSC-029
JUNE 1971

OMS LO2
TANKS (2)

OMS LH2
TANKS (2)

ON-ORBIT
ENGINES

Again, the main engines were in the booster, and all that remained in the MSC-029 were small on-orbit maneuvering engines. Propellants were carried under the payload bay.

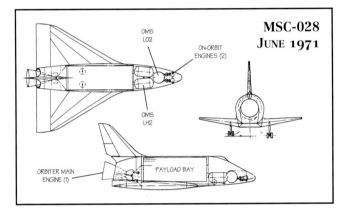

MSC-028
JUNE 1971

OMS
LO2

ON-ORBIT
ENGINES (2)

OMS
LH2

ORBITER MAIN
ENGINE (1)   PAYLOAD BAY

Providing a 15 by 40 foot payload bay, the MSC-028 returned to having a single large boost engine in the orbiter, although all propellants were in the external tank.

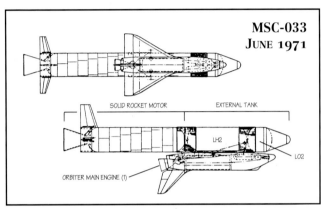

MSC-033
JUNE 1971

SOLID ROCKET MOTOR   EXTERNAL TANK

LH2   LO2

ORBITER MAIN ENGINE (1)

The MSC-033 reversed the trend towards larger payloads with a 12 by 40 foot bay and 20,000 pounds capacity. It was significant in that it could return with an equal payload.

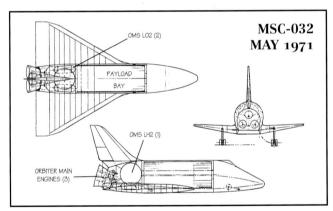

MSC-032
MAY 1971

OMS LO2 (2)

PAYLOAD
BAY

OMS LH2 (1)

ORBITER MAIN
ENGINES (3)

The MSC-032 was converging on the design that would be built. It had three J-2S engines and the boost propellants moved into an external tank, mounted ahead of the single SRM.

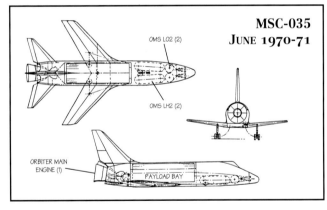

MSC-035
JUNE 1970-71

OMS LO2 (2)

OMS LH2 (2)

ORBITER MAIN
ENGINE (1)   PAYLOAD BAY

Two versions of this design were proposed, one with a 12 by 40 foot, 20,000 pound payload capacity (above) and the other with a 12 by 60 foot, 40,000 pound capacity (below).

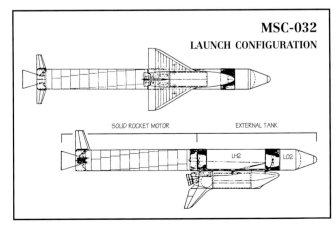

MSC-032
LAUNCH CONFIGURATION

SOLID ROCKET MOTOR   EXTERNAL TANK

LH2   LO2

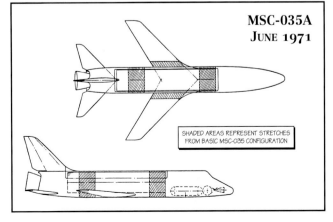

MSC-035A
JUNE 1971

SHADED AREAS REPRESENT STRETCHES
FROM BASIC MSC-035 CONFIGURATION

MSC-034
JUNE 1971

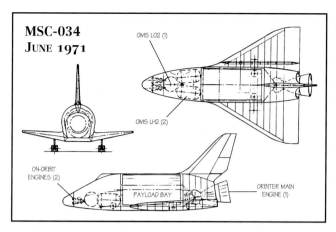

A small 15 by 30 foot payload bay was provided by the MSC-034 which was essentially a shortened MSC-023. A single large solid rocket motor was used as a first stage.

MSC-036
JUNE 1971

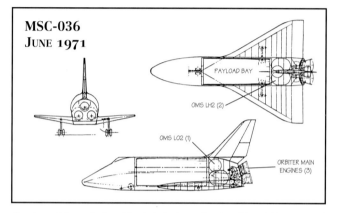

The shape of the production orbiter is beginning to show here. Three J-2S engines and a 15 by 40 foot payload bay capable of carrying 20,000 pounds were provided.

MSC-036A
JUNE 1971

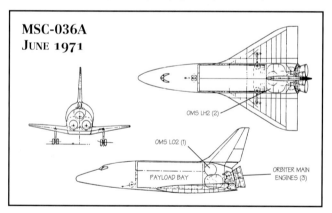

Continued refinement of the basic MSC-036 shape resulted in the MSC-036A and MSC-036B, with most of the changes being limited to the forward fuselage and cockpit.

MSC-036B
JULY 1971

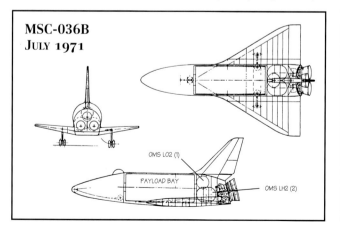

MSC-037
JUNE 1971

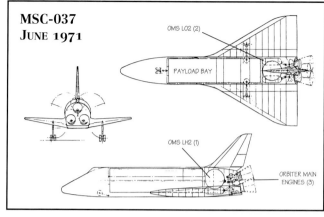

A 15 by 60 foot payload bay and three J-2S engines were features of the MSC-037. The main engines were to be operated at 15 percent thrust for on-orbit maneuvering.

MSC-038
JUNE 1971

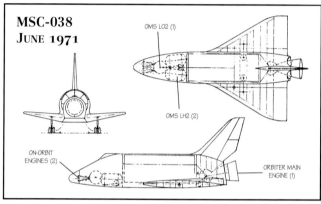

Essentially an improved MSC-036 with a single 550,000-lbf main engine, the MSC-038 still had a payload bay limited to 15 by 40 feet with a 20,000 pound capacity.

MSC-036C
AUGUST 1971

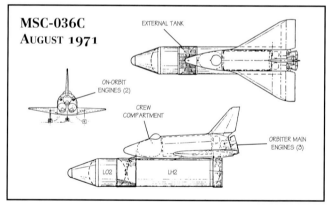

A further refinement of the MSC-036B, this orbiter greatly resembled the MSC-040 that became the basis for the Phase C/D proposals. The payload bay measured 15 by 40 feet.

MSC-036C
LAUNCH CONFIGURATION

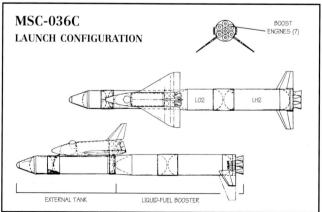

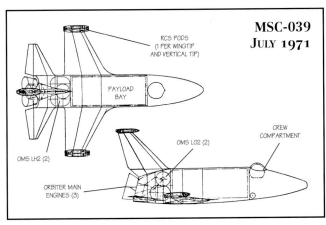

MSC-039
JULY 1971

A compromise to the straight-wing, the MSC-039 used a slightly swept leading edge, along with the fuselage contours of the MSC-036C. Note the RCS pods on the wing-tips.

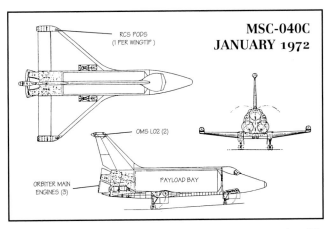

MSC-040C
JANUARY 1972

The MSC-040C represented an aerodynamic refinement of the MSC-040A, and would be the orbiter ordered into production, although other designs were still being investigated.

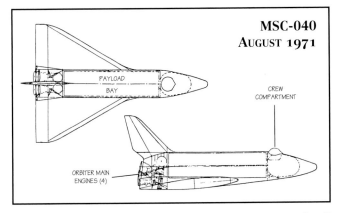

MSC-040
AUGUST 1971

The beginning of the end. The basic MSC-040 was penned on 30 August 1971, and would form the basis for the production orbiters. A 15 by 60 foot payload bay was provided.

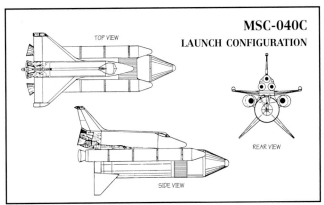

MSC-040C
LAUNCH CONFIGURATION

Looking much like the final product, the MSC-040C with an external tank and two 156-inch solid rocket motors. Note the large stabilizing fins on the external tank.

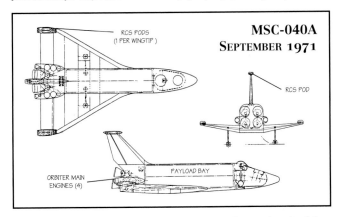

MSC-040A
SEPTEMBER 1971

The MSC-040A added RCS pods to the wing and tail-tips, and OMS pods to the aft fuselage just above the wing. The use of a single pressure-fed booster was envisioned (below).

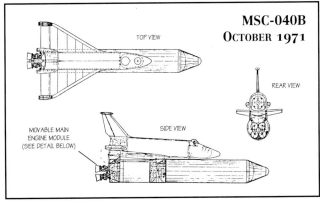

MSC-040B
OCTOBER 1971

The MSC-040B was an unsuccessful attempt to explore new propulsion concepts, such as the 'propulsion unit transfer concept' detailed below. This design was quickly abandoned.

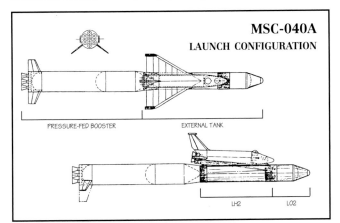

MSC-040A
LAUNCH CONFIGURATION

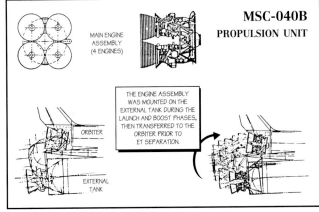

MSC-040B
PROPULSION UNIT

THE ENGINE ASSEMBLY WAS MOUNTED ON THE EXTERNAL TANK DURING THE LAUNCH AND BOOST PHASES, THEN TRANSFERRED TO THE ORBITER PRIOR TO ET SEPARATION.

MSC-040C
FEBRUARY 1972

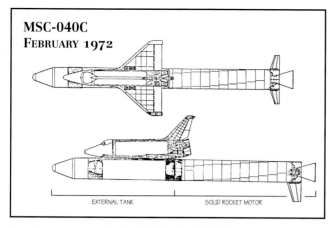

EXTERNAL TANK   SOLID ROCKET MOTOR

Other boosters were considered by NASA for the MSC-040C orbiter, such as this single large solid rocket motor attached to the aft end of the external tank.

PROPOSED
CIRCA 1970-72

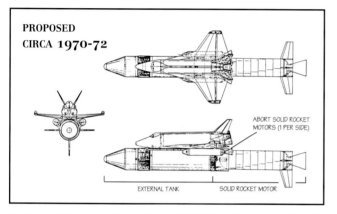

ABORT SOLID ROCKET MOTORS (1 PER SIDE)

EXTERNAL TANK   SOLID ROCKET MOTOR

Bearing absolutely no resemblance at all to the MSC-040C, this design was the first of the 'bat' shapes that were inspired by aerodynamic research at NASA Langley.

MSC-041
AUGUST 1971

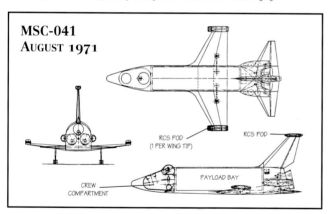

RCS POD (1 PER WING TIP)   RCS POD

CREW COMPARTMENT   PAYLOAD BAY

One of the last attempts by the straight-wing advocates was the MSC-041. Although the payload bay had grown in size to 15 by 60 feet, its capacity was only 15,000 pounds.

MSC-041A
SEPTEMBER 1971

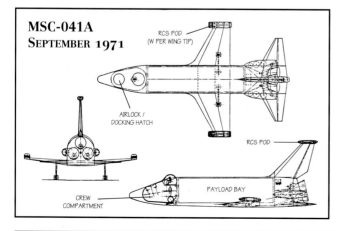

RCS POD (W PER WING TIP)

AIRLOCK / DOCKING HATCH

RCS POD

CREW COMPARTMENT   PAYLOAD BAY

MSC-043
DECEMBER 1971

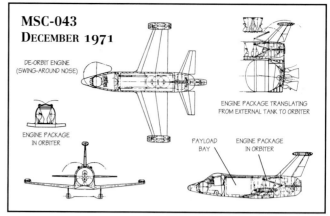

DE-ORBIT ENGINE (SWING-AROUND NOSE)

ENGINE PACKAGE TRANSLATING FROM EXTERNAL TANK TO ORBITER

ENGINE PACKAGE IN ORBITER

PAYLOAD BAY   ENGINE PACKAGE IN ORBITER

Similar to the MSC-040B in concept, this design used a complicated retracting mechanism to pull the main engine package from the back of the ET into the payload bay.

MSC-042B
NOVEMBER 1971

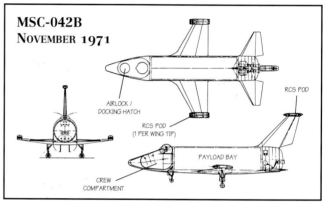

AIRLOCK / DOCKING HATCH

RCS POD

RCS POD (1 PER WING TIP)

CREW COMPARTMENT   PAYLOAD BAY

A straight wing derivative of the MSC-042A (below), this orbiter also contained no propulsion and used a Titan IIIL6 booster. A 25,000 pound payload could be carried.

MSC-042A
NOVEMBER 1971

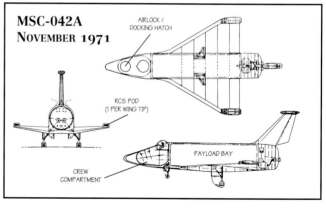

AIRLOCK / DOCKING HATCH

RCS POD (1 PER WING TIP)

CREW COMPARTMENT   PAYLOAD BAY

Basically an MSC-040A minus the main engines, this design relied entirely on an uprated Titan IIIL6, essentially a larger Titan III with six strap-on solids, to get it to orbit.

MSC-042A
LAUNCH CONFIGURATION

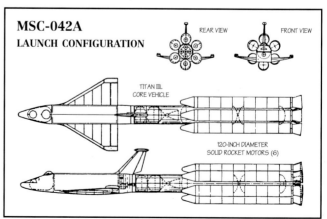

REAR VIEW   FRONT VIEW

TITAN IIIL CORE VEHICLE

120-INCH DIAMETER SOLID ROCKET MOTORS (6)

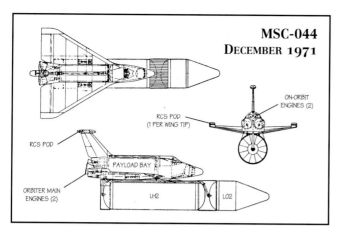

MSC-044
DECEMBER 1971

ON-ORBIT ENGINES (2)

RCS POD (1 PER WING TIP)

RCS POD

PAYLOAD BAY

ORBITER MAIN ENGINES (2)

LH2   LO2

Two 550,000-lbf engines of the same type baselined by MSFC for the ASSC study were used by the MSC-044, which could carry 25,000 pounds in a 10 by 30 foot bay.

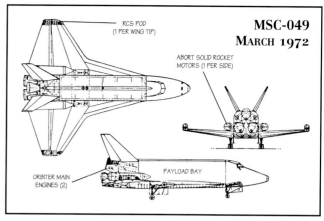

RCS POD (1 PER WING TIP)

MSC-049
MARCH 1972

ABORT SOLID ROCKET MOTORS (1 PER SIDE)

ORBITER MAIN ENGINES (2)

PAYLOAD BAY

Essentially an MSC-048 with three engines, the MSC-049 could still carry 40,000 pounds in its 15 by 60 foot payload bay. Two 62-inch diameter abort SRMs were provided.

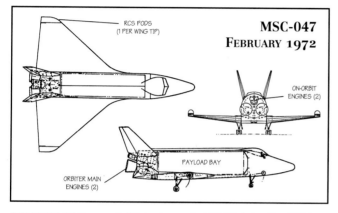

RCS PODS (1 PER WING TIP)

MSC-047
FEBRUARY 1972

ON-ORBIT ENGINES (2)

ORBITER MAIN ENGINES (2)

PAYLOAD BAY

Twin vertical stabilizers marked their return on the MSC-047 design. Two ASSC baseline 550,000-lbf engines and a 40,000 pound payload capacity were featured.

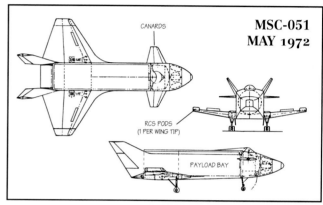

CANARDS

MSC-051
MAY 1972

RCS PODS (1 PER WING TIP)

PAYLOAD BAY

The MSC-051 had no main engines, and could carry 25,000 pounds in its 15 by 60-foot payload bay. The vertical stabilizers were smoothly faired into the sides of the fuselage.

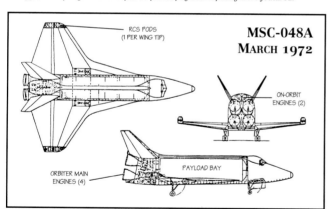

RCS PODS (1 PER WING TIP)

MSC-048A
MARCH 1972

ON-ORBIT ENGINES (2)

ORBITER MAIN ENGINES (4)

PAYLOAD BAY

The MSC-048A used four 550,000-lbf engines, and could lift 40,000 pounds in its 15 by 60-foot payload bay. Several different boosters were investigated (below).

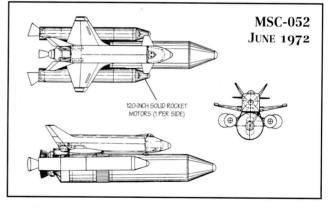

MSC-052
JUNE 1972

120-INCH SOLID ROCKET MOTORS (1 PER SIDE)

Part of the continuing research to find an method of handling the engines, this design again attempted to retract the engines into the orbiter when the ET was jettisoned.

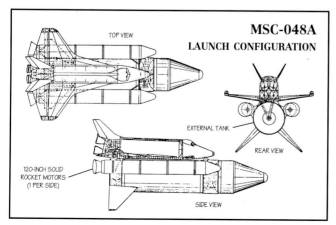

TOP VIEW

MSC-048A
LAUNCH CONFIGURATION

EXTERNAL TANK

REAR VIEW

120-INCH SOLID ROCKET MOTORS (1 PER SIDE)

SIDE VIEW

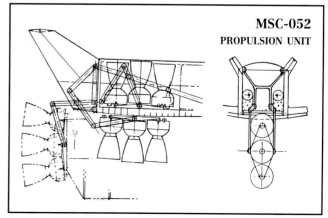

MSC-052
PROPULSION UNIT

PHASE B PRIME

On 12 September 1971 both Phase B contractors, along with Grumman/Boeing and Lockheed, were told to reevaluate their studies using the MSC-040C orbiter and external tank concept. The Chrysler ASSC proposal had been too far out of the mainstream, and the company elected not to continue participating in the airframe competition. Officially termed the Phase B Extension (even for the ASSC contractors), this was more often called Phase B Prime. This represented a major change from the original approach of reducing the number of contractors as the phases progressed – four contractors would now compete on a theoretically equal basis for the Phase C/D development and production contract.

During Phase B, NASA had not specified a landing speed for the space shuttle components, leaving this up to the contractors to propose. But to a large extent, the desired landing speed is a major weight driver since it determines how large a wing is necessary and was another item that made comparing the various designs difficult. Phase B Prime specified a 'subsonic design velocity' to be used to size the orbiter wing defined as the "trimmed velocity at an angle-of-attack equivalent to tail scrape angle at touchdown" – usually called landing velocity by those involved. A landing velocity of 190 mph was chosen by NASA since man-in-the-loop simulations indicated that this produced actual touchdown speeds of 205–220 mph, well within the existing state-of-the-art for landing gear systems. This criteria was used for the remainder of the program.[18]

Although NASA never seriously considered it for production, the MSC-040B design showed the creative concepts discussed by engineers. In an effort to lower the weight of the orbiter during abort scenarios, the MSC-040B moved two of the four J-2S engines into pods that retracted from the orbiter's belly to nestle between a pair of solid rockets and the external tank. The theory was that the orbiter could jettison the pods in the event of an engine failure during ascent, thereby lowering the weight of the returning vehicle. In the nominal case, the engines would be retracted back into the orbiter for refurbishment and reuse on the next mission. There were other variations to this idea, including one where the 'engine pack-

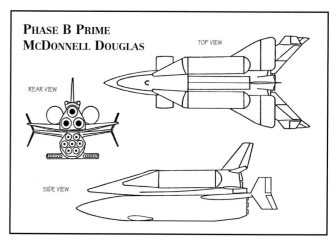

McDonnell Douglas also investigated external tank concepts using the booster designed by Martin Marietta. Again, only LH2 was carried externally. (NASA)

age' was installed on the back of an expendable booster during ascent, then physically transferred to the orbiter just prior to staging. The complexity of this installation outweighed any possible advantages it could have had during aborts, and the design was quickly abandoned in favor of the MSC-040C, although similar concepts were explored as late as the MSC-054 during June 1972.[19]

For Phase B Prime, the choice of possible boosters was left largely to the airframe contractors, but probable alternatives included modified Saturn IB or Saturn V stages, clusters of Titan-derived stages, or the development of new liquid or solid-propellant boosters. The choice to make the boosters expendable or reusable also rested with the contractors, with economics and safety being the drivers.

In contrast to the orbiter, there was no clear consensus on what the booster should look like. At the end of the original Phase B study, both contractors were proposing boosters that were significantly larger and heavier than a Boeing 747. Each needed a dozen boost engines to propel it to over 6,000 mph, then another 10–12 air-breathing engines to let it fly back to its launch site. Although the two designs took very different approaches to achieving their performance, particularly in their thermal protection systems, either would have cost a fortune to develop.[20]

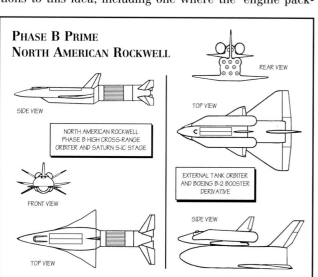

North American studied various concepts under the Phase B Prime extension. These included the Phase B high cross-range orbiter using a modified Saturn S-IC first stage, and the Grumman-based external LH2 orbiter using a derivative of the ASSC Boeing B-2 booster. (NASA)

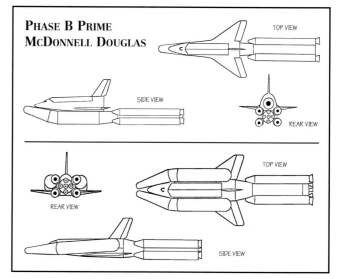

McDonnell Douglas explored several variations on the external tank concept, including these two variations on a delta-wing external LH2 tank orbiter – one that used two drop-tanks, and another that used a single large tank. Both of these designs used solid rocket motors instead of liquid boosters. (NASA)

Then along came the Grumman study, with its proposal to lower the staging velocity to only 4,750 mph. To design a booster for their orbiter, Grumman had teamed with The Boeing Company, which had built the S-IC (first) stage for the Saturn V, and was in the process of flight testing the world's largest commercial airliner, the 747. Lowering the staging velocity brought a disproportionate reduction in the size and weight of the booster, which now required far less propellant. In addition, the booster would be closer to the launch site when it released the orbiter, and would need less jet fuel for the return flight. All of this translated into a booster about 40 percent smaller than the one required for a fully reusable two-stage system.[21]

Boeing also discovered that the lower staging velocity effectively eliminated the need for thermal protection on the booster. By using a hot-structure (called a heat-sink by Grumman/Boeing), the booster could withstand its brief high-speed flight in much the same manner that the X-15 did. Although some particularly hot areas (the wing leading edges, etc.) would require a titanium or Inconel-X, most of the airframe could be manufactured from aluminum, greatly reducing the cost of the structure.[22]

McDonnell Douglas went even further. It reduced the staging velocity to only 4,200 mph, resulting in a booster that had a structure composed of 82 percent aluminum. Although engineers investigated lowering the staging velocity yet further by moving more of the ascent propellants into the orbiter, they found that total system weight began to creep upward again because the orbiter rapidly got larger. It seemed that the optimum staging velocity was somewhere around 4,500 mph.[23]

All four contractors agreed that the external tank concept appeared to lower peak annual funding requirements from the $2,200 million of the fully reusable con-

cept to about $1,800 million – a significant reduction, but still a long way from the $1,000 being imposed by the OMB. Perhaps the Air Force could help.

BUT NO DoD MONEY

That the space shuttle had survived into 1971 despite a general restructuring of the national space program, and the budget axe of OMB, was due in no small part to the Air Force. At the same time, DoD needs drove the size and capacity of the payload bay, and the general planform via their high cross-range requirement. To a lesser degree, DoD requirements also drove the decision to construct the airframe primarily of aluminum alloys, and the attendant problems this forced on the thermal protection system.

It has been popular to describe the United States as having two distinct space programs – one military and one civilian. But this has never truly been the case. The first NASA astronauts were selected from active duty military personnel, and almost every space shuttle study group, including the SSTG, included members from the Department of Defense. What is interesting is the fact that the Air Force had almost no money that it was willing to expend on space shuttle development. Other than ongoing research at the Flight Dynamics Laboratory, and participation in the various study groups, the Air Force contributed little financially to the early design efforts.

However, the Air Force was willing to support NASA at congressional and OMB budget hearings and this gave the appearance of a united front, without which space shuttle would surely have been killed. NASA briefly toyed with asking the Air Force for financial assistance during 1971 when the OMB imposed their $1,000 million annual funding restriction on the program. Although high-level dis-

Grumman continued to champion the external tank concept, and soon developed designs in addition to those proposed during the ASSC study. This straight-wing concept uses three liquid rocket boosters, in addition to two small boost engines on the orbiter itself. Each LRB contained all of its own propellants, and the center LRB also furnished propellants to the orbiter's boost engines during ascent. (Grumman Aerospace)

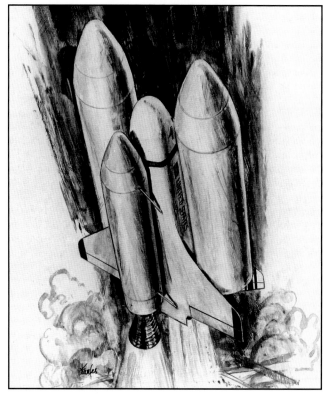

This Grumman delta-wing orbiter carried all of its propellants externally in two much enlarged drop tanks. A single liquid rocket booster was carried under the orbiter and provided additional first-stage thrust. It could be recovered after it was jettisoned and reused several times, further reducing the cost of the program. This particular orbiter used three new-development main engines. (Grumman Aerospace)

cussions undoubtedly took place, the idea never matured sufficiently to become an official request. This was probably just as well since the Air Force would likely have declined to invest any significant amount of money into a program they were somewhat skeptical of in any case.

PHASED DEVELOPMENT

Regardless, NASA still needed to lower the projected annual funding for shuttle development. During June 1971 the new NASA Administrator, James C. Fletcher, embraced what the agency had previously rejected – a phased development approach. This called for the extensive use of interim systems, mostly derived from existing Air Force and Apollo components, that would allow the development of an initial space shuttle to continue within the budget. A follow-on vehicle, using the same basic airframe, would include more advanced systems at some later date.[24]

It was widely recognized that such an approach was wasteful and carried potentially serious political consequences. Wasteful because money would need to be spent on finding, evaluating, procuring, and testing the interim systems. Although no development, *per se*, would be required on these systems, it had often been found that adapting them to new uses could be almost as expensive as developing new systems – and you still do not end up with what you wanted in the first place. The idea was politically dangerous simply because, in Washington, few things are so permanent as a temporary solution. If NASA could begin flying any sort of space shuttle, even an interim design, the agency might face stiff opposition to funding the follow-on system it really wanted. The attractiveness of the idea was that a phased approach would spread the program over a longer period of time, thereby reducing the peak annual funding level – although the total program would ultimately cost more. This concept became public knowledge on 16 June 1971 when Fletcher released a statement saying "We now believe a 'phased approach' is feasible and may offer significant advantages."[25]

Now somebody had to determine what systems would fall into the interim category. The space shuttle main engine (SSME) was a likely candidate, mainly because it had been slated for early development since it was going to power both the booster and orbiter stages. But if the choice was made to use a smaller booster, or an existing expendable stage instead, then the SSME might be able to be deferred. An alternative existed for the orbiter in the form of an improved Rocketdyne J-2, called the J-2S, that required only minimal development and would be sufficient to power an orbiter with a greatly reduced payload (15,000–20,000 pounds). Max Faget and MSC had already been using J-2S engines in some designs. This was not an ideal answer, but it would allow the postponement of a large – and expensive – development effort.[26]

But most of the cuts in peak funding would come from postponing plans to develop the fully-reusable fly-back booster. Even the smaller aluminum vehicles* proposed during Phase B Prime carried significant development costs and risk. Instead, NASA would use a derivative of an existing expendable booster to propel the orbiter to staging velocity. A three year delay in beginning booster development would cut the peak funding requirement to $1,300 million, still over the OMB mandate, but getting close.

There were three leading candidates for an expendable shuttle booster. The most obvious was the Boeing S-IC stage from the Saturn V – Boeing had already been investigating this idea as part of their involvement with Grumman during the ASSC studies. Although there were no pending orders for additional Saturn Vs, the Boeing production line remained mothballed in place and could turn out modified S-ICs with minimal changes. The second candidate was the Titan IIIM, developed (but never built) for MOL. Although the IIIM itself was too small to carry the proposed interim orbiter, it could be scaled up into the Titan IIIL ('large') that used a 16-foot diameter core instead of the normal 10-foot unit, and would have up to six 120-inch diameter solid rocket boosters arranged around the core to provide additional thrust. The solid boosters were already in production for the Titan IIIC, and could be lengthened by adding more segments[†] if necessary. The third possibility was a derivative of the Saturn IB booster, although this vehicle was out of production and did not have the lift capacity to boost the orbiter without modification. Nevertheless, there was a significant body of NASA officials that appreciated the simplicity of the smaller Saturn IB. Costs estimates for each flight ranged from $30 million for the Titan IIIL to $73 million for the S-IC stage, with the Saturn IB somewhere in between.[27]

Engineers at MSFC also had an idea – a pressure-fed expendable booster. This amounted to returning the liquid-fueled rocket to its earliest origins, using concepts selected for their simplicity and low cost. The most expensive items on the modern liquid rocket engine are the turbopumps that are required to feed propellants into the combustion chamber under tremendously high pressure. These turbopumps are also the main reason that liquid rocket engines are so efficient at producing power from relatively small and lightweight packages.[28]

The concept from MSFC eliminated the turbopumps and proposed to 'push' the propellants into the engine using high pressure in the propellant tanks. The propellant tank structure would need to be much heavier to withstand the required pressures, and the resulting engine would be less efficient than the high-pressure turbopump-driven engines used on most liquid rockets, but the booster and engine would be fairly inexpensive. This was exactly the opposite philosophy from that driving the development of the SSME with its high-technology turbopumps and almost unbelievable operating pressures.

The heavy structure of the pressure-fed booster did have one advantage over the other concepts – it lended itself easily to being recovered and reused. The thick skin used to beef up the propellant tanks could also serve as an efficient hot-structure during reentry. A set of parachutes could be fitted, gently lowering the booster into the ocean about 200 miles downrange from the launch site. Ships would recover the booster, which could then be refurbished and reused. Not terribly elegant, perhaps, but workable.[29]

There was another idea. Adding wings and stabilizers to various boosters to make recoverable vehicles had been investigated during the Air Force SR-89774 studies of 1957–62, and Boeing had participated in studies of recoverable Saturn boosters under NASA contract as early as 1962 (see page 52). Boeing dusted off these ideas and suggested a winged version of the S-IC stage. Although the S-IC had not been designed to reenter the atmosphere, Boeing proposed making the skin thicker to provide a sufficient hot-structure. Many began to get carried away with this concept, believing this would offer a fully-reusable system with a minimal development cost.[30]

---

* Referred to in the press, and even official correspondence, as 'aluminum boosters.'
† The Titan IIIC used 5-segment SRBs, while the Titan 34D used 5.5-segment boosters. The IIIM would have used 7-segment boosters if it had been built.

Another possibility was the use of a solid rocket motor. There were four major manufacturers of solid rockets in the United States – Aerojet General, Lockheed Propulsion, Thiokol, and the United Technology Center (UTC). All were sure that some combination of solid rockets could be used as an interim booster for the space shuttle.

When the concept of using solid-propellant rocket stages for space shuttle began to be openly discussed, many questioned the idea. Solid rockets had never been used for a piloted vehicle* – in the U.S. or the Soviet Union – and NASA had little experience with the technology. But solids were far from new. Several small boosters were based on them, as were all Navy submarine launched ballistic missiles (SLBM) and the Air Force Minuteman ICBM. What was relatively new were large segmented solids – previously, solid rockets had been manufactured as a monolithic cases with the appropriate bolt-on nose caps, nozzles, etc., limiting their practical size.

Aerojet appears to have been the first contractor to actively develop segmented solids. During 1957 Aerojet cut a 20-inch-diameter Regulus II booster into three segments, then put them back together using bolted flange joints. Aerojet subsequently tested the motor, achieving nearly identical performance to the single-piece unit.[31]

It was becoming obvious that solid motors would soon become too large to be transported (except by barge), so segmenting them clearly made sense. No doubt encouraged by Aerojet's preliminary successes, in April 1959 the Air Force awarded the company a $495,000 contract for a series of firings using 100-inch-diameter motors. The first step was testing a 65-inch-diameter Minuteman first stage that was cut in half and rejoined with a lock-ring (also called lock-strip) joint. This joint had outer and inner lock strips held together by a key in the center and an O-ring inboard – both key and O-ring holding portions of the joint together and preventing leakage. This motor yielded 160,000 lbf during a 60 second test on 5 May 1961.[32]

Next, Aerojet developed a 100-inch-diameter test-weight motor (TW-1) consisting of three segments – a rolled and welded barrel center segment plus two welded and spun hemispherical sections for the head and the nozzle ends. Aerojet successfully tested the motor on 3 June 1961, yielding 450,000 lbf for over 45 seconds.

Aerojet then tested two flight-weight motors – FW-1 had two center segments, while FW-2 had three. One of the center segments on FW-1 used the refurbished segment case from the TW-1 motor to demonstrate it could be reused. The FW-1 motor was tested on 26 August 1961 and subsequent inspection showed that most parts could be used again after refurbishing. All of the motor parts were thought to be reusable after a 88.1-second test of FW-2 on 17 February 1962.

This concluded the first phase of the Air Force Large Segmented Solid Rocket Motor Program, although Aerojet subsequently tested two additional motors – FW-3 had five center segments while the FW-4 motor had only two. The Aerojet 100-inch-diameter tests were not completely successful since there had been anomalies with both the

thrust-vector-control system and ablative nozzles, but they had demonstrated that a segmented design was a viable solution for the transportation of large rocket motors.

By this time both Lockheed and UTC[†] had also tested large segmented motors. UTC, for example, had tested motors measuring 87 and 90 inches in diameter and featuring either three or four segments (in those orders) during 1961. Lockheed followed with three motors measuring at least 100 inches in diameter from 1962 to 1964.[33]

Subsequently, in May 1962 the Air Force awarded UTC a contract to produce the strap-on solid rocket motors for the Titan III launch vehicle. Only a year later, on 20 July 1963, UTC successfully test fired a five-segment prototype of the motor. The 5-segment Titan SRM had a propellant configuration consisting of an 8-point star in the forward closure of the motor, with the internal perforation becoming cylindrical through the rest of the motor including the aft closure. The motor was 84.65 feet long and 10 feet in diameter, with a case made of D6AC carbon-bearing steel. The segments were joined by tongue-and-groove clevis joints with o-ring seals. This development was a major advance in large solid-rocket technology, eventually paving the way for the space shuttle boosters.[34]

In late 1962 the Air Force Rocket Propulsion Laboratory (AFRPL) at Edwards AFB and NASA jointly initiated a program to develop even larger solid-propellant motors. The Air Force provided funding for 120- and 156-inch-diameter segmented motors and for further work on TVC systems, while NASA provided funding for part of the 156-inch research and a follow-on 260-inch program. TVC was integral to the effort because fixed nozzles were thought to offer the lowest design risk, but methods had to be found to provide directional control during ascent. Liquid thrust vector control,[‡] jet tabs, and jet vanes were all tested with various degrees of success. In the process, Lockheed developed a Lockseal® mounting structure that allowed the nozzle itself to gimbal to provide thrust vector control, much like a liquid-fueled rocket. Thiokol later succeeded in scaling this concept up to the size necessary for the large motors (becoming known as Flexseal).

Another critical technology to be developed was case material for the 156-inch (and later, 260-inch) motors. A new material that seemed promising was maraging steel – a tough, strong steel that was low in carbon and high in nickel, and that formed hardening precipitates as it aged. Applications for the maraging steel varied from submarines and hydrofoils to rocket cases. Maraging steels with 18-percent nickel were the primary focus of the testing because they demonstrated greater fracture toughness than conventional steels with the same level of strength, permitting thinner (thus lighter) cases.[35]

Meanwhile, AFRPL awarded Lockheed a contract for a 120-inch motor with a single center segment using a case made from H-11 steel that was rolled and welded by Exelco of Silver Creek, New York. The motor contained 164,000 pounds of propellant and used a fixed nozzle with two separate liquid TVC systems. The motor was successfully tested for 123 seconds, yielding 350,000 lbf, on 12 May 1962. In a demonstration of the possibility of refurbishing case segments, this one was subsequently reused by Aerojet in a subscale motor for a 260-inch test and then by Thiokol to evaluate a hot-gas-injection TVC system.

Lockheed also conducted the first 156-inch-diameter motor test, on 28 May 1964. This motor featured a single center segment plus fore and aft closures and used a case made of rolled and welded maraging steel manufactured by Exelco. The motor was 70.4 feet long and produced

---

* The escape rockets on Mercury and Apollo had used solid rocket motors.

† United Aircraft Corporation purchased increasingly large shares of UTC, and eventually assumed the name. At that point the solid rocket division changed its name to the Chemical Systems Division.

‡ In liquid TVC, metering valves inject a liquid (usually Freon 114, strontium perchlorate, or nitrogen tetroxide) into the exit cone of the nozzle, creating a shock pattern that causes the exhaust stream to deflect in the desired direction.

949,000 lbf during a test lasting 108 seconds. After refurbishing, Lockheed reloaded the same case and conducted a 142.8-second test on 30 September 1964 that yielded 1,094,110 lbf, a record at that time.

Meanwhile, AFRPL had awarded Thiokol a contract to develop a 156-inch motor employing a gimballing nozzle. The motor and nozzle operated successfully except for a failure of the silica phenolic cloth exit cone 30 seconds into the test. Nevertheless, the motor yielded 1,471,000 lbf and burned for 126 seconds. A follow-on test took place on 27 February 1965 using a motor that had a 100.5-foot maraging steel case with two center segments. It yielded 3,250,000 lbf in a test lasting 58.7 seconds, far exceeding any previous motor.[36]

The next step in the 156-inch program was a 75.1-foot long Lockheed flight-weight case that used a fixed nozzle submerged in the motor itself with a liquid TVC system. On 14 December 1965 the motor was tested for 55.25 seconds, producing 3,107,000 lbf. Lockheed followed this with a monolithic (unsegmented) flight-weight motor that was 33.3 feet long with a deeply-submerged ablative nozzle and a nitrogen-tetroxide TVC system. On 15 January 1966 this motor was tested for 65.0 seconds, yielding a maximum of 1,025,000 lbf.

The three final motors in the 156-inch series were developed by Thiokol. Two of these had fiberglass cases – a monolithic 21.1-foot motor and a segmented 58.3-foot motor with one center segment and fore and aft closures. The first was tested on 13 May 1966 in a diffuser chamber to provide altitude simulation, but was not completely successful. The second motor was tested on 25 June 1968 for 118 seconds and yielded 1,089,000 lbf. Following the firing, a hydroburst test caused the case to fail just short of 1,100 pounds per square inch, meeting the minimum safety factor of 1.25, the firing having produced 806 psi.

The third motor had a maraging steel case and featured a TVC nozzle that used a Flexseal bearing. Lockheed had also bid on this contract, but Thiokol won despite Lockheed having developed the Lockseal® concept. This 34.5-foot long motor used a monolithic case and yielded 983,000 lbf during a 77-second test on 26 May 1967. The Flexseal bearing performed as expected.[37]

For the final phase of the program, during 1963 the AFRPL awarded parallel $25 million contracts to Aerojet and Thiokol to develop monolithic 260-inch motors man-

ufactured out of rolled plates of maraging steel. NASA Lewis took over responsibility for the 260-inch program on 1 March 1965. Meanwhile, Thiokol contracted with the Newport News Shipbuilding Corporation to build its case, which was larger than any previous rocket motor – in fact, it was about two-thirds the diameter of a Polaris submarine hull. Apparently Thiokol incorrectly specified the heat-treatment for the case, and on 11 April 1965 the case failed during hydrotesting at 540 pounds per square inch (psi) of pressure, slightly more than half of the 1,040 psi design strength. The cause was determined to have been a pre-existing flaw in the 0.75-inch-thick steel next to a longitudinal weld. This effectively eliminated Thiokol from the 260-inch program, ending its long series of successes in scaling up rocket motors.[38]

Aerojet set more conservative standards with the Sun Shipbuilding and Dry-dock of Chester, Pennsylvania, and the Aerojet cases passed their hydrotests. Aerojet had the first 260-inch case floated by barge from Pennsylvania to its facility in Dade County, Florida, where it arrived during a hurricane but was undamaged when winds beached the barge. The motor was 80.7 feet long, about half that needed to replace a Saturn booster. Aerojet mixed the 1,676,000 pounds of PBAN propellant using both batch and continuous mixers, resulting in no discernible difference in physical properties or performance between the two procedures. A 150-foot-deep caisson sunk into the ground provided a pit for casting and curing the propellant in the case, and then served as a test stand, with the motor firing vertically. The SL-1 test took place on the night of 25 September 1965, with the firing lasting 113.7 seconds and producing 3,567,000 lbf – the flame was visible 35 miles away in Miami. A second identical motor (SL-2) was tested on 23 February 1966 with comparable results. A follow-on 77-foot long motor was tested on 17 June 1967, yielding 5,884,000 lbf during a 77-second firing – achieving a record for a single rocket propulsion unit that stood until at least 1994.[39]

There was no direct application of the technologies from the 260-inch program, but the Large Segmented Solid Rocket Motor Program provided experience, designs, materials, methods of fabrication, and test results eventually contributed to the development of the space shuttle solid rocket boosters when the program finally abandoned thoughts of liquid boosters.

The plume from the SL-2 motor during the 260-inch program. The 3,500 million candlepower plume was visible in Miami, 35 miles away. The only part of the motor that is visible is the nozzle – the remaining 60-feet is below ground. The firing produced 3,567,000 lbf and burned for almost two minutes. (Miami-Metro News Bureau photo via Aerojet)

A technician suspended on a rope inspects the 500,000 square inches of propellant surface inside the SL-1 test motor for the 260-inch program. Note the 3-point star shape. This area of propellant would burn at over 5,500 degF, and propellant was expended at a rate of six tons per second during the firing. (Aerojet-General photo 2-66-SP-000)

Choosing a Launch Site

Almost all of the early studies included evaluations of possible launch sites other than the obvious ones at Cape Canaveral and Vandenberg AFB. By 1970 when it became obvious that a space shuttle most probably would actually be built, the choosing of a launch site became a political issue. It gained even more importance when acting NASA Administrator George M. Low told a congressional committee that shuttle was the "keystone of the whole space program during the coming decades."[40]

During the summer of 1970 the NASA Office of Facilities made it known it was looking at possible alternate launch sites. This led to widespread competition for the facility, with NASA eventually receiving over 100 unsolicited bids from most every state in the Union. The most serious came from delegations representing Florida (Cape Canaveral), California (Vandenberg AFB, and also Edwards AFB), New Mexico (White Sands Missile Range and Holloman AFB), and Utah (Wendover AFB on the Dugway Proving Grounds). To eliminate possible charges of favoritism, NASA awarded a $380,000 contract to the Ralph M. Parsons Company in Los Angeles to perform reviews on potential locations, and also established a 14-member Space Shuttle Facilities Group to select the final site. The group was chaired by Robert H. Curtin of the NASA Office of Facilities, and included ten other NASA members and three Air Force representatives.[41]

The potential importance of the launch site to a local economy was tremendous, although not as great as it had been during Apollo. During the peak years of the Moon program, employment at KSC had peaked at just over 20,000 with an annual budget of almost $90 million. Regardless of some of the fantasies contained in the various space shuttle studies, the Office of Facilities was estimating that shuttle would employ about 6,000 people with an annual budget of $20–30 million.[42]

The fact that NASA had already invested over $1,000 million in launch facilities at KSC obviously made it a logical choice. However, much of the rationale that had always dictated a coastal location – namely having a large ocean to drop expended stages in – was not necessarily applicable to space shuttle, at least initially. KSC had always presented concerns over corrosion and hurricanes, and potential inland sites would eliminate those. However, the best launch sites are located close to the equator in order to take advantage of the Earth's rotation during easterly launches, and KSC was as near the equator as anywhere in the U.S. mainland that was likely to host a launch site. Estimates to modify facilities at KSC ran $200–400 million, while the cost of a new facility was double that amount.[43]

Many states and local governments committed funds for studies, ad campaigns, and frequently to purchase land or improvements. Leading these, of course, were California, Florida, and New Mexico. Senator Alan Cranston (D-California) formed a statewide task force of business, labor, and political leaders to ensure the launch site ended up at either Edwards or Vandenberg. In some cases, local groups used non-government funds, such as the $50,000 raised by private subscriptions in Alamogordo, New Mexico (White Sands). Congressional hearings were held, and in early 1970 Representative Olin E. Teague (D-Texas) had the House Science and Space Committee issue a directive ordering NASA to favor existing facilities, or have very good explanations otherwise.[44]

The entire issue, normally very serious, did have its lighter moments. For instance, Mill Creek, Oklahoma,

wrote NASA suggesting that it's municipal airport would make a good launch site – Brownsville, Texas, had a similar suggestion. A group in Boardman, Oregon, thought a under-utilized local industrial park would be appropriate. A gentleman in Maine offered to sell NASA 150 acres. And finally, there was a bid submitted for Mackinack and Chippewa counties on the upper Michigan peninsula. Richard Lyons of the *New York Times* finally tracked down the source of the bid. It had come from an unemployed truck driver who needed work, and explained: "Some of my friends and I were drinking it up a little at the town bar, and this guy came in who had just read about the space base competition ..."[45]

Other than the established launch locations at KSC and Vandenberg, the White Sands Missile Range stood the best chance of selection. This was the largest military reservation in the United States (4,000 square miles), and already had a small NASA installation (the White Sands Test Facility) that was administratively controlled by MSC in Houston. Since it was a missile range, much of the infrastructure (radar, telemetry, etc.) already existed, although it would need significant augmentation for shuttle duties. The site was at an elevation of 4,000 feet, providing a small increase in payload capacity since the first mile of atmosphere was already behind it, although it was much further north than KSC. Even as expendable boosters began to be considered, White Sands did not necessarily loose its appeal since it controlled a large land area that could – potentially – be used as an impact area for the boosters. And since it could theoretically launch in all directions, it would allow military and NASA flights from a single location, reducing funding requirements.[46]

As long as a fully-reusable system was being developed, the location of the launch site could be decided on other than purely technical or safety issues. But in the end, as it became obvious that some sort of partially expendable system would be selected for space shuttle, the choice of launch sites became much easier – particularly with the unveiled threats from Congressman Teague. The last site to be eliminated was White Sands. NASA announced on 14 April 1972 that the primary space shuttle launch site would be located at KSC and a smaller Air Force launch site would be built at Vandenberg AFB.[47]

KSC would be used for easterly launches, accounting for the vast majority of NASA missions. Vandenberg would be used for polar launches, accounting for most Air Force missions, plus a few NASA science missions. A single launch site was not possible since it is political infeasible to launch polar missions from KSC (they would overfly Cuba, and also pose safety problems with southern Florida and Mexico, or overfly portions of the northeast U.S. and Canada). At the same time it is not possible to launch toward the east from Vandenberg because of safety concerns, and the fact there is no place to drop expendable stages (or propellant tanks). Even given its large land area, White Sands had proven a bit too small to accommodate all-azimuth launches with expendable stages. In announcing the selection, NASA's George Low estimated it would cost $150 million to modify facilities at KSC, with another $500 million being required to modify the former MOL facility at Vandenberg.[48]

The day before the public announcement the DoD had written a letter to James Fletcher concurring with the selection. In this letter it was stated that "a second operational site for missions requiring high inclination launches not feasible from KSC is planned at Vandenberg Air Force Base toward the end of the 1970's." As it turned out, a very optimistic forecast.[49]

SAFETY IN EARTH ORBIT

On 12 July 1972, North American Rockwell released the final report from the Safety in Earth Orbit Study. This 12-month study had examined five specific safety issues associated with low-earth orbit operations of manned spacecraft. Three different vehicles were considered during the study, including the fully-reusable two-stage shuttle vehicle proposed by North American during Phase B, the MSC-040A-derived orbiter being investigated during Phase B Prime, and the various space station concepts also undergoing separate Phase B studies at the time. The five specific issues included (1) hazardous payloads, (2) docking, (3) on-board survivability, (4) tumbling spacecraft, and (5) escape and rescue.[50]

The main concern of the study was personnel safety, and a lesser emphasis was placed on preventing damage to, or loss of, the vehicles. The analysis was confined to the manned on-orbit phase of the missions, and launch, ascent, deorbit, reentry, and landing were not considered. The study drew on earlier analyses conducted by The Boeing Company on space station safety, Lockheed Missiles and Space Company on space shuttle safety, and ongoing studies by The Aerospace Corporation on shuttle safety and possible escape and rescue systems. Pennsylvania State University provided information on tumbling dynamics.[51]

During the investigations of hazardous payloads, North American concluded that either orbiter design was extremely sensitive to even small explosions in the payload bay. Uncontained explosions equivalent to as little as 0.01 pound of TNT could exceed the structural limits of the payload bay structure. By comparison, a normal hand grenade was equivalent to approximately 0.025 pound of TNT, and a fully fueled Centaur upper stage was equivalent to approximately 6,000 pounds of TNT.[52]

The study also concluded that any structural failure of a loaded liquid-fueled upper stage while in the payload bay which resulted in large leaks of both oxidizer and fuel would almost certainly be catastrophic to the orbiter and crew. The energy content of even the smallest liquid upper stage, if released suddenly, would be far more than could be tolerated by the payload bay structure. It was discovered that vehicle accelerations caused by the reaction to leaking fluids would ensure mixing, and an ignition source would undoubtedly be present during the process

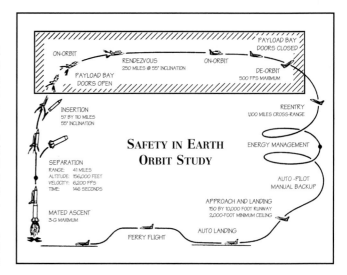

Typical mission profile examined by the Safety in Earth Orbit Study. Note the use of a pressure-fed booster concept. (North American Rockwell)

of structural failure. The study noted that the chemical and physical behavior of gases, liquids, and cryogenic fluids was not well understood in zero-g, and that the possibilities could not be fully explored due to this. Other hazardous payloads examined included pressurized space station modules, since the rupture of a 5,000 cubic foot module pressurized to one atmosphere would release an energy equivalent to 22 pounds of TNT.[53]

The main recommendations from the hazardous payload portion of the study was that the payload bay doors should be opened as early as feasible and closed as late as possible while on-orbit. This would minimize the potential hazards from leakage and explosions. Also, the liquid content of any payload being returned to Earth should be dumped while in space to avoid the possibility of an uncontrolled increase in payload-internal pressure during reentry. Nevertheless, the orbiter should have the capability to return with a fully-fueled upper stage if required. The study recommended that all upper stages carried by the orbiter should be man-rated, that is, subjected to the same safety criteria applied to a manned vehicle. Other recommendations included strengthening the orbiter payload bay structure to the maximum extent possible, and increasing the venting capability of the payload bay with the doors closed, to better withstand accidents.[54]

The portion of the study concerned with docking concentrated on examining the relative merits of three different docking modes. The first was direct docking the shuttle orbiter to the space station, similar to the concept used by Project Apollo. The second involved the use of manipulator arms, either on the space station or the orbiter, to provide a more mechanically determinate berthing maneuver at a lower contact velocity than was thought practical with direct docking. The last option was an extendable soft-dock system that used a flexible tunnel to provide a large distance between the co-located vehicles.[55]

The study concluded that each of the three docking techniques could be made adequately safe. All three systems had the potential of damaging the vehicle or the docking system, and the damage could conceivably be critical enough to result in the loss of the vehicle and crew. The manipulator docking system had the minimum potential for inadvertent collision between vehicles, but had more failure modes than the other systems. The study noted that the hazards and risks of the three systems were not equally well understood, mainly because of the differ-

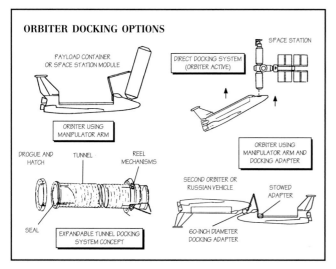

Possible orbiter docking options examined in the study. All of these assumed a separate airlock on the nose of the orbiter. (North American Rockwell)

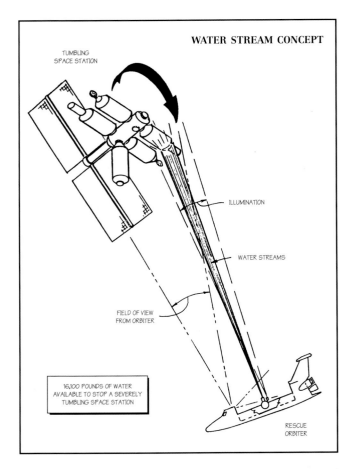

WATER STREAM CONCEPT

TUMBLING
SPACE STATION

ILLUMINATION

WATER STREAMS

FIELD OF VIEW
FROM ORBITER

16,100 POUNDS OF WATER
AVAILABLE TO STOP A SEVERELY
TUMBLING SPACE STATION

RESCUE
ORBITER

One of the more interesting ideas to come out of the study was this concept to control a tumbling space station. A stream of high pressure water would be directed at the tumbling space station from a direction opposite of its tumble. introducing a high drag coefficient on the space station, slowly stopping it. (North American Rockwell)

ent development status of the systems. The direct docking system was relatively well understood from Gemini and Apollo experience; the manipulator system had been defined somewhat during the Phase B shuttle studies, but had not been tested or extensively simulated; and the extendable tunnel system was only in a conceptual stage.[56]

The on-board survivability analysis was used to evaluate the personnel traffic patterns, escape routes, and other on-board survivability issues from a safety standpoint, without particular regard to mission effectiveness. After examining these issues, the study recommended that quick-donning pressure suits that did not require pre-breathing be provided for all on-board personnel. The pressurized compartment of the orbiter should be divided into two sections by a partition which could exclude smoke and flames. These compartments should be capable of being separately pressurized, and offer protection against excessive heat from a fire. The study noted that a separate airlock was not required* if the above recommendations were met, however it pointed out that sizing the airlock to accommodate all possible passengers eliminated the need for two compartments and pressure suits (except for the flight crew, who would need to remain on the flight deck regardless of the situation).[57]

Uncontrolled tumbling of a spacecraft following the loss of its attitude control capability was another area of study. For a vehicle with reentry capability, such as n orbiter, neither deorbit nor reentry would be possible

under these conditions. Also, a rescue vehicle might be unable to assist since it would be incapable of docking to a rapidly tumbling spacecraft. Such a situation could, therefore, be catastrophic and result in the loss of both the vehicle and crew.[58]

North American studied some novel ways for arresting the motion of an out-of-control tumbling spacecraft by means <u>external</u> to the vehicle in order to save on-board personnel and, if possible, the vehicle. The tumbling spacecraft considered in the study included shuttle orbiters, space stations, and individual small space vehicles (SSV). Sortie modules or space station modules were considered typical of SSVs in size, mass properties, and geometry. The tumbling vehicle was considered to be disabled and non-cooperative, so that use of on-board systems or personnel to assist in arresting tumbling was not feasible.[59]

The rescue vehicle was always considered to be an orbiter with an appropriate payload. All the concepts considered for arresting tumbling could, however, equally well be used by remote control mode from an unmanned tug – this might have been more acceptable where the tumbling vehicle posed an unacceptable risk to the orbiter. Where the crew of a tumbling spacecraft needed to evacuate, an orbiter was considered to be near enough to provide refuge within the scope of the disabled crew's life support suits.[60]

Two separate concepts were investigated, the first involving a water stream directed at the tumbling spacecraft, the other using 'stick-on' rockets that attached themselves to the out-of-control vehicle. Both concepts were considered feasible, low development cost, operationally practical concepts. Each concept could be operated with an adequate margin for the worst case tumbling orbiter, space station, or SSV with only one rescue orbiter.[61]

The water stream concept appeared to be most attractive. It could be developed from off-the-shelf, non-spacecraft hardware having large factors of safety to minimize qualification testing. It was felt that the system could be stored at the launch site for long periods of time without any upkeep cost, and be made ready for launch within a couple of hours. It could deal with any configuration and size of spacecraft, was simple to use, tolerant of errors, and required little training. The water streams themselves posed little danger to the tumbling spacecraft, other than to delicate appendages such as solar panels and radio antennas. In this concept, the rescue vehicle

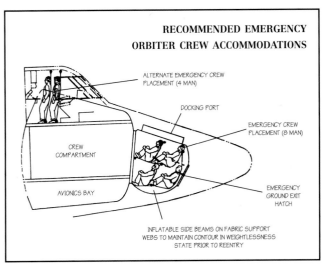

RECOMMENDED EMERGENCY
ORBITER CREW ACCOMMODATIONS

ALTERNATE EMERGENCY CREW
PLACEMENT (4 MAN)

DOCKING PORT

EMERGENCY CREW
PLACEMENT (8 MAN)

CREW
COMPARTMENT

AVIONICS BAY

EMERGENCY
GROUND EXIT
HATCH

INFLATABLE SIDE BEAMS ON FABRIC SUPPORT
WEBS TO MAINTAIN CONTOUR IN WEIGHTLESSNESS
STATE PRIOR TO REENTRY

The separate forward-located airlock could also be used as a safe-haven in the event of emergencies in the main compartment. (North American Rockwell)

---

* This was because one of the compartments could be utilized as an airlock as required.

contained a water tank, a pump, and a remotely controlled nozzle.[62]

The jet of water would be directed by positioning and orienting the rescue vehicle so that the water stream impinged on the tumbling spacecraft to reduce its angular momentum. Since the stream would be visible, and be moving in a straight line, the operator could easily adjust the rescuing vehicle orientation so that the stream impacted the tumbling vehicle at the desired point. The rescue orbiter would position itself roughly 200 feet from the tumbling spacecraft, and the water jet would be delivered at a pressure of 1,100 psi. The worst case scenario would require delivering 16,100 pounds of water to arrest the tumbling of the late-1970s space station design. The reaction of the water jet on the rescuing vehicle could be counteracted by the attitude control system, potentially requiring a great deal of propellant.[63]

The stick-on rocket concept also appeared feasible. It could use existing small solid rocket motors to minimize development costs, was light enough so that a single rescue orbiter could still achieve a degree of 'over-kill' to ensure success, and had a long, low-maintenance storage life. Like the water stream, it could deal with any anticipated size or configuration spacecraft, was relatively simple to use, and tolerant of errors. However, compared to the water stream concept it did have a number of disadvantages. Because it contained propellants, it posed a hazard during storage, to the orbiter during ascent, and to the tumbling vehicle. It would have required the development of some attachment mechanism, and this mechanism could also result in damage to the tumbling spacecraft, possibly breaching the pressure vessel. This concept involved firing many small rockets that would attach themselves to the tumbling spacecraft, to be ignited upon command to counter the rotational momentum of the vehicle. The trick would be to get the rockets to attach themselves to the correct side of the tumbling vehicle. It was expected to take approximately 200 rockets of 40 pounds each for the space station.[64]

The study went on to discuss other possible rescue schemes, including large nets that could be snag the tumbling spacecraft. It also investigated possible actions that could cause a spacecraft to begin tumbling, as well as the dynamics of a tumbling spacecraft, and proposed some means of escaping from them.[65]

The purpose of another portion of the study was to review and analyze different rescue and survivability concepts defined in earlier studies, and to determine the adaptability of these concepts to the space shuttle orbiter and space station. The most applicable of these concepts were to be identified and studied for adaptation to crew sizes of up to 6 or 12 men. Three different terms were defined for the purpose of the study:[66]

- Escape, meaning to egress from a distressed vehicle and return to Earth without outside assistance,
- Rescue, the use of outside assistance to effect the return of personnel from a distressed vehicle, and
- Survivability, the use of a vehicle or equipment to separate from a distressed spacecraft and provide a safe haven in orbit until rescue could be effected.

The groundrules of the study were that both the space shuttle and space station would use this capability at some time. However, no attempt was made in the study to determine the possible causes for requiring abandonment from the spacecraft. Also, no determination as to the statistical probability, relative or absolute, of such causes, or as to how much time might be available to the crew, was made. In all the evaluations, it was assumed that the potential escape, rescue, or survivability devices were accessible to the crew, that they were able to operate them, and that they had sufficient time to do so.[67]

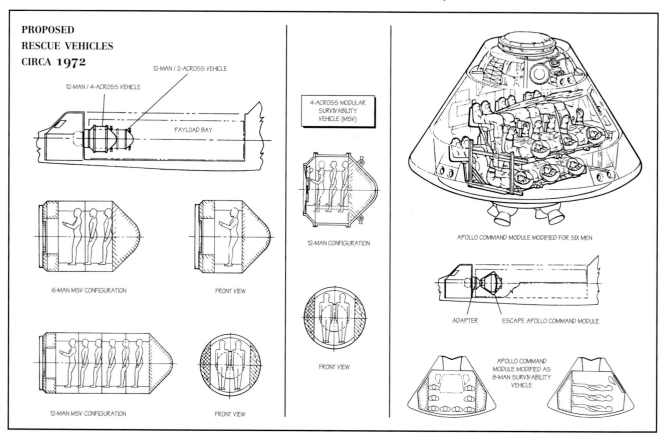

PROPOSED
RESCUE VEHICLES
CIRCA 1972

12-MAN / 2-ACROSS VEHICLE
12-MAN / 4-ACROSS VEHICLE
PAYLOAD BAY

4-ACROSS MODULAR SURVIVABILITY VEHICLE (MSV)

12-MAN CONFIGURATION

APOLLO COMMAND MODULE MODIFIED FOR SIX MEN

6-MAN MSV CONFIGURATION   FRONT VIEW

FRONT VIEW

ADAPTER   ESCAPE APOLLO COMMAND MODULE

12-MAN MSV CONFIGURATION   FRONT VIEW

APOLLO COMMAND MODULE MODIFIED AS 8-MAN SURVIVABILITY VEHICLE

The study concluded that the space shuttle should be the primary vehicle for dealing with emergencies involving manned spacecraft in low-earth orbit. It recommended that a shuttle orbiter be available for rapid emergency rescue whenever manned orbital flight was in progress. The study pointed out that this did not necessarily mean a dedicated rescue shuttle, but could involve normal operational vehicles on which a variety of rescue kits could rapidly replace the planned payload in an emergency. It should be noted that this was based on the early estimate that roughly one shuttle flight a week would take place during the operational era.[68]

There were, however, several alternatives discussed for possible use during the early period of the shuttle program when a shuttle rescue was not possible because an orbiter was not available. The recommended approach was to keep a modified Apollo CM stacked on a Saturn IB or Titan IIIC booster available at all times – it was expected that several surplus or refurbished CMs could be made available during the 1974–78 time period (when shuttle was expected to become operational) at minimal cost. These would be modified to seat between six and ten people in an extremely uncomfortable, but nevertheless survivable, configuration. The service module was not required, but a retro-package of some sort would have needed to be developed for deorbit.[69]

Ten additional evacuation concepts were also investigated, although most of them had been originated by various other manufacturers. The ideas ranged from small lifting-body lifeboats to individual seats protected by an undeveloped foam insulator that would allow astronauts to reenter the atmosphere individually. None had been tested to any great degree, and indeed, until the space shuttle was available to ferry test versions of the concepts into space, it was not economically feasible to test them.[70]

Under the survivability category, five different concepts were investigated. All of these would have provided some sort of life support system to allow crew members (of a shuttle or space station) to evacuate their vehicle and survive until a rescue flight arrived. One of these, 'cocoon,' consisted of encapsulating individual suited crew members in foamed plastic to afford thermal protection. The device was relatively simple and light, but depended upon the crew member's space suit for all life support. Another idea was to have a 'sortie module' (Spacelab or equivalent) attached to either the orbiter or space station that the crew could retreat into, then separate from their crippled vehicle. This module would then provide the necessary life support functions for a limited time period.[71]

The study concluded that the modified Apollo CM was the preferred* escape method, while the sortie module offered the most reasonable chances of surviving until help arrived. Although this study, and several others, pointed out the desirability of having equipment available for rescue, escape and survivability, the reality of economics has resulted in little actual procurement for the concept. At a maximum flight rate of eight per year, it is extremely unlikely that an orbiter could ever be made flight ready quickly enough to effect a rescue of a crippled orbiter in space. The year 2000-version of the International Space Station is again wrestling with how to offer the permanent crew some options in the event of a catastrophic failure, and is currently testing the X-38 prototype of an assured crew return vehicle.[72]

_____

* This was not totally surprising since North American had designed and built the Apollo Command Module (CM), and desired to identify a further use for the vehicle.

TYPE	NAME	CHARACTERISTICS	ESCAPE CONCEPTS
DEPLOYABLE	AIRMAT (GOODYEAR)	• TWO MAN • SUITS REQUIRED • INFLATABLE • EJECTION SEAT • 1,140 POUNDS • NEW TECHNOLOGY REQUIREMENTS • FLEXIBLE HEAT SHIELD • MATERIAL	
	RIB-STIFFENED EXPANDABLE (NORTH AMERICAN ROCKWELL)	• THREE MAN • SHIRTSLEEVE ENVIRONMENT • MECHANICALLY RIGID • CANISTER STORED • 1,452 POUNDS • NEW TECHNOLOGY REQUIREMENTS • ARTICULATING RIB-TRUSS STRUCTURE • MATERIAL	
	PARACONE (MCDONNELL DOUGLAS)	• ONE MAN • SUIT REQUIRED • INFLATABLE • 425 POUNDS • NEW TECHNOLOGY REQUIREMENTS • LARGE INFLATABLE AND DEPLOYABLE STRUCTURE • MATERIAL	
	MOOSE "COCOON" (GENERAL ELECTRIC)	• ONE MAN • SUIT REQUIRED • HAND-HELD RETRO • ALL EQUIPMENT CARRIED EVA • FOAM-IN-PLACE • 475 POUNDS • NEW TECHNOLOGY REQUIREMENTS • FOAM IN SPACE • FOLDABLE HEAT SHIELD	
	ENCAP (LOCKHEED)	• ONE MAN • SUIT REQUIRED • ALL EQUIPMENT CARRIED EVA • MECHANICALLY RIGID • 588 POUNDS • NEW TECHNOLOGY REQUIREMENTS • MECHANICAL DEPLOYMENT MECHANISM • FOLDABLE HEAT SHIELD	
RIGID CONCEPTS	EGRESS (MARTIN MARIETTA)	• ONE MAN • SHIRTSLEEVE ENVIRONMENT • EJECTION SEAT • 820 POUNDS • NEW TECHNOLOGY REQUIREMENTS • MOVABLE CANOPY • NEW HEAT SHIELD • MODIFIED B-58 CAPSULE	
	LIFE RAFT (GENERAL ELECTRIC)	• THREE MAN • SUITS REQUIRED • PERSONAL CHUTES REQUIRED • 936 POUNDS • NEW TECHNOLOGY REQUIRED • NEW HEAT SHIELD • FOAM MATERIAL	
	LIFTING BODY (NORTHROP)	• THREE MAN • SHIRTSLEEVE ENVIRONMENT • 4,330 POUNDS • NEW TECHNOLOGY REQUIREMENTS • NEW HEAT SHIELD • REENTRY TECHNIQUE • HIGH-SPEED PILOT TECHNIQUES	
	EEOED* (NASA / LOCKHEED)	• THREE MAN • SHIRTSLEEVE ENVIRONMENT • 2,769 POUNDS • NEW TECHNOLOGY REQUIREMENTS • NEW HEAT SHIELD	
	APOLLO ESCAPE COMMAND MODULE (NORTH AMERICAN ROCKWELL)	• TWO TO SIX MAN • SHIRTSLEEVE ENVIRONMENT • 10,000 POUNDS • NEW TECHNOLOGY REQUIREMENTS • NONE	

One of the primary concerns of researchers was how to protect any space shuttle orbiter from the thermal effects of reentry. Earlier experience with the X-15A-2, ASSET, and PRIME vehicles had demonstrated that ablative heat shields had some serious operational drawbacks, and in fact, the ablator on the X-15A-2 had failed in flight, forcing the premature retirement of that vehicle. On 17 May 1972 NASA Langley awarded the Denver Division of Martin Marietta a 15 month contract (NAS1-11592) to conduct technical and cost trade-off studies between an ablative heat shield and reusable surface insulation (RSI) techniques of protecting the future orbiter. It should be noted that by this time the reusable surface insulation technique had already been baselined for the orbiter and that the continuing studies were more to ensure that there were alternate approaches if the RSI development effort should fail or be seriously delayed. As Eugene S. Love noted during the 10th von Kàrmàn Lecture "... ablators offer a confident fall-back solution for both leading edges and large surface areas should development of the baseline approaches lag ..."[73]

The studies used mission profiles that were bounded by a 2.5-g acceleration limit and a 26 BTU per square-foot per second heating rate at a reference point 50 feet aft of the fuselage nose on the bottom centerline. Heating rates and total heats were generated for the total orbiter, which was based on the Model 619 orbiter proposed by Grumman for the Phase C contract. The three mission profiles that were examined included:[74]

- Mission 1: This mission was to deliver a 65,000 pound payload to a 115-mile due-east circular orbit. The orbiter would perform an on-orbit translation ΔV of 950 fps. A reentry cross-range of 920 miles was expected;
- Mission 2: This was a resupply mission to a 310-mile 55 degree orbit with a 25,000 pound payload. The orbiter on-orbit translation ΔV was estimated at 1,400 fps. Again, a reentry cross-range of 920 miles was expected; and,
- Mission 3: This was to be a 40,000-pound DoD payload delivery mission launched from Vandenberg into a 100 mile polar orbit and returning to the launch site after one revolution. The on-orbit translation ΔV was 500 fps, and this mission included the maximum expected operational cross-range reentry, 1,265 miles.

The history of ablators in the space program dated to the earliest attempts to protect reentry vehicles, and the materials had been used successfully on Gemini, Apollo, *Viking*, and others. However, the need for labor-intensive refurbishment following each thermal usage all but ruled out their application to an operational vehicle such as space shuttle. One of their primary disadvantages resulted from the fact that early ablators had all been bonded directly to the airframe they were designed to protect. In an attempt to overcome this limitation, Martin Marietta devised several approaches that mechanically attached the ablator either directly or on standoffs that allowed additional batt-type insulation between the ablator and the airframe.

Three different ablators were defined to cover the orbiter – SLA-561 was used for large surface areas, such as the upper wings and fuselage; and ESA-3560HF or ESA-5500M3 were specified for the wing leading edges and other surface areas where heating was too severe for SLA-561. The SLA-561 ablator had a nominal density of 14.5 pounds per cubic foot, while ESA-3560HF and ESA-5500M3 were rated at 30.0 and 58.1 pounds per cubic foot, respectively. It was felt that all of these ablators would be reusable for an unlimited number of missions in locations were the temperature did not exceed 800 degF. In all cases the ablator would not fail during one mission, and was designed for exposure periods 300 percent in excess of those expected during any single mission. The SLA-561 ablator had been developed by Martin for the *Viking* aeroshell which successfully entered the Martian atmosphere in 1976. The other two ablators, also developed by Martin, had been successfully flown on the PRIME test vehicles.[75]

The reusable surface insulation (RSI) concept used for the various comparison studies was the Lockheed-developed LI-1500 which weighed 15 pounds per cubic foot. These tiles were 99.8 percent pure amorphous fibers that were derived from common sand. The study assumed that ESA-3560HF and ESA-5500M3 would continue to be used in extreme high-heat areas such as the wing leading edges and nose cap. The RSI used strain isolators between the tiles and the orbiter airframe. It was noted that the LI-1500 tiles would require a thin outer coating "... for better emissivity characteristics and moisture-proofing..." – a fact that would later haunt the Space Shuttle Program during the orbital flight test series

The studies concluded that an ablative heat shield was indeed possible, and that it would weigh some ten percent less than an equivalent RSI thermal protection system. In the ablator designs it was estimated that approximately 75 percent of the TPS operational cost involved the fabrication of the ablator slab. The other 25 percent encompassed assembly, installation, removal, tooling, repair, and inspection of the ablator. It was further estimated that a typical TPS would account for approximately ten percent of total space shuttle operational costs. Not surprisingly, since Martin Marietta was one of the leading ablator researchers and manufacturers, the study thought that a reusable ablative coating would make a preferable heat shield for an operational shuttle.[77]

Another Martin study completed in 1971 had determined that the cost of fabricating an ablative heat shield panel varied from a high of $296.66 per square foot for a double-contoured panel made of 67 percent silicone resin and 33 percent phenolic microballoons, to a low of $48.96 per square foot for a flat panel made of 90 percent microballoons and ten percent silicone resin. Various other compositions of ablative coatings were investigated, as were possible fabrication techniques in an effort to lower the cost of ablative heat shields, but the study pointed out the continued difficulties in achieving a consistent product.[78]

McDonnell Douglas also studied possible ablative heat shields in an attempt to identify the least expensive thermal protection system for an orbiter. Their report noted that "...a significant fraction of the total operational cost is the vehicle's thermal protection system (TPS) cost ..." The study concluded that the externally removable heat shield panel concept was the most efficient for near optimum system reusability/refurbishability. The panel concept offered minimum weight and shorter turnaround times, primarily since the entire vehicle need not be involved in the TPS refurbishment cycle. McDonnell Douglas suggested that refurbishment techniques be a significant factor in the selection of a baseline TPS for the shuttle, but noted that "... not all aerospace companies agree as to the magnitude of the maintenance costs since there is no historical data to use as a reference ..." The report concluded that further research should be undertaken as a high priority to further define the problems of using a reusable TPS for the space shuttle. NASA Langley continued to study the problem and was firmly convinced that a removable ablative heat shield was the best solution.[79]

But others in NASA remembered the headaches associated with the Martin-developed MA-25S ablator used on the X-15A-2, and had serious doubts if the material could be trusted over the long operational life expected of the space shuttle. Martin countered these criticisms by stating that MA-25S had not been given a fair chance, since it had been used on an extremely limited number of X-15A-2 flights. It was further argued that after sufficient experience had been gained with its use and possible refurbishment techniques had been perfected, that costs would decrease substantially. Nevertheless, although NASA would continue to fund ablative heat shield development and testing at a low level, all bets were on the Lockheed-developed RSI tiles.

The baseline thermal protection system at this point differed somewhat from the system ultimately used on the production Orbiters. The leading edges of the wings and the nose cap where temperatures were expected to reach above 2,300 degF were protected by a composite composed of pyrolyzed carbon fibers in a pyrolyzed carbon matrix with a silicon carbide coating, commonly known as 'carbon-carbon.' This material had been perfected by the Vought Corporation (later LTV, now part of Lockheed Martin) based upon work initially conducted during the Dyna-Soar program.[80]

The bottom of the fuselage and wings, the entire vertical stabilizer, and most of the forward fuselage were protected by LI-2200 high-temperature reusable surface insulation (HRSI) tiles. These areas were expected to experience temperatures less than 2,300 degF, but more than 650 degF at some point during ascent or reentry. The tile thickness varied as necessary to limit the aluminum airframe's maximum temperature to 350 degF. Actually, earlier studies showed two variations of tiles,

LI-2200 and LI-3000 would be required, although this was later reduced to entirely LI-2200, then to a combination of LI-900 and LI-2200. The tiles were to be coated with a silicon carbide waterproof emissivity coating whose thickness depended on the tile's location on the airframe. This coating was to prevent water absorption into the HRSI tiles, and to protect the tiles from erosion due to rain, ice, or dust.

The upper surfaces of the wings and aft fuselage, along with the entire payload bay door area, was to be covered with an ablative elastomeric reusable surface insulation (ERSI). These areas were expected to be subjected to temperatures of less than 650 degF during the entire flight. The ERSI was to be bonded directly to the aluminum airframe and consisted of a silicone resin with titanium dioxide and carbon-black pigments for thermal control.

Further study would show some drawbacks to this protection system, and the ERSI would subsequently be replaced by a combination of lower-temperature versions of the Lockheed RSI tiles, and by Nomex felt blankets when *Columbia* was manufactured. The HRSI tiles were used over a smaller percentage of the orbiter than originally anticipated when further analysis showed that reentry heating would not be as severe as expected, mainly because of improved flight profiles.[81]

Even as late as 1982, studies were being performed on possible alternatives to the silica-based RSI used on the Orbiters. Most of these studies had the goals of reducing the life-cycle costs of the vehicle, and also to improve its operational characteristics. A 1982 study was performed by NASA Langley into various metallic, ablator, and carbon-carbon concepts. The study was conducted by an industry team led by Rockwell, with Rohr Industries being assigned metallic concepts, Vought Corporation looking into carbon-carbon concepts, and General Electric investigating ablators. These assignments were logical, based on the fact that Rohr was a leading metallic alloy manufacturer, Vought built the carbon-carbon TPS used on the Orbiters, and General Electric having provided various

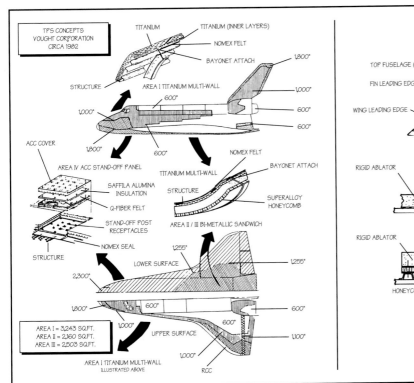

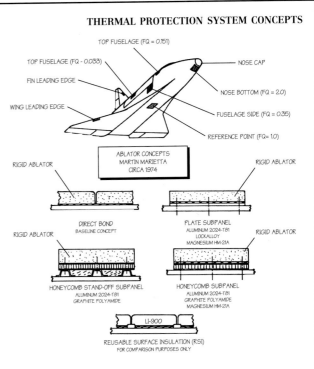

ablators for spacecraft in the past. Battelle Columbus Laboratories provided materials technology data, and Rockwell, by now the prime vehicle contractor, assisted in evaluating the various concepts, coordinating the work effort, and writing the final report.[82]

This particular study was limited to replacing areas of the Orbiter protected with the Lockheed-developed tiles. Other areas, namely those protected by blanket insulation, hot-structure, and the carbon-carbon leading edges and nose cap were not investigated since they were already considered to be the optimum configuration. This study gave an indication of the general uneasiness NASA felt with the technologically advanced, but extremely fragile, RSI tiles. Although these tiles have worked well, they force a high amount of refurbishment work after every flight.

Only metallic, ablator, and carbon-carbon concepts were considered. The concepts investigated fell into three general categories – prepackaged, stand-off, and ablators. In the prepackaged concepts, the insulation was completely encapsulated to form a single unit which would be installed on the vehicle. In the stand-off concepts, insulation was secured between the vehicle skin and a heat shield panel (metallic or carbon-carbon) which was attached to the vehicle via stand-off supports. All the ablator concepts investigated involved applying the ablator directly to the vehicle skin. Interestingly, the prepackaged ablators investigated by Martin Marietta and McDonnell Douglas in earlier studies did not resurface. As a result of their operational limitations, ablators were not seriously considered during the study, and no in-depth evaluations were performed.[83]

In addition to normal thermal and structural analysis for each concept design, a series of trade studies, including panel geometry, insulation packaging and attachment, panel-to-panel interface, concept-to-concept interface, and penetration and close-out requirements was conducted during the detailed engineering investigation. Potential operational impacts to the orbiter were assessed throughout the analysis and design efforts in terms of mass, payload capability, changes in the outer mold line (the aerodynamic shape), turnaround time, and flight tra-

jectory. Some changes in the mold lines were considered inevitable, but efforts were made to minimize them.

The prepackaged family of concepts included titanium multi-wall panels for protection up to 1,000 degF, and an Inconel-617/titanium bimetallic sandwich for applications up to 2,000 degF. A new Columbium/titanium bimetallic sandwich for protection above 2,000 degF was also briefly investigated. All of the concepts were similarly packaged in one-foot squares, and attached to the orbiter by means of a bayonet and clip arrangement. The bayonets and clips were diffusion-bonded to the TPS panels, and the clips were mechanically attached to the vehicle. Dyna-Flex or Micro-Quartz insulation was placed between the TPS panels and the orbiter's skin to minimize heat transfer.[84]

The stand-off concept used panels approximately 20 inches square attached by four posts located five inches from each corner. Seventeen additional posts were provided in the panel for support, but did not physically attach to the vehicle. Various materials, including reinforced carbon-carbon and Haynes-188 superalloy, were used to construct the panels depending on their location. A layer of Micro-Quartz insulation was placed between the panel and the orbiter to minimize heat transfer.[85]

The 'best' alternative was considered a hybrid configuration consisting of prepackaged metallic concepts for application in areas where the temperatures ranged from 700 degF to 1,800 degF, and a carbon-carbon stand-off concept for areas where the temperature exceeded 1,800 degF. The current Nomex felt blankets would continue to be used in the lower temperature areas. The estimated total installed cost for a fleet of five vehicles was approximately $425 million, with five years being required for development and manufacture.[86]

PHASE B DOUBLE PRIME

The Phase B Prime studies had shown that a space shuttle could be developed with significantly lower RDT&E costs and reduced peak annual funding than those that had been defined during the initial Phase B studies. However, the costs for these vehicles still exceeded the acceptable peak annual funding figures desired by NASA and the OMB.

In mid-1971 the president's new scientific advisor, Edward David, convened a panel chaired by Alexander Flax to review the progress made so far on the space shuttle and recommend how NASA should proceed. Flax was the president of the Institute for Defense Analyses, a Pentagon think-tank. The first panel meeting took place on 3 August 1971, and the panel met approximately every four weeks for the rest of the year.[87]

In the meantime, NASA was facing another, but short-lived, budget crisis. Proposals had surfaced within the OMB to further reduce the NASA budget to $2,800 million – $400 million less than had been expected only a few months before. Fortunately when the deputy director of the OMB, Caspar Weinberger, was informed that these numbers would mean the end to manned space flight, he responded by assuring NASA that sufficient funds would be found to provide at least $3,200 million for FY73 and subsequent years. Weinberger's boss, George Shultz, concurred with this, and President Nixon signaled his agreement during late August 1971.[88]

But the President's concurrence did not specifically mention the space shuttle program, and for several weeks afterwards James Fletcher and George Low debated cancelling space shuttle and pursuing some less ambitious

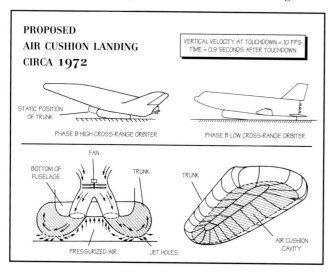

One of the more interesting ideas explored during the space shuttle definition studies, was the concept of air cushion landing. Bell Aerospace Company (formerly Bell Aircraft) investigated the possibility of deleting the landing gear on the orbiter and using an air cushion device for landing on hard surfaces. This was similar to studies and trials conducted on small transport aircraft around the same time. Although the MSC-040C-derived orbiter had been selected by this time, Bell's work centered around the two original baseline Phase B orbiter designs since the work had been on-going for some time. (NASA) [200]

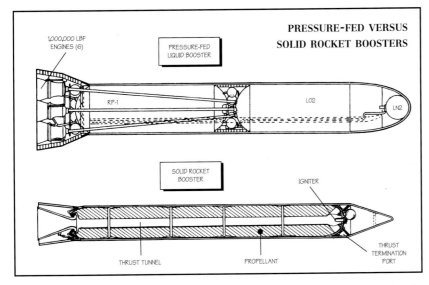

PRESSURE-FED VERSUS SOLID ROCKET BOOSTERS

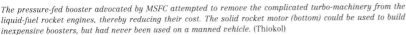

The pressure-fed booster advocated by MSFC attempted to remove the complicated turbo-machinery from the liquid-fuel rocket engines, thereby reducing their cost. The solid rocket motor (bottom) could be used to build inexpensive boosters, but had never been used on a manned vehicle. (Thiokol)

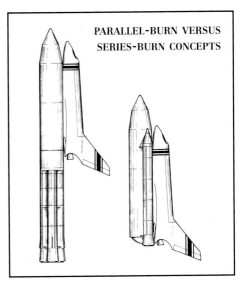

PARALLEL-BURN VERSUS SERIES-BURN CONCEPTS

The serial-burn configuration (left) used conventional staging as all previous piloted vehicles. In the parallel-burn design, all engines were ignited on the ground prior to lift-off. (Thiokol)

manned space program in its place. By the end of September, however, Fletcher had decided to continue with shuttle and it was included in his $3,225 million FY73 budget request. Of this amount, $200 million was earmarked for shuttle development, and $28 million for building shuttle facilities.[89] By this time NASA had abandoned the idea of postponing the development of the booster, preferring to develop the entire space shuttle system at one time. What had fallen by the wayside was the concept of a fully-reusable two-stage system – now the booster would be either a winged reusable S-IC derivative, the MSFC-proposed pressure-fed liquid booster, or a new solid rocket booster.[90]

Nevertheless, the phased approach was not totally dead. Although it had received slight attention previously, now the orbiter was the center of attention – the beginning of the so-called Mark I/Mark II orbiter concept. Much of the phased orbiter concept was developed in-house at MSC, with the Phase B Prime/Double Prime contractors playing supporting roles. The MSC position was that the Mark I orbiter would feature a 15 by 60-foot payload bay, although four J-2S engines instead of the planned three SSMEs would severely limit the total payload weight that could be carried to orbit. This would allow SSME development to be deferred several years, greatly reducing peak funding. The advanced thermal protection system – either metallic shingles or ceramic tiles – would be deferred, and a spray-on ablative heat shield would be used initially. Much of the advanced avionics would also fall by the wayside, with modified Apollo systems being substituted. The follow-on Mark II orbiter would utilize the same basic airframe with SSMEs, the advanced TPS, and state-of-the-art avionics.

The Flax Committee had its own ideas – an interim report issued on 19 October outlined a set of manned space flight alternatives to the large general-purpose space shuttle being developed by NASA. The preferred concept was a small glider launched by a Titan IIIL – but this configuration could only carry 10,000 pounds of payload, hardly acceptable to either NASA or the Air Force. Other alternatives included continuing to fly refurbished Apollo command modules on top of Saturn IBs, or the development of the Big-G Gemini launched by a Titan IIIM.[91]

The Flax Committee also disputed the potential cost effectiveness of the proposed space shuttle. Surprisingly,

the report did not question NASA's estimated cost per flight ($9 million for Mark I; $5.5 million for Mark II), but suggested that given the potential overruns likely to be incurred during development, there was no way the shuttle could pay for itself over its expected 13 year service life. Essentially the Flax Committee was saying that there was no economic justification for the space shuttle, something the OMB had been stating for the past year.[92]

During October 1971, Alexander Flax wrote Edward David that "most members of the panel doubt that a viable shuttle program can be undertaken without a degree of national commitment over the long term analogous to that which sustained the Apollo program. Such a degree of political and public support may be attainable, but it is certainly not now apparent. Planning a program as large and as risky (with respect to both technology and cost) as the shuttle, with a long-term prospect of fixed ceiling budgets for the program and NASA as a whole does not bode well for the future."[93]

Phase B Prime was scheduled to end on 30 October 1971, concurrent with the scheduled release of the RFP for Phase C/D. But the timing was not right since NASA still not know exactly what it could afford to build. Because vehicle definition was still ongoing, NASA awarded $2.8 million extensions to Grumman/Boeing, Lockheed, McDonnell Douglas/Martin Marietta, and North American Rockwell to continue reporting the results of various study efforts. These were collectively known as Phase B Double Prime and ran from 1 November 1971 through 15 March 1972, the revised Phase C/D RFP release date.[94]

The primary open technical issue in Phase B Double Prime was the booster configuration, although a low cross-range and a high cross-range orbiter were still being carried by each contractor. Emphasis shifted to three alternate booster configurations: (1) a fully recoverable pressure-fed booster, (2) a fly-back pump-fed liquid fueled booster based on the S-IC, (3) and several configurations of solid rocket boosters. For each of the booster concepts, there were two further variations to be studied – whether the system should use parallel-burn or series-burn staging. Parallel-burn was where all the engines are ignited on the ground – both the booster and orbiter engines would be firing during the initial ascent, similar to the McDonnell Douglas Triamese concepts presented during the ILRV and

PHASE B DOUBLE PRIME
McDONNELL DOUGLAS
PRESSURE-FED BOOSTER

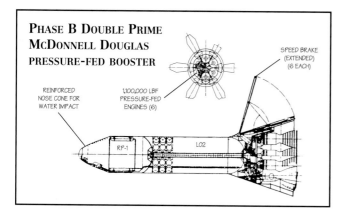

REINFORCED
NOSE CONE FOR
WATER IMPACT

1,100,000 LBF
PRESSURE-FED
ENGINES (6)

SPEED BRAKE
(EXTENDED)
(6 EACH)

RP-1

L02

Based on the preferences of engineers at MSFC, all of the Phase B Double Prime contractors designed pressure-fed boosters in an attempt to drive down development costs. The great number of unknowns with this type of technology resulted in the booster becoming too expensive to compete with solid rocket motors. (McDonnell Douglas)

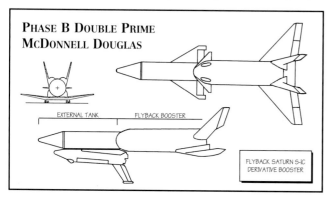

Three possible space shuttles. At left is the OMB-recommended 'glider' that used a Titan IIIL (with four SRMs in this case – others used six solids). The middle configuration has a pair of pressure-fed liquid boosters flanking the external tank. The version at the right could be mounted on the nose of an S-IC stage or other booster. (NASA)

PHASE B DOUBLE PRIME
McDONNELL DOUGLAS

EXTERNAL TANK

FLYBACK BOOSTER

FLYBACK SATURN S-IC
DERIVATIVE BOOSTER

PHASE B DOUBLE PRIME
NORTH AMERICAN ROCKWELL

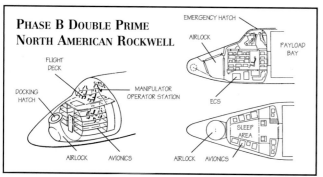

EMERGENCY HATCH

AIRLOCK

PAYLOAD
BAY

FLIGHT
DECK

MANIPULATOR
OPERATOR STATION

DOCKING
HATCH

ECS

AIRLOCK   AVIONICS

SLEEP
AREA

AIRLOCK   AVIONICS

PHASE B DOUBLE PRIME
NORTH AMERICAN ROCKWELL

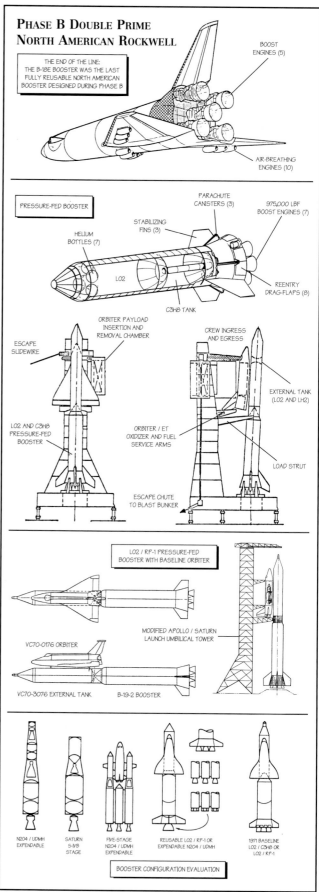

THE END OF THE LINE:
THE B-18E BOOSTER WAS THE LAST
FULLY REUSABLE NORTH AMERICAN
BOOSTER DESIGNED DURING PHASE B

BOOST
ENGINES (5)

AIR-BREATHING
ENGINES (10)

PRESSURE-FED BOOSTER

PARACHUTE
CANISTERS (3)

975,000 LBF
BOOST ENGINES (7)

STABILIZING
FINS (3)

HELIUM
BOTTLES (7)

L02

C3H8 TANK

REENTRY
DRAG-FLAPS (8)

ESCAPE
SLIDEWIRE

ORBITER PAYLOAD
INSERTION AND
REMOVAL CHAMBER

CREW INGRESS
AND EGRESS

L02 AND C3H8
PRESSURE-FED
BOOSTER

ORBITER / ET
OXIDIZER AND FUEL
SERVICE ARMS

EXTERNAL TANK
(L02 AND LH2)

LOAD STRUT

ESCAPE CHUTE
TO BLAST BUNKER

L02 / RP-1 PRESSURE-FED
BOOSTER WITH BASELINE ORBITER

MODIFIED APOLLO / SATURN
LAUNCH UMBILICAL TOWER

VC70-0176 ORBITER

VC70-3076 EXTERNAL TANK

B-19-2 BOOSTER

N204 / UDMH
EXPENDABLE

SATURN
S-IVB
STAGE

FIVE-STAGE
N204 / UDMH
EXPENDABLE

REUSABLE L02 / RP-1 OR
EXPENDABLE N204 / UDMH

1971 BASELINE
L02 / C3H8 OR
L02 / RP-1

BOOSTER CONFIGURATION EVALUATION

Rockwell expended considerable company funds on the Phase B Double Prime effort, and undertook a detailed development of the MSC-040A orbiter (left). Various boosters were also studied in detail, including the launch and recovery aspects of operating them (above). (North American Rockwell)

Phase A studies. Series-burn was the traditional staged-rocket approach where the upper stages were ignited in flight as the stage below them burned out. There were advantages and disadvantages for each method, and the contractors were to determine which was the most desirable. A new groundrule imposed by NASA was that the staging velocity could not exceed 4,100 mph, a significant reduction from earlier projections.

Five mid-period reviews were held during Phase B Double Prime, with each concentrating on a different issue pertaining to vehicle definition. The first of these, held on 15 December 1971 in Washington DC, was a discussion of the external tank orbiter using both the pressure-fed and pump-fed liquid boosters. The next review was on 3 February 1972 at MSFC in Huntsville, and was primarily concerned with various subsystems and ground handling techniques. Another meeting took place in Huntsville on 15 February 1972 to discuss all of the candidate booster systems. A review the following day in Houston concentrated on the issue of parallel-burn versus series-burn systems, as well as a philosophical discussion of the relative merits of liquid versus solid rocket boosters. The last review, again on booster issues, was held on 22 February 1972 in Washington DC.

Several different configurations were advanced for the fly-back pump-fed liquid booster concept. The majority concentrated on the maximum use of components developed for the S-IC stage, including the basic tank structure. Attached to this were either a moderately swept wing with wing tip mounted vertical stabilizers, or a delta wing with canard surfaces and a central vertical stabilizer. In both cases the wing was mounted low on the booster's fuselage, and approximately 7,000 square feet of wing area was provided in order to achieve the desired landing speed. Air-breathing engines were provided to enable the booster to fly back to the launch site for refurbishment and reuse. These designs used the 1,522,000 lbf sea level (1,748,000 lbf vacuum) Rocketdyne F-1 engine originally used on the S-IC, although these were not reusable and would have needed replaced after every flight. The majority of the designs utilized the same five engine layout as the original Saturn stage, but several explored four

engine configurations. These were eventually eliminated from further study because of inadequate safety margins and the same gas buildup concerns that had accompanied the original S-IC development.[95]

Both manned and unmanned versions of the fly-back S-IC configuration were investigated, and it was expected that the booster alone would weigh in excess of 3.5 million pounds fully loaded. The orbiter and external tank were mounted on the nose of the booster, and a staging velocity of 4,000 mph was expected. This all resulted in a truly enormous vehicle, rivaling some of the early Phase A/B designs in sheer size. The inability to reuse the relatively expensive F-1 engines was a serious economic drawback. Interestingly, the estimated development costs of this booster differed drastically among the contractors. For instance, Boeing – who built the original S-IC stages and should have had a good handle on it – estimated the cost of making the stage reusable at $4,500 million, with a peak requirement for $1,110 million per year. Lockheed generally concurred with this estimate. McDonnell Douglas, on the other hand, believed it would cost over $7,500 million with a peak requirement for $1,300 million.[96]

The MSFC-advocated pressure-fed liquid booster was an attempt to reduce costs by employing a relatively simple design approach with a minimum number of components. In the pressure-fed design all of the engine turbo machinery was eliminated and replaced by a large pressurization system to provide the propellants at a usable pressure to the engines. Two different configurations were studied in detail. The first consisted of a 32.6-foot diameter booster that was 159 feet long with the orbiter and ET mounted on the top (nose). This stage had seven 1,350,000 lbf engines and a gross lift-off weight of 5,791,000 pounds. The other design used three stages side-by-side with the orbiter and external tank mounted on the top of the middle booster (although Grumman had a unique variation of this concept). Each of the three stages had four 1,048,000 lbf engines, and the integrated vehicle weighed 5,279,000 pounds at lift-off. Possible propellants were LO2 and RP-1 and a staging velocity of 3,300 mph was specified. Both designs were fitted with parachute recovery systems for limiting the impact speed to 150 fps.

Three of the competitors show their designs during Phase B Double Prime. All of these used an MSC-040-based orbiter with both LO2 and LH2 moved into the external tank. (1) McDonnell Douglas with two parallel-burn pressure-fed liquid boosters, (2) McDonnell Douglas with two parallel-burn 156-inch solid boosters, (3) McDonnell Douglas with a large series-burn pressure-fed booster, (4) North American with two parallel-burn 156-inch solid boosters, (5) North American with two parallel-burn pressure-fed liquid boosters, (6) North American with a series-burn S-IC derivative, (7) Grumman with two parallel-burn pressure-fed liquid boosters, (8) Grumman with two parallel-burn 120-inch solid boosters, (9) Grumman / Boeing with a new-design series-burn pump-fed booster. For some reason, none of the Lockheed designs were shown in this lineup. (NASA)

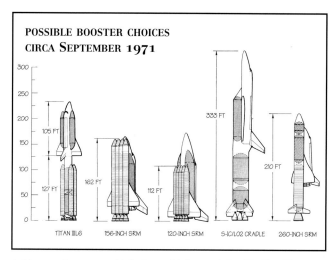

In late 1971, these appeared to be the possible booster choices. The Titan IIIL was not really up to the job, but was advocated by the OMB as an inexpensive alternative. Only the 260-inch SRM on the extreme right moved all the orbiter propellants externally. Oddly, the S-IC derivative only moved the LO2 externally, resulting in a large orbiter. The other three moved the LH2 outside per Grumman's advice, and in fact all of the orbiters except the S-IC appear to be based on Grumman's ASSC design. (NASA)

A dual 156-inch solid rocket booster configuration featured a gross lift-off weight of 4,898,000 pounds with a staging velocity of 3,600 mph. The exhaust nozzles contained a noticeable precant to ensure the thrust vector went through the integrated vehicle's center-of-gravity. The nozzles could not be gimbaled in flight, and some other method to provide thrust vector control during ascent would need to be found. Thrust termination ports were provided to simplify abort scenarios. Each booster contained 2,740,000 pounds of propellant and burned for 130 seconds, providing 2,939,000 lbf at sea level.

Cost comparisons showed that the liquid systems had the lowest per flight costs, while the 120-inch solid had the highest cost. The 156-inch solid rocket boosters would cost $3,700 million to develop, with a total system cost of $10,800 million – resulting in a cost of $228 per pound to orbit. The pressure-fed liquid booster would cost $4,600 million to develop, but the total system would only cost $9,400 million, with a pound costing $127. The S-IC pump-fed liquid booster would put a pound into orbit for $125, after costing $4,200 million to develop and $8,900 million to build. It should be noted that the solids, although they had a higher total program cost, would require less early-year funding than the liquid systems, but would yield higher per-flight costs.[97]

It was felt that the separation dynamics related to the series-burn systems were significantly more straight-forward than those for a parallel-burn system. Experience and relative risk were in favor of the series-burn systems because of the long history of successful sequential staged launch vehicles (Titan, Saturn, etc.). The parallel-burn system was thought to have some advantages in ground handling, and since the main engines were started prior to lift-off it gave some assurance that they were indeed startable, and running well prior to committing to flight. It was felt that the liquid propellant system was more flexible because of the ability to tailor the thrust at almost any point in the ascent phase, however, the development risk appeared to be in favor of the solids because of their relative simplicity. Recovery of the liquid boosters presented more of a challenge since they were more fragile, and ground handling of the solids was considered easier. It seemed there was no clear consensus.[98]

But politics within NASA came into play during the eval-

uation of the boosters. The work split between the three major NASA space centers had taken the same mold as Apollo – MSC was responsible for the manned portion (the orbiter in this case; the Apollo spacecraft previously), while MSFC was responsible of the booster. The third space center, KSC, was responsible for launching whatever MSC and MSFC gave them. But MSFC had no experience with solid rockets and was not particularly eager to gain any. This was evident when a series of directives were issued to the contractors instructing them to limit their investigations to the pressure-fed booster and the winged S-IC – effectively locking the solids out of the running. For now.[99]

Regardless of the Air Force experience, MSFC did not like solid rockets, and none had ever been used on a manned spacecraft except the small escape rockets on the Mercury and Apollo capsules. One of the drawbacks MSFC saw to solids was that they could never be tested as thoroughly as liquid engines. A liquid system could be tested over and over, "literally thousands of times," according to Bill Brown at MSFC, who had experience with both liquids and solids. "The cost of testing large [solid] rocket motors repeatedly is very, very high … They have … maybe tens of tests rather than hundreds or thousands of tests such as you would have in a liquid system. So, there has to be much more extrapolation of the data." According to LeRoy Day "There were people who were not enthusiastic about them. Von Braun was one who didn't think we should go with solids."[100]

But MSFC had to be careful, especially after NASA Headquarters began to endorse the possibility of using solid rocket motors. If MSFC protested too loudly, or too long, it was likely that booster development would simply be moved to MSC – and MSFC would be left out.

On the other hand, the winged S-IC appeared to be everything NASA could want. It promised most of the same capabilities as the original fully-reusable boosters, yet had significantly lower peak development costs. It also kept MSFC happy – comfortable with a technology they had largely developed and nurtured. But the OMB was not convinced. The more Donald Rice and the OMB looked into the cost estimates flowing from NASA and the contractors, the more they began to question whether any of them were truly accurate. Nobody made any public accusations as to whether any discrepancies were intentional – designed to get the program approved – or simply because nobody had ever actually built anything even remotely as complex as the space shuttle was proving to be.[101]

What was really questionable was that the estimated development cost for the shuttle had been cut in half – from $10,000 million to $5,000 million – with little attendant loss in capability. True, the estimated cost per flight had risen slightly, but it was still low enough to validate the Mathematica economic analysis and let NASA point to space shuttle as a good investment. Rice questioned whether NASA was being totally honest with itself, or Congress. He particularly noted that NASA never seemed to change its basic requirements, yet had managed to significantly cut costs. Something was being missed.

Still, progress had been made towards defining a much more reasonable space shuttle system. The two-stage fully-reusable concepts were dead – and at least Charles Donlan would not miss them: "It wasn't till the Phase B's came along and we had a hard look at the reality of what we meant by fully reusable that we shook our heads saying, 'No way you're going to build this thing in this century.' As I say, 'Thank God for all the pressures that were brought to bear to not go that route.'"[102]

The winged S-IC stage would follow by the end of 1971. Further investigation by the contractors, NASA, and the OMB pointed out that converting a stage that had originally be designed to be expendable would prove much more difficult and expensive than originally thought. In essence, it would be similar to developing a large hypersonic aircraft – the same pitfall as the original Phase B concepts.

ANOTHER IDEA

The original economic analysis from Mathematica had drawn widespread criticism because NASA had not allowed Oskar Morgenstern and Klaus Heiss to examine concepts other than the baseline fully-reusable two-stage Phase B designs. This began to change during mid-1971. Independently, Heiss and the Flax Committee had reached many of the same conclusions. In studying alternatives, designers traded a reduction in development costs against an increase in per flight cost – if the per-flight costs went too high, shuttle could not be competitive with expendable boosters; but if the development costs stayed too high, there would not be a shuttle. The estimated mission cost for the Mark I orbiter was $9 million, comfortably under the $12 million average for expendables. Heiss believed that this cost per mission could go as high as $10 million and still be competitive, especially if it meant a significant reduction in development cost.[103]

But Heiss could not design a space shuttle – he was an economist, not an engineer. Nevertheless, while reviewing the contractor reports from Phase B, ASSC, and the various extensions, he began to see a concept that promised to do just what he wanted. He coined a new term – thrust assisted orbiter shuttle (TAOS) – to describe the concept. What interested Heiss was a Mark I/Mark II orbiter with an external propellant tank large enough to permit the orbiter's engines to operate in parallel from lift-off to orbital insertion. Other rocket boosters, either solid or liquid, would provide additional thrust for the first couple of minutes of ascent, then be jettisoned. These boosters might, or might not, be reused.

The two designs that had sparked this interest were from Grumman and McDonnell Douglas. Grumman called the idea TAHO (thrust assisted hydrogen-oxygen), and used a single external tank mounted on the belly of the orbiter with two large solid rocket boosters flanking it. McDonnell Douglas called the concept RATO (rocket assisted take-off) and used a single 180-inch diameter solid booster mounted to the belly of the orbiter with the propellants in two large external tanks mounted on the sides of the fuselage.

The major disadvantage of either TAOS concept was that the boosters were unpiloted, and NASA still wanted a manned, fully-reusable system. A second drawback was that both concepts used solid boosters instead of the pressure-fed liquid configurations preferred by MSFC. As late as 15 October 1971 NASA officials did not include either of the TAOS designs in a presentation of possible designs to the Flax Committee. Regardless, on 28 October 1971, Morgenstern and Heiss wrote a 15-page report to James Fletcher that strongly endorsed the TAOS concept.[104]

Specifically, the economists wrote that the non-recurring costs of the TAOS were estimated at $6,000 million over a six year period, within the OMB guidelines. Development risks for the TAOS development were lower than the baseline two-stage system, but were still considered substantial. The concepts that had been examined promised the same performance as the larger system – a

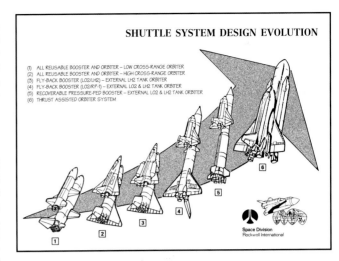

At the end of the Phase B Extension, North American included this chart in their final report showing the conceptual evolution of the space shuttle from Phase A through Phase C. (North American Rockwell)

40,000 pound payload in a 15 by 60-foot payload bay. Mathematica also believed that any orbiter should use the advanced SSMEs from the beginning, and recommended abandoning the phased Mark I/Mark II orbiter concept entirely. Surprisingly, Heiss thought that the TAOS designs might be able to achieve launch costs equal to those of the early fully-reusable two-stage system after the external tanks and the boosters came down in price due to manufacturing efficiencies.[105]

Grumman and McDonnell Douglas soon supported the Mathematica analysis. Slightly modified designs appeared that used pressure-fed boosters, and both contractors assumed that the boosters could be successfully recovered from the Atlantic and refurbished at reasonable cost. But continued analysis indicated that the lowest cost system appeared to use solid rocket boosters.[106]

GETTING CLOSER

On 22 October 1971 the senior OMB staff met to determine the future of several NASA programs including the space shuttle. The OMB staff recommendation was blunt – cancel the space shuttle program – but Caspar Weinberger was not eager to accept this recommendation. NASA still supported the Mark I/Mark II orbiter concept with a recoverable liquid booster, although Weinberger believed some of the alternatives from the Flax Committee and Mathematica held a better chance of approval. The following month saw a flurry of letters and reports between the OMB and NASA, trying to find a middle ground acceptable to everybody. Periodically the Air Force chimed in to remind everybody of its requirements, although only NASA appeared to be listening, and then only occasionally.[107]

In advance of this meeting, James Fletcher had written privately to Caspar Weinberger, presenting NASA's arguments for the space shuttle. "The United States urgently needs the space shuttle to provide 'routine' access to near-earth space ... The Shuttle provides the capability for a continuing U.S. manned space flight program, a capability we believe to be essential – without flying men just for their own sake ... The aerospace industry will be hurt by continuing indecision ... a firm go-ahead, on the other hand, will quickly create jobs in this industry ... It will not be possible to sustain the momentum now built up in the shuttle program much longer."[108]

Despite Mathematica's suggestions, NASA did not abandon the Mark I/Mark II concept and, ignoring the Air Force, the baseline orbiter still included a 15 by 60-foot payload bay with only a 30,000 pound capacity. On 22 November, George Low wrote to Donald Rice: "The most promising candidate configuration today is the Mark I/Mark II orbiter with the parallel-staged pressure-fed booster."[109]

This letter also contained the current definitions for some of the concepts still being carried. The 'baseline' was listed as a two-stage fully-reusable system with an orbiter that carried its LH2 in external tanks and had a payload capacity of 65,000 pounds in a 15 by 60-foot payload bay. However, most of the letter described the Mark I/Mark II orbiter concept where the LO2, as well as LH2, was moved to external tanks. "Some of the subsystems would be phased, starting out in the Mark I model with more nearly existing technology in areas such as the heat shield and avionics, and phasing in more advanced versions later in Mark II." Four different boosters were still being considered in conjunction with the Mark I/Mark II orbiter: (1) a series-burn fly-back booster based on a winged S-IC stage, (2) a series-burn pressure-fed booster that would be recovered via parachutes, (3) parallel-burn pressure-fed booster that would also be recovered via parachute, and (4) expendable parallel-burn solid rocket boosters. Although it was fairly well determined that any liquid booster would use LO2/RP-1 propellants, the contractors also investigated alternate fuels such as propane (C3H8) and LH2.[110]

Low also listed the OMB's preference for a smaller glider with a 30,000-pound capacity in a 12 by 30-foot payload bay – but NASA stacked this option to make sure it was not cost competitive. For instance, NASA moved the engines and avionics from the orbiter into the booster, then expended the booster on every flight. This ensured that

the cost per mission was abnormally high, although the development cost remained the lowest of the options.[111]

The letter described projected development costs and recurring flight costs for each option, with the original Phase B fully-reusable two-stage system continuing to have the lowest per flight cost, but the highest development cost. The OMB-supported glider was at the other end of the scale, with the lowest development cost but by far the highest per flight cost. All of the different Mark I/Mark II options fell somewhere in the middle.[112]

While NASA and OMB were haggling over shuttle size and costs during November and December 1971, the space agency had also been attempting to solicit political support for its proposals, particularly at DoD and the White House. In October, James Fletcher met with Deputy Secretary of Defense David Packard, who told him that the agency's approach to selling the shuttle "was all wrong." Packard apparently did not believe that "the shuttle should ever be sold on the basis of the cost savings that might result, or even the flexibility in payloads." Rather, Packard noted "the real point has to do with national security and an intangible thing which might be called 'man's presence in space.'" He also said "that it is not surprising that it is the Defense Department and the State Department together with Henry Kissinger who offer the most support for the shuttle." He suggested that NASA and DoD assemble a top-level team to develop a rationale for the shuttle and communicate it to the President and Congress. Fletcher agreed, but later told Low that it was important that the rationale "doesn't become unduly military in its flavor."[113]

Conversations between NASA and DoD continued into December, and George Low suggested to the Assistant Secretary of Defense for Research and Development, Johnny Foster, that what was needed was "an imaginative military space program taking advantage of the new capabilities that the shuttle would represent."[114]

The Pentagon never came forward with such a program, but NASA had its own list of military missions that the shuttle might perform. For example, one suggestion was: " ... the shuttle could be maintained on ready alert, making possible rapid responses to foreseeable and expected situations and greatly increase flexibility and timeliness of responses to military or technological surprises, such as: (a) rapid recovery and replacement of a faulty or failed spacecraft essential to national security; (b) examination of unidentified or suspicious orbiting objectives; (c) capture, disablement, or destruction of unfriendly spacecraft; (d) vapid examination of crucial situations developing on Earth or in space whenever such events are observable from an orbiting spacecraft; and (e) rescue or relief of stranded or ill astronauts."[115]

It apparently was shuttle capabilities such as these that were attractive to President Nixon. Chief of staff John Erlichman remembers that "a strong influence was what the military could do with the larger bay in terms of the uses of satellites" and "the capability of capturing satellites, or recovering them." These factors, said Erlichman, "weighed into my attitude toward the larger shuttle, and ... also weighed into Nixon's."[116]

Another argument in favor of the space shuttle was its potential employment impacts, particularly in view of the then-depressed state of the aerospace industry and the upcoming 1972 presidential election. Fletcher told the White House that "an accelerated start on the shuttle would lead to a direct employment of 8,800 by the end of 1972, and 24,000 by the end of 1973." This was "a very

This straight-wing concept from Grumman used three liquid boosters in a rather unique arrangement. The two long extensions that attached to the orbiter's sides are propellant tanks that were retained after the three lower boosters have been jettisoned, although the tanks were also jettisoned later in ascent. (Grumman Aerospace)

important consideration in Nixon's mind," according to Erlichman. During a review of the employment impacts of various federal programs in states crucial to the President's reelection, the White House found that "when you look at employment numbers and key them to battleground states, the space program has an importance out of proportion to its budget."[117]

Another aspect in gaining Nixon's support was the leadership aspect of maintaining a vigorous U.S. manned space program. Fletcher argued that "the United States cannot forego its responsibility – to itself and to the free world – to have a part in manned space flight ... For the U.S. not to be in space, while others do have men in space, is unthinkable, and a position which America cannot accept." This argument reportedly appealed to Nixon, who saw astronauts as representing the very best of American values, and because of "a commitment that had to do with chauvinism. We had to be at the leading edge of this kind of applied technological development."[118]

It was considerations such as these that would eventually lead to Nixon approving the space shuttle over the 1971–72 New Years weekend. According to Erlichman, "it was Nixon's decision. During that time there wasn't anyone else making those final decisions. In defense, space, certain kinds of domestic problems, he was the final arbiter."[119]

In the meantime, on 2 December 1971 the OMB sent a memorandum to Nixon recommending "a smaller reduced cost version of a manned reusable Shuttle" with a total development cost of under $5,000 million. Interestingly, the OMB wrote that "for national security purposes, we may not want all of eggs in one basket." and that the nation should "...retain the reliable Titan III expendable booster to launch the few largest payloads that would not fit the smaller Shuttle. These include space telescopes and large intelligence satellites."[120]

Although the memorandum to the president had not included a payload bay size or capacity, the OMB had provided these privately to NASA. The vehicle OMB envisioned could carry 30,000 pounds in a 10 by 30-foot payload bay – this did not meet NASA or, especially, Air Force requirements. All of this threw the configuration of the shuttle back into turmoil. The OMB recommendation allowed NASA to develop new engines, avionics, and the thermal protection system, completely bypassing the Mark I phase. Also included in the OMB numbers was the development of the pressure-fed booster advocated by MSFC. But the recommended payload capacity neglected Air Force requirements, making the basic design of a delta-wing orbiter unnecessary since it was the Air Force that drove the 1,265-mile cross-range requirement. If the space shuttle could not satisfy the Air Force payload requirements, perhaps it would be best to fall back to one of Max Faget's simple straight-wing orbiters instead.[121]

Charles Donlan quickly rejected this tactic. Although the Air Force had originally drove a 1,726-mile cross-range requirement (later relaxed to 1,265 miles), subsequent studies by NASA had indicated that a high cross-range was also required for certain abort scenarios that were deemed essential. For instance, substantial cross-range was required to return to the launch site during an abort-once-around, or to land at contingency sites during a downrange abort.* NASA had also determined that a substantial degree of aerodynamic maneuvering capabil-

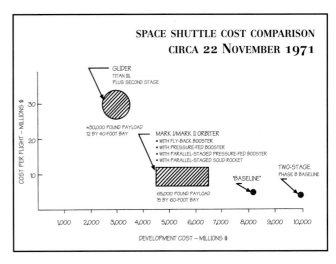

SPACE SHUTTLE COST COMPARISON
CIRCA 22 NOVEMBER 1971

This chart was attached to 22 November 1971 correspondence from NASA Deputy Administrator George M. Low to Donald B. Rice, Assistant Director of the Office of Management and Budget. (NASA)

ity at hypersonic and supersonic speeds was fundamental to the safe operation of the orbiter. This would afford frequent opportunities to return to the launch site during due-east missions from KSC, and allow the selection of reentry routes "over sparsely populated land mass or water in the event that sonic boom over pressures are judged to be objectionable levels ...."[122]

NASA engineers determined that the minimum cross-range requirement was 1,265 miles to support the abort-once-around option – interestingly, this exactly matched the single-orbit profile desired by the Air Force. It was the consensus among NASA and the contractors that this cross-range could best be achieved by a delta-wing orbiter, finally putting to rest any chance that Max Faget's straight-wing orbiter would ever be built. Apart from the operational requirements for cross-range, the delta-wing configuration was preferred for basic aerodynamic considerations. The critical periods during reentry were all in the hypersonic and high-supersonic flight, and the delta-wing configuration was stable throughout a wide range of angles of attack in this regime. By modulating angle-of-attack and bank angle, the delta-wing orbiter could take full advantage of trajectory shaping and cross-range options, whereas the straight-wing orbiter was largely ballistic until it reached a very low-speed and comparatively low altitude.[123]

The flow field over the delta-wing vehicle tended to be smooth, producing predictable aerodynamic heating gradients and relatively low and uniform temperatures (600–800 degF) over the sides and upper surfaces of the vehicle. The flow over the sides of the fuselage was smoothly blended with no shock interactions and few, if any, hot spots. These conditions were favorable to straightforward heating predictions, something that was important to the engineers given the lack of real-world experimental data on very-high-speed atmospheric flight, and the limitations of existing simulation tools. On the other hand, the straight-wing configuration suffered from unsteady flow and buffeting in the transonic regime. During hypersonic flight, the straight-wing vehicle had flow-fields that were complex, with strong bow and wing shock interactions with the fuselage. These strong interference flow fields resulted in high local temperatures and severe temperature gradients on the wing, fuselage, and empennage. Vortex flows in the wing-fuselage and empennage-fuselage junctures tended to result in local hot spots,

---

* Now called a trans-oceanic abort landing (TAL).

† NASA still wanted a 15-foot diameter bay to accommodate space station modules, while the Air Force need a 60-foot long bay for intelligence satellites.

complicating heating predictions. Temperatures on the side of the fuselage were higher (900–1,300 degF), requiring a more robust thermal protection system. Overall, NASA believed the delta-wing orbiter would be the simpler configuration to design and build, and that it would result in an operationally superior vehicle.[124]

NASA also attempted to explain the complications in selecting the size of the payload bay, besides the operational constraints it would cause.[†] During evaluations of payload bays ranging in size from 12 by 40 feet to 15 by 60 feet, NASA found that the vehicle dry weight varied by as little as 15 percent. Some of this was because many of the orbiter systems (crew module, auxiliary power, etc.) were not directly related to payload size and were fixed by other requirements (orbital duration, safety, etc.). There were also physical reasons reducing the payload bay size did not necessarily mean a smaller orbiter. For instance, reducing the diameter from 15 to 12 feet did not reduce the size of the aft fuselage because a 15-foot diameter boat-tail was required to house the main engines.[125]

At one point the OMB had suggested that NASA initially build a smaller orbiter than it desired, then when funds became available 'stretch' it much as was done on civilian airliners such as the DC-8. NASA pointed out that subsonic transport aircraft were not indicative of the problems associated with hypersonic flight. The stretch of an orbiter would significantly alter the hypersonic flow field, resulting in greatly different stability, control, and thermal characteristics for the orbiter, as well as the complex launch configuration of the orbiter and booster combined. Extensive aerodynamic, static, and dynamic ground tests, plus additional flight test development would undoubtedly be required, possibly approaching the level of an entirely new development project. NASA concluded that the "most cost effective system is one sized properly at the outset for its intended use."[126]

By the end of November 1971, NASA management had firmly embraced the TAOS configuration, at least in concept. It was still undecided if the vehicle would use liquid or solid boosters, and the question of parallel-burn or series-burn staging was still open. On 13 December 1971 George Low asked Dale Myers, head of the Office of Manned Space Flight, to evaluate four alternate space shuttles – the vehicle OMB was suggesting (30,000 pound payload in a 10 by 30-foot bay), the baseline NASA vehicle (65,000 pounds in a 15 by 60-foot bay), a design with a 30,000-pound capacity in a 12 by 40-foot payload bay, and an odd-ball case with the full 65,000-pound capacity in a 14 by 50-foot bay. Subsequently, a 45,000-pound capacity in a 14 by 45-foot payload bay was added to the analysis.[127]

The results of this evaluation were sent from James Fletcher to Caspar Weinberger on 29 December 1971. NASA concluded "… that the 16´ by 60´ – 65,000# payload shuttle still represents a 'best buy,' and in ordinary times should be developed. However, in recognition of the extremely severe near-term budgetary problems, we are recommending a somewhat smaller vehicle – one with a 14´ by 45´ – 45,000# payload capacity, at a somewhat reduced overall cost."[128]

Fletcher went on to state that the recommended vehicle would be useful for continued manned flight into space, as well as being able to deliver a variety of unmanned payloads. It would, however, be unable to accommodate many DoD payloads and some NASA scientific payloads. It also could not reasonably accommodate the proposed Space Tug. Specifically, NASA indicated that at the 50-foot length many DoD and NASA science payloads were eliminated, as was the capability to fly a Space Tug and payload combination. A 30-foot length eliminated "nearly all" DoD payloads, more NASA science payloads, most "application payloads" (i.e., communication satellites), all planetary payloads, and "useful manned modules" (Spacelab – still called a sortie module at this point – and space station modules).[129]

The minimum acceptable diameter was set at 14 feet. The OMB-proposed 10-foot diameter led to an outside module diameter of only 9 feet (one foot was required for clearance between the payload bay sides and the module). This left an inside diameter of only 8 feet, which could not accommodate being 'squared off' to provide cabling, plumbing, consoles, cabinets, etc. in a manned module. NASA pointed out that Skylab was 22 feet in diameter, and the Apollo capsule was 13 feet. Many science, planetary, and application payloads were also better accommodated in a 14-foot diameter bay.[130]

Some interesting facts came to light in the discussion on payload weight. Although NASA conceded that the initial 65,000-pound requirement[*] came from the DoD, it pointed out that in order to carry an application payload and Space Tug on a single mission, the NASA requirement was over 60,000 pounds. In addition, carrying manned modules required at least a 45,000-pound capability. This was because, although the modules only weighed 15,000-20,000 pounds, they needed launched into a highly-inclined 55-degree 270 mile orbit, equating to a 45,000-pound due-east capability.[131]

Finally, A Decision

Monday, 3 January 1972, dawned like many before it. NASA was still unsure what sort of shuttle would be authorized for development, although they knew that the current Director of the OMB, George Shultz, privately supported the baseline NASA vehicle. Fletcher and Low had agreed that the minimum they could accept was a vehicle with a 45,000-pound capacity in a 14 by 45-foot payload bay. At 6 p.m., Fletcher and Low met in Shultz's office with Weinberger, Rice, and David. This would be the first time that Shultz would publicly side with NASA and agree that a 65,000-pound 15 by 65 payload bay shuttle was the only one that made sense to develop. Despite the objections of Rice and David, Shultz confirmed that he would agree to $200 million in startup funds in the FY73 NASA budget. NASA was asked to pre-

SPACE SHUTTLE COST ESTIMATES CIRCA DECEMBER 1971					
CASE	1	2	2A	3	4
Payload Bay (feet):	10 by 30	12 by 40	14 by 45	14 by 50	15 by 60
Payload (pounds):	30,000	30,000	45,000	65,000	65,000
Development (millions $):	4,700	4,900	5,000	5,200	5,500
Mission Cost (millions $):	6.6	7.0	7.5	7.6	7.7
Payload Cost ($/pound):	220	223	167	115	118

Dale Meyers provided these cost estimates for various space shuttle concepts in December 1971, and James Fletcher forwarded them to Caspar Weinberger at the OMB as the basis for Nixon's decision to approve the program. (NASA)

---

[*] DoD actually required a 40,000-pound into polar orbit capability, which translated to a 65,000-pound capability into a due-east orbit.

pare a draft of a presidential statement announcing the approval of the space shuttle program.[132]

Fletcher and Low flew to the Western White House in California to meet with President Nixon before the announcement. Although the president had intended to spend only 15 minutes of a hectic day with the representatives from NASA, he ended up spending over half-an-hour with them. Nixon was "fascinated" by model of a TAOS space shuttle that Fletcher had brought along, according to Erlichman. "He held it and I wasn't sure that Fletcher was going to be able to get it away from him." Nixon told Fletcher and Low that NASA "should stress civilian applications, but not to the exclusion of the military applications …" Nixon "liked the fact that ordinary people would be able to fly in the shuttle." Fletcher continued to assure the President that the space shuttle was economically viable, but Nixon "indicated that even if it were not a good investment, we would have to do it anyway, because space flight is here to stay. Men are flying in space now and will continue to fly in space, and we'd best be a part of it."[133]

On 5 January Nixon released his formal statement:[134]

"I have decided today that the United States should proceed at once with the development of an entirely new type of space transportation system designed to help transform the space frontier of the 1970s into familiar territory, easily accessible for human endeavor in the 1980s and 90s.

"This system will center on a space vehicle that can shuttle repeatedly from earth to orbit and back. It will revolutionize transportation into near space, by routinizing it. It will take the astronomical cost out of astronautics. In short, it will go a long way toward delivering the rich benefits of practical space utilization and the valuable spinoffs from space efforts into the daily lives of Americans and all people. However, all these possibilities, and countless others with direct and dramatic bearing on human betterment, can never be more than fractionally realized so long as every single trip from Earth to orbit remains a matter of special effort and staggering expense. This is why the commitment to the Space Shuttle program is the right next step for American to take, in moving out from our present beachhead in the sky to achieve a real working presence in space – because Space Shuttle will give us routine access to space by sharply reducing costs in dollars and preparation time.

"Views of the earth from space have shown us how small and fragile our home planet truly is. We are learning the imperatives of universal brotherhood and global ecology – learning to think and act as guardians of one tiny blue and green island in the trackless oceans of the universe.

"'We must sail sometimes with the wind and sometimes against it,' said Oliver Wendell Holmes, 'but we must sail, and not drift, nor lie at anchor.' So with man's epic voyage into space – a voyage the United States of America has led and still shall lead."

Before the announcement, NASA had prepared a list of possible names for the new program, including Pegasus, Hermes, Astroplane, and Skylark. The White House favored Space Clipper, and this name appeared in several early drafts of the statement. But in the end, Nixon himself decided it would be better to refer to the vehicle simply as Space Shuttle.[135]

This statement largely stopped the bickering over the configuration of the vehicle, except within NASA. Nixon did not care if the vehicle had a 45-foot or 60-foot payload bay. As long as the development cost did not seriously exceed the numbers approved by George Shultz and the OMB –

$5,000 million – nobody in the White House was interested in the details. This was as it should be, and why the independent agencies, such as NASA, are staffed with competent engineers and administrators. There would still be debates in the press, and on the floor of Congress, but the program was approved and began to move ahead quickly.

Almost immediately, the Air Force authorized expending approximately $15 million to set up a small program office to work with NASA in defining the final space shuttle requirements. It was decided against setting up a formal System Program Office (SPO) within the Air Force Systems Command, and instead a less formal group was established within the Space and Missile Systems Office (SAMSO) to support NASA through the preliminary design review. A formal SPO would eventually be established in late 1974, mainly to oversee the development of the Vandenberg launch site.[136]

THE FINAL CONFIGURATION

With program approval in hand, NASA now had to decide exactly what they wanted to build. Although the issue of payload bay size and capacity was still open, the basic orbiter shape, based on the MSC-040C, was nearly settled. The overall vehicle was becoming more refined, helped by many hours in various wind tunnels courtesy of the Phase B Extension contractors, in-house testing at NASA Ames and Langley, and many hours at the Air Force Arnold Engineering Development Center. The Mark I/Mark II phasing concept had fallen completely by the wayside.[137]

Decisions were finally made about what materials to construct the orbiters out of – based largely on Air Force preferences, aluminum was chosen for the primary airframe structure. This did not overly disappoint NASA since all the contractors knew how to build aluminum air-

A Grumman concept using four 120-inch solid rockets. These appear to be largely unmodified Titan SRMs, and the small nitrogen tetroxide tanks used for thrust vector control can be seen nestled between the SRMs and the external tank Given that the boosters were already developed, this should have been a low-cost concept. (Grumman Aerospace)

craft, but only a few had large-scale experience with titanium. The choice of aluminum, however, dictated that the Lockheed-developed silica tiles would almost certainly need to be used as a thermal protection system, adding a substantial amount of risk to the program. Various studies into alternate heat shields, mostly ablative concepts, continued as possible contingencies, but the operational future of the orbiter was tied to the success of the tiles.[138]

As late as the end of November 1971, NASA had still been undecided on whether the orbiter's engines should be ignited simultaneously with the booster's (the parallel-burn approach), or if a more traditional 'staged' (series-burn) approach. In December 1971, NASA formally adopted the parallel-burn design, but left open the decision of liquid or solid boosters. Total development costs were now estimated by NASA at $5,800 million using the MSFC-preferred pressure-fed booster.[139]

The winged S-IC had finally been dropped at the end of 1971 when even Boeing could not drive its development costs low enough to stay within the OMB guidelines. And the more people looked at the idea, the more potential risk it seemed to have – after all, it was not really much different than the original fly-back boosters envisioned during Phase B. But the Saturn V would not go away. Boeing and MSFC proposed another alternative – a pump-fed booster based largely on the S-IC stage, but this time without wings. Instead, it was fitted with large parachutes and clamshell doors over the engines, and was to be recovered at sea. It was never seriously in the running.[140]

Originally, many concepts had been based around a pair of pressure-fed liquid boosters, one on each side of the external tank. But by the end of 1971 this configuration had also largely disappeared as estimated development costs continued to soar. The primary reason for the fall of the pressure-fed booster, MSFC's future employment not withstanding, would be that nobody had ever developed a large pressure-fed booster before, and that significantly increased program risk. On the other hand, the Air Force had 120-inch solids in production for the Titan IIIC, and various Air Force and NASA experimental programs had demonstrated 156-inch, and indeed 260-inch, solids. There appeared to be little, if any, development risk for a solid rocket booster. The most promising concept appeared to be solid rocket boosters mounted beside the external propellant tank. Several variations to this theme existed – a pair of 156-inch diameter solids; four 120-inch solids; two very tall 120-inch solids. To verify this, in mid January NASA awarded $150,000 one-month contracts to each of the four manufacturers of solid rocket motors asking for technical details and cost estimates.[141]

In the end, development cost estimates for the solids were approximately $1,000 million less than for the pressure-fed booster, but both were within the realm OMB had given NASA. However, on 9 February 1972 Caspar Weinberger wrote James Fletcher reminding him that the NASA budget was unlikely to exceed $3,200 million, and that "We [OMB] also fully expect NASA to develop a shuttle system within the $5.5 billion* estimate." If solid boosters were selected, the difference in development costs would create a comfortable management reserve to pay for unexpected problems with the SSMEs or thermal protection system. In retrospect, a very wise decision.[142]

There was also a constant exchange of data between NASA and the contractors during the final gasps of the Phase B Double Prime studies. Many of these meetings concerned the possible choice of boosters, and George Low commented that these briefings "yielded the recom-

mendations for each contractor that were most predicable based on vested interests."[143]

Boeing, still part of the Grumman team, had developed the S-IC and continued to champion it. As early as September 1971 Boeing had proposed using unmodified S-IC stages for the first 30 shuttle flights while the final booster was being developed. By February 1972 Boeing was supporting a new-development pump-fed booster, although Grumman was proceeding with designs utilizing all three major booster concepts – pump-fed and pressure-fed liquids, as well as solids.[144]

In contrast, Lockheed was a major manufacturer of solid rockets and had proposed a variety of sizes and shapes during September. By February it had expanded its offerings to include liquid boosters, although internal trade-studies continued to show solids were the most economical.[145] North American did not appear to have a favorite, somewhat surprising since their Rocketdyne division might possibly provide the engine for any liquid-powered booster. But then, Rocketdyne was thoroughly engaged in the SSME competition at the time.[146]

McDonnell Douglas liked solids. As a company, it had long experience with liquid rocket engines, having built the Thor and Delta launch vehicles, and the S-IVB stage for Saturn. But the company was also familiar with solids, and used them extensively to augment the thrust of the Delta. In its February final briefing, McDonnell Douglas summarized 2,128 solid-rocket motor firings that covered Delta and Titan strap-on boosters, Minuteman ICBMs, and the small four-stage Scout light launch vehicle. Of this number, 13 boosters had failed in ways that mattered to shuttle, and McDonnell Douglas carefully detailed the failure mode and what could be done to prevent each failure in the future. The report noted that for most of these failures it would be possible to safely abort a space shuttle launch.[147]

Most – an important word. McDonnell Douglas noted there was one case that appeared to offer no reasonable abort scenario. Large solid rocket motors, such as the Titan III strap-on, were manufactured in segments that were bolted together at field joints. This was the only way the boosters could be transported to their launch sites, and also eased manufacturing. There was a concern, however, that the propellant could 'burn through' one of the joints. McDonnell Douglas noted that if this occurred near the external tank, particularly the hydrogen tank, "sensing may not be feasible and abort not possible."[148]

Between January and March 1972, NASA constantly revised its cost estimate for developing the shuttle system. During that period, the cost of developing the orbiter itself increased almost $700 million, and the estimate for the SSMEs increased $130 million. In the final review on 13 March the estimates for the booster were: pressure-fed – $1,400 million; pump-fed – $1,080 million; solid – $350 million. There was only one way NASA stood a chance to stay within the OMB-mandated $5,500 million total.[149]

The booster decision was announced on 15 March 1972 – 156-inch diameter solids. James Fletcher reported to the OMB that solid boosters could be developed faster and for $700 million less than an equivalent liquid booster, lowering estimated total development costs to $5,150 million in FY71 dollars. There was another side to the economics that is seldom mentioned – and it also favored the solids. If one of the liquid boosters could not be recovered, the loss to the program would be substantially greater

---

* The OMB had added a ten percent overrun contingency on top of the $5,000 million estimate.

than if a solid booster was lost – simply because the boosters themselves were far more expensive and NASA was unlikely to have as many of them.[150]

The booster decision placed a large share of the burden of paying for the space shuttle program on its future users. The 1972 mission model for NASA, DoD, and other users called for some 580 flights over a twelve year period (1979–90), an average of almost 50 flights per year. Thirty-eight percent of these flights were to be scientific research aboard Spacelab modules, 31 percent were to be in support of the DoD, and the remaining 31 percent were to be commercial satellite deliveries and space station support missions.[151]

Later (1974) models would call for over 60 flights per year, and as late as 1985, just prior to the *Challenger* accident, NASA was still predicting 20 flights per year. Using the 1972 model, launch and direct launch-related costs using existing expendable vehicles were estimated at $13,200 million, while the total Space Shuttle launch costs were estimated at $8,100 million, excluding all research and development costs. The total cost per shuttle flight was estimated at $10.5 million, or $175 per pound* for a full payload bay.

ORBITER PHASE C/D

Unlike the previous two phases of the space shuttle competition, Phase C/D would not involve government funding.† The RFP was issued on 17 March 1972 to Grumman/Boeing, Lockheed, McDonnell Douglas/Martin Marietta, and North American Rockwell for the production and initial operations of the Space Shuttle System. The technical proposals were due on 12 May 1972, with the associated cost and contractual data due 19 May 1972.[152]

The RFP broke the program into two phases – Shuttle System Development and Production, and Shuttle Operations. The selected contractor was to be responsible for overall systems engineering and integration, but the solid rocket motor, external tank, SSME, and air-breathing engine (ABE) were to be government-furnished equipment provided under separate contracts. Of these, the SSME had already been defined and was in the process of being procured by MSFC, but the other three elements would be defined by the space shuttle contractor via the writing of procurement specifications – these would be provided to NASA, which would then competitively procure the items and provide them back to the prime contractor.[153]

Within the two phases, four increments had been defined. The first increment was the initial design effort. This was followed by detailed design and the production of two orbiters in the second increment. These two increments made up the design, development, test, and evaluation (DDT&E) effort leading to the first manned orbital flight (FMOF). The third increment consisted of producing three additional orbiters and refurbishing the two DDT&E vehicles to an operational configuration. The fourth increment was operating the shuttle system during the operational phase of the contract.[154]

The proposal would consist of nine volumes totaling 1,700 pages. The contractor needed to deliver up to 200 copies of some volumes to the government, with all volumes requiring at least 100 copies. Of this total, only 425 pages were devoted to technical subjects – 1,050 were devoted to cost data, and the remaining were management information or the executive summary.[155]

The RFP stated that each orbiter should have a useful life of ten years and be capable of up to 500 mission each. However, the RFP asked each contractor to provide information on the impact of lowering this to only 100 missions per orbiter, a figure that was subsequently adopted. Three reference missions were specified: (1) 65,000 pounds into a 310-mile due-east orbit from KSC, (2) 25,000 pounds into a 310-mile 55-degree orbit from KSC while carrying the air-breathing engines, and (3) 40,000 pounds into a 115-mile circular polar orbit from Vandenberg. The first and last missions excluded the use of the air-breathing engines. The orbiter was to have the capability to return to its launch site after a single orbit, although an actual cross-range was not specified.[156]

Interestingly, the RFP did not specifically state the bidders needed to use the MSC-040C design, but this was widely assumed by everybody involved. The payload bay was to have a clear volume 15-feet in diameter and 60 feet long, and be able to accommodate payloads up to 50 feet long. The vehicle was to be capable of being launched within two hours from a standby status, and be able to hold in a standby status for up to 24 hours. The crew cabin needed to accommodate four astronauts and be able to support them for up to a week on-orbit. An additional six astronauts were to be accommodated for shorter periods of time as needed. Maximum acceleration during ascent or reentry was to be limited to 3-g.[157]

For ferry flights, the orbiter was to be cable of flying across the United States using its own air-breathing engines (ABE). This range could be achieved using either multiple refueling stops, or aerial refueling – the bidders' choice. Surprisingly, the RFP stated that the ABE system for ferry flights could be separate and different from the ABE provided for space missions (reference mission 2). While returning from orbital missions, the ABE was to provide 15 minutes of loiter time at 10,000 feet "to allow operational assessment of conditions prior to landing." During atmospheric flight the orbiter was to be capable of +2.5/–1-g maneuvering.[158]

An unspecified reusable surface insulative thermal protection system was to be used, although ablative material or other special forms of TPS were allowed "where beneficial to the program." The orbiter thermal protection system was to be capable of surviving an abort scenario from a 500-mile circular orbit.[159]

Two 156-inch solid rocket boosters were to be used as the baseline, and they were to be ignited on the ground in parallel with the orbiter main engines. The SRBs were to include a thrust termination capability, and the entire shuttle system was to be capable of intact aborts even while the SRBs were thrusting. The SRBs were to be designed for water recovery, refurbishment, and subsequent reuse.[160]

Although the prime contractor was to be responsible for avionics and software development, the RFP contained an interesting clause "... because of the unique operational interfaces which will affect NASA, a separate direct technical working relationship with the Contractor or subcontractor providing software will be established with NASA. This direct interface will provide for requirements, formulation, development, and verification control between the software development division or subcontractor ... and NASA."[161]

---

* In 1988, the OMB found it cost $300 million per flight, or $5,000 per pound and that development had cost $6,651 million in FY71 dollars – $1,500 million more than estimated.

† Sort of, anyway. Aerospace companies receive something called 'B&P' (bid and proposal) funds from the Government based on a percentage of existing government contracts. This money may only be used to prepare proposals for future government contracts. Companies often supplement this with internal funds.

Grumman / Boeing

Over the course of the ASSC and various extensions, Grumman/Boeing had investigated 162 different orbiter configurations. The baseline Phase C proposal used an orbiter 128.1 feet long that varied from the MSC-040C design in some details – the Model 619 orbiter had a distinctive 'humpback' when viewed from the side. Grumman and North American were the only competitors that designed the orbiter for the baseline 500 mission service life – the other two opted for the 100 mission alternate.[162]

The Grumman payload bay was actually almost 16 feet in diameter and 61 feet long in order to accommodate payloads measuring 15 by 60 feet. The payload bay doors were divided into two sections on each side, allowing the bay to be divided to carry different payloads. This also allowed the air-breathing propulsion system (ABPS) to be installed in the aft section of the bay with swing-out engines deployed from the aft portions of the payload bay doors. On missions when there was no anticipated need for the ABPS, this area could be used as additional payload volume.

Two 500,000 lbf, 35-foot long ASRMs were located along the fuselage sides immediately above the wing juncture. The 92.5-foot span wing had a severe delta planform with slightly forward-swept trailing edges and rounded wingtips. The reaction control system was mounted in blended pods on each wing-tip and in the nose, and the on-orbit engines were carried in pods on the fuselage shoulder beside the vertical stabilizer.

The SRBs were 138.2 feet long and used exhaust nozzles with a 13.5 degree precant angle venting aft of the wing. The SRBs were mounted above the ET centerline and were 38.2 feet aft of the ET nose. The external tank was 31.8 feet in diameter and 163 feet long. Two large stabilizers were mounted on the bottom (in relation to the orbiter) of the ET, mainly to provide additional directional stability during ascent.

The crew compartment could accommodate six passengers and included a galley and 'hygiene facility.' A crew of two was seated on the flight deck that had 19-degree over-the-nose visibility, and a stand-up aft station supported payload operations while on-orbit. The docking port and airlock were located in the nose of the orbiter since Grumman thought this simplified the docking maneuver.

Grumman did not completely trust the Lockheed-developed reusable surface insulation, and proposed to add a layer of ablative material around the crew compartment and the OMS/RCS modules during the early development flights to guard against a possible RSI burn-through.

Over 93 percent of the orbiter structure was composed of various aluminum alloys, and 94 percent of the major items could be manufactured from flat or single-curvature pieces. Interestingly, Grumman proposed assembling the orbiters at KSC from major subassemblies manufactured at other locations (mainly the Grumman New York facility).

A major strength of the Grumman proposal was that the orbiter was designed for ease of maintenance, and most major components and systems could be accessed from inside the orbiter without requiring large work stands or disassembly. Items requiring frequent access were located in the cabin lower deck or the payload access passageway immediately aft of the cabin. All four APUs could be easily accessed while the vehicle was on the pad.

Grumman estimated that the maintenance features of the orbiter would permit a ground turn-around time of 154 hours. An Automated Ground Checkout System was integrated with the On-Board Checkout System to allow many

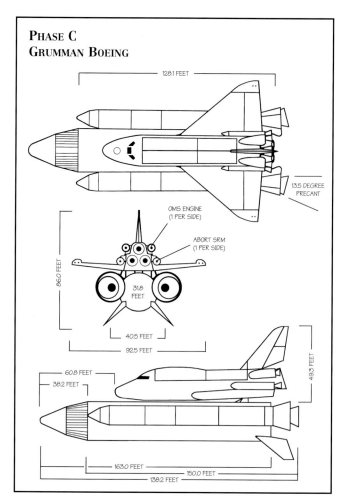

PHASE C
GRUMMAN BOEING

The Grumman Model 619 sitting on the pad at Launch Complex 39 at the Kennedy Space Center. Note the relatively simply tower and the complete lack of weather protection for the vehicle – this would prove to be a major problem at KSC. (Grumman Aerospace)

level I and level II maintenance tasks to be automated. After landing at KSC, the orbiter would be purged and safed, then towed to a new Reusable Surface Insulation refurbishment facility where the TPS was repaired as necessary, then "sprayed to maintain the thermal protection properties and prevent water absorption. The orbiter would then be towed to the Vehicle Assembly Building (VAB) for maintenance and mating with the ET and SRBs.

After the pyrotechnics were installed the stack would be transported to the pad. All of this would be monitored by the computerized checkout systems.

The Grumman proposal was by far the most detailed of the four submittals, and Grumman had gone as far as constructing full-scale mockups of the solid rockets to test their splash-down characteristics to ensure they would be recoverable and reusable.

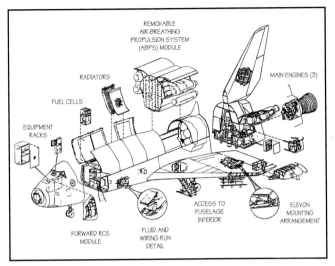

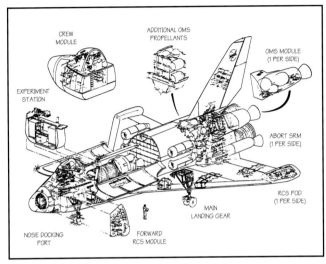

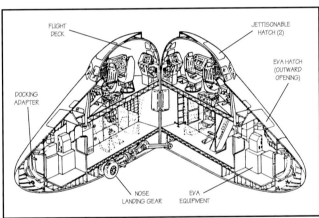

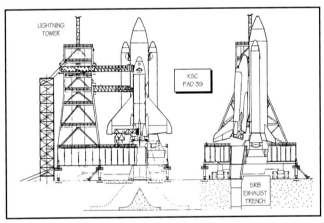

The Model 619 orbiter returning home to KSC. Note the deployed air-breathing engines in the rear of the payload bay. Most competitors opted for a removable engine pack that could be used in this manner. (Grumman Aerospace)

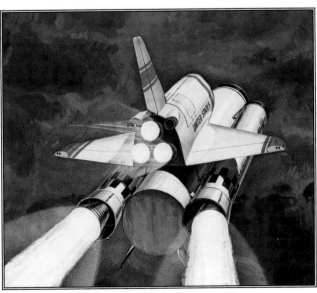

Not looking very much different than the real thing. Note the RCS pods on the wingtips and the two stabilizing fins on the external tank. This drawing does not show the abort solid rocket motors. (Grumman Aerospace)

Lockheed

The Lockheed Phase C/D proposal used an orbiter 126.1 feet long with a 78.3-foot wingspan that also took some liberties with the basic MSC-040C design. Specifically the wing planform showed a decided leaning towards a double-delta design, long before the program officially sanctioned this change. The OMS engines were co-located with the reaction control system in pods faired into the fuselage shoulders at the base of the vertical stabilizer, another feature not yet adopted by the program as a whole. A 56-foot long ASRM was mounted at the wing-fuselage juncture with the nozzles canted outward at approximately ten degrees. The orbiter was mounted 62.1 feet aft of the ET nose.[163]

The 31.7-foot diameter external tank was 173.2 feet long with a tapered nose and conical tail. The solid rocket boosters were 162.4 feet long and were mounted 50.4 feet aft of the ET nose and slightly above the ET centerline. Each SRB nozzle exhausted well aft of the orbiter wing and had a five degree precant angle to ensure the primary thrust vector was through the mated vehicle's center of gravity. The SRB centerlines were 41.6 feet apart.

One of the more controversial aspects of the Lockheed proposal was that they intended to act as a systems integrator instead of a manufacturer. More than 70 percent of the orbiter would be subcontracted out, then assembled by Lockheed. The rationale was "Rather than incur the expense and delay inherent in developing the system within a single company and the resulting relocation of thousands of people ... Each section contractor will be given full responsibility for the technical management associated with developing, producing, and delivering the hardware, ready to be mated with adjacent sections." Lockheed pointed to their successful track record on the Navy fleet ballistic missile program where the company managed over 2,000 subcontractors and suppliers in 40 states. In the end, NASA did not buy the argument, believing instead that

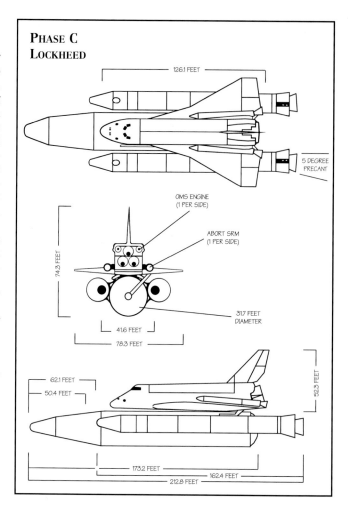

allowing a multitude of companies to design and build the majority of orbiter would substantially increase the risk of things not working when it all came together.

The forward fuselage section consisted of a carbon-carbon nose cap, a modular forward reaction control system, a crew compartment, nose landing gear, and an equipment bay. The RCS module was installed above the nose gear in a manner that allowed it to be easily removed in either horizontal or vertical maintenance. An in-flight refueling receptacle module could replace the forward RCS module for ferry flights. The equipment bay, located behind the pressurized crew compartment, contained the fuel cells and environmental control system. A door in the side of the fuselage permitted maintenance access.

The main cabin provided over 2,400 cubic feet of space for a four-man crew, equipment, and passengers. The cockpit windows were of three-pane construction and were considered fail-safe and bird-proof. Interestingly, Lockheed provided a windshield wiper for the forward windscreen.

The aft fuselage section was comprised of the main load-carrying structures that supported the payload attach points, payload bay actuator system, center wing structure, and both the main engine and abort rocket thrust structures. Unlike most of the designs, the Lockheed payload bay doors formed an integral part of the fuselage load-carrying shell. Lockheed claimed this resulted in a weight savings of 4,800 pounds over non-load-carrying doors.

For missions that required air-breathing propulsion (reference mission 2 in the RFP), Lockheed provided two pairs of General Electric GE101/F12A3 turbofan engines in the rear portion of the payload bay. This installation required a fuselage kit that replaced the aft sections of the

The artist concept of launch shows how far aft the solid rocket nozzles were located, eliminating both thermal and acoustic loads on the orbiter wings and aft fuselage. Note the small Lockheed stars on the OMS pods. (Lockheed)

payload bay doors. In addition to the engines and their extension mechanisms, this special kit also included JP-4 tanks that provided sufficient fuel for 15 minutes of loiter time after reentry. For ferry flights these engines were supplemented by two additional engines mounted in pods under the fuselage between the main landing gear wells. The main landing gear used three wheels per side and was derived from the SR-71 design.

The external tank proposed by Lockheed was an all-aluminum vessel containing separate LO2 and LH2 tanks that used weld-bond construction. The tank structure was insulated with a spray-on foam insulation and cork ablator for protection during ascent. The external tank was attached to the orbiter at three points in a tripod arrangement.

Orbiter ascent boost was provided by two 3,520,000 lbf 156-inch diameter solid motors. The SRMs were positioned to thrust through the orbiter center-of-gravity and were equipped with a thrust vector control system based on the Lockseal design. The composite PBAN propellant was contained within D6AC steel case segments. Thrust termination stacks were installed at the forward closure/nose cone area, and arranged to avoid impingement of the plume and debris on the orbiter. A set of laterally separation solid rockets was installed forward and aft on each SRM to provide direct translation of the boosters away from the stack when they were jettisoned. A parachute recovery system was installed in the nose cone of each SRM.

Long before the program as a whole adopted it, Lockheed was proposing to use a double-delta wing. Surprisingly, the proposal did not specifically address why Lockheed had chosen this design, other than it slightly simplified the thermal protection system. For instance, Lockheed proposed LI-1500 tiles for the inboard part of the leading edge, using the more expensive carbon-carbon system only on the outboard sections. The overall orbiter thermal protection system was designed to provide 100 'normal operational' reentries at 2,500 degF, or a single contingency reentry at 3,000 degF.

Lockheed calculated that their orbiter had a cross-range of 1,575 miles. Automatic landings would use the Lockheed Autoland System developed for the L-1011 Tristar airliner that had recently been certified by the FAA for Category III landings. The design landing speed was 172 mph, with a FAR field length of just under 10,000 feet based on the use of brakes and drag chutes.

A new operating entity, the Space Shuttle Program Division, was going to be created within the Lockheed Corporation specifically to manage the NASA effort. Interestingly the proposed division manager, J. F. Milton, had been the chief systems engineer on Dyna-Soar when he worked at Boeing.

Calculated in 1971 dollars, Lockheed estimated that

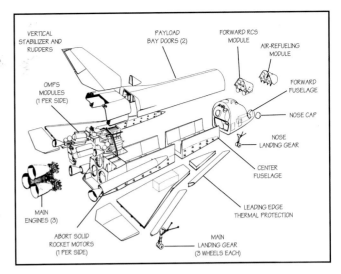

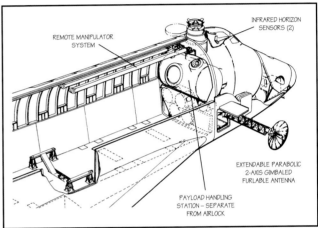

development would cost $2,695 million (in addition to $956 million in GFE development for the SSMEs, etc.). Production of five orbiters would cost $572 million, with operations for the first two years coming to $1,259 million (plus an additional $3,331 for government-furnished operational costs such as propellants).

Lockheed also proposed a rather unique in-house design that borrowed heavily from the MSC-048 series design. This twin-tail vehicle had a 'bat-shaped' double-delta planform and used three main engines. Reaction control system pods were mounted on the wing-tips, and the OMS engines were located on top of the fuselage between the vertical stabilizers. A spacious cockpit with somewhat limited visibility was provided. The orbiter was 113 feet long with an 81-foot span. Although interesting, only the MSC-040-derived vehicle received serious consideration from NASA.

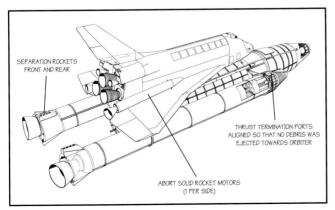

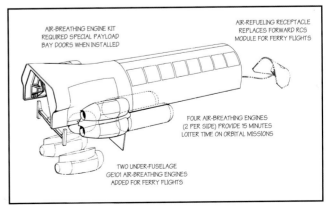

McDonnell Douglas

Since Martin Marietta had been responsible for the fly-back booster, which was no longer required, McDonnell Douglas bid Phase C with TRW instead. The company remained fairly true to the original MSC-040C design. The 126.7-foot long orbiter had a severely swept delta-wing that spanned 70.6 feet over the ends of its wing-tip mounted reaction control system pods – the trailing edge was also slightly swept. The orbiter was 51 feet high and contained three SSMEs in the aft of the fuselage, along with two OMS engines. As with all the Phase C/D designs, the payload bay doors extended slightly aft of the leading edge of the vertical stabilizer. The orbiter was mounted 60.8 feet aft of the nose of the 32.8-foot diameter external tank.[164]

The three main engines were spaced as far apart as feasible to preclude a catastrophic failure of one from affecting the others. The SSMEs would be throttled down to only 68 percent power to limit maximum dynamic pressure. The engines could be gimbaled ±10 degrees in pitch and ±6.5 degrees in yaw at rates up to 8 degrees per second.

The thermal protection system proposed by McDonnell Douglas was not the expected Lockheed-developed RSI silica tiles. Instead, McDonnell proposed using Mullite hardened-compacted fibrous (HCF) tiles that were glued to the bottom and side of the orbiter over a thin layer of foam for strain relief. The top of the orbiter would use ESA-3560 or SLA-561 ablator, as would the leading edges of the wing and vertical stabilizer.

The flight deck had provisions for a pilot and mission commander, both with full controls and displays. Both orbital and aerodynamic maneuvers were controlled via a single side-stick controller, similar to the one first used on the X-15. Provisions were made to incorporate a conventional center stick for aerodynamic control during flight testing if required. Interestingly, McDonnell Douglas proposed to use instruments "… of conventional design to eliminate development risk and enhance crew acceptance and confidence." Two separate airlocks were proposed, one used for docking at the front of the lower cabin, and one used for EVAs and payload bay access at the rear of the lower cabin. Each airlock had a volume of 210 cubic feet and could accommodate two men in space suits. One of the more interesting items to be proposed was a food preparation area that used a 'laminar flow bench' that prevented small trash and food scraps from circulating within the cabin. A trash compactor was also installed.

The air-breathing engine arrangement was unique among the bidders. Two retractable engine pods, each

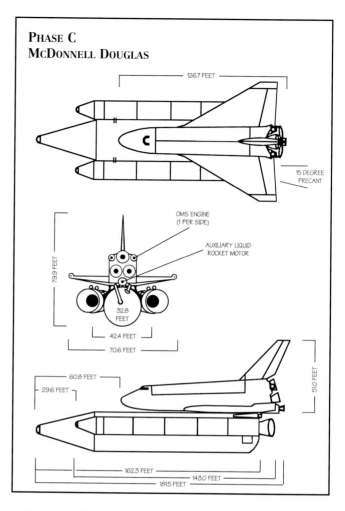

with a pair of 16,000-lbf General Electric F101/F12A3 turbofans, were located in the belly of the orbiter between the main landing gear wells. At subsonic speeds these pods would swivel out of the fuselage. Fuel was carried in dedicated tanks in the bottom of the payload bay. McDonnell Douglas estimated the weight of this system at 16,150 pounds (dry). The retractable pods were located such that they could be removed without affecting the orbiter center-of-gravity. For ferry flights, two twin-engine pods containing 30,000-lbf F101-GE-100 afterburning engines were bolted to the bottom of the fuselage, and a larger fuel tank was installed in the payload bay.

The external tank was 162.3 feet long and had an extreme conical shape to the forward LO2 tank and a con-

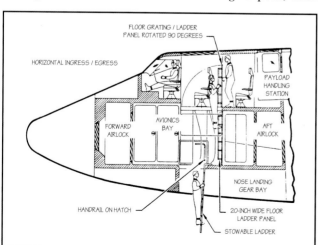

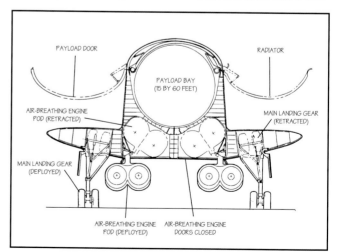

vex aft end formed by the trailing bulkhead of the LH2 tank. The 143-foot long SRBs were mounted well above the ET centerline, almost flush with the 'top' of the external tank. The SRBs were mounted 29.6 feet aft of the ET nose. The SRB nozzles had a 15 degree precant angle and exhausted ahead of the trailing edge of the orbiter wing, the only one of the four designs to do so. This would probably have led to an unacceptable heating of the wing trailing edge from the SRB exhaust plume.

McDonnell Douglas did not include provisions for mounting an abort solid rocket motor system on their orbiter, again the only one of the four teams not to do so. However, a single 931,000-lbf liquid-fueled rocket engine could be mounted low on the orbiter aft fuselage to assist in abort situations, along with thrust from the orbital maneuvering

system engines. This motor had sufficient propellants for a single 7.4-second burn, considered sufficient to allow the orbiter to separate from the ET and SRBs. The total inert weight of the installation was 7,793 pounds, and 29,500 pounds of propellants were carried. When installed, this motor would be jettisoned prior to reaching orbit.

Like Lockheed, McDonnell Douglas also proposed a twin-tail 'bat-wing' orbiter as a possible alternate design. In this case the orbiter would use three main engines arranged side-by-side across the fuselage. This engine arrangement resulted in "reduced cant losses, shorter body length, shorter thrust structure, and a direct thrust load path into the wing carry-through structure." A second alternate design was similar to the baseline but used a double-delta planform much like that used by Lockheed.

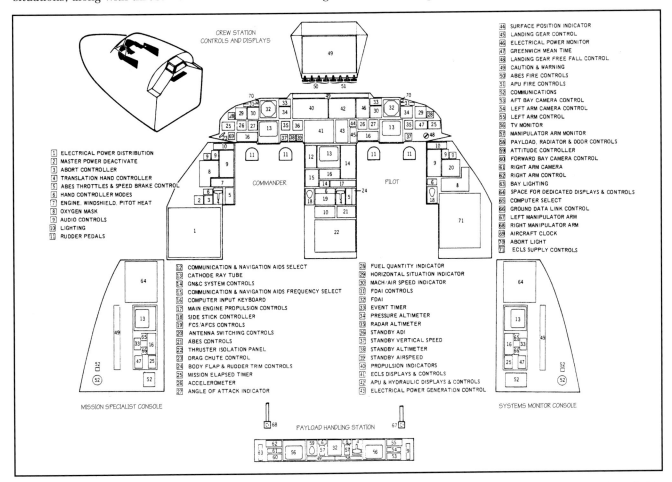

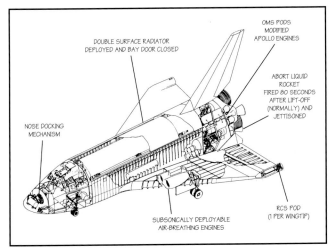

North American Rockwell

North American also took some liberties with the basic MSC-040C design, both in the wing planform and in the shape of the top of the orbiter fuselage. Since the NASA design had not identified a location for the manipulator arm, North American created a raised fairing along the top of the fuselage immediately aft of the crew station that continued along the top of the payload bay doors to contain the arm. A 'bubble' style canopy was provided over the flight deck to improve rearward visibility during on-orbit payload operations, although this had been seen previously on various MSC designs, including versions of the MSC-040A. The wing planform had a gracefully rounded leading edge and wingtips that terminated in a straight-cut trailing edge. The wing also featured increased incidence versus the original twist to improve the vehicle's aerodynamic trim capability. The main gear was retracted into the wing instead of the fuselage to preclude the reduction in fuselage depth available to carry payloads. An ASRM was located on each side of the fuselage, slightly above the wing and immediately below the OMS pod, which was mounted lower on the fuselage sides than either the Grumman or Lockheed orbiter. The orbiter was 124.25 feet long and had a wing span of 79.6 feet.[165]

The external tank was 32.8 feet in diameter and 210.1 feet long including the retro-rocket contained in the extreme nose. The SRBs were mounted 59.9 feet behind the nose of the ET and slightly above the ET centerline. The SRB nozzles exhausted behind the wing trailing edge and had an 11 degree precant angle. Four long, but narrow, fins were mounted on the aft end of each SRB to provide stability immediately after separation until their recovery parachutes could deploy.

North American used two 385,000-lbf abort solid rocket motors mounted between the OMS pod and the wing next to the fuselage. The ASRMS provided meaningful

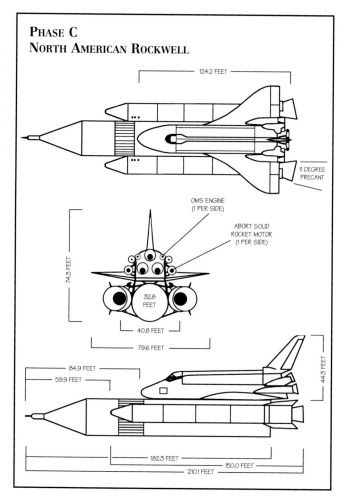

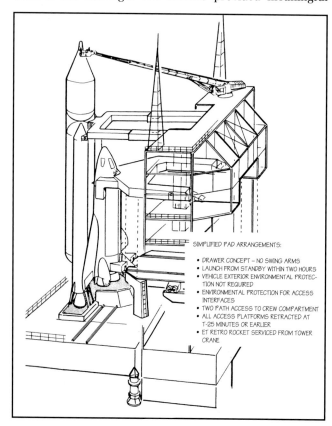

SIMPLIFIED PAD ARRANGEMENTS:

- DRAWER CONCEPT – NO SWING ARMS
- LAUNCH FROM STANDBY WITHIN TWO HOURS
- VEHICLE EXTERIOR ENVIRONMENTAL PROTECTION NOT REQUIRED
- ENVIRONMENTAL PROTECTION FOR ACCESS INTERFACES
- TWO PATH ACCESS TO CREW COMPARTMENT
- ALL ACCESS PLATFORMS RETRACTED AT T-25 MINUTES OR EARLIER
- ET RETRO ROCKET SERVICED FROM TOWER CRANE

escape for the first 30 seconds of ascent, and would be jettisoned soon afterwards. The combined weight of the two ASRMs and their structure was 92,700 pounds. Two 30,000 lbf Pratt & Whitney F401-PW-400 air-breathing engines could be carried in the rear of the payload bay on swing-out doors to provide a limited loiter capability after reentry. Two additional engines and a larger fuel tank were added for ferry flights.

The basic airframe was constructed from 2024 aluminum alloy, and the 'floating' crew compartment featured a circular cross-section to provide additional strength while pressurized. All of the major thrust loads were carried within the fuselage – the wings and vertical stabilizer were bolt-on units to simplify manufacture and maintenance.

The primary avionics system consisted of a fail-operational/fail-safe computer network with manual backup. Interestingly, North American proposed "a removable kit for the unmanned first vertical flight." Each payload would be monitored via controls and displays hardwired to the payload specialist console on the aft flight deck.

The landing gear made maximum use of off-the-shelf equipment. The wheels, tires, and brakes were identical to those used by North American on the B-1A bomber prototypes, while the drag chute came from a B-52 and the drogue chute and mortar were Apollo units.

Like Grumman, North American projected a 500 mission life for the orbiter. To accomplish this, the company selected a thermal protection system designed to subject the primary structure to no more than 350 degF. North American was not completely convinced that the Lockheed-developed silica tiles would mature quickly enough for the early orbital flights, so the baseline used

Mullite HCF tiles instead. North American continued to monitor the progress being made on the silica-tiles since they were lighter weight. The leading edges and nose cap were carbon-carbon, and no substantial use of ablators was made anywhere on the orbiter.

Interestingly, North American proposed using SSMEs that were not equipped with computerized engine control systems, predicting a $10 million cost savings over the life of the program. Instead the engines would use a simple closed-loop mixture control system.

Although not as detailed in many respects as the Grumman proposal, North American had constructed many full-size mock-ups of various parts of the orbiter to ensure the design was optimized for maintenance. Many of these mock-ups had first been constructed during the Phase A and Phase B studies, and had been continually modified (or rebuilt) as the design of space shuttle progressed through its myriad of changes.

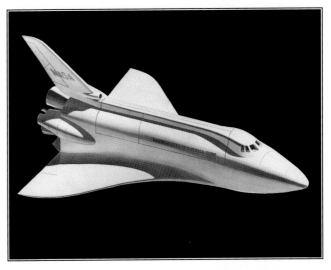

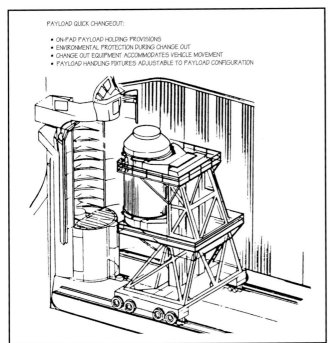

PAYLOAD QUICK CHANGEOUT:

• ON-PAD PAYLOAD HOLDING PROVISIONS
• ENVIRONMENTAL PROTECTION DURING CHANGE OUT
• CHANGE OUT EQUIPMENT ACCOMMODATES VEHICLE MOVEMENT
• PAYLOAD HANDLING FIXTURES ADJUSTABLE TO PAYLOAD CONFIGURATION

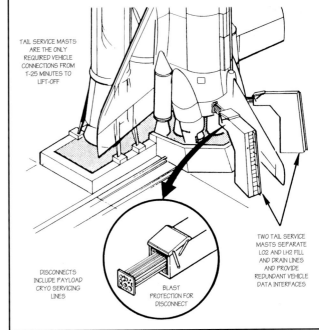

TAIL SERVICE MASTS ARE THE ONLY REQUIRED VEHICLE CONNECTIONS FROM T-25 MINUTES TO LIFT-OFF

DISCONNECTS INCLUDE PAYLOAD CRYO SERVICING LINES

BLAST PROTECTION FOR DISCONNECT

TWO TAIL SERVICE MASTS SEPARATE LO2 AND LH2 FILL AND DRAIN LINES AND PROVIDE REDUNDANT VEHICLE DATA INTERFACES

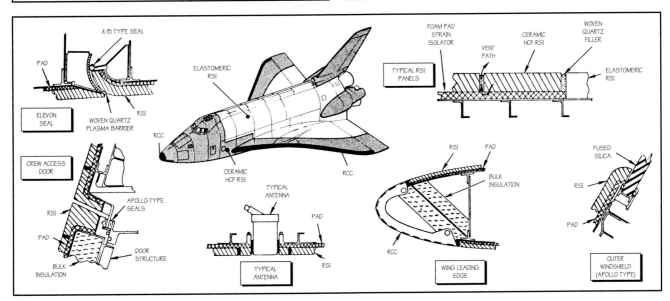

North American also decided against using the Lockheed-developed silica tiles, opting instead to baseline Mullite ceramic HCF (hardened-compacted fibrous) tiles because "test data show Mullite to be superior in material stability and coating compatibility to higher temperatures." North American noted, however, that "silica has made marked improvements with high-purity fibers" and indicated they would readdress this issue at the first major requirements review. (North American Rockwell)

The source selection report provided some interesting insights. In the preliminary evaluations, the four competitors ranked (1) North American, (2) Grumman, (3) McDonnell Douglas, and (4) Lockheed. All four companies were deemed to be within the 'competitive range,' meaning that the government felt that any of them were capable of performing the task at hand within the budget available. Each company was invited to participate in oral discussions with the government and to answer any questions that the government had regarding their proposals. After these discussions the relative rankings did not change.[166]

North American received the highest score in mission suitability and had the lightest-weight design. The source evaluation board liked the proposed guidance, navigation, and control system, which was considered to be a very good simple design with minimum interfaces. The board was less thrilled by the North American docking scheme, which used a male-female concept instead of the preferred androgynous method. North American also presented an excellent analysis of maintainability and proposed optimum turnaround concepts. But overall, the largest North American advantage was considered to be their proposed management structure. The board regarded North American's proposed key personnel as the best of any offeror. The only weakness noted, and a minor one at that, was the company's "lack of recent experience with large operational airframes."[167]

Grumman received the second highest mission suitability score, very close to North American. In general the Grumman proposal went into significantly more technical detail than any of the others, and these details were judged to be very good. Because of this, Grumman proposed weight summary was deemed to be the most accurate. The board was less thrilled, however, with Grumman's complex approach to the guidance, navigation, and control system (although it was very close to the one actually built). Grumman was rated outstanding in their maintainability concept, providing easy access to all major components. The board also liked Grumman's proposed external tank, but was concerned about placing the LO2 and LH2 fill lines in the same umbilical plate. The Grumman management proposal was considered strong, but lacking in large cryogenic system experience.[168]

McDonnell Douglas was third, but ranked significantly behind the first two. The primary technical strength was the design of the air-breathing propulsion system, and the fact that the proposed orbiter had the largest internal fuselage volume. In retrospect, however, this might not have worked well, since the air-breathers could not be easily deleted. Other commendable features included a good design for the reaction control system, and the best radiator/payload bay door system of the offerors. The board did not feel that McDonnell Douglas had taken full advantage of their maintainability experience with commercial airliners, and also felt that the proposed management structure was overly complex, especially the work split between the California and Missouri locations.[169]

Lockheed received, by far, the lowest score, being rated only 'fair' (compared to 'good' for McDonnell Douglas and 'very good' for the other two). The board liked the fact that Lockheed had extended the solid rocket boosters well past the trailing edge of the orbiter wing, and felt that the proposed thrust vector control system on the SRBs would provide better ascent control. However, the Lockheed design was the heaviest, and also had a lack of consistent technical depth. A major weakness was that Lockheed had a 65-second gap during ascent where there were no abort options for the crew. Minor weaknesses included the vehicle having a higher-than-specified landing speed, and that most subsystems were considered overly complex. Lockheed did, however, propose to use two tail service masts that separated the LO2 and LH2 fill lines, and the board thought this was a technical strength. Interestingly, Lockheed proposed to subcontract all of the major components of the orbiter instead of building them in-house. Even worse, from the board's perspective, was that Lockheed intended to allow the subcontractors to perform the actual design work, generating very complex organizational interfaces and exposing final assembly to a multitude of potential risks. The Lockheed key personnel team was thought to be lacking in overall strength and balance.[170]

Cost estimates contained in the proposals varied widely. North American was the lowest, followed closely by Lockheed – Grumman and McDonnell Douglas were significantly higher. The primary differences stemmed not from the cost of doing work in different locations, but the fact that the two groups (North American and Lockheed, versus Grumman and McDonnell Douglas) had widely divergent estimates on the total amount of work required to do the job. Although the board attempted to adjust some of these differences (as is often the case during source selection), in the end they decided to allow each proposal to stand largely unchanged and to evaluate the management processes further. This favored North American, mostly because the board believed the management techniques proposed by the company would "provide earlier identification of cost problems." Grumman and McDonnell Douglas received essentially neutral ratings, with the board believing that each company had proposed a workable management structure, and that their proposed costs reflected this structure. The board, however, had far less confidence in the Lockheed approach, stating that "its estimating techniques, its management plans, and its technical approaches all were set forth in its proposal with a lack of depth which contributed to an impression that many unforeseen problems might arise to jeopardize the company's control over its costs." In the end, the board rated the cost proposals (1) North American Rockwell, (2) Grumman, (3) McDonnell Douglas, and a distant (4) Lockheed.[171]

In an attempt to make the space shuttle seem more like an airplane, most early artist concepts showed a minimum of maintenance equipment. Compare this to the complex work stands and special equipment in the OPFs. (NASA)

James Fletcher, George Low, and Richard McCurdy met in the morning of 26 July for the final review of the selection board results.* They noted that McDonnell Douglas and Lockheed ranked significantly below the other two competitors, so concentrated instead on North American and Grumman. After carefully reviewing the mission suitability scores, the three men decided that the advantage went to North American. Since North American also presented the lowest probable cost, they were deemed the winner.

After the stock market closed on 26 July 1972, it was announced that the Space Transportation Systems Division of North American Rockwell† had been awarded the $2,600 million contract (NAS9-14000) to design and build the Space Shuttle Orbiter.[172] The design was loosely based on the final MSC-040C configuration, and two space-rated Orbiters, a full-scale structural test article (STA), and a main propulsion test article (MPTA) were included in the contract. Due-east flight from KSC would carry up to 65,000 pounds in a payload bay measuring 15 by 60 feet. During its return to Earth, the vehicle would be capable of a 1,265 mile reentry maneuver on either side of its flight path. The contract also included the systems integration function, where Rockwell would ensure all elements of the space shuttle system worked together, and included supporting the first two years of operations at KSC. Almost immediately there were charges of favoritism levied against Nixon for selecting a 'home-state' company. The Democratic National Committee reported that five of the members of the Rockwell board had contributed to Nixon's campaign – further investigation showed that the contributions totaled a whopping $6,000 between all five men.[173]

What was not talked about was the potential conflict of interest of Dale D. Meyers at the Office of Manned Space Flight. Meyers had spent a significant portion of his career at North American, and had also been responsible for selecting the members of the source evaluation board that reviewed the space shuttle proposals. Careful selection of a source evaluation board can skew the results of a competition one way or another, and Meyers, Low, and Fletcher all knew it. Meyers had divested himself of all connections with North American when he came to NASA, and had no monetary interest in the company. He selected the members of the board from senior NASA management based on their responsibilities within the agency. The only non-NASA members were two officers from the Air Force.‡ The NASA legal counsel concurred with the selections. Meyers himself did not participate in the source selection, and did not review the results until after they were decided.[174]

MAIN ENGINE PHASE C/D AND AWARD

As originally intended by MSFC, the SSME contract would have followed an unmodified four phase development plan. At the end of the SSME Phase B, MSFC had intended to select two contractors that would be funded during Phase C to build and test prototypes of the SSME. A single contractor would be selected at the end of Phase C to perform final design and manufacturing of the winning engine. But funding constraints changed this – a single contractor would need to be selected at the end of Phase B – with C/D combined, as with the Orbiter.[175]

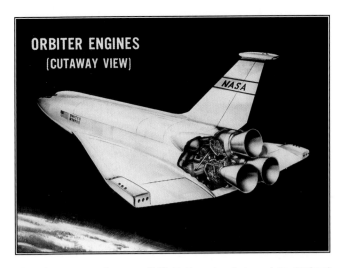

This February 1972 artist concept highlights the engines that were being developed under the direction of MSFC. This is an early MSC-040 orbiter – note the RCS pod on top of the vertical stabilizer. (NASA)

Officials at MSFC worried that "once we choose a company and a configuration, we are locked in." Nor was the approach expected to be less costly in the long run. Richard L. Brown, who helped evaluate the SSME proposals, claimed "there were economic studies that indicated it would actually be cheaper to run the competition because of its influence on price" and to arrive at "a better definition of cost, and therefore less overrun." Nevertheless, only a single contractor would proceed into a combined Phase C/D.[176]

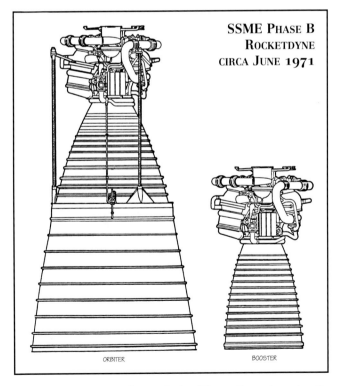

When the space shuttle was still supposed to be a fully-reusable two-stage system, the space shuttle main engine was supposed to power both stages. But the some of the requirements for an engine to power a vehicle within the lower atmosphere – where the booster would mainly fly – are decidedly different than the requirements of a vehicle in the extreme upper atmosphere – where the orbiter would fly. The majority of these differences surround the exhaust nozzle, and the expansion ratio necessary for optimum performance. The other differences are in propellant mixture, but that can be handled by the engine management system. This drawing shows the large difference in nozzle size between the 550,000 lbf booster engine and the 632,000 lbf orbiter engine that were baselined at the end of the SSME Phase B studies. The power head, turbopumps, and most of the mechanical items were identical on these engines. (Rocketdyne)

---

*   At this time, Fletcher was Administrator, Low was Deputy Administrator, and McCurdy was Associate Administrator for Organization and Management.
†   North American Rockwell became Rockwell International on 16 February 1973.
‡   The Air Force also had 24 members on various technical panels.

An RFP was released by MSFC on 1 March 1971 for the design, development, and delivery of 36 SSMEs, all to be completed by the end of 1978. The same three companies that had been participating in the Phase B SSME studies – Aerojet, Pratt & Whitney, and Rocketdyne – submitted proposals on 21 April 1971. Each proposal consisted of multiple volumes of technical, management, and cost data.[177]

The Rocketdyne Division of North American Rockwell was selected on 13 July 1971 as the winner of the SSME competition. The decision to select Rocketdyne was heavily influenced by the company-funded construction and testing of a nearly full-scale model of the combustion devices for their proposed SSME powerhead. However, shortly after the award, Pratt & Whitney filed a protest with the General Accounting Office to block the contract award (NAS8-40000). The protest charged that the selection was "manifestly illegal, arbitrary and capricious, and based upon unsound, imprudent procurement decisions." Both senators from Alabama joined seven colleagues from the southeast protesting the selection of a California company over one based in Florida. The GAO ordered that Rocketdyne could not receive any funds under the contract until the dispute was settled, so MSFC issued a level-of-effort contract to Rocketdyne to support to the still-competing airframe contractors.[178]

The GAO ultimately decided the protest in favor of Rocketdyne on 31 March 1972, and the $202,766,000 40-month contract was signed on 14 August 1972. The finalization of the engine requirements began in May 1972 and over 250 separate issues were identified and resolved over a two month period. But the actual definition of the physical, electrical, and functional interfaces could not begin until MSC selected an Orbiter contractor – a decision that was finally made on 26 July 1972

The long-term ramifications of the protest were more serious than the slight delay in SSME development that had actually occurred. NASA still needed Congressional support for the space shuttle, and this meant they also needed the support of the entire aerospace industry, which was suffering from a rather severe economic depression at the time. One of the ways to ensure this was to spread out space shuttle development among as many aerospace contractors as possible. But sound politics does not necessarily translate to sound engineering. This would become very evident when the solid rocket motor contracts were awarded.[180]

SRB Phase C/D and Award

Negotiations for the solid rocket motor contract were laden with controversy from the very beginning. The first disagreement was internal, as MSFC began to prepare an RFP for release to industry. MSFC envisioned the solid rocket booster (SRB) as a system comprised of the steel case loaded with propellant (the SRM) and several non-propulsive elements such as the forward and aft skirts, nose cone, attachment structure, thrust vector control, and recovery devices. Rather than award a single contract that covered the entire SRB, James Fletcher decided to contract only for the SRM, then to give MSFC responsibility to integrate the other components to create the SRB.[181]

On 27 January 1972 MSFC issued one-year $150,000 study contracts to Aerojet General, Lockheed Propulsion, Thiokol, and United Technology to study what the final configuration of the solid rocket motors should be. NASA then used this information to formulate the request for proposals for the production contract.

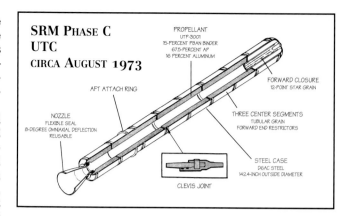

The segmented design from UTC was typical of the proposals from Lockheed, Thiokol, and UTC. All three contractors dismissed the potential risks of the segmented design, pointing to the Titan IIIC record – outstanding at the time, but a very limited sampling. (UTC)

But NASA would make a key decision affecting the booster configuration before MSFC had finished writing the RFP. During April 1973 MSC, supported by Rockwell, decided to eliminate the baseline requirement for the SRB thrust termination system. As envisioned at the time, the thrust termination system consisted of pyrotechnic charges that would blow a hole in the forward dome of the SRB, negating much of the normal thrust of the booster. But there were two significant problems with this concept – it produced potentially unsurvivable dynamic loads due to its rapid onset, and it also produced forward debris that could impinge on the Orbiter and External Tank.[182]

There was, and is, debate over the relative risks of large segmented solid rockets. But there was no question that the SRB thrust termination would be expensive. It would require a much more robust Orbiter to survive the sudden dynamic forces, possibly adding as much as 8,000 pounds to the empty weight of the vehicle. The External Tank would also need more structure, increasing its weight. In turn, this would require higher-performing SSMEs. It was a vicious cycle. MSFC conceded to the change, but argued for an option to implement it later if necessary. Headquarters declined this request, not wanting to incur the 'scar penalties' (extra weight) that might have made later addition of the thrust termination system possible.[183]

The RFP was finally released on 16 July 1973, involving only the motor segments, and lacking any requirement for thrust termination. The baseline design had evolved somewhat from that used in the orbiter competition, and instead of being 156-inches in diameter, the RFP requested motors that were somewhat smaller. The same four companies that had participated in the 1972 studies all submitted technical and cost proposals on 27 August and 30 August 1973, respectively. Oral and written discussions were held with the competitors between 24 September and 10 October 1973, and all four companies submitted their 'best and final' offers on 15 October.[184]

All of the bidders concluded that the drivers for the space shuttle SRMs were different from any previous motors. Primarily this was because the shuttle motors were to be reusable and man-rated. All previous motors had been expendable, resulting in cases and systems designed for minimal acquisition costs, and no thought given to refurbishment. For shuttle it would be possible to spend more money on the case and subsystems, but they would also need to be more robust to support multiple reuses.[185]

Three of the four proposals were generally similar, offering segmented weld-free cases. Where the motors were manufactured varied widely, with one offeror (UTC)

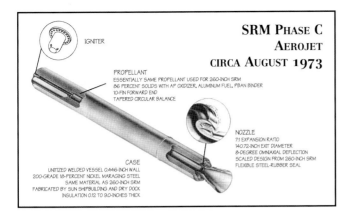

SRM PHASE C
AEROJET
CIRCA AUGUST 1973

IGNITER

PROPELLANT
ESSENTIALLY SAME PROPELLANT USED FOR 260-INCH SRM
86 PERCENT SOLIDS WITH AP OXIDIZER, ALUMINUM FUEL, PBAN BINDER
10-FIN FORWARD END
TAPERED CIRCULAR BALANCE

NOZZLE
7:1 EXPANSION RATIO
140.72-INCH EXIT DIAMETER
8-DEGREE OMNIAXIAL DEFLECTION
SCALED DESIGN FROM 260-INCH SRM
FLEXIBLE STEEL-RUBBER SEAL

CASE
UNITIZED WELDED VESSEL 0.446-INCH WALL
200-GRADE 18-PERCENT NICKEL MARAGING STEEL
SAME MATERIAL AS 260-INCH SRM
FABRICATED BY SUN SHIPBUILDING AND DRY DOCK
INSULATION 0.12 TO 9.0-INCHES THICK

Aerojet's monolithic case eliminated all of the o-rings used to seal the segmented cases proposed by the other competitors and preferred by NASA. As later events would show, this might have been a worthy design. (Aerojet)

suggesting it would be least expensive to have two separate plants – one on the east coast (Michoud) and one on the west coast (Coyote Center) – to support a 60 per year flight rate. But it was the fourth proposal that sparked the most controversy.[186]

Aerojet General elected to forego the weld-free segmented case proposed by everybody else – and required by the RFP – and proposed a welded monolithic case. The primary rationale for this decision was "the attendant low cost for its production, refurbishment, and operational support. The conventional [monolithic] design permits a proof test of each completed case to provide positive verification of the integrity of each flight motor." Aerojet continued "We also pioneered making 100- and 120-inch diameter segmented motors, but for a specific and unique military application ... a first-stage ICBM booster capable of overland transportation to remote missile sites. The inherent reliability of one-piece construction was **compromised** to satisfy a transportation limitation. But no such limitation exists for the Space Shuttle Program ..." (emphasis in the original).[187]

Aerojet presented an argument that the monolithic case was lighter weight, less expensive, and much safer than a segmented case. The primary drawback was handling a single unit weighing over 1,000,000 pounds. But Aerojet pointed out that both launch sites were located on the coast, as was the Aerojet production site. This would facilitate using barges for transportation, with only very short over-land routes required. The case proposed by Aerojet was 200-grade 18-percent nickel maraging steel 0.452-inch thick that was rolled and welded by the Sun Shipbuilding and Dry Dock Company, the same fabricator that was used in the experimental 260-inch program. The monolithic case had an outside diameter of 141.21 inches and an overall length of 110.12 feet. Internal insulation would be installed at the Aerojet facility using premolded sections. The propellant was ANB-3412, essentially similar to the ANB-3105 used in the 260-inch program with a one percent increase in oxidizer to boost performance. The propellant would use a tapered circular bore with ten fins at the forward end to provide additional surface area for the initial high trust requirement.[188]

After an evaluation of the proposals by teams involving 289 people representing five NASA Centers, NASA Headquarters, and all three military services, the source evaluation board ranked Lockheed, Thiokol, and UTC as 'very good' with mission suitability scores of 714, 710, and 710, respectively. Aerojet was ranked last, with a score of 655 and a subjective rating of 'good.' Apparently the gov-

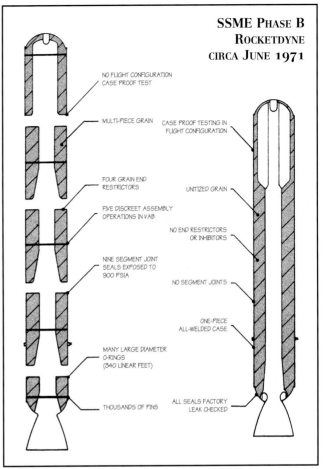

SSME PHASE B
ROCKETDYNE
CIRCA JUNE 1971

NO FLIGHT CONFIGURATION
CASE PROOF TEST

MULTI-PIECE GRAIN

CASE PROOF TESTING IN
FLIGHT CONFIGURATION

UNITIZED GRAIN

FOUR GRAIN END
RESTRICTORS

FIVE DISCREET ASSEMBLY
OPERATIONS IN VAB

NO END RESTRICTORS
OR INHIBITORS

NINE SEGMENT JOINT
SEALS EXPOSED TO
900 PSIA

NO SEGMENT JOINTS

ONE-PIECE
ALL-WELDED CASE

MANY LARGE DIAMETER
O-RINGS
(340 LINEAR FEET)

THOUSANDS OF PINS

ALL SEALS FACTORY
LEAK CHECKED

Drawing from the Aerojet proposal showing the perceived benefits of a monolithic case. Aerojet believed that the monolithic design had significantly fewer single-point failure paths than the segmented case and could also be subjected to more effective pre-flight functional verifications. Aerojet pointed out that there had been thousands of firings of large monolithic cases, but that the total flight experience with segmented cases consisted of 25 Titan IIIC launches. The experimental base for large segmented cases was not much greater, with fewer than 100 total having been fired. (Aerojet)

ernment did not believe Aerojet's claim that the monolithic case provided additional safety margins without compromising operations. Given that Aerojet's cost proposal was not substantially different than the other three competitors, Aerojet was eliminated from the competition. Lockheed's main strengths were technical, and they had the second-lowest cost proposal. Thiokol had the lowest cost proposal, and was rated second technically. UTC was third in both categories, but not by much. The board could find no significant technical drivers between the three proposals, so they chose the winner based on cost.[189]

On 20 November 1973, NASA announced that Thiokol had won the development contract for the solid rocket motors. Immediately there were charges that James Fletcher had somehow rigged the results to award business in his home state of Utah. Fletcher denied the charge, and others on the source selection board defended him. Nevertheless. the next day Lockheed filed a protest with the General Accounting Office (GAO), charging that "the entire NASA evaluation was marred by plain mistakes, inconsistency, arbitrary judgements, and improper procedures." The GAO would investigate these charges for the next six months, but finally, on 24 June 1974, concluded that there was "... no reasonable basis to question ..." the NASA selection. As is normally the case in the these matters, the GAO left NASA the option to reconsider their award, but NASA declined the opportunity.[190]

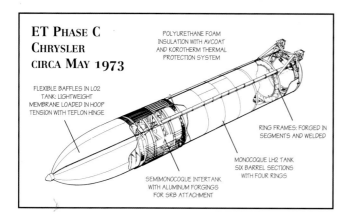

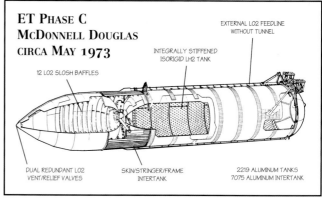

Lockheed's protest had centered, seemingly, around transportation costs. Lockheed had proposed using the Michoud Assembly Facility to refurbish the solids, then ship them via barge to KSC. The other contractors planned to ship their segments via rail, much like the Titan SRMs – in fact, this was one of the major drivers to the maximum segment size on the Thiokol and UTC designs. Lockheed pointed out that the Air Force was charged $8 per 100 pounds to ship the Titan segments; Lockheed had proposed charging only $2.50 per 100 pounds. What was usually left out of this discussion, as far as Lockheed care to say, was that Thiokol already had a functioning plant where they would perform the work – Lockheed had to build one largely from scratch at Michoud, at a cost of many millions of dollars.[191]

In the meantime, on 15 February 1974 MSFC had awarded a 90-day interim contract to Thiokol for studies, analysis, planning, and design in support of the SRM in the overall space shuttle system. Lockheed also protested this interim award, but the contract was allowed to move forward. On 20 May NASA extended this interim contract by 45-days, and again Lockheed protested. This extension was also allowed to stand so that work on space shuttle could continue. On 26 June 1974, two days after the GAO decision, MSFC finally awarded Thiokol a final contract for SRB design. This was followed on 15 May 1975 by a six year $136,546,674 contract covering development, testing, and production.[192]

The design of the solid rocket motors and the SRB as a whole was conservative, reflecting the prevailing approach of MSFC and the need to rate solid boosters for manned flight. The case was made of the same D6AC steel used on the Minuteman ICBM and Titan IIIC solid rocket motors rather than the 18-percent nickel maraging steel used on the experimental 156- and 260-inch motors. The final SRB was 146 inches in diameter and 125.3 feet long, with each of four segments being 164 inches long and the fore and aft sections making up the rest of the length. The Ladish Company in Cudahy, Wisconsin, made the cases for each segment without welding by a process called rolled ring forging – a hole was punched in a hot ingot of metal and then rolled to the 146-inch diameter.[193]

A separate firm, Cal Doran near Los Angeles, provided heat treatment to the case segments, after which they went further south to Chula Vista near San Diego where tang-and-clevis joints were added mechanically. The joint was significantly different from the one used on the Titan motors that had used a single O-ring with the outside portion of the joint pointed downward, discouraging the for-

mation of moisture in the joint. For the Shuttle SRB, a tang was added to the bottom of the casing, and it fit into the clevis on top of the adjoining section. Rohr milled 177 pins for each joint, and they were inserted through holes drilled in both the tang and clevis to hold adjacent segments together but still permit disassembly and reuse. To seal the joints, UTC used a careful application of putty to prevent high-pressure gas 'blow by' due to improper sealing of the O-rings – to man-rate the SRBs, the designers added a second O-ring.[194]

For thrust vector control Thiokol used the Flexseal design it had scaled up from Lockheed's design for the experimental 156-inch test program. It was capable of eight degrees of deflection, which was necessary among other reasons for the Shuttle to perform its now familiar roll downrange after lifting off the launch pad, a maneuver for which the gimballing of the SSMEs was not sufficient. A liquid-injection thrust-vector-control system such as that used on the Titan motors was also not feasible because of the level of thrust and the sheer size of the SRBs would have required too much fluid.[195]

The decision to separate the SRM from the rest of the SRB led to some interesting contractual arrangements at MSFC. On 21 December 1973, United Space Boosters, Inc. (USBI) – a part of United Technologies – was selected to manufacture most of the non-motor elements of the SRBs. But what was particularly unusual was that the MSFC Science and Engineering Directorate (S&E) performed as a third prime contractor responsible for the recovery system, booster separation motors, and the integrated electronic assembly. This arrangement not only gave MSFC more business than it would have had if all SRB work had been given to a single contractor, but it required less funding during the early years of development.[196]

ET PHASE C/D AND AWARD

Although the external tank appeared to be the least demanding of the space shuttle components, in reality it was a fairly advanced piece of engineering. Of primary importance was keeping weight and cost at a minimum, and this drove several aspects of the design. In addition, the ET was an extremely critical structural component of the integrated vehicle since all thrust loads between the elements were transferred through it.[197]

An RFP was released on 2 April 1973 to Boeing, Chrysler, McDonnell Douglas, and Martin Marietta, all of whom responded with proposals submitted on 17 May. Rockwell was explicitly prohibited from bidding on the ET contract since they were the prime Orbiter contractor, but nevertheless the company teamed with Chrysler for a joint bid. All four proposals were generally similar and considered satisfactory.

The source selection board rated Martin Marietta and McDonnell Douglas as the top technical competitors. The arguments made by Martin that it was the only company with similar experience rang true – the external tank was conceptually similar to the Titan core stage in many respects (mainly, having two large solid rockets on either side of a large propellant tank). And Martin's total costs were far lower than any of the others. The board recognized that Martin was 'buying in' – bidding a lower cost than could probably be met. But the board still determined that Martin's most-likely costs were still the lowest of the bidders. It was felt that McDonnell Douglas, although presenting a good design, had shaved the weight margins too tightly. Interestingly, two of the original 'booster' contractors that found themselves without work on the revised partially-expendable space shuttle teamed up. The board found this marriage awkward, and the Boeing/General Dynamics team placed a distant third in the scoring. Despite the association with Rockwell, Chrysler was rated an even more distant fourth, being considered weak during both the technical and management scoring.[198]

On 16 August 1973 Martin Marietta was selected to design, develop, and test the ET. This contract included three ground test tanks and six development flight test tanks. For a change, the award was not protested.[199]

EVOLUTIONARY CHANGES

Through the end of FY73, a total of $438.9 million had already been spent by NASA directly on the space shuttle system, mainly for studies and early facility work. This included $12.5 million in FY70, $80 million in FY71 ($78.5 million for R&D studies), $118.5 million the following year ($100 million in R&D), and $227.9 million in FY73 ($198.6 million for R&D). With the award of the various element contracts, the division of responsibilities within NASA shifted slightly. Instead of the 'lead center' concept that had been in place during the study efforts, NASA reverted to using a Space Shuttle Program Office under the Office of Manned Space Flight. This office was responsible for the detailed assignment of responsibilities, basic performance requirements, control of major milestones, and funding allocations to the various NASA field centers. In essence however, little had changed, and MSC continued to lead the shuttle effort for all intents and purposes.[201]

MSC was officially responsible for program control (cost accounting), overall systems engineering, systems integration, and the "overall responsibility and authority for definition of those elements of the total system which interact with other elements, such as total configuration and combined aerodynamic loads." In other words, if it somehow touched the Orbiter, MSC had control. MSC was also responsible for the development, production, and delivery of the Orbiter, and as such, managed the contract with North American Rockwell.[202]

MSFC was responsible for the development, production, and delivery of the Space Shuttle Main Engine, the Solid Rocket Boosters, and the External Tank. KSC was responsible for the design and construction of the launch and landing facilities in Florida, and KSC personnel would also play a major role in assisting the Air Force with the design and construction of the facilities at Vandenberg.[203]

At this point, several major studies were performed in an attempt to define the lightest possible design. These studies had been identified as a result of the different Phase B final reports, and were performed by various contractors and in-house NASA engineering groups. For example, it was shown that a space-frame concept for the thrust structure would save approximately 1,730 pounds compared with a plate-girder concept. Similarly, the integration of the aft wing carry-through spar and the aft payload bay bulkhead could save approximately 450 pounds compared to a floating carry-through spar. A single-point drag attachment between the Orbiter and external tank was studied since it achieved a statically determinate interface, but was found to be heavier and more difficult to integrate with the natural load paths from the thrust structure.

Results from other studies confirmed the weight effectiveness of the separate crew module compared to a crew module that was manufactured integrally with the fuselage. Weight and cost studies led to the selection of the composite material used in the payload bay doors, the OMS pod external shells, and as a reinforcement in the thrust structure. All totaled, these design choices led to the savings of over 1,000 pounds from the baseline MSC-040C orbiter design.

During this same period, a decision had already been made to remove the air-breathing landing engines after the first couple of years of operations to recover the 20,000 pound* payload penalty. NASA also decided to provide an ascent abort capability for the Orbiter by incorporating abort solid rocket motors (ASRM) mounted on the aft portion of the Orbiter fuselage. It was subsequently determined that the weight penalty of the ASRM system could be offset by the performance gained through their use during a nominal ascent. Specifically, when the vehicle had achieved sufficient velocity to reach low-orbit without the abort motors, they would be fired to provide the additional thrust needed to reach a higher orbit. Careful sequencing of the ASRM firing was required together with throttling of the Orbiter main engines to avoid over-accelerating the Orbiter, but using this technique allowed less liquid propellants to be carried, recovering most of the weight penalty imposed by the ASRMs. The rational for requiring the ASRMs was that in order to keep the SRBs simple and inexpensive, no thrust vector control system would be incorporated on them. This would require another method of ensuring the Orbiter could abort early in the ascent phase, hence the ASRMs.

At the time of the contract award to North American Rockwell, NASA had estimated the space shuttle would require some 32,000 hours of wind tunnel time during its early development effort. This was confirmed by the four contractor proposals, which had called for between 27,000 and 50,000 hours. Between contract award and the final assembly of the first Orbiter, approximately 46,000 hours were spent in NASA, Air Force, and university wind tunnels around the country.[204]

Space shuttle represented the first time engineers would actually be able to design a reusable lifting-reentry spacecraft. It would also represent the largest hypersonic vehicle ever designed, providing the first real test of experimental and theoretical knowledge of high-speed flight. No real precedents existed to help establish the design requirements for such a vehicle, especially after the decision was made to abandon the hot-structure airframe

---

* This included the air-breathing engines, their mounts, and dedicated propellant tanks.

approach that had been used on the X-15. A primary challenge for the designers were the preflight prediction of the aerodynamic characteristics of the vehicle with an accuracy consistent with establishing sufficient confidence to conduct the first orbital flight with a human crew. This also required overcoming the unknowns involved in hypersonic wind tunnel testing, and developing reliable simulation techniques without having an extensive data base to compare the results against. An additional challenge, given the magnitude of the project and the number of agencies and contractors involved, was careful configuration management of a continuously evolving aerodynamic data base top ensure that at any one time all developers were using the same set of results.[205]

The two years immediately after contract award brought many changes, as problems were better understood and resolved. The modified MSC-040C design that was authorized in March 1972 at the time of 'authority to proceed' (ATP) evolved considerably before the Program Readiness Review (PRR)* in October 1972. These designs, somewhat after the fact, were known as Vehicle 1 and Vehicle 2, respectively. The next 18 months would bring Vehicles 2A, 3, 4, 5, and finally Vehicle 6, which would be the final configuration.

The ATP Orbiter (Vehicle 1) had a blended delta wing with a 50-degree leading edge sweep that was sized to provide a 172 mph landing speed when returning with a 40,000 pound payload. The trailing-edge elevons were sized to trim at hypersonic speeds over an angle-of-attack range from 20 degrees to 50 degrees with a center-of-gravity travel of three percent of body length. The payload

---

* This was also known as the Preliminary Requirements Review.

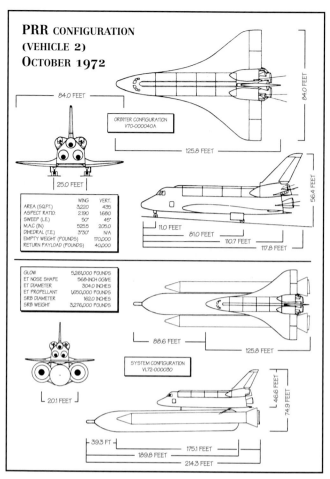

PRR CONFIGURATION
(VEHICLE 2)
OCTOBER 1972

ORBITER CONFIGURATION
V70-000040A

	WING	VERT.
AREA (SQ.FT.)	3,220	435
ASPECT RATIO:	2.190	1.680
SWEEP (L.E.)	50°	45°
M.A.C. (IN)	525.5	205.0
DIHEDRAL (T.E.)	3°30'	N/A
EMPTY WEIGHT (POUNDS)	170,000	
RETURN PAYLOAD (POUNDS)	40,000	

GLOW	5,261,000 POUNDS
ET NOSE SHAPE	568-INCH OGIVE
ET DIAMETER	304.0 INCHES
ET PROPELLANT	1,650,000 POUNDS
SRB DIAMETER	162.0 INCHES
SRB WEIGHT	3,276,000 POUNDS

SYSTEM CONFIGURATION
VL72-000030

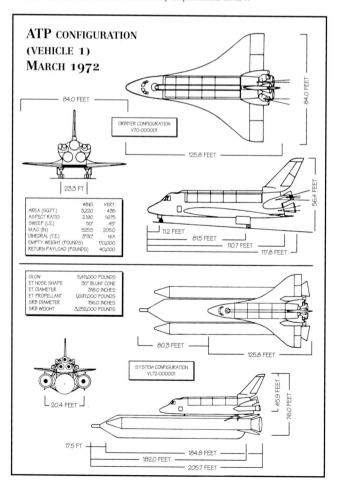

ATP CONFIGURATION
(VEHICLE 1)
MARCH 1972

ORBITER CONFIGURATION
V70-000001

	WING	VERT.
AREA (SQ.FT.)	3,220	435
ASPECT RATIO:	2.190	1.675
SWEEP (L.E.)	50°	45°
M.A.C. (IN)	525.5	205.0
DIHEDRAL (T.E.)	3°30'	N/A
EMPTY WEIGHT (POUNDS)	170,000	
RETURN PAYLOAD (POUNDS)	40,000	

GLOW	5,411,000 POUNDS
ET NOSE SHAPE	30° BLUNT CONE
ET DIAMETER	318.0 INCHES
ET PROPELLANT	1,697,000 POUNDS
SRB DIAMETER	156.0 INCHES
SRB WEIGHT	3,252,000 POUNDS

SYSTEM CONFIGURATION
VL72-000001

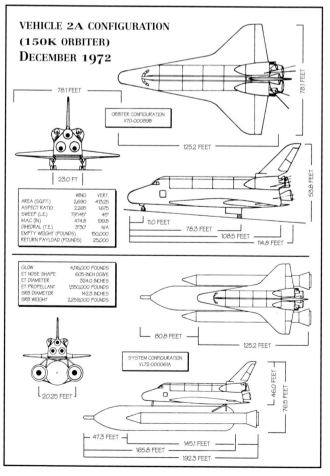

VEHICLE 2A CONFIGURATION
(150K ORBITER)
DECEMBER 1972

ORBITER CONFIGURATION
V70-000898

	WING	VERT.
AREA (SQ.FT.)	2,690	413.25
ASPECT RATIO:	2.265	1.675
SWEEP (L.E.)	79°/45°	45°
M.A.C. (IN)	474.8	199.8
DIHEDRAL (T.E.)	3°30'	N/A
EMPTY WEIGHT (POUNDS)	150,000	
RETURN PAYLOAD (POUNDS)	25,000	

GLOW	4,116,000 POUNDS
ET NOSE SHAPE	605-INCH OGIVE
ET DIAMETER	324.0 INCHES
ET PROPELLANT	1,550,000 POUNDS
SRB DIAMETER	142.3 INCHES
SRB WEIGHT	2,258,000 POUNDS

SYSTEM CONFIGURATION
VL72-000061A

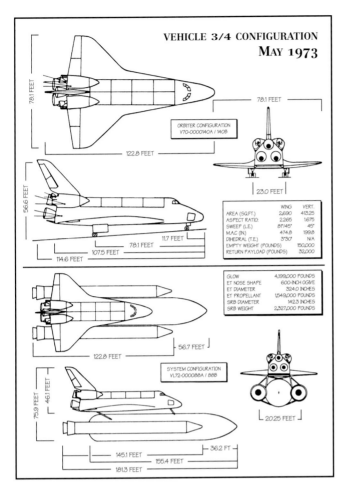

VEHICLE 3/4 CONFIGURATION
MAY 1973

ORBITER CONFIGURATION
V70-0000140A / 140B

78.1 FEET
78.1 FEET
122.8 FEET
23.0 FEET

56.6 FEET
11.7 FEET
78.1 FEET
107.5 FEET
114.6 FEET

	WING	VERT.
AREA (SQ.FT.)	2,690	413.25
ASPECT RATIO:	2.265	1.675
SWEEP (L.E.)	81°/45°	45°
MAC (IN)	474.8	199.8
DIHEDRAL (T.E.)	3°30'	N/A
EMPTY WEIGHT (POUNDS)	150,000	
RETURN PAYLOAD (POUNDS)	32,000	

GLOW	4,199,000 POUNDS
ET NOSE SHAPE	600-INCH OGIVE
ET DIAMETER	324.0 INCHES
ET PROPELLANT	1,549,000 POUNDS
SRB DIAMETER	142.3 INCHES
SRB WEIGHT	2,327,000 POUNDS

122.8 FEET
56.7 FEET

SYSTEM CONFIGURATION
VL72-0000088A / 88B

75.9 FEET
46.1 FEET
20.25 FEET

145.1 FEET
36.2 FT
155.4 FEET
181.3 FEET

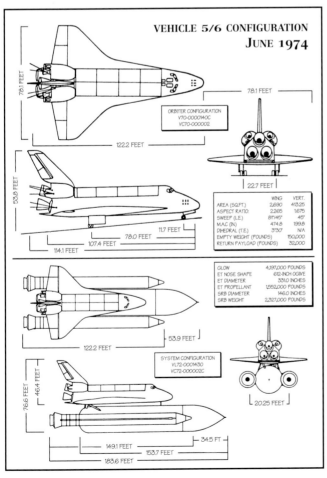

VEHICLE 5/6 CONFIGURATION
JUNE 1974

ORBITER CONFIGURATION
V70-0000140C
VC70-000002

78.1 FEET
78.1 FEET
122.2 FEET
22.7 FEET

53.8 FEET
11.7 FEET
78.0 FEET
107.4 FEET
114.1 FEET

	WING	VERT.
AREA (SQ.FT.)	2,690	413.25
ASPECT RATIO:	2.265	1.675
SWEEP (L.E.)	81°/45°	45°
MAC (IN)	474.8	199.8
DIHEDRAL (T.E.)	3°30'	N/A
EMPTY WEIGHT (POUNDS)	150,000	
RETURN PAYLOAD (POUNDS)	32,000	

GLOW	4,197,000 POUNDS
ET NOSE SHAPE	612-INCH OGIVE
ET DIAMETER	331.0 INCHES
ET PROPELLANT	1,552,000 POUNDS
SRB DIAMETER	146.0 INCHES
SRB WEIGHT	2,327,000 POUNDS

122.2 FEET
53.9 FEET

SYSTEM CONFIGURATION
VL72-0001430
VC72-000002C

76.6 FEET
46.4 FEET
20.25 FEET

149.1 FEET
34.5 FT
153.7 FEET
183.6 FEET

---

bay was 15 by 60 feet, and the remote manipulator arms were stored in a dorsal fairing along the top center of the payload bay doors. The size of the base (aft body) of the fuselage was dictated by the packaging requirements of the SSMEs. Nose camber and cross-section, along with the upward sloping forebody sides, were selected to improve hypersonic pitch trim and directional stability – along with wing-fuselage blending that also helped to reduce the reentry heating effects on the fuselage sides. The lower body flap, originally proposed simply to shield the SSMEs from excessive heating during reentry, was found to provide excellent pitch control and became the primary pitch trim device.[206]

Three main engines were located in the extreme aft of the fuselage, and the orbital maneuvering systems (OMS) were installed in pods attached to the sides of the lower aft fuselage. For ferry operations and reentry assist, two air-breathing engines were installed in the payload bay using an air intake at the base of the vertical stabilizer, with the jet exhausts immediately below the three main engines. The cockpit location was 17 feet aft of the extreme nose, and provided a view 20 degrees up and 24.5 degrees down from the Orbiter centerline. The nose radius was 25 inches and blended smoothly into the low fineness ratio body. Provisions for two ASRMs were located on the aft end of the Orbiter fuselage between the OMS pods and the upper surface of the wing.[207]

The ATP integrated vehicle had the Orbiter located 80.3 feet aft of the ET's nose with the Orbiter's fuselage reference line canted down at 1.2 degrees negative incidence with respect to the ET centerline. To facilitate deorbit, the External Tank was fitted with a retro-rocket package on its nose – housed inside a 10.33-foot long hemispherical-

---

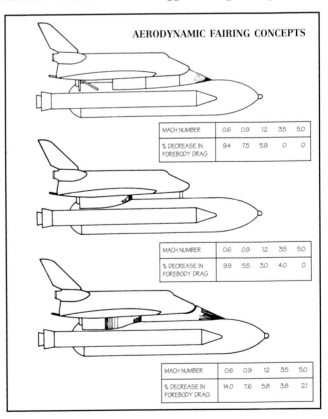

AERODYNAMIC FAIRING CONCEPTS

MACH NUMBER	0.6	0.9	1.2	3.5	5.0
% DECREASE IN FOREBODY DRAG	9.4	7.5	5.9	0	0

MACH NUMBER	0.6	0.9	1.2	3.5	5.0
% DECREASE IN FOREBODY DRAG	9.9	5.5	3.0	4.0	0

MACH NUMBER	0.6	0.9	1.2	3.5	5.0
% DECREASE IN FOREBODY DRAG	14.0	7.6	5.8	3.8	2.1

Several ideas have been considered to reduce the parasitic drag between the external tank and the orbiter. Most centered around various fairings, some of which resulted in a significant decrease in subsonic or low supersonic drag. However, since the vehicle spends a relatively short time at subsonic speeds during ascent, and the fairings were not particularly effective at higher speeds, the concept was not economical. (NASA)

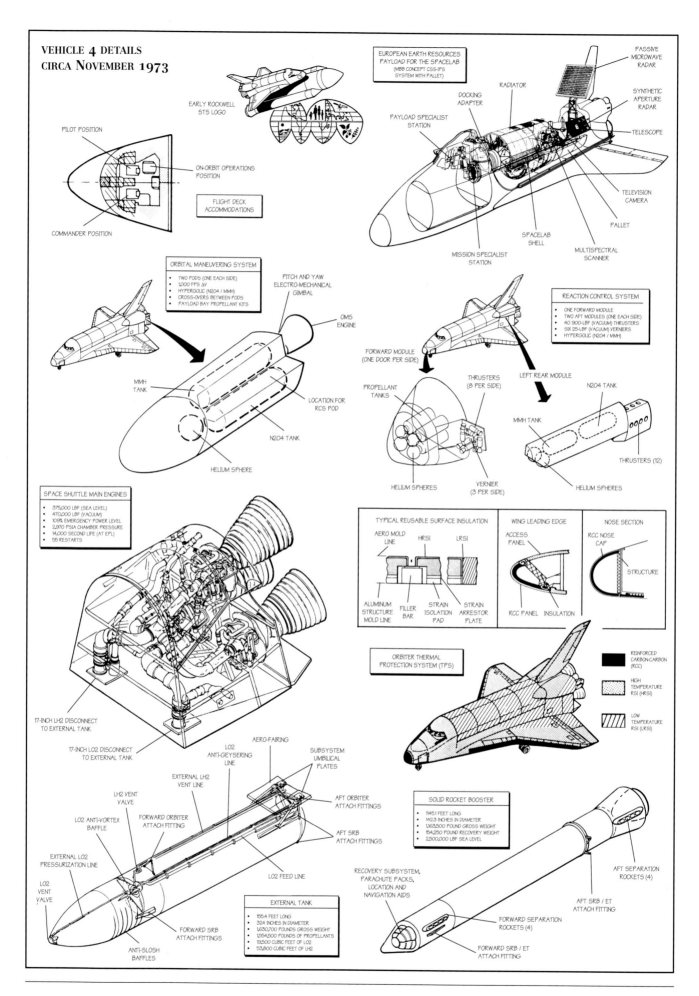

VEHICLE 4 DETAILS
CIRCA NOVEMBER 1973

EARLY ROCKWELL STS LOGO

FLIGHT DECK ACCOMMODATIONS
- PILOT POSITION
- ON-ORBIT OPERATIONS POSITION
- COMMANDER POSITION

EUROPEAN EARTH RESOURCES PAYLOAD FOR THE SPACELAB
(MBB CONCEPT C55-IPS SYSTEM WITH PALLET)
- PASSIVE MICROWAVE RADAR
- RADIATOR
- DOCKING ADAPTER
- SYNTHETIC APERTURE RADAR
- PAYLOAD SPECIALIST STATION
- TELESCOPE
- TELEVISION CAMERA
- PALLET
- SPACELAB SHELL
- MULTISPECTRAL SCANNER
- MISSION SPECIALIST STATION

ORBITAL MANEUVERING SYSTEM
- TWO PODS (ONE EACH SIDE)
- 1,000 FPS ΔV
- HYPERGOLIC (N204 / MMH)
- CROSS-OVERS BETWEEN PODS
- PAYLOAD BAY PROPELLANT KITS

- PITCH AND YAW ELECTRO-MECHANICAL GIMBAL
- OMS ENGINE
- MMH TANK
- LOCATION FOR RCS POD
- N204 TANK
- HELIUM SPHERE

REACTION CONTROL SYSTEM
- ONE FORWARD MODULE
- TWO AFT MODULES (ONE EACH SIDE)
- 40 900-LBF (VACUUM) THRUSTERS
- SIX 25-LBF (VACUUM) VERNIERS
- HYPERGOLIC (N204 / MMH)

- FORWARD MODULE (ONE DOOR PER SIDE)
- LEFT REAR MODULE
- THRUSTERS (8 PER SIDE)
- N204 TANK
- PROPELLANT TANKS
- MMH TANK
- THRUSTERS (12)
- HELIUM SPHERES
- VERNIER (3 PER SIDE)
- HELIUM SPHERES

SPACE SHUTTLE MAIN ENGINES
- 375,000 LBF (SEA LEVEL)
- 470,000 LBF (VACUUM)
- 109% EMERGENCY POWER LEVEL
- 2,970 PSIA CHAMBER PRESSURE
- 14,000 SECOND LIFE (AT EPL)
- 55 RESTARTS

TYPICAL REUSABLE SURFACE INSULATION
- AERO MOLD LINE
- HRSI
- LRSI
- ALUMINUM STRUCTURE MOLD LINE
- FILLER BAR
- STRAIN ISOLATION PAD
- STRAIN ARRESTOR PLATE

WING LEADING EDGE
- ACCESS PANEL
- RCC PANEL
- INSULATION

NOSE SECTION
- RCC NOSE CAP
- STRUCTURE

ORBITER THERMAL PROTECTION SYSTEM (TPS)
- REINFORCED CARBON-CARBON (RCC)
- HIGH TEMPERATURE RSI (HRSI)
- LOW TEMPERATURE RSI (LRSI)

- 17-INCH LH2 DISCONNECT TO EXTERNAL TANK
- 17-INCH LO2 DISCONNECT TO EXTERNAL TANK
- LO2 ANTI-GEYSERING LINE
- AERO-FAIRING
- SUBSYSTEM UMBILICAL PLATES
- EXTERNAL LH2 VENT LINE
- LH2 VENT VALVE
- LO2 ANTI-VORTEX BAFFLE
- FORWARD ORBITER ATTACH FITTING
- AFT ORBITER ATTACH FITTINGS
- AFT SRB ATTACH FITTINGS
- EXTERNAL LO2 PRESSURIZATION LINE
- LO2 FEED LINE
- LO2 VENT VALVE
- FORWARD SRB ATTACH FITTINGS
- ANTI-SLOSH BAFFLES

SOLID ROCKET BOOSTER
- 1145.1 FEET LONG
- 142.3 INCHES IN DIAMETER
- 1,163,500 POUND GROSS WEIGHT
- 154,250 POUND RECOVERY WEIGHT
- 2,500,000 LBF SEA LEVEL

- RECOVERY SUBSYSTEM, PARACHUTE PACKS, LOCATION AND NAVIGATION AIDS
- AFT SEPARATION ROCKETS (4)
- AFT SRB / ET ATTACH FITTING
- FORWARD SEPARATION ROCKETS (4)
- FORWARD SRB / ET ATTACH FITTING

EXTERNAL TANK
- 155.4 FEET LONG
- 324 INCHES IN DIAMETER
- 1,630,700 POUNDS GROSS WEIGHT
- 1,554,500 POUNDS OF PROPELLANTS
- 19,500 CUBIC FEET OF LO2
- 53,800 CUBIC FEET OF LH2

cylinder with a nose radius of 20.5 inches. The conical nose portion of the ET had a semi-vertex angle of 30 degrees which blended smoothly into the 13.25-foot diameter cylindrical section of the tank. The ET had an overall length of 182.0 feet. The SRBs were located 17.5 feet aft of the ET nose, and were mounted 3.1 feet above the ET centerline, with the SRB and ET centerlines parallel to each other. The nose radius of the SRBs was 13 inches using a semi-vertex cone with 18 degree angles. The 156-inch diameter SRBs had an overall length of 184.8 feet and used fixed nozzles that were canted outward 11 degrees in the yaw plane to thrust through the approximate center-of-gravity.[208]

The PRR configuration (Vehicle 2) featured OMS pods that were moved from the sides of the aft fuselage to the fuselage shoulders, and were also slightly longer, now overlapping the payload doors. The cockpit windscreen was moved aft approximately 4.3 feet, and the view from the cockpit was reduced to 7 degrees up and 18 degrees down. The forebody was redesigned to accommodate internal packaging revisions and to improve the aerodynamic transition to the mid-body. Wing refinements included an increased thickness ratio, a slight leading-edge droop and minor wing/body fillet modifications. The Orbiter incidence when mated to the External Tank was changed to a positive 0.5 degrees to improve ET separation characteristics. The ASRMs and air-breathing engines were deleted, although this was controversial and both would return in later configurations. The ET nose was changed to an ogive shape to reduce parasitic drag, and all four vehicle elements (two SRBs, ET, and Orbiter) were repositioned slightly to improve the element-to-element interference drag. The SRBs were shortened to 175.1 feet, but increased in diameter to 162 inches, and moved slightly aft in relation to the ET to eliminate the SRB plume effects on the Orbiter base. Thrust vector control (TVC) was added to the SRBs along with a reduced precant, and the four strakes on each SRB were deleted.[209]

The largest changes were shown by Vehicle 2A, which was also referred* to as the '150K Orbiter.' Many of the changes were proposed at the PRR in an effort to reduce the overall cost of the program. To accommodate this, both the Orbiter empty weight and the payload return weight were reduced significantly, leading to a complete resizing of the Orbiter. A higher design landing velocity (196 mph) was also adopted in order to make the wing smaller, further reducing weight. As part of this effort, NASA directed Rockwell to use a double-delta wing planform with the forward portion being swept 79 degrees, and the outer portions 45 degrees. There was also a slight forward sweep on the trailing edge, and the wing included twist, camber, and incidence revisions for improved subsonic performance based on continuing wind tunnel studies. NASA had been studying the double-delta planform, proposed by Lockheed in its Phase C proposal, for over a year and concluded that it offered exceptional landing performance. In addition, aerodynamic stability and trim could be adjusted by modifying the lightly-loaded forward delta (glove). This simple control of aerodynamic features allowed the design of the main delta wing box to be frozen so that manufacturing could begin. Any center-of-gravity or aerodynamic problems that might surface during program development could then be corrected by modifying the forward glove – a relatively easy and inexpensive solution.[210]

Nose camber and radius, body cross-section and upward sloping forebody slab sides were selected to improve hypersonic pitch trim and directional stability. By combining these with wing-body blending, reentry heating of the fuselage sides was theoretically lowered. At the same time, it was decided to attempt to use as many uniform dimension tiles (i.e., all the same size) tiles as possible, resulting in a vehicle whose surface was composed of large flat areas, limiting curvatures to smaller areas between flat ones. The center-of-gravity travel requirement was reduced from three percent of body length, to two percent. The air-breathing engines made their return, this time as the air-breathing propulsion system (ABPS) with five TF33-P7A turbofans in three removable pods located under the Orbiter. Fuel for the engines would be carried in a special tank installed in the payload bay. It was intended to use the ABPS for ferry flights only, and no provisions for taking it into space were made. Provisions were again included for an ASRM on each side of the Orbiter's aft fuselage, but it was extremely unlikely by this point if they would ever be installed. The SRBs were shortened again, this time to 145.1 feet, and their diameter was significantly reduced to only 142.3 inches, reflecting the generally lighter vehicle and the need for less thrust. The SRBs were moved further aft in relation to the ET, the yaw gimbal setting precant was reduced to zero degrees, and the nozzle flare angle was also reduced. Due to the smaller Orbiter and SRBs, the ET was also shortened, and the relative position of the Orbiter was changed.[211]

But wind tunnel testing revealed that this configuration was not workable. Aerodynamic tests showed difficulty in providing trim capability at the forward center-of-gravity in supersonic flight, and aerothermodynamic tests indicated the blunt fuselage nose resulted in early transitional flow and high temperatures along the lower body surfaces. Also, the wing incidence, camber, and thickness – selected for maximum subsonic performance – created unacceptable local hot spots in some locations.[212]

Vehicles 3 and 4 were developed during mid-1973 to overcome these problems, and were essentially the same from an external view, but differed somewhat in internal layout. When compared to Vehicle 2A, the Orbiter was 3.17 feet shorter with a smaller nose radius and smoother nose-body area. The wing glove leading edge sweep was increased to 81 degrees and the incidence was decreased

A substantially evolved orbiter from the one that North American originally bid is depicted here. Note the early forward RCS 'doors' and the OMS pods that overlap the payload bay doors. Nevertheless, this is substantially similar to what was built. (NASA)

---

* In reference to the Orbiter's empty weight of 150,000 pounds.

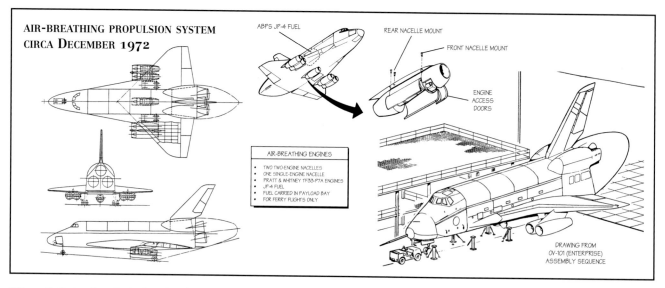

AIR-BREATHING PROPULSION SYSTEM
CIRCA DECEMBER 1972

ABPS JP-4 FUEL
REAR NACELLE MOUNT
FRONT NACELLE MOUNT
ENGINE ACCESS DOORS

AIR-BREATHING ENGINES
• TWO TWO-ENGINE NACELLES
• ONE SINGLE-ENGINE NACELLE
• PRATT & WHITNEY TF33-P7A ENGINES
• JP-4 FUEL
• FUEL CARRIED IN PAYLOAD BAY
• FOR FERRY FLIGHTS ONLY

DRAWING FROM
OV-101 (ENTERPRISE)
ASSEMBLY SEQUENCE

This was the final configuration for the air-breathing propulsion system. Two of these illustrations come from the 'Final Assembly, Systems Installation and Checkout: Building 294 – Palmdale' drawings dated 8 January 1974. The configuration shown here was only going to be used for ferry flights – there were no provisions for an orbital system. The boat-tail would cover the SSMEs on all flights since without it the Orbiter had much more aerodynamic drag. (North American Rockwell)

from 3 degrees to 0.5 degree. Some minor airfoil changes were made, including thickening the wing six inches at the elevon hingeline. The wing was also lowered four inches on the fuselage and the lower body contours smoothed. In addition, the body flap span was reduced slightly. The primary purpose of these changes was to improve the overall aerodynamic and aerothermo-dynamic performance of the Orbiter. The center-of-gravi-ty range requirement was increased to 2.5 percent of body length to allow at least 0.5 degree for aerodynamic trim uncertainties. A major identifying change was the moving of the remote manipulator arm from its dorsal fairing to a location inside either side of the payload bay. The SRBs and the Orbiter were moved forward relative to the External Tank, but remained the same size. The ET was shortened again, and the retro-package (spike) was removed, resulting in a slightly shorter overall vehicle.[213]

By early 1973 several other design changes had also been approved. Continued studies of abort profiles indi-cated the emergency thrust termination system planned for the SRBs could be dropped. The method that had been selected to accomplish thrust termination was to blow a hole in the nose bulkhead of the top motor segment, allow-ing thrust to escape from both ends. In theory, with the SRB thrust terminated additional options would be avail-able for aborts or crew escape – but thrust termination occurred so suddenly that it introduced dynamic loads that could cause various Orbiter structural components to fail. Detailed analyses showed that to design the Orbiter to withstand the stresses caused by rapid thrust termination would add an additional, and prohibitive, 8,000 pounds to the projected empty weight of the Orbiter, so the thrust termination requirement was dropped on 27 April 1973.[214]

Vehicle 5 resulted from an early-1974 decision to shorten and re-fair the OMS pods not to overlap the pay-load bay doors. This simplified the pod/door interface, and allowed the use of a one-piece (per side) door instead of the four-piece door originally envisioned. Other changes included modified wingtips and increased elevon gaps. This design again deleted the provisions for any type of air-breathing engines.[215]

The ASRMs were finally deleted from Vehicle 5 as a cost and weight saving measure since they would cost over $300 million to design and build, and could provide

meaningful assistance during just 30 seconds of each flight. A decision was made to equip the Orbiter with a parachute braking system during flight tests to reduce its landing roll-out. As construction of the first Orbiter con-tinued, NASA deleted the parachute braking system* as unnecessary since it was reasoned that Edwards' lake beds were more than long enough. The hydraulic system was also redesigned to incorporate three redundant sys-tems, instead of the earlier four.[216]

Changes in mid-1974 added recessed thermal glass in the windscreens, observation windows, and hatch win-dows. At this time, Rockwell's design had a forward reac-tion control system that was contained behind doors that opened after the Orbiter reached orbit. Further study indi-cated it would be significantly less complex (and less expensive) to design a system capable of withstanding the rigors of ascent and reentry while exposed, leading to the system in use on the current Orbiters. The covers were also removed from the umbilical doors on the aft fuselage sides. The ET and SRB lengths were changed slightly and an ascent air data system was added to the nose of the ET. The SRB diameter was increased slightly, to 146 inches, to accommodate the slightly higher-than-expected Orbiter empty weight. These resulted in the Vehicle 6 configura-tion, and this design was presented at the critical design review (CDR) in February 1975 – and subsequently built.[217]

When North American Rockwell had submitted their proposal, they had committed to subcontracting 53 per-cent of the work (by dollars) on the Orbiter. Given that the aerospace industry as a whole was in the middle of a severe recession, this work would be welcomed by all recipients. Although NASA and Rockwell were hesitant to being subcontracting too early – it would lead to expensive and time consuming changes as the vehicle evolved – there was political pressure to spread the wealth around as soon as possible. On 29 March 1973 Rockwell formally award-ed four major subcontracts: Fairchild-Republic would build the vertical stabilizer, Grumman Aerospace the wings, General Dynamics/Convair the mid-fuselage, and McDonnell Douglas would manufacture the OMS pods. Space shuttle was well on its way to becoming a reality.[218]

---

* The parachute braking system made a return on OV-105, *Endeavour*, and was retro-fitted to the remaining production Orbiters by the end of 1992.

Design data from selected concepts from each ILRV phase and contract award are presented below for comparison. In most cases, many versions were presented for each design concept, so the data below is not necessarily representative of all possible variants. Also, it should be noted that many contractors submitted multiple concepts during each phase. Only the primary design is presented here.

PHASE:	A	A	A	A	A	B	B	B
Date :	01 Nov 69	11 Nov 69	22 Dec 69	31 Oct 69	22 Dec 69	11 Dec 69	11 Dec 69	13 Nov 70
Contractor / Team :	MDC	MMC	Lockheed	Convair	NAR	MDC / MMC	MDC / MMC	NAR / GD
Design :	—	SpaceMaster	LS-112	FR-3A	—	LCR	HCR	NAR-134-C

ORBITER:

Length Overall (feet) :	107.00	181.20	163.00	179.00	202.00	147.60	171.00	192.30
Wing Span (feet) :	75.70	107.00	92.00	146.00	146.00	113.80	97.50	126.60
Height (feet) :	48.30	30.00	36.75	45.00	51.30	62.40	57.60	49.30
Wing Area (sq ft) :	—	1,812	—	—	2,830	1,375	—	—
Weight (empty – pounds) :	—	—	183,830	287,000	256,515	188,600	203,500	217,732
No. of Rocket Engines :	2	2	3	3	2	2	2	2
Thrust (Sea Level – lbf) :	415,000	415,000	415,000	415,000	510,000	415,000	415,000	415,000
No. of Air-Breathers :	4	none	4	2	4	4	4	2
Payload (pounds) :	25,000	25,000	50,000	50,000	50,000	15,000	15,000	20,000
Payload Bay Size (feet) :	15 by 30	22 by 30*	22 by 30*	15 by 60	15 by 60	15 by 30	15 by 30	15 by 60
Cross-Range (miles) :	—	1,700	1,725	1,500	1,240	230	1,726	1,726
External Tanks ? :	none	none	none	none	none	none	none	none

* 15 by 60-foot payloads could also be accommodated

BOOSTER:

Type of Booster :	Flyback	Flyback	Flyback	Flyback	Flyback	Flyback	Flyback	Flyback
No. of Boosters :	1	1	1	1	1	1	1	1
Length Overall (feet) :	195.00	197.00	237.00	235.50	280.00	220.20	232.20	257.00
Wing Span (feet) :	151.00	148.00	200.50	203.10	244.00	151.00	151.00	142.00
Height (feet) :	69.50	58.00	56.90	47.50	—	56.90	56.90	50.00
Wing Area (sq ft) :	—	4,850	—	—	8,050	3,970	—	2,001
Weight (empty – pounds) :	—	448,369	322,550	517,000	587,522	452,800	483,900	—
No. of Rocket Engines :	10	14	13	15	10	13	14	12
Thrust (Sea Level – lbf) :	415,000	415,000	415,000	415,000	510,000	415,000	415,000	415,000
No. of Air-Breathers :	4	8	4	4	4	10	10	4

PHASE:	—	ASSC	ASSC	ASSC	**Baseline**	**Award**	**Production**	**Production**
Date :	27 Apr 70	06 Jul 71	04 Jun 71	30 Jun 71	30 Aug 71	18 Feb 72	08 Mar 79	07 May 91
Contractor / Team :	MSC	Grumman	Lockheed	Chrysler	MSC	MSC	RIC	RIC
Design :	DC-3	H-33	LS-200	SERV	MSC-040	MSC-040C	OV-102	OV-105

ORBITER:

Length Overall (feet) :	122.67	157.00	156.50	90.00 (dia)	121.88	123.75	122.17	122.17
Wing Span (feet) :	90.83	97.00	92.00	n/a	75.00	80.20	78.06	78.06
Height (feet) :	41.66	61.25	49.00	66.50	46.88	48.13	56.58	56.58
Wing Area (sq ft) :	1,175	1,060	1,229	n/a	—	2,500	2,690	2,690
Weight (empty – pounds) :	120,000	197,000	294,399	—	—	150,000	158,289	151,205
No. of Rocket Engines :	2	3	9	1	4	3	3	3
Thrust (Sea Level – lbf) :	297,000	415,000	550,000	5,800,000	325,000	375,000	375,000	375,000
No. of Air-Breathers :	6	4	6	28	0	0	0	0
Payload (pounds) :	15,000	60,000	65,000	116,439	60,000	60,000	60,000	55,250
Payload Bay Size (feet) :	8 by 30	15 by 60	15 by 60	23 by 60	15 by 60	15 by 60	15 by 60	15 by 60
Cross-Range (miles) :	200	1,265	1,500	—	1,500	1,500	975	1,085
External Tanks ? :	none	LH2	LO2/LH2	none	LO2/LH2	LO2/LH2	LO2/LH2	LO2/LH2

BOOSTER:

Type of Booster:	Flyback	Flyback	none	none	SRB	SRB	SRB	RSRM
No. of Booster :	1	1	n/a	n/a	2	2	2	2
Length Overall (feet) :	203.00	245.00	n/a	n/a	—	184.80	149.16	149.16
Wing Span (feet) :	141.00	177.50	n/a	n/a	n/a	n/a	n/a	n/a
Height (feet) :	74.00	88.30	n/a	n/a	13.00 (dia)	13.00 (dia)	12.17 (dia)	12.17 (dia)
Wing Area (sq ft) :	2,840	—	n/a	n/a	n/a	n/a	n/a	n/a
Weight (empty – pounds) :	290,000	220,135	n/a	n/a	—	150,000	154,023	150,023
No. of Rocket Engines :	11	12	n/a	n/a	1	1	1	1
Thrust (Sea Level – lbf) :	297,000	415,000	n/a	n/a	2,500,000	2,750,000	2,940,000*	3,310,000
No. of Air-Breathers :	4	12	n/a	n/a	n/a	n/a	n/a	n/a

NOTES:

— = data not available

n/a = not applicable to this configuration

George M. Low. (NASA photo S65-39212)

SHUTTLE CARRIER AIRCRAFT

When the air-breathing engines were finally deleted from the Orbiter design in early 1974, NASA was faced with the question of how to conduct the atmospheric flight tests, as well as how to ferry the Orbiter from a remote landing location back to the launch site.

Independently, NASA Langley conducted some preliminary design work on a new aircraft – called VIRTUS – with a 472-foot wing span and twin 293-foot long fuselages. In concept, this was generally similar to several ideas proposed by Robert Salkeld as early as 1965 (see page 73). The Orbiter would be suspended under the high-mounted wing between the two fuselages, or alternately, other over-sized payloads such as external tanks could be carried in the same manner. Four 40,000-lbf JT9D-class turbofan engines would be mounted in single nacelles under the outer wing panels. A 1/34-scale model of the proposed aircraft was constructed and tested in a low-speed wind tunnel with encouraging results. The primary drawbacks of the design were its sheer size, and the long development and testing cycle necessary to gain certification. A quicker and less expensive alternative would need to be found.[1]

John Conroy* suggested modifying one of the 'jumbo-jet' type aircraft into a Shuttle Carrier Aircraft (SCA). Proposals were subsequently solicited and received from both jumbo-jet manufacturers. Lockheed initially proposed a twin-fuselage variation of the C-5 military transport that carried the Orbiter suspended between the fuselages – again, remarkably similar to a design by Salkeld. Boeing wanted to carry the Orbiter on top of a slightly modified 747 airliner. At the time, thought was also given to carrying the External Tanks to Vandenberg on top of the SCA – but wind-tunnel tests later indicated the drag of the ET would pose a serious safety hazard, and the tanks were shipped by barge through the Panama Canal.[2]

Although the atmospheric flight tests would be easier from the Lockheed aircraft, it presented the same problems as the earlier Langley proposal – it was far too wide for any known runway, and it would be very expensive to develop and certify. Lockheed subsequently submitted a proposal to carry the Orbiter on top of a slightly modified C-5A. This second Lockheed proposal and the one from Boeing were

deemed roughly equal in the technical evaluation, and as late as March 1974 no final decision had been made.

By mid-April 1974 NASA had apparently decided on using the C-5 and, on 24 April, George Low wrote a letter to the Secretary of the Air Force, John L. McLucas, outlining the plan. Interestingly, according to the letter there were four primary landing sites – KSC, Vandenberg, Edwards, and Palmdale. A comparison of the C-5 to the 747 showed that "the C-5 is considered to have some attractive advantages. It would involve the least acquisition/capitalization expense, and would require less structural rework than the 747." NASA requested that the Air Force make 3–5 C-5As available, with the final number to be determined after a firm manifest was developed. The modifications were expected to add between 400 and 600 pounds to the overall weight of each aircraft, and have "miniscule" effect on its cargo-carrying ability. NASA would pay[+] for the modifications, and then lease the aircraft as-needed "under normal Military Airlift Command industrial fund procedures." NASA would also be responsible in the event of the loss or serious damage to a C-5. One aircraft would be needed full time to develop the modifications and the removable ferry kit, and also to conduct the atmospheric flight tests. A week later the Air Force agreed in principle with this request.[3]

There was some lingering concern over the possible impact of the Orbiter turbulence impinging on the tee-tail of the C-5A, but it was the availability of low-cost used 747 aircraft, and the relative scarcity of C-5As, that finally drove the selection of the Boeing aircraft. After being informed of possible Air Force restrictions on the use of the C-5As, NASA decided that it was easier to have complete control over their destiny and own the SCA than it was to compete with military priorities.[4]

The technical community expressed calm assurance that a mated ferry flight program was a feasible undertaking. The separation of the two vehicles in flight did not produce the same response. The space shuttle already had two parallel separations to contend with – separating

---

* Conroy was the co-developer of the Guppy series of over-sized cargo carrying aircraft.

+ This is not unlike what actually happened with the two C-5C aircraft that have been modified to carry full-size shuttle payloads.

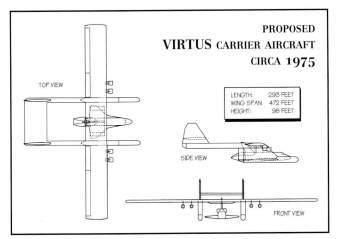

PROPOSED
VIRTUS CARRIER AIRCRAFT
CIRCA **1975**

TOP VIEW

LENGTH:	293 FEET
WING SPAN:	472 FEET
HEIGHT:	98 FEET

SIDE VIEW

FRONT VIEW

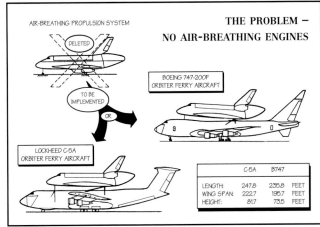

AIR-BREATHING PROPULSION SYSTEM

DELETED

TO BE IMPLEMENTED

OR

**THE PROBLEM –
NO AIR-BREATHING ENGINES**

BOEING 747-200F
ORBITER FERRY AIRCRAFT

LOCKHEED C-5A
ORBITER FERRY AIRCRAFT

	C-5A	B747	
LENGTH:	247.8	235.8	FEET
WING SPAN:	222.7	195.7	FEET
HEIGHT:	81.7	73.5	FEET

the SRBs from the ET, and jettisoning the ET from the Orbiter. Both required knowledge of the aerodynamic effects when the vehicles were in proximity, and a great deal of wind tunnel time had been spent studying the separation maneuvers. Additional tests now had to be accomplished to understand the separation effects of the 747 and Orbiter, although the fact that this would happen at subsonic speeds greatly simplified matters.[5]

Structural clearance was the initial concern, but computer simulations revealed that the vortex wake of the SCA might present a larger issue so additional wind tunnel tests were scheduled. Separation was to be accomplished by the mated vehicles entering a dive to increase airspeed, followed by the 747 reducing power and deploying spoilers to reduce lift and increase drag. Such a configuration was necessary to create the relative motion required to aerodynamically drive the two vehicles apart. Unfortunately, this also resulted in a near maximum vortex wake condition since the SCA was now configured, essentially, in a landing configuration.[6]

No vortex wake testing was scheduled, but a Boeing 747 model was available in an ongoing NASA Langley wind tunnel test, and JSC was allowed to test the separation configuration to gather wake vortex information. Additionally, Boeing was conducting tests on the prototype 747 at Edwards, and allowed the aircraft to be fitted with smoke generators on the wingtips and aft fuselage. A Cessna T-37 was then flown in close proximity to measure the actual severity of the wake vortex.[7]

On 18 July 1974 NASA purchased a used 747-123

N905NA shortly after being procured by NASA. During August 1974 the aircraft flew 30 wake vortex research flights with the NASA Ames LearJet and an Air Force T-37 as chase aircraft. (NASA photo EC74-4243)

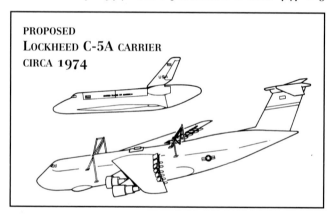

**PROPOSED
LOCKHEED C-5A CARRIER
CIRCA 1974**

Studies into using a C-5A progressed quite far, with the Air Force and Lockheed-Georgia carrying out limited wind tunnel testing of the configuration. Note the large truss-structures selected to attach the Orbiter to the C-5A and the close proximity of the Orbiter's aft boat-tail fairing to the carrier aircraft's tee-tail.(NASA photo 74-H-425 below)

(N9668, msn 20107) from American Airlines. The aircraft was the 86th plane off the 747 production line, and had been delivered to American on 29 October 1970. It had logged 8,999 hours during 2,985 flights, mostly on long-haul flights between New York and Los Angeles. The aircraft received the new civil registration' N905NA.[8]

Before the aircraft was modified into an SCA, NASA Dryden used the 747 as part of the overall NASA study of trailing wake vortices. Trailing vortices are the invisible flow of spiraling air that trails from the wings of large aircraft and can 'upset' smaller aircraft flying behind them. The data gathered in the 747 studies complemented data from a previous (1973–74) joint NASA–FAA wake vortices study that had used a Boeing 727.[9]

Six smoke generators were installed under the wings of the 747 to provide a visual image of the trailing vortices. The objective of the experiments was to determine how to break up or lessen the strength of the vortices. These tests were not directly connected to space shuttle in any way, but were intended to help the FAA draft new regulations for civil aircraft. Approximately 30 flights were made using various combinations of wing spoilers in an attempt to reduce wake vortices. To evaluate the effectiveness of the different configurations, chase aircraft were flown into the vortex sheets to probe their strengths and patterns at different times – a T-37 and a Learjet were used since they represented the types of smaller business jets and other small aircraft that might encounter large ('heavy' in FAA terms) passenger aircraft on approach or landings around major airports. The results of the tests led to shorter spacing between landings and take-offs, which, in turn, helped alleviate air-traffic congestion.[10]

Tests without the 747's wing spoilers deployed produced violent 'upset' problems for the T-37 at a distance of approximately 3 miles, and minor disturbances could be felt at distances as great as 10 miles. With two spoilers deployed on the 747's wing panels, the T-37 could fly at a distance of 3 miles and not experience difficulties.[11]

Subsequently, the 747 was used for additional space shuttle tests – a Lockheed F-104 from Dryden was positioned near the 747 wing and both vehicles flew a simulated separation maneuver. When the 747 reached the

---

* The civil registration numbers ('N' numbers) assigned to NASA aircraft also indicate the center responsible for operating them. In this case the leading '9' indicates the Johnson Space Center. NASA contract numbers also abide by this convention, with all 'NAS1' contracts being issued by Langley, 'NAS8' by MSFC, 'NAS9' by JSC, and so forth.

† This included $15.6 million for the aircraft itself.

appropriate conditions for separation, the F-104 pulled away and replicated the planned Orbiter maneuver following separation. The test confirmed that adequate clearance between the 747 vortex wake and the Orbiter flight path would be maintained. The tests revealed that although the separation maneuver would need to be flown very precisely, there were no major technical reasons not to proceed with the atmospheric flight tests.[12]

Under a $30 million contract[+] from Rockwell, a Boeing Aerospace–Boeing Commercial Airplane Company team began modifying the 747 on 2 August 1976. The Orbiter would be carried on top of the 747 fuselage, mildly reminiscent of the two-stage Phase A/B concepts, mounted to struts at three points – one forward and two aft – that matched the socket fittings intended for attaching the External Tank during ascent.[13]

The modifications performed by Boeing involved permanent (type-one) modifications that included the installation of bulkheads to strengthen the fuselage, skin reinforcements at critical stress areas, reinforcement of the horizontal stabilizer structure, adding L-band telemetry and C-band transponders, installing fittings for the Orbiter attach struts, and the installation of a 747-200 rudder ratio-changer.[14]

In addition to these permanent changes, Boeing also developed a series of removable structures that were only intended to be installed when needed. These type-two changes included:[15]

- a telescopic Orbiter forward support assembly to be used during the atmospheric flight tests. This was a bipod consisting of two tubes, each 13 feet long, with an adjustable drag strut on the aft-side of each tube to support the bipod after Orbiter release. The Orbiter would be mounted at a six degree angle-of-attack using this support, making separation a little easier.
- a fixed Orbiter forward support assembly for use during ferry missions. This is also a bipod, consisting of two 8.5-foot long tubes, but without the drag struts. The Orbiter would be mounted at a three degree angle-of-attack using this support, providing less drag during ferry flights.
- aft support assemblies, each consisting of a drag strut 12 feet long and a vertical strut 4.5 feet long. The right-aft support is fitted with a non-adjustable 4.8-foot side strut, and the left-aft support is fitted with dual non-load-bearing adjustable side snubbers.
- vertical endplates, each measuring 10 by 20 feet, on the tip of each horizontal stabilizer. Each endplate has a drag strut connecting to the upper surface of the horizontal stabilizer. These endplates provide additional aerodynamic stability when the 747 is carrying the Orbiter. In practice, these are never removed from the SCA.

The 747 longitudinal trim system was modified to permit two degrees more trim in order to counteract a nose-up tendency caused by the downwash off the Orbiter wing onto the 747 horizontal stabilizer. Additionally, most of the lower (main) deck interior was stripped, although a number of seats were retained to transport support personnel on ferry missions. In all, the modifications added 11,500 pounds to the empty weight of the 747. Under a separate contract, the four Pratt & Whitney JT9D-3A engines were converted to the JT9D-7AHW configuration, increasing available power from 43,500 lbf to 46,950 lbf each. As of 1996, both SCAs used 50,000-lbf JT9D-7J engines.[16]

A separation monitoring and control system (SMCS) was installed on the main deck of the 747 for the atmospheric

N905NA during 1977 after modification into the Shuttle Carrier Aircraft. The American Airline logos can still be seen on the aluminum fuselage surfaces. The special access stands around the forward and aft Orbiter supports are noteworthy. (Dennis R. Jenkins)

flight tests. Displays and controls for the SMCS were added at the pilot stations, and an Orbiter-to-ground S-band relay link and SCA-to-Orbiter intercom were also carried. A load measurement system was installed to record the attach forces between the two vehicles during the mated portion of each flight. Load cells instrumented to measure axial and shear forces were located on each of the three attach struts. This data was evaluated in real-time and used as part of the criteria for approving separation during the first tailcone-off flight. All of this equipment was removed following the completion of the ALT program.[17]

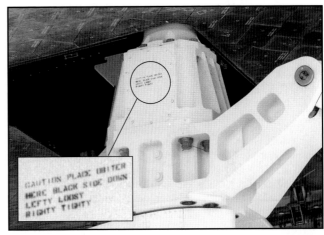

Space humor. Note the small writing on N905NA's aft Orbiter support strut. (Tony Landis)

The forward (left) and aft (right) Orbiter support struts on N905NA during September 1981. The forward strut is the higher ALT configuration. (Dennis R. Jenkins)

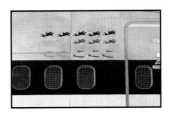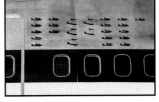

Following the tradition of the NB-52s at Dryden, the SCA crew added 'mission marks' to symbolize each ALT and ferry mission flown. The photo at left was taken in June 1978, while the one at right was taken in September 1981. (Dennis R. Jenkins)

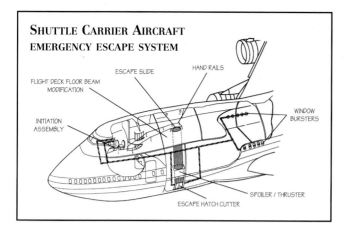

SHUTTLE CARRIER AIRCRAFT
EMERGENCY ESCAPE SYSTEM

FLIGHT DECK FLOOR BEAM
MODIFICATION

ESCAPE SLIDE   HAND RAILS

INITIATION
ASSEMBLY

WINDOW
BURSTERS

SPOILER / THRUSTER

ESCAPE HATCH CUTTER

The Orbiter mated location on the SCA was selected based on static stability and control, required structural modification, weight, and mission performance. Center-of-gravity limits for the SCA while carrying the Orbiter are 15 percent of the SCA mean aerodynamic chord (MAC) for the forward limit, and 33 percent MAC for the aft limit. Ballast is carried by the SCA in standard 747 cargo containers in the forward cargo compartment to ensure center-of-gravity limits are not exceeded. This ballast is adjusted for each ferry flight based on which Orbiter (each has a different empty weight) is being carried, and any payloads that might be installed in the Orbiter.[18]

On 14 January 1977 Boeing turned the modified 747 over to Rockwell for acceptance testing. Upon completion of weight and balance checks and a limited flight test series, the aircraft was delivered to NASA.[19]

Concerns over 747 crew safety during the atmospheric flight tests prompted NASA to award a contract to Teledyne-McCormick-Selph to develop an escape system for the 747 crew (the Orbiter had ejection seats). The resulting design bore a great resemblance to the system used by Boeing on the first prototype 747 during the initial flight test series conducted during 1969. The system allowed crew members to safely abandon the aircraft in the event of an emergency, and consisted of a 16-foot long escape slide connecting the flight deck with an egress port located in the forward cargo compartment. In the event of an emergency, a crew member would activate the system by pulling an initiation handle, one of which was located on either side of the autopilot control pedestal. Upon activation, 30 fuselage windows would be explosively fractured to allow rapid aircraft decompression – 3 seconds later the emergency egress hatch would be explosively severed and

blown clear of the fuselage and a spoiler extended into the airstream to allow crew members to clear the engine nacelles, landing gear, and empennage. Crew members would descend using individual parachutes. Tests showed that each crew member required only 11 seconds to clear the aircraft from the decision to abandon the 747.[20]

In February 1988 NASA announced plans to acquire a second 747 to serve as a backup to N905NA. This was, in part, due to the realization that a substantial number of flights would continue to land at Edwards AFB, and also in response to a recommendation made by the Rogers Commission after the *Challenger* accident. The aircraft selected was the first 747-100SR (JA8117, msn 20781) which had been declared surplus by Japan Air Lines (JAL) as they updated their fleet with more modern 747s. Boeing Military Airplane Company purchased the aircraft for NASA in April 1988, and began modifications during early 1990. The aircraft – reregistered N911NA – was delivered to NASA on 20 November 1990 and was first used to ferry OV-105 from Palmdale to KSC in May 1991. In late-1995 the N911NA vertical stabilizer was repainted with a stylized new-old NASA logo, and N905NA followed during early 1996.[22]

During the original SCA feasibility assessment it was found the blunt aft end of the Orbiter would adversely affect the mated vehicle performance, substantially limiting both climb speed and range. It was also thought that the Orbiter wake impinging upon the SCA vertical stabilizer would cause excessive structural fatigue. A drag-reducing tailcone was baselined, and Boeing was awarded a $4.2 million contract to build a single unit. The tailcone significantly reduces aerodynamic drag by smoothing over the SSME and OMS engines. The mated vehicle cruising speed and fuel economy are both improved by the fairing, as is the tolerance to cross-wind situations. The tailcone is 36 feet long, 25 feet wide, and 22 feet high. It is constructed primarily of aluminum, weighs 5,927 pounds, and attaches to the Orbiter with eight adjustable steel fittings. The fairing can be disassembled, placed in special shipping containers, and transported by truck as needed.[21]

The new drag-chute installation on OV-105 required a modified tailcone fairing. It was not possible to simply modify the existing fairing since it would be several years before the entire Orbiter fleet was equipped with drag chutes, necessitating keeping both old and new style tailcones available until the last Orbiter was modified. In 1990 Boeing was awarded a $5.3 million contract to build the new fairing, and it was first used to ferry *Endeavour* from Palmdale to KSC in May 1991. The most distinctive feature

Although most of the interior has been removed from the SCAs, N905NA has this attractive royal-blue carpeted mural on the forward bulkhead. (Tony Landis)

The original tailcone being removed from OV-101 at NASA Dryden in preparation for the fourth free-flight. (U.S. Air Force/AFFTC photo 77-1674)

Two views of N905NA carrying Challenger. These views give a good idea of exactly where the Orbiter is carried in relation to the SCA wing and empennage. (NASA via Tony Landis)

of the new fairing was a large oval vent in the extreme rear. After the last Orbiter was equipped with a drag chute, Boeing modified the original tailcone to the current configuration, giving NASA a fairing to go with each SCA.

N905NA and N911NA were originally maintained by a commercial contractor who provided both routine and depot-level maintenance, as well as normal pre-flight and post-flight servicing. This freed NASA from having to recruit and train maintenance personnel to service a one-of-a-kind (within NASA) aircraft. Actually, this type of arrangement is not terribly unusual even in the commercial airline industry, and many smaller airlines frequently contract out at least the depot-level maintenance on their aircraft. However, during 1993 it became more economical for personnel from NASA Dryden to provide routine maintenance for the SCAs. This changed a few years later, and contractor (DynCorp and Evergreen) personnel again maintain the aircraft under contract from the JSC Aircraft Support Division. Engine maintenance is provided by Pratt & Whitney. One SCA is typically kept at Evergreen Air Center facilities in Marana, Arizona, while the other is at NASA Dryden – they are switched out approximately every three months for maintenance.[23]

The maximum airspeed of an SCA is Mach 0.6 (250 KIAS), typical cruise altitude during a ferry mission is 13,000–15,000 feet, and the maximum ferry range is 1,150 miles. Without an Orbiter attached, the aircraft can attain altitudes of 24,000–26,000 feet and have a range of 6,300 miles. The minimum crew carried by the SCA during ferry

missions is two pilots and two flight engineers, although only one flight engineer is required when not carrying an Orbiter. N905NA has an empty weight of 318,053 pounds, while N911NA weighs 323,034 pounds. The maximum take-off weight of both aircraft is 710,000 pounds, and maximum landing weight is 600,000 pounds.[24]

NASA has examined the possibility of using aerial refueling to extend the range of the SCAs. The equipment is readily available since Boeing has delivered a handful of 747s (VC-25A and E-4s) to the Air Force that are equipped with flying-boom-type receptacles. Analysis indicated there was no particular reason the idea would not work,

One of the few external differences between the SCAs is the number of upper deck windows. N911NA (left) has five – N905NA has two. (NASA photo EC95-43339-1)

The new and the old. In late 1995 N911NA received a new stylized NASA tail logo – shown here with N905NA and its old 'worm' logo. The first SCA would receive its new logo a few months later and the two aircraft again appear virtually identical. (NASA photo EC95-43339-4 by Tony Landis)

Enterprise and N905NA lift-off on the first of the Approach and Landing Test captive-inert flights. The remains of the American Airlines logo, as well as some structural patches on the aft fuselage, can be seen on the 747. In theory the small auxiliary end-plate vertical stabilizers are removable, but this has never been done. Despite its ungainly appearance, the Orbiter weighs a relatively small fraction of the 747's normal payload capacity, providing a considerable power margin. (Dennis R. Jenkins)

Discovery and N905NA at Dryden. The black tiles on the OMS pods shows up well here, and is the easiest way way to tell if the OMS pods are real. (Tony Landis Collection)

The first use of N911NA was to deliver Endeavour to KSC, arriving on 7 May 1991 – a shiny new SCA with a shiny new Orbiter. (NASA photo KSC-91PC-864)

The mate-demate devices at KSC and Dryden are generally similar, and much larger structures than the lifting frame at Vandenberg/Palmdale. (NASA photo EC91-187-01)

The lifting frame from Vandenberg was moved to Palmdale where it is now used during OMDPs – here is Atlantis and N911NA on 26 May 1994. (Tony Landis Collection)

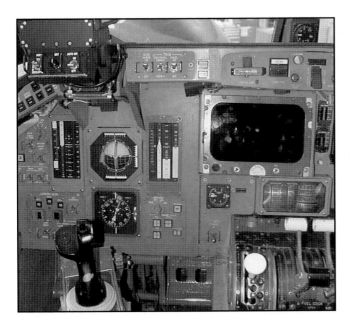

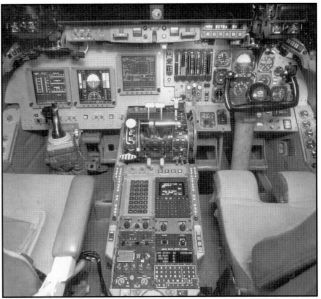

The Shuttle Training Aircraft (STA) cockpit before (left) and after the MEDS installation. The right seat instrumentation is generally similar to a stock Gulfstream II, including the normal control wheel. But the left seat is almost pure Orbiter, including the center hand controller and HUD. Note that in the original STAs only a single CRT was installed, along with several dedicated displays (aero instruments) beside it. In the MEDS installation, three displays are included. (NASA photos [no number for left] and S98-05111)

Since the original ventral fins were removed, the STAs do not look much different than standard Gulfstream II business jets on the outside. Here one of the aircraft is shown (at two different times) making approaches to Northrup Strip at White Sands, New Mexico. (NASA photos S96-06240 and S97-15629)

Originally, the STAs used large ventral fins under the wing to generate side forces in order to aerodynamically simulate the Orbiter. As more experience was gained, and the software-driven flight control system became more sophisticated, these were removed. Note the early markings in this photo compared with the two above. (NASA via Tony Landis)

The second SCA taxies to the runway at Edwards AFB with Endeavour. Pilots report that the SCA handles remarkably well with its large external load, but range is adversely affected, usually necessitating several stops between Edwards and KSC. (Christian Ledet via Tony Landis)

although there was some concern about fuel impinging on the Orbiter tiles after disconnect. The idea went as far as actually flying N905NA in formation with a KC-135 for a simulated refueling. The next step was to install OV-101 on top of the SCA and repeat the tests, but this never happened. A routine inspection of the SCA showed some minor cracking around the base of the vertical stabilizer – ground crews indicated the cracks had been there for a while, and were most probably caused by flow off the Orbiter impinging on the area during ferry flights. An alternate explanation was that the cracks had been caused by wake turbulence from the tanker. In the absence of more information, or a pressing need to extend the range of the SCA, the test program was cancelled.[25]

In addition to the SCA, another pair of aircraft were being readied in preparation for the atmospheric flight tests. Two G1159 Gulfstream II business jets (N946NA and N947NA) were purchased by NASA in 1974–75 and modi-

At one time, NASA had evaluated using a barge to transport Orbiters to the Kennedy Space Center instead of the 747. The Orbiter model is in the Approach and Landing Test (ALT) configuration with an air data probe and boat-tail tailcone fairing. Of interest is the configuration of the Reaction Control System exhaust ports on the forward fuselage. (NASA photo 108-KSC-75PC-95)

fied into Shuttle Training Aircraft (STA). The STAs provide Orbiter pilots with a realistic simulation of Orbiter cockpit motion, cues, and handling qualities, while simultaneously matching the Orbiter's atmospheric decent trajectory from 35,000 feet to the actual Orbiter cockpit height above the runway at touchdown. As originally configured the aircraft flaps were modified to move up as well as down, large vertical surfaces on the underside of the fuselage generated side forces, the thrust reversers could be activated in flight, and the landing gear was modified so that the mains could be extended separately from the nose gear and at higher-than-normal speeds. With the main gear, flaps, and thrust reversers deployed, the STA was capable of duplicating the high-drag landing profile of the Orbiter. An advanced digital avionics system (ADAS) controlled the direct lift control (DLC) and the in-flight reverse thrust, as well as the conventional aircraft controls.[26]

On the flight deck the left side instrument panel was modified with a set of Orbiter displays and controls, while the right side contained the normal Gulfstream instruments as a safety measure. The ADAS computer system translated the pilot's inputs into control movements that largely mimicked an Orbiter. Subsequent changes to the software provided more accurate simulations of the Orbiter, and also allowed the removal of the large ventral fins. This software was largely developed by NASA Ames and included one of the first airborne applications of 'adaptive fuzzy-logic.' Beginning in 1997, the STAs were modified with MEDS displays in the cockpit, replacing the older DEUs and dedicated displays.[27]

Two additional Shuttle Training Aircraft (N943NA and N944NA) were acquired during the 1980s, and another Gulfstream II (N948NA) was purchased in 1992 but has not yet been converted into an STA.

The NASA White Sands Test Facility (WSTF) operates the White Sands Space Harbor (WSSH), the primary training area for space shuttle pilots flying practice approaches and landings in the STA and T-38 chase aircraft. WSSH is located approximately 30 miles west of Alamogordo, New Mexico, on the White Sands Missile Range. The runways are located at Akalai Flats, a site formerly known as Northup Strip. NASA selected Northrup Strip as the site for pilot training in early 1976. A second lakebed runway was subsequently marked, and in 1979 both runways were lengthened, allowing the facility to serve as a back-up landing site. The only time a shuttle has used the facility was for the landing of STS-3 in March 1982. After the STS-3 landing, the strip became an emergency landing site, and and act of Congress renamed the facility the White Sands Space Harbor.[28]

IT'S A BIRD; IT'S A PLANE ...

As development progressed, an interesting question was posed – was the space shuttle an aircraft or a space vehicle. This distinction is not totally random, as there are various U.S. laws and international treaties that come into play depending how a vehicle is categorized. In cooperation with the Federal Aviation Administration (FAA) the NASA Counsel for General Law, Gerald J. Mossinghoff, attempted to determine whether shuttle should be classified as an aircraft within the meaning of the Federal Aviation Act of 1958, as amended. Several interesting arguments were put forth by Mossinghoff:[29]

"(1) The National Aeronautics and Space Act of 1958 authorizes NASA to develop, test, and operate both 'aeronautical and space vehicles.' The legislative history makes clear that aeronautical vehicles are those designed for operation 'within the atmosphere' whereas space or astronautical vehicles are designed for operation 'primarily outside the atmosphere, although often passing through the atmosphere on the way to outer space.' Based on this history, although there is legally no precise dividing line between the atmosphere and outer space, it is clear that the Space Shuttle is a space vehicle under our act, and not an aeronautical vehicle.

"(2) Although the definition of aircraft in the Federal Aviation Act is quite broad ('any contrivance now known or hereafter invented, used, or designed for navigation of or flight through the air'), the fact that something falls within this literal definition does not mean that it legally will be considered an aircraft even by the FAA, which recognizes 'non-aircraft' airborne objects, for example, surface-effects (air-cushion) vehicles.

"(3) NASA's authorizing committees, when describing the Space Shuttle in reports accompanying our annual authorization acts, have consistently characterized it as a 'reusable [sic] manned space vehicle.

"(4) Under our interpretation of the Convention on International Liability for Damage Caused by Space Objects, the Space Shuttle would clearly be a 'space object' so as to impose absolute liability upon the United States for 'damage caused on the surface of the earth or to aircraft in flight.'

"(5) Similarly, Space Shuttle flights would be registrable [sic] under the Convention on Registration of Objects Launched into Outer Space ..."

The FAA accepted these arguments and agreed that the space shuttle was not an aircraft under existing laws. This decision also applied to similar international treaties, and therefore exempted the space shuttle from the various noise, environmental, and safety regulations otherwise imposed on aircraft. By agreeing that the shuttle was a 'space object' it became the subject of different international rules that govern how virtually every country on the planet treats spacecraft and astronauts.[30]

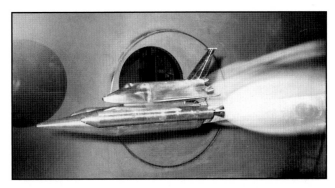

Shuttle exhaust plume studies were conducted using this 1.9-percent scale model in the 9 by 7 foot supersonic wind tunnel at NASA Ames. Note the early configuration of the vehicle in this 21 May 1973 photo – the SRBs are mounted above the ET centerline, the ET has an extremely pointed nose, and the early MSC-040 orbiter configuration. The exhaust plumes were simulated by pumping compressed air through the model. (NASA photo 73-H-342)

A 36-percent-scale model of the Orbiter in the Ames 40 by 80-foot wind tunnel. This January 1976 test was gathering data for the Approach and Landing Tests. (NASA)

STRUCTURAL DYNAMICS

Up to this point, a launch vehicle had been subjected to an extensive flight test program before being 'man rated.' The shuttle program philosophy, on the other hand, was to test key hardware elements and use verified analytical models to certify the space shuttle launch configuration. The Orbiter was to carry a crew of two on the first manned orbital flight (FMOF), tentatively scheduled for 1979.

Design engineers made extensive use of analytical models to investigate the structural dynamics, pogo, and flutter characteristics of the space shuttle. These were partially verified using data gained during laboratory testing to ensure that the models were accurate. The philosophy was to correlate theoretical analysis with element testing and wind tunnel results to an acceptable degree of accuracy, and to infer that the actual vehicle dynamics could be predicted with the same amount of accuracy.[31]

Analytical studies during the Phase B extensions had pointed out some unique dynamic characteristics for any parallel-burn configuration. In particular, a very high modal density – with 200 structural modes below 20 hertz encountered during a wide spectrum of conditions – presented a large variety of dynamic problems. Studies conducted at NASA Langley on a 1/8-scale structural model reinforced these concerns, and indicated the substantial influence of element interface stiffness on the primary low frequency modes of the integrated vehicle. Although the 1/8-scale model was coarse, the overall configuration was representative and the results were cause for concern.[32]

Immediately after ATP an emphasis was placed on developing a plan that would ensure early resolution of

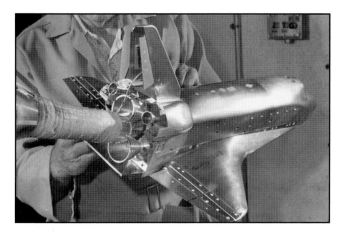

This was the highest-fidelity model of the Orbiter tested at AEDC, and incorporated such details as windows, retractable air data probes, and a rocket-powered reaction control system. This model was tested up to 1,500 mph. (U.S. Air Force/AEDC photo 78-441)

A 0.0175-percent model in the AEDC von Karman Gas Dynamics Facility Wind Tunnel A during 1980. This was the definitive configuration of for the first manned orbital flight. (U.S. Air Force/AEDC photo 80-1550)

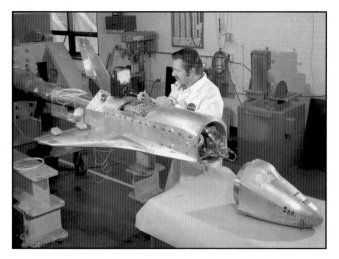

A highly-instrumented model of the Orbiter being prepared for testing in the 16-foot transonic wind tunnel at NASA Langley in 1978. Note that the entire forward fuselage is removable to allow access to the instrumentation. (NASA photo EL-1997-00090)

A 1/8-scale model of the space shuttle, held aloft by a maneuverable tower lift, is tested in the anechoic chamber at JSC. The chamber walls are completely covered with foam pyramids for absorbing stray electromagnetic energy during antenna radiation pattern tests. A model of the entire stack was also tested. (NASA photo S73-38424)

the key structural dynamics issues. A series of tests were initiated at various NASA, Air Force, and contractor facilities to ensure that the dynamics were fully understood before committing the space shuttle to flight. The shuttle vibration tests (SVT) eventually consisted of three major parts: (1) a 1/4-scale structural model program to provide early design data to validate other analyses, (2) testing full-scale flight elements, including OV-101, and (3) a full-scale mated vertical ground vibration test.[33]

The plan also established procedures for generating and updating the structural dynamic math models. Each of the 'element' contractors* (Rockwell, USBI, and Martin Marietta) was responsible for generating and updating element-unique models, and the systems integration contractor (Rockwell) was responsible for maintaining the complete model. The structural dynamic models were derived from a 'stress model' that was a detailed finite-element model of the space shuttle structure. The stress model had approximately 50,000 degrees of freedom – this was reduced by various techniques to about 1,000 for each of the dynamic models.[34]

The final verification of the models would require an assessment of flight data, the bulk of which eventually came from the orbital flight tests (OFT). Approximately 1,000 measurements were part of the developmental flight instrumentation (DFI) package that was carried in the payload bay, and these were used to correlate the earlier model data with actual flight test results. This final correlation of data was one of the decisive factors used to declare the shuttle 'operational' after STS-4.[35]

The 1/4-scale structural model program was initiated in early 1975 by a JSC–Rockwell team to acquire high-quality structural dynamic data on the integrated vehicle flight configuration. The structural model was a high-fidelity replication of OV-102, a standard-weight ET, and three different SRB flight configurations. The 1/4-scale program was the most comprehensive SVT model, and was structured most like the operational flight hardware. The 1/4-scale vibration test configuration included: (1) an integrated vehicle configuration with a 45,000 pound payload during lift-off, the period of maximum dynamic pressure, and just prior to SRB separation, (2) an Orbiter–ET configuration with a 45,000 pound payload immediately after SRB separation, during mid-boost, and just prior to orbital insertion, (3) the orbiter alone, both with and with-

---

* These became known as the 'RUM' contractors, taking the first letter of each company name.

Wind tunnel tests have continued through the life of the program, as shown in this 1988 photo taken in the AEDC 16-foot transonic wind tunnel. This test was to study the aerodynamic effects generated by various engine-out configurations. Exhaust from the SSMEs was simulated by short-duration burn of gaseous hydrogen and oxygen. The SRB plumes were generated by burning a mixture of gases designed to match the thermal and propulsive properties of the actual SRB plumes. (U.S. Air Force/AEDC photo 88-31616)

out a 45,000 pound payload, (4) an ET, just after jettison from the Orbiter, and (5) an SRB, just after separation.[36]

During the testing, water was used to simulate LO2 in the External Tank, and the weight and hydroelastic effects of the LH2[*] were neglected, a procedure that would also be used in the later mated vertical ground vibration tests conducted on OV-101. To complement the vibration test program, load-deflection tests were conducted separately on a pre-production SRB and ET in order to provide data that could be used to resolve anomalous or unexplained vibration test data. At the completion of the test program, the 1/4-scale model was stored at JSC until mid-2000 when the orbiter model was loaned to the Calgary, Canada, airport for public display. The ET and SRBs were transferred to the National Air and Space Museum, which also hopes to obtain the orbiter from Calgary at a later date.[37]

One of the major problems encountered during the early Saturn V program[†] had been a phenomena called 'pogo' – caused by a partial vacuum being created in the propellant lines by the pumping rocket engines. This condition produced a hydraulic resonance that usually matched the natural frequency of the vehicle, creating a severe shaking. In addition, the engine 'burped' when the bubbles caused by the partial vacuum reaching the thrust chamber. Luckily, the first few Saturn V flights had been unmanned, and engineers developed a solution prior to its first manned mission. Because of this, coupled with the fact that the first shuttle mission would be manned, strict pogo requirements[‡] were placed on the space shuttle. Therefore, the hydroelastic model of the ET was of particular concern.[38]

A general lack of correlation between earlier tests and analysis with the Langley 1/8-scale model data indicated that the same deficiency could be expected from the

1/4-scale hydroelastic analysis of the External Tank. As a consequence, several configurations were selected for testing. In parallel, Martin Marietta developed a new hydroelastic analysis technique that became available before the 1/4-scale testing commenced. The quality of correlation between the new analysis and the ET vibration data was judged to be excellent, and provided confidence that allowed a reduction in the scope of the ET vibration testing. Generally, the analysis frequencies were higher than the actual test frequencies, and these differences were attributed to pressure effects in the LO2 and LH2 tanks.[39]

The SRB tests identified several areas in the model that required additional study. These included the ET/SRB interface, which required additional detail in the finite element model, and the incorporation of a representative shear modulus for the solid propellant. The post-test analysis resulted in several changes being incorporated into the model that corrected some of these deficiencies.[40]

SPACE, THE FINAL FRONTIER ...

On 4 June 1974 workers at Air Force Plant 42 in Palmdale, California, started structural assembly of the first space shuttle orbiter. This initial vehicle was designated[§] OV-101 (Orbiter Vehicle 101) by NASA, and was to carry the name *Constitution* in honor of the U.S. Constitution Bicentennial. However, some people had other ideas about the name for the first reusable manned spacecraft – nearly 100,000 fans of the TV science-fiction show *Star Trek* staged a write-in campaign urging the White House to rename OV-101. The 'Trekkers' had their way, and when the doors opened on 17 September 1976, the name on the side of the Orbiter was *Enterprise*.[41]

But before OV-101 was rolled out publicly, the vehicle had participated in the second phase of the shuttle vibration tests. During the summer of 1976, OV-101 was used in the horizontal ground vibration test (HGVT) conducted at Palmdale. This was the first opportunity for engineers to obtain structural dynamic data from an actual flight vehicle for math model verification. Although OV-101 was not identical to the configuration planned for *Columbia*, the differences were well understood and accounted for in the model. For instance, OV-101 did not have provisions for mounting 'real' OMS pods (but used structural boilerplate replicas) and the vertical stabilizer was built-

OV-101 coming together on 29 March 1976 at Palmdale. Note that the forward fuselage is constructed in upper and lower halves, with only the lower portion shown here. The crew module has not been installed yet. (NASA photo 108-KSC-76P-49)

---

* LH2 is very light, and even large quantities have little effect on vehicle dynamics. Besides, there is no reasonable substitute since LH2 is lighter than any other liquid, and testing with LH2 usually represents an unreasonable danger.
† Actually this problem had been present on nearly all launch vehicles, and in fact, the Gemini program had almost been postponed when an early Titan II ICBM suffered a catastrophic failure attributed to pogo effects.
‡ Pogo suppressors for the engines had been baselined early in the SSME program.
§ It is not unusual for aircraft programs to assign numbers above 100 (or 1,000) to production aircraft and numbers below that to test articles.

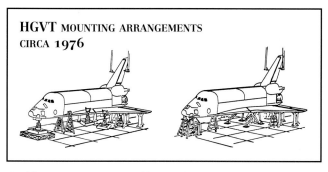

Two different arrangements were used during the horizontal ground vibration test (HGVT) series. The 'soft' free-free mount (left) and the 'rigid' mounting arrangement. (NASA)

up using skin and stringers as opposed to the integrally machined structure of OV-102.* The payload installed in the Orbiter during HGVT testing was the 10,000-pound developmental flight instrumentation (DFI) package that would be used during the atmospheric flight tests.[42]

There were two test configurations, one with the Orbiter supported in a 'free-free' condition to simulate reentry and landing, and the other with the Orbiter rigidly attached to the ground at the Orbiter–ET interface to simulate the ascent configuration. Tests were also conducted with the payload bay doors opened to simulate an on-orbit configuration. Ferry locks were used to secure the aerodynamic control surfaces during the testing, mainly to prevent unexpected damage. The test objectives were to determine selected mode shapes, frequencies and modal damping in

the range of 0.5 to 50.0 hertz, and to acquire frequency response data at the Orbiter guidance and control sensor locations. The structural mode shapes were predicted to be extremely complex and not generally amenable to classic model descriptions, a theory the testing confirmed. Following the completion of the tests, minor modifications were made to the vehicle prior to the public roll out.[43]

For the atmospheric flight tests, OV-101 would not carry all of the equipment and systems required for space travel. *Enterprise* contained no main propulsion system plumbing, internal propellant lines or tankage, and the main engines and their nozzles were mock-ups, as were the orbital maneuvering system pods beside the vertical stabilizer. Lead ballast was used to maintain the vehicle weight and balance within the limits anticipated for space-rated Orbiters. There were ballast provisions in the nose wheel well that could accommodate up to 1,350 pounds, on the X_O378 bulkhead (just behind the nose wheel well) that could handle 2,660 pounds, and in the payload bay for an additional 18,000 pounds. The Orbiter fuel cells were fed hydrogen and oxygen from high-pressure tanks instead of cryogenic dewars. The payload bay did not contain structural supports or mounting points for payloads, and the payload bay doors did not have radiators or use hydraulic actuators. The crew module did not contain an airlock, mid-deck lockers, galley, shower, or other crew support items. The fragile thermal protection system tiles were simulated by white and black polyurethane foam, while the exotic carbon-carbon nose-cap and wing leading edges were made of fiberglass.[44]

On the flight deck, most of the navigation, guidance, and propulsion controls were missing, along with the star tracker controls. The heads-up displays were not

---

* Most public reports also indicate that the thrust structure did not use the boron-epoxy reinforcements found in the other Orbiters, but this appears to be incorrect based on a visual examination of OV-101 by JSC engineers at the Smithsonian.[251]

Although destined never to fly in space, Enterprise looked the part during rollout ceremonies at the Rockwell Palmdale facility. Here Fred Haise poses with the new Orbiter. (NASA photo S76-29561)

OV-101 during the roll-out ceremonies at Palmdale. Notice the different black-tile pattern at the base of the vertical stabilizer compared to the photo (top) showing the left side. Also note the Bicentennial '76' on the tow tractor. (Dennis R. Jenkins)

Enterprise was greeted by 2,000 well-wishers when she was rolled out. Note the Apollo capsule boilerplate near the right wing. At this point, NASA still expected the first orbital flight in 1979. (Boeing North American photo B02832-0004)

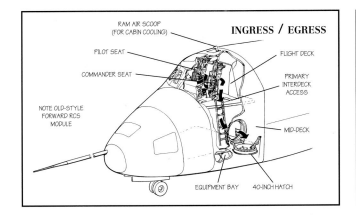

INGRESS / EGRESS

RAM AIR SCOOP (FOR CABIN COOLING)
PILOT SEAT
COMMANDER SEAT
NOTE OLD-STYLE FORWARD RCS MODULE
FLIGHT DECK
PRIMARY INTERDECK ACCESS
MID-DECK
EQUIPMENT BAY
40-INCH HATCH

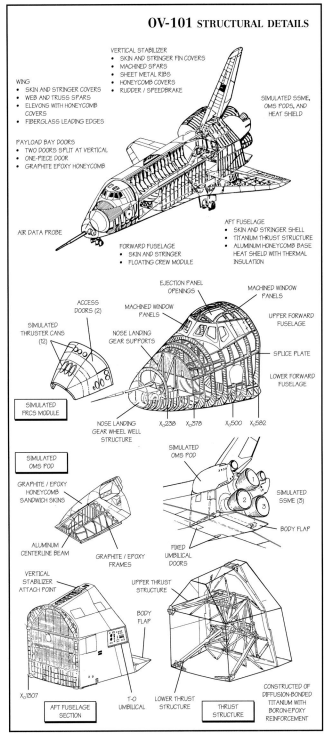

ALT-UNIQUE ITEMS

ORBITAL SYSTEMS NOT INSTALLED FOR ALT
- OMS / RCS
- AIRLOCK
- SSME (3)
- PAYLOAD
- RADIATORS
- TPS
- CREW STATIONS
- EPS CRYO TANKS
- STAR TRACKERS
- UNIFIED S-BAND
- RENDEZVOUS RADAR AND KU-BAND

EJECTION SEATS AND OVERHEAD PANELS
DISPLAY AND CONTROL KITS
AIR DATA PROBE
BALLAST
BACKUP FLIGHT CONTROL SYSTEM
RAM AIR PROVISIONS
DEVELOPMENTAL FLIGHT INSTRUMENTATION PALLET
SIMULATED TPS
C-BAND BEACON
SCA SEPARATION SYSTEM
FIBERGLASS LEADING EDGES
FUEL CELL REACTANTS (HI-PRESSURE O2 AND H2)
SIMULATED SSME, OMS PODS, AND HEAT SHIELD
BALLAST
FLIGHT TEST UMBILICAL TO SCA

OV-101 STRUCTURAL DETAILS

WING
- SKIN AND STRINGER COVERS
- WEB AND TRUSS SPARS
- ELEVONS WITH HONEYCOMB COVERS
- FIBERGLASS LEADING EDGES

PAYLOAD BAY DOORS
- TWO DOORS SPLIT AT VERTICAL
- ONE-PIECE DOOR
- GRAPHITE EPOXY HONEYCOMB

VERTICAL STABILIZER
- SKIN AND STRINGER FIN COVERS
- MACHINED SPARS
- SHEET METAL RIBS
- HONEYCOMB COVERS
- RUDDER / SPEEDBRAKE

SIMULATED SSME, OMS PODS, AND HEAT SHIELD

AIR DATA PROBE

FORWARD FUSELAGE
- SKIN AND STRINGER
- FLOATING CREW MODULE

AFT FUSELAGE
- SKIN AND STRINGER SHELL
- TITANIUM THRUST STRUCTURE
- ALUMINUM HONEYCOMB BASE HEAT SHIELD WITH THERMAL INSULATION

EJECTION PANEL OPENINGS
ACCESS DOORS (2)
SIMULATED THRUSTER CANS (12)
MACHINED WINDOW PANELS
NOSE LANDING GEAR SUPPORTS
MACHINED WINDOW PANELS
UPPER FORWARD FUSELAGE
SPLICE PLATE
LOWER FORWARD FUSELAGE
SIMULATED FRCS MODULE
NOSE LANDING GEAR WHEEL WELL STRUCTURE
X_0238 X_0378 X_0500 X_0582

SIMULATED OMS POD
GRAPHITE / EPOXY HONEYCOMB SANDWICH SKINS
ALUMINUM CENTERLINE BEAM
GRAPHITE / EPOXY FRAMES

SIMULATED OMS POD
SIMULATED SSME (3)
BODY FLAP
FIXED UMBILICAL DOORS

VERTICAL STABILIZER ATTACH POINT
X_01307
AFT FUSELAGE SECTION
T-O UMBILICAL
UPPER THRUST STRUCTURE
BODY FLAP
LOWER THRUST STRUCTURE
THRUST STRUCTURE
CONSTRUCTED OF DIFFUSION-BONDED TITANIUM WITH BORON-EPOXY REINFORCEMENT

---

installed, nor were all the panels, switches, and indicators for the ET and SRB systems. A small amount of additional instrumentation was added just below the left display electronics unit for the air data system installed on the long flight test boom on the nose. Three cameras recorded the pilot's actions and Lockheed SR-71 zero-zero ejection seats were provided in the event escape was necessary. Two blow-out panels were installed above the pilots to facilitate ejection, or rapid egress on the ground. The two windows looking from the aft flight deck into the payload bay were absent, covered by aluminum panels, as were the overhead rendezvous windows. Aluminum braces were installed where the mid-deck airlock should have been. The landing gear system was lowered by explosive bolts and gravity, and there were no provisions for hydraulic retraction.[45]

While Rockwell was building OV-101, NASA was defining a flight test series to verify the subsonic airworthiness of the Orbiter. Once the 747 was selected as the SCA, a great deal of wind tunnel data needed to be gathered to prove the vehicle's could be separated in flight – it was these tests that led to the addition of the endplates to the horizontal stabilizers. These tests were also the first to evaluate if the Orbiter could be carried without fear of damaging the 747 vertical stabilizer. The tests indicated that the buffeting, although having a negative impact on performance, did not seriously exceed recommended structural loads, giving confidence that a limited number of flights without the tailcone would not damage the SCA.[46]

With wind-tunnel data in hand, NASA and Rockwell engineers were now confident that the atmospheric flight tests could be safely accomplished. Interestingly, although the engineers were comfortable with the idea, upper management was not. To better demonstrate the separation maneuver, the engineers coupled various simulations with computer graphics to produce a movie of the maneuver – very state-of-the-art for the mid-1970s. This convinced everybody that separation was possible, and plans moved forward rapidly. As initially envisioned, the newly-named Approach and Landing Tests (ALT) would take four forms:[47]

- Taxi Tests. These would be used to verify the very low-speed dynamics of the mated vehicles. The taxi tests would use the concrete runway (04/22) at Edwards, and be conducted early in the morning to minimize problems associated with heat build-up in the SCA tires and brakes. The first run would take the mated vehicles to 86 mph, then be stopped using normal braking. The second test would get up to 138 mph and also stop using normal braking. The last run would be at 155 mph and use full braking, thrust reversers, and speedbrakes to simulate an aborted take-off.

- Captive-Inert Flights. Six flights with an unmanned inert Orbiter were planned. These tests would verify the performance, stability and control, flutter margin, and buffet characteristics of the mated configuration in flight patterns similar to the manned free flights. The Orbiter would weigh 150,000 pounds, resulting in a combined weight of from 585,000 to 630,000 pounds depending on the 747 fuel load. Flights would be conducted up to 25,000 feet and 316 mph.

- Captive-Active Flights. A crew would be aboard OV-101 during the six captive-active flights which were designed to determine the optimum separation profile based on the captive-inert test results, refine and finalize Orbiter and SCA crew procedures, and evaluate Orbiter integrated systems operations. Five of the flights were scheduled with the Orbiter tailcone attached, and one flight without if data indicated it as prudent. The mated combination would fly at altitudes up to 24,000 feet and speeds between 270 to 300 mph.

- Free-Flights. A series of up to eight free-flights were scheduled to follow the first five captive-active flights. The free-flights were designed to verify Orbiter subsonic airworthiness, integrated system operations, and both piloted and automated landing capabilities. The first five free-flights would be conducted with the tailcone on. Following the successful completion of these, the last captive-active flight (tailcone-off) would be conducted. If this flight was successful, the last three free-flights with the tailcone off would be flown.

The tailcone-on free flights would involve the separation of the Orbiter from the SCA at 287 mph at 22,000 feet (AGL). *Enterprise* would glide for about 25 miles on its way to a landing, which would be either on the lakebed or the concrete runway at Edwards AFB. The tailcone-off flights would be conducted at 18,000 feet and, due to the increased drag resulting from the lack of the tailcone, the Orbiter would only glide about 12 miles. Landings would all be made on the lakebed since it had more overrun available.[48]

On 31 January 1977, *Enterprise* was towed 36 miles through the California desert from Palmdale to Edwards. The trip took the better part of the day, using mostly public roads that had been cleared of traffic, and a few specially constructed gravel roads* to provide a shortcut around one of the dry lakebeds.[49]

Enterprise was mated to N905NA on 7–8 February 1977 using the mate/demate device at Dryden. The mated vehicles underwent weight, balance, and vibration checks during the morning of 15 February 1977, followed by

---

* All of the Orbiters except OV-105 would initially be moved between Palmdale and Edwards in this manner. The Orbiter Lifting Frame (portable mate/demate device) originally constructed for the Vandenberg Launch Site was installed at Palmdale in late 1990 to allow Orbiters to be flown directly to/from Palmdale as required.

The cockpit of Enterprise while under construction in Palmdale during April 1976. The CRTs and keyboards are missing from the instrument panel, as are the seats. Note the large ejection hatch above each pilot, and the simple control panels over the windshields (at least compared to later Orbiters – see page 374). (NASA photo 108-KSC-76PC-236)

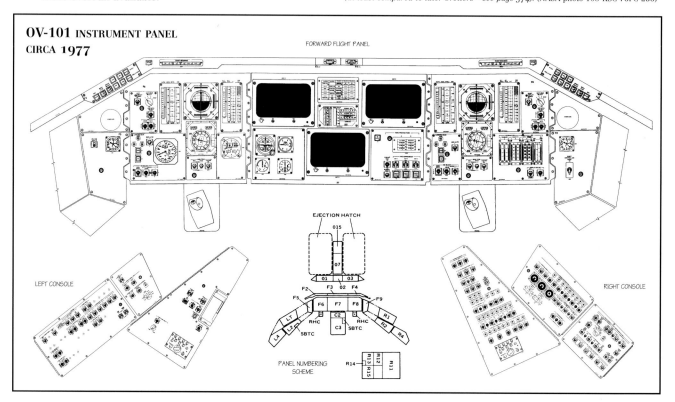

OV-101 INSTRUMENT PANEL
CIRCA 1977

FORWARD FLIGHT PANEL

LEFT CONSOLE

RIGHT CONSOLE

EJECTION HATCH

PANEL NUMBERING SCHEME

The overland move of Enterprise through the streets and desert between Palmdale and Edwards AFB presented a strange sight. Power lines along part of the route had to be removed and/or relocated to allow the Orbiter to pass, while many street signs had a hinge permanently installed to allow them to 'fold' under the Orbiter's wings. This would be the first journey for all of the Orbiters except Endeavour. Much of what was learned here was later applied to the overland move at Vandenberg. (Dennis R. Jenkins)

three taxi-tests. During the taxi tests, the 747 weighed approximately 400,000 pounds with an Orbiter weight of about 144,000 pounds. These test validated the 747 steering and braking systems, and no unexpected problems were uncovered.[50]

On 18 February 1977, the Orbiter and SCA took-off for the first captive-inert flight. Four additional flights would be accomplished by 2 March 1977, and the results were sufficiently encouraging that NASA cancelled the sixth flight. The pilots of the SCA were pleasantly surprised to find the Orbiter had little adverse effect on the handling of the 747. Test objectives included engine-out simulations, control and stability evaluations, and air data calibration. The fourth and fifth flights practiced the short-field landing that would be required when the Orbiter was ferried to the 7,500-foot long runway at MSFC – the SCA demonstrated that it could successfully stop in less than 6,000 feet. None of the five captive-inert flights revealed any stability or flight control problems.[51]

The first captive-active flight was scheduled for 17 June 1977, but during a pre-flight check one of the shuttle on-board computers was 'voted out' by the other three. The flight was postponed 24 hours while the computer was replaced. On 18 June veteran astronaut (Apollo 13) Fred W. Haise and test pilot Charles Gordon Fullerton were aboard *Enterprise* when it and the SCA lifted off from Edwards at just past 08:06 hours. The flight was once around an oval that measured 78 miles on the straight sections and 10 miles through the curves. This

test was the first in which the Orbiter was powered up, and Haise and Fullerton practiced moving the aerosurfaces and rudder/speedbrake (to 60, 80, and 100 percent) for the first time. No problems were uncovered during the 55 minute and 46 second flight. Fullerton later said that the positioning of the Orbiter on the SCA resulted in the Orbiter crew not being able to see any part of the 747, which made it feel as if they were flying alone.[52]

The next flight, with Joe H. Engle and Richard H. Truly at the controls of the Orbiter, was designed to simulate the separation maneuver that would be used for the first free flight. Take-off was at 08:52 hours and during the initial climb-out various low-speed flight control system tests were performed by both the Orbiter and SCA. A higher-speed (310 mph) set of flight control system tests were then performed to assess the accuracy of predicted control surface responses and structural characteristics with respect to aerodynamic vibration. The vehicles then climbed to 22,030 feet for a separation maneuver test. The 747 crew accomplished a pushover and descended at approximately 3,000 feet per minute. The Orbiter elevons were positioned as they would be during an actual separation. Following the separation test, the vehicles climbed to 19,300 feet and established a six degree glide slope to simulate an Orbiter AUTOLAND mode fly-through. The Orbiter crew monitored the MSBLS (microwave scanning bean landing system) to ensure it was performing as expected (it was). After waving off the AUTOLAND approach, the vehicles performed a normal landing.[53]

MATED PROFILE FREE-FLIGHT ONE

FLIGHT SEQUENCE

ITEM	TIME (MIN)	ALTITUDE (FEET AGL)	RANGE (NM)	EVENT
1	0	0	0	SCA TAKEOFF
2	12	16,400	8	INTERSECT RACETRACK
3	22	22,000	32	IN-FLIGHT FCS CHECKS
4	28	24,100	14	REACH MCT 200 FPM CEILING
5	38	24,100	31	SCA BEGIN SRT
6	45	26,000	9	SCA PUSHOVER
7	46	22,800	7	SCA ORBITER SEPARATION
8	51	0	0	ORBITER TOUCHDOWN

NOTE: THERE IS ONLY 1 FULL GO AROUND IF THE ORBITER DOES NOT SEPARATE ON THE FIRST ATTEMPT, THE FLIGHT WILL BE TERMINATED

GROUND TRACK

ALTITUDE PROFILE

NOTE: ALTITUDES ARE ABOVE GROUND LEVEL (AGL) AND ARE REFERENCED TO ORBITER GROUND AIM POINT ON THE RUNWAY (ADD 2300 FEET TO OBTAIN MEAN SEA LEVEL)

Enterprise and N905NA depart Edwards for one of the ALT flights. The nose-high attitude of the Orbiter is obvious if compared with a photo of a ferry flight later in this book. Note the T-38 chase aircraft. (Dennis R. Jenkins)

The Shuttle Carrier Aircraft flight crew during the Approach and Landing Tests, from left: Fitzhugh L. Fulton Jr., Thomas C. McMurtry, Victor W. Horton, and Louis E. Guidry, Jr. (Dennis R. Jenkins)

The last free-flight of Enterprise. The fifth free flight was the first to land on the concrete runway at Edwards, and resulted in a pilot-induced oscillation by Fred Haise just as he touched down. Fortunately, the flight ended successfully. (Dennis R. Jenkins)

At the end of the third captive-active flight, Fred Haise deployed the landing gear of Enterprise while still attached to the SCA just to make sure everything would work on the first free-flight. The cherry-picker was used by the Orbiter crews to ingress/egress while the Orbiter was on top of the SCA. Note the DC-3 to the right. (Dennis R. Jenkins)

The table below lists the tests conducted as part of the ALT at Edwards AFB during the first half of 1977. In all cases the carrier aircraft was N905NA, a Boeing 747-123, and the Orbiter was OV-101, Enterprise. These tests confirmed that the aerodynamic handling of the basic Orbiter shape was satisfactory, and confirmed preliminary weight and balance information for the landing configuration.

TAXI TESTS:

FLIGHT #	DATE	OV CREW	MAXIMUM SPEED	BRAKING SPEED
1	15 FEB 77	—	89	27
2	15 FEB 77	—	140	23
3	15 FEB 77	—	157	57

CAPTIVE-INACTIVE FLIGHTS:

FLIGHT #	DATE	OV CREW	DURATION	TAILCONE	SPEED	ALTITUDE
1	18 FEB 77	—	2 HR 05 MIN	ON	287	16,000
2	22 FEB 77	—	3 HR 13 MIN	ON	328	22,600
3	25 FEB 77	—	2 HR 28 MIN	ON	425	26,600
4	28 FEB 77	—	2 HR 11 MIN	ON	425	28,565
5	02 MAR 77	—	1 HR 39 MIN	ON	474	30,000

CAPTIVE-ACTIVE FLIGHTS:

FLIGHT #	DATE	OV CREW	DURATION	TAILCONE	SPEED	ALTITUDE	COMBINED WEIGHT
1	18 JUN 77	H/F	55 MIN 46 SEC	ON	208	14,970	508,000
2	28 JUN 77	E/T	62 MIN 00 SEC	ON	310	22,030	557,080
3	26 JUL 77	H/F	59 MIN 53 SEC	ON	311	30,292	565,600

FREE-FLIGHTS:

FLIGHT #	DATE	OV CREW	DURATION	TAILCONE	LAUNCH SPEED	LAUNCH ALTITUDE	LANDING SPEED	OV INERT WEIGHT	OV LANDING WEIGHT
1	12 AUG 77	H/F	5 MIN 21 SEC	ON	310	24,100	213	127,144	149,574
2	13 SEP 77	E/T	5 MIN 28 SEC	ON	310	26,000	225	127,144	149,574
3	23 SEP 77	H/F	5 MIN 34 SEC	ON	290	24,700	221	127,144	149,971
4	12 OCT 77	E/T	2 MIN 34 SEC	OFF	278	22,400	230	127,459	150,876
5	26 OCT 77	H/F	2 MIN 01 SEC	OFF	283	19,000	219	127,459	150,846

NOTES:

OV CREW H/F WAS FRED W. HAISE JR. AND CHARLES G. 'GORDON' FULLERTON.
OV CREW E/T WAS JOSEPH H. ENGLE AND RICHARD H. TRULY.

SCA CREW FOR ALL FLIGHTS INCLUDED FITZHUGH L. FULTON, JR. (P), THOMAS C. MCMURTRY (P),
LOUIS E. GUIDRY, JR. (FE) AND VICTOR W. HORTON (FE).

SPEEDS ARE IN MPH, ALTITUDES IN FEET AGL, WEIGHTS IN POUNDS.

Upon the successful completion of the first two captive-active tests, NASA revised its plans to stand down for 15 days to modify *Enterprise*. The modifications included connecting the reserve hydraulic reservoir system, replacing the #1 and #3 APUs, swapping the #5 (spare) GPC, and various minor updates to other on-board systems. The revised plan accomplished this work while the Orbiter and SCA remained mated in the mate/demate device, instead of unloading the Orbiter from the SCA, performing the mods, then remating the vehicles. This change saved four days.[54]

The third captive-active flight was flown on 26 July 1977 with Haise and Fullerton in the Orbiter for a 07:47 take-off. This was a 'dress-rehearsal' for the first free flight and consisted of once around an oval measuring 84 miles on the straights and 24 miles through the curves. During the climb-out various avionics checks and flight control surface deflections were conducted. Shortly thereafter the Orbiter caution and warning system detected a malfunction in the #1 APU, which Fullerton shut down. The trouble was later traced to a faulty sensor, and the shut-down had no impact on the remainder of the flight. The 747 pitched down from 30,292 feet, performing the maneuver that would be flown during an actual separation. The mated vehicle then continued to descend on essentially the same profile that would be flown by the Orbiter in free-flight. The pair landed on runway 22 with no problems. After landing, while still mated to the SCA, Haise deployed the Orbiter landing gear as a final check prior to the first separation. These flights revealed no reason not to proceed with the free flights, and the last two tailcone-on captive-active flights were cancelled.[55]

During the week of 8 August 1977, a two-day Shuttle Readiness Review was completed – all conditions were 'go.' Shortly after 08:00, 12 August, Fitzhugh L. Fulton, Jr. and Thomas C. McMurtry guided N905NA down runway 22 with Haise and Fullerton back at the controls of *Enterprise*. At 08:48, while flying at 310 mph, Fulton

Enterprise just after release from the SCA for free-flight #4 on 12 October 1977. This was the first flight without the tailcone covering the Orbiter main engines. Just over 2.5 minutes later the Orbiter would land on the lakebed runway at Edwards. (NASA)

A flying brick. Enterprise returns from free-flight #4 without the tailcone. Note that the rudder/speedbrake is partially open. The extreme descent angle would become commonplace as shuttle became operational. (NASA)

pushed the nose of the 747 down seven degrees, and Haise detonated the seven explosive bolts holding the pair together. *Enterprise* separated cleanly and Haise put the Orbiter into a right-hand turn and pitched up. Haise held two degrees of pitch for three seconds, then banked 20 degrees to the right and towards runway 17. The Orbiter maintained a 9-degree nose down attitude as Haise and Fullerton executed two 90-degree turns. On final, Haise pointed the Orbiter down the centerline and opened the speedbrake. Descending at one foot per second, the main gear tires hit the lakebed at 213 mph, and the Orbiter rolled 11,000 feet with minimal usage of the wheel brakes. The first flight of *Enterprise* had lasted just over five minutes, with no serious problems.[56]

Engle and Truly were aboard when *Enterprise* flew again on 13 September 1977 for a series of more extensive maneuvers. *Enterprise* was stable as it began a wide 55 degree turn with its nose angled up three degrees. Truly took control and flew the vehicle down to 2,000 feet, handing off to Engle for a final series of maneuvers with the speedbrake open 40 percent. Five minutes and 28 seconds after leaving Fulton and McMurtry in the 747, the Orbiter settled onto the lakebed, rolling out 10,037 feet in just over a minute. The most serious problem during the flight was not in either of the vehicles – the radar at Dryden failed 28 minutes into the captive portion of the flight, and almost caused an abort before it was brought back on-line. The Orbiter pilots reported that the use of elevons after nosewheel touchdown was not as effective in steering the Orbiter as predicted, but other than that, the vehicle had performed well in flight.[57]

During the third free flight, on 23 September 1977, Haise and Fullerton tested the autoland system, letting the general purpose computers (GPC) fly the Orbiter down to the 900-foot pre-flare. The crew hard-braked the vehicle on lakebed runway 17 without any problems. The first three free flights were deemed so successful that the final tailcone-on flight was cancelled. The original plan of conducting a mated flight with the tailcone off was also scrubbed – if the buffeting from the mated vehicles became excessive, they could simply abort the flight and land on the lakebed. The time had come to see how *Enterprise* would fly in a return-from-space configuration.[58]

The fourth free flight took place on 12 October 1977, and this time the Orbiter flew like the brick it was, with a flight time under half of what it was with the tailcone on. The buffeting in the 747 as the Orbiter pulled away was moderate but acceptable, and the handling qualities of the Orbiter were basically the same, with the exception of the more rapid descent. Engle and Truly experienced some problems with the TACAN system, and some minor error messages from the GPCs, but the flight cleared the way for the final test – landing *Enterprise* on the concrete runway.[59]

The fifth, and last, free flight was on 26 October 1977 with Haise and Fullerton at the controls of *Enterprise*. The Orbiter separated 51 minutes after take-off at an altitude of 19,000 feet. As before, the crew put the Orbiter through a series of maneuvers, finding again that the gliding performance of the Orbiter was better than predicted. This time, however, as the crew set up their final approach, trouble began. Coming out of the pre-flare, *Enterprise* was dropping at 334 mph, considerably quicker than planned. In an attempt to slow the Orbiter, Haise opened the speedbrake early, but instead of slowing down, speed increased, so Haise deployed the landing gear and pitched the nose down to make the desired touchdown point. Since *Enterprise* was unpowered, there was no option of making a second pass. The wings dipped and Haise struggled to correct as the rear wheels hit the tarmac hard. Instead of dropping the nose, *Enterprise* suddenly took off again, bouncing 20 feet into the air. After a few anxious seconds, the Orbiter smoothed out and stabilized for the remainder of the roll-out, with the nose finally dropping onto the runway – a classic case of pilot-induced oscillation (PIO).[60]

Regardless, NASA announced that the ALT program had met its objectives, and no further flights would be necessary. The ALT flights had also largely validated the various aerodynamic data bases and wind tunnel predictions, giving engineers greater confidence that they truly understood how the Orbiter would fly in all flight regimes. The end result was that the Orbiter was aerodynamically cleared for flight into space. What remained was to finish testing other systems, such as the SSMEs, that were beginning to lag far behind their development schedules, and were putting the anticipated 1979 first manned orbital flight in jeopardy.[61]

Following the ALT flights, technicians reinstalled the tailcone on *Enterprise*, drained and purged the fluid systems, removed the flight-test air-data probe from the nose, and installed locks on the elevons to secure them in place. The forward attachment strut on N905NA was replaced with the lower ferry unit, and during mid-November 1977, Fulton and the 747 crew completed a series of four test flights with the Orbiter in the ferry configuration. OV-101 was then returned to the NASA hanger at Dryden and modified for upcoming tests at the Marshall Space Flight Center in Alabama. On 13 March 1978, the SCA ferried *Enterprise* to Redstone Arsenal Next to MSFC for the mated vertical ground vibration tests (MVGVT) – as predicted, landing on the short runway produced no surprises. These tests used a set of exciters and sensors placed on the skin of the mated elements to create and monitor vibrations and resonances similar to those that would later be encountered during powered ascent.[62]

A sensor system, known as the shuttle modal test and analysis system (SMTAS), provided automatic control of up to 24 preselected exciter channels from the available thirty-six 150-pound and twenty 1,000-pound exciters that applied precise pressure at preselected points on the vehicle. The mostly-enclosed test facility had been constructed in 1964 to conduct similar tests on the Saturn V. It was modified during 1975–77, enlarging it from 98 by 98 by 360 feet to 98 by 122 by 360 feet to accommodate the winged space shuttle instead of the cylindrical Saturn V. The total cost of the modifications was $2,880,000.[63]

The first test configuration included *Enterprise* and an ET to simulate the high altitude portion of ascent prior to orbital insertion, but after SRB separation. During these tests, the LO2 tank contained between 3,450 and 101,000 gallons of deionized water, and the LH2 tank was pressurized but empty. The combined Orbiter–ET weighed 1,200,000 pounds and was suspended by a combination of air bags and cables attached to the top of the test stand.[64]

The second test configuration added a set of SRBs containing inert propellant to simulate lift-off conditions. This marked the first time that a complete set of dimensionally correct elements of the Space Transportation System had been assembled together. The LO2 tank was filled with 140,600 gallons of deionized water, and the test article weight of 4,000,000 pounds was supported by four hydrodynamic stands, two under each SRB. These stands contained 1,000 gallons of special oil pressurized to 1,500 psi, and bearings on top of the stands created the 'floating' characteristics desired for the tests. The stands were originally used for the Saturn V tests, and were refurbished by the Denver Division of Martin Marietta. The test in the lift-off configuration was completed on 15 September 1978, and in the burn-out configuration on 5 December 1978.[65]

The final test configuration was similar to the second except that the SRBs were empty and the LO2 tank held 101,000 gallons. These tests simulated the period of flight just prior to SRB separation, and the vehicle was supported using the same system as in the second test series. These tests were completed on 23 February 1979. The engineer in charge of the tests, Eugene Cagle, said that the testing "provided a detailed understanding of the structural dynamic characteristics of the vehicle. This gives us confidence that the Shuttle will withstand the vibratory forces it will encounter during powered flight …" One of the changes that resulted from the test was the addition of strengthening brackets in the forward portion of the SRBs where the rate gyros are mounted.[66]

The modal data extracted from these tests compared favorably with the data derived from the 1/4-scale model testing when the known configuration differences were considered. The major result of these tests was the iden-

A stiff-leg derrick system was designed by KSC for use at MSFC and was essentially a collection of commercially-rented derricks positioned so the Orbiter could be lifted while N905NA was backed from under it. This is similar to what would be used at any of the contingency landing sites around the world. (NASA photo 108-KSC-78PC-33)

Enterprise is prepared for hoisting into position for the mated vertical ground vibration tests at MSFC. Noteworthy is the fact that the simulated OMS pods have been replaced by mass weights and the cockpit windows are covered to prevent accidental damage. (NASA photo S78-35589)

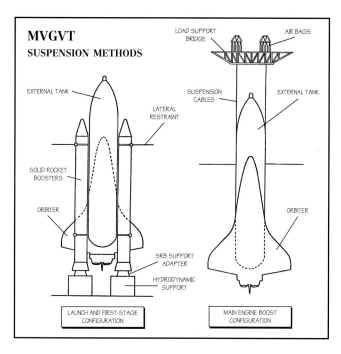

MVGVT
SUSPENSION METHODS

LOAD SUPPORT BRIDGE

AIR BAGS

EXTERNAL TANK

LATERAL RESTRAINT

SUSPENSION CABLES

EXTERNAL TANK

SOLID ROCKET BOOSTERS

ORBITER

ORBITER

SRB SUPPORT ADAPTER

HYDRODYNAMIC SUPPORT

LAUNCH AND FIRST-STAGE CONFIGURATION

MAIN ENGINE BOOST CONFIGURATION

tification of local resonances in the area of the SRB rate gyro locations that had the effect of corrupting the sensor signals. If occurring in flight, this would have had the same result as a malfunctioning sensor. Similar anomalies were noted on the Orbiter side-mounted rate gyros. The only major issue not resolved concerned the SRB propellant stiffness. Because of the nonlinear viscoelastic properties of the SRB propellant, the eventual resolution of this problem was to adjust the propellant shear stiffness

in the analysis to agree with test data. Fortunately, the structural modes with significant SRB propellant motion had relatively high damping, and were not significant in the performance of the vehicle.[67]

At this point, *Enterprise* was supposed to have gone back to Palmdale to be refitted as a flight Orbiter. But several things convinced NASA to change its plan. As OV-101 was being built, numerous lessons were learned regarding the design and the materials used in its construction. All subsequent Orbiters (from STA/OV-099 on) would use wings and a mid-fuselage significantly stronger than those installed on *Enterprise*, and some aluminum castings in other areas of the fuselage were changed to titanium as a weight saving measure. The wings would need to be returned to Grumman in New York, the mid-fuselage to Convair in San Diego, and the aft-fuselage to Downey – all expenses the program did not have funds to cover.

The need to retrofit these changes on *Enterprise*, as well as the time and expense required to disassemble her, led to a decision to modify the structural test article (STA-099) instead. Since STA-099 was never completely assembled, there would not be as much rework required, and most of the structural and materials changes had been incorporated during production. This decision had apparently been made during late 1977, and the STA test sequence was modified to ensure the airframe was not damaged. Some estimates have indicated this may have saved the program as much as $100 million.[68]

So instead of heading to California, *Enterprise* was loaded aboard the SCA for the one hour and 52 minute flight to the Kennedy Space Center on 10 April 1979. While at KSC the Orbiter would be used to check out facilities and procedures that would support *Columbia* during STS-1.

Enterprise being mated to the ET and SRBs for the MVGVT at MSFC. This was the first time that a dimensionally correct stack had been assembled. (NASA photo MSFC-00687)

The bottom of Enterprise shows that there were not doors over the aft ET attach points. (NASA photo 108-KSC-78PC-412)

On 3 April 1978 the space shuttle 'orbiter weight simulator,' more commonly known as Pathfinder, arrived at the KSC turn basin. The vertical stabilizer and the out wing sections have been removed. Note the 'NASA MSFC' markings on the aft fuselage – these would quickly be repainted at KSC. (NASA photo 108-KSC-78P-63)

The orbiter simulator did not look like much, but it served its purpose. Note that the 'NASA MSFC' has been replaced by 'NASA' and that a set of simulated OMS pods has been added. The simulator was used to check out the OPF, VAB, and convoy operations at the Shuttle Landing Facility. (NASA photo 108-KSC-379-104)

Pathfinder had been built in order to validate MSFC facilities being used for the MVGVT tests on Enterprise, such as the MSFC Dynamic Test Stand. (NASA photo MSFC-00395)

Fit checks were performed in OPF-1 to ensure the work platforms were positioned correctly and would not hit the Orbiter when used. (NASA photo 108-KSC-78P-300)

Prior to the arrival of *Enterprise*, a 75 ton 'orbiter weight simulator' had been used to practice lifting and handling the Orbiter. This simulator was a steel structure roughly resembling the shape of an Orbiter, and was more or less dimensionally correct. Dubbed *Pathfinder*, it was originally constructed at MSFC during 1977 as a stand-in for the *Enterprise* to fit-check the roads and facilities that were used during the mated vertical ground vibration tests, as well as the hoisting system that was used to lift OV-101 into the test facility. During mid-1978, the *Pathfinder* mock-up was shipped by ocean barge to KSC where it was used to fit-check the mate/demate device, OPF and VAB work platforms, and for ground crew training. In early 1979, *Pathfinder* was used to rehearse post-landing procedures at the shuttle landing facility, allowing convoy operations to be simulated with some accuracy. Later during 1979 *Pathfinder* was returned to MSFC for storage.[69]

Several years later, a group of Japanese businessmen expressed an interest in obtaining a full-scale orbiter model for display during the 'Great Space Shuttle Exposition.' The Japanese readily agreed to provide almost $1,000,000 to have Teledyne-Brown Engineering modify the ungainly orbiter simulator to more closely resemble an Orbiter. Following display in Tokyo, *Pathfinder* was returned to MSFC for permanent display at the nearby Space and Rocket Center in Huntsville. It is mounted on a special platform, mated to the External Tank (MPTA-ET) used during the main propulsion tests, and a pair of inert

Pathfinder is now on display at the Space and Rocket Center just outside the gates of MSFC, mounted on the MPTA-ET, along with a pair of filament wound SRBs from the attempt to create lightweight SRBs for Vandenberg. The markings are generally duplicates of those found on Columbia, including the black wing chines. (NASA)

Enterprise went through an entire 'flow' at KSC, including the trip to Pad 39A on 1 May 1979. Photos of OV-101 on the pad in 1979 are easily distinguished from Columbia in 1980 by the lack of black chines. (NASA photo KSC-79PC-0203 and KSC-79PC-0247)

A newly-refurbished Pathfinder at the Great Space Shuttle Exposition in Tokyo. (NASA)

Solid Rocket Boosters. The MPTA-ET had been delivered to NASA on 2 March 1977 for a nine-month test series on test stand B-2 at the NSTL – it ended up remaining there for over ten years and conducting 12 three-engine cluster test firings totaling 3,810 seconds. It was finally removed on 16 February 1988 and sent via Michoud to MSFC where it was visually refurbished by Martin Marietta and readied for display. The SRBs were assembled from composite casings produced during the attempt to build filament-wound motors for Vandenberg, as well as various engineering and structural tests pieces, such as nose segments and aft skirts. Interestingly, in early 1999 NASA asked that the nose segments and aft skirts be returned. These items were refurbished for use by the Space Shuttle Program, and a set of mockups was provided to replace them.[70]

During April 1979, OV-101 was mated to a pair of inert Solid Rocket Boosters and the ET (serial number 2) scheduled to be used on STS-1. The stack was transported atop MLP-1 to Pad 39A on 1 May 1979. During almost three months at the pad, *Enterprise* would help verify that maintenance platforms were in the correct locations, and that crew escape procedures worked properly. On 23 July 1979, *Enterprise* was rolled back to the VAB and demated from the SRBs and ET. *Enterprise* arrived at Vandenberg AFB on 15 August 1979 to let the workers building the launch complex there have a look at an Orbiter. The next day OV-101 was flown to Edwards, demated from the SCA on 23 August, and moved overland to Palmdale on 30 October. While at Palmdale, selected parts of *Enterprise* were removed and refurbished for use on later Orbiters as a cost saving measure. For the most part these items were controls, displays, and avionics from the flight deck, but a few other parts were removed also. Wearing a new coat of paint, OV-101 returned to Edwards on 6 September 1981.[71]

During May and June 1983, *Enterprise* was ferried to France for an appearance at the Paris Air Show, and also visited Germany, Italy, England, and Canada. It marks the only time to date that an Orbiter has been outside the United States. During the trip, the SCA was fitted with infrared countermeasures equipment to guard against possible terrorist missile attack. Immediately prior to this

Enterprise and N905NA returned to Edwards from Vandenberg on 16 August 1979 – the vehicle had spent less than a day at the western launch site. (Dennis R. Jenkins)

For two months during 1985, OV-101 sat alongside the Saturn V fit-check vehicle in the parking lot of the VAB at KSC. Tourists loved it. (NASA photo KSC-85PC-0541)

trip, both OV-101 and N905NA received new paint jobs and slightly modified markings. On OV-101 this included substantially more black areas on the vertical stabilizer and forward fuselage in an attempt to look more like *Columbia*. The new paint also finally eliminated all traces of the 747's previous association with American Airlines.[72]

In late March 1984, OV-101 was ferried to Mobile, Alabama, and on 5 April 1984, the Orbiter was floated down the Mississippi River on a barge to the World's Fair in New Orleans, Louisiana. Cranes carefully lifted the Orbiter onto a display stand where *Enterprise* remained through the end of the Fair.

During late 1984 and early 1985, OV-101 was again called to be a pathfinder. Mated to a pair of inert SRBs, and the ET scheduled for STS-3V, *Enterprise* performed in a series of flight vehicle verification (FVV) tests at the Vandenberg Launch Site (VLS).* These tests gave preliminary indications that all of the facilities at Space Launch Complex Six (SLC-6) and in the orbiter maintenance and checkout facility would support the planned launch of OV-103 during 1986. Although on a somewhat smaller scale than KSC, VLS was a fully-equipped launch and landing site. During early 1986, OV-102 was scheduled to arrive to conduct tanking tests and a flight readiness firing (FRF) to complete validation of the launch complex. This was cancelled after the loss of *Challenger* as "… too risky

---

* The official name was the Vandenberg Launch and Landing Site, but this was always abbreviated VLS. Regardless, most documentation, but official and unofficial, simply used Vandenberg AFB or VAFB.

…", and also to save the $60 million cost. *Discovery* was to fly the first mission from Vandenberg and be permanently based there, but all launch preparations were suspended before this occurred. See Appendix C for more details.[73]

After completion of the Vandenberg tests, OV-101 was ferried to Dryden on 24 May 1985. Following another ferry flight on 20 September 1985, OV-101 was put on display next to the Saturn V SA-500F fit-check vehicle at KSC for several months while a place at Dulles Airport was prepared. *Enterprise* was officially transferred to the National Air and Space Museum (NASM) on 18 November 1985.

But this was hardly the end of NASA's use of OV-101. During the week of 8 June 1987 a landing barrier, similar to ones used by the military to catch damaged aircraft, was erected at Dulles and *Enterprise* was slowly winched into it to determine if an Orbiter could successfully use one. The tests of the shuttle orbiter arresting system (SOAS) were successful, and barriers are kept at the three primary TAL sites (Moron and Zaragota, Spain, and Banjul, The Gambia), although none are installed at any of the primary landing sites (KSC or Edwards). Later in 1987, OV-101 was used for tests of the various crew bail-out concepts being investigated in response to the *Challenger* accident.[74]

OV-101 was used to test the shuttle amateur radio experiment (SAREX) during 1990 when the team needed to check-out a new antenna design. None of the operational Orbiters were available for the tests, and none of the mock-up windows used the proper glass materials. So the team went to Dulles and mounted the antenna on the inside of one of the windows on *Enterprise* to test the

Enterprise and N905NA, flown by pilots Joe Algranti and A. J. Roy, returned from the Paris Air Show on 15 June 1983 – here the pair is shown at Dulles Airport. Noteworthy is the '376' marking on the fuselage of the 747 – its exhibit number at Paris. A NASA 25th anniversary logo is just behind the forward 747 door. (NASA photo 83-H-453)

OV-101 sits on a display stand at the 1984 World's Fair in New Orleans during April 1984. The bi-pod supporting the nose would find later use while Enterprise was in storage at the Smithsonian – interestingly, the bipod was theoretically capable of securing the Orbiter to the ground during a hurricane. (NASA)

Enterprise at the Vandenberg Launch Site. The 76-wheel transporter was designed to keep the Orbiter level while traveling over the hills between the runway facility (located on North VAFB) and the SLC-6 launch site (located on South VAFB). (Dennis R. Jenkins)

NASA flew Enterprise and the SCA in for a second visit to Vandenberg in October 1984. A year later the Orbiter would spend a considerable amount of time at the launch site for the flight vehicle verification tests. (Dennis R. Jenkins)

radio gear. SAREX has subsequently flown several times, and STS-63 used an antenna based on the SAREX design for communications with *Mir*.[75]

In April 1990 NASA 'borrowed' the main landing gear from *Enterprise* for use in a series of tests at NASA Dryden involving the Convair 990 landing gear research aircraft. The main gear was sent to NASA Dryden in anticipation of attaching it to the test fixture on the 990 in order to conduct 'full up' landing gear, wheel, and tire tests. But to use the OV-101 main gear would have required cutting off the upper two feet of the main gear trunion attach fittings in order to clear the available height/space in the belly of the 990. This was considered a drastic thing to do to a $500,000 part knowing that it would have to be rewelded before it was returned for use on OV-101 – NASA always agrees not to 'visually alter' a part borrowed from *Enterprise*. Fortunately, NASA held off performing this modification to the main gear because sufficient data was obtained from the 990 by conducting single tire tests using the existing *Challenger*-based test fixture. It was finally determined that full-up main gear tests were not necessary. The main gear was returned to NASM during March 1996, and was reinstalled on OV-101 in June 1997, except for the wheels and tires. The wheels are currently being refurbished by KSC technicians who will also install newer tires.[76]

A growing concern is the effect of small objects hitting the Orbiter while on-orbit. In 1993 NASA contracted with Rockwell to study the potential effects of these impacts

using the hypervelocity test facility at JSC. To be as accurate as possible in their assessment, Rockwell wanted to test actual flight hardware. Most of the test articles were obtained from discarded flight hardware or excess spare inventory, including various test pieces that had never flown but had been built to flight specifications. But one item could not be obtained from these sources – a flipper door used to cover the elevon actuators. The search even went so far as to look through the records to see if an intact flipper door had been recovered from the

After being retired to the Smithsonian, Enterprise was used to determine if Orbiters could successfully use an arresting barrier in the event of a brake failure. During June 1987, OV-101 was slowly winched into the shuttle orbiter arresting system (SOAS) barrier temporarily erected at Dulles Airport in Washington, D.C. (NASA photo S87-35883)

Enterprise seen from the top of the Access Tower at Vandenberg. The Crew Cabin Access Arm is in place, allowing technicians to enter the Orbiter's crew cabin. Unlike KSC where the arm 'swings' away from the vehicle, at Vandenberg the arm retracted straight back away from the Orbiter. (Dennis R. Jenkins)

Challenger debris, but none had. JSC therefore decided to borrow one from OV-101 for the tests. Door number 4 from the starboard wing was borrowed in July 1993, refurbished at Rockwell in Downey, then sent to JSC for the impact tests. At the conclusion of the tests and subsequent analysis, the door was returned to Downey, restored to its original visual condition, then reinstalled on *Enterprise* during March 1994.[77]

On 5–9 February 1996 a team from JSC inspected *Enterprise* to determine its structural condition since there were proposals that involved refurbishing the airframe for use as an additional flight vehicle. Although the team could not conduct a thorough inspection due to the way OV-101 was positioned in its temporary storage location at Dulles, a fairly complete visual inspection was accomplished. Overall, the team found the vehicle to be very well preserved given the minimum amount of environmental control over the years. The cavities around the body flap and elevons did not contain any significant corrosion, but the interior of the aft fuselage had numerous areas of corrosion, primarily at the floor, X_O1307 bulkhead, and the base heat shield – the team associated this with fasteners corroding in earlier standing water, but felt the damage was repairable. The wings were found to be in acceptable condition, with only minor corrosion on the aluminum-honeycomb leading edge spar, an area that has been a problem on other Orbiters. The crew module and FRCS

The payload bay of Enterprise photographed in 1997 at the Smithsonian from the crew module hatch. Note the high-pressure gas bottles (many of which had already been removed) along the side of the bay. (NASA photo KSC-387C-1458.22 via Chris Hansen)

cavity were satisfactory with the exception of significant corrosion around the nose wheel well that had probably been caused by entrapped and standing water at some point. The only area with potentially significant corrosion was the lower forward fuselage, but limited access did not allow the team to define the full extent of the problem. In all the team felt the vehicle was in "fairly good condition" considering its long-term exposure to the elements.[78]

For two weeks at the end of June 1997, JSC engineers conducted an evaluation of the structural integrity of the payload bay doors on OV-101 to determine if the composite construction had degraded due to environmental exposure over the years. JSC was interested in the conditions of the doors in case a replacement is needed in the future for an operational Orbiter, and the non-destructive evaluation was accomplished using state-of-the-art shearography inspection techniques. The inspection showed that the OV-101 doors are still in serviceable condition and could be used after refurbishment if required.[79]

Subsequently, *Enterprise* would also contribute to the Shuttle Upgrades Program. One of the areas of interest is improving the landing and deceleration systems on the Orbiter – primarily relating to the tires. Due to weight growth during development, the Orbiter landing gear are undersized for the vehicle weight and landing speed, and these effects are amplified by the negative 4.5-degree angle-of-attack after nose pitch-down. As the nose pitches over, the aerodynamic loads apply a tremendous pressure on the tires and gear – in fact, the tires experience well in excess of their 'rated' loads.* The available test data combined with analysis indicate there is very little margin in the tires. Depending on exactly when a failure might occur, it is possible that one tire failing could also damage the tire next to it on the strut, essentially eliminating the ability to roll on that side of the Orbiter. Because of this, a great deal of time and effort is spent to protect the tires – they undergo pre-flight X-ray inspections and the mains are only used for one landing (the nose gear tires get used twice). The Shuttle Upgrades Program has awarded Michelin a contract to develop a tire with a higher rated load.[80]

Another solution is to lengthen the nose gear, thus lowering aerodynamic loads after nose pitch-over. This concept had been examined during the late 1970s, and again

This was as close as Vandenberg would get – Enterprise, a pair of inert SRBs, and a real ET. Unlike KSC where the vehicle is stacked in the VAB and transported to the pad, at Vandenberg everything was stacked in-place. (Dennis R. Jenkins)

* The rated load doesn't mean much, particularly the way the Shuttle program uses tires, but is the value at which the manufacturer (now Michelin) certifies that the tires have infinite life (at least in terms of structural failure, not tread wear).

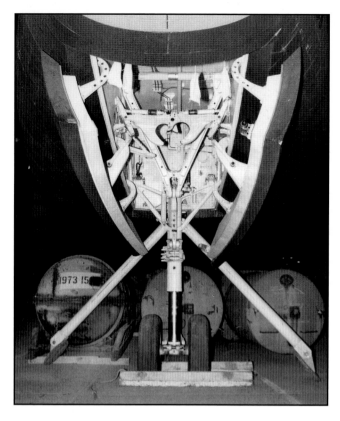

This is the nose gear that was on Enterprise in June 1997. The bipod is the same one that was used at the World's Fair, and would be secured to the concrete before all of the Orbiter's weight was put on it. Note the uplocks in the wheel well and the lack of hydraulic retraction devices. (NASA photo KSC-397C-1592.26 via Chris Hansen)

The modified nose gear in the test rig at JSC in 1999. The silver corrugated-looking part is the 24-inch extension. When the nose gear is retracted into the wheel well, the extension is compressed so that the nose strut is no longer than normal. As the gear is lowered the extension expands and locks into place. (NASA photo via Chris Hansen)

as recently as 1992, but all of the proposals suffered from some complication that made implementation too difficult. The problem is that it is not possible to make the nose wheel well larger, and the existing well can not accommodate a larger gear.[81]

Engineers at JSC began by using the landing simulators at JSC and Ames to evaluate the effect that a longer nose gear had on the handling qualities of the vehicle and the loads on the gear itself. Lengths from 17 inches to 52 inches were studied, and it was determined that 24 inches was the optimal length, and provided substantial loads relief for the main gear tires. The concept developed by the engineers at JSC was to remove the nose gear strut and replace it with one of the same length but a somewhat larger diameter. Inside the larger strut is a second, telescoping strut that contains all of the same energy absorbing devices already in use. The new telescoping strut would automatically extend 24 inches and lock in place when the nose gear was deployed prior to landing.[82]

Building a functioning laboratory prototype was the next step. In order to keep costs down, it was decided to use the original test fixture that had cycle-tested the nose gear to certify the design for flight, and to use the nose strut assembly from OV-101. The test fixture was an exact replica of the wheel well, and the gear on *Enterprise* was even better than a replica, it was flight-like gear. This was important, because the new design relied on using as many parts from the old gear as possible to save cost – the pistons, axles, wheels, drag links, and all of the uplock hardware would be integrated with a newly-designed strut.[83]

In June 1997 a JSC–KSC team went to Dulles to borrow the nose gear from *Enterprise*, and spent three days removing every piece of hardware from the wheel well. Fortunately, the NASM had also received the bipod that

was used to support OV-101 during its barge trip to the World's Fair in New Orleans. The nose gear shock strut was used as a jack to lift the vehicle, allowing the bipod to be placed under the nose. Holes were drilled in the cement floor to secure the bipod, and the nose strut was removed, packed, and shipped to JSC.[84]

Once the hardware arrived at JSC, it was installed on the test fixture to verify everything fit correctly. As it turned out, the nose gear that had flown on *Enterprise* during the ALT flights had been removed and used as spares[*] for the operational vehicles. The gear that had originally been used to certify the cycle-life had been installed on OV-101 before the Orbiter was delivered to the museum. After carefully checking, cleaning, repainting, and assembling all of the components to make sure they worked, engineers began modifying the strut. Modifications complete, the new strut was installed in the test rig and evaluated – the idea worked, fit in the nose well, and was economical. But the modification potentially introduces several new failure modes, and program management is committed to allowing Michelin to develop new main gear tires instead.[85]

Subsequently, another use was found for this hardware. JSC became aware that the cycle-life of the operational nose gears was rapidly approaching. The gear are cycled more than originally anticipated, and NASA needed to establish a new cycle-life. The only option seemed to be to remove a gear from a flight vehicle and re-test it, an action that would impact operations. Instead, the engineers at JSC suggested using the refurbished test rig to re-certify the gear (after all, it had been used to certify it originally).

---

[*] This was not true of the main gear since it had not had been flight certified, despite its usage on the ALT flights.

The OV-101 nose strut gear was demodified (nothing permanent had been done) and refurbished to flight standards. As of late-2000 the test rig had surpassed 4,000 cycles, and probably saved several million dollars. As soon as the recertification is complete, it is planned to return the nose gear to NASM and reinstall it on OV-101.[86]

Perhaps the most intrusive use of *Enterprise* as a test-bed was during April–May 1999. A JSC–Boeing team used OV-101 for a proof-of-concept demonstration for a proposed FRCS interconnect system (FICS) modification. This change (since approved) will allow propellant to be moved between tanks in the OMS pods and the Forward RCS (FRCS) module by running transfer lines between them. The team removed the simulated TPS tiles from the right side of the forward fuselage. The splice plate that connects the upper and lower fuselage halves together was removed by releasing 300 fasteners, and the thermal control system blankets under the splice plate were also removed. The FRCS mockup (called a simulator) was lifted from the nose to gain access to the X_O376 bulkhead. Three holes were drilled in this bulkhead – two for the propellant lines and one for access while working. The propellant lines would run through the forward fuselage – under the skin but outside the crew module.[87]

The team temporarily installed brackets along the frame end caps and fabricated the propellant lines using seven sections of flight-type tubing. The lines were installed, welded, and inspected to ensure that the entire procedure could be accomplished without damage to the flight vehicles, and that sufficient access could be gained in all locations to perform the Astro Arc welding and X-ray inspection. The temporary brackets were then removed, and Korpon primer was applied to all the holes that had been drilled to protect against corrosion. All but two of the gold TCS blankets were reinstalled – one was retained by the NASM for display purposes, and one slipped between the outer skin and crew module and fell to the lower vehicle centerline where it could not be retrieved. The splice plate and simulated TPS were reinstalled and painted to match the rest of *Enterprise*.

After problems were discovered at KSC with the wiring used in the operational Orbiters, NASA returned to Dulles in October 1999 to remove some samples since "the wiring in *Enterprise* represents the oldest specimens of Kapton wiring available to NASA." The intent was to determine how Kapton wiring degrades as it ages. Eight samples were removed from OV-101 – 3 feet from each the crew module, mid-deck, and avionics bay; 6 feet from the mid-fuselage; and four samples of 6 feet each from the aft-fuselage. Unlike most of the other modern uses of *Enterprise*, the wiring removal was permanent, although NASA made "every attempt to select harnesses which can be removed in the least intrusive fashion and that minimize the effect on the appearance of the vehicle."[88]

Other parts were removed from OV-101 prior to the Orbiter being turned over to the NASM, including: the ejection seats, most of the cockpit instrumentation and consoles, the control sticks, most of the avionics, many of the high-pressure bottles in the mid-bay, the flight-test boom that had been installed on the nose, and the movable water ballast (although the nose ballast is still installed). Serviceable parts were returned to the program for use as spares for the flight vehicles.[89]

In late 2000 the Smithsonian National Air and Space Museum began construction on the new $153 million Steven F. Udvar-Hazy Center, south of the main terminal at Dulles Airport. The 711,636-square-foot facility will sit on 176.5 acres and include exhibit hangars, an observation tower to watch activity at Dulles, collections storage, classrooms, archives, a large-format theater, restaurants, and gift shops. The main purpose of the new facility is to display hundreds of aircraft and space artifacts that have been in storage because the Museum did not have sufficient room in the main building in downtown Washington DC. *Enterprise* will be one of the showcase exhibits in the new facility when it opens in 2003.[90]

N905NA and Enterprise circa 1980. Noteworthy are the windshield wipers on the 747, and the patch on the nosecone of OV-101 where the flight-test air data probe was originally attached. (Dennis R. Jenkins)

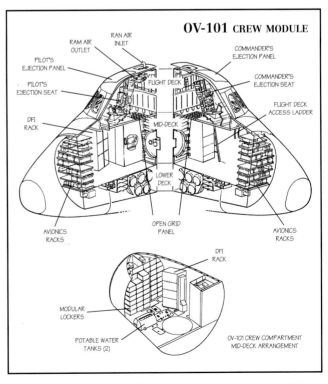

The crew module of Enterprise was tailored for its initial role of an atmospheric flight test vehicle. Many of the components needed for space flight were missing. (NASA)

When Enterprise was transported to the Paris Air Show in 1983, the 747 was fitted with infrared countermeasures (IRCM) on each engine pylon – note the small canisters at the upper aft edge of each pylon – to protect against possible missile attacks by terrorists. Fortunately, in the end the precaution was unnecessary. (NASA photo ECN-24314)

OV-101 being worked on in at Edwards in 1983 – note the ballast cradle being lifted out of the payload bay. (NASA photo EC83-22740)

While at Edwards, Enterprise shows its payload bay. Note the lack of windows on the aft flight deck, and the unfinished payload bay interior. (Tony Landis Collection)

A partially disassembled tailcone fairings at NASA Dryden during 2000. The tailcone can be packed into crates and trucked to wherever it is needed after the Orbiter lands. Current procedures state that it is transported by ship to the TAL sites, but in actuality it would most probably be transported by a C-5 or C-17. (NASA photos by Tony Landis)

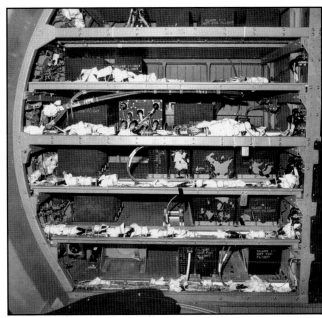

NASA used Enterprise to test-fit the FICS modification during early 1999. At top, one of the fuel or oxidizer lines running from the OMS pods to the FRCS is supported by a bracket in the forward fuselage. The lower photos shows both lines installed. (NASA)

Avionics bay number one on Enterprise during the 1996 inspection. All of the avionics boxes that remain are labeled 'Class III – Not For Flight' and were of no particular use to the program, explaining why they are still there. (NASA photo S96-01934)

The flight deck of Enterprise during the 1996 inspection. Everything that was serviceable had been previously removed, and little remains. Compare with the photo on page 208. Note the rudder pedals and rails for the ejection seats. (NASA photo S96-01956)

A view of the payload bay of Enterprise in 1996 reveals two T-shaped steel supports used to carry ballast. The supports attached to the keel and sill longerons just like any other payload carried by the Orbiters. (NASA photo S96-02000 via Julie Kramer)

An artist concept of the Nation Air and Space Museum's new Steven F. Udvar-Hazy Center at the Dulles International Airport, Virginia, outside Washington D.C. Naturally, Enterprise will be one of the star attractions when the new center opens in 2003. (NASM/Visualization by Interface Multimedia via Dr. Valerie Neal)

OTHER TESTING

Previous manned space programs had used complete spacecraft during vibration and acoustic testing, as well as complete reentry vehicle thermal testing. Initial proposals for space shuttle involved subjecting forward and aft sections of the Orbiter to vehicle-level acoustic tests in addition to the MVGVT conducted on OV-101. After detailed technical and programmatic examination, the differences with the past precedents became clear. Secondary structure and installations on the Orbiter were designed in accordance with the life-cycle requirements of a reusable vehicle, as opposed to previous spacecraft which were essentially throw-away vehicles. Generic installation concepts were used throughout the Orbiter, and required the development of additional verification procedures. The prevention and detection of acoustic fatigue in the primary structure emerged as a serious requirement to ensure reusability. Detection was complicated by the fact that major portions of the exterior of the Orbiter were covered with thermal protection, along with extensive use of internal insulation, which made regular inspection extremely costly and impractical.[91]

As a result, *Enterprise* was not the only test article built, and two other partial airframes were constructed – the main propulsion test article (MPTA-098) and a structural test article (STA-099) that was later modified into OV-099. During October 1973 a 20-foot long mid-fuselage section was used to measure internal stress distribution and heat transfer, allowing Rockwell to finalize the requirements for the flight articles. An entire upper forward fuselage also was constructed and sent to Holloman AFB on 26 October 1976 for rocket-sled testing of the shuttle ejection seats. Initial testing commenced on 18 November 1976 and several unmanned ejections were successfully accomplished at speeds between zero and 450 mph beginning on 11 January and finishing on 5 May 1977. This cleared the use of the ejection seats in *Enterprise* during the ALT flights. Further testing was

accomplished at Holloman during April 1980 to clear *Columbia*'s seats for the orbital flight test series.[92]

Additionally, a right-side OMS pod (designated RPST, later rebuilt into RP03), a vertical stabilizer, a liquid oxygen tank, and an Orbiter nose cap test article were built and tested to ensure their design integrity, and also to feed data for verification of the structural models.

The Thermostructures Research Facility at NASA Dryden was utilized to apply mechanical and thermal loads to a test portion of an Orbiter wing and elevon. This test was intended to verify proper functioning of the elevon seals which were designed to prevent free stream air from entering the gap between the aluminum wing structure and the elevon during movement of the control surfaces. The Dryden tests verified that the design was adequate.

At the NASA White Sands Test Facility, located outside Alamogordo, New Mexico, tests were being conducted on the smaller engines that used hypergolic propellants. WSTF had been established to test the reaction control system on the Apollo spacecraft, and was performing a similar role on shuttle. Development testing of the forward reaction control system (FRCS) began on 4 November 1977, and was completed on 13 July 1978. The first orbital maneuvering system (OMS) pod arrived at WSTF on 3 July 1978, and development tests began on 21 July, finishing on 15 November 1978. The OMS pod was built by McDonnell Douglas with the OMS engines manufactured by Aerojet-General, while the RCS thrusters were made by Marquardt. The development testing of the RCS thrusters, which were uprated versions of a unit originally designed for the Air Force MOL program, was extensive – the primary thrusters were fired over 14,000 times, mostly in bursts of less than a second, and the verniers were fired over 100,000 times.[93]

Formal qualification testing of the OMS engine began on 9 December 1978 and was completed on 18 February 1980. Aft RCS qualifications began on 15 December 1978 and were completed on 1 February 1980, while the FRCS tests began on 4 May 1979 and finished on 20 February 1980. In addition to engine testing, WSTF also performed various acoustic, thermal, and vibration tests on the OMS pod and FRCS module to qualify them for flight.[94]

There were other tests, of course. The APU and fuel cells both presented challenges for their developers and each had prolonged test programs, although both had to be ready in time for the atmospheric flight tests. The APU, in particular, had more than its share of problems considering it did not attempt to be truly state-of-the-art (a lesson learned from the X-15 APU, which pushed the art a bit too far and was a maintenance nightmare). Interestingly, both the APU and fuel cell would prove to be troublesome in operational service, leading the Orbiter Upgrades program to initiate programs to replace them.[95]

A test dummy is ejected from an Orbiter crew cabin mock-up during tests at Holloman AFB during April 1980. (NASA photo S80-33102)

This unusual space shuttle crash rescue trainer was in front of the Holloman AFB flight line fire station on 13 June 1981. (1Lt. M. J. Kasiuba via the Terry Panopalis Collection)

SSME Qualification

When the SSME design had been selected on 13 July 1971, it was still intended to use derivatives of the same engine on both the fly-back booster and the Orbiter. As later events played out, the fly-back boosters were abandoned in favor of the SRBs. The parallel-burn concept meant the Orbiter SSMEs would need to operate at sea level, something that had not been considered in their original design since they were meant to be ignited at altitude. This meant that Rocketdyne and NASA needed to redefine the SSME and limit the nozzle expansion area to 77.5:1 or less. This was held up by the protest concerning the Rocketdyne contract and it was not until May 1972 that work began in earnest on the final SSME configuration. Once the Orbiter contractor was selected, definition progressed rapidly, and an Interface Control Document (ICD) was released on 9 February 1973.[96]

The original requirement for the fully-reusable two-stage SSME had been 100 missions or 27,000 seconds, including the ability to operate six times at an 'emergency power level' of 109 percent. With the parallel-burn shuttle, 27,000 seconds equated to only 55 missions, so this became the baseline. Further analysis indicated that if the total number of missions was held to 55, then the time spent operating at 109 percent could be raised to 14,000 seconds – the terminology was subsequently changed to 'full power level.' After the definitive ascent trajectory was defined it became clear that the SSMEs did not need to be capable of throttling back to 50 percent, and the 'minimum power level' was raised to 65 percent.[97]

Main engine development testing was planned to be conducted at the NASA rocket test site in Mississippi beginning in late 1974. The Mississippi Test Facility (MTF)* had been used for static testing various stages of the Saturn launch vehicle, and these facilities were modified during 1973 to accommodate SSME testing. While the modifications were underway, early SSME component testing was to be conducted at the Coca facility at Rocketdyne's Santa Susana Field Laboratory in Chatsworth, California. Various test stands at this facility needed to be modified to accommodate the turbopumps and combustion devices required for the SSME, and this work quickly fell behind, leading to a six month schedule slip. A program realignment in the summer of 1974 led to the decision to abandon testing at the Santa Susana facility, and to conduct all tests at the NSTL. This decision would later be reversed again, and testing at Santa Susana resumed after the first engines were flight qualified.[98]

The SSME operated at much high chamber pressures than any previous liquid-fuel rocket engine – interestingly, it ran at the same chamber pressure that Eugen Sänger had predicted would be necessary for the *Silverbird*. The new engine also incorporated 'pre-burning' of the propellants to help generate a higher specific impulse. To begin obtaining test results before a complete SSME was available, Rocketdyne elected to build an integrated subsystem test bed (ISTB). This was essentially a complete engine assembly that was not built to flight-hardware specification, i.e.; it was larger and heavier, but would provide for proof-of-concept and component testing. The first full thrust chamber test of the ISTB was conducted on 23 July 1975 on Test Stand A-1 at NSTL.[99]

A space shuttle main engine (SSME) during a test firing at the National Space Technology Laboratory in Mississippi. (NASA/NSTL photo 81-201-1)

A great many problems were identified and corrected during the component testing using the ISTB, and on 12 March 1976 the engine successfully demonstrated a 65 percent power level for 42.5 seconds. This test had been scheduled to last 50 seconds, but a failure in the high-pressure fuel turbopump caused it to be terminated early. On 24 March 1977, a failure of a high-pressure oxidizer turbopump caused a one month delay in testing while the cause was investigated. No concise cause was identified, so several general modifications were made and additional instrumentation was added to the test cell. On 27 April 1977 testing was resumed, and 25 tests were run on two engines at the NSTL with no serious difficulties. However, problems with the new advanced-design turbopumps would continue to plague the engine program throughout the development process. Another critical failure involved the use of an incorrect welding wire, forcing an inspection of all welds. The next two years would see a series of failures and fires involving the SSMEs.[100]

As part of the Orbiter contract Rockwell constructed MPTA-098 to allow Rocketdyne to test the SSMEs in a realistic structural environment. MPTA-098 consisted of an aft-fuselage, a truss arrangement which simulated the mid-fuselage, and a complete thrust structure including all main propulsion system plumbing and electrical systems. MPTA-098 was shipped from Palmdale to the NSTL on 24 June 1977 and mated with an External Tank (MPTA-ET) and three prototype SSMEs for propellant loading and static firing tests. The first static firing of a non-flight rated engine took place on 21 April 1978 and lasted for 2.5 seconds. By the summer of 1979, flight hardware had been installed on MPTA-098, and testing resumed, but not without problems.[101]

On 2 July 1979, the main fuel valve on engine #2002 developed a major fracture that allowed hydrogen to leak

---

* The NASA Mississippi Test Facility (MTF) was renamed the National Space Technology Laboratory (NSTL) in 1973. The NSTL was renamed the Stennis Space Center (SSC) during 1989.

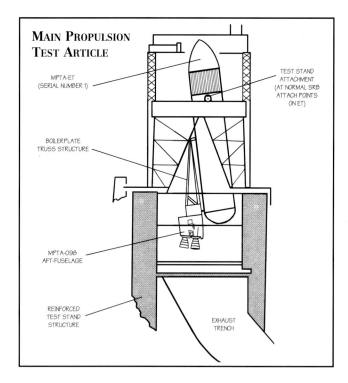

MAIN PROPULSION TEST ARTICLE

MPTA-ET
(SERIAL NUMBER 1)

TEST STAND
ATTACHMENT
(AT NORMAL SRB
ATTACH POINTS
ON ET)

BOILERPLATE
TRUSS STRUCTURE

MPTA-098
AFT-FUSELAGE

REINFORCED
TEST STAND
STRUCTURE

EXHAUST
TRENCH

engine ground test program (excluding the MPTA). A goal of 65,000 seconds was established by John Yardley as representing a sufficient level of confidence to consider the engine flight worthy, and this goal was soon established as a flight constraint. This requirement was achieved on 24 March 1980 during a test on engine #2004.[103]

Then, on 16 April, temperature limits on a high-pressure fuel turbopump were exceeded, causing an engine to shut down 4.6 seconds into a 544 second test. Another success was followed by another failure. The tenth static firing, this time with flight-design nozzles, was shut down 105 seconds into the test due to a burn-through in a hydrogen preburner. During November 1980, a weak brazing section on a nozzle failed during a 581-second test, causing a hole in the nozzle several inches in diameter. Thanks to the automated engine control system, and the use of highly-instrumented test stands, most of the SSME failures resulted in damage to components, not to the entire engine. This was because the failure could be detected and the engine shut down quickly enough to prevent the kind of catastrophic failures that had characterized the early space program, when even entire test stands were frequently destroyed. Nevertheless, the SSMEs were rapidly becoming the pacing item towards the first manned orbital flight.[104]

After the engines were installed in *Columbia*, significant changes were made to the engine design as a result of problems discovered during the ongoing test program. Because of the number and complexity of the changes, it was decided to repeat the final engine acceptance tests after the flight engines were modified. During June 1980 the engines were removed from *Columbia* and returned to the NSTL where they all satisfactorily completed a 520-second flight mission demonstration test. The engines were shipped back to KSC and reinstalled in OV-102.[105]

But the program had largely given up on the 109 percent 'full power level' capability, being willing to settle for 100 percent, at least for the orbital flight tests. Of the 65,000 seconds of test time, just over 1,000 seconds had been at 109 percent power. According to Bob Biggs, a Rocketdyne manager in charge of the ground test program, "We felt like we should at least demonstrate that we could do it without the engine blowing up. ... But we tested very little at 109 percent before the first flight. The reason was obvious; We didn't want an engine to blow at 109 percent a month before the first flight. ... We stopped blowing up engines at 109 because we stopped testing at 109."

into the enclosed aft compartment. The engine was commanded to shut-down, but before this process could be completed the pressure in the aft compartment exceeded the structural capability of the heat shield supports, and MPTA-098 sustained major structural damage. Testing was resumed in September, but on 4 November 1979, a high-pressure oxidizer turbopump failed after 9.7 seconds of a scheduled 510 second test of a three engine cluster. A completely successful static firing (the sixth 6 of the series) finally took place on 17 December 1979 involving three non-flight rated engines cycling between 70 and 100 percent power for 554 seconds.[102]

The first of four flight configuration engines were assembled and acceptance tested during the first half of 1979. Engine acceptance testing included a 1.5 second start verification, a 100-second calibration firing, and a 520-second flight demonstration test. Engine #2004 was allocated to the preliminary flight certification program, and engines #2005, 2006, and 2007 were installed on OV-102. But perhaps the most important test milestone was established in terms of total accumulated test duration of the single-

The space shuttle main engine test facility at the National Space Transportation Laboratory, what is now the Stennis Space Center. In the shot at left the External Tank can be seen just to the right of the crane, complete with the white FRL that was also used on the first two flight tanks. Barges parked at the lower right contain the cryogenic propellants that will be loaded into the External Tank. The engine exhaust is located on the back side, as shown in the photograph at left. (NASA photos S80-29627 and S80-32740)

Low performance, combined with the engine failures during the summer of 1977, had led to concerns by NASA and contractor management that developing the 109 percent capability in the time allotted would be too difficult to achieve, and some felt it might jeopardize the ability to support the flight program. In February 1978, Mike Malkin, the NASA program manager, declared a moratorium on testing at 109 percent until after the first manned orbital flight. Testing began as soon as STS-1 was launched. Within a month, 109 percent test programs were initiated with three engines on three different test stands. Within five months, all three engines experienced major failures. In the following year, three more engines were lost. A second moratorium was established in February 1983 and rescinded in August when the SSME program was restructured as two programs – Flight and Development.[106]

Although 109 percent power was desirable for use at Vandenberg, problems continued to delay it. There was even talk of trying for 120 percent, but given the difficulties reaching 109 percent, this seemed unreasonable. It would not be until the introduction of the Block II engine in 2001 that 109 percent became available on a routine basis.

The preliminary flight certification (PFC) program began in early 1980, and was defined in terms of a unit of tests that were called 'cycles.' Each cycle consisted of 13 tests and 5,000 seconds of test exposure which emulated normal and abort flight profiles. It was required to complete two PFC cycles on each of two engines of the flight configuration to certify that configuration for ten missions. All of the tests in each cycle had to be completely successful – if any test was less than perfect, the entire cycle did not count. Eventually, eight PFC cycles were completed prior to STS-1. On 17 January 1981, just three months prior to the first launch date, MPTA-098 successfully demonstrated a 625-second* firing, complete with simulated abort profiles. This test also included loading an ET without the LO2 anti-geyser line to evaluate the feasibility of removing it on future 'lightweight' External Tanks. All totaled, by the launch of STS-1 the SSME program had accumulated 110,253 seconds during 726 tests.[107]

On 20 February 1981 engineers and technicians at KSC loaded the ET with propellants and, without the astronauts in the cockpit, conducted a 20 second flight readiness firing (FRF) of the three SSMEs installed in OV-102. This test, in addition to verifying the main engines operated correctly, checked the auxiliary power units and the SSME gimbal system. It also allowed engineers to measure the 'twang' – the amount the vehicle stack moved when the main engines were ignited. *Columbia* moved forward 25.5 inches, slightly more than the 19 inches that had been predicted, but still well within expected limits. The FRF also allowed engineers to practice how to shut down and safe an Orbiter after an engine start (but without a launch). This was the final major test, and essentially cleared the way for the first flight of *Columbia*.[108]

---

* Two engines completed the 625 second demonstration. The #1 engine shut down at 239 seconds

† The Titan-IIIM (or Titan 34D-7) version of the venerable Titan booster had been designed with man-rating in mind for the Manned Orbiting Laboratory (MOL) program, but according to most sources the test sequence was not completed before the program was cancelled. However, UTC claims that both the UA-1205 (5-segment) and UA-1207 (7-segment) motors were "man-rated designs." Also, no Soviet booster had used solids prior to 1993, and none had ever been man-rated.

‡ The term Solid Rocket Motor (SRM) refers to the motor segments that make up a Solid Rocket Booster (SRB), which is a complete stack including the non-motor forward and aft sections, electronics, hydraulics, etc. Thiokol is responsible for the development and manufacture of the SRM/RSRM; USBI and MSFC for the remainder of the SRB hardware (now accomplished by United Space Alliance).

SRB QUALIFICATION

Since the Solid Rocket Boosters were relatively new technology for NASA, and none had ever been man-rated† before the shuttle program began, an extensive test program was planned. The first hydroburst test of an empty case was successfully completed on 30 September 1977 at the Morton Thiokol facility in Utah. This provided the data required to verify fracture mechanics and crack growth analysis, and to demonstrate the case's cyclic pressure load capability. A second hydroburst was successfully conducted on 19 September 1980 using the aft dome, two cylindrical segments, and the forward dome.[109]

Four development motor static firings were successfully conducted at Thiokol – DM-1 on 18 July 1977, DM-2 on 18 January 1978, DM-3 on 19 October 1978, and DM-4 on 17 February 1979. Three qualification motor static firings also were successfully completed – QM-1 on 13 June 1979, QM-2 on 27 September 1979, and QM-3 on 13 February 1980. These seven tests provided the data required to certify the solid rocket motor‡ design. The data obtained included ballistic performance, ignition system performance, case structural integrity, nozzle structure integrity, internal insulation, thrust performance, thrust reproducibility, dynamic thrust vector alignment, nozzle performance and the flight readiness of the SRM. Interestingly, none of these tests attempted to duplicate the loads or conditions anticipated during ascent – an oversight that would gain particular importance after the *Challenger* accident in 1986. Compare these 7 tests with the 726 tests required to certify the main engines.[110]

SRB decelerator subsystem (DSS) drop tests were conducted to evaluate the DSS under conditions simulating actual recovery conditions. The test program consisted of a series of six air drops using full-scale, flight-type pilot, drogue, and main parachutes. The tests provided data for evaluating the design, deployment process, parachute performance, and parachute structural integrity. The drop tests were performed using the NB-52 that had been used during the X-15 and lifting-body programs. The tests were conducted at the National Parachute Test Range at NAS El Centro, California. The drop test vehicle weight was approximately one-third the recovery weight of the flight SRB. Drop tests were conducted on 15 June, 4 August, and 14 December 1977; and 10 May, 26 July, and 12 September 1978.[111]

The SRB made a very effective snow remover! This is qualification test QM-8 at test bay T-97 in Utah on 20 January 1989. (Morton Thiokol photo 8903738)

Various structural tests on complete SRB assemblies simulated static loads on the launch pad, wind loads, flight loads, forward and aft attach point loads, parachute attach loads, water impact loads, and ultimate internal case pressure loads. All of these tests were conducted in facilities located at MSFC and were completed by the end of May 1980, well in advance of the first flight.[112]

After stacking on the mobile launch platform, but prior to ET mating, the two SRBs destined for use on STS-1 were connected at the forward ET attach points by a stiffener beam while two hydraulic beams at the aft attach points induced bending motions measured by sensors located at various points along the boosters. The purpose of this test was to determine how much the boosters were able to flex under the cold-induced contraction caused by filling the External Tank with cryogenics. These tests were conducted in the VAB at KSC during late 1980, and the contraction forces were arrived at via model analysis. No External Tank or cryogenic propellants were used during the tests.[113]

Even before the launch of STS-1, it was obvious that additional performance would be needed, especially for launches from Vandenberg. Since the ET was already on a diet, the program next turned to the SRBs looking for ways to shave weight. It should be noted that there was no serious consideration given to increasing the thrust of the SRBs, or increasing the performance of the SSMEs past the 109 percent level already baselined (but never achieved). A program was initiated to develop a new filament-wound case, constructed from a composite material of plastic reinforced with graphite fibers, for the SRB. Each of the new cases weighed 70,000 pounds – approximately 28,000 pounds less than the metal cases. This did not provide nearly the payload increase that would seem logical, mainly because the SRBs are dropped early in ascent and are not as weight critical as the Orbiter or ET. Nevertheless, the new cases would have provided an increase of 4,600 pounds in payload to polar orbit from Vandenberg.[114]

The SRB decelerator subsystem test vehicle after the first drop on 10 June 1977 at the National Parachute Test Range, El Centro, California. (via the Terry Panopalis Collection)

The solid rocket motors were tested laying on their sides at the Thiokol Wasatch facility, near Brigham City Utah. This test method was criticized after the Challenger accident as being unrealistic since it did not accurately reflect flight loads on the booster. (Morton Thiokol)

The three main chutes were representative of those that would be used on the operational SRBs. Note the recovery truck standing by. (via the Terry Panopalis Collection)

The black stripe around the nose identifies this as the ET destined for STS-1 (serial number 2) Here it is being raised to the vertical in the vehicle assembly building – soon it will be lifted into high bay #3 for mating with the SRBs. (NASA photo 108-KSC-81P-207)

ET QUALIFICATION

The design of the External Tank contained some significant differences from the Saturn stages it so closely resembled in theory. The Saturn had its engines at the bottom of the stage, and the LH2 tank was on top of the LO2 tank, eliminating the need for the hydrogen tank to support the heavy mass of oxygen. This also reduced the moments of inertia, making the stage more responsive to steering inputs. But with the ET, the LO2 tank went on top. The reason was that the SSMEs did not thrust vertically through the centerline of the tank – they were mounted on the Orbiter, well off to one side. In order for the SSME thrust vector (after the SRBs had separated) to pass through the ET center-of-gravity, the LO2 tank had to be on top. This necessitated a much more robust LH2 tank and intertank than had been used on Saturn.[115]

At MSFC, three structural components of the External Tank were tested beginning in March 1977 – an intertank structural test article, an LO2 tank, and an LH2 tank. The intertank STA and LO2 tank were tested in a fixture that consisted of 28 hydraulic jacks that could exert up to 4,350,000 pounds of pressure on the forward portion of the intertank STA, while ten other jacks applied shearing forces to the mid-section. This represented 140 percent of worse case predictions for actual flight loads. Over 2,800 stress measurement channels were monitored during the tests. The LH2 tank was tested at the former S-IC static firing test stand, and was subjected to internal pressures of 42 psi for 14 hours while various structural loads as high as 1,200,000 pounds were applied at the attach points. Following the tests, the LO2 and LH2 tanks were carefully examined, then assembled with a new intertank to form a complete ET that was used at NSTL to support engine testing – it was delivered as a complete unit on 26 August 1977. This was separate from the MPTA-ET that had been delivered on 2 March 1977[116]

MSFC also conducted 13 tests using 'mini-tanks' to evaluate the spray-on foam insulation that covered the outside of the LH2 tank. Each of these small aluminum tanks was subjected to sound levels averaging about 170 decibels to ensure that the insulation would not be cracked when subjected to the intense noise at launch. Pressure tests also evaluated the boil-off rate of LH2 in the tank to determine the effectiveness of the insulation to protect the LH2 during 7-hour 'holds' during countdown.[117]

The first 'tanking' of an ET – loading it with LO2 and LH2 – was conducted at the NSTL on 21 December 1977 using the ET contracted from the structural test components. The tanking test included flowing propellants as far the main engine inlet valves. A few days earlier the ET had been 40 percent filled and vibrated with three large shakers to provide information on its natural frequencies.[118]

But even as the first ETs were being manufactured, the program was embarking on some major changes. In February 1975 the Grumman Corporation had released a study that had been commissioned by MSFC to investigate possible alternatives to using aluminum for the production of the External Tank. The goal was to lower the recurring cost of the only expendable component of the space shuttle system. The concept investigated by Grumman consisted of building a 1/4-scale model of an ET constructed from a 0.25-inch Nomex honeycomb core bonded to a 0.040-inch aluminum liner with a film adhesive. This was then overwrapped with a cloth and wet-wound in the hoop direction with a 0.015-inch thick layer of E-glass (electrical grade glass) and epoxy.[119]

The conclusion of the study was that the tank would be less expensive to produce, but somewhat heavier than the baseline External Tank. Since the Orbiter was already over its weight goals, this construction technique was dropped from further investigation, although MSFC continued to pursue other alternatives that would lower the weight, and hopefully the recurring costs, of the ET.[120]

In fact, even as Martin Marietta was building the first External Tank (MPTA-ET), NASA was asking if it was possible to reduce the weight of the ET. The weight reductions would need to be accomplished with no net increase in recurring cost, and NASA let it be known that it would be beneficial if a cost reduction could be realized. A small weight savings (between 1,110 and 2,300 pounds each) was realized during the first production batch of 'standard-weight tanks' (SWT), but Martin Marietta undertook a concerted effort to reduce the empty weight of the second production batch of ETs by at least 6,000 pounds. The first of this batch was tentatively scheduled for delivery in June 1982, and would be used to support anticipated launches from Vandenberg AFB.[121]

The SWTs weight savings had been accomplished by slightly shaving the tolerances for the thickness of various metal and thermal protection system parts to the minimum allowed by the specification. Any further weight savings was going to require redesigning significant portions of the tank. Because the ET is the structural backbone of the integrated vehicle, load paths are complex, making weight reduction a difficult task. Also, the dual goals of lowering weight and also lowering production costs were often at odds.[122]

Martin Marietta identified an initial list of 30 candidates with a potential total savings of 7,500 pounds, providing a 25 percent contingency to ensure meeting the 6,000 pound reduction goal. A screening process was established using various recurring and non-recurring production costs as discriminators. The recurring cost screen was selected at $75 per pound and the non-recurring cost was $15,000 per pound, based on removing the same weight from the Orbiter. It was discovered that it would be difficult to mix the standard-weight tank and the light-weight tank (LWT) across the same production tooling, so it was decided that only a single production line would be maintained to minimize total costs. This one line would produce only LWTs after the initial production run of SWTs.[123]

Additionally, all excessive safety margins were reduced and a unique approach to factors of safety was applied that was tailored to the repeatability and predictability of loads. The standard safety factor was 140 percent (i.e.; an item with an expected load of 1.00 was designed to withstand a load of 1.40) for all aerodynamic and dynamic loads (which varied for each flight), whereas a factor of safety of 1.25 was applied to all well understood and fairly non-variable loads such as thrust loads, internal pressures, etc. Most of the significant design optimization candidates not only saved weight (2,748 pounds), but also resulted in lower recurring costs.[124]

The anti-geyser line was deleted* and its function replaced by direct helium injection into the main feedline to prevent geysers – a modification that took four years to develop through extensive testing of the MPTA-ET at the NSTL. Main feedline injection is possible because helium rising in the feedline provides transpiration cooling to keep the liquid below its saturation temperature, thus precluding the formation of vapor which can cause pogo. The change eliminated expensive hardware and a TPS ablator strip along most of the length of the LH2 tank, and allowed a more efficient packaging of the remaining propulsion lines. The total weight savings resulting from this change was 666 pounds.[125]

---

* A gaseous hydrogen pressurization line was moved to the location formerly occupied by the anti-geyser line, so it is difficult to tell the difference between tanks equipped with the anti-geyser line and without.

The ETs arrive at KSC via barge from the Michoud Assembly Facility. The tanks are carried on special trailers inside the covered barge – the empty trailers are also returned to the MAF via the barge. This is a late lightweight tank (serial number 67) that arrived in May 1993 and was used on STS-66. (NASA photo KSC-93PC-1060)

The ET-Orbiter aft crossbeam height was deepened and the chord thickness was reduced, providing an increase in structural stiffness and a resulting weight savings of 91 pounds. This component was constructed of 7050-T73 aluminum, and was initially considered a prime candidate for composite construction as that technology matured. However, the decision to pursue a new-design aluminum-lithium ET eliminated the need to incorporate new technology into the existing LWT. A net savings of 235 pounds was achieved by deleting some of the stringers and Z-frames in the LH2 tank. Actually, several hundred more pounds were realized by the stringer removal, but certain other components (the aft bulkhead among them) had to be strengthened to ensure that the LH2 tank did not buckle.[126]

Surprisingly, the thermal protection system on the ET is more complex than it appears. On Saturn, the insulation acted merely to reduce the boiloff rates to acceptable levels. If ice formed on the tank, it simply acted as additional insulation, and nobody much cared if the ice fell off during launch. But on shuttle, engineers did care – ice could damage the Orbiter's fragile tiles. In addition, the ET needed more than simple insulation since the complex configuration of the shuttle stack ensured that the ET would suffer shock-impingement heating during ascent. The top of the tank would be safe – it sat out in front of everything else and would only have to deal with the high-speed airflow. But a little further back the SRBs and Orbiter would be creating shock waves that would impinge on the sides of the ET, resulting in heating rates of over 40 BTUs per square foot per second. In addition, the attachment hardware for the Orbiter and SRBs would create complex flow fields, resulting in high local heating rates.[127]

After extensive testing, the spray-on CPR-488 thermal protection system was deemed adequate for ascent protection, and the top-coat of fire-retardant latex (FRL) was deleted, saving 595 pounds. This coating was also deleted from the last four standard-weight tanks, and only STS-1 and STS-2 used the distinctive white-colored tanks protected with the FRL.[128]

Other weight savings were achieved by optimizing various design details included using 6A1-4V titanium alloy for the construction of various fittings and interface hardware, using an integrated receiver/decoder in the range safety system, miscellaneous electrical wiring changes and the elimination of the External Tank Developmental Flight Instrumentation (DFI). In all, later LWTs achieved a weight reduction in excess of 10,000 pounds compared with the first production SWTs.[129]

ADVANCED VEHICLE AUTOMATION

Events effecting the choice of computer systems for the emerging space shuttle can be traced back to the Apollo era. Before the first manned Apollo flight, NASA was at work on defining an orbiting laboratory ultimately known as Skylab. Even on Earth, Skylab was impressive – basically a gutted and modified Saturn S-IVB upper stage, the laboratory was a fully equipped orbiting research facility that encompassed more than 10,000 cubic feet. At one end, beyond an airlock module that would allow Apollo Command Modules to ferry crews and supplies to the laboratory, was the Apollo telescope mount – a 12 ton observatory that held instruments that allowed the crew to make detailed studies of the Sun. Not only would the Sun be a primary object of study during the Skylab missions, it would also supply all of the laboratory's power, a first for the U.S. manned space program.[130]

About 10 percent of this power was dedicated to a revolutionary computer system. Called the attitude and pointing control system (APCS), the system was primarily responsible for controlling the attitude of the laboratory along all three axes, particularly important to the Sun-observing experiments. The APCS represented the first opportunity to employ a truly dual-redundant digital computer system aboard a spacecraft. While Gemini and Apollo used complex redundant circuitry inside their single computers, and were backed-up by ground-based systems at Mission Control and hand-calculated solutions by the astronauts, the APCS included two identical on-board computers. One functioned as the primary unit, the second as a backup – each machine, however, was fully capable of performing all the necessary functions. If system redundancy management software detected deviations from preselected criteria, it could automatically command the primary computer to relinquish control to the backup. Such a switchover could also be commanded by the astronauts, or from the ground.[131]

Unlike the special computers devised for the cramped quarters aboard Gemini and Apollo, the computers for Skylab had not been custom built. The central processing unit (CPU) of each computer was from IBM's 4Pi series of processors, effectively a miniaturized and hardened version of the System/360 computers developed by IBM in the early 1960s. The IBM 4Pi computers were also used in several military aircraft programs, such as the Republic F-105 Thunderchief and the Boeing B-52 Stratofortress, amongst others. The 4Pi model chosen for Skylab was the TC-1, adapted for use on Skylab by the addition of a custom input/output assembly to communicate with the unique sensors and equipment aboard the laboratory.[132]

The use of tested, off-the-shelf technology proved so successful with Skylab that it became a standard NASA policy for subsequent programs. But in many respects, the choice was somewhat problematic. As had been the case with Apollo (and would occur again with space shuttle), the hardware decisions were made first, long before the software specifications had been finalized. For Skylab, the hardware choice was made in October 1969, but the software specification document was not issued until the following July. In trying to meet the program specifications, IBM produced software modules that ranged from 9,000 to 20,000 words of code – but the maximum capacity of the ferrite-core memory used in the TC-1 was only 16,000 words. As a result, IBM engineers had to request numerous deviations from the specifications in order to make the software fit into the available memory without losing any critical functionality. Despite these obstacles, IBM delivered the final release of the software on 20 March 1973, two months before the scheduled launch of the laboratory.[133]

As a further guarantee of reliability, an auxiliary system called the memory load unit (MLU) was added in 1971 to provide a backup. Should the software be inadvertently erased from the core memory, the entire package could be reloaded from read-only tape stored on the MLU. In theory this could also give Skylab the capability to reload only selected parts of memory, allowing different programs to be swapped in and out of memory as required to support various experiments or other tasks. This capability was apparently never exploited, but served as a model for the future space shuttle.[134]

The success of the APCS aboard Skylab helped set the stage for a major expansion of the computer's role in manned space flight. Building on Skylab's foundation of redundancy management software and dual computers,

NASA had already decided that space shuttle would be the most automated vehicle in history. George Mueller had made this clear early during the Phase A studies (see page 80). That the expanded use of computers would have inevitably happened with or without the experience of APCS is undeniable. However, computers were still new and largely unknown devices in the early 1970s, before the advent of relatively inexpensive microprocessors and personal computers. IBM had traditionally been NASA's only spacecraft computer supplier, and was at the heart of these decisions. Early on, during the Phase A studies, IBM had teamed with several of the airframe contractors, most notably North American Rockwell to supply computers and software for the upcoming space shuttle. Although it was not cast in stone, it was largely assumed that IBM would produce the hardware and software regardless of the ultimate selection of an airframe contractor.

The fully-reusable Phase B vehicles were more aircraft than spacecraft, although they combined the most demanding requirements of both worlds. Not only would the flight control system have the usual responsibilities for figuring out how to make a high-speed* aircraft stable at high speeds and controllable during landing, it would also have the additional need to control the vehicle on-orbit and during reentry. In an era of mainly mechanical flight control systems, this was asking for quite an advancement.[135]

But a computer-based fly-by-wire (FBW) control system could probably do this. Fly-by-wire systems were not necessarily new – the German A-4 (V2) rocket used one during World War II, the Avro CF-105 Arrow used one in 1958, the X-20 Dyna-Soar would have used one, and to some degree, each of the U.S. manned capsules and the lunar lander had used them. But these were all analog systems, inherently limited in what they could do by a lack of processing power, and usually (but not always) backed-up by a mechanical system of some description. In 1972 NASA Dryden began research flights with the first aircraft equipped with a digital fly-by-wire (DFBW) control system.† This system eliminated the direct mechanical or hydraulic control linkage from the pilot's control stick to each control surface (although the aerosurface was still moved by a hydraulic actuator). Instead, the stick provided inputs to a computer that determined how each control surface should be moved to generate the required motion. To demonstrate this system, NASA Dryden modified a Vought F-8 Crusader, at first using a digital com-

puter from the Apollo Command Module, and later using three AP-101s, the same computer selected for space shuttle. The first flight with the AP-101 was on 27 August 1976 and 169 missions were flown before the last flight on 16 December 1985. The F-8 was also used to validate the control laws used in the space shuttle flight control system before the atmospheric flight tests.[36]

Choosing a computer for shuttle was an interesting task in itself. The logical choice at first appeared to be the Autonetics D-216, a 16-bit computer with 16 kilowords of memory that cost approximately $85,000 each. This was the computer being used in the North American B-1A prototypes, but the small word size and limited memory led space shuttle officials to continue their search. A meeting on 5 October 1972 to discuss proposed modifications to the DFBW F-8 brought to light two computers that could possibly be used on shuttle. The 32-bit Singer-Kearfott SKC-2000 had floating-point arithmetic (unusual in those days), and could be expanded to 24 kilowords of memory.‡ Its major drawbacks were that it used 430 watts of power, weighed 90 pounds, and cost $185,000 each. During November 1972, the shuttle program 'discovered' the IBM AP-101, a variant of the same 4Pi computer that had flown on Skylab. This 32-bit machine had 32 kilowords of memory, a floating-point instruction set, consumed 370 watts of power, weighed slightly less than 50 pounds, and cost $87,000 each.[137]

It should be noted that no off-the-shelf microprocessors (no Z80s, 80x86s, 680x0s, etc.) were then available, and large-scale integrated circuit technology was emerging, but years away from maturity. Since little, if anything, was known about the effects of lightning or radiation on high-density solid-state circuitry, ferrite-core memory was the only reasonably available choice for the Orbiter computers. Therefore, memory size was limited by the power, weight, and heat constraints associated with core memory.[138]

---

* Actually, the flight velocities proposed by the Phase A and B designs had not been achieved by any aircraft, including the research rocket-planes at Edwards, making total guesswork out of the flight software requirements.

† A month earlier the Air Force had begun testing a modified F-4 that is usually called the first FBW aircraft. The F-4's claim-to-fame was that it was the first FBW aircraft without a mechanical backup system, but it used an analog FBW system as opposed to the NASA F-8's digital system.

‡ In addition to having much more memory in modern computers, memory is much less expensive today. Adding 8,000 words (32 kilobytes) of memory to the SKC-2000 cost $15,000, or $1.88 per word. Today it is not even possible to buy such a small quantity, but 100 times as much (32 megabytes) costs about $25.

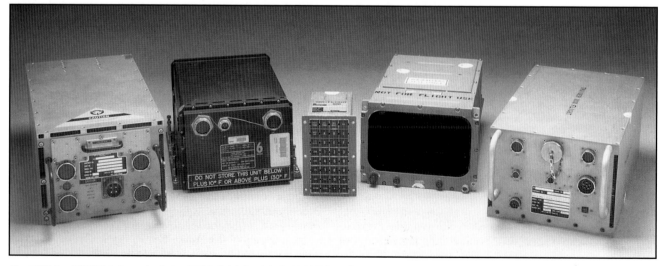

The mid-life (pre-MEDS) Data Processing System (DPS) included, from left, the AP-101S General Purpose Computer (GPC); the Mass Memory Unit (MMU); and the keyboard, Display Unit (DU) and Display Electronics Unit (DEU) that made up the Multifunction CRT Display System (MCDS). The original AP-101B used two boxes similar to the single AP-101S. (IBM)

However modest the final space shuttle was compared to the early Phase A/B studies, the role of its on-board computers was the most ambitious yet for any aircraft or spacecraft. When the RFP for the Phase C/D orbiter contract was released, it contained language that essentially removed the flight software from the prime contract and created a direct link between NASA and the software developer. NASA felt that it needed to keep tighter control on the software development effort than would have been possible in a once-removed subcontractor arrangement.[139]

A separate contract was issued to IBM for the development and verification of the most complicated computer software yet attempted. The initial $6,618,500 18-month contract was awarded on 25 September 1973, and covered the design and maintenance of the primary avionics system software (PASS), the development of appropriate tools for developing the software, and the design and development of the Avionics Development Laboratory (ADL) and the Shuttle Avionics Integration Laboratory (SAIL) at JSC.[140]

Interestingly, although NASA had removed the software, the Orbiter contract still contained provisions for Rockwell to supply the computer hardware. Rockwell subsequently awarded IBM a $15 million contract for a version of the AP-101B coupled with a new-design input-output processor (IOP). The computer was already being used in the Dryden DFBW F-8 program, and in fact, the shuttle program had provided Dryden with over $1 million in funding for the DFBW program to assist in perfecting the machine. As in Skylab, each CPU was coupled with a uniquely-developed IOP that provided communications with the various systems aboard the space shuttle.[141]

Data bus technology for real-time avionics systems was emerging but could not be considered operational, and the first digital data bus standard, Mil-Std-1553 would not be officially approved by the Air Force until 1975. All previous on-board systems had used bundles of wires, each dedicated to a specific signal. The sheer weight of the wiring would be unacceptable, and space shuttle elected to adopt a sort-of preliminary version of the -1553 data bus during late 1974.[142]

The use of mass storage (magnetic tape) for software programs in a dynamic environment was limited and suspect, especially for program overlays while in flight (the MLU had flown on Skylab, but was not used operationally to load overlays). Software design was evolving rapidly with the emergence of top-down, structured techniques, but no large-scale flight-critical systems had yet been developed and qualified using this methodology.[143]

It had been decided as early as Phase A that a computer system would be required to control the vehicle during all phases of flight. In fact, the final Orbiter configuration was an inherently unstable flight vehicle, incapable of even rudimentary control without the help of the on-board computer systems. Therefore, even more so than on Skylab, the computer system had to be completely reliable. That ideal was achieved through a novel approach to full redundancy, an advancement of the dual system used on Skylab. On the shuttle, four identical AP-101Bs would function simultaneously as a quadruple-redundant set during critical mission phases such as ascent and reentry, processing the same information, derived from completely separate data busses, in precise synchronization. If a conflict arose among the four primary computers, the majority* ruled, voting the conflicting unit out of

The Digital Fly-by-Wire F-8 poses with Enterprise. The Phase II flights made during the DFBW F-8 program tested the AP-101B computers that were initially used on space shuttle. the flights also proved many of the concepts that would be used in the software, and some flights tested aspects of the BFS. (NASA via the Tony Landis Collection)

the loop. None of the computers, singly or en masse, could turn off any other – that step was left to the crew. An errant machine would announce itself to the crew with warning lights, audio signals, and display-screen messages – all suggesting that the crew might want to isolate (i.e.; turn-off) the offending computer from the system.[144]

With the principle of redundancy solidly established, the remaining challenge was to actually write the software. The huge expense and difficulties suffered in the development of software for Apollo caused NASA to take a hard look at what language to use for shuttle's software. Most of the software for Apollo had been meticulously coded in assembly language, which required the programmer to pay close attention to the machine-instruction set and to sequences of addresses in memory. To give programmers greater flexibility, NASA urged the use of a high-order computer language for space shuttle.[145]

A variation of a the HAL language, called HAL/S, was created specifically for the shuttle project by Intermetrics, Incorporated, a Massachusetts company founded in 1969 by five M.I.T. programmers who had worked on developing software for Apollo. The name was a tribute to computer pioneer J. Halcombe Laning, who had invented an algebraic compiler in 1952 that ran on the M.I.T. Whirlwind – the first real-rime computer. The 'S' appended to the name has been variously explained as meaning 'subset' (of what is unclear), or 'shuttle,' and it is left to the reader to determine which is most fitting.[146]

Looking a lot like FORTRAN syntactically, what set HAL/S apart from other languages of the era was its ability to schedule tasks according to priority levels defined by the programmer. This capability would be the source of constant battles between IBM, who used the capability extensively, and Rockwell, who shunned its use for the backup software. Distilled from the best programming techniques of the era, HAL/S defined strict access rules that allowed various segments of the program to interact only as the program's author intended. But the language's greatest distinction may have been that it featured specific statements for real-time computing. Several programs could appear to share computer resources simultaneously, allowing both the crew and the many automated systems controlled by the computer to respond to rapidly changing events inside and outside the spacecraft without any knowledge of each other.[147]

---

* If the system ends up with a 2-to-2 vote, it displays warning messages to the crew advising the activation of the Backup Flight System (BFS).

The capabilities of HAL/S, specifically the ability to schedule tasks on a priority basis, caused the infamous scrub of the first space shuttle launch. At approximately T-20 minutes, the backup computer alerted the ground-based Launch Processing System that it had not been able to synchronize itself with the redundant-set of primary computers. It ended up that a minor programming error that had survived thousands of hours of testing allowed the redundant-set to begin processing information 40 milliseconds earlier than it should have. The basic problem stemmed from the difference in programming techniques between IBM and Rockwell. The Rockwell-authored back-up software was time-sliced, or synchronous – it dedicated a specific slice of CPU time to each processing task, and executed those tasks in a fixed order. The IBM software was priority-interrupt-driven, or asynchronous – it performed computations on demand and in strict observance to a predefined order of importance. Rockwell and IBM had disagreed over the relative merits of the two techniques for two years, although in the end, it really did not matter. Once the bug had been discovered, it became a rather trivial matter to fix it. In fact, it was rather ironic that the problem occurred on STS-1, since engineers later estimated that the problem would have occurred on just one out of every 67 launches.[148]

The STS-1 software development effort had released 17 interim versions during a 31-month period starting in October 1977 and culminating in December 1980 with the STS-1 flight release. Although full software functionality was provided after the ninth release in December 1978, an additional eight releases of the software were necessary to accommodate continued requirements changes, and the discrepancy corrections inherent in a large, complex, first-of-a-kind software system.[149]

As anyone who has ever used a computer knows, software is seldom error free. A statistical average for software used in critical systems (flight control, air traffic control, etc.) shows that programs average 10–12 errors for every 1,000 lines of software code. This was clearly unacceptable to NASA for use on space shuttle. As a result, NASA forced one of the most stringent test and verification processes ever undertaken on IBM for the primary avionics system software.[150]

The result achieved by the 300 IBM programmers, analysts, engineers, and subcontractors was impressive. An analysis accomplished after the *Challenger* accident showed that the IBM-developed PASS software had a latent defect rate of just 0.11 errors per 1,000 lines of code – for all intents and purposes, it was considered error free. But this remarkable achievement did not come easily or cheap. In an industry where the average line of code cost (at the time of the report) the government approximately $50 (written, documented, and tested), the Primary Avionics System Software cost NASA slightly over $1,000 per line. A total of $500 million was paid to IBM for the initial development and support of PASS.[151]

The Backup Flight System

The initial design of the orbiter flight control system was comprised of a quad-redundant computer complex that handled flight-critical functions. Systems management and most non-avionics functions were contained in a fifth computer that was not considered flight critical. This concept was well into development when a blue ribbon panel was asked to review all aspects of the upcoming atmospheric flight tests. One of the conclusions reached by the panel was that an unnecessary risk was being taken by not providing a backup flight control system for the first atmospheric flights. This decision was based on the relative complexity of the redundant-set computer synchronization scheme being implemented by IBM, and the lack of a direct manual flight control capability in the Orbiter.[152]

To protect against a latent programming error existing in the primary avionics system software that would render the space shuttle uncontrollable during a critical flight phase, NASA decided to develop a backup flight system (BFS). This system would have its own set of requirements, and would not reuse any of the code being developed for PASS. To ensure that the BFS was as independent as possible, NASA contracted with Rockwell to write it, and even different development environments and configuration management systems were specified. Nevertheless, like IBM, Rockwell elected to use HAL/S as the programming language, although the BFS tends to use a different programming approach than PASS.[153]

The BFS was chartered to protect against a software fault in the most sophisticated flight software system ever implemented. One of the drivers towards using a BFS was the fact that it would take an estimated 10,000 years to test every possible branch in the primary flight software's 500,000 words of code. This, obviously, was not feasible. And at this point, the concept of 'error free' software did not exist since this declaration came only after extensive analysis of the performance of PASS during 25 flights and tens of thousands of hours of ground-based testing.[154]

Initially BFS was to be a very simple system installed for ALT only, capable of bringing the Orbiter in for a landing on one of the lakebed runways. The BFS would be deleted once confidence had been gained in the primary flight system. The word 'simple' is very important because one of the primary concerns that was driving the development of the BFS was NASA's inability to properly verify the large amount of software being developed for PASS. The approach taken for the BFS was to develop a very simple and straightforward software program and then test it in every conceivable manner. The result was a program that contained only 12,000 words of executable instructions, including the ground checkout and built-in test for the computer. The actual flight control portion of the software consisted of approximately 6,000 words. The remainder was for the systems management functions that still had to be performed by the fifth computer, along with the backup autopilot functions.[155]

While the BFS was being written for the ALT flights, NASA was also considering extending its use to the orbital flight tests. Although some of the PASS software would have been demonstrated during the ALT flights, a great deal more of it would not have been exercised. Therefore NASA extended the Rockwell contract to include a rudimentary ability to operate during the critical flight phases (ascent and reentry) of the OFT flights. No on-orbit capability would be included except for the ability set up a de-orbit burn and fire the OMS engines.

After this decision, it was expected that the BFS would be deleted after the completion of the OFT flights. The expectation was that after OFT, the entire shuttle design, including the PASS, would be proven safe for operational use and, therefore, the BFS would no longer be needed. Close to the end of the OFT, however, an examination of the number of changes being included into the PASS resulted in a decision to keep the BFS for the foreseeable future. It was concluded that the quantity of changes in PASS (for mission enhancements, correction of known errors, etc.) would continue to present a risk for unknown PASS software errors, and that therefore, BFS should be available if needed.[156]

The BFS has never been required to demonstrate its capability. A proposed on-orbit engagement and orbital maneuvering system engine burn was deleted in 1982 from the list of flight test requirements. This change was mainly the result of a busy test schedule during the orbital flight test series, but also indicates the level of confidence that NASA has in the BFS. However, between 18 March and 15 April 1977 the DFBW F-8 conducted eight flights at NASA Dryden that used the BFS software package during practice landing approaches.[157]

AN IMPORTANT MISSION – RESCUE SKYLAB

The concept of an orbiting laboratory based on readily available components used in the Apollo missions had been discussed for several years at MSFC. Finally, at 13:30 EDT on 14 May 1973, a two-stage Saturn V (SA-513) boosted the Skylab orbital cluster into low-earth orbit. For almost two years the laboratory would be nearly continually manned by a total of three crews, the last setting a mission record of 84 days 1 hour and 14 minutes.[158]

Before undocking from the Skylab on 8 February 1974, astronaut Gerald Carr had fired the Apollo CSM-118 service module's attitude-control thrusters for three minutes, nudging the cluster 6.8 miles higher, into an orbit 269 by 283 miles. After the crew had returned to Earth, and the end-of-mission engineering tests were finished, flight controllers at JSC vented the atmosphere from the workshop, oriented the cluster with the docking adapter pointed away from Earth, and shut down most of its systems. This ended the operational use of the first American space station.[159]

Calculations made during the mission, based on 1974 values for solar activity and expected atmospheric density, gave the workshop just over nine years in orbit. Slowly at first – dropping just 18 miles by 1980 – and then faster – another 62 miles by the end of 1982 – Skylab would reenter the Earth's atmosphere. It was predicted that sometime in March 1983, Skylab would burn up during reentry. If, as planners hoped, development of the space shuttle went smoothly, one of its first missions would be to boost the workshop into a higher orbit. Later missions would attempt to repair the space station, and possibly to re-activate it. If not, the 240,000 pound workshop would probably attract a great deal more attention than NASA wanted when it returned to Earth, and flight controllers could do little to alter the course of its reentry.[160]

The nine-year lifetime of the orbiting laboratory seemed ample in 1974, and in any case, NASA had more pressing problems to worry about. The space shuttle development, although not going badly for such a complex effort, was not progressing at the rate originally envisioned. Funding constraints had forced NASA to spend considerably longer in Phase B than had been planned, and economics was continuing to haunt space shuttle by stretching out the development program. By 1977 *Enterprise* was being prepared for the atmospheric flight tests, and planners were busy thinking about possible payloads and missions for the first few orbital flights. Early in 1977, NASA Headquarters directed JSC and MSFC to outline schedules and funding requirements for a space shuttle mission to boost Skylab into a higher orbit. Engineers at JSC were not optimistic. Rendezvous and docking with the inert workshop had not been thoroughly studied, and preliminary analysis showed that a Skylab visit could not be accomplished earlier than the fifth test flight of the Orbiter. This event was tentatively scheduled for late 1979, although it was beginning to look like the schedule was slipping rapidly. As the next

One of the first missions envisioned for Space Shuttle was to boost the falling Skylab into a higher orbit to permit its continued use. This mission was originally scheduled for early 1979, but was cancelled due to delays in achieving the first flight. Skylab eventually reentered Earth's atmosphere on 11 July 1979, with fragments falling on western Australia. (NASA)

solar maximum period approached (1980–81), it was also becoming clear that the Sun was considerably more active than previously anticipated. This created problems for Skylab, since the increased solar activity heated the Earth's upper atmosphere, increasing its density at orbital altitudes, and dragging the workshop down faster than had been predicted in 1974.[161]

In March 1977, MSFC told Headquarters that a study contract to define a booster stage for the Skylab mission should be awarded no later than midyear. Headquarters then set the fifth space shuttle flight as the target mission, and 1 September 1977 as the latest date to commit to the rescue attempt. This would allow just over two years for the necessary hardware development, planning, and training for the first orbital rescue mission.[162]

NASA Headquarters formally approved the mission in September, and in November 1977 MSFC awarded a $1.75 million letter contract to Martin Marietta to proceed with the analysis and design studies necessary to conduct the mission. Since time was critical, as much already designed and qualified hardware as possible was to be used – this was in keeping with the overall philosophy that had originally been used to build Skylab. The first design review was scheduled for March 1978, but new analysis by the National Oceanic and Atmospheric Administration (NOAA) pointed out that the solar activity in this period was looking to be the second most intense of the century, and that Skylab might tumble back to Earth even earlier than NASA predicted.[163]

Early in 1978, Skylab was rudely thrust into the glare of publicity when the unmanned Soviet Cosmos 954 reentered over northern Canada, scattering pieces of it's nuclear-fueled electrical power module over a wide area. The Russian satellite caused no injuries, and little property damage, although the nuclear clean-up effort consumed a great deal of effort. NASA assured the public that the Skylab workshop contained no nuclear fuel, and that current predictions showed it would not drop below 170 miles before October 1979.[164]

By far the most stimulating reaction to the Cosmos incident was from the U.S. State Department. In view of the worldwide interest in Cosmos 954, State wanted to know what, if anything, was NASA planning to do about Skylab? Diplomatic repercussions were possible almost anywhere in the world if a piece of Skylab fell on a citizen somewhere, since the laboratory's orbit took it over the heads of 90 percent of the world's population. Although NASA's studies had shown that the risk to humans was small, it was not zero – a fact that was very important to an agency sensitive to public opinion.[165]

NASA immediately got to work to determine the condition of Skylab's systems. If the laboratory was to be rebooted for later use or brought out of orbit at a site of NASA's choosing, it was necessary to determine just how much control could be exercised from the ground. In the most favorable circumstances this was limited to controlling the workshop's attitude, thereby increasing or decreasing atmospheric drag. It was impossible to increase its altitude. Analysis showed that if everything worked well, Skylab's orbital lifetime might be extended by as much as five months. However, the problem was compounded by a general update to NASA's tracking system, leaving the Bermuda tracking station as the only site capable of communicating with the obsolete UHF telemetry equipment aboard Skylab.[166]

Martin Marietta, in the meantime, had designed a teleoperator unit, mostly fuel tanks and engines, that would be guided by an astronaut in the space shuttle to attach to the workshop's multiple docking adapter. Once attached, the unit would fire its engines to boost Skylab to a higher orbit. The design was mostly complete, and fabrication and assembly were due to begin shortly. The completed unit was scheduled for delivery to KSC in August 1979, for an September 1979 launch on what was now the third test flight. It was an ambitious schedule, especially considering that the first Orbiter had not yet been launched.[167]

The effort to save Skylab was also becoming costly. Not counting expenditures for hardware development and planning, NASA had spent $750,000 on the rescue attempt by 1 June 1978, and expected to spend another $3 million by the end of the year. At least one NASA official, Chris Kraft, Director of JSC, publicly expressed his opinion that this money was largely wasted. He expected neither the laboratory's systems to last long enough, or the space shuttle to be flight qualified soon enough, to successfully complete a reboost mission. Nevertheless, JSC began to staff up a Skylab rescue group under the direction of Charles Harlan. By October 1978 Harlan had sufficient people and resources to keep five flight control teams working three shifts a day. A few had sat behind the same consoles during Skylab's three operational missions, but most were new with little operational experience in monitoring on-orbit operations of any sort, let alone ones as complex as Skylab's.[168]

Having started with little confidence in the aging systems aboard Skylab, the group soon discovered that they were in remarkably good shape. The group regained the use of the on-board computer complex, and engineers from MSFC devised new procedures to control the laboratory's attitude without using the limited propellants available for the reaction thrusters – the propellants would be necessary to regain control after the reboost to higher orbit.[169]

But it was not to be, and the year-long effort to save Skylab ended abruptly in December 1978. Although the teleoperator propulsion unit was approaching final assembly, problems with the SSMEs had convinced planners that a space shuttle mission was not likely before the end of the decade. NASA Administrator Robert Frosch informed the President on 15 December 1978 that Skylab could not reasonably be saved, but that NASA would attempt to command reentry over unpopulated areas.[170]

At 03:45 EDT on 11 July 1979, on its 34,981st orbit, controllers at JSC commanded Skylab to tumble. Unfortunately, Skylab did not break up as thoroughly as expected, and at 12:37 EDT, pieces of the space station showered an area around Perth, Australia, causing no injuries and remarkably little property damage. It would be 20 years before another American space station would be put into orbit.[171]

DEPARTMENT OF DEFENSE SUPPORT

Although it appears that the Air Force continued to harbor some misgivings about the space shuttle, they continued to publicly support the effort and even provided funds to ensure some items were developed or procured in a timely manner. For instance, on 7 August 1974 the Deputy Secretary of Defense, William P. Clements, Jr., wrote to NASA's James Fletcher:" The Department of Defense is planning to use the Space Shuttle ... to achieve more effective and flexible military space operations in the future. Once the Shuttle's capabilities and low operating costs are demonstrated we expect to launch essentially all of our military space payloads on this new vehicle and phase out of inventory our current expendable launch vehicles." The letter continued: "Recent budget actions assure that adequate funding will be available to develop a low cost modified upper stage [the IUS] for use with the Shuttle. This stage will be ready for operational use at Kennedy Space Center concurrently with the Shuttle in 1980. Funding is also included now in our budget for establishing a minimum cost Shuttle launch capability at Vandenberg AFB consistent with realistic DoD and NASA needs. This addition should be available around December 1982 ..."[172]

In late 1976, the definitive 'NASA/DoD Memorandum of Understanding on Management and Operation of the Space Transportation System' was finally signed. This document listed the specific responsibilities of each party with regards to space shuttle. For instance, it listed the NASA responsibilities as including: (1) the development of the space shuttle, including the vehicle and general-purpose ground support equipment and facilities, (2) including DoD requirements into shuttle where possible, (3) developing the launch site at KSC, (4) using the Interim Upper Stage (IUS) being developed by the Air Force for all planetary and earth-orbital missions not more economically served by the SSUS, (5) planning and integrating all shuttle flights and users, including DoD missions, (6) providing integrated logistics and training for the shuttle, and (7) establishing a pricing policy for all non-DoD users. The MoU read: "Because of DoD's heavy investment, large usage, and the operation of VAFB, the DoD pricing and reimbursement arrangements will be jointly negotiated between NASA and DoD ..."[173]

The Department of Defense responsibilities included: (1) using the STS as the primary vehicle for placing payloads in orbit, (2) developing the IUS and general-purpose ground support equipment, and (3) developing the launch site at Vandenberg. The MoU went on to detail that flight control for all missions would be the responsibility of the Mission Control Center in Houston, and that DoD would advise NASA of any security measures that needed to be taken at the Mission Control for classified flights.

Concurrently with formalizing the MoU, NASA and the Air Force were negotiating a price schedule for the military use of shuttle. The final pricing schedule showed a dedicated flight costing $12.2 million (based on FY75 dollars, with an inflation clause) for the first six years, with the price adjusted based of the actual cost of operations beginning in the seventh year. The final MoA also prohibited NASA from providing shuttle services to foreign military payloads "unless the payloads are officially designated as DoD Cooperative Agreement payloads..."[174]

But the "capabilities and low operating costs" expected in 1974 never materialized, and official DoD commitment to shuttle began to waiver in the early 1980s. Two years before the *Challenger* accident, testimony before the Senate Appropriations Committee, Subcommittee on Defense, included "DoD remains fully committed to shuttle. However, total reliance upon the STS for access to space represents an 'unacceptable national security risk.' Consequently, DoD has identified a requirement for a complementary system to the Shuttle/Centaur that will deliver 10,000 pounds to geosynchronous orbit. The requirement is for two launches per year for five years out of KSC ..."[175]

THOSE PESKY TILES

During the original studies of lifting-reentry vehicles during the late 1950s and 1960s, there had been a great debate over the relative merits of active cooling systems versus passive systems for the vehicle structure. The active systems were attractive – on paper – but nobody could quite figure out how to make them work. Therefore the choices were largely narrowed to either a hot-structure like that used on the X-15, or a more conventional structure protected by some sort of insulation. The hot-structure approach required the use of rare and expensive superalloys, and there was always a great deal of doubt if it would have worked on a vehicle as large as the shuttle. Generally most contractors seemed to prefer a

The most visible sign of Air Force support for the space shuttle program was the construction of the western launch site at Vandenberg AFB, shown here in about 1982 before the shuttle assembly building (SAB) was built. But DoD support before Congress and the OMB was probably more significant in the long run. (Dennis R. Jenkins)

fairly conventional structure made of titanium and protected by a series of metallic shingles with a thick layer of insulation in between the two. There was some investigation into ablative coatings, but the unhappy X-15A-2 experience made just about everybody shy away from this technology except as a last resort.

Things began to change as the Lockheed Missiles and Space Company made quick progress with the development of the ceramic reusable surface insulation (RSI) concept. This work had begun during the late 1950s, and by December 1960 Lockheed had applied for a patent for a reusable insulation material made of ceramic fibers. The first use for the material came in 1962 when Lockheed developed a 32-inch diameter radome for the Apollo spacecraft made from a filament-wound shell and a lightweight layer of internal insulation cast from short silica fibers. But the Apollo design changed, and the radome never flew.[176]

But the experience led to the development of a fibrous mat that had a controlled porosity and microstructure called Lockheat®. The mat was impregnated with organic fillers such as methyl methacylate (Plexiglas) to achieve a structural quality. These composites were not ablative – they did not char to provide protection. Instead Lockheat evaporated, producing an outward flow of cool gas. Lockheed investigated a number of fibers – silica, alumina, and boria – during the Lockheat development effort. By 1965 this had led to the development of LI-1500, the first of what became the shuttle tiles. This material was 89 percent porous, had a density of 15 pounds per cubic foot, appeared to be truly reusable, and was capable of surviving repeated cycles to 2,500 degF. A test sample was flown on the Air Force Pacemaker reentry test vehicle during 1968, reaching 2,300 degF with no apparent problems.[177]

Lockheed decided to continue the development of the silica RSI, but decided to produce the material in two different densities to protect different heating regimes – 9 pounds per cubic foot (designated LI-900) and 22 pounds per cubic foot (LI-2200). The ceramic consisted of silica fibers bound together and sintered with other silica fibers, and then glaze coated by a reaction-cured glass consisting of silica, boron oxide, and silicon tetraboride. Since this mixture was not waterproof, a silicon polymer was coated over the undersurface (i.e., non-glazed) side. This material was very brittle, with a low coefficient of linear thermal expansion, and therefore Lockheed could not cover an entire vehicle with it. Rather the material would have to be installed in the form of small tiles, generally 6 by 6 inch squares. The tiles would have small gaps between them (averaging about 0.01-inch) to permit relative motion and allow for the deformation of the metal structure under them due to thermal effects. A second concern was the movement of the metal skin directly under an individual tile – since a tile would still crack under this loading,

engineers decided to isolate the skin from the tile by bonding the tile to a felt pad, then bonding the felt pad to the skin. Both of these bonds were done with a room-temperature vulcanizing (RTV) adhesive.[178]

In their Phase C response, Rockwell had proposed using mullite tiles made from aluminum silicate instead of the Lockheed-developed tiles since the technology was better understood and more mature. But the mullite tiles were heavier, and potentially not as durable. Given the progress Lockheed had made subsequently, Rockwell and NASA asked the Battelle Memorial Institute to evaluate both candidate systems – an evaluation that the Lockheed product won. But the Lockheed material was not appropriate for all applications. Very high temperature areas of the Orbiter – the nose cap and wing leading edges – would use a reinforced carbon-carbon (RCC) material originally developed by LTV for the Dyna-Soar program. The RCC would provide protection above 2,700 degF, yet keep the aluminum structure of the Orbiter comfortably below its 350 degF maximum. Tiles were used for the entire underside of the vehicle and for most of the fuselage sides and vertical stabilizer. Black tiles could protect up to 2,300 degF, while white tiles protected up to 1,200 degF. Flexible reusable surface insulation (FRSI) protected areas not expected to exceed 750 degF.[179]

The magnitude of the potential TPS problem can be imagined by looking at the bottom of an Orbiter – in this case Columbia just prior to STS-3. The entire bottom surface of the vehicle is covered by tiles, generally measuring no larger than 6 by 6 inches. Early in the program most of the fuselage sides were also covered with tiles (blankets have since replaced the tiles). In all, over 30,000 tiles were on the early Orbiters. (NASA photo 108-KSC-82P-44)

Interestingly, NASA and Rockwell originally believed that the leeward side (top) of the vehicle would not require any thermal protection. But in March 1975 the Air Force Flight Dynamics Laboratory conducted a briefing for space shuttle engineers on the classified results of the ASSET, PRIME, and boost-glide reentry vehicle (BGRV) programs that indicated leeward side heating was a serious consideration. The thermal environment was not particularly severe, but easily exceeded the 350 degF capability of the aluminum skin. FRSI blankets were subsequently baselined for this area.

But in the meantime another problem had developed – with the tiles themselves. As flight profiles were refined and aero-loads better understood, engineers began to question whether the tiles could survive the punishment. By mid-1979 it had become obvious that in certain areas the tiles "did not have sufficient strength to survive the tensile loads of a single mission." NASA immediately began a massive search for a solution that eventually involved outside blue-ribbon panels, government agencies, academia, and most of the aerospace industry. As LeRoy Day recalled: "... there is a case [the tile crisis] where not enough engineering work, probably, was done early enough in the program to understand the detail – the mechanical properties – of this strange material that we were using ..."[180]

The final solution to the tile problem (at least this one) involved strengthening the bond between the tiles and the felt strain-isolation pads (SIP). Analysis indicated that while each individual component – a tile, the SIP under the tile, and the two layers of adhesives – each had satisfactory tensile strength, when combined as a system the components lost about 50 percent of their combined strength. This was largely attributed to stiff spots in the SIP (caused by needling) that allowed the system strength to decline as far as 6 psi instead of the baseline 13 psi. In October 1979 NASA decided on a 'densification' process that involved filling voids between fibers at the inner moldline (the part next to the SIP pad) with a special slurry mixture consisting of Ludox (a colloidal silica made by DuPont) and a mixture of silica and water. Since the tiles had been waterproofed during manufacture, the process began by applying isopropyl alcohol do dissolve the water-proofing, then painting the back of the tile with the Ludox. After air-drying for 24 hours the tiles were baked in an oven at 150 degF for two hours. After a visual and weight check, each tile was re-waterproofed using Dow-Corning's standard Z-6070 product (methyltrimethoxysilane). The densified layer acted as a 'plate' on the bottom of the tile, eliminating the effect of the local stiff spots in the SIP, bring the total system strength back up to 13 psi.[181]

But the installation presented its own problems. Rockwell quickly ran out of time to install tiles while OV-102 was in Palmdale – NASA needed to present the appearance of maintaining a schedule, and *Columbia* moving to KSC was a very visible milestone. So in March 1979 *Columbia* was flown from Palmdale to KSC on the SCA, and quickly moved into the Orbiter Processing Facility (OPF). Just over 24,000 tiles had been installed in Palmdale, with 6,000 left to go. But it now appeared that all of the tiles would need to be removed so that they could be densified.

The challenge became to salvage as many of the installed tiles as possible while ensuring sufficient structural margin for a safe flight. The approach developed to overcome this almost insurmountable challenge was called the tile proof test. This involved the application of

This F-104 (N826NA) was primarily involved with testing the elevon trailing edge tiles at maximum dynamic pressures up to 1,125 psf. The NASA Dryden aircraft provided a valuable resource for the shuttle program since they allowed real-world testing without endangering a shuttle crew. (NASA via the Tony Landis Collection)

a load to the installed tile so as to induce a stress over the entire footprint equal to 125 percent of the maximum flight stress experienced at the most critical point on the tile footprint. This approach could potentially salvage thousands of installed tiles.[182]

The device used for the proof test employed a vacuum chuck to attach to the tile, a pneumatic cylinder to apply the load, and six pads attached to surrounding tiles to react the load. Since any appreciable tile load might cause some internal fibers to break, acoustic sensors placed in contact with the tiles were used to monitor the acoustic emissions for any internal fiber breakage. The proof testing not only salvaged tens of thousands of installed tiles but also revealed those tiles (13 percent failed the proof

The second F-15 (71-0281) was used to test the windshield tiles (shown here) as well as several other areas. The F-15 subjected the windshield tiles to a maximum dynamic pressure of 1,140 psf at speeds up to Mach 1.4. (NASA via the Tony Landis Collection)

test) with inadequate flight strengths. The tiles that failed would be replaced with densified tiles.[183]

Two other techniques were developed to strengthen tiles while they were still on the vehicle. The first involved 'thick' tiles – usually on the underside of the Orbiter – that were relatively small. As shock waves swept air over these tiles, they tended to rotate, inducing high stresses at the SIP bonds. The solution was to install a 'gap filler' that prevented the tile from rotating. But this solution would not be very effective for small thin tiles, so a technique was developed where the filler bar surrounding the SIP was bonded to the tiles. This was done by inserting a crooked needle into the tile-to-tile gap and depositing RTV on top of the filler bar. This significantly increased the total bonded footprint and decreased the effects of a shock-imposed overturning moment.[184]

For the next 20 months, technicians worked three shifts per day, six days per week testing and installing 30,759 tiles. By the time the tiles were installed, proof-tested, often removed and reinstalled, then re-proof-tested, the technicians averaged 1.3 tiles per man per week. Sometimes it seemed like the workers were making no progress at all. During June 1979, Rockwell estimated that 10,500 tiles needed to be replaced – by January 1980 over 9,000 of these had been installed, but the number remaining had ballooned to 13,100 as additional tiles failed their proof tests or were otherwise damaged. By September 1980 only 4,741 tiles remained to be installed, and by Thanksgiving the number was below 1,000. It finally appeared that the end was in sight.[185]

Between April 1978 and January 1979 a team from the AFFDL conducted a review of potential Orbiter heating concerns, and concluded that the OMS pod might have unanticipated problems. To support the conclusion, the Air Force ran a series of tests at the Arnold Engineering Development Center (AEDC) between May and November 1979, and further tests were run at the Naval Surface Weapons Center shock tunnel in May 1980. It was discovered that the OMS pod structure would deflect considerably more than originally anticipated – the thin 8 by 8-inch tiles were relatively weak under bending loads, and it was feared that the tiles might fracture and separate from the vehicle. Since the tiles had already been installed, a unique solution was developed where each tile was 'diced' while still attached to the pod. This involved carefully cutting each of the 8-inch tiles into nine equal parts. Each cut was carefully monitored to ensure neither the tile nor the underlying structure was damaged. The technique proved so successful that it has subsequently been used on other areas of the Orbiter when needed.[186]

The thermal protection system is subjected to numerous loadings from the severe aerodynamic environment, including shocks and pressure gradients. It was comparatively easy to model these loads for the majority of the tiles since they were on flat surfaces on the Orbiter. However, many tiles adjacent to 'boundaries' (such as the wing leading edge, the windshield, etc.) did not have a simply geometry and were difficult to analyze because they were located in very complex flow fields. To better understand the problem, NASA initiated a combination of flight and wind tunnel testing. The flight testing was undertaken during 1980 at NASA Dryden with a total of 60 flights by a McDonnell Douglas F-15 Eagle and a Lockheed F-104 Starfighter. The F-15 was used to evaluate tiles from the wing leading-edge closeout, the wing glove, windshield closeout, and vertical stabilizer leading edge; the F-104 tested tiles from the elevon trailing edge

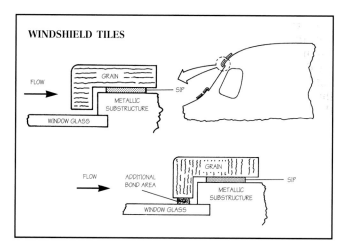

The test program on the NASA Dryden F-15 showed that the windshield tiles needed to be changed. The tile grain was reversed, running perpendicular to the surface instead of parallel, and the leading edge of the tile was bonded directly to the windscreen. (NASA)

and wing cove/elevon. The tiles were subjected to a variety of aerodynamic load conditions including flights up to Mach 1.4 and dynamic pressures of 1,140 pounds per square foot. Three wind tunnels were used for tile tests – the 16-foot tunnel at AEDC, the 11-foot supersonic tunnel at NASA Ames, and the 8-foot tunnel at NASA Langley. Various anomalies were uncovered that resulted in at least some of the tiles being redesigned and retested.[187]

The tiles around the windshield posed some unique problems since they were subjected to high stagnation pressures that tried to lift the tiles. All Orbiter tiles are machined from blocks of RSI so that the layers of silica material run in a direction generally parallel to the skin of the Orbiter. This grain orientation is a thermal requirement to minimize the conduction of heat from the outer moldline to the inner moldline (i.e., through the tile to the skin). However, this grain orientation also causes a reduction in strength because of the relatively low number of fibers that run vertically (perpendicular to the Orbiter skin) – the vertical fibers are what a transfer loads to the SIP. Because of the unusual aero-loads around the windshield, researchers decided to machine the tiles with the grain perpendicular to the Orbiter skin to provide twice the strength. But this ran the risk of overheating the aluminum structure. A thermal analysis revealed that the heavy framing around the windows acted as a large heat sink that prevented unacceptable temperatures. Further structural analysis, however, indicated that an adequate margin of safety was still not achieved, so it was decided to bond the portion of the tile that overhung the window directly to the glass (previously it had not been bonded). This provided acceptable safety margins.[188]

IN PRODUCTION AT LAST

While *Enterprise* and MPTA-098 were undergoing tests, Lockheed was busy trying to verify the structural integrity of the Orbiter structural test article (STA-099). On 4 February 1978, Rockwell had delivered the nearly-complete airframe named *Challenger* to the Lockheed-California Company located across the Plant 42 runway. Detailed planning of the structural test conditions was well underway for a classical static test program followed by fatigue investigations when new requirements emerged. The Orbiter structure had evolved under such weight-saving pressure that virtually all the primary structure had a significant structural stress component.

Attempts to factor mechanical loading into equivalent thermal loadings resulted in inconsistent stress distributions, and were not considered meaningful. Thus, it became clear that the classical demonstration of ultimate design strength was not feasible without an accompanying detailed thermal simulation.[189]

Twelve months of testing would take place in a 430-ton steel rig built especially for the space shuttle test program, finally concluding on 4 October 1979. The special test rig contained 256 hydraulic jacks that distributed loads across 836 application points to simulated various stress levels under control of a computer. These stress levels duplicated the launch, ascent, on-orbit, reentry, and landing phases of flight. Three 1,000,000 pound-force hydraulic cylinders were used to simulate the thrust from the SSMEs. Heating and cooling simulations were conducted along with the stress tests – gaseous nitrogen being used to simulate the cold of space flight, while heating blankets developed for the Lockheed A-12/SR-71 test program were used to simulate ascent and reentry heating. Twenty-five different heating and cooling zones were established along the length of the fuselage, and the thermal loads were applied directly to the airframe's metal structure (i.e.; no thermal protection system was installed).[190]

During late 1977 a decision had been made that it would be less expensive to modify the STA-099 airframe to a flight vehicle than it would be to modify *Enterprise*. Traditionally, manned spacecraft had been tested to 140 percent of their design strength – a practice that was common the aerospace industry. The SRB, ET, and various small components of the Orbiter had already been tested at this level. But testing STA-099 to 140 percent might result in damage that would make it unadvisable or uneconomical to modify it into a flight vehicle. Engineers led by Thomas L. Moser at JSC thought they had an alternative. Before testing began they developed an analytical model that simulated all of the 3,000+ measurement points on the airframe. The model showed that the Orbiter could easily withstand 140 percent loads. But still, nobody wanted to risk the airframe. The engineers convinced management that if the actual data from the tests matched the model at 120 percent, then there was no need to go to 140 percent – at that point you had confidence the model was accurate. If the model was not matching reality at 120 percent, then the tests would continue to 140 percent. Although there were some skeptics, the idea was approved.[191]

Moser was right – the stress distributions in the critical test regions compared within ten percent of the pretest model results with few exceptions. These exceptions included the ET forward and aft fittings and support structure, the distribution of reaction between the forward and aft spars of the vertical stabilizer, and in the y-load distribution of the main landing gear reaction. Supplementary instrumentation and component test articles were defined to fully instrument and certify these regions. The thermal tests measured stresses that compared well with preliminary analysis. The skin stresses in the circumferential directions were considerably less (30 percent) than predicted analytically, and in generally good agreement longitudinally. Hardware modifications that resulted from the testing included the forward reaction control system tank support structure, which failed because of inadequate lateral stiffness, and the vertical stabilizer which was redesigned to redistribute the loads between the front and rear spars.[192]

A conscious decision was made early in the program to limit the Orbiter's maximum dynamic pressure (max-q) to

The structural test article (STA-099) is positioned in the test rig at Lockheed-Palmdale. The three large hydraulic cylinders used to simulate the SSMEs are visible at the rear. Note the wing leadings edges are attached in this photo (Rockwell International)

The test rig is closed around STA-099 on 15 March 1978. For all the hi-tech equipment used to test space shuttle, the flooring of the test stand was made of redwood planks. (Lockheed-California Company)

STA-099 completely engulfed in the Lockheed test rig. Noteworthy are the actuators and beams attached to the wing (visible at center left) necessary to induce flight loads. Note that the wing leading edges are missing in this photo. (Lockheed-California Company)

ORBITER MILESTONES	MPTA-098	Challenger STA-099	Challenger OV-099	Enterprise OV-101	Columbia OV-102	Discovery OV-103	Atlantis OV-104	Endeavour OV-105
Contract award	26 Jul 72	26 Jul 72	05 Jan 79	26 Jul 72	26 Jul 72	29 Jan 79	29 Jan 79	31 Jul 87
Start structural assembly crew module *	n/a	21 Nov 75	28 Jan 79	21 Jun 74	04 Jun 74	28 Jun 76	03 Mar 80	15 Feb 82
Start structural assembly aft-fuselage	24 Jun 75	14 Jun 76	n/a	26 Aug 74	13 Sep 76	10 Nov 80	23 Nov 81	28 Sep 87
Wings arrive at Palmdale from Grumman	n/a	16 Mar 77	n/a	23 May 75	26 Aug 77	30 Apr 82	13 Jun 83	22 Dec 87 •
Start final assembly	12 Jul 76	30 Sep 77	03 Nov 80	25 Aug 75	07 Nov 77	03 Sep 82	02 Dec 83	01 Aug 87
Complete final assembly	27 May 77	10 Feb 78	23 Oct 81	12 Mar 76	23 Apr 78	12 Aug 83	10 Apr 84	06 Jul 90
Rollout (at Palmdale)	n/a	14 Feb 78	30 Jun 82	17 Sep 76	08 Mar 79	16 Oct 83	06 Mar 85	25 Apr 91
Overland transport from Palmdale to Edwards	n/a	n/a	01 Jul 82	31 Jan 77	12 Mar 79	05 Nov 83	09 Apr 85	n/a
Delivery to Kennedy Space Center **	24 Jun 77	n/a	05 Jul 82	10 Apr 79	25 Mar 79	09 Nov 83	03 Apr 85	07 May 91
Flight Readiness Firing +	21 Apr 78	n/a	19 Dec 82	n/a	20 Feb 81	02 Jun 84	05 Sep 85	06 Apr 92
First space flight	n/a	n/a	04 Apr 83	n/a	12 Apr 81	30 Aug 84	03 Oct 85	07 May 92
Orbiter empty weight (without SSMEs) at rollout	——	——	155,400	150,000	158,289	151,419	151,315	151,205
Leave KSC for first OMDP ‡	——	——	——	——	08 Oct 94	17 Feb 92 ++	17 Oct 92	29 Jul 96
Arrive KSC after first OMDP	——	——	——	——	14 Apr 95	03 Nov 92	29 May 94	27 Mar 97
Leave KSC for second OMDP	——	——	——	——	24 Sep 99	27 Sep 95	11 Nov 97	
Arrive KSC after second OMDP	——	——	——	——	26 Feb 01 ‡‡	29 Jun 96	27 Sep 98	

Notes:
*	=	simulated crew module on STA-099
**	=	delivery to NSTL for MPTA-098
+	=	first full-up static firing for MPTA-098
‡	=	*Columbia* had several 'mini' modifications prior to OMDP
•	=	structural spares in storage at Downey
++	=	Performed at KSC instead of Palmdale
‡‡	=	Estimate for OV-102 return from OMDP-2
n/a	=	not applicable

Orbiter empty weight is in pounds (see page 438 for current weight data)
Enterprise empty weight includes ballast.

This list of Orbiter assembly data details the significant milestones in the assembly of each Orbiter. OV-105, *Endeavour*, is hard to break out since the major parts of the airframe were initially procured as 'structural spares,' with no intention of assembling them into an Orbiter. This explains the significant gap between the structural assembly of the crew module, and the award of a production contract.

650 pounds per square foot nominal, and the maximum longitudinal acceleration to 3-g. The desired effect was to keep the q-sensitive parameters such as peak differential pressure, buffet intensity, aerodynamic noise and loading, control authority, and flutter requirements within known and manageable bounds. The constant battle to maintain the empty weight of the Orbiter to 150,000 pounds, and the mission flexibility built into the vehicle subsequently resulted in changes to the ascent loading requirements late in the Orbiter design cycle.[193]

Changes to the detailed structural design criteria were necessary to minimize the impact of the updated ascent configurations, and this resulted in essentially two different loads criteria being applied to the various Orbiters. MPTA-098, STA-099, OV-101, and OV-102 were designed to a so-called 5.1 loads data base, even though it was clear during construction that later certification to a 5.4 loads data base would be needed. All future Orbiters would be designed to the newer 5.4 loads data base, and the early flight Orbiters would eventually be modified to this level.[194]

As it ended up, MPTA-098 was certified to the new level by analysis, *Enterprise* was placarded* to 80 percent of the load limit, *Columbia* was restricted to an 80 percent load during the orbital flight test phase, and modified after STS-9 to the new limits. STA-099 was modified to the 5.4 loads data base during her conversion to OV-099.[195]

Part of the new loads data base analysis also allowed Rockwell to relax the requirements for the wing design on OV-103 and OV-104 in order to achieve a slight weight reduction. But subsequent flight data acquired by instrumentation installed on *Columbia* during early operational missions raised questions about the structural integrity of the relaxed standards, and the wings were subsequently modified by the addition of doublers and stiffeners during the *Challenger* stand-down. The wings for OV-105 were constructed to the modified design.[196]

During 1992, in order to raise the maximum allowable landing weight of the Orbiter to 250,000 pounds, NASA implemented a 6.0 loads data base, and each Orbiter is being modified as it goes through its OMDP cycle.[197]

Rockwell's original $2,600 million Orbiter production contract (NAS9-14000) had authorized the building of a pair of test articles (MPTA-098 and STA-099) and two initial flight-test Orbiters (OV-101 and OV-102). The final

assembly of OV-102 started on 7 November 1977, with rollout from the Palmdale plant on 8 March 1979. *Columbia* arrived at the KSC on 25 March 1979. In the meantime, on 5 January 1979, NASA ordered Rockwell to modify STA-099 into a space-rated Orbiter (OV-099), and followed this on 29 January 1979 with an order to construct two additional Orbiters (OV-103 and OV-104). This $1,900 million contract extension also covered modifying OV-102 following the orbital flight-test series. The eventual addition of OV-105, plus a significant amount of modification to all the Orbiters, have resulted in this contract being valued at $5,815,778,189 by early 1996.[198]

After testing was completed, STA-099 was returned to Rockwell on 7 November 1979 for conversion into OV-099. This conversion, while easier than it would have been to convert *Enterprise*, still involved a major disassembly of the vehicle. Within a month of arriving back from Lockheed, the payload bay doors, elevons, and body flap had been removed so they could be returned to the original vendors for modifications. By 18 January 1980 the vertical stabilizer had been removed and shipped back to Fairchild-Republic in New York for rework. *Challenger* had been built with a simulated crew module, and the forward fuselage halves had to be separated to remove it – this occurred on 15 February 1981, and the upper forward fuselage was subsequently sent to Downey for rework (the lower fuselage was modified in Palmdale). Additionally, the wings were modified to the 5.4 loads data base in Palmdale. The entire aft fuselage was removed and sent to Downey for modifications, returning to Palmdale on 21 July 1981. *Challenger* would end up some 2,889 pounds lighter than OV-102, in spite of having additional operational equipment installed and the more robust structure.[199]

Discovery (OV-103) and *Atlantis* (OV-104) would benefit from lessons learned and weighed approximately 6,870 pounds less than OV-102 at delivery. *Discovery* was rolled out on 16 October 1983 and was delivered to the KSC on 9 November. *Atlantis'* was rolled-out on 6 March 1985, with delivery to KSC following on 3 April. The experience gained during the assembly process enabled

---

* Installing a placard in the cockpit/flight manual of a vehicle is common practice in the aviation industry, placing the burden on the pilots not to exceed certain known limitations of the vehicle.

OV-104 to be completed with a 49.5 percent reduction in man-hours compared to OV-102. A rather significant part of the decrease can probably be traced to the greater use of thermal protection blankets on *Atlantis*, which required less manpower to install than LRSI tiles.[200]

Interestingly, the total number of man-hours required to assemble OV-105 was considerably greater than those required for the final assembly of OV-104. This can be directly attributed to the fact that the Orbiter assembly line had been shut down for almost two years when *Endeavour* was ordered, and many of the skilled and experienced workers had found other jobs, mostly on the Rockwell B-1B bomber program (which was also built at Air Force Plant 42 in Palmdale).[201]

During the construction of OV-103 and OV-104, NASA had opted to have the various contractors manufacture a set of 'structural spares' to facilitate the repair of an Orbiter if one was damaged during an accident. This $389 million contract amendment was issued in April 1983 and was seen as a way to keep the industrial base alive to preserve an option for the fifth Orbiter. The spares consisted of an aft-fuselage, mid-fuselage, forward fuselage halves, vertical stabilizer and rudder, wings, elevons, and a body flap. Following the destruction of *Challenger*, NASA authorized Rockwell to assemble these spares into the fifth Orbiter with a contract (NAS9-17800) awarded on 31 July 1987 valued at $1,371,599,318, plus award fee.[202]

OV-105 was built with all the improvements and modifications incorporated on the other Orbiters since the beginning of the program. In addition, a few new items were aboard, including a 39-foot braking parachute, much like that deleted in 1974. The drag chute was meant to relieve the stress on the brakes, and also reduces the landing rollout distance by 1,000–2,000 feet.[203]

A series of eight drag chute deployment tests had been carried out at NASA Dryden using the NB-52 during 1990 to validate the design. The NB-52 landed at speeds ranging from 160 to 230 mph on one of the lakebed runways, and also on the 15,000-foot concrete runway at Edwards AFB. Instrumentation on the NB-52 obtained data during the deployments to validate predicted drag loads that an operational Orbiter would sustain using a drag chute during landing and roll-out.[204]

Endeavour was also equipped as the first extended duration orbiter (EDO), capable of remaining in orbit up to 28 days. This capability has since been removed during OMDP to save weight for ISS missions. Updated avionics systems installed on OV-105 included AP-101S general purpose computers (GPC), improved inertial measurement units and tactical air navigation (TACAN) systems, enhanced master events controller (EMEC), enhanced multiplexer-demultiplexers (EMDM), a solid-state star tracker, and an improved nose wheel steering system. An improved version of the auxiliary power unit (IAPU) was also installed. These modifications would be incorporated into the rest of the Orbiter fleet as funds permit. *Endeavour* was delivered to KSC on 7 May 1991, and her maiden flight, STS-49, occurred a year later on 7 May 1992.[205]

During 1988, the NASA Space Station program office studied the purchase of an additional Orbiter for use as a space station rescue vehicle. It was felt that 'piggy-backing' the purchase of another Orbiter during the construction of OV-105 might be more economical that designing a new vehicle for the rescue function. Because of various economic and political problems, this did not occur, and the Orbiter production line was shut down after the completion of OV-105. It would be partially opened again later in 1991 to perform OMDP modifications to the Orbiters, an effort that is expected to continue for the life of the program.[206]

On 6 October 1989 NASA let contracts for another set of structural spares, similar to the ones used to construct *Endeavour*. Unfortunately, budgetary constraints caused NASA to terminate the contract before any significant spares were completed, raising doubts about the future ability to repair a seriously damaged Orbiter.[207]

During 1993 Rockwell proposed the development of an automated orbiter kit (AOK) that would allow an Orbiter to fly without a crew. The AOK maintained all current mission capabilities, while adding automated rendezvous, docking, and landings. A weight reduction of 12,500 pounds would have been achieved by eliminating all crew support equipment such as seats, lithium hydroxide canisters, stowage lockers, galley, hygiene facilities, breathing air cryogenic storage tanks, the aft radiator panels, and most flight deck windows. These items would be replaced by approximately 1,000 pounds of automation equipment installed on the Orbiter mid-deck. Automated sequences could be commanded from Mission Control, or stored on-board and executed autonomously. Rockwell estimated it would take three years to develop the system and retrofit an existing Orbiter, most likely *Columbia*. NASA declined the offer.[208]

The Orbiter 'assembly line' at the Rockwell facility in Palmdale during mid-1983. The major components of Atlantis are in the foreground. The final assembly areas are to the extreme left, with a small sign pointing to Discovery. A few months after this photo was taken, both OV-103 and OV-104 would be side-by-side in final assembly. (NASA photo 108-KSC-83PC-702)

As part of the original space shuttle decision, NASA committed to providing a Space Tug that would move payloads into and out of orbits higher than the Orbiter could reach. This reusable tug was to be capable of reaching geosynchronous orbits, and some versions of it were to be able to carry satellites to lunar orbit. The tug was to be an autonomous vehicle, and some of the more advanced versions proposed were capable of servicing or refueling satellites without human intervention. As part of the work sharing agreement, MSFC was assigned the responsibility of developing the Space Tug.[209]

Unfortunately, the tug became one of the first casualties of the budget dilemma that faced NASA, and development was postponed. MSFC began looking at existing upper stages that might be able to serve until the budget crisis passed, and also at other possible funding sources for the Space Tug. The source that immediately sprang to mind was the Air Force. George Low met with Malcom R. Currie, Director of Defense Research and Development (DDR&E) during August 1973 to discuss various aeronautic facility issues, but the subject of the Space Tug was also broached. According to Low, "Currie is not at all adverse to this idea ..."[210]

Low and Currie met again a month later and tentatively agreed that the Air Force would develop an interim upper stage for the space shuttle. The preliminary concept involved modifying an existing upper stage at a cost not to exceed $100 million. But Currie did not have the authority to commit the Air Force to such a program, and it would not be until 11 July 1974 that a formal endorsement came. Four existing upper stages were considered as a basis for the newly christened, logically, Interim Upper Stage (IUS) – Agena, Centaur, Delta, and the Transtage from Titan. Interestingly, all of these stages used liquid propellants. NASA preferred the Centaur mainly because it had sufficient capacity for planetary missions. The Air Force preferred the Transtage since it would make it easy to switch payloads between the shuttle and Titan III if the need arose. Boeing had another idea – a variation of the Solid Burner II. This solid-propellant upper stage was too small for consideration, but Boeing believed they could easily modify it to meet the projected requirements.[211]

At the beginning of October 1974 the Air Force awarded $635,000 nine-month study contracts to Boeing (Solid Burner), General Dynamics/Convair (Centaur), Lockheed (Agena), McDonnell Douglas (Delta), and Martin Marietta (Transtage). The Air Force made it clear that the IUS they were intending to build would meet DoD requirements, but possibly not all of NASA's – particularly the planetary ones. George Low later commented that "We [NASA] would either like to get the high performance of the Centaur stage or the simplicity of the solid stages. The Transtage, which may well be the Air Force's number one candidate, has the disadvantages of both and the advantages of neither."[212]

On 4 September 1975 the Air Force decided that the IUS would be expendable and use solid propellants. In the grander scheme of things this was considered the wisest move – solids were less expensive to develop, and were potentially much safer to carry in the confined payload bay of the Orbiter. They were also easier to handle, both on the ground and in orbit, and generally simpler to integrate with a satellite prior to launch. But although the Air Force had selected the approach advocated by Boeing, it had not yet selected a contractor. An RFP was released for the solid-motor expendable IUS and four contractors responded – Boeing, General Dynamics, Lockheed, and Martin Marietta. In August 1976 the Boeing Aerospace Company in Seattle was selected to develop the IUS under contract to the Space and Missile Systems Center of the Air Force Materiel Command.[213]

Boeing proposed a rather unique two-stage design. The smaller stage used 6,000 pounds of propellant and could boost Delta-class payloads from the shuttle to low-earth orbit. The larger stage used 21,400 pounds of propellant and could do the same for Atlas–Centaur-class payloads. By using both stages together, most geosychronous payloads could be launched. Boeing also devised a scheme that used two large stages topped by a single small one that could potentially boost most planetary missions, but by this time the payload bay was getting full and it was questionable if the 3-stage IUS would actually fit with a payload attached. Various other combinations could also be used as needed for any payload, a seemingly versatile arrangement.[214]

But interestingly, the IUS did not satisfy NASA's potentially largest customer – the communication satellite industry. Most communication satellites are spin stabilized, a technique adopted because it is much less expensive than a three-axis stabilization system. Hughes Aircraft and The Aerospace Corporation conducted studies that concluded that a turntable mounted in the payload bay could impart the necessary rotation on a satellite, point it in the correct direction, eject it a short distance, then allow the Orbiter to slowly back away. After a safe distance was achieved, an attached solid motor would ignite and carry the satellite to geosynchronous orbit. Several small existing solid motors appeared capable of performing this role.[215]

McDonnell Douglas picked up on this concept, and offered it to NASA for use on the TDRS satellites being designed for the shuttle program. NASA declined the offer, but McDonnell Douglas management seized on the opportunity to use company funds to develop the idea and offer it for sale to commercial, NASA, and DoD users. This was the beginning of the spinning solid upper stage (SSUS), which later was renamed the payload assist module (PAM). The SSUS was derived from the Thiokol TU-844 solid motor used as the third stage of the Minuteman III ICBM with the TVC and roll control systems removed. Two versions were designed, the SSUS-A that could accommodate 4,400-pound Atlas–Centaur payloads, and the SSUS-D that could accommodate Delta-class payloads up to 2,750 pounds.[216]

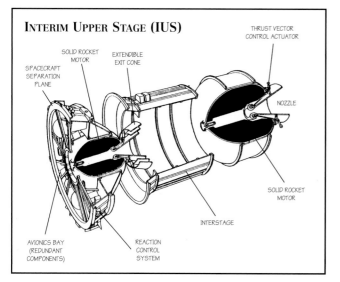

INTERIM UPPER STAGE (IUS)

THRUST VECTOR CONTROL ACTUATOR

SOLID ROCKET MOTOR

EXTENDIBLE EXIT CONE

SPACECRAFT SEPARATION PLANE

NOZZLE

SOLID ROCKET MOTOR

INTERSTAGE

AVIONICS BAY (REDUNDANT COMPONENTS)

REACTION CONTROL SYSTEM

Studies of the Space Tug continued at a low level at MSFC and several contractors, but it was becoming increasingly obvious that it would never be built. A similar concept would reappear as the Orbital Maneuvering Vehicle (OMV) and the Orbital Transfer Vehicle (OTV) during the mid-1980s (see page 455). Symbolically, in December 1977 the IUS was renamed the Inertial Upper Stage (thereby keeping the same abbreviation), and was extended to the Titan family of launchers as well.

The IUS development did not proceed as quickly or as smoothly as anybody expected. There were difficulties coming up with an agreeable avionics design, and the solid motors themselves experienced quality-control problems. By mid-1979 the first launch had slipped either one year (Boeing estimate) or two years (Air Force estimate), depending whom you believe. It did not matter much since the first launch of shuttle was slipping even faster. The delays were adding to the cost however – the initial development contract had been for $263 million – by the end of 1979 this had risen to $430 million, and six months later topped $500 million. Boeing did finally succeed – the first IUS was launched on a Titan 34D from Cape Canaveral on 30 October 1982.

But the IUS being developed by Boeing had a significant limitation – it was not powerful enough to launch a payload to a distant planet like Jupiter in a direct shot. To get to its destination the payload would have to slingshot around several planets to gain a gravity-assisted speed boost. Most engineers considered this approach inelegant and too time consuming. In addition, the planetary model of the IUS had been placed on the back burner while Boeing tried to get the standard Air Force version developed. During this time its costs had risen more than 50 percent to $179 million – more than NASA was willing to pay. Robert Frosch argued that the only other alternative was Centaur: "No other alternative upper stage is available on a reasonable schedule or with comparable costs."[217]

The Centaur had been developed by the Convair Division of General Dynamics under an MSFC contract during the 1950s and was the first successful use of LO2 and LH2 as propellants. Management of the project moved from MSFC to NASA Lewis during 1962. Fittingly, given its Convair heritage, the first use of the Centaur was as an upper stage for the Atlas booster – a successful November 1963 launch signaled the first in-flight ignition of a LO2/LH2 engine.[218]

During the 1960s engineers had explicitly rejected Centaur for manned missions – in 1961 NASA Langley considered the possibilities of using Centaur for lunar and planetary missions but came to the conclusion that it was "out of the question." Centaur was not 'man-rated' and most predicted that it never would be. But by the early 1970s this opinion began to change, and during 1973 General Dynamics/Convair conducted the Reusable Centaur Study that analyzed safety, cost, schedule, weight, reliability, reusability, and flight complexity. The final assessment was that the "Centaur programs are extremely low risk" and that no technology breakthroughs were required to achieve a reusable Centaur.[219]

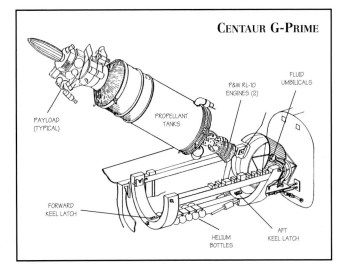

CENTAUR G-PRIME

PAYLOAD (TYPICAL)

P&W RL-10 ENGINES (2)

FLUID UMBILICALS

PROPELLANT TANKS

FORWARD KEEL LATCH

HELIUM BOTTLES

AFT KEEL LATCH

There were a number of reasons that Centaur was more attractive than the IUS. First and most important was that Centaur was more powerful and had the ability to propel a payload directly to another planet. Second, Centaur was 'gentler' – solid rockets had a harsh initial thrust that had the potential to damage the sensitive instruments aboard a planetary payload. Liquid rockets increase their thrust more slowly, thus reducing this threat. The perceived advantage of the IUS over the Centaur was safety – LH2 presented a significant challenge. Nevertheless, NASA decided to accept the risk and go with the Centaur.[220]

The Lewis Research Center seemed the logical choice to manage the new Centaur program due to its decade of experience with the upper stage. But, somewhat unexpectedly, Lewis encountered strong opposition from within NASA, specifically the manned space centers. The leadership at KSC, JSC, and MSFC opposed a Lewis-led Centaur program for space shuttle, and they recommended that MSFC should take the lead. In May 1981, Alan Lovelace decided to allow Lewis to keep the lead role. This altered the politics surrounding Centaur, and probably guaranteed that it would ultimately fail.[221]

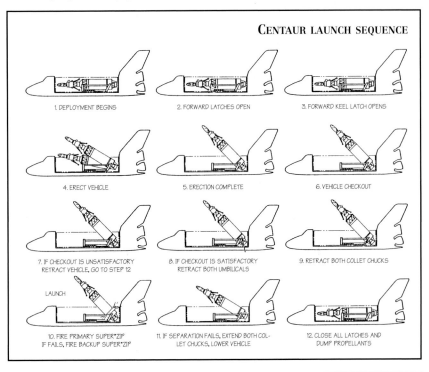

CENTAUR LAUNCH SEQUENCE

1. DEPLOYMENT BEGINS

2. FORWARD LATCHES OPEN

3. FORWARD KEEL LATCH OPENS

4. ERECT VEHICLE

5. ERECTION COMPLETE

6. VEHICLE CHECKOUT

7. IF CHECKOUT IS UNSATISFACTORY RETRACT VEHICLE, GO TO STEP 12

8. IF CHECKOUT IS SATISFACTORY RETRACT BOTH UMBILICALS

9. RETRACT BOTH COLLET CHUCKS

LAUNCH

10. FIRE PRIMARY SUPER*ZIP IF FAILS, FIRE BACKUP SUPER*ZIP

11. IF SEPARATION FAILS, EXTEND BOTH COLLET CHUCKS, LOWER VEHICLE

12. CLOSE ALL LATCHES AND DUMP PROPELLANTS

Many of the technical features of the Shuttle/Centaur were similar to the Atlas/Centaur configuration, but with two major differences. The first was a change to the stage itself to allow it to fit into the payload bay – both the Centaur and its payload had to fit within roughly 60 feet of length. Engineers developed two different configurations of the Centaur that met this requirement. The Centaur G was 20 feet long, carried 29,000 pounds of propellants and a 40-foot payload, and was designed to place satellites into geosynchronous orbit. The other* configuration, Centaur G-Prime, was nearly 30 feet long and capable of carrying 30.4-foot payloads. This version was optimized for planetary missions and the payload space was smaller since the stage used longer tanks carrying 45,000 pounds of propellants. The Air Force funded half of the initial $269 million required to develop the Centaur G as a backup to the IUS, while NASA funded the G-Prime configuration since the Air Force had little interest in going to other planets. Overall, the two versions shared over 80 percent of the basic design.[222]

The second new technical development was the launching mechanism – the centaur integrated support structure (CISS). This 10-foot diameter aluminum structure supported the upper stage in the payload bay, and enabled the Centaur to fly with a limited number of design modifications to the Orbiter itself. The CISS was responsible for all of the mechanical, electrical, and fluid interfaces between the Centaur and the Orbiter, and could be refurbished and reused for up to ten missions.[223]

Two Orbiters, *Challenger* and *Discovery*, were modified at KSC to enable them to carry the Centaur. The modifications included extra plumbing to load and vent the cryogenic propellants, and controls on the aft flight deck for loading and monitoring Centaur. These modifications cost approximately $5 million per Orbiter. *Atlantis* was initially delivered from Palmdale with a Centaur capability. Modifications to the launch pads at KSC and Vandenberg were also completed.

The LO2 was to be loaded from plumbing tapped into the MPS lines inside the aft compartment. The LH2 fill and drain was serviced with a second umbilical plate located just in front of the left side aft compartment access door. The hardware on the FSS to mate to the flight umbilical was called the RBUS (Rolling Beam Umbilical System) and extended from a platform on the FSS out to the Orbiter. At lift-off it was to be pulled straight back into the platform by a massive drop weight powering retract cables and pulleys.

The Centaur dump system vented LO2 from the right side of the aft fuselage and LH2 from the left. There was also a relief valve port between the left OMS pod and the vertical stabilizer The relief valve presented some concerns since there were some abort scenarios where the crew would not have enough time to finish the cryo dumps and there was the possibility of hydrogen burping from the relief valve after landing and being ignited by the hot APU exhaust ports. Late in the project there was some consideration of installing a vent line that ran inside the aft fuselage, up through the vertical stabilizer and out through the tip of the stabilizer, but this did not materialize prior to the project being cancelled.

The destinations for the first two Shuttle/Centaur launches were both extremely important to the scientific community. *Galileo* was NASA's only planned planetary mission under development during the 1980s. Originally this mission was scheduled to be launched with a two-stage booster (IUS), but Centaur was ultimately chosen to provide faster trip. Initially the *Galileo* team was pleased with this

decision, never suspecting that soon they would have to search for an alternative. The second Shuttle/Centaur mission was a joint venture with the European Space Agency. Originally called the International Solar Polar Mission, the probe was renamed *Ulysses* in 1984.[224]

In May 1985 KSC was busy making plans for the *Galileo* and *Ulysses* launches – more complicated than usual since there were only 6 days between the launches, with each having only a one hour window. Therefore, there would be a shuttle on both Pad 39A and 39B at the same time, each with a Centaur G-Prime in the payload bay. But before either could be launched, *Challenger* exploded, and immediately the Shuttle/Centaur program was in jeopardy – suddenly the upper stage appeared to be an unacceptable risk.

During May 1986 NASA Lewis held a series of meetings regarding Centaur safety in light of the *Challenger* accident. Representatives from the Air Force, Analex, Boeing, General Dynamics, JSC, KSC, Lockheed, Martin Marietta, MSFC, and TRW attended. The goal was to prove that the Centaur was safe – but the participants could not agree that the stage presented an acceptable risk situation.[225]

In late June 1986, James Fletcher, now serving a second term as NASA Administrator, gave the order: "You are directed to terminate the Shuttle/Centaur upper stage program." After spending almost $1,000 million, the program ended – the official reason was labeled "safety concerns." One report stated, "The final decision was made on the basis that even following certain modifications identified by the ongoing reviews, the resultant stage would not meet safety criteria being applied to other cargo or elements of the space shuttle system." Fletcher stated that "although the Shuttle/Centaur decision was very difficult to make, it is the proper thing to do…"[226]

This decision was in keeping with numerous safety studies that had been made as early as 1972 (see page 156), all of which had recommended against carrying a liquid-propellant upper stage in the payload bay of a manned spacecraft. No Centaur flights were ever flown from shuttle.

Fifth Orbiter Debate

The joint NASA-DoD position paper of 1973 had anticipated 581 shuttle flights between 1979 and 1999, and based on anybody's best guess of operational usage, derived a requirement for a five Orbiter fleet. And since the beginning of Phase C/D, NASA had envisioned a five Orbiter fleet to meet the anticipated use of space shuttle. The plan was to dedicate three Orbiters to KSC, and two to Vandenberg in the following order:

Orbiter	Delivery Date	Launch Site
2	August 1978	KSC
3	December 1979	KSC
1 (OV-101)	December 1980	KSC
4	December 1981	VLS
5	December 1982	VLS

But as the development cost of space shuttle rose, available funding became tight, and in the end only three Orbiters were authorized – and only two had been ordered (OV-101 and OV-102 on 26 June 1972). As early as 1976 NASA began lobbying to get funds for a five Orbiter fleet, usually enlisting the assistance of the Department of Defense to present a united front. For instance, on 23

---

* Centaur G-Prime was also known as Centaur F within General Dynamics/Convair in San Diego, at least for the *Galileo* mission.

January 1976 George Low (NASA Deputy Administrator) and Malcolm Currie (Director of Defense Research and Engineering) released a joint position statement on space shuttle procurement that read in part: "The National Aeronautics and Space Administration and the Department of Defense agree that five Space Shuttle Orbiters are needed to meet our national traffic model requirements." But the statement went on to recognize that NASA had only funded three Orbiters so far, and that neither DoD nor NASA had funds for any further procurement.[227]

On 22 October 1976, James Fletcher wrote a letter to the Office of Management and Budget presenting the NASA and DoD case for a five Orbiter fleet. This was a follow-up to various other letters between NASA and OMB, and expanded upon a briefing given to OMB on 10 September. The three page letter was very eloquent in pleading the case for a five Orbiter fleet, waving the flag of national prestige, and warning of possible defense shortcomings if such a fleet was not available. The letter also touched upon a crucial economic point: "An FY 1978 start on the procurement of additional Shuttle Orbiters will be required to maintain reasonable schedules and to avoid the severe cost penalties of a break in production." At this point the national traffic model was still predicting 60 flights per year, and that a five Orbiter fleet was more cost effective than a mixed vehicle fleet using 3–4 shuttle Orbiters and some expendable launch vehicles.[228]

Of significance in the letter was a mention that "Attrition of an Orbiter from the three Orbiter fleet would significantly worsen this posture [space launch capacity]." The letter noted that since 1972 NASA and the DoD had funded $11,000 million in shuttle development and procurement, and that it was "prudent to add the approximate additional ten percent ... to provide the fleet size we believe is the minimum essential ..." The projected cost for two addition Orbiters was listed as $1,177 million in FY78 dollars spread over seven years. Although DoD agreed with the procurement of additional Orbiters, it stated that funding "for the additional Orbiters should be placed where the responsibility for management and performance now rests: with NASA."

The OMB, on the other hand, was firmly convinced that three Orbiters were sufficient. The analysts at the budget agency seriously doubted that demand would ever materialize for the ambitious 581-flight manifest that NASA and DoD advertised, and wanted to defer the production of the last two Orbiters until such time that it was more obvious the users would materialize.

What finally convinced the OMB, and indeed, the Carter Administration that more than three Orbiters were needed was an excellent appeal by Secretary of Defense Harold Brown. Sidestepping the issue of flight rates and user demand, Brown pointed out that the first two Orbiters (*Columbia* and *Challenger*) had come in above their weight targets, and were too heavy to carry the largest reconnaissance satellites. While the third operational Orbiter (*Discovery*) would be able to do so easily, if it was lost in an accident, national security would be severely impacted. Brown recommended the procurement of a fourth Orbiter – using NASA funds – to ensure national security considerations were met. This argument weighed heavily on Jimmy Carter, and ultimately was approved.

By late 1977 the issue of a fifth Orbiter had been broached several times with Carter, and during early December Robert Frosch apparently thought he had approval to begin negotiations for an option for a fifth Orbiter. But James T. McIntyre, Jr. the acting director of

the OMB, was quick to point out that Carter had clearly supported a four Orbiter fleet and agreed that NASA should negotiate an 'option for an option' for a fifth Orbiter. McIntyre wrote: "The option for a fifth Orbiter should be kept open for <u>future</u> Presidential consideration and it is NASA's obligation to assure that no actions, contractual or otherwise, are taken that might tend to preempt the President's future decision on a fifth Orbiter." The White House position was that plans should be made for an early transition from expendable launch vehicles to the shuttle, for both military and civilian payloads, from launch sites on both coasts by 1984. The Administration believed that a total fleet of four operational Orbiters would meet civilian and military shuttle flight requirements, and intended to fund the production of a four-Orbiter fleet beginning in the FY79 budget.[229]

On 5 January 1979 Rockwell was authorized to modify STA-099 into OV-099, and three weeks later NASA ordered the two 'production' Orbiters (OV-103 and OV-104), along with the 'option-for-an-option.'

As part of this option-for-an-option, some interesting language was written into the Rockwell Orbiter production contract: "NASA may be authorized to procure a fifth Space Shuttle Orbiter vehicle. Should that authorization materialize, the parties [NASA and Rockwell] agree to enter into negotiations for the procurement of such vehicle." The contract went on to require that Rockwell submit budget and schedule estimates for the fifth Orbiter to NASA twice a year – on 24 January and 24 July. These reports continued for several years.[230]

This essentially ended the debate over the fifth Orbiter. Although the DoD and NASA repeatedly briefed the Carter and Reagan administrations, no further budget authority would be forthcoming. Eventually the option-for-an-option for the fifth Orbiter was partially exercised, resulting in the procurement of a set of 'structural spares' as a means of keeping the Orbiter supplier-base active while a final last-ditch effort was made to secure permission to build a fifth Orbiter. This request was turned down, and the structural spares were stored and later used as the basis for constructing OV-105 after the *Challenger* accident.

THE FINAL PUSH

As with any major NASA program, especially one involving human operations, NASA maintained an active safety program that constantly reviewed the risks associated with the program. For shuttle, these were documented in a document called JSC-09990 that was maintained by the Safety Division at JSC. As of late November 1976 there were 24 open safety concerns, 18 closed safety concerns, and 9 accepted risks.[231]

These issues and risks ran a wide variety of subjects from major high-level concepts, to very specific technical details. For instance, one concerned the subject of on-orbit rescue during early flights. Potential mitigation included retaining the capability to launch an Apollo capsule if needed, carrying an Apollo capsule in the payload bay of the Orbiter during the orbital flight tests, or delaying the FMOF until a second Orbiter was ready. In the end the program decided to accept the risk with no further action. Other concerns included the ability of the windshield to withstand bird strikes (analysis concluded it could, although no testing was ever accomplished), the decision to use unpowered landings (closed based on the X-15 and lifting-body results at NASA Dryden), the placement of a smoke detector in the crew compartment of OV-101 (flights were believed short enough to warrant not having one), and the lack of a thrust termination system for the SRBs (closed since the SRBs "are inherently reliable").[232]

As late as 1978 NASA still envisioned a six flight orbital flight test (OFT) program. Each of these flights would be limited to 100 percent SSME power – a criteria that became increasingly important as the SSME program lagged behind and experienced severe difficulties with power levels above 100 percent. Only two pilots would be carried on each flight, and OV-102 would be used for all six. Ejection seats would be fitted to *Columbia* that allowed meaningful escape during ascent and reentry at speeds under Mach 2.7 and altitudes below 75,000 feet.[233]

But the expected 1979 first manned orbital flight was in serious jeopardy. Problems with the SSME and TPS were conspiring to push the flight into early 1980 at the earliest. During the fall of 1979, the OMB recommended canceling the space shuttle program based on the delays and problems being encountered during its development. Fortunately, the Department of Defense again came to the rescue and convinced the Carter Administration to allow the program to continue.[234]

In late 1979 Robert Frosch met with President Carter to review the status of the shuttle program. At this time the major hurdle remaining to be overcome was the successful certification of the SSME. Frosch estimated there was only a ten percent likelihood of the first manned orbital flight occurring by July 1980, with a 50 percent chance that it might occur by September. Both of these dates were predicated on the SSME being flight certified by June 1980. By this time the orbital flight test series had been cut to four flights, meaning that 'operational' flights might begin by December 1981 if the first manned orbital flight happened by September 1980. Vandenberg was expected to be operational by December 1983 assuming the "timely delivery" of the three production Orbiters – OV-099 in June 1982, OV-103 in September 1983, and OV-104 in December 1984.[235]

Probably the most critical item on the road to first flight – the Space Shuttle Main Engine. This is serial number 0003, shown on 29 June 1976. (Rocketdyne)

Frosch noted that 75 percent of all shuttle testing had been successfully completed, and that the key risk at this point was final certification of the SSME. Just prior to this meeting, two engine problems had been discovered that needed additional analysis. Assuming that testing resumed by mid-December 1979, Frosch stated that at least seven additional successful tests were required, with a minimum of three weeks between the tests. Other potential risks included the thermal protection system (something that, in fact, did materialize), and the fact that all of the flight and ground pieces had never played together and there was some concern regarding software compatibility.[236]

During an independent review of the shuttle program, former astronaut William Anders pointed out that the Orbiter had not been tested to the same standards that had been applied to all other manned spacecraft. Specifically, a choice had been made to subject the STA-099 to only 120 percent of its designed loads instead of the more-normal 140 percent. The 20 percent difference was made up by analysis. (The ET and SRBs had been tested to 140 percent.) In addition, Apollo had several unmanned test flights that had further reduced risk. NASA management countered that the FMOF was a very "benign mission" that was designed to limit loads to 80 percent of design – effectively increasing the factor of safety to 180 percent of the design (based on the 120 percent testing, 20 percent analysis, and the margin by flying a benign mission). Interestingly, NASA touted the FMOF as a risk-reduction since the presence of astronauts aboard the Orbiter opened additional abort scenarios and contingency plans.[237]

Frosch also defended the cost of the shuttle program. Originally planned as a six-year development effort, the program was running approximately three years late. At least one year of this could be attributed to a lack of early-year funding due to OMB budget cuts, but the other two years were the direct result of technical problems with the SSME and TPS. By the end of 1979 the estimate at completion had increased nearly $2,900 million in FY81 dollars, or a total of about 20 percent over the original $5,150 million estimate in constant FY71 dollars. This was considered about normal for a high-technology project. In FY81 dollars, NASA now estimated the total shuttle program would cost $17,098 million, including $2,234 million of DoD funds (this did not include the cost of the Vandenberg launch site). Congress again supported the project, appropriating $826 million more than NASA had originally asked for in their FY81 budget request to cover the overruns. This sweeping support reflected a general (but not unanimous) appreciation that the space shuttle was something that America needed, and was a program that served the national interest.[238]

By the end of 1979 NASA had firm commitments from 15 different users who planned to fly 60 payloads during the first three years of shuttle operations. NASA had manifested 38 flights beginning in late 1981 to launch these payloads. Of these, 27 non-DoD payloads were being manufactured such that they could only be launched on shuttle. This put additional pressure on the program since it was not possible for these payloads to be remanifested on expendable launchers in the event of serious shuttle delays. NASA recognized their responsibility for potential launch delays, and agreed to help fund the development of an uprated version of Delta if shuttle continued to be delayed. Nevertheless, each payload customer would be responsible for paying the $17–24 million additional that an uprated Delta would cost versus launching on shuttle.

NASA also noted that Europe was developing the Ariane "as a potential backup for Delta class payloads …"[239]

A year later shuttle had still not flown, and Robert Frosch again released a review of the program. "1. The extraordinary attention which has been given to the Thermal Protection System over the past 18 months has greatly enhanced out knowledge of the system's requirements and our confidence in its capabilities. This has allowed us to … plan the first flight without including the Manned Maneuvering Unit and the Tile Repair Kit … 2. The formal certification process of the Shuttle main engine is about 70% complete. Despite recent problems, it is clear that the basic design has been proven. 3. The Flight Certification Assessment is essentially complete. Overall it has provided a strong endorsement to the manner in which the program is being carried out."[240]

Frosch continued that the plan was to roll OV-102 out of the OPF on 23 November 1980, with the intent "to launch by the end of March 1981, although we recognize that this is a tight schedule." It was a schedule that could not, politically, slip again. Fortunately, everything finally came together and the schedule (mostly) held.

WHAT'S IN A NAME ?

Sometime during the mid-1970s NASA decided upon a name for at least the first Orbiter, but any further list has apparently been lost to time. All that is known is that *Constitution* was proposed as the name for OV-101 in recognition of the U.S. Constitution Bicentennial that was occurring the same year the vehicle was rolled out. But the public (i.e. Trekker) fervor to change the name of the first Orbiter to *Enterprise* finally convinced Washington to do so. In response, and to keep a similar incident from happening again, NASA decided to formalize its naming convention.

On 26 May 1978 John F. Yardley, the NASA Associate Administrator for Space Transportation Systems, sent the Public Affairs Office a list of potential Orbiter names. "In accordance with paragraph 4 of [NASA Management Instruction] 7620.1, I convened and chaired a meeting of an ad hoc committee to recommend names for the Space Shuttle Orbiters. The committee consisted of Mike Malkin, Roy Day, Chet Lee, Dave Garrett, and Dan Nebrig. We elected to recommend names having significant relationship to the heritage of the United States or to the Shuttle's mission of exploration. The attached list of names is recommended in descending order of preference. The committee further recommends that the name *Enterprise* be reserved for Orbiter Five, assuming that there is a fifth Orbiter, to carry on the name assigned to Orbiter 101 during the [Approach and Landing Test] Program."[241]

The attached list included 15 names: *Constitution, Independence, America, Constellation, Enterprise, Discoverer, Endeavour, Liberty, Freedom, Eagle, Kitty Hawk, Pathfinder, Adventurer, Prospector,* and *Peace.*[242]

This apparently did not satisfy everybody, and the issue remained open. Another Orbiter naming committee met three times through the end of November 1978 to determine a strategy for naming the new spacecraft. The committee was chaired by Arnold W. Frutkin (Associate Administrator for External Relations) and included seven permanent members, plus numerous 'visitors' that also contributed ideas. Several themes were proposed: bright stars, constellations, American flight history, American his-

An unlikely combination – a Jumbo Jet and a spaceship. Enterprise and N905NA wait in the mate/demate facility at NASA Dryden on 18 June 1977. (NASA photo 77-P-427)

tory, explorer vessels, and American Indian tribes. The final recommendations were sent to Acting Administrator Dr. Alan M. Lovelace on 11 December 1978. "Three categories of names were selected by the committee by their vote in the following order of preference: (I) Explorers' Vessels – *Enterprise, Endeavor,* Discovery, Resolution, Adventure*; (II) American Tradition and Spirit – *Enterprise, Independence, Constitution, Freedom, Liberty, Republic, Columbia*; and (III) Stars and Constellations – [*Enterprise*], *Orion, Arcturus, Polaris, Pegasus, Canopus, Capella, Alpha Centauri.*" It was noted in earlier memos that *Enterprise* did not fit well into the third category, and that there was some concern over the name *Polaris* because of its association with the Navy fleet ballistic missile program.[243]

ORBITER NAMES

In the end, NASA chose explorer's ships – but did not use the suggested names. A subsequent decision was made not to reuse the name *Enterprise* after the vehicle was retired to the National Air and Space Museum.[244]

In March 1988, NASA Headquarters announced a contest to name OV-105. As a tribute to school teacher Christa McAuliffe, who was on *Challenger*, the contest was open to all kindergarten through 12th grade students in the United States. In honor of the crew, the name *Challenger* was permanently retired. During the fall of 1988, 71,652 students formed 6,154 teams which submitted 422 different names.[245]

The official list of finalist names included: *Adventure, Calypso, Chatham, Deepstar, Desire, Dove, Endeavour, Godspeed, Hôkûle'a* (a Polynesian heritage name), *Horizon, Nautilus, North Star, Pathfinder, Phoenix, Resolution, Trieste, Victoria,* and *Victory.*[246]

Other public reports have indicated there were 29 names on the final list, including: *Adventure, Blake, Calypso, Deepstar, Desire, Dove, Eagle, Endeavor* (American spelling), *Endeavour* (British spelling), *Endurance, Godspeed, Griffin, Gulf Stream, Chatham, Hôkûle'a, Horizon, Investigator, Meteor, Nautilus, North Star, Pathfinder, Phoenix, Polar Star, Resolution, Rising Star, Royal Tern, Trieste, Victoria,* and *Victory.*[247]

The names were evaluated by two separate NASA committees, and the results forwarded to the NASA Administrator, and then to President George Bush. Continuing with the tradition of naming the Orbiters after famous sailing ships, OV-105 was christened *Endeavour* in May 1989.[248]

---

* Note the American spelling, not the British spelling that was ultimately adopted for OV-105, *Endeavour.*

OV-099, *Challenger*, after a British naval research vessel which from 1872 to 1876 made a prolonged exploration of the Atlantic and Pacific oceans. It was also the name of the Apollo 17 Lunar Module that carried Eugene A. Cernan and Harrison H. Schmitt to the surface of the Moon. The name was doubly appropriate since the basic airframe* was rebuilt from the structural test article (STA-099).[249]

---

* For some period of time during 1978, the 'rebuilt' STA-099 was known as OV-101M.

Challenger was the first Orbiter delivered with a name on the upper right wing (above), and the 'USA' and American flag on the left. The previous two Orbiters (OV-101 and OV-102) had the American flag on the left and the 'USA' on the right wing (left). Challenger did not change appearance substantially during its operational life. Note the white tiles on the outer portions of the wings. (NASA)

OV-099 immediately prior to the official rollout ceremonies at the Rockwell/Palmdale assembly facility. Challenger was the first Orbiter to have its name painted ahead of the payload doors instead of on the forward portion of the door. This was done to facilitate reading the name while the payload doors were open on-orbit. The T–0 disconnect panel located below the orbital maneuvering system (OMS) pod is unpainted. The use of black tiles on the vertical stabilizer also differed considerably from the other Orbiters (notice that the entire top of the vertical is black and the extended black area ahead of the lower portion of the rudder), and the black tiles around the forward crew hatch were also somewhat different. The demarcation line between areas protected by tiles and blankets is clearly visible in a 'stair' pattern progressing up the fuselage from the wing's leading edge.

OV-101, *Enterprise*, after the starship in the science-fiction television show Star Trek. This airframe was originally to be named *Constitution* in honor of the U.S. Constitution's Bicentennial. Several naval vessels have also carried the name *Constitution*. A write-in campaign during 1976 produced over 100,000 requests for the White House to override NASA, which it did. *Enterprise* was still appropriate since several naval ships, including the a famous WWII aircraft carrier and the world's first nuclear-powered aircraft carrier, have carried the name.

OV-101 during the Approach and Landing Tests at Edwards AFB. Only Enterprise and Columbia had the Orbiter name painted on the forward payload doors instead of ahead of the door. The early markings were decidedly non-production, with a minimum of simulated black tiles on the vertical stabilizer, the forward RCS module, and around the crew hatch. (Dennis R. Jenkins)

Enterprise in the markings originally painted for the Paris Air Show, although this is at VLS in October 1984. Note is the increased use of black around the rudder/speed-brake leading and trailing edges, and the black stripe on the aft edge of the forward RCS module. Also, the frames around the cockpit glass have been painted to more closely resemble the later production Orbiters. (Dennis R. Jenkins)

Enterprise was used as a facility verification vehicle at the Kennedy Space Center before STS-1. OV-101 can be distinguished from OV-102 by the black chines used exclusively by Columbia, and the lack of black forward reaction control system (RCS) module on Enterprise. Noteworthy are the American flag and 'USA' markings on the wings, also shared with OV-102. (NASA photo 108-KSC-79P-202)

OV-101 during the facility verification at the Vandenberg Launch Site. The markings on the wing surfaces have been changed to reflect the production configuration introduced by Challenger ('USA' and American flag on left wing, 'NASA' and Orbiter name on right). The FRCS module has been repainted, although still not correctly, and additional black markings have been added around the windshield to simulate tiles. (Dennis R. Jenkins)

<u>OV-102</u>, *Columbia*, after a sailing frigate launched in 1836 that was one of the first Navy ships to circumnavigate the globe, under the command of Robert Gray. *Columbia* was also the name of the Apollo 11 Command Module which carried Armstrong, Collins, and Aldrin to the moon for the first lunar landing mission, 20 July 1969.

OV-102 during the overland move from Palmdale to Edwards AFB following rollout. The Orbiter followed the same route as Enterprise before it. Schedule considerations prevented the thermal protection system from being completed at Palmdale, and the missing tiles were subsequently installed at KSC prior to STS-1. (NASA photo 108-KSC-79P-69)

Other than the tile pattern on the vertical stabilizer (and the SILTS pod), the exterior appearance of Columbia changed very little over the last 20 years. The photo at left is STS-83 on Pad 39A, while the photo below is STS-87 heading towards Pad 39B. (NASA photos KSC-97PC-0481 and KSC-97PC-1576)

Columbia was extensively modified during the Challenger stand-down. Here OV-102 is being transferred from OPF-1 to the OMRF (what is now OPF-3) in order to make room for Discovery upon her return from the STS-26R mission. Most of OV-102's thermal protection system has been removed, exposing her aluminum skin. Note the SILTS pod on the vertical stabilizer. (NASA photo KSC-88P-1179)

<u>OV-103</u>, *Discovery*, after two ships; Henry Hudson's which in 1610–11 attempted to search for a northwest passage between the Atlantic and Pacific oceans and instead discovered Hudson Bay, and James Cook's which discovered the Hawaiian Islands and explored southern Alaska and western Canada. *Discovery* was also the name of two ships used by The Royal British Geographical Society to explore the North Pole and Antarctica.

Discovery gets a lift home after STS-64. The black tiles on the forward edge of the OMS pod shows up well, making it easy to discern between real OMS pods and the aerodynamic dummies used on some ferry flights. (NASA photo EC94-42750-6)

Discovery at Pad 39A before her maiden voyage on STS-14/41-D in August 1984. OV-103 and OV-104 were as close to a 'production series' as the Orbiter project came, and the two vehicles were nearly identical except for the names painted on the fuselage and upper wing surface. Note the outer wing panels use blankets instead of the white tiles used on OV-102 (see page 250). (NASA)

Discovery lands at Edwards at the conclusion of STS-64. These were the markings that most of the Orbiters used for the first 20 years of the program. The NASA worm on the aft payload bay doors would fall victim to Dan Goldin's edicts against it and be replaced by a small meatball logo on the aft fuselage. (NASA photo EC94-42750-4 by Dennis Taylor)

<u>OV-104</u>, *Atlantis*, after a two-masted ketch operated for the Woods Hole Oceanographic Institute from 1930 to 1966 and traveled more than half a million miles during her research trips. The 460-ton ketch was the first U.S. vessel to be used for oceanographic research. She featured a crew of 17 and room for five scientists.

A banner on the gate to Pad 39A advertises the current resident, being prepared for STS-101. The only Orbiter that is easily identifiable from this distance would be Columbia – the black chines are readily identifiable and are unique in the fleet. (NASA photo KSC-00PP-0636)

Atlantis returns from its first flight after the Challenger accident. A look at the base of the vertical stabilizer shows that the drag chute modification has not yet been installed. (NASA photo EC88-0247-1 by Ken Wiersma)

OV-104 displays the new meatball logo and rearranged wing markings while sitting on pad 39A for STS-101. The flag and Orbiter name are now left-justified, not centered on each other as the worm and name had been. (NASA photo KSC-00PP-0411)

OV-105, *Endeavour*, continues the famous sailing ship tradition, being the first ship commanded by James Cook. In August 1768, on *Endeavour*'s maiden voyage, Cook sailed to the South Pacific on a mission to observe a seldom occurring event when the planet Venus passes between the Earth and Sun. *Endeavour* was also the name of the Apollo 15 Command Module which carried Scott, Worden, and Irwin to the Moon in July 1971.

Endeavour launches as STS-99 from Pad 39A. The new side markings are evident, with the meatball on the aft fuselage instead of the NASA worm on the payload bay doors. (NASA photo KSC-00PC-0223)

Workers put the final touches on Endeavour's markings prior to roll-out. The NASA worm was dark gold while the Orbiter names was black. The different texture of the TPS blanket may be seen at the extreme right of the photo. (Rockwell International)

The roll-out ceremonies in palmdale were well attended by both official visitors and invited members of the public. Note the SCA (N911NA), T-38s, and various business aircraft (two Gulfstream IIs and a Gulfstream I) on the ramp in the background. The extensive use of blankets on the forward fuselage shows up well here. (DVIC photo DF-ST-99-05011)

APOLLO / SATURN VERSUS SHUTTLE COSTS

Apollo/Saturn

Fiscal Year	RY $M	CPI	Inflation Factor	2000 $M	# Flights	Mission Numbers
1962	75.6	90.6	5.673	429		
1963	1,184.0	91.7	5.605	6,637		
1964	2,273.0	92.9	5.533	12,576		
1965	2,614.6	94.5	5.439	14,221		
1966	2,992.2	97.2	5.288	15,823		
1967	3,002.6	100.0	5.140	15,433		
1968	2,556.3	104.2	4.933	12,610		
1969	2,025.0	109.8	4.681	9,480	5	7 thru 11
1970	1,684.4	116.3	4.420	7,444	2	12, 13
1971	913.7	121.3	4.237	3,872	2	14, 15
1972	601.2	125.3	4.102	2,466	1	16
1973	56.7	133.1	3.862	219	1	17
Totals	19,979.3			101,210	11	

Average cost of each Apollo/Saturn launch was
  $9,201,000,000 in FY00 dollars
  including amortizing all R&D and CoF costs
Based on 253,000 pounds to LEO equals
  $36,368 per available pound in FY00 dollars

> Sources
> *   Henry C. Dethloff,
>   *Suddenly, Tomorrow Came. . . , A History of the Johnson Space Center,*
>   NASA SP-4307 (Washington DC: NASA, 1993), page 286.
> **   Average of 1990 and 1992 (no actual data available)
> ***   Wayne Littles Briefing to Congress, 9 November 1995
> ****   Actual Costs according to OMB
> ++   http://ifmp.nasa.gov/cobeb/budget2000/html/shuttle.htm
> CPI   All Urban Consumers (1967 = 100) at the Bureau of Labor Statistics
>   2000 CPI is estimate based on first 11 months
>
> Launch data is per Fiscal Year, not calender year (hence why only 98 shuttle flights, not 100)

> Non-amortized year 2000 costs: based on $2,500 million authorization per year, with a 7 flight per year manifest – 5 flights to ISS (maximum payload of 40,000 pounds) and 2 low-inclination launches (55,000 pounds each).
>
> $357 million per launch = $8,000 per available pound

Space Shuttle (NASA only)

Fiscal Year	RY $M	Source	CPI	Inflation Factor	2000 $M	# Missions
1970	12.5	*	116.3	4.420	55	
1971	78.5	*	121.3	4.237	333	
1972	100.0	*	125.3	4.102	410	
1973	198.6	*	133.1	3.862	767	
1974	475.0	*	147.7	3.480	1,653	
1975	797.5	*	161.2	3.189	2,543	
1976	1,206.0	*	170.5	3.015	3,636	
1977	1,413.1	*	181.5	2.832	4,002	
1978	1,349.2	*	195.4	2.631	3,549	
1979	1,707.8	*	217.4	2.364	4,038	
1980	2,054.9	*	246.8	2.083	4,280	
1981	2,301.8	*	272.4	1.887	4,343	1
1982	2,689.2	*	289.1	1.778	4,781	3
1983	3,357.5	*	298.4	1.723	5,783	4
1984	3,068.9	*	311.1	1.652	5,070	4
1985	2,786.7	*	322.2	1.595	4,446	8
1986	2,987.9	*	328.4	1.565	4,677	5
1987	5,138.3	*	340.4	1.510	7,759	
1988	2,917.9	*	354.3	1.451	4,233	1
1989	3,500.3	*	371.3	1.384	4,846	4
1990	3,818.2	*	391.4	1.313	5,014	5
1991	4,011.7	**	408.0	1.260	5,054	8
1992	4,205.2	***	420.3	1.223	5,143	7
1993	4,044.1	***	432.7	1.188	4,804	7
1994	3,558.7	***	444.0	1.158	4,120	8
1995	3,155.1	***	456.5	1.126	3,553	6
1996	3,081.0	****	469.9	1.094	3,370	8
1997	3,001.0	****	480.8	1.069	3,208	8
1998	2,912.8	++	488.3	1.053	3,066	4
1999	2,998.3	++	499.0	1.030	3,088	4
2000	2,986.2	++	514.0	1.000	2,986	3
Totals	69,948.6				114,609	98

Average cost of each Space Shuttle launch was
  $1,169,500,000 in FY00 dollars
  including amortizing all R&D and CoF costs
Based on 55,000 pounds to LEO equals
  $21,300 per available pound in FY00 dollars

This table shows the overall costs of the Apollo/Saturn program and the Space Shuttle program converted into year 2000 dollars. The numbers are not exact due to government accounting methods, but are deemed generally reliable and representative. The costs listed include research and development, construction of facilities, and operations. These figures do not include Air Force expenditures, support from the other armed forces (particularly the Navy), or (apparently) the cost of programs such as Shuttle/Centaur, ASRM, or Shuttle-C.

A few of the mockups in Building 9-A at JSC during 1993. Note the Spacelab module in the mockup in the foreground, and the open crew hatch on the one at right. The Manipulator Development Facility in the background has a functional crane that operates similar to the RMS to allow payload handling. (NASA photo S9348140)

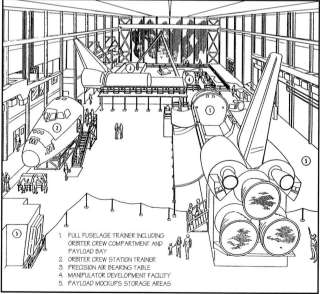

1. FULL FUSELAGE TRAINER INCLUDING ORBITER CREW COMPARTMENT AND PAYLOAD BAY
2. ORBITER CREW STATION TRAINER
3. PRECISION AIR BEARING TABLE
4. MANIPULATOR DEVELOPMENT FACILITY
5. PAYLOAD MOCKUPS STORAGE AREAS

A different perspective – and a slightly different arrangement since the simulators have been moved around a bit – of the same area in Building 9-A at JSC as seen in the photo at left. The Orbiter Crew Station Trainer (number 2 above) can be rotated 90 degrees to simulate the Orbiter in either the horizontal or vertical position. (NASA)

Various muck-ups of the Orbiter have been built to support many different purposes. Some of these mock-ups are used in engineering evaluations by NASA or the contractors. Others serve strictly as public relation functions for NASA or business. A few of the mock-ups that have been used are illustrated on these two pages. Many others have existed at various points in time.

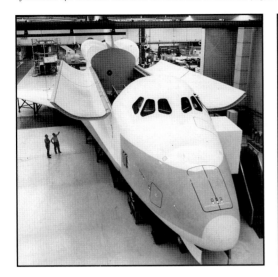

Many mock-ups of the space shuttle Orbiter have been constructed to serve many purposes. The mock-up shown above was a full-size representation of the Orbiter minus the left wing, and at various times, the vertical stabilizer. It was located in the Rockwell engineering and manufacturing facility in Downey, California. Noteworthy in these 12 April 1974 shots are the OMS pods that continue onto the payload bay doors and the RCS thrusters on the nose that are protected by doors. Both items were changed before the first Orbiter was completed in 1975. This mock-up was continually updated, and was in use by Boeing until the Downey facility was closed in 1999. (NASA photos 108-KSC-74PC-254 and S74-19403)

The Egress Trainer at JSC – similar unit exists at KSC. Noteworthy are the wheels on the right side, necessary since the trainer can be oriented in either launch (vertical) or landing (horizontal) positions. The trainer at KSC is periodically placed in a nearby lake for ditching egress training. In this shot, Astronaut John Young is exiting the trainer. (NASA)

U.S. Senator John H. Glenn, Jr. (D.-Ohio), uses a device called a Sky Genie to simulate rappelling at JSC in April 1998 during training for STS-95. This exercise trains the crew members in procedures to follow while egressing a troubled Orbiter on the ground. (NASA photo S98-06945 by Joe McNally, National Geographic)

The Astronaut Hall of Fame in Titusville, Florida (right outside the KSC gate) erected the Shuttle to Tomorrow. This mock-up has a movie theater installed in its payload bay. The outside of this full-size mock-up is extremely detailed and realistic. (Dennis R. Jenkins)

The Visitor Complex at the Kennedy Space Center built the Explorer, a full-size replica of an Orbiter. Visitors can walk through the vehicle which has a simulated flight deck and payload. This is a 'don't miss' attraction in Central Florida. (Dennis R. Jenkins)

"The actual reliability of the Shuttle system is unknown, but may lie between 97 percent and 98 percent. If reliability is 98 percent, the Nation faces a 50–50 chance of losing an additional Orbiter in the next 34 flights …

"… if the United States wishes to send people into space on a routine basis, the Nation will have to accept the risk these activities entail. If such risks are perceived to be too high, the Nation may wish to reduce its emphasis on placing humans in space."

Access to Space: The Future of U.S. Space Transportation Systems
The Office of Technology Assessment, 1989

The crew of Challenger on STS-33/51-L. (NASA photo S85-44253)

Broken In Mid-Stride

Flight Profile

A nominal Space Shuttle flight follows basically the same routine every time. It starts with the Solid Rocket Boosters being stacked on the Mobile Launch Platform (MLP) in the Vehicle Assembly Building (VAB) while the Orbiter is being serviced in the Orbiter Processing Facility (OPF). The External Tank is then mated to the stacked SRBs, and the Orbiter is moved* from the OPF to the VAB (called 'roll-over') and mated to the ET. Electrical and fluid connections are made, the integrated vehicle is checked out, and the Range Safety System ordnance is installed while still in the VAB.[1]

A Crawler-Transporter moves the MLP with the entire Space Shuttle stack to one of the two launch pads approximately 3.5 miles away (called 'roll-out'), where servicing and checkout activities begin. If the payload was not installed in the OPF, it will be installed on the launch pad. On launch day the Orbiter and ET are filled with propellants, and the crew arrives and boards approximately three hours prior to launch. At T-20 (20 minutes prior to launch) there is a built-in 10 minute hold that allows the launch team and flight crew to catch up on any outstanding preparations and to make final consultations before proceeding. There is another built-in hold at T-9 minutes.[2]

At launch, the three SSMEs are ignited and verified to be operating at the proper thrust level. A signal is then sent to ignite the SRBs, and shortly thereafter the eight hold-down bolts are explosively released. Compared to the lumbering lift-off of the Saturn V, the Space Shuttle fairly much jumps off the pad due to the rapid thrust buildup of the solid boosters. As soon as the velocity exceeds 127 fps (about T+7 seconds), the vehicle rolls to a heads-down attitude to align the stack's velocity vector

with the desired orbital plane. During first stage (SRB) ascent, heads-down makes it easier for the stack to maintain a negative angle-of-attack that reduces wing loading during the period of maximum dynamic pressure (max-q). Max-q is reached early in the ascent, nominally 30–60 seconds after lift-off.[3]

Kennedy Space Center (KSC) has responsibility for all mating, prelaunch testing, and launch control ground activities until the vehicle clears the launch pad tower. Responsibility is then turned over to the Mission Control Center (MCC) at the Johnson Space Center (JSC) in Houston, Texas. Mission Control's responsibilities include ascent, on-orbit operations, reentry, approach and landing until run-out completion. Once the Orbiter comes to a stop, the post-landing operations team from KSC takes over to safe the vehicle and tows it to the OPF or prepares it for the ferry flight to KSC. This team also deploys to Edwards if that is the planned landing site, and would also deploy to a TAL or contingency landing site.[4]

Approximately 120 seconds into the ascent, the two SRBs are separated from the ET – this is also accomplished explosively, and eight small rocket motors fire to ensure the boosters separate cleanly from the vehicle. At a predetermined altitude the SRBs deploy parachutes which lower the boosters into the Atlantic approximately 141 miles downrange from the launch site. The boosters are recovered by a pair of ships operating from the Cape Canaveral Air Force Station† (CCAFS) adjacent to KSC.[5]

Meanwhile, the Orbiter and ET continue to ascend using the thrust provided by the three SSMEs ('second stage'). At this point the stack is flying above most of the atmosphere, so wing loading is less critical. Originally, the vehicle stayed heads-down to allow S-band communications through the Bermuda tracking station. That station has since been retired, and beginning with STS-87 most flights have done a roll-to-heads-up (RTHU) during second stage to allow communications through the TDRS satellites. This occurs at approximately T+6 minutes when the vehicle is traveling at 12,200 fps. The roll maneuver is performed at about 5 degrees per second.

---

* Originally the Orbiter was towed on its landing gear; after Vandenberg was closed, the special 76-wheel carrier designed for use there was transferred to KSC and is used to carry the Orbiter from the OPF to the VAB.

† This facility has spent the 1990s trying to figure out if it wants to be an Air Force Station (AFS) or just an Air Station (AS). As of late 2000 it is Cape Canaveral Air Force Station.

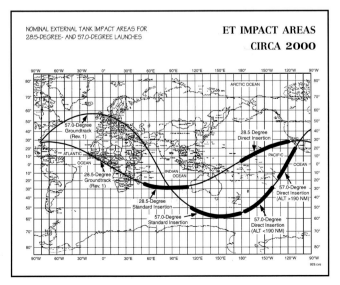

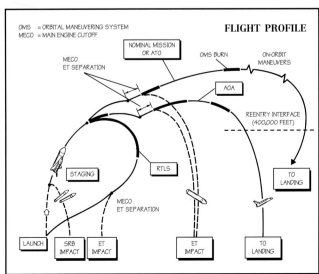

— 259 —

CHAPTER VIII – BROKEN IN MID-STRIDE

Approximately 8 minutes 30 seconds after launch, and just short of orbital velocity, the main engines are shut down (main engine cutoff – MECO), and the ET is jettisoned. The forward and aft Reaction Control System (RCS) thrusters provide the translation away from the ET at separation and return the Orbiter to the proper attitude prior to the Orbital Maneuvering System (OMS) burn. Nominal velocity at MECO should be approximately 25,650 fps (17,489 mph) for a standard insertion trajectory, or 25,850 fps (17,625 mph) for a direct insertion mission.[6]

The ET continues on a ballistic trajectory and enters the atmosphere where it disintegrates, assisted by a tumbling motion caused by unused LO2 vaporizing and venting through the gaseous oxygen vent in the nose of the ET. On the original standard insertion 28.5-degree missions, the ET impacted in the Indian Ocean – for direct insertion flights this has moved into the Pacific near Hawaii. On 57-degree standard insertion flights the impact area was south of Australia, with direct insertion tanks impacting in the south or mid-Pacific.[7]

For the early missions, two thrusting maneuvers were made using the OMS engines. The first OMS burn (OMS 1) raised the maximum altitude and occurred a few minutes after MECO. The second burn (OMS 2) raised the lowest point of the orbit (perigee) and was usually used to circularize the orbit. This is now referred to as the 'standard insertion' mission profile since it was standard practice early in the program when there was less information on the exact performance of the SSMEs and some lack of targeting precision. Later missions have usually flown 'direct-insertion' trajectories and use a single OMS burn to circularize the orbit (the OMS 1 burn is omitted). The last mission to use a standard-insertion profile was STS-30R. In the year 2000, a standard insertion profile would only be flown to low orbital altitude where the ET impact point was outside safe limits using a direct-insertion profile.[8]

The orbital altitude of a mission is highly dependent upon the objectives of the mission. The nominal altitude can vary between 130 to 350 miles. The lowest orbit during the first 100 flights was 138 miles during STS-68, while the highest was 385 miles during STS-82.[9]

The forward and aft RCS thrusters provide attitude (roll, pitch, and yaw) control of the Orbiter as well as any minor translation maneuvers along a given axis on-orbit. The OMS engines are used on-orbit for any major velocity changes. At the completion of orbital operations the Orbiter is oriented in a tail first attitude by the RCS and an OMS burn reduces the Orbiter's velocity 200–500 fps, depending upon orbital altitude, in preparation for deorbit.[10]

The RCS turns the Orbiter's nose forward for reentry which occurs at 400,000 feet, slightly over 5,000 miles from the landing site. The velocity at reentry is approximately 17,000 mph with a 40-degree angle-of-attack. The FRCS thrusters are inhibited by the GPCs immediately prior to reentry, and the aft RCS thrusters maneuver the vehicle until a dynamic pressure of 10 pounds per square foot, which is when the Orbiter's ailerons become effective. The aft RCS roll thrusters are then deactivated. At a dynamic pressure of 2 pounds per square foot the Orbiter's elevators become effective and the aft RCS pitch thrusters are deactivated. The Orbiter's speedbrake is used below Mach 10 to induce more positive downward elevator trim. At approximately Mach 5 the rudder is activated, and the aft RCS yaw thrusters are deactivated at Mach 1.[11]

The guidance system must dissipate the tremendous amount of energy the Orbiter possesses when it reenters the Earth's atmosphere to ensure the Orbiter does not either burn up (entry angle too steep) or skip back out of the atmosphere (entry angle too shallow). During reentry, excess energy is dissipated by atmospheric drag on the Orbiter – a steep trajectory gives higher atmospheric drag levels which results in faster energy dissipation. Normally, the angle-of-attack and roll angle enable the atmospheric drag of any flight vehicle to be controlled. However, for the Orbiter, angle-of-attack variation was rejected because it creates exterior surface temperatures in excess of the insulation capabilities of the thermal protection system. The angle-of-attack schedule is loaded into the GPCs as a func-

LANDING SITES
CIRCA 2000

Site	EOM AOA	TAL	ECAL	ELS Low (28°-37°)	ELS Mid (38°-49°)	ELS High (50°-53.5°)	ELS High (53.6°-63.5°)
Amberley, Australia				X	X	X	X
Amilcar Cabral, Cape Verde Islands				X	X		
Andersen AFB, Guam				X	X	X	X
Arlanda, Sweden							O
Ascension Island							
Banjul, The Gambia		X		X	X		
Beja, Portugal						X	
Ben Guerir, Morocco		X		X	X	X	X
Bermuda International, Bermuda				X	X		
Cherry Point MCAS, NC, USA			X			X	X
Dakar, Senegal				X			
Darwin, Australia				X	X	X	X
Diego Garcia, Chagos Islands, UK				X	X	X	X
Dover AFB, DE, USA			X				X
Dyess AFB, TX, USA				X	X	X	X
Edwards AFB, CA USA	X			X	X	X	X
Ellsworth AFB, SD, USA						O	O
Elmendorf AFB, AK, USA							
Esenboga, Turkey						X	X
Fairford, England						X	X
Falmouth, MA, USA			X			X	X
Gander, Newfoundland, Canada			X			X	X
Goose Bay, Newfoundland, Canada			X				X
Grant County, WA, USA						O	O
Grissom AFB, IN, USA							
Halifax, Nova Scotia, Canada			X			X	X
Hao Atoll, Society Islands, France				X	X	X	X
Hickam AFB, HI, USA				X	X	X	X
Hoedspruit, South Africa				X	X		
King Khalid (Riyadh), Saudi Arabia				X	X	X	X
Kinshasa, Zaire				X			
Koln.Bonn, Germany						O	X
Kennedy Space Center, FL, USA	X			X	X	X	X
Lajes AB, Azores, Portugal				X	X	X	
Las Palmas, Canary Islands, Spain				X	X		
Lincoln Municipal, NE, USA						O	O
Mataveri, Easter Island, Chile							
Miramar NAS, CA, USA							
Moron AB, Spain		X		X	X	X	X
Mountain Home AFB, ID, USA						O	O
Myrtle Beach, SC, USA			X				X
Nassau, Bahamas				X	X	O	O
Northrup Strip (WSSH), NM, USA	X			X	X	X	X
Oceana NAS, VA, USA			X			X	X
Orlando, FL, USA				X	X	O	O
Portsmouth, HN, Usa			X			X	X
Plattsburgh AFB, NY, USA							
Roberts Field, Liberia				X	X		
Rota NS, Spain							
St. John's, Newfoundland, Canada			X				
Santiago, Spain							
Shannon, Ireland						X	X
Shearwater, Nova Scotia, Canada							
Souda Bay, Crete, Greece				X	X		
Stephenville, Newfoundland, Canada			X				
Tamanrasset, Algeria				X	X		
Upper Hayford, UK							
Wake Island AAF, Wake Island							
Westover AFB, MA, USA							
Wright-Patterson AFB, OH, USA							
Yokota, Tokyo, Japan							
Zaragota AB, Spain		X			X	X	X

Sites with no annotations are not currently used. An 'O' indicates that the coordinates are not carried in the onboard software and the approach would have to be flown manually. Dakar, Senegal was a TAL site early in the program, but no longer has State Department concurrence. The Vandenberg TAL sites would have been Hoa and Easter Islands in the South Pacific, and construction of facilities on both islands was completed prior to cancellation. EOM is end of mission; ELS is emergency launch site. (NASA)[135]

tion of relative velocity, leaving only roll angle for energy control. Increasing the roll angle decreases the vertical component of lift, causing a higher sink rate and a higher energy dissipation rate. Increasing the roll angle raises the surface temperature of the Orbiter, but not nearly as drastically as an equivalent angle-of-attack variation.[12]

If the Orbiter is low on energy (current range-to-go greater than nominal at current velocity), reentry guidance will command lower than nominal drag levels. If the Orbiter has too much energy (current range-to-go less than nominal), reentry guidance will command higher than nominal levels to dissipate the extra energy, within the limits of the Orbiter to withstand the additional surface heating. The goal is to maintain a constant heating rate until the Orbiter is below 13,000 mph (19,000 fps) – this is the temperature control phase.[13]

Roll angle is also used to control cross-range. Azimuth error is the angle between the plane containing the Orbiter's position vector and the heading alignment circle tangency point, and the plane containing the Orbiter's position vector and the Orbiter's velocity vector. When the azimuth error exceeds a predetermined number, the Orbiter's roll angle is reversed.[14]

The equilibrium glide phase transitions the Orbiter from the rapidly increasing drag levels of the temperature control phase to the constant drag phase. Equilibrium glide flight is defined as flight in which the flight path angle – the angle between the local horizontal and the local velocity vector – remains constant. Equilibrium glide flight provides the maximum downrange capability, and continues until the drag acceleration reaches 33 feet per second squared.[15]

The constant drag phase begins at this point, with an initial 40 degree angle-of-attack, ramping down to 36 degrees by the end of the phase. In the transition phase, the angle-of-attack continues to ramp down, reaching about 14 degrees as the Orbiter reaches the terminal area energy management (TAEM) interface, at approximately 83,000 feet altitude, 2,500 fps (1,700 mph), and 69 miles from the runway. Control is then transferred to TAEM guidance.[16]

The TAEM guidance steers the Orbiter to the nearest of two heading alignment circles (HAC), whose radii are approximately 18,000 feet and which are located tangent to, and on either side of, the approach end of the runway centerline. In TAEM guidance, excess energy is dissipated with S-turns, increasing the ground track range as the Orbiter turns away from the nearest HAC until sufficient energy is dissipated to allow a normal approach and landing. The Orbiter may also be flown near the velocity for maximum lift-over-drag or wings level to increase range. The vehicle slows to subsonic velocity at approximately 49,000 feet about 30 miles from the runway.[17]

The approach and landing phase begins at 10,000 feet at an airspeed of 345 mph, either 6.7 or 7.3 miles from touchdown (for –20 or –18-degree glideslopes,* respectively). Autoland guidance is initiated at this point to guide the Orbiter to the –20 or –18-degree glideslope aimed at a target 7,500 feet short of the runway. The descent rate is greater than 10,000 feet per minute.† The speedbrake is used to control velocity, and at 2,000 feet altitude a pre-flare maneuver is started to position the vehicle for a 1.5-degree glideslope in preparation for landing. The final phase reduces the sink rate to less than 9 fps, and the crew deploys the landing gear at 300 feet altitude. A final flare maneuver lowers the sink rate to 3 fps and touchdown

Recovery personnel surround OV-103 (Discovery) in April 1990 on runway 22 at Edwards AFB after STS-31R. Trucks and tubing at right prevent hazardous vapors from escaping. (NASA photo EC90-129-6)

occurs approximately 2,500 feet past the runway threshold at a speed of roughly 225 mph (for lightweight Orbiters; 235 mph for heavy vehicles).[18]

Spacecraft recovery operations at the nominal end-of-mission landing are supported by the KSC post-landing operations team. Ground team members wearing self-contained atmospheric protective ensemble (SCAPE) equipment that protects them from toxic chemicals approach the spacecraft as soon as it stops rolling. The ground team takes measurements to ensure the atmosphere in the vicinity of the vehicle is not hazardous – in the event of a propellant leak, a wind machine truck carrying a large fan will be moved into the area to create a turbulent airflow that will break up gas concentrations and reduce the potential for an explosion.[19]

An air conditioning purge unit is attached to the right-hand T-0 umbilical so that cool air can be directed through the aft-fuselage, payload bay, forward fuselage, wings, vertical stabilizer, and OMS pods to dissipate the heat of reentry. A second ground cooling unit is connected to the left-hand T-0 umbilical Freon-21 coolant loops to provide cooling for the flight crew and avionics during the post-landing checks. The Orbiter fuel cells remain powered up until after the flight crew exits the vehicle, then a ground crew powers-down the Orbiter.[20]

If the landing was made at KSC, the Orbiter and its support convoy is moved to the OPF. If the landing was made at Edwards, the Orbiter is safed on the runway and then moved to the mate/demate device at Dryden. After a detailed inspection of the vehicle, the tailcone is installed and the Orbiter is lifted and mounted on the SCA for the ferry flight back to KSC.[21]

In the event of a return to an alternate landing site, a team from KSC will deploy to the landing site to assist the astronaut crew in preparing the Orbiter for loading aboard the SCA for transport back to KSC. For landings outside the United States, personnel at contingency landing sites are provided minimum training on safe handling of the Orbiter with an emphasis on crash rescue training, how to safe the vehicle and its propellants, and procedures on how to safely tow the Orbiter to a parking area.[22]

Upon its return to the Orbiter Processing Facility, the Orbiter is safed (ordnance devices removed), the payload (if any) is removed, and the Orbiter payload bay is reconfigured for the next mission. Any required maintenance is also performed in the OPF. A payload for the next mission may be installed in the OPF, or it may be installed when the Orbiter is at the launch pad.[23]

---

* This is over seven times as steep as that normally used by commercial airliners.
† The FAA standard descent rate is 500 feet per minute for commercial air traffic.

ORBITAL MECHANICS

As originally envisioned, Space Shuttle missions to equatorial orbits would be launched from the Kennedy Space Center (KSC), and those requiring polar orbit were to be launched from the Vandenberg Launch Site (VLS). Orbital mechanics, the complexities of mission requirements, plus safety constraints and the possibility of infringing on foreign air space, prohibit polar orbit launches from KSC. In 1986 plans to activate Vandenberg were cancelled, and KSC is used for all Space Shuttle launches – leaving the United States without a manned polar launch capability.

A 35-degree launch azimuth from KSC places the Orbiter in an equatorial orbital inclination of 57 degrees – the vehicle will never exceed an Earth latitude higher or lower than 57 degrees north or south of the equator. A launch azimuth of 90 degrees (due east) will place the vehicle in an orbital inclination of 28.5 degrees (it will be above or below 28.5 degrees north or south of the equator).[24]

These two azimuths – 35 and 90 degrees – represent the upper and lower launch limits from KSC. Any azimuth angles further north or south would launch the shuttle over a habitable land mass, adversely affect safety for abort or vehicle separation conditions, or potentially present the politically undesirable possibility that the SRBs or ET could land on foreign land or sea space. Flights into slightly higher (62.0 degrees) and lower (28.35 degrees) orbits have been accomplished using inefficient dog-leg profiles that initially used the approved launch azimuths, then turn toward the desired orbit during second-stage ascent.[25]

Launches from Vandenberg would have had allowable launch limits of 201 and 158 degrees. At a 201-degree launch azimuth, the vehicle would orbit at a 104-degree inclination. Zero degrees would be due north of the launch site, and the orbital trajectory would be within 14 degrees east or west of the north-south pole meridian. With a launch azimuth of 158 degrees, the Orbiter would be at a 70 degree inclination, and the trajectory would be within 20 degrees east or west of the polar meridian.

Mission requirements and payload weight penalties also are major factors in selecting a launch site. The Earth rotates from west to east at approximately 1,035 mph. An easterly launch uses the Earth's rotation somewhat as a springboard, increasing maximum payload and altitude capabilities. This rotation was also the reason the Orbiter was designed with a cross-range capability of 1,265 miles, to provide an abort-once-around capability.[26]

Attempting to launch and place a spacecraft in polar orbit from KSC and still avoid habitable land mass would be uneconomical because the Orbiter's payload would be reduced severely – to approximately 17,000 pounds. A northerly launch into polar orbit of 8 to 20 degrees azimuth would necessitate a path over a land mass; and most safety, abort, and political constraints would have to be waived. It should be noted, however, that NASA has accomplished the necessary technical analysis and planning to conduct such launches, and the flight software accommodates these unique flight profiles, so a polar launch from KSC could be accomplished if the national objective outweighed the economic and political considerations.

The current (1999) assessment of Orbiter ascent and landing weights assumes a maximum SSME throttle of 104 percent. The first five flights after the *Challenger* accident required more conservative flight design criteria and additional instrumentation, reducing these basic capabilities by approximately 1,600 pounds. NASA earlier (1986) had calculated the limits for Vandenberg, even though there are no plans to use the launch site.[27]

1. Kennedy Space Center satellite deployment missions. The basic payload-lift capability for a due-east (28.5-degree orbit) launch is 55,250 pounds to a 126-mile orbit using OV-103, OV-104, or OV-105 to support a four-day mission. This capability is reduced approximately 115 pounds for each additional mile of altitude.
   The payload capability for the same mission with a 57 degree orbit is 41,000 pounds. Performance for intermediate inclinations may be estimated by subtracting 500 pounds per degree between 28.5 and 57 degrees.
   If OV-102 (*Columbia*) is used, the payload capability must be decreased by approximately 6,550 pounds. This weight difference is attributed to an approximately 5,000 pound difference in inert weight.
2. Vandenberg Launch Site satellite deployment missions. For all Orbiters except OV-102, the payload capability was 29,600 pounds for a 98 degree launch inclination and 126-mile polar orbit. Again, an increase in altitude cost approximately 115 pounds per mile.
   This same mission at a 68 degree inclination (minimum inclination based on range safety limitations) could carry 49,400 pounds. Performance for intermediate inclinations may be estimated by subtracting 660 pounds for each degree between of 68 and 98 degrees.
3. Landing weight limits. Since the 6.0 loads structural modifications, the Orbiters have been limited to a total landing weight of 250,000 pounds for abort landings and 240,000 pounds for nominal end-of-mission landings.

It should be noted that each additional crew member beyond the five-person standard is chargeable to the payload weight allocation and reduces the payload capability by approximately 500 pounds, including life support and crew escape equipment.

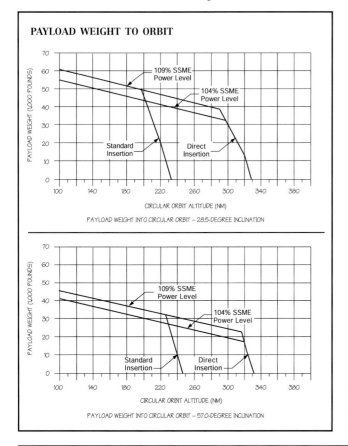

PAYLOAD WEIGHT TO ORBIT

109% SSME Power Level

104% SSME Power Level

Standard Insertion

Direct Insertion

PAYLOAD WEIGHT INTO CIRCULAR ORBIT – 28.5-DEGREE INCLINATION

PAYLOAD WEIGHT INTO CIRCULAR ORBIT – 57.0-DEGREE INCLINATION

CIRCULAR ORBIT ALTITUDE (NM)

PAYLOAD WEIGHT (1,000 POUNDS)

ABORTS

A Space Shuttle launch may be scrubbed or aborted up to the ignition of the SRBs. Normally, scrubs prior to SSME start are followed by an orderly safing and crew egress, assisted by the closeout crew. An abort after main engine ignition is controlled by the ground launch sequencer – this occurred during the second launch attempt for STS-16/41-D. The most serious hazard is the presence of excess hydrogen that could result in a fire which is invisible to the eye – special cameras have been installed around the launch pad that can detect these fires.[28]

Depending upon the severity of the situation on the launch pad, an emergency egress/escape may be required to evacuate personnel. The shuttle program has classified four launch pad escape modes that are preplanned and rehearsed by the flight crew, closeout crew, and all launch pad personnel. These modes ensure standard procedures are followed and minimize the risk of additional injury and damage. Launch pad personnel and each flight crew practice these procedures prior to every mission.[29]

After lift-off, aborts fall broadly into three categories – performance aborts, systems aborts, and range safety actions. A performance abort occurs when the shuttle loses one or more SSMEs during ascent. The performance penalty when an engine loses thrust or completely fails is directly related to the time of the problem. Early engine degradation or failure while the vehicle is heavy with propellant may preclude achieving a safe orbit – late engine problems may result in no underspeed at all. Fortunately, it is not necessary to fly all the way to MECO to find out how large the underspeed will be. By using a computer program called the abort region determinator (ARD), engineers at Mission Control can predict the underspeed that will result from any performance problem. Thus, it can be determined immediately whether a safe orbit can be achieved – if not, some type of abort will be required. A systems abort occurs when one or more critical systems aboard the Orbiter – such as the crew module pressurization system – fails during ascent. In most cases the crew must rely on the insight provided by Mission Control using vehicle telemetry to fully understand the magnitude of any systems problem during ascent.[30]

Despite all planning, design, and tests, the possibility that a malfunction could cause a Space Shuttle to fly out of control toward a populated area must be considered. The primary function of Range Safety is to protect the public at large, particularly the cities immediately around the launch area and downrange. During first stage, this is accomplished by the Air Force remotely detonating explosive charges aboard the SRBs. During second stage ascent, the responsibility for range safety is on Mission Control and the flight crew, with inputs from Range Safety. To protect populated areas, impact limit lines are drawn around them, and no potentially lethal piece of a vehicle can be allowed to land behind an impact limit line. Since any flight termination needs to occur well before an impact

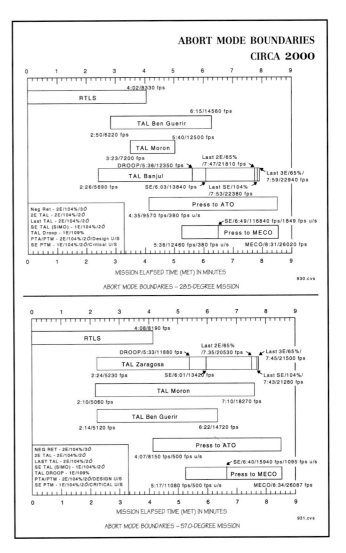

ABORT MODE BOUNDARIES – 28.5-DEGREE MISSION

ABORT MODE BOUNDARIES – 57.0-DEGREE MISSION

limit line is reached, destruct lines are drawn inside the boundaries of the impact limit lines. A vehicle passing outside the boundaries of the destruct line would be subject to termination.[31]

During first-stage ascent, trajectory deviations may lead to a violation of a destruct line by a Space Shuttle that is still under control. It may be possible to return the vehicle toward its nominal trajectory or to safely execute an abort. Therefore, the Flight Director (FD) and Flight Dynamics Officer (FDO) at Mission Control are in voice communication with the Eastern Range Flight Control Officer (FCO) (formerly called the range safety officer) during ascent. If the FCO detects a violation, the FDO and FD are immediately informed. The FD must determine whether the shuttle is controllable or uncontrollable, and inform the FCO. As long as the FD declares the vehicle controllable, the FCO takes no action to terminate the flight for trajectory deviations alone. It should be noted that a destruct command will not be sent until after the flight crew has attempted an abort to separate the Orbiter from the rest of the stack.[32]

Within the three broad abort categories, there are two basic abort modes – intact aborts and contingency aborts. Intact aborts are designed to provide a safe return of the Orbiter to a planned landing site. Contingency aborts are designed to permit flight crew survival following more severe failures when an intact abort is not possible. A contingency abort generally results in either a crew bailout or an Orbiter ditch* operation.[33]

---

* There have been numerous studies conducted by NASA on the survivability of ditching an Orbiter. Generally it has been determined that a light (i.e.; almost empty) Orbiter could be successfully ditched in calm seas if the impact was at a 12 degree nose-up attitude, gear up, body flap up. This would result in a fairly smooth run-out of about six fuselage lengths, approximately 6-g in longitudinal acceleration, and approximately 9-g in braking deceleration. It is thought that the Orbiter would be damaged beyond repair, but that the crew could survive. Unfortunately, the possible survivability of such a ditching is greatly reduced by the additional mass of even a small payload, so the general consensus is that a reentry ditching (without returning payload) is possibly survivable, but a ditching resulting from an ascent abort (with a payload) is most likely not survivable.[136]

There are four basic intact aborts – abort-to-orbit (ATO), abort-once-around (AOA), transoceanic abort landing[*] (TAL), and return-to-launch-site (RTLS). In addition there is a fifth intact abort scenario that is sort of a variation of TAL and is generally not considered separately.

- Abort-to-Orbit: The ATO mode is designed to allow the vehicle to achieve a temporary orbit that is lower than nominal. This mode requires less performance and allows time to evaluate problems and then choose either an early deorbit maneuver or an OMS thrusting maneuver to raise the orbit and continue the mission. If an SSME fails in a region that results in a MECO under the desired velocity, Mission Control will determine that an abort mode is necessary and will inform the crew. On one occasion (mission STS-26/51-F) this method was been successfully used and the mission concluded successfully. ATO may be selected for either a performance shortfall or certain systems failures. Reentry is handled as a nominal case.[34]

- Abort-Once-Around: The AOA is designed to allow the vehicle to fly once around the Earth and make a normal reentry and landing. The AOA mode is used in cases where vehicle performance has been lost to such an extent that it is impossible to achieve a viable orbit, or not enough OMS propellants remain for orbit circularization and/or deorbit. In addition, an AOA is used in cases where a major subsystem failure makes it desirable to land as quickly as possible. The reentry sequence is similar to normal, although the Orbiter has slightly less energy to dissipate.[35]

  In the AOA abort mode, one OMS thrusting sequence is made to adjust the post-MECO orbit so a second OMS thrusting will result in the vehicle deorbiting and landing at the AOA landing site (White Sands, Edwards, or KSC). Thus, an AOA results in the Orbiter circling the Earth once and landing approximately 105 minutes after lift-off. The actual landing is handled as a nominal case.

- Transoceanic Abort Landing: The TAL mode is designed to permit an intact landing, usually on the other side of the Atlantic Ocean. This mode results in a ballistic trajectory that does not require an OMS maneuver for deorbit. The TAL mode was developed to improve the options available when an SSME fails after the last RTLS opportunity but before an AOA can be accomplished, or when a major system failure (i.e.; cabin leak) makes an AOA undesirable.[36]

  In a TAL abort, the vehicle continues on a ballistic trajectory across the Atlantic to land at a predetermined runway approximately 45 minutes after launch. The landing site is selected near the nominal ascent ground track to make the most efficient use of SSME propellants. The landing site also must have U.S. State Department approval, and have some minimal equipment.

  To select the TAL abort mode, the crew must place the ABORT MODES rotary switch in the TAL/AOA position and depress the ABORT push-button. It should be noted that depressing the ABORT push-button after MECO automatically engages the AOA mode. The TAL abort mode begins sending commands to steer the vehicle toward the plane of the pre-selected contingency landing site. It also rolls the vehicle heads-up (if necessary) before MECO and sends commands to begin an OMS propellant dump by thrusting at a low level through the OMS and RCS systems. This dump is necessary to decrease weight, and also to place the vehicle's center of gravity in the correct position for landing. The actual landing operations are handled similarly to a nominal reentry.[37]

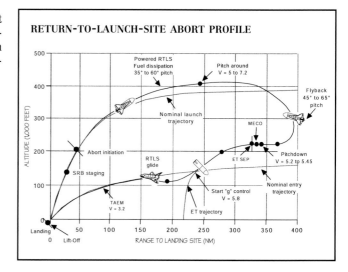

RETURN-TO-LAUNCH-SITE ABORT PROFILE

- East-Coast Abort Landing: The option for an ECAL abort was added during the early 1990s for high-inclination launches, although it is available for a very small period of time during ascent. NASA considers this a variation of TAL, but it is broken out here for clarity. The ECAL capability replaced certain bailout regions with landing capability for 2 engine-out scenarios prior to a single-engine TAL capability. Small ECAL capability regions also exist for 2 engine-out scenarios prior to RTLS powered pitch around and for some 3 engine-out trajectories. The theory is that it is always better to try and fly towards a land mass, even if you can't make a landing site, rather than fly wings level straight ahead and bailout (which is what regular contingency abort logic does).[38]

- Return-to-Launch-Site: The RTLS mode involves flying downrange to dissipate propellants and then turning around under power to return directly to a landing at, or near, the launch site approximately 25 minutes after lift-off. The RTLS abort mode is designed to allow the return of the Orbiter, flight crew, and payload to the launch site approximately 25 minutes after lift-off. The RTLS profile is optimized to accommodate the loss of thrust from one SSME between SRB separation and approximately 4 minutes 20 seconds, when there is insufficient main propulsion system propellants remaining to return to the launch site.[39]

  An RTLS consists of three major phases – a powered phase, during which the SSMEs are still thrusting; an ET separation phase; and the glide phase, during which the Orbiter glides to a landing. The powered RTLS phase begins with the crew selecting an RTLS abort after SRB separation. The crew selects the abort mode by positioning the rotary ABORT MODES switch to RTLS and depressing the ABORT push-button. The time at which the RTLS is selected depends on the reason for the abort. For example, a three engine RTLS is selected at the last minute, approximately 3 minutes 34 seconds into the mission, whereas an RTLS chosen due to an engine out at lift-off is selected at the earliest possible time, approximately 2 minutes 20 seconds into the mission (immediately after SRB separation).

  After RTLS is selected, the vehicle continues downrange to expend excess propellants from the ET. The goal is to leave only enough propellant to turn around, fly back towards KSC, and achieve the proper main engine cutoff

---

[*] TAL originally stood for trans-atlantic landing, but was changed early in the program because Vandenberg would not abort across the Atlantic Ocean. It was also known as a TLA (transatlantic landing abort) in some documentation.

conditions so the vehicle can glide back to the KSC after ET separation. During the downrange phase, a pitch-around maneuver is initiated to orient the Orbiter–ET configuration to a heads-up attitude, pointing towards the launch site. At this time the vehicle is still moving away from the launch site, but the SSMEs are now thrusting to nullify the downrange velocity. In addition, excess OMS and RCS propellants are dumped by continuous thrustings to improve Orbiter weight and center-of-gravity for the glide phase and landing.

Ideally the vehicle will reach the desired MECO point with less than two percent excess propellant remaining in the ET. At MECO minus 20 seconds, a pitch-down maneuver takes the mated vehicle to the required ET separation attitude and pitch rate. After MECO has been commanded, the ET separation sequence begins, including an RCS translation that ensures that the Orbiter does not re-contact the ET and that the Orbiter has achieved the necessary pitch attitude to begin the glide phase of the RTLS.

After the RCS translation maneuver has been completed, the glide phase of the RTLS begins. From this point through landing, the RTLS is handled similarly to a nominal reentry.

The type of failure and when the failure occurs during ascent largely determines which type of abort is selected. But there is a definite order of preference for the various abort modes. In cases where performance loss is the only factor, the preferred modes would be ATO, AOA, TAL, and RTLS, in that order. The mode chosen is the highest one that can be completed with the remaining vehicle performance. In the case of some system failures, such as cabin leaks, the preferred mode might be the one that would end the mission most quickly, thus assuring the survival of the crew. In these cases, TAL or RTLS might be preferable to AOA or ATO. The priorities associated with choosing an abort mode are: not endangering the general public; saving the flight crew; saving the vehicle; and completing the mission, even partially. A contingency abort is never chosen if another abort option exists.[40]

Mission Control is responsible for calling these aborts because it has a more precise knowledge of the Orbiter's position than can be obtained from the on-board systems. Before MECO, Mission Control makes periodic calls to the crew to tell them which abort mode is (or is not) available. If ground communications are lost, the flight crew has on-board methods to determine the current abort region, although not as accurately as the ground could. It should be noted that the flight crew – and only the crew – can initiate an abort by positioning an ABORT MODES switch and depressing the ABORT push-button.

FINAL PREPARATIONS

ON 8 March 1979 the first flight-rated Orbiter, *Columbia*, was rolled-out of the Rockwell International facility in Palmdale, and transported to Edwards AFB following the same overland route used by *Enterprise*. In order to make the scheduled arrival date at KSC, many of the tiles were missing, as were several other items of equipment. The Orbiter was mated to the SCA and made a short test flight on 9 March, but several of the simulated tiles that had been glued and taped to the outside of the Orbiter fell off during take-off. Hasty repairs were made and a second test flight took place on 20 March, this time without incident. The following day the SCA ferried *Columbia* as far as Biggs Army

Columbia stacked and ready for the FRF prior to STS-1. Note the cargo net on the ET where some insulation had worked loose during an earlier tanking test. It was decided to wait until after the FRF to repair the damage. (NASA photo 81-HC-159)

Air Field, near El Paso, Texas. A short hop to Kelly AFB in San Antonio was made on 22 March, and on 23 March the pair made it as far as Eglin AFB, Florida. The next day the Orbiter arrived at KSC and was placed inside OPF-1 to begin the painstaking process of getting the thermal protection system ready for the first flight.[41]

The first flight ready ET had arrived at KSC on 29 June 1979, and on 3 November 1980 it was mated with the pair of SRBs destined for use on STS-1. Finally, after spending a record 613 days in the OPF, *Columbia* was rolled-over to VAB High Bay #3 on 24 November 1980, and mounted to the ET on 26 November. The entire stack was powered up in the VAB for the first time on 4 December. MLP-1, carrying the first complete flight-rated Space Shuttle stack, was rolled-out to Launch Pad 39A on 29 December 1980, making the trip in just over ten hours. Propellant loading tests were performed on 22 January 1981 (LH2) and 24 January (LO2) to verify that the ground and flight system would work together correctly, and allowed ground personnel in the Firing Room to gain some real-life experience. However, during the propellant loading tests several pieces of the thermal insulation on the ET debonded. After investigating the problem, it was decided not to attempt repairs until after the Flight Readiness Firing (FRF), so a cargo net was draped over the affected area. The FRF was conducted on 20 February 1981 – running *Columbia*'s main engines for 20 seconds at 100 percent thrust while the vehicle remained firmly secured to the launch pad as a final check of all systems.[42]

After the FRF, technicians from Martin Marietta began repairing the insulation on the ET. On 19 March 1981 a dress rehearsal was conducted with the astronauts in the cockpit, but tragedy struck later in the morning when two technicians were asphyxiated in a nitrogen-purged portion of the aft fuselage. Two additional propellant loading tests were conducted on 25 and 27 March to verify repairs to the ET insulation. Subsequently, a launch readiness review set the date for the first flight – 10 April.[43]

THE FIRST TWENTY-FIVE FLIGHTS

Prior to the Challenger stand-down, there were 25 flights flown, and the popular way to list them is by these consecutive sequence numbers (Flight 1 thru Flight 25). The original official numbering scheme was to use sequential numbers preceded by 'STS,' hence Flight 1 was known as STS-1. Under this system, the numerically highest mission flown was STS-33; eight flights that were assigned numbers having been cancelled. After STS-9, NASA instituted a system where each flight carried a two number/one letter designation. The first digit indicated the fiscal year of the scheduled launch (4 for FY84); the second digit identified the launch site (1 = KSC, 2 = VLS); and the letter was the sequence for the fiscal year ('A' being the first mission of the year, 'B' being second, etc.).[133]

Seq. #	Mission #	Manifest #	Orbiter Flight #	Dates Launch Landing	Time Launch Landing	Sites Launch Landing	Crew	Orbit Altitude Inclination	Weights Lift-Off / OV Launch / Payload / OV Landing	Payloads	Mission Duration	Reentry Velocity Cross-Range Touchdown Roll-out	Notes
1	STS-1		OV-102 1	10 Apr 81 (pd) 12 Apr 81 14 Apr 81	102:12:00:04 104:18:20:55 195,472	KSC, Pad A EAFB, 23	John W. Young (C) Robert L. Crippen (P)	166 40.30	4,457,111 219,440 10,823 8,993	DFI Pallet ACIP RMS = none	36 Orbits 54 hrs 20 mins 32 secs 1,074,567 miles	25,731 262 183	Lift-off at 07:00:0398.010 EST First orbital flight by a winged spacecraft 16 tiles lost, 148 damaged Orbiter returned to KSC on 28 Apr 81
2	STS-2		OV-102 2	09 Oct 81 (s) 04 Nov 81 (s) 12 Nov 81 14 Nov 81	316:15:09:59 318:21:23:14	KSC, Pad A EAFB, 23	Joe H. Engle (C) Richard H. Truly (P)	157 38.00	4,470,308 230,938 18,778 204,262	OFT Pallet DFI Pallet ACIP IECM OSTA-1 RMS = s/n 201	36 Orbits 54 hrs 13 mins 13 secs 1,074,757 miles	25,726 73 197 7,711	First re-use of a manned spacecraft First test of Remote Manipulator System Flight shortened due to Fuel Cell #1 12 tiles damaged Orbiter returned to KSC on 25 Nov 81
3	STS-3		OV-102 3	22 Mar 82 30 Mar 82	081:16:00:59 089:16:04:48	KSC, Pad A WSMR, 17	Jack R. Lousma (C) Charles Gordon Fullerton (P)	147 38.00 22,710	4,468,755 235,556 ACIP 207,072	OSS-1 Pallet DFI Pallet 3,334,904 miles RMS = s/n 201	129 Orbits 192 hrs 04 mins 45 secs 220	25,659 318 13,732	EAFB landing site flooded by heavy rains 36 tiles lost, 19 damaged First unpainted External Tank Orbiter returned to KSC on 06 Apr 82
4	STS-4		OV-102 4	27 Jun 82 04 Jul 82	178:14:59:59 185:16:09:45	KSC, Pad A EAFB, 22	Thomas K. 'Ken' Mattingly II (C) Henry 'Hank' W. Hartsfield, Jr. (P)	197 28.45	4,481,935 241,772 11,644 208,946	IECM DFI Pallet DoD 82-1 RMS = s/n 201	112 Orbits 169 hrs 09 mins 40 secs ≈2,900,000 miles	25,797 669 204 9,878	First DoD Mission (non-dedicated) SRBs not recovered due to chute failure First concrete runway landing Orbiter returned to KSC on 15 Jul 82
5	STS-5		OV-102 5	11 Nov 82 16 Nov 82	315:12:19:00 320:14:33:26	KSC, Pad A EAFB, 22	Vance D. Brand (C) Robert F. Overmyer (P) Joseph P. Allen (MS1) William B. Lenoir (MS2)	184 28.45	4,487,268 247,112 20,830 202,480	ANIK-C3/PAM-D SBS-C/PAM-D DFI Pallet GAS (test) RMS = none	81 Orbits 122 hrs 14 mins 26 secs 2,110,849 miles	25,758 667 198 9,553	First operational flight Largest US crew flown to date Orbiter returned to KSC on 22 Nov 82
6	STS-6		OV-099 1	20 Jan 83 (s) 04 Apr 83 09 Apr 83	094:18:30:00 099:18:53:47	KSC, Pad A EAFB, 22	Paul J. Weitz (C) Karol J. Bobko (P) Donald H. Peterson (MS1) F. Story Musgrave (MS2)	178 28.45	4,487,255 256,928 46,662 190,330	TDRS-A/IUS CBSA GAS (3) RMS = none	80 Orbits 122 hrs 14 mins 26 secs 2,094,293 miles	25,755 435 190 7,244	First *Challenger* flight First Space Shuttle-based EVA First use of lightweight SRBs First use of lightweight ET Orbiter returned to KSC on 16 Apr 83
7	STS-7		OV-099 2	09 Jun 83 (s) 18 Jun 83 24 Jun 83	169:11:33:00 175:13:57:08	KSC, Pad A EAFB, 15	Robert L. Crippen (C) Frederick H. Hauck (P) Sally K. Ride (MS1) John M. Fabian (MS2) Norman E. Thagard (MS3)	195 28.45	4,482,241 249,362 31,893 204,043	ANIK-C2/PAM-D PALAPA-B1/PAM-D SPAS-01 OSTA-2 CBSA GAS (7) RMS = s/n 201	97 Orbits 146 hrs 23 mins 59 secs 2,530,567 miles	25,771 849 202 10,450	First U.S. woman in space First Shuttle rendezvous Orbiter returned to KSC on 29 Jun 83
8	STS-8		OV-099 3	04 Aug 83 (s) 20 Aug 83 (pd) 30 Aug 83 05 Sep 83	242:06:32:00 248:07:40:43	KSC, Pad A EAFB, 22	Richard H. Truly (C) Daniel C. Brandenstein (P) Dale A. Gardner (MS1) Guion S. Bluford, Jr. (MS2) William E. Thornton (MS3)	191/139 28.45	4,492,074 242,912 25,790 203,945	INSAT-1B/PAM-D PFTA DFI Pallet CBSA SPAS-01 GAS (12) RMS = s/n 201	97 Orbits 145 hrs 08 mins 43 secs 2,514,478 miles	25,649 597 195 9,371	First night launch Fight night landing Orbiter returned to KSC on 09 Sep 83
9	STS-9	(41-A)	OV-102 6	30 Sep 83 (pd) 28 Nov 83 08 Dec 83	332:16:00:00 342:23:47:24	KSC, Pad A EAFB, 17L	John W. Young (C) Brewster H. Shaw, Jr. (P) Owen K. Garriott (MS1) Robert A. Parker (MS2) Bryon K. Lichtenberg (PS1) Ulf D. Merbold (PS2)	155 57.00	4,503,361 247,807 33,131 220,027	Spacelab-1 Cryo Sets 4 & 5 RMS = s/n 201	166 Orbits 247 hrs 47 mins 24 secs 4,295,853 miles	25,696 79 185 8,456	First Spacelab mission Orbiter returned to KSC on 15 Dec 83
10	STS-11	41-B	OV-099 4	29 Jan 84 (pd) 03 Feb 84 11 Feb 84	034:13:00:00 042:12:15:55	KSC, Pad A KSC, 15	Vance D. Brand (C) Robert L. 'Hoot' Gibson (P) Bruce McCandless II (MS1) Robert L. Stewart (MS2) Ronald E. McNair (MS3)	202 28.45	4,498,443 250,482 28,252 201,238	PALAPA-B2/PAM-D WESTAR-VI/PAM-D SPAS-01 IMAX Camera GAS (5) MMU RMS = s/n 201	127 Orbits 191 hrs 15 mins 55 secs 3,311,380 miles	25,752 603 196 10,807	First MMU test First KSC landing
11	STS-13	41-C	OV-099 5	04 Apr 84 (pd) 06 Apr 84 13 Apr 84	097:13:58:00 104:13:38:06	KSC, Pad A EAFB, 17L	Robert L. Crippen (C) Francis R. 'Dick' Scobee (P) George D. Nelson (MS1) James D. van Hoften (MS2) Terry J. Hart (MS3)	313/288 28.45	4,508,234 254,554 33,831 196,975	LDEF-1 SMM SMRM IMAX Camera MMU (2) RMS = s/n 302	107 Orbits 167 hrs 40 mins 07 secs ≈2,870,000 miles	25,998 438 213 8,716	First satellite repair (*Solar Maximum*) First operational use of MMUs First 'direct insertion' trajectory Orbiter returned to KSC on 18 Apr 84
12	STS-16	41-D	OV-103 1	22 Jun 84 (s) 25 Jun 84 (pd) 26 Jun 84 (abort) 29 Aug 84 (pd) 30 Aug 84 05 Sep 84	243:12:41:50 249:13:37:55	KSC, Pad A EAFB, 17L	Henry 'Hank' W. Hartsfield, Jr. (C) Michael L. Coats (P) Judith A. Resnik (MS1) Steven A. Hawley (MS2) Richard M. Mullane (MS3) Charles D. Walker (PS1)	205 28.45	4,517,534 263,477 41,382 201,674	SBS-D/PAM-D TELSTAR-3C/PAM-D Leasat-2 OAST-1 SSIP IMAX Camera RMS = s/n 301	96 Orbits 144 hrs 56 mins 04 secs ≈2,490,000 miles	25,776 545 200 10,275	First *Discovery* flight First on-pad abort of program First 3-satellite deployment Orbiter returned to KSC on 10 Sep 84
13	STS-17	41-G	OV-099 6	01 Oct 84 (s) 05 Oct 84 13 Oct 84	279:11:03:00 287:16:26:39	KSC, Pad A KSC, 33	Robert L. Crippen (C) Jon A. McBride (P) Kathryn D. Sullivan (MS1) Sally K. Ride (MS2) David C. Leestma (MS3) Marc Garneau (PS1) Paul D. Scully-Power (PS2)	218 57.00	4,493,317 242,790 17,592 202,266	ERBS ASTA-3 LFC/ORS APE CANEX SASSE GAS (8) Galley RMS = s/n 302	132 Orbits 197 hrs 23 mins 33 secs 3,434,444 miles	25,684 707 208 10,565	EVA refueling simulation
14	STS-19	51-A	OV-103 2	07 Nov 84 (pd) 08 Nov 84 16 Nov 84	313:12:15:00 321:12:00:01	KSC, Pad A KSC, 15	Frederick H. Hauck (C) David M. Walker (P) Anna L. Fisher (MS1) Dale A. Gardner (MS2) Joseph P. Allen (MS3)	224 28.45	4,519,901 263,324 38,003 207,505	ANIK-D2/PAM-D Leasat-1 DMOS RME MMU (2) RMS = s/n 301	126 Orbits 191 hrs 44 mins 56 secs 3,289,406 miles	25,869 559 186 9,454	First on-orbit satellite retrieval (WESTAR-VI and PALAPA-B2)
15	STS-20	51-C	OV-103 3	23 Jan 85 (pd) 24 Jan 85 27 Jan 85	024:19:50:00 027:21:23:29	KSC, Pad A KSC, 15	Thomas K. 'Ken' Mattingly II (C) Loren J. Shriver (P) Ellison S. Onizuka (MS1) James F. Buchli (MS2) Gary E. Payton (PS1)	220 28.45	— 250,981 — —	DoD (ORION-1/IUS) IOCM OASIS CLOUDS RMS = s/n 301	48 Orbits 73 hrs 33 mins 23 sec ≈1,250,000 miles	25,855 437 185 7,352	First dedicated DoD mission Payload codenamed MAGNUM
16	STS-23	51-D	OV-103 4	19 Mar 85 (s) 28 Mar 85 12 Apr 85 19 Apr 85	102:13:59:05 109:13:54:33	KSC, Pad A KSC, 33	Karol J. Bobko (C) Donald E. Williams (P) Margaret Rhea Seddon (MS1) Jeffrey A. Hoffman (MS2) S. David Griggs (MS3) Charles D. Walker (PS1) Senator E. J. 'Jake' Garn (PS2)	289 28.45	4,505,245 250,891 28,747 198,014	Leasat-3 ANIK-C1/PAM-D CFES-III AFE PPE SSIP GAS (2) RMS = s/n 301	109 Orbits 167 hrs 55 mins 23 secs 2,889,785 miles	25,955 596 200 10,298	Mission extended two days
17	STS-24	51-B	OV-099 7	29 Apr 85 (s) 06 May 85	119:16:00:00 119:16:02:19	KSC, Pad A EAFB, 17L	Robert F. Overmeyer (C) Frederick D. Gregory (P) Don Leslie Lind (MS1) Norman E. Thagard (MS2) William E. Thornton (MS3) Lodewijk van den Berg (PS1) Taylor G. Wang (PS2)	222 57.00	4,512,009 247,291 30,748 213,499	Spacelab-3 GAS (1) Airlock Long Tunnel Galley RMS = none	110 Orbits 168 hrs 08 mins 46 secs 2,890,383 miles	25,857 315 204 8,317	First crosswind landing Orbiter returned to KSC on 11 May 85

Seq. #	Mission #	Manifest #	Orbiter Flight #	DATES Launch Landing	TIME Launch Landing	SITES Launch Landing	Crew	ORBIT Altitude Inclination	WEIGHTS Lift-Off OV Launch Payload OV Landing	Payloads	Mission Duration	REENTRY Velocity Cross-Range Touchdown Roll-out	Notes
18	STS-25	51-G	OV-103 5	17 Jun 85 24 Jun 85	168:11:33:00 175:13:12:00	KSC, Pad A EAFB, 23	Daniel C. Brandenstein (C) John O. Creighton (P) Shannon W. Lucid (MS1) John M. Fabian (MS2) Steven R. Nagel (MS3) Patrick Baudry (PS1) Salman Abdul aziz Al-Sàud (PS2)	240 28.45	4,516,613 256,421 38,258 204,169	MORELOS-A/PAM-D ARABSAT-1B/PAM-D TELSTAR-3D/PAM-D SPARTAN-101 GAS (3) RMS = s/n 301	111 Orbits 169 hrs 38 mins 53 secs 2,916,127 miles	25,850 799 198 7,433	Orbiter returned to KSC on 28 Jun 85
19	STS-26	51-F	OV-099 8	12 Jul 85 (abort) 29 Jul 85 06 Aug 85	210:21:00:00 218:19:45:27	KSC, Pad A EAFB, 23	Charles Gordon Fullerton (C) Roy D. Bridges, Jr. (P) F. Story Musgrave (MS1) Anthony W. England (MS2) Karl G. Henize (MS3) Loren W. Acton (PS1) John-David F. Bartoe (PS2)	207/165 49.50	4,515,554 252,628 33,012 216,735	Spacelab-2 EPDP RMS = s/n 302	126 Orbits 190 hrs 45 mins 26 secs 3,283,543 miles	25,813 694 199 8,569	Second on-pad abort of program In-flight abort-to-orbit successful after one SSME shutdown Orbiter returned to KSC on 11 Aug 85
20	STS-27	51-I	OV-103 6	24 Aug 85 (pd) 25 Aug 85 (pd) 27 Aug 85 03 Sep 85	239:10:55:10 246:13:15:46	KSC, Pad A EAFB, 23	Joe H. Engle (C) Richard O. Covey (P) James D. van Hoften (MS1) John M. Lounge (MS2) William F. Fisher (MS3)	278 28.45	4,512,130 262,309 38,884 196,856	ASC-1/PAM-D AUSSAT-1/PAM-D Leasat-4 PVTOS Galley RMS = s/n 301	111 Orbits 170 hrs 18 mins 29 secs 2,919,576 miles	25,829 796 191 6,100	On-orbit repair of Leasat-2 Orbiter returned to KSC on 08 Sep 85
21	STS-28	51-J	OV-104 1	03 Oct 85 07 Oct 85	276:15:15:31 280:17:00:12	KSC, Pad A EAFB, 23	Karol J. Bobko (C) Ronald J. Grabe (P) David C. Hilmers (MS1) Robert L. Stewart (MS2) William A. Pailes (MS3)	320 28.50	— — — —	DoD (2 x DSCS-III/IUS) OASIS-II AMOS CLOUDS-II RMS = none	65 Orbits 97 hrs 44 mins 38 secs ≈1,725,000 miles	— — — 8,056	First Atlantis flight Second dedicated DoD mission Orbiter returned to KSC on 11 Oct 85
22	STS-30	61-A	OV-099 9	30 Oct 85 06 Nov 85	303:17:00:00 310:17:44:53	KSC, Pad A EAFB, 17L	Henry 'Hank' W. Hartsfield, Jr. (C) Steven R. Nagel (P) James F. Buchli (MS1) Guion S. Bluford, Jr. (MS2) Bonnie J. Dunbar (MS3) Reinhard Furrer (PS1) Ernst Messerschmid (PS2) Wubbo J. Ockels (PS3)	207 57.00	4,508,496 243,762 30,519 214,171	Spacelab-D1 GLOMAR Airlock Long Tunnel Galley RMS = s/n 302	111 Orbits 168 hrs 44 mins 51 secs 2,909,352 miles	25,830 79 203 8,304	Largest flight crew in history Orbiter returned to KSC on 11 Nov 85
23	STS-31	61-B	OV-104 2	26 Nov 85 03 Dec 85	331:00:29:00 337:21:33:54	KSC, Pad A EAFB, 22	Brewster H. Shaw, Jr. (C) Bryan D. O'Conner (P) Mary L. Cleave (MS1) Sherwood C. Spring (MS2) Jerry L. Ross (MS3) Rudolfo Neri-Vela (PS1) Charles D. Walker (PS2)	240 28.45	4,514,530 261,610 42,788 205,732	MORELOS-B/PAM-D SATCOM-Ku2/PAM-D2 AUSSAT-2/PAM-D EASE/ACCESS IMAX Camera GAS (1) RMS = s/n 303	108 Orbits 165 hrs 04 mins 49 secs 2,838,972 miles	25,882 613 189 10,759	First construction of structures in orbit Second night launch Orbiter returned to KSC on 07 Dec 85
24	STS-32	61-C	OV-102 7	18 Dec 85 (s) 19 Dec 85 (pd) 06 Jan 86 (pd) 07 Jan 86 (pd) 09 Jan 86 (s) 10 Jan 86 (pd) 12 Jan 86 18 Jan 86	012:11:55:00 018:13:58:53	KSC, Pad A EAFB, 22	Robert L. 'Hoot' Gibson (C) Charles F. Bolden, Jr. (P) Franklin R. Chang-Diaz (MS1) Steven A. Hawley (MS2) George D. Nelson (MS3) Robert J. Cenker (PS1) Congressman C. William 'Bill' Nelson (PS2)	213 28.45	4,509,360 256,003 28,625 210,161	SATCOM-Ku1/PAM-D2 MSL-2 Hitchhiker G-1 IR-IE GAS (13) Galley RMS = none	97 Orbits 146 hrs 03 mins 51 secs 2,528,658 miles	25,815 761 217 10,202	First flight of OV-102 following mods First flight of SEADS, SILTS, and SUMS flight test package Orbiter returned to KSC on 23 Jan 86
25	STS-33	51-L	OV-099 10	22 Jan 86 (s) 23 Jan 86 (s) 26 Jan 86 (pd) 27 Jan 86 (pd) 28 Jan 86	028:16:38:00	KSC, Pad B	Francis R. 'Dick' Scobee (C) Michael J. Smith (P) Judith A. Resnik (MS1) Ellison S. Onizuka (MS2) Ronald E. McNair (MS3) Gregory B. Jarvis (PS1) Sharon Christa McAuliffe (PS2)	150 * 28.5 * * planned	4,526,583 268,829 48,633 n/a	TDRS-B/IUS SPARTAN-203/Halley FDE CHAMP Galley RMS = s/n 302	— 01 min 13 secs	n/a n/a n/a n/a	First Shuttle launch from Pad 39B Vehicle exploded 73 seconds after launch due to O-ring failure in right SRB Crew, vehicle and payload lost

Launch (s) is scheduled date (no launch attempt)
Launch (pd) is delay on pad during launch attempt
Launch (abort) is an abort after SSME ignition
n/a indicates not applicable
— indicates classified data or data not available
RMS = Remote Manipulator System serial number

Edwards AFB (EAFB) Runways 05R, 17L, 23L, and 33 are lakebed
Edwards AFB (EAFB) Runways 04 and 22 are concrete
White Sands Missile Range (WSMR) Runway 17 is lakebed
Kennedy Space Center (Shuttle Landing Facility) Runways 15 and 33 are concrete
Orbital altitude is highest apogee
Times are GMT (day:hour:minute:second)

(C) = Shuttle Commander
(P) = Shuttle Pilot
(PC) = Payload Commander
(MS) = Mission Specialist
(PS) = Payload Specialist

Weights in pounds
Altitude and cross-range in statue miles
Inclination in degrees from Equator
Velocity in feet per second
Touchdown in knots indicated airspeed
Roll-out in feet after main gear touchdown

John W. Young and Robert L. Crippen give a 'thumbs-up' during a photo session inside OV-102 during October 1980 while the Orbiter was in the OPF. (NASA photo S80-40384)

Columbia being mated to the ET (SWT-2) in preparation for STS-2. This ET did not have the black lightning protection strip around the top of the LO2 tank – one of the few visible external differences between the the first and second missions. (NASA)

The launch of STS-2. This was the last flight to use the white ET. Note the black wing chines on Columbia, the only Orbiter in the fleet so marked. This has always made it easy to identify OV-102 from a distance. (NASA)

Flight 1 – (STS-1) – A launch attempt on 10 April was scrubbed due to a timing slew when the Backup Flight System failed to synchronize with the Primary Avionics Software System – the trouble was quickly diagnosed, and a software patch was installed. The flight was launched 12 April 1981 as a two-day demonstration of *Columbia's* ability to fly into orbit, conduct on-orbit operations, and return safely. Due to a late engineering reassessment of the 'twang' motion after main engine start, STS-1 lifted off at T+4 seconds on countdown clock instead of T–0 (this was corrected for subsequent countdowns). The main payload was a developmental flight instrumentation (DFI) pallet containing equipment for recording temperatures, pressures, and acceleration levels at various points on the vehicle. This marked the first use of solid rockets on a manned vehicle, and the first time astronauts had piloted a new type of spacecraft on it's maiden flight. All the Orbital Flight Test missions were scheduled to land at Edwards AFB, and *Columbia* settled onto lakebed Runway 23 at 10:21 on 14 April. Post-flight inspection showed *Columbia* suffered minor damage from an overpressure created by the SRBs at ignition, and had lost 16 tiles, with 148 more being damaged. Otherwise, OV-102 was in good condition for a 'used' spaceship, and returned to KSC on 28 April 1981 atop the SCA. The backup crew for this flight were Joseph H. Engle (Cdr) and Richard H. Truly (Plt).[44]

Flight 2 – (STS-2) – The second Orbital Flight Test was scheduled seven months after the first in order to give engineers time to evaluate the data collected on the first flight, and to make minor changes to the vehicle and launch complex. The original launch date of 9 October 1981 was rescheduled when a nitrogen tetroxide spill occurred during the loading of the FRCS module – necessitating replacing or repairing 379 tiles. A second launch attempt on 4 November was delayed by an apparent low reading on a fuel cell oxygen tank pressure transducer, then aborted at T–31 seconds when clogged APU fuel filters caused an overtemp condition. *Columbia* was successfully launched on 12 November after a two hour and 40 minute delay to replace a multiplexer/demultiplexer (MDM). It was the first manned flight of a 'used' spacecraft. Richard Truly became the first person launched on his birthday, and, to date, this was the last all-rookie American crew. This flight marked the first test of the Remote Manipulator System (RMS), and carried a payload of Earth survey instruments, as well as the same DFI pallet carried on STS-1. The failure of fuel cell #1 shortened the planned five day flight by three days, although over 90 percent of the original mission objectives were still accomplished. Modifications to the launch pad sound suppression water system to absorb the SRB overpressure were successful, and only 12 damaged tiles were noted when OV-102 returned to the Edwards lakebed. The backup crew for STS-2 were Jack R. Lousma (Cdr) and Charles G. Fullerton (Plt).[45]

Flight 3 – (STS-3) – Practice was helping lower the processing time between flights, and *Columbia* spent only 68 days in the OPF after STS-2 being prepared for the third Orbital Flight Test. The launch was delayed one hour by a failure of a heater on a ground-based nitrogen gas purge line. *Columbia* stayed on-orbit for eight days, making this the longest OFT flight. Activities included a special test of the manipulator arm which removed a package of instruments from the payload bay, but did not release it. The flight included experiments in materials processing, and thermal testing of the Orbiter. The latter was accomplished by exposing the tail, nose, and top of the Orbiter to the sun for varying periods of time, rolling it in between

The payload for STS-3 included the Office of Space Science pallet (OSS-1) that contained a variety of thermal, electrical, and gassing experiments, as well as the same DFI pallet that was carried on all four orbital flight tests. The remote manipulator arm may be seen along the far payload bay sill. (NASA photo S82-26912)

tests to stabilize temperatures over the entire body. For the first time a number of experiments were carried in the mid-deck lockers, including a Continuous Flow Electrophoresis System to study separation of biological components and a Monodisperse Latex Reactor to produce uniform micron-sized latex particles. The first Shuttle Student Involvement Project (SSIP) – the study of insect motion – also was carried in a mid-deck locker. This marked the first flight of an unpainted ET, saving 595 pounds. During the flight, both crew members experienced some space sickness, the toilet malfunctioned, one APU overheated (but worked properly during descent), and three communications links were lost on 26 March. Heavy rains at Edwards caused the landing site to be changed to Northrup Strip at the White Sands Missile Range in New Mexico, although high winds at this site still caused the mission to be extended one day. Some brake damage occurred during landing, and an unexpected dust storm caused extensive contamination of the Orbiter. *Columbia* returned to KSC on 6 April 1982 with a total of 36 tiles missing and another 19 slightly damaged. The backup crew for this flight was Thomas K. Mattingly (Cdr) and Henry W. Hartsfield (Plt).[46]

Flight 4 – (STS-4) – This was the first Space Shuttle flight to be launched on schedule. The payload consisted of the first get-away specials (GAS), including nine scientific experiments provided by students from Utah State University, and a classified Air Force payload. In the mid-deck, the Continuous Flow Electrophoresis System and Monodisperse Latex Reactor were flown for the second time. The crew conducted a lightning survey with hand-held cameras, performed medical experiments on themselves for two student projects, and operated the RMS with the Induced Environment Contamination Monitor instrument mounted on its end designed to obtain information on gases or particles being released by the Orbiter in flight. The only major problem on this flight was the

loss of the two SRBs when the main parachutes failed to deploy properly and the two casings impacted the water at too high a velocity and sank. They were later found and examined by remote camera, but were not recovered. Although classified as a research flight, STS-4 carried the first commercial experiment (the CFES electrophoresis test). With the landing of STS-4 on the concrete runway at Edwards, the Orbital Flight Test program came to an end with 95 percent of its objectives completed. Since more than the two pilots would be carried on future flights, the ejection seats were disabled after this flight.[47]

The Crawler-Transporter arrives at the launch pad. In 1963 the Marion Power Shovel Company of Marion, Ohio, was awarded the contract to build two crawler transporters which were assembled in 1965–66. The cost of both transporters was under $15 million. As the largest land vehicles ever built, in 1977 the transporters were designated as National Historic Mechanical Engineering Landmarks by the American Society of Mechanical Engineers. Each transporter weighs 6,000,000 pounds. The flat top is 131 feet long and 114 wide, and is adjustable by hydraulic jacks, from 20 feet to 26 feet in height. The transporter averages about 127 gallons of diesel fuel to the mile. As of 1990, each transporter had traveled about 1,000 miles, total. (NASA)

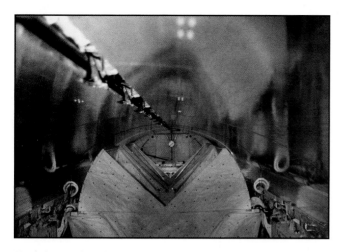

View looking into the payload bay of Columbia on STS-5. The two satellite containers are empty after the 'We Deliver' team successfully deployed the first commercial satellites – ANIK-C3 and SBS-C – to be launched by Space Shuttle. (NASA photo STS-5-10-370)

Flight 5 – (STS-5) – This was the second on-time launch of the Space Transportation System. On the first 'operational' Space Shuttle flight, *Columbia* launched two commercial communications satellites (ANIK-C3 and SBS-C), that were later boosted into geosynchronous orbit by attached payload assist modules (PAM). This was the largest U.S. space crew yet, featuring four astronauts, and they advertised themselves as the 'We Deliver' team. This was also the first launch with the astronauts taking advantage of the 'shirt-sleeve' environment in the crew module, marking the first U.S. crew not to wear pressure suits during ascent. In addition to the first commercial satellite cargo, the flight carried a West German-sponsored microgravity GAS experiment canister in the payload bay. A

planned extra-vehicular activity (EVA) by Allen and Lenoir was cancelled when a ventilation motor in one suit and a pressure regulator in the other malfunctioned.[48]

Flight 6 – (STS-6) – The first launch of *Challenger* was originally set for 20 January 1983, but a 20-second flight readiness firing (FRF) on 19 December 1982 revealed a hydrogen leak in a fuel line to SSME #1. A second FRF conducted on 25 January 1983 revealed leaks in all three SSMEs – all three engines were removed from the Orbiter while the vehicle was on the launch pad, and the cracked fuel lines were repaired. SSMEs #2 and #3 were then reinstalled following extensive failure analysis and testing, while SSME #1 (s/n 2011) was replaced with a spare (2017) after analysis revealed the possibility of further problems. Meanwhile, as the engine repairs were underway, a severe storm caused contamination of the TDRS-A payload while it was in the payload changeout room (PCR) on the rotating service structure (RSS) at the launch pad. The satellite had to be taken back to its checkout facility where it was cleaned and rechecked, and the PCR and Orbiter payload bay also had to be cleaned. The first flight of *Challenger* was finally launched on 4 April 1983, with no unexpected holds during the countdown. This was the first use of the lightweight External Tank and lightweight SRB casings. The flight was highlighted by the first Space Shuttle-based EVA, lasting a little over four hours and 17 minutes. The crew successfully deployed TDRS-A but the IUS shut down early, stranding the satellite in a low elliptical orbit. Nearly two months of short bursts of the satellite's attitude control thrusters eventually nudged it into the correct orbit.[49]

Flight 7 – (STS-7) – This was the third flight to be launched with no delays, and *Challenger* delivered a pair of commercial communications satellites. It was the first flight of an American woman astronaut – Sally K. Ride – and also the largest crew to fly in a single spacecraft up to that time, five persons. Additional tests of the remote

At left is the payload for STS-7 while still in the Vertical Processing Facility (VPF) at KSC. From top to bottom (forward to aft): SPAS-01, OSTA-2, PALAPA-B, and ANIK-C2. Above is what the payload bay looked like on-orbit. Multiple GAS canisters can be seen to the right, and a single one to the left. SPAS-01 was the first object to be retrieved from free-flight in space. The closed ANIK-C2 canister can also be seen behind SPAS-01. (NASA photos S83-34035 and S07-19-0937)

manipulator system were performed, releasing and recapturing the SPAS-01 (shuttle pallet satellite) test article. SPAS carried 10 experiments to study formation of metal alloys in microgravity, the operation of heat pipes, instruments for remote sensing observations, and a mass spectrometer to identify various gases in the payload bay. While SPAS was being held by the arm, Bob Crippen fired the RCS thrusters to observe the effect this had on the extended manipulator arm. The first communication tests using the TDRS-A satellite deployed on STS-6 were made on this flight. A KSC landing was waved off due to bad weather, but a subsequent problem with APU #3 forced a quick landing at Edwards two orbits later.[50]

Flight 8 – (STS-8) – *Challenger* made the first night launch of the Space Shuttle program, providing a spectacular view for several thousand spectators. The time of launch was dictated by the tracking requirements of the INSAT-1B primary payload, although lift-off was delayed 17 minutes due to weather. The 5-member crew, included the first black American to fly in space, Guion S. Bluford Jr. In addition to INSAT, the payload bay carried 12 GAS canisters. Four canisters contained experiments while the remaining eight contained special STS-8 postal covers. Two other boxes of covers were mounted on an instrument panel, bringing the total number of the special philatelic covers on board to 260,000. These were later sold to collectors by the U.S. Postal Service. Additional tests of the TDRS system were conducted in preparation for the upcoming Spacelab mission. *Challenger*'s nose was held away from the sun for 14 hours to test the flight deck's performance in the extreme cold. An additional experiment had OV-099 lower its orbit to 139 miles as part of an effort to identify the cause of the glow that tends to surround the Orbiter at night. The flight ended in the first night landing, conducted just to make sure it would work.[51]

Flight 9 – (STS-9/41-A) – The ninth flight was originally scheduled for 30 September 1983, but was delayed for two months because of a suspect exhaust nozzle on the right SRB. The problem was discovered after the vehicle was moved to the launch pad, and necessitated a return to the VAB to replace the nozzle. *Columbia*, in its last flight prior to an extended modification period, carried the first Spacelab. The crew was divided into two teams, each working 12-hour shifts for the duration of the mission. Seventy-two scientific experiments were conducted in atmospheric and plasma physics, astronomy, solar physics, material sciences, technology, life sciences, and Earth observations. The effort went so well that the mission was extended an additional day to 10 days, making it the longest Shuttle flight to date. This marked the first operational use of the TDRS system. Two of the quad-redundant GPCs (#1 and #2) and IMU #1 failed during this flight, causing a 7.5 hour delay in landing. Upon landing, several small fires were discovered in the aft fuselage, caused by hydrazine leaking from the APUs.[52]

Flight 10 – (STS-11/41-B) – Launch was originally scheduled for 29 January 1984, but was delayed five days to replace the Orbiter APUs as a precautionary measure after test data revealed possible discrepancies. This *Challenger* mission was highlighted by the introduction of the manned maneuvering unit (MMU), an untethered backpack propulsion unit that allowed astronauts to maneuver in space independent of the Orbiter. Astronaut Bruce McCandless – the first human Earth-orbiting satellite – ventured out 320 feet from the Orbiter, while Robert Stewart tested the 'work station' foot restraint at the end of the RMS. An inflated balloon was to be released from

Columbia being hoisted up from the transfer isle in preparation for mating with the ET and SRBs for STS-9 Note the open attach-point/disconnect doors on the underside of the Orbiter. (NASA Photo 108-KSC-83P-317)

the payload bay as a rendezvous target, but it burst upon inflation. However, it still formed a large enough target, and rendezvous operations in preparation for the *Solar Maximum* retrieval scheduled for STS-13/41-C were completed successfully. The mission also launched two communications satellites, but both PAM upper stages failed to ignite, leaving them stranded in low-earth orbit. (Both satellites were retrieved successfully the following November during STS-19/51-A). The SPAS pallet originally carried by STS-7 was flown again, marking the first time a satellite had been refurbished and returned to space. This time, however, SPAS remained in the payload bay due to a failure of the RMS. For the first time, a Space Shuttle landed on the concrete runway at KSC.[53]

Flight 11 – (STS-13/41-C) – After an on-time launch, this was the first flight to use a 'direct insertion' trajectory, where the SSMEs carried it all the way to its planned operational altitude of 288 miles (later raised to 313 miles to rendezvous with *Solar Max*) – the OMS engines were only used to circularize the orbit. The ET was photographed breaking-up during reentry near Hawaii. Other activities included deployment of the Long Duration Exposure Facility (LDEF-1) – a retrievable, 21,300-pound, 12-sided cylinder that was 14 feet in diameter and 30 feet long. The LDEF carried 57 experiments. *Challenger* demonstrated an important capability of Space Shuttle – the retrieval, repair, and redeployment of the malfunctioning *Solar Maximum* spacecraft with the help of the MMUs. The Orbiter was maneuvered within 200 feet of *Solar Max* while Nelson and van Hoften made several unsuccessfully attempts to capture the spacecraft. The next morning, they were successful on the first attempt, capturing, and securing the satellite in the payload bay for repairs that ultimately took two separate space walks. *Solar Max* was deployed back into orbit the next day, thus concluding one of the most unique rescue and repair missions in the history of the space program.[54]

<u>Flight 12</u> – (STS-16/41-D) – *Discovery* conducted a successful FRF on 2 June 1984. The planned launch date of 22 June was later changed to 25 June, and the STS-14/41-D launch attempt on that day was scrubbed during the T-9 minute hold due to a failure in the backup GPC. The offending computer was replaced, but another launch attempt the next day was scrubbed by the on-board computers at T-4 seconds when redundant control over the SSME #3 main fuel valve was lost immediately after ignition. SSME #2 had run for two seconds and #1 had not

ignited when the shutdown occurred. This was the first on-pad abort-after-ignition of the Space Shuttle program. After this abort, the STS-14/41-D mission was remanifested to include the most important items from both the originally payload, and that intended for STS-16/41-F, which was then cancelled. This required returning the stack to the VAB for disassembly and reconfiguration. A 29 August launch attempt failed when a discrepancy was noted in the flight software for the Master Events Controller. *Discovery* was finally launched for her maiden flight on 30 August, after a delay of 6 minutes 50 seconds when a private aircraft intruded into a warning area off the coast of KSC. The mission demonstrated repeated deployment and retraction of a large, foldable solar array to investigate

At Bottom, STS-13/41-C Astronaut James D. van Hoften poses with the Solar Maximum spacecraft that was retrieved, repaired, and redeployed. The RMS is visible at the extreme right. Notice the Orbiter's elevons are not locked, and free float while on-orbit. This was the highest U.S. manned space flight flown to-date, excluding the Apollo moon missions and Gemini XI. At left Nelson and van Hoften work on Solar Max in the payload bay. The shot at right shows the loneliest man in the world – and the smallest manned satellite in history (in this case Bruce McCandless practicing with the MMU on STS-11/41-B). (NASA photos 108-KSC-84PC-359, and S13-52-2646, and DVIC photo DF-SC-84-10567)

the practicality of using them as power sources for extended Space Shuttle missions and the space station, and launched three communications satellites. One of these, Leasat-2, was the first large communications satellite designed specifically to be deployed from the Space Shuttle. The landing was intentionally planned for Edwards since this was *Discovery*'s first flight.[55]

Flight 13 – (STS-17/41-G) – A perfect countdown resulted in an on-time launch of *Challenger* and the NASA Earth Radiation Budget Explorer. The major mission activity was operating the Shuttle Imaging Radar-B (SIR-B) that was part of the OSTA-3 experiment package in the payload bay. This package included the large format camera (LFC) to photograph Earth, another camera called MAPS that measured air pollution, and a feature identification and location experiment called FILE which consisted of two TV cameras and two 70-mm still cameras. The record-sized crew of seven included two women and a Canadian, and focused on scientific research. A rehearsal EVA was conducted by Sullivan (the first American woman to walk in space) and Leestma of the orbital refueling system (ORS) that would be used to refuel Landsat-4 on a later mission. The only problem came when the Ku-band antenna lost the ability to track, requiring the crew to 'point' the Orbiter at TDRS instead of the usual procedure of pointing the antenna. Because of this, much of the data from the OSTA-3 package had to be recorded onboard the Orbiter rather than transmitted to Earth in real-time as originally planned.[56]

Flight 14 – (STS-19/51-A) – Less than a month after the STS-17/41-G flight, the second *Discovery* mission was launched. A launch attempt the day before was scrubbed at the T–20-minute built-in hold because of high shear winds in the upper atmosphere. *Discovery* deployed two communications satellites, and retrieved two others that had been sent into unusable orbits after deployment on STS-11/41-B. This was the first time an object placed into orbit by one vehicle had been recovered by another. The retrieval of PALAPA-B2 and WESTAR-VI involved considerable determination by astronauts Gardner and Allen during extended EVAs. On day five, *Discovery* rendezvoused with PALAPA and Allen and Gardner performed an EVA to capture the satellite with a device known as a 'stinger' that was inserted into the apogee motor nozzle by Allen. The satellite's rotation was slowed to one rpm and Fisher, operating from a position on the end of the RMS, attempted unsuccessfully to grapple the satellite. However, Allen was able manually to maneuver the satellite into its cradle with help from Gardner and aided by the RMS which was operated by Fisher. The improvised – but ultimately successful – rescue effort took two hours. The recovery of WESTAR-VI was not as difficult and took place a day later. This time Gardner, using the same muscle power technique that Allen had used, captured the satellite. With Allen's help, he placed it in a cradle in the cargo bay.[57]

Flight 15 – (STS-20/51-C) – Mission 51-C had been manifested for *Challenger*, but problems with the Thermal Protection System required the substitution of *Discovery*. The original launch date of 23 January 1985 was postponed one day due to freezing weather. This was the first dedicated DoD flight (small DoD payloads had flown on nine previous flights). The payloads that have been acknowledged to have flown on this flight include the Interim Operational Contamination Monitor (IOCM) and the OEX Autonomous Supporting Instrumentation System (OASIS) in the payload bay, and Cloud Logic to Optimize Use of Defense Systems (CLOUDS – also flown on STS-16/41-D) – and OCEANS (photography of ocean phenomena) in the mid-deck. Speculation is that the primary payload was the National Security Agency (NSA) ORION-1 signals intelligence satellite (code-named MAGNUM) that was boosted into higher orbit by an attached IUS. All the official Air Force statements ever said was that the flight "… successfully met its mission objectives …"[58]

Flight 16 – (STS-23/51-D) – This *Discovery* flight was a composite mission, carrying part of its original manifest, and part of that from STS-22/51-E, which had been cancelled. The original launch date had been set for 19 March 1985, but was delayed to allow the payloads and crews to be shuffled. The crew was entirely from the cancelled mission, except for Charles Walker, who substituted for Patrick Baudry because the latter's flight experiments were no longer on the manifest. STS-22/51-E was to have been a *Challenger* flight, and the crew patch selected for the remanifested STS-23/51-D was that of the cancelled flight with the substitution of *Discovery*'s name for *Challenger*'s on the patch. Two communication satellites were deployed, but one, Leasat-3, did not successfully fire its booster stage. The flight was extended for two days to permit the crew to try and work-around the problem using a 'fly-swatter' to trigger a malfunctioning switch. Despite repeated efforts, the spacecraft remained lifeless. (It was eventually repaired on STS-27/51-I.) The mission also featured the first flight of an elected official, Senator E.J. 'Jake' Garn (R-Utah), chairman of the Senate committee with oversight responsibilities for the NASA budget. A tire was badly shredded upon landing at KSC, causing extensive brake damage, and it was decided that all future landings would be at Edwards AFB until the inactive nose-wheel steering system could be activated and tested.[59]

Atlantis at Pad 39A for STS-31/61-B This would be the second flight of Atlantis and carried three communications satellites to orbit. (NASA photo 108-KSC-85P-258)

Flight 17 – (STS-24/51-B) – After a delay of 2 minutes and 18 seconds, the second Spacelab mission was carried aloft by *Challenger*. This was the first flight of a fully operational Spacelab configuration, and a variety of multi-disciplinary microgravity research was successfully conducted. The gravity gradient attitude of the Orbiter proved quite stable, allowing the delicate experiments in materials processing and fluid mechanics to proceed successfully. Two monkeys and 24 rodents were flown in special cages, the first time American astronauts have flown with live mammals aboard. Spacelab 3 carried a large number of experiments, including 15 primary ones, of which 14 were successfully performed. There were five basic discipline areas – materials sciences, life sciences, fluid mechanics, atmospheric physics, and astronomy – with numerous experiments in each. Two get-away special experiments required that they be deployed from their canisters, a first for the GAS program – NUSAT (Northern Utah Satellite) deployed successfully, but GLOMR (global low orbiting message relay satellite) did not deploy and was returned to Earth.[60]

Flight 18 – (STS-25/51-G) – No launch delays were experienced and an international flavor was added to *Discovery* with the addition of a Saudi Arabian Prince and a French 'spatialnaut' to the crew. Three communications satellites, one Saudi, one Mexican, and one belonging to AT&T were successfully deployed. Also flown were the 2,223-pound free-flying SPARTAN-101 astronomy experiment, six GAS canisters, a High Precision Tracking Experiment (HPTE) for the 'Star Wars' Strategic Defense Initiative, a materials processing furnace, and French biomedical experiments. SPARTAN-101 was deployed and operated successfully, before being retrieved. The materi-

als furnace, French biomedical experiments, and GAS experiments were all successfully performed, although the GO34 get-away special shut down prematurely. The SDI experiment failed during the first attempt on orbit 37 because the Orbiter was not at the correct attitude, but successfully worked on orbit 64.[61]

Flight 19 – (STS-26/51-F) – During a 12 July launch attempt a hydrogen coolant valve in SSME #2 failed to close, causing an on-pad abort at T-3 seconds – after main engine ignition. Launch was rescheduled for 29 July, but was delayed 1 hour and 37 minutes due to an Orbiter communications problem. At T+5 minutes 45 seconds, the on-board GPCs shut down SSME #1 after detecting an overtemp condition – resulting in an abort-to-orbit (AOA). The engine failure occurred after the SSMEs had throttled back from 104 to 65 percent. The remaining two engines were throttled up to 91 percent and fired for an extra 70 seconds to provide sufficient energy to reach orbit. While this procedure was underway, another sensor indicated one of the remaining SSMEs was overheating and was about to be shut down by the GPCs. Had this engine also failed, the Orbiter would have attempted a TAL abort to an emergency landing at Zaragoza, Spain. Fortunately, a flight controller suspected the sensor was malfunctioning and ordered the crew to disable the sensor prior to the computer commanding an engine shutdown. *Challenger* entered an initial orbit of 124 by 165 miles that was later raised via a series of OMS burns to 196 by 207 miles. This flight carried a Spacelab payload consisting of an igloo and three pallets that contained scientific instruments dedicated to life sciences, plasma physics, astronomy, high-energy astrophysics, solar physics, atmospheric physics and technology research.[62]

Flight 20 – (STS-27/51-I) – This *Discovery* launch was originally planned for 24 August, but was delayed at T–5 minutes due to thunderstorms and lightning in the pad area. Launch was rescheduled for 25 August, but was further delayed until 27 August to replace a failed GPC (#5), and to inspect the MPS engine ducts. The countdown on 27 August was trouble-free except for a 3-minute 1-second delay caused by an unauthorized ship in the SRB recovery area. The primary mission was to deploy three commercial communications satellites and repair Leasat-3. In addition, a mid-deck materials processing experiment was flown. The mission was shortened by one day when the AUSSAT-1 sunshield hung-up on the RMS, and AUSSAT had to be deployed before scheduled. Leasat-3 which had failed to activate after deployment on STS-23/51-D was retrieved, repaired, and successfully redeployed during EVAs that totaled 11 hours 27 minutes.[63]

Flight 21 – (STS-28/51-J) – After a 20-second FRF on 5 September 1985, *Atlantis*' first flight was devoted to the second dedicated classified DoD mission. The flight was delayed 22 minutes 30 seconds because a main engine LH2 prevalve indicator showed a faulty reading. Acknowledged payloads included OASIS-II in the payload bay, and AMOS (Air Force Maui Optical Site Calibration Test), CLOUDS-II, CST (Contrast Sensitivity Test), MARC-DN (Measurement of Atmospheric Radiance Camera-Day/Night), and OCEANS on the mid-deck. It has since been learned that the primary payload was a pair of DSCS-III satellites (F-2 and F-3) carried by a single IUS. At the time, the Air Force again issued a statement calling the flight "... successful ..."[64]

Flight 22 – (STS-30/61-A) – After a countdown with no delays, *Challenger* was launched on a Spacelab mission devoted to materials processing experiments. Some of these experiments had predecessors which had returned

Details of Columbia during launch. A careful look at the SRB will reveal 'loaded' stencils on each segment. Note the star tracker locations on the forward fuselage. (NASA)

data obtained on earlier flights, making it possible to pre-pare experiment regimens that were 'second generation' with respect to technical concept and experiment installa-tion. Almost all of the experiments took advantage of the microgravity environment to perform work not possible, or very much more difficult to do, on Earth. This was the first Space Shuttle mission largely financed and operated by another nation, West Germany, and for the first time a ground station other than Mission Control in Houston con-trolled non-Orbiter portions of the mission – scientific operations were controlled by the German Space Operations Center located in Oberpfaffenhofen, near Munich. The GLOMR satellite that had not worked proper-ly on STS-24/51-B was successfully deployed during the mission, and five experiments mounted on the separate structure behind the Spacelab module obtained good data. One unusual item of equipment was the Vestibular Sled, an ESA contribution consisting of a seat that could be moved backward and forward with precisely controlled accelerations and stops. Various physiological measure-ments were made while crew members rode the sled. This mission carried the largest flight crew to date – eight.[65]

Flight 23 – (STS-31/61-B) – The second night launch of the program was accomplished with no unplanned delays. This *Atlantis* flight was highlighted by astronaut assembly of structures in orbit, and attendant study of extravehicular dynamics and human factors. The EASE/ACCESS experiment was a 'high-rise' tower com-posed of many small struts and nodes to form a geomet-ric structure shaped like an inverted pyramid. This exper-iment demonstrated the feasibility of assembling large preformed structures in space, a critical technology for the eventual space station. The assembly activities were filmed by high resolution IMAX cameras mounted in the payload bay. The mission also deployed three communi-cations satellites that were successfully boosted into geo-synchronous orbit by their payload assist modules (SATCOM Ku-1 used a PAM-D2, the first time this booster had flown on Shuttle). One GAS canister in the cargo bay carried an experiment by Canadian students to fabricate mirrors in microgravity with higher performance than ones made on Earth. The flight was shortened by one orbit due to weather conditions at Edwards, where *Atlantis* landed on the concrete runway since the lakebeds were wet.[66]

Flight 24 – (STS-32/61-C) – Fresh from a major refit in Palmdale, *Columbia*'s launch would be delayed several times from a planned date of 18 December 1985. On that date it was delayed because it was decided that addition-al time was needed to close out the Orbiter aft compart-ment. On 19 December, the count was halted at T–14 sec-onds due to an out-of-tolerance reading in the right SRB hydraulic system. Another launch attempt on 6 January was halted at T–31 seconds when a problem with the fill and drain valve in the LO2 system accidentally drained 14,000 pounds of oxidizer – the launch window ended before the problem could be resolved. On 7 January the launch team tried again, but marginal weather for an emergency return to KSC, plus bad weather at the con-tingency landing sites at Dakar and Moron forced a post-ponement at T–9 minutes. Launch was delayed 48 hours so the launch team could rest, but the planned 9 January launch date was delayed an extra day to permit removal of a LO2 sensor that had broken and become lodged in the SSME #2 prevalve. Another launch attempt was made on 10 January, but eventually delayed another 48 hours due to heavy rains in the launch area. The actual lift-off on 12

Dale Gardner uses an MMU to retrieve WESTAR VI during STS-19/51-A. This marked the first commercial satellite retrieval. (NASA Photo 84-H-655)

January was achieved without major incident, but this record five attempts to get a mission off the pad caused serious work-load problems at the various NASA Centers and contractor sites. *Columbia* deployed a communica-tions satellite, conducted experiments in infrared imag-ing, and acquired photos and spectral images of Comet Halley. Florida Congressman Bill Nelson was among the crew of seven. Mission controllers decided to shorten the planned flight by one day to provide more processing time on the ground for the next flight of *Columbia*, however the early landing attempt at KSC on 16 January had to be waved off due to unfavorable weather, and was waved off the next day as well. The mission was extended for one more day for a third attempt at KSC, but when that was again cancelled due to weather, *Columbia* landed at Edwards after a one orbit extension.[67]

Flight 25 – (STS-33/51-L) – *Challenger* was destroyed 73 seconds after launch on 28 January 1986.

Challenger being rolled-over from the OPF to the VAB. Before the 76-wheel transporter was brought from Vandenberg, the Orbiters were moved on their landing gear, which was then retracted and secured while in the VAB. (NASA)

Challenger (OV-099) lifts-off on the ill-fated STS-33/51-L mission from Pad 39B at the Kennedy Space Center, Florida. (NASA photo 108-KSC-86PC-158)

THE ACCIDENT

The Flight Readiness Review (FRR) for STS-33/51-L was held on 15 January 1986 and certified that *Challenger* was "... flight ready ..." The launch was postponed three times and scrubbed once from the planned date of 22 January 1986. The first postponement was announced on 23 December 1985, adding an extra day (until 23 January) to accommodate the final integrated simulation that had slipped one day due to the late launch of STS-32. On 22 January 1986, the date was slipped to 26 January, primarily because of KSC work requirements. The third postponement occurred on 25 January when forecasts indicated the launch site weather would be unacceptable. The new launch date was set for 27 January.[68]

The launch attempt on 27 January began with loading the ET at 00:30 hours, Eastern Standard Time. The crew was awakened at 05:07, and events proceeded normally with the crew strapped into the Orbiter at 07:56. At 09:10, however, the countdown was halted when the ground crew reported a problem with the exterior handle on the crew hatch. By the time the hatch problem was solved at 10:30, the KSC recovery site designated for an RTLS abort had exceeded the allowable velocity for crosswinds. The launch attempt was scrubbed at 12:35, and rescheduled for the following morning.[69]

The weather on 28 January was forecast to be clear and very cold, with temperatures dropping into the low twenties overnight. The management team directed engineers to assess the possible effects of temperatures on the launch. No critical issues were identified to NASA or contractor management officials, and while the evaluation continued, it was decided to proceed with the countdown and load propellants into the External Tank.[70]

A significant amount of ice had accumulated on Pad 39B overnight, creating considerable concern for the launch team. In reaction, the ice inspection team was sent to the pad at 01:35, and returned to the launch control center at 03:00. After meeting to consider the team's report, the Space Shuttle program manager continued the countdown, and scheduled another ice inspection for T-3 hours.[71]

At 08:44 the ice team completed its second inspection, and after hearing the team's report, the program manager decided to allow additional time for the ice to melt. He also elected to send the ice team to perform one final inspection during the scheduled hold at T-20 minutes. At this point, the launch had been delayed two hours past the original 09:38 scheduled time.[72]

At 11:15 the final ice inspection was completed, and during the scheduled hold at T-9 minutes, the flight crew and all members of the launch team, gave a 'go' for launch. The ambient air temperature at launch was 36 degF measured at ground level approximately 1,000 feet from the vehicle. This was 15 degrees colder than any previous launch. The final flight of *Challenger* began at 11:38:00.010, Eastern Standard Time, 28 January 1986.[73]

From lift-off until telemetry from the vehicle was lost, no flight controller observed any indications of a problem, although post-flight analysis showed telemetry had uncovered some anomalies. The SSMEs throttled down to limit the maximum dynamic pressure, then throttled up to full thrust as expected. Voice communications were normal. The crew called to indicate the vehicle had begun its roll to head due east, and fifty-five seconds later, Mission Control informed the crew that the engines had successfully throttled up, and that all other systems were satisfactory. Dick Scobee's acknowledgment of this call was the last voice communication from *Challenger*.[74]

Challenger lifts-off on 28 January 1986. The failure of the SRB seal is evident even at this early point in the flight as a small puff of smoke (arrow) escapes from the field joint between the two lower segments. (NASA photo 108-KSC-386C-774/36)

When the External Tank exploded, the two Solid Rocket Boosters continued to fly unguided momentarily until the Air Force range safety officer (RSO) at Cape Canaveral destroyed them. (NASA photo 108-KSC-386C-3560/38)

A front-end loader holds a large piece of debris from the Challenger after it washed ashore in Cocoa Beach during 1996. The piece, about 15 by 6 feet, was part of an elevon – one of the largest pieces to wash ashore to date. A smaller piece was found several blocks south. Other pieces have washed ashore both before and since this incident. Combined with the wreckage originally recovered during the search operations, approximately 50 percent of the Orbiter has been recovered. (NASA photo KSC-96PC-1357)

There were no alarms sounded in the cockpit, and the crew apparently had no indication of a problem before the rapid break-up of the vehicle. The first evidence of the accident came from live video coverage and when radar began tracking multiple targets. The flight dynamics officer at Mission Control confirmed to the flight director that "… RSO [range safety officer] reports the vehicle has exploded …," and thirty seconds later added that the Air Force range safety officer had sent the destruct signal to the SRBs.[75]

During the period of flight while the Solid Rocket Boosters are thrusting, there are no survivable abort options. There was nothing that the crew, or the flight controllers, could have done to avert the catastrophe.[76]

A combined Coast Guard/NASA/Air Force/Navy search team spent the next three months searching the Atlantic for the remains of *Challenger* and her crew. The wreckage was found in water between 100 and 1,200 feet deep. During the initial recovery operations, approximately thirty percent of the Orbiter was retrieved, including all three

A large section of Challenger is lowered into an abandoned Minuteman III test silo at Complex 31 on the Cape Canaveral Air Force Station adjacent to KSC for permanent storage. Waymon 'Cotton' Higgins is on the left. (NASA photo KSC-87PC-22)

SSMEs, the forward fuselage with the crew module, the right inboard and outboard elevons, a large portion of the right wing, the lower portion of the vertical stabilizer, three rudder/speedbrake panels, and portions of the mid-fuselage. Portions of the SRBs were also recovered. The debris was evaluated by NASA and the NTSB in typical accident investigation fashion.[77]

The forward frustum from the left SRB is prepared for storage at Complex 31. Both frustums were recovered, with the right unit showing significant damage from where it impacted the ET. (NASA photo via Waymon 'Cotton' Higgins)

Approximately 30-percent of Challenger was initially recovered, and was taken to a large logistics facility at KSC where a team of NASA and NTSB specialists examined it in an attempt to find out what happened. (NASA photo 108-KSC-386C-855/1)

SEQUENCE OF MAJOR EVENTS

GMT (hr:min:sec)	MET (sec)	Event
16:37:53.444	-6.566	ME-3 ignition command
16:37:53.564	-6.446	ME-2 ignition command
16:37:53.684	-6.326	ME-1 ignition command
16:38:00.010	0.000	SRB ignition command (T=0)
16:38:00.018	0.008	Holddown post 2 PIC firing
16:38:00.260	0.250	First continuous vertical motion
16:38:00.688	0.678	Confirmed smoke above field joint on RH SRM
16:38:00.846	0.836	Eight puffs of smoke (from 0.836 thru 2.500 sec MET)
16:38:02.743	2.733	Last positive evidence of smoke above right aft SRB/ET attach ring
16:38:03.385	3.375	Last positive visual indication of smoke
16:38:04.349	4.339	SSME 104 percent command
16:38:05.684	5.674	RH SRM pressure 11.8 psi above nominal
16:38:07.734	7.724	Roll maneuver initiated
16:38:19.869	19.859	SSME 94 percent command
16:38:21.134	21.124	Roll maneuver completed
16:38:35.389	35.379	SSME 65 percent command
16:38:37.000	36.990	Roll and yaw attitude response to wind shear (36.990 to 62.990 seconds)
16:38:51.870	51.860	SSME 104 percent command
16:38:58.798	58.788	First evidence of flame on RH SRM
16:38:59.010	59.000	Reconstructed max-q (720 psf)
16:38:59.272	59.262	Continuous well defined plume on RH SRM
16:38:59.763	59.753	Flame from RH SRM in +Z direction (seen from south side of vehicle)
16:39:00.014	60.004	SRM pressure divergence (RH vs. LH)
16:39:00.248	60.238	First evidence of plume deflection, intermittent
16:39:00.258	60.248	First evidence of SRB plume attaching to ET ring frame
16:39:00.998	60.988	First evidence of plume deflection, continuous
16:39:01.734	61.724	Peak roll rate response to wind
16:39:02.094	62.084	Peak TVC response to wind
16:39:02.414	62.404	Peak yaw response to wind
16:39:02.494	62.484	RH outboard elevon actuator hinge moment spike
16:39:03.934	63.924	RH outboard elevon actuator delta pressure change
16:39:03.974	63.964	Start of planned pitch rate maneuver
16:39:04.670	64.660	Change in anomalous plume shape (LH2 tank leak near 2058 ring frame)
16:39:04.715	64.705	Bright sustained glow on sides of ET
16:39:04.947	64.937	Start SSME gimbal angle large pitch variations
16:39:05.174	65.164	Beginning of transient motion due to changes in aero forces due to plume
16:39:06.774	66.764	Start ET LH2 ullage pressure deviations
16:39:12.214	72.204	Start divergent yaw rates (RH vs. LH SRB)
16:39:12.294	72.284	Start divergent pitch rates (RH vs. LH SRB)
16:39:12.488	72.478	SRB major high-rate actuator command
16:39:12.507	72.497	SSME roll gimball rates 5 degrees per second
16:39:12.535	72.525	Vehicle max +Y lateral acceleration (+0.227-g)
16:39:12.574	72.564	SRB major high-rate actuator motion
16:39:12.574	72.564	Start of LH2 tank pressure decrease with 2 flow control valves open
16:39:12.634	72.624	Last state vector downlinked
16:39:12.974	72.964	Start of sharp MPS LO2 inlet pressure drop
16:39:13.020	73.010	Last full computer frame of TDRS data
16:39:13.054	73.044	Start of sharp MPS LH2 inlet pressure drop
16:39:13.055	73.045	Vehicle max -Y lateral acceleration (-0.254-g)
16:39:13.134	73.124	Circumferential white pattern on ET aft dome (LH2 tank failure)
16:39:13.134	73.124	RH SRM pressure 19 psi lower than LH SRM
16:39:13.147	73.137	First hint of vapor at intertank E207 Camera
16:39:13.153	73.143	All engine systems start responding to loss of fuel and LOX inlet pressure
16:39:13.172	73.162	Sudden cloud along ET between intertank and aft dome
16:39:13.201	73.191	Flash between Orbiter and LH2 tank
16:39:13.221	73.211	SSME telemetry data interference from 73.211 to 73.303
16:39:13.223	73.213	Flash near SRB forward attach point and brightening of flash between Orbiter and ET
16:39:13.292	73.282	First indication intense white flash at SRB forward attach point
16:39:13.337	73.327	Greatly increased intensity of white flash
16:39:13.387	73.377	Start of RCS jet chamber pressure fluctuations
16:39:13.393	73.383	All engines approaching HPFT discharge temperature redline limits
16:39:13.492	73.482	ME-2 HPFT discharge temperature Channel A vote for shutdown; 2 strikes on Channel B
16:39:13.492	73.482	ME-2 controller last time word update
16:39:13.513	73.503	ME-3 in shutdown due to HPFT discharge temperature redline exceeded
16:39:13.513	73.503	ME-3 controller last time word update
16:39:13.533	73.523	ME-1 in shutdown due to HPFT discharge temperature redline exceedance
16:39:13.553	73.543	ME-1 last telemetered data point
16:39:13.628	73.618	Last validated Orbiter telemetry measurement
16:39:13.641	73.631	End of last reconstructed data frame with valid synchronization and frame count
16:39:14.140	74.130	Last radio frequency signal from Orbiter
16:39:14.597	74.587	Bright flash in vicinity of Orbiter nose
16:39:16.447	76.437	RH SRB nose cap separation chute deployment
16:39:50.260	110.250	RH SRB RSS destruct
16:39:50.262	110.252	LH SRB RSS destruct

THE ANALYSIS

President Ronald Reagan, seeking to ensure a thorough and unbiased investigation of the *Challenger* accident, formed a Presidential Commission with the issuance of Executive Order 12546 on 3 February 1986. The Commission was chaired by William P. Rogers, a former Secretary of State under President Nixon, and other members included astronauts Neil A. Armstrong and Sally K. Ride; Robert B. Hotz, former editor-in-chief of <u>Aviation Week & Space Technology</u>; Brigadier General Charles Yeager, USAF (Retired); and several distinguished scientists and engineers including a Nobel laureate.[78]

The report of the Commission concluded:[79]

> "The consensus of the Commission and participating investigative agencies is that the loss of the Space Shuttle Challenger was caused by a failure in the joint between the two lower segments of the right Solid Rocket Motor. The specific failure was the destruction of the seals that are intended to prevent hot gases from leaking through the joint during the propellant burn of the rocket motor. The evidence assembled by the Commission indicates that no other element of the Space Shuttle system contributed to this failure."

More than 160 individuals were interviewed and more than 35 formal panel investigative sessions were held, generating almost 12,000 pages of transcript. Nearly 6,300 documents, totaling more than 122,000 pages, and hundreds of photographs were examined and made a part of the Commission's permanent record.[80]

Foot-long icicles on a lower level of the Fixed Service Structure frame the attachment point where the Orbiter is connected to the External Tank (arrow). The icing was even more severe at upper levels. (NASA)

The series of events leading up to the destruction of *Challenger* was determined to have started with a leak in an O-ring seal in the right SRB. Just 0.678 seconds into the flight, photographic data showed a strong puff of gray smoke was spurting from the vicinity of the aft field joint on the right SRB, but the two Pad 39B cameras that would have recorded the precise location of the puff were inoperative. Computer graphic analysis of film from other cameras indicated the initial smoke came from the 270 to 310-degree sector of the aft field joint of the right SRB. This area of the booster faces the ET. The vaporized material streaming from the joint indicated there was not complete sealing action within the joint.[81]

Eight more distinctive puffs of increasingly black smoke were recorded between T+0.836 and T+2.500 seconds. The smoke appeared to puff upwards from the joint. While each smoke puff was being left behind by the upward flight of the vehicle, the next fresh puff could be seen near the level of the joint. The multiple smoke puffs in this sequence occurred about four times per second, approximating the frequency of the structural load dynamics and resultant joint flexing. Computer graphics applied to NASA photos from a variety of cameras in this sequence again placed the smoke puffs origin in the 270- to 310-degree sector of the SRB. The last smoke was seen above the field joint at 2.733 seconds. The black color and dense composition of the smoke puffs suggest that the grease, joint insulation, and rubber O-rings in the joint seal were being burned and eroded by the hot propellant gases.[82]

At approximately T+37 seconds, *Challenger* encountered the first of several high-altitude wind shear conditions, which lasted about 27 seconds. The wind shear created forces on the vehicle with relatively large fluctuations. These were immediately sensed and countered by the ascent flight control system.[83]

The SRB thrust vector control system responded to all commands and wind shear effects, causing the steering system to be more active than on any previous flight. The SSMEs and SRBs were operated at reduced thrust while passing through the area of maximum dynamic pressure (720 psf) and the SSMEs had throttled up to 104 percent and the SRBs were increasing their thrust when the first flickering flame appeared on the right SRB in the area of the aft field joint. This first very small flame was detected on image enhanced film at T+58.788 seconds appeared to originate at about 305 degrees around the booster circumference at or near the aft field joint.[84]

One film frame later from the same camera, the flame was visible without image enhancement. It grew into a continuous, well-defined plume at T+59.262 seconds. At about the same time, telemetry showed a pressure differential between the chamber pressures in the right and left boosters – the right booster chamber pressure was lower, confirming the growing leak in the area of the field joint.[85]

As the flame plume increased in size, it was deflected rearward by the aerodynamic slipstream and circumferentially by the protruding structure of the upper ring attaching the booster to the ET. These deflections directed the flame plume onto the surface of the ET, and this sequence of flame spreading was confirmed by analysis of the recovered wreckage.[86]

The first visual indication that swirling flame from the right SRB breached the ET was at 64.660 seconds when there was an abrupt change in the shape and color of the plume – indicating it was mixing with leaking hydrogen from the ET. LH2 pressurization telemetry confirmed the leak. Within 45 milliseconds of the breach of the ET, a bright sustained glow developed on the black-tiled underside of the *Challenger* between it and the ET.[87]

At about T+72 seconds, a series of events began that terminated the flight. Telemetered data indicate a wide variety of flight system actions that support the visual evidence of the photos as the vehicle struggled futilely against the forces that were destroying it.[88]

The lower strut linking the SRB and ET was severed or pulled away from the weakened LH2 tank at T+72.20 seconds, permitting the right SRB to rotate around the upper attachment strut. This rotation was indicated by divergent yaw and pitch rates between the left and right boosters.[89]

At T+73.124 seconds, a circumferential white vapor pattern was observed blooming from the side of the ET bottom dome. This was the beginning of the structural failure of hydrogen tank that culminated in the entire aft dome dropping away. This released massive amounts of LH2 from the tank and created a sudden forward thrust of about 2.8 million pounds, pushing the hydrogen tank upward into the intertank structure. At about the same time, the rotating right SRB impacted the intertank structure and the lower part of the LO2 tank. These structures failed at T+73.137 seconds as evidenced by the white

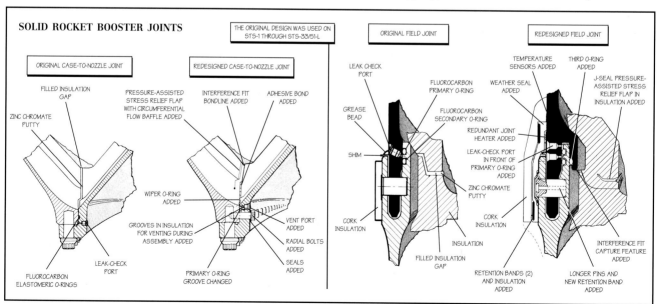

SOLID ROCKET BOOSTER JOINTS

vapors appearing in the intertank region. Within milliseconds there was massive, almost explosive, burning of the hydrogen streaming from the failed tank bottom and liquid oxygen breach in the area of the intertank.[90]

At this point in its trajectory, while traveling at Mach 1.92 and an altitude of 46,000 feet, *Challenger* was totally enveloped in the explosive burn. The reaction control system ruptured and a hypergolic burn of its propellants occurred as it exited the oxygen-hydrogen flames. The reddish brown colors of the hypergolic fuel burn are visible on the edge of the main fireball. The Orbiter, under severe aerodynamic loads, broke into several large sections which emerged from the fireball. Separate sections that can be identified on film include the aft-fuselage with the engines still burning, one wing, and the forward fuselage trailing a mass of umbilical lines pulled loose from the payload bay.[91]

It should be noted that although post-flight analysis showed that indications of the disaster were present in the telemetry stream, it appears that this relatively obscure data was not being actively monitored. Much of the telemetry data on each flight is simply recorded for subsequent engineering analysis, not evaluated in real-time. Even if a flight controller had been monitoring the specific data, there was less than six seconds between the first indication of a problem and the destruction of the vehicle – hardly sufficient time to recognize a problem, check the data, and take any meaningful action.

All fractures and material failures examined on recovered sections of the Orbiter, with the exception of the main engines, were the result of overload forces, and they exhibited no evidence of internal burn damage, or exposure to explosive forces. This indicated the destruction of the Orbiter occurred predominantly from aerodynamic and inertial forces that exceeded design limits. Additionally, chemical analysis indicated that the right side of the Orbiter was sprayed by hot propellant gases exhausting from the hole in the right SRB. Evaluation of the SSMEs showed extensive internal thermal damage as a consequence of an oxygen-rich shutdown. The supply of fuel to the main engines would have been abruptly terminated when the LH2 tank disintegrated.[92]

The wreckage of the crew module was found submerged in about 90 feet of water concentrated in an area of about 20 feet by 80 feet. Portions of the forward fuselage outer shell structure were found among the pieces of the crew module. There was no evidence of an internal explosion, heat or fire damage on the forward fuselage or crew module. The crew module had mostly disintegrated, with the heaviest fragmentation and crash damage on the left side. The fractures were typical of overload forces resulting from the impact with the ocean. The consistency of damage to the left side of the fuselage and the crew module indicated these structures remained attached to each other and intact until impact with the water. The entire ET range safety system was recovered intact, and the SRBs were observed to have destructed upon command, thus ruling out any possibility that the RSS malfunctioned and caused the accident.[93]

Following the completion of the accident investigation, the remains of *Challenger* were permanently sealed underground at the former Minuteman III test location at Complex 31 on Cape Canaveral AFS.

O-ring anomalies had been detected to varying degrees on 12 previous flights. Erosion of either the primary or secondary O-rings had been seen on Flights 2, 10, 11, 12, 15, 16, 17, 18, 20, 22, 23, and 24. A more serious problem,

the actual blow-by of exhaust gases past an O-ring had occurred on Flights 11, 12, 15, 16, 17, 18, 22, 23, and 24. All of these anomalies were recorded upon occurrence, and this data was known at the Flight Readiness Review for STS-33/51-L. Failure analysis conducted as early as 1979 on the Space Shuttle had concluded that one in fifty flights would encounter a catastrophic accident during ascent, and one in 100 would fail to land successfully. The failure analysis have been updated repeatedly, and until recently, had always reached much the same conclusion.[94]

THE FINDINGS

The genesis of the *Challenger* accident – the failure of a joint in the right Solid Rocket Motor – began with decisions made during the design of the joint and in the failure by both Thiokol and NASA's SRB project office to understand and respond to facts obtained during testing.[95]

The Commission concluded that neither Thiokol nor NASA responded adequately to internal warnings about the faulty seal design. Furthermore, Thiokol and NASA did not make a timely attempt to develop and verify a new seal after the initial design was shown to be deficient. Neither organization developed a solution to the unexpected occurrences of O-ring erosion and blow-by even though this problem was experienced frequently during earlier flights. Instead, Thiokol and NASA management came to accept erosion and blow-by as unavoidable and an acceptable flight risk. Specifically, the Commission found that:[96]

1. The joint test and certification program was inadequate. There was no requirement to configure the qualifications test motor as it would be in flight, and the motors were static tested in a horizontal position, not in the vertical flight position.

2. Prior to the accident, neither NASA nor Thiokol fully understood the mechanism by which the joint sealing action took place.

3. NASA and Thiokol accepted escalating risk apparently because they "got away with it last time." As Commissioner Feynman observed, the decision making was: "a kind of Russian roulette. ... [Shuttle] flies [with O-ring erosion] and nothing happens. Then it is suggested, therefore, that the risk is no longer so high for the next flights. We can lower our standards a little bit because we got away with it last time. ... You got away with it, but it shouldn't be done over and over again like that."

4. NASA's system for tracking anomalies for Flight Readiness Reviews failed in that, despite a history of persistent O-ring erosion and blow-by, flight was still permitted. It failed again in the strange sequence of six consecutive launch constraint waivers prior to 51-L, permitting it to fly without any record of a waiver, or even of an explicit constraint. Tracking and continuing only anomalies that are "outside the data base" of prior flight allowed major problems to be removed from and lost by the reporting system.

5. The O-ring erosion history presented to Level I at NASA Headquarters in August 1985 was sufficiently detailed to require corrective action prior to the next flight.

6. A careful analysis of the flight history of O-ring performance would have revealed the correlation of O-ring damage and low temperature. Neither NASA nor Thiokol carried out such an analysis; consequently, they were unprepared to properly evaluate the risks of launching the 51-L mission in conditions more extreme than they had encountered before.

The Recommendations

Although the Commission found the *Challenger* accident was a direct result of the SRM seal failure, numerous other areas of concern were uncovered during the investigation. These areas included everything from the Orbiter braking system, to a decision making process described as "flawed." It was found that in its efforts to produce an 'operational' system, NASA had abandoned many of the procedures that had made it successful during it's first 25 years.[97]

The Commission provided NASA with nine major recommendations to help ensure a safe return to flight:[98]

1. Redesign the SRB joint and seal. The recertification process should more closely resemble flight conditions and loads. Alternate SRB designs should be evaluated. Procedures for assembly and inspection of SRBs prior to launch operations should be evaluated and refined.

2. Review the Space Shuttle management structure. An effort should be taken to consolidate decision making and review processes into a Space Shuttle program organization instead of as an individual Center effort. The earlier NASA practice of utilizing astronauts in management positions should be reinstituted. An independent flight safety panel should be established.

3. A complete review of all safety critical items should be undertaken and the procedures for granting waivers should be tightened.

4. NASA should establish an Office of Safety, Reliability, and Quality Assurance reporting directly to the NASA Administrator.

5. NASA should improve communications between Centers, and the designated flight commander (or his representative) should participate in the Flight Readiness Review process.

6. Actions should be taken to improve landing safety, especially the Orbiter brakes and nosewheel steering. A second Shuttle Carrier Aircraft should be purchased.

7. Efforts should be made to provide a crew escape system for use during controlled, gliding flight.

8. NASA must establish a flight rate that is consistent with its resources. The nation's reliance on a single launch system should be avoided in the future.

9. Installation, test, and maintenance procedures must be especially rigorous for critical items. An improved system for tracking maintenance intensive items should be implemented.

During the thirty-two month stand-down after the *Challenger* accident, NASA made great strides towards meeting most of these recommendations. By the beginning of September 1988 (just prior to STS-26R), 76 mandatory Orbiter modifications had been completed, including the crew escape system, increasing the thermal protection on the chin panel, new brakes, and recertifying the 17-inch propellant disconnects between the Orbiter and the ET. One hundred eighty five ground system modifications had also been accomplished. But by far the most extensive work had been to the Solid Rocket Boosters.

SRB Recertification

Following the *Challenger* accident NASA developed a plan to provide a Redesigned* Solid Rocket Motor (RSRM). The primary objective of the plan was to produce an SRM that was safe to fly, but an important secondary objective was to minimize the impact on the schedule by using existing hardware to the maximum extent possible without compromising safety. An SRM Redesign Project Plan was developed to formalize the methodology for the motor redesign and requalification. The plan provided an overview of the organizational responsibilities and relationships, design objectives, verification approach, and a master schedule.[99]

All aspects of the existing SRM were assessed, and design changes were required in the field joint, case-to-nozzle joint, nozzle, factory joint, propellant grain shape, ignition system, and ground support equipment. No changes were made in the propellant, liner, or castable inhibitor formulations. Design criteria were established for each component to ensure a safe design with an adequate margin of safety. The criteria were converted into specific design requirements which were evaluated during the Engineering Design Review and Preliminary Requirements Review held on 10 July and 19 September 1986, respectively. The design developed from these requirements was assessed at the Preliminary Design Review (PDR) held on 10 October 1986, and the final design was approved at the Critical Design Review (CDR) completed on 4 February 1988. Manufacture of the RSRM test hardware and the first flight hardware began prior to the CDR and continued in parallel with the hardware certification program.[100]

A comprehensive test program was developed by Morton Thiokol to evaluate and verify all changes incorporated into the Redesigned Solid Rocket Motors. The tests were designed to thoroughly evaluate design changes under all relevant environments and loading conditions.

Two full-scale, short-duration, 'simulators' were built. The first was the joint environment simulator (JES), which was tested seven times between 14 August 1986 and 28 July 1988. The JES was designed to evaluate field joint hardware, insulation and seal performance. Next was the nozzle joint environment simulator (NJES), which was tested nine times between 8 February 1987 and 14 August 1988. This simulator was designed to evaluate case-to-nozzle joint hardware, insulation and seal performance. A transient pressure test article (TPTA) was designed to evaluate both the field joint and case-to-nozzle joint performance, and was fired six times between 3 October 1987 and 1 September 1988. A structural test article (STA) of the entire booster was also constructed, and tests were conducted on 18 December 1987 and 1 April 1988 to evaluate the structural margins of the redesigned hardware.[101]

A booster (engineering test motor – ETM-1A) with STS-33/51-L configuration hardware was static tested on 27 May 1987. The two forward field joints incorporated the original Viton O-rings and external graphite reinforcing bands. The aft field joint had one Viton and one silicone O-ring. All three joints were equipped with wrap-around electrical joint heaters. In addition to evaluating a new O-ring seal material in the unmodified hardware, ETM-1A was used to test the effectiveness of the external graphite composite stiffener rings in reducing joint rotation. It was also used to evaluate the performance of field joint heaters, measure baseline hardware joint deflections in both the field and case-to-nozzle joints, and define the effectiveness of new nozzle nose-inlet rings.[102]

The first test of the redesigned configuration was demonstration motor eight (DM-8). The DM-8 static test on 30 August 1987 was used to evaluate the performance

---

* In 1992 the name was changed to Reusable SRM, but still abbreviated RSRM.

of major design features of the redesigned motor, including the prototype bonded J-seal insulation design. It was also used to evaluate the redesigned structural backup nozzle outer boot ring and evaluate the effects of the ET attachment ring on the redesigned hardware. The capability of the joint heaters to maintain seal temperatures of 75 degF was also demonstrated. DM-9, the second full-scale test of the redesigned motor, was fired on 23 December 1987 under ambient temperatures of 20 degF. The test was designed to further study the performance of the major redesign features and the joint heaters.[103]

On 20 April 1988, qualification motor six (QM-6) started the process of recertifying the SRBs for manned flight. The center field joint was assembled with an intentional manufacturing-type flaw in the surface of the bonded insulation J-seal. The defect was designed to allow hot gasses to penetrate the bonded insulation as far as the capture O-ring. Also incorporated into QM-6 was an intentional flaw in the case-to-nozzle joint which allowed gases to penetrate the bonded insulation to the new wiper

O-ring. QM-6 and all subsequent boosters included the structural backup configuration outer boot ring. A static test of QM-7 on 14 June 1988 was used to evaluate the redesign without intentional manufacturing or assembly flaws in the joints. Three hydraulically actuated struts were programmed to simulate ignition, lift-off and flight loads were externally attached to the QM-7 test article.[104]

Production verification motor one (PVM-1) was fired on 18 August 1988 and was the final qualification test prior to the launch of STS-26R. PVM-1 was the most seriously flawed motor ever tested, containing 14 distinct flaws in all but one of the major joints. Post-test disassembly and inspection that all joints performed as predicted.[105]

All of the static tests were full duration (122 seconds) with the booster in a horizontal position. Each booster was fitted with 400 to 600 channels of instrumentation to measure pressure, temperature, deflection, thrust, strain, and other conditions. The hydraulic loads system could apply simulated flight loads on the cases, at least mostly addressing on of the Presidential Committee's concerns.[106]

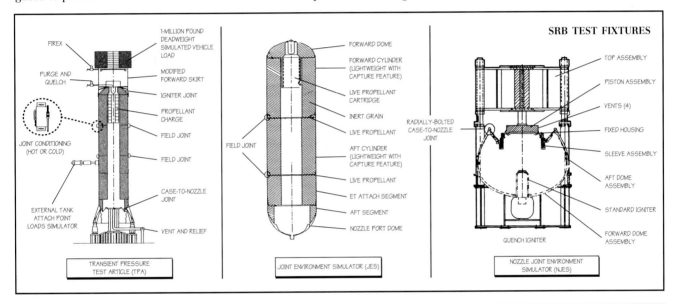

A Redesigned Solid Rocket Motor is prepared at the Morton Thiokol Wasatch facility in Utah. This was the first full-scale firing of an RSRM during the recertification process after the Challenger accident. (NASA photo KSC-87P-573)

As recommended by the Presidential Commission report, and at the request of the NASA Administrator, the National Research Council (NRC) established an Independent Oversight Panel chaired by Dr. H. Guyford Stever, who reported directly to the NASA Administrator. Initially, the panel was given introductory briefings on Space Shuttle requirements, implementation and control, the original design and manufacturing of the SRM, *Challenger* accident analyses, and preliminary plans for the solid motor redesign. The panel frequently reviewed the RSRM design criteria, engineering analyses and design, and certification program planning. Panel members continuously reviewed the design and testing for safe operation, selection and specifications for material, and quality assurance procedures. The panel continued to review the design as it progressed through certification and also reviewed the manufacturing and assembly of the first flight RSRM. Six written reports were provided by the panel to the NASA Administrator.[107]

Since the Return-to-Flight, full-scale flight support motors (FSM) have been tested approximately every 18 months through 1999. These tests were designed to confirm the performance and safety of RSRM systems, components, materials, and processes. The FSMs also provided the opportunity to evaluate and/or certify design, material, or process changes. Beginning with the firing of FSM-8 on 17 February 2000, these tests will be conducted once per year. In addition, a new series of sub-scale test motors (designated MNASA) provide a ballistic environment similar to the full-scale RSRM using components that are approximately one-fifth scale. These motors use 10,000 pounds of propellants, burn for 28 seconds, and have been fired in Utah and at MSFC. The motors have been used as test beds for evaluating case insulation and nozzle phenolic changes. During 1999 a 24-inch solid rocket test motor (SRTM) was developed as another testbed for the RSRM to provide a relatively low-cost fast-turnaround representation of the RSRM internal environment. Two configurations were designed – a one-segment version to test nozzle components, and a two-segment version to test insulation.[108]

RETHINKING ABORTS AND ESCAPES

Between 1973 and 1983, first stage (while the SRBs are thrusting) abort provisions were assessed numerous times by all levels of NASA and contractor management. Many methods of saving the Orbiter and/or the crew have been proposed – and rejected. Any type of ejection seat or separable crew module were continually rejected due to their limited utility and significant cost and weight impacts. Because of these factors, NASA adopted the philosophy that the reliability of the first stage (SRB) must be assured, and that design and testing must preclude time critical failures that would require emergency action before the normal SRB burnout.

If a problem arose that required the Orbiter to get away from the SRBs, the separation would have to be performed extremely quickly. Time is of the essence for two reasons. First, as *Challenger* demonstrated, if a problem develops in the SRBs, it can escalate very suddenly. Second, the ascent trajectory is carefully designed to control aerodynamic loads on the vehicle – a very small deviation from the programmed path will produce excessive loads, so if the vehicle begins to diverge from its path there is very little time (seconds) before structural breakup occurs.

The normal separation sequence to free the Orbiter from

SOLID ROCKET STATIC FIRINGS[137]					
Designation	Configuration	Date		Results	Test Bay
DM-1	SRM	18 Jul	77	Success	T-24
DM-2	SRM	18 Jan	78	Success	T-24
DM-3	SRM	19 Oct	78	Success	T-24
DM-4	SRM	17 Feb	79	Success	T-24
QM-1	SRM	13 Jun	79	Success	T-24
QM-2	SRM	27 Sep	79	Success	T-24
QM-3	SRM	13 Feb	80	Success	T-24
DM-5	HPM	21 Oct	82	Success	T-24
QM-4	HPM	21 Mar	83	Success	T-24
DM-6	FWC	25 Oct	84	Success	T-24
DM-7	FWC	09 May	85	Success	T-24
QM-5	FWC	Feb	86	Cancelled	T-24
ETM-1A	HPM (FM)	27 May	87	Success	T-24
DM-8	RSRM	30 Aug	87	Success	T-24
DM-9	RSRM	23 Dec	87	Success	T-24
QM-6	RSRM	20 Apr	88	Success	T-24
QM-7	RSRM	14 Jun	88	Success	T-97
PVM-1	RSRM	18 Aug	88	Success	T-24
TEM-1	HPM (FM)	08 Nov	88	Success	T-24
QM-8	RSRM	20 Jan	89	Success	T-97
TEM-2	HPM (FM)	24 Feb	89	Success	T-24
TEM-3	HPM (FM)	23 May	89	Success	T-97
TEM-4	HPM (FM)	19 Aug	89	Success	T-24
TEM-5	HPM (FM)	23 Jan	90	Success	T-97
TEM-6	HPM (FM)	16 Mar	90	Success	T-24
FSM-1	RSRM	15 Aug	90	Success	T-24
TEM-7	HPM (FM)	11 Dec	90	Success	T-97
TEM-8	HPM (FM)	31 Jul	91	Success	T-97
FSM-2	RSRM	20 Nov	91	Success	T-24
TEM-9	HPM (FM)	19 Mar	92	Success	T-97
FSM-3	RSRM	29 Jul	92	Success	T-24
TEM-10	HPM (FM)	27 Apr	93	Success	T-24
TEM-11	HPM (FM)	28 Sep	93	Success	T-97
FSM-4	RSRM	10 Mar	94	Success	T-24
FSM-5	RSRM	16 Nov	95	Success	T-24
FSM-6	RSRM	24 Apr	97	Success	T-97
FSM-7	RSRM	24 Jun	98	Success	T-97
FSM-8	RSRM	17 Feb	00	Success	T-97
FSM-9	RSRM	26 Apr	01*		T-97
ETM-2	RSRM	13 Sep	01*		T-24
FSM-10	RSRM	11 Apr	02*	* = Scheduled	T-97

the rest of the system takes 18 seconds, far too long to be of use during a first stage (SRB) emergency. Therefore a capability called 'fast-separation' was built into the flight software for use at any time. Fast-separation bypasses or reduces the normal built-in delays in order to achieve separation in approximately three seconds. Some risk was accepted by NASA to obtain this contingency capability. Unfortunately, subsequent analysis has shown that if fast separation is attempted while the SRBs are thrusting, the Orbiter will 'hang-up' on its aft attach points and pitch violently, with the probable destruction of the vehicle. Therefore fast-separation does not provide a meaningful way to escape unless some form of SRB thrust-termination system is implemented, and this was rejected for sound technical and safety reasons in 1973 and again in 1986.[109]

The current concept of fast-separation does have some uses. Contingency aborts resulting from the loss of two or three main engines immediately after SRB separation require fast separation from the ET so the Orbiter can attain a reentry attitude quickly. Unfortunately, nearly all contingency aborts of this type culminate in water impact, and it is extremely unlikely that the Orbiter, or the crew, could survive a ditching. It is therefore desirable to have available to the crew a reasonable 'bail-out' option during controlled, gliding flight.[110]

After the loss of *Challenger*, the issue of crew escape was also reexamined, and after again rejecting various proposals for ejection seats/capsules, escape pods, etc., two very different options were proposed – a telescopic slide-pole and a tractor rocket extraction system. Both methods were considered useful only below 230 mph and 20,000 feet during controlled gliding flight. The tractor rocket system provided small rockets that could be attached to the individual parachutes and fired from the crew access hatch to carry the crew member clear of the Orbiter. The chosen system is a telescopic pole that is a spring loaded and extends from the crew access hatch.[111]

The in-flight crew escape system is provided for use only when the Orbiter is in controlled gliding flight and unable to reach a runway. This condition would normally lead to ditching, and the crew escape system provides the flight crew with an alternative to water ditching or to landing on terrain other than a usable runway. Studies have shown that there is little, if any, probability of the flight crew surviving a ditching while a payload is installed.[112]

The hardware changes performed to the Orbiters during the *Challenger* stand-down enable the crew to equalize the pressure in the crew module with the outside atmosphere, pyrotechnically jettison the side hatch, and bail out from the mid-deck through the side hatch after manually deploying the escape pole. One by one, each crew member attaches a lanyard hook assembly, which surrounds the deployed escape pole, to his or her parachute harness and egresses through the side hatch opening. Attached to the escape pole, the crew member slides down the pole and off the end on a trajectory that takes them below the Orbiter's left wing.

Changes were also made to the PASS and BFS software for the TAL and RTLS abort modes. These changes provide the Orbiter with an automatic-mode which provides the Orbiter with a stable gliding flight for crew bailout. This software change, which was required to allow the flight crew commander's departure, automatically controls the Orbiter's velocity and angle-of-attack to the desired bailout conditions while the commander escapes.[113]

The crew needs to make the escape decision at an altitude of approximately 60,000 feet and engage the flight control system autopilot mode. As the Orbiter descends to 30,000 feet, its airspeed is decreased to approximately 230 mph. At 25,000 feet, a crew member in the mid-deck (referred to as the jump master and seated in the forward left seat in the mid-deck) raises a cover on the left side of the mid-deck at floor level and pulls a T-handle that activates the pyrotechnics for the depressurization valve, equalizing the crew cabin and outside pressure before the side hatch is jettisoned.[114]

At approximately 25,000 feet, the autopilot mode changes the Orbiter's angle-of-attack to approximately 15 degrees. This angle-of-attack must remain nearly con-

stant for approximately three minutes while the crew prepares to leave, and actually abandons the Orbiter. The jump master then jettisons the side hatch. Pyrotechnics separate the hatch assembly by severing the side hatch hinge, then three pyrotechnic thrusters propel the hatch from the Orbiter at a velocity of approximately 50 fps. The jump master then pulls the 'pip' pin on the escape pole and pulls the ratchet handle down, which permits the two telescoping sections of the escape pole to be deployed through the side hatch opening by spring tension.[115]

A magazine assembly located near the side hatch contains a lanyard assembly, consisting of a hook attached to a Kevlar strap, for every crew member. Each crew member attaches a lanyard hook assembly to the escape pole, and jumps out the hatch opening. The escape pole extends 9.8 feet downward from the side hatch and provides the crew member with a trajectory beneath the Orbiter's left wing.

Practice escapes have shown that it takes approximately 90 seconds for a crew of eight to bail out. After the first crew member leaves the mid-deck, the remaining crew members follow at 12-second intervals until all are out by approximately 10,000 feet altitude.[116]

This escape system was tested on a Lockheed C-141B Starlifter during the spring of 1988. William A. Chandler, crew system escape manager at JSC, noted that although both the slide-pole and tractor rocket systems provided adequate escape clearance for the crew, the slide-pole system was considered safer and would prove more cost effective over the life of the Orbiters. Tractor rockets would have necessitated carrying pyrotechnics in the Orbiter cabin, and would also have required additional support during the turnaround periods between flights.

The modifications required to install the new system in *Discovery* were completed on 15 April 1988, and the system was operational in time for the launch of STS-26R. The escape pole, as well as an inflatable escape slide for use during ground evacuations, was subsequently installed on *Columbia* and *Atlantis* before their return to flight, and were production features on *Endeavour*.[117]

Studies into alternate methods to ensure a safe recovery were also made. These involved looking at arresting barriers at the end of landing site runways (except lake bed runways), installing a skid on the landing gear that could preclude the potential for a second blown tire on the same gear after the first tire has blown, providing a 'roll on rim' capability for a predictable roll if both tires are lost on a single or multiple gear, and adding a drag chute.

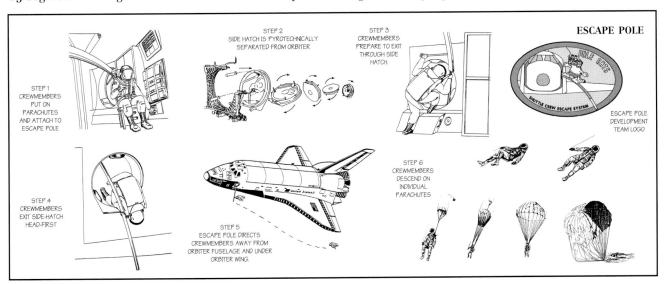

STEP 1
CREWMEMBERS PUT ON PARACHUTES AND ATTACH TO ESCAPE POLE

STEP 2
SIDE HATCH IS PYROTECHNICALLY SEPARATED FROM ORBITER

STEP 3
CREWMEMBERS PREPARE TO EXIT THROUGH SIDE HATCH.

ESCAPE POLE

POLE CATS
SHUTTLE CREW ESCAPE SYSTEM

ESCAPE POLE DEVELOPMENT TEAM LOGO

STEP 4
CREWMEMBERS EXIT SIDE-HATCH HEAD-FIRST

STEP 5
ESCAPE POLE DIRECTS CREWMEMBERS AWAY FROM ORBITER FUSELAGE AND UNDER ORBITER WING.

STEP 6
CREWMEMBERS DESCEND ON INDIVIDUAL PARACHUTES

WESTAR VI (near camera) and PALAPA-B2 in the payload bay of Challenger during STS-11/41-B. (NASA photo S84-26062)

PRE-CHALLENGER COMMERCIAL PAYLOADS[123]

Mission	Orbiter	Satellite	Type	Operator
STS-5	*Columbia*	ANIK-C3	Hughes HS-376	Telesat (Canada)
STS-5	*Columbia*	SBS-3	Hughes HS-376	SBS (USA)
STS-7	*Challenger*	ANIK-C2	Hughes HS-376	Telesat (Canada)
STS-7	*Challenger*	PALAPA-B1	Hughes HS-376	Perumtel (Indonesia)
STS-8	*Challenger*	INSAT-1B	Ford	ISRO (India)
STS-11/41-B	*Challenger*	PALAPA-B2	Hughes HS-376	Perumtel (Indonesia)
STS-11/41-B	*Challenger*	WESTAR-VI	Hughes HS-376	Western Union (USA)
STS-16/41-D	*Discovery*	SBS-4	Hughes HS-376	SBS (USA)
STS-16/41-D	*Discovery*	TELSTAR-3C	Hughes HS-376	AT&T (USA)
STS-16/41-D	*Discovery*	Leasat-2	Hughes HS-381	Hughes (US Navy)
STS-19/51-A	*Discovery*	ANIK-D2	Hughes HS-376	Telesat (Canada)
STS-19/51-A	*Discovery*	Leasat-1	Hughes HS-381	Hughes (US Navy)
STS-23/51-D	*Discovery*	Leasat-3	Hughes HS-381	Hughes (US Navy)
STS-23/51-D	*Discovery*	ANIK-C3	Hughes HS-376	Telesat (Canada)
STS-25/51-G	*Discovery*	MORELOS-1	Hughes HS-376	SCT (Mexico)
STS-25/51-G	*Discovery*	ARABSAT-1B	Aerospatiale	ASCO (Arab League)
STS-25/51-G	*Discovery*	TELSTAR-3D	Hughes HS-376	AT&T (USA)
STS-27/51-I	*Discovery*	ASC-1	RCA S-3000	ASC (USA)
STS-27/51-I	*Discovery*	AUSSAT-A1	Hughes HS-376	Aussat (Australia)
STS-27/51-I	*Discovery*	Leasat-4	Hughes HS-381	Hughes (US Navy)
STS-31/61-B	*Atlantis*	MORELOS-2	Hughes HS-376	SCT (Mexico)
STS-31/61-B	*Atlantis*	SATCOM-K2	RCA S-4000	RCA (USA)
STS-31/61-B	*Atlantis*	AUSSAT-A2	Hughes HS-376	Aussat (Australia)
STS-24/61-C	*Columbia*	SATCOM-K1	RCA S-4000	RCA (USA)

The last idea was judged to offer the best overall improvement, and *Endeavour* was delivered with a drag chute, similar to the original concept developed in 1974 prior to the construction of *Enterprise*. The other three Orbiters were modified to incorporate the drag chute during normal maintenance periods. Arresting barriers were subsequently installed at each of the primary TAL sites.[118]

WHAT WAS LAUNCHED BEFORE 51-L

Although it is hard to imagine in this day of widespread use of commercial satellites and such ambitious private systems as Iridium and Globalstar, the age of the communication satellite did not really begin until the mid-1970s. Prior to 1975, essentially all satellites belonged to the government or to government-chartered corporations such as Comsat or Intelsat. But as the Space Shuttle was being developed, this began to change, and for the first time it was financially justifiable and legal for individual corporations to purchase their own communication satellites for private use instead of buying services from Comsat or Intelsat. Large corporations and telephone companies around the world were beginning to order their own satellites in order to increase profits and provide a wider range of services – all of these satellites would need launched.[119]

Space Shuttle was well positioned to provide launch services for this new generation of satellites. The most popular was the Hughes HS-376 – and the Space Shuttle payload bay could, in theory, accommodate five of them on each flight. Shuttle only had to carry the satellites into low-earth orbit since attached upper stages would boost them to their final orbit. Each HS-376 used essentially the same support equipment, allowing NASA to predict a quick turnaround between missions. It appeared that launching satellites would be a lucrative market.[120]

By legislation Space Shuttle had acquired a monopoly in the domestic market, and NASA hoped to launch an Orbiter once per week to capitalize on this. To ensure a full payload bay for each of these flights, NASA was marketing the unique capabilities of Space Shuttle overseas, hoping to attract foreign satellites as well. As with any good marketing effort, NASA offered several introductory deals where customers could launch a number of satellites for a substantially discounted rate during the first five years of Shuttle operations. A similar deal had been made with the U.S. Department of Defense* to guarantee their continued support. In every case, NASA was actually losing money on each launch, but was sure that once customers became accustomed to the services provided by Shuttle, they would return at the full-price rate. Nine major telecommunications providers would take NASA up on this introductory offer – Aussat in Australia, Telesat in Canada, Perumtel in Indonesia, SCT in Mexico, and AT&T, RCA, and SBS in the United States. In addition, Hughes took advantage of the offer to manifest five of its HS-381 SYNCOM-IV satellites that were being leased to the U.S. Navy as Leasats.

The Space Shuttle was declared operational after its fourth orbital flight test. The fifth flight (STS-5) on 11 November 1982 carried the first commercial payload ever delivered to space by a manned spacecraft – the ANIK-C3 and SBS-C communication satellites. The satellites were mounted vertically in separate cradles in the rear of the payload bay. Each HS-376 was essentially a drum covered with solar cells and, on station, the satellites would spin

---

* During FY82/83, the DoD had paid NASA a total of $268 million for nine military Space Shuttle launches.

[124]PRE-CHALLENGER TDRS AND SCIENCE PAYLOADS

Mission	Orbiter	Satellite	Type	Operator
STS-6	*Challenger*	TDRS-A	Tracking & Relay	NASA
STS-13/41-C	*Challenger*	LDEF-1	Science	NASA
STS-33/51-L	*Challenger*	TDRS-B	Tracking & Relay	NASA

Workers in the Vertical Processing Facility lower an Inertial Upper Stage (IUS) booster into a workstand for preflight processing After the IUS is checked out, it will be mated to a TDRS satellite and loaded aboard shuttle. (NASA photo KSC-95PC-0515)

constantly to provide stability, and also to mitigate the thermal stresses encountered in space. But sitting in their launch cradles they would be baked on one side and frozen on the other as soon as the payload bay doors opened. To protect the satellites, clamshell covers were closed over them and electric heaters turned on until they were ready to be released. When it was time to release each satellite, *Columbia* faced its payload bay in the desired direction, opened the clamshell doors, and the satellite began spinning on a turntable in its cradle. Eight hours into the mission, SBS-C became the first commercial satellite to be released by the Space Shuttle. The release was videotaped and relayed to the ground, showing the spinning satellite being ejected from its cradle and slowly drifting past the vertical stabilizer. The Shuttle crew advertised themselves as the 'We Deliver' team.[121]

Over the next three years, 24 commercial satellites would be deployed from Shuttle, including four of the Leasats that were owned by Hughes and leased to the Navy. Each was boosted to geostationary orbit to serve as a communications relay. Although the Shuttle payload bay was technically capable of carrying up to five HS-376-class satellites, in realty, three was the maximum number carried on any given mission. This was partially a factor of the marketplace – seldom were there more than three satellites ready at any given time – but also partially because NASA did not yet fully understand the effect of a forward center-of-gravity movement on the Orbiter's emergency landing characteristics. Limiting the load to three satellites gave a comfortable performance margin.[122]

But things do not always go as planned. Although in each case *Challenger* performed its deployment mission perfectly, on STS-11/41-B both PAMs failed to fire correctly, leaving PALAPA-B2 and WESTAR-VI in useless

low-earth orbits. Despite the fact that the failure had nothing to do with the deployment, NASA felt its image had been tarnished. But the Space Shuttle offered the unique capability – one that can never be matched by any expendable vehicle – of being able to retrieve objects from space and return them to the Earth. *Columbia* had first demonstrated this during STS-7 when the SPAS-01 experiment pallet had been released on-orbit, retrieved, and returned. In November 1984, under contract to the insurance companies, *Discovery* retrieved PALAPA-B2 and WESTAR-VI during the STS-19/51-A mission. The insurance companies had already paid Perumtel and Western Union for the loss of the satellites, so both satellites would be refurbished by Hughes and sold again to new customers, with the proceeds going to the insurers.

On STS-27/51-I, NASA demonstrated another unique capability of the Space Shuttle. After deploying three communications satellites (which, incidentally, were of three different designs), *Discovery* maneuvered next to Leasat-3. This satellite had been deployed during STS-23/51-D, but had failed to activate correctly after deployment. During EVAs that totaled 11 hours and 27 minutes, astronauts successfully retrieved, repaired, and redeployed the satellite. All of this was televised live around the world – the publicity was priceless.

Government Payloads

Commercial satellites were not all that was deployed from Space Shuttle before the *Challenger* accident. Immediately after the last Skylab mission, NASA had begun decommissioning many of its ground tracking and communications stations in order to save money. The plan for Space Shuttle had always included the use of space-based Tracking and Data Relay System (TDRS) satellites – ironically, Space Shuttle would have to deploy the constellation before it could use them. The first TDRS satellite was deployed by *Challenger* on STS-6 – *Challenger* was to be the workhorse for these satellites since *Columbia* was too heavy to effectively carry them. Unfortunately, the second satellite of the constellation was lost in the *Challenger* accident.

Another government mission (STS-13/41-C) involved the Long-Duration Exposure Facility (LDEF) – a large science payload designed to measure the effects of being in space for a prolonged period of time. On the same flight, astronauts conducted the first on-orbit repair of a satellite when they successfully (after several failed attempts) repaired the *Solar Maximum* science satellite. There were also a variety of missions that did not release payloads into space. Four Spacelab flights carried a large scientific module in the payload bay that allowed research into a variety of subjects – Spacelab modules were not released, and were brought back to Earth with the Orbiter landed. The two most important science missions, *Ulysses* and *Galileo*, were just getting ready for launch at the time of the *Challenger* accident. Two classified DoD flights were also conducted and will be further discussed in Chapter IX.

The publicity during some of these flights had been tremendous – people around the world watched in awe as astronauts maneuvered themselves and the Orbiter into position to launch or retrieve satellites. The Shuttle had successfully demonstrated that it was the carrier of choice for the current generation of communications satellites. This was significant since fully half the commercial satellites deployed were for foreign customers – contracts which had been won against competition from Europe's Ariane. It appeared that NASA had successfully demonstrated that there was indeed a market for Shuttle's services.[125]

But this apparent commercial success was deceiving. It was widely known that NASA had cut some good deals on multiple-satellite launch services, and this was somewhat expected for a new vehicle. What was not fully appreciated was that these 'deals' were even better than they seemed. NASA's standard fee for deploying a satellite had been calculated from early estimates of flight rates, turn-around times, and operating costs. By the end of 1985, it was apparent that these estimates had been unrealistically low. The best annual flight rate Shuttle had managed so far was nine missions in 1985 – even if you looked at the best floating 12-month window (late January 1985 to late January 1986), the best that had been managed was 11 missions, and one of those was the ill-fated *Challenger*. It was a far cry from 60 missions per year.

At this lower flight rate, each mission was costing substantially more than had been envisioned. In fact, the cost of each mission was continuously increased as NASA invested more money in the Space Shuttle infrastructure, and the Orbiters demonstrated they required substantially more maintenance between flights than had been expected. But the fee being charged to most customers was fixed – and would be until 1988 at the earliest based on the introductory prices NASA had agreed with to attract customers. Unwittingly, the U.S. government was subsidizing the satellite launch business.

Commercial Consequences of the Accident

The *Challenger* accident would change the face of the Space Shuttle program forever. Of course, shuttle was not unique in losing a vehicle. But it was unique in that it was a *piloted* vehicle – and the loss occurred live on national television. Of the five major launch vehicles then in service, four of them experienced a launch failure during the first 38 months Space Shuttle was operational.[126]

During the time NASA was actively launching commercial satellites on Space Shuttle, eight Arianes successfully deployed 13 commercial satellites while a single launch failure destroyed two satellites – interestingly, only one HS-376 was launched on Ariane. In addition, an Ariane was used to launch the *Giotto* probe to observe Halley's

Comet. Of the 18 Atlas boosters launched, 17 were successful, but only Intelsat used Atlas commercially – the rest of the payloads were for the Department of Defense, National Oceanic and Atmospheric Administration, or NASA. All 13 Titan III missions carried classified DoD payloads, and one was lost due to a launch failure. The Delta had long been the launch vehicle of choice for small communications satellites, and probably suffered most from Shuttle entering the market. Only 12 Deltas were launched – all successfully – delivering six commercial satellites and six government payloads.[128]

Even discounting PALAPA-B2 and WESTAR-VI that were left in useless orbits (not Shuttle's fault), the Shuttle managed to deploy nearly as many commercial satellites as all of its competitors combined. This all would change.

Although the Presidential Commission found the *Challenger* accident was a direct result of the SRB O-ring failure, numerous other areas of concern were uncovered during the investigation. It found that in its efforts to produce an "operational" system, NASA had abandoned many of the procedures that had made it successful during it's first 25 years. The Commission provided NASA with nine major recommendations to help ensure a safe return to flight, with the primary one being to redesign the SRB joint and seal. But perhaps the most far-reaching recommendation was that "The nation's reliance on a single launch system should be avoided in the future."[129]

This last recommendation resulted in a change to U.S. policy and law. No longer was the Space Shuttle to be the Nation's only launch vehicle – in fact, its use to deliver commercial satellites was essentially forbidden. The new law read:[130]

U.S. Code, Section 42 – The Public Health and Welfare, Chapter 26 – National Space Program, Section 2465a – Space Shuttle Use Policy.
(a) Use policy (1) It shall be the policy of the United States to use the Space Shuttle for purposes that (i) require the presence of man, (ii) require the unique capabilities of the Space Shuttle or (iii) when other compelling circumstances exist. (2) The term "compelling circumstances" includes, but is not limited to, occasions when the Administrator determines, in consultation with the Secretary of Defense and the Secretary of State, that important national security or foreign policy interests would be served by a Shuttle launch. (3) The policy stated in subsection (a)(1) of this section shall not preclude the use of available cargo space, on a Space Shuttle mission otherwise consistent with the policy described under subsection (a)(1) of this section, for the purpose of carrying secondary payloads (as defined by the Administrator) that do not require the presence of man if such payloads are consistent with the requirements of research, development, demonstration, scientific, commercial, and educational programs authorized by the Administrator.

Other sections of the law pertained to the NASA Administrator implementing the law and reporting on adherence to the law are not reproduced here.

Subsidies ?

There were other reasons to remove commercial payloads from Shuttle. First, it was decidedly non-cost effective in a full-cost recovery environment to use the Orbiters to launch communication satellites. This was particularly obvious when compared with Arianespace.

LAUNCH VEHICLE FAILURES[127]

| Launch Vehicle | Launches | | Satellites | | |
	Total	Failures	Total +	Commercial	Lost
Shuttle	25	1	33	20 *	1
Ariane	10	1	16	13	2
Delta	12	0	12	6	0
Atlas	18	1	18	5	1
Titan III	13	1	Classified	0	Classified

* Excluding the Leasats, which were closer to Government payloads.
† All satellites; but in the case of shuttle, not free-flyers returned on the same mission.

Created in March 1980 as a private stock company by European aerospace firms, banks, and the French space agency, Arianespace soon took over operation of the multinational European Space Agency's Ariane launch vehicle. This included overseeing and financing of Ariane production, organizing worldwide marketing of launch services, and managing launch operations at Kourou, French Guyana.

Ariane launches began in December 1979, and the initial series of missions was conducted under ESA responsibility. The first fully commercial mission under Arianespace control was the launcher's ninth flight in May 1984, when an Ariane I successfully lifted the GTE SPACENET-1 satellite into orbit. By the spring of 1985, Arianespace held firm orders for thirty satellites and had options to launch twelve more – representing a combined order book value of about $750 million. Of those orders, half were from satellite customers outside the European home market. Arianespace marketing combined the best of both worlds – the marketing freedom of a private company, plus the direct support of government agencies.

The Space Shuttle was not the only U.S. response to Arianespace competition. Transpace Carriers, a private U.S. firm created to provide McDonnell Douglas Delta launch services, attempted to halt trading by Arianespace in the United States on grounds of unfair subsidy pricing. Arianespace used a two-tier pricing policy, charging higher prices to the European Space Agency (ESA) and its member states. The French space agency subsidized Arianespace launch and range services, as well as administrative and technical personnel, and ESA member states subsidized Arianespace insurance rates. Transpace Carriers filed a petition on 25 May 1984, with the Office of the United States Trade Representative (USTR) under Section 301 of the Trade Act of 1974 alleging that the European Space Agency was engaged in predatory pricing and other unfair trade practices in the sale of Ariane launch services. The USTR accepted the case on 9 July 1984, and meetings between European and U.S. government officials began in November 1984. Talks soon turned to pricing and subsidy comparisons of the Space Shuttle and Ariane. On 17 July 1985, President Reagan determined that the pricing and subsidy practices of Arianespace were neither unreasonable nor a restriction on U.S. commerce, because Arianespace practices were not sufficiently different from those of the Space Shuttle to warrant action under the Trade Act of 1974.[131]

Then came the *Challenger* accident. National Security Decision Directive 254, released shortly after the *Challenger* accident, took NASA and the Space Shuttle out of competition with U.S. launch providers for commercial and foreign spacecraft payloads. This, in turn, was the impetus for U.S.C 42 cited previously. NSDD 254 bore fruit three years later when the Office of Commercial Space Transportation issued four commercial launch licenses, and had five more license applications pending. As the OCST reported to Congress in 1990:[132]

"Fiscal Year 1989 was a turning point for the Office of Commercial Space Transportation (OCST) and the U.S. commercial launch industry." The OCST had licensed the first U.S. commercial launches, thus marking "the beginning of a new era in the history of U.S. space endeavors."

And marking an end in the history of Space Shuttle.

Atlantis is rolled-over from the OPF to the VAB in preparation for the STS-31/61-B mission. This would be the second flight for Atlantis and would deliver three communications satellites into orbit. This mission was also the second night launch for the program, and resulted in some beautiful on-orbit IMAX footage. (Dennis R. Jenkins)

Astronauts Joe Allen and Dale Gardner on STS-19/51A after retrieving WESTAR-VI and PALAPA-B2. Note the 'For Sale' sign. (NASA)

A LONG 32 MONTHS

In January 1986 *Atlantis* was in the OPF being prepared for the *Galileo* mission and was ready to be mated to the SRBs and ET in the VAB. *Columbia* had just completed the STS-32/61-C mission and was also in the OPF undergoing post-flight deconfiguration and maintenance. *Discovery* was in temporary storage in the VAB awaiting transfer to the OPF for preparation for the first flight from Vandenberg scheduled for later during 1986. The next flight of *Challenger* was scheduled to carry the *Ulysses* probe.[1]

Various manifest options were being considered, and it was subsequently decided that *Atlantis* would be rolled out to Pad 39B for fit checks of new weather protection system and take part in emergency egress exercises and count-down demonstration tests. It also was decided that *Columbia* would be flown to Vandenberg for more exten-sive fit checks than *Enterprise* could provide – including a flight readiness firing – prior to *Discovery* being transferred there for the first flight.[2]

After the *Challenger* accident, all plans for launches from Vandenberg were abandoned, so neither *Columbia* nor *Discovery* made the trip west. *Atlantis* did eventually take part in the Pad 39B tests, although they were slight-ly later than anticipated. In the months that followed the accident, *Discovery* was selected for the first Return-to-Flight mission and was moved from VAB High Bay 2 into OPF-1 the last week of June 1986. The Orbiter was pow-ered up for systems checks until mid-September 1986 when it was transferred back to the VAB while the OPF-1 facility underwent modifications.[3]

Preparations for the Return-to-Flight began on 30 October 1986 when OV-103 was moved from the VAB back to OPF-1 where most of the Orbiter's major systems and components were removed and sent to the respective ven-dors for modifications or to be rebuilt. After an extensive powered-down period that began in February 1987, *Discovery* was powered up on 3 August 1987 and remained in the OPF while workers implemented over 200 modifica-tions and outfitted the payload bay for the TDRS satellite.[4]

Flight processing began in mid-September during which the major components of the vehicle were rein-stalled and checked out, including the OMS pods and FRCS module. In January 1988, *Discovery*'s three main engines arrived at KSC and were installed – engine number 2019 was delivered on 6 January 1988, and was installed in the number 1 position on 10 January; engine 2022 arrived on 15 January and was installed in the number 2 position nine days later; engine 2028 arrived 21 January and was installed in the number 3 position, also on 24 January.[5]

Discovery in VAB High Bay 2 on 18 June 1986 during the Challenger stand-down. Note the main engines and OMS pods are missing, and protective mats cover the top of the body flap to protect it while the vehicle is being worked on. Also note the 'protective canopy' over the Orbiter to keep things from dropping onto it. (NASA photo 108-KSC-86P-128)

The RSRM segments began arriving at KSC on 1 March 1988 and the first segment was stacked on MLP-2 in VAB High Bay 3 on 29 March. Technicians started with the left aft booster and continued stacking the four left hand segments before beginning the right hand segments on 5 May. The forward assemblies/nose cones were attached 27–28 May, and the SRB field joints were closed out prior to mating the ET to the boosters on 10 June. A functional interface test between the SRBs and ET was conducted in the VAB a few days later to verify the connections.[6]

The OASIS payload was installed in the payload bay on 19 April while the Orbiter was still in the OPF. The TDRS arrived at the Vertical Processing Facility on 16 May and its Inertial Upper Stage (IUS) on 24 May. The TDRS/IUS mechanical mating was accomplished on 31 May, and the payload was kept in the VPF since it would be installed while the Orbiter was on the pad.[7]

Discovery was rolled-over from the OPF to the VAB on 21 June and mated to the rest of the stack on 24 June. While still in the VAB, an interface test was conducted to check out the mechanical and electrical connections between the various elements of the stack and the function of the onboard flight systems. The vehicle was rolled-out to Pad 39B on top of a crawler-transporter on 4 July for a few major tests and final launch preparations.[8]

On 15 July 1988 a leak was found in one OMS pod, but this did not affect a wet countdown test conducted on 1 August – although several hydrogen leaks were discovered in and around the vehicle during the tanking. On

A full-scale replica of a TDRS satellite is displayed suspended from wires at the TRW facility in Redondo Beach, California. The actual satellite – at 5,000 pounds and 57 feet across – can not be deployed in Earth's gravity. (TRW photo 202261-85)

4 August an FRF was attempted to verify the three main engines, but a sluggish valve in one engine aborted the test prior to SSME start (т–6 seconds). After repairs to the engine a 20-second FRF was successfully completed on 10 August. Nine days later the OMS leak was repaired while the vehicle was on the pad, and the TDRS-C payload was installed on 29 August. A full-up TCDT dress rehearsal for launch – with the flight crew in the Orbiter – was conducted on 8 September. Thirty-two months after the *Challenger* accident, all was deemed ready for launch.[9]

In the early morning of 9 October 1986, Atlantis was rolled-out to Pad 39B for seven weeks of testing. While at the pad, OV-104 supported the checkout of the new weather protection system, launch team proficiency exercises, and emergency egress simulations. This was the first time a vehicle had been on the pad since Challenger. Interestingly, Atlantis did not have any SSMEs installed during the tests – the engines were in the process of being overhauled. (NASA photo S86-38627)

The Return-to-Flight launch of Discovery) as STS-26R on 29 September 1988. (NASA photo KSC-88PC-1033)

The confusing alphanumeric designation scheme was dropped after the Challenger accident in 1986, the program returning to the original 'STS' numbering system. This, however, also caused some problems since NASA decided that the first flight after the accident would be STS-26 (Challenger was the 25th flight launched). To overcome the problem of having two 'STS-26' missions (Flight 19/51-F before the accident, one after), the mission was carried internal to NASA as STS-26R, the 'R' signifying 'reflight.' The 'R' suffix was added to missions through STS-33R, the number carried internally by Challenger.[120]

Seq. #	Mission #	Orbiter Flight #	DATES Launch Landing	TIME Launch Landing	SITES Launch Landing	Crew	ORBIT Altitude Inclination	WEIGHTS Lift-Off OV Launch Payload OV Landing	Payloads	Mission Duration	REENTRY Velocity Cross-Range Touchdown Roll-out	Notes
26	STS-26R	OV-103 7	29 Sep 88 03 Oct 88	273:13:37:00 277:16:37:12	KSC, Pad B EAFB, 17L	Frederick H. Hauck (C) Richard O. Covey (P) John M. Lounge (MS1) George D. Nelson (MS2) David C. Hilmers (MS3)	205 28.50	4,522,411 254,606 44,601 194,184	TDRS-C/IUS Galley RMS = none	63 Orbits 97 hrs 00 mins 04 secs ≈1,680,000 miles	28,790 441 187 7,451	Return-to-Flight after Challenger accident Orbiter returned to KSC on 08 Oct 88
27	STS-27R	OV-104 3	01 Dec 88 (pd) 02 Dec 88 06 Dec 88	337:14:30:33 341:23:36:11	KSC, Pad B EAFB, 17L	Robert L. 'Hoot' Gibson (C) Guy S. Gardner (P) Richard M. Mullane (MS1) Jerry L. Ross (MS2) William M. Shepherd (MS3)	— 57.00 190,956	4,505,773	DoD (LACROSSE-1) CRUX-A OASIS CLOUDS RMS = s/n 201	68 Orbits 105 hrs 05 mins 35 secs ≈1,820,000 miles	25,121 598 194 7,123	Third dedicated DoD flight Classified payload Orbiter returned to KSC on 13 Dec 88
28	STS-29R	OV-103 8	18 Feb 89 (s) 13 Mar 89 18 Mar 89	072:14:57:00 077:14:35:49	KSC, Pad B EAFB, 22	Michael L. Coats (C) John E. Blaha (P) James P. Bagian (MS1) James F. Buchli (MS2) Robert C. Springer (MS3)	187 28.45	4,524,261 256,357 45,316 194,789	TDRS-D/IUS SHARE OASIS-1 IMAX Camera RMS = none	79 Orbits 119 hrs 38 mins 52 secs ≈2,000,000 miles	25,787 442 205 9,339	Orbiter returned to KSC on 24 Mar 89
29	STS-30R	OV-104 4	28 Apr 89 (s) 04 May 89 08 May 89	124:18:46:59 128:19:43:25	KSC, Pad B EAFB, 22	David M. Walker (C) Ronald J. Grabe (P) Norman E. Thagard (MS1) Mary L. Cleave (MS2) Mark C. Lee (MS3)	185 28.85	4,527,426 261,118 45,823 192,459	Magellan/IUS RMS = none	64 Orbits 96 hrs 57 mins 31 secs 1,681,997 miles	25,788 403 196 10,295	Long delayed probe to Venus Last 'standard insertion' trajectory Orbiter returned to KSC on 15 May 89
30	STS-28R	OV-102 8	08 Aug 89 13 Aug 89	220:12:37:00 225:13:37:08	KSC, Pad B EAFB, 17L	Brewster H. Shaw, Jr. (C) Richard N. Richards (P) James C. Adamson (MS1) David C. Leestma (MS2) Mark N. Brown (MS3)	191 57.00 200,214	4,510,019	DoD (SDS-B1) CRUX-B HEIN-LO CLOUDS RMS = none	80 Orbits 121 hrs 00 mins 09 secs ≈2,100,000 miles	25,803 214 155 6,015	Fourth dedicated DoD flight Classified payload Orbiter returned to KSC on 21 Aug 89
31	STS-34	OV-104 5	12 Oct 89 (pd) 17 Oct 89 (pd) 18 Oct 89 23 Oct 89	291:16:53:40 296:16:33:01	KSC, Pad B EAFB, 23L	Donald E. Williams (C) Michael J. McCulley (P) Franklin R. Chang-Diaz (MS1) Shannon W. Lucid (MS2) Ellen S. Baker (MS3)	204 34.30	4,524,224 257,569 45,905 195,954	Galileo/IUS SSBUV-01 IMAX Camera RMS = none	79 Orbits 119 hrs 39 mins 24 secs ≈2,000,000 miles	25,784 571 195 9,677	Probe to Jupiter Orbiter returned to KSC on 29 Oct 89
32	STS-33R	OV-103 9	20 Nov 90 (s) 22 Nov 89 27 Nov 89	327:00:23:29 332:00:30:18	KSC, Pad B EAFB, 04	Frederick D. Gregory (C) John E. Blaha (P) F. Story Musgrave (MS1) Manley L. Carter (MS2) Kathryn C. Thornton (MS3)	347 28.45 194,282	4,529,160	DoD (ORION-2) RMS = none	79 Orbits 120 hrs 06 mins 49 secs ≈2,100,000 miles	25,998 260 199 7,764	Fifth dedicated DoD flight Classified payload Orbiter returned to KSC on 04 Dec 89
33	STS-32R	OV-102 9	18 Dec 89 (s) 08 Jan 90 (pd) 09 Jan 90 20 Jan 90	009:12:35:00 020:09:35:36	KSC, Pad A EAFB, 22	Daniel C. Brandenstein (C) James D. Wetherbee (P) Bonnie J. Dunbar (MS1) G. David Low (MS2) Marsha S. Ivins (MS3)	205 28.50	4,519,487 255,994 18,317 228,335	Syncom IV-5 IMAX Camera Galley RMS = s/n 201	171 Orbits 261 hrs 00 mins 37 secs 4,509,972 miles	25,328 428 207 10,096	Retrieval of Long-Duration Exposure Facility (LDEF) from orbit Longest Shuttle flight to date Orbiter returned to KSC on 26 Jan 90
34	STS-36	OV-104 6	22 Feb 90 (pd) 23 Feb 90 (pd) 25 Feb 90 (pd) 26 Feb 90 (pd) 28 Feb 90 04 Mar 90	059:07:50:22 063:18:08:44	KSC, Pad A EAFB, 23L	John O. Creighton (C) John H. Casper (P) Richard M. Mullane (MS1) David C. Hilmers (MS2) Pierre J. Thuot (MS3)	152 62.00 187,200	4,507,283	DoD (MYSTY) RME-III VFT-1 VFT-2 RMS = none	72 Orbits 106 hrs 18 mins 23 secs 1,868,588 miles	25,713 293 199 7,900	Sixth dedicated DoD flight Classified payload Orbiter returned to KSC on 13 Mar 90
35	STS-31R	OV-103 10	18 Apr 90 (s) 12 Apr 90 (s) 10 Apr 90 (s) 24 Apr 90 29 Apr 90	114:12:33:51 119:13:49:57	KSC, Pad B EAFB, 22	Loren J. Shriver (C) Charles F. Bolden, Jr. (P) Steven A. Hawley (MS1) Bruce McCandless II (MS2) Kathryn D. Sullivan (MS3)	380 28.45	4,514,665 249,109 25,517 189,118	Hubble Telescope IMAX Camera APM-01 Galley RMS = s/n 301	79 Orbits 121 hrs 16 mins 05 secs 2,068,213 miles	26,120 483 177 8,889	Hubble Space Telescope deployment First use of carbon brakes on landing Orbiter returned to KSC on 07 May 90
36	STS-41	OV-103 11	06 Oct 90 10 Oct 90	279:11:47:15 283:13:57:19	KSC, Pad B EAFB, 22	Richard N. Richards (C) Robert D. Cabana (P) William M. Shepherd (MS1) Bruce E. Melnick (MS2) Thomas D. Akers (MS3)	184 28.45	4,544,024 259,593 46,173 196,869	Ulysses/IUS+PAM-S SSBUV-02 ISAC Galley RMS = s/n 301	65 Orbits 98 hrs 10 mins 03 secs 1,707,445 miles	25,762 566 192 8,532	Braking test on Runway 22 Heaviest payload to date Orbiter returned to KSC on 16 Oct 90
37	STS-38	OV-104 7	15 Jul 90 (s) 09 Nov 90 (s) 15 Nov 90 20 Nov 90	319:23:48:15 324:21:42:46	KSC, Pad B KSC, 33	Richard O. Covey (C) Frank L. Culbertson (P) Robert C. Springer (MS1) Carl J. Meade (MS2) Charles D. 'Sam' Gemar (MS3)	163 28.50 191,091	4,531,909	DoD (SDS-B2) AMOS APE-B SPADVOS RMS = ???	78 Orbits 117 hrs 54 mins 22 secs ≈2,030,000 miles	25,729 3 199 9,003	Launch delayed from July 1990 by hydrogen leaks at 17-inch disconnect Seventh dedicated DoD flight First KSC landing since Apr 85 (STS-23)
38	STS-35	OV-102 10	16 May 90 (pd) 30 May 90 (pd) 01 Sep 90 (pd) 18 Sep 90 (s) 02 Dec 90 10 Dec 90	336:06:49:01 345:05:54:09	KSC, Pad B EAFB, 22	Vance D. Brand (C) Guy S. Gardner (P) Jeffrey A. Hoffman (MS1) John M. Lounge (MS2) Robert A. Parker (MS3) Samual T. Durrance (PS1) Ronald A. Parise (PS2)	218 28.45	4,600,228 256,385 27,760 225,329	ASTRO-1 BBXRT ACIP Galley RMS = none	143 Orbits 215 hrs 05 mins 08 secs 3,728,636 miles	25,858 490 201 10,566	Launch delayed from May 1991 by hydrogen leaks at 17-inch disconnect Orbiter returned to KSC on 20 Dec 90
39	STS-37	OV-104 8	05 Apr 91 11 Apr 91	095:14:22:45 101:35:55:29	KSC, Pad B EAFB, 33	Steven R. Nagel (C) Kenneth D. Cameron (P) Jerry L. Ross (MS1) Jerome 'Jay' Apt (MS2) Linda M. Godwin (MS3)	285 28.45	4,519,158 255,824 36,800 190,098	GRO CETA APM-02 RMS = s/n 303	93 Orbits 143 hrs 32 mins 44 secs 2,456,263 miles	24,612 432 168 6,364	Gamma Ray Observatory deployment First U.S. Spacewalk in five years Landing delayed one day due to weather Orbiter returned to KSC on 18 Apr 91
40	STS-39	OV-103 12	09 Mar 90 (s) 28 Apr 91 (pd) 28 Apr 91 06 May 91	118:11:33:14 126:18:55:37	KSC, Pad A KSC, 15	Michael L. Coats (C) L. Blaine Hammond, Jr. (P) Guion S. Bluford, Jr. (MS1) Gregory J. Harbaugh (MS2) Richard J. Hieb (MS3) Donald R. McMonagle (MS4) Charles Lacy Veach (MS5)	161 57.00 211,512	4,512,698 247,373 21,413	SPAS-II IBSS/SPAS-II MEPC CRO AFP-675 STP-1 RMS = s/n 301	133 Orbits 199 hrs 22 mins 22 secs ≈3,470,000	25,765 709 218 9,235	Eighth dedicated DoD flight Unclassified primary payload
41	STS-40	OV-102 11	21 May 91 (pd) 01 Jun 91 (pd) 05 Jun 91 14 Jun 91	156:13:24:51 165:15:39:11	KSC, Pad B EAFB, 22	Bryan D. O'Connor (C) Sidney M. Gutierrez (P) Margaret Rhea Seddon (MS1) James P. Bagian (MS2) Tamara E. Jernigan (MS3) F. Drew Gaffney (PS1) Millie Hughes-Fulford (PS2)	181 39.00	4,518,801 251,970 28,114 226,535	Spacelab SLS-01 GAS Bridge Tunnel RMS = none	145 Orbits 218 hrs 14 mins 20 secs 3,779,940 miles	25,772 243 203 9,438	Fifth dedicated Spacelab mission Last flight of SEADS and SUMS Orbiter returned to KSC on 21 Jun 91
42	STS-43	OV-104 9	23 Jul 91 (pd) 24 Jul 91 (pd) 01 Aug 91 (pd) 02 Aug 91 11 Aug 91	214:15:02:00 223:12:23:25	KSC, Pad A KSC, 15	John E. Blaha (C) Michael A. Baker (P) Shannon W. Lucid (MS1) James C. Adamson (MS2) G. David Low (MS3)	200 28.45	4,522,828 259,374 46,712 196,088	TDRS-E/IUS SSBUV-03 SHARE-II GAS (1) RMS = none	141 Orbits 213 hrs 21 mins 22 secs 3,700,400 miles	25,794 207 197 9,890	
43	STS-48	OV-103 13	12 Sep 91 18 Sep 91	255:23:11:04 261:07:38:42	KSC, Pad A EAFB, 22	John O. Creighton (C) Kenneth S. Reightler, Jr. (P) James F. Buchli (MS1) Charles D. 'Sam' Gemar (MS2) Mark N. Brown (MS3)	360 57.00	4,503,424 240,062 17,144 192,780	UARS GAS Bridge RMS = s/n 301	80 Orbits 128 hrs 27 mins 34 secs 2,193,670 miles	26,077 794 203 9,384	Orbiter returned to KSC on 26 Sep 91
44	STS-44	OV-104 10	19 Nov 91 (pd) 24 Nov 91 01 Dec 91	328:23:44:00 335:22:34:44	KSC, Pad A EAFB, 05R	Frederick D. Gregory (C) Terence T. Henricks (P) F. Story Musgrave (MS1) Mario Runco, Jr. (MS2) James S. Voss (MS3) Thomas J. Hennen (PS1)	226 28.50	4,520,641 259,904 44,637 194,818	DoD (DSP F-16/IUS) IOCM GAS Bridge AMOS RMS = none	109 Orbits 166 hrs 50 mins 42 secs 2,890,067 miles	25,868 436 189 11,191	Ninth dedicated DoD flight Last of the original DoD missions Unclassified payload Orbiter returned to KSC on 08 Dec 91

Seq. #	Mission #	Orbiter Flight #	Launch Landing	Launch Landing	Launch Landing	Crew	Altitude Inclination	Orbit Lift-Off OV Launch Payload OV Landing	Payloads	Mission Duration	Velocity Cross-Range Touchdown Roll-out	Notes
45	STS-42	OV-103 14	22 Jan 92 30 Jan 92	022:14:52:33 030:16:07:17	KSC, Pad A EAFB, 22	Ronald J. Grabe (C) Stephen S. Oswald (P) Norman E. Thagard (MS1) David C. Hilmers (MS2) William F. Readdy (MS3) Roberta L. Bondar (PS1) Ulf D. Merbold (PS2)	187 57.00	4,518,872 243,494 28,663 218,089	Spacelab IML-01 IMAX Camera GAS Bridge GAS (12) RMS = none	128 Orbits 193 hrs 14 mins 44 secs 2,921,153 miles	257,885 617 196 9,841	Mission extended one day for continued scientific experimentation Orbiter returned to KSC on 16 Feb 92
46	STS-45	OV-104 11	23 Mar 92 (pd) 24 Mar 92 02 Apr 92	084:13:13:40 093:11:23:08	KSC, Pad A KSC, 33	Charles F. Bolden, Jr. (C) Brian K. Duffy (P) Kathryn D. Sullivan (MS1) David C. Leestma (MS2) C. Michael Foale (MS3) Byron K. Lichtenberg (PS1) Dirk D. Frimout (PS2)	184 57.00	4,495,720 233,652 17,683 205,588	Spacelab ATLAS-01 GAS (1) RMS = none	142 Orbits 214 hrs 09 mins 28 secs 3,238,177 miles	257,885 781 192 9,227	Mission extended one day for continued scientific experimentation
47	STS-49	OV-105 1	04 May 92 (s) 07 May 92 16 May 92	128:23:40:00 137:20:57:38	KSC, Pad B EAFB, 22	Daniel C. Brandenstein (C) Kevin P. Chilton (P) Pierre J. Thuot (MS1) Kathryn C. Thornton (MS2) Richard J. Hieb (MS3) Thomas D. Akers (MS4) Bruce E. Melnick (MS5)	224 28.35	4,518,872 256,392 32,809 201,235	INTELSAT-VI F3/PKM RMS = s/n 303	140 Orbits 213 hrs 17 mins 38 secs 3,696,019 miles	25,841 472 194 9,490	First *Endeavour* flight Recovery and relaunch of INTELSAT-VI First use of braking parachute (test) First three astronaut EVA Longest EVA to date (8 hrs 29 mins) First Shuttle mission to feature four EVAs Orbiter returned to KSC on 30 May 92
48	STS-50	OV-102 12	25 Jun 92 09 Jul 92	177:16:12:23 191:11:42:27	KSC, Pad A KSC, 33	Richard N. Richards (C) Kenneth D. Bowersox (P) Bonnie J. Dunbar (PC) Ellen S. Baker (MS2) Carl J. Meade (MS3) Lawrence J. DeLucas (PS1) Eugene H. Trinh (PS2)	184 28.45	4,520,103 257,338 24,305 225,615	USML-01 RMS = none	220 Orbits 331 hrs 30 mins 04 secs 5,758,000 miles	25,186 448 203 10,674	First EDO flight Longest Shuttle flight to date Diverted to KSC due to weather at EAFB First *Columbia* landing at KSC
49	STS-46	OV-104 12	31 Jul 92 08 Aug 92	213:13:56:48 221:13:11:51	KSC, Pad B KSC, 33	Loren J. Shriver (C) Andrew M. Allen (P) Jeffrey A. Hoffman (PC) Franklin R. Chang-Diaz (MS2) Claude Nicollier (MS3) Marsha S. Ivins (MS4) Franco Malerba (PS1)	264 28.45	4,516,789 256,026 28,585 209,532	EURECA-1L TSS-01 LDCE PHCF UVPI IMAX Camera RMS = s/n 201	126 Orbits 191 hrs 15 mins 03 secs 3,321,007 miles	25,698 574 195 10,860	EURECA deployment delayed one day Anomalies with TSS deployment system restricted tether to 860 feet instead of planned 12 miles Mission extended one day to complete scientific objectives
50	STS-47	OV-105 2	12 Sep 92 20 Sep 92	256:14:23:00 164:12:53:23	KSC, Pad B KSC, 33	Robert L. 'Hoot' Gibson (C) Curtis L. Brown, Jr. (P) Mark C. Lee (PC) Jerome 'Jay' Apt (MS2) N. Jan Davis (MS3) Mae C. Jemison (MS4) Mamoru Mohri (PS1)	191 57.00	4,506,804 244,645 27,607 220,195	Spacelab-J GAS Bridge GAS (12) RMS = s/n 303	125 Orbits 190 hrs 31 mins 11 secs 3,271,844 miles	25,803 770 202 8,567	First professional Japanese astronaut First married couple (Lee and Davis) First operational use of drag chute (opened before nose gear touchdown)

Launch (s) is scheduled date
Launch (pd) is delay on pad
Launch (abort) is an abort after SSME ignition
n/a indicates not applicable
— indicates classified data or data not available
RMS = Remote Manipulator System serial number

Edwards AFB (EAFB) Runways 05R, 17L, 23L, and 33 are lakebed.
Edwards AFB (EAFB) Runways 04 and 22 are concrete.
White Sands Space Harbor (WSSH) Runway 17 is lakebed.
Kennedy Space Center (Shuttle Landing Facility) Runways 15 and 33 are concrete.
Orbital altitude is highest apogee
Times are GMT (day:hour:minute:second)

(C) = Shuttle Commander
(P) = Shuttle Pilot
(PC) = Payload Commander
(MS) = Mission Specialist
(PS) = Payload Specialist

Weights in pounds
Altitude and cross-range in statue miles
Inclination in degrees from Equator
Velocity in feet per second
Touchdown in knots indicated airspeed
Roll-out in feet after main gear touchdown

Discovery touches down on lakebed runway 17L at Edwards AFB on 3 October 1988. Note the original shape of the vertical stabilizer under the rudder, and compare this to later photos of the Orbiters after they were equipped with drag chutes. (NASA photo S26-(S)139)

Flight 26 – (STS-26R) – The Return-to-Flight of *Discovery* was delayed for 1 hour and 38 minutes to replace fuses in the cooling system of two of the crew's flight pressure suits, and also because of higher than expected upper wind conditions. After the suits were repaired, a waiver was obtained for the upper winds, and the flight was launched with no further delays. The TDRS-C satellite was deployed six hours after launch. Two anomalies were experienced by *Discovery* during the flight – a problem with the Ku-band telemetry system resulting in some flight data being lost, and the Orbiter's Flight Evaporator System iced up, causing the crew compartment temperature to rise above 80 degF.[10]

Flight 27 – (STS-27R) – This *Atlantis* launch was originally set for 1 December during a window lasting from 06:32 to 09:32 EDT, but was postponed due to unacceptable cloud cover at KSC. A second attempt during the same window the following day was successful although there are reports that lift-off was delayed by 1 minute and 11 seconds due to weather at the TAL site. This was the third mission dedicated to the Department of Defense, and was termed "... successful ..." by the Air Force. Acknowledged payloads include the Cosmic Ray Upset Experiment (CRUX-A) and OASIS in the payload bay, and AMOS, APE-A, CLOUDS-1A-04, RME-III-09, and VFT-2-02 on the mid-deck. Speculation is that the primary payload consisted of the LACROSSE-1 radar imaging satellite that weighed nearly 40,000 pounds and did not use an upper stage.[11]

Flight 28 – (STS-29R) – The original manifest date of 18 February was reassessed during late-January to allow the replacement of suspected faulty LO2 turbopumps on all three *Discovery* main engines. At the time it was felt a late-February date could be achieved, but difficulties with

a Master Events Controller postponed the flight an additional two weeks. The actual launch was delayed 1 hour and 50 minutes by morning ground fog and upper winds. The TDRS-D payload was deployed approximately six hours after launch.[12]

Flight 29 – (STS-30R) – A launch attempt on 28 April was scrubbed at T–31 seconds due to a problem with the LH2 recirculation pump on SSME #1 and a vapor leak in the four-inch LH2 recirculation line between the Orbiter and ET. Repairs were accomplished on the pad, and the 4 May countdown was delayed until the last five minutes of a 64 minute window due to cloud cover and high winds at KSC, violating return-to-launch-site limits. The long-delayed *Magellan* Venus mapping probe was deployed 6 hours and 14 minutes after launch. The probe's IUS fired as planned, sending *Magellan* on a 15 month journey to Venus.[13]

Flight 30 – (STS-28R) – The fourth dedicated DoD mission was launched with no announced delays within a window extending from 07:30 to 11:30 EDT. *Columbia* entered a 57-degree high-inclination orbit and deployed the classified payload. Observers originally believed this was an advanced reconnaissance satellite, but subsequent observations confirmed that the satellite was spin-stabilized (it 'flashed' as it spun) – something not consistent with a reconnaissance satellite. Current speculation was that this payload was the National Reconnaissance Office SDS-B1 communication relay satellite that was later transferred to a highly-elliptical orbit. The SDS-B satellites are used to relay data from reconnaissance satellites to the ground, and are reportedly modified versions of the Hughes HS-389 commercial communications satellite. Acknowledged payloads include CRUX-B, HEIN-LO, IOCM, and MPEC in the payload bay, and AMOS, CLOUDS-1A, RME-III, SAM, and VFT-2 on the mid-deck. The Air Force again termed this mission "... successful ..."[14]

Flight 31 – (STS-34) – The first flight since *Challenger* not to carry the 'R' suffix was originally scheduled for 12 October, but was postponed until 17 October because of a faulty Main Engine Controller on SSME #2. Weather constraints forced a further one day delay, and reconfiguring the GPCs at T–5 minutes delayed the final count by 3 minutes and 40 seconds. *Atlantis*' primary payload was the *Galileo* probe, which was successfully deployed 6.5 hours into the mission to begin its 6 year journey to Jupiter. This mission had been designed to use the liquid-fueled Centaur upper stage that was later cancelled for safety reasons. Because the substitute IUS was not as powerful as the Centaur, *Galileo* was propelled on a trajectory known as Venus-Earth-Earth Gravity Assist (VEEGA) and swung around Venus, the Sun, and Earth before making it's way toward Jupiter. Landing was brought forward by two orbits because of an approaching weather front at Edwards.[15]

Flight 32 – (STS-33R) – The scheduled launch of *Discovery* on 20 November was delayed to allow the replacement of suspect integrated electronics assemblies on both SRBs. No delays were announced during the countdown for the fifth dedicated DoD mission on 22 November, although some reports indicate a hold of 1 minute 30 seconds at T–5 minutes due to pressurization system adjustments. Most speculation is that the primary payload was the 38,500-pound ORION-2 National Security Agency (NSA) signals intelligence satellite that was boosted into higher orbit by an attached IUS. Acknowledged payloads included CRUX-B in the payload bay, and AMOS, APE-B, CLOUDS, RME-III, and VFT-1 on the mid-deck. Typically, the Air Force simply made their standard "... successful ..." announcement.[16]

Endeavour fires her main engines for 22 seconds during the 7th FRF of the shuttle program. The test began at 11:12 on 6 April 1992 on Pad 39B. (NASA photo KSC-92PC-733)

The TDRS-D satellite photographed by a 70-mm camera from the payload bay windows of Discovery on STS-29R. (NASA photo S29-71-029)

Flight 33 – (STS-32R) – A delay in completing facility modifications to Pad 39A forced a postponement from the planned 18 December 1989 date. This was the first flight to use 39A since the original STS-32/61-C in January 1986. A launch attempt on 8 January 1990 was scrubbed due to weather conditions at KSC, and the attempt the following day proceeded with no delays. This *Columbia* mission deployed a Navy communications satellite, and also retrieved the Long Duration Exposure Facility (LDEF) that had been left in orbit by STS-13. This was the longest flight to-date (almost 11 days). The planned landing on 19 January was waved off due to fog, and was delayed for one orbit the following day due to a computer malfunction.[17]

Flight 34 – (STS-36) – A combination of weather and the illness of commander John Creighton forced the postponement of this *Atlantis* mission from 22 February, to the 23rd, the 24th, and finally the 25th – NASA had long since stopped assigning primary and backup crews to all but the most important Space Shuttle flights. This was the first time since Apollo 13 in 1970 that a manned space mission had been affected by the illness of a crew member. The launch attempt on 25 February was scrubbed due to the failure of an Air Force range safety computer, and an attempt on the 26th was cancelled due to weather. Launch was reset for 28 February in a classified window lying within launch period extending from 00:00 to 04:00 EST. At T–9 minutes there was reportedly a 1 hour, 56 minute, 52 second hold due to rain at KSC and TAL site weather. A second hold lasted 2 minutes at the T–5 minute mark. This flight marked the highest-inclination orbit flown by shuttle – 62 degrees – and required a very inefficient 'dog-leg' ascent over the North Atlantic due to range safety constraints. This was the sixth "… successful …" DoD flight. Acknowledged payloads included only RME-III-12, VFT-1, and VFT-2 on the mid-deck with speculation that the primary payload was an advanced (possibly 'stealth') reconnaissance satellite (named MYSTY).

The satellite was apparently deployed using the Stabilized Payload Deployment System (SPDS) that essentially dumped the payload over the left payload bay sill.[18]

Flight 35 – (STS-31R) – This flight carried the *Hubble* Space Telescope (HST), perhaps the most significant payload carried to date. *Discovery* had originally been scheduled for launch on 18 April, a date which was moved forward to 12 April and finally 10 April following the Flight Readiness Review. This was the first time that a Space Shuttle launch was attempted ahead of schedule. The 10 April launch attempt was scrubbed at T–4 minutes due to a faulty valve in APU #1, and the launch was rescheduled for two weeks later to allow the batteries in the HST to be recharged. The 24 April countdown was held for 2 minutes and 52 seconds at T–31 seconds when computer software failed to shut the ET LO2 fill and drain valve but otherwise proceeded smoothly. The HST was deployed approximately 24 hours after launch. *Discovery* landed on runway 22 at Edwards, marking the first use of the new carbon brakes.[19]

Flight 36 – (STS-41) – The European *Ulysses* (formerly known as Solar-Polar) spacecraft was the primary payload for *Discovery*. The count was held for 10 minutes and 43 seconds at T–9 minutes due to rain over SLF, then at T–5 minutes for 10 seconds due to RSLS fault adjustment, then at T–31 seconds for 1 minute and 22 seconds for further RSLS adjustment – total hold time was 12 minutes and 15 seconds. Two upper stages, a two-stage Inertial Upper Stage (IUS) and a mission-specific payload assist module (PAM-S), combined together to send *Ulysses* into a polar orbit of the Sun – this was the substitute for the cancelled Centaur upper stage that had been designed to launch *Ulysses*.[20]

A National Reconnaissance Office SDS-B satellite in launch configuration. The 'drop skirt' is stowed, covering the cylindrical main body and its heat radiator. One of the two side-mounted antennas is also visible. The folded omni antenna is visible at left. Similar satellites were carried on STS-28R, STS-38 and STS-53. (NRO)

Things had seemed to be going well for Space Shuttle – 11 missions had been launched in just over 20 months since the Return-to-Flight, and the manifest was fairly stable. The 37th flight was scheduled to be STS-35 using *Columbia* on 16 May 1990 from Pad 39A. The Flight Readiness Review (FRR) had mandated a Freon-21 cooling loop proportioning valve on the Orbiter be replaced, so the announcement of a firm launch date was delayed. A subsequent FRR was conducted after the faulty valve had been changed, and the launch was set for 30 May. During LH2 tanking, a minor hydrogen leak was discovered near the tail service mast (TSM) on the mobile launch platform (MLP), but subsequent investigation revealed a major leak at the Orbiter–ET 17-inch disconnect. Gaseous hydrogen was also found in the sealed aft compartment of *Columbia*, and was believed to be associated with the leak at the 17-inch disconnect. The launch attempt was scrubbed, and the ET was drained and inerted.[21]

A mini-tanking test was conducted on 6 June to try and further define the location of the leak at the 17-inch disconnect. It was decided that repairs could not be accomplished at the launch pad, so *Columbia* was rolled back to the VAB on 12 June, demated from the ET–SRB stack, and returned to the OPF on 15 June. The Orbiter side of the hydrogen 17-inch disconnect was replaced with one 'borrowed' from the yet-uncompleted OV-105, and the ET side was fitted with new hardware sent from Michoud.[22]

While NASA and contractor engineers were trying to dis-

Columbia (left) as STS-35 and Atlantis as STS-38 pass each other in the early morning of 9 August 1990 as NASA tried to deal with the hydrogen leaks that plagued both launches. This is one of the few times full Shuttle stacks have passed so near each other. (NASA)

cover what had caused the problem with the 17-inch disconnects on *Columbia*, the launch team proceeded with plans to launch *Atlantis* on the STS-38 mission in mid-July 1990. *Atlantis* was rolled-out to Pad 39A on 18 June, and NASA decided to conduct a precautionary tanking test to ensure the problems that plagued *Columbia* were not present on the *Atlantis*. Liquid hydrogen was loaded into the

Toxic fumes are a major concern after landing, as is cooling on-board systems. A full set of equipment is kept at Edwards since it is always the alternate end-of-mission landing site, and was the primary site for most missions early in the program. This is Columbia, evidenced by the SILTS pod on the vertical stabilizer and black chines. (Tony Landis Collection)

STS-38 ET on 29 June, and hydrogen gas was quickly detected around the 17-inch disconnect seal. A great deal of additional instrumentation was fitted around the disconnect in an attempt to pinpoint the exact location of the leak, and an additional tanking test was conducted on 13 July. Special sealant was added to the disconnect in an attempt to stop the leaks, and another mini-tanking test was performed on 25 July. Hydrogen gas was again detected, and it was decided to roll STS-38 back to the VAB on 9 August to further investigate the problem. During the roll-back, *Atlantis* was parked outside the VAB for most of a day while *Columbia* was transferred to Pad 39A. This resulted in some memorable photographs of the two vehicle stacks passing each other in the early morning. Unfortunately, while sitting outside the VAB *Atlantis* suffered some minor damage to the thermal protection system tiles from a passing thunderstorm that included small hail.[23]

NASA officials continued to believe that the hydrogen leaks affecting STS-35 and STS-38 were unrelated, and *Columbia* was again rolled out to Pad 39A on 9 August in anticipation of a 1 September launch date. Two days before launch, an avionics box on the ASTRO-1 payload malfunctioned and had to be changed. Launch was rescheduled to 6 September to allow time for replacement and retesting.[24]

Propellant loading began as normal on 5 September, and hydrogen gas was again detected in the Orbiter aft compartment. The launch team concluded that *Columbia* had experienced two separate hydrogen leaks from the beginning – one at the 17-inch disconnect, and a second in the Orbiter aft compartment that had resurfaced. The investigation concentrated on a package of three hydrogen recirculation pumps in the aft compartment which were replaced and retested. A damaged Teflon seal in SSME #2 was also discovered and replaced. Launch was rescheduled for 18 September, but hydrogen gas was again detected in the aft compartment during the tanking. The STS-35 mission was put on hold until the problem could be investigated by a special 'tiger team' assigned by Space Shuttle Program manager Bob Crippen.[25]

Meanwhile, *Atlantis* had been repaired in the OPF, and was rolled back to the VAB on 2 October. While the Orbiter was being lifted to the vertical position in preparation for mating, a temporary work platform beam that should have been removed from the aft compartment fell and caused minor damage inside the compartment. The damage was repaired in the VAB, and stacking operations continued.

Columbia was moved from Pad 39A to Pad 39B (termed 'roll-around' by workers – coined for this event to supplement roll-over and roll-out) on 8 October to make room for *Atlantis* at Pad 39A, but Tropical Storm Klaus took an unexpected turn and forced NASA to roll *Columbia* back to the VAB on 9 October – this was the first time a vehicle had been rolled-back to the VAB twice on a single mission (although it would happen again on STS-79). The STS-38 stack was rolled to Pad 39A on 12 October, and *Columbia* was transferred back to Pad 39B on 14 October.[26]

By this time NASA had significantly modified the rules concerning the amount of hydrogen leakage that was acceptable. It had been determined that the original threshold levels were unrealistically low, and had been unknowingly violated on several occasions earlier in the flight program without incident. More sophisticated monitoring equipment had recently been installed, adding to the confusion. A fourth and final mini-tanking test on *Atlantis* was conducted on 24 October, and no leakage was discovered that violated the new rules. A Flight Readiness Review set the STS-38 launch date for 9 November.

Special sensors and video cameras (including a plexiglass aft compartment door) were installed in *Columbia* and next to the 17-inch disconnects. A final mini-tanking test was conducted on 30 October, and no leakage in excess of the new rules was detected, so the STS-35 launch was scheduled for 2 December.[27]

NASA continued to explain the leaks in *Columbia* and *Atlantis* as being mostly unrelated, although there were changes to the procedures used to test the 17-inch disconnects after manufacturing. The leaks also prompted NASA to raise the priority on development of a new 14-inch disconnect valve that had been under development as an answer to numerous design deficiencies with the existing 17-inch units. The new 14-inch disconnects were eventually cancelled in 1993, although some of the technology was subsequently used to upgrade the existing 17-inch valves. The new rules regarding hydrogen leakage were made a part of the normal launch commit criteria, and have been used on all subsequent Space Shuttle flights.

The newest Orbiter, Endeavour, being stacked in the VAB in preparation for its maiden flight. OV-105 had just spent seven months in the OPF being prepared for flight. Note how the lifting frame supports the Orbiter during mating. (NASA photo KSC-92PC-523)

Flight 37 – (STS-38) – After solving the problems with hydrogen leaks, *Atlantis* was scheduled for a 9 November launch. Problems with the classified DoD payload forced a further delay to 15 November. Lift-off occurred during classified launch window lying within a period extending from 18:30 to 22:30 EST. This marked the seventh mission dedicated to the DoD, which "... successfully ..." met all of its mission objectives. Acknowledged payloads included AMOS, APE-B, RME-III-13, SPADVOS, and VFT-1 on the mid-deck. Most speculation is that the primary payload was the SDS-B2 communications relay satellite that was later transferred to geosynchronous orbit – interestingly, it appears this was accomplished without the use of an IUS. Landing was delayed one day from 19 November due to high winds at Edwards, and finally diverted to KSC on 20 November when the situation did not improve at Edwards.[28]

Flight 38 – (STS-35) – The 2 December date set after the resolution of the hydrogen leak problems held, and *Columbia* was launched following a hold of 21 minutes and 1 second due to cloud cover at KSC. The flight crew experienced trouble dumping waste water because of a clogged drain, but managed using spare containers. The ASTRO-1 payload had been serviced regularly and remained in *Columbia*'s payload bay during the repairs and reprocessing forced by the hydrogen leaks. The primary objective of the mission was the round-the-clock observation of the celestial sphere in ultraviolet and X-ray astronomy using the ASTRO-1 observatory that consisted of four telescopes. The loss of both data display units (used for pointing the telescopes and operating experiments) during the mission forced ground teams at MSFC to aim the ultraviolet telescopes via remote-control with some fine-tuning by the flight crew. The mission was cut short by one day due to impending bad weather at Edwards, which was still designated the primary landing site. Nevertheless, the science teams at NASA Goddard and MSFC estimated that 70 percent of the planned science objectives were achieved.[29]

Columbia roars aloft from Pad 39A into the Florida skies. During the ten-day STS-32 mission, the five-member crew deployed the SYNCOM IV-5 military communications satellite and retrieved the LDEF experiment. (NASA photo KSC-90PC-0035)

Flight 39 – (STS-37) – The 5 April launch attempt proceeded with only a 4-minute and 45-second hold due to low cloud cover at KSC. *Atlantis* carried the long-delayed Gamma Ray Observatory (GRO) into low-earth orbit. The GRO high-gain antenna failed to deploy correctly, and was freed by astronauts Ross and Apt during an unscheduled EVA – the first since April 1985. The next day the same astronauts performed a scheduled spacewalk to test concepts to move themselves and equipment while assembling and maintaining Space Station Freedom. The initial landing opportunity at Edwards was waved off, and landing was rescheduled for KSC the next day. That attempt was also waved off due to fog and the landing was shifted to Edwards once again, one orbit later. *Atlantis* landed 600 feet short of the threshold on dry lakebed runway 33.[30]

Flight 40 – (STS-39) – *Discovery* was originally scheduled for launch on 9 March 1991, but during processing activities at Pad 39A, significant cracks were discovered in all four hinges on the two ET umbilical door drive mechanisms. NASA opted to roll the vehicle back to the VAB on 7 March, and then to the OPF on 15 March for repair. Hinges from *Columbia* were also found to have minor cracks, but were removed, reinforced, and installed on *Discovery*. Cracks were also found on the hinges on *Atlantis*. The stack was returned to Pad 39A on 1 April for a scheduled launch on 23 April. The mission was again postponed during ET tanking when a transducer on the high-pressure oxidizer turbopump on SSME #3 showed out-of-tolerance readings. The transducer and its wiring harness were replaced, and a new launch date of 28 April was set. The unclassified DoD mission was launched after a hold of 32 minutes and 14 seconds due to a false start of payload recorders (not a GPC problem).[31]

Flight 41 – (STS-40) – This *Columbia* mission was postponed less than 48 hours before launch when it became known that a LH2 transducer in the MPS was defective. The sensor had been replaced during leak testing the previous year, but had unknowingly failed testing at the vendor. Engineers feared that if the transducer broke, pieces of it could cause a catastrophic failure of the SSME turbopump. In addition, one of the Orbiter's GPCs and an MDM failed while the vehicle was on the pad. A new computer and MDM were installed and tested, one LH2 and five LO2 transducers were replaced, and three other LH2 transducers were removed all together. A launch attempt on 1 June was scrubbed at 07:15 after several attempts to calibrate IMU #1 failed – the IMU was replaced, and launch proceeded smoothly on 5 June. This was the fifth Spacelab mission, and the first dedicated to life sciences. The mission featured the most detailed and interrelated physiological measurements in space since the three Skylab missions of 1973–74. The measurements focused on the crew, 30 rodents, and several thousand tiny jellyfish.[32]

Flight 42 – (STS-43) – Launch was originally scheduled for 23 July, but was moved to the 24th to allow time to replace a faulty electronics unit that controls Orbiter–ET separation. The launch was postponed again about five hours before lift-off because of a faulty Main Engine Controller on SSME #3 – the MEC was replaced and launch was rescheduled for 1 August. The planned 11:01 lift-off was delayed because of a cabin pressure vent valve reading, and postponed until 2 August due to unacceptable RTLS weather conditions. *Atlantis* successfully deployed the fourth Tracking and Data Relay Satellite (TDRS-E) approximately six hours after launch.[33]

Flight 43 – (STS-48) – The countdown for *Discovery* went smoothly, and the Orbiter was launched on 12

September with only a 14-minute and 4 second delay due to communication problems at T–5 minutes. The Upper Atmosphere Research Satellite (UARS) primary payload was deployed on the third day of the mission. During its planned 18-month mission the 14,500-pound observatory would make the most extensive study ever conducted of the Earth's troposphere, the upper level of the planet's life-sustaining gases which also include the protective ozone layer. This flight was scheduled to land at KSC, but weather conditions in central Florida forced a diversion to the alternate landing site at Edwards.[34]

Flight 44 – (STS-44) – This launch of *Atlantis* was originally scheduled for 19 November, but was postponed prior to tanking operations because of a faulty inertial measurement unit in the IUS upper stage. The unit was replaced and launch rescheduled for 24 November. Liftoff was delayed for 13 minutes to allow adequate time to replenish the LO2 level in the ET following minor repairs on the LO2 plumbing in the MLP. The primary payload on this unclassified DoD flight was the Defense Support Program (DSP) F-16 ('Liberty') early warning satellite. This mission was also scheduled to land at KSC, but was diverted to Edwards following the on-orbit failure of one of the Orbiter's three IMUs. This failure also shortened the planned ten day mission by three days.[35]

Flight 45 – (STS-42) – The launch of *Discovery* was delayed 59 minutes and 33 seconds due to a fuel cell anomaly and cloud cover at KSC. Once on-orbit, the mission was extended for one day to continue experiments on the International Microgravity Laboratory (IML) to explore in depth the complex effects of weightlessness on living organisms and materials processing. The international crew, divided into Red and Blue teams, conducted experiments in the pressurized Spacelab module on the human nervous system's adaptation to low gravity and the effects of microgravity on other life forms such as shrimp eggs, lentil seedlings, fruit fly eggs, and bacteria.[36]

Flight 46 – (STS-45) – The launch of *Atlantis* was originally scheduled for 23 March, but was delayed one day because of higher than allowable concentrations of LO2 and LH2 in the Orbiter's aft compartment during tanking operations. During troubleshooting the leaks could not be reproduced, leading engineers to theorize they were the result of the plumbing not being properly conditioned to the cryogenic propellants. No repairs were deemed necessary, and the launch was rescheduled for the following day. This attempt was delayed by 13 minutes and 40 seconds due to weather concerns at KSC and the TAL sites. *Atlantis* carried the Atmospheric Laboratory for Applications and Science (ATLAS-1) experiment on Spacelab pallets in the payload bay. The non-deployable payload – equipped with 12 instruments from the U.S., France, Germany, Belgium, Switzerland, The Netherlands, and Japan – conducted studies in atmospheric chemistry, solar radiation, space plasma physics, and ultraviolet astronomy. The mission was extended one day for continued scientific research.[37]

Flight 47 – (STS-49) – A 22-second flight readiness firing (FRF) was conducted at Pad 39B on 6 April 1992 in preparation for OV-105's maiden flight. This *Endeavour* launch was originally scheduled for 20:34 on 4 May, but was moved to 19:06 on 7 May to allow more daylight for photographic documentation of vehicle behavior during the launch phase. The launch on 7 May was delayed 34 minutes due to weather conditions at the TAL site. An ambitious plan called for *Endeavour* to rescue the INTELSAT-VI (603) satellite that was left stranded in an unusable orbit since its launch on a Titan III in March

1990. The capture of the satellite required three separate EVAs, including the first three-astronaut (Thout, Hieb, and Akers) EVA of the U.S. manned space program. Mission commander Daniel Brandenstein maneuvered *Endeavour* within a couple of feet of the 4.5-ton Intelsat on several occasions. The astronauts managed to attach a new perigee kick motor which later boosted the satellite to its proper orbit, marking the first time that a crew had attached a live rocket motor in space. The crew also practiced assembly techniques for the upcoming Space Station Freedom, although the amount of time spent on this activity was reduced due to the extended EVAs required to rescue INTELSAT-VI. Upon landing at Edwards AFB, *Endeavour* tested the new braking parachute, deploying it after nose-gear touchdown. This flight included the first and second longest EVAs accomplished to date – 8 hours 29 minutes; and seven hours 45 minutes.[38]

Flight 48 – (STS-50) – Launch was delayed 5 minutes and 23 seconds at T–5 minutes to wait for clouds around KSC to disperse. This *Columbia* flight set a record for the longest Shuttle flight, and was the longest U.S. manned space flight with the exception of the three Skylab missions. A failure of an air recycling device (RCRS) installed on the vehicle as a part of the extended duration orbiter (EDO) modifications did not affect the mission since *Columbia* still carried the older, standard hardware also. The flight was extended one day to allow the weather at Edwards to clear for landing. When this did not occur, *Columbia* was diverted to KSC and was the second Orbiter to use a braking parachute upon landing.[39]

Flight 49 – (STS-46) – *Atlantis* was launched 48 seconds late because the flight crew fell behind configuring switches on the flight deck, but this countdown was nevertheless regarded as the 'cleanest' since the Return-to-Flight after the *Challenger* stand-down. This long-awaited mission tested a Tethered Satellite System (TSS) – an Italian satellite was to be deployed on a 12-mile long tether connected to the Orbiter. The effect of the tether passing through space was expected to generate 5,000 volts of electricity and conduct this current to the Orbiter. Problems with the deployment apparatus did not allow the satellite to deploy further than 860 feet, although it still managed to generate approximately 40 volts of power. *Atlantis* also released the 9,424-pound European Space Agency EURECA (European Retrievable Carrier) satellite. This was the largest satellite produced in Europe and carried 15 major science experiments, mostly in microgravity sciences. The EURECA satellite was recovered by STS-57 in June 1993.[40]

Flight 50 – (STS-47) – The second *Endeavour* mission marked the 'golden flight' of the Space Shuttle program with the first on-time launch since the original STS-31 in 1985. Several firsts were marked with the STS-47 mission, including the first black woman astronaut, the first professional Japanese astronaut (a Japanese press reporter had flown previously with the Soviets), and the first married couple on the same mission. The Spacelab-J experiments focused on life sciences and material processing, and included a cargo of hornets, frogs, and Japanese Carp. Stanley N. Koszelak was the alternate for Mae Jemison – the first U.S. mission specialist to be assigned a back-up. The mission was extended one day to allow additional science to be accomplished. Weather threatened to divert the landing, but cleared enough to return to KSC one orbit later than originally planned. This was the first operational use of the new braking parachute, with it being deployed slightly before nose-gear touchdown.[41]

The Space Shuttle finally began to fulfill some of its promise during the mid-1990s. A steady schedule of seven or eight flights per year took some of the pressure of the launch and landing crews, giving the appearance at least, of 'routine' operations. Flights began to the Russian Mir space station in preparation for the assembly of the International Space Station. Other than a few U.S. Government satellites, Space Shuttle no longer delivered commercial payloads to space, but a highly successful Hubble repair mission illustrated the usefulness of allowing men to travel into space.[120]

Seq. #	Mission #	Orbiter Flight #	DATES Launch / Landing	TIME Launch / Landing	SITES Launch / Landing	Crew	ORBIT Altitude Inclination	WEIGHTS Lift-Off OV Launch Payload OV Landing	Payloads	Mission Duration	REENTRY Velocity Cross-Range Touchdown Roll-out	Notes
51	STS-52	OV-102 13	15 Oct 92 (s) / 22 Oct 92 / 01 Nov 92	296:17:09:39 / 306:14:05:52	KSC, Pad B / KSC, 33	James D. Wetherbee (C); Michael A. Baker (P); Charles Lacy Veach (MS1); William M. Shepherd (MS2); Tamara E. Jernigan (MS3); Steven G. MacLean (PS1)	188 / 28.45	4,515,380 / 250,399 / 20,132 / 215,979	LAGEOS; USMP-1; RMS = s/n 301	158 Orbits; 236 hrs 56 mins 13 secs; 4,120,972 miles	25,666 / 257 / 211 / 10,708	
52	STS-53	OV-103 15	02 Dec 92 / 09 Dec 92	337:13:24:00 / 344:20:43:47	KSC, Pad A / EAFB, 22	David M. Walker (C); Robert D. Cabana (P); Guion S. Bluford, Jr. (MS1); James S. Voss (MS2); Michael R. 'Rich' Clifford (MS3)	200 / 57.00	4,507,750 / 243,944 / 26,118 / 193,851	DoD (SDS-B3); ODERACS; GLO; RMS = none	115 Orbits; 175 hrs 19 mins 47 secs; 3,011,688 miles	25,813 / 910 / 212 / 10,165	Final DoD payload; First flight of Discovery after OMDP; First Discovery use of drag chute; Orbiter returned to KSC on 18 Dec 92
53	STS-54	OV-105 3	13 Jan 93 / 19 Jan 93	013:13:59:30 / 019:13:37:49	KSC, Pad B / KSC, 33	John H. Casper (C); Donald R. McMonagle (P); Mario Runco, Jr. (MS1); Gregory J. Harbaugh (MS2); Susan J. Helms (MS3)	190 / 28.45	4,523,381 / 259,764 / 46,540 / 197,353	TDRS-F/IUS; DXS; RMS = none	95 Orbits; 143 hrs 38 mins 19 secs; 2,484,183 miles	25,780 / 368 / 212 / 8,724	Extensive EVA to practice Space Station assembly
54	STS-56	OV-103 16	06 Apr 93 (pd) / 08 Apr 93 / 17 Apr 93	098:05:29:00 / 107:11:37:24	KSC, Pad B / KSC, 33	Kenneth D. Cameron (C); Steven S. Oswald (P); C. Michael Foale (MS1); Kenneth D. Cockrell (MS2); Ellen Ochoa (MS3)	184 / 57.00	4,502,299 / 237,213 / 16,439 / 207,946	SPARTAN-201-01; ATLAS-2; SAREX II; RMS = s/n 201	147 Orbits; 222 hrs 08 mins 24 secs; 3,831,267	25,797 / 7 / 206 / 9,530	
55	STS-55	OV-102 14	25 Feb 93 (s) / 14 Mar 93 (s) / 21 Mar 93 (s) / 22 Mar 93 (abort) / 24 Apr 93 (pd) / 26 Apr 93 / 06 May 93	116:14:50:00 / 126:14:29:59	KSC, Pad A / EAFB, 22	Steven R. Nagel (C); Terence T. Henricks (P); Jerry L. Ross (MS1); Charles J. Precourt (MS2); Bernard A. Harris, Jr. (MS3); Ulrich Walter (PS1); Hans W. Schlegel (PS2)	188 / 28.45	4,518,969 / 255,441 / 26,881 / 227,209	Spacelab-D2; MAUS; AOET; GAUSS; MOMS; GAS (1); RMS = none	159 Orbits; 239 hrs 39 mins 59 secs; 4,146,971 miles	25,779 / 737 / 217 / 10,125	At conclusion of mission, combined Shuttle fleet had accumulated a total of 365 days, 23 hours and 48 minutes in flight; Orbiter returned to KSC on 14 May 93
56	STS-57	OV-105 4	18 May 93 (s) / 03 Jun 93 (s) / 20 Jun 93 (pd) / 21 Jun 93 / 01 Jul 93	172:13:07:22 / 182:12:52:16	KSC, Pad B / KSC, 33	Ronald J. Grabe (C); Brian K. Duffy (P); G. David Low (MS1); Nancy J. Sherlock (MS2); Peter J. K. Wisoff (MS3); Janice E. Voss (MS4)	290 / 28.45	4,518,566 / 252,710 / 19,630 / 224,906	EURECA Retrieval; Spacehab-01; GAS (10); RMS = s/n 303	154 Orbits; 239 hrs 44 mins 54 secs; 4,115,989 miles	25,988 / 676 / 207 / 9,954	First flight of Spacehab module
57	STS-51	OV-103 17	17 Jul 93 (pd) / 24 Jul 93 (pd) / 04 Aug 93 (s) / 12 Aug 93 (abort) / 10 Sep 93 (s) / 12 Sep 93 / 22 Sep 93	255:11:45:00 / 265:07:56:11	KSC, Pad B / KSC, 15	Frank L. Culbertson (C); William F. Readdy (P); James H. Newman (MS1); Daniel W. Bursch (MS2); Carl E. Walz (MS3)	184 / 28.45	4,532,125 / 261,486 / 42,637 / 206,931	ACTS/TOS; ORFEUS-SPAS; IMAX Camera; RMS = s/n 201	156 Orbits; 236 hrs 11 mins 11 secs; 4,052,051 miles	25,794 / 102 / 194 / 8,271	
58	STS-58	OV-102 15	14 Oct 93 (pd) / 15 Oct 93 (pd) / 18 Oct 93 / 01 Nov 93	291:14:53:10 / 305:15:05:42	KSC, Pad B / EAFB, 22	John E. Blaha (C); Richard A. Searfoss (P); Margaret Rhea Shedon (MS1); William S. McArthur, Jr. (MS2); David A. Wolf (MS3); Shannon W. Lucid (MS4); Martin J. Fettman (PS1)	178 / 39.00	4,517,138 / 256,097 / 23,127 / 229,368	Spacelab SLS-2; RMS = none	224 Orbits; 336 hrs 12 mins 32 secs; 5,809,886 miles	25,755 / 166 / 198 / 9,640	Second flight of Spacelab Life Sciences module; Orbiter returned to KSC on 08 Nov 93
59	STS-61	OV-105 5	01 Dec 93 (pd) / 02 Dec 93 / 13 Dec 93	336:09:27:00 / 347:05:25:37	KSC, Pad B / KSC, 33	Richard O. Covey (C); Kenneth D. Bowersox (P); Kathryn C. Thornton (MS1); Claude Nicollier (MS2); Jeffrey A. Hoffman (MS3); F. Story Musgrave (MS4); Thomas D. Akers (MS5)	369 / 28.45	4,511,794 / 250,279 / 17,401 / 212,835	HST Repair; IMAX Camera; RMS = s/n 303	162 Orbits; 259 hrs 58 mins 37 secs; 4,396,207 miles	26,096 / 3 / 201 / 7,922	First Hubble servicing mission; Second night landing at KSC
60	STS-60	OV-103 18	03 Feb 94 / 11 Feb 94	034:12:10:00 / 042:19:19:22	KSC, Pad A / KSC, 15	Charles F. Bolden, Jr. (C); Kenneth S. Reightler, Jr. (P); N. Jan Davis (MS1); Ronald M. Sega (MS2); Franklin R. Chang-Diaz (MS3); Sergei K. Krikalev (MS4)	220 / 57.00	4,442,490 / 245,767 / 22,311 / 216,594	Wake Shield; Bremen; Spacehab-02; GAS (4); RMS = s/n 201	129 Orbits; 199 hrs 09 mins 22 secs; 3,379,914 miles	25,858 / 433 / 205 / 7,771	First flight of Russian cosmonaut on Shuttle; Second flight of Spacehab module
61	STS-62	OV-102 16	03 Mar 94 (pd) / 04 Mar 94 / 18 Mar 94	063:13:53:00 / 077:13:09:41	KSC, Pad B / KSC, 33	John H. Casper (C); Andrew M. Allen (P); Peirre J. Thout (MS1); Charles D. 'Sam' Gemar (MS2); Marsha S. Ivins (MS3)	165/188 / 39.00	4,519,801 / 256,584 / 19,972 / 228,250	USML-2; RMS = s/n 301	223 Orbits; 335 hrs 16 mins 41 secs; 5,797,960 miles	25,708 / 133 / 207 / 10,151	Orbit lower 23 miles for last half of mission
62	STS-59	OV-105 6	07 Apr 94 (s) / 08 Apr 94 (pd) / 09 Apr 94 / 20 Apr 94	099:11:05:00 / 110:16:54:30	KSC, Pad A / EAFB, 22	Sidney M. Gutierrez (C); Kevin P. Chilton (P); Jerome 'Jay' Apt (MS1); Michael R. 'Rich' Clifford (MS2); Linda M. Godwin (PC); Thomas D. Jones (MS4)	139 / 57.00	4,511,411 / 246,851 / 27,447 / 221,865	Space Radar Lab-1; GAS (3); RMS = s/n 303	182 Orbits; 269 hrs 49 mins 30 secs; 4,675,934 miles	25,660 / 830 / 215 / 10,691	SRL-1 surveyed 20% of Earth's surface; First flight of TUFI tiles; Orbiter returned to KSC on 02 May 94
63	STS-65	OV-102 17	08 Jul 94 / 23 Jul 94	189:16:43:00 / 204:10:38:00	KSC, Pad A / KSC, 33	Robert D. Cabana (C); James D. Halsell, Jr. (P); Richard J. Hieb (MS1); Carl E. Walz (MS2); Leroy Chiao (MS3); Donald A. Thomas (MS4); Chiaki Mukai (PS1)	184 / 28.45	4,523,441 / 258,585 / 24,159 / 229,307	Spacelab IML-2; RMS = none	234 Orbits; 353 hrs 55 mins 01 secs; 6,078,077 miles	25,720 / 207 / 199 / 10,211	Longest Shuttle flight to date; First Japanese woman in space; Longest flight to date by female astronaut
64	STS-64	OV-103 19	09 Sep 94 / 20 Sep 94	252:22:22:55 / 263:21:12:52	KSC, Pad B / EAFB, 04	Richard N. Richards (C); L. Blaine Hammond, Jr. (P); Jerry M. Linenger (MS1); Susan J. Helms (MS2); Carl J. Meade (MS3); Mark C. Lee (MS4)	161 / 57.00	4,503,921 / 240,884 / 19,260 / 212,141	SPARTAN-201-02; ROMPS; LITE; GAS (10); RMS = s/n 201	175 Orbits; 262 hrs 49 mins 57 secs; 4,520,281 miles	25,727 / 127 / 198 / 9,656	First flight of Lidar Experiment; Orbiter returned to KSC on 27 Sep 94
65	STS-68	OV-105 7	18 Aug 94 (abort) / 30 Sep 94 / 11 Oct 94	273:11:16:00 / 284:17:02:08	KSC, Pad A / EAFB, 22	Michael A. Baker (C); Terrence W. Wilcutt (P); Steven L. Smith (MS1); Daniel W. Bursch (MS2); Peter J. K. Wisoff (MS3); Thomas D. Jones (PC)	138 / 57.00	4,510,613 / 247,136 / 27,582 / 221,571	Space Radar Lab-2; GAS (3); RMS = s/n 303	181 Orbits; 269 hrs 46 mins 08 secs; 4,649,105 miles	25,658 / 858 / 193 / 8,495	Orbiter returned to KSC on 20 Oct 94
66	STS-66	OV-104 13	27 Oct 94 (s) / 03 Nov 94 / 14 Nov 94	307:16:59:43 / 318:15:33:45	KSC, Pad B / EAFB, 22	Donald R. McMonagle (C); Curtis L. Brown, Jr. (P); Ellen Ochoa (PC); Joseph R. Tanner (MS2); Jean-Francois Clervoy (MS3); Scott E. Parazynski (MS4)	189 / 57.00	4,508,715 / 243,089 / 18,001 / 211,327	CRISTA-SPAS; Spacelab ATLAS-3; RMS = s/n 202	173 Orbits; 262 hrs 34 mins 02 secs; 4,499,056 miles	25,798 / 357 / 193 / 7,642	Landing diverted to Edwards due to Tropical Storm Gorden; Orbiter returned to KSC on 22 Nov 94
67	STS-63	OV-103 20	02 Feb 95 (pd) / 03 Feb 95 / 11 Feb 95	034:05:22:04 / 042:11:50:19	KSC, Pad B / KSC, 15	James D. Wetherbee (C); Eileen M. Collins (P); Bernard A. Harris, Jr. (PC); C. Michael Foale (MS1); Janice Voss (MS2); Vladimir G. Titov (MS3)	249 / 51.60	4,511,630 / 247,555 / 19,108 / 212,646	SPARTAN-204; ODERACS; Spacehab-3; RMS = s/n 201	128 Orbits; 198 hrs 28 mins 15 secs; 3,373,819 miles	25,903 / 469 / 212 / 11,002	First approach and flyaround of Mir; Second flight of Russian cosmonaut on Shuttle; First female Shuttle pilot; OV-103 first to complete 20 missions

Seq. #	Mission #	Orbiter Flight #	DATES Launch Landing	TIME Launch Landing	SITES Launch Landing	Crew	ORBIT Altitude Inclination	WEIGHTS Lift-Off OV Launch Payload OV Landing	Payloads	Mission Duration	REENTRY Velocity Cross-Range Touchdown Roll-out	Notes
68	STS-67	OV-105 8	02 Mar 95 18 Mar 95	061:06:38:13 077:21:47:01	KSC, Pad A EAFB, 22	Stephen S. Oswald (C) William G. Gregory (P) John M. Grunsfeld (MS1) Wendy B. Lawrence (MS2) Tamara E. Jernigan (PC) Samual T. Durrance (PS1) Ronald A. Parise (PS2)	215 28.45	4,519,114 256,293 20,250 217,481	ASTRO-2 GAS (2) RMS = s/n 303	261 Orbits 399 hrs 08 mins 48 secs ≈6,900,000 miles	25,839 202 209 11,617	Longest Shuttle flight to date Orbiter returned to KSC on 27 Mar 95
69	STS-71	OV-104 14	23 Jun 95 (pd) 24 Jun 95 (pd) 27 Jun 95 07 Jul 95	178:19:32:19 188:14:54:36	KSC, Pad A KSC, 15	Robert L. 'Hoot' Gibson (C) Charles J. Precourt (P) Ellen S. Baker (PC) Gregory J. Harbaugh (MS2) Bonnie J. Dunbar (MS3) Anatoly Y. Solovyev (Mir-19) (up) Nikolai M. Budarin (Mir-19) (up) Vladimir N. Dezhurov (Mir-18) (down) Gennady M. Strekalov (Mir-18) (down) Norman R. Thagard (Mir-18) (down)	195/249 51.60	4,511,586 248,857 18,299 214,879	Mir/ODS Spacelab RMS = none	152 Orbits 235 hrs 22 mins 17 secs ≈4,100,000 miles	25,913 645 207 8,364	First Mir docking 100th Manned launch from Cape Canaveral Largest spacecraft (Shuttle–Mir) ever on-orbit First shuttle to land with different crew than it was launched with
70	STS-70	OV-103 21	22 Jun 95 (s) 08 Jun 95 (s) 13 Jul 95 22 Jul 95	194:13:41:55 203:12:02:02	KSC, Pad B KSC, 33	Terence T. Henricks (C) Kevin R. Kregel (P) Nancy J. Currie (MS1) Donald A. Thomas (MS2) Mary Ellen Weber (MS3)	184 28.45	4,520,654 258,798 44,232 196,575	TDRS-G/IUS MSX-01 PARE RMS = none	143 orbits 214 hrs 20 mins 07 secs ≈3,700,000 miles	25,789 430 211 8,465	First use of new Mission Control Center Last flight of Discovery before overhaul Quickest turnaround (STS-71 to STS-70) First Block I SSME flight Final TDRS deployment mission
71	STS-69	OV-105 9	05 Aug 95 (pd) 31 Aug 95 (pd) 07 Sep 95 18 Sep 95	250:15:09:00 261:11:37:56	KSC, Pad A KSC, 33	David M. Walker (C) Kenneth D. Cockrell (P) James S. Voss (PC) James H. Newman (MS2) Michael L. Gernhardt (MS3)	190 28.45	4,514,647 256,645 25,352 219,377	SPARTAN-201-03 Wake Shield RMS = s/n 303	170 Orbits 260 hrs 28 mins 56 secs ≈4,500,000 miles	25,799 220 215 10,230	First deployment and retrieval of two satellites on same mission
72	STS-73	OV-102 18	28 Sep 95 (pd) 05 Oct 95 (pd) 06 Oct 95 (pd) 07 Oct 95 (pd) 15 Oct 95 (pd) 19 Oct 95 (s) 20 Oct 95 05 Nov 95	293:13:53:00 309:11:45:22	KSC, Pad B KSC, 33	Kenneth D. Bowersox (C) Kent V. Rominger (P) Catherine G. 'Cady' Coleman (MS1) Michael E. Lopez-Alegria (MS2) Kathryn C. Thornton (PC) Fred W. Leslie (PS1) Albert Sacco, Jr. (PS2)	172 39.00	4,521,518 257,017 25,167 230,469	USML-2 OARE-06 STABLE RMS = none	255 orbits 381 hrs 52 mins 22 secs ≈6,600,000 miles	25,744 231 210 9,117	
73	STS-74	OV-104 15	11 Nov 95 (pd) 12 Nov 95 20 Nov 95	316:12:30:43 324:17:01:29	KSC, Pad A KSC, 33	Kenneth D. Cameron (C) James D. Halsell, Jr. (P) Chris A. Hadfield (MS1) Jerry L. Ross (MS2) William S. McArthur, Jr. (MS3)	249 51.60	4,512,395 274,560 13,525 202,951	Mir/ODS Russian Docking Module RMS = s/n 301	128 orbits 196 hrs 30 mins 46 secs ≈3,400,000 miles	25,840 612 203 8,691	Second Mir docking Russian Docking Adapter left at Mir
74	STS-72	OV-105 10	10 Jan 96 20 Jan 96	011:09:41:00 020:07:41:45	KSC, Pad B KSC, 15	Brian Duffy (C) Brent W. Jett, Jr. (P) Leroy Chiao (MS1) Winston E. Scott (MS2) Koichi Wakata (MS3) Daniel T. Barry (MS4)	288 28.45	4,518,376 258,391 19,351 222,934	SFU Retrieval SPARTAN/OAST RMS = s/n 303	142 orbits 214 hrs 00 mins 45 secs ≈3,700,000 miles	25,697 424 208 8,729	Retrieval of Japanese SFU launched 18 March 95 aboard an H-2
75	STS-75	OV-102 19	22 Feb 96 09 Mar 96	053:20:18:00 069:13:58:21	KSC, Pad B KSC, 33	Andrew M. Allen (C) Scott J. Horowitz (P) Jeffrey A. Hoffman (MS1) Maurizio Cheli (MS2) Claude Nicollier (MS3) Franklin R. Chang-Diaz (PC) Umberto Guidoni (PS1)	184 28.45	4,523,663 261,927 18,071 229,031	TSS-1R USMP-03 RITSI RMS = none	251 orbits 377 hrs 40 mins 21 secs ≈6,500,000 miles	25,823 679 212 8,460	Tether broke: TSS-1R lost

Launch (s) is scheduled date
Launch (pd) is delay on pad
Launch (abort) is an abort after SSME ignition
n/a indicates not applicable
—— indicates classified data or data not available
RMS = Remote Manipulator System serial number

Edwards AFB (EAFB) Runways 05R, 17L, 23L, and 33 are lakebed.
Edwards AFB (EAFB) Runways 04 and 22 are concrete.
White Sands Space Harbor (WSSH) Runway 17 is lakebed.
Kennedy Space Center (Shuttle Landing Facility) Runways 15 and 33 are concrete.
Orbital altitude is highest apogee
Times are GMT (day:hour:minute:second)

(C) = Shuttle Commander
(P) = Shuttle Pilot
(PC) = Payload Commander
(MS) = Mission Specialist
(PS) = Payload Specialist

Weights in pounds
Altitude and cross-range in statue miles
Inclination in degrees from Equator
Velocity in feet per second
Touchdown in knots indicated airspeed
Roll-out in feet after main gear touchdown

Endeavour at Edwards after STS-68 on 11 October 1994. Columbia is overhead on her way to Palmdale for OMDP-1. (NASA photo EC94-42789-4 by Jim Ross via Tony Landis)

Flight 51 – (STS-52) – The original target launch date in mid-October slipped when it was decided to replace the SSME #3 (2034 replaced 2038), prompted by concerns about possible cracks in the LH2 coolant manifold on the nozzle. Changing the engine at the pad was less complex than continued X-ray analysis of the suspect area. Lift-off was delayed 1 hour and 53 minutes due to crosswinds at KSC and clouds at the Banjul TAL site. Primary objectives were the deployment of Laser Geodynamic Satellite II (LAGEOS II) and operation of the U.S. Microgravity Payload-1 (USMP-1). LAGEOS II, a joint effort between NASA and the Italian Space Agency, was deployed on flight day two and boosted into an initial elliptical orbit by the Italian Research Interim Stage (IRIS), flying for first time.[42]

Flight 52 – (STS-53) – Lift-off was originally set for 06:59, but was delayed to allow sunlight to melt ice that had formed on the ET after tanking due to overnight temperatures in the upper-40s. This was the first flight of *Discovery* after an OMDP. In a small switch from normal procedures, after the classified DoD payload was deployed on first day, the remaining flight activities became unclassified. As usual, DoD had little comment. Acknowledged payloads included BLAST-01, CLOUDS-1A-03, CREAM-03, GCP, HERCULES-01, MIS-I-01, RME-III-09, STL-02, and VFT-2-02 on the mid-deck. The primary payload was probably the third SDS-B to be deployed by Space Shuttle (others on STS-28R and STS-38). The landing was diverted to Edwards due to clouds at KSC, but delayed one revolution because of high winds in the desert. Even the DoD can have a little fun – Navy Captain David Walker had been better known through his years of military service as 'Red Dog.' The pack of flight controllers, astronaut trainers, and astronauts involved in STS-53 eventually became known as the 'Dogs of War,' and the first 'Dog Crew' was born, although never officially acknowledged by NASA.[43]

Flight 53 – (STS-54) – Lift-off was delayed 7 minutes and 30 seconds due to unacceptable upper winds. The primary payload was the fifth Tracking and Data Relay Satellite (TDRS-F), deployed about six hours after lift-off. The attached IUS fired one hour later to propel TDRS-F into an intermediate checkout orbit. On flight day five, Runco and Harbaugh spent nearly five hours working in the open payload bay, performing a series of EVAs designed to increase NASA's knowledge of working in space. The two astronauts

tested their abilities to move about freely in payload bay, climb into foot restraints without using their hands, and simulated carrying large objects in the microgravity environment. Landing was rescheduled one orbit earlier due to an approaching weather front, eventually being waved off one orbit due to fog.[44]

Flight 54 – (STS-56) – The first launch attempt of *Discovery* on 6 April was halted at T–11 seconds by the onboard GPCs when a LH2 high-point bleed valve in the main propulsion system indicated closed instead of open. Later analysis indicated valve was properly configured and 48-hour scrub turnaround procedures were implemented. The final countdown on 8 April proceeded smoothly. The primary payload was the second Atmospheric Laboratory for Applications and Science (ATLAS-2), designed to collect data on relationship between Sun's energy output and Earth's middle atmosphere and how these factors affect the ozone layer. Six instruments were mounted on the Spacelab pallet in the payload bay, with a seventh mounted in two GAS canisters. On 11 April the crew used the remote manipulator arm to deploy the Shuttle Point Autonomous Research Tool for Astronomy (SPARTAN-201), a free-flying science instrument platform designed to study velocity and acceleration of solar wind and observe Sun's corona. The crew also made numerous radio contacts to schools around world using the second Shuttle Amateur Radio Experiment (SAREX II) and a brief radio contact with Russian *Mir* space station – the first such contact between Shuttle and *Mir* using amateur radio equipment. The landing was postponed one day due to weather conditions at KSC. This was the first use of a revised drag chute that incorporated a second reefing line to provide additional stability after deployment.[45]

Flight 55 – (STS-55) – This *Columbia* launch was first scheduled for 25 February, but slipped to early March after questions arose about turbine blade tip seal retainers in the high-pressure oxidizer turbopumps on all three SSMEs. When engineers could not verify whether old or new retainers were on *Columbia*, NASA opted to replace all three turbopumps at the pad as a precautionary measure. A revised launch date of 14 March slipped again after a hydraulic flex hose burst in the aft compartment. All 12 hydraulic lines were removed and inspected – nine lines were re-installed and three new lines put in. Launch was set for 21 March but pushed back 24 hours due to Eastern Range conflicts caused by a one-day Delta II launch delay. The launch attempt on 22 March was aborted at T–3 seconds by the on-board GPCs when SSME #3 failed to ignite completely – the LO2 preburner check valve leaked internally, causing an over-pressurization of the purge system, which in turn precluded full engine ignition. The was the first on-pad main engine abort since the return-to-flight, and only the third in program history (STS-26/51-F and STS-16/41-D were the others). The valve leak was later traced to contamination during manufacturing, and NASA decided to replace all three main engines on *Columbia*. Launch was rescheduled for 24 April, but scrubbed early on launch morning when one of three IMUs gave a faulty reading. Lift-off was postponed 48 hours to allow the replacement of the IMU, and the final launch countdown on 26 April proceeded smoothly. This was the last launch scheduled from Pad 39A until February 194 to allow for pad refurbishment and modification. The Spacelab D-2 payload was the second under German mission control. All communications with *Columbia* were lost for about hour and a half on 4 May due to an errant command from Mission Control, but no serious problems were encoun-

Discovery begins the 3.5-mile trip to Pad 39A for STS-60. The Launch Control Center (LCC) may be seen in the background to the left of the MLP. Note that the VAB high bay doors may be opened in seemingly endless combinations, and the upper sections are frequently open for ventilation. (NASA photo KSC-94PC-0108)

tered. On 2 May, NASA determined that sufficient electrical power remained to extend the flight by one day to allow additional experiments to be conducted. The landing was diverted to Edwards because of low cloud ceiling and moisture in the air at KSC.[46]

Flight 56 – (STS-57) – Launch was originally targeted for 18 May, but rescheduled to 3 June to allow both lift-off and landing to occur in daylight. The 3 June launch was slipped when NASA decided to replace the high-pressure oxidizer turbopump on SSME #2 after concerns arose over a misplaced penetration verification stamp. The launch attempt on 20 June was scrubbed at T–5 minutes due to low clouds and rain at KSC, and weather concerns at all three TAL sites. STS-57 marked the first flight of the commercially-developed Spacehab pressurized laboratory – altogether 22 experiments were flown, covering materials, life sciences, and wastewater recycling for space station. On 24 June the crew captured the EURECA satellite deployed on STS-46. However, EURECA ground controllers were unable to stow the spacecraft's two antennas, and on 25 June, Low and Wisoff spent the beginning of a scheduled EVA manually folding the antennas. The remainder of the 5-hour, 50-minute EVA was spent on planned tasks. Landing attempts on 29 and 30 June were waved off due to unacceptable cloud cover and rain showers at KSC – STS-32/61-C in 1986 was last time there were two landing wave-offs. *Endeavour* burned 330 pounds more hypergolic propellant during reentry than planned, prompting an increase in Orbiter reserves for all future missions. After landing, the STS-57 crew in *Endeavour* talked to the STS-51 crew in *Discovery* at Pad 39B, marking the first orbiter-to-orbiter crew conversation since the orbiting STS-23/51-D crew talked to the STS-24/51-B crew at KSC in 1985.[47]

Flight 57 – (STS-51) – The first launch attempt on 17 July was scrubbed during the T–20 minute hold due to premature and unexplained charging of the pyrotechnic initiator controllers (PIC) located on the MLP that controlled the T–0 LH2 vent arm umbilical and SRB holddown bolts. The problem was traced to a faulty circuit card in the PIC rack on MLP. The second launch attempt on 24 July was halted at T–19 seconds due to problems with the APU turbine assembly for one of two hydraulic power units (HPU) on the right SRB. The APU was removed and replaced at the pad. Launch was rescheduled for 4 August then changed to 12 August to avoid the Perseid meteor shower, which was expected to peak on 11 August. The 12 August launch attempt was halted at T–3 seconds (after main engine ignition) due to a faulty sensor monitoring the fuel flow on SSME #2. This was the fourth pad abort in Space Shuttle history and the second during 1993. All three main engines were changed at the pad. Launch was rescheduled to 10 September, then slipped to 12 September to allow time to complete review of Advanced Communications Technology Satellite (ACTS) design, production, and testing history following loss of contact with the *Mars Observer* and NOAA-13 satellite (which used a similar design). The 12 September launch attempt proceeded smoothly and ACTS was deployed on flight day one. On flight day two, the crew deployed the second primary payload, the Orbiting and Retrievable Far and Extreme Ultraviolet Spectrograph-Shuttle Pallet Satellite (ORFEUS-SPAS), first in a series of ASTRO-SPAS astronomical missions. After six days of data collection, ORFEUS-SPAS was retrieved with the manipulator arm and returned to payload bay. On 16 September, Newman and Walz performed an EVA lasting 7 hours, 5 minutes,

Atlantis on the launch pad. Note the protective covers over the Orbiter windscreens and the black photo-reference markings on the SRBs. The proximity to the Atlantic Ocean is more evident here than usual – the pads are very close to the beach. (NASA)

and 28 seconds to evaluate tools, tethers, and foot restraint platforms intended for the upcoming *Hubble Space Telescope* servicing mission. The two landing opportunities that were available on 21 September were waved off due to isolated rain showers in central Florida. The landing on the 22nd marked the first KSC night landing; after landing, plumes were visible from the venting of APUs #1 and #2.[48]

Flight 58 – (STS-58) – The first launch attempt on 14 October was delayed 2 hours by bad weather. When it cleared and the count resumed, a failure in an Air Force range safety command message encoder verifier cancelled the launch at the T–31 second mark. This system is used to transmit a vehicle destruct signal if it should become necessary. The second launch attempt on 15 October was scrubbed at T–9 minutes when one of the two S-Band communication transponders failed onboard the Orbiter – flight rules require that both transponders be functional for launch. An 18 October countdown proceeded smoothly to lift-off, delayed only by several seconds because of an unauthorized aircraft in the launch area. During recovery of the SRBs, engineers observed one of the four forward booster separation motor covers was missing from the right-hand booster. These covers protect the motors that are used to separate the SRBs from the ET after the boosters have been expended. Past occurrences of missing forward separation motor covers (STS-28R and STS-48) had been found to occur during SRB descent, frustum water impact, or frustum retrieval from the ocean when parachute lines become entangled with the doors and are not considered flight safety issues. This was the second dedicated Spacelab Life Sciences mission (SLS-2). Fourteen experiments were conducted in four primary areas – regulatory physiology, cardiovascular/cardiopulmonary, musculoskeletal, and neuroscience.[49]

Flight 59 – (STS-61) – Launch of the first *Hubble* Space Telescope service mission was originally scheduled to occur from Pad 39A, but after roll-out, contamination was found in the Payload Changeout Room and a decision was made to move to Pad 39B. The contamination appeared to have been caused by sandblasting grit from recent Pad A modifications, internal HST payload package was not affected because it was tightly sealed. Roll-around occurred on 15 November, but on 18 November *Endeavour* experienced a failure of a delta-p transducer on an elevon hydraulic actuator. Changing the actuator would have required a roll-back to the OPF because access to the actuator is through the main landing gear wheel well. Since there are 4 delta-P transducers and only 3 are required for launch, the failed transducer was disconnected and not used during the flight. The first launch attempt on 1 December was scrubbed due to weather conditions at KSC. Just before the scrub the range was also in a no-go situation due to a ship in restricted area. The final flight of 1993 was one of the most challenging and complex manned missions ever attempted. During a record five EVAs totaling 35 hours and 28 minutes, two teams of astronauts completed the first servicing of the HST. In many instances, tasks were completed sooner than expected and the few contingencies that did arise were handled smoothly. The repairs to *Hubble* were generally praised as restoring the telescope to better than anticipated condition, leading to several exciting discoveries.[50]

Flight 60 – (STS-60) – This *Discovery* mission marked the first flight of a Russian cosmonaut on Shuttle as one of the first elements in implementing the Agreement on

NASA/Russian Space Agency Cooperation in Human Space Flight. Sergei Krikalov's back-up was Vladimir G. Titov. This mission also marked the second flight of Spacehab pressurized module and the 100th GAS payload to fly in space. Also on board was the first Wake Shield Facility (WSF-1), making its first in a planned series of flights. The experiment took advantage of the near vacuum of space to grow innovative thin film materials for use in electronics. The WSF was deployed by the remote manipulator arm, flew in formation with *Discovery* at a distance of up to 46 miles from the Orbiter for 56 hours, then was retrieved and stowed in the payload bay.[51]

Flight 61 – (STS-62) – Launch was originally set for 3 March, but postponed at T–11 hours because of predicted unfavorable weather in the KSC area. The countdown on 4 March proceeded smoothly, and the only deviation to normal was a delay in deploying the SRB recovery ships because of high seas. The two recovery ships left port on launch day, and recovered the boosters and their parachutes on 6 March. USMP-2 included five experiments investigating materials processing and crystal growth in microgravity, while OAST-2 featured six experiments focusing on space technology and space flight. USMP-2 experiments received emphasis at the beginning of the flight, and later in the mission *Columbia*'s orbit was lowered about 230 miles to facilitate OAST-2 experiments. The crew also conducted a number of biomedical activities aimed at better understanding and countering the effects of prolonged space flight.[52]

Flight 62 – (STS-59) – A launch attempt originally set for 7 April was postponed at T–27 hours to allow inspections of metallic vanes in the SSME high pressure oxidizer preburner pumps – similar pumps were found to have anom-

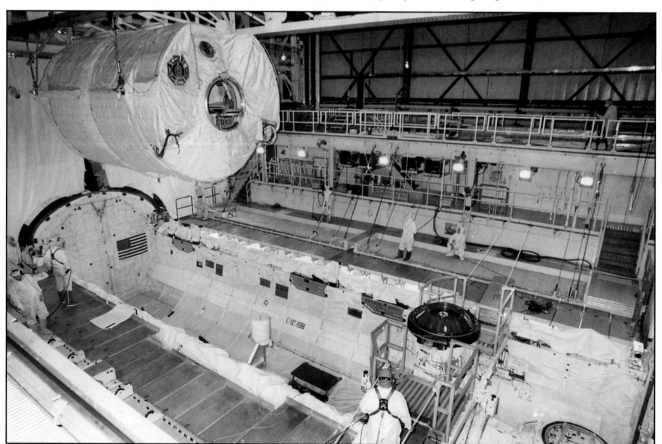

The STS-71 Spacelab/Mir module being lowered into the payload bay of Atlantis in OPF-3. The bright-red cover on the Russian-built androgynous peripheral docking system (APDS) on top of the Orbiter Docking System (ODS) is at the lower right-center of the photo. Spacelab would serve as a medical laboratory during the mission. (NASA photo KSC-95PC-453)

alies at Rocketdyne. *Endeavour*'s launch on 8 April was scrubbed at T–9 minutes due to weather, but the count proceeded smoothly the following day. The German Space Agency (DARA) and the Italian Space Agency (ASI) SIR-C/X-SAR imaged approximately 38.5 million square miles of the Earth, the equivalent of 20 percent of the planet. This mission also marked the first flight of toughened uni-piece fibrous insulation (TUFI), an improved thermal protection tile – several test tiles were placed on the Orbiter's base heat shield between the three main engines. The mission was extended one day following payload technical difficulties on flight day one, and on 19 April the mission was extended another day due to low cloud ceiling at KSC – two attempts had already been waved off. On 20 April, the first attempt was waved off from KSC due to weather concerns, and the landing was diverted to Edwards.[53]

<u>Flight 63</u> – (STS-65) – This *Columbia* mission marked the second flight of the International Microgravity Laboratory (IML-2), carrying more than twice the number of experiments and facilities as IML-1. More than 80 experiments, representing more than 200 scientists from 6 space agencies, were located in the Spacelab module. Chiaki Mukai became the first Japanese woman to fly in space, and she also set a record for the longest flight to date by a female astronaut. Mukai's back-up was Jean Jacques Favier from France. The first landing attempt on 22 July was waved off due to low clouds, winds, and heavy offshore rains within 30 miles of KSC. STS-65 was *Columbia*'s last mission before scheduled modification and refurbishment at Rockwell's Palmdale facility. OV-102 departed for California atop the SCA on 8 October 1994.[54]

<u>Flight 64</u> – (STS-64) – This *Discovery* mission marked the first flight of Lidar In-space Technology Experiment (LITE) and the first untethered EVA in 10 years. The LITE payload employed a laser optical radar (lidar) to study Earth's atmosphere as part of the Mission to Planet Earth. The 2.5-hour launch window opened at 16:30 EDT, and the late afternoon launch was scheduled to permit night-time operation of the LITE laser early in the mission. The hold at T–9 minutes was extended due to heavy cloud cover, and a there was a delay at T–5 minutes because of potential rain over the SLF – total hold time was 1 hour, 52 minutes, 55 seconds. The LITE instrument operated for 53 hours, yielding more than 43 hours of high-rate data. Unprecedented views were obtained of cloud structures, storm systems, dust clouds, pollutants, forest burning, and surface reflectance. The instrument studied the atmosphere above northern Europe, Indonesia, the south Pacific, Russia, and Africa. Sixty-five groups from 20 countries subsequently made validation measurements with ground- and aircraft-based instruments to verify the LITE data. The mission was extended by one day to allow extra LITE operations. Four landing attempts were waved-off, two each on 19 and 20 September due to bad weather at KSC – the landing was finally diverted to Edwards.[55]

<u>Flight 65</u> – (STS-68) – The first launch attempt of *Endeavour* on 18 August was halted at T–1.9 seconds when the on-board GPCs shut down all three main engines after detecting an unacceptably high discharge temperature in the SSME #3 high-pressure oxidizer turbopump turbine. This was the fifth on-pad abort of the program. *Endeavour* was returned to the VAB and all three engines were replaced. The countdown for the second launch attempt proceeded smoothly to an on-time lift-off. This marked the second flight in 1994 of the Space Radar Laboratory (first flight was STS-59 in April). Flying SRL during different seasons allowed comparing of changes between the first and

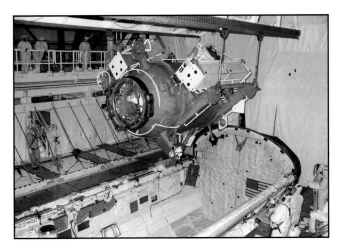

The Russian-built Docking Module (DM) is lowered for installation into the payload bay of Atlantis while in OPF-2. The docking module was left permanently attached to Mir for use during future missions.The white structures attached to the module's sides are solar panels that were attached to Mir during the mission. (NASA photo KSC-95PC-1324)

second flights. On flight day 6 the mission was extended one day to gather additional science data. The maneuvering capability of Orbiter was demonstrated as *Endeavour* was piloted to within 30 feet of where it had flown on the first Space Radar mission in April. A KSC landing attempt was waved off prior to the deorbit burn due to low clouds and impending rain, and a successful landing was made at Edwards later the same day.[56]

<u>Flight 66</u> – (STS-66) – This was the first *Atlantis* flight since her OMDP. The launch was originally scheduled for 27 October but the need to refurbish three engines after the RSLS abort of the initial launch attempt of STS-68 caused a one week delay. During launch processing one of *Columbia*'s overhead windows was used to replace one on *Atlantis* that was found to have a tiny scratches. Other concerns included a check of *Atlantis*' plumbing after a water leak onboard *Endeavour* during the landing of STS-68. Lift-off was delayed slightly waiting for weather to clear at the TAL sites. CRISTA-SPAS was released from the Orbiter on the second day of the mission. Flying at a distance of 25–44 miles behind *Atlantis*, the payload collected data for more than eight days before being retrieved and returned to the payload bay. The CRISTA instruments gathered the first global information about medium- and small-scale disturbances in trace gases in the middle atmosphere, which could lead to better models of the atmosphere and Earth's energy balance. Landing was diverted to Edwards due to high winds, rain, and clouds at KSC caused by Tropical Storm Gordon. This was the fourth diverted landing of 1994, and the third in a row.[57]

<u>Flight 67</u> – (STS-63) – Minor adjustments were made to the countdown sequence to better accommodate the extremely short five-minute launch window required for the first rendezvous with *Mir* in a 51.6-degree orbit. The first launch attempt on 2 February was postponed at L–1 day when one of three IMUs failed. This mission included the first flight of a female Shuttle pilot, the second flight of a Russian cosmonaut on Shuttle, and the first approach and fly-around of *Mir*. Beginning on flight day one, a series of RCS burns brought *Discovery* in line with *Mir* – the original plan called for *Discovery* to approach no closer than 32.8 feet from *Mir*, and then complete a fly-around of the Russian space station. However, 3 of 44 RCS thrusters sprang leaks prior to rendezvous. Shortly after MECO, two leaks occurred in the aft primary thrusters, one of which – R1U – was key to the rendezvous. The third

leak occurred later on-orbit in a forward primary thruster, but the crew was able to fix the problem. After extensive negotiations and technical information exchanges between U.S. and Russian space teams, the Russians concluded a close approach could still be safely achieved, and the STS-63 crew given a 'go' to proceed with rendezvous. The R1U thruster manifold was closed and a backup thruster selected for the approach. After station-keeping at a distance of 400 feet from *Mir*, and with James Wetherbee manually controlling the Orbiter, *Discovery* flew to within 37 feet of the Russian space station. "As we are bringing our spaceships closer together, we are bringing our nations closer together," Wetherbee said after *Discovery* was at the point of closest approach. "The next time we approach, we will shake your hand and together we will lead our world into the next millennium." After landing, the cosmonauts aboard *Mir* radioed their congratulations to the *Discovery* crew. Titov's backup for the mission was Sergei Krikalov.[58]

Flight 68 – (STS-67) – Lift-off was delayed for about a minute due to concerns about a heater system on the flash evaporator system – a backup heater was selected instead. *Endeavour* logged 6.9 million miles in completing longest Shuttle flight to date, allowing sustained examination of the 'hidden universe' of ultraviolet light. The primary payload had flown once before – on STS-35 in December 1990 – but this second flight had almost twice the duration. Planned ASTRO-2 observations built on discoveries made by ASTRO-1, as well as seeking answers to other questions. STS-67 became the first advertised Shuttle mission connected to Internet. Users of more than 200,000 computers from 59 countries logged onto the ASTRO-2 home page at MSFC, and more than 2.4 million requests were recorded during the mission, many answered by the crew on-orbit. On 17 March, the mission was extended by one day due to poor weather conditions at KSC and Edwards. The planned landing at KSC the following day was waved off due to winds, clouds, and potential rain and the Orbiter diverted to Edwards. Scott D. Vangen was the alternate payload specialist.[59]

Flight 69 – (STS-71) – This launch of *Atlantis* was originally targeted for late May, but slipped into June to accommodate Russian activities necessary for the first Space Shuttle–Mir docking, including a series of spacewalks to reconfigure *Mir* for docking and the launch of the new *Spektr* module containing U.S. research hardware. The launch set for 23 June was scrubbed when rain and lightning prevented loading the ET earlier that day. The second try on 24 June was scrubbed at T-9 minutes, again due to persistent stormy weather in central Florida, coupled with the short 10-minute launch window. Lift-off was rescheduled for 27 June, and the final countdown proceeded smoothly. STS-71 marked a number of firsts – the 100th U.S. human space launch conducted from Cape Canaveral, the first Shuttle–Mir docking and joint on-orbit operations, and the first on-orbit changeout of a Space Shuttle crew. Docking occurred at 09:00 EDT on 29 June using an R-Bar (radius vector; i.e., from above or below), approach with *Atlantis* closing in on *Mir* from directly below. R-bar approaches allow natural forces to brake the Orbiter more than would occur with a V-bar (velocity vector; i.e., from in front or behind) approach directly in front of space station. The manual phase of docking began with *Atlantis* about a half-mile below *Mir* – Gibson was at the controls on the aft flight deck. Stationkeeping was performed when the Orbiter was about 250 feet from *Mir*, pending approval from Russian and U.S. flight directors to proceed. Gibson then maneuvered the Orbiter to a point about 30 feet from *Mir* before beginning his final approach to the station. The closing rate was targeted at 0.1 fps, and the actual closing velocity was approximately 0.107 fps at contact. Interface contact was nearly flawless – less than one inch lateral misalignment and an angular misalignment of less than 0.5-degrees per axis. Docking occurred about 249 miles above the Lake Baykal region of the Russian Federation. The Orbiter docking system (ODS) androgynous port served as the actual connection to a similar interface on the docking port on the *Krystall* module. When linked, *Atlantis* and *Mir* formed largest spacecraft ever in orbit, with a total mass of almost 500,000 pounds. The spacecraft undocked on 4 July, but just prior to the undocking, the Mir-19 crew temporarily flew away in their Soyuz spacecraft so they could photograph *Atlantis* and *Mir* separating. The Soyuz unlatched at 06:55, and Gibson undocked *Atlantis* from *Mir* at 07:10. The returning crew of eight equaled the largest crew (STS-30/61-A) in shuttle history. To ease their reentry into gravity environment after more than 100 days in space, Mir-18 crew members Thagard, Dezhurov, and Strekalov lay supine in custom-made Russian seats installed in the Orbiter mid-deck. Thagard had just logged the longest American space flight.[60]

Flight 70 – (STS-70) – Lift-off was first targeted for 22 June, however, due to various Russian delays affecting STS-71, NASA opted to swap the STS-70 and STS-71 launch dates, and accelerated processing to ready *Discovery* and her payloads for lift-off no earlier than 8 June, with *Atlantis* to follow as STS-71 later in June. This schedule was thrown off following an extended Memorial Day weekend, when Northern Flicker Woodpeckers poked about 200 holes in the foam insulation on *Discovery*'s External Tank at Pad 39B. Attempts to repair the damage at pad were unsuccessful, and the stack returned to the VAB on 8 June, with a new launch date set for 13 July. The holes ranged in size from large excavations about four inches in diameter to single pecks and claw marks. The launch on 13 July proceeded smoothly, with only a 55 second hold at T-31 seconds while a possible ET range safety system problem was looked into. The primary objective was the deployment of the seventh Tracking Data and Relay Satellite (TDRS), which became the sixth placed in operational use (the second TRDS was lost with *Challenger*). This marked the maiden flight of a Block I SSME (2036 in position 1) that featured a new high-pressure liquid oxygen turbopump, two-duct powerhead, baffleless main injector, single-coil heat exchanger, and start sequence modifications. Both KSC landing opportunities on 21 July were waved off due to ground fog, as was the first opportunity the following day. The weather finally cooperated on the second opportunity on 22 July and *Discovery* landed from her final flight prior to being shipped to California for OMDP.[61]

Flight 71 – (STS-69) – The original launch date for STS-69 was 5 August, but would be delayed by several events. On 1 August *Endeavour* was rolled-back to the VAB for protection from Hurricane Erin which was threatening the central Florida coast. Subsequently, the vehicle was returned to Pad 39A on 8 August, but NASA became concerned over minor SRB O-ring erosion that had been detected during STS-70 and STS-71 and delayed the flight until the issue could be resolved. After engineers decided the erosion was not a concern, the launch date was set for 31 August, but was scrubbed prior to tanking due to a higher than allowable temperature in fuel cell #2, which was subsequently replaced. Lift-off on 7 September was preceded by a smooth countdown. STS-69 marked first time two different payloads were retrieved during same mission, and also featured an EVA to practice for space station activities and to evaluate space suit design modifications. This was second flight of a 'dog crew,' a flight crew tradition that began on STS-53, on which both Walker and Voss flew. Each STS-69 crew member adopted a dogtag or nickname: Walker was 'Red Dog;' Cockrell was 'Cujo;' Voss, 'Dog Face;' Newman, 'Pluto;' and Gernhardt, 'Under Dog.'[62]

Flight 72 – (STS-73) – The launch of *Columbia* was originally scheduled for 28 September, but was scrubbed prior to the crew ingressing when the main fuel valve on SSME #1 began to leak. The attempt on 5 October 1995 was scrubbed due to bad weather from Hurricane Opal, and one the next day because of a hydraulic problem with the nose wheel steering system. The count on 7 October was halted by a failure in a Main Engine Controller. An attempt on 15 October was scrubbed due to weather, and another attempt could not be made until the 20th since an Atlas had priority for the Eastern Range. On launch day, a fire alarm went off on the 155-foot level of the launch pad while the crew was on the pad. A quick investigation showed no fire, and the crew boarded the vehicle and the count continued, resulting in a launch three minutes later than scheduled. The second U.S. Microgravity Laboratory (USML-2) mission built upon a foundation begun on the original USML-1 aboard STS-50. The mission provided new insights into theoretical models of fluid physics, the role of gravity in combustion and flame spreading, and how gravity affects the formation of semiconductor crystals.[63]

Flight 73 – (STS-74) – An initial launch attempt on 11 November was scrubbed due to poor weather at the TAL sites. *Atlantis* lifted-off the next day at the beginning of a 10 minute window. STS-74 was the second of seven scheduled *Mir* missions, and delivered the Russian Docking Module and a pair of solar arrays. The mission marked the first time astronauts from ESA, Canada, Russia, and the U.S. were in space on the same complex at the same time. The Russian Docking Module was lifted from the payload bay and attached to the Orbiter Docking System at the forward end of the payload bay. *Atlantis* then docked to the *Krystall* module for three days. When OV-104 departed, the Russian Docking Module remained attached to *Krystall* to provide additional clearance between Shuttle and *Mir*'s solar arrays on future docking flights. *Atlantis* also delivered water, supplies, equipment, and two solar arrays to *Mir*. Various experiment samples, products manufactured aboard *Mir*, and some equipment that needed repaired were returned to Earth.[64]

Flight 74 – (STS-72) – *Endeavour* was launched 23 minutes late due to minor communication link difficulties. The launch window was 49 minutes long, based primarily on the position of the Japanese Space Flyer Unit that Japan had paid NASA $50 million to recover. This mission marked the first reflight of a Block I SSME (2036 in position 3), and was also the first flight with two of the new engines. Two extravehicular activities (EVA) were conducted as part of a continuing series to prepare for on-orbit construction of the International Space Station. On flight day five the first EVA lasted 6 hours and 9 minutes – Chiao and Barry evaluated a new portable work platform and a structure known as the rigid umbilical, which might be used on the station to hold various fluid and electrical lines. The evaluation was continued during a 6-hour and 53-minute second EVA on flight day seven.

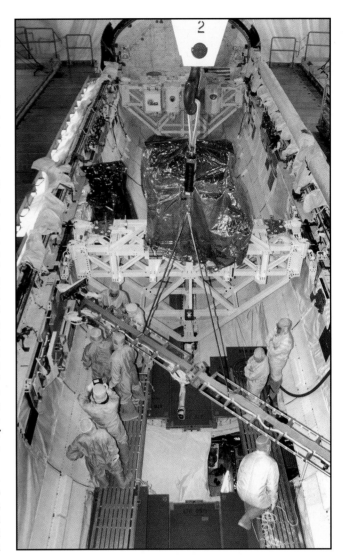

Members of the STS-72 flight crew are assisted by technicians as they look over payload bay hardware during the Crew Equipment Interface Test (CEIT) in the OPF. The structure suspended above Endeavour's payload bay is an international space station-type rigid umbilical that the astronauts would test during an EVA. (NASA photo KSC-95PC-1726)

Scott also evaluated the spacesuit's warmth in severe cold up to minus 104 degF.[65]

Flight 75 – (STS-75) – About 4 seconds after lift-off, instruments showed that one of *Columbia*'s SSMEs was operating at only 45 percent of nominal power. It was quickly determined that the problem was in the instrumentation and not in the engine and the ascent proceeded smoothly. On the fourth flight day, operation of the Tethered Satellite System (TSS) began. After TSS-1R had deployed 12.24 of an expected 13 miles, the 0.1-inch diameter tether broke, resulting in the loss of the Italian satellite. Before the tether broke, electrical voltages of 3,500 volts and 500 milliamps were recorded, largely validating the experiment. NASA decided against trying to intercept and retrieve the satellite based primarily on OV-102's propellant reserves and possible dangers to the Orbiter from the electrically-charged satellite. This flight represented the second attempt to conduct the TSS experiment, the first ending in failure aboard STS-46 in 1992. Four landing opportunities (2 at KSC and 2 at Edwards) were waved off due to weather on 8 March, forcing a one day extension. On 9 March, the first available KSC opportunity was waved off due to cloud cover over the SLF, but the second opportunity was successful.[66]

As the 100th Space Shuttle flight approached, NASA was finally realizing its dream of building a space station – in fact, STS-92, the 100th mission, was an International Space Station Assembly flight. The dismal failure of all attempts to replace shuttle, due as much to a lack of political will than any particular technology, guarantees that Space Shuttle will continue to fly until at least 2010, and more probably 2012 – at least barring any major accidents. Recent and planned upgrades are making the vehicle much safer, and recent budget increases have allowed more personnel to be hired at the launch site, in theory increasing safety.[120]

Seq. #	Mission #	Orbiter Flight #	DATES Launch Landing	TIME Launch Landing	SITES Launch Landing	Crew	ORBIT Altitude Inclination	WEIGHTS Lift-Off OV Launch Payload OV Landing	Payloads	Mission Duration	REENTRY Velocity Cross-Range Touchdown Roll-out	Notes
76	STS-76	OV-104 16	22 Mar 96 31 Mar 96	082:08:13:04 091:13:28:57	KSC, Pad B EAFB, 22	Kevin P. Chilton (C) Richard A. Searfoss (P) Ronald M. Sega (MS1) Michael R. 'Rich' Clifford (MS2) Linda M. Godwin (MS3) Shannon W. Lucid (MS4) (up)	249 51.60	4,510,934 246,337 14,152 209,993	Mir/ODS Spacehab-SM Mir logistics RMS = none	145 Orbits 221 hrs 15 mins 53 secs ≈3,800,000 miles	25,898 878 204 8,460	Third Mir docking First flight of pressurized Spacehab to Mir Orbiter returned to KSC on 12 Apr 96
77	STS-77	OV-105 11	19 May 96 29 May 96	140:10:30:00 150:11:09:24	KSC, Pad B KSC, 33	John H. Casper (C) Curtis L. Brown. Jr. (P) Andrew S. W. Thomas (MS1) Daniel W. Bursch (MS2) Mario Runco, Jr. (MS3) Marc Garneau (MS4)	177 39.00	4,517,872 254,891 27,393 221,828	Spacehab-4 IAE SPARTAN-207 RMS = s/n 301	160 Orbits 240 hrs 39 mins 24 secs ≈4,100,000 miles	25,763 361 215 9,290	
78	STS-78	OV-102 20	20 Jun 96 07 Jul 96	172:14:49:00 189:12:36:36	KSC, Pad B KSC 33	Terence T. Henricks (C) Kevin R. Kregel (P) Richard M. Linnehan (MS1) Susan J. Helms (PC) Charles E. Brady, Jr. (MS3) Jean-Jacques Favier (PS1) Robert Brent Thirsk (PS2)	173 39.00	4,516,637 256,145 23,666 228,545	Spacelab LMS-1 RMS = none	271 Orbits 405 hrs 47 mins 36 secs ≈7,000,000 miles	25,749 105 214 9,335	Longest Shuttle flight to date Last flight with ET range safety package First live video during ascent First live video during reentry
79	STS-79	OV-104 17	31 Jul 96 (s) 14 Sep 96 (s) 16 Sep 96 26 Sep 96	260:08:54:49 271:12:13:13	KSC, Pad A KSC, 15	William F. Readdy (C) Terrence W. Wilcutt (P) Jerome 'Jay' Apt (MS1) Thomas D. Akers (MS2) Carl E. Walz (MS3) John E. Blaha (MS4) (up) Shannon W. Lucid (MS4) (down)	245 51.60	4,511,714 249,328 19,039 215,027	Mir/ODS Spacehab-DM Mir logistics RMS = none	159 Orbits 243 hrs 18 mins 24 secs ≈3,900,000 miles	25,872 894 217 10,981	Fourth Mir docking Lucid set U.S. record for time in space First flight of Spacehab double module Two roll-backs SRBs replaced prior to launch First flight without ET range safety package
80	STS-80	OV-102 21	30 Oct 96 (s) 07 Nov 96 (s) 15 Nov 96 (s) 19 Nov 96 07 Dec 96	324:19:55:47 342:11:49:06	KSC, Pad B KSC, 33	Kenneth D. Cockrell (C) Kent V. Rominger (P) Tamara E. Jernigan (PC) Thomas D. Jones (MS2) F. Story Musgrave (MS3)	219 28.45	4,522,305 260,935 21,422 227,383	ORFEUS-SPAS-II WSF-3 RMS = s/n 202	277 Orbits 423 hrs 53 mins 19 secs ≈7,000,000 miles	25,877 83 210 8,705	Longest Shuttle flight to date Musgrave sets two spaceflight records (oldest person and most shuttle flights) First time two free-flyers have been deployed and retrieved
81	STS-81	OV-104 18	12 Jan 97 22 Jan 97	012:09:27:23 022:14:22:46	KSC, Pad B KSC, 33	Michael A. Baker (C) Brent W. Jett, Jr. (P) Peter J.K. 'Jeff' Wisoff (MS1) John M. Grunsfeld (MS2) Marsha S. Ivins (MS3) Jerry M. Linenger (MS4) (up) John E. Blaha (MS4) (down)	245 51.60	4,511,710 249,936 19,156 214,711	Mir/ODS Spacehab-DM Mir logistics RMS = none	160 Orbits 244 hrs 55 mins 23 secs ≈4,100,000 miles	25,891 39 199 9,417	Fifth Mir docking First crop of wheat grown in space
82	STS-82	OV-103 22	13 Feb 97 (s) 11 Feb 97 21 Feb 97	042:08:55:17 052:08:32:24	KSC, Pad A KSC, 15	Kenneth D. Bowersox (C) Scott J. Horowitz (P) Joseph R. Tanner (MS1) Steven A. Hawley (MS2) Gregory J. Harbaugh (MS3) Mark C. Lee (PC) Steven L. Smith (MS5)	368/385 28.45	4,513,793 251,238 16,497 213,486	HST service RMS = s/n 301	149 Orbits 239 hrs 37 mins 07 secs ≈4,000,000 miles	26,120 557 184 7,073	Second HST servicing mission Highest Shuttle flight to date
83	STS-83	OV-102 22	03 Apr 97 (s) 04 Apr 97 08 Apr 97	094:19:20:32 098:18:33:11	KSC, Pad A KSC, 33	James D. Halsell, Jr. (C) Susan L. Still (P) Janice E. Voss (PC) Michael L. Gernhardt (MS2) Donald A. Thomas (MS3) Roger K. Crouch (PS1) Gregory T. Linteris (PS2)	184 28.45	4,521,943 259,144 25,082 230,276	Spacelab MSL-1 EXPRESS RMS = none	63 Orbits 95 hrs 12 mins 39 secs ≈1,500,000 miles	25,791 64 193 8,623	First mission to end early since 1991 (STS-2 and STS-44 were others)
84	STS-84	OV-104 19	15 May 97 24 May 97	135:08:07:48 144:12:23:33	KSC, Pad A KSC, 33	Charles J. Precourt (C) Eileen M. Collins (P) Jean-Francois Clervoy (PC) Carlos I. Noriega (MS2) Edward Tsang Lu (MS3) Elena V. Kondakova (MS4) C. Michael Foale (MS5) (up) Jerry M. Linenger (MS5) (down)	245 51.60	4,511,290 249,462 28,329 213,547	Mir/ODS Spacehab-DM Mir logistics RMS = none	143 Orbits 221 hrs 19 mins 56 secs ≈3,700,000 miles	25,906 36 210 8,202	Sixth Mir docking
85	STS-94	OV-102 23	01 Jul 97 17 Jul 97	182:18:02:00 198:10:46:36	KSC, Pad A KSC, 33	James D. Halsell, Jr. (C) Susan L. Still (P) Janice E. Voss (PC) Michael L. Gernhardt (MS2) Donald A. Thomas (MS3) Roger K. Crouch (PS1) Gregory T. Linteris (PS2)	184 28.45	4,519,811 260,249 25,556 230,515	Spacelab MSL-1R EXPRESS RMS = none	250 Orbits 370 hrs 44 mins 36 secs ≈6,200,000 miles	25,792 94 208 8,910	First reflight of same vehicle, crew, and payload Fastest turnaround of mission to date Was known as STS-83R during planning Reflight of MSL-1 from STS-83
86	STS-85	OV-103 23	07 Aug 97 19 Aug 97	219:14:41:00 231:11:08:00	KSC, Pad A KSC, 33	Curtis L. Brown, Jr. (C) Kent V. Rominger (P) N. Jan Davis (PC) Robert L. Curbeam, Jr. (MS2) Stephen K. Robinson (MS3) Bjarni V. Tryggvason (PS1)	184 57.00	4,511,117 249,696 24,933 219,077	CRISTA-SPAS-02 TAS-01 IEH-02 Shuttle Glow 5/6 RMS = s/n 301	189 Orbits 284 hrs 27 mins 00 secs ≈4,700,000 miles	25,755 398 185 8,745	Jeffrey S. Ashby had been replaced as pilot due to his promotion
87	STS-86	OV-104 20	25 Sep 97 06 Oct 97	269:02:34:19 279:21:55:12	KSC, Pad A KSC, 15	James D. Wetherbee (C) Michael J. Bloomfield (P) Vladimir Georgievich Titov (MS1) Scott E. Parazynski (MS2) Jean-Loup J.M. Chrétien (MS3) Wendy B. Lawrence (MS4) David A. Wolf (MS5) (up) C. Michael Foale (MS5) (down)	245 51.60	4,513,814 252,035 21,293 214,509	Mir/ODS Spacehab-DM Mir logistics RMS = none	169 Orbits 259 hrs 20 mins 53 secs ≈4,400,000 miles	25,989 432 198 11,947	Seventh Mir docking First joint U.S.-Russian EVA Last flight of Atlantis prior to OMDP
88	STS-87	OV-102 24	19 Nov 97 05 Dec 97	323:19:46:00 339:12:20:02	KSC, Pad B KSC, 33	Kevin R. Kregel (C) Steven W. Lindsey (P) Kalpana Chawla (MS1) Winston E. Scott (MS2) Takao Doi (MS3) Leonid K. Kadenyuk (PS2)	172 28.45	4,521,645 260,799 22,130 230,110	USMP-4 SPARTAN-201-04 CUE (GAS) RMS = s/n 301	251 Orbits 376 hrs 34 mins 02 secs ≈6,500,000 miles	25,670 76 188 8,003	First 'roll-to-heads-up' trajectory First Japanese citizen to EVA (Doi)
89	STS-89	OV-105 12	22 Jan 98 31 Jan 98	023:02:48:15 031:22:35:10	KSC, Pad A KSC, 15	Terrence W. Wilcutt (C) Joe Frank Edwards, Jr. (P) James F. Reilly (MS1) Michael P. Anderson (MS2) Bonnie J. Dunbar (PC) Shakirovich Sharipov Salizhan (MS4) Andrew S. W. Thomas (MS5) (up) David A. Wolf (MS5) (down)	238 51.60	4,512,426 252,316 21,849 216,802	Mir/ODS Spacehab-DM Mir logistics RMS = none	138 Orbits 211 hrs 46 mins 55secs ≈3,600,000 miles	25,900 690 202 9,770	Eighth Mir docking First Endeavour flight after OMDP Had been manifested as Discovery flight Changed to OV-105 on 22 May 97
90	STS-90	OV-102 25	16 Apr 98 (pd) 17 Apr 98 03 May 98	107:18:19:00 123:16:08:58	KSC, Pad B KSC, 33	Richard A. Searfoss (C) Scott D. Altman (P) Richard M. Linnehan (PC) Kathryn P. 'Kay' Hire (MS2) Dafydd 'Dave' Rhys Williams (MS3) Jay C. Buckey (PS1) James A. 'Jim' Pawelczyk (PS2)	171 39.00	4,523,791 262,357 26,150 231,343	Neurolab Shuttle Vibration Forces Bioreactor Demo 4 RMS = none	255 Orbits 381 hrs 49 mins 58 secs 6,375,000 miles	25,758 283 224 9,949	Final Spacelab mission module flight First Astronaut from KSC (Hire)

Seq. #	Mission #	Orbiter Flight #	Launch Landing (Dates)	Launch Landing (Time)	Launch Landing (Sites)	Crew	Altitude Inclination	Lift-Off OV Launch Payload OV Landing (Weights)	Payloads	Mission Duration	Velocity Cross-Range Touchdown Roll-out (Reentry)	Notes
91	STS-91	OV-103 24	02 Jun 98 12 Jun 98	153:22:06:24 163:18:00:24	KSC, Pad A KSC, 15	Charles J. Precourt (C) Dominic L. Pudwill-Gorie (P) Franklin R. Chang-Diaz (MS1) Wendy B. Lawrence (MS2) Janet Lynn Kavandi (MS3) Valery Viktorvitch Ryumin (MS4)	235 51.60	4,514,378 259,653 26,237 226,017	Mir/ODS Spacehab-SM Mir logistics AMS RMS = s/n 201	154 Orbits 235 hrs 54 mins 00 secs ≈3,800,000 miles	25,889 365 214 10,730	Ninth and final *Mir* docking First use of SLWT External Tank First docking mission for *Discovery*
92	STS-95	OV-103 25	29 Oct 98 07 Nov 98	302:19:19:34 311:17:03:31	KSC, Pad B KSC, 33	Curtis L. Brown, Jr. (C) Steven W. Lindsey (P) Stephen K. Robinson (PC) Scott E. Parazynski (MS2) Pedro Duque (MS3) Chiaki Mukai (PS1) John H. Glenn, Jr. (PS2)	349 28.45	4,521,292 263,987 28,804 227,783	Spacehab-SM HOST PANSAT SPARTAN-201-05 IEH RMS = s/n 201	134 Orbits 213 hrs 43 mins 57 secs ≈3,600,000 miles	26,063 200 199 9,511	First launch attended by U.S. President First flight of the Block II SSME John Glenn's first flight in 36 years, 8 months, and 9 days Drag chute not used on landing
93	STS-88	OV-105 13	04 Dec 98 15 Dec 98	338:08:35:34 350:03:53:32	KSC, Pad A KSC, 15	Robert D. Cabana (C) Frederick W. 'Rick' Sturckow (P) Jerry L. Ross (MS1) Nancy J. Currie (MS2) James H. Newman (MS3) Sergei Konstantinovich Krikalev (MS4)	247 51.60 31,363	4,519,508 239,059 SAC-A 200,296	ISS-Unity (Node 1) ICBC SAC-A MightySat-1 SEM-07 RMS = s/n 202	185 Orbits 283 hrs 17 mins 03 secs ≈4,600,000 miles	25,889 176 197 8,322	ISS assembly flight 2A Ross' seventh EVA (44 hrs 9 mins total)
94	STS-96	OV-103 26	20 May 99 (pd) 27 May 99 06 Jun 99	147:10:49:42 157:06:02:45	KSC, Pad B KSC, 15	Kent V. Rominger (C) Rick D. Husband (P) Tamara E. Jernigan (MS1) Ellen Ochoa (MS2) Daniel T. Barry (MS3) Julie Payette (MS4) Valery Ivanovich Tokarev (MS5)	236 51.60	4,514,454 262,035 20,056 220,980	ISS-OTD crane ISS-STRELA crane ISS logistics STARSHINE IVHM-HEDS Shuttle Vibration Forces RMS = s/n 303	153 Orbits 235 hrs 13 mins 03 secs ≈3,600,000 miles	25,915 712 210 8,866	ISS assembly flight 2A.1 First flight to dock with ISS
95	STS-93	OV-102 26	20 Jul 99 (pd) 22 Jul 99 (pd) 23 Jul 99 27 Jul 99	204:04:31:00 209:03:20:36	KSC, Pad B KSC, 33	Eileen M. Colins (C) Jeffrey S. Ashby (P) Catherine G. 'Cady' Coleman (MS1) Steven A. Hawley (MS2) Michel Tognini (MS3)	168/174 28.45	4,524,727 270,142 50,212 219,980	Chandra/IUS SWUIS RMS = none	79 Orbits 118 hrs 49 mins 36 secs ≈1,800,000 miles	25,762 82 201 6,851	First woman Shuttle commander Shortest mission since 1990 Wiring problems shorted two MECs off-line LO2 post-pin ejected during ascent Heaviest payload to-date
96	STS-103	OV-103 27	15 Oct 99 (s) 19 Nov 99 (s) 02 Dec 99 (s) 09 Dec 99 (s) 18 Dec 99 (pd) 19 Dec 99 27 Dec 99	354:00:50:00 362:00:00:47	KSC, Pad B KSC, 33	Curtis L. Brown, Jr. (C) Scott J. Kelly (P) Steven L. Smith (PC) Jean-Francois Clervoy (MS2) John M. Grunsfeld (MS3) C. Michael Foale (MS4) Claude Nicollier (MS5)	374 28.45	4,506,759 248,000 29,471 211,129	HST service RMS = s/n 301	118 Orbits 191 hrs 10 mins 47 secs 3,267,000 miles	26,114 155 187 7,075	Third HST servicing mission 'call-up' mission Second time is U.S. history a crew has been in orbit for Christmas
97	STS-99	OV-105 14	31 Jan 00 11 Feb 00 22 Feb 00	042:17:43:40 053:23:22:24	KSC, Pad A KSC, 33	Kevin R. Kregel (C) Dominic L. Pudwill-Gorie (P) Gerhard P.J. Thiele (MS1) Janet L. Kavandi (MS2) Janice E. Voss (MS3) Manoru Mohri (MS4)	150 57.00	4,520,415 256,560 29,000 225,669	SRTM EarthKAM RMS = s/n none	180 Orbits 269 hrs 38 mins 44 secs ≈4,064,000 miles	25,714 244 206 9,964	Acquired 12 terabytes of digital data Covered 99.98 percent of planned mapping area 13-knot cross-wind landing test
98	STS-101	OV-104 21	10 Feb 00 (s) 24 Apr 00 (pd) 25 Apr 00 (pd) 26 Apr 00 (pd) 18 May 00 (s) 19 May 00 29 May 00	140:11:11:10 150:07:21:19	KSC, Pad A KSC, 15	James D. Halsell, Jr. (C) Scott J. Horowitz (P) Mary Ellen Weber (MS1) Jeffrey N. Williams (MS2) James S. Voss (MS3) Susan J. Helms (MS4) Yuri Vladimirovich Usachev (MS5)	230/250 51.60	4,519,645 262,565 33,894 221,271	Spacehab-DM ISS logistics ISS-ICC RMS = s/n 301	154 Orbits 236 hrs 09 mins 51 secs 4,076,241 miles	26,021 743 205 9,253	ISS assembly flight 2A.2a Reboost station from 230 to 250 miles First MEDS flight
99	STS-106	OV-104 22	19 Aug 99 (s) 08 Sep 00 19 Sep 00	252:13:45:47 263:08:58:01	KSC, Pad B KSC, 15	Terrence W. Wilcutt (C) Scott D. Altman (P) Daniel C. Burbank (MS1) Edward Tsang Lu (MS2) Richard A. 'Rick' Mastracchio (MS3) Yuri Ivanovich Malenchenko (MS4) Boris V. Morukov (MS5)	250 51.60	4,519,645 254,099 22,528 221,271	Spacehab-DM ISS logistics ISS-ICC RMS = s/n 301	184 Orbits 283 hrs 12 mins 15 secs ≈4,900,000 miles	25,973 698 211 8,989	ISS assembly flight 2A.2b
100	STS-92	OV-103 28	05 Oct 00 (pd) 06 Oct 00 (s) 09 Oct 00 (pd) 10 Oct 00 (pd) 11 Oct 00 24 Oct 00	286:00:17:00 298:22:00:47	KSC, Pad A EAFB, 22	Brian K. Duffy (C) Pamela A. Melroy (P) Koichi Wakata (MS1) Leroy Chiao (MS2) Peter J.K. Wisoff (MS3) Michael E. Lopez-Alegria (MS4) William S. McArthur, Jr. (MS5)	250 51.60	4,520,596 253,807 41,952 204,455	ISS-Z-1 Truss CMG Ku/S-Band antenna PMA-3/SLP DDCU RMS = s/n 303	203 Orbits 309 hrs 43 mins 47 secs ≈5,300,000 miles	25,996 728 203 10,349	ISS assembly flight 3A Orbiter returned to KSC on 03 Nov 00

Launch (s) is scheduled date
Launch (pd) is delay on pad
Launch (abort) is an abort after SSME ignition
n/a indicates not applicable
— indicates classified data or data not available
RMS = Remote Manipulator System serial number

Edwards AFB (EAFB) Runways 05R, 17L, 23L, and 33 are lakebed.
Edwards AFB (EAFB) Runways 04 and 22 are concrete.
White Sands Space Harbor (WSSH) Runway 17 is lakebed.
Kennedy Space Center (Shuttle Landing Facility) Runways 15 and 33 are concrete.
Orbital altitude is highest apogee
Times are GMT (day:hour:minute:second)

(C) = Shuttle Commander
(P) = Shuttle Pilot
(PC) = Payload Commander
(MS) = Mission Specialist
(PS) = Payload Specialist

Weights in pounds
Altitude and cross-range in statue miles
Inclination in degrees from Equator
Velocity in feet per second
Touchdown in knots indicated airspeed
Roll-out in feet after main gear touchdown

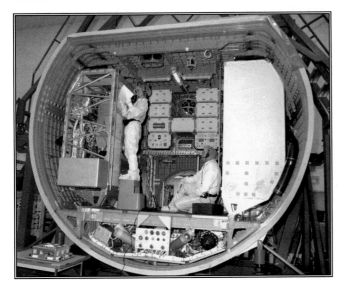

Technicians from the National Space Development Agency of Japan (NASDA) test the real-time radiation monitoring device on the Spacehab module destined for STS-89. (NASA photo KSC-97PP-1592)

Discovery is ventilated via the hose running underneath at the end of STS-96 at the KSC Shuttle Landing Facility. Note that the left nose landing gear door is slightly wider than the right door. (NASA photo KSC-99PP-0630)

Flight 76 – (STS-76) – A one day scrub was called prior to tanking on 21 March because of high winds and rough seas at the launch site. The countdown on 22 March proceeded smoothly and resulted in an on-time launch. The *Atlantis* mission was launched with a hydraulic leak in APU #3, which did not impact the mission except in disallowing flight day extensions. Post-launch inspections of MLP-3 revealed a 63-foot long crack on one of the steel plates running from the north end of the left-hand flame hole to the north end of the MLP surface. Cracks are sometimes found and easily repaired on the MLPs following launch operations, but this one was particularly large. Shannon Lucid stayed on *Mir*, returning on STS-79. John Blaha served as Lucid's back-up. On 28 March the flight was cut short by one day due to weather concerns at KSC on 31 March and 1 April. But the weather deteriorated sooner than expected and the two morning opportunities at KSC on 30 March were waved off. On 31 March the first KSC opportunity was waved off due to thunderstorm activity and the mission landed at Edwards. Unusually, there was a problem on the ferry flight back to KSC. On 6 April *Atlantis* departed Edwards on top of N905NA, but a fire warning indicator light for the right inboard engine on the 747 convinced Gordon Fullerton and Tom McMurtry to shut down the engine and return to Edwards on the remaining three engines The engine was replaced and the SCA carrying *Atlantis* arrived at KSC on 12 April.[67]

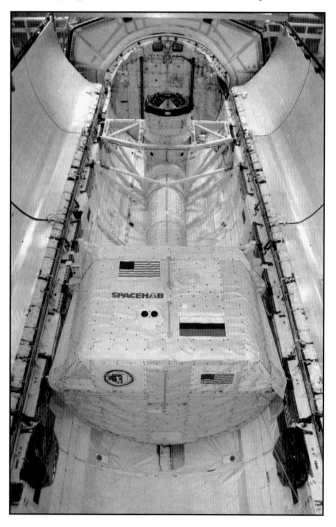

The Spacehab-SM (single module) in the payload bay of Atlantis for STS-76. This was the third Shuttle–Mir docking mission. Note the tunnel connecting Spacehab and the ODS (at the top of the photo). (NASA photo KSC-96PC-0379)

Flight 77 – (STS-77) – The planned 16 May launch date was reset to the 19th because the original date was not available on the Eastern Range. There were no unscheduled holds in the count, and the *Endeavour* mission launched on time. This was the first flight to use three Block I main engines. The Spacehab single module carried nearly 3,000 pounds of experiments and support equipment for 12 commercial space product development payloads in the areas of biotechnology, electronic materials, polymers, and agriculture as well as several experiments for other NASA payload organizations. One of these, the Commercial Float Zone Facility (CFZF) was developed through international collaboration between the U.S., Canada, and Germany. The NASA Goddard SPARTAN-207 satellite was used to test the Inflatable Antenna Experiment (IAE) which laid the groundwork for future technology development on inflatable space structures. The IAE tested the performance of a large inflatable antenna during a ninety-minute mission – the antenna structure was jettisoned and the SPARTAN spacecraft recovered and stowed in the payload bay. This was the first shuttle mission controlled from the new Mission Control Center at JSC, finally replacing the Apollo-era complex used for the first 76 missions.[68]

Flight 78 – (STS-78) – On 18 June NASA decided to X-ray the power drive units (PDU) for the Orbiter–ET disconnect doors on *Columbia*. The units were suspected of having possible loose screws in the PDU terminal circuitry boards since this had been discovered during inspections of the PDUs aboard *Atlantis*. The X-rays verified that all suspect screws were secure and properly installed, and the aft compartment was closed again for flight. The count proceeded smoothly with no unscheduled holds. The Life and Microgravity Spacelab (LMS) module was used to study the effects of long-duration space flight on human physiology. The were 13 life sciences and six microgravity experiments in the laboratory and mid-deck lockers. This marked the eighth use of the EDO pallet carried in the payload bay.[69]

Flight 79 – (STS-79) – *Atlantis* was rolled-out to Pad 39A on the night of 30 June, with a planned 31 July launch. But concerns about a hot gas penetration of the STS-78 SRBs caused the vehicle to be rolled-back to the VAB on 10 July – technicians disassembling the booster from STS-78 discovered that hot gases had seeped through the field joints between segments. An investigation into the seepage identified the most probable cause was the use of a new adhesive and cleaning fluid that had been adopted to comply with Environmental Protection Agency (EPA) regulations to reduce ozone depleting substances. The STS-79 booster set included the same adhesive so NASA decided to unstack the SRBs and to assemble a new set of boosters using the old adhesives. On 29 July *Atlantis* remained in the VAB while stacking the replacement set of SRBs continued in the adjacent high bay. On 25 July a leak check of the field joint between the right aft center and right forward center segments of the replacement SRB failed, and the forward center segment was destacked and cleaned. During inspections of the secondary O-rings, an applicator brush bristle was found and was believed to be the reason for the field joint leakage. New O-rings were installed and the segment was restacked and retested without incident. On 3 August *Atlantis* was demated from the original stack and transported to OPF-3 – the original SRBs would be destacked, cleaned, inspected, and restacked for use on STS-81. The original ET was destacked from the first set of SRBs and mated to the new SRBs on 8 August. *Atlantis* returned to the VAB on 13 August, mated with the new stack, and rolled-out to the pad on 21 August. The new launch date was 14 September. However, Hurricane Fran interrupted these plans, and *Atlantis* was rolled-back for the second time on 4 September. The storm passed overnight, and the vehicle was again rolled-out to Pad 39A the following day. The countdown proceeded smoothly, but APU #2 failed during ascent, although it would not affect the overall mission. The fourth *Mir* docking proceeded without incident, and a variety of food, clothing, experiment supplies, and spare equipment was transferred from the double Spacehab module to *Mir*. Jerry M. Linenger was John Blaha's back-up. Shannon Lucid returned to Earth, holding the record for both a woman in space and longest time by an American.[70]

Flight 80 – (STS-80) – The launch of *Columbia* scheduled for 30 October was delayed 8 days to replace the forward windows (#3 and 4) after an engineering analysis suggested that windows with a high number of flights could tend to fracture more easily. One of the windows had flown eight times and the other seven times. On 4 November, a further 7 day delay was called for to complete analysis of unusual erosion of the SRB nozzles recovered from STS-79. A new launch date of 15 November was set, but two days beforehand the mission was rescheduled for 19 November because of conflicts on the Eastern Range. A 2-minute and 47-second delay at T–31 seconds was due to a minor hydrogen leak in the aft compartment, but engineers determined it did not represent a threat and continued. During the mission, the flight was extended one day, but a revised weather forecast later cancelled the extension. Landing was waved-off from all opportunities on 5 and 6 December due to weather at KSC and Edwards, finally landing at KSC two days later than planned. This was the longest shuttle mission to date.[71]

Flight 81 – (STS-81) – Due to a decision on 12 July to replace the boosters on STS-79 with the set intended for STS-81, then the series of frustrations experienced by STS-79, the launch date for this *Atlantis* mission was delayed several times. The countdown on 12 January went smoothly and the mission carried another double Spacehab module full of supplies to *Mir*. While the vehicles were docked, 1,400 pounds of water, 1,137.7 pounds of U.S. science equipment, 2,206.1 pounds of Russian logistics, and 268.2 pounds of miscellaneous material were transferred to *Mir*. The first landing opportunity was waved-off due to cloud cover at KSC. Jerry Linengers' back-up was C. Michael Foale. This flight used the SRBs originally stacked for STS-79, after they had been refurbished with the older adhesive.[72]

Flight 82 – (STS-82) – The date for this *Discovery* launch was originally set for 13 February, but despite some minor glitches, the flight was launched two days early. This was the second in a series of planned servicing missions to the orbiting *Hubble* Space Telescope that had been placed in orbit during STS-31R, and first serviced by STS-61. Beginning on the third day of the mission, four different astronauts conducted five EVAs to repair the HST. The NASA Goddard High Resolution Spectrometer and Faint Object Spectrograph were replaced by the functionally similar Space Telescope Imaging Spectrograph (STIS) and Near Infrared Camera and Multi-Object Spectrometer (NICMOS). A new Solid State Recorder (SSR) also replaced one of original reel-to-reel tape recorders.[73]

Flight 83 – (STS-83) – On 1 April the decision was made to slip this *Columbia* launch one day, after it was determined that a water coolant line in the payload bay was

Atlantis is rolled-over to the VAB for STS-79. The roads surrounding the OPFs and VAB are used for daily traffic, and are blocked-off to allow the Orbiter to pass. Many employees usually come out to watch the Orbiter during roll-over. (NASA photo KSC-96PC-0849)

not properly insulated and might possibly freeze during the 16-day flight. A 20 minute and 23 second delay was added to the T–9 minute hold to replace a seal on the orbiter hatch. The primary payload was the Microgravity Science Laboratory (MSL), a collection of microgravity experiments housed inside a European Spacelab long module. This mission built upon foundation of the International Microgravity Laboratory (IML-1 on STS-42 and IML-2 on STS-65), the United States Microgravity

The External Tank (s/n 80) from the STS-79 stack in the transfer isle of the VAB after being removed from the original SRBs. (NASA photo KSC-96PC-0969

The Manipulator Flight Demonstration (MFD) payload is lowered into the payload bay of Discovery in OPF-2. The MFD is one of several payloads that flew on STS-85. This payload was designed to test the operational capability of the Japanese Experiment Module Remote Manipulator System (JEM RMS) Small Fine Arm (SFA), which may be seen atop its Multi-Purpose Experiment Support Structure (MPESS) carrier. (NASA photo KSC-97PP-0815)

The first super lightweight external tank (SLWT) is lifted in the VAB for STS-91 pre-flight processing. The improved tank is 7,500 pounds lighter than its predecessors and was developed to increase the Shuttle payload capacity on International Space Station assembly flights. The new liquid oxygen and liquid hydrogen tanks are constructed of aluminum lithium, and the redesigned walls of the liquid hydrogen tank are machined to provide additional strength and stability. (NASA photo KSC-98PC-0282)

Laboratory (USML-1 on STS-50 and USML-2 on STS-73), the Japanese Spacelab (Spacelab-J on STS-47), the Spacelab Life and Microgravity Science mission (LMS on STS-78), and the German Spacelab (D-1 on STS-30/61-A and D-2 on STS-55). Unfortunately, the mission was cut over 11 days short out of 15 scheduled due to failure of fuel cell #2. Since the space station assembly schedule was encountering significant delays, leaving shuttle with little to do, it was subsequently decided to refly the MSL payload on STS-94 (known as STS-83R during planning).[74]

Flight 84 – (STS-84) – As with all the *Mir* missions, the exact time of launch was only determined about 90 minutes before lift-off based on the location of the *Mir* space station. The countdown was smooth and the mission was launched on schedule. The sixth Shuttle–Mir docking mission carried 7,314 pounds of water and supplies to and from *Mir*. During the docked phase, 1,025 pounds of water, 844.9 pounds of U.S. science equipment, 2,576.4 pounds of Russian logistics along with 392.7 pounds of miscellaneous material were transferred to *Mir*. Returning to Earth aboard *Atlantis* was 897.4 pounds of U.S. science material, 1,171.2 pounds of Russian material, 30 pounds of ESA material, and 376.4 pounds of miscellaneous material. James S. Voss was Foale's back-up. The first landing opportunity was waved off due to cloud cover at KSC.[75]

Flight 85 – (STS-94) – This was a reflight of the STS-83 Microgravity Science Laboratory (MSL) mission that was

originally launched on 4 April 1997. The first mission had been intended to be on-orbit for 15 days and 16 hours, but was cut short due to a problem with fuel cell #2 – *Columbia* landed only 3 days and 23 hours after lift-off. During the extremely quick planning phase (by far the shortest for any shuttle flight) this mission was called STS-83R, but was subsequently assigned a number in the normal mission sequence. On 30 June, hoping to preempt expected afternoon thunderstorms, NASA management decided to move up the launch time for *Columbia* by 47 minutes. The decision to launch early removed one end-of-mission daylight landing opportunity at Edwards, but still allowed two daylight landing opportunities at KSC. Weather still impacted the launch, causing a 12 minute delay due to clouds and rain. Paul D. Ronney served as the alternate payload specialist, as he had on STS-83. This mission proved much more successful than the first attempt, and all objectives were accomplished.[76]

Flight 86 – (STS-85) – Unusually, Jeff Ashby was originally assigned as pilot for this *Discovery* mission, but was transferred to a managerial post in March 1997, and was replaced as pilot by Kent Rominger. The launch occurred within the scheduled window, and the primary mission objective was the deployment and retrieval of a satellite designed to study Earth's middle atmosphere along with a test of potential International Space Station hardware. The primary payload for the flight was the Cryogenic Infrared

Spectrometers and Telescopes for the Atmosphere-Shuttle Pallet Satellite (CRISTA-SPAS-2) on its second flight (previously on STS-66 in 1994). During the flight, Davis used the manipulator arm to deploy the CRISTA-SPAS payload for nine days of free-flight. Landing was waved off one day due to the possibility of ground fog.[77]

Flight 87 – (STS-86) – The seventh Shuttle–Mir mission carried a double Spacehab module, and included five days docked with *Mir* out of the ten day mission. The 6–10-minute launch window opened at 22:29 EDT, but instead of launching at the opening of this period, NASA decided to target the most optimum launch time of 22:34 for vehicle performance reasons. Wendy B. Lawrence was scheduled to replace Foale onboard *Mir*. However, due to concerns about the minimum size restrictions of the Russian Orlan EVA spacesuit, her backup David A. Wolf was launched in her place. Wolf was originally scheduled to fly on the STS-89 mission to *Mir*. Landing attempts on 5 October were waved off due to bad weather at KSC. The returning Foale had spent 144 days, 13 hours, 47 minutes, and 17 seconds in space.[78]

Flight 88 – (STS-87) – This *Columbia* mission carried the United States Microgravity Payload (USMP-4) and the SPARTAN-201-04 as the primary payloads. After a nominal launch, this was the first mission to perform a 'roll-to-heads-up' (RTHU) maneuver approximately six minutes into flight. This procedure would be used on all future low inclination (due-east) launches and allowed the Orbiter to communicate with the TDRS satellites 2.5 minutes sooner, eliminating the need for the Bermuda tracking station. Two EVAs were conducted to demonstrate space station assembly techniques, and the first EVA was also used to

retrieve the errant SPARTAN satellite that had begun a slow rotation after being released from the RMS. This rotation made it impossible for the RMS to grapple the satellite and retrieve it without assistance. Yaroslav Pustovoy, of Ukraine, served as Kadenyuk's back-up.[79]

Flight 89 – (STS-89) – The eighth of nine planned Shuttle–Mir flights, this *Endeavour* mission carried more than 7,000 pounds of experiments, supplies, and hardware to *Mir*. Dave Wolf, who had been on *Mir* since late September 1997, was replaced by Andrew Thomas. James S. Voss was the alternate for Thomas. During launch a problem with the ground-based Launch Processing System (LPS) required extending the hold at the T–20 minute mark. The problem was corrected and the count was resumed. The scheduled 46 minute T–9 hold was reduced to 25 minutes and 15 seconds to ensure lift-off within the short launch window. Unusually, the end-of-mission weather forecast for KSC was so good that NASA did not even activate the backup landing site at Edwards.[80]

Flight 90 – (STS-90) – A launch attempt on 16 April was scrubbed due to a faulty Orbiter network signal processor – before the crew had boarded the Orbiter. The launch attempt the following day proceeded without incident. *Columbia* carried the Neurolab, a Spacelab mission focusing on the effects of microgravity on the nervous system. A Neurolab reflight was considered, but was eventually dropped to preserve September launch opportunities for STS-88. Alexander W. Dunlap and Chiaki Mukai were back-ups for Buckey and Pawelczyk.[81]

Flight 91 – (STS-91) – This was the last Shuttle–Mir flight, and also carried the Alpha Magnetic Spectrometer Investigation (AMS) to search for anti-matter and dark

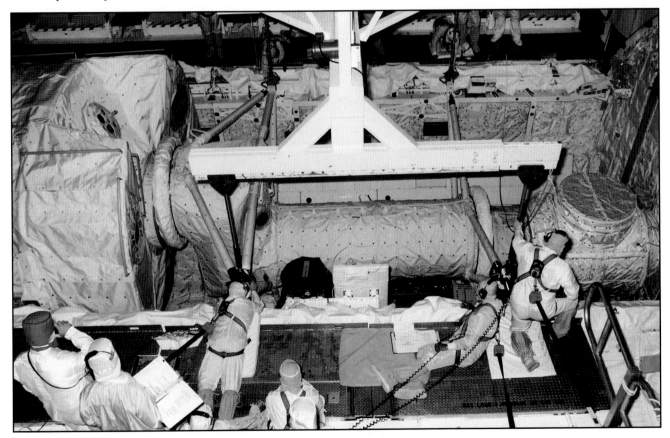

The long transfer tunnel that leads from the Columbia airlock to the Microgravity Science Laboratory (MSL-1) Spacelab module in the payload bay is removed in OPF-1 following STS-83. The tunnel was taken out to allow better access to the MSL-1 module during servicing operations to prepare it for its reflight as MSL-1R (STS-94). This was the first time that a Spacelab payload was serviced without removing it from the payload bay. The Spacelab module was scheduled to fly again with the full complement of STS-83 experiments after that mission was cut short due to a faulty fuel cell. (NASA photo KSC-97PC-0670)

With its drag chute deployed, Columbia touches down on runway 33 at the Shuttle Landing Facility to complete the nearly 16-day STS-90 mission. This was Columbia's 13th landing at KSC and the 43rd landing at KSC. (NASA photo KSC-98PC-0566)

matter in space and to study astrophysics. This *Discovery* flight was the first to use the new aluminum-lithium SLWT External Tank. Andrew Thomas was returned from *Mir* after spending 140 days, 16 hours, 11 minutes, and 45 seconds in space.[82]

Flight 92 – (STS-95) – The primary objectives of this *Discovery* mission included conducting a variety of science experiments in the pressurized Spacehab module, the deployment and retrieval of the SPARTAN-201-05 free-flyer, and operations with the Hubble Space Telescope Orbiting Systems Test (HOST) and the International Extreme Ultraviolet Hitchhiker (IEH) payloads carried in the payload bay. The single-module Spacehab flew in the forward portion of the payload bay with the crew gaining access to the module through the airlock tunnel system. The HOST platform carried experiments to validate components planned for installation during the third *Hubble* Space Telescope servicing mission and to evaluate new technologies in an Earth-orbiting environment. Despite its science objectives, most of the publicity surrounding the flight came from one payload specialist – John H. Glenn. Because the door that covered the drag-chute compartment had fallen off during lift-off, it was decided not to use the drag chute during landing since it might have been damaged during ascent or reentry.[83]

Flight 93 – (STS-88) – The first Space Shuttle flight to the International Space Station was launched exactly on

time. This seven-day *Endeavour* mission was highlighted by the mating of the U.S.-built *Unity* module (also known as Node 1) to the Functional Energy Block (FGB) and two spacewalks to connect power and data transmission cables between the Node and the FGB. *Unity* had originally been scheduled for launch in December 1997 but was rescheduled after repeated delays on the Russian module. *Unity* has two pressurized mating adapters (PMA), one attached to either end – one PMA is permanently mated to the FGB and the other used for orbiter dockings and crew access to the station. The mission was originally scheduled to land on SLF runway 33, but two hours before landing NASA switched to runway 15 (North to South). This mission featured yet another 'Dog Crew' – this time 'Stealth Dog Crew III' (stealth because NASA management did not necessarily share in the spirit of fun). The crew featured Jim 'Pluto' Newman, Robert 'Mighty Dog' Cabana (a veteran of the original STS-53 Dog Crew litter), Rick 'Hooch' Sturckow, Jerry 'Devil Dog' Ross, Nancy 'Laika' Currie, and cosmonaut Sergei 'Spotnik' Krikalev.[84]

Flight 94 – (STS-96) – Hail damage forced a roll-back on 8 May 1999 for repairs at the VAB – the stack returned to Pad 39B on 20 May. *Discovery* conducted the first logistics and resupply mission to the International Space Station. A Spacehab double module contained cargo for station outfitting. The Integrated Cargo Carrier (ICC) carried the Russian STRELA cargo crane that was mounted to the exterior of the Russian station segment, the Spacehab Oceaneering Space System Box (SHOSS), and a U.S. built crane called the ORU Transfer Device. Other payloads were STARSHINE (student tracked atmospheric research satellite for heuristic international networking equipment), the Shuttle Vibration Forces Experiment (SVF), and the Orbiter Integrated Vehicle Health Monitoring – HEDS Technology Demonstration (IVHM HTD). Again, weather at KSC was good enough to allow NASA to not activate Edwards as a backup landing site.[85]

Flight 95 – (STS-93) – The original launch attempt for this *Columbia* mission was scheduled for 20 July, but a hydrogen buildup (640 parts per million) was detected in the aft engine compartment, and the launch was aborted just prior to the beginning of main engine ignition at T–6 seconds. A subsequent hydrogen reading at T–8 seconds (after the decision to abort) registered normal levels, indicating the earlier reading was false, but by then it was too late. The launch was rescheduled for 22 July, but a thunderstorm caused it to be cancelled at T–5 minutes. This

Discovery at the end of STS-92, being towed in the early evening of 24 October 2000. The lights allow the ground crew to monitor activities around the purge and air conditioning units while the Orbiter is under tow. (NASA photo EC00-311-6 by Tom Tschida)

The end of the 100th mission – STS-92 at Edwards AFB on 24 October 2000. Pitchover (nose coming down) puts an unusual aero-load on the main gear, contributing to the tire problems that have plagued the Orbiters. (NASA photo EC00-311-3 by Jim Ross)

gave NASA just one more opportunity to launch before the Air Force closed the Eastern Range for modifications. The launch attempt just after midnight on 23 July proceeded with only minor problems, although the ascent experienced two major anomalies that will be discussed later. The primary objective of this *Columbia* mission was to deploy the $1,550-million *Chandra* X-Ray Observatory. The third of the 'great observatories' was the most sophisticated X-ray observatory ever built, and was designed to observe high energy regions of the universe, such as hot gas in the remnants of exploded stars. *Chandra* was originally called the Advanced X-Ray Astrophysics Facility (AXAF), but was renamed in honor of the late Indian-American Nobel Laureate Subrahmanyan Chandrasekhar. 'Chandra' also means 'Moon' or 'luminous' in Sanskrit.[86]

GROUNDED, YET AGAIN

The launch of *Columbia* on STS-93 may have turned out to have been the most hazardous to-date – two serious in-flight anomalies occurred during ascent. About five seconds after lift-off, flight controllers noted a voltage drop on one of the Orbiter electrical buses. This caused the primary Main Engine Controllers on two of the three SSMEs to shut down. The redundant controllers on those two engines (#1 and 3) continued to function normally, and *Columbia* made it to orbit without difficulty. The left engine was apparently unaffected by the problem.[87]

Post-flight inspection revealed a single 14-gauge polyimide wire had arced to a burred screw head in the payload bay, where all the major fore-aft wiring bundles run. This had resulted in a short that caused the two MECs to drop off-line. The second anomaly was a LO2 low-level cutoff 0.15 seconds before the planned MECO. Post-flight inspection of the affected engine indicated that an LO2 post pin had been ejected and had penetrated three nozzle coolant tubes, causing a fuel leak and premature engine shut-off. Flight controllers knew something was amiss when an extra 5,000 pounds of LO2 was consumed and the engines shut down one second early, leaving the Orbiter with a 16 fps underspeed and an orbit 8 miles lower than the planned altitude of 176 miles. The damage to the nozzle cooling tubes caused SSME #3 to leak 2,500 pounds of hydrogen during ascent, resulting in a lower pressure in the engine combustion chamber. The engine controller detected the pressure loss and attempted to compensate by opening LO2 valves to consume more oxidizer. That essentially caused *Columbia* to run out of LO2, triggering the early engine shutdown.[88]

These problems seemed to be a continuation of a recent run of bad luck experienced by the program. During STS-95 the drag chute door released prematurely about 2 seconds after main engine ignition during lift-off. Still another incident occurred on the ferry flight of OV-102 to Palmdale, for which washers on several attachment bolts were not installed. System design and redundancy successfully handled each anomaly and allowed safe flight of the vehicle and mission completion. However, the occurrence of the anomalies raised concerns over the adequacy of Shuttle operations and maintenance procedures, particularly in light of the age and projected extended life (to the year 2012) of the Space Shuttle.[89]

In the STS-93 post pin incident, three of the 1,080 nozzle coolant tubes in engine 2019 were ruptured and showed evidence of impact damage. There was also evidence of slight impact on the main combustion chamber, although no penetration of the coolant channels occurred.

It had been common practice to deactivate main injector LO2 posts when they were determined to be life-limited because of manufacturing or operational damage. When a post life-limit was reached, a pin was inserted in the LO2 post supply orifice that shut off the LO2 flow through the post, reducing high-cycle fatigue loading. The tapered 0.1-inch diameter pin was about 1-inch long, gold coated, and pressed with interference fit into the orifice. There have been 212 pins used during the program, and 20 prior instances of pin loss during ground testing, with no impact damage. The practice was to insert the pin and perform a vacuum leak check. If there was no leak, an engine firing and subsequent successful vacuum leak check were required to ascertain that the pin would not be ejected. It is significant to note that 19 of 20 pins were ejected on the first engine firing. The one exception was engine 0220 that had a pin expelled after 31 hot fires. In November 1990, STS-38 was flown immediately after pin insertion, and this practice was repeated for nine other pin installations on five missions (on STS-40, 42, 52, 56, and 75) before STS-93, with no pin losses. During the recent engine modifications (Block I, II, and IIA), the main injector manufacturing processes were improved to preclude LO2 post damage, and currently, there are no pinned posts in the fleet. All future Space Shuttle flights, starting with STS-103, will use either Block II or Block IIA main engines, and none of these have deactivation pins in any of the LO2 injector posts at this time and it is not planned to fly any more pinned posts.[90]

Prior to the launch of STS-91, on 18 May 1998, a tanking test was performed while Discovery was on Pad 39A to verify the new SLWT External Tank performed as expected – it did. (NASA photo KSC-98PC-0621)

The initial assessment of the wiring problem indicated that this was an isolated anomaly caused by a technician inadvertently pressing the wires against the screw during maintenance. Further investigation would prove otherwise.

An evaluation of payload bay wiring aboard *Columbia* revealed the potential for damaged wire to exist in the other Orbiters, and as a precaution NASA decided to check the wiring in the payload bay of all four Orbiters. On 12 August 1999 NASA delayed the roll-over of *Endeavour* from OPF-2 to the VAB to conduct wiring inspections and preventative wire maintenance in the payload bay. The additional work was expected to delay the STS-99 launch to early October technicians had to remove the Shuttle Radar Topography Mission (SRTM) payload to gain access to the lower cable trays that run the length of the Orbiter mid-fuselage. Technicians soon noticed more cracked wiring – 26 instances on OV-102 and 38 on OV-105 – ranging from small nicks in the insulation to bare wiring where the insulation had fallen (or rubbed) off. The situation was not unlike one also being experienced by commercial jetliners, where aging wiring is becoming and increasing concern. NASA quickly grounded the fleet for an inspection of all 100 miles of wiring in each of the four Orbiters. It would take a while.[91]

The 16 September *Endeavour* radar mapping mission and an early October *Discovery* mission to service the *Hubble* Space Telescope were indefinitely postponed. Shuttle Program Manager Ronald D. Dittemore hoped the manifest could be back on track by mid-October, but nobody knew for sure.[92]

As damaged wires were found on *Endeavour* and *Discovery*, they were repaired at KSC. In addition, technicians took measures to further protect the wiring from future damage by installing flexible plastic tubing over some wires, smoothing and coating rough edges near wires, and installing other protective shields in selected locations. Inspections of *Atlantis* were postponed since she was third on the manifest. In-depth inspections were not conducted on *Columbia* at KSC since the Orbiter was scheduled to go to Palmdale for an OMDP where it would be much easier to inspect. Eventually, while in Palmdale

Discovery nears touchdown on runway 15 at the Shuttle Landing Facility to complete STS-91. Note the open speedbrake and the position of the body flap under the main engines. (NASA photo KSC-98PC-0743)

some 7,000 wiring anomalies were found in OV-102, significantly delaying *Columbia*'s return from OMDP.[93]

Because of the problems, the Associate Administrator for Space Flight, Joseph Rothenberg, convened a Shuttle Independent Assessment Team (SIAT) led by NASA Ames director Dr. Henry McDonald. The team was comprised of representatives from NASA, industry, academia, and the military and was chartered to evaluate Shuttle sub-systems and maintenance practices. The team was to issue a preliminary report within 60 days[93]

By early October, the wiring inspections of *Discovery* were almost finished, and those of *Endeavour* were 90 percent complete. Inspections on *Atlantis* were just beginning at KSC, and *Columbia* was at Palmdale. NASA had decided to again push back the *Discovery* HST servicing mission on STS-103 to the early morning hours of 2 December. The mission had previously been scheduled for no earlier than 19 November. The STS-99 radar mapping mission of *Endeavour was* postponed to 13 January 2000. No firm launch date had been officially planned for STS-99, although it had appeared unlikely before the end of December. The following mission, the STS-101 launch of *Atlantis* to the International Space Station, was pushed back from late January to no earlier than 10 February — the continuing series of delays caused by the wiring inspections has pushed the launch back from its original early December date.[95]

"Our number one priority for the Space Shuttle is to fly safely, and that is why we delayed our launch preparations and have performed comprehensive wiring inspections and repairs," Space Shuttle Program manager Ron Dittemore said. "As a result of our inspections, we've made significant changes in how we protect electrical wiring. We believe those changes, along with changes to the work platforms and procedures we use in the shuttle's payload bay, will prevent similar wire damage from recurring."[96]

Delays in the shuttle schedule also contributed to the 1 October decision by Russian and American officials to delay the launch of the next major segment of the International Space Station. With no shuttle mission to the station planned before February, officials saw no need to rush and launch the Zvezda service module as had been planned. After years of waiting for the Russians, now it was the Americans' turn to delay construction. On 29 October NASA announced that the launch of *Atlantis* on STS-101 would occur no earlier than 16 March to finish the wiring inspections and repair efforts, along with the unplanned replacement of the ammonia boiler after inspections revealed corrosion.[97]

The problems seemed to continue during the stand-down, however. In early December, damage was found on a 4-inch hydrogen recirculation line in the aft compartment of *Discovery*. The STS-103 mission had taken on a new urgency because *Hubble* desperately needed serviced – enough of its gyroscopes had failed to take the telescope off-line, causing a minor uproar among scientists. Compounding the problem, however, was a decision by NASA to not fly over the dreaded 'Y2K' threshold – all the Orbiters were to be safely on the ground at midnight 31 December. NASA had checked all of the software (none of which really cares about the date anyway since it is mostly based on GMT that does not have a year in it, let alone a century), but decided it wanted to be completely safe, and the easiest way to do that was not to fly. Therefore *Discovery* needed to be launched before 17 December in order to complete the 10-day servicing mission.[98]

Although the SIAT team would not publish its final report until 7 March 2000, preliminary versions of the report were released to support decisions concerning the launch of STS-103 and STS-99. The team documented many positive elements during the course of their interviews – particularly noteworthy were the observations dealing with the skill and dedication of the workforce.[99]

But the SIAT had nine major issues with how the Space Shuttle Program (SSP) was managed. The first sentence of the first issue was particularly true, and is something that must always be remembered: "Human rated spaceflight implies significant inherent risk." With this in mind, the nine issues were:[100]

1. NASA must support the Space Shuttle Program with the resources and staffing necessary to prevent the erosion of flight-safety critical processes.
2. The past success of the Shuttle program does not preclude the existence of problems in processes and procedures that could be significantly improved.
3. The SSP's risk management strategy and methods must be commensurate with the 'one strike and you are out' environment of Shuttle operations.
4. SSP maintenance and operations must recognize that the Shuttle is not an 'operational' vehicle in the usual meaning of the term.
5. The SSP should adhere to a 'fly what you test / test what you fly' methodology.
6. The SSP should systematically evaluate and eliminate all potential human single-point failures.
7. The SSP should work to minimize the turbulence in the work environment and its effects on the workforce.
8. The size and complexity of the Shuttle system and of the NASA/contractor relationships place extreme importance on understanding, communication, and information handling.
9. Due to limitations in time and resources, the SIAT could not investigate some Shuttle systems and/or processes in depth.

Independent assessments, like the SIAT, have been used repeatedly throughout the history of the Space Shuttle program. NASA's goal for these independent assessments has been to identify opportunities to improve safety. This report brought the Shuttle maintenance and operations processes a perspective from the best practices of the external aviation community. The SIAT focused their activities on eleven technical areas: avionics, human factors, hydraulics, hypergols and auxiliary power unit, problem reporting and tracking process, propulsion, risk assessment and management, safety and mission assurance, software, structures, and wiring.

"The SIAT was asked by the SSP for its views on the return to flight of STS-103. The SIAT had earlier considered this question and had concluded that a suitable criterion would be that STS-103 should possess less risk than, for example, STS-93. In view of the extensive wiring investigation, repairs and inspections that had occurred this condition appeared to have been satisfied. Furthermore, none of the main engines scheduled to fly have pinned Main Injector liquid oxygen posts. The SIAT did suggest that prior to the next flight the SSP make a quantitative assessment of the success of the visual wiring inspection process. In addition, the SIAT recommended that the SSP pay particular attention to inspecting the 76

areas of local loss of redundancy and carefully examine OV-102 being overhauled at Palmdale for wiring damage in areas that were inaccessible on OV-103. Finally, the team suggested that the SSP review in detail the list of outstanding waivers and exceptions that have been granted for OV-103. The SSP is in the process of following these specific recommendations and so far has not reported any findings that would cause the SIAT to change its views."[101]

Overall, the SIAT was impressed with the state of the shuttle program, but nevertheless documented 81 recommendations. Four of these were identified as 'immediate' (solutions required prior to return to flight):[102]

1. The reliability of the wire visual inspection process should be quantified (success rate in locating wiring defects may be below 70% under ideal conditions).
2. Wiring on the Orbiter *Columbia* [at Palmdale for OMDP] should be inspected for wiring damage in difficult-to-inspect regions. If any of the wires checked are determined to be especially vulnerable, they should be re-routed, protected, or replaced.
3. The 76 Crit 1 [criticality 1 safety] areas should be reviewed to determine the risk of failure and ability to separate systems when considering wiring, connectors, electrical panels, and other electrical nexus points. Each area that violates system redundancy should require a program waiver that outlines risk and an approach for eliminating the condition. The analysis should assume arc propagation can occur and compromise the integrity of all affected circuits. Another concern is that over 20% of this wiring can not be inspected due to limited access; these violation areas should, as a minimum, be inspected during heavy maintenance and ideally be corrected.
4. The SSP should review all waivers or deferred maintenance to verify that no compromise to safety or mission assurance has occurred.

Unique problems require unique solutions. Northern Flicker Woodpeckers poked holes in the External Tank on STS-71, so NASA installed plastic owls around Pad 39B to scare them away. The rough surface in the background is the spray-on insulation on an ET. As of late-2000 the Owls are still in place, along with a couple of yellow balloons with 'meanie' faces painted on them. And even with all that, NASA still has people sit up on the 215-foot level of the service structure and watch for birds roosting on the ET. One of the hazards of launching from a wildlife refuge. (NASA photo KSC-95PC-769)

These recommendations were reviewed and disposi-
tioned prior to the Flight Readiness Review for STS-103
(the first flight following the stand-down of the Orbiter
fleet for wire inspections). Most of the other recommendations made by the SIAT concerned process and procedure improvement rather than specific technical items.
Thirty-seven recommendations were identified as 'short-
term' (solutions required prior to making more than four
more flights), 30 as 'intermediate' (required prior to
1 January 2001), and ten more as 'long-term' (required
prior to 1 January 2005).[103]

Preparing to Fly

One of the items addressed in light of the SIAT report
was the declining workforce at KSC. In response, the
Space Shuttle Program hired additional personnel and
conducted practice vehicle flows to find any deficiencies
that could hamper the coming resurgence in launch operations necessary to assemble the International Space
Station. Nearly 500 people were hired into the program at
the three human spaceflight centers (JSC, KSC, and
MSFC); about 200 of these were at KSC which had an
overall workforce of nearly 15,000, including all government and contractor personnel. "We are embarking on a

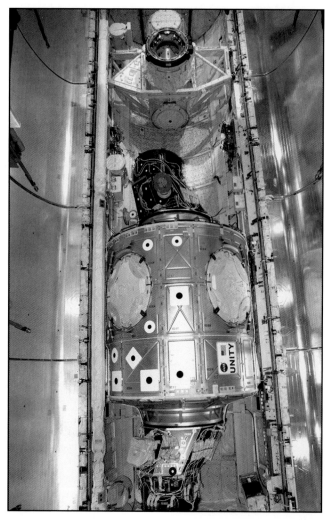

The Unity connecting module rests inside the payload bay of Endeavour at Pad 39A for
STS-88. At the top of bay is the ODS mechanism first used with launches to Mir. Unity
is the first U.S. element of the International Space Station (ISS), and is a connecting
passageway to the living and working areas of ISS. While on-orbit, the flight crew
deployed Unity from the payload bay and attached it to the Russian-built Zarya control module. (NASA photo KSC-98PC-1731)

set of activities as complex as anything we have ever done
in the space business, including landing on the Moon,"
said Dittemore. Between August 2000 and the end of
2001 the schedule showed 11 shuttle missions – nine to
the International Space Station – and a sustained flight
rate of 7–8 missions per year downstream.[104]

"What we get done on one flight is absolutely critical to
the next flight. They all build on each other," Dittemore
said. "It will be a real challenge for us." This was unlike
the previous 20 years where some years had eight or
more flights – then, every mission was a distinct sortie,
usually not dependent upon the flight that had gone
before it. Now with space station assembly being the primary mission for shuttle, each flight had to be flown in
order and close to its original schedule.[105]

In addition to processing Atlantis for her STS-101
flight, during the spring of 2000 Dittemore had KSC
process Endeavour and Discovery through simultaneous
flows to demonstrate the kind of rates KSC would be
required to sustain for the ISS assembly push. By doing
that "we identified several weaknesses in critical skills,
both in how we process vehicles and manage facilities,",
Dittemore said. "We were able to identify those and correct them this summer." The early processing was done
on Discovery for its STS-92 mission and Endeavour for its
STS-97 mission. It showed that the restructured workforce should be able to maintain a constant 80–85-day
OPF flow capability.[106]

Also, during late 1999 KSC awarded contracts to create
a 'safe haven' in the 130-million-cubic-foot Vehicle
Assembly Building (VAB), a project that was completed in
mid-2000. External changes involved unburying a section
of the crawlerway that had existed during the Apollo era,
but had long since been covered over. The crawlerway
again extends around the north to the west side of the VAB,
allowing access to High Bay 2 . For the past 20 years, these
high bays have been used primarily for ET checkout and to
store SRM segments and ground support equipment. In
High Bay 2, crews removed a 125-ton crane, modified the
steel frame, and completed significant floor and foundation
work – including replacement of the existing mobile launch
platform (MLP) mounts. The utility infrastructure for electrical, potable water, gaseous helium, nitrogen, and compressed air was also installed. In High Bay 4, a protective
canopy was added above the Orbiter storage area and communication equipment was installed. The changes allow
High Bay 2 to store a fully assembled stack and High Bay 4
to accommodate an Orbiter in horizontal storage. The
1,250-foot crawlerway extension was topped-off with
about 3,000 tons of river rock. Crews also constructed a
new Orbiter tow-way into High Bay 4.[107]

Every year from June through November, KSC insti-
tutes a plan designed to protect the Space Shuttle during
hurricane season. When a vehicle is at the launch pad
and winds of 69 mph or greater are forecast, a decision
can be made to roll back into the VAB for protection. The
VAB can withstand winds up to 125 mph. The changes to
the VAB allow up to three complete or partial shuttle
stacks to be parked in the VAB, and the horizontal storage
of the fourth Orbiter (the VAB offers more protection than
the OPFs). The modifications were completed in early
August 2000, and the first fully stacked vehicle was
moved into High Bay 2 on 12 August when Atlantis and
the stack for STS-106 was rolled-out of High Bay 1 and
transported around the north side of the VAB for a fit
check. After this was successfully completed, Atlantis was
rolled-out to Pad 39B in preparation for STS-106.[108]

Atlantis being rolled-over to the VAB for STS-106. Note the revised markings on the side of the fuselage, the result of NASA Administrator Dan Goldin's banning of the 'worm' logo. The lighter-colored leading edges are reinforced carbon-carbon panels, similar to the nose cap. (NASA photo KSC-00PP-1072)

Flight 96 – (STS-103) – The scheduled launch on 18 December was scrubbed before propellants were loaded due to bad weather. The count on 19 December proceeded without incident, but by this time the planned 10-day mission had been cut to 8 days to ensure *Discovery* was back on the ground before the Y2K deadline. This was only the third shuttle flight of the year, marking the lowest launch rate since the return-to-flight in 1988. This *Discovery* flight was the third *Hubble* Space Telescope servicing mission. NASA decided to move up part of the servicing mission that had been scheduled for June 2000 after three of the telescope's six gyroscopes failed. Three gyroscopes must be working to meet the telescope's very precise pointing requirements, and the telescope's flight rules dictated that NASA consider a 'call-up' mission before a fourth gyroscope failed – unfortunately, a fourth gyro had failed while the fleet was grounded for wiring inspections. Four new gyroscopes were installed during the first servicing mission (STS-61) in December 1993 and all six gyroscopes were working during the second servicing mission (STS-82) in February 1997. But a gyroscope failed later during 1997, another in 1998 and two more in 1999. In addition to replacing all six gyroscopes, the crew replaced a guidance sensor, the telescope's computer, and installed a voltage/temperature kit for the batteries. A new transmitter, solid state recorder, and thermal insulation blankets were also installed.[109]

Flight 97 – (STS-99) – The countdown was held at T–9 minutes to resolve three minor technical issues relating to the cabin leak check supply pressure, a hydraulic recirculation pump, and an LH2 manifold tank heater. These were corrected and the launch of *Endeavour* proceeded smoothly. The Shuttle Radar Topography Mission (SRTM) was an international project spearheaded by the National Imagery and Mapping Agency and NASA, with participation of the German Aerospace Center (DLR). The primary objective was to obtain the most complete high-resolution digital topographic database using a specially modified radar system to gather data that produced unrivaled 3D images of the Earth's surface. The only significant in-flight problem occurred as the 197-foot boom that carried the C- and X-band receivers was retracted into its canister – the 440 pounds of copper and fiber cables attached to it did not fold correctly, preventing the lid of the container from closing completely. If the lid could not be closed and latched, the entire system would have been jettisoned and abandoned in space. Fortunately, the crew switched the motor to a 'full torque' setting and completed the retraction. Nearly 12 terabytes of digital data were gathered during the flight. The first landing opportunity was waved off due to weather at KSC. On the next opportunity, they flew into a direct 12–13 knot crosswind, enabling the pilots to fulfill a crosswind development test objective that had been planned for 57 previous missions.[110]

Flight 98 – (STS-101) – On 18 February 2000 NASA confirmed plans to fly an additional Space Shuttle mission to the ISS during the year 2000. The plan distributed the original STS-101 mission objectives between two flights – STS-101 and STS-106 – both using *Atlantis*. STS-101 remained targeted for launch no earlier than 13 April and the STS-106 launch was expected no earlier than 19 August. A firm launch date was set for 24 April, and on that day the attempt was scrubbed at T–5 minutes due to excessive crosswinds at KSC. An attempt the following day was also scrubbed for high winds, this time at T–38 minutes. On 26 April everything looked good at KSC, but weather at the TAL sites caused yet another delay. A crowded Eastern Range schedule caused the next attempt to be delayed until 18 May, but the scrub of an Atlas III launch on 16 May and its need to recycle caused STS-101 to be delayed another day to 19 May. This time all went well, and *Atlantis* was launched at the beginning of a five

The payload of Atlantis for STS-101 prior to door closure. In the center is the Spacehab-DM (double module), and on the lower right end are two GAS canisters (see photo below). (NASA photo KSC-00PP-0492)

Two Get-Away Special (GAS) canisters in the payload bay of Atlantis for STS-101 – MARS (left) and SEM-06. The MARS payload had 20 tubes filled with materials for various classroom investigations designed by the MARS schools. The SEM program is student-developed, focusing on the science of zero-gravity and microgravity. Selected student experiments on this sixth venture are testing the effects of space on Idaho tubers, seeds, paint, yeast, film, liquids, electronics, and magnetic chips. (NASA photo KSC-00PP-0494)

minute window. This was the first flight with the new MEDS glass cockpit. The primary mission objective of STS-101 was to deliver supplies to the International Space Station, perform an EVA at the station, and reboost the station from 230 to 250 miles. Detailed objectives included ISS ingress/safety to take air samples, monitor carbon dioxide, deploy portable, personal fans, measure air flow, rework/modify ISS ducting, replace air filters, and replaced the fire extinguishers and smoke detectors on Zarya.[111]

Flight 99 – (STS-106) – Out for a small scenic tour, Atlantis took a brief detour to VAB High Bay 2 for a fit check of the 'safe haven' modifications after being rolled out of High Bay 1. On 5 September, while *Atlantis* was being prepared for flight, Pad 39B was struck by lightning – subsequent checks confirmed that the lightning protection system performed as expected with no damage to the vehicle or ground support equipment. The countdown on 8 September went smoothly and *Atlantis* was launched at the opening of the window. *Atlantis* used the Spacehab double module and the Integrated Cargo Carrier (ICC) to bring 1,300 pounds of supplies and 4,000 pounds of new equipment to the station. This flight found the ISS to be a growing outpost – with the addition of Zvezda and a Progress resupply ship, the station measured 143 feet long and weighed 67 tons. The mission also included two EVAs to connect various electrical, communications, and telemetry cables between Zvezda and the Zarya Control Module.[112]

Flight 100 – (STS-92) – The 100th Space Shuttle flight – in 1979 this had been expected to occur before 1983, not in the year 2000. The primary objectives the *Discovery* mission was to bring the Z-1 Truss (mounted on a Spacelab pallet), control moment gyros, Pressurized Mating Adapter (PMA-3), and two DDCU (heat pipes) to the International Space Station. The first launch attempt on 5 October was scrubbed early because of concerns over the forward bolt that holds the ET to the Orbiter. Film of the same bolt during the STS-106 mission showed that it did not retract as completely as expected after the frangible nut was detonated, leading to concerns that the Orbiter could 'hang' on the exposed part of the bolt and not jettison the ET correctly. Launch was tentatively rescheduled for the following day after engineers dismissed the concern, but was subsequently set for 9 October. Weather on that day caused a 24-hour scrub. A 10 October attempt was cancelled at the T–3 hour mark when a metal pin – typically used to secure removable handrails to work platforms – was observed lodged on a strut connecting the Orbiter and ET. NASA briefly considered the consequences of launching without removing the pin, but concluded that it might damage the thermal protection system during launch and elected to retrieve it instead. Due to safety considerations the pin could not be removed with propellants loaded on the vehicle, causing a one day delay. *Discovery* was finally launched without further delay on 11 October. Just prior to beginning the rendezvous with the ISS, the Ku-band antenna on *Discovery* failed, taking the rendezvous radar with it. This forced the crew to accomplish the docking maneuver without the aid of radar – a task they completed successfully. Perhaps more disappointing to those on the ground, the lack of the Ku antenna meant there was no live video feed during the docking. Persistent high winds at KSC forced NASA to divert landing to Edwards – for the first time in four years (STS-76). Post flight inspections of *Discovery*'s thermal protection system revealed a total of 127 debris hits of which 24 measured one-inch or larger. The return flight aboard N905NA was delayed one day at Edwards because one of the tailcone bolts broke.

Inside the VAB, Atlantis is lowered onto the ET and SRBs. Also visible are the three SSME nozzles, each measuring 7.8 feet across and 9.4 feet high. This stacking was in preparation for STS-101 to the International Space Station.(NASA photo KSC-00PP-0365)

After the Rotating Service Structure at Pad 39A is retracted, Atlantis awaits a fourth launch attempt as STS-101. This view shows the large flame trench under the MLP – and the curved flame deflector in the trench. (NASA photo KSC-00PP-0637)

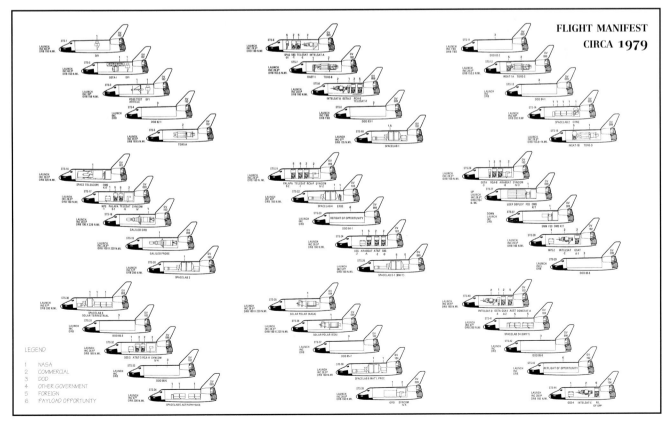

FLIGHT MANIFEST
CIRCA 1979

LEGEND

1   NASA
2   COMMERCIAL
3   DOD
4   OTHER GOVERNMENT
5   FOREIGN
6   PAYLOAD OPPORTUNITY

POST-CHALLENGER MISSIONS

Commercial satellites were essentially banned from Space Shuttle when it returned to flight in 1986. In its place, the Reagan administration reverted to a 'mixed fleet' strategy of using the shuttle for missions that required the presence of humans in space, primarily for science, and of using expendable launch vehicles to place most satellites – commercial or government – in orbit. The DoD immediately bailed out of the Space Shuttle program, closing the mostly-completed launch site at Vandenberg AFB, ordering additional Atlas and Titan III boosters, and initiating the development of the Titan IV with the same lift capabilities as shuttle. Nevertheless, given the lead times for the new boosters, and the fact that several classified payloads had been designed specifically for Shuttle, the DoD would fly six major missions* after the Return-to-Flight.

The commercial users were left in more of a lurch. The few remaining Atlas and Titan boosters were all reserved for government use, and there were only five remaining Deltas in the inventory – two were reserved for the Strategic Defense Initiative Office, two for NOAA, and the last was used to compensate Indonesia for the embarrassing stranding of PALAPA-B2. McDonnell Douglas had closed the Delta production line in the early 1980s when it became clear that Space Shuttle would be the launch vehicle of choice. General Dynamics has scaled back Atlas production considerably, and Martin Marietta only had a few Titan IIIs rolling down the line as contingency vehi-

SPACELAB AND SPACEHAB MISSIONS [118]

Mission	Date			Orbiter	Spacelab	Purpose
STS-9	28 Nov	83		Columbia	Spacelab-1	General research demo
STS-24	29 Apr	85		Challenger	Spacelab-3	Microgravity and life sciences
STS-26	12 Jul	85		Challenger	Spacelab-2	Solar physics
STS-30	30 Oct	85		Challenger	Spacelab-D1	Microgravity and life sciences
STS-35	02 Dec	90		Columbia	ASTRO-1	Astronomy
STS-40	05 Jun	91		Columbia	Spacelab SLS-01	Space life sciences
STS-42	22 Jan	92		Discovery	Spacelab IML-01	Microgravity
STS-45	24 Mar	92		Atlantis	ATLAS-1	Atmospheric studies
STS-50	25 Jun	92		Columbia	USML-1	Microgravity
STS-47	12 Sep	92		Endeavour	Spacelab-J1	Microgravity and life sciences
STS-56	08 Apr	93		Discovery	ATLAS-2	Atmospheric studies
STS-55	26 Apr	93		Columbia	Spacelab-D2	Microgravity
STS-57	21 Jun	93		Endeavour	Spacehab-1	Materials and life sciences
STS-58	18 Oct	93		Columbia	Spacelab SLS-02	Life sciences
STS-60	03 Feb	94		Discovery	Spacehab-2	Material sciences
STS-65	08 Jul	94		Columbia	Spacelab IML-02	Microgravity
STS-66	03 Nov	94		Atlantis	ATLAS-3	Atmospheric studies
STS-63	03 Feb	95		Discovery	Spacehab-3	Materials and life sciences
STS-67	02 Mar	95		Discovery	ASTRO-2	Astronomy
STS-71	27 Jun	95		Atlantis	Spacelab-Mir	Life sciences
STS-73	20 Oct	95		Columbia	USML-2	Microgravity
STS-77	19 May	96		Endeavour	Spacehab-4	Materials and life sciences
STS-78	20 Jun	96		Columbia	LMS-1	Life and microgravity sciences
STS-83	04 Apr	97		Columbia	MSL-1	Materials sciences
STS-94	01 Jul	97		Columbia	MSL-1R	Materials sciences
STS-90	17 Apr	98		Columbia	Neurolab	Neurological life sciences
STS-95	29 Oct	98		Discovery	Spacehab-5	Life sciences

---

* The two TDRS missions were also – sort of – in support of the DoD since the Air Force is a major user of TDRS capacity.

On the 195-foot level of the Fixed Service Structure on Pad 39A, STS-92 Pilot Pamela Ann Melroy (left) sits in the slidewire basket while Commander Brian Duffy reaches for the lever to release the basket. All crew members are take part in emergency egress training during the Terminal Countdown Demonstration Test activities. The slidewire baskets provide a means for the crew to escape from the launch pad – at the other end are underground bunkers that provide safe haven during an emergency. (NASA photo KSC-00PP-1380)

POST-CHALLENGER **TDRS** AND SCIENCE MISSIONS

Mission	Date	Orbiter	Satellite	Type
STS-26R	29 Sep 88	*Discovery*	TDRS-C	Tracking & Relay
STS-29R	13 Mar 89	*Discovery*	TDRS-D	Tracking & Relay
STS-30R	04 May 89	*Atlantis*	*Magellan*	Venus Probe
STS-34	18 Oct 89	*Atlantis*	*Galileo*	Jupiter Probe
STS-32R	09 Jan 90	*Columbia*	LDEF Retrieval	Long-Delayed
STS-31R	24 Apr 90	*Discovery*	*Hubble* Space Telescope	Great Observatory
STS-41	06 Oct 90	*Discovery*	*Ulysses*	ex-Solar-Polar
STS-37	05 Apr 91	*Atlantis*	GRO	Great Observatory
STS-43	02 Aug 91	*Atlantis*	TDRS-E	Tracking & Relay
STS-48	12 Sep 91	*Discovery*	UARS	Earth observatory
STS-54	13 Jan 93	*Endeavour*	TDRS-F	Tracking & Relay
STS-61	02 Dec 93	*Endeavour*	HST Service Mission #1	HST Service
STS-70	13 Jul 95	*Discovery*	TDRS-G	Tracking & Relay
STS-82	11 Feb 97	*Discovery*	HST Service Mission #2	HST Service
STS-93	23 Jul 99	*Columbia*	*Chandra*	Great Observatory
STS-103	19 Dec 99	*Discovery*	HST Service Mission #3	HST Service

cles, all for the Air Force. Only Ariane remained in series production in 1986 and was able to increase it operations immediately. Arianespace quickly won contracts to launch three commercial satellites that had been scheduled for launch on Shuttle – AUSSAT-A3, INSAT-1C, and SBS-5. The nine Ariane launches conducted during the 32 months that Shuttle was grounded successfully orbited 16 satellites, and destroyed one other.[113]

Nevertheless, the Space Shuttle had a surprisingly busy launch schedule ahead of it when it returned to flight in 1988. There were still six DoD flights to be flown, and the TDRS constellation needed to be completed. The TDRS satellites had been designed specifically to be launched from Shuttle, and could not economically be reconfigured for another launch vehicle. Spacelab missions also continued to be flown, and NASA was looking forward to the

The third HST servicing mission – STS-103. Astronauts Steven L. Smith and John M. Grunsfeld appear as small figures in this wide scene photographed during an EVA to replace gyroscopes contained in rate sensor units (RSU). (NASA photo STS103-713-048)

Ulysses and its attached IUS/PAM-S in the payload bay of Discovery prior to STS-41. The unusual IUS/PAM-2 combination was an innovative solution to the cancellation of planned Centaur upper stage after the Challenger accident in 1986. Note the RMS stowed at the bottom of the photo. (NASA photo KSC-90P-1474)

The Long-Duration Exposure Facility had been left in orbit by STS-13 in April 1984, but the Challenger accident and stand-down resulting in it becoming a little 'longer duration' than planned. STS-32R in January 1990 finally retrieved LDEF, a little worse for wear but full of data for the scientists. (NASA photo EL-1994-00473)

Mission	Date		Orbiter	Purpose
STS-63	03 Feb	95	*Discovery*	Mir Rendezvous and fly-by
STS-71	27 Jun	95	*Atlantis*	1st Shuttle-Mir Docking
STS-74	12 Nov	95	*Atlantis*	2nd Shuttle-Mir Docking
STS-76	22 Mar	96	*Atlantis*	3rd Shuttle-Mir Docking
STS-79	16 Sep	96	*Atlantis*	4th Shuttle-Mir Docking
STS-81	12 Jan	97	*Atlantis*	5th Shuttle-Mir Docking
STS-84	15 May	97	*Atlantis*	6th Shuttle-Mir Docking
STS-86	25 Sep	97	*Atlantis*	7th Shuttle-Mir Docking
STS-89	22 Jan	98	*Endeavour*	8th Shuttle-Mir Docking
STS-91	02 Jun	98	*Discovery*	9th and final Shuttle-Mir Docking
STS-88	04 Dec	98	*Endeavour*	ISS Assembly Flight 2A
STS-96	27 May	99	*Discovery*	ISS Assembly Flight 2A.1
STS-101	19 May	00	*Atlantis*	ISS Assembly Flight 2A.2a
STS-106	08 Sep	00	*Atlantis*	ISS Assembly Flight 2A.2b
STS-92	11 Oct	00	*Discovery*	ISS Assembly Flight 3A

construction of the space station in the not too distant future. But perhaps more importantly, there were five major scientific payloads that required the unique capabilities of shuttle – and Space Shuttle missions were needed to service the *Hubble* Space Telescope.

The *Challenger* accident greatly compounded the problems being faced by the probes *Galileo* and *Ulysses*. The accident happened just before both probes were due to be launched using Centaur upper stages – in fact, *Atlantis* was being readied for the *Galileo* mission and *Challenger*'s next payload was to be *Ulysses*. There was no question that both probes would wait for shuttle to

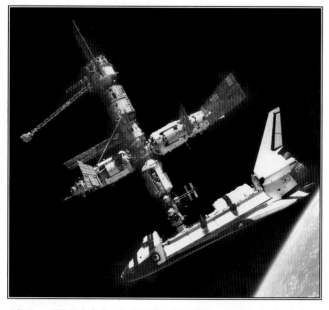

Atlantis as STS-71 docked to the Kristall module of Mir on the first shuttle mission to the Russian space station. The photo was taken by Cosmonaut Nikolai M. Budarian on 4 July 1995 from a Soyuz spacecraft. Atlantis had docked to Mir on 29 June and undocked shortly after this photo was taken on 4 July. At the time the mated vehicle formed the largest spacecraft ever, with a mass of over 500,000 pounds. (Russian Space Agency photo courtesy of NASA photo KSC-00PP-0494)

come back into service – both had been designed for a shuttle launch and there was no reasonable way to transfer them to another vehicle, even if a heavy-lift expendable had been available.

In the immediate aftermath of the accident, the Jet Propulsion Laboratory (JPL – in charge of both probes) was told to prepare *Galileo* for the next launch window in June 1987 on the assumption that shuttle would be back in service by then. But it was not to be. Not only would the stand-down last much longer than initially anticipated, one of the conclusions NASA reached was that the program would not continue to certify the SSMEs at 109 percent power. *Galileo* had become very overweight – a fully loaded Centaur with the probe required 109 percent. With only 104 percent available, JPL came up with a workaround where the Centaur would only carry a partial propellant load and make a 'fly-by' trajectory around Earth to pick up the additional energy needed for the trip to Jupiter.[114]

But the program was dealt a more serious blow when the Centaur itself was cancelled. Now the only choice was to use the IUS that had been rejected years earlier – but the three-stage version of the IUS had been cancelled when the two probes had elected to go with the Centaur. The two-stage IUS was simply not powerful enough to propel the heavy-weight *Galileo* into the far solar system, and it appeared the project might be cancelled entirely. In an ironic way, the extended stand-down of shuttle worked to JPL's advantage. It gave the lab time to develop a complex multiple-encounter flyby trajectory that used gravity assist once from Venus and twice from the Earth. It would take a long time for the probe to get to Jupiter using this method, but a standard two-stage IUS could handle the task.[115]

Another interesting work-around was developed for the launch of *Ulysses*. Like *Galileo*, the solar probe had been designed to use the now-cancelled Centaur upper stage, forcing designers to figure out another way to propel the probe into its polar orbit of the Sun. The solution was unique. When *Discovery* launched *Ulysses*, the satellite sat atop both an IUS and a PAM-S. The first stage of the IUS fired for 148 seconds, shut down, and was jettisoned – pretty much like all other two-stage IUS launches. After a 125-second coast, the second stage motor ignited and burned for 108 seconds before releasing its payload. At this point *Ulysses* was traveling just under 7 miles per second (24,607 mph), barely sufficient to escape the gravitation pull of Earth. Four small solid rockets on the outside of the PAM-S burned for a few seconds to get the probe spinning at 70 rpm to enhance stability when the PAM motor fired. The solid-fueled motor burned for 88 seconds, and *Ulysses* was now traveling 9.5 miles per second (34,450 mph), making it the fastest probe ever launched.[116]

Only one flight resembled the commercial satellite deployment flights that had characterized the early shuttle manifest. In January 1990 *Columbia* deployed the Leasat-5 satellite, much like the four Leasats that had gone before. Again, the HS-381 bus had been designed for launch from Shuttle, and could not easily be reconfigured for an expendable launch vehicle. What was different about this mission happened after the satellite was deployed – the Long Duration Exposure Facility was retrieved. LDEF had been left in orbit by STS-13/41-C in April 1984 for a one year mission – it ended up being almost six years.

During *Endeavour*'s first flight, the Space Shuttle again proved its unique capabilities. The second launch of a Commercial Titan had left INTELSAT-VI (603) stranded in a useless low orbit in March 1990 when the attached kick motor failed to start. After much soul-searching, NASA agreed to try and rescue it. Fortunately, the satellite was a Hughes HS-393, which was a slightly larger version of the HS-381 that had been designed to be launched by Shuttle – only the grounding of the Shuttle fleet had forced Intelsat to use Ariane and Titan launchers instead. The April 1992 STS-49 mission of *Endeavour* successfully captured INTELSAT-VI, installed a new kick motor, and released the satellite which later boosted itself into the proper orbit.

The Spacelab missions conducted by Space Shuttle simply could not have been accomplished on any other launch vehicle, and at least partially vindicated NASA's choice of payload bay sizes. The Spacelab and privately-funded Spacehab modules were sufficiently large to permit meaningful scientific work to be accomplished before a Space Station was completed. Ironically, as the assembly for the International Space Station (ISS) kept getting delayed, NASA looked to Spacelab/Spacehab flights to keep the Shuttle launch rate at a somewhat economical level.

Space Station related flights began with a series of nine missions to the Russian *Mir* space station. In each case the Space Shuttle carried passengers to and from *Mir*, along

Atlantis as STS-74 docked to Mir in November 1995. Visible in the photo are the GLO-4 and PSADE experiments in GAS canisters in the payload bay, and the remote manipulator arm extending forward from behind the solar panel. (NASA photo STS074-341-012)

with needed supplies. Finally, beginning with STS-88, crews from Space Shuttle began the construction of the International Space Station on-orbit. The dream NASA had been chasing for over 30 years was becoming a reality.

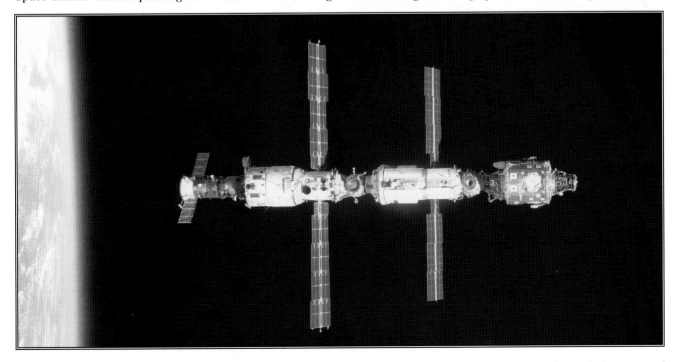

Backdropped against black space above Earth's horizon, the International Space Station (ISS) is seen following its undocking with Atlantis during STS-106. This frame was one of an extensive series of photos acquired after Atlantis separated from the ISS. When Atlantis was at a safe distance from the station, about 450 feet, pilot Scott D. Altman performed a 90-minute, double-loop fly around to enable the crew to document the station's exterior. (NASA photo STS-106-E-5329)

DoD Missions

Although the Department of Defense had always seemed a reluctant partner with NASA during the development of Space Shuttle, by the time STS-1 made the first flight there were several 'national security' payloads that had been designed specifically to be launched by Space Shuttle. Even if the Air Force had not agreed to slow down (and eventually end) the Titan production line, the existing Titan boosters were not capable of launching these satellites (either because of size or weight, or because the satellites had been designed around the Orbiter payload bay instead of as tail-sitters for an expendable vehicle).

The Air Force had purchased nine dedicated Shuttle flights in 1982 for the bargain price of $268 million, mostly reflecting the discount NASA had promised for continued DoD support during the early years of Shuttle development and the tremendous underestimation of actual launch costs. This was in addition to the anticipated dedicated use of OV-103 at the Vandenberg Launch Site beginning in 1985 (later delayed to 1986). The Air Force used all nine of their missions (STS-20/51-C, 28/51-J, 27R, 28R, 33R, 36, 38, 39, and 44), then subsequently added a tenth (STS-53), mainly to launch classified payloads. The effect of the *Challenger* accident on DoD priorities can be seen by the heavy concentration of DoD flights immediately after the Return-to-Flight – although NASA had several high-priority payloads of their own, 6 of the first 15 flights after *Challenger* were in support of the Air Force (it can also be argued that the two TDRS launches were also in support of the DoD, at least partially).

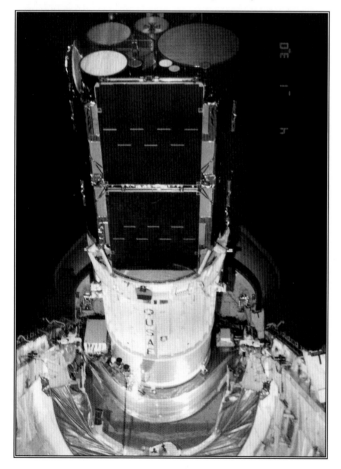

Two DSCS III satellites on top of an Inertial Upper Stage (IUS) during deployment from the payload bay on STS-28/51-J. Another view of this payload is in the color section. (U.S. Air Force via Dwayne A. Day)

But DoD also flew many payloads on other Shuttle flights. In fact, the first DoD payload was carried on the very first 'operational' mission of *Columbia*, STS-4. Called DOD-82-01, the classified payload consisted of several experiments mounted on the Engineering Support Structure (ESS) in the payload bay. The basic ESS structure consisted of three parts – the payload support structure (PSS), long experiment section (LES), and short experiment section (SES). The PSS was a simple truss structure that provided the mounting locations for the other two sections and transferred the loads to the Orbiter through four sill longeron trunnions and one keel trunnion. A total of seven experiments were mounted on the ESS, mostly to gather data on the effect of the Orbiter on the surrounding space (optical and plasma contamination, etc.). Several of these experiments would also fly on future missions.[127]

Many of the planned DoD flights (which were managed by the Air Force, so the organizations are used interchangeably here) needed the use of an upper stage to reach geosynchronous Earth orbits (GEO). In response to this requirement, and with NASA's support following the delay and eventual cancellation of the planned Space Tug, the Air Force developed the Interim Upper Stage (IUS). The IUS was designed to be compatible with both the Shuttle and Titan, and the first IUS boosted a pair of classified military communications satellites into GEO in October 1982 after being launched on a Titan III. This successful IUS flight, combined with the completion the Orbital Flight Test series using *Columbia*, would finally allow the Air Force to begin launching the backlog of missions that had accumulated during the protracted Space Shuttle development period.[128]

But first both NASA and the DoD needed at least part of the TDRS constellation in place to provide communications with the Orbiter. NASA planned to use the IUS to launch TDRS satellites on STS-6, 8, 12, and 15 – as soon as the first two were in place, STS-9 would carry the first dedicated Spacelab flight to ease relations with the Europeans, who were trying to justify the $500 million development of the formerly-named 'sortie module.' The Air Force quickly reserved STS-10 for its first classified payload, and booked five additional flights over the next two years (STS-13, 22, 25, 27, and 30).

Unfortunately, the launch of the first TDRS did not go according to plan. The first-stage of the IUS boosted the satellite to the top of the geostationary transfer orbit (GTO) with no problems, but after the IUS second stage fired to circularize the orbit, trouble began. A simple hydraulic actuator for the thrust vector control system failed, causing the stack to begin tumbling – one of the few contingencies that had not been planned for in the development of the IUS. Fortunately, controllers at the Air Force Satellite Operations Center in Sunnyvale, California, who were controlling the IUS, quickly realized what was happening and released the TDRS from the IUS. Eventually the satellite reached the proper orbit by using its station-keeping thrusters (and thereby greatly reducing its on-orbit life), but the reliability of the IUS was suddenly in doubt, and all payloads that used the upper stage were put on hold. The Air Force promptly cancelled STS-10, and returned the other five manifest slots to NASA.[129]

Eighteen months later, it was decided to try the IUS again. The first classified DoD flight was STS-20/51-C, carrying the same crew and payload that had been booked for STS-10. This mission had been manifested for *Challenger*, but problems with the thermal protection system forced the substitution of *Discovery* instead. At the time of the flight, all the Air Force would reveal about the primary payload was its unremarkable DoD-85-01 designation.

The payload, initially code-named MAGNUM, was apparently a large signals intelligence satellite later called ORION-1 along with its associated IUS. Development of this series of satellites was reportedly ordered after the National Security Agency (NSA) lost its listening post in Iran after the overthrow of the Shah in 1979. The Air Force would not reveal the weight of the payload, but if the Orbiter launch weight of 250,981 pounds released by NASA is correct, it would allow about 40,000 pounds of payload to be carried, slightly less than a TDRS/IUS stack. A similar payload would be carried on STS-33R in November 1989.[130]

The next dedicated DoD mission was STS-28/51-J, and again the Air Force would not reveal the payload. Subsequently it was revealed that the payload was a pair of Defense Satellite Communications Systems (DSCS-III) satellites carried on a single IUS. This is the same payload configuration that was launched on the very first IUS flight on a Titan III in 1982. This should not have come as a surprise to anybody since the Air Force had announced as early as 1981 that launching the DSCS-III constellation would be a primary task assigned to Shuttle once it was operational. Another pair was manifested for 61-N in September 1986, but the *Challenger* accident would cause these to be off-loaded onto a Titan 34D. Each DSCS satellite uses six super high frequency transponder channels capable of providing secure voice and high rate data communications. DSCS-III satellites also carry a single-channel transponder for disseminating emergency action and force direction messages to nuclear-capable forces. Each satellite weighs 2,716 pounds and is roughly six-feet on a side before the 38-foot solar arrays are extended.[131]

The next DoD flight remains a mystery, although it is widely speculated by unofficial sources that the payload was a LACROSSE radar imaging satellite. A year prior to the launch of STS-27R in December 1988, Aviation Week & Space Technology had noted that a large radar imaging satellite was ready for launch. Like all good spy satellites during the Cold war, LACROSSE needed placed into an orbit inclined as highly as possible – *Atlantis* flew the maximum inclination available* from KSC with a heavy payload, 57 degrees. The fact that this payload was assigned to the second flight after the *Challenger* accident emphasizes its importance, and also the probable fact that it could not be reconfigured easily for launch on another vehicle. Subsequently, the National Reconnaissance Office (NRO) released a photograph of a similar satellite (see page 330) that was likely launched by a Titan IV in 2000.[132]

The fourth DoD flight originally presented somewhat of a puzzle for experienced space watchers. Given the secrecy surrounding the launch, and the 57 degree orbit, it was widely assumed that the payload was a reconnaissance satellite of some description, perhaps the much-rumored 'KH-12.' However, certain behavior of the satellite was incongruent with this assumption. For one, observers noted that the satellite flashed at regular intervals – suggesting it was spin stabilized – something not consistent with a spy-satellite. Then, after *Columbia* was well clear, the satellite boosted itself into a highly-inclined 'Molniya' orbit. The puzzle would continue for a while.[133]

Four months later, in November 1989, *Columbia* was again tasked with a DoD mission on STS-33R. This was a

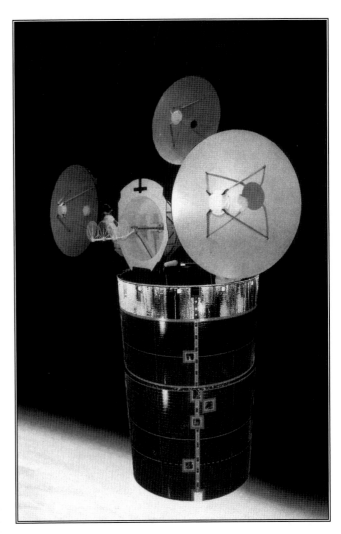

Composite image of an SDS-B in orbit. The satellite is based on the INTELSAT-VI bus. Three of these satellites have been deployed by shuttle to Molniya orbits and one to geosynchronous orbit. Note the low-data-rate omni antenna to the right. (NRO)

repeat of the STS-20/51-C mission, and delivered the second ORION signals intelligence satellite.[134]

STS-36 posed an interesting problem for the Air Force and NASA – the orbit that the payload was going into violated the launch rules on the Eastern Range. The northerly-most launch azimuth approved for shuttle results in an orbit inclined at 57 degrees. This satellite needed 62 degrees. The flight analysts knew how to launch into such an orbit – they have accomplished similar tricks in the past for expendable launch vehicles. The reason the rules exist is to protect people and property along the Eastern seaboard from any catastrophic failure of the launch vehicle, or from falling expended stages (or SRBs in this case). The answer was to launch the shuttle as if it was heading towards a 57-degree orbit, then after the SRBs had separated and sufficient altitude had been gained, turn and 'dog-leg' into a higher inclination. The problem with this, besides the compilations in mission planning and tracking, is that the maneuver carries a severe penalty on the payload capability of the vehicle.

Apparently this did not matter, and on 4 March 1990 *Atlantis* lifted off from Pad A carrying the AFP-731 payload, whatever that may have been. The orbit is generally consistent with some sort of surveillance platform, and most speculation seems to center on it being a new generation 'stealth' reconnaissance satellite called MYSTY, but nobody that knows for sure is talking. Four days after

---

* STS-36 later proved the Shuttle could safely fly a dog-leg over the Atlantic and reach a 62-degree orbit, but this comes at a fairly large penalty in payload weight and would not have been practical – apparently – for STS-27R and LACROSSE.

the satellite – reportedly deployed over the payload bay sill by the Stabilized Payload Deployment System – was left in orbit, observers on the ground reported tracking four distinct objects. The first reports indicated the satellite may have exploded, but subsequent observations showed it had maneuvered into an even higher 65-degree orbit and that the debris was most likely covers that were jettisoned after deployment. The 65-degree orbit would give excellent coverage of the Soviet Union.[135]

The STS-38 mission of *Atlantis*, the seventh dedicated DoD flight, flew into a normal 28.5-degree orbit, and its payload remained a mystery until April 1996. At this point a legal case against the Hughes Electronics Corporation revealed that a military communications satellite had been launched in the winter of 1990. The only classified DoD flight on any vehicle that was unaccounted for was the STS-38 mission. The payload deployed on this flight had remained in low Earth orbit for several months, then slowly made its way to GEO. The legal papers indicated that the 'B-2' satellite had suffered a failure resulting in "a much shorter lifespan and diminished work capacity for the satellite." The two fit.[136]

A further piece of the puzzle feel into place in early 1998 when the NRO released a videotape that showed, among other things, two previously unacknowledged satellites. One of those was a Satellite Data System (SDS) spacecraft built by Hughes. The NRO followed-up in July 1998 with photos of the spacecraft. These photos have led specialists to conclude that the SDS-B satellites are built on the same Hughes bus used by INTELSAT-VI, based mainly on physical appearance and the development timeline. Nevertheless, there are some physical differences between the satellite types, and the evidence is not conclusive. In most probability the SDS-B2 carried on STS-38 used a perigee kick motor (PKM), and this is what failed during the deployment.[137]

The eighth dedicated DoD flight was STS-39, and was the first dedicated flight to carry a completely announced payload, this time for the Strategic Defense Initiative Office (SDIO – better known as 'Star Wars'). The unclassified portion of the payload included the Air Force Program-675 (AFP-675) space vehicle that carried five experiments in the payload bay (CIRRIS-1A, FAR-UV, URA, QINMS, and HUP), Infrared Background Signature Survey (IBSS), Critical Ionization Velocity (CIV), Chemical Release Observation (CRO), Shuttle Pallet Satellite (SPAS-II)

An advanced radar imaging satellite during assembly at Lockheed Martin in late 1997. The code name was changed to ONYX after the original LACROSSE code name became too well known. This particular satellite was likely launched by a Titan IV in 2000, but an earlier version was carried by STS-28R in 1988. (National Reconnaissance Office)

experiments, and Space Test Payload (STP-1). A classified payload was inside a multi-purpose release canister (MPEC) in the payload bay (essentially a GAS canister). Also on board was Radiation Monitoring Equipment (RME-III) and Cloud Logic to Optimize Use of Defense Systems-(CLOUDS-IA). Many of these experiments had flown on unclassified Shuttle flights as secondary payloads.[138]

The last of the original nine DoD flights carried an unclassified Defense Support Program (DSP) satellite (F-16, called 'DSP Liberty' by the crew). This payload had been designed for launch on either the Space Shuttle or the larger Titan family and, as it happened, this was the only one carried to orbit by Space Shuttle.

The *Discovery* flight on STS-53 in December 1992 marked an additional, and final, dedicated DoD flight. The mission profile was essentially identical to that flown by STS-28R, leading many to assume it carried an similar payload. In fact, these two flights along with STS-38 had all carried the same payload – SDS-B satellites. Two of these satellites (from STS-28R and STS-53) later entered highly-inclined Molniya orbits, while the third (from STS-38) eventually reached a geosynchronous orbit. The primary function of these satellites is to provide a relay for imaging data from low-flying reconnaissance satellites. The recce satellites send data up to the SDS satellites, which then relay it to the ground station at Fort Belvoir, Virginia. The highly inclined orbit allows at least one of these satellites to be in line-of-sight with both the polar-orbiting reconnaissance satellites and the ground station. The correlation between the first and third of these missions was apparent fairly early, but it was not until after the Hughes legal revelations that observers concluded the middle mission was also an SDS-B.[139]

At the time of the *Challenger* accident, preparations were well underway for the first launch (STS-1V/62-A) from Vandenberg. This first-ever manned launch from the west coast was scheduled to carry a surveillance satellite called Teal Ruby. The spacecraft was designed to detect and track aircraft – particularly cruise missiles – passively from space. The CIRRIS (cryogenic infrared radiance instrument in space) experiment carried on STS-4 gathered calibration data for Teal Ruby. The satellite was to use a 10-foot focal length telescope with a matrix of 250,000 individual infrared detectors chilled by liquid helium; considered a significant advancement in the state-of-the-art at the time. Before being launched as an independent satellite into a 375-mile orbit, Teal Ruby was to be tested in the payload bay of of an early Shuttle mission from KSC. But the development of the advanced IR detectors proved almost as difficult as developing Space Shuttle, and neither the satellite or the Shuttle was ready for the projected 1980 launch.[140]

It was decided to move the Teal Ruby launch to Vandenberg so that the satellite could realistically be tested in polar orbit, and it was manifested on STS-1V, originally scheduled for 1985, then various dates in 1986. The $500 million test was also going to carry Under Secretary of the Air Force Edward C. 'Pete' Aldridge, Jr., sort of the Air Force's version of carrying Senators and Congressmen. But *Challenger* changed all the plans – *Discovery* stayed at KSC, and Vandenberg was closed.

Teal Ruby was subsequently remanifested on STS-39 from KSC, but the innovative sensor technology that had been so advanced in the late 1970s was virtually obsolete in the early 1990s. It was decided there was little point in flying Teal Ruby, so the payload was withdrawn. The Air Force substituted a variety of experiments – including an

[142] CIVILIAN MISSIONS THAT DEPLOYED DoD PAYLOADS

Mission	Date	Orbiter	Satellite	Type
STS-16	30 Aug 84	*Discovery*	Leasat-2 (Syncom IV-1)	Communications
STS-19	08 Nov 84	*Discovery*	Leasat-1 (Syncom IV-2)	Communications
STS-23	12 Apr 85	*Discovery*	Leasat-3 (Syncom IV-3)	Communications
STS-24	29 Apr 85	*Challenger*	GLOMR*	Communications
STS-27	27 Aug 85	*Discovery*	Leasat-4 (Syncom IV-4)	Communications
STS-30	30 Oct 85	*Challenger*	GLOMR	Communications
STS-32R	09 Jan 90	*Discovery*	Leasat-5 (Syncom IV-5)	Communications
STS-66	03 Nov 94	*Atlantis*	MAHRSI (on SPAS)	Experimental
STS-85	07 Aug 97	*Discovery*	MAHRSI (on SPAS-02)	Experimental
STS-95	29 Oct 98	*Discovery*	MightySat-1	Technology Demo
STS-99	11 Feb 00	*Endeavour*	SRTM	Radar Topography

List only includes deployed payloads; items in the payload bay or mid-deck are not listed
(small DoD payloads have been carried on at least 73 of the first 100 missions)

* = Failed to deploy

GLOMR = GLobal Low Orbiting Message Relay Satellite
MAHRSI = Middle Atmosphere High Resolution Spectrograph Investigation
SRTM = Shuttle Radar Topographer Mapper

[141] DEDICATED DoD MISSIONS

Mission	Date	Orbiter	Satellite	Type
STS-4	27 Jun 82	*Columbia*	DoD-82-01 (ESS)	Experimental
STS-20	24 Jan 85	*Discovery*	ORION-1	Signals Intelligence
STS-28	03 Oct 85	*Atlantis*	DSCS-III F-2 & F-3	Communications
STS-27R	02 Dec 88	*Atlantis*	LACROSSE-1	Radar Imaging
STS-28R	08 Aug 89	*Columbia*	SDS B-1	Communications
STS-33R	22 Nov 89	*Columbia*	ORION-2	Signals Intelligence
STS-36	28 Feb 90	*Atlantis*	MYSTY	Reconnaissance ?
STS-38	15 Nov 90	*Atlantis*	SDS-B2	Communications
STS-39	28 Apr 91	*Discovery*	SFP-675	Experimental
STS-44	24 Nov 91	*Atlantis*	DSP F-16 ('Liberty')	Early Warning
STS-53	02 Dec 92	*Discovery*	SDS B-3	Communications

updated CIRRIS payload on STS-39. The SPAS free-flying pallet, which had flown twice on early missions, was purchased from Germany and fitted with a variety of sensors for an infrared background signature survey in support of advanced Teal Ruby follow-ons.

During the late 1970s, despite the apparent misgivings about totally committing to Space Shuttle, the DoD was nevertheless pressing ahead with plans to launch all of their satellites using the Orbiter. This included the GPS navigation constellation, Defense Meteorological Satellite Program (DMSP), and various classified payloads. However, regardless of this commitment, the Air Force was cautious and waited to see just how the Space Shuttle would develop – after all, first flight was over three years behind the original schedule. This caution kept a small quantity of expendable launch vehicles available for immediate needs such as launching the original GPS demonstration constellation. It was not until the early 1980s that planning began in earnest to development payloads specifically for Space Shuttle, and it is unclear exactly how far programs like GPS went in their planning before the *Challenger* accident.

Discovery getting ready to roll-over from OPF-2 to the Vehicle Assembly Building in preparation for STS-85. The relationship between the three main engines and the two smaller OMS engines can be seen here. Note that the aerodynamic control surfaces droop when the vehicle is not powered up. (NASA photo KSC-97EC-0996)

EXTERNAL TANK SUMMARY

The following chart summarizes the type of External Tank, its serial number, the quantity of LO2 and LH2, the amount of each remaining at significant mission milestones, and the External Tank impact location with its deviation from preflight predictions. ET serial number 1 was assigned to the MPTA-ET. The first six flight tanks (numbers 1 thru 6) were contracted on 16 August 1973, along with three ground test tanks. A production contract for 54 tanks (7 thru 60) was awarded to Martin on 17 February 1978, with an additional 60 (61 thru 120) being contracted on 2 November 1984. A further 60 tanks (121 thru 179) were ordered in early 1999. In the chart, the 'delivered' date is when the ET arrived at KSC. Separation MET is seconds from lift-off, velocity is in feet per second, and altitude is in feet.[121]

		EXTERNAL TANK			LIQUID OXYGEN QUANTITY (pounds)				LIQUID HYDROGEN QUANTITY (pounds)				SEPARATION			REENTRY IMPACT POSITION		
Seq.	Mission	ET Type	Serial #s	Delivered	Loaded	At T-0	At MECO	Residual	Loaded	At T-0	At MECO	Residual	MET	Velocity	Altitude	Latitude	Longitude	From Prediction
1	STS-1	Standard	2 / SWT-1	29 Jun 79	1,356,020	1,346,260	18,335	16,879	227,348	225,528	5,914	5,222	532.1	25,600	365,100	28.40°S	82.75°E	150 nm Down Range
2	STS-2	Standard	3 / SWT-2	27 Feb 81	1,356,914	1,347,238	17,267	15,968	228,424	226,617	4,989	4,277	537.2			28.38°S	81.45°E	312 nm Down Range
3	STS-3	Standard	4 / SWT-3	28 Sep 81	1,352,359	1,342,796	12,254	10,791	228,741	226,747	4,988	4,248	531.5			31.20°S	94.40°E	150 nm Down Range
4	STS-4	Standard	5 / SWT-4	17 Jan 82	1,355,660	1,346,256	9,874	8,459	228,385	226,656	4,268	3,507	530.4		354,800	28.41°S	83.07°E	— Down Range
5	STS-5	Standard	6 / SWT-5	26 May 82	1,356,698	1,347,091	7,200	5,739	227,980	226,301	3,980	3,210	528.8		366,193	28.30°S	82.44°E	170 nm Down Range
6	STS-6	Light	8 / LWT-1	08 Sep 82	1,359,741	1,350,244	10,534	9,283	229,077	227,358	3,824	3,145	517.6		366,139	28.34°S	82.99°E	150 nm Down Range
7	STS-7	Standard	7 / SWT-6	26 Jul 82	1,366,400	1,345,936	10,000	8,606	229,088	227,191	4,300	3,580	518.2	25,668	366,506	28.39°S	83.73°E	75 nm Down Range
8	STS-8	Light	9 / LWT-2	12 Jan 83	1,373,679	1,364,099	22,800	21,347	231,009	229,253	5,500	4,769	539.7	25,669	365,878	28.35°S	81.49°E	127 nm Up Range
9	STS-9	Light	11 / LWT-4	06 May 83	1,378,418	1,368,926	7,800	6,460	231,341	229,608	3,600	2,885	527.3	25,732	378,580	56.96°S	149.90°E	300 nm Down Range
10	STS-11	Light	10 / LWT-3	01 Mar 83	1,378,612	1,365,168	14,000	12,607	231,168	229,458	5,200	4,480	539.6	25,667	366,061	28.29°E	80.63°E	120 nm Up Range
11	STS-13	Light	12 / LWT-5	22 Jul 83	1,378,350	1,368,776	5,000	3,596	231,692	229,956	4,280	3,552	528.9	26,013	367,011	18.94°N	149.91°W	22 nm Up Range
12	STS-16	Light	13 / LWT-6	13 Sep 83	1,377,535	1,367,999	6,400	5,050	231,439	229,691	3,976	3,264	533.0	25.666	365,877	28.25°S	80.02°E	86 nm Up Range
13	STS-17	Light	15 / LWT-8	15 Dec 83	1,378,090	1,368,487	10,300	9,014	231,121	229,379	4,100	3,400	549.4	25,719	378,859	57.06°S	150.04°E	70 nm Up Range
14	STS-19	Light	16 / LWT-9	27 Jan 84	1,377,918	1,368,282	8,300	6,800	230,856	229,108	4,000	3,230	531.3	25,669	366.591	27.69°S	81.99°E	146 nm Down Range
15	STS-20	Light	14 / LWT-7	03 Nov 83	1,378,474	1,368,669	6,000	4,568	231,455	229,652	4,600	3,879	—	—	—	28.10°S	78.30°E	90 nm Down Range
16	STS-23	Light	18 / LWT-11	24 Apr 84	1,378,156	1,368,544	9,200	7,801	231,040	229,290	2,700	1,972	550.0	26,014	369,603	20.24°N	149.37°W	48 nm Down Range
17	STS-24	Light	17 / LWT-10	16 Mar 84	1,378,626	1,369,045	10,760	9,341	231,520	229,772	3,500	2,772	533.1	25,723	378.806	57.09°S	150.83°E	153 nm Down Range
18	STS-25	Light	20 / LWT-13	05 Jul 84	1,378,428	1,368,830	6,400	5,016	231,411	229,668	3,400	2,677	533.9	25.914	368.228	14.89°N	159.50°W	58 nm Down Range
19	STS-26	Light	19 / LWT-12	24 May 84	1,378,020	1,368,475	4,259	3,111	231,666	229,924	2,430	1,868	599.3	25,756	381,932	48.90°S	159.00°E	40 nm Up Range
20	STS-27	Light	21 / LWT-14	25 Jul 84	1,377,882	1,368,249	6,800	5,401	231,295	229,532	3,600	2,855	525.8	25,912	368,863	11.50°N	157.60°W	50 nm Down Range
21	STS-28	Light	25 / LWT-18	20 Dec 84	1,382,233	1,372,639	9,325	8,022	231,267	229,510	2,530	1,827	—	—	—	20.63°N	148.26°W	194 nm Down Range
22	STS-30	Light	24 / LWT-17	16 Nov 84	1,382,740	1,373,141	13,000	11,512	232,030	230,302	4,500	3,776	533.0	25,721	379,121	56.97°S	147.96°E	50 nm Down Range
23	STS-31	Light	22 / LWT-15	24 Aug 84	1,377,926	1,368,562	9,995	8,686	231,424	229,709	3,028	2,312	529.5	25,915	367,881	17.31°N	156.69°W	68 nm Down Range
24	STS-32	Light	30 / LWT-23	18 Jun 85	1,381,446	1,371,963	17,693	16,310	231,760	230,052	5,170	4,435	519.8	25,670	366,430	28.30°S	81.30°E	10 nm Down Range
25	STS-33	Light	26 / LWT-19	15 Mar 85	1,381,361	1,372,007	n/a	n/a	231,671	229,962	n/a	n/a	n/a	n/a	n/a	n/a	n/a	n/a n/a
26	STS-26R	Light	28 / LWT-21	16 Apr 85	1,387,981	1,372,690	8,435	7,114	231,993	230,146	4,155	3,433	530.5	25,869	367,122	12.58°N	164.04°W	17 nm Up Range
27	STS-27R	Light	23 / LWT-16	24 Sep 84	1,387,161	1,370,430	7,248	6,052	231,779	229,855	4,184	3,477	—	—	—	02.86°S	123.48°W	— Classified
28	STS-29R	Light	36 / LWT-29	10 Dec 85	1,388,645	1,373,305	10,063	8,695	231,944	230,080	4,542	3,807	528.0	25,868	367,500	13.26°N	162.65°W	83 nm Down Range
29	STS-30R	Light	29 / LWT-22	15 May 85	1,381,861	1,372,358	9,767	8,419	231,745	229,996	3,497	3,437	526.7	25,677	365,579	28.85°S	86.89°E	137 nm Up Range
30	STS-28R	Light	31 / LWT-24	16 Jul 85	1,382,851	1,373,393	7,629	6,195	231,820	230,082	3,703	2,971	—	—	—	38.64°S	149.65°W	— Classified
31	STS-34	Light	27 / LWT-20	29 Jan 85	1,381,656	1,372,219	7,383	6,009	231,784	230,044	4,084	3,345	530.1	25,869	371,779	3.40°N	147.60°W	7 nm Down Range
32	STS-33R	Light	38 / LWT-31	15 Apr 86	1,382,286	1,372,813	8,057	6,704	231,814	230,074	3,180	2,438	—	—	—	28.57°S	86.42°E	— Classified
33	STS-32R	Light	32 / LWT-25	14 Aug 85	1,382,321	1,372,817	8,348	7,094	231,741	229,975	3,902	3,186	530.0	25,912	368,882	10.44°N	157.22°W	31 nm Up Range
34	STS-36	Light	33 / LWT-26	15 Oct 85	1,381,385	1,311,898	8,252	6,974	231,657	229,925	4,277	3,560	—	—	—	61.46°S	145.11°E	— Classified
35	STS-31R	Light	34 / LWT-27	23 Sep 85	1,378,556	1,369,038	8,616	7,260	231,568	229,825	4,091	3,352	528.0	26,134	372,084	19.95°N	150.00°W	85 nm Up Range
36	STS-41	Light	39 / LWT-32	15 Apr 86	1,380,562	1,371,063	7,322	6,028	231,659	229,913	3,710	2,984	528.4			12.52°N	164.18°W	52 nm Up Range
37	STS-38	Light	40 / LWT-33	20 Jun 86	1,381,562	1,372,056	7,861	6,421	231,689	229,913	4,007	3,253	—	—	—	28.52°S	84.91°W	— Classified
38	STS-35	Light	35 / LWT-28	08 Nov 85	1,381,403	1,371,917	10,725	9,336	231,664	229,900	3,669	2,923	529.7			15.09°N	159.01°W	63 nm Up Range
39	STS-37	Light	37 / LWT-30	15 Jan 86	1,381,727	1,372,232	7,906	6,554	231,688	229,950	3,857	3,116	530.2	26,000	369,797	20.23°N	149.32°W	64 nm Up Range
40	STS-39	Light	46 / LWT-39	09 Sep 87	1,382,470	1,372,994	9,828	8,493	231,829	230,070	2,465	2,718	531.4	25,793	367,696	43.82°S	156.29°W	81 nm Up Range
41	STS-40	Light	41 / LWT-34	25 Jul 86	1,383,242	1,373,758	11,130	9,859	231,869	230,116	4,252	3,523	527.9	25,869	366,809	01.05°N	146.06°W	77 nm Up Range
42	STS-43	Light	47 / LWT-40	18 Dec 87	1,381,725	1,372,239	9,734	8,295	231,778	230,011	3,380	3,070	526.0	25,689	366,906	13.47°N	162.24°W	16 nm Down Range
43	STS-48	Light	42 / LWT-35	19 Sep 86	1,383,048	1,373,542	7,034	5,618	231,828	230,097	3,640	2,887	534.5	26,704	382,935	00.26°N	121.93°W	56 nm Up Range
44	STS-44	Light	53 / LWT-46	31 Jul 89	1,308,419	1,370,955	8,124	6,778	231,758	229,993	3,461	2,717	528.1	25,924	367,944	17.01°N	154.05°W	72 nm Up Range
45	STS-42	Light	52 / LWT-45	03 May 89	1,381,971	1,372,471	9,563	8,224	231,724	229,980	4,243	3,502	528.6	24,994	382,889	44.70°S	157.92°W	88 nm Up Range
46	STS-45	Light	44 / LWT-37	11 Mar 87	1,382,579	1,373,082	16,465	15,111	231,764	229,995	5,253	4,507	528.6	25,231	369,432	42.74°S	154.99°W	115 nm Up Range
47	STS-49	Light	43 / LWT-36	10 Nov 86	1,382,912	1,737,400	10,265	9,053	231,838	230,074	3,557	2,829	527.6	24,446	368,365	12.17°S	163.62°W	130 nm Up Range
48	STS-50	Light	50 / LWT-43	31 Oct 88	1,382,444	1,372,955	10,326	9,043	231,811	230,041	3,848	3,104	526.5	25,868	366,831	13.28°N	162.64°W	1 nm Up Range
49	STS-46	Light	48 / LWT-41	22 Apr 88	1,382,117	1,372,638	9,183	7,898	237,753	230,016	3,991	3,248	527.7	25,978	369,300	17.85°N	153.04°W	55 nm Up Range
50	STS-47	Light	45 / LWT-38	08 Jun 87	1,381,187	1,371,677	9,992	8,667	231,753	229,992	3,818	3,051	532.0	25,822	381,731	43.99°S	156.84°W	17 nm Up Range
51	STS-52	Light	55 / LWT-48	01 Feb 90	1,382,616	1,373,115	15,925	14,595	231,663	229,903	4,786	4,042	530.4	25,866	367,142	12.90°S	163.39°W	103 nm Up Range
52	STS-53	Light	49 / LWT-49	08 Aug 88	1,383,320	1,373,842	9,323	7,997	231,834	230,073	3,444	2,685	531.6	25,876	383,692	40.95°S	152.60°W	17 nm Up Range
53	STS-54	Light	51 / LWT-51	03 Feb 89	1,382,560	1,373,067	9,818	8,439	231,739	229,956	3,731	2,965	528.3	25,869	366,833	12.92°N	163.34°W	32 nm Up Range
54	STS-56	Light	54 / LWT-54	30 Oct 89	1,382,200	1,373,260	16,077	14,790	231,780	230,003	4,562	3,822	532.7	25820	381,364	42.42°N	154.36°W	74 nm Up Range
55	STS-55	Light	56 / LWT-56	10 May 90	1,382,324	1,372,836	13,366	12,046	231,762	230,000	4,054	3,305	528.6	25,867	367,076	12.75°N	163.68°W	83 nm Up Range
56	STS-57	Light	58 / LWT-58	16 Oct 90	1,381,214	1,371,694	8,738	7,405	231,745	229,969	3,741	2,982	531.4	26,019	370,822	16.09°N	142.98°W	16 nm Down Range
57	STS-51	Light	59 / LWT-59	27 Mar 91	1,382,560	1,373,033	7,976	6,696	231,746	229,992	3,734	2,997	528.8	25,867	367,409	12.89°N	163.41°W	47 nm Up Range
58	STS-58	Light	57 / LWT-57	26 Jul 90	1,382,386	1,372,904	7,836	6,405	231,787	230,006	3,550	2,788	534.4	24659	373,247	3.92°N	137.78°W	19 nm Down Range
59	STS-61	Light	60 / LWT-60	27 Jun 91	1,381,727	1,372,162	7,400	6,068	231,689	229,920	3,731	3,060	530.5	26,108	372,736	16.44°N	142.15°W	74 nm Up Range
60	STS-60	Light	61 / LWT-61	15 Oct 91	1,382,073	1,372,521	7,188	5,876	231,635	229,874	3,997	3,330	531.1	25,911	380,432	2.69°N	123.15°W	58 nm Up Range
61	STS-62	Light	62 / LWT-62	21 Jan 92	1,382,569	1,373,068	8,293	6,982	231,758	230,004	3,910	3,245	529.7	25,882	372,880	8.13°N	132.93°W	81 nm Up Range
62	STS-59	Light	63 / LWT-63	21 Apr 92	1,381,864	1,372,315	8,412	7,097	231,697	229,926	4,717	4,005	532.4	25,775	379,629	45.02°N	158.06°W	111 nm Up Range
63	STS-65	Light	64 / LWT-64	20 Jul 92	1,381,891	1,372,383	9,955	8,622	231,687	229,930	4,166	3,525	529.9	25,872	367,744	13.63°S	163.26°W	38 nm Up Range
64	STS-64	Light	66 / LWT-66	26 Jan 93	1,382,565	1,373,131	15,189	13,858	231,777	230,102	4,007	3,336	533.8	25,799	381,528	43.31°S	155.54°W	98 nm Up Range
65	STS-68	Light	65 / LWT-65	02 Nov 92	1,382,548	1,372,994	6,701	7,389	231,782	230,032	3,730	3,058	532.7	25,771	379,587	43.91°S	156.35°W	17 nm Up Range
66	STS-66	Light	67 / LWT-67	17 May 93	1,381,794	1,372,301	9,456	8,145	231,705	229,926	3,813	3,152	533.1	25,822	383,039	44.18°S	156.94°W	45 nm Up Range
67	STS-63	Light	68 / LWT-68	06 Aug 93	1,382,490	1,372,970	9,734	8,427	231,757	229,959	3,339	2,681	530.1	25,884	377,632	0.04°S	125.61°W	63 nm Up Range
68	STS-67	Light	69 / LWT-69	03 Nov 93	1,382,268	1,372,760	12,539	11,211	231,773	229,999	3,848	3,161	526.3	25,910	368,592	15.50°N	159.45°W	78 nm Up Range
69	STS-71	Light	70 / LWT-70	17 Feb 94	1,382,416	1,372,882	7,952	6,620	231,737	229,971	3,433	2,753	529.6	24,868	377,285	0.08°S	125.38°W	2 nm Down Range
70	STS-70	Light	71 / LWT-71	05 May 94	1,381,477	1,371,966	11,298	9,973	231,685	229,899	3,769	3,090	529.8	24,432	367,140	13.75°N	162.99°W	12 nm Up Range
71	STS-69	Light	72 / LWT-72	01 Jul 94	1,382,585	1,373,086	13,584	12,274	231,795	230,038	3,833	3,181	529.3	24,464	368,783	18.82°N	151.91°W	62 nm Up Range
72	STS-73	Light	73 / LWT-73	04 Oct 94	1,382,857	1,373,333	10,955	9,601	231,788	229,992	3,231	2,544	528.6	24,638	373,585	2.84°N	138.97°W	44 nm Up Range
73	STS-74	Light	74 / LWT-74	17 Nov 94	1,381,943	1,372,410	9,916	8,591	231,718	229,957	3,325	2,653	532.8	24,930	377,862	0.31°S	125.58°W	18 nm Up Range
74	STS-72	Light	75 / LWT-75	03 Feb 95	1,381,774	1,373,213	16,458	15,103	231,703	229,932	4,660	3,983	529.6	24,868	377,285	18.43°N	145.55°W	158 nm Up Range
75	STS-75	Light	76 / LWT-76	29 Mar 95	1,382,373	1,374,669	7,300	5,992	231,764	229,982	3,162	2,494	527.3	24,511	367,741	13.60°N	163.29°W	25 nm Down Range
76	STS-76	Light	77 / LWT-77	24 May 95	1,383,107	1,373,565	9,808	8,507	231,752	229,978	3,104	2,453	531.8	24922	377,247	0.47°N	125.85°W	29 nm Up Range
77	STS-77	Light	78 / LWT-78	17 Jul 95	1,381,730	1,372,204	3,467	12,104	231,715	229,898	3,559	2,871	527.0	24,605	373,450	2.97°N	138.82°W	7 nm Up Range
78	STS-78	Light	79 / LWT-79	13 Sep 95	1,382,922	1,373,438	13,004	11,668	231,762	229,975	2,799	2,121	528.5	24,653	373,616	2.86°N	138.95°W	13 nm Up Range
79	STS-79	Light	82 / LWT-82	12 Mar 96	1,383,260	1,373,759	8,974	7,682	231,271	229,499	2,489	1,812	533.4	25,866	377,680	0.65°S	125.96°W	45 nm Up Range
80	STS-80	Light	80 / LWT-80	17 Nov 95	1,379,557	1,370,058	7,119	5,813	231,399	229,397	3,767	3,105	535.3	25,907	370,959	15.46°S	159.56°W	74 nm Up Range
81	STS-81	Light	83 / LWT-83	25 Apr 96	1,382,580	1,373,072	7,989	6,667	231,203	229,416	2,999	2,315	532.0	25,877	377,525	0.38°S	125.59°W	8 nm Up Range
82	STS-82	Light	81 / LWT-81	19 Jan 96	1,382,070	1,372,253	9,808	8,419	231,189	229,377	3,819	3,139	528.7	26,226	372,748	17.46°N	141.08°W	68 nm Up Range
83	STS-83	Light	84 / LWT-84	17 Jun 96	1,382,156	1,372,653	9,383	8,090	231,185	229,410	3,518	2,853	530.0	25,867	367,486	13.67°N	163.15°W	27 nm Up Range
84	STS-84	Light	85 / LWT-85	07 Aug 96	1,382,730	1,373,218	6,917	5,654	231,257	229,498	3,159	2,496	532.2	25,866	377,380	0.95°S	126.00°W	63 nm Up Range
85	STS-94	Light	86 / LWT-86	08 Oct 96	1,383,318	1,373,797	9,857	8,544	231,271	229,483	3,297	2,653	527.9	25,867	371,149	13.52°N	163.46°W	49 nm Up Range
86	STS-85	Light	87 / LWT-87	22 Nov 96	1,381,938	1,372,413	8,888	7,582	231,231	229,438	2,989	2,315	531.4	26,820	372,586	42.77°S	154.86°W	88 nm Up Range
87	STS-86	Light	88 / LWT-88	17 Jan 97	1,382,899	1,373,335	6,329	5,026	231,303	229,528	3,333	2,661	530.1	25,869	377,982	0.52°S	126.53°W	80 nm Up Range
88	STS-87	Light	89 / LWT-89	26 Jun 97	1,382,591	1,373,070	11,406	10,107	231,202	229,403	3,490	2,822	529.2	25,864	366,604	20.28°N	147.99°W	103 nm Up Range
89	STS-89	Light	90 / LWT-90	08 Aug 97	1,382,408	1,373,463	9,394	8,091	231,241	229,454	3,268	2,580	528.4	25,873	378,254	0.68°N	125.67°W	66 nm Up Range
90	STS-90	Light	91 / LWT-91	14 Nov 97	1,381,517	1,371,996	7,875	6,577	231,223	229,445	3,328	2,651	527.6	25,855	373,938	1.88°N	139.90°W	77 nm Up Range
91	STS-91	SLWT	96 / SLWT-1	12 Jan 98	1,383,846	1,374,094	7,105	5,818	231,382	229,569	3,322	2,636	529.3	25,924	347,352	2.68°S	127.20°W	35 nm Down Range
92	STS-95	SLWT	98 / SLWT-3	04 Jun 98	1,383,512	1,374,545	9,110	7,830	231,071	229,296	3,243	2,559	520.5	26,093	373,176	20.80°N	147.23°W	77 nm Up Range
93	STS-88	SLWT	97 / SLWT-2	30 Mar 98	1,382,502	1,373,338	7,654	6,287	231,825	230,037	4,111	3,420	521.3	25,927	367,414	1.72°S	127.16°W	48 nm Up Range
94	STS-96	SLWT	100 / SLWT-5	25 Nov 98	1,383,439	1,374,102	8,123	7,290	231,326	229,468	3,412	2,530	522.4	25,790	369,923	0.67°N	142.39°W	30 nm Up Range
95	STS-93	SLWT	99 / SLWT-4	27 Jul 98	1,383,215	1,373,874	10,402	8,954	231,168	229,521	3,209	2,890	528.4	25,834	376,249	19.53°N	132.35°W	57 nm Up Range
96	STS-103	SLWT	101 / SLWT-6	25 Nov 98	1,383,485	1,372,908	9,476	8,932	231,680	229,125	3,326	2,832	523.5	25,893	373,217	20.32°N	145.43°W	65 nm Up Range
97	STS-99	Light	92 / LWT-92	19 Apr 99	1,383,302	1,374,235	5,230	4,239	231,431	230,010	3,530	2,630	531.9	25,830	371,490	43.02°S	153.25°W	23 nm Up Range
98	STS-101	SLWT	102 / SLWT-7	01 Feb 99	1,383,843	1,374,823	9,923	7,849	231,225	229,989	3,762	3,103	525.3	26,001	373,492	2.23°S	127.50°W	95 nm Up Range
99	STS-106	SLWT	103 / SLWT-8	22 Jun 99	1,383,594	1,374,129	10,350	9,128	231,317	229,256	3,420	3,000	521.7	25,909	371,287	1.98°S	127.02°W	62 nm Up Range
100	STS-92	SLWT	104 / SLWT-9	01 Jun 99	1,383,701	1,373,871	7,609	6,870	231,245	229,555	3,301	2,893	522.2	25.899	368,945	2.54°S	127.16°W	44 nm Up Range

Each Orbiter was originally delivered with a set of OMS pods and a FRCS module: OV-099 (LP01, RP01, FRC9), OV-102 (LV01, RV01, FRC2), OV-103 (LP03, RP03, FRC3), OV-104 (LP04, RP04, FRC4), and OV-105 (LP05, RP05, FRC5). Two OMS pods (LV01 and RV01) and one FRCS module (FRC9) were lost with Challenger. A set of aerodynamic dummies (ALTA – Approach and Landing Test Articles) were also delivered and are normally used for OMDP ferry flights. The apogee and perigee are the initial ones resulting from each OMS burn – on some flights the Orbiter subsequently maneuvered into other orbits. MET is hours:minutes:seconds after lift-off, duration is in seconds, apogee and perigee are in miles, and propellant is in pounds.[123]

ORBITAL MANEUVERING SYSTEM SUMMARY

			SERIAL NUMBERS			OMS-1 BURN				OMS-2 BURN				FUEL (MMH)		OXIDIZER (N204)	
Seq.	Mission	Orbiter	Left Pod	Right Pod	FRCS	MET	Duration	Apogee	Perigee	MET	Duration	Apogee	Perigee	Loaded	Used	Loaded	Used
1	STS-1	OV-102	LV01	RV01	FRC2	00:10:34.0	87.0			00:44:02.0	75.0			6,706	4,983	11,048	8,017
2	STS-2	OV-102	LV01	RV01	FRC2	00:10:33.9	77.6			00:41:41.7	71.0			6,687	5,115	11,017	8,598
3	STS-3	OV-102	LV01	RV01	FRC2	00:10:34.3	86.0			00:40:50.4	88.0			6,650	4,741	10,954	7,806
4	STS-4	OV-102	LV01	RV01	FRC2	00:10:32.6	88.0	149.6	39.1	00:37:40.6	105.0	149.6	149.6	8,420	6,353	13,612	10,610
5	STS-5	OV-102	LV01	RV01	FRC2	00:10:30.8	137.8	184.4	59.6	00:44:40.8	117.0	184.1	184.1	7,430	6,177	12,252	10,466
6	STS-6	OV-099	LP01	RP01	FRC9	00:10:19.6	135.3	177.5	58.2	00:43:37.6	117.0	178.1	177.2	7,266	6,437	12,000	10,733
7	STS-7	OV-099	LP01	RP01	FRC9	00:10:20.2	139.5	184.4	59.7	00:44:30.2	117.5	184.6	184.4	7,870	6,547	12,976	10,925
8	STS-8	OV-099	LP01	RP01	FRC9	00:10:41.7	138.1	184.1	59.3	00:44:51.7	115.6	184.7	184.4	8,250	7,100	13,596	11,926
9	STS-9	OV-102	LV01	RV01	FRC2	00:10:29.3	63.8	155.4	51.3	00:40:37.3	101.5	155.2	154.7	5,952	4,592	9,802	7,824
10	STS-11	OV-099	LP01	RP01	FRC9	00:10:41.6	150.2	190.0	58.8	00:45:24.6	125.0	189.9	189.9	9,288	7,136	15,262	11,916
11	STS-13	OV-099	LP03	RP01	FRC9	—	—	—	—	00:42:54.0	95.2	289.5	132.8	9,424	7,298	15,486	12,715
12	STS-16	OV-103	LP03	RP03	FRC3	00:10:36.9	153.5	185.0	58.6	00:44:52.2	126.2	185.0	185.0	8,970	6,954	14,740	11,800
13	STS-17	OV-099	LP01	RP01	FRC9	00:10:50.4	134.5	219.2	61.5	00:46:30.4	143.3	220.6	219.1	9,424	7,914	15,486	13,386
14	STS-19	OV-103	LP03	RP03	FRC3	00:10:33.3	151.0	184.4	59.5	00:44:43.0	114.8	184.2	173.1	9,424	7,709	15,488	12,922
15	STS-20	OV-103	LP03	RP03	FRC3	DoD	DoD	DoD	DoD	DoD	DoD	212.9	212.9	DoD	DoD	DoD	DoD
16	STS-23	OV-103	LP03	RP03	FRC3	—	—	—	—	00:43:15.0	142.6	286.1	184.6	8,020	6,173	13,292	10,413
17	STS-24	OV-099	LP01	RP04	FRC9	00:10:35.0	133.4	219.3	61.7	00:46:15.0	145.2	219.3	219.1	8,562	7,883	14,186	12,979
18	STS-25	OV-103	LP04	RP03	FRC3	—	—	—	—	00:40:29.0	177.6	221.1	219.2	7,011	5,804	11,510	9,711
19	STS-26	OV-099	LP01	RP04	FRC9	00:11:41.0	106.4	164.1	60.9	00:33:00.0	119.4	164.4	125.1	9,424	7,408	15,488	12,519
20	STS-27	OV-103	LP04	RP03	FRC3	—	—	—	—	00:40:28.0	183.2	219.0	218.3	9,310	7,691	15,184	12,826
21	STS-28	OV-104	LP03	RP01	FRC4	DoD	DoD	DoD	DoD	DoD	DoD	322.2	322.2	DoD	DoD	DoD	DoD
22	STS-30	OV-099	LP01	RP03	FRC9	00:10:35.0	121.2	204.8	60.3	00:44:40.0	131.6	204.5	202.3	6,875	6,082	11,322	10,090
23	STS-31	OV-104	LP03	RP01	FRC4	—	—	—	—	00:40:25.0	180.4	220.0	219.3	7,515	5,751	12,423	9,588
24	STS-32	OV-102	LP04	RP04	FRC2	00:10:22.0	164.2	201.5	62.0	00:46:06.0	134.6	203.1	201.5	8,496	7,796	14,002	13,132
25	STS-33	OV-099	LV01	RV01	FRC9	n/a	n/a	n/a	n/a	n/a	n/a	n/a	n/a	8,057	n/a	13,357	n/a
26	STS-26R	OV-103	LP04	RP03	FRC3	—	—	—	—	00:39:56.0	141.6	204.8	186.4	5,231	4,755	8,700	7,854
27	STS-27R	OV-104	LP01	RP01	FRC4	DoD	DoD	DoD	DoD	DoD	DoD	280.8	275.0	6,981	6,387	11,595	10,628
28	STS-29R	OV-103	LP04	RP03	FRC3	—	—	—	—	00:39:58.0	141.6	191.7	187.0	5,248	4,792	8,710	7,900
29	STS-30R	OV-104	LP01	RP01	FRC4	00:10:29.0	141.8	184.7	58.3	00:44:27.0	125.6	191.4	184.1	7,054	6,539	11,708	10,882
30	STS-28R	OV-102	LP03	RP04	FRC2	DoD	DoD	DoD	DoD	DoD	DoD	191.0	184.1	5,290	DoD	8,871	DoD
31	STS-34	OV-104	LP01	RP03	FRC4	—	—	—	—	00:39:55.0	140.6	194.1	185.7	5,165	4,622	8,781	7,744
32	STS-33R	OV-103	LP04	RP01	FRC3	DoD	DoD	DoD	DoD	DoD	DoD	347.5	145.0	8,707	DoD	14,380	DoD
33	STS-32R	OV-102	LP03	RP04	FRC2	—	—	—	—	00:40:25.6	139.5	204.8	199.1	9,421	6,799	15,422	11,345
34	STS-36	OV-104	LP01	RP03	FRC4	DoD	DoD	DoD	DoD	DoD	DoD	151.9	132.3	4,639	DoD	7,841	DoD
35	STS-31R	OV-103	LP04	RP01	FRC3	—	—	—	—	00:42:35.9	304.4	383.2	376.3	9,379	8,893	15,512	14,449
36	STS-41	OV-103	LP04	RP01	FRC3	—	—	—	—	00:39:53.4	143.6	184.4	183.4	5,394	4,820	8,960	7,960
37	STS-38	OV-104	LP01	RP03	FRC4	DoD	DoD	DoD	DoD	DoD	DoD	163.4	132.3	5,761	DoD	9,447	DoD
38	STS-35	OV-102	LP03	RP04	FRC2	—	—	—	—	00:40:24.7	179.2	218.9	216.0	7,022	6,379	11,596	10,824
39	STS-37	OV-104	LP01	RP01	FRC4	—	—	—	—	00:41:43.1	234.7	284.2	275.0	7,494	6,625	12,403	11,088
40	STS-39	OV-103	LP04	RP03	FRC3	—	—	—	—	00:36:07.5	129.0	161.1	158.8	8,595	5,807	13,803	9,539
41	STS-40	OV-102	LP03	RP04	FRC2	—	—	—	—	00:42:17.6	124.1	180.7	168.0	4,899	4,236	8,121	7,155
42	STS-43	OV-104	LP01	RP01	FRC4	—	—	—	—	00:39:51.0	142.7	200.2	185.3	5,257	4,645	8,683	7,743
43	STS-48	OV-103	LP04	RP03	FRC3	—	—	—	—	00:43:40.0	266.2	360.2	347.5	8,481	7,717	13,876	12,672
44	STS-44	OV-104	LP01	RP01	FRC4	—	—	—	—	00:40:48.0	183.8	226.7	223.3	6,138	5,337	10,025	9,186
45	STS-42	OV-103	LP04	RP03	FRC3	—	—	—	—	00:36:08.1	159.2	187.6	185.3	5,405	4,789	8,828	8,048
46	STS-45	OV-104	LP01	RP01	FRC4	—	—	—	—	00:36:20.0	145.6	186.4	183.0	6,331	5,391	10,425	8,944
47	STS-49	OV-105	LP03	RP04	FRC5	—	—	—	—	00:39:57.8	124.3	209.4	170.3	7,420	5,549	12,341	9,241
48	STS-50	OV-102	LP05	RP05	FRC2	—	—	—	—	00:39:50.7	141.3	184.1	184.1	6,202	5,484	10,352	9,056
49	STS-46	OV-104	LP01	RP01	FRC4	—	—	—	—	00:41:23.4	222.4	264.7	262.4	9,290	8,070	15,441	13,493
50	STS-47	OV-105	LP03	RP04	FRC5	—	—	—	—	00:39:50.7	158.7	184.1	184.1	5,420	4,478	8,974	7,496
51	STS-52	OV-102	LP05	RP05	FRC2	—	—	—	—	00:39:55.5	137.4	187.6	184.1	6,488	5,473	10,665	9,151
52	STS-53	OV-103	LP04	RP03	FRC3	—	—	—	—	00:36:53.8	204.0	346.3	231.0	6,931	5,968	11,541	9,858
53	STS-54	OV-105	LP03	RP04	FRC5	—	—	—	—	00:38:53.4	143.8	188.8	184.5	5,312	4,702	8,814	7,885
54	STS-56	OV-103	LP01	RP03	FRC3	—	—	—	—	00:37:18.2	148.8	184.9	182.9	6,535	5,241	10,847	8,658
55	STS-55	OV-102	LP05	RP05	FRC2	—	—	—	—	00:39:54.9	140.2	187.5	180.9	5,837	4,551	9,607	7,568
56	STS-57	OV-105	LP03	RP04	FRC5	—	—	—	—	00:42:12.4	198.6	290.2	242.2	9,446	7,707	15,546	12,942
57	STS-51	OV-103	LP01	RP03	FRC3	—	—	—	—	00:39:53.8	145.2	184.6	184.1	6,268	4,732	10,322	7,756
58	STS-58	OV-102	LP05	RP05	FRC2	—	—	—	—	00:41:54.9	124.9	178.6	174.9	5,396	4,229	8,878	7,092
59	STS-61	OV-105	LP03	RP04	FRC5	—	—	—	—	00:43:30.2	201.4	354.9	247.3	9,425	8,621	15,410	14,429
60	STS-60	OV-103	LP01	RP03	FRC3	—	—	—	—	00:42:16.3	163.5	220.3	218.2	7,012	5,799	10,880	9,572
61	STS-62	OV-102	LP05	RP05	FRC2	—	—	—	—	00:42:20.0	132.2	187.6	184.1	6,252	4,935	10,291	8,157
62	STS-59	OV-105	LP04	RP01	FRC5	—	—	—	—	00:35:10.3	100.2	139.6	138.7	4,904	3,400	8,229	5,708
63	STS-65	OV-102	LP05	RP05	FRC2	—	—	—	—	00:39:55.2	141.3	187.5	184.5	6,099	5,128	10,042	8,592
64	STS-64	OV-103	LP01	RP03	FRC3	—	—	—	—	00:36:04.8	125.4	162.0	161.0	6,317	4,713	10,318	7,683
65	STS-68	OV-105	LP04	RP01	FRC5	—	—	—	—	00:35:09.0	99.0	137.8	137.4	4,961	3,431	8,206	5,749
66	STS-66	OV-104	LP03	RP04	FRC4	—	—	—	—	00:36:13.1	160.0	190.0	188.7	7,746	5,669	12,990	9,452
67	STS-63	OV-103	LP01	RP03	FRC3	—	—	—	—	00:42:10.0	155.7	211.6	194.3	8,996	7,038	14,813	11,411
68	STS-67	OV-105	LP04	RP01	FRC5	—	—	—	—	00:40:21.5	177.6	219.1	215.5	9,054	6,896	14,945	11,612
69	STS-71	OV-104	LP03	RP04	FRC4	—	—	—	—	00:42:57.7	47.1	184.1	96.7	8,221	6,545	13,850	11,049
70	STS-70	OV-103	LP01	RP03	FRC3	—	—	—	—	00:38:49.4	143.2	184.8	184.5	5,654	5,086	9,302	8,389
71	STS-69	OV-105	LP04	RP05	FRC5	—	—	—	—	00:41:44.4	187.1	231.1	229.2	9,377	7,490	15,461	12,337
72	STS-73	OV-102	LP05	RP01	FRC2	—	—	—	—	00:41:28.7	118.4	172.6	168.7	4,628	3,596	7,789	7,213
73	STS-74	OV-104	LP03	RP04	FRC4	—	—	—	—	00:41:51.9	131.1	187.0	186.8	9,471	7,486	15,529	12,714
74	STS-72	OV-105	LP04	RP05	FRC5	—	—	—	—	00:43:30.0	71.6	286.4	109.6	9,407	8,710	15,478	14,444
75	STS-75	OV-102	LP05	RP01	FRC2	—	—	—	—	00:39:52.4	114.2	185.3	183.3	7,092	6,516	11,771	10,901
76	STS-76	OV-104	LP03	RP04	FRC4	—	—	—	—	00:42:21.9	47.4	183.4	98.2	8,484	7,140	13,028	11,785
77	STS-77	OV-105	LP04	RP05	FRC5	—	—	—	—	00:41:47.1	126.1	176.3	175.8	7,280	6,745	12,048	11,057
78	STS-78	OV-102	LP05	RP01	FRC2	—	—	—	—	00:41:28.6	117.5	176.8	168.8	4,848	4,166	8,101	6,959
79	STS-79	OV-104	LP03	RP04	FRC4	—	—	—	—	00:42:53.3	47.4	183.0	98.3	8,033	6,716	13,296	11,130
80	STS-80	OV-102	LP05	RP01	FRC2	—	—	—	—	00:40:24.4	182.0	219.1	216.2	7,637	5,177	12,636	10,441
81	STS-81	OV-104	LP03	RP04	FRC4	—	—	—	—	00:43:00.1	47.3	184.2	98.0	8,083	5,853	13,336	13,940
82	STS-82	OV-103	LP01	RP03	FRC3	—	—	—	—	00:44:33.6	172.0	360.2	214.4	9,446	8,521	15,394	14,196
83	STS-83	OV-102	LP05	RP05	FRC2	—	—	—	—	00:39:54.7	142.8	356.7	214.4	9,236	8,303	5,552	4,938
84	STS-84	OV-104	LP03	RP04	FRC4	—	—	—	—	00:44:04.1	47.3	185.2	98.5	8,118	6,257	13,402	10,381
85	STS-94	OV-102	LP05	RP05	FRC2	—	—	—	—	00:39:53.0	142.7	188.0	183.6	5,559	4,671	9,252	7,865
86	STS-85	OV-103	LP01	RP03	FRC3	—	—	—	—	00:37:04.0	160.0	184.9	183.7	6,362	4,884	10,572	8,040
87	STS-86	OV-104	LP03	RP04	FRC4	—	—	—	—	00:41:51.0	108.0	185.5	159.8	8,111	6,571	13,418	10,817
88	STS-87	OV-102	LP05	RP05	FRC2	—	—	—	—	00:41:08.0	128.0	177.8	173.5	5,997	4,270	9,936	7,168
89	STS-89	OV-105	LP04	RP01	FRC5	—	—	—	—	00:41:48.4	136.0	188.0	118.0	7,723	6,464	12,801	10,638
90	STS-90	OV-102	LP05	RP05	FRC2	—	—	—	—	00:41:27.0	112.0	177.0	158.5	5,841	5,074	9,676	5,860
91	STS-91	OV-103	LP01	RP03	FRC3	—	—	—	—	00:44:10.8	104.8	203.9	148.8	8,197	6,820	13,580	11,270
92	STS-95	OV-103	LP01	RP03	FRC3	—	—	—	—	00:41:57.0	305.4	339.9	348.9	9,401	8,645	15,477	14,124
93	STS-88	OV-105	LP04	RP01	FRC5	—	—	—	—	00:43:41.0	66.8	201.4	100.3	9,566	8,317	15,248	13,991
94	STS-96	OV-103	LP01	RP03	FRC3	—	—	—	—	00:42:51.2	64.2	198.5	97.5	9,375	7,802	15,477	12,807
95	STS-93	OV-102	LP05	RP05	FRC2	—	—	—	—	00:41:18.0	126.4	175.8	173.0	5,482	4,854	9,086	8,081
96	STS-103	OV-103	LP01	RP03	FRC3	—	—	—	—	00:42:51.2	321.2	333.5	370.5	9,374	8,680	15,460	14,287
97	STS-99	OV-105	LP04	RP01	FRC5	—	—	—	—	00:37:25.3	129.2	142.6	138.8	7,425	5,209	12,025	8,563
98	STS-101	OV-104	LP03	RP04	FRC4	—	—	—	—	00:44:09.5	104.5	185.3	149.6	9,580	7,562	15,362	13,856
99	STS-106	OV-104	LP03	RP04	FRC4	—	—	—	—	00:44:12.4	106.3	180.8	143.4	9,436	7,950	15,428	14,012
100	STS-92	OV-103	LP01	RP03	FRC3	—	—	—	—	00:43:56.9	110.0	191.1	162.9	9,965	8,123	15,224	13,995

SOLID ROCKET BOOSTER SUMMARY

During the first 25 missions, several different motor and case types were flown as the optimal design was sought – the HPM/HP was slightly more powerful than the original standard types. All motors flown since the Challenger accident have been of the Redesigned SRM (RSRM) type, and all cases have been of the Light-Weight Case Redesign (LWC-R) type. Twelve early flights used a serial number scheme that assigned an individual serial number to each SRB, while later flights assigned a common serial number to the SRB pair. In August 1999 NASA awarded Thiokol a $1,730 million contract for 73 RSRMs – 35 flight sets plus three test motors. This brings the total purchased to 230 flight motors and 11 test motors, not counting the FWC motors intended for use at Vandenberg. Separation MET is seconds from lift-off, velocity is in feet per second, and altitude is in feet. [122]

		LEFT SOLID ROCKET BOOSTER								RIGHT SOLID ROCKET BOOSTER								SEPARATION		
Flight	STS	Serial Number	Ignition Weight	Separation Weight	Usable Propellant	Maximum Thrust	Burn Time (Seconds)	Motor Type	Case Type	Serial Number	Ignition Weight	Separation Weight	Usable Propellant	Maximum Thrust	Burn Time (Seconds)	Motor Type	Case Type	MET	Velocity	Altitude
1	STS-1	A07	1,295,940	182,022	1,105,970	2,813,700	131.826	STD	STD	A08	1,298,160	182,738	1,107,563	2,803,200	131.828	STD	STD	131.7	4,230	150,150
2	STS-2	A09	1,296,747	181,590	1,107,967	2,831,300	130.047	STD	STD	A10	1,296,782	181,814	1,107,951	2,812,200	130.046	STD	STD	130.0		
3	STS-3	A11	1,296,697	182,537	1,106,118	2,889,400	127.888	STD	STD	A12	1,296,915	182,225	1,106,600	2,887,000	127.889	STD	STD	127.9		
4	STS-4	A13	1,298,213	182,937	1,107,102	2,824,000	129.979	STD	STD	A14	1,298,253	182,651	1,108,784	2,820,000	129.981	STD	STD	130.0		156,024
5	STS-5	A15	1,298,013	183,299	1,106,744	2,876,700	129.206	STD	STD	A16	1,298,506	183,906	1,107,663	2,873,300	129.207	STD	STD	129.1		155,206
6	STS-6	A17	1,295,519	179,315	1,108,215	2,872,600	129.486	STD	LTWT	A18	1,296,328	179,621	1,108,727	2,836,440	129.487	STD	LTWT	129.4		151,453
7	STS-7	A51	1,295,752	178,177	1,109,552	2,900,300	126.323	STD	LTWT	A52	1,294,403	177,930	1,108,527	2,900,800	126.323	STD	LTWT	126.2	4,319	149,357
8	STS-8	A53	1,297,016	179,183	1,110,544	3,102,600	124.410	HPM	STD	A54	1,297,509	179,050	1,111,248	3,091,100	124.410	HPM	STD	124.3	4,235	152,110
9	STS-9	A55	1,298,366	182,763	1,107,609	3,037,030	126.326	HPM	STD	A60	1,297,983	182,365	1,107,621	3,062,000	126.326	HPM	STD	126.2	4,289	161,689
10	STS-11	A57	1,295,569	180,675	1,107,385	3,054,700	128.010	HPM	MWC	A58	1,296,187	180,729	1,107,483	3,074,600	128.010	HPM	MWC	127.9	4,330	152,605
11	STS-13	BI-012	1,295,903	179,723	1,108,408	3,108,200	125.560	HPM	MWC	BI-012	1,296,386	179,600	1,108,831	3,108,200	125.560	HPM	MWC	125.6	4,137	169,426
12	STS-16	BI-013	1,296,101	178,516	1,110,396	3,096,400	124.590	HPM	LWC	BI-013	1,298,244	180,878	1,110,178	3,082,600	124.590	HPM	LWC	124.5	3,990	162,535
13	STS-17	A63	1,296,572	178,688	1,110,046	3,154,000	124.200	HPM	LWC	A64	1,296,481	179,182	1,109,342	3,112,000	124.144	HPM	LWC	124.1	4,157	157,374
14	STS-19	A65	1,299,609	181,984	1,109,692	3,092,000	125.769	HPM	LWC	A66	1,299,831	181,506	1,110,640	3,110,000	125.769	HPM	LWC	125.6	4,095	156,242
15	STS-20	BI-015	1,294,714	179,014	1,108,850	3,088,000	127.850	HPM	LWC	BI-015	1,295,660	178,574	1,109,157	3,084,000	127.850	HPM	LWC	—	—	—
16	STS-23	BI-018	1,297,460	179,079	1,109,976	3,048,000	126.888	HPM	LWC	BI-018	1,296,665	179,720	1,109,194	3,068,000	126.888	HPM	LWC	126.8	4,077	153,102
17	STS-24	BI-016	1,296,246	179,890	1,108,525	3,094,000	127.403	HPM	LWC	BI-016	1,296,969	179,979	1,109,242	3,077,000	127.403	HPM	LWC	125.9	4,195	156,700
18	STS-25	BI-019	1,297,968	182,237	1,108,553	3,054,000	124.723	HPM	MWC	BI-019	1,298,704	182,379	1,109,897	3,066,000	124.723	HPM	MWC	124.7	3,973	164,000
19	STS-26	BI-017	1,300,211	182,568	1,109,705	3,102,000	125.100	HPM	MWC	BI-017	1,300,031	182,521	1,109,573	3,090,000	125.100	HPM	MWC	125.2	4,284	157,308
20	STS-27	BI-020	1,297,697	180,146	1,109,602	3,172,000	121.114	HPM	LWC	BI-020	1,298,536	180,136	1,110,451	3,158,000	121.114	HPM	LWC	121.0	4,235	154,479
21	STS-28	BI-021	1,298,230	180,647	1,109,788	3,185,000	124.600	HPM	LWC	BI-021	1,299,053	180,761	1,110,152	3,144,000	124.600	HPM	LWC	—	—	—
22	STS-30	BI-022	1,298,021	180,712	1,109,426	3,110,000	124.900	HP	LWC	BI-022	1,297,886	180,698	1,109,226	3,100,000	124.900	HP	LWC	125.1	4,190	162,256
23	STS-31	BI-023	1,296,606	180,172	1,108,582	3,129,000	123.600	HP	LWC	BI-023	1,296,018	180,584	1,107,516	3,146,000	123.600	HP	LWC	123.6	4,275	147,782
24	STS-32	BI-024	1,295,611	180,243	1,107,463	3,058,000	128.397	HP	LWC	BI-024	1,295,702	180,300	1,107,565	3,037,000	128.397	HP	LWC	127.2	4,442	152,555
25	STS-33	BI-026	1,297,828	n/a	n/a	n/a	n/a	HP	LWC	BI-026	1,297,849	n/a	n/a	n/a	n/a	HP	LWC	n/a	n/a	n/a
26	STS-26R	BI-029	1,301,513	188,343	1,106,197	3,060,000	124.806	RSRM	LWC-R	BI-029	1,301,428	188,407	1,106,045	3,060,000	124.846	RSRM	LWC-R	124.8	4,127	151,816
27	STS-27R	BI-030	1,302,581	188,905	1,106,653	3,060,000	126.300	RSRM	LWC-R	BI-030	1,301,490	188,597	1,105,920	3,060,000	126.300	RSRM	LWC-R	—	—	—
28	STS-29R	BI-031	1,300,254	189,207	1,104,157	3,040,000	126.040	RSRM	LWC-R	BI-031	1,300,917	189,030	1,104,804	3,050,000	126.080	RSRM	LWC-R	124.5	4,200	154,800
29	STS-30R	BI-027	1,300,247	187,240	1,106,021	3,080,000	125.120	RSRM	LWC-R	BI-027	1,300,881	188,168	1,105,784	3,070,000	125.160	RSRM	LWC-R	125.3	4,190	155,000
30	STS-28R	BI-028	1,301,088	187,739	1,106,380	3,090,000	124.280	RSRM	LWC-R	BI-028	1,300,644	187,873	1,105,771	3,080,000	124.320	RSRM	LWC-R	—	—	—
31	STS-34	BI-032	1,300,813	186,802	1,106,943	3,060,000	124.880	RSRM	LWC-R	BI-032	1,300,165	187,441	1,105,654	3,070,000	124.920	RSRM	LWC-R	125.0	5,277	156,990
32	STS-33R	BI-034	1,299,924	187,837	1,105,634	3,030,000	126.750	RSRM	LWC-R	BI-034	1,299,919	188,803	1,105,629	3,050,000	126.790	RSRM	LWC-R	—	—	—
33	STS-32R	BI-035	1,299,175	187,026	1,105,393	3,090,000	125.120	RSRM	LWC-R	BI-035	1,299,405	187,301	1,105,355	3,100,000	125.160	RSRM	LWC-R	125.0	5,281	157,254
34	STS-36	BI-036	1,299,251	187,304	1,105,247	3,070,000	125.800	RSRM	LWC-R	BI-036	1,299,095	187,503	1,104,854	3,060,000	125.840	RSRM	LWC-R	—	—	—
35	STS-31R	BI-037	1,300,241	187,649	1,105,792	3,070,000	125.760	RSRM	LWC-R	BI-037	1,300,214	188,020	1,105,479	3,070,000	125.800	RSRM	LWC-R	125.0	5,324	155,453
36	STS-41	BI-040	1,301,372	188,362	1,106,358	3,060,000	124.120	RSRM	LWC-R	BI-040	1,301,388	188,408	1,106,404	3,060,000	124.160	RSRM	LWC-R	126.1	4,113	156,553
37	STS-38	BI-039	1,301,957	188,149	1,107,535	3,100,000	123.840	RSRM	LWC-R	BI-039	1,301,066	188,044	1,106,533	3,080,000	123.840	RSRM	LWC-R	—	—	—
38	STS-35	BI-038	1,300,088	187,520	1,106,120	3,050,000	125.760	RSRM	LWC-R	BI-038	1,300,124	187,617	1,106,020	3,060,000	125.760	RSRM	LWC-R	125.7	4,200	152,000
39	STS-37	BI-042	1,300,130	187,533	1,106,148	3,070,000	125.040	RSRM	LWC-R	BI-042	1,299,254	187,021	1,105,628	3,080,000	125.040	RSRM	LWC-R	125.3	5,295	156,097
40	STS-39	BI-043	1,299,733	187,310	1,105,908	3,090,000	124.720	RSRM	LWC-R	BI-043	1,301,485	187,426	1,107,422	3,090,000	124.720	RSRM	LWC-R	124.8	5,036	152,683
41	STS-40	BI-044	1,301,303	188,417	1,106,653	3,010,000	124.840	RSRM	LWC-R	BI-044	1,301,723	187,698	1,106,785	3,030,000	124.840	RSRM	LWC-R	124.7	5,275	153,002
42	STS-43	BI-045	1,299,661	187,674	1,105,687	3,030,000	125.560	RSRM	LWC-R	BI-045	1,299,220	188,188	1,105,064	3,010,000	125.560	RSRM	LWC-R	125.6	4,265	149,261
43	STS-48	BI-046	1,298,959	188,011	1,104,717	3,060,000	125.480	RSRM	LWC-R	BI-046	1,298,580	187,915	1,104,372	3,070,000	125.480	RSRM	LWC-R	125.0	4,149	155,365
44	STS-44	BI-047	1,298,356	187,449	1,104,705	3,030,000	126.560	RSRM	LWC-R	BI-047	1,300,086	188,306	1,105,506	3,010,000	126.560	RSRM	LWC-R	126.6	4,229	152,899
45	STS-42	BI-048	1,299,484	188,159	1,105,779	3,020,000	127.840	RSRM	LWC-R	BI-048	1,298,658	188,235	1,104,672	3,020,000	127.840	RSRM	LWC-R	127.8	4,253	155,716
46	STS-45	BI-049	1,298,567	187,441	1,104,893	2,990,000	128.120	RSRM	LWC-R	BI-049	1,298,967	187,791	1,105,018	2,980,000	128.120	RSRM	LWC-R	128.0	4,222	155,044
47	STS-49	BI-050	1,299,195	187,572	1,105,396	3,010,000	127.240	RSRM	LWC-R	BI-050	1,298,790	187,154	1,105,364	3,010,000	127.240	RSRM	LWC-R	127.2	4,216	156,553
48	STS-50	BI-051	1,298,413	187,826	1,103,799	3,010,000	126.320	RSRM	LWC-R	BI-051	1,299,050	188,868	1,103,835	3,030,000	126.320	RSRM	LWC-R	126.2	4,216	157,456
49	STS-46	BI-052	1,297,746	187,985	1,103,370	3,050,000	125.480	RSRM	LWC-R	BI-052	1,298,292	187,548	1,104,470	3,040,000	125.480	RSRM	LWC-R	125.1	4,233	152,815
50	STS-47	BI-053	1,298,225	187,635	1,104,207	3,070,000	124.120	RSRM	LWC-R	BI-053	1,299,291	188,125	1,104,912	3,070,000	124.120	RSRM	LWC-R	124.0	4,122	157,413
51	STS-52	BI-054	1,299,188	187,816	1,105,037	3,110,000	123.120	RSRM	LWC-R	BI-054	1,300,396	189,026	1,105,115	3,100,000	123.120	RSRM	LWC-R	123.1	4,108	149,168
52	STS-53	BI-055	1,299,174	188,013	1,104,863	3,040,000	126.360	RSRM	LWC-R	BI-055	1,298,532	187,840	1,104,434	3,030,000	126.360	RSRM	LWC-R	126.3	4,216	153,648
53	STS-54	BI-056	1,299,818	187,796	1,105,869	3,040,000	125.960	RSRM	LWC-R	BI-056	1,299,187	187,782	1,105,225	3,050,000	125.960	RSRM	LWC-R	125.7	4,212	152,107
54	STS-56	BI-058	1,299,765	187,991	1,105,515	3,060,000	125.880	RSRM	LWC-R	BI-058	1,300,514	187,559	1,106,723	3,070,000	125.880	RSRM	LWC-R	125.8	4,179	151,248
55	STS-55	BI-057	1,298,514	187,524	1,104,746	3,050,000	125.480	RSRM	LWC-R	BI-057	1,300,561	188,338	1,105,941	3,060,000	125.480	RSRM	LWC-R	125.5	4,172	154,326
56	STS-57	BI-059	1,300,547	187,659	1,106,672	3,040,000	124.760	RSRM	LWC-R	BI-059	1,300,983	188,795	1,105,973	3,040,000	124.760	RSRM	LWC-R	124.7	4,224	146,296
57	STS-51	BI-060	1,298,328	188,343	1,104,376	3,050,000	124.680	RSRM	LWC-R	BI-060	1,298,670	188,413	1,104,651	3,040,000	124.680	RSRM	LWC-R	124.7	4,203	154,673
58	STS-58	BI-061	1,298,197	187,672	1,104,316	3,050,000	123.640	RSRM	LWC-R	BI-061	1,298,503	188,397	1,103,797	3,070,000	123.640	RSRM	LWC-R	123.6	4,074	152,511
59	STS-61	BI-063	1,298,560	187,757	1,104,316	3,010,000	126.680	RSRM	LWC-R	BI-063	1,298,435	187,742	1,104,442	3,030,000	126.680	RSRM	LWC-R	126.5	4,229	154,032
60	STS-60	BI-062	1,298,776	187,973	1,104,531	3,090,000	125.080	RSRM	LWC-R	BI-062	1,299,783	188,075	1,105,413	3,100,000	125.080	RSRM	LWC-R	125.0	4,216	154,650
61	STS-62	BI-064	1,299,668	188,010	1,105,414	3,070,000	126.240	RSRM	LWC-R	BI-064	1,299,184	188,085	1,104,783	3,060,000	126.240	RSRM	LWC-R	126.3	4,227	155,483
62	STS-59	BI-065	1,300,061	188,268	1,105,551	3,080,000	126.120	RSRM	LWC-R	BI-065	1,300,299	188,593	1,105,452	3,080,000	126.120	RSRM	LWC-R	124.1	4,196	150,886
63	STS-65	BI-066	1,299,578	189,271	1,104,118	3,060,000	123.480	RSRM	LWC-R	BI-066	1,300,094	188,764	1,105,009	3,070,000	123.480	RSRM	LWC-R	123.4	4,138	150,801
64	STS-64	BI-068	1,298,944	188,348	1,104,375	3,090,000	122.480	RSRM	LWC-R	BI-068	1,299,118	188,603	1,104,301	3,090,000	122.480	RSRM	LWC-R	122.5	4,156	148,952
65	STS-68	BI-067	1,299,293	188,278	1,104,738	3,060,000	123.960	RSRM	LWC-R	BI-067	1,299,523	188,598	1,104,657	3,060,000	123.960	RSRM	LWC-R	123.6	4,158	151,793
66	STS-66	BI-069	1,299,860	188,147	1,105,521	3,050,000	124.720	RSRM	LWC-R	BI-069	1,300,230	188,573	1,105,412	3,050,000	124.720	RSRM	LWC-R	124.7	4,197	149,107
67	STS-63	BI-070	1,299,711	188,588	1,105,002	3,120,000	125.000	RSRM	LWC-R	BI-070	1,300,127	188,177	1,105,683	3,100,000	125.000	RSRM	LWC-R	124.9	4,242	155,257
68	STS-67	BI-071	1,299,857	188,545	1,105,059	3,100,000	125.040	RSRM	LWC-R	BI-071	1,299,392	187,696	1,105,424	3,080,000	125.040	RSRM	LWC-R	125.0	4,264	152,923
69	STS-71	BI-072	1,299,083	187,089	1,105,811	3,060,000	122.500	RSRM	LWC-R	BI-072	1,298,854	187,456	1,105,138	3,070,000	122.500	RSRM	LWC-R	123.5	4,280	150,973
70	STS-70	BI-073	1,299,219	187,499	1,105,492	3,100,000	122.870	RSRM	LWC-R	BI-073	1,299,339	188,111	1,105,472	3,090,000	122.870	RSRM	LWC-R	122.7	4,157	152,664
71	STS-69	BI-074	1,299,384	187,291	1,105,836	3,090,000	122.495	RSRM	LWC-R	BI-074	1,299,175	187,529	1,105,327	3,100,000	122.495	RSRM	LWC-R	122.4	4,206	151,749
72	STS-73	BI-075	1,299,553	187,895	1,105,653	3,070,000	122.600	RSRM	LWC-R	BI-075	1,300,509	188,139	1,106,099	3,070,000	122.600	RSRM	LWC-R	123.5	4,217	153,975
73	STS-74	BI-076	1,299,872	187,704	1,105,959	3,100,000	122.010	RSRM	LWC-R	BI-076	1,299,878	187,612	1,106,113	3,090,000	122.010	RSRM	LWC-R	122.9	4,205	151,230
74	STS-72	BI-077	1,302,278	188,337	1,107,691	3,120,000	124.400	RSRM	LWC-R	BI-077	1,301,220	188,239	1,106,629	3,090,000	124.400	RSRM	LWC-R	124.3	4,217	156,819
75	STS-75	BI-078	1,300,541	188,241	1,106,063	3,050,000	126.300	RSRM	LWC-R	BI-078	1,300,635	188,362	1,106,374	3,070,000	126.300	RSRM	LWC-R	126.2	4,244	155,143
76	STS-76	BI-079	1,299,896	187,989	1,105,812	3,070,000	125.980	RSRM	LWC-R	BI-079	1,299,353	187,856	1,105,368	3,070,000	125.980	RSRM	LWC-R	125.8	4,248	153,959
77	STS-77	BI-080	1,300,764	188,814	1,105,737	3,070,000	124.240	RSRM	LWC-R	BI-080	1,299,175	187,015	1,106,937	3,070,000	124.240	RSRM	LWC-R	124.2	4,236	154,408
78	STS-78	BI-081	1,297,867	187,194	1,104,460	3,100,000	123.690	RSRM	LWC-R	BI-081	1,298,253	187,056	1,104,980	3,110,000	123.690	RSRM	LWC-R	123.7	4,227	155,249
79	STS-79	BI-083	1,297,828	187,273	1,104,318	3,140,000	122.100	RSRM	LWC-R	BI-083	1,298,848	188,197	1,104,363	3,313,000	122.100	RSRM	LWC-R	121.9	4,178	152,913
80	STS-80	BI-084	1,299,137	187,273	1,105,080	3,110,000	124.020	RSRM	LWC-R	BI-084	1,299,854	188,197	1,105,867	3,120,000	124.020	RSRM	LWC-R	124.7	4,213	154,621
81	STS-81	BI-082	1,298,753	187,581	1,104,886	3,060,000	125.700	RSRM	LWC-R	BI-082	1,298,211	187,764	1,104,146	3,060,000	125.700	RSRM	LWC-R	125.1	4,222	153,816
82	STS-82	BI-085	1,299,604	188,656	1,104,736	3,110,000	123.700	RSRM	LWC-R	BI-085	1,298,386	186,753	1,105,358	3,130,000	123.700	RSRM	LWC-R	123.7	4,214	155,372
83	STS-83	BI-086	1,299,459	187,990	1,105,263	3,120,000	123.220	RSRM	LWC-R	BI-086	1,298,204	186,710	1,104,298	3,120,000	123.220	RSRM	LWC-R	123.0	4,178	151,302
84	STS-84	BI-087	1,298,206	187,298	1,104,651	3,090,000	124.000	RSRM	LWC-R	BI-087	1,298,123	187,379	1,104,430	3,100,000	124.000	RSRM	LWC-R	123.8	4,234	152,341
85	STS-94	BI-088	1,297,078	187,632	1,103,239	3,070,000	123.500	RSRM	LWC-R	BI-088	1,297,346	187,716	1,103,362	3,070,000	123.500	RSRM	LWC-R	123.6	4,195	154,302
86	STS-85	BI-089	1,298,435	187,922	1,103,997	3,110,000	123.600	RSRM	LWC-R	BI-089	1,299,129	187,962	1,104,611	3,110,000	123.600	RSRM	LWC-R	123.9	4,225	156,447
87	STS-86	BI-090	1,297,660	187,903	1,103,490	3,070,000	123.000	RSRM	LWC-R	BI-090	1,298,078	188,051	1,103,811	3,070,000	123.000	RSRM	LWC-R	123.0	4,231	153,405
88	STS-87	BI-092	1,297,838	187,080	1,104,549	3,130,000	123.700	RSRM	LWC-R	BI-092	1,298,044	187,852	1,104,055	3,110,000	123.700	RSRM	LWC-R	124.2	4,220	151,750
89	STS-89	BI-093	1,298,227	187,382	1,104,736	3,120,000	123.300	RSRM	LWC-R	BI-093	1,298,526	187,636	1,104,628	3,120,000	123.300	RSRM	LWC-R	123.3	4,232	155,386
90	STS-90	BI-094	1,298,901	188,501	1,104,673	3,070,000	125.100	RSRM	LWC-R	BI-094	1,298,520	187,815	1,104,482	3,070,000	125.100	RSRM	LWC-R	125.4	4,234	153,998
91	STS-91	BI-091	1,298,618	187,756	1,104,680	3,080,000	123.000	RSRM	LWC-R	BI-091	1,297,292	186,855	1,104,263	3,090,000	123.000	RSRM	LWC-R	123.1	4,238	154,520
92	STS-95	BI-096	1,297,332	187,565	1,104,201	3,080,000	122.600	RSRM	LWC-R	BI-096	1,298,008	187,893	1,104,549	3,070,000	122.360	RSRM	LWC-R	122.6	4,402	153,662
93	STS-88	BI-095	1,297,827	187,463	1,104,798	3,040,000	123.600	RSRM	LWC-R	BI-095	1,297,944	187,967	1,104,412	3,050,000	123.760	RSRM	LWC-R	123.9	4,376	154,436
94	STS-96	BI-098	1,299,251	187,295	1,105,836	3,090,000	123.470	RSRM	LWC-R	BI-098	1,299,184	187,652	1,105,325	3,100,000	123.500	RSRM	LWC-R	123.4	4,253	154,526
95	STS-93	BI-097	1,300,620	187,958	1,105,653	3,070,000	123.600	RSRM	LWC-R	BI-097	1,300,562	188,236	1,106,052	3,070,000	123.100	RSRM	LWC-R	124.6	4,512	153,285
96	STS-103	BI-099	1,299,752	187,705	1,105,952	3,100,000	123.000	RSRM	LWC-R	BI-099	1,299,842	187,632	1,105,883	3,090,000	122.950	RSRM	LWC-R	123.2	4,169	152,895
97	STS-99	BI-100	1,302,120	188,350	1,107,691	3,120,000	124.500	RSRM	LWC-R	BI-100	1,301,220	188,285	1,106,974	3,110,000	124.300	RSRM	LWC-R	124.9	4,352	153,854
98	STS-101	BI-102	1,300,594	188,254	1,106,064	3,050,000	126.300	RSRM	LWC-R	BI-102	1,300,659	188,362	1,106,125	3,070,000	126.100	RSRM	LWC-R	123.2	4,222	154,126
99	STS-106	BI-101	1,299,539	187,989	1,105,385	3,080,000	125.950	RSRM	LWC-R	BI-101	1,299,583	187,658	1,105,628	3,050,000	125.800	RSRM	LWC-R	123.8	4,284	153,529
100	STS-92	BI-104	1,299,128	188,852	1,106,091	3,100,000	124.500	RSRM	LWC-R	BI-104	1,299,539	187,219	1,105,985	3,090,000	124.650	RSRM	LWC-R	123.6	4,324	153,764

MAIN ENGINE SUMMARY

SSME #1 is installed in the top-center position; SSME #2 is at the lower left; and SSME #3 is at the lower right (as viewed from the rear of the Orbiter). The maximum throttle listed in the table at right is regardless of whether it was before or after the throttle change made at max-q.[124]

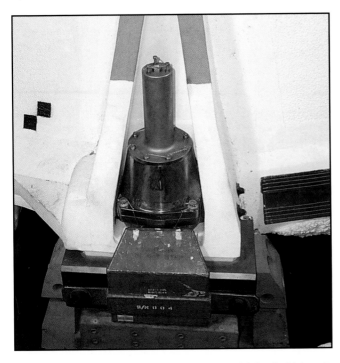

The SRB hold down. The photo below shows the hold-down bolt before the debris catcher ('blast can') has been installed. Although difficult to discern in a monochrome photo, there is a red cap at the corner of the nut closest to the bottom – this is where the pyro will be installed, with a similar one installed 180 degrees opposite. The photo above shows the hold-down after the pyro and blast can have been installed. The hold down bolt is 28 inches long and 3.5 inches in diameter. Each SRB is secured to the MLP using four of these bolts, and this is all that attaches the entire Space Shuttle stack to the MLP. Note the safety wire on the various bolts. (Kim Keller)

Seq. #	Mission	Serial Number (Pos #1)	# of Flights	Duration (secs)	Serial Number (Pos #2)	# of Flights	Duration (secs)	Serial Number (Pos #3)	# of Flights	Duration (secs)	Max Throttle
	FRF-1	2007	–	21.86	2006	–	23.83	2005	–	23.97	100
1	STS-1	2007	1	519.42	2006	1	519.56	2005	1	519.68	100
2	STS-2	2007	2	520.13	2006	2	520.24	2005	2	520.36	100
3	STS-3	2007	3	519.67	2006	3	519.80	2005	3	519.91	100
4	STS-4	2007	4	519.03	2006	4	519.13	2005	4	519.31	100
5	STS-5	2007	5	517.04	2006	5	517.21	2005	5	517.29	100
	FRF-2	2011	–	21.80	2015	–	23.76	2012	–	23.88	100
	FRF-3	2011	–	21.80	2015	–	23.80	2012	–	23.92	100
6	STS-6	2017	1	505.76	2015	1	505.87	2012	1	505.99	104
7	STS-7	2017	2	506.50	2015	2	506.60	2012	2	506.71	104
8	STS-8	2017	3	527.98	2015	3	528.11	2012	3	528.19	100
9	STS-9	2011	1	515.52	2018	1	515.65	2019	1	515.78	104
10	STS-11	2109	1	527.79	2015	4	527.89	2012	4	528.01	100
11	STS-13	2109	2	517.14	2020	1	517.25	2012	5	517.40	104
	FRF-4	2021	–	17.60	2018	–	19.60	2017	–	19.72	100
	STS-14	2109	–	0.00	2018	–	2.00	2017	–	0.22	Abort
12	STS-16	2109	3	521.53	2018	2	521.67	2021	1	521.78	104
13	STS-17	2023	1	536.69	2020	2	536.83	2021	1	536.95	100
14	STS-19	2109	4	519.53	2018	3	519.67	2012	6	519.79	104
15	STS-20	2109	5	517.02	2018	4	517.15	2012	7	517.27	104
16	STS-23	2109	6	538.23	2018	5	538.33	2012	8	538.47	100
17	STS-24	2023	2	521.63	2020	3	521.46	2021	2	521.57	104
18	STS-25	2109	7	522.12	2018	6	522.24	2012	9	522.36	104
	STS-26	2023	–	3.52	2020	–	1.74	2021	–	3/76	Abort
19	STS-26	2023	3	349.75	2020	4	587.51	2021	3	587.85	104
20	STS-27	2109	8	513.92	2018	7	514.04	2012	10	514.16	104
	FRF-5	2011	–	19.20	2019	–	20.50	2017	–	20.50	104
21	STS-28	2011	2	518.28	2019	1	518.40	2017	2	518.52	104
22	STS-30	2023	4	521.32	2020	5	521.46	2021	4	521.58	104
23	STS-31	2011	3	517.65	2019	2	517.77	2017	3	517.88	104
24	STS-32	2015	1	508.00	2018	8	508.12	2109	1	508.24	104
25	STS-33	2023	5	79.40	2020	6	79.56	2021	5	79.62	104
	FRF-6	2019	–	19.20	2022	–	20.52	2028	–	21.92	100
26	STS-26R	2019	4	518.71	2022	1	518.83	2028	1	518.95	104
27	STS-27R	2027	1	519.92	2030	1	520.04	2029	1	520.16	104
28	STS-29R	2031	1	515.52	2022	2	515.64	2028	2	515.76	104
29	STS-30R	2027	2	515.32	2030	2	515.44	2029	2	515.56	104
30	STS-28R	2019	5	521.68	2022	3	521.84	2028	3	521.92	104
31	STS-34	2027	3	518.40	2030	3	518.56	2029	3	518.68	104
32	STS-33R	2011	4	512.36	2031	2	512.48	2107	1	512.56	104
33	STS-32R	2024	1	518.48	2022	4	518.56	2028	4	518.72	104
34	STS-36	2019	6	515.92	2030	4	516.04	2027	4	516.16	104
35	STS-31R	2011	5	516.92	2031	3	517.08	2107	2	517.16	104
36	STS-41	2011	6	516.40	2031	4	516.52	2107	3	516.60	104
37	STS-38	2019	7	516.20	2022	5	516.32	2027	5	516.40	104
38	STS-35	2024	2	517.64	2012	11	517.80	2028	5	517.92	104
39	STS-37	2019	8	519.04	2031	5	519.12	2107	4	519.24	104
40	STS-39	2026	1	520.44	2030	5	520.56	2029	4	520.68	104
41	STS-40	2015	6	516.16	2022	6	516.24	2027	6	516.40	104
42	STS-43	2024	3	513.92	2012	12	514.04	2028	6	514.16	104
43	STS-48	2019	9	522.44	2031	6	522.56	2107	5	522.72	104
44	STS-44	2015	7	516.04	2030	6	516.12	2029	5	516.28	104
45	STS-42	2026	2	516.44	2022	7	516.56	2027	7	516.64	104
46	STS-45	2024	4	516.76	2012	13	516.87	2028	7	517.00	104
	FRF-7	2035	–	19.19	2033	–	20.52	2034	–	21.91	100
47	STS-49	2030	7	515.70	2015	8	515.84	2017	6	515.94	104
48	STS-50	2019	10	513.56	2031	7	513.68	2011	7	513.81	104
49	STS-46	2032	1	515.72	2033	1	515.84	2027	8	515.96	104
50	STS-47	2026	3	519.97	2022	8	520.08	2029	6	520.21	104
51	STS-52	2030	8	518.40	2015	9	518.52	2034	1	518.64	104
52	STS-53	2024	5	519.64	2012	14	519.77	2017	7	519.87	104
53	STS-54	2019	11	516.28	2033	2	516.41	2018	9	516.52	104
54	STS-56	2024	6	520.64	2033	3	520.77	2018	10	520.89	104
	STS-55	2030	–	3.84	2034	–	2.70	2011	–	1.50	Abort
55	STS-55	2031	8	516.60	2109	16	516.72	2029	7	516.85	104
56	STS-57	2019	12	518.52	2034	2	518.65	2017	8	518.77	104
	STS-51	2030	–	3.84	2033	–	1.50	2032	–	2.80	Abort
57	STS-51	2031	9	515.84	2034	3	515.97	2029	8	516.08	104
58	STS-58	2024	7	521.92	2109	11	522.04	2018	11	522.16	104
59	STS-61	2019	13	517.64	2033	4	517.77	2017	9	517.89	104
60	STS-60	2012	15	518.35	2034	4	518.49	2032	2	518.60	104
61	STS-62	2031	10	516.77	2109	12	516.88	2029	9	517.01	104
62	STS-59	2028	8	519.37	2033	5	519.48	2018	12	519.60	104
63	STS-65	2019	14	516.93	2030	9	517.06	2017	10	517.19	104
64	STS-64	2031	11	520.95	2029	13	521.10	2029	10	521.23	104
	STS-68	2012	–	6.96	2034	–	5.80	2032	–	4.72	Abort
65	STS-68	2028	9	519.71	2033	6	519.85	2026	4	519.97	104
66	STS-66	2030	10	520.12	2034	5	520.24	2017	11	520.37	104
67	STS-63	2035	1	517.49	2109	14	517.60	2029	11	517.70	104
68	STS-67	2012	16	513.33	2033	7	513.43	2031	12	513.57	104
69	STS-71	2028	10	516.83	2034	6	516.96	2032	3	517.06	104
70	STS-70	2036	1	516.97	2019	15	517.06	2017	12	517.20	104
71	STS-69	2035	2	516.49	2109	15	516.60	2029	12	516.73	104
72	STS-73	2037	1	515.68	2031	13	515.81	2038	1	515.92	104
73	STS-74	2012	17	519.01	2026	5	519.11	2032	4	519.22	104
74	STS-72	2028	11	513.01	2039	1	513.13	2036	2	513.25	104
75	STS-75	2035	3	514.35	2034	7	514.23	2017	13	514.61	104
76	STS-76	2035	3	518.92	2109	16	519.04	2019	16	519.17	104
77	STS-77	2037	2	514.02	2040	1	514.20	2038	2	514.33	104
78	STS-78	2041	1	515.63	2039	2	515.76	2036	3	515.89	104
79	STS-79	2012	18	520.36	2031	14	520.48	2033	8	520.61	104
80	STS-80	2032	5	516.41	2026	6	516.52	2029	14	516.63	104
81	STS-81	2041	2	518.12	2034	8	518.25	2042	1	518.37	104
82	STS-82	2037	3	515.68	2040	2	515.81	2038	3	515.93	104
83	STS-83	2012	19	516.76	2109	17	516.90	2019	17	517.00	104
84	STS-84	2032	6	519.30	2031	15	519.30	2029	15	519.45	104
85	STS-94	2037	4	515.06	2034	9	515.22	2033	9	515.33	104
86	STS-85	2041	3	518.47	2039	3	518.60	2042	2	518.73	104
87	STS-86	2012	20	517.20	2040	3	517.33	2019	18	517.44	104
88	STS-87	2031	16	516.00	2039	4	516.13	2037	5	516.25	104
89	STS-89	2043	1	515.38	2044	1	515.52	2045	1	515.64	104.5
90	STS-90	2041	4	514.32	2032	7	514.45	2012	21	514.56	104
91	STS-91	2047	1	516.00	2040	4	516.12	2042	3	516.25	104/104.5
92	STS-95	2048	1	507.56	2043	2	507.67	2045	2	507.79	104.5
93	STS-88	2050	1	508.49	2044	2	508.60	2041	3	508.73	104/104.5
94	STS-96	2047	2	508.48	2051	1	508.61	2049	1	508.73	104.5
95	STS-93	2012	22	513.99	2031	17	514.25	2019	19	514.25	104
96	STS-103	2053	1	511.71	2043	4	511.85	2049	2	511.97	104.5
97	STS-99	2052	1	509.80	2044	4	509.92	2047	3	510.05	104.5
98	STS-101	2043	5	511.21	2054	1	511.31	2049	3	511.45	104.5
99	STS-106	2052	2	511.46	2044	4	511.57	2047	4	511.68	104.5
100	STS-92	2045	4	511.95	2053	2	512.09	2048	1	512.21	104.5

FLIGHT READINESS FIRINGS

	Orbiter		Pad	Date	Duration (SSME1/2/3)			%
FRF-1	OV-102	Columbia	39A	20 Feb 81	21.86	23.83	23.97	100
FRF-2	OV-099	Challenger	39A	18 Dec 82	21.80	23.76	23.88	100
FRF-3	OV-099	Challenger	39A	25 Jan 83	21.80	23.80	23.92	100
FRF-4	OV-103	Discovery	39A	02 Jun 84	17.60	19.60	19.72	100
FRF-5	OV-104	Atlantis	39A	12 Sep 85	19.20	20.50	20.50	100
FRF-6	OV-103	Discovery	39B	10 Aug 88	19.20	20.52	21.92	100
FRF-7	OV-105	Endeavour	39B	06 Apr 92	19.19	20.52	21.91	100
	OV-102	Columbia	VLS	Cancelled				

The following chart summarizes the location and number of days required to process each vehicle flow. It should be noted that the days listed are calendar days, not work days, and include storage time if it occurred in the middle of a flow. The number of ground power-on hours were not available for all flows. Where vehicles were rolled back, the pertinent data is shown.[120]

Seq. #	Mission	Orbiter	OPF In-Date	OPF Bay	OPF # Days	VAB Roll-Over	VAB High Bay	VAB # Days	MLP Stacked	MLP	PAD Roll-Out	PAD Pad	PAD # Days	# Hours Ground Power-On	Notes
1	STS-1	OV-102	22 Mar 79	1	613	24 Nov 80	3	37	26 Nov 80	1	31 Dec 80	39A	102	4,508	
2	STS-2	OV-102	29 Apr 81	1	103	10 Aug 81	3	21	26 Aug 81	1	31 Aug 81	39A	74	2,755	
3	STS-3	OV-102	26 Nov 81	1	68	03 Feb 82	3	14	12 Feb 83	1	16 Feb 82	39A	34	1,814	
4	STS-4	OV-102	07 Apr 82	1	42	18 May 82	3	7	25 May 82	1	26 May 82	39A	33	1,318	
5	STS-5	OV-102	16 Jul 82	1	57	09 Sep 82	3	11	17 Sep 82	1	21 Sep 82	39A	52	1,647	
6	STS-6	OV-099	06 Jul 82	2	141	22 Nov 82	3	7	29 Nov 82	2	30 Nov 82	39A	126	4,556	
7	STS-7	OV-099	17 Apr 83	1	34	21 May 83	3	5	26 May 83	1	26 May 83	39A	24	1,143	
8	STS-8	OV-099	30 Jun 83	1	27	27 Jul 83	3	6	30 Jul 83	2	02 Aug 83	39A	29	998	
–	STS-9	OV-102	02 Jul 83	2	81	23 Sep 83	3	5	28 Sep 83	1	28 Sep 83	39A	18	–	Recycle due to SRB nozzle problem
9	STS-9	OV-102	19 Oct 83	2	14	03 Nov 83	3	3	07 Nov 83	1	08 Nov 83	39A	21	3,116	
10	STS-11/41-E	OV-099	11 Sep 83	1	67	06 Jan 84	3	6	12 Jan 84	2	12 Jan 84	39A	22	1,439	
11	STS-13/41-C	OV-099	11 Feb 84	2	32	14 Mar 84	3	4	18 Mar 84	1	18 Mar 84	39A	19	1,001	
–	STS-14/41-D	OV-103	10 Jan 84	1	124	12 May 84	3	6	12 May 84	2	19 May 84	39A	56	–	Launch aborted, returned to VAB
12	STS-16/41-D	OV-103	17 Jul 84	1	16	14 Jul 84	3	7	01 Aug 84	2	09 Aug 84	39A	20	2,330	41-D (STS-14) with 41-F (STS-16) payloads
13	STS-1741-G	OV-099	18 Apr 84	2	69	08 Sep 84	1	5	08 Sep 84	1	13 Sep 84	39A	23	1,440	
14	STS-19/51-A	OV-103	11 Sep 84	2	37	18 Oct 84	3	5	19 Oct 84	2	12 Oct 84	39A	17	1,042	
15	STS-20/51-C	OV-103	16 Nov 84	2	35	21 Dec 84	1	14	21 Dec 84	1	05 Jan 85	39A	20	999	
16	STS-23/51-D	OV-103	27 Jan 85	2	46	23 Mar 85	3	13	23 Mar 85	1	28 Mar 85	39A	16	990	
–	STS-22/51-E	OV-099	12 Oct 84	1	120	10 Feb 85	3	5	10 Feb 85	2	15 Feb 85	39A	17	–	STS-22/51-E cancelled due to TDRS payload problems
17	STS-24/51-B	OV-099	14 Mar 85	1	30	10 Apr 85	1	5	10 Apr 85	2	15 Apr 85	39A	15	2,532	
18	STS-25/51-G	OV-103	19 Apr 85	2	38	28 May 85	1	7	29 May 85	1	04 Jun 85	39A	14	927	
19	STS-26/51-F	OV-099	11 May 85	1	43	24 Jun 85	3	5	24 Jun 85	2	29 Jun 85	39A	31	1,482	
20	STS-27/51-I	OV-103	29 Jun 85	2	30	30 Jul 85	1	7	30 Jul 85	1	06 Aug 85	39A	22	1,002	
21	STS-28/51-J	OV-104	31 Apr 85	1/2/1	27/51/12	12 Aug 85	3	18	12 Aug 85	2	30 Aug 85	39A	35	2,590	*Atlantis* in VAB High Bay #2 storage from 10 May to 28 May 85
22	STS-30/61-A	OV-099	11 Aug 85	1	40	12 Oct 85	1	4	12 Oct 85	1	16 Oct 85	39A	15	1,092	
23	STS-31/61-B	OV-104	12 Oct 85	1	27	08 Nov 85	3	4	08 Nov 85	2	12 Nov 85	39A	15	854	
24	STS-32/61-C	OV-102	18 Jul 85	2	108	22 Nov 85	1	10	22 Nov 85	1	02 Dec 85	39A	42	2,155	*Columbia* in VAB High Bay #2 storage from 08 Sep to 26 Sep 85
25	STS-33/51-L	OV-099	11 Nov 85	1	35	17 Dec 85	3	5	17 Dec 85	1	22 Dec 85	39B	38	1,213	
26	STS-26R	OV-103	02 Oct 87	1	263	21 Jun 88	1	13	21 Jun 88	2	04 Jul 88	39B	88	5,900	*Discovery* stored in OPF-1 during Challenger stand-down
27	STS-27R	OV-104	22 Feb 86	1	243	22 Oct 88	3	10	24 Oct 88	1	02 Nov 88	39B	31	1,853	*Atlantis* stored in OPF-2 during Challenger stand-down
28	STS-29R	OV-103	12 Oct 88	1	106	02 Mar 89	1	11	29 Jan 89	2	04 Feb 89	39B	39	2,204	
29	STS-30R	OV-104	14 Dec 88	2	87	11 Mar 89	3	11	11 Mar 89	1	22 Mar 89	39B	44	1,725	
30	STS-28R	OV-102	08 Sep 88	2	208	03 Jul 89	1	12	04 Jul 89	2	15 Jul 89	39B	25		
31	STS-34	OV-104	16 May 89	2	97	21 Aug 89	1	8	22 Aug 89	1	29 Aug 89	39B	51		
32	STS-33R	OV-103	25 Mar 89	1	134	05 Oct 89	3	21	06 Oct 89	2	27 Oct 89	39B	27		
33	STS-32R	OV-102	22 Aug 89	1	87	16 Nov 89	1	11	17 Nov 89	3	28 Nov 89	39A	43		
34	STS-36	OV-104	19 Jan 90	1	89	19 Jan 90	1	6	19 Jan 90	1	25 Jan 90	39A	35		
35	STS-31R	OV-103	05 Dec 89	2	88	05 Mar 90	1	11	05 Mar 90	2	16 Mar 90	39B	40		
36	STS-41	OV-103	08 May 90	1	111	27 Aug 90	3	9	27 Aug 90	2	04 Sep 90	39B	32		
–	STS-38	OV-104	14 Mar 90	2	87	08 Jun 90	1	9	08 Jun 90	2	18 Jun 90	39A	52		Hydrogen leaks at pad forced recycle
37	STS-38	OV-104	14 Aug 90	2	49	02 Oct 90	1	7	04 Oct 90	1	12 Oct 90	39A	34		
–	STS-35	OV-102	27 Jan 90	1	79	16 Apr 90	3	6	16 Apr 90	3	22 Apr 90	39A	50		Hydrogen leaks at pad forced recycle
–	STS-35	OV-102	15 Jun 90	2	48	02 Aug 90	3	3	02 Aug 90	3	09 Aug 90	39B	59		Roll-around to Pad 39B to make room for *Atlantis*
–	STS-35	OV-102	–		–	–		–	–		08 Oct 90	39B	1		Tropical Storm Klaus forced second roll-back
38	STS-35	OV-102				09 Oct 90	3	5	–		14 Oct 90	39B	48		
39	STS-37	OV-104	20 Nov 90	2	108	08 Mar 91	3	6	08 Mar 91	1	14 Mar 91	39B	22	1,238	
–	STS-39	OV-103	17 Oct 90	1	113	09 Feb 91	1	6	09 Feb 91	2	15 Feb 91	39A	20		Cracked ET door hinges forced roll-back
40	STS-39	OV-103	14 Mar 91	2	11	25 Mar 91	1	6	26 Mar 91	3	01 Apr 91	39A	28	1,499	
41	STS-40	OV-102	09 Feb 91	1	75	26 Apr 91	3	6	26 Apr 91	3	01 May 91	39B	35	2,053	
42	STS-43	OV-104	19 Apr 91	2	61	19 Jun 91	1	6	19 Jun 91	1	25 Jun 91	39A	39	1,442	
43	STS-48	OV-103	07 May 91	1	80	25 Jul 91	1	11	02 Aug 91	3	12 Aug 91	39A	32	1,502	
44	STS-44	OV-104	11 Aug 91	2	68	18 Oct 91	1	5	19 Oct 91	1	23 Oct 91	39A	32	1,609	
45	STS-42	OV-103	27 Sep 91	3	77	12 Dec 91	1	5	12 Dec 91	3	19 Dec 91	39A	23	1,232	
46	STS-45	OV-104	09 Dec 91	2	55	31 Jan 92	3	7	13 Feb 92	1	19 Feb 92	39A	23	1,316	
47	STS-49	OV-105	25 Jul 91	1	225	07 Mar 92	1	6	07 Mar 92	2	13 Mar 92	39B	56	2,940	
48	STS-50	OV-102	10 Feb 92	2	109	29 May 92	3	7	29 May 92	3	03 Jun 92	39A	23	1,897	
49	STS-46	OV-104	02 Apr 92	2	63	04 Jun 92	1	6	05 Jun 92	1	11 Jun 92	39B	51	1,279	
50	STS-47	OV-105	31 May 92	2	78	17 Aug 92	3	8	20 Aug 92	2	25 Aug 92	39B	19	1,136	
51	STS-52	OV-102	09 Jul 92	1	73	20 Sep 92	1	5	20 Sep 92	3	26 Sep 92	39B	27	1,058	
52	STS-53	OV-103	17 Feb 92	3	251	03 Nov 92	3	5	03 Nov 92	1	08 Nov 92	39A	25		
53	STS-54	OV-105	20 Sep 92	1	63	23 Nov 92	1	10	23 Nov 92	1	03 Dec 92	39B	42		
54	STS-56	OV-103	19 Dec 92	3	74	02 Mar 93	3	12	03 Mar 93	1	15 Mar 93	39B	24	1,258	
55	STS-55	OV-102	01 Nov 92	1	93	02 Feb 93	3	5	02 Feb 93	3	07 Feb 93	39A	79	2,079	On pad abort caused all three main engines to be replaced
56	STS-57	OV-105	20 Jan 93	1	63	24 Mar 93	1	34	25 Mar 93	2	28 Apr 93	39B	55		
57	STS-51	OV-103	17 Apr 93	3	61	16 Jun 93	1	8	18 Jun 93	3	26 Jun 93	39B	79	1,869	
58	STS-58	OV-102	14 May 93	2	89	12 Aug 93	3	36	12 Aug 93	1	17 Sep 93	39B	32	1,504	
–	STS-61	OV-105	01 Jul 93	1	111	21 Oct 93	1	7	21 Oct 93	2	28 Oct 93	39A	18	–	Changed pads due to contamination at Pad A
59	STS-61	OV-105	–		–	–		–	–		15 Nov 93	39B	18	1,610	
60	STS-60	OV-103	22 Sep 93	3	105	05 Jan 94	3	5	05 Jan 94	3	10 Jan 94	39A	25	1,306	
61	STS-62	OV-102	09 Nov 93	2	86	03 Feb 94	1	7	03 Feb 94	1	10 Feb 94	39A	23	1,097	
62	STS-59	OV-105	12 Dec 93	1	91	14 Mar 94	3	5	14 Mar 94	2	19 Mar 94	39A	22	1,143	

Table Continued at the Top of the Following Page

The power drive unit inside the vertical stabilizer – this is what powers the rudder/speedbrake. The Orbiter is in a vertical position at the pad. (Kim Keller)

The left wing shows the flipper doors that provide access to the elevon actuators. Note the tile configuration around the wing tip at left. (Kim Keller)

Seq. #	Mission	Orbiter	OPF In-Date	Bay	# Days	VAB Roll-Over	High Bay	# Days	MLP Orbiter Stacked	MLP	PAD Roll-Out	Pad	# Days	# Hours Ground Power-On	Notes
63	STS-65	OV-102	18 Mar 94	2	82	08 Jun 94	1	6	08 Jun 94	3	14 Jun 94	39A	24	1,030	
64	STS-64	OV-103	08 Jun 94	3/2	103/63	11 Aug 94	1	8	11 Aug 94	2	18 Aug 94	39B	22	1,617	
–	STS-68	OV-105	03 May 94	1	79	21 Jul 94	3	6	21 Jul 94	1	27 Jul 94	39A	23		On pad abort caused all three main engines to be replaced
65	STS-68	OV-105	–	–	–	24 Aug 94	1	20			13 Sep 94	39A	18	1,489	
66	STS-66	OV-104	30 May 94	3	126	04 Oct 94	3	6	04 Oct 94	3	10 Oct 94	39A	25	2,290	
67	STS-63	OV-103	28 Sep 94	2	99	04 Jan 95	1	5	05 Jan 95	2	10 Jan 95	39B	25	1,237	
68	STS-67	OV-105	21 Oct 94	1	108	03 Feb 95	3	5	03 Feb 95	1	08 Feb 95	39A	23	1,171	
69	STS-71	OV-104	23 Nov 94	3	148	19 Apr 95	3	6	20 Apr 95	3	26 Apr 95	39A	63	3,888	
–	STS-70	OV-103	11 Feb 95	2	80	03 May 95	3	6	03 May 95	2	11 May 95	39B	27		Woodpecker damage to ET insulation forced roll-back
70	STS-70	OV-103	–	–	–	08 Jun 95	3	11			15 Jun 95	39B	28	1,426	
–	STS-69	OV-105	28 Mar 95	1	92	27 Jun 95	1	8	28 Jun 95	1	06 Jul 95	39A	26		Hurricane Erin forced Roll-back
71	STS-69	OV-105	–	–	–	01 Aug 95	1	6			08 Aug 95	39A	31	1,633	
72	STS-73	OV-102	21 Apr 95	3	121	21 Aug 95	1	7	21 Aug 95	3	28 Aug 95	39B	54	1,883	
73	STS-74	OV-104	07 Jul 95	1	86	03 Oct 95	1	9	03 Oct 95	2	12 Oct 95	39A	32	1,274	
74	STS-72	OV-105	19 Sep 95	3	73	01 Dec 95	3	5	01 Dec 95	1	06 Dec 95	39B	37	1,121	
75	STS-75	OV-102	06 Nov 95	2	78	23 Jan 96	1	5	24 Jan 96	3	29 Jan 96	39B	25	1,244	
76	STS-76	OV-104	21 Nov 95	1	90	19 Feb 96	3	8	20 Feb 96	2	28 Feb 96	39B	22	991	
77	STS-77	OV-105	20 Jan 96	3	78	08 Apr 96	1	7	16 Apr 96	3	19 Apr 96	39B	34	1,368	
78	STS-78	OV-102	09 Mar 96	2	74	21 May 96	3	9	23 May 96	3	30 May 96	39B	22	1,072	
–	STS-79	OV-104	13 Apr 96	1	72	24 Jun 96	1	9	25 Jun 96	1	01 Jul 96	39A	9		SRB concerns and Hurricane Bertha forced roll-back
–	STS-79	OV-104	03 Aug 96	3	10	13 Aug 96	3	8	13 Aug 96	1	21 Aug 96	39A	13		Hurricane Fran forced second roll-back
79	STS-79	OV-104	–	–	–	04 Sep 96	3	1			05 Sep 96	39A	11	1,129	
80	STS-80	OV-102	07 Jul 96	3	93	10 Oct 96	3	6	10 Oct 96	3	16 Oct 96	39B	33	1,344	
81	STS-81	OV-104	27 Sep 96	3	70	05 Dec 96	1	5	06 Dec 96	2	10 Dec 96	39B	33	913	
82	STS-82	OV-103	30 Jun 96	2	196	11 Jan 97	3	5	12 Jan 97	1	17 Jan 97	39A	26	1,612	
83	STS-83	OV-102	08 Dec 96	1	93	05 Mar 97	1	6	05 Mar 97	3	11 Mar 97	39A	25	1,178	
84	STS-84	OV-104	23 Jan 97	3	86	09 Apr 97	3	5	20 Apr 97	2	24 Apr 97	39A	22	1,175	
85	STS-94	OV-102	09 Apr 97	1	56	04 Jun 97	1	7	04 Jun 97	1	11 Jun 97	39A	21	1,134	
86	STS-85	OV-103	21 Feb 97	2	136	07 Jul 97	3	7	08 Jul 97	1	14 Jul 97	39A	25	1,010	
87	STS-86	OV-104	25 May 97	1	78	11 Aug 97	1	7	11 Aug 97	2	18 Aug 97	39A	39	894	
88	STS-87	OV-102	18 Jul 97	2	101	24 Oct 97	3	5	24 Oct 97	1	29 Oct 97	39B	22	1,292	
89	STS-89	OV-105	05 Jun 97	1	182	12 Dec 97	3	7	12 Dec 97	3	19 Dec 97	39A	32	1,958	
90	STS-90	OV-102	06 Dec 97	3	100	16 Mar 98	3	7	16 Mar 98	2	23 Mar 98	39B	25	1,204	
91	STS-91	OV-103	30 Oct 97	2	180	27 Apr 98	1	4	28 Apr 98	2	02 May 98	39A	31	1,835	
92	STS-95	OV-103	15 Jun 98	2	93	14 Sep 98	1	7	14 Sep 98	2	21 Sep 98	39B	39	1,184	
93	STS-88	OV-105	01 Feb 98	1	256	15 Oct 98	3	6	19 Oct 98	2	21 Oct 98	39A	45	1,741	
–	STS-96	OV-103	07 Nov 98	1	155	12 Apr 99	3	11	19 Apr 99	2	23 Apr 99	39B	15		Hail damage forced roll-back – repaired in VAB w/o destacking
94	STS-96	OV-103	–	–	–	16 May 99	3	4			20 May 99	39B	7		
95	STS-93	OV-102	04 May 98	3	393	02 Jun 99	1	5	04 Jun 99	1	07 Jun 99	39B	46		
96	STS-103	OV-103	06 Jun 99	1	148	04 Nov 99	1	9	05 Nov 99	2	13 Nov 99	39B	36		
97	STS-99	OV-105	15 Dec 98	2	355	02 Dec 99	1	11	03 Dec 99	3	13 Dec 99	39A	59		
98	STS-101	OV-104	28 Sep 98	2	535	17 Mar 00	3	8	17 Mar 00	1	25 Mar 00	39A	55		
99	STS-106	OV-104	29 May 00	3	69	07 Aug 00	1	6	08 Aug 00	2	13 Aug 00	39B	25		
100	STS-92	OV-103	27 Dec 99	1/3	235/2	23 Aug 00	3	17	24 Aug 00	3	11 Sep 00	39A	30		

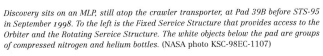

Discovery sits on an MLP, still atop the crawler transporter, at Pad 39B before STS-95 in September 1998. To the left is the Fixed Service Structure that provides access to the Orbiter and the Rotating Service Structure. The white objects below the pad are groups of compressed nitrogen and helium bottles. (NASA photo KSC-98EC-1107)

Another view of Pad 39B with Discovery as STS-95. The track over the flame trench that allows the Rotating Service Structure to roll around the Orbiter is visible here. Above it is the 80-foot fiberglass lightning mast that provides protection from lightning strikes. (NASA photo KSC-98PC-1108)

LANDING AND FERRY FLIGHT SUMMARY

Early in the program the preferred landing site was Edwards AFB because the long runways eliminated a small amount of risk during the landing phase. Landings then switched to KSC to ease post-flight processing, but problems with the Orbiter braking system caused the program to switch back to Edwards. The Challenger accident prompted the program to remain at Edwards for safety reasons until confidence was regained and landings again switched to KSC. Ferry flight constraints include no flight through clouds; no flight in air cooler than 15 degF; no flight in air with an ambient pressure less than eight pounds per square inch (equal to 16,000 feet); no flight at night; no flight within 25 nautical miles of thunderstorms; no flight through moderate or greater turbulence; and finally, there can be no precipitation greater than 'light' at overnight stopping locations.[126]

			LANDING SITE		RUNWAY			FERRY FLIGHT					
Seq. #	Mission	Orbiter	Planned	Actual	No.	Type	Winds	SCA	Departure	Stop #1 (nights)	Stop #2 (nights)	Arrival	Remarks
1	STS-1	OV-102	Edwards AFB	Edwards AFB	23	Lakebed	000 / 00	N905NA	27 Apr 81	Tinker AFB, OK (1)	—	28 Apr 81	
2	STS-2	OV-102	Edwards AFB	Edwards AFB	23	Lakebed	220 / 08	N905NA	24 Nov 81	Bergstrom AFB, TX (1)	—	25 Nov 81	
3	STS-3	OV-102	Edwards AFB	Northrup Strip	17	Gypsum	220 / 13	N905NA	06 Apr 82	Barksdale AFB, LA	—	06 Apr 81	Wet lakebeds at Edwards
4	STS-4	OV-102	Edwards AFB	Edwards AFB	22	Concrete	240 / 12	N905NA	14 Jul 82	Dyess AFB, TX (1)	—	15 Jul 82	
5	STS-5	OV-102	Edwards AFB	Edwards AFB	22	Concrete	012 / 02	N905NA	21 Nov 82	Kelly AFB, TX (1)	—	22 Nov 82	
6	STS-6	OV-099	Edwards AFB	Edwards AFB	22	Concrete	210 / 18	N905NA	14 Apr 83	Kelly AFB, TX (2)	—	16 Apr 83	
7	STS-7	OV-099	KSC	Edwards AFB	15	Lakebed	190 / 07	N905NA	28 Jun 83	Kelly AFB, TX	—	29 Jun 83	Poor visibility at KSC
8	STS-8	OV-099	Edwards AFB	Edwards AFB	22	Concrete	210 / 06	N905NA	09 Sep 83	Sheppard AFB, TX	—	09 Sep 83	First night landing
9	STS-9	OV-102	Edwards AFB	Edwards AFB	17L	Lakebed	010 / 03	N905NA	14 Dec 83	Kelly AFB, TX (1)	Eglin AFB, FL	15 Dec 83	
10	STS-11	OV-099	KSC	KSC	15	Concrete	010 / 03	—	— — —	—	Changed from 33 to 15		
11	STS-13	OV-099	KSC	Edwards AFB	17L	Lakebed	000 / 00	N905NA	17 Apr 84	Kelly AFB, TX (1)	—	18 Apr 83	Storms at KSC
12	STS-16	OV-103	Edwards AFB	Edwards AFB	17L	Lakebed	220 / 04	N905NA	09 Sep 84	Altus AFB, OK (1)	—	10 Sep 84	
13	STS-17	OV-099	KSC	KSC	33	Concrete	320 / 08	—	—	—	—	—	
14	STS-19	OV-103	KSC	KSC	15	Concrete	330 / 05	—	— — —	—			
15	STS-20	OV-103	KSC	KSC	15	Concrete	160 / 08	—	—	—	—	—	
16	STS-23	OV-103	KSC	KSC	33	Concrete	090 / 09	—	—	—	—	—	
17	STS-24	OV-099	Edwards AFB	Edwards AFB	17L	Lakebed	210 / 05	N905NA	10 May 85	Kelly AFB, TX (1)	—	11 May 85	
18	STS-25	OV-103	Edwards AFB	Edwards AFB	23	Lakebed	160 / 11	N905NA	28 Jun 85	Bergstrom AFB, TX	—	28 Jun 85	
19	STS-26	OV-099	Edwards AFB	Edwards AFB	23	Lakebed	220 / 10	N905NA	10 Aug 85	Davis Monthan AFB, AZ	Kelly AFB, TX (1)	11 Aug 85	Fuel also at Eglin AFB, FL
20	STS-27	OV-103	Edwards AFB	Edwards AFB	23	Lakebed	240 / 18	N905NA	07 Sep 85	Kelly AFB, TX (1)	—	08 Sep 85	
21	STS-28	OV-104	Edwards AFB	Edwards AFB	23	Lakebed	270 / 12	N905NA	11 Oct 85	Kelly AFB, TX	—	11 Oct 85	
22	STS-30	OV-099	Edwards AFB	Edwards AFB	17L	Lakebed	170 / 01	N905NA	10 Nov 85	Davis-Monthan AFB, AZ	Kelly AFB, TX (1)	11 Nov 85	Fuel also at Eglin AFB, FL
23	STS-31	OV-104	Edwards AFB	Edwards AFB	22	Concrete	010 / 05	N905NA	07 Dec 85	Kelly AFB, TX	—	07 Dec 85	
24	STS-32	OV-102	KSC	Edwards AFB	22	Concrete	260 / 01	N905NA	22 Jan 86	Davis-Monthan AFB, AZ	Kelly AFB, TX (1)	23 Jan 86	Fuel also at Eglin AFB, FL
25	STS-33	OV-099	KSC	—	—	—	—	—	—	—	—	—	Vehicle destroyed at T+73 seconds
26	STS-26R	OV-103	Edwards AFB	Edwards AFB	17L	Lakebed	010 / 02	N905NA	08 Oct 88	Kelly AFB, TX	—	08 Oct 88	
27	STS-27R	OV-104	Edwards AFB	Edwards AFB	17L	Lakebed	110 / 02	N905NA	11 Dec 88	Davis-Monthan AFB, AZ	Kelly AFB, TX (2)	13 Dec 88	
28	STS-29R	OV-103	Edwards AFB	Edwards AFB	22	Concrete	220 / 06	N905NA	23 Mar 89	Kelly AFB, TX (1)	—	24 Mar 89	
29	STS-30R	OV-104	Edwards AFB	Edwards AFB	22	Concrete	280 / 08P16	N905NA	13 May 89	Biggs AAF, TX (2)	Dallas-Ft. Worth, TX	15 May 89	Fuel also at Robbins AFB, GA
30	STS-28R	OV-102	Edwards AFB	Edwards AFB	17L	Lakebed	160 / 06	N905NA	20 Aug 89	Kelly AFB, TX (1)	Robbins AFB, GA	21 Aug 89	
31	STS-34	OV-104	Edwards AFB	Edwards AFB	23L	Lakebed	170 / 04	N905NA	23 Oct 89	Biggs AAF, TX (1)	Columbus AFB, MS	29 Oct 89	
32	STS-33R	OV-103	Edwards AFB	Edwards AFB	04	Concrete	070 / 08P19	N905NA	03 Dec 89	Kelly AFB, TX	Eglin AFB, FL (1)	04 Dec 89	
33	STS-32R	OV-102	Edwards AFB	Edwards AFB	22	Concrete	300 / 04P05	N905NA	25 Jan 90	Davis-Monthan AFB, AZ	Kelly AFB, TX (1)	26 Jan 90	Fuel also at Eglin AFB, FL
34	STS-36	OV-104	Edwards AFB	Edwards AFB	23L	Lakebed	260 / 16P18	N905NA	11 Mar 90	Biggs AAF, TX (2)	Columbus AFB, MS	13 Mar 90	Weather delay at Biggs AAF
35	STS-31R	OV-103	Edwards AFB	Edwards AFB	22	Concrete	180 / 07P10	N905NA	05 May 90	Sheppard AFB, TX (1)	Robbins AFB, GA (1)	07 May 90	
36	STS-41	OV-103	Edwards AFB	Edwards AFB	22	Concrete	279 / 03	N905NA	15 Oct 90	Sheppard AFB, TX	Eglin AFB, FL (1)	16 Oct 90	Weather delay at Eglin AFB
37	STS-38	OV-104	Edwards AFB	KSC	33	Concrete	020 / 05P07	—	—	—	—	—	Weather at Edwards AFB
38	STS-35	OV-102	Edwards AFB	Edwards AFB	22	Concrete	014 / 01P02	N905NA	18 Dec 90	Biggs AAF, TX	Kelly AFB, FL (1)	20 Dec 90	Overnight also at Eglin AFB, FL (1)
39	STS-37	OV-104	Edwards AFB	Edwards AFB	33	Lakebed	008 / 17P21	N905NA	16 Apr 91	Columbus AFB, MS (1)	MacDill AFB, FL (1)	18 Apr 91	Weather delay at MacDill AFB
40	STS-39	OV-103	Edwards AFB	KSC	15	Concrete	155 / 12P16	—	—	—	—	—	Weather at Edwards AFB
41	STS-40	OV-102	Edwards AFB	Edwards AFB	22	Concrete	227 / 12P17	N911NA	19 Jun 91	Biggs AAF, TX (1)	Columbus AFB, MS (1)	21 Jun 91	
42	STS-43	OV-104	KSC	KSC	15	Concrete	240 / 07	—	—	—	—	—	
43	STS-48	OV-103	KSC	Edwards AFB	22	Concrete	200 / 08	N911NA	24 Sep 91	Biggs AAF, TX	Tinker AFB, OK (1)	26 Sep 91	Overnight also at Columbus AFB (1)
44	STS-44	OV-104	KSC	Edwards AFB	05R	Lakebed	074 / 13P15	N911NA	07 Dec 91	Sheppard AFB, TX (1)	—	08 Dec 91	
45	STS-42	OV-103	Edwards AFB	Edwards AFB	22	Concrete	300 / 02	N905NA	14 Feb 92	Biggs AFB, TX (1)	Kelly AFB, TX (1)	16 Feb 92	
46	STS-45	OV-104	KSC	KSC	33	Concrete	290 / 08P12	—	—	—	—	—	
47	STS-49	OV-105	Edwards AFB	Edwards AFB	22	Concrete	232 / 02P06	N911NA	27 May 92	Biggs AAF, TX (1)	Kelly AFB, TX (1)	30 May 92	
48	STS-50	OV-102	Edwards AFB	KSC	33	Concrete	258 / 05P08	—	—	—	—	—	
49	STS-46	OV-104	KSC	KSC	33	Concrete	330 / 01	—	—	—	—	—	
50	STS-47	OV-105	KSC	KSC	33	Concrete	270 / 02P04	—	—	—	—	—	
51	STS-52	OV-102	KSC	KSC	33	Concrete	080 / 06P09	—	—	—	—	—	
52	STS-53	OV-103	KSC	Edwards AFB	22	Concrete	274 / 14P19	N911NA	15 Dec 92	Kelly AFB, TX (3)	—	18 Dec 92	
53	STS-54	OV-105	KSC	KSC	33	Concrete	360 / 04P06	—	—	—	—	—	
54	STS-56	OV-103	KSC	KSC	33	Concrete	320 / 06P08	—	—	—	—	—	
55	STS-55	OV-102	KSC	Edwards AFB	22	Concrete	220 / 14P19	N905NA	11 May 93	Biggs AAF, TX (1)	Kelly AFB, TX (2)	14 May 93	
56	STS-57	OV-105	KSC	KSC	33	Concrete	320 / 07P10	—	—	—	—	—	
57	STS-51	OV-103	KSC	KSC	15	Concrete	000 / 00	—	—	—	—	—	
58	STS-58	OV-102	Edwards AFB	Edwards AFB	22	Concrete	029 / 02P03	N911NA	07 Nov 93	Columbus AFB, MS	—	08 Nov 93	
59	STS-61	OV-105	KSC	KSC	33	Concrete	330 / 05P08	—	—	—	—	—	
60	STS-60	OV-103	KSC	KSC	15	Concrete	150 / 08P18	—	—	—	—	—	
61	STS-62	OV-102	KSC	KSC	33	Concrete	190 / 05P08	—	—	—	—	—	
62	STS-59	OV-105	KSC	Edwards AFB	22	Concrete	354 / 00P02	N911NA	29 Apr 94	Biggs AAF, TX (1)	Little Rock AFB, AR (2)	02 May 94	
63	STS-65	OV-102	KSC	KSC	33	Concrete	150 / 03P04	—	—	—	—	—	
64	STS-64	OV-103	KSC	Edwards AFB	04	Concrete	040 / 10P13	N905NA	26 Sep 94	Kelly AFB, TX (1)	—	27 Sep 94	
65	STS-68	OV-105	KSC	Edwards AFB	22	Concrete	200 / 08P10	N911NA	10 Oct 94	Biggs AAF, TX	Dyess AFB, TX (1)	20 Oct 94	
66	STS-66	OV-104	KSC	Edwards AFB	22	Concrete	360 / 04	N911NA	21 Nov 94	Biggs AAF, TX	Eglin AFB, FL (1)	22 Nov 94	
67	STS-63	OV-103	KSC	KSC	15	Concrete	170 / 05P07	—	—	—	—	—	
68	STS-67	OV-105	KSC	Edwards AFB	22	Concrete	231 / 15P22	N905NA	26 Mar 95	Dyess AFB, TX (1)	Columbus AFB, MS	27 Mar 95	
69	STS-71	OV-104	KSC	KSC	15	Concrete	030 / 07P11	—	—	—	—	—	
70	STS-70	OV-103	KSC	KSC	33	Concrete	200 / 05	—	—	—	—	—	
71	STS-69	OV-105	KSC	KSC	33	Concrete	220 / 05	—	—	—	—	—	
72	STS-73	OV-102	KSC	KSC	33	Concrete	034 / 05P07	—	—	—	—	—	
73	STS-74	OV-104	KSC	KSC	33	Concrete	010 / 07P10	—	—	—	—	—	
74	STS-72	OV-105	KSC	KSC	15	Concrete	315 / 06P08	—	—	—	—	—	
75	STS-75	OV-102	KSC	KSC	33	Concrete	326 / 13P20	—	—	—	—	—	
76	STS-76	OV-104	KSC	Edwards AFB	22	Concrete	142 / 01P04	N905NA	06 Apr 96	—	—	—	Engine fire on 747; return to EAFB
								N905NA	11 Apr 96	Davis-Monthan AFB, AZ	Dyess AFB, TX (1)	12 Apr 96	Refuel also at Eglin AFB
77	STS-77	OV-105	KSC	KSC	33	Concrete	260 / 08P11	—	—	—	—	—	
78	STS-78	OV-102	KSC	KSC	33	Concrete	180 / 03P05	—	—	—	—	—	
79	STS-79	OV-104	KSC	KSC	33	Concrete	120 / 04P09	—	—	—	—	—	
80	STS-80	OV-102	KSC	KSC	33	Concrete	200 / 06P09	—	— — —	—			
81	STS-81	OV-104	KSC	KSC	33	Concrete	140 / 04P07	—	—	—			
82	STS-82	OV-103	KSC	KSC	15	Concrete	140 / 07P15	—	—	—	—	—	
83	STS-83	OV-102	KSC	KSC	33	Concrete	020 / 09P18	—	—	—	—	—	
84	STS-84	OV-104	KSC	KSC	33	Concrete	111 / 09P13	—	—	—	—	—	
85	STS-94	OV-102	KSC	KSC	33	Concrete	150 / 01	—	—	—	—	—	
86	STS-85	OV-103	KSC	KSC	33	Concrete	200 / 06P09	—	—	—	—	—	
87	STS-86	OV-104	KSC	KSC	33	Concrete	075 / 09P14	—	—	—	—	—	
88	STS-87	OV-102	KSC	KSC	33	Concrete	078 / 06P12	—	—	—	—	—	
89	STS-89	OV-105	KSC	KSC	15	Concrete	020 / 05P11	—	—	—	—	—	
90	STS-90	OV-102	KSC	KSC	33	Concrete	230 / 04P11	—	—	—	—	—	
91	STS-91	OV-103	KSC	KSC	15	Concrete	040 / 07P11	—	—	—	—	—	
92	STS-95	OV-103	KSC	KSC	33	Concrete	060 / 09P14	—	—	—	—	—	
93	STS-88	OV-105	KSC	KSC	15	Concrete	310 / 5P9	—	—	—	—	—	
94	STS-96	OV-103	KSC	KSC	15	Concrete	090 / 4P7	—	—	—	—	—	
95	STS-93	OV-102	KSC	KSC		Concrete	249 / 5P6	—	—	—	—	—	
96	STS-103	OV-103	KSC	KSC	33	Concrete	235 / 7P8	—	—	—	—	—	
97	STS-99	OV-105	KSC	KSC	33	Concrete	050 / 9P13	—	—	—	—	—	
98	STS-101	OV-104	KSC	KSC	15	Concrete		—	—	—	—	—	
99	STS-106	OV-104	KSC	KSC	15	Concrete		—	—	—	—	—	
100	STS-92	OV-103	KSC	Edwards AFB	22	Concrete	130 / 02	N905NA	02 Nov 00	Altus AFB, OK	Whiteman AFB, MO (1)	03 Nov 00	

'Roll-back' is the term used when the entire stack must be returned to the VAB because of the threat of severe weather or the the need to repair flight hardware or payloads that cannot be performed at the launch pad.[125]

Seq. #	Mission	Orbiter	Date	Reason
9	STS-9	*Columbia*	19 Oct 83	*Columbia* was rolled back to the VAB, de-stacked, and the Orbiter returned to the OPF on because of a suspect exhaust nozzle on the right Solid Rocket Booster.
12	STS-16/41-D	*Discovery*	11 Jul 84	*Discovery* was rolled back to the VAB following a pad abort on 26 June. The vehicle was returned to the VAB, de-stacked, and the Orbiter returned to the OPF to remove and replace SSME #3.
—	STS-22/51-E	*Challenger*	05 Mar 85	*Challenger* was rolled back due to a timing problem with the TDRS primary payload. The vehicle was de-stacked in VAB and the Orbiter returned to OPF. This mission, 51-E, was cancelled and the Orbiter re-manifested with STS-24/51-B payloads.
38	STS-35	*Columbia*	12 Jun 90	1st rollback. *Columbia* was rolled back after a hydrogen leak was detected in the Orbiter–ET 17-inch umbilical.
37	STS-38	*Atlantis*	09 Aug 90	*Atlantis* was rolled back to the VAB after tests confirmed a hydrogen leak on the ET side of the 17-inch quick disconnect umbilical. The vehicle was de-stacked.
38	STS-35	*Columbia*	09 Oct 90	2nd rollback. *Columbia* was rolled back a second time due to the threat of severe weather from Tropical Storm Klaus.
40	STS-39	*Discovery*	07 Mar 91	*Discovery* was rolled back to the VAB after significant cracks were found on all four lug hinges on the two ET umbilical door drive mechanisms. The vehicle was de-stacked.
65	STS-68	*Endeavour*	24 Aug 94	*Endeavour* was rolled back to the VAB due to a pad abort due to an unacceptably high discharge temperature in the high-pressure oxidizer turbopump on SSME #3. All three engines were replaced in the VAB.
70	STS-70	*Discovery*	08 Jun 95	*Discovery* was rolled back to the VAB after Northern Flicker Woodpeckers poked about 200 holes in the ET foam insulation.
71	STS-69	*Endeavour*	01 Aug 95	*Endeavour* was rolled back to the VAB due to the threat of severe weather from Hurricane Erin.
79	STS-79	*Atlantis*	10 Jul 96	*Atlantis* was rolled back to the VAB due to the threat of severe weather from Hurricane Bertha and SRB concerns.
79	STS-79	*Atlantis*	04 Sep 96	*Atlantis* was rolled back to the VAB a second time due to the threat of severe weather from Hurricane Fran.
94	STS-96	*Discovery*	16 May 99	*Discovery* rolled-back due to hail damage to the ET insulation. Damage repaired in VAB without destacking the vehicle.

A number of Shuttle missions have been canceled during the planning process for various reasons. Even more were cancelled following the Challenger accident in January 1986. This partial list includes the first two Vandenberg missions (62-x).

Manifest	Date	Payload	Crew	Manifest	Date	Payload	Crew
STS-10	Nov 83	DoD	Thomas K. Mattingly (C) Loren J. Shriver (P) Ellison S. Onizuka (MS) James F. Buchli (MS) Gary E. Payton (PS)	61-H	24 Jun 86	Com. Sat.	Michael L. Coats (C) John E. Blaha (P) Robert C. Springer (MS) James F. Buchli (MS) Anna L. Fisher (MS) Pratiwi Sudarmono (PS) Nigel R. Wood (PS)
STS-12	Mar 84	TDRS/IUS	Henry W. Hartsfield (C) Michael L. Coats (P) Richard M. Mullane (MS) Steven A. Hawley (MS) Judith A. Resnik (MS)	62-A	29 Jan 86 (15 Jul 86) (20 Mar 86) (15 Jul 86) (15 Oct 86)	Teal Ruby	Robert L. Crippen (C) Guy S. Gardner (P) Dale A. Gardner (MS) Jerry L. Ross (MS) Richard M. Mullane (MS) Edward C. Aldridge (PS) John B. Watterson (PS)
41-E	Jul 84	DoD	Same as STS-10 except Jeffrey E. Detroye (PS)				
41-F	Aug 84	Com. Sat.	Karol J. Bobko (C) Donald E. Williams (P) M. Rhea Seddon (MS) S. David Griggs (MS) Jeffrey A. Hoffman (MS)	61-M	22 Jul 86	TDRS/IUS	Loren J. Shriver (C) Bryan D. O'Connor (P) Mark C. Lee (MS) Sally K. Ride (MS) William F. Fisher (MS) Robert J. Wood (PS)
41-H	Sep 84	DoD	Frederick H. Hauck (C) David M. Walker (P) Joseph P. Allen (MS) Anna L. Fisher (MS) Dale A. Gardner (PS) Frank J. Casserino (PS) Gary E. Payton (PS)	61-J	Aug 86	*Hubble*	John W. Young (C) Charles F. Bolden (P) Bruce McCandless (MS) Steven A. Hawley (MS) Kathryn D. Sullivan (MS)
51-E	Mar 85	TDRS/IUS	Same as STS-41F plus Patrick Baudry (PS) E. Jacob Garn (Senator)	61-N	04 Sep 86	DoD	Brewster H. Shaw (C) Michael J McCulley (P) David C. Leestma (MS) James C. Adamson (MS) Mark N. Brown (MS) Frank Casserino (PS)
51-D	Apr 85	Com. Sat.	Daniel C. Brandenstein (C) John O. Creighton (P) Steven R. Nagel (MS) John M. Fabian (MS) Shannon W. Lucid (MS) Gregory B. Jarvis (PS) Charles D. Walker (PS)	61-I	27 Sep 86	LDEF(R)	Donald E. Williams (C) Michael J. Smith (P) Bonnie J. Dunbar (MS) Manley L. Carter (MS) James P. Bagian (MS) Nagapathi C. Bhat (PS) ? (Journalist)
51-G	Jul 85	TDRS/IUS	Brewster H. Shaw (C) Bryan D. O'Connor (P) Jerry L. Ross (MS) Mary L. Cleave (MS) Sherwood C. Spring (MS)	62-B	29 Sep 86	DoD	No Crew Listed
51-K	Aug 85	Spacelab-D	No crew listed. Remanifested as 61-A	61-K	Oct 86	EOM-1/2	Vance D Brand (C) S. David Griggs (P) Robert L. Stewart (MS) Claude Nicollier (MS) Owen K. Garriott (PS) Byron K. Lichtenberg (PS) Michael L. Lampton (PS)
51-H	Nov 85	EOM-1	No crew listed. Combined with 61-K				
61-E	06 Mar 86	ASTRO-1	Com. John A. McBride Richard N. Richards (P) David C. Leestma (MS) Jeffrey A. Hoffman (MS) Robert A. Parker (MS) Samuel T. Durrance (PS) Ronald A. Parise (PS)	61-L	Nov 86	Com. Sat.	No crew listed
				71-B	Dec 86	Com. Sat.	No crew listed
				71-A	Jan 87	ASTRO-2	No crew listed
				71-C	Jan 87	Com. Sat.	No crew listed
61-F	15 May 86	*Ulysses*	Frederick H. Hauck (C) Roy D. Bridges (P) John M. Lounge (MS) David C. Hilmers (MS)	71-D	Feb 87	Com. Sat.	No crew listed
				71-E	Mar 87	SLS-1	No crew listed
61-G	20 May 86	*Galileo*	David M. Walker (C) Ronald J. Grabe (P) Norman E. Thagard (MS) James D. van Hoften	71-F	Mar 87	Com. Sat.	No crew listed
				71-K	Jul 87	Antelsat	No crew listed
				71-M	Aug 87	ASTRO-3	No crew listed

These two photos show the change of markings that replaced the NASA 'worm' logo with the 'meatball' insignia. Endeavour (left) on 24 May 1997 shows the old markings; Discovery on 4 November 1999 shows the new. Note that the name is considerably lower on the wing with the new markings. (NASA photos KSC-97PC-0830 and KSC-99PC-1280)

FLIGHT CREW MEMBER SUMMARY

As of the 100th mission, 262 people had flown on Space Shuttle, including representatives of several different countries, members of Congress, and a teacher. Many have only flown once – others have flown as many as six times.[143]

NOTES:
a = Payload Commander b = does not include time aboard Mir
c = ascent only d = descent only
e = Other flights listed under Nancy J. Sherlock or Nancy J. Currie

NAME					1ST MISSION		2ND MISSION		3RD MISSION		4TH MISSION		5TH MISSION		6TH MISSION		TOTAL TIME		
Last	First	Middle	Rank / Title	Affiliation	STS	Pos.	STS	Pos.	STS	Pos.	STS	Pos.	STS	Pos.	STS	Pos.	hrs	min	sec
Acton	Loren	W.	Ph.D.	Civ	51-F	PS1											190	45	27
Adamson	James	C.	Lt.Col.	USA	28R	MS1	43	MS2									334	21	33
Akers	Thomas	D.	Col.	USAF	38	MS3	49	MS4	61	MS5	79	MS2					834	29	10
Al-Saúd	Salman	Abdul aziz		Civ	51-G	PS2											169	39	00
Allen	Andrew	M.	Lt.Col.	USAF	46	Plt	62	Plt	75	Cdr							904	12	05
Allen	Joseph	P.	Ph.D.	Civ	5	MS1	51-A	MS3									313	59	27
Altman	Scott	D.	LCDR	USN	90	Plt	106	Plt									665	02	13
Anderson	Michael	P.	Maj.	USAF	89	MS2											211	46	35
Apt	Jerome	'Jay'	Ph.D.	Civ	37	MS2	47	MS2	59	MS1	79	MS1					847	11	01
Ashby	Jeffrey	S.	CAPT	USN	93	Plt											118	49	36
Bagian	James	P.	M.D.	Civ	29R	MS1	40	MS2									337	53	09
Baker	Ellen	S.	M.D.	Civ	34	MS3	50	MS2	71	MS1							686	32	42
Baker	Michael	A.	CAPT	USN	43	Plt	52	Plt	68	Cdr	81	Cdr					964	59	09
Barry	Daniel	T	M.D. Ph.D.	Civ	72	MS4	96	MS3									449	13	48
Bartoe	John-David	F.	Ph.D.	Civ	51-F	PS2											190	45	27
Baudry	Patrick		Lt.Col.	FAF	51-G	PS1											169	39	00
Blaha	John	E.	Col.	USAF (Ret.)	29R	Plt	33R	Plt	43	Cdr	58	Cdr	79 b, c	MS4	81 b d	MS4	1,036	08	57
Bloomfield	Michael	J	Maj.	USAF	86	Plt											259	20	53
Bluford, Jr.	Guion	S.	Col.	USAF	8	MS2	61-A	MS2	39	MS1	53	MS1					688	35	46
Bobko	Karol	J.	Col.	USAF	6	Plt	51-D	Cdr	51-J	Cdr							387	03	56
Bolden, Jr.	Charles	F.	Col.	USMC	61-C	Plt	31R	Plt	45	Cdr	60	Cdr					680	38	49
Bondar	Roberta	L.	Ph.D.	Civ.	42	PS1											193	14	44
Bowersox	Kenneth	D.	CDR	USN	50	Plt	61	Plt	73	Cdr	82	Cdr					1,212	58	10
Brady, Jr.	Charles	E.	CDR	USN	78	MS3											405	47	364
Brand	Vance	D.		Civ.	5	Cdr	41-B	Cdr	35	Cdr							528	35	28
Brandenstein	Daniel	C.	CAPT	USN	8	Plt	51-G	Cdr	32R	Cdr	49	Cdr					789	05	57
Bridges, Jr.	Roy	D.	Col.	USAF	51-F	Plt											190	45	27
Brown	Curtis	L.	Lt.Col.	USAF	47	Plt	66	Plt	77	Plt	85	Cdr	95	Cdr	103	Cdr	1,383	05	32
Brown	Mark	N.	Col.	USAF	28R	MS3	48	MS3									249	27	46
Buchli	James	F.	Col.	USMC	51-C	MS2	61-A	MS1	29R	MS2	48	MS1					490	24	48
Buckey	Jay	Clark	M.D.	Civ.	90	PS1											381	49	58
Budarin	Nikolai	M.		Civ	71 c	Mir-19											44	36	00
Burbank	Daniel	C.	Lt.Cdr.	USCG	106	MS1											283	12	15
Bursch	Daniel	W.	CDR	USN	51	MS2	68	MS2	77	MS2							746	36	43
Cabana	Robert	D.	Col.	USMC	41	Plt	53	Plt	65	Cdr	88	Cdr					910	42	50
Cameron	Kenneth	D.	Col.	USMC	37	Plt	56	Cdr	74	Cdr							562	11	53
Carter	Manley	L.	CDR	USN	33R	MS2											120	06	48
Casper	John	H.	Col.	USAF	36	Plt	54	Cdr	62	Cdr	77	Cdr					825	52	46
Cenker	Robert	J.		Civ.	61-C	MS3											146	03	53
Chang-Diaz	Franklin	R.	Ph.D.	Civ.	61-C	MS1	34	MS1	46	MS2	60	MS3	75	MS4 a	91	MS1	1,269	41	58
Chawla	Kalpana		Ph.D.	Civ.	87	MS1											376	34	02
Cheli	Maurizio		Lt.Col.	Italian AF	75	MS2											377	40	21
Chiao	Leroy		Ph.D.	Civ.	65	MS3	72	MS1	92	MS2							877	39	32
Chilton	Kevin	P.	Col.	USAF	49	Plt	59	Plt	76	Cdr							704	23	01
Chrétien	Jean-Loup	J.M.	BGen.	FAF	86	MS3											259	20	53
Cleave	Mary	L.	Ph.D.	Civ.	61-B	MS2	30R	MS2									262	01	20
Clervoy	Jean-Francois			ESA	66	MS3	84	MS1 a	103	MS2							675	04	45
Clifford	Michael	R. 'Rick'		Civ.	53	MS3	59	MS2	76	MS2							666	25	10
Coats	Michael	L.	CAPT	USN	41-D	Plt	29R	Cdr	39	Cdr							463	57	17
Cockrell	Kenneth	D.		Civ.	56	MS2	69	Plt	80	Cdr							906	30	38
Coleman	Catherine	G. 'Cady'	Capt.	USAF	73	MS1	93	MS1									500	41	58
Collins	Eileen	M.	Col.	USAF	63	Plt	84	Plt	93	Cdr							538	37	47
Covey	Richard	O.	Col.	USAF	51-I	Plt	26R	Plt	38	Cdr	61	Cdr					645	11	4
Creighton	John	O.	CAPT	USN	51-G	Plt	36	Cdr	48	Cdr							404	25	00
Crippen	Robert	L.	CAPT	USN	1	Plt	7	Cdr	41-C	Cdr	41-G	Cdr					565	48	44
Crouch	Roger	K.	Ph.D.	Civ	83	PS1	94	PS1									465	57	15
Culbertson	Frank	L.	CAPT	USN	38	Plt	51	Cdr									354	05	42
Curbeam, Jr.	Robert	L.	Lt.Cdr.	USN	85	MS2											284	27	00
Currie	Nancy	J.	Maj.	USA	88 e	MS2											737	23	00
Davis	N.	Jan	Ph.D.	Civ.	47	MS3	60	MS1	85	MS1 a							674	06	45
DeLucas	Lawrence	J.	Ph.D.	Civ.	50	PS1											331	31	04
Dezhurov	Vladimir	N.	Lt.Col.	Russian AF	71 d	Mir-18											190	46	19
Doi	Takao		Ph.D.	NASDA	87	MS3											376	34	02
Duffy	Brian	K.	Lt.Col.	USAF	45	Plt	57	Plt	72	Cdr	92	Cdr					977	38	54
Dunbar	Bonnie	J.	Ph.D.	Civ.	61-A	MS3	32R	MS1	50	MS1	71	MS3	89	MS3 a			1,208	25	25
Duque	Pedro			Civ.	95	MS3											213	43	57
Durrance	Samuel	T.	Ph.D.	Civ.	35	PS1	67	PS1									614	13	56
Edwards, Jr.	Joe	Frank	CDR	USN	89	Plt											211	46	35
England	Anthony	W.	Ph.D.	Civ.	51-F	MS2											190	45	27
Engle	Joe	H.	Col.	USAF	2	Cdr	51-I	Cdr									224	30	58
Fabian	John	M.	Col.	USAF	7	MS2	51-G	MS2									316	03	08
Favier	Jean-Jacques	J.	Ph.D.	FSA	78	PS1											405	47	36
Fettman	Martin	J.	DVM, Ph.D.	Civ.	58	PS1											336	12	32
Fisher	Anna	L.	M.D.	Civ.	51-A	MS1											191	45	01
Fisher	William	F.	M.D.	Civ.	51-I	MS3											170	17	44
Foale	C.	Michael	Ph.D.	Civ.	45	MS3	56	MS1	63	MS2	84 c	MS5	86 b, d	MS5	103	MS4	1,077	30	46
Frimout	Dirk	D.	Ph.D.	Civ.	45	PS2											214	9	28
Fullerton	Charles	Gordon	Col.	USAF	3	Plt	51-F	Cdr									382	50	15
Furrer	Reinhard		Ph.D.	Civ.	61-A	PS1											168	44	53
Gaffney	F.	Drew		Civ.	40	PS1											218	14	20
Gardner	Dale	A.	Capt.	USN	8	MS1	51-A	MS2									336	53	44
Gardner	Guy	S.	Col.	USAF	27R	MS3	35	Plt									320	10	45
Garn	E.	J. 'Jake'	Senator	Civ.	51-D	PS2											167	55	28
Garneau	Marc		Ph.D.	Civ.	41-G	PS1	77	MS4									438	03	03
Garriott	Owen	K.	Ph.D.	Civ.	9	MS1											247	47	24
Gemar	Charles	D. 'Sam'	Lt.Col.	USA	38	MS3	48	MS2	62	MS2							581	38	50
Gernhardt	Michael	L.	Ph.D.	Civ	69	MS3	83	MS2	94	MS2							726	26	11
Gibson	Robert	L. 'Hoot'	CAPT	USN	41-B	Cdr	61-C	Cdr	27R	Cdr	47	Cdr	71	Cdr			868	17	59
Glenn, Jr.	John	H.	Senator	Civ.	95	PS2											213	43	56
Godwin	Linda	M.	Ph.D.	Civ.	37	MS3	59	MS3 a	76	MS3							634	38	07
Grabe	Ronald	J.	Col.	USAF	51-J	Plt	30R	Plt	42	Cdr	57	Cdr					628	40	45
Gregory	Frederick	D.	Col.	USAF	51-B	Plt	33R	Cdr	44	Cdr							455	06	26
Gregory	William	G.	Lt.Col.	USAF	67	Plt											399	8	48
Griggs	S.	David		Civ.	51-D	MS3											167	55	28
Grunsfeld	John	M.	Ph.D.	Civ.	67	MS1	81	MS2	103	MS3							835	14	58
Guidoni	Umberto			Civ.	75	PS1											377	40	21
Gutierrez	Sidney	M.	Col.	USAF	40	Plt	59	Cdr									488	03	50
Hadfield	Chris	A.	Maj.	CAF	74	MS1											196	30	46
Halsell, Jr.	James	D.	Lt.Col.	USAF	65	Plt	74	Plt	83	Cdr	94	Cdr	101	Cdr			1,252	32	52
Hammond, Jr.	L.	Blaine	Lt.Col.	USAF	39	Plt	64	Plt									462	12	20
Harbaugh	Gregory	J.		Civ.	39	MS2	54	MS2	71	MS2	82	MS3					818	00	06
Harris, Jr.	Bernard	A.	M.D.	Civ.	55	MS3	63	MS1 a									438	08	14
Hart	Terry	J.		Civ.	41-C	MS3											167	40	06

NAME					1ST MISSION		2ND MISSION		3RD MISSION		4TH MISSION		5TH MISSION		6TH MISSION		TOTAL TIME		
Last	First	Middle	Rank / Title	Affiliation	STS	Pos.	STS	Pos.	STS	Pos.	STS	Pos.	STS	Pos.	STS	Pos.	hrs	min	sec
Hartsfield, Jr.	Henry	W. 'Hank'	Col.	USAF (Ret)	4	Plt	41-D	Cdr	61-A	Cdr							482	50	43
Hauck	Frederick	H.	CAPT	USN	7	Plt	51-A	Cdr	26R	Cdr							435	09	21
Hawley	Steven	A.	Ph.D.	Civ.	41-D	MS2	61-C	MS2	31R	MS1	82	MS2	93	MS2			770	42	47
Helms	Susan	J.	Lt.Col.	USAF	54	MS3	64	MS2	78	MS2	101	MS4					1,048	25	43
Henize	Karl	G.	Ph.D.	Civ.	51-F	MS3											190	45	27
Hennen	Thomas	J.	CWO-3	USA	44	PS1											166	50	44
Henricks	Terence	T.	Col.	USAF	44	Plt	55	Plt	70	Cdr	78	Cdr					1,020	38	26
Hieb	Richard	J.		Civ.	39	MS3	49	MS3	65	MS1							766	35	11
Hilmers	David	C.	Col.	USMC	51-J	MS1	26R	MS3	36	MS3	42	MS2					495	17	59
Hire	Kathryn	P. 'Kay'	CDR	USN	90	MS2											381	49	58
Hoffman	Jeffrey	A.	Ph.D.	Civ.	51-D	MS2	35	MS1	46	MS1 a	61	MS3	75	MS1			1,211	54	37
Horowitz	Scott	J.	Lt.Col.	USAF	75	Plt	82	Plt	101	Plt							853	27	19
Hughes-Fulford	Millie		Ph.D.	Civ.	40	PS2											218	14	20
Husband	Rick	D.	Lt.Col.	USAF	96	Plt											235	13	03
Ivins	Marsha	S.		Civ.	32R	MS3	46	MS4	62	MS3	81	MS3					1,032	27	43
Jarvis	Gregory	B.		Civ.	51-L	PS1											0	1	13
Jemison	Mae	C.	M.D.	Civ.	47	MS4											190	30	23
Jernigan	Tamara	E.	Ph.D.	Civ.	39	MS3	52	MS3	67	MS3 a	80	MS1 a	96	MS1			1,494	33	46
Jett, Jr.	Brent	W.	LCDR	USN	72	Plt	81	Plt									458	56	08
Jones	Thomas	D.	Ph.D.	Civ.	59	MS4	68	MS4 a	80	MS2							963	28	57
Kadenyuk	Leonid	K.	Col.	Russian AF	87	PS2											376	34	02
Kavandi	Janet	Lynn	Ph.D.	Civ	91	MS2	99	MS2									505	32	44
Kelly	Scott	J.	CDR	USN	103	Plt											191	10	47
Kondakova	Elena	V		Russia	84 b	MS4											221	19	56
Kregel	Kevin	R.		Civ	70	Plt	78	Plt	87	Cdr	99	Cdr					1,266	20	29
Krikalev	Sergei	K.		Russia	60	MS4	88	MS4									482	27	21
Lawrence	Wendy	B.	CDR	USN	67	MS2	86	MS4	91	MS2							894	23	41
Lee	Mark	C.	Col.	USAF	30R	MS3	47	MS1 a	64	MS4	82	MS4 a					789	53	53
Leestma	David	C.	CDR	USN	41-G	MS3	28R	MS2	45	MS2							532	33	15
Lenoir	William	B.	Ph.D.	Civ.	5	MS2											122	14	26
Leslie	Fred	W.	Ph.D.	Civ.	73	PS1											381	52	22
Lichtenberg	Bryon	K.	Ph.D.	Civ.	9	PS1	45	PS1									461	56	52
Lind	Don	Leslie	Ph.D.	Civ.	51-B	MS1											168	08	54
Lindsey	Steven	W.	Lt.Col.	USAF	87	Plt	95	Plt									590	17	58
Linenger	Jerry	M.	CAPT	USN	64	MS1	81 c	MS3	84 b, d	MS5							502	13	53
Linnehan	Richard	M.	DVM	Civ	78	MS1	90	MS1 a									787	37	34
Linteris	Gregory	T.	Ph.D.	Civ.	83	PS2	94	PS2									465	57	15
Lopez-Alegria	Michael	E.	LCDR	USN	73	MS2	92	MS4									691	36	09
Lounge	John	M.		Civ.	51-I	MS2	26R	MS1	35	MS2							482	23	04
Lousma	Jack	R.	Col.	USMC	3	Cdr											192	04	48
Low	G.	David	Ph.D.	Civ.	32R	MS2	43	MS3	57	MS1							714	06	55
Lu	Edward	Tsang	Ph.D.	Civ.	84	MS3	106	MS2									504	32	11
Lucid	Shannon	W.	Ph.D.	Civ.	51-G	MS1	34	MS2	43	MS1	58	MS4	76 b, c	MS4	79 b, d	MS4	1,063	59	32
MacLean	Steven	G.	Ph.D.	Civ.	52	PS1											236	56	13
Malenchenko	Yuri	I.	Col.	Russian AF	106	MS4											283	12	15
Malerba	Franco		Ph.D.	Civ.	46	PS1											191	15	03
Mastracchio	Richard	A. 'Rick'		Civ.	106	MS3											283	12	15
Mattingly II	Thomas	K. 'Ken'	CAPT	USN	4	Cdr	51-C	Cdr									242	43	13
McArthur, Jr.	William	S.	Lt.Col.	USA	58	MS2	74	MS3	92	MS5							842	27	05
McAuliffe	Sharon	Christa		Civ.	51-L	PS2											0	1	13
McBride	Jon	A.	CDR	USN	41-G	Plt											197	23	39
McCandless II	Bruce		CAPT	USN	41-B	MS1	31R	MS2									312	32	00
McCulley	Michael	J.	CDR	USN	34	Plt											119	39	21
McMonagle	Donald	R.	Lt.Col.	USAF	39	MS4	54	Plt	66	Cdr							605	34	44
McNair	Ronald	E.	Ph.D.	Civ.	41-B	MS3	51-L	MS3									191	17	07
Meade	Carl	J.	Col.	USAF	38	MS2	50	MS3	64	MS3							712	15	32
Melnick	Bruce	E.	CDR	USCG	41	MS4	49	MS5									311	27	42
Melroy	Pamela	A.	Lt.Col.	USAF	92	Plt											309	43	47
Merbold	Ulf	D.	Ph.D.	Civ.	9	PS2	42	PS2									441	2	8
Messerschmid	Ernst		Ph.D.	Civ.	61-A	PS2											168	44	53
Mohri	Mamoru		Ph.D.	Civ.	47	PS1	99	MS4									460	09	07
Morukov	Boris	V.	M.D., Ph.D.	Civ.	106	MS5											283	12	15
Mukai	Chiaki		M.D., Ph.D.	NASDA	65	PS1	95	PS1									567	38	56
Mullane	Richard	M.	Col.	USAF	41-D	MS3	27R	MS1	36	MS1							356	20	04
Musgrave	F.	Story	M.D., Ph.D.	Civ.	6	MS2	51-F	MS1	33R	MS1	44	MS1	61	MS4	80	MS3	1,285	58	42
Nagel	Steven	R.	Col.	USAF	51-G	MS3	61-A	Plt	37	Cdr	55	Cdr					721	36	36
Nelson	C. 'Bill'	William	Congressman	Civ.	61-C	PS2											146	03	53
Nelson	George	D.	Ph.D.	Civ.	41-C	MS1	61-C	MS3	26R	MS2							410	44	11
Neri-Vela	Rodolfo		Ph.D.	Civ.	61-B	PS1											165	04	54
Newman	James	H.	Ph.D.	Civ.	51	MS1	69	MS2	88	MS3							779	58	06
Nicollier	Claude		Capt.	Swiss AF/Civ.	46	MS3	61	MS2	75	MS3	103	MS5					1,021	04	48
Noriega	Carlos	I.	Maj.	USMC	84	MS2											221	19	56
O'Connor	Bryan	D.	Col.	USMC	61-B	Plt	40	Cdr									383	19	14
Ochoa	Ellen		Ph.D.	Civ.	56	MS3	66	MS1 a	96	MS2							719	55	28
Ockels	Wubbo	J.	Ph.D.	Civ.	61-A	PS3											168	44	53
Onizuka	Ellison	S.	Lt.Col.	USAF	51-C	MS1	51-L	MS2									73	34	41
Oswald	Steven	S.		Civ.	42	Plt	56	Plt	67	Cdr							814	31	55
Overmyer	Robert	F.	Col.	USMC	5	Plt	51-B	Cdr									290	23	20
Pailes	William	A.	Maj.	USAF	51-J	PS1											98	44	41
Parazynski	Scott	E.	M.D.	Civ.	66	MS4	86	MS2	95	MS2							735	38	51
Parise	Ronald	A.	Ph.D.	Civ.	35	PS2	67	PS2									614	13	56
Parker	Robert	A.	Ph.D.	Civ.	9	MS2	35	MS3									462	52	32
Pawelczyk	James	A. 'Jim'	Ph.D.	Civ.	90	PS2											381	49	58
Payette	Julie			Canada	96	MS4											235	13	03
Payton	Gary	E.	Maj.	USAF	51-C	PS1											73	33	28
Peterson	Donald	H.		USAF (Ret)	6	MS1											120	23	47
Precourt	Charles	J.	Col.	USAF	55	MS2	71	Plt	84	Cdr	91	Cdr					932	16	12
Pudwill-Gorie	Dominic	L.	CDR	USN	91	Plt	99	Plt									505	32	44
Readdy	William 'Reads'	F.		USN	42	MS3	51	Plt	79	Cdr							672	34	19
Reightler, Jr.	Kenneth	S.	CAPT	USN	48	Plt	60	Plt									327	37	00
Reilly	James	F.	Ph.D.	Civ.	89	MS1											211	46	35
Resnik	Judith	A.	Ph.D.	Civ.	41-D	MS1	51-L	MS1									144	57	18
Richards	Richard	N.	CAPT	USN	28R	Plt	41	Cdr	50	Cdr	64	Cdr					813	31	13
Ride	Sally	K.	Ph.D.	Civ.	7	MS1	41-G	MS2									343	47	47
Robinson	Stephen	K.	Ph.D.	Civ.	85	MS3	95	MS1 a									498	10	56
Rominger	Kent	V.	CDR	USN	73	Plt	80	Plt	85	Plt	96	Cdr					1,325	25	44
Ross	Jerry	L.	Col.	USAF	61-B	MS3	27R	MS2	37	MS1	55	MS1	74	MS2	88	MS1	1,133	11	57
Runco, Jr.	Mario		Lt.Cdr.	USN	44	MS2	54	MS1	77	MS3							551	8	27
Ryumin	Valery	Victorvitch		Russia	91	MS4											235	54	00
Sacco, Jr.	Albert		Ph.D.	Civ	73	PS2											381	52	22
Salizhan	Shakirovich	Sharipov		Russia	89	MS4											211	46	35
Schlegel	Hans	W.		Civ.	55	MS2											239	39	59
Scobee	Francis	R. 'Dick'		USAF (Ret)	41-C	Plt	51-L	Cdr									167	41	19
Scott	Winston	E.	CAPT	USN	72	MS2	87	MS3									590	34	47
Scully-Power	Paul	D.		Civ.	41-G	PS2											197	23	39
Searfoss	Richard	A.	Lt.Col	USAF	58	Plt	76	Plt	91	Cdr							939	18	23
Seddon	Margaret	Rhea	M.D.	Civ.	51-D	MS1	40	MS1	58	MS1							722	22	20
Sega	Ronald	M.	Ph.D.	Civ.	60	MS2	76	MS1									420	25	15
Shaw, Jr.	Brewster	H.	Col.	USAF	9	Plt	61-B	Cdr	28R	Cdr							533	52	26
Shepherd	William	M.	CAPT	USN	27R	MS3	38	MS1	52	MS2							459	56	21
Sherlock	Nancy	J.	Maj.	USA	57	MS2	70	MS3									454	05	01
Shriver	Loren	J.	Col.	USAF	51-C	Plt	31R	Cdr	46	Cdr							386	04	37
Smith	Michael	J.	CDR	USN	51-L	Plt											0	01	13
Smith	Steven	L.		Civ.	68	MS1	82	MS5	103	MS1							700	34	02

<table>
<thead>
<tr><th colspan="5">NAME</th><th colspan="2">1ST MISSION</th><th colspan="2">2ND MISSION</th><th colspan="2">3RD MISSION</th><th colspan="2">4TH MISSION</th><th colspan="2">5TH MISSION</th><th colspan="2">6TH MISSION</th><th colspan="3">TOTAL TIME</th></tr>
<tr><th>Last</th><th>First</th><th>Middle</th><th>Rank / Title</th><th>Affiliation</th><th>STS</th><th>Pos.</th><th>STS</th><th>Pos.</th><th>STS</th><th>Pos.</th><th>STS</th><th>Pos.</th><th>STS</th><th>Pos.</th><th>STS</th><th>Pos.</th><th>hrs</th><th>min</th><th>sec</th></tr>
</thead>
<tbody>
<tr><td>Solovyev</td><td>Anatoly</td><td>Y.</td><td></td><td>Russia</td><td>71 c</td><td>Mir-19</td><td></td><td></td><td></td><td></td><td></td><td></td><td></td><td></td><td></td><td></td><td>44</td><td>36</td><td>00</td></tr>
<tr><td>Spring</td><td>Sherwood</td><td>C.</td><td>Lt.Col</td><td>USA</td><td>61-B</td><td>MS2</td><td></td><td></td><td></td><td></td><td></td><td></td><td></td><td></td><td></td><td></td><td>165</td><td>04</td><td>54</td></tr>
<tr><td>Springer</td><td>Robert</td><td>C.</td><td>Col.</td><td>USMC</td><td>29R</td><td>MS3</td><td>38</td><td>MS1</td><td></td><td></td><td></td><td></td><td></td><td></td><td></td><td></td><td>237</td><td>33</td><td>20</td></tr>
<tr><td>Stewart</td><td>Robert</td><td>L.</td><td>Col.</td><td>USA</td><td>41-B</td><td>MS2</td><td>51-J</td><td>MS2</td><td></td><td></td><td></td><td></td><td></td><td></td><td></td><td></td><td>290</td><td>00</td><td>35</td></tr>
<tr><td>Still</td><td>Susan</td><td>L.</td><td>LCDR</td><td>USN</td><td>83</td><td>Plt</td><td>94</td><td>Plt</td><td></td><td></td><td></td><td></td><td></td><td></td><td></td><td></td><td>465</td><td>57</td><td>15</td></tr>
<tr><td>Strekalov</td><td>Gennady</td><td>M.</td><td></td><td>Russia</td><td>71 d</td><td>Mir-18</td><td></td><td></td><td></td><td></td><td></td><td></td><td></td><td></td><td></td><td></td><td>190</td><td>46</td><td>19</td></tr>
<tr><td>Sturckow</td><td>Frederick</td><td>W. 'Rick'</td><td>Maj.</td><td>USMC</td><td>88</td><td>Plt</td><td></td><td></td><td></td><td></td><td></td><td></td><td></td><td></td><td></td><td></td><td>283</td><td>17</td><td>50</td></tr>
<tr><td>Sullivan</td><td>Kathryn</td><td>D.</td><td>Ph.D.</td><td>Civ.</td><td>41-G</td><td>MS1</td><td>31R</td><td>MS3</td><td>45</td><td>MS1</td><td></td><td></td><td></td><td></td><td></td><td></td><td>532</td><td>49</td><td>33</td></tr>
<tr><td>Tanner</td><td>Joseph</td><td>R.</td><td></td><td>Civ.</td><td>66</td><td>MS2</td><td>82</td><td>MS1</td><td></td><td></td><td></td><td></td><td></td><td></td><td></td><td></td><td>502</td><td>11</td><td>09</td></tr>
<tr><td>Thagard</td><td>Norman</td><td>E.</td><td>M.D.</td><td>Civ.</td><td>7</td><td>MS3</td><td>51-B</td><td>MS2</td><td>30R</td><td>MS1</td><td>42</td><td>MS1</td><td>71 d</td><td>Mir-18</td><td></td><td></td><td>795</td><td>30</td><td>32</td></tr>
<tr><td>Thiele</td><td>Gerhard</td><td>P. J.</td><td>Ph.D.</td><td>ESA</td><td>99</td><td>MS1</td><td></td><td></td><td></td><td></td><td></td><td></td><td></td><td></td><td></td><td></td><td>269</td><td>38</td><td>44</td></tr>
<tr><td>Thirsk</td><td>Robert</td><td>Brent</td><td>M.D.</td><td>CSA</td><td>78</td><td>PS2</td><td></td><td></td><td></td><td></td><td></td><td></td><td></td><td></td><td></td><td></td><td>405</td><td>47</td><td>36</td></tr>
<tr><td>Thomas</td><td>Andrew</td><td>S.W.</td><td>Ph.D.</td><td>Civ</td><td>77</td><td>MS1</td><td>89 c</td><td>MS5</td><td>91 d</td><td>MS5</td><td></td><td></td><td></td><td></td><td></td><td></td><td>497</td><td>32</td><td>18</td></tr>
<tr><td>Thomas</td><td>Donald</td><td>A.</td><td>Ph.D.</td><td>Civ.</td><td>65</td><td>MS4</td><td>70</td><td>MS1</td><td>83</td><td>MS3</td><td>94</td><td>MS3</td><td></td><td></td><td></td><td></td><td>1,034</td><td>12</td><td>22</td></tr>
<tr><td>Thornton</td><td>Kathryn</td><td>C.</td><td>Ph.D.</td><td>Civ.</td><td>33R</td><td>MS3</td><td>49</td><td>MS2</td><td>61</td><td>MS1</td><td>73</td><td>MS3 a</td><td></td><td></td><td></td><td></td><td>975</td><td>15</td><td>25</td></tr>
<tr><td>Thornton</td><td>William</td><td>E.</td><td>M.D.</td><td>Civ.</td><td>8</td><td>MS3</td><td>51-B</td><td>MS3</td><td></td><td></td><td></td><td></td><td></td><td></td><td></td><td></td><td>313</td><td>17</td><td>37</td></tr>
<tr><td>Thuot</td><td>Pierre</td><td>J.</td><td>CDR</td><td>USN</td><td>36</td><td>MS3</td><td>49</td><td>MS1</td><td>62</td><td>MS1</td><td></td><td></td><td></td><td></td><td></td><td></td><td>654</td><td>52</td><td>41</td></tr>
<tr><td>Titov</td><td>Vladimir</td><td>G.</td><td>Col.</td><td>Russian AF</td><td>63</td><td>MS4</td><td>86</td><td>MS1</td><td></td><td></td><td></td><td></td><td></td><td></td><td></td><td></td><td>457</td><td>49</td><td>08</td></tr>
<tr><td>Tognini</td><td>Michel</td><td></td><td>Col.</td><td>French AF</td><td>93</td><td>MS3</td><td></td><td></td><td></td><td></td><td></td><td></td><td></td><td></td><td></td><td></td><td>118</td><td>49</td><td>36</td></tr>
<tr><td>Tokerav</td><td>Valery</td><td>Ivanovich</td><td>Col.</td><td>Russian AF</td><td>96</td><td>MS5</td><td></td><td></td><td></td><td></td><td></td><td></td><td></td><td></td><td></td><td></td><td>235</td><td>13</td><td>03</td></tr>
<tr><td>Trinh</td><td>Eugene</td><td>H.</td><td>Ph.D.</td><td>Civ.</td><td>50</td><td>PS2</td><td></td><td></td><td></td><td></td><td></td><td></td><td></td><td></td><td></td><td></td><td>331</td><td>31</td><td>04</td></tr>
<tr><td>Truly</td><td>Richard</td><td>H.</td><td>RADM</td><td>USN (Ret)</td><td>2</td><td>Plt</td><td>8</td><td>Cdr</td><td></td><td></td><td></td><td></td><td></td><td></td><td></td><td></td><td>199</td><td>21</td><td>57</td></tr>
<tr><td>Tryggvason</td><td>Bjarni</td><td>V.</td><td></td><td>CSA</td><td>85</td><td>PS1</td><td></td><td></td><td></td><td></td><td></td><td></td><td></td><td></td><td></td><td></td><td>284</td><td>27</td><td>00</td></tr>
<tr><td>Usachev</td><td>Yuri</td><td>V.</td><td></td><td>Russia</td><td>101</td><td>MS5</td><td></td><td></td><td></td><td></td><td></td><td></td><td></td><td></td><td></td><td></td><td>236</td><td>09</td><td>51</td></tr>
<tr><td>van den Berg</td><td>Lodewijk</td><td></td><td>Ph.D.</td><td>Civ.</td><td>51-B</td><td>PS1</td><td></td><td></td><td></td><td></td><td></td><td></td><td></td><td></td><td></td><td></td><td>168</td><td>08</td><td>54</td></tr>
<tr><td>van Hoften</td><td>James</td><td>D.</td><td>Ph.D.</td><td>Civ.</td><td>41-C</td><td>MS2</td><td>51-I</td><td>MS2</td><td></td><td></td><td></td><td></td><td></td><td></td><td></td><td></td><td>377</td><td>57</td><td>50</td></tr>
<tr><td>Veach</td><td>Charles</td><td>Lacy</td><td></td><td>Civ.</td><td>39</td><td>MS5</td><td>52</td><td>MS1</td><td></td><td></td><td></td><td></td><td></td><td></td><td></td><td></td><td>436</td><td>18</td><td>36</td></tr>
<tr><td>Voss</td><td>James</td><td>S.</td><td>Lt.Col</td><td>USA</td><td>44</td><td>MS3</td><td>53</td><td>MS2</td><td>69</td><td>MS1</td><td>101</td><td>MS3</td><td></td><td></td><td></td><td></td><td>838</td><td>49</td><td>18</td></tr>
<tr><td>Voss</td><td>Janice</td><td>E.</td><td>Ph.D.</td><td>Civ.</td><td>57</td><td>MS4</td><td>63</td><td>MS3</td><td>83</td><td>MS1 a</td><td>94</td><td>MS1 a</td><td>99</td><td>MS3</td><td></td><td></td><td>1,173</td><td>49</td><td>08</td></tr>
<tr><td>Wakata</td><td>Koichi</td><td></td><td></td><td>NASDA</td><td>72</td><td>MS3</td><td>92</td><td>MS1</td><td></td><td></td><td></td><td></td><td></td><td></td><td></td><td></td><td>532</td><td>44</td><td>32</td></tr>
<tr><td>Walker</td><td>Charles</td><td>D.</td><td></td><td>Civ.</td><td>41-D</td><td>PS1</td><td>51-D</td><td>PS1</td><td>61-B</td><td>PS2</td><td></td><td></td><td></td><td></td><td></td><td></td><td>477</td><td>56</td><td>27</td></tr>
<tr><td>Walker</td><td>David</td><td>M.</td><td>CAPT</td><td>USN</td><td>51-A</td><td>Plt</td><td>30R</td><td>Cdr</td><td>53</td><td>Cdr</td><td>69</td><td>Cdr</td><td></td><td></td><td></td><td></td><td>724</td><td>30</td><td>10</td></tr>
<tr><td>Walter</td><td>Ulrich</td><td></td><td></td><td>Civ.</td><td>55</td><td>PS1</td><td></td><td></td><td></td><td></td><td></td><td></td><td></td><td></td><td></td><td></td><td>239</td><td>39</td><td>59</td></tr>
<tr><td>Walz</td><td>Carl</td><td>E.</td><td>Lt.Col.</td><td>USAF</td><td>51</td><td>MS3</td><td>65</td><td>MS2</td><td>79</td><td>MS3</td><td></td><td></td><td></td><td></td><td></td><td></td><td>833</td><td>24</td><td>36</td></tr>
<tr><td>Wang</td><td>Taylor</td><td>G.</td><td>Ph.D.</td><td>Civ.</td><td>51-B</td><td>PS2</td><td></td><td></td><td></td><td></td><td></td><td></td><td></td><td></td><td></td><td></td><td>168</td><td>08</td><td>54</td></tr>
<tr><td>Weber</td><td>Mary</td><td>Ellen</td><td>Ph.D.</td><td>Civ.</td><td>70</td><td>MS3</td><td>101</td><td>MS1</td><td></td><td></td><td></td><td></td><td></td><td></td><td></td><td></td><td>450</td><td>29</td><td>58</td></tr>
<tr><td>Weitz</td><td>Paul</td><td>J.</td><td></td><td>USN (Ret)</td><td>6</td><td>Cdr</td><td></td><td></td><td></td><td></td><td></td><td></td><td></td><td></td><td></td><td></td><td>120</td><td>23</td><td>47</td></tr>
<tr><td>Wetherbee</td><td>James</td><td>D.</td><td>CDR</td><td>USN</td><td>32R</td><td>Plt</td><td>52</td><td>Cdr</td><td>63</td><td>Cdr</td><td>86</td><td>Cdr</td><td></td><td></td><td></td><td></td><td>955</td><td>45</td><td>57</td></tr>
<tr><td>Wilcutt</td><td>Terrence</td><td>W.</td><td>Maj.</td><td>USMC</td><td>68</td><td>Plt</td><td>79</td><td>Plt</td><td>89</td><td>Cdr</td><td>106</td><td>Cdr</td><td></td><td></td><td></td><td></td><td>1,172</td><td>50</td><td>49</td></tr>
<tr><td>Williams</td><td>Dafydd</td><td>R. 'Dave'</td><td>M.D.</td><td>Civ.</td><td>90</td><td>MS3</td><td></td><td></td><td></td><td></td><td></td><td></td><td></td><td></td><td></td><td></td><td>381</td><td>49</td><td>58</td></tr>
<tr><td>Williams</td><td>Donald</td><td>E.</td><td>CAPT</td><td>USN</td><td>51-D</td><td>Plt</td><td>34</td><td>Cdr</td><td></td><td></td><td></td><td></td><td></td><td></td><td></td><td></td><td>287</td><td>34</td><td>49</td></tr>
<tr><td>Williams</td><td>Jeffrey</td><td>N.</td><td>Lt.Col.</td><td>USA</td><td>101</td><td>MS2</td><td></td><td></td><td></td><td></td><td></td><td></td><td></td><td></td><td></td><td></td><td>236</td><td>09</td><td>51</td></tr>
<tr><td>Wisoff</td><td>Peter</td><td>J. K.</td><td>Ph.D.</td><td>Civ.</td><td>57</td><td>MS3</td><td>68</td><td>MS3</td><td>81</td><td>MS1</td><td>92</td><td>MS3</td><td></td><td></td><td></td><td></td><td>1,064</td><td>10</td><td>12</td></tr>
<tr><td>Wolf</td><td>David</td><td>A.</td><td>M.D.</td><td>Civ.</td><td>58</td><td>MS3</td><td>86 c</td><td>MS5</td><td>89 b, d</td><td>MS5</td><td></td><td></td><td></td><td></td><td></td><td></td><td>524</td><td>59</td><td>07</td></tr>
<tr><td>Young</td><td>John</td><td>W.</td><td></td><td>USN (Ret)</td><td>1</td><td>Cdr</td><td>9</td><td>Cdr</td><td></td><td></td><td></td><td></td><td></td><td></td><td></td><td></td><td>302</td><td>8</td><td>15</td></tr>
</tbody>
</table>

Discovery while in the mate/demate device at Dryden after STS-92. (Tony Landis)

The flight crew goes over preflight checklists on the flight deck of Atlantis during the final phase of the Terminal Countdown Demonstration Test (TCDT) exercises for STS-81, a simulated final countdown that goes through all the motions until just before main engine ignition. The crew are in their flight positions with the Orbiter in a vertical attitude at Pad 39B – note all the checklists and placards around the cockpit. The angle at right creates the illusion that astronaut support assistant Pam Melroy (second from right) is floating in space. Seated on the left is Pilot Brent W. Jett, Jr., while Mission Commander Michael A. Baker is on the right. From the left in the back row are Mission Specialists Peter J. K. 'Jeff' Wisoff and John M. Grunsfeld. (NASA photos KSC-96PC-1362 and KSC-96PC-1363)

A technician closes an access panel on the flight deck while the Orbiter is vertical at the pad. The forward instrument panel is at the top of the photo, with the lightweight commander's seat back shown horizontal at the bottom right. (Kim Keller)

Discovery shows the rain covers installed on the forward-firing thrusters and dessicant packs on the other FRCS thrusters. The FRCS service panels are still open. Note the covers on all the cabin windows, and the location of the star trackers. (Kim Keller)

Details of the aft fuselage as an Orbiter is ready to be lifted in the VAB. A bit of the lifting sling may be seen above the NASA meatball. The cover just behind the meatball is the 50-1 door penetration used to gain access to the aft fuselage. (Kim Keller)

A close up of Discovery at the pad. Red 'remove-before-flight' covers are installed on the OMS engine nozzles, and dessicant assemblies are on the RCS thrusters. Note the shape of the vertical stabilizer base for the drag chute installation. (Kim Keller)

This is a view looking into the aft compartment through the 50-1 door on the left side of the Orbiter – view is towards the belly. A NASA inspector is laying just below avionics bay 6, and an insulated LH2 feedline is in the foreground. (Kim Keller)

This is the LH2 vent arm (actually, the gaseous umbilical carrier plate) that vents the gaseous hydrogen that vaporizes from the LH2 tank while the vehicle is on the pad. The intertank access hatch is to the left. (photo by the Ice Team via Kim Keller)

Details of the SRB mounting to the MLP showing two of the hold-down bolts. Note the separation rockets on the far SRB. The piping and water bags (painted red in real life) are part of the sound suppression water system. (Kim Keller)

Gaseous hydrogen vents through the SSME as the main propulsion system is chilled down for launch. The Ice Team are the only people allowed on the pad during this time, and then only for short, well-controlled periods to inspect the vehicle for ice build-up prior to launch. (photo by the Ice Team via Kim Keller)

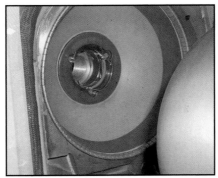

Another view through the 50-1 door showing the LH2 manifold plumbing leading to the three SSMEs. A technician in the upper right corner lends some scale to the scene. The Orbiter belly is to the left. (Kim Keller)

One of the two sockets in the bottom of the Orbiter where the ET attaches. The green circular area is called the 'salad bowl' and will receive the silver ET aft attach post, visible at right. (Kim Keller)

The forward Orbiter-ET 'bipod' attach point – the Orbiter is vertical to the left. (Kim Keller)

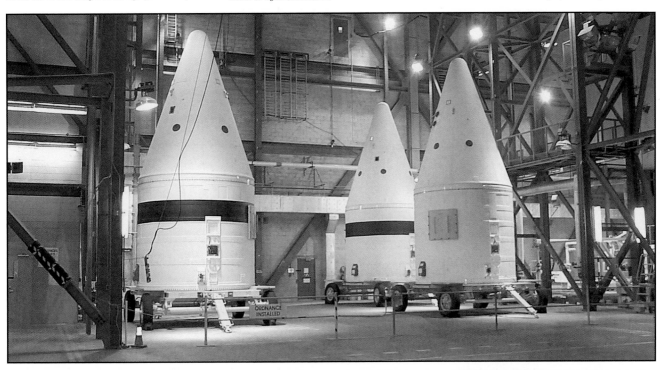

Three SRB forward skirt assemblies stored in the low bay of the VAB. (Kim Keller)

A technician works on wiring inside the aft compartment of the Orbiter. (via Kim Keller)

Two views of the aft ET–Orbiter attach points. The technician at left gives an indication of the size of the fittings. The protective covers (held on by black bands – the covers are blue) are removed before launch. (Kim Keller)

The last Orbiter, OV-105 Endeavour, under construction at Rockwell's Palmdale final assembly facility. (NASA)

OV-101 during her overland move from Palmdale to Edwards. Although she looked the part, in fact much of Enterprise was merely for show, such as the fiberglass wing leading edges and nose cap, and the styrofoam TPS tiles. (Dennis R. Jenkins)

As usual, a T-38 is close by as Enterprise comes in for a landing on one of the dry lakebed runways at Edwards. Although not equipped for spaceflight, OV-101 was very representative of future Orbiters aerodynamically. (Dennis R. Jenkins)

Unlike KSC and Dryden that were equipped with large Mate-Demate Devices, Vandenberg had a much simpler lifting frame arrangement. In theory the frame was transportable to contingency landing sites. The frame is now installed at Palmdale. (Dennis R. Jenkins)

A brand new Space Shuttle Main Engine being installed on a test stand at the National Space Transportation Laboratory, now called the Stennis Space Center. This was – and still is – the most efficient rocket engine ever developed. (Rocketdyne)

This was the classified payload aboard Atlantis on STS-28/51-J in 1985 – a pair of DSCS-III communications satellites riding on a single IUS. (Air Force via Dwayne Day)

Two shots of Enterprise being off-loaded from N905NA at the Redstone Arsenal adjacent to MSFC on 14 March 1978. (NASA photos MSFC-00677 and MSFC-00676)

A brand-new Atlantis at her roll-out ceremony in Palmdale. The OMS pods are aerodynamic dummies since the pods destined for OV-104 had been borrowed to support launches of Discovery at KSC. In fact, it would be several years before Atlantis flew with the OMS pods manufactured for her. (Rockwell International)

Am interesting view of Atlantis aboard the MLP as she leaves the VAB. Note how fragile the forward attach strut between the Orbiter and ET looks. (NASA photo KSC-96-0997)

The Space Shuttle shows one of its unique capabilities – repairing satellites on-orbit. This is the Hubble Space Telescope during STS-103. Note the built-in locations for the RMS effector to grab – one on each side of the satellite. (NASA photo S103E-5162)

The Kennedy Space Center is located within a wildlife sanctuary, and many different species of birds call the area home – although launch tends to alarm them somewhat. This is Endeavour during STS-57. (NASA photo NASA KSC-93PC-0879)

Columbia shows her unique black chines – these make it very easy to identify OV-102. Many models of the Space Shuttle incorrectly have black chines regardless of which Orbiter they are supposed to be replicating. The SILTS installation on top of the vertical stabilizer and the number of black tiles around it are also unique. (NASA photo KSC-98PC-0371)

The payload bay of Atlantis being prepared for STS-44. The primary payload on this mission was a DSP F-16 – called 'DSP Liberty' by the crew. This was the last of the original nine flights purchased by the Air Force. (Air Force photo PL91C-11927.06)

The new meatball marking on the left wing is evident as Discovery rolls-back to VAB High Bay 1 to repair damage to the ET's foam insulation caused by hail. The repairs could not be performed at the pad due to access limitations. (NASA photo KSC-99PC-0533)

Atlantis shows her new Multifunction Electronic Display System (MEDS) during April 1999. The 'glass cockpit' represents a significant upgrade to the Orbiter fleet, and will enable many future upgrades to provide increased situational awareness to the crew. For now most of the displays are duplicates of those that had been shown on the MCDS and dedicated displays and do not take any real advantage of the increased display capabilities of the new system. (NASA photo KSC-99PP-0440)

Enterprise being lowered onto the ET (below the load support bridge at left) during the MVGVT tests at MSFC on 21 April 1978. (NASA photo MSFC-00684)

A forklift hoists the first Block I Space Shuttle Main Engine (#2036) into the number one position on Discovery in preparation for STS-70. (NASA photo KSC-95PC-0585)

OV-104 – Atlantis – at Palmdale during an OMDP. Note the strong backs that are attached to the payload bay doors to provide additional support during ground operations. Essentially, one of the four operational Orbiters is always at Palmdale being overhauled and modified. (Boeing North American)

Sometimes it is easy to forget how close the Space Shuttle launch pads are to the ocean. – the pad perimeter is only a few hundred feet from the beach. Although this was planned for safety reasons, the result is a constant corrosion problem. Here a camera near Pad 39A records Challenger lifting-off for STS-6. (NASA photo KSC-83PC-0137)

The curved crawlerway leading into VAB High Bay 2 was part of the Safe Haven project, enabling the storage of Orbiters during severe weather. The road circles around OPF-3 at left center. OPF-1 and OPF-2 are just below the curve at lower left. The crawlerway also extends from the east side of the VAB out to the two launch pads – one visible in the distance to the left of the VAB, the other hidden behind the building. The horizon is the Atlantic Ocean. To the right of the crawlerway leading to the pads is the turn basin, where ships tow the barges to off-load new ETs from Louisiana. A corner of the Launch Control Center projects from the right of the VAB. (NASA photo KSC-00PP-0727)

A Super Light-Weight Tank (SLWT) just prior to completion at Michoud. Note that the nose cap and gaseous oxygen vent valve have not been installed yet. (Lockheed Martin Space Systems Company, Michoud Operations)

Sunrise and sunset launches are particularly beautiful, and the beaches around Cape Canaveral and Cocoa Beach are often filled with locals and visitors watching the spectacle. This is Discovery during STS-64. (NASA photo KSC-94PC-1137)

All of the major structures at SLC-6, except the LCC, are visible here. At the extreme left is the Payload Changeout Room (PCR), which contrary to its name was not a room but a mobile building that translated between the Payload Preparation Room (PPR) even further to the left, and the launch mount where Enterprise is standing. The Shuttle Assembly Building (SAB) is at center-left (with USAF written on it) and moved together with the Mobile Service Tower (MST) at the extreme right to completely enclose the launch mount during stacking operations and checkout. The low building in the background is the SRB storage facility. Note the oil platform off the coast to the right. (Dennis R. Jenkins)

The large white tube protruding from the top of the service structure is part of the lightning protection system that is installed on both lunch pads – necessary in the lightning capital of the nation. This is Atlantis on STS-106. (NASA photo KSC-00-PP-1269)

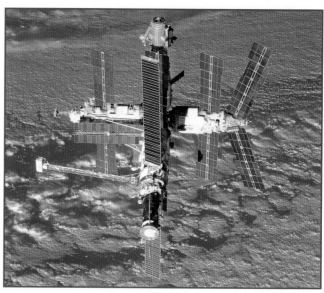

The Russian Mir space station. The Orbiter usually docked at the port in the upper center of the photo – the adapter was added to provide additional clearance between the Orbiter and the solar arrays. (NASA)

With its drag chute deployed, Discovery touches down on Runway 15 at KSC to complete the STS-91 mission on 12 June 1998. This was the 44th KSC landing and the 15th consecutive one. (NASA photo NASA 98EC-0742)

Atlantis rolls-over to the VAB for STS-106. The bright red covers are installed on the SSMEs and OMS engines to prevent possible foreign object damage (FOD) during ground handling and will be removed once the vehicle is vertical. (NASA photo KSC-00PP-1074)

Atlantis at Pad 39A for STS-101. Note the new markings – the NASA meatball on the left wing and both sides of the fuselage – instead of the stylized logo that most other NASA aircraft received. (NASA photo KSC-00PP-0634)

The International Space Station as photographed against Earth's horizon during a fly-around by Atlantis during STS-106 in September 2000. (NASA photo STS106-712-028)

Discovery is silhouetted against the early morning sky as it is makes the 3.5 mile trek to Pad 39A in preparation for STS-91. (NASA photo KSC-98PC-0568)

Space Shuttle Program, Launch Site, and Other Non-Mission Insignia

1979-1987

Vandenberg Launch Site

1991

Tenth Anniversary of STS-1

Modern

Space Shuttle Program Insignia

Traditional

Space Shuttle Program Insignia

STS-1V / 62-A

Go For Stack For First VLS Mission

1999–2000

Integrated Vehicle Health Management Insignia

Flight 25

STS-33/51-L Teacher in Space Insignia

STS-1

Go For Pad For First KSC Mission

Flight 1

STS-1
Columbia

Flight 2

STS-2
Columbia

Flight 3

STS-3
Columbia

Flight 4

STS-4
Columbia

Flight 5

STS-5
Columbia

Flight 6

STS-6
Challenger

Flight 7

STS-7
Challenger

Flight 8

STS-8
Challenger

Flight 9

STS-9
Columbia

Flight 10

STS-11 / 41-B
Challenger

Flight 11

STS-13 / 41-C
Challenger

Flight 12

STS-16 / 41-D
Discovery

Flight 13

STS-17 / 41-G
Challenger

Flight 14

STS-19 / 51-A
Discovery

Flight 15

STS-20 / 51-C
Discovery

Flight 16

STS-23 / 51-D
Discovery

Flight 17

STS-24 / 51-B
Challenger

Flight 18

STS-25 / 51-G
Discovery

Flight 19

STS-26 / 51-F
Challenger

Flight 20

STS-27 / 51-I
Discovery

Flight 21

STS-28 / 51-J
Atlantis

Flight 22

STS-30 / 61-A
Challenger

Flight 23

STS-31 / 61-B
Atlantis

Flight 24

STS-32 / 61-C
Columbia

Flight 25

STS-33 / 51-L
Challenger

Flight 26

STS-26R
Discovery

Flight 27

STS-27R
Atlantis

Flight 28

STS-29R
Discovery

Flight 29

STS-30R
Atlantis

Flight 30

STS-28R
Columbia

Flight 31

STS-34
Atlantis

Flight 32

STS-33R
Discovery

Flight 33

STS-32R
Columbia

Flight 34

STS-36
Atlantis

Flight 35

STS-31R
Discovery

Flight 36

STS-41
Discovery

Flight 37

STS-38
Atlantis

Flight 38

STS-35
Columbia

Flight 39

STS-37
Atlantis

Flight 40

STS-39
Discovery

Flight 41

STS-40
Columbia

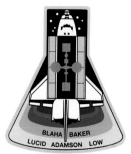

Flight 42

STS-43
Atlantis

Flight 43

STS-48
Discovery

Flight 44

STS-44
Atlantis

Flight 45

STS-42
Discovery

Flight 46

STS-45
Atlantis

Flight 47

STS-49
Endeavour

Flight 48

STS-50
Columbia

Flight 49

STS-46
Atlantis

Flight 50

STS-47
Endeavour

Flight 51

STS-52
Columbia

Flight 52

STS-53
Discovery

Flight 53

STS-54
Endeavour

Flight 54

STS-56
Discovery

Flight 55

STS-55
Columbia

Flight 56

STS-57
Endeavour

Flight 57

STS-51
Discovery

Flight 58

STS-58
Columbia

Flight 59

STS-61
Endeavour

Flight 60

STS-60
Discovery

Flight 61

STS-62
Columbia

Flight 62

STS-59
Endeavour

Flight 63

STS-65
Columbia

Flight 64

STS-64
Discovery

Flight 65

STS-68
Endeavour

Flight 66

STS-66
Atlantis

Flight 67

STS-63
Discovery

Flight 68

STS-67
Endeavour

Flight 69

STS-71
Atlantis

Flight 70

STS-70
Discovery

Flight 71

STS-69
Endeavour

Flight 72

STS-73
Columbia

Flight 73

STS-74
Atlantis

Flight 74

STS-72
Endeavour

Flight 75

STS-75
Columbia

Flight 76

STS-76
Atlantis

Flight 77

STS-77
Endeavour

Flight 78

STS-78
Columbia

Flight 79

STS-79
Atlantis

Flight 80

STS-80
Columbia

Flight 81

STS-81
Atlantis

Flight 82

STS-82
Discovery

Flight 83

STS-83
Columbia

Flight 84

STS-84
Atlantis

Flight 85

STS-94
Columbia

Flight 86

STS-85
Discovery

Flight 87

STS-86
Atlantis

Flight 88

STS-87
Columbia

Flight 89

STS-89
Endeavour

Flight 90

STS-90
Columbia

Flight 91

STS-91
Discovery

Flight 92

STS-95
Discovery

Flight 93

STS-88
Endeavour

Flight 94

STS-96
Discovery

Flight 95

STS-93
Columbia

Flight 96

STS-103
Discovery

Flight 97

STS-99
Endeavour

Flight 98

STS-101
Atlantis

Flight 99

STS-106
Atlantis

Flight 100

STS-92
Discovery

Cancelled Missions, Humorous, and Miscellaneous Insignia

Humorous

STS-8
('Spooky' patch)
(Courtesy of Liem Bahneman)

Cancelled

STS-22/51-E

Cancelled

61-E

Cancelled

61-F
(Ulysses)
(Courtesy of Liem Bahneman)

Cancelled

61-G
(Galileo)
(Courtesy of Liem Bahneman)

Humorous

STS-70
(With Woody Woodpecker)
(Courtesy of Liem Bahneman)

Original

STS-85
(With Pilot Jeff Ashby)
(Courtesy of Liem Bahneman)

Original

STS-88
(Before Sergei Krikalev)
(Courtesy of Liem Bahneman)

Shuttle-Mir
Insignia

First (STS-61)
Hubble Space Telescope
Servicing Mission
Insignia

Second (STS-82)
Hubble Space Telescope
Servicing Mission
Insignia

Flight 60

STS-60
Russian Insignia

THE PLAYERS

The Space Shuttle system consists of three major flight elements – the Orbiter, External Tank (ET), and Solid Rocket Boosters (SRB). Of these, only the External Tank is not recovered and reused.

The basic Orbiter Vehicle (OV) was developed and manufactured by the Space Systems Division of Rockwell International's, Downey, California, under contract to the NASA Lyndon B. Johnson Space Center (JSC) in Houston, Texas. The Orbiters were assembled at the Rockwell assembly facility located at Air Force Plant 42, Site 1, in Palmdale, California. With the consolidations and mergers within the aerospace industry, the company now responsible for the Orbiters is the Reusable Space Systems Division* of Boeing North American. Major mid-life overhauls (both OMDPs and OMMs) of the Orbiters are also accomplished at Palmdale.

The Solid Rocket Motors (SRM) are built by the Thiokol Propulsion division+ of Cordant Technologies, Inc., Brigham City, Utah. The non-motor parts of the SRBs (forward and aft skirts, separation motors, frustrum, parachutes and the SRB nose cap) were originally manufactured by United Space Boosters Incorporated, (USBI) in Huntsville, Alabama, with other portions being made in-house by the MSFC Science and Engineering Directorate. At first, MSFC was listed as 'prime contractor' for this effort, with USBI being a subcontractor. Eventually, these roles reversed, and finally MSFC stopped producing parts and became a customer only. On 1 October 1999 the functions performed by USBI were rolled into the Space Flight Operations Contract under United Space Alliance (USA).

The External Tank is built‡ by Lockheed Martin Space Systems Company, Michoud Operations Division at the NASA Michoud Assembly Facility in Louisiana. The Space Shuttle Main Engines (SSME) are manufactured by the Rocketdyne§ Division of Boeing North American, Canoga Park, California. All of these propulsion-related contracts are under the direction of the NASA George C. Marshall Space Flight Center (MSFC) in Huntsville, Alabama.

For the first two years of Space Shuttle operations, most aspects were handled by Rockwell as part of the original Space Shuttle systems contract awarded in 1972. The three other major contractors (Martin Marietta, Rocketdyne, and USBI) also participated in operations at KSC and JSC to support their specific flight elements. In January 1984 all ground operations at KSC and VLS were

consolidated into the Shuttle Processing Contract (SPC) that was awarded to Lockheed Space Operations Company (LSOC). The flight operations at JSC were performed under the Space Operations Contract (SOC) by the Rockwell Shuttle Operations Company (RSOC).

In August 1995, NASA expressed a desire to consolidate the large number of Space Shuttle Program contracts under a single prime contractor. Although originally to be procured as a competitive bid, in the end the new Space Flight Operations Contract (SFOC) was awarded sole-source to a joint venture of Lockheed and Rockwell called United Space Alliance, LLC (USA). On 7 November 1995, NASA Administrator Daniel Goldin announced in his 'Determination and Findings' report that the space agency would pursue a sole-source agreement with United Space Alliance. In April 1996, USA assumed management responsibility for both the SPC and SOC efforts, establishing the company as the prime contractor for the Space Shuttle Program. Other contracts, such as the USBI effort at MSFC and the former-IBM flight software effort at JSC were subsequently brought into the SFOC contract. Eventually the ET and SRM manufacturing contracts will also be consolidated within SFOC.[1]

SPACE SHUTTLE COORDINATE SYSTEM

The coordinate reference system provides a means of locating specific points on the Space Shuttle. The system is generally similar to ones used by most aircraft manufacturers, and is measured in inches and decimal places – X_O designates the longitudinal (forward and aft) axis, Y_O the lateral (side to side) axis, and Z_O the vertical (up and down) axis. The subscript '$_O$' indicates Orbiter; similar reference systems are used for the External Tank – '$_T$', Solid Rocket Booster – '$_B$', and the integrated Space Shuttle system – '$_S$'.[2]

In each coordinate system, the X-axis zero point is located forward of the nose – that is, the Orbiter nose tip location is 236 inches aft of the zero point (at X_O236), the ET nose cap tip location is at X_T322.5, and the SRB nose

---

* In December 1996 The Boeing Company purchased the space and defense divisions of Rockwell International and renamed them Boeing North American. The Space Systems Division was renamed the Reusable Space Systems Division (to differentiate it from other space divisions within Boeing that build Delta, etc.).

† When the SRM contract was awarded, the company was called the Wasatch Division of Morton Thiokol, Inc. On 1 July 1989, Morton Thiokol, Inc. was split into two separate companies – Morton International, Inc. and Thiokol Corporation. In 1998, Thiokol Corporation changed its name to Cordant Technologies, Inc. – the word Cordant is derived from 'concordant,' defined as harmonious or in agreement. Cordant is presently owned by Alcoa.

‡ The original contract was awarded to Martin Marietta Corporation. Lockheed Corporation and Martin Marietta Corporation merged in 1995.

§ Rocketdyne had been part of Rockwell International at contract award, and was bought by Boeing in December 1996 with the rest of the Rockwell space and defense businesses. It is now the Rocketdyne Division of Boeing North American.

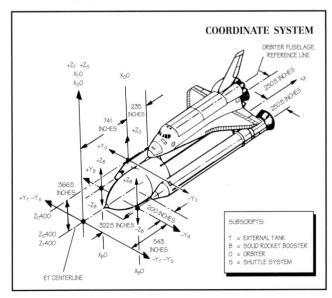

COORDINATE SYSTEM

tip location is at X_B200. In the Orbiter, the horizontal X_O / Y_O reference plane is located at Z_O400, which is 336.5 inches above the ET horizontal X_T / Y_T reference plane located at Z_T400. The SRB horizontal X_B / Y_B reference plane is located at Z_B0 and coincident with the ET horizontal plane at Z_T400. The vertical X_B and Z_T planes are located at $+Y_S250.5$ and $-Y_S250.5$. Also, the Orbiter, ET, and Shuttle system center X and Z planes coincide.

From the X-zero point, aft is positive and forward is negative for all coordinate systems. Looking forward, each Space Shuttle element Y-axis point right of the center plane (starboard) is positive and each Y-axis point left of center (port) is negative. The Z axis of each point within all elements of the Space Shuttle except the SRBs is positive, with Z-zero located below the element. In the SRBs, each Z-coordinate point below the X_B / Y_B reference plane is negative and each point above is positive.

LOCATION CODES

Orbiter location codes enable crewmembers to locate displays and controls, stowage compartments and lockers, access panels, and wall-mounted equipment in the Orbiter crew compartments (flight deck, mid-deck, and airlock). A fourth compartment became part of the configuration when the Spacelab was flown. Because of compartment functions and geometry, each has a unique location coding format.

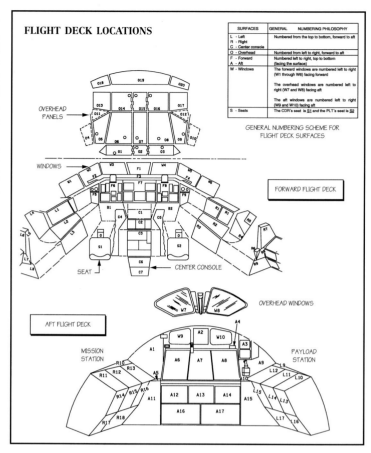

FLIGHT DECK LOCATIONS

SURFACES	GENERAL NUMBERING PHILOSOPHY
L - Left	Numbered from the top to bottom, forward to aft
R - Right	
C - Center console	
O - Overhead	Numbered from left to right, forward to aft
F - Forward	Numbered left to right, top to bottom (facing the surface)
A - Aft	
W - Windows	The forward windows are numbered left to right (W1 through W6) facing forward
	The overhead windows are numbered left to right (W7 and W8) facing aft
	The aft windows are numbered left to right (W9 and W10) facing aft
S - Seats	The CDR's seat is S1 and the PLT's seat is S2

GENERAL NUMBERING SCHEME FOR FLIGHT DECK SURFACES

OVERHEAD PANELS

FORWARD FLIGHT DECK

WINDOWS

SEAT

CENTER CONSOLE

AFT FLIGHT DECK

OVERHEAD WINDOWS

MISSION STATION

PAYLOAD STATION

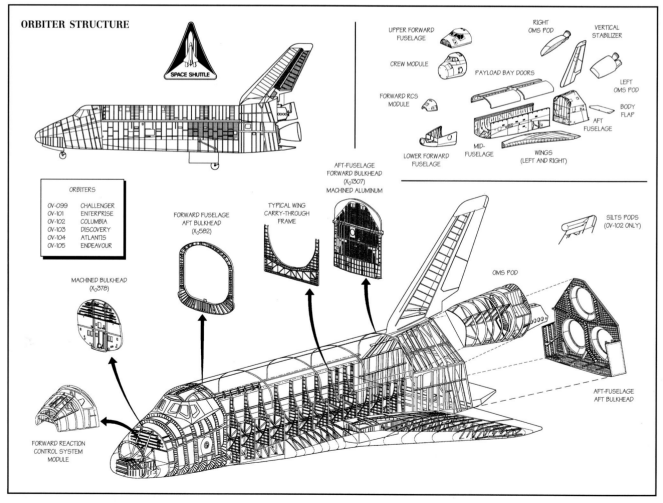

ORBITER STRUCTURE

SPACE SHUTTLE

ORBITERS	
OV-099	CHALLENGER
OV-101	ENTERPRISE
OV-102	COLUMBIA
OV-103	DISCOVERY
OV-104	ATLANTIS
OV-105	ENDEAVOUR

MACHINED BULKHEAD (X_O378)

FORWARD FUSELAGE AFT BULKHEAD (X_O582)

TYPICAL WING CARRY-THROUGH FRAME

AFT-FUSELAGE FORWARD BULKHEAD (X_O1307) MACHINED ALUMINUM

UPPER FORWARD FUSELAGE

CREW MODULE

FORWARD RCS MODULE

LOWER FORWARD FUSELAGE

MID-FUSELAGE

PAYLOAD BAY DOORS

RIGHT OMS POD

VERTICAL STABILIZER

LEFT OMS POD

BODY FLAP

AFT FUSELAGE

WINGS (LEFT AND RIGHT)

SILTS PODS (OV-102 ONLY)

OMS POD

AFT-FUSELAGE AFT BULKHEAD

FORWARD REACTION CONTROL SYSTEM MODULE

Orbiter

The Orbiter is a double-delta winged reentry vehicle capable of carrying both passengers and cargo to low-earth orbit and back to a controlled gliding landing. The Orbiter is constructed primarily of aluminum alloys covered by reusable surface insulation in tile and blanket form, with limited use of advanced composites. Each Orbiter was required to be capable of 100 flights, but this is a theoretical number. At present each Orbiter undergoes an overhaul (OMDP) every 8 missions, and no major structural issues have been identified to date. It is, therefore, possible that the Orbiters have a nearly indefinite structural life given the extensive maintenance philosophy currently used.

Forward Fuselage

The Orbiter forward fuselage section is of conventional aircraft manufacture, contains the pressurized crew module, and supports the forward RCS module, nose cap, nose gear wheel well, nose landing gear, and nose gear doors. The forward fuselage is manufactured as an upper and lower section to allow the pressurized crew module to be inserted during final assembly. The primary structure is constructed of 2024-T81 aluminum alloy stringer panels, frames, and bulkheads. The panels are single curvature with stretched-formed skins and riveted stringers spaced 3 to 5 inches apart. Frames are riveted to stringer panels, with 30 to 36 inch spacing between the frames. The upper

_____

* Early in the program this was 2,660 pounds.

portion of the X_0378 forward bulkhead is constructed of flat aluminum and formed sections riveted and bolted together, while the lower portion is a machined section. The bulkhead provides structure for the nose section, which contains large machined beams and struts. The forward fuselage carries the basic body-bending loads (a tendency to change the radius of a curvature of the body) and reacts to the nose landing gear loads.[3]

The nose landing gear well structure consists of two support beams, two upper closeout webs, hydraulic cylinder attachment points, drag-link support struts, and the landing gear door fittings. The nose landing gear doors are constructed of aluminum alloy honeycomb, and the left-hand door is slightly wider than the right-hand door.

The forward fuselage skin has structural provisions for installing antennas, deployable air data probes, and has eyelet openings for the two star trackers. Each opening has a door for environmental control.

The forward Orbiter–ET attach fitting is at the X_0378 bulkhead structure just aft of the nose gear wheel well. Lead ballast in the nose wheel well and on the X_0378 bulkhead provides weight and center-of-gravity control. The nose wheel well can accommodate up to 1,350 pounds of ballast, while an additional 1,971 pounds* can be attached to the X_0293 bulkhead. Purge and vent control is provided by flexible boots between the forward fuselage and crew module around the windshield, overhead observation windows, crew hatch window, and star tracker openings. The forward fuselage is isolated from the payload bay by a flexible membrane between the forward fuselage and crew compartment at X_0582.[4]

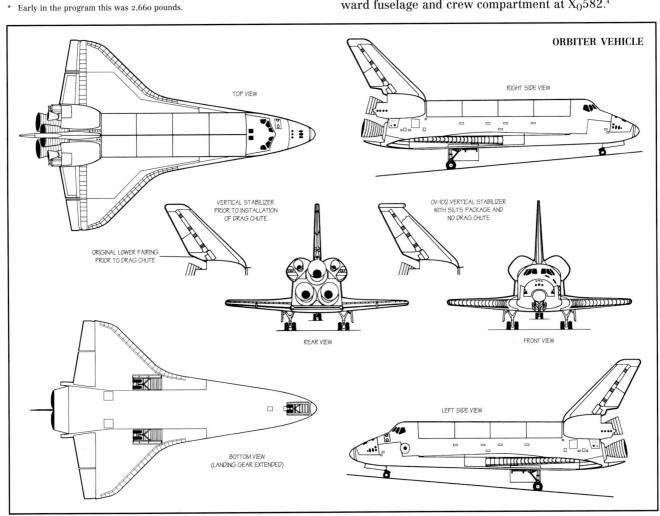

ORBITER VEHICLE

TOP VIEW

RIGHT SIDE VIEW

VERTICAL STABILIZER
PRIOR TO INSTALLATION
OF DRAG CHUTE

OV-102 VERTICAL STABILIZER
WITH SILTS PACKAGE AND
NO DRAG CHUTE

ORIGINAL LOWER FAIRING
PRIOR TO DRAG CHUTE

REAR VIEW

FRONT VIEW

LEFT SIDE VIEW

BOTTOM VIEW
(LANDING GEAR EXTENDED)

The lower forward fuselage for OV-105 in Building 200 at Downey. The hole just below the forward corner of the FRCS cutout is for the air data probes. The two nose landing gear support struts may be seen inside the FRCS area. (NASA photo S88-27847)

The upper forward fuselage for Endeavour. The outer pane of glass for all of the windscreens will be attached to this outer shell, while the two inner panes will be attached to the pressurized crew module that goes inside the shell. (NASA photo S88-55911)

The forward fuselage of Columbia being prepared for STS-28R. Note the holes in the nose cap for the SEADS experiment. The Orbiter is suspended by a lifting frame in the VAB, and the technicians are installing some last-minute gap filler between tiles. By this time the carbon-carbon chin panel had been installed on the Orbiter. The insert shows a deployed air data probe (not related to this Orbiter) with an arrow pointing to its location on the forward fuselage. (NASA photo KSC-89P-639; air data probe via Kim Keller)

The nose cap is made of reinforced carbon-carbon for protection from peak reentry heating. The upper and lower forward fuselage, and the Forward Reaction Control System (FRCS) modules were built by Rockwell at Downey, California. The carbon-carbon nose cap was manufactured by LTV (now part of Lockheed Martin) in Grand Prairie, Texas.

Crew Module

The crew module is a pressurized compartment housing flight control stations, living areas, and equipment bays. It contains three levels and has a volume of 2,325 cubic feet if the internal airlock is installed and 2,625 otherwise. The upper level is the flight deck and provides normal seating for up to four crewmembers. The mid-deck contains passenger seating, a living area, and an airlock into the payload bay (if installed), along with avionics equipment compartments. Originally, each Orbiter was equipped with the internal airlock on the mid-deck; as of late 2000, only OV-102 retains the internal airlock, the others have the airlock in the payload bay as part of the Orbiter Docking System (ODS). Up to three passengers can be accommodated on the mid-deck, except for OV-102 that accommodates four. The lower deck contains environmental control equipment and is accessible through removable floor panels on the mid-deck. The crew module is pressurized to 14.7 ±0.2 psia and provides a 'shirt sleeve' environment for the crew.[5]

Basically conical in shape, the crew module is composed of 2219 aluminum alloy plates with integral stiffening stringers and internal framing welded together to create a pressure-tight vessel. The module has a side hatch for normal ingress and egress, a hatch into the airlock from the mid-deck, and a hatch from the airlock through the aft bulkhead into the payload bay. Approximately 300 penetrations in the pressure shell are sealed with plates and fittings. The crew module is supported within the forward fuselage at only four attach points to minimize thermal conductivity. The two major attach points are located at the aft end of the crew module at the flight deck floor level. The vertical load reaction link is on the centerline of the forward bulkhead and the lateral load reaction link is on the lower segment of the aft bulkhead. Gold-coated multi-layer insulation (MLI) blankets are used to insulate the crew module. On OV-101 and OV-102 the MIL is attached to the interior frames and skins of the forward fuselage; on OV-103 and later Orbiters the MLI is attached directly to the outside of the crew module. Rockwell in Downey, California, built the crew modules.[6]

Redundant pressure window panes are provided in the six forward windshields, two overhead viewing windows, two aft viewing windows, and the side hatch window. The six windshields on the flight deck provide pilot visibility, and are the thickest pieces of optical quality glass ever produced for 'see-through' viewing. Each of the outer windscreens on the forward fuselage is 42 inches diagonal, and the two center windscreens are 35 inches diagonal. The innermost pane is constructed of tempered alumino-silicate glass to withstand the cabin pressure with a high degree of reliability, and is 0.625-inch thick. Alumino-silicate glass is a low-expansion glass that can be tempered to provide maximum mechanical strength. The outer surface of this pane is coated with a red-reflector that reflects the infrared (heat) while transmitting the visible spectrum.[7]

Due to its high optical quality and excellent thermal shock resistance, the center pane in all of the crew com-

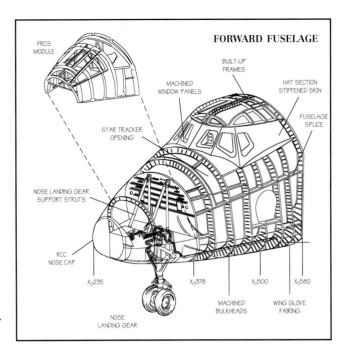

FORWARD FUSELAGE

The forward bulkhead for the OV-104 crew module being manufactured in Downey. (Rockwell photo A821018G-10)

The OV-105 crew module is fit-checked with the lower forward fuselage. Note the large opening in the aft bulkhead for the mid-deck airlock. This is the only pressurized area on the Orbiter. (NASA photo S88-31449)

Commander Kenneth D. Cockrell peers out the window of Columbia after a successful landing at KSC, completing STS-80. Note how far inset the windows are from the moldline of the fuselage. Part of the star tracker opening may be seen in the extreme lower left corner of the photo. (NASA photo KSC-96PC-1340)

Eight of the ten flight deck windows are visible here. The six forward windscreens provide about the same visibility as a 747, while the two overhead windows on the aft flight deck are finally being used for the purpose they were designed – docking with a space station. This is Atlantis as seen from Mir on STS-71. (NASA)

partment windows is constructed of 1.3-inch thick low-expansion fused silica glass. The surfaces of this pane are coated with a high efficiency anti-reflection coating to improve visible light transmission. These windows must successfully withstand a stress pressure of 8,600 psi in a flexure test performed at 240 degF and 0.017-percent relative humidity.[8]

The outer panes are fused silica glass 0.625-inch thick designed for a reentry temperature of 800 degF. Although the exterior of these panes are uncoated, the interior uses the same anti-reflective coating as the center pane.

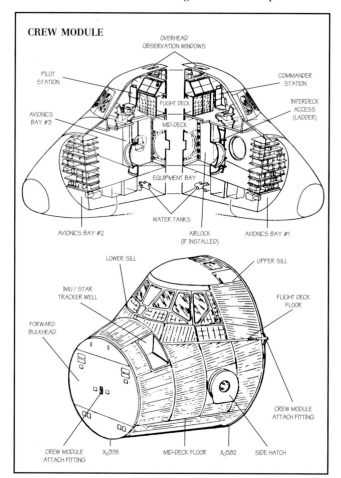

The outer panes of the six forward windows are mounted to the forward fuselage, while the center and inner panes are mounted to the crew module. The windshields are held in the crew module by glass retainers. For the innermost panel (pressure panel), grooves are provided in the glass retainer for redundant seals that attach the inner pane to the window frame. The tolerances are so tight that no sealing or bonding compounds are used.

There are also two 20 by 20 inch overhead windows directly above the aft flight station. These windows are identical in construction to the windscreens except for thickness. The inner and center panes are 0.45-inch and the outer pane is 0.68-inch thick. Again, the outer pane is attached to the forward fuselage structure, while the center an dinner panes are attached to the crew module. The left-hand overhead window also provides the crew with a secondary emergency egress route. The inner and center panes open into the crew compartment and the outer pane is jettisoned outward, providing a 20 by 20 inch opening.[9]

On the aft flight deck, the two windows for viewing the payload bay consist of only two panes of glass, which are identical to the forward windshields' inner and center panes. The outer thermal panes are not installed since the windows are protected by the payload bay. The windows are 14.5 by 11 inches and each pane is 0.3 inch thick.[10]

During orbital operations, the large window areas expose the flight crew to sun glare; therefore, window shades and filters are provided to preclude or minimize exposure. Shades are provided for all windows, and filters are supplied for the aft and overhead viewing windows. The window shades and filters are stored on the mid-deck.

The forward window shades (W-1 through W-6) are rolled up and stowed at the base of the windows. Solid metal shades are used on overhead windows (W-7 and W-8). The overhead window filters are fabricated from Lexan and used interchangeably with the shades. The aft window shades (W-9 and W-10) are the same as the overhead window shades except that a 0.63-inch-wide strip of Velcro has been added around the perimeter of the shade. The shade is attached to the window by pressing the Velcro strip to the pile strip around the window opening. The aft window filters are the same as the overhead window filters except for the addition of a Velcro hook strip. The side hatch window cover is permanently attached to the window frame and is hinged to allow opening and closing.[11]

The side hatch viewing window consists of three panes identical to the forward windscreens. The inner pane is 11.4 inches in diameter and 0.25 inch thick, while the center pane is also 11.4 inches in diameter but has a thickness of 0.5-inch. The outer pane is 15 inches in diameter and 0.3 inch thick. The effective viewing area is 10-inches in diameter. The side hatch glass for OV-099, OV-101, and OV-102 was of slightly different construction than the later Orbiters. *Columbia* has since been retrofitted with the later configuration; *Challenger* was lost prior to being modified, and there was no reason to modify OV-101. The contractor for all Orbiter windows was Corning Glass Company, Corning, New York.[12]

Smoke detection and fire suppression capabilities are provided in the crew module avionics bays, the crew module, and pressurized payload modules (i.e., Spacelab). Ionization detection elements sense levels of smoke concentrations or rate of concentration change. If an ionization detection element senses a smoke particle concentration of 2,000 (±200) micrograms per cubic meter for at least 5 seconds and/or a rate of smoke increase of 22 micrograms per cubic meter per second for eight consecutive counts in 20 seconds, the crew is warned. Fire suppression in the avionics bays is provided by three permanently-mounted Halon fire bottles. Three hand-held fire extinguishers are available in the crew module – one on the flight deck and two on the mid-deck.[13]

Flight Deck

The flight deck is on the uppermost level of the crew module. The commander and pilot work stations are positioned side-by-side in the forward portion of the flight deck. These stations have controls and displays for maintaining autonomous control of the vehicle throughout all mission phases. Directly behind and to the sides of the commander and pilot centerline are the mission specialist seats. The forward flight deck, which includes the center console and seats, contains approximately 24 square feet. However, the side console controls and displays add an additional 3.5 square feet, and the center console occupies 5.2 square feet. The aft flight deck has approximately 40 square feet of usable area.[14]

The commander and pilot seats have two shoulder harnesses and a lap belt for restraints, with the shoulder harnesses having an inertia reel lock/unlock feature. The commander and pilot can move their seats along the Orbiter Z (vertical) and X (longitudinal) axes so that they can reach and see controls better during the ascent and reentry phases of flight. Seat movement for each axis is provided by a single electric motor, and the total travel distance for the Z and X axes is 10 and 5 inches, respectively. If the seat motors fail, the seat can be adjusted manually, however, manual seat adjustment can only take place on-orbit and is accomplished with a special seat adjustment tool provided in the in-flight maintenance tool kit. The seats accommodate stowage of in-flight equipment and have removable seat cushions and mounting provisions for oxygen and communications connections to the crew altitude protection system.

Each specialist seat also has two shoulder harnesses and a lap belt for restraints. The seats have controls to manually lock and unlock the tilt of the seat back. Each seat has removable seat cushions and provisions for communications and oxygen connections. The seats are removed and stowed in the mid-deck while on-orbit. No tools are required since the legs of each seat have quick-disconnect fittings. The seats are 25.5 inches long, 15.5 inches wide, and 11 inches high when folded for stowage.

In 1995 NASA initiated a program to design lightweight seats for the Orbiter to increase its payload capacity on International Space Station missions. These seats use advanced construction, but otherwise are interchangeable with the original seats. The lightweight seats are made primarily of 7075 aluminum, with some other metal alloys such as 2024 aluminum, 6061 aluminum, titanium, Inconel-718, and some MP-35N used as appropriate. The new seats were designed to significantly higher loads, including a 20-g load applied simultaneously with a 15-degree floor warp, and a 16-g dynamic crash load. The original seats were designed only to a 10-g static load. The mission specialist seats were originally about 110 pounds, while the new seats are only 49 pounds. Weight was cut out

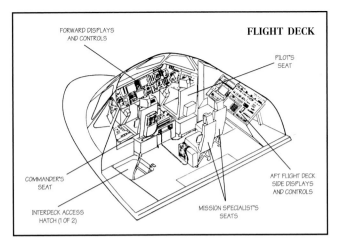

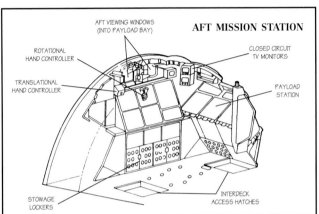

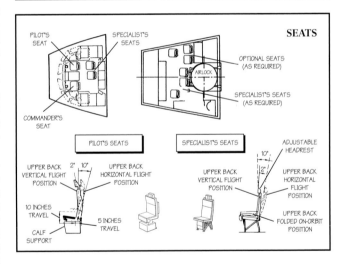

of the seats primarily with some very aggressive analysis, testing, and by using the seat design in a unique way. The standard floor fittings in the Orbiter were replaced with a spherical ball connection, which greatly reduces the load at the attach points. As the floor warps, the seat behaves very much like a mechanism and accommodates the movement by rotating about its joints instead of bending. This led to a seat that is very strong, and yet flexible enough to tolerate the floor warping. The floor warping requirement had been developed by the Federal Aviation Administration (FAA) due to some common failures that were being seen in commercial airline accidents (i.e. seats were too stiff and commonly failed at the floor connection points). Interestingly NASA subjected the seats to an actual dynamic crash test at the FAA crash test facility in Oklahoma City. To demonstrate the complication that the floor warping requirement caused – it is equivalent to holding three of the four seat attach points in place, and moving the fourth attach point almost 3 inches vertically ... then proceeding to crash the seat with an occupant at 20-g.[15]

After the specialists seats were designed, the pilot seats were replaced with similar units. The new pilot seats used all of the old drive mechanisms (for seat adjustability) but replaced the rest with components similar to the MS seats. These seats were not subjected to the floor warping requirement. The seats have been installed in all Orbiters as they pass through their OMDP cycles.

During the Approach and Landing Tests (ALT) phase *Enterprise* was equipped with Lockheed zero-zero ejection seats (modified A-12/SR-71 units) for the two pilot positions, and no other seating was provided. During the Orbital Flight Test (OFT) phase (STS-1 through STS-4), *Columbia* was also equipped with Lockheed zero-zero ejection seats for use on the launch pad and during landing operations. The ejection seats were disabled after STS-4, and removed at Palmdale following STS-9. No crew escape system was provided during STS-5 through STS-33/51-L.

After the loss of *Challenger*, the issue of crew escape systems was reexamined, and after again rejecting various proposals for ejection seats, escape pods, etc., a telescop-

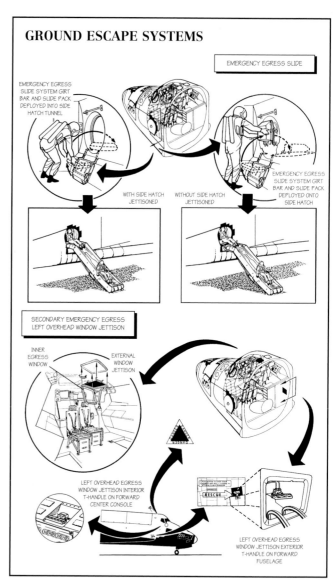

GROUND ESCAPE SYSTEMS

The shot above shows one of the new lightweight pilot seats with the back in the raised position. The back folds forward almost flush with the seat bottom while on-orbit. At right are the three major styles of mid-deck seats. Furthest from the camera is one of the original specialist seats – note there is no strut connecting the seat back to the seat bottom. This is a major distinguishing feature of the new seats, shown in the middle. In the foreground is a lightweight specialist seat in the recumbent position used to bring de-conditioned astronauts back from Mir and the ISS. The astronaut lays with their back to the floor and their legs tucked inside the open mid-deck locker just behind the seat. This attitude is much easier for the astronaut to tolerate during reentry and landing after a long stay in zero-g. (NASA photos S96-07476 and S96-16208 via Chris Hansen)

ic slide-pole was selected. This method is considered useful only below 200 knots and 30,000 feet during controlled gliding flight. The 9.8-foot long telescopic pole extends from the Orbiter side hatch, and during evacuation, crew members slide down the pole with special attachments on the parachute harness. The escape pole curves sharply downward and slightly aft, directing the crew members under the Orbiter's left wing. A crew of eight can evacuate the Orbiter in approximately 90 seconds.[16]

The escape pole is constructed of aluminum alloy and steel. The arched housing for the pole is 126.75 inches long and is attached to the mid-deck ceiling above the airlock hatch and at the two o'clock position at the side hatch for deployment during launch and reentry. The escape pole telescopes from its 3.5-inch diameter mid-deck housing through the side hatch in two sections. The primary extension is 73 inches long, and the end extension is 32 inches long. The two telescoping sections are slightly smaller in diameter. The escape pole itself weighs approximately 241 pounds, for a total 248 pounds with attachments.

Modifications required to install the escape system in OV-103 were completed on 15 April 1988, and it was operational in time for the Return-to-Flight of STS-26R. The system was subsequently installed on *Columbia* and *Atlantis*, and was a production feature on *Endeavour*.

The side hatch jettison thruster contractor was OEA of Denver, Colorado. The pyrotechnics contractor for the hatch tunnel, hinge and the energy transfer system lines was Explosive Technology, Incorporated, of Fairfield, California. The escape pole was government-furnished equipment supplied by the Johnson Space Center.

The first four Shuttle flights used David Clark Company Model S1030A Ejection Escape Suits (EES). These full-pressure suites satisfied the OFT emergency egress requirements up to Mach 2.7 and 80,000 feet. The S1030A was a variant of the Air Force S1030 Pilot's Protective Assembly (PPA) worn by SR-71 crew members – the primary change was an integrated anti-g suit. When the ejection seats were disarmed after STS-4, ascent was considered a 'shirt sleeve' environment, and no protective suits were worn, although a clamshell-type helmet was worn to provide supplemental oxygen in the event of a cabin depressurization during ascent or reentry below 50,000 feet.[17]

After the *Challenger* accident, NASA abandoned the shirt-sleeve concept for ascent and initiated an effort to develop a pressure suit that would provide hypobaric protection during ascent and reentry, as well as cold water immersion protection in the event of a bailout over water. The David Clark Company was selected for a fast-track development effort that took advantage of previous pressure suit developments, with each major LES component being selected from previously qualified systems. The new David Clark Company S1032 Launch Entry Suit (LES) was ready in time for STS-26R and were used on all subsequent missions. It should be noted that the LES was a partial-pressure suit, not a full-pressure garment.[18]

It was always realized that the LES was not the optimum suit, and in 1990 a new development effort was initiated that resulted in the Model S1035 Advanced Crew Escape Suit (ACES). This development effort drew heavily from an earlier Air Force-sponsored effort that had resulted in the S1034 PPA. The new S1035 suit was designed to be a simplified, lightweight, low-bulk full-pressure suit that facilitated self-donning/doffing, and provided enhanced overall performance. Production of the new suit commenced in 1993 and the first units were delivered to NASA in May 1994. The ACES first flew on

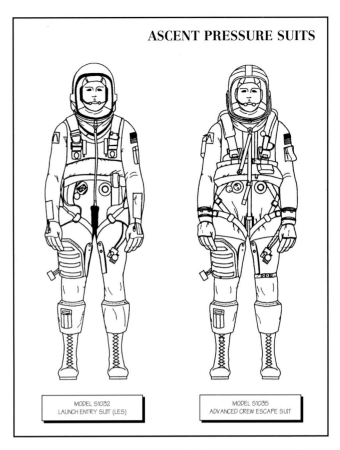

The partial-pressure LES suit (left) was used beginning on the Return-to-Flight of STS-26R. Starting with STS-68 the program began transitioning to the full-pressure ACES suit (right). (David Clark Company)

STS-68 in August 1994 and all flights since mid-1995 have used the ACES.[19]

An emergency egress slide is also provided to provide the Orbiter crew members with a rapid emergency egress through the Orbiter mid-deck side hatch after a normal opening of the hatch, or after explosively jettisoning the hatch. The emergency egress slide was fitted to all of the remaining Orbiters during the *Challenger* stand-down, and was a production feature on *Endeavour*. The slide replaced an emergency side hatch escape bar that allowed crew members to hang from the side hatch feet first. This required crew members to drop approximately 10.5 feet to the ground, and it was felt that this drop could cause injury to the crew members and prevent an injured crew member from moving to a safe distance from the Orbiter. The crew emergency egress slide is generally similar to the ones in use by all commercial airliners, and was manufactured by Inflatable Systems Incorporated, a division of OEA of Denver, Colorado.

Over 2,100 displays and controls are located on the flight deck instrument panels, and many additional functions are controlled through keypads and CRTs that interface with the five on-board general purpose computers (GPC). Dual Kaiser Electronics heads-up displays (HUD) were included on *Challenger* and subsequent Orbiters, and were retrofitted to OV-102 after STS-9. Rotational hand controllers (RHC) are used to control vehicle rotation about all three axes. During ascent the controllers may be used to gimbal the SSMEs and SRBs. For orbital insertion and deorbit, the RHCs may be used to gimbal the OMS engines and to command thrusting of the RCS. On-orbit, the controllers are used to command the RCS thrusters. During reentry the hand controllers provide

ROTATIONAL HAND CONTROLLER

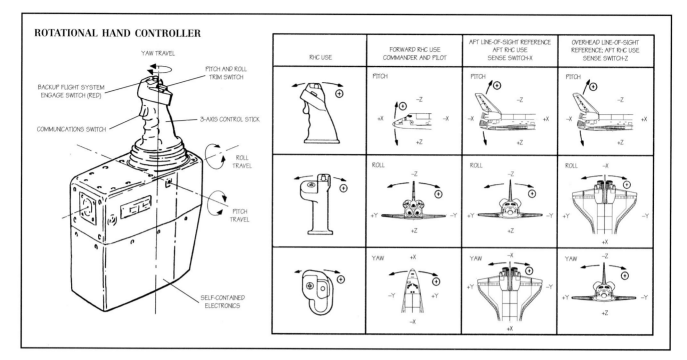

normal flight control-type inputs, commanding either the RCS thrustings or aerodynamic flight controls as appropriate. A controller is also provided at the aft mission station for control of the remote manipulator system (RMS).[20]

Normal rudder pedals are provided at the commander and pilot stations, controlling the rudder during atmospheric flight, and the nose wheel steering system and main wheel brakes during ground operations. There are also two speedbrake/thrust controllers on the flight deck that can be used to vary the thrust level of the SSMEs during ascent, and speedbrake operations during descent. A dedicated caution-and-warning display is provided for Orbiter systems and payloads consisting of four master caution lights, a 40-light enunciator on panel F7 and a separate 120-light enunciator on panel R13U. In addition, an audio cue is sent over the communications system to the headsets or speaker boxes to alert the crew. Aft of the pilot's seat is the mission station, which is operated while standing and contains no seating. Displays and controls for the RMS, and payloads contained in the payload bay, are located on the aft bulkhead wall.[21]

The flight deck, bunk sleep stations, and mid-deck modular stowage lockers were built by Rockwell's Space Transportation Systems Division, Downey, California. The original crew seat contractor was AMI of Colorado Springs, Colorado, but later seats were constructed by Rockwell's Space Systems Division.

Multifunction CRT Display System

As originally delivered, and flown for almost 20 years, the Orbiters were equipped with the Multifunction CRT Display System (MCDS). In 1998 the system began to be replaced by the Multifunction Electronic Display System (MEDS), which is described later. The MCDS is composed of three types of hardware – 4 display electronics units

The side panels contain a wide variety of switches, and occupy an abnormally large amount of floor space. This is OV-102 before the ejection seats were removed. Note the lack of HUDs, and tight fit of the Lockheed zero-zero seats. Also notice the small cameras outboard and above each position. (NASA photos S84-43422 and S84-43428)

The three bags on the side of the compartment at upper left contain escape devices to allow the crew to egress through the overhead windows. Note the opening to the mid-deck at the lower left. These two photos are of the Full Fuselage Trainer at JSC, not an actual Orbiter – although it is hard to tell. (NASA photo S87-41940)

The aft station contains a hand controller for the manipulator arm, and a DU/keypad for controlling payloads. The blank panels are used for payload-unique instrumentation as required. Note the two CCTV monitors in the upper right corner, and the ADI instrument in the upper left corner. (NASA photo S87-41942)

(DEU); 4 display units (DU) that include the CRTs, and 3 keyboard units, which communicate with the GPCs over the display/keyboard data bus network.[22]

Three DUs and two keyboards are mounted in the main instrument panel, while a single keyboard and DU are mounted at the side aft station. The keyboards each consist of 32 momentary double-contact push button keys. A switch on each of the forward keyboards determines which DU it communicates with (the center DU is shared while the outboard ones are not). There are ten numeric keys, six letter keys (for hexadecimal input), two algebraic keys, a decimal key, and 13 special key functions.

Each of the four DEUs responds to computer commands, transmits data, executes its own software to process keyboard inputs, and sends signals to drive displays on the CRTs. The four DEUs store display data, generate the GPC/keyboard unit and GPC/display unit interface displays, update and refresh on-screen data, check keyboard entry errors, and echo entries to the CRT.

The display unit uses a 5 by 7-inch magnetic-deflected, electrostatic-focused CRT. When supplied with deflection signals and video input, the CRT displays alphanumeric characters, graphic symbols, and vectors on a green-on-green phosphorous screen activated by a magnetically controlled beam. The CRTs are not capable of displaying true free-form raster graphics, being limited to a symbol library and vector lines. Each CRT has a brightness control for ambient light adjustment. If one of the CRTs (number 1 or 2 only) on the main panels fails, the crew can swap the CRT from the aft station with the failed unit for reentry.[23]

Dedicated Displays

When the MCDS is installed, the dedicated displays provide the primary flight instrumentation required to fly the vehicle manually or to monitor automatic flight control system performance, and are located in front of the commander and pilot seats and on the aft flight deck

panel. The data on the dedicated displays may be generated by the navigation or flight control system software, or more directly by one of the navigation sensors.[24]

The dedicated displays include: attitude director indicators (ADI), two horizontal situation indicators (HSI), alpha Mach indicators (AMI), altitude/vertical velocity indicators (AVVI), flight control position indicators, reaction control system activity lights, and a g-meter. Not all of the dedicated displays are available in every operational sequence or major GPC mode.

The display driver unit (DDU) interfaces between the GPCs and the primary flight displays. The DDU receives data signals from the computers and decodes them to

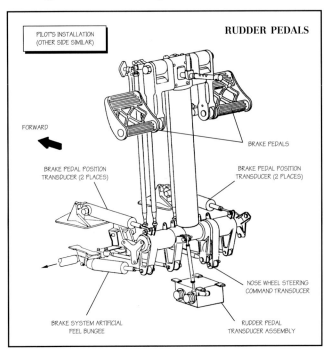

Columbia during final assembly. The CRTs are missing, but otherwise the instrument panel is complete. Note the different style of rotary switches on the overhead panel – compared to page 375 – and the different configuration of the panels themselves. The switch configuration on the sides of the glare shield is also different. (Rockwell)

drive the dedicated displays. The unit also provides dc and ac power for the ADIs and the rotational and translational hand controllers. It contains logic for setting flags on the dedicated instruments for such items as data dropouts and failure to synchronize. When the MEDS is installed, the DDU is retained since it continues to provide power to the RHCs, but no longer performs any display functions.

All display parameters, regardless of their origin, are ultimately processed through the GPC dedicated display processor software (except for the g-meter, which is totally self-contained). The display parameters are then routed to the respective displays through either a DDU or multiplexer/demultiplexer; DDUs send data to the ADI, HSI, AMI, and AVVI displays, while MDMs provide data for the RCS activity lights. There are three display driver units – one interfaces with the ADI, HSI, AVVI, and AMI displays on panel F6 at the commander's station; the second interfaces with the same instruments on panel F8 at the pilot's station; and the third unit interfaces with the ADI at the aft flight station. The DDU contractor was Rockwell International, Collins Radio Group, Cedar Rapids, Iowa.

The heads-up display (HUD) is an optical device that cues the commander and/or pilot during the final phase of reentry and during the final approach to the runway. The HUDs use the same data that drive the dedicated displays. The HUD uses a CRT to create the image, which is then

projected through a series of lenses onto a combining glass located above the glareshield in the direct line of sight of the commander and the pilot. HUDS were not originally installed on OV-101 or OV-102 – *Columbia* was modified after STS-9 to include them. All subsequent Orbiters were equipped with HUDs during manufacture.[25]

Multifunction Electronic Display System

Although the Orbiter had been greatly ahead of its time with the MCDS and its associated CRTs, they were rapidly limiting future enhancements by their lack of color and true graphics capabilities. In 1988, Rockwell began investigating updating the Orbiter flight deck instrumentation. By this time it was common within the commercial aircraft industry to utilize 'glass cockpits' consisting of multiple color displays instead of dedicated instrumentation. By 1992 this concept had progressed to the point of a definitive design and an engineering model for the Multifunction Electronic Display System (MEDS).

MEDS utilizes 11 identical color multifunction display units (MDU), 4 integrated display processors (IDP), and 4 analog-to-digital converters (ADC) per Orbiter. Nine display units are installed at the forward flight station where they replace the three CRTs and the dedicated displays (ADI, HIS, AMI, etc.). The remaining two display units are

The cockpit of Endeavour is representative of the final configuration before the installation of MEDS. The HUDs are missing, but their location is evident. The three switches on top of the glare panel on each side of the HUD openings are for the drag chute – a release switch, a redundant release switch, and an indicator. (Rockwell photo A901207R-15C)

located at the aft station replacing the CRT and the ADI instrument. The DDUs from the MCDS are retained since they provide power for the rotational hand controllers (although this is all they do now). The MEDS units are interconnected via four Mil-Std-1553B serial data buses operating in half-duplex (1553B simplex) mode. The four-string system meets all functional and fault-tolerant requirements imposed on the flight system. The existing keyboards are maintained, providing an operational transparency between the original and upgraded systems. Additional crew interaction with the MEDS is provided by software-controlled edge keys mounted along the bottom of the MDU bezels.[26]

The IDPs are provided by Honeywell, and use 16 MHz 80386DX processors. The IDP is a direct form-fit replacement for the original DEUs, and performs all functions of the DEU and DDU (except powering the RHCs). The IDP controls the MEDS operation, and provides the interface to the on-board GPCs. Each IDP has 2.5 MB of volatile static RAM memory, and 1 MB of non-volatile EEPROM memory that provides storage for critical functions, display formats, and associated programs downloadable to the MDUs. The required mass storage is provided by an off-the-shelf Raymond Engineering 300MB 5.25-inch SCSI magnetic hard disk repackaged to fit the IDP envelope and modified to meet radiation requirements.

The MDUs are based on the flat panel displays developed for the Boeing 777, modified to utilize an LCD produced in the United States. The screen is 6.71 inches square, provides a resolution of 99.3 color pixels per inch, produces 16 shades of each primary color, and has a allowable horizontal viewing angles of ±60 degrees and +45/-10 degrees of vertical viewing. The active matrix design uses amorphous silicon thin film transistors, and has a high ambient luminous of 10,000 foot-candles. Display application software written in Ada by Rockwell replicates the original 94 GPC displays and additional software handles replicating the dedicated flight instrumentation and menu functions.

Honeywell ADCs are provided to convert 32 analog flight instrument signals into digital transmissions for use by MEDS. The analog data is continually converted and stored in memory, then placed on the 1553B data buses.

Originally MEDS was going to be installed at KSC in two phases that allowed the dedicated display instrumentation to be retained for a period after MEDS was installed. As confidence was gained during testing this approach was abandoned and it was decided to install MEDS in its entirety during OMDPs. The MEDS installation increases the empty weight of the Orbiter by approximately 475 pounds. The total electrical load and heat dissipation of the new system is roughly equivalent (1,600 watts versus

Atlantis shows off the new MEDS installation during late 1998. For now, most of the displays shown on the color displays are simple recreations of the original green GPC displays and the analog dedicated display instruments. Eventually NASA plans of implementing more elaborate displays to increase the crew's situational awareness. (NASA)

1,640 watts) to the original system. *Atlantis* was the first Orbiter to receive MEDS, during the 1998 OMDP, and made the initial flight on STS-98 in May 2000. *Columbia* is receiving MEDS during her late-2000 OMDP, and OV-103 will receive the system during her OMDP-3; *Endeavour* will receive MEDS during her OMDP-2.[27]

Closed-Circuit TV System and Cameras

The closed circuit television (CCTV) system is used on-orbit to provide support to Orbiter and payload activities. This support includes transmitting real-time and recorded video from the Orbiter to Mission Control through either the S-band, FM, or Ku-band communications systems. Mission requirements for CCTV and camera configurations are somewhat unique for each shuttle flight.

The CCTV system consists of video processing equipment, TV cameras (and lens assemblies), pan/tilt units (PTU), camcorders, video tape recorders, color television monitors, and all the cabling and accessories required to make these components work together. The crew can control all CCTV operations, although most CCTV configuration commands can also be executed by the

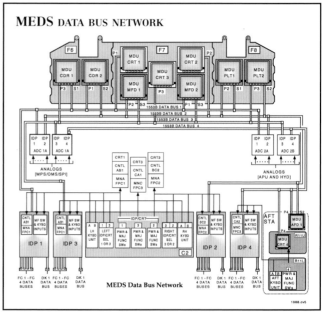

On the GPC side, each IDP is connected to flight-critical data buses 1 through 4 and one display/keyboard (DK) bus, in addition to the panel switches and the keyboards. Each IDP is a functional replacement for a single DEU/CRT (e.g., IDP1 replaces CRT1, etc.). In general, an IDP can display flight instrument and subsystem data on more than one crew station (left, right, aft). On the MEDS side, each IDP controls a Mil-Std-1553B data bus that allows the IDP to interface with the MDUs and a pair of the ADCs. (NASA)

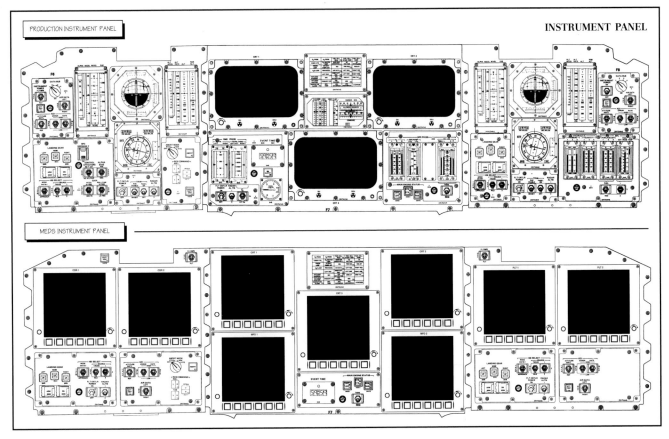

PRODUCTION INSTRUMENT PANEL

INSTRUMENT PANEL

MEDS INSTRUMENT PANEL

Although various phased approaches to implementing MEDS were discussed during the development program, in the end the program elected to install the entire system at once. Remarkably little of the main instrument panel was left unchanged – the 40-light caution and warning panel at center top, and the abort modes rotary switch were both left in their original spots at the request of the Astronaut Office to ease training requirements and preserve reaction time during critical situations.

Instrumentation and Communications Officer (INCO) at Mission Control. In addition, the crew frequently carries handheld video camcorders (originally Cannon A1 Hi-Ban 8-mm units; more recently Cannon L1 units with interchangeable lenses). The hand-held units cannot interface to the Orbiter CCTV system due to incompatible formats.[28]

Various still and motion picture cameras are used on-orbit to document crew activities, meet payload requirements, and record Earth observations. Two Hasselblad 500EL/S 70-mm cameras and two slightly modified Nikon 35-mm F4S autofocus SLR cameras are normally manifested for each mission. Interchangeable autofocus lenses of various focal lengths, including 28-mm f/2.8, 24–50-mm f/4.5 and 35–70-mm f/2.8 zoom are available for the Nikons. A 300mm telephoto and 2x converter can be flown for photographing the ET post-jettison. A Kodak/Nikon electronic still camera stores digital still images on removable hard drives, and images can be downlinked to Mission Control. An Arriflex 16-mm camera is used for motion picture photography. When stowage and weight provisions allow, an Aero Linhof 4 by 5 aerial photography camera or a NASA-developed electronic still camera are also carried. Additional camera systems may be flown for specific experiments or as part of an ongoing NASA evaluation of new cameras and technology. In addition, on flights with a likelihood of an EVA, two factory-modified Nikon F3 35-mm cameras are carried. These cameras use a special thermal cover and have controls and eyepieces modified for use with spacesuits. Before the EVA each camera is loaded with ISO-100 negative film, and one of two special lenses (28-mm f/2 or 35-mm f/1.4) are fitted. Periodically, the IMAX Corporation manifests its 70mm motion picture camera as a commercial payload.[29]

Mid-Deck

The mid-deck is located directly beneath the flight deck. Access to the mid-deck is through two 26 by 28-inch interdeck openings. Normally, the right interdeck opening is closed and the left is open. A ladder attached to the left interdeck access allows passage in 1-g conditions. The mid-deck provides crew accommodations and contains three avionics equipment bays. The two forward avionics bays utilize the complete width of the cabin and extend into the mid-deck 39 inches from the forward bulkhead. The aft bay extends into the mid-deck 39 inches from the aft bulkhead on the right side of the airlock. On the Orbiter centerline, just aft of the forward avionics equip-

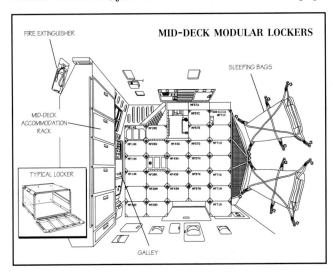

ment bay, an opening in the ceiling provides access to the inertial measurement units. The completely stripped mid-deck is approximately 160 square feet while the gross mobility area is nominally 100 square feet.[30]

The side hatch on the port side of the mid-deck is used for normal crew ingress/egress and may be operated from within the crew module or externally. Modifications made during the *Challenger* stand-down allow the hatch to be explosively jettisoned. The side hatch is attached to the crew compartment by hinges, a torque tube, and support fittings. The hatch opens outward 90 degrees down with the Orbiter horizontal, or 90 degrees sideways with the Orbiter vertical. The side hatch is 40 inches in diameter, has a window in its center, and weighs 294 pounds. A pressure seal is compressed by the side hatch latch mechanisms when the hatch is locked closed. A thermal barrier of Inconel wire mesh spring with a ceramic fiber braided sleeve is installed between the reusable surface insulation tiles on the forward fuselage and the side hatch.

Located on the port side wall of the mid-deck are lockers for stowage, a waste management facility (toilet) and personal hygiene station (shower), privacy screen, and galley. Located on the forward wall are additional stowage space, a work or dining table, and lockers for experiments. The starboard wall contains three sleep stations, a vertical sleep restraint station, and modular stowage space. The sleep stations and galley may be omit-

How do you go to the bathroom in Space? Much the same as on Earth. The unisex urinal and seatbelt are missing from this photo. (NASA photo S83-32562)

ted depending upon mission requirements. In addition, three (all except OV-102) or four (OV-102) seats of the same type as the mission specialists' seats on the flight deck can be installed in the mid-deck. An additional three seats could be installed in the mid-deck if the bunk sleep stations are removed. The mid-deck also provides a stowage volume of 140 cubic feet.[31]

The mid-deck floor contains removable panels that provide access to the Environmental Control and Life Support System (ECLSS) equipment. An equipment bay below the mid-deck floor houses the major components of the waste management and air revitalization systems, such as pumps, fans, lithium hydroxide absorbers (used in purging contaminants from the breathing air), heat exchangers, and ducting. This compartment has space for stowing used lithium hydroxide canisters and trash, and five separate spaces for crew equipment stowage with a total volume of 29.92 cubic feet.

Modular stowage lockers are used to store the flight crew's personal gear, mission-necessary equipment, personal hygiene equipment, and experiments. The current lockers are made of sandwich panels of Kevlar-epoxy with a non-metallic core. The use of composite construction reduced weight to 295 pounds, compared to 495 pounds for the original all-aluminum lockers. There are 42 identical lockers that are 11 by 18 by 21 inches.[32]

Three types of exercisers are available for use in flight – a treadmill, cycle ergometer, and a rower. Only one type is flown per mission, with selection dependent on crew preference, availability of on-orbit space, and science objectives. All exercisers are designed to minimize muscle loss in the legs and maintain cardiovascular fitness in a zero-g environment.[33]

In addition to time scheduled for sleep periods and meals, each crewmember has housekeeping tasks that require from 5 to 15 minutes of time at intervals throughout the day. These include cleaning of the waste management compartment, the dining area and equipment, floors and walls (as required), and the cabin air filters; trash collection and disposal; and changeout of the crew module lithium hydroxide canisters. The materials and equipment available for cleaning operations are biocidal cleanser, disposable gloves, general-purpose wipes, and a vacuum cleaner. The vacuum is stowed in a mid-deck locker, and the remaining hardware is stowed primarily in the waste management compartment.

A limited amount of medical equipment is carried on all flights to provide in-flight care for minor illnesses and

An Orbiter mid-deck – from left to right: the airlock, side hatch, and galley. (NASA photo S86-26812)

Mid-deck lockers and galley (left) Note the fire extinguisher to the left of the galley. (NASA photo S86-26825)

The other side of the mid-deck. Two sleeping stations are on the wall. (NASA photo S86-26822)

injuries. Two kits are carried – the medications and bandages kit (MBK) and the emergency medical kit (EMK). The kits also provide support for stabilizing severely injured or ill crewmembers until they are returned to Earth. A resuscitator, breathing oxygen, various medicines, and bandages are included in the kits. Interestingly, an EKG is built-in to the Orbiter as part of the operational bioinstrumentation system. This provides an amplified EKG signal to the avionics system that downlinks it to the ground for the flight surgeon in Mission Control. The system was designed to provide information on astronaut heart rates during ascent, but can be used on-orbit if necessary.[34]

Airlock

Two different airlocks have been used on the program. Originally, each Orbiter except OV-101 was equipped with an internal airlock in the rear of the mid-deck. This airlock has a hatch into the crew module, and another into the payload bay (or to a Spacelab module) and was intended for EVAs to service payloads while preserving the maximum payload bay volume. To facilitate docking with *Mir* and the ISS, the internal airlock was replaced by a 185-cubic-foot airlock in the forward portion of the payload bay. This airlock has three hatches – one into the crew module, one aft into the payload bay (or into a pressurized Spacehab module), and one upward used for docking. As of late 2000, only OV-102 still retains the internal airlock – the other three Orbiters use the external airlock to permit ISS operations. During the late-2000 OMDP, OV-102 received structural and wiring provisions for the external airlock and ODS, although there are no plans to install such systems on *Columbia*.[35]

The internal airlock has an inside diameter of 63 inches, is 83 inches long, and has two 40-inch diameter D-shaped openings that are 36 inches across. The interior volume of the airlock is approximately 150 cubic feet. The airlock is constructed of machined aluminum sections welded together to form a cylinder with hatch mounting flanges. The upper cylindrical section and bulkheads are constructed of aluminum honeycomb sandwich. Two semi-cylindrical aluminum sections are welded to the airlock's primary structure to house the ECLSS and electrical support equipment. Each semi-cylindrical section has three feedthrough plates for plumbing and cable routings from the Orbiter to the airlock. Normally, two extravehicular mobility units are stowed in the airlock.[36]

The inner (Orbiter side) airlock hatch is mounted on the exterior of the airlock, opens into the mid-deck, and isolates the airlock from the Orbiter crew module. The outer (payload bay side) hatch is mounted inside the airlock and opens into the airlock. The outer hatch isolates the airlock from the unpressurized payload bay when closed and permits the crew members performing extravehicular activities (EVA) to exit from the airlock to the payload bay. Note that the two D-shaped airlock hatches open toward the primary pressure source to achieve pressure-assisted sealing.

Airlock repressurization is controllable from the crew module or inside the airlock. Valves mounted on the inner hatch equalize the airlock pressure with that of the crew module, resulting in a slight drop of crew module pressure until the ECLSS an make it up. The airlock is depressurized from inside the airlock by venting pressure overboard.

Each hatch has six interconnected latches and a gearbox/actuator, a hinge mechanism and hold-open device, a differential pressure gauge on each side, and two equalization valves. There is a 4-inch diameter window in each airlock hatch. The dual window panes are made of polycarbonate plastic and mounted directly to the hatch by means of bolts fastened through the panes, and each hatch window has dual pressure seals, with seal grooves located in the hatch. Each airlock hatch has dual pressure seals to maintain pressure integrity. One seal is mounted on the airlock hatch and the other on the airlock structure. A leak check quick disconnect is installed between the hatch and the airlock pressure seals to verify hatch pressure integrity.

A gearbox with latch mechanisms on each hatch allows the crew to open and close the hatch during transfers and

The mid-deck airlock for Endeavour prior to completion. This is the mid-deck side. The size of the airlock can be seen compared to the woman working on – consider that two fully-suited astronauts use it at once for EVAs. (NASA photo S88-55913)

Looking into the mid-deck airlock through the open mid-deck hatch. The closed hatch leads to the payload bay. The relative placement of this opening can be seen at the left side of the left photo on page 378. (Rockwell)

The Mir/ODS in Atlantis for the STS-106 mission. This shows the androgynous peripheral docking system device on top of the ODS. The relative merits of an androgynous (where both sides are the same) versus a traditional American male-female docking device had long been debated within the Shuttle program. (NASA photo KSC-00PP-0997)

The Mir/ODS in Atlantis prior to the STS-79 mission. The Orbiter crew module is at the top while a Spacehab-SM is at the bottom. The placement of the ODS is based largely on clearance issues with Mir and the ISS. Later ODS installations cover the truss that connects the ODS to the sill longeron with blanket insulation. (NASA photo KSC-96PC-1005)

EVA operations. The gearbox and the latches are mounted on the low-pressure side of each hatch, with a gearbox handle installed on both sides to permit operation from either side of the hatch. The gearbox is used to provide the mechanical advantage to open and close the latches. The hatch actuator lock lever requires a force of eight to ten pounds through an angle of 180 degrees to unlatch the actuator. A minimum rotation of 440 degrees with a force of 30 pounds applied to the actuator handle is required to operate the latches to their fully unlatched positions.

The airlock air circulation system provides conditioned air to the airlock during non-EVA periods. The airlock revitalization system duct is attached to the outside airlock wall at launch. Upon airlock hatch opening in flight, the duct is rotated by the flight crew through the inner airlock hatch, installed in the airlock and held in place by a strap holder. The duct has a removable air diffuser cap, installed on the end of the flexible duct, which can adjust the air flow. The duct must be rotated out of the airlock before the inner airlock hatch is closed to depressurize the airlock. During the EVA preparation period, the duct is rotated out of the airlock and can be used for supplemental air circulation in the mid-deck.

To assist crew members before and after EVA operations, the airlock incorporates various handrails and foot restraints. Handrails are located alongside the avionics and ECLSS panels. Aluminum alloy hand holds mounted on the side of each hatch have oval configurations 0.75 by 1.32 inches, and are painted yellow. They are bonded to the airlock walls with an epoxyphenolic adhesive. Each handrail has a clearance of 2.25 inches between the airlock wall and the handrail to allow the astronauts to grip it while wearing a pressurized glove. A ceiling hand hold is installed near the cabin side of the airlock. Foot restraints are

installed on the airlock floor near the payload bay side. The foot restraints can be rotated 360 degrees by releasing a spring-loaded latch and lock in every 90 degrees. A rotation release knob on each foot restraint is designed for shirt-sleeve operation and, therefore, must be positioned before the suit is donned. The foot restraints are sized for the EMU boot. There are also four floodlights in the airlock.

Normally, two extravehicular mobility units (EMU) are stowed in the airlock, although the internal airlock can accommodate storage of up to four. The EMU is an integrated space suit assembly and life support system that allows crew members to leave the pressurized crew compartment and work near the Orbiter. The airlock is sized to accommodate two fully-suited crew members at the same time, and individual EVAs are procedurally prohibited by NASA. The first flight of *Endeavour* proved that three suited Astronauts could successfully fit in the airlock during a last-ditch attempt to rescue INTELSAT-VI.[37]

For most Spacelab pressurized module missions, the airlock remained in the mid-deck, and a tunnel adapter connected the airlock with the Spacelab was installed in the payload bay. The airlock tunnel adapter, hatches, tunnel extension and tunnel permitted the crew to transfer from the mid-deck to Spacelab's pressurized shirt-sleeve environment. The tunnel and Spacelab were accessed via the tunnel adapter, which was located in the payload bay and was attached to the airlock at Orbiter station X_0576 and the tunnel extension at X_0660. The tunnel adapter had an inside diameter of 63 inches at its widest section and tapered in the cone area at each end to two 40-inch diameter D-shaped openings 36 inches across. A similar opening was also located at the top of the tunnel adapter. Two pressure-sealing hatches were located in the tunnel adapter, one in the upper area of the tunnel adapter and

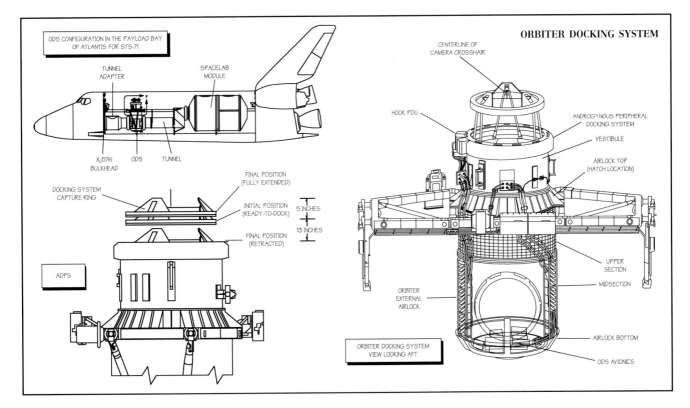

one at the aft end. The tunnel adapter was a welded structure constructed of 2219 aluminum with 2.4 by 2.4 inch exposed structural ribs on the exterior surface and external waffle skin stiffening.

The hatch located in the aft end of the tunnel adapter opened into the tunnel adapter and isolated the airlock/tunnel adapter from the tunnel extension and Spacelab. The hatch located in the tunnel adapter at the upper D-shaped opening isolated the airlock/tunnel adapter from the unpressurized payload bay when closed and permitted the crew to exit from the airlock/tunnel adapter to the payload bay when open. This hatch also opened into the tunnel adapter.

The airlock was constructed by Rockwell's Space Transportation Systems Division, Downey, California. The Spacelab pressurized module tunnel adapter and tunnel contractor was McDonnell Douglas Astronautics, of Huntington Beach, California.

Orbiter Docking System

In July 1992 NASA initiated the development of the Orbiter Docking System (ODS) to support flights to the Russian *Mir* space station. This interim ODS design was also intended to be easily modified to a final configuration to support later International Space Station (ISS) operations. The $95,200,000 Mir/ODS consisted of an external airlock, a supporting truss structure, a docking base, avionics required to operate the system, and a 632-pound Russian-built docking mechanism called the Androgynous Peripheral Docking System* (APDS), which is mounted on top of the airlock and docking base. The ODS preliminary design review (PDR) was held on 9 March 1993. Two U.S. critical design reviews (CDR) occurred on 3 November 1993 and 4 February 1994, while the Russian CDR was held on 10 December 1993.[38]

The Mir/ODS was installed near the forward end of the payload bay in *Atlantis*. The docking unit measured approximately 15 feet wide, 6.5 feet long, 13.5 feet high, and weighed 4,016 pounds. It was connected by short tunnels to the existing airlock inside the Orbiter mid-deck and a pressurized Spacelab module that could be installed aft of the airlock in the payload bay. The Spacelab module did not need to be installed, and several mission were flown without it.

The APDS was manufactured by RSC-Energia in Kaliningrad, Russia, and was a variation of the docking system used by the Soviets in 1975 for the Apollo-Soyuz Test Project (ASTP). The new unit differed from its predecessor in several key respects – it was much more compact, with an overall external diameter of 60 inches compared to 80 inches on ASTP, although the inner egress tunnel diameter remains approximately the same; the APAS docking mechanism had 12 structural latches, compared to eight on the ASTP; the APDS guide ring and its extend/retract mechanism were packaged inside the egress tunnel rather than being outside of the mechanism as they were on ASTP; and the three guide petals on the APDS pointed inboard rather than outboard. The APDS was based on a design originally intended for the Soviet *Buran* space shuttle.

Both the Orbiter and *Mir* were equipped with an APDS, which consisted of a three-petal androgynous capture ring mounted on six interconnected ball screw shock absorbers that operated like a sophisticated car suspension system. The absorbers arrested the relative motion of the two vehicles and prevented them from colliding.

The requirement that the ODS be at least dual fault tolerant to allow demating (and subsequent Orbiter payload bay door closure) was achieved with two independent (no common cause failure mode) methods. If the primary separation method failed, pyrotechnic charges could be used to shear each of the Orbiter's 12 active latch retention bolts, permitting the springs to separate the vehicles. When fired, the pyrotechnic charges fractured the bolts, allowing the latch hook to rotate about an independent

---

* Originally called the Androgynous Peripheral Assembly System (APAS).

rotation point to release the mating hook. Hook rotation was powered by both the pyrotechnic charge and the off-center force created by the Belleville washers on the mating passive hook. A spring-loaded pin prevented hook rebound. A third method was available if the second also fails. This method required two astronauts to perform an EVA to remove the 96 bolts that held the docking base on the upper flange of the external airlock.

In June 1994, Rockwell began the development of the ODS for the International Space Station, derived from the semi-prototype *Mir* unit carried by *Atlantis*. Two new units were manufactured for OV-103 and OV-105, and the early OV-104 unit was modified to the final configuration. Each of the Orbiters was modified during OMDP, and *Discovery* became the second Orbiter to dock with *Mir* during STS-91 in June 1998.[39]

Since the ISS uses a derivative of the *Mir* docking adapter, the new ODS does not differ significantly from the one initially used on *Atlantis*. Like the original unit, it consists of a forward ring, bellows, forward adapter, and aft adapter. The ODS may be installed in either the 2nd or 3rd bay of the payload bay. The airlock provides the necessary hardware and power distribution to pressurize and light the docking base and tunnel. The largest change between the original Mir/ODS and the permanent ISS/ODS installation is the removal of the mid-deck airlock from the Orbiter. This compensates for the additional weight of the ODS, and also provides additional room on the mid-deck for experiments and other equipment. Minor structural modifications were also required to the mid-deck, payload bay liner channels, and the payload bay doors.[40]

Mid-Fuselage

The mid-fuselage was manufactured by the Convair Aerospace Division of General Dynamics Corporation in San Diego, California. It contains the payload bay and connects with the forward fuselage, aft-fuselage, wings, and payload bay doors. The assembly is 60 feet long, 17 feet wide, 13 feet high and weighs 13,502 pounds.

The forward and aft ends of the mid-fuselage are open, with reinforced skin and longerons interfacing with the bulkheads on the forward and aft fuselage. Twelve main frame assemblies provide stabilization of the mid-fuselage structure. Each frame consists of machined vertical sides and horizontal braces constructed of boron/aluminum tube

trusses with bonded titanium end fittings. This construction technique saved approximately 305 pounds. The upper portion consists of door and sill longerons that absorb the bending loads of the vehicle and support payloads contained in the payload bay. The mid-fuselage skins were integrally machined by numerical control, an innovative technique at the time they were built. The panels above the wing glove and wings are composed of 13 bays – the forward eight bays have longitudinal T-stringers, while the five aft bays use aluminum honeycomb panels. The side skins in the shadow of the wing were also machined by numeric control, but use vertical stiffeners. The sill longerons contain the hinges for the payload bay doors, and also provide the base support for the Remote Manipulator System (if installed), the Ku-band rendezvous antenna, and the payload bay door actuation system.[41]

The side wall forward of the wing carry-through structure provides the inboard support for the main landing gear. The total lateral landing gear loads are reacted by the mid-fuselage structure. The mid-fuselage also supports the two electrical wire trays that contain the wiring between the crew module and aft-fuselage. Plumbing and wiring in the lower portion of the mid-fuselage are supported by fiberglass 'milk stools.'

Since analysis of actual flight data from the orbital flight test series showed that descent-stress thermal gradient loads were greater than expected, torsional straps were added to tie all the lower mid-fuselage stringers in bays 1 through 11 together in a manner similar to a box section. This eliminated rotational (torsional) tendencies, and provided additional positive margins of safety. Also, to bring OV-102, OV-103, and OV-104 up to the 6.0 loads data base, room-temperature vulcanizing silicone rubber material was bonded to the lower mid-fuselage from bays 4 through 12 to act as a heat sink, distributing temperatures more evenly across the bottom of the mid-fuselage structure.[42]

Payloads and their carriers are attached to structural supports along the Orbiter's mid-fuselage longerons. Attachment points in the payload bay are in 3.933-inch increments along the left and right-side longerons, and along the bottom centerline. Of the potential 172 attach points on the longerons, 48 are unusable because of their proximity to Orbiter hardware. The remaining 124 may be used for deployable payloads. Along the Orbiter's centerline keel, 89 attach points are available, 75 of which may be used for deployable payloads.

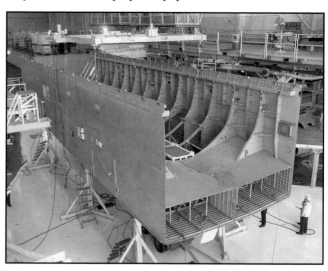

The structural spares mid-fuselage on 15 October 1987 as it was being prepared to be assembled into Endeavour. This view is from the back looking forward. (Rockwell)

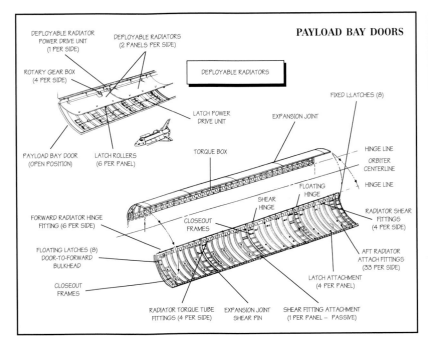

DEPLOYABLE RADIATORS

An engineering development article of a radiator panel at the LTV factory in Grand Prairie, Texas. (LTV Corporation)

Payload Bay Doors

The payload bay doors consist of left and right doors hinged at each side of the mid-fuselage. Each door is made up of five segments that are interconnected by circumferential expansion joints. Each door rests on 13 Inconel-718 external hinges that attach to the mid-fuselage sill longeron. There are five shear (non-moving) hinges, and eight floating hinges that allow fore and aft movement of the door panels for thermal expansion. Thermal seals on the doors provide a relatively air-tight payload compartment when the doors are closed and latched, although it should be noted that the payload bay is not a pressurized area. The right door must be opened first and closed last because of the arrangement of the centerline latch mechanism and the structural and seal overlap. The doors are held closed by 32 latches – 16 centerline latches that secure the doors to each other on the centerline, and 8 forward and 8 aft bulkhead latches that secure the doors to the forward and aft-fuselage bulkheads. The doors were manufactured by Rockwell's Tulsa, Oklahoma, division.[43]

Each payload bay door is 60 feet long, has a mean chord of approximately* 10 feet and a gross area of approximately 800 square feet. The payload bay doors are constructed of a graphite-epoxy/Nomex composite that is about 23 percent lighter than a equivalent aluminum honeycomb sandwich. The left-hand door with attached systems weighs approximately 2,375 pounds and the right door weighs about 2,535 pounds. The right door contains the centerline latching mechanism, which accounts for the weight difference. These weights do not include the radiator system, which adds approximately 833 pounds per door without the aft radiator kit installed.[44]

The payload bay doors also serve as a strongback for the Orbiter radiator panels. While on-orbit the payload bay doors are opened to allow heat rejection of the Orbiter's systems via the radiators. A radiator is always installed on the forward end of the aft payload bay door, and an additional radiator may be installed on the aft end of the aft door as required for a specific mission. When deployed, the two forward radiator panels on each side are unlatched and tilted to allow heat to be radiated from both the front and back sides of the panels; the fixed aft panels dissipate heat only from the outer side. When the forward radiators are deployed, they are angled 35.5 degrees from the payload bay doors.[45]

The total heat rejection capability of the baseline configuration is 21,000 BTU per hour, and the addition of the aft radiator on each side raises this to 30,550 BTU per hour. Each panel is 15.1 feet long and 10.5 feet wide constructed of 0.11-inch 2024-T81 aluminum face-sheets and 5056-H39 aluminum honeycomb cores. Each of the forward deployable panels contains 68 tubes spaced 1.9 inches apart, and the aft panels contain 26 longitudinal tubes spaced 4.96 inches apart. The tubes are 6061-T6 aluminum and the panel surfaces are coated with gloss silver Teflon tape. The radiator panels on the left and right sides are configured to flow in series, while flow

The right aft payload bay door segment is lowered into place on Endeavour at Palmdale during late 1990 The left door is already complete and in place. (Rockwell International)

---

* The doors have a mean chord of 10 feet, but are curved to match the fuselage mold line, resulting in a greater area than would seem correct.

The remote manipulator arm stowed in the payload bay of Columbia. The black squares on the side of the payload bay are part of the vent system. (NASA photo S82-26911)

Columbia's arm in action during STS-2. The box on the elbow (middle) joint is a CCTV camera. Newer cameras have been installed since 1981. (NASA photo S81-39557)

* For a while this was officially called the Shuttle Remote Manipulator System (SRMS) to differentiate it from a similar system being designed for the space station.

within each panel is parallel through a bank of tubes connected by an inlet and outlet connector manifold. The radiator panels on the left side are connected in series with Freon coolant loop 1, while the right side are connected to loop 2. The radiators were manufactured by LTV (now Lockheed Martin) in Grand Prairie, Texas.[46]

After the radiators were delivered, NASA implemented a series of upgrades to minimize the chance of an impact causing damage to a cooling tube and potentially affecting a mission. The primary upgrade consisted of redesigning the radiators to include cooling tube shielding – thin aluminum strips bonded over the tubes – to protect the tubes from impacts. A series of isolation valves and check valves was also installed that allows the crew to isolate a leaking radiator from the rest of the cooling loop (previously, the entire loop had to be shut down). The modifications were first incorporated in OV-104 during her 1997 OMDP. *Endeavour* was modified at KSC in early 2000, and *Columbia* is being modified during OMDP-2. OV-103 will be modified during a future OMDP.[47]

Payload Deployment and Retrieval System

The Payload Deployment and Retrieval System (PDRS) consists of the Remote* Manipulator System (RMS), the manipulator positioning mechanisms (MPM), the manipulator retention latches (MRL), the manipulator controller interface unit (MCIU), and various displays and indicators.[48]

The RMS is the mechanical arm portion of the PDRS. The arm can perform several tasks, such as deploying and retrieving a payload, providing a stable platform for EVA crew-member foot restraints or work stations, mating space station components, and taking payload bay surveys. The RMS arm is normally mounted on the left main longeron of the payload bay upper wall. Provisions were incorporated in each Orbiter to mount a similar device along the right side, although this would have required the removal of the Orbiter's Ku-band antenna and only one arm could be operated at a time since there was only one control location at the aft mission station, although wiring existed for two. By 1999 the idea of ever flying two RMS units on the same Orbiter had been abandoned, although some controls and displays still reflect the possibility.[49]

As originally installed, the RMS was capable of deploying or retrieving payloads weighing up to 65,000 pounds. Changes implemented beginning in 1998 have raised this to 586,000 pounds, and all four of the in-service RMS units had been modified by the end of 1999. So far, the heaviest load handled by the RMS is the *Zarya* ISS module at 42,637 pounds on STS-88. The RMS can also provide a mobile extension ladder for extravehicular activity

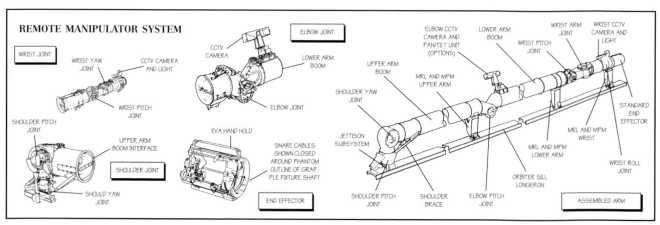

REMOTE MANIPULATOR SYSTEM

and be used as an inspection aid to allow crew members to view the Orbiter through a television camera mounted on its end. The arm was built via an international agreement between the National Research Council of Canada and NASA. Spar Aerospace, Limited, designed and built the arm, and CAE Electronics, Limited, in Montreal, provided the electronic interfaces, servo-amplifiers, and power conditioners.[50]

The basic RMS configuration consists of the manipulator boom, an RMS display and control panel, and a manipulator interface unit that interfaces with the onboard GPCs. The arm is 50.25 feet long and 15 inches in diameter. It weighs 905 pounds, and the total system weighs 994 pounds including controllers. The arm has six joints that correspond roughly to the joints in the human arm, with shoulder yaw and pitch joints, an elbow pitch joint, and wrist pitch, yaw and roll joints. The boom, constructed in two pieces (upper and lower), is made of reinforced graphite-epoxy, and weighs only 93 pounds.

A complete RCA Astro-Electronics closed-circuit TV system includes a color camera in the crew compartment and several black-and-white or color cameras in the payload bay and on the manipulator arm. These facilitate payload handling and provide TV coverage for engineers and the general public on Earth. The video from these cameras can be viewed, in monochrome only, on two monitors located on the flight deck. The video can also be recorded by on-board video recorders on-board for later viewing, or sent to the ground via the S-band FM or Ku-band telemetry systems.

Purge, Vent, and Drain System

The Purge, Vent, and Drain (PVD) System is designed to provide unpressurized compartments with gas purge for thermal conditioning, prevent accumulation of hazardous gases, vent the unpressurized compartments during ascent and reentry, drain trapped fluids (water and hydraulic fluid), and condition window cavities to maintain visibility.

Three purge circuits are connected by the T-0 umbilical to ground support equipment during the pre-flight countdown and post-landing phases. Purge gas (cool, dry air and gaseous nitrogen) is provided to three sets of distribution plumbing: the forward fuselage, OMS/RCS pods, wings and vertical stabilizer; the mid-fuselage; and the aft-fuselage. The purge gas acts to inert all the unpressurized volumes, maintains constant humidity and temperature, forces out any hazardous gases, and ensures that external contaminants cannot enter.

There are 18 active vent doors in the Orbiter fuselage, 9 on each side. Each vent has a door that can be positioned for a specific purpose at various phases of flight. For identification, each door is numbered, starting with number 1 at the front of the Orbiter. Each compartment has a dedicated vent on the left and right side of the Orbiter for redundancy. Fourteen of the doors are functional – an engineering analysis showed that vent doors 3, 5, and 6 provide sufficient venting of the payload bay and mid-fuselage. Because of this, doors 4 and 7 were permanently capped shut on both sides of *Columbia*, *Discovery*, and *Atlantis* during each Orbiter's first OMDP. The associated actuators and linkages were also removed. *Endeavour* did not have the equipment installed, although provisions exist to do so in the future if necessary.[51]

All vent doors are driven by electromechanical actuators. Vent doors located near each other share common actuators and controls. Vent doors 1 and 2, 4 and 7, and 8

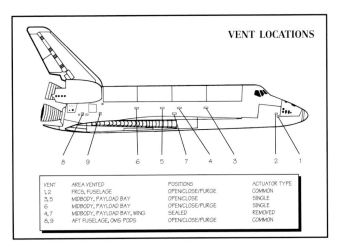

VENT LOCATIONS

VENT	AREA VENTED	POSITIONS	ACTUATOR TYPE
1, 2	FRCS, FUSELAGE	OPEN/CLOSE/PURGE	COMMON
3, 5	MIDBODY, PAYLOAD BAY	OPEN/CLOSE	SINGLE
6	MIDBODY, PAYLOAD BAY	OPEN/CLOSE/PURGE	SINGLE
4, 7	MIDBODY, PAYLOAD BAY, WING	SEALED	REMOVED
8, 9	AFT FUSELAGE, OMS PODS	OPEN/CLOSE/PURGE	COMMON

and 9 share drive mechanisms on the left and right side – however, when doors 4 and 7 were disabled, the actuators and linkages were removed. The 18 doors are divided into six groups of four ac motors each and are staggered so that all 24 motors do not run at the same time. All vent doors are driven inward, and each door has a pressure seal and thermal seal. The normal opening or closing time of a door with two motors operating is five seconds. Vent doors 1, 2, 8, and 9 have purge positions that control flow from the forward and aft volumes, respectively. Vent door 6 has two purge positions and a closed position that accommodates the different purge flow rates available to the payloads and payload bay. These doors are placed in the purge position immediately prior to launch.

The purge and vent ducting was originally made of fiberglass and aluminum alloy, but has since been replaced with ducts made of Kevlar-epoxy (115 pieces, up to 11 inches in diameter), which reduced the weight of the ducts 33 percent, or approximately 200 pounds.

The PVD system is controlled through the Orbiter GPCs in both PASS and BFS. The active vents are positioned by software on the basis of mission time or mission events during ascent, reentry and aborts, and can also be controlled by crew inputs.

Aft-Fuselage

Consisting of an outer shell, primary thrust structure, and internal secondary structure, the aft-fuselage provides the primary support for the Main Propulsion System, Orbital Maneuvering System pods, and vertical stabilizer. The structure – 18 feet long, 22 feet wide and 20 feet high – was built by Rockwell International in Downey.

The aft-fuselage provides the load path to the mid-fuselage main longerons, and provides main wing spar continuity across the forward bulkhead of the aft-fuselage. It also provides structural support for the body flap and is a structural housing around all internal systems, providing protection from the operational environment (pressure, thermal, and acoustic), as well as controlled internal pressure venting during flight. The forward bulkhead closes off the aft-fuselage from the mid-fuselage and is composed of machined and beaded aluminum sheet segments. The upper portion of the bulkhead provides attachment for the front spar of the vertical stabilizer.[52]

The internal thrust structure supports the three Space Shuttle Main Engines (SSME) and includes the SSME load reaction truss structure, engine interface fittings, and the SSME gimbal actuator support structure. The internal thrust structure is composed of 28 machined and diffu-

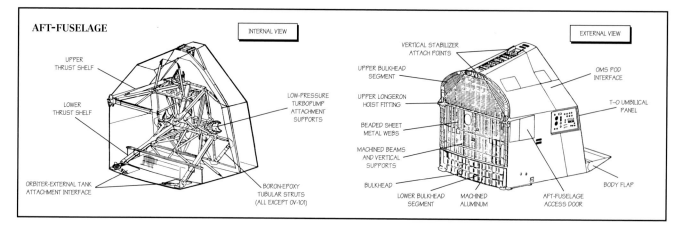

AFT-FUSELAGE   INTERNAL VIEW

UPPER
THRUST SHELF

LOWER
THRUST SHELF

LOW-PRESSURE
TURBOPUMP
ATTACHMENT
SUPPORTS

ORBITER-EXTERNAL TANK
ATTACHMENT INTERFACE

BORON-EPOXY
TUBULAR STRUTS
(ALL EXCEPT OV-101)

VERTICAL STABILIZER
ATTACH POINTS   EXTERNAL VIEW

UPPER BULKHEAD
SEGMENT

UPPER LONGERON
HOIST FITTING

BEADED SHEET
METAL WEBS

MACHINED BEAMS
AND VERTICAL
SUPPORTS

BULKHEAD

LOWER BULKHEAD
SEGMENT

MACHINED
ALUMINUM

OMS POD
INTERFACE

T-0 UMBILICAL
PANEL

BODY FLAP

AFT-FUSELAGE
ACCESS DOOR

sion-bonded titanium truss members in OV-099 through OV-104, and of built-up titanium forgings in OV-105. In diffusion bonding, titanium strips are bonded together under heat and pressure. This fuses the titanium strips into a single hollow, homogeneous mass that is lighter and stronger than a forged part. In selected areas the titanium structure is reinforced with boron-epoxy tubular struts to minimize weight and add stiffness. The use of composite materials reduced the weight of the structure approximately 900 pounds compared to an all metal piece of equal stiffness.[53]

The upper thrust structure supports the OMS pods, and is of integral machined aluminum construction with aluminum frames. An exception is the vertical stabilizer support frame, which is constructed of titanium. The skin panels are integrally machined aluminum and attach to each side of the vertical stabilizer to react to drag and torsion loading. A drag chute compartment is attached to the upper thrust structure. The drag chute was a production feature in OV-105, and was subsequently added to OV-102, OV-103, and OV-104.

The outer shell of the aft-fuselage is constructed of integrally machined aluminum and various penetrations are provided in the outer shell for access to installed systems. The secondary structure is of conventional aluminum construction, except that some titanium and fiberglass is utilized for thermal isolation of selected equipment. The structure consists of brackets, built-up webs, truss members, and machined fittings as required by subsystem loading and physical support constraints. Certain system components, such as avionics shelves, are shock-mounted to the secondary structure. Support provisions are also included for the auxiliary power units, hydraulics, electrical wire runs, and environmental control and life support systems.

Two ET umbilicals interface with the LO2 and LH2 feed lines and electrical wire runs. The umbilicals are retracted and the umbilical areas are closed off after ET separation by an electromechanically operated beryllium door at each umbilical. Thermal barriers are employed at each umbilical door and the exposed area of each closed door is covered with reusable surface insulation.

The aft-fuselage heat shield provides a closeout of the Orbiter aft base area. The aft heat shield consists of machined aluminum and attached domes of honeycomb construction that support flexible and sliding seal assemblies. Each of three engine-mounted heat shields are of Inconel honeycomb construction and is removable for access to the main engine power heads. The heat shield is covered with a reusable thermal protection tiles, except for the Inconel segments, which are left uncovered.

Body Flap

Pitch control trim during atmospheric flight is provided by the Orbiter's body flap, which also shields the main engines from extreme thermal conditions during reentry. The body flap is capable of being articulated during all portions of reentry and atmospheric flight.[54]

The body flap is an aluminum structure consisting of ribs, spars, skin panels, and a trailing edge assembly. The main upper and lower forward honeycomb skin panels are joined to the ribs, spars and honeycomb trailing edge with structural fasteners. The forward upper skin consists of five removable access panels attached to the ribs with quick-release fasteners. The four integral-machined aluminum actuator ribs provide the aft-fuselage interface through self-aligning bearings.

Two bearings are located in each rib for mechanical attachment to four rotary actuators located in the aft-fuselage that are controlled by the flight control system. The remaining ribs consist of eight stability ribs and two closeout ribs constructed of chemically-milled aluminum webs bonded to aluminum honeycomb core. The forward spar web is of chemically-milled sheets with flanged holes and stiffened beads. The spar web is riveted to the ribs. The trailing edge includes the rear spar, which is composed of piano-hinge half-cap angles, chemically milled skins, honeycomb aluminum core, closeouts, and plates. Two moisture drain lines and one hydraulic fluid drain line penetrate the trailing edge honeycomb core for horizontal and vertical drainage.[55]

The body flap's upper and lower surfaces are covered with high-temperature RSI tiles. An articulating pressure and thermal seal on the lower surface of the body flap blocks heat and air flow from entering the aft-fuselage structure, and also protects the hinges and actuators from thermal damage. The body flap was built by Rockwell International's Structures Division in Columbus, Ohio.

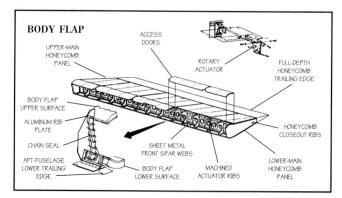

BODY FLAP

ACCESS
DOORS

UPPER-MAIN
HONEYCOMB
PANEL

BODY FLAP
UPPER SURFACE

ALUMINUM RIB
PLATE

CHAIN SEAL

AFT-FUSELAGE
LOWER TRAILING
EDGE

ROTARY
ACTUATOR

FULL-DEPTH
HONEYCOMB
TRAILING EDGE

HONEYCOMB
CLOSEOUT RIBS

SHEET METAL
FRONT SPAR WEBS

BODY FLAP
LOWER SURFACE

MACHINED
ACTUATOR RIBS

LOWER-MAIN
HONEYCOMB
PANEL

Wings

The blended double-delta wings of the Orbiter provide aerodynamic lift and control of the vehicle during atmospheric flight. The left and right wings consist of a wing glove, an intermediate section that contains the main landing gear well, a torque box, the forward spar for mounting the leading edge, and trailing edge elevons. The wings use a modified NACA-0010 section, with an 81 degree sweep on the inner leading edge, and 45 degrees on the outer. The wing trailing edge has a 3° 31′ dihedral. Each of the main wing assemblies are 60 feet long and have a maximum thickness of approximately five feet at the fuselage intersection. The main wing assemblies were constructed by Grumman in Bethpage, Long Island, New York. The completed wings were shipped via the Panama Canal to Long Beach, California, and then transported overland to Rockwell's assembly facility in Palmdale.

The forward wing box is an extension of the basic wing that aerodynamically blends the wing leading edge into the mid-fuselage wing glove. It is of conventional design with aluminum ribs, aluminum tubes, and tubular struts. The leading edge spar is constructed of corrugated aluminum, except for OV-101 and OV-102 which use a non-corrugated aluminum honeycomb sandwich construction. The upper and lower wing skin panels are stiffened aluminum alloy.[56]

The intermediate wing section consists of the same conventional aluminum design, except that the upper and lower skins are of aluminum honeycomb construction. A portion of the lower wing skin panel is made up of the main landing gear door. The intermediate wing section houses the main landing gear compartment, and absorbs a portion of the main landing gear loads. A structural rib provides support for the outboard main landing gear door hinges, and the outboard main gear trunnion and drag link. The support for the inboard main gear trunnion and drag link attachment is provided by the mid-fuselage. The main landing gear doors are constructed of aluminum honeycomb with machined aluminum hinge beams and hinges. The remainder of the door is conventional aluminum beams and multi-stringers with stiffened aluminum skin. A recessed area in the door is provided for additional tire clearance.

The wing torque box incorporates a conventional eleven rib truss arrangement with four corrugated aluminum spars on OV-101 and OV-102, and four graphite composite spars on OV-103 and subsequent Orbiters. The ribs on OV-101 were designed on 60-inch centers, but this was later shown to be inadequate and an intermediate rib was installed. The extra stiffness provided by this rib was built into a new lightweight stiffener on OV-102 and subsequent vehicles. The forward spar on all vehicles is of aluminum honeycomb and provides the attachment of the RCC leading edge thermal protection system (the leading edge of OV-101 is fiberglass instead of RCC). The rear spar provides the attach points for the elevons, hinged upper seal panels (flipper doors), and associated hydraulic and electrical systems. The upper and lower wing skin panels are stiffened aluminum covered with TPS tiles and blankets.[57]

The elevons provide flight control during atmospheric flight, and are of conventional aluminum multi-rib and beam construction with aluminum honeycomb skins. Each elevon is divided into two segments, and each segment is supported by three hinges. Attachments for the flight control systems are provided along the forward extremity of the elevons, and all hinge moments are at these points. The upper leading edge of each elevon panel incorporates

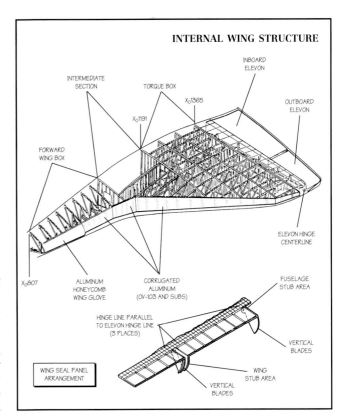

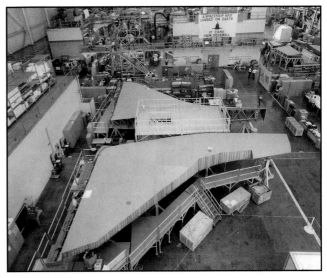

The wings for Atlantis after their arrival in Palmdale from Grumman in New York. The wings were transported by ship through the Panama Canal to Long Beach, then trucked to Palmdale. Discovery can be seen in her final assembly position at top left, and the elevons for Atlantis are in her final assembly area at top right. The corrugated leading edge spar was used on OV-103 and later Orbiters – OV-101 and OV-102 used a non-corrugated aluminum honeycomb sandwich instead. (NASA photo 108-KSC-83PC-703)

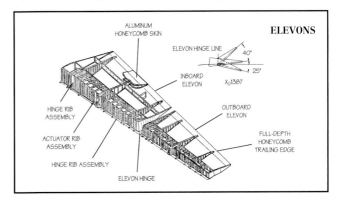

titanium rub strips. Hinged 'flipper doors' on the upper wing are used to seal the wing-elevon gap, and originally were not covered by the thermal protection system. All of the flipper doors on OV-101 were constructed from titanium – subsequent vehicles used Inconel for the outer elevon doors. The three ISS Orbiters are migrating towards aluminum flipper doors covered with FRSI in order to save 520 pounds. Each elevon is capable of deflecting 40 degrees upward and 25 degrees down, with a maximum deflection rate of 20 degrees per second – however, software limits travel to only 33 degrees up and 18 degrees down to avoid over-stressing the airframe.[58]

During 1981 the trailing edge of the right inboard elevon on OV-102 was badly damaged and was deemed not to be repairable. Instead of constructing a new piece, NASA elected to remove the full-depth honeycomb trailing edge wedge from OV-101. In its place a wooden trailing edge was attached, and this is the configuration of OV-101 as of late-2000. Unfortunately, the repair is beginning to come apart, and will undoubtedly be a challenge for the excellent restoration facility at the Smithsonian before *Enterprise* is placed on display in 2003.[59]

The wing is attached to the fuselage is accomplished with a tension bolt splice along the upper surface. A shear bolt splice along the lower surface of the fuselage carry-through completes the fuselage attachment interface.

Prior to building the wings for OV-103 and OV-104, a weight reduction program resulted in a redesign of certain areas of the wing structure. An assessment of wing air loads was made from actual flight data that indicated greater loads on the wing structure than previously anticipated, resulting in the 6.0 loads data base. To maintain positive margins of safety during ascent, five (OV-103 and OV-104) or seven (OV-102) truss tubes in the wing box and three in the wing glove were replaced with stronger units. In addition, aluminum doublers and stiffeners were added to high-stress areas, and other portions of the wing spars were reinforced in OV-103 and OV-104. The wing for OV-105 was modified to the revised design prior to assembly. During 1992, a decision was made to further strengthen the wing to allow higher landing weights and the Orbiters were all modified during normal maintenance at KSC or Palmdale.[60]

Vertical Stabilizer

The vertical stabilizer has a leading edge sweep of 45 degrees and consists of a structural fin surface, the rudder/speedbrake surface, a tip, and a lower trailing edge. The rudder splits into two halves to serve as a speedbrake. The vertical stabilizers for the original Orbiters were constructed by Fairchild Republic in Farmingdale, Long Island, New York. With the demise of that company, the vertical stabilizer for OV-105 was constructed by Grumman Aerospace, also of Long Island, which took over engineering support for many of Republic's contracts.

The vertical stabilizer (except for OV-101) consists of integral aluminum ribs, webs, stringers and machined aluminum spars to make a torque box for primary load carrying. OV-101 used conventional skin and stringer construction. The lower trailing edge area that houses the rudder/speedbrake power drive unit has aluminum honeycomb skins. The stabilizer is attached using two tension tie bolts at the root of the front spar to the forward bulkhead of the aft-fuselage and by eight shear bolts at the root of the rear spar to the upper structure of the aft-fuselage.[61]

The rudder/speedbrake control surface consists of con-

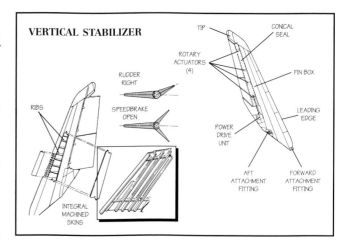

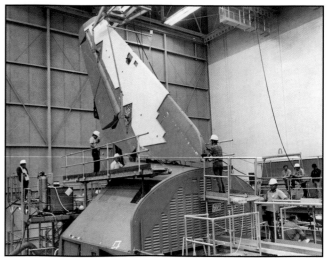

The vertical stabilizer of OV-101 is attached to the aft-fuselage during final assembly. The white material is simulated (foam) thermal protection. (NASA photo KSC-76P-50)

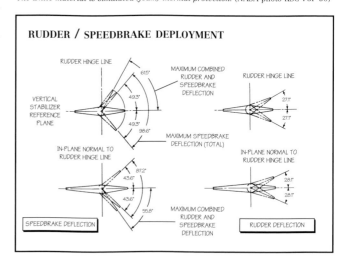

ventional aluminum ribs and spars with aluminum honeycomb skin panels for the primary load carrying members, and is attached through rotating hinge parts to the vertical stabilizer. An Inconel honeycomb aerodynamic seal is used between the rudder/speedbrake and the vertical stabilizer, and is not covered by the thermal protection system.

Mission requirements call for the rudder/speedbrake to be locked during ascent, on-orbit, and high-speed reentry. Speedbrake control is provided from Mach 10 to 5. Below Mach 5, speedbrake and rudder controls are provided as required. The movable surfaces have a maximum deflec-

OMS Pod RP03 is towed to the Hypergolic Maintenance Facility at KSC on 11 February 1987. The pod would be used on Discovery for STS-26R. (NASA photo KSC-87PC-89)

tion rate of 14 degrees per second as a rudder, and 10 degrees per second as a speedbrake. The maximum combined rudder/speedbrake deflection is ±61.5 degrees. When acting solely as a speedbrake, the maximum deflection is 98.6 degrees, or ±27.1 degrees when used as a rudder only. The unit is capable of limited rudder operations (±12.2 degrees) even when fully (98.6 degrees) deployed as a speedbrake.

The vertical stabilizer is designed for a 163 decibel acoustic environment with a maximum temperature of 350 degF. The surface of the vertical stabilizer is covered by the thermal protection system. Inconel honeycomb conical seals house the rotary actuators and provide a pressure and thermal seal that withstands a maximum of 1,200 degF. The split halves of the rudder panels and trailing edge contain a thermal barrier seal. A thermal barrier is also employed at the interface between the vertical stabilizer and the aft-fuselage.

Orbital Maneuvering System

The Orbital Maneuvering System (OMS) provides the thrust to perform orbital insertion, orbit circularization, orbit transfer, rendezvous, and deorbit. The OMS may also be used to provide small amounts of thrust above 70,000 feet during atmospheric flight. The system consists of two pods, one on each side of the upper aft-fuselage flanking the vertical stabilizer of the Orbiter. These pods also contain the aft Reaction Control System (RCS). The pods and propellant tanks were fabricated by McDonnell Douglas Astronautics, and the OMS engines were manufactured by Aerojet (a GenCorp company).[62]

Each pod is 21.8 feet long, 11.37 feet wide at its aft end, and 8.41 feet wide at the forward end, with a surface area of approximately 435 square feet. Attachment to the aft-fuselage is by 11 bolts per pod. Each pod is divided into two compartments – one for the OMS and one for the aft RCS. The forward and aft bulkhead aft tank support bulkhead and floor truss beam are machined aluminum 2124 while the centerline beam is 2024 aluminum sheet with titanium stiffeners and graphite-epoxy frames. The OMS thrust structure is conventional 2124 aluminum, the cross braces are aluminum tubing, and the attach fittings at the forward and aft fittings are 2124 aluminum. The intermediate fittings are corrosion-resistant steel. The pod skin panels are graphite-epoxy honeycomb sandwich that reduced the weight by approximately 225 pounds per OMS pod.[63]

The first three pods (RV01 and LV01 delivered with OV-102, and the static test article – RPST) were built to the 5.1 loads data base. Like the Orbiter itself, later pods (RP01 and LP01) were built to the 5.4 loads data base (Increment II), then still later to an Increment III standard (RP02 and LP02 and subsequent). LV01 and RV01 were lost with *Challenger*, and RPST (right pod, structural test) became RP03 after replacing all of the composite skins

The OMS engines are extremely compact, especially when the nozzle is not installed. The engine has a dry weight of only 305 pounds, yet produces 6,000 lbf. (Aerojet)

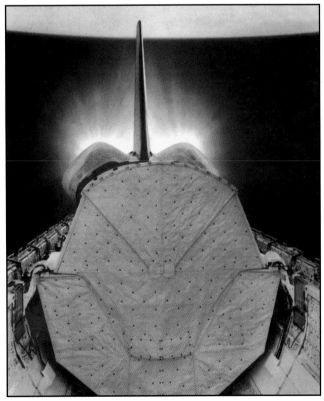

The OMS engines light up the night on STS-5. The OMS/RCS is a mixed blessing – the hypergolic propellants guarantee reliability, but are very toxic. (courtesy of Aerojet)

ORBITAL MANEUVERING SYSTEM

NOTE: COMPONENTS NOT SHADED ARE PART OF RCS

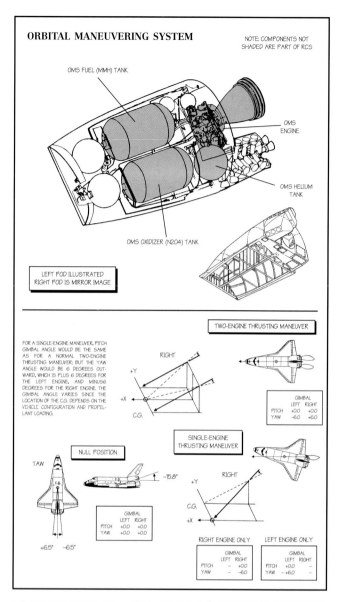

OMS FUEL (MMH) TANK

OMS ENGINE

OMS HELIUM TANK

OMS OXIDIZER (N2O4) TANK

LEFT POD ILLUSTRATED
RIGHT POD IS MIRROR IMAGE

FOR A SINGLE-ENGINE MANEUVER, PITCH GIMBAL ANGLE WOULD BE THE SAME AS FOR A NORMAL TWO-ENGINE THRUSTING MANEUVER; BUT THE YAW ANGLE WOULD BE 6 DEGREES OUTWARD, WHICH IS PLUS 6 DEGREES FOR THE LEFT ENGINE, AND MINUS6 DEGREES FOR THE RIGHT ENGINE. THE GIMBAL ANGLE VARIES SINCE THE LOCATION OF THE C.G. DEPENDS ON THE VEHICLE CONFIGURATION AND PROPELLANT LOADING.

TWO-ENGINE THRUSTING MANEUVER

RIGHT

+Y

+X

C.G.

GIMBAL
	LEFT	RIGHT
PITCH	+0.0	+0.0
YAW	-6.0	+6.0

SINGLE-ENGINE THRUSTING MANEUVER

NULL POSITION

YAW

-15.8°

GIMBAL
	LEFT	RIGHT
PITCH	+0.0	+0.0
YAW	+0.0	+0.0

+6.5° -6.5°

RIGHT

+Y

C.G.

+X

RIGHT ENGINE ONLY

GIMBAL
	LEFT	RIGHT
PITCH	–	+0.0
YAW	–	-6.0

LEFT ENGINE ONLY

GIMBAL
	LEFT	RIGHT
PITCH	+0.0	–
YAW	+6.0	–

The aft RCS module on the right OMS pod as Discovery was prepared for STS-103. Technicians were replacing SSME #3 in the VAB. (NASA photo KSC-99PC-1294)

and the entire housing. All current pods have mostly Increment III configuration structure; however, the aft closeout assemblies of RP01 and LP01 are beefed up Increment II structure since they were long lead items and production began prior to the Increment III redesign.[64]

The RCS housing, which attaches to the OMS pod structure, contains the RCS thrusters and associated propellant feed lines. The RCS housing is constructed of aluminum sheet metal, including flat outer skins. The curved outer skin panels are graphite-epoxy honeycomb sandwich. Twenty-four doors in the skins provide access to the OMS and RCS components and the pod attach points.

The pods can withstand 162 decibel acoustic noise and a temperature range from -170 to +135 degF. The exposed areas of the OMS/RCS pods are covered by the thermal protection system, and a pressure and thermal seal is installed at the OMS/RCS pod aft-fuselage interface. Thermal barriers are installed at each RCS thruster.[65]

Each pod contains a high-pressure helium storage bottle, four tank pressure regulators and controls, a fuel tank, an oxidizer tank, and a pressure-fed, regeneratively cooled, rocket engine. A maximum of 7,773 pounds of nitrogen tetroxide (N2O4) oxidizer and 4,718 pounds of monomethyl hydrazine (MMH) fuel can be carried in each pod. The propellants are the same as used by the RCS.

Each engine produces 6,000 lbf in a vacuum and a specific impulse of 313 seconds at a chamber pressure of 125 psia, and a nozzle expansion ratio of 55:1. Each engine is designed to be reusable for 100 missions, and capable of sustaining 1,000 starts and 15 hours of cumulative firing. Two engines firing together create an acceleration of approximately 2 fps/s or 0.06-g. The total dry weight of each engine is about 305 pounds. Aerojet has developed an uprated OMS engine that raises the chamber pressure from 125 psia to 350 psia. Although this does not result in a thrust increase, the specific impulse improves from 313 to 334 seconds. There are no firm plans to use the uprated engines operationally. The OMS engines are gimbaled by electromechanical actuators during firing for thrust vector control. The actuation system provides multi-axis gimballing of ±8 degrees in pitch and ±7 degrees in yaw.

The OMS propellants are nitrogen tetroxide (N2O4) as the oxidizer and monomethyl hydrazine (MMH) as the fuel. The design mixture ratio of 1.65 to 1 (oxidizer to fuel weight) permits the use of identically-sized tanks for both propellants. The OMS tankage is sized to provide propellant capacity for a velocity change of 1,000 fps at maximum gross weight – a portion of this (usually about 500 fps) is used during ascent. The minimum firing of an OMS engine is two seconds. Early in the program there were plans to carry up to three additional OMS propellant kits in the payload bay; each kit providing an additional 500 fps velocity capability. In support of this planned capability, all the structural provisions, wiring, and controls/displays were installed in the Orbiters. As of 2000 NASA indicates it is unlikely this capability will ever be used, and the controls/displays are inoperative and likely to remain so.[66]

For velocity changes less than 6 fps, the RCS is used. For changes greater than 6 fps, a single OMS engine is used in order to minimize the number of engine starts. Both OMS engines are used for large velocity changes. Propellant from one pod can be cross-fed to the other pod. OMS propellants may also be used by the aft RCS as needed.

During the *Challenger* stand-down, each of the 64 valves operated by electric motors in the OMS and RCS were modified to incorporate a 'sniff' line to permit monitoring for propellants in the electrical portion of the valves during ground operations. This line reduced the probability of particles in the electrical microswitch portion of each valve. By the end of 1997, a new-design motor valve was being introduced that eliminated the need for the sniff line, which was subsequently removed during OMDPs beginning with *Endeavour*'s OMDP-1.

Reaction Control System

The Reaction Control System (RCS) has 38 primary thrusters and six vernier thrusters to provide attitude control and three-axis translation during the orbit insertion, on-orbit, and reentry phases of flight. The RCS is used as the primary flight control when the vehicle is above 70,000 feet. Manufactured by the Marquardt Division of CCI Corporation, the primary thrusters are type R-40A, and the vernier are R-1E-3 models. The RCS consists of three propulsion modules, one in the forward fuselage, and one in each OMS pods.

The forward RCS (FRCS) module is constructed of conventional 2024 aluminum alloy skin-stringer panels and frames. The panels are composed of single-curvature, stretch-formed skins with riveted stringers, and the frames are riveted to the skin-stringer panels. The FRCS is secured to the forward fuselage nose section and forward bulkhead of the forward fuselage with 16 fasteners, that permit the installation and removal of the module.[67]

The forward RCS module contains 14 primary and 2 vernier thrusters, and each of the aft RCS modules have 12 primary and two vernier thrusters. Each of the three RCS modules also contain two helium tanks that pressurize the oxidizer and fuel tanks. Like the OMS, the RCS propellants are used in a 1.65:1 ratio. The capacity of each tank is 1,464 pounds of N2O4 or 923 pounds of MMH. Since the propellants are hypergolic, there is no ignition system.

Each RCS jet is identified by the propellant manifold that supplies the jet and by the direction of the jet plume. The first identifier designates a jet as forward (F), left aft (L), or right aft (R). The second identifier, number 1 through 5, designates the propellant manifold. The third identifier designates the direction of the jet plume: A (aft), F (forward), L (left), R (right), U (up), D (down). For example, jets F2U, F3U, and F1U are forward RCS jets receiving propellants from forward RCS manifolds 2, 3, and 1, respectively; the jet plume direction is up.[68]

The R-40A primary thrusters produce 870 lbf in a vacuum and a specific impulse of 289 seconds. They are operable in a maximum steady-state thrusting mode of 1 to 150 seconds, with a maximum single-mission contingency of 800 seconds for the aft RCS plus-X engines, and 300 seconds maximum for the forward RCS plus-X engines. The primary thrusters can also operate in an impulse mode above 125,000 feet, with a minimum impulse time of 0.08 second. Each thruster is designed to be reusable for 100 missions, and is capable of sustaining 50,000 starts and 20,000 seconds of cumulative firing.

The R-1E-3 vernier thrusters produce 25 lbf and a specific impulse of 228 seconds. They are operable in a steady-state thrusting mode of 1 to 125 seconds, as well as in an impulse mode with a minimum impulse time of 0.08 second. Each vernier is also designed for 100 missions, capable of being started 330,000 times for a total of 125,000 seconds of cumulative firing. The nominal chamber pressure for the primary thruster is 152 psia, and 110 psia for the vernier thrusters.

The RCS is also used during off-nominal situations. In the case of loss of two SSMEs on ascent, the OMS-to-RCS

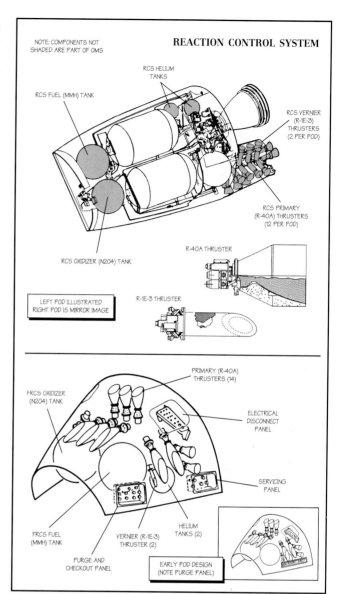

A FRCS module being removed from Discovery prior to the Return-to-Flight. Note the open purge and checkout panel at the lower edge. (NASA photo KSC-87P-693)

interconnect is automatically commanded, and the RCS provides single-engine roll control. If the OMS gimballing system is not performing adequately to control vehicle attitude during an OMS burn, RCS jets are used to help maintain attitude – this is also known as 'RCS wraparound.' The RCS is also used to adjust an orbit if the OMS fails prematurely. During aborts, the RCS may be used to assist with ascent propellant dumps to decrease vehicle weight, improve performance, and the control center-of-gravity.[69]

When the Orbiter is used to reboost the International Space Station, under some conditions it is preferable to use a FRCS thruster rather than an aft RCS thruster to avoid gas impingement on the station. The FRCS Interconnect System (FICS) modification will ensure there is adequate propellant to use whichever thruster is most appropriate. OV-103 will be the first Orbiter equipped with the new system during her second OMDP in August 2001, followed by *Atlantis*, *Endeavour*, and maybe, *Columbia*. The system allows propellants from the OMS pods to be transferred to the Forward RCS module using a new set of propellant transfer lines. These lines tie into the existing cross-connect lines between the OMS pods, run down the $X_0 1307$ bulkhead, under the liner along the right wall of the payload bay, into the right wing outboard of the main landing gear well, back into the forward payload bay, then along the forward fuselage splice on the right side to the FRCS module. Originally the lines were going to run under the liner for the entire length of the payload bay, but it was decided that there was less potential for damage to the lines (while working in the payload bay during turnaround ops) if they were routed through the wing instead.[70]

The excess OMS propellants that are normally loaded in case the OMS-1/2 burn during ascent needs to be extended will now be able to be used by the FRCS. Originally, these same lines were to interface with the Orbiter Propellant Transfer System that would have been used to refuel the United States Propulsion Module (USPM) on the ISS. Propellants from the OMS pods would have been transferred using FICS plumbing to additional plumbing on the ODS, through a set of quick disconnects to the USPM attached to the ISS. However, the cancellation of the USPM in March 2000 in favor of Russian hardware has eliminated the need for this capability. The OI-29 software will be the first to feature support for the FICS modifications.[71]

Electrical Power

The Orbiter Electrical Power System (EPS) consists of three primary subsystems – power reactant storage and distribution (PRSD), electrical power distribution and control (EPDC), and the fuel cell power plants. The EPS provides all vehicle power when not connected to ground power, and operates during all flight phases. Electrical power is generated by the three fuel cells using cryogenically stored oxygen and hydrogen reactants. Each fuel cell is connected to one of three independent electrical buses. During peak and average power loads, all three fuel cells and buses are used; during minimum power loads, only two fuel cells are used, but they are interconnected to all three buses. The third fuel cell is put into a standby mode, and can be brought back on-line instantly. Excess heat from the fuel cells is transferred to a Freon-21 cooling loop through heat exchangers. The combined fuel cell system provides 14 kW continuous and 24 kW peak at 28 Vdc. The fuel cells are located under the payload bay in the forward portion of the mid-fuselage. Each fuel cell is overhauled after 2,400 hours* of operation.[72]

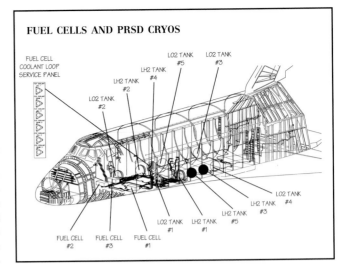

FUEL CELLS AND PRSD CRYOS

Each fuel cell consists of a power section, where the chemical reaction occurs, and an accessory section that controls and monitors the power section's performance. The power section, where oxygen and hydrogen are transformed into electrical power, water and heat, consists of 96 cells contained in three substacks. Manifolds run the length of these substacks and distribute oxygen, hydrogen, and coolant to the cells. The cells contain electrolyte consisting of potassium hydroxide and water, an oxygen electrode (cathode) and a hydrogen electrode (anode).

A fuel cell redesign effort was undertaken during the *Challenger* stand-down, and numerous components of the fuel cells were improved in an effort to increase the reliability and maintainability of the system. Among the changes: end-cell heaters on each fuel cell power plant were deleted because of potential electrical failures and replaced with Freon-21 coolant loop passages; the hydrogen pump and water separator in each fuel cell were improved to minimize excessive hydrogen gas entrapped in the power plant by-product water; several new sensors were added to provide more visibility into possible hydrogen pump overload and thermal conditions; and the by-product water from all three fuel cells now flows to a single water relief control panel to simplify ECLSS operations.

The PRSD subsystem stores the reactants (cryogenic oxygen and hydrogen) under pressure (731 psia for the oxygen and 188 psia for the hydrogen) and supplies them to the three fuel cells. In addition, oxygen is supplied to the Environmental Control and Life Support System for crew compartment pressurization. The tanks are grouped in sets consisting of one LO2 and one LH2 tank. The number of tank sets installed depends on specific mission requirements, and up to five sets can be accommodated in the mid-fuselage under the payload bay liner on all four Orbiters (and OV-099 when it was in service). The oxygen tanks are 36.8 inches in outside diameter, weigh 215 pounds, and store 781 pounds of oxygen each. The oxygen tanks consist of inner pressure vessels made of Inconel-718 and outer shells of 2219 aluminum alloy. The hydrogen tanks are 45.5 inches in outside diameter, weigh 227 pounds, and store 92 pounds of hydrogen each. On the hydrogen tanks, both the inner pressure vessel and the outer shell are constructed of 2219 aluminum.

In order to better support Spacelab missions, NASA contracted with Rockwell to develop the Extended

---

* This has usually been reported as 5,000 hours, but that was the original design goal, not what has been realized in operation.

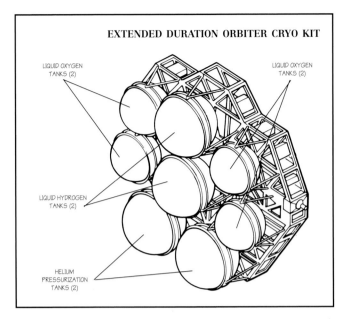

EXTENDED DURATION ORBITER CRYO KIT

LIQUID OXYGEN TANKS (2)

LIQUID OXYGEN TANKS (2)

LIQUID HYDROGEN TANKS (2)

HELIUM PRESSURIZATION TANKS (2)

buses are used to supply power to flight controls. Two pre-flight buses are used only during ground operations.

The hydrazine-fueled, turbine-driven auxiliary power unit (APU) generates mechanical shaft power to drive a pump which produces pressure for the Orbiter's hydraulic system. Each system provides hydraulic pressure to position actuators for thrust vector control of the main engines by gimbaling the three SSMEs, actuation of various control valves on the SSMEs, movement of the Orbiter aerosurfaces (elevons, body flap, rudder/speedbrake), retraction of the Orbiter–ET 17-inch LO2 and LH2 disconnect umbilicals within the Orbiter at ET jettison, main/nose landing gear deployment, main landing gear brakes and anti-skid, and nose wheel steering.[74]

There are three separate Sunstrand APUs, and three 3,000 psi hydraulic pumps, one for each of the triple redundant hydraulic systems. The APU systems are located in the aft-fuselage and consist of a fuel tank, a fuel feed system, an APU and controller, an exhaust duct, and a lube oil cooling system. The fuel tanks have a maximum capacity of 350 gallons of hydrazine (N2H4). Each tank is a 28-inch diameter sphere, and two (tanks #1 and #2) are located on the left and one (tank #3) on the right side of the aft-fuselage. Each APU is rated at 135 hp at normal speed and weighs just under 88 pounds including its controller. The APU exhaust ports are located on the top of the aft-fuselage, beside the vertical stabilizer's leading edge.

The typical pre-launch fuel load for each tank is approximately 325 pounds. The fuel supply supports the nominal power unit operating time of 90 minutes in a mission – as well as all defined abort modes where the APUs run continuously for approximately 110 minutes. Under operating load conditions, an APU consumes approximately 3–3.5 pounds of fuel per minute.[75]

The APUs that were used for the first 25 missions had a limited life, each unit being refurbished after 25 hours of operations because of cracks in the turbine housing. Improved APUs (IAPU) were delivered in late 1988 with a new turbine housing that increased the operating life to 75 hours (roughly 50 typical missions). The program procured two modification kits for use during qualification tests, nine modification kits for flight units, and three new flight-rated IAPUs (delivered with OV-105).[76]

The water spray boiler (WSB) system consists of three identical independent water spray boilers, one for each APU/hydraulic system. The boilers are located in the aft-fuselage and cools the corresponding APU lube oil system and hydraulic system by spraying water onto their lines –

Duration Orbiter (EDO) cryo kit. A 3,500 pound, 15-foot diameter wafer pallet provides a support structure for tanks and associated control panels and avionics equipment. The tanks store 3,124 pounds of LO2 and 368 pounds of LH2. When filled with cryogens, the pallet weighs approximately 7,000 pounds. The EDO cryo kit attaches vertically to the payload bay rear bulkhead on OV-102 and OV-105. Combined with the cryogens normally carried, the EDO kit allows an Orbiter to stay on-orbit for up to 16 days. There was some consideration being given to allow OV-105 to carry a second EDO pallet immediately in front of the first (for a total of eight additional tank sets, or 13 sets total). This, however, was abandoned as the International Space Station began to be assembled, and in fact, the EDO capability has been removed from OV-105 in order to save weight. OV-104 was also modified with many of the provisions necessary to make her EDO-capable, but NASA elected not to proceed with the final changes and *Atlantis* is not capable of EDO missions.[73]

The EPDC subsystem consists of 3-bus system that generates and distributes 115-Vac, 3-phase, 400-hertz electrical power and distributes 28-Vdc electrical power to the Orbiter, ET, and both SRBs. There are 3 main ac buses, and 3 main dc buses. Three essential buses supply control power to selected flight controls and operational power to electrical loads that are deemed essential. Nine control

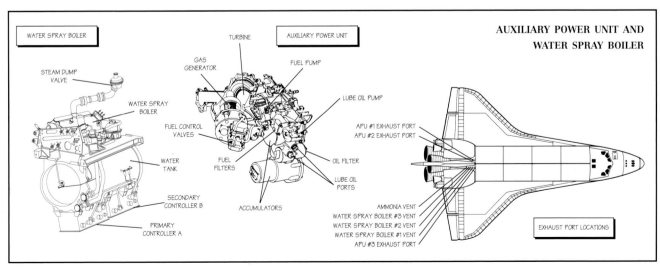

AUXILIARY POWER UNIT AND WATER SPRAY BOILER

WATER SPRAY BOILER

STEAM DUMP VALVE

WATER SPRAY BOILER

TURBINE

GAS GENERATOR

AUXILIARY POWER UNIT

FUEL PUMP

LUBE OIL PUMP

FUEL CONTROL VALVES

WATER TANK

FUEL FILTERS

OIL FILTER

LUBE OIL PORTS

SECONDARY CONTROLLER B

ACCUMULATORS

PRIMARY CONTROLLER A

APU #1 EXHAUST PORT

APU #2 EXHAUST PORT

AMMONIA VENT

WATER SPRAY BOILER #3 VENT

WATER SPRAY BOILER #2 VENT

WATER SPRAY BOILER #1 VENT

APU #3 EXHAUST PORT

EXHAUST PORT LOCATIONS

as the water boils off, the lube oil and hydraulic fluid are cooled. The steam that boils off in each water spray boiler exits through its own exhaust duct, located on the starboard side of the vertical stabilizer. Each WSB is 45 inches long, 31 inches high, 19 inches wide, and weigh 18.1 pounds including controller and vent nozzle. Each boiler contains 142 pounds of water.[77]

Environmental Control and Life Support

The Environmental Control and Life Support System (ECLSS) is made up of three major subsystems: the atmosphere revitalization and pressure control subsystem (ARPCS) that controls the atmospheric environment for the crew and the thermal environment for electronics; the food, water and waste subsystem to provide hygiene and other life support functions; and the active thermal control subsystem (ATCS).[78]

The ARPCS controls crew module pressure at 14.7 ±0.2 psia, with an average mixture of 80 percent nitrogen and 20 percent oxygen. Oxygen partial pressure is maintained between 2.95 psia and 3.45 psia, with sufficient nitrogen pressure of 11.5 psia added to achieve the desired cabin total pressure. Relief valves are activated if the compartment pressure rises above 16 psia. The ARPCS receives oxygen from two PRSD cryogenic oxygen systems and two nitrogen systems (consisting of two nitrogen tanks for each) located in the mid-fuselage of the Orbiter. An optional mission kit consists of an emergency gaseous oxygen tank located in the mid-fuselage of the Orbiter. The gaseous nitrogen system is also used to pressurize the potable and waste water tanks located below the mid-deck floor.[79]

Based on the crew module volume of 2,325 cubic feet (plus 150 for the airlock) and 330 cubic feet of air movement per minute, one air change occurs in approximately 7 minutes, or approximately 8.5 air changes every hour.

Air leaves the cabin fan at a rate of about 1,400 pounds per hour, and an orifice in the duct directs approximately 120 pounds per hour to each of two lithium hydroxide (LiOH) canisters where carbon dioxide is removed. Activated charcoal canisters remove odors and trace contaminants. The canisters are changed periodically on a predetermined schedule, generally once or twice per day, through an access door. (For larger crews, the canisters are changed more frequently.) Each canister is rated at 48 man-hours, and up to 30 spare canisters are stored under the mid-deck floor in a locker between the cabin heat exchanger and water tanks. For non-EDO missions, the lithium hydroxide canisters are the primary means of CO2 control. OV-102 and OV-105 were modified to use a regenerable carbon dioxide removal system (RCRS) for EDO missions while on-orbit. The system was subsequently removed from OV-105 as a weight reduction during OMDP-1, and the normal LiOH canisters were reinstalled. OV-103 and OV-104 are not configured for RCRS and must use the LiOH system for carbon dioxide removal.[80]

The ability to use the RCRS in the EDO Orbiters solved a major weight and volume stowage problem encountered when attempting to conduct 10–16 day missions with large crews. Carbon dioxide removal is accomplished by passing cabin air through one of two identical solid amine resin beds consisting of a polyethylenimine (PEI) sorbent coating on a porous polymeric substrate. Upon exposure to carbon dioxide laden cabin air, the resin combines with water vapor in the air to form a hydrated amine which reacts with carbon dioxide to form a weak bicarbonate bond. Water is required for the process since dry amine cannot

react with the carbon dioxide directly. While one bed adsorbs carbon dioxide, the other bed regenerates with thermal treatment and vacuum venting. This latter requirement prevents the use of the RCRS during ascent or reentry. The adsorption/regeneration process runs continuously with the beds automatically alternating every 13 minutes. A full cycle is made up of two 13 minute processes. An RCRS configured vehicle uses a single LiOH canister for launch and another for entry. Activated charcoal canisters are still used to remove odors.

Potable water produced by the three fuel cells is stored in potable water tanks for crew consumption and personal hygiene. The potable water is also used to supply to the flash evaporator system where it is used to cool the Freon-21 coolant loops and by the RCRS, if installed. A waste water tank is located below the mid-deck floor to collect waste water from the crew compartment heat exchanger and crew waste water. Solid waste remains in the waste management system in the mid-deck until the Orbiter is serviced during ground turnaround operations.

The Waste Collection System (WCS) is an integrated, multifunctional system used primarily to collect and process biological wastes from crew members in a zero-gravity environment. The WCS is located on the mid-deck in a 29 inch wide area immediately aft of the side hatch. The commode is 27 by 27 by 29 inches and is used like a standard toilet. The system collects, stores, and dries fecal wastes and associated tissues; processes urine and transfers it to the waste water tank; processes EMU condensate water from the airlock and transfers it to the waste water tank if an EVA is required on a mission; provides an interface for venting trash container gases overboard; provides an interface for dumping waste water overboard in a contingency situation; and transfers waste water to the waste water tank.[81]

A door on the waste management compartment, and two privacy curtains attached to the inside of the compartment door, provide privacy for crew members while they are using the facilities. One curtain is attached to the top of the door and interfaces with the edge of the interdeck access, and the other is attached to the door and interfaces with the galley, if installed. The door also serves as an ingress platform during prelaunch (vertical) operations since the crew must enter the flight deck over the waste management compartment. The door has a friction hinge and must be open to gain access to the WCS.

The WCS consists of a commode, urinal, fan separators, odor and bacteria filter, vacuum vent quick disconnect, and waste collection system controls. The commode contains a single multilayer hydrophobic porous bag liner for collecting and storing solid waste. When the commode is in use, it is pressurized, and transport air flow is provided by the fan separator. When the commode is not in use, it is depressurized for solid waste drying and deactivation. The urinal is essentially a funnel attached to a hose and provides the capability to collect and transport liquid waste to the waste water tank. The urinal can accommodate both males and females. It can be used in a standing position or can be attached to the commode by a pivoting mounting bracket for use in a sitting position. The fan separator provides air flow to transport the urine and to separate the waste liquid from the air flow. The liquid is drawn off to the waste water tank, and the air returns to the crew cabin through the odor and bacteria filter.[82]

The ATCS provides Orbiter heat rejection during all phases of the mission. It consists of two Freon-21 coolant loops, cold plate networks for cooling electronic avionics units, liquid-to-liquid heat exchangers for cooling various Orbiter systems, and four heat sink systems for rejecting excess heat outside the Orbiter. The Freon-21 coolant loops transport excess heat from the fuel cell power plant heat exchangers, payload heat exchangers and mid-body and aft avionics electronic units; heat the hydraulic systems; and deliver heat to the heat sinks. When the Orbiter is on-orbit, the payload bay doors are opened, exposing radiator panels on the underside of the doors to provide heat rejection to the cold of space.[83]

Flash evaporators are used for heat rejection during ascent and reentry, and at other times when the payload bay doors are closed and the radiators are inoperative. They can also supplement the radiators if needed while on-orbit. The flash evaporators are located in the aft-fuselage.

Thermal Protection System

The Thermal Protection System (TPS) protects the aluminum structure of the Orbiter during the atmospheric portion of flight, maintaining skin temperatures at less than 350 degF. These materials must perform in temperature ranges from minus 250 degF in the cold soak of space to reentry temperatures approaching 3,000 degF. The TPS must also sustain the forces induced by deflections of the Orbiter airframe as it responds to the various external environments. Because the TPS is installed on the outside of the Orbiter, it establishes the aerodynamics of the vehicle and must be capable of withstanding all aero-loads as well.[84]

The peak heating rates and longest exposure occur during high cross-range reentry maneuvers when equilibrium surface temperatures range from 3,000 degF at stagnation points on the nose and leading edges of the wing and tail, down to 600 degF on leeward surfaces of the wing.

The TPS is composed of four types of reusable surface insulation (RSI) tiles, reusable surface insulation blankets, thermal window panes, and thermal seals. The TPS is a passive system consisting of materials selected for

The carbon-carbon nose cap of Endeavour is mated at the Palmdale assembly facility after being fabricated by LTV in Grand Prairie, Texas. LTV used experience gained during the Dyna-Soar program to develop the RCC for Shuttle. (Rockwell International)

stability at high temperatures and weight efficiency. These materials are:[85]

1. Reinforced carbon-carbon (RCC) is used on the wing leading edges; the nose cap, including an area immediately aft of the nose cap on the lower surface (chin panel); and the immediate area around the forward Orbiter–ET structural attachment. RCC protects areas where temperatures exceed 2,300 degF during reentry.

2. Black high-temperature reusable surface insulation (HRSI) tiles are used on the entire underside of the vehicle where RCC is not used, the base heat shield, and in selected areas on the upper forward fuselage, including around the forward fuselage windows, portions of the OMS/RCS pods, the leading and trailing edges of the vertical stabilizer, wing glove areas, elevon trailing edges, adjacent to the RCC on the upper wing surface, the interface with wing leading edge RCC, the upper body flap surface, and around the SILTS pod on OV-102. The HRSI tiles protect areas where temperatures are between 1,200 and 2,300 degF. These tiles have a black surface coating necessary for thermal emittance.

3. Black fibrous refractory composite insulation (FRCI) tiles have replaced some of the HRSI tiles in selected areas of the Orbiter.

4. Toughened uni-piece fibrous insulation (TUFI) tiles were developed in 1993, and are used on limited areas of the Orbiters, mainly on the base heat shield around the SSMEs. The black-colored TUFI tiles are more resistant to damage than the HRSI tiles they replace.

5. White low-temperature reusable surface insulation (LRSI) tiles are used in selected areas of the forward fuselage, mid-fuselage, aft-fuselage, vertical stabilizer, upper wing, and OMS/RCS pods. These tiles protect areas where temperatures are below 1,200 degF and have a white surface coating to provide better thermal characteristics while on-orbit.

6. White advanced flexible reusable surface insulation (AFRSI) blankets were developed after the initial delivery of *Columbia*. The AFRSI blankets provide improved producibility and durability, reduced fabrication and installation time and costs, and a weight reduction over that of the LRSI tiles. The AFRSI blankets protect areas where temperatures are below 1,200 degF and aerodynamic flight loads are minimal.

7. White blankets made of coated Nomex felt reusable surface insulation (FRSI) are used on the upper payload bay doors, portions of the mid-fuselage and aft-fuselage sides, portions of the upper wing surface, and parts of the OMS/RCS pods. The FRSI blankets protect areas where temperatures are below 700 degF.

8. Other materials are used in other special areas. These materials include thermal panes for the windows; a combination of white and black-pigmented silica cloth for thermal barriers and gap fillers around operable penetrations; metal for the FRCS and elevon seal panels on the upper wing to elevon interface; and room-temperature vulcanizing (RTV) material for the heavy aluminum T-0 umbilicals on the sides of the Orbiter aft-fuselage.

Where temperatures reach above 2,300 degF, such as the wing leading edge and the nose cap, a composite composed of pyrolyzed carbon fibers in a pyrolyzed carbon matrix with a silicon carbide coating, commonly known as carbon-carbon, is used. The carbon-carbon nose cap and wing leading edges were manufactured by LTV (now Lockheed Martin) of Grand Prairie, Texas.

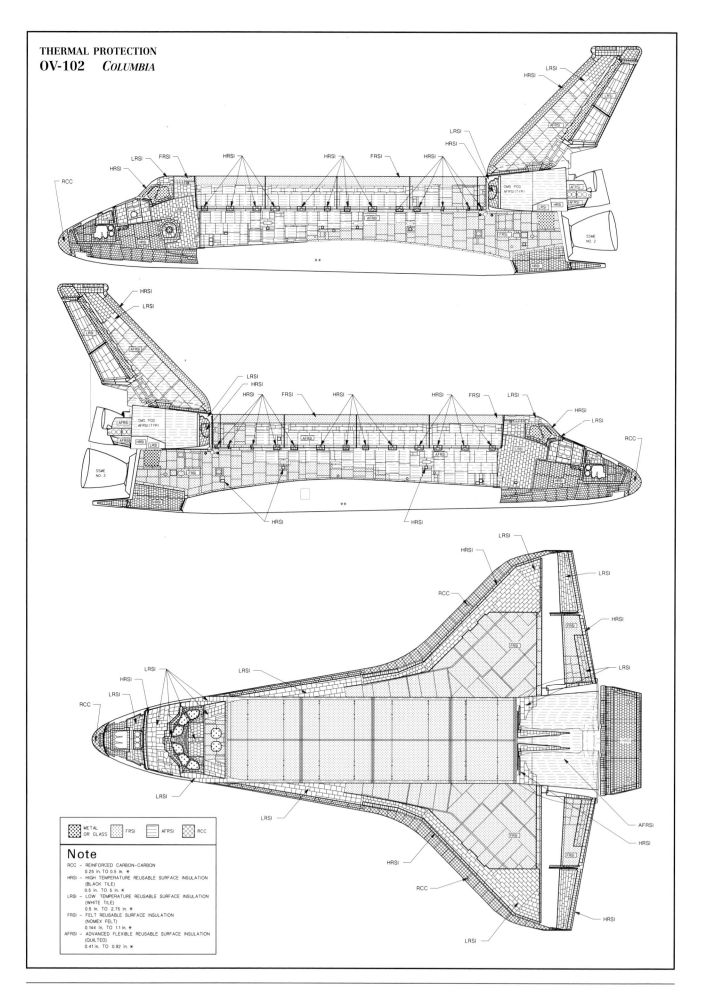

METAL
OR GLASS   FRSI   AFRSI   RCC

Note
RCC – REINFORCED CARBON–CARBON
    0.25 in. TO 0.5 in. ✳
HRSI – HIGH TEMPERATURE REUSABLE SURFACE INSULATION
    (BLACK TILE)
    0.5 in. TO 5 in. ✳
LRSI – LOW TEMPERATURE REUSABLE SURFACE INSULATION
    (WHITE TILE)
    0.5 in. TO 2.75 in. ✳
FRSI – FELT REUSABLE SURFACE INSULATION
    (NOMEX FELT)
    0.144 in. TO 1.1 in. ✳
AFRSI – ADVANCED FLEXIBLE REUSABLE SURFACE INSULATION
    (QUILTED)
    0.41 in. TO 0.92 in. ✳

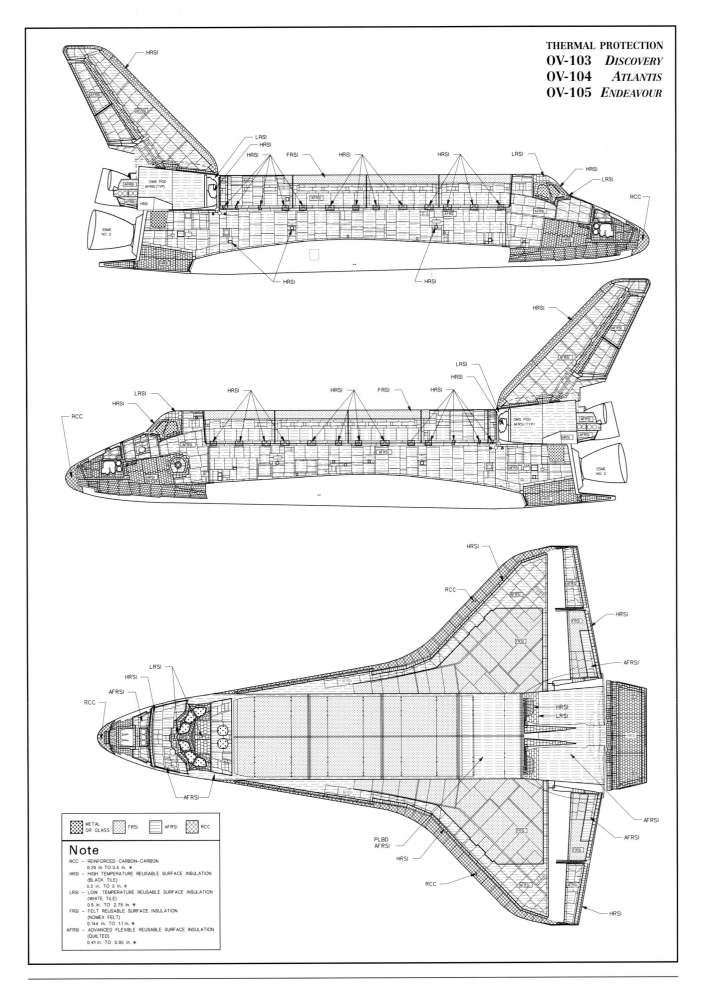

Note

RCC − REINFORCED CARBON–CARBON
  0.25 in. TO 0.5 in. ∗
HRSI − HIGH TEMPERATURE REUSABLE SURFACE INSULATION
  (BLACK TILE)
  0.5 in. TO 5 in. ∗
LRSI − LOW TEMPERATURE REUSABLE SURFACE INSULATION
  (WHITE TILE)
  0.5 in. TO 2.75 in. ∗
FRSI − FELT REUSABLE SURFACE INSULATION
  (NOMEX FELT)
  0.144 in. TO 1.1 in. ∗
AFRSI − ADVANCED FLEXIBLE REUSABLE SURFACE INSULATION
  (QUILTED)
  0.41 in. TO 0.92 in. ∗

Legend: METAL OR GLASS | FRSI | AFRSI | RCC

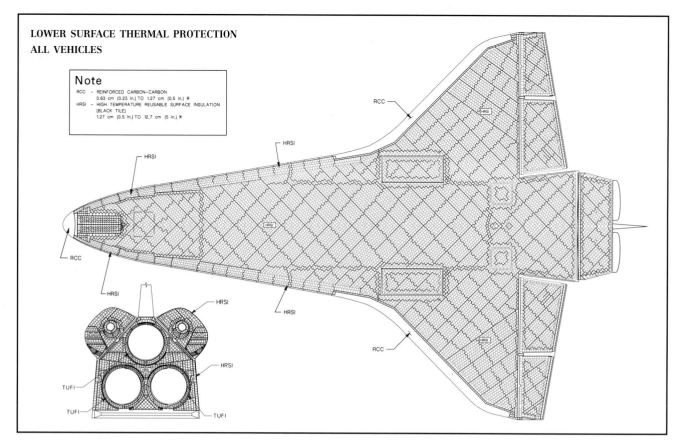

LOWER SURFACE THERMAL PROTECTION
ALL VEHICLES

Note
RCC - REINFORCED CARBON-CARBON
0.63 cm (0.25 in.) TO 1.27 cm (0.5 in.) ✳
HRSI - HIGH TEMPERATURE REUSABLE SURFACE INSULATION
(BLACK TILE)
1.27 cm (0.5 in.) TO 12.7 cm (5 in.) ✳

RCC fabrication begins with a rayon cloth graphitized and impregnated with a phenolic resin. This impregnated cloth is layed up as a laminate and cured in an autoclave. After being cured, the laminate is pyrolyzed to convert the resin to carbon. This is then impregnated with furfural alcohol in a vacuum chamber, then cured and pyrolyzed again to convert the furfural alcohol to carbon. This process is repeated three times until the desired carbon-carbon properties are achieved.[86]

To provide sufficient oxidation resistance to enable reuse, the outer layers of the RCC are converted to silicon carbide. The RCC is packed in a retort with a dry-pack material made up of a mixture of alumina, silicon, and silicon carbide. The retort is placed in a furnace, and the coating conversion process takes place in argon with a stepped-time-temperature cycle up to 3,200 degF. A diffusion reaction occurs between the dry pack and carbon-carbon in which the outer layers of the carbon-carbon are converted to silicon carbide (whitish-gray color) with no thickness increase. It is this silicon-carbide coating that protects the carbon-carbon from oxidation. The silicon-carbide coating develops surface cracks caused by differential thermal expansion mismatch. This requires further oxidation resistance, provided by impregnation of a coated RCC part with tetraethyl orthosilicate. The part is then sealed with a glossy overcoat. The RCC laminate is superior to a sandwich design because it is light weight and rugged; and it promotes internal cross-radiation from the hot stagnation region to cooler areas, thus reducing stagnation temperatures and thermal gradients around the leading edge. The operating range of RCC is from -250 degF to +3,000 degF. The RCC is highly resistant to the fatigue loading experienced during ascent and reentry.

The RCC panels are mechanically attached to the wing with a series of floating joints to reduce loading on the panels caused by wing deflections. The seal between each wing leading edge panel is referred to as a T-seal. The T-seals allow for lateral motion and thermal expansion differences between the RCC and the Orbiter wing. In addition, they prevent the direct flow of hot boundary layer gases into the wing leading edge cavity during reentry. The T-seals are also constructed of RCC.

Since carbon is a good thermal conductor, the adjacent aluminum and metallic attachments must be protected from exceeding temperature limits by internal insulation. Inconel-718 and A-286 fittings are bolted to flanges on the RCC components and are attached to the aluminum wing spars and nose bulkhead. Inconel-601-covered Dyna-Flex (Cerachrome) bulk insulation (also called 'Incoflex') protects the metallic attach fittings and spar from the heat radiated from the inside surface of the RCC wing leading-edge panels.

Beginning in 1998, the program began developing a modification to the leading edge RCC to enable it to withstand larger punctures. The original design allowed a 1-inch hole in the upper surface of any panel. But on the lower surface, no penetrations were allowed on panels 5-13 because any hole generated by orbital debris (or other causes) would allow heat from the plasma flow during reentry to quickly erode the 0.004-inch Inconel foil on the Incoflex insulators. This would expose the leading edge attach fittings and wing front spar to the direct blast of hot plasma. The upgrade included additional insulation that is able to withstand a penetration of up to 0.25-inch diameter in the lower surface of panels 9–12, and up to 1-inch on panels 5–8 and 13. To achieve this, Nextel 440 fabric insulation was wrapped around the Incoflex insulators – one layer for panels 5 through 7 and 11 through 13, and two fabric layers for panels 8 through 10 (which have the highest potential heating environment). The overall weight increase was 50 pounds. *Atlantis* was the first vehicle to be modified during May 1998, and

Discovery followed during January 1999. OV-102 is being modified during OMDP-2, and no schedule has yet been set for *Endeavour* although it will likely happen during her next OMDP.[87]

The nose cap thermal insulation uses an AB312 blanket made from ceramic fibers and filled with silica fibers. HRSI or FRCI tiles are used to protect the forward fuselage from the heat radiated from the hot inside surface of the RCC. An AB312 blanket is placed on the forward fuselage in the immediate area surrounding the forward Orbiter–ET attach point. RCC is placed over the blanket and is attached by metal standoffs for additional protection from the forward Orbiter–ET attach point pyrotechnics.

The HRSI tiles originally used on the lower forward fuselage ahead of the nose gear well of all Orbiters was replaced by the RCC 'chin' panel during 1988. The tiles in this area were constantly being damaged by impacts during ascent and overheating during reentry.

The HRSI tiles are made of a low-density, high-purity, 99.8 percent amorphous silica fibers derived from common sand, one to two mils thick, that are made rigid by ceramic bonding. Because 90 percent of the tile is void and the remaining ten percent is material, the tiles weigh approximately 9 pounds per cubic foot. A slurry containing fibers mixed with water is frame-cast to form soft, porous blocks to which a colloidal silica binder solution is added. When it is sintered, a rigid block is produced that is cut into quarters and then machined to the precise dimensions required for individual tiles. HRSI tiles can vary in thickness from one inch to five inches with the thickness determined by the heat load encountered during reentry. Generally, the HRSI tiles are thicker at the forward areas of the Orbiter and thinner toward the aft end. The HRSI tiles are nominally 6 by 6 inch squares, but vary in size and shape in the closeout areas. The tiles have a notable 'grain' (all the silica fibers run in generally one direction) and are machined such that the grain runs parallel to the surface the tile protects. An exception to this are the tiles around the windshield. These are machined so that the grain is perpendicular to the surface to provide additional strength to resist aerodynamic loads. The HRSI tiles withstand on-orbit cold soak conditions, repeated heating and cooling thermal shock, and extreme acoustic environments (163 db) at launch.

The HRSI tiles are coated on the top and sides with a mixture of powdered tetrasilicide and borosilicate glass with a liquid carrier. This material is sprayed on the tile to coating thicknesses of 16 to 18 mils. The coated tiles then are placed in an oven and heated to a temperature of 2,300 degF. This results in a black, waterproof glossy coating that has a surface emittance of 0.85 and a solar absorptance of about 0.85. After the ceramic coating heating process, the remaining silica fibers are treated with a silicon resin to provide bulk waterproofing.

As an aside, an HRSI tile taken from a 2,300 degF oven can be immersed in cold water without damage. Also, surface heat dissipates so quickly that an uncoated tile can be held by its edges with an ungloved hand seconds after removal from the oven while its interior still glows red.

As the higher density HRSI tiles are replaced, new fibrous refractory composite insulation (FRCI) tiles are usually used instead. This newer insulation was developed by NASA Ames later in the program and is considered more durable and resistant to cracking than the earlier HRSI. The FRCI-12 tiles have a density of 12 pounds per cubic foot. The tiles are essentially similar to the original HRSI tiles with the addition of a 3M Company additive called

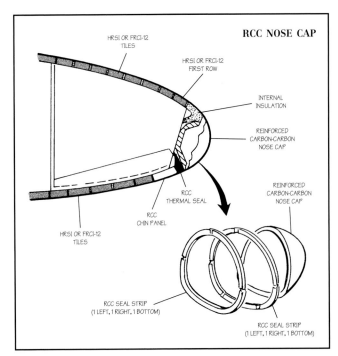

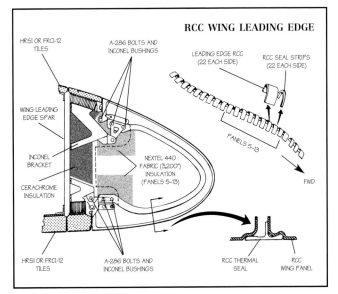

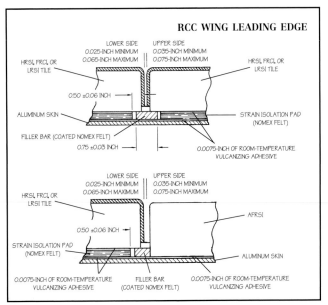

Nextal (AB312 – alumina-borosilicate fiber). Nextal, with an expansion coefficient ten times that of the 99.8-percent-pure silica, acts like a preshrunk concrete reinforcing bar in the fiber matrix. This additive activates a boron fusion which effectively 'welds' the pure silica fibers into a rigid structure during sintering. The resulting composite fiber refractory material, composed of 20 percent Nextal and 80 percent pure silica, has entirely different physical properties from the original 99.8 percent pure silica tiles.

The reaction-cured glass (black) coating of the FRCI-12 tiles is compressed as it is cured to reduce the coating's sensitivity to cracking during handling and operations. In addition to the improved coating, the FRCI-12 tiles are about 10 percent lighter than the HRSI tiles for a given thermal protection. FRCI-12 tiles have demonstrated a tensile strength at least three times greater than that of the HRSI tiles, and a use temperature approximately 100 degF higher than that of HRSI tiles.

The FRCI-12 manufacturing process is essentially the same as that for the 99.8-percent-pure silica HRSI tiles, the only change being in the wet-end pre-binding of the slurry before it is cast. It also requires a higher sintering temperature. When the material is dried, a rigid block is produced which is cut into quarters and then machined to the precise dimensions required for each tile. FRCI-12 tiles are the same 6 by 6 inch size as HRSI tiles and vary in thickness from 1 to 5 inches. They vary also in size and shape at the closeout areas and are bonded to the Orbiter in essentially the same way as the HRSI tiles.

LRSI tiles are of the same construction and have the same basic characteristics as the HRSI tiles, but are cast thinner (0.2 to 1.4 inches). The 99.8-percent-pure silica LRSI tiles are manufactured in the same manner as the HRSI tiles, except that the tiles are produced in 8 by 8 inch squares and have a white optical and moisture-resistant coating applied 10 mils thick to the top and sides. The coating is made of silica compounds with shiny aluminum oxide to obtain optical properties, and provides on-orbit thermal control for the Orbiter. The LRSI tiles are treated with bulk waterproofing and are installed on the Orbiter in the same manner as the HRSI tiles. The LRSI tile has a surface emittance of 0.8 and a solar absorptance of 0.32.

Because of evidence of plasma flow on the lower wing trailing edge and elevon leading edge tiles (wing/elevon cove) at the outboard elevon tip and inboard elevon, the original LRSI tiles were replaced with FRCI-12 and HRSI-22 tiles along with gap fillers, except on *Columbia*, where only gap fillers are installed in this area.

A new high-temperature tile, known as toughened uni-piece fibrous insulation (TUFI), was developed by NASA Ames and first flown on STS-59. The initial use was on *Endeavour*'s base heat shield, between the three SSMEs. The results were encouraging enough for TUFI tiles to be installed on an attrition basis to replace 304 HRSI tiles on the base heat shield and lower body flap surface. Depending upon operational experience with this application, future use could include around the main landing gear doors and on the forward RCS module.

TUFI is the first of a new type of composite known as 'functional gradient materials' where the density of the material varies from relatively high on the outer surface and increasingly lower within the insulation. Unlike the HRSI tiles, where the tetrasilicide and borosilicate glass coating receives little support from the underlying tile, TUFI's outer surface is fully integrated into the insulation, resulting in a more damage resistant tile. TUFI tends to 'dent' instead of 'shatter' when hit. This tile material is

also known as an alumina-enhanced thermal barrier (AETB-8), or as TUFI/RCG-coated tiles.

Regardless of type, every tile is unique to a location on the Orbiter, being cut to size, thickness, and shape to conform to the contours of the vehicle and the necessary insulative properties. An identification code is painted in yellow using a commercial product called Spearex, which does not burn off during reentry.

Note that the tiles cannot withstand airframe load deformation and therefore, stress isolation is necessary between the tiles and the Orbiter structure. This is provided by strain isolation pads (SIP) made of Nomex felt in thicknesses of 0.090, 0.115, or 0.160 inch. SIP isolates the tiles from the Orbiter's structural deflections, expansions, and acoustic excitation, thereby preventing stress failure in the tiles. The SIP is bonded to the tiles, and then bonded to the Orbiter skin using a room-temperature vulcanizing (RTV) silicon adhesive.

The RTV silicon adhesive is applied to the Orbiter surface in a layer 0.008 inch thick. The very thin bond line reduces weight and minimizes the thermal expansion at temperatures of 500 degF during reentry and below -170 degF on-orbit. The tile/SIP bond is cured at room temperature under pressure applied by vacuum bags.

However, the SIP introduces stress concentrations at the needled fiber bundles. This results in localized failure in the tile just above the RTV bond line. To solve this problem, the inner surface of the tile is densified to distribute the load more uniformly. The densification process was developed from a Ludox ammonia-stabilized binder. When mixed with silica slip particles, it becomes a cement, and when mixed with water, it dries to a finished hard surface. A silica-tetraboride coloring agent is mixed with the compound for penetration identification. Several coats of the pigmented Ludox slip slurry are brush-painted on the SIP/tile bond interface and allowed to air-dry for 24 hours and then heat treated. The densification coating penetrates the tile to a depth of 0.125 inch, and the strength and stiffness of the tile and SIP system are increased by a factor of two.

Since the tiles thermally expand or contract very little compared to the Orbiter structure, it is necessary to leave gaps of 25 to 65 mils between them to prevent tile-to-tile contact. Nomex felt material insulation is required in the bottom of the gap between tiles. These 'filler bars' are supplied in thicknesses corresponding to the SIPs, cut into strips 0.75 inch wide, and bonded to the Orbiter structure. The filler bar is waterproof and temperature-resistant up to approximately 800 degF.

Nomex felt is an aramid fiber that is 2 deniers in fineness, 3 inches long, and crimped. These are loaded into a carding machine that untangles the clumps of fibers and combs them to make a tenuous mass of lengthwise-oriented, relatively parallel fibers called a web. The cross-lapped web is fed into a loom, where it is lightly needled into a batt. Generally, two such batts are placed face-to-face and needled together to form felt. The felt then is subjected to a multi-needle pass process until the desired strength is reached. The needled felt is calendered to stabilize at a thickness of 0.16 inch to 0.40 inch by passing through heated rollers at selected pressures. The calendered material is heat-set at 500 degF to thermally stabilize the felt.

Where temperatures are less than 700 degF, and aerodynamic loads are minimal, such as the upper payload bay doors and lower aft-fuselage, a FRSI blanket consisting of coated Nomex felt is used. FRSI is the same Nomex material as the SIP used under the HRSI and LRSI tiles. The FRSI

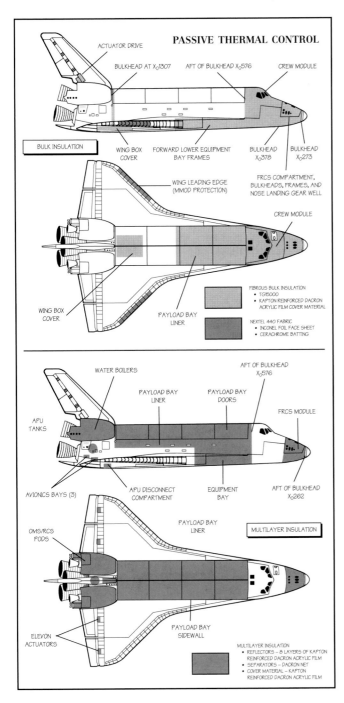

PASSIVE THERMAL CONTROL

ACTUATOR DRIVE
BULKHEAD AT X₀1307 AFT OF BULKHEAD X₀576 CREW MODULE
BULK INSULATION
WING BOX COVER FORWARD LOWER EQUIPMENT BAY FRAMES BULKHEAD X₀378 BULKHEAD X₀273

WING LEADING EDGE (MMOD PROTECTION)
FRCS COMPARTMENT, BULKHEADS, FRAMES, AND NOSE LANDING GEAR WELL
CREW MODULE
WING BOX COVER PAYLOAD BAY LINER

FIBROUS BULK INSULATION
• TG15000
• KAPTON REINFORCED DACRON ACRYLIC FILM COVER MATERIAL

NEXTEL 440 FABRIC
• INCONEL FOIL FACE SHEET
• CERACHROME BATTING

WATER BOILERS AFT OF BULKHEAD X₀576
PAYLOAD BAY LINER PAYLOAD BAY DOORS
APU TANKS FRCS MODULE
AVIONICS BAYS (3) APU DISCONNECT COMPARTMENT EQUIPMENT BAY AFT OF BULKHEAD X₀262
PAYLOAD BAY LINER
MULTILAYER INSULATION

OMS/RCS PODS
PAYLOAD BAY SIDEWALL
ELEVON ACTUATORS PAYLOAD BAY SIDEWALL

MULTILAYER INSULATION
• REFLECTORS – 8 LAYERS OF KAPTON REINFORCED DACRON ACRYLIC FILM
• SEPARATORS – DACRON NET
• COVER MATERIAL – KAPTON REINFORCED DACRON ACRYLIC FILM

After the initial delivery of *Columbia*, an advanced flexible reusable surface insulation (AFRSI) was developed. AFRSI consists of a low-density fibrous silica batting that is made up of high-purity silica and 99.8-percent amorphous silica fibers one to two mils thick. This batting is sandwiched between an outer woven silica high-temperature fabric and an inner woven glass lower temperature fabric. After the composite is sewn with silica thread, it has a quilt-like appearance. The AFRSI blankets are coated with a ceramic colloidal silica and high-purity silica fibers (referred to as C-9) that provide endurance. The AFRSI composite density is approximately eight to nine pounds per cubic foot and varies in thickness from 0.45 to 0.95 inch. The AFRSI thickness is determined by the heat load the blanket encounters during reentry. The blankets are cut to the planform shape required and bonded directly to the Orbiter by RTV silicon adhesive 0.20 inch thick. The sewn quilted fabric blanket is manufactured in 3 by 3 foot squares of the proper thickness.[88]

AFRSI blankets were used on *Discovery* and *Atlantis* to replace the majority of the LRSI tiles. The LRSI on *Columbia*'s mid-fuselage, payload bay doors, and vertical stabilizer were replaced with AFRSI blankets during the *Challenger* stand-down. *Endeavour* was delivered with an even greater use of AFRSI, and the other Orbiters (except OV-102) initially migrated towards this configuration during normal maintenance, as well as major overhauls. However, in an effort to save as much weight as possible, much of the AFRSI on the mid- and aft-fuselage, payload bay doors, and upper wing surfaces of the three ISS Orbiters is being replaced by the lighter FRSI during the OMDP cycle. During *Endeavour*'s OMDP-1, 1,472 pounds were saved by this modification. A change to the flipper doors that included FRSI saved another 520 pounds.[88A]

The HRSI, LRSI, and FRCI are manufactured by Lockheed Martin in Sunnyvale, California. Boeing North American is responsible for making the AFRSI blankets from material supplied by various other companies. As of 1999, *Atlantis'* is protected with 501 HRSI-22 tiles, 19,725 HRSI-9 tiles, 2,945 FRCI-12 tiles, 322 TUFI-8 tiles, 77 LRSI-12 tiles, 725 LRSI-9 tiles, 2,277 AFRSI blankets, and 977 FRSI blankets. Approximately 20 tiles are damaged on each flight, and about 70 are usually replaced between flights for various reasons.

During the Return-to-Flight modifications, many of the unique aspects of the thermal protection system on *Columbia* were modified to more closely resemble the newer Orbiters. Larger protective tiles were installed in the elevon leading edge and wing trailing edges to improve flight durability and decrease turnaround maintenance. *Columbia*'s payload bay doors and fuselage were originally covered with small white 'diced' tiles. During the stand-down, technicians replaced more than 2,300 of these diced tiles with thermal protective blankets. In addition, the RCC chin panel replaced about 40 tiles between the nose cap and nose landing gear doors. A significant amount of the worn interior thermal control blankets were also replaced. Nevertheless, even after her OMDP-2, *Columbia* will still be unique within the fleet.[89]

After each flight, the thermal protection system is re-waterproofed. Dimethylethoxysilane (DMES) is injected into each tile through an existing hole in the surface coating with a needleless gun, and the AFRSI blankets are injected with DMES from a needle gun. The need for re-waterproofing had been identified early in the development of the TPS, but was largely ignored for the first couple of flights. When NASA finally decided to perform the opera-

blankets generally consist of sheets 3 by 4 feet, and 0.16 to 0.40 inch thick, depending on location. A white-pigmented silicon elastomer coating is used to waterproof the felt and provide the required thermal and optical properties. The FRSI has an emittance of 0.8 and solar absorptance of 0.32 and covers nearly half of the Orbiter's upper surfaces.

The FRSI is bonded directly to the Orbiter by RTV silicon adhesive 0.20-inch thick. The very thin glue line reduces weight and minimizes the thermal expansion during temperature changes. This direct application of the blankets to the Orbiter results in weight reduction, improved producibility and durability, reduced fabrication and installation cost, and reduced installation time.

---

* This varies slightly for each vehicle, with *Columbia* having additional HRSI-22 tiles on the vertical stabilizer (to protect the SILTS pod) and additional HRSI-9 tiles on the upper wing leading edge chines, and *Endeavour* having fewer tiles due to the extensive use of AFRSI.

tion, a commercially available product (3M ScotchGuard™) was used until the definitive mixture was developed.

Thermal barriers are used in the closeout areas between various components of the Orbiter and TPS, such as the forward and aft RCS, rudder/speedbrake, nose and main landing gear doors, crew module side hatch, vent doors, External Tank umbilical doors, vertical stabilizer/aft-fuselage interface, payload bay doors, wing leading edge RCC–HRSI interface, and nose cap-HRSI interface. The various materials used are white AB312 ceramic alumina-borosilicate fibers or black-pigmented AB312 ceramic fiber cloth braided around an inner tubular spring made from Inconel-750 wire with silica fibers within the tube, alumina mat, quartz thread, and Macor machinable ceramic.

An internal passive thermal control system helps maintain the temperature of the Orbiter systems and components within their temperature limits. This system uses available Orbiter heat sources and heat sinks supplemented by insulation blankets, thermal coatings, and thermal isolation methods. Heaters are provided on components and systems in areas where passive thermal control techniques are not adequate.[90]

The internal insulation blankets are of two basic types – fibrous bulk and multi-layer. The bulk blankets are fibrous materials with a density of 2 pounds per cubic foot and a sewn cover of reinforced Kapton acrylic film . The cover material has 13,500 holes per square foot for venting. Acrylic film tape is used for cutouts, patching and reinforcements. Tufts throughout the blankets minimize billowing during venting. The multi-layer blankets are constructed of alternate layers of perforated Kapton acrylic film reflectors and Dacron net separators. There are 16 reflector layers in all, the two cover halves counting as two layers. Covers, tufting and acrylic film tape are similar to that used for the bulk blankets.[91]

Flags and letters are painted on the Orbiter with a Dow Corning-3140 silicon-base material colored by adding pigments. It is basically the same paint used to paint automobile engines and will break down in temperature ranges between 800 to 1,000 degF. Because of this, almost all markings are painted in relatively low-temperature areas of the Orbiter.

Avionics

The Space Shuttle avionics systems consists of more than 300 major electronic units located throughout the vehicle, connected by more than 300 miles of electrical wiring. There are approximately 120,400 wire segments and 6,500 connectors – the wiring and connectors weigh approximately 7,000 pounds, wiring alone weighing over 4,600 pounds. This is considerably less than the system would have weighed if extensive use of data bus technology not been employed. Total weight of the Orbiter avionics, including wiring and connectors, is over 6,000 pounds.[92]

Orbiter avionics consist of a fail-operational/fail-safe Guidance, Navigation and Control (GNC) system including three inertial measuring units (IMU); triple AIL manufactured Ku-band microwave scanning beam landing system; three Northrop rate-gyro assemblies; three Hoffman Electronics Corporation L-band TACANs; three Bendix accelerometer assemblies; two Honeywell C-band radar altimeters; four AiResearch air data transducer assemblies; three Lear Siegler attitude direction indicators; two Collins horizontal situation displays; two Sperry Mach indicators; two Bendix barometric altimeters; and two Sperry 4096 Mode-C ATC transponders.[93]

A High-Accuracy Inertial Navigation System (HAINS) began to augment the original Singer-Kearfott KT-70 IMUs during 1991 and were the primary units in service by 1994. A total of 17 flight and two non-flight units were purchased. The HAINS offered improved performance, lower failure rates, and were smaller and lighter. The HAINS units also contain a dedicated microprocessor with memory for processing and storing compensation factors from the vendor's factory calibration, thereby reducing the need for extensive initial load data from the Orbiter's computers. The HAINS is physically and functionally interchangeable with the original Singer KT-70 IMUs. During 1997 Honeywell began the development of a new Space Integrated Global Position System / Inertial Navigation System (SIGI) receiver to replace the IMU, HAINS, TACAN, and GPS systems.[94]

Two Ball Brothers star trackers are located just forward and to the left of the commander's windscreen. The

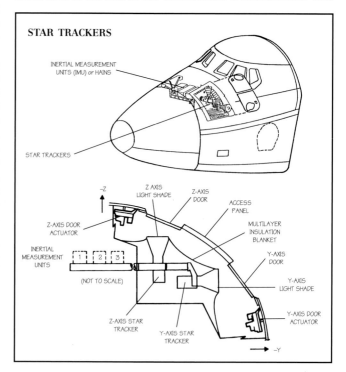

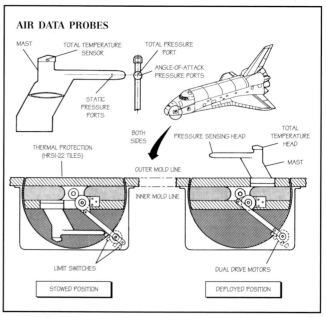

star trackers are used to align the Orbiter's IMUs, as well as to track targets and provide line-of-sight vectors for rendezvous calculations. The star tracker cannot be used if the IMU alignment error is greater than 19 degrees because the angles the star tracker is given for searching are based on current knowledge of the orbiter attitude, which is based on IMU gimbal angles. If that attitude is greatly in error, the star tracker may acquire and track the wrong star. In this case, either a special HUD mode or the crew optical alignment sight (COAS) may be used to orient the star trackers. This is done by the crew orienting the Orbiter until a known star appears in the HUD or COAS, then queuing the GPCs when the star is sighted. The COAS is an optical device with a reticule focused at infinity projected on a combining glass.[95]

Each star tracker has a door to protect it during ascent and reentry that is normally left open on-orbit to permit use of the star trackers. The GPCs contain inertial infor-

mation for 50 stars chosen for their brightness and ability to provide complete sky coverage. There is no redundancy management for the star trackers – they operate independently and either unit can accomplish the entire task. New solid-state star trackers were delivered with OV-105, and additional units have been purchased as funds have been available. The new start trackers are physically and functionally interchangeable with the originals and, in fact, the two designs may be mixed on a single flight.[96]

Two air data probes are located on the left and right sides of the Orbiter's lower forward fuselage. During ascent, on-orbit, deorbit, and initial reentry the probes are stowed inside the fuselage. At approximately Mach 5, the probes are deployed into the airstream to provide information on the flight environment. Each probe assembly is 4.87 inches high, 21.25 inches long, 4.37 inches wide, and weighs 19.2 pounds. Each air data probe has four pressure-port sensors and two temperature sensors. These measure static pressure, total pressure, angle-of-attack upper pressure, and angle-of-attack lower pressure. The extension of the air data probes must be initiated manually by the flight crew, and is one reason the Orbiter can not land autonomously.

Communications

Direct communication for NASA missions takes place through Space Flight Tracking and Data Network (STDN) ground stations. For military missions, Air Force Satellite Control Network (AFSCN) remote tracking station sites – also known as space-ground link system (SGLS) ground stations – are used. Signals from the ground to the Orbiter are referred to as uplinks (or forward links), and signals from the Orbiter to the ground are downlinks (or return links).[97]

Orbiter communications equipment includes two S-band phase modulation (PM) transceivers, two S-band frequency modulation (FM) transmitters, one or two Ku-band rendezvous radar/satellite communicators on the starboard side of the payload bay, and two P-band UHF radios for EVA and ATC communications.

The S-band PM system provides two-way communication between the Orbiter and the ground, either directly or through a relay satellite (generally TDRS). S-band is a 'line-of-sight' system, meaning that the receiver must be visible from the transmitter. With the availability of the TDRS satellite system, over 80 percent of any given orbit is covered. The remaining 20 percent (called a 'zone of exclusion') is generally located over the Indian Ocean where TDRS coverage is limited. Certain DoD satellites may also provide relay services under some circumstances.[98]

The S-band PM uplink is phase modulated on a center carrier frequency of either 2,106.4 MHz or 2,041.9 MHz for NASA missions and operates through TDRS. The two frequencies are provided to prevent interference in the event two Orbiters are in operation at the same time. For DoD missions the forward link carrier frequencies are 1,831.8 MHz or 1,775.5 MHz through the AFSCN. The uplink transfers data at a rate of 72 kbs, consisting of two 32 kbs voice channels and an 8 kbs command channel.

NASA and DoD both use an S-band PM downlink center carrier frequency of 2,287.5 MHz or 2,217.5 MHz. The downlink operates at 192 kbs consisting of two 32 kbs voice channels and a 128 kbs data channel. Alternately, a low data rate forward link of 32 kbs (one 24 kbs voice and one 8 kbs command channel) and a low data rate downlink of 96 kbs (one 32 kbs voice and one 64 kbs data channel), are available. Two four-quadrant S-band PM

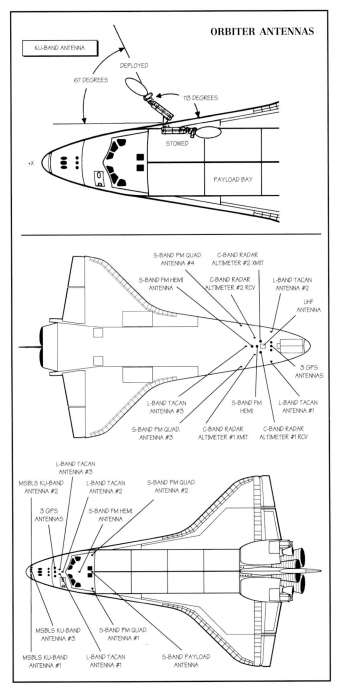

ORBITER ANTENNAS

KU-BAND ANTENNA

DEPLOYED

67 DEGREES

113 DEGREES

STOWED

+X

PAYLOAD BAY

S-BAND PM QUAD. ANTENNA #4

C-BAND RADAR ALTIMETER #2 XMIT

S-BAND FM HEMI ANTENNA

C-BAND RADAR ALTIMETER #2 RCV

L-BAND TACAN ANTENNA #2

UHF ANTENNA

3 GPS ANTENNAS

L-BAND TACAN ANTENNA #3

S-BAND FM HEMI

L-BAND TACAN ANTENNA #1

S-BAND PM QUAD. ANTENNA #3

C-BAND RADAR ALTIMETER #1 XMIT

C-BAND RADAR ALTIMETER #1 RCV

L-BAND TACAN ANTENNA #3

MSBLS KU-BAND ANTENNA #2

L-BAND TACAN ANTENNA #2

S-BAND PM QUAD. ANTENNA #2

3 GPS ANTENNAS

S-BAND FM HEMI ANTENNA

MSBLS KU-BAND ANTENNA #3

S-BAND PM QUAD. ANTENNA #1

MSBLS KU-BAND ANTENNA #1

L-BAND TACAN ANTENNA #1

S-BAND PAYLOAD ANTENNA

antennas are located on the forward fuselage outer skin just aft of the cockpit windows and just forward of the wing leading edge, and are covered by RSI tiles.

The S-band frequency modulation (FM) system cannot receive information, and is used to downlink data from eight different sources (the three SSME EIUs, the video switching unit, the operations recorder for recorder dumps, the payload recorder for recorder dumps, payload analog, or payload digital), one at a time. The S-band FM return link can originate from two S-band transmitters aboard the Orbiter, both tuned to 2,250.0 MHz. The S-band FM return link can be transmitted simultaneously with the S-band PM return link. This link was used primarily during the dedicated DoD missions, and had a maximum data rate of 1,024 kbs. The Orbiters are equipped with communications security (COMSEC) equipment that provide a capability to encrypt and decrypt data. This capability was used on classified DoD flights, and is also used selectively on some NASA flights.[99]

The Ku-band antenna aboard the Orbiter is located in the payload bay, so the Ku-band system can only be used when the Orbiter is on-orbit and the payload bay doors are open. The Ku-band carrier frequencies are 13,775 GHz from the TDRS to the Orbiter and 15,003 GHz from the Orbiter to the TDRS. The Ku-band system transmits at a much higher data rate than the S-band PM system, with a downlink of up to 50 Mbs and an uplink of up to 216 kbs. The Ku-band system can handle more data than the S-band systems, and transmits three channels of data, one of which is the same interleaved voice and telemetry processed by the S-band PM system. The Ku-band antenna assembly is 7 feet long and 1 foot wide when stowed in the payload bay, and weighs 304 pounds. When deployed the antenna is 3 feet in diameter. If the antenna cannot be stowed after operations, there is a capability to jettison it. The antenna dish is edge-mounted on a two-axis gimbal. The alpha gimbal provides a 360-degree roll movement around the pole or axis of the gimbal. The beta gimbal provides a 162-degree pitch movement around its axis. The alpha gimbal has a stop at the lower part of its movement to prevent wraparound of the beta gimbal control cable. Since the beta gimbal has only a 162-degree movement, there is a 4-degree non-coverage zone outboard around the pole and a 32-degree non-coverage zone toward the payload bay. At times the Ku-band system, in view of a TDRS, is interrupted because the Orbiter blocks the Ku-band antenna's view to the TDRS, because of Orbiter attitude requirements, or because payload radiation sensitivities prohibit its use. Currently, the Ku-band system is one of the most maintenance intensive systems aboard the Orbiter but a replacement is elusive.[100]

The Hughes Ku-band system can also be used as a rendezvous radar to track nearby objects. Radar search uses a wide spiral scan of up to 60 degrees. Objects may be detected by reflecting the radar beam off the surface of the target (skin track or passive mode), or by using the radar to trigger a transponder beacon on the target (active mode). Angle tracking can be accomplished automatically or manually, but range tracking is always automatic. The on-board GPCs are capable of fully automatic target acquisition and rendezvous. During a rendezvous operation, the radar system is used as a sensor that provides target angle, angle rate, and range rate information for updating the rendezvous navigation data in the GNC computer. The operation is similar to using the COAS or star trackers, except that the radar provides range data in addition to angle data. The Ku-band system can not be used for com-munications and radar operations simultaneously.

The ultra-high frequency (UHF) system is used as a backup for the S-band and Ku-band voice communications, primarily during EVAs. For communications with the STDN ground stations, the UHF system operates in a simplex mode, which means that the Orbiter can transmit or receive, but cannot do both simultaneously. Two frequencies are available for transmitting to the ground, 296.8 MHz and 259.7 MHz, and these plus 243.0 MHz (guard) are available for receiving. Additional frequencies are available for EVA activities and other local activities.

The payload communication system is used to transfer information between the Orbiter and its payload or payloads. It supports hardline and radio frequency communications with a variety of payloads. The system is used to activate, check out, and deactivate attached and detached payloads. Communication with an attached payload takes place through the payload patch panel at the flight deck aft station, which is connected to payloads in the payload bay. All command and telemetry signals that meet the payload communication system specifications can be processed onboard. Incompatible signals can be sent through the Ku-band system directly to payload ground stations. This method of transmission, referred to as bent-pipe telemetry, means that no onboard signal processing occurs before the telemetry is sent to the Ku-band system. The S-band payload antenna is located on the top of the outer skin of the Orbiter's forward fuselage, just aft of the S-band FM upper hemispherical antenna and is covered with reusable thermal protection system. This antenna is used as the radiating element for S-band transmission and reception to and from the Orbiter to detached payloads through the forward and return links.[101]

Three independent Ku-band Microwave Scanning Beam Landing Systems* (MSBLS) are carried onboard the Orbiter. MSBLS is used during the terminal area energy management and the approach and landing flight phases, and also during a return-to-launch-site abort. The system is fully capable of landing an Orbiter without human+ intervention, although this capability is not currently implemented, and there are no plans to do so in the future. The systems were provided by Eaton Corporation, AIL Division of Farmingdale, New York.

Triple-redundant Gould TACAN (tactical air navigation) units were initially installed on each Orbiter. These were replaced by improved units manufactured by Rockwell Collins during normal maintenance. The new units are based on a commercial TACAN, and are significantly more reliable than the original units.

The use of the Global Position System (GPS – NavStar) on the Orbiters has been discussed since 1982. *Endeavour* was delivered with wiring for a possible GPS installation, but the equipment was never integrated into the basic GPC flight control system. The use of GPS took on a new urgency in the early 1990s when the Air Force announced plans to decommission the TACAN system used by the Orbiter (and many aircraft) prior to the year 2000. A serious effort to fit GPS to the Orbiter fleet was begun during 1995, and all of the Orbiters were equipped with a prototype GPS units as part of the OI-26B development test objectives (DTO). This resulted in each vehicle being

---

* NASA has begun referring to these as a Microwave Landing System (MLS), but the technology is slightly different than the commercial MLS installations operated by the FAA and are not interoperable.

+ Although MSBLS is capable of controlling the Orbiter all the way through landing, as currently configured the GPCs can not lower the landing gear or deploy the air data probes – these steps still require human intervention.

equipped with a single-string GPS and three TACANs. The OI-27 software release (first flown on STS-96) added the necessary software to support an all-TACAN configuration, a single-string GPS with a three-string TACAN, or an all-GPS configuration.[102]

Atlantis received the wiring and control panels for a three-string GPS system during her first OMDP (November 1997–September 1998), with the GPS receivers scheduled to be installed at KSC afterwards. But before the receivers were installed a decision was made to delay the implementation of the three-string GPS system for two primary reasons – the results of early testing of the single-string system using OI-26B were not terribly encouraging, and the Air Force decided to delay the final decommissioning of TACAN, eliminating the urgency for the new system.

The reason the test results were not encouraging was that GPS was never intended to be used by an orbital client. The primary difficulty were that the relative velocity between the NAVSTAR satellites and the Orbiter were much greater than any existing GPS receiver was able to compute accurately, mostly because the Orbiter is traveling at 25,000 mph, significantly faster than anything else GPS is used by. Another problem was that no existing receiver was able to deal with having up to 12 satellites in view – something that never happens on the ground. There have also been problems with large velocity noise, probably caused by ionospheric interference.

Each GPS receiver is equipped with two antennas, one on the lower forward fuselage and one on the upper forward fuselage. All are covered by thermal protection system tiles. The convection-cooled GPS receivers are located in the Orbiter forward avionics bays. Each receiver is 6.78 inches high, 3.21 inches wide, 12.0 inches long, and weighs 12.3 pounds. It should be noted that there are four slots available for navigation devices – currently, all Orbiters carry one GPS and three TACAN. In addition, the control panels on the flight deck are tailored for either GPS or TACAN, depending which is installed.[103]

For the time being the Orbiters will retain a single-string GPS and three-string TACAN, although efforts to incorporate the software for GPS continues, with OI-28 (scheduled for STS-98 offering full support for both GPS and TACAN, depending upon what hardware is installed.

Although the GPS system will be fully integrated into the guidance, navigation, and control (GNC) function of PASS and BFS, it will not be used for landings – the MSBLS will continue to be used for the foreseeable future.

The position and velocity of the Orbiter must be uniquely defined in terms of an inertial coordinate system – a reference frame that is neither rotating nor accelerating. Such a system ensures that the normal equations of motion are valid. The inertial coordinate system used by the onboard navigation system is the Aries Mean of 1950 Cartesian method, generally called the M50 system. The M50 system is used to establish and maintain the inertial position and velocity of the Orbiter during all flight phases. The X axis points toward the mean vernal equinox (the apparent point on the celestial sphere where the Sun crosses the Earth's equator on its northward journey) of the year 1950. The Z axis points along Earth's mean rotational axis of 1950 with the positive direction toward the North Pole. The Y axis completes the right-handed system, with the corresponding X–Y plane lying in the plane of the equator. Note that this is an inertial system and, although its origin is at the Earth's center, the M50 system is completely independent of the Earth's rotation.[104]

However, some computations are considerably simpler if performed in other coordinate systems. For example, when landing the Orbiter, the position of the vehicle relative to the runway is more meaningful than its position in a coordinate system fixed in space. Therefore, the onboard navigation system uses a number of different coordinate systems to simplify the various inputs, outputs, and computations required.

Data Processing System

Orbiter operational instrumentation (OI) is used to collect, route, and process information from transducers and sensors located throughout the Orbiter and payloads. This system also interfaces with the ET and SRBs when they are attached to the Orbiter. Over 2,000 data points are monitored, and the data are routed through 20 MDMs (multiplexer/demultiplexers). The instrumentation system consists of transducers, signal conditioners, two non-redundant pulse code modulation master units (PCMMU), transmission encoding equipment, two operational recorders, one payload recorder, and on-board checkout equipment.

The Data Processing System (DPS) hardware consists of five general purpose computers (GPC) for computation and control, two 134-megabit magnetic tape mass memory units (MMU) for large-volume bulk storage, a modified Mil-Std-1553 digital data bus, 20 Orbiter and 4 SRB MDMs, 3 SSME engine interface units (EIU), 4 display system processors (or the MEDS equivalent), 2 data bus isolation amplifiers to interface with the ground-based Launch Processing System (LPS), 2 master events controllers, and a master timing unit (MTU). Early in the program a sixth GPC was generally carried aboard each Orbiter (in avionics bay 1 after the AP-101Ss were installed), but was not connected to the data busses. This GPC could be manually installed if one of the other GPCs fails during flight. This is not always the case any longer.[105]

GPCs 1 and 4 are located in forward mid-deck avionics bay 1, GPCs 2 and 5 are located in forward mid-deck avionics bay 2, and GPC 3 is located in aft mid-deck avionics bay 3. The GPCs receive forced-air cooling from an avionics bay fan. If both fans in an avionics bay fail, the computers will overheat within 25 minutes (at 14.7 psi cabin pressure or 17 minutes at 10.2 psi) after which their operation cannot be relied upon. An operating GPC may or may not survive for up to an additional 30 minutes beyond the certifiable thermal limits.

The IBM 4Pi/AP-101S incorporated semiconductor memory in place of the AP-101B's ferrite-core memory. The CMOS memory has built-in error detection and correction (ECC) and is supplied with a battery backup in case of a momentary power loss. (IBM/FSD)

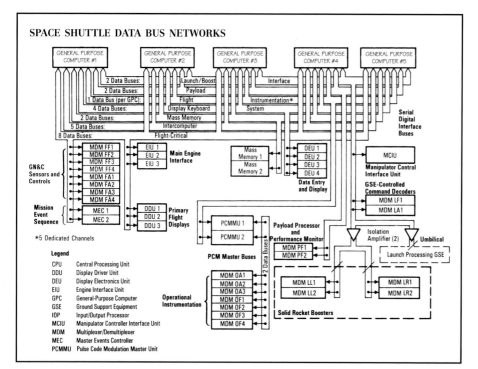

SPACE SHUTTLE DATA BUS NETWORKS

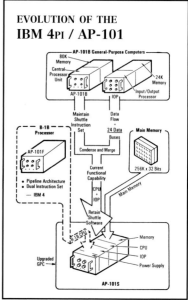

EVOLUTION OF THE
IBM 4PI / AP-101

The new AP-101S computers are hybrids using the Central Processing Unit (CPU) from the Air Force AP-101F, but optimized to run the existing PASS/BSS software without modification. (NASA)

Each of the five original IBM 4Pi/AP-101B general purpose computers was composed of two separate units, a central processor unit (CPU) and an input/output processor (IOP). Each GPC was equipped with 106,496 32-bit words (81,920 words in the CPU and 24,576 in the IOP) of main memory in random-access, nonvolatile, destructive-readout ferrite core. Ferrite core memory was originally selected for the GPCs because little was known about the effects of radiation on the high-density solid-state RAM that is now used in almost all digital computer hardware. The fact that the AP-101B used non-volatile memory was put to use after the *Challenger* accident; once the GPCs were recovered from the ocean floor, NASA was able to recover a significant percentage of the memory contents, helping to confirm the events that transpired during the accident.

Each AP-101B CPU was 7.62 inches high, 10.2 inches wide, 19.55 inches long, and weighed 57 pounds. The IOPs were the same size and weight as the CPUs. Thirty-eight AP-101Bs were built by IBM, and 24[*] of those remained flight worthy when they were retired.

The original AP-101Bs began to be replaced by updated AP-101S units during 1991, and the AP-101S first flew on STS-37 in April 1991. A total of 1 prototype, 1 qualification, 8 pre-production, and 26 flight-rated AP-101S computers were procured from IBM by Rockwell, and the new computers were installed on all Orbiters by late 1993. The upgraded computers are derived from the AP-101F developed for the Air Force B-1B bomber, and have allowed NASA to incorporate more capabilities into the Orbiters, and apply advanced computer technologies that were not available when the Orbiter was first designed.

Development of the new computer began in January 1984, twelve years after the original computers were designed. The new GPCs take the basic CPU designed for the AP-101F, and mate it with the unique data bus interfaces originally developed for the Orbiter, along with a newly developed 256k by 32-bit hardened solid-state memory module. New machine-level software was written for the CPU that enables it to run the same instruction-set as the AP-101B, allowing NASA to 'port' the Shuttle software.[106]

The upgraded GPCs provide almost two-and-a-half times the original memory capacity, and three times the processor speed (1,200,000 instructions per second, as opposed to the AP-101B's 400,000), with minimal impact to flight software. The new computers are half the size (the CPU and IOP are one unit), weigh approximately half as much (64 pounds), and require less power to operate. The AP-101S uses 262,144 32-bit words of solid state CMOS memory with battery backup. Because the CMOS memory can be altered by the level of radiation which the Shuttle encounters in orbit, the memory was designed with error detection and correction (EDAC) logic. Whenever the GPC is powered on, a hardware process continually checks the GPC's memory for alterations. This process is capable of detecting and correcting all single-bit errors; multi-bit errors can be detected, but not corrected. All of the GPC's memory is checked every 1.7 seconds, and each memory location is also checked prior to being executed.[107]

The AP-101S has a mean-time-between-failure (MTBF) of approximately 6,000 hours, with growth potential to 10,000 hours. The older GPCs averaged 5,200 hours between failures, more than five times better than the 1,000 hours called for in the original contract.

During flight critical flight phases, four of the computers are assigned to perform guidance, navigation and control (GNC) tasks acting as a cooperative, redundant set. The computers run in sync with each other and the computations of each computer are verified by the others. In this way, the computer complex is able to achieve a very high system reliability, as well as a 'fault tolerant' operational capability that lets the system endure component or connection failure in one or more of the computers without a loss of operational capability. The Primary Avionics Software System (PASS) for the GPCs, designed and developed by the Federal Systems Division of IBM (now part of

---

[*] Of the other 14, six were aboard OV-099 (s/n 0009, 0011, 0012, 0019, 0026, and 0027), two had non-isolated failures that could not be repaired (0008 and 0033), two were dedicated to tests (0001 and 0013), and four others (0003, 0004, 0016, and 0030) had 'hard' failures, but could have been repaired and returned to flight status if required.

Lockheed Martin/United Space Alliance), remains the most sophisticated and complex set of programs ever written for aerospace use (although, at only 500,000 words, far from the largest). The software is written in a Shuttle-unique language known as HAL/S developed by Intermetrics, Incorporated, of Cambridge, Massachusetts. During non-critical flight periods on-orbit, one computer is used for GNC and control tasks, and another for system management. The remaining three can be used either for payload management, or deactivated as standby replacements.

The PASS – also referred to as primary flight software – contains all the programming needed to fly the vehicle through all phases of the mission and manage all vehicle and payload systems. Since the ascent and reentry phases of flight are so critical, four of the five GPCs are loaded with the same PASS software and perform all GNC functions simultaneously and redundantly.

The DPS software is divided into two major groups – system software and applications software – that are combined to form a memory configuration for a specific mission phase. System software is the operating system software that controls the interfaces among the computers and the rest of the DPS. It is loaded into the computer when it is first initialized (called IPL – initial program load), always resides in the GPC main memory, and is common to all memory configurations. The system software controls GPC input and output, loads new memory configurations, keeps time, and consists of three sets of programs. The flight computer operating system (FCOS) (the executive) controls the processors, monitors key system parameters, allocates computer resources, provides orderly program interrupts for higher priority activities, and updates computer memory. The user interface processes flight crew commands or requests and displays messages and data to the crew. The system control program initializes each GPC and arranges for multi-GPC operation during flight-critical phases. One of the system software functions is to manage the GPC input and output operations that includes assigning computers as commanders and listeners on the data buses and exercising the logic involved in sending commands to these data buses at specified rates and upon request from the applications software.[108]

The applications software performs the functions required to fly and operate the vehicle. To conserve main memory, the applications software is divided into three major functions:

- Guidance, navigation, and control (GNC): specific software required for launch, ascent to orbit, maneuvering on-orbit, reentry, and landing. This is the only major function where redundant set synchronization can occur.
- Systems management (SM): tasks that monitor various Orbiter systems, such as life support, thermal control, communications, and payload operations. SM is a simplex major function – only one GPC at a time can actively process an SM memory configuration.
- Payload (PL): this major function currently contains mass memory utility software. The PL major function is usually unsupported in flight, which means that none of the GPCs are loaded with PL software. It is only used in vehicle preparation at KSC, and is also a simplex major function. Software to support payload operations is included as part of the SM GPC memory configuration.

Major functions are divided into mission phase oriented blocks called operational sequences (OPS). Each OPS is associated with a particular memory configuration that must be loaded separately into a GPC from the MMUs. Therefore, all the software residing in a GPC at any given time consists of system software and an OPS major function; i.e., one memory configuration. Except for memory configuration 1, each memory configuration contains one OPS. Memory configuration 1 is loaded for GNC at launch and contains both OPS 1 (ascent) and OPS 6 (RTLS), since there would be no time to load in new software for a return to launch site (RTLS) abort.

During the transition from one OPS to another – called an OPS transition – the flight crew requests a new set of applications software to be loaded in from the MMU. Every OPS transition is initiated by the flight crew. When an OPS transition is requested, the redundant OPS overlay contains all major modes of that sequence. Major modes (MM) are further subdivisions of an OPS that relate to specific portions of a mission phase. As part of one memory configuration, all major modes of a particular OPS are resident in GPC main memory at the same time. The transition from one major mode to another may be automatic (e.g., in GNC OPS 1 from pre-count MM 101 to first stage MM 102 at lift-off) or manual (e.g., in SM OPS 2 from on-orbit MM 201 to payload bay door MM 202 and back). Each major mode has an associated CRT display, called a major mode display or OPS display, that provides the crew with information concerning the current portion of the mission phase and allows crew interaction. There are three levels of CRT displays. Certain portions of each OPS display can be manipulated by flight crew keyboard input (or ground link) to view and modify system parameters and enter data. The specialist function (SPEC) of the OPS software is a block of displays associated with one or more operational sequences and enabled by the flight crew to monitor and modify system parameters through keyboard entries. The display function (DISP) of the OPS software is a group of displays associated with one or more OPS. These displays are for parameter monitoring only (no modification capability) and are called from the keyboard.

To guard against the possibility of a software failure ('bug') affecting all four primary computers, the fifth computer is loaded with Backup Flight System (BFS) written by Intermetrics, Rockwell International in Downey, California, and the Charles Stark Draper Laboratory (CSDL) in Cambridge, Massachusetts. The BFS is capable of all necessary ascent and reentry operations, and maintaining vehicle control on-orbit, but can not support an operational mission. During orbital operations, the fifth computer is usually assigned to system management functions. The BFS can be engaged at any time the fifth GPC is running by depressing a button located on the commander's or pilot's rotational hand controller.

In order to be prepared to take over the vehicle at any time, the BFS also performs the SM functions during ascent and reentry. BFS software is always loaded into GPC #5 before flight, but any of the five GPCs could run the BFS if necessary. Since the BFS is intended to be used only in a contingency, its programming is much simpler than that of the PASS. Only the software necessary to complete ascent or reentry safely, maintain vehicle control on-orbit, and perform SM functions during ascent and reentry is included. Thus, all the software used by the BFS can fit into one GPC and never needs to access mass memory. For added protection, however, the BFS software is loaded into the MMUs in case of a BFS GPC failure and the need to IPL a new BFS GPC.[109]

The GPC complex requires a stable, accurate time source because its software uses Greenwich Mean Time (GMT) to schedule processing. Each GPC uses the master timing unit (MTU) to update its internal clock. The MTU provides precise frequency outputs for various timing and synchronization purposes to the GPC complex and many other Orbiter subsystems. Its three time accumulators provide GMT and mission elapsed time (MET), which can be updated by external control. The accumulator's timing is in days, hours, minutes, seconds, and milliseconds up to one year. The MTU is a stable, crystal-controlled frequency source that uses two oscillators for redundancy. The MTU is located in crew compartment mid-deck avionics rack 3B and is cooled by a water coolant loop cold plate.[110]

The portable* general support computer (PGSC) is a Pentium-based laptop portable computer used either as a standalone computer or as a terminal device for communicating with other electronic systems. Other Intel and Macintosh computers have been used on earlier flights. PGSCs are used in the mid-deck or flight deck to interface with flight-specific experiments that may be located in the cabin or payload bay. The PGSCs are used to monitor experiment data and/or issue commands to payloads or experiments in the payload bay. An Orbiter communications adapter allows the PGSC to interface to the S-band PM or Ku-band communications system. The OCA essentially acts as a modem operating up to 4 Mbs, enabling the crew to transfer files and receive email from the ground.[111]

Landing Gear

The Orbiter landing gear is arranged in a conventional tricycle configuration consisting of a nose landing gear, and left and right main gear. Each landing gear includes a shock strut and two wheel and tire assemblies. The nose landing gear is located in the lower forward fuselage, and the main landing gear are located in the lower left and right wing area adjacent to the mid-fuselage. When the nose landing gear is retracted, two doors cover the landing gear well. The main gear each have a single door that enclose the gear well.[112]

The left and right nose landing gear doors are attached by hinge fittings in the nose section. The doors are constructed of aluminum alloy honeycomb, and although the doors are the same length, the left door is slightly wider than the right. Each door has an up-latch fitting at the forward and aft ends to lock the door closed when the gear is retracted, and each has a pressure seal in addition to a thermal barrier. Lead ballast in the nose wheel well and on the X_0293 bulkhead provides weight and center-of-gravity control. The nose wheel well will accommodate 1,350 pounds of ballast, and the X_0293 bulkhead will accommodate a maximum of 1,971 pounds.[113]

The landing gear is constructed of stress and corrosion resistant high strength steel and aluminum alloys, stainless steel, and aluminum-bronze. Urethane paint and cadmium-titanium plating is applied to all exposed steel surfaces, and conventional anodizing plus urethane paint is applied on exposed aluminum surfaces.[114]

The nose and main landing gear each contain a shock strut which is the primary source of shock attenuation during Orbiter landing. The shock struts are conventional pneudraulic (gaseous nitrogen/hydraulic fluid) shock absorbers. A departure from convention is the separation of the gaseous nitrogen from the hydraulic fluid by a floating diaphragm to preclude the possibility of the nitrogen dispersing throughout the hydraulic fluid during zero-g

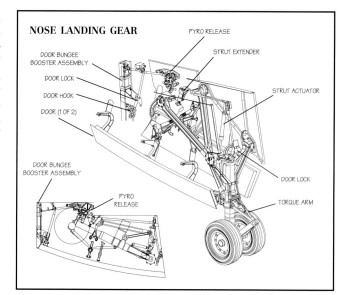

NOSE LANDING GEAR

DOOR BUNGEE BOOSTER ASSEMBLY
DOOR LOCK
DOOR HOOK
DOOR (1 OF 2)
DOOR BUNGEE BOOSTER ASSEMBLY
PYRO RELEASE
PYRO RELEASE
STRUT EXTENDER
STRUT ACTUATOR
DOOR LOCK
TORQUE ARM

This is the nose landing gear of Atlantis. Each nose gear tire may be used twice if it passes a post-flight inspection. The ET attachment point may be seen immediately ahead of the nose gear. (NASA photo 00-SPN-043)

---

* This was originally known as the payload and general support computer since it was first proposed by the payload community. After it began being used for other things, its name was changed to more accurately identify it.

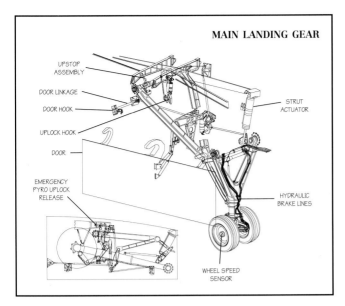

MAIN LANDING GEAR

UPSTOP ASSEMBLY

DOOR LINKAGE

DOOR HOOK

UPLOCK HOOK

DOOR

EMERGENCY PYRO UPLOCK RELEASE

STRUT ACTUATOR

HYDRAULIC BRAKE LINES

WHEEL SPEED SENSOR

The right main gear on Discovery while at Dryden after STS-92. Note the highly reflective interior surface on the gear door, and the latches along its lower (inboard) edge. The main gear tires are only used once. (Tony Landis)

---

* There was nothing wrong with the Goodrich tires – they were simply designed to an early specification. As the Orbiter weight grew and more data was attained on the landing stresses, new tires were competitively purchased, and Michelin won.

operations. This separation is required to maintain proper shock strut performance at touchdown.

All of the landing gear retract forward. During retraction, each gear is hydraulically rotated forward and up by ground support equipment until it engages an uplock hook for each gear in its respective wheel well. The uplock hooks onto a roller on each strut. A mechanical linkage driven by each landing gear mechanically closes each landing gear door. The gear may not be retracted in flight.[115]

When deployment of the landing gear is commanded by the crew, the uplock hook for each gear is unlocked by hydraulic pressure. Once the hook is released from the roller on the strut, the gear is driven down and aft by springs, hydraulic actuators, aerodynamic forces, and gravity. The doors free-fall open when unlocked, sometimes aided by the wheels and tires pushing on them. The landing gear reach the full-down and extended position within 10 seconds and are locked in the down position by spring-loaded downlock bungees. If hydraulic pressure is not available to release the uplock hook, a pyrotechnic initiator at each landing gear uplock hook automatically releases the uplock hook on each gear one second after the crew has commanded gear down. The landing gear is deployed at 300 ± 100 feet (AGL) altitude and at a maximum of 312 knots equivalent air-speed (KEAS).

The nose and main landing gear strut actuators assist in the landing gear deployment through hydraulic pressure. The strut actuators include an oil snubber to control the rate of extension of the gear free fall, and thereby prevent damage to the landing gear and downlink linkage. The strut actuators also serve as the landing gear retract actuators, although this can not take place in flight. The nose landing gear also contains a pyro boost system to further assure nose gear door and gear extension in case high aerodynamic forces on the nose gear door are present. This pyro system is fired each time the landing gear is deployed.

The Michelin nose gear tires are 32 by 8.8 with a rated static load of 45,000 pounds (this was only 22,300 pounds early in the program with the B.F. Goodrich* tires) and an inflation pressure of 350 psi (300 psi early in the program). Gaseous nitrogen is used to inflate the tires, and the nose gear tire life is nominally five landings, although in practice they are only reused once. The nose landing gear shock strut has a 22 inch stroke and the maximum allowable vertical sink rate is approximately 11.5 fps. The maximum allowable landing speed is 217 knots (250 mph) and the maximum allowable derotation rate is approximately 9.9 degrees per second.[116]

The nose landing gear is electro-hydraulically steerable. *Columbia* and *Challenger* were delivered with a nose wheel steering system that proved to be ineffectual at controlling the Orbiter during rapid transient maneuvers at high speeds. While a solution was investigated, the nose wheel steering system was deactivated on OV-102 following STS-4 and never activated on OV-099. *Discovery* and *Atlantis* were delivered with the plumbing, wiring, and fittings for a nose wheel steering system, but none was actually installed pending the design of a workable system.

An improved nose wheel steering system was incorporated on *Columbia* in time for STS-32/61-C, and the system has since been fitted to *Discovery* and *Atlantis*. *Challenger* had not been modified prior to her loss, and *Endeavour* was delivered with the new system already installed. The modification allows a safe high-speed engagement of the nose wheel steering system, and provides positive lateral directional control of the Orbiter during roll-out in the presence of high crosswinds and blown tires. Nose wheel steering is accomplished either through the on-board GPCs, or through the rudder pedals.[117]

Each of the main landing gear wheels has a brake assembly equipped with an anti-skid system. The Michelin main gear tires are 44.5 by 16 with an inflation pressure of 370 psi. (315 psi early in the program with the Goodrich tires) using gaseous nitrogen. The main gear tires are currently rated at 225 knots (260 mph), and each tire is only used once. Each tire is inspected by X-ray at KSC to ensure its structural integrity prior to use.[118]

The maximum allowable load per main landing gear tire is 132,000 pounds (123,000 pounds early in the program with Goodrich tires). If the Orbiter touches down with a 60/40 percent load distribution on a strut's two tires, with one tire supporting the maximum load, then the other tire can support a load of only 82,410 pounds. Therefore, the maximum load on a strut is 220,000 pounds (205,410 pounds early in the program). The shock strut stroke is 16 inches. The maximum allowable main gear sink rate for a 211,000-pound Orbiter is 9.6 fps; for a 240,000-pound Orbiter it is 6.0 fps. With a 20 knot crosswind, the maximum allowable sink rate for a 211,000-pound Orbiter is 6.0 fps; for a 240,000-pound Orbiter, it is approximately 5.0 fps. (Note that the current allowable cross-wind limit is 15 knots.)

An on-going program to improve the landing gear of the Orbiters has resulted in the incorporation of thicker main axles to provide a stiffer configuration that reduces brake-to-axle deflections and minimizes potential brake damage. This change has significantly improved the Orbiter's resistance to landing gear damage. Strain gauges have also been added to each nose and main wheel to monitor tire pressure before launch, on-orbit, reentry, and landing.

Brakes

There are four brake assemblies, one for each main gear wheel. The brakes installed on the Orbiters for the first 25 flights were designed for the original (1973) predicted weight of the Orbiter, and since the 'as-built' Orbiters are somewhat heavier, the brakes had little or no margin. Three early missions, STS-5, STS-23/51-D, and STS-32/61-C experienced severe thermal stator damage, and all 24 missions prior to the *Challenger* accident experienced some brake damage.[119]

Each of the original B. F. Goodrich brake assemblies used four beryllium rotors and three carbon lined stators. The brakes were designed to absorb 36.5 million foot-pounds of energy for normal stops, or one emergency stop of 55.5 million foot-pounds. Each brake was fitted with a Hydro-Aire anti-skid system. Due to concerns of extreme brake damage, the maximum speed at which the original brakes could be applied was limited to 205 mph.

Because of damage sustained during early landings, the program restricted landing weights and dictated that all landings would occur at Edwards AFB, where there was less stress on the brakes due to the long lakebed runways. This caused an unacceptable impact on the program, and during the *Challenger* stand-down, improved brakes were fitted to OV-103 and OV-104. The revised brakes were essentially the same as the original design, except that the carbon-lined beryllium stator discs in each main landing gear brake were replaced with thicker discs to increase the braking energy available (up to 65 million foot-pounds). The 205 mph speed limit was carried-over to the improved beryllium design.

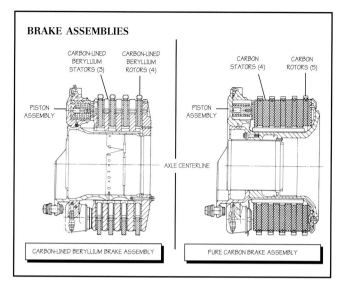

BRAKE ASSEMBLIES

A long-term structural carbon brake program was also initiated to replace the carbon-lined beryllium stator discs with an all-carbon configuration to provide higher braking capacity by increasing maximum energy absorption. These brakes were developed by B. F. Goodrich and NASA, with significant assistance by the Air Force.

The pure carbon brakes are similar in many respects to the older carbon-lined beryllium brakes, but have design features that significantly increase their performance and life-expectancy. The carbon brakes have a maximum operating temperature of 2,100 degF, compared to 1,750 degF for the older design, and their nominal temperature operating range is 1,200 degF. Under normal operations, the carbon brakes are expected to be used for 20 missions, compared to a single mission for the older brakes. The brakes weigh 180 pounds per assembly, approximately 30 pounds more than the original design.

Additionally, the carbon brakes can be applied at landing speeds as high as 260 mph, and actually slow the Orbiter at a rate much greater than the older beryllium brakes. This is particularly important for abort landings, since many of the airfields used as contingency sights have comparatively short runways, and the Orbiter will land heavier than usual during an abort situation.

Similar in size and design to the improved beryllium brakes, the carbon brakes have five rotors (instead of four),

One of the reasons that Orbiters were not allowed to land at KSC after early missions is evident here. Severe tire wear, caused partially by an incorrect surface grooving of the shuttle landing facility runway, and partially by braking problems, occurred on several flights. This is Discovery after STS-23/51-D. (NASA photo S23-9110)

The shuttle 'test' wheel and tire was installed between the Convair's main landing gear. A very fuzzy digital photo (below) shows a tremendous trail of sparks and fire as the test wheel is driven into the concrete runway at Edwards, eventually destroying the tire and wheel. (NASA photos EC93-41018-17 and ED95-43234-1)

LANDING SYSTEM RESEARCH AIRCRAFT

As the empty weight of the Orbiter kept increasing, the landing gear and brakes were becoming marginal to handle the extra load. A series of improvements were initiated to take care of immediate concerns, but a long-range plan was also needed. The shortcomings of the original design also pointed out a need for better research and development tools to assist designers of future large high-speed aircraft.

To test any new Shuttle landing gear systems, NASA investigated the concept of a Landing System Research Aircraft (LSRA) during 1985. Initial research showed that a large commercial transport could be modified to handle the high touch-down speeds (≈200 knots) experienced by Shuttle. Approval for the program was given by the Johnson Space Center in January 1989, and a Convair 990 (then called N710NA) that had been used for various research projects at the Ames Research Center was acquired by Dryden FRC and extensively modified beginning in May 1990.

A truss was installed between the 990's main landing gear to provide a support for mounting test landing gear and the structure was significantly strengthened around this area. A computer-controlled actuator system was installed that could apply up to 250,000 pounds of force on the test landing gear to drive it downward into the runway at precise speeds and loads. A total of 48 nitrogen bottles and 16 hydraulic accumulators were installed in the aft cargo hold to power the actuators. Four test consoles were installed in the forward passenger compartment.

Other modifications included installing heavy metal plates to protect the aircraft's belly from debris during tests; two 100-gallon water tanks for fire suppression; high-speed video and film cameras to record tire reactions; and over 200 sensors to record loads, pressures, temperatures, slip angles, and other data. This added over 40,000 pounds to the empty weight of the aircraft.

The LSRA modifications were completed in the fall of 1992, and DFRC reregistered the 990 as N810NA. Checkout of Shuttle test components began in April 1993, and concluded in August 1995 after 155 flights. These included 25 landings during the fall of 1993 on the 15,000-foot Shuttle Landing Facility at KSC. All of the tests consisted of landing on the LSRA's main landing gear, establishing the desired speed and slip angles, then forcing the test landing gear down in a precise manner.

Several of the components tested, including portions of the landing gear struts, wheels, and brakes, were items recovered from *Challenger* after the accident. NASA also 'borrowed' the main landing gear from OV-101 in the National Air and Space Museum during April 1990, intending to use it on the LSRA. But this never came to pass, and the gear was subsequently returned to the NASM. Unfortunately, the LSRA came too late to assist with most of the brake and landing gear trouble experienced by Shuttle during its early service life.[208]

and four stators (instead of three). The stators are attached to a torque tube inside the wheels, and remain stationary while the rotors revolve around them. The carbon brakes have been successfully tested for energy absorption of up to 100 million foot-pounds, although the nominal operating value is approximately 82 million foot-pounds.

Discovery was chosen to be the first Orbiter equipped with carbon brakes since strain gauges and accelerometers to monitor brake performance had been fitted previously. The first flight to use the new brakes was STS-31R, and the brakes performed satisfactorily. The other three Orbiters have since been fitted with carbon brakes during normal maintenance cycles.

The brakes are applied using conventional toe-action of the rudder pedals from either (or both) flight positions. The anti-skid system will not allow the brakes to be activated until approximately 1.9 seconds after weight on the main gear has been sensed.[120]

Drag Chute

The Orbiter drag chute was designed to assist in safely stopping the vehicle on the runway at end of mission (EOM) and abort weights. Design requirements included the ability to stop a 248,000 pound TAL abort Orbiter in 8,000 feet with a 10 knot tail-wind on a hot (103 degF) day and maximum braking at 140 knots ground speed or one half runway remaining. The drag chute, housed at the base of the vertical stabilizer, is manually deployed by redundant commands from the flight crew prior to derotation. The drag chute is jettisoned at 60 (±20) knots ground speed to prevent damage to the main engine bells. The drag chute can be used on lakebed and concrete runways except with crosswinds greater than 15 knots or in the presence of main engine bell repositioning problems. The drag chute may be deployed without engine bell repositioning if landing/rollout control problems exist.[121]

During reentry, at a velocity between 8,000 to 3,500 fps, the main engine bells are repositions 10 degrees below the nominal to preclude damage during drag chute deployment. Although the drag chute may be deployed at speeds up to 230 knots, current procedures call for it to be deployed at 195 knots. If the chute is deployed at more than 230 knots, the drag chute pivot pin is designed to fail, resulting in the chute being jettisoned. The drag chute is also designed to be deployed after the main gear has touched down. In fact, if the drag chute is deployed at an altitude between 135 and 40 feet, a loss of control will most likely result in the loss of the Orbiter.[122]

When drag chute deployment is initiated, the door is blown off of the chute compartment by pyros and a mortar fires deploying a 9-foot pilot chute. The pilot chute in turn extracts the 40 foot, partially reefed conical main chute. The main chute is reefed to 40 percent of its total diameter for about 3.5 seconds to lessen the initial loads on the vehicle. The main chute trails the vehicle by 89.5 feet on a 41.5 foot riser.[123]

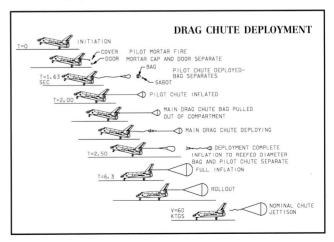

DRAG CHUTE DEPLOYMENT

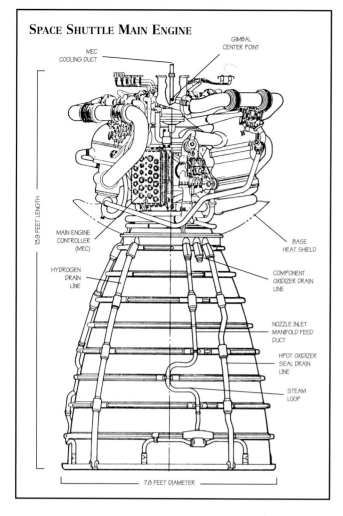

SPACE SHUTTLE MAIN ENGINE

MEC COOLING DUCT

GIMBAL CENTER POINT

13.9 FEET LENGTH

MAIN ENGINE CONTROLLER (MEC)

HYDROGEN DRAIN LINE

BASE HEAT SHIELD

COMPONENT OXIDIZER DRAIN LINE

NOZZLE INLET MANIFOLD FEED DUCT

HPOT OXIDIZER SEAL DRAIN LINE

STEAM LOOP

7.8 FEET DIAMETER

MAIN PROPULSION SYSTEM

The Main Propulsion System (MPS) consists of three Space Shuttle Main Engines (SSME), three main engine controllers (MEC), the External Tank, the Orbiter MPS propellant management system and helium subsystem, four ascent thrust vector control units, and six SSME hydraulic TVC servo-actuators. Most of the MPS is located in the aft fuselage beneath the vertical stabilizer.

Space Shuttle Main Engines

Three Space Shuttle Main Engines (SSME), built by Boeing Rocketdyne, are located in the extreme rear of the Orbiter's aft-fuselage. The SSMEs provide thrust in conjunction with the two SRBs during ascent.

Each SSME consumes 1,035 pounds of propellants per second (889 pounds of LO2 and 146 pounds of LH2) in a 6:1 oxidizer-to-fuel ratio – at the rated 100 percent power level, each engine develops 375,000 lbf at sea level and 470,000 lbf in a vacuum. The 104 percent power level corresponds to 393,800 lbf at sea level and 488,800 lbf in a vacuum, consuming 933 pounds of liquid oxygen and 155 pounds of liquid hydrogen per second. The nozzle area ratio is 69:1 and a specific impulse of 452 seconds is achieved. The main engines are 13.9 feet long, 7.8 feet in diameter at the nozzle exit, and each weighs approximately 7,480 pounds, including the main engine controller. Ancillary systems, including the gimbal system,

hydraulic supply, installation hardware, etc., weigh an additional 1,776 pounds per engine. The propellant feed system plumbing adds a further 1,674 pounds per engine, for a total weight of 10,930 pounds per engine.[124]

Each engine is throttleable, in one-percent increments, from 67 (originally 65) to 104 percent of rated thrust under normal circumstances. After the *Challenger* accident, an on-going certification to 109 percent was halted only a month prior to being completed, and the engines used for most of the program were limited to 104 percent power. Prior to the *Challenger* accident, the upper limit was to be increased to 120 percent, but this was at first postponed, then cancelled. Each SSME was contractually required to operate for 27,000 seconds consisting of 55 starts at 8 minutes per flight. However, current estimates are that each engine is good for 15,000 seconds of operation over a life span of 30 starts. The engines are certified to 754 seconds of continuous power (67–104 percent) in any one flight.

All three engines receive the same throttle command at the same time through the on-board GPCs. During certain contingency situations, manual control of engine throttling is possible through the speedbrake/thrust controller handle in the cockpit. The ability to throttle the SSMEs reduces vehicle loads during maximum aerodynamic pressure, and limits vehicle acceleration to 3-g maximum during second stage ascent. The engines are gimbaled by hydraulic actuators for thrust vector control during ascent. The maximum gimballing capability is ±10.5 degrees in pitch, and ±8.5 degrees in yaw. The gimbal bearing is bolted to the main injector and dome assembly and is the thrust interface between the engine and Orbiter. The bearing assembly is approximately 11.3 by 14 inches.[125]

In theory, the engines are capable of a full power level of 109 percent that produces 512,250 lbf in a vacuum – however, the performance of the entire MPS has not been certified beyond a power level of 104 percent, so the use of higher power levels is not allowed under normal circumstances. The ability to use 106 percent power under some emergency circumstances was added in the OI-27 software release, and the flight crew can manually throttle Block IIA engines up to 109 percent power if necessary to avoid ditching. The Block II engines are being certified for normal operations at 109 percent, finally achieving a level that has been a goal of the program for over 20 years.

The crew of STS-93 gather in and around an SSME – giving an excellent indication of the relative size of the engine. A major improvement program during the last decade has resulted in what is essentially a completely new engine compared to what flew on STS-1 in 1981. (NASA photo KSC-99PC-0179)

---

* The numbers in this section reflect the Block IIA engine configuration.

Disconnect Valves

Liquid oxygen (LO2) and liquid hydrogen (LH2) are used as oxidizer and fuel, respectively, for the Main Propulsion System, and are supplied from the External Tank. The LO2 and LH2 tanks are pressurized with gaseous helium while on the launch pad, which forces the propellants from the ET through 17-inch feed lines in the aft underside of the Orbiter. Normal boil-off and heated gas recirculation provides pressurization during ascent.

Each mated pair of 17-inch disconnects contains two flapper valves, one on the Orbiter side and one on the ET side. Both valves in the disconnect pair are opened to permit propellant flow between the Orbiter and External Tank. Prior to separation, both valves in each mated pair are closed to prevent propellant discharge during separation. Valve closure on the Orbiter side also prevents contamination of the Orbiter main propulsion system during landing and subsequent ground operations.

Inadvertent closure of either valve in each 17-inch disconnect during main engine thrusting would abruptly stop propellant flow from the ET to all three main engines – catastrophic failure of all three SSMEs and both ET feed lines would most probably result. A latch mechanism was added during the *Challenger* stand-down in the Orbiter side of each disconnect to prevent inadvertent closure by providing a mechanical backup to the fluid-induced forces originally used. The latch is mounted on a shaft in the flow stream so that it overlaps both flappers and obstructs closure.

In preparation for ET separation, both valves in each 17-inch disconnect are commanded closed. Pneumatic pressure from the MPS causes the latch actuator to rotate the shaft in each Orbiter 17-inch disconnect 90 degrees, thus freeing the flapper valves to close as required. A backup mechanical separation capability is provided in case a latch pneumatic actuator fails.

In 1989, Rockwell International was contracted to develop and build an improved disconnect system for the Orbiters. The new system was to use more robust 14-inch valves, and was expected to be certified for operational use in the 1994–95 timeframe. Incentive for this change was increased due to the adverse publicity received during the 'hydrogen leak' episodes during 1990 when the Space Shuttle fleet was grounded for six months. However, on 1 November 1991 the 14-inch valve program was cancelled for funding reasons, although several of the innovations developed for it have been retrofitted to the existing 17-inch disconnect valves.[126]

Power Heads

Oxidizer enters the Orbiter at the liquid oxygen feed line disconnect valve and flows into the Orbiter's MPS liquid oxygen feed line. There it branches out into three parallel paths, one 12-inch feed line to each engine. In each branch, a liquid oxygen prevalve must be opened to permit flow to the low-pressure oxidizer turbopump (LPOT).

The LPOT is an axial-flow pump driven by a six-stage turbine powered by liquid oxygen and boosts the LO2 pressure from 100 psia to 422 psia. The flow from the LPOT is supplied to the high-pressure oxidizer turbopump (HPOT). The LPOT operates at approximately 5,150 rpm, and is roughly 18 by 18 inches in size. The LPOT is flange-mounted to the Orbiter propellant ducting. During engine operation, the pressure boost permits the HPOT to operate at high speeds without cavitating.[127]

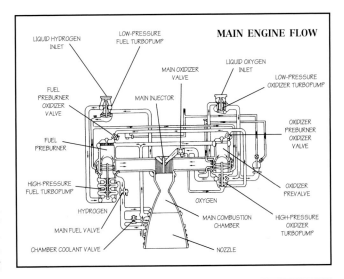

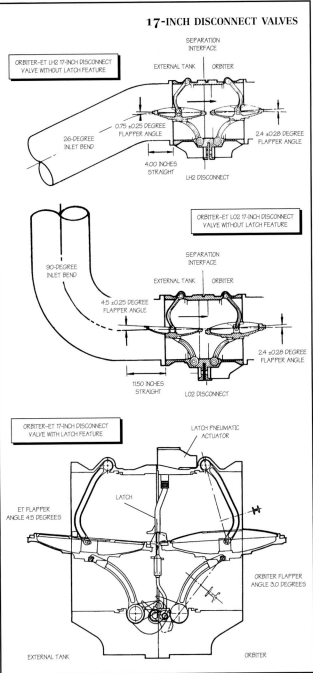

The HPOT consists of two single-stage centrifugal pumps (a main pump and a preburner pump) mounted on a common shaft and driven by a three-stage, hot-gas turbine. The main pump boosts the LO2 pressure from 422 psia to 4,030 psia while operating at approximately 21,150 rpm. The HPOT discharge flow splits into several paths, one of which is routed to drive the LPOT turbine. Another path is routed through the main oxidizer valve and enters into the main combustion chamber. Another small flow path is tapped off and sent to the oxidizer heat exchanger. The LO2 flows through an anti-flood valve that prevents it from entering the heat exchanger until sufficient heat is present to convert the liquid oxygen to gas. The gas is routed to the External Tank to pressurize the liquid oxygen tank. Another path enters the HPOT second-stage preburner pump to boost the LO2 pressure from 4,030 psia to 6,940 psia. The HPOT is approximately 24 by 36 inches and is attached by flanges to the hot-gas manifold on the SSME itself.[128]

Fuel enters the Orbiter at the liquid hydrogen feed line disconnect valve, then flows into the hydrogen feed line manifold and branches out into three parallel paths to each engine. In each liquid hydrogen branch, a prevalve permits LH2 to flow to the low-pressure fuel turbopump (LPFT) when the prevalve is open.

The LPFT is an axial-flow pump driven by a two-stage turbine powered by gaseous hydrogen. It boosts the pressure of the LH2 from 30 psia to 303 psia and supplies it to the high-pressure fuel turbopump (HPFT). During engine operation, the pressure boost provided by the LPFT permits the HPFT to operate at high speeds without cavitating.

The LPFT operates at approximately 15,670 rpm, and is roughly 18 by 24 inches in size. The LPFT is flange-mounted to the SSME at the inlet to the low-pressure fuel duct.[129]

The HPFT is a three-stage centrifugal pump driven by a two-stage, hot-gas turbine. It boosts the pressure of the liquid hydrogen from 303 psia to 5,945 psia, and operates at approximately 34,290 rpm. The discharge flow from the turbopump is routed through the main fuel valve and then splits into three flow paths. One path is through the jacket of the main combustion chamber, where the hydrogen is used to cool the chamber walls. The resulting gaseous hydrogen is then routed from the main combustion chamber to the LPFT where it is used to drive the LPFT turbine. A small portion of the flow from the LPFT is directed to a common manifold from all three engines to form a single path to the External Tank to maintain liquid hydrogen tank pressurization. The remaining hydrogen passes between the inner and outer walls of the hot-gas manifold as a coolant, and is discharged into the main combustion chamber. The second hydrogen flow path from the main fuel valve is through the engine nozzle where it is used to cool the nozzle. It then joins the third flow path from the chamber coolant valve and the combined flow is directed to the fuel and oxidizer preburners. The HPFT is approximately 22 by 44 inches, and is attached by flanges to the hot-gas manifold.[130]

The HPOT turbine and HPOT pumps are mounted on a common shaft. Mixing of the fuel-rich hot gas in the turbine section and the liquid oxygen in the main pump could create a hazard. To prevent this, the two sections are separated by a cavity that is continuously purged by

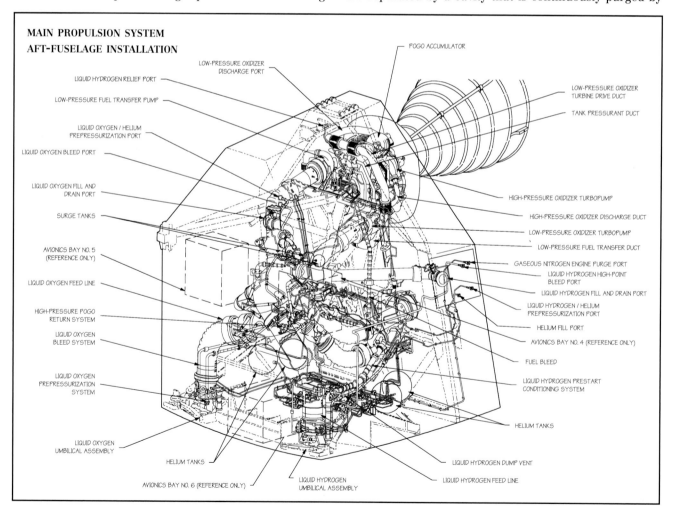

MAIN PROPULSION SYSTEM
AFT-FUSELAGE INSTALLATION

POGO ACCUMULATOR

LOW-PRESSURE OXIDIZER DISCHARGE PORT

LIQUID HYDROGEN RELIEF PORT

LOW-PRESSURE FUEL TRANSFER PUMP

LIQUID OXYGEN / HELIUM PREPRESSURIZATION PORT

LIQUID OXYGEN BLEED PORT

LIQUID OXYGEN FILL AND DRAIN PORT

SURGE TANKS

AVIONICS BAY NO. 5 (REFERENCE ONLY)

LIQUID OXYGEN FEED LINE

HIGH-PRESSURE POGO RETURN SYSTEM

LIQUID OXYGEN BLEED SYSTEM

LIQUID OXYGEN PREPRESSURIZATION SYSTEM

LIQUID OXYGEN UMBILICAL ASSEMBLY

HELIUM TANKS

AVIONICS BAY NO. 6 (REFERENCE ONLY)

LIQUID HYDROGEN UMBILICAL ASSEMBLY

LOW-PRESSURE OXIDIZER TURBINE DRIVE DUCT

TANK PRESSURANT DUCT

HIGH-PRESSURE OXIDIZER TURBOPUMP

HIGH-PRESSURE OXIDIZER DISCHARGE DUCT

LOW-PRESSURE OXIDIZER TURBOPUMP

LOW-PRESSURE FUEL TRANSFER DUCT

GASEOUS NITROGEN ENGINE PURGE PORT

LIQUID HYDROGEN HIGH-POINT BLEED PORT

LIQUID HYDROGEN FILL AND DRAIN PORT

LIQUID HYDROGEN / HELIUM PREPRESSURIZATION PORT

HELIUM FILL PORT

AVIONICS BAY NO. 4 (REFERENCE ONLY)

FUEL BLEED

LIQUID HYDROGEN PRESTART CONDITIONING SYSTEM

HELIUM TANKS

LIQUID HYDROGEN DUMP VENT

LIQUID HYDROGEN FEED LINE

The first Block I SSME (#2036) waits to be installed in Discovery in preparation for STS-70. The program was very conservative and only flew a single Block I engine along with two Phase II engines on STS-71. Interestingly, when the Block IIA engines came on-line for STS-89, all three positions used the new engine. (NASA photo KSC-95PC-0586)

the MPS engine helium supply during operation. Two seals minimize leakage into the cavity – one seal is located between the turbine section and the cavity; the other is between the pump section and cavity. Loss of helium pressure in this cavity results in an engine shutdown.

The low-pressure oxygen and low-pressure fuel turbopumps are mounted 180 degrees apart on the Orbiter's aft-fuselage thrust structure for safety. The ducts from the low-pressure turbopumps to the high-pressure turbopumps contain flexible bellows that enable the low-pressure turbopumps to remain stationary while the rest of the engine is gimbaled for thrust vector control. The liquid hydrogen line from the LPFT to the HPFT is insulated to prevent the formation of liquid air.[131]

The speed of the HPOT and HPFT turbines depends on the position of the corresponding oxidizer and fuel preburner oxidizer valves. These valves are positioned by the main engine controller (MEC), which uses them to throttle the flow of liquid oxygen to the preburners and, thus, control engine thrust. The oxidizer and fuel preburner oxidizer valves increase or decrease the liquid oxygen flow, thus increasing or decreasing preburner chamber pressure, HPOT and HPFT turbine speed, and liquid oxygen and gaseous hydrogen flow into the main combustion chamber, thus throttling the engine. The oxidizer and fuel preburner valves operate together to throttle the engine and maintain a constant 6:1 propellant mixture ratio.

Hot Gas Manifold and Combustion Chamber

The hot-gas manifold is the structural backbone of the engine. It supports the two preburners, the high-pressure turbopumps, and the main combustion chamber. Hot gas generated by the preburners, after driving the high-pressure turbopumps, passes through the hot-gas manifold on the way to the main combustion chamber.[132]

The oxidizer and fuel preburners are welded to the hot-gas manifold. Liquid oxygen and liquid hydrogen from the high-pressure turbopumps enter the preburners and are mixed so that efficient combustion can occur. The preburners produce the fuel-rich hot gas that passes through the turbines to generate the power to operate the high-pressure turbopumps. The oxidizer preburner's outflow drives a turbine that is connected to the high-pressure oxidizer turbopump and the oxidizer preburner

boost pump. The fuel preburner's outflow drives a turbine connected to the high-pressure fuel turbopump.

The oxidizer heat exchanger converts liquid oxygen to gaseous oxygen for tank pressurization and pogo suppression. The heat exchanger receives its liquid oxygen from the high-pressure oxidizer turbopump discharge flow.

Each engine main combustion chamber receives fuel-rich hot gas from the fuel and oxidizer preburners. The high pressure oxidizer turbopump supplies liquid oxygen to the combustion chamber where it is mixed with fuel-rich gas by the main injector. A small augmented spark igniter chamber is located in the center of the injector. The dual-redundant igniter is used during the engine start sequence to initiate combustion. The igniters are turned off after approximately 3 seconds because the combustion process is self-sustaining. The main injector and dome assembly are welded to the hot-gas manifold. The main combustion chamber is bolted to the hot-gas manifold.

Nozzle

The inner surface of each combustion chamber, as well as the inner surface of the nozzle, is cooled by gaseous hydrogen flowing through 1,080 coolant passages. The nozzle assembly is a bell-shaped extension bolted to the main combustion chamber. The nozzle is 113 inches long, and the outside diameter of the exit is 94 inches – creating a 69:1 expansion ratio. A support ring welded to the forward end of the nozzle is the engine attach point to the Orbiter-supplied heat shield. Thermal protection is provided for the nozzles to protect them from the high heating rates experienced during ascent and reentry. The insulation consists of four layers of metallic batting covered with a metallic foil and screening.[133]

MPS Propellant Management Subsystem

Liquid hydrogen and liquid oxygen pass from the ET to the propellant management subsystem (PMS). The PMS consists of manifolds, distribution lines, and valves. It also contains lines needed to transport gases from the engines to the External Tank for pressurization.[134]

During prelaunch activities, the MPS propellant management subsystem is used to control the loading of LO2 and LH2 into the ET. During SSME thrusting periods, propellants from the ET flow into this subsystem and to the three SSMEs. The subsystem also provides the path that allows gases tapped from the three main engines to flow back to the ET through two gas umbilicals to maintain pressure in the ET propellant tanks. After MECO, this subsystem controls MPS dumps, vacuum inerting, and MPS repressurization for reentry.

All the valves in the MPS are either electrically or pneumatically operated. Pneumatic valves are used where large loads are encountered, such as in the control of liquid propellant flows. Electrical valves are used for lighter loads, such as in the control of gaseous flows. The pneumatically actuated valves are divided into two basic types – those that require pneumatic pressure to open and close the valve and those that are spring loaded to one position and require pneumatic pressure to move to the other position.

There are two 17-inch diameter MPS propellant feed line manifolds in the Orbiter aft-fuselage, one for liquid oxygen and one for liquid hydrogen. Each manifold has an outboard and inboard fill and drain valve in series that interface with the respective right and left T-0 umbilical. The right T-0 umbilical is for liquid oxygen; the left for

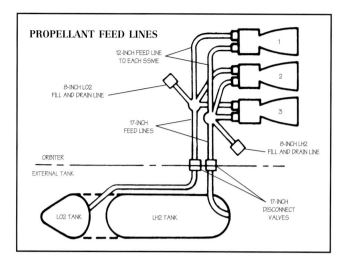

PROPELLANT FEED LINES

12-INCH FEED LINE
TO EACH SSME

8-INCH LO2
FILL AND DRAIN LINE

17-INCH
FEED LINES

ORBITER

EXTERNAL TANK

8-INCH LH2
FILL AND DRAIN LINE

17-INCH
DISCONNECT
VALVES

LO2 TANK

LH2 TANK

liquid hydrogen. In addition, each manifold connects the Orbiter to the External Tank in the lower aft-fuselage through the 17-inch disconnect valves.

There are three outlets in both the LO2 and LH2 17-inch manifolds between the Orbiter–ET umbilical disconnect valves and the inboard fill and drain valve. The outlets in the manifolds provide liquid oxygen and liquid hydrogen to each SSME in 12-inch diameter feed lines. The prevalve in each of the three LO2 and LH2 12-inch feed lines to each engine isolates liquid oxygen and liquid hydrogen from each engine or permits liquid oxygen and liquid hydrogen to flow to each engine. Each 17-inch LO2 and LH2 manifold has a one inch-diameter line that is routed to a feed line relief isolation valve and feed line relief valve in the respective liquid hydrogen and liquid oxygen system. When a feed line relief isolation valve is opened, the corresponding manifold can relieve excessive pressure overboard through its relief valve.

The LH2 feed line manifold has another outlet directed to the two liquid hydrogen RTLS dump valves in series. When opened, these valves enable the liquid hydrogen dump during RTLS aborts or provide a backup to the normal LH2 dump after a nominal main engine cutoff. In an RTLS abort dump, LH2 is dumped overboard through a vent at the outer aft-fuselage left side between the OMS pod and the upper surface of the wing.

The MPS propellant management subsystem also contains two 2-inch diameter manifolds, one for gaseous oxygen and one for gaseous hydrogen. Each manifold individually permits ground servicing with helium through the respective T-0 umbilical and provides initial pressurization of the External Tank LO2 and LH2 Orbiter–ET disconnect umbilicals. Self-sealing quick disconnects are provided at the T-0 umbilical and the Orbiter–ET umbilical.

Pogo Suppression Subsystem

A pogo suppression subsystem prevents the transmission of low-frequency flow oscillations into the high-pressure oxidizer turbopump and, ultimately, prevents main combustion chamber pressure (engine thrust) oscillation. Flow oscillations transmitted from the Space Shuttle vehicle are suppressed by a partially filled gas accumulator, which is attached by flanges to each high-pressure oxidizer turbopump inlet duct.

The system consists of a 0.6 cubic foot accumulator with an internal standpipe, helium precharge valve, gaseous oxygen supply valve, and two recirculation isolation valves (one located on the Orbiter). During engine

start, the accumulator is charged with helium 2.4 seconds after the start command to provide pogo protection until the engine heat exchanger is operational and gaseous oxygen is available. The accumulator is partially chilled by LO2 during the engine chill-down operation. It fills to the overflow standpipe line inlet level, which is sufficient to preclude gas ingestion at engine start. During engine operation, the accumulator is charged with a continuous gaseous oxygen flow maintained at a rate governed by the current engine operation point.[135]

Main Engine Controllers

Each SSME has an engine control computer that monitors parameters such as pressure and temperature, and automatically adjusts engine operation for the required thrust. The main engine controllers (MEC) provide responsive control of engine thrust and propellant mixture ratio, updating instructions to the engine control elements 50 times per second. Each MEC is attached to the forward end of its SSME with shock-mounted fittings and contains two digital computers. Each MEC* receives commands transmitted by the Orbiter's GPCs through its own engine interface unit (EIU). Engine data are sent to the engine controller which maintains a record of engine operating history for maintenance purposes. This data is sent to the EIU where it is routed to the GPCs and also relayed in real-time to the ground through a 60 kbs PCM telemetry link per engine.[136]

The MEC is a general-purpose digital computer that provides the computational capability necessary for all engine control functions. The original MEC had a program storage capacity of 16,384 data and instruction words (17-bit words; 16 bits for program use, one bit for parity).

A Block-II main engine controller was certified for use during early 1991, and provides greatly increased computing power, as well as additional telemetry for monitoring engine performance during ascent. The first flight of a Block-II MEC was STS-49, and all flight-worthy SSMEs were upgraded to the Block-II MEC by the middle of 1994. It should be noted that the Block-II MEC has no direct relationship to the Block II SSME discussed later.

Thrust Vector Control

Two actuators per engine provide attitude control and trajectory shaping by gimbaling the SSMEs in conjunction with the Solid Rocket Boosters during first-stage and without the SRBs during second-stage ascent. Each SSME servoactuator receives hydraulic pressure from two of the three Orbiter hydraulic systems.

The three pitch actuators gimbal the engine up or down a maximum of 10.5 degrees from the installed null position. The three yaw actuators gimbal the engine left or right a maximum of 8.5 degrees from the installed position. The installed null position for the left and right main engines is 10 degrees up from the X axis in a negative Z direction and 3.5 degrees outboard from the engine centerline parallel to the X axis. The center engine installed null position is 16 degrees above the X axis for pitch and on the X axis for yaw. When any engine is installed in the null position, the other engines do not come in contact with it. The minimum gimbal rate is 10 degrees per second – the maximum rate is 20 degrees per second.[137]

---

* Please note this acronym is used twice on Space Shuttle – on the Orbiter it stands for Master Events Controller; on the engines it is the Main Engine Controller.

Engine Start and Ascent

At T–4 minutes, the system purge begins. It is followed at T–3 minutes 25 seconds by the beginning of the engine gimbal tests, during which each gimbal actuator is operated through a set profile of extensions and retractions. If all actuators function satisfactorily, the engines are gimbaled to a predefined position at T–2 minutes 15 seconds. The engines remain in this position until engine ignition.[138]

At T–2 minutes 55 seconds, the launch processing system closes the liquid oxygen tank vent valve, and the tank is pressurized to 21 psig with ground support equipment-supplied helium. The 21-psig pressure corresponds to a LO2 engine manifold pressure of 105 psia. At T–1 minute 57 seconds, the Launch Processing System closes the LH2 vent valve, and the tank is pressurized to 42 psig with ground-supplied helium. At T–31 seconds, the onboard redundant set launch sequencer is enabled by the launch processing system. From this point on, all sequencing is performed by the Orbiter GPCs in the redundant set. The GPCs still respond, however, to hold, resume count, and recycle commands from the Launch Processing System.

At T–16 seconds, the GPCs begin to issue arming commands for the SRB ignition pyro initiator controllers, the hold-down release pyro initiator controllers, and the T–0 umbilical release pyro initiator controllers. At T–9.5 seconds the engine chill-down sequence is complete and the GPCs open the LH2 prevalves (the LO2 prevalves are open during loading to chill-down the engines).

At T–6.6 seconds the GPCs issue the engine start command and the main fuel and oxidizer valves in each engine opens. If all three SSMEs reach 90 percent of their rated thrust by T–3 seconds, then at T–0, the GPCs will issue the commands to fire the SRB ignition pyro initiator controllers, the hold-down release pyro initiator controllers, and the T–0 umbilical release pyro initiator controllers. If one or more of the three main engines do not reach 90 percent of their rated thrust by T–3 seconds, all SSMEs are shut down, the SRBs are not ignited, and a pad abort condition exists.

At T–0 the SSME gimbal actuators that were locked in their special preignition position, are first commanded to

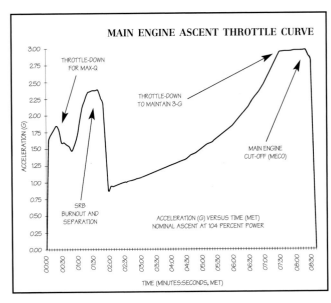

MAIN ENGINE ASCENT THROTTLE CURVE

'Generic' SSME throttle profile – i.e., this does not represent any particular mission. In general there are two periods where the SSMEs throttle down. The first is at about 30 seconds after lift-off – the SSMEs throttle down to 65 percent power to help limit the maximum dynamic pressure on the vehicle. The SRB propellant is also shaped to lower thrust during this period. The second SSME throttle-down is late in the ascent as the vehicle picks up speed since it is much lighter. The SSMEs begin to throttle down to limit the maximum acceleration to 3-g. Shortly after this is main engine cut-off. (NASA)

their null positions for SRB start and then are allowed to operate as needed for thrust vector control. Lift-off occurs almost immediately because of the extremely rapid thrust buildup of the solid boosters. Starting the main engines 6.6 seconds before T–0 allows the vehicle base bending loads ('twang') to return to minimum by T–0. Between lift-off and MECO, as long as the SSMEs perform nominally, all MPS sequencing and control functions are executed automatically by the GPCs. During this period, the crew monitors MPS performance, backs up automatic functions, if required, and provides manual inputs in the event of MPS malfunctions.

During ascent, the SSME thrust level depends on mission requirements – it is usually 104 percent, but higher settings may be used for emergency situations. As dynamic pressure rises, the GPCs throttle the engines to a lower power level (usually 67 percent) to minimize structural loading while the Orbiter is passing through the region of maximum dynamic pressure (max-q). This is called the 'thrust bucket' because of the way the thrust plot appears on the graph. Although the bucket duration and thrust level vary, a typical bucket runs from about T+30 to T+65 seconds MET. The SRB propellant is also shaped to reduce thrust during this period.

At approximately T+65 seconds MET, the engines are once again throttled up to the appropriate power level (usually 104 percent) and remain at that setting until 3-g throttling is initiated. Beginning at approximately T+7 minutes 40 seconds, the engines are throttled back to maintain vehicle acceleration at 3-g or less. Approximately 6 seconds before main engine cutoff, the engines are throttled back to 67 percent in preparation for shutdown.

Although MECO is based on the attainment of a specified velocity, the engines may also be shut down due to the depletion of LO2 or LH2 before the specified velocity is reached. Liquid oxygen depletion is sensed by four sensors in the Orbiter LO2 feedline manifold. Liquid hydrogen depletion is sensed by four sensors in the bottom of the LH2 tank. If any two of the four sensors in either system indicate a dry condition, the GPCs will shut down the engines.

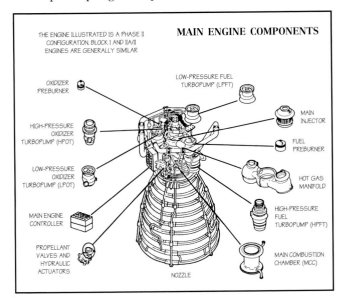

MAIN ENGINE COMPONENTS

THE ENGINE ILLUSTRATED IS A PHASE II CONFIGURATION. BLOCK I AND IIA/II ENGINES ARE GENERALLY SIMILAR

OXIDIZER PREBURNER

LOW-PRESSURE FUEL TURBOPUMP (LPFT)

HIGH-PRESSURE OXIDIZER TURBOPUMP (HPOT)

LOW-PRESSURE OXIDIZER TURBOPUMP (LPOT)

MAIN ENGINE CONTROLLER

PROPELLANT VALVES AND HYDRAULIC ACTUATORS

MAIN INJECTOR

FUEL PREBURNER

HOT GAS MANIFOLD

HIGH-PRESSURE FUEL TURBOPUMP (HPFT)

MAIN COMBUSTION CHAMBER (MCC)

NOZZLE

The major components of a Phase II SSME are shown here. The hot gas manifold is the backbone of the engine, providing the mounting location for most of the other major components. Block II engines use a large throat main combustion chamber (LTMCC) and Pratt & Whitney high-pressure turbopumps, but are otherwise similar. (Rocketdyne)

SSME Enhancements

The engineers at Rocketdyne had begun to think about possible improvements to the SSME even before the engine received its flight certification in 1981. In fact, during September 1980 Rocketdyne had briefed Space Shuttle Program Manager John Yarldey on various potential upgrades, the most promising one being a 'big throat' main combustion chamber. It was a complex proposal since, although the big throat chamber promised increased thrust, it also lowered the specific impulse, negating most of the gain. The engineers had an answer for this, namely eliminating some of the 'stability aids' that ensured stable combustion. It was believed that the big throat combustion chamber would provide sufficient stability by itself. Nevertheless, the successful flight of STS-1 focused attention on operational issues, and no formal development project was initiated.[139]

By the end of 1982, however, the amount of maintenance required between each flight and a series of ground test failures during full-power level (109 percent) testing again focused attention on SSME upgrades. In August 1983 a formal program to improve the SSME was initiated. There were three goals: Phase I would decrease the maintenance required between flights; Phase II would provide the 109 percent power level required for Vandenberg launches; and Phase III would provide long-term margin for flight at full power (109 percent).

In order to reliably achieve the 109 percent thrust rating, Phase II focused on improving durability with changes to the high-pressure turbopumps, main combustion chamber, hydraulic actuators, and high-pressure turbine discharge temperature sensors. Improvements made to the turbopumps were aimed primarily at extending the time between overhauls. Changes to the high-pressure fuel turbopump included reducing the operating temperature, applying a surface texture to the fuel turbine blades, and minor changes to the rotor dynamics (whirl) and bearings.

An SSME is removed from the number one position on Atlantis prior to STS-101 while still in the VAB. An inventory review concerning suspect main engine fuel pump tip seals indicated that defective seals might have been be present on the fuel pump for the engine. (NASA photo KSC-00PP-0380)

The turbine blades in the high-pressure oxidizer turbopump received an additional damper to reduce vibration.

The main combustion chamber service life was increased by plating the welded outlet manifold with nickel. Margin improvements were also incorporated into five hydraulic actuators to preclude a loss of redundancy on the launch pad. Improvements in quality were incorporated into the servo-component coil design along with modifications to increase margin. To address the in-flight temperature sensor anomaly, the sensor was redesigned, and minor changes were incorporated into the main engine controller software to improve performance.

At the time of the *Challenger* accident in January 1986, there had been 25 flights of the SSMEs, accumulated by 13 different engines. One engine (#2012) had flown ten times, and ten other engines had flown between five and nine times each. Two off-nominal conditions were experienced on the launch pad (STS-12 and STS-26/51-F), and one during ascent (again, STS-26). The two launch pad failures both occurred after engine start, but before SRB ignition. In each case, the main engine controller detected a loss of redundancy in the hydraulic actuator system and commanded an engine shutdown in keeping with the launch commit criteria. During the STS-26/51-F ascent a sensor indicated an overtemperature condition in the SSME #1. The engine was commanded to shut down, but almost immediately a temperature sensor in a second engine indicated an anomaly. Only quick thinking by flight controllers advising the crew to bypass the sensor kept the second engine from shutting down.

After the *Challenger* accident, the 109 percent full-power certification was cancelled as was the Phase III proposal, but almost all the improvements developed during Phase II were subsequently incorporated into the flight engines. To certify these various improvements, an aggressive test program was initiated during December 1986, running until December 1987. During these evaluations, 151 individual tests totaling 52,363 seconds of operations (equivalent to 100 flights) were performed. These hot-fire ground tests were conducted at the single-engine test stands at the NSTL in Mississippi, and at the Rocketdyne Santa Susana Field Laboratory in California. After certification, it was necessary to return all flight engines to Rocketdyne where they were completely disassembled and rebuilt with the phase II improvements.[140]

Other improvements to the SSMEs were also initiated as the result of studies performed after the *Challenger* accident. Several of these could not be adequately funded immediately after the accident, while others could not be completely integrated into the operational engines for several years after the Return-to-Flight of STS-26R, due primarily to the complexity of the changes and the lengthy recertification programs involved. None of these changes were considered essential to the safety of flight, being designed instead to improve the operational efficiency of the engine, or to provide a performance growth.

While the Phase II test and certification program was underway, NASA finally admitted that the Space Shuttle Main Engines, advanced as they might have been when designed, could benefit from major improvements in design and manufacturing. Therefore, a handful of components that had historically experienced the highest failure and/or maintenance rates were evaluated for possible modifications. Five major components were selected for advanced development – the high-pressure fuel turbopump; high-pressure oxidizer turbopump; powerhead; heat exchanger; and main combustion chamber.

Rocketdyne defined a two-step approach to incorporating much of the new technology originally envisioned for the Phase III upgrade into the SSMEs. Separately, in August 1986 NASA awarded Pratt & Whitney a $300 million contract to develop and produce 44 'alternate' turbopumps that would be supplied to Rocketdyne as government-furnished equipment. The alternate HPOT, a new two-duct powerhead, and a single-coil heat exchanger would be incorporated into a new configuration known as the Block I engine. A follow-on Block II configuration would add the alternate HPFT and a large throat main combustion chamber.[141]

High-Pressure Turbopumps

The newly-designed Pratt & Whitney turbopumps have 50 percent fewer rotating parts, although they have slightly more piece parts. The original design for the turbopumps featured welded construction, which required meticulous inspections during the manufacturing process, and also between major overhauls. The new design uses a unique casting process to eliminate all but six of the previous 300 welds. In addition, the original design required special coatings on the turbine blades for thermal protection. These coatings are no longer necessary since the new turbopump is constructed of materials which are more heat tolerant and less sensitive to the hydrogen environment.[142]

On the original Rocketdyne turbopump the bearings on the impeller shaft had to be replaced after two flights. The Pratt & Whitney turbopump design incorporates a new bearing with rotating surfaces made of silicon nitrate. These ceramic bearings are 30 percent harder and 40 percent lighter than the original steel bearings, and have an ultra-smooth finish that produces significantly less friction during pump operation. The new turbopump operates at 23,700 rpm, generates 25,850 shp, and has a discharge pressure of 7,250 psi. The new HPOT was first fired in May 1990, and flight certification was completed on 15 March 1995, marking the end of a test program equivalent to 40 flights.[143]

The new HPFT was first fired in February 1990, but was put on hold in 1991 due to budget constraints. During 1993 the new HPFT was once again been given the go-ahead to be incorporated into the Block II SSME. However, the new HPFT ran into trouble during a 25 January 1996 test when test engine 0523 had a catastrophic failure of a new HPFT during a 754-second test firing, destroying the engine. Up until this point Pratt &

Whitney had accumulated a total of 3,593.9 seconds of testing on two new HPFTs. A total of 60,000 seconds of successful testing is required for certification. The new HPFT operates are 36,200 rpm, generating 72,900 shp, and an outlet discharge pressure of 6,400 psi.

Further investigation lead to the discovery of fatigue cracks in the turbine second-stage vane assembly on other high-cycle fuel turbopumps. Similarities between the turbines on the fuel turbopumps and the new Pratt & Whitney oxidizer turbopump used on the Block I engines led NASA to temporarily ground the new engines effective in February 1996 pending further investigations. The grounding only lasted several weeks until engineers confirmed the problem was not present on the HPOT. Pratt & Whitney continued testing the new turbopump, working through various problems as they were uncovered. But certification seemed elusive.

Two-Duct Powerhead and Heat Exchanger

The Block I engine also contained a new two-duct powerhead. The powerhead contains the preburners that generate the gas to drive the turbopump turbines, and also collects the hot gases downstream of the turbines and ducts them into the main injector. The original powerhead had five ducts – three on one side of the engine to collect gases from the fuel turbine, and two on the other for the oxidizer turbine. By replacing the three smaller fuel ducts in the original design with two enlarged ducts, the new design significantly improved hot gas flows within the engine. Pressure and loads were decreased, turbulence reduced, maintenance was mostly eliminated, and inspections were minimized. New production processes were also used, resulting in 52 fewer detailed parts and the elimination of over 80 welds.[144]

Another enhancement to the main engine powerhead improved the engine's heat exchanger. The main engines must supply pressure to the ET that in turn provides propellants at the correct pressures to the engine turbopumps. This pressure is produced by the engine's heat exchanger, a 41-foot long piece of coiled stainless steel alloy tubing. Liquid oxygen is routed through the tubing, passes through the engine's hot gas manifold, and is heated by the hot exhaust from the high-pressure oxidizer turbopump turbine. The oxygen turns to a superheated gas as it reaches about 500 degF, and supplies pressure to the ET.

The original heat exchanger had seven welds in the 41-foot tube. Welding can change the properties of a metal and leave flaws, impairing its ability to withstand repeated thermal cycles and loads. The new single coil heat exchanger is a continuous piece of stainless steel alloy with no welds. The new design also uses thicker walls to reduce wear on the tube and lessen the chances of damage, resulting in reduced maintenance and postflight inspections. Surprisingly, the new single coil design was also 40 percent less expensive to fabricate.[145]

Large Throat Main Combustion Chamber

This was one of the original changes identified in 1980, but it would take almost 20 years to actually make it into operational engines. Although the LTMCC was seen to offer a way to improve margins and gain a slight thrust advantage, it was deemed too risky due to combustion stability concerns and the anticipated loss of specific impulse.[146]

The main combustion chamber is where the LO2 and LH2 are mixed and burned to provide thrust. The older

Three SSMEs in the old engine shop in VAB room 1M4 during November 1990. (NASA photo KSC-390C-7727.05 via Boeing Rocketdyne/Fred H. Jue)

chamber design requires frequent inspections and maintenance due to its welded construction. The new LTMCC is cast from several large pieces of metal rather than being made from many smaller pieces welded together. In addition to eliminating over 50 welds, casting also reduces the assembly time and labor required to build and maintain the hardware. Inspections are significantly reduced since there are fewer welds to check during maintenance periods. The throat of the new chamber is also 11 percent larger than the original design, resulting in reduced pressure in the chamber and throughout the engine. This lowers turbine temperatures, improves engine cooling, reduces overall maintenance, and extends the life of the hardware.

Testing on the first LTMCC began during 1988 at Santa Susanna and soon demonstrated that chamber pressure was reduced by 270 psi, while pressures and temperatures in both high-pressure turbopumps dropped up to ten percent at the 109-percent power level. But it was not until late 1992 that testing finally put all concerns about combustion instability to rest. As expected, the LTMCC penalized specific impulse by 1.4 seconds, but test data showed that lower pressures also reduced the power demand from the HPFT by over 3,000 hp. Overall, the engine margin improved by over seven percent, significantly improving the safety of the engine. With this discovery, the LTMCC became the single most important SSME upgrade.[147]

Block I

The Block I engines incorporated the 'alternate' Pratt & Whitney HPOT, new two-duct powerhead, and the single-coil heat exchanger. In addition, a new Kevlar jacket was fitted on the LPFT housing to eliminate liquid air formation. The previous design often trapped moisture, leading to corrosion that required extensive maintenance. The Kevlar jacket worked so well, in fact, that its use is being extended to the low-pressure fuel duct. A new HPFT turbine inlet fairing was also introduced, fabricated from a single sheet of metal that eliminated hundreds of high-maintenance welds. The improved HPFT also featured improved turbine blades, assembly processes that reduced a persistent rotor imbalance, and a revised main turbine housing.[148]

Seven new SSMEs (2036–2042) were manufactured to the interim Block I configuration. Block I engine 2036 was the first to fly, in the center position on STS-70 during July 1995. Since this was an evaluation flight, the other two SSMEs were of the older Phase II design. STS-77 in May 1996 was the first flight to use three Block I engines.

The incorporation of the Block I engine offered a significant decrease in the risk of losing a vehicle during ascent. With three Phase II engines installed, the computed loss-of-vehicle probability was one in 262 flights; the Block I engine reduced this to one in 335 flights.[149]

Block IIA

The continued difficulties in certifying the Pratt & Whitney high-pressure fuel turbopump was delaying the introduction of the Block II engine. But the tests of the LTMCC proved so successful at lowering the risk during ascent that NASA decided not to wait until the alternate HPFT was ready. An interim configuration, dubbed Block IIA, included all the improvements originally intended for Block II – except the alternate HPFT. This was considered a very low-risk change – the LTMCC had successfully passed many tests totaling nearly 80,000 seconds, and the existing HPFT was the veteran of over 775,000 seconds of

firing. Less than a year passed from the decision to create the interim Block IIA configuration and the first flight.[150]

The new LTMCC required other changes as well. The LPFT turbine nozzle flow area was recalibrated to maintain the proper downstream pressure margins. The LPOT performance was increased to accommodate the reduced oxidizer system pressure required by the LTMCC. To accomplish this, the inducer's blade incidence profile was optimized to increase the cavitation margin. Engineers also took the opportunity to replace the metallic thrust bearing with a ceramic material that had already been incorporated in the new HPOT. The main injectors were also refined in an attempt to gain back some of the lost specific impulse. A positive leak-free faceplate seal that eliminated the boundary-layer coolant holes and installed coolant orifices in the injector fuel sleeves was also added. These changes recovered about 0.4 seconds of specific impulse.[151]

In order to compensate for the remaining loss of specific impulse, the Block IIA engine was certified to operate at 104.5 percent power (although it is still called 104 percent). This also offset the slightly heavier weight of the Block IIA engines. The Block IIA engine is capable of 109 percent power for contingency aborts. The maintenance goal for the Block IIA engine is to fly for ten missions without the need to replace any major components or tear down the engine for inspection.

Fourteen engines (2043–2056) were built to the Block IIA configuration, and all will be modified to the Block II configuration when the new HPFT becomes available. These engines were not completely new-builds, and used many components from the earlier Block I and Phase II engines. The Block IIA first flew on STS-89 in January 1998, with all three engines being the new design. The Block IIA engine further reduced the ascent risk of the vehicle, to one in 438.[152]

Block II

The Block II configuration was scheduled to enter certification testing in late 1996, and was tentatively scheduled to make its first flight in September 1997. Obviously, the January 1996 test failure of the alternate HPFT delayed this. Finally, towards the end of 2000 it finally appears that all the problems have been overcome, and the Block II configuration is in the final process of being qualified for flight. NASA currently expects to fly a single Block II engine (with two Block IIAs) during early 2001.

The Block II engine incorporates all of the Block IIA improvements plus the new Pratt & Whitney 'alternate' high-pressure fuel turbopump (HPFT). All existing Block IIA engines will eventually be brought up to the Block II standard, resulting in 14 of the new engines being available. However, the Space Shuttle Program has a requirement for 15 engines – at any given time this would allow 12 engines to support launch operations at KSC, plus three engines undergoing periodic maintenance or overhaul. Unfortunately, the last engine is currently unfunded and it is uncertain when, or if, the engine will be manufactured.[153]

And the full-up Block II main engine finally provides a capability that has been sought almost since the beginning of the program – the ability to routinely operate at a full-power level of 109 percent. The increase in power will further boost the payload capacity into high-inclination orbits. Based on preliminary data, the Block II engine will again reduce the ascent risk for the vehicle. Current estimates are that with three Block II engines, the chances of losing a vehicle during ascent will be one in 483.[154]

EXTERNAL TANK

The External Tank (ET) contains the propellants for the Space Shuttle Main Engines, and is built by Lockheed Martin Space System Company in Michoud, Louisiana. Each ET is 153.8 feet long and 27.6 feet in diameter. The ET consists of the forward LO2 tank, an unpressurized intertank, and an LH2 tank.[155]

Liquid Oxygen Tank

The LO2 tank is a 2219 aluminum (early tanks) or 2915 aluminum-lithium (SLWT) alloy monocoque structure composed of a fusion-welded assembly of preformed, chemically-milled gores, panels, machined fittings, and ring chords. It operates in a pressure range of 20–22 psig. The tank contains anti-slosh and anti-vortex provisions to minimize liquid residuals and damp fluid motion. The tank feeds into a 17-inch diameter feed line that conveys the liquid oxygen through the intertank, then outside the ET to the aft right-hand ET–Orbiter disconnect umbilical. The right aft umbilical assembly consists of an electrical disconnect, the 2-inch gaseous oxygen (O2 or GOX) pressurization disconnect used for pressurizing the ET LO2 tank, and the 17-inch liquid oxygen disconnect.[156]

The 17-inch diameter feed line permits LO2 to flow at approximately 2,787 pounds per second with the SSMEs operating at 104 percent or permits a maximum flow of 17,592 gallons per minute. The double-wedge nose cone of the LO2 tank reduces aerodynamic drag and heating, contained the ascent air data system (for the first nine tanks only), and serves as a lightning rod for the Shuttle assembly. Total capacity of the LO2 tank is 19,182 cubic feet, which translates to 143,351 gallons or 1,391,936 pounds of LO2. The tank is 27.58 feet in diameter, 49.33 feet long, and weighs 12,000 pounds empty.

intertank

The intertank is 22.5 feet long, 27.58 feet in diameter, and weighs 12,100 pounds. The cylindrical aluminum alloy semi-monocoque structure has flanges on each end for joining to the LO2 and LH2 tanks. The intertank houses the ET instrumentation components and provides an umbilical plate that interfaces with the launch pad for purge gas supply, hazardous gas detection, and hydrogen gas boiloff during ground operations. It consists of mechanically joined skin, stringers, and machined panels of 2219 aluminum (early tanks) or 2915 aluminum-lithium (SLWT) alloy. The intertank is vented in flight. The SRB thrust beam and fittings that distribute the SRB thrust loads to the LO2 and LH2 tanks are contained in the intertank.[157]

Liquid Hydrogen Tank

The LH2 tank is a 96.66 foot long fusion-welded 2219 aluminum (early tanks) or 2915 aluminum-lithium (SLWT) structure designed to operate in a pressure range of 32–34 psia. It consists of barrel sections, five major ring frames, and forward and aft ellipsoidal domes. The tank contains an anti-vortex baffle and siphon outlet to transfer LH2 from the tank through a 17-inch line to the left aft umbilical. The left aft umbilical assembly consists of an electrical disconnect, the 2-inch gaseous hydrogen pressurization disconnect used for pressurizing the LH2 tank, the four inch recirculation disconnect used during prelaunch to precondition the SSMEs, and the 17-inch LH2 disconnect.[158]

The LH2 tank does not contain the complex anti-slosh baffles that are installed in the LO2 tank for the simple reason that LH2 is so light that sloshing does not induce significant forces. The LH2 feed line flow rate is 465 pounds per second with the SSMEs at 104 percent or a maximum flow of 47,365 gallons per minute. At the forward end of the liquid hydrogen tank is the ET–Orbiter forward attachment pod strut, and at its aft end are the two ET–Orbiter aft attachment ball fittings as well as the aft SRB–ET stabilizing strut attachments. The LH2 tank weighs 29,000 pounds when empty, and can hold 385,265 gallons of LH2 weighing 237,641 pounds. It has an internal volume of 51,737 cubic feet.

ET Thermal Protection System

The ET is thermally protected with a nominal one inch CPR-488 spray-on foam insulation, and charring ablators (MA-25S and SLA-220) to withstand localized high heating during ascent. The thermal protection system weighs approximately 4,823 pounds. The insulation was protected by a white-colored Fire-Retardant Latex (FRL) coating for the first two flights (STS-1 and STS-2), but has been left bare on subsequent flights to save 595 pounds and $15,000. Thermal isolators are required for the LH2 tank attachments to preclude the liquefaction of air-exposed metallic attachments and to reduce heat flow into the liquid hydrogen.[159]

Systems

Each propellant tank has a vent and relief valve at its forward end. This dual-function valve can be opened by ground support equipment for the vent function during prelaunch and can open during flight when the ullage pressure of the LH2 tank reaches 36 psig, or the LO2 tank reaches 31 psig.[160]

There are eight propellant-depletion sensors, four each for oxidizer and fuel. The LO2 depletion sensors are mounted in the Orbiter LO2 feed line manifold, while the fuel depletion sensors are mounted in the bottom of the LH2 tank. The locations of the LO2 sensors allow the maximum amount of oxidizer to be consumed in the engines, while allowing sufficient time to shut down the engines before the oxidizer pumps cavitate (run dry). In addition, 1,100 pounds of LH2 are loaded over and above that required by the 6:1 oxidizer to fuel engine mixture ratio. This assures that MECO from the depletion sensors is fuel-

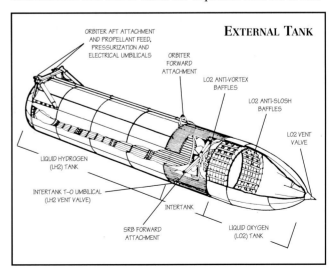

EXTERNAL TANK

ORBITER AFT ATTACHMENT AND PROPELLANT FEED, PRESSURIZATION AND ELECTRICAL UMBILICALS

ORBITER FORWARD ATTACHMENT

LO2 ANTI-VORTEX BAFFLES

LO2 ANTI-SLOSH BAFFLES

LO2 VENT VALVE

LIQUID HYDROGEN (LH2) TANK

INTERTANK T-0 UMBILICAL (LH2 VENT VALVE)

INTERTANK

SRB FORWARD ATTACHMENT

LIQUID OXYGEN (LO2) TANK

rich; oxidizer-rich engine shutdowns can cause burning and severe erosion of engine components. Normally, main engine cutoff is based on a predetermined velocity; however, if any two of the oxidizer or fuel sensors indicate dry, the engines will shut down to prevent catastrophic damage.[161]

The ET is separated from the Orbiter 18 seconds after main engine cutoff, just prior to reaching orbital velocity, and proceeds on a ballistic path for impact in the Pacific or Indian Ocean. Separation is normally commanded by the on-board GPCs, but can also be initiated manually by the flight crew. Gaseous oxygen is vented from the LO2 vent valve in the nose of the ET to induce a tumbling motion (at a minimum of ten degrees per second end-over-end) that ensures the ET breaks up in flight. The ET is the only major component of the Space Transportation System that is not recovered and reused.

The forward structural attachment from the ET to the Orbiter consists of a shear bolt mounted in a spherical bearing. The bolt separates at a break point when two pressure cartridges are initiated – one or both cartridges drives one of a pair of pistons to shear the bolt. A centering mechanism rotates the unit from the displacement position to a centered position, aligning the bearing flush with the adjacent thermal protection system mold line.

The ET has five propellant umbilical valves that interface with Orbiter umbilicals – two for the LO2 tank and three for the LH2 tank. One of the LO2 tank umbilical valves is for liquid oxygen, the other for gaseous oxygen. A swing-arm-mounted beanie cap on the fixed service structure covers the gaseous oxygen vent valve on top of the ET during the countdown and vents oxygen vapor that threatens to form ice on the ET, thus protecting the Orbiter's TPS during launch. The LH2 tank umbilical has two valves for liquid and one for gas. The intermediate-diameter liquid hydrogen umbilical is a recirculation umbilical used only during the LH2 chill-down sequence during prelaunch. The ET also has two electrical umbilicals that carry electrical power from the Orbiter to the ET and the two SRBs and provide information from the SRBs and ET to the Orbiter.

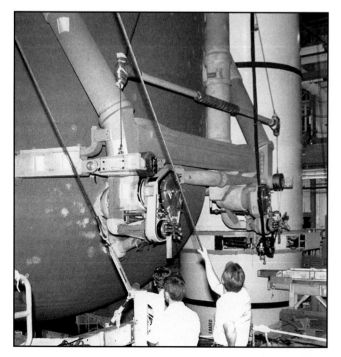

The ET for STS-30R shows the two umbilicals that carry all power, fluids, and gases between the ET and the rest of the Space Shuttle stack. (NASA photo KSC-90P-2014)

Each of the two ET umbilical plates mate with a corresponding plate on the bottom of the Orbiter. The plates help maintain alignment among the umbilicals. Physical strength at the umbilical plates is provided by bolting corresponding umbilical plates together. The aft structural attachment consists of a pair of bolts and pyrotechnically actuated frangible nuts that attach the ET struts to the Orbiter attach points. When the GPCs command ET separation, the bolts are severed by pyrotechnic devices. At separation, the frangible nuts are split by a booster cartridge initiated by a detonator cartridge and the attach bolts are driven by the separation forces and a spring into a cavity within the tank strut. On the Orbiter side, each frangible nut and its detonators is enclosed in a debris container that captures the nut fragments.

After the release of each of the aft attach struts, three lateral support arms at each Orbiter umbilical plate hold the plate in the lateral position while the ET separates from the umbilical plates. The Orbiter umbilical plates are then retracted approximately 2.5 inches inside the Orbiter aft-fuselage by three hydraulic actuators and locked to permit closure of the umbilical doors in the bottom of the aft-fuselage. A closeout curtain is installed at each umbilical to prevent hazardous gases (gaseous oxygen and hydrogen) from entering the Orbiter aft-fuselage through the umbilical openings before the umbilical doors are closed. Each umbilical door, which is approximately 50 square inches, is covered with reusable thermal protection system tiles. Additionally, each door has an aerothermal barrier that requires approximately 6 psi to seal it with adjacent thermal protection tiles. There are two types of latches associated with the ET doors, center-line latches and uplock latches. The centerline latches hold the ET doors open during ascent, and the uplock latches secure the doors when they are closed. The centerline latches must be disengaged from the doors before both ET doors can be closed. In the latched position, each centerline latch fits into a notch on the outer edge of the left and right ET doors. When stowed, the two centerline latches rotate and retract into the body of the Orbiter so that they are flush with the thermal protection system mold line. After the ET doors are closed, uplock latches pull the doors fully closed, compressing them against aerothermal seals and positioning them flush with the Orbiter body. These latches secure each door closed and prevent them from vibrating or opening during reentry. Three uplock latches are located inside each umbilical cavity. The latches engage three uplock rollers located on each ET door. The latches can engage the rollers as long as the ET door is within 2 inches of being closed.[162]

Light-Weight Tank

A 'light-weight' ET entered production in time for the first flight (STS-6) of *Challenger*, weight being saved primarily by eliminating portions of the stringers in the hydrogen tank, and milling various skin panels to a reduced thickness. Weight was also saved by redesigning the anti-slosh baffling and eliminating the anti-geyser line used to circulate liquid oxygen during fill. Material changes to the ET included converting all 5A1-2.5 titanium alloy fittings to the more common 6A1-4V titanium alloy, and all 7075-T73 aluminum hardware to 7050-T73, allowing thinner sections because of a ten percent increase in strength. The initial light-weight tanks (LWT) had an empty weight of 66,800 pounds, 10,300 pounds lighter than the first standard-weight tanks (SWT).[163]

Composite Nose Cone

In June 1989, MSFC and Martin Marietta began development of a new advanced composite nose cone for the ET. The development was in response to a report that indicated that debris from the ET nose cone thermal protection system had damaged several Orbiter TPS tiles on STS-27R.

The new nose cone is constructed from a high-temperature resistant polymeric-matrix composite. The piece is manufactured by hand lay-up using 18-21 plies (sheets) of graphite phenolic cloth inside a graphite mold, then placed in an autoclave at 350 degF for 13 hours. An addition 16 hours is spent at 415 degF to ensure the chemical reaction of the raw materials is complete. The finished product can withstand temperatures in excess of 900 degF, eliminating the need for the thermal protection system.

The new design also results in a weight savings of 21 pounds. The original design used over 1,100 fasteners to assemble multiple sheet metal pieces – the new composite nose cone is fabricated as a single piece, resulting in a significant reduction in assembly time.

The new nose cone is manufactured by the MSFC Productivity Enhancement Center. Two non-production units were delivered for testing in January 1994, and the first production article was delivered to the ET production facility at Michoud for use on ET #81.

Super Light-Weight Tank

In 1991 MSFC awarded Martin Marietta a contract to further reduce the weight of the External Tank by applying new material science to its construction. Improvements in the manufacturing of the last production batch of light-weight tanks had decreased their weight even further, to approximately 65,000 pounds. However, the cancellation of the Advanced Solid Rocket Motor (ASRM), and the decision to construct the International Space Station (ISS) in an extremely high-inclination (51.6-degree) orbit, again levied a requirement to increase the payload capacity of Shuttle.[164]

Martin Marietta Labs in Baltimore developed and patented an 2195 aluminum-lithium (Al-Li) alloy that is being manufactured by the Mill Products Division of Reynolds Aluminum in McCook, Illinois. The alloy is 40 percent stronger and 10 percent less dense than the 2219 aluminum alloy it replaces. And like 2219, the new alloy actually becomes slightly stronger at cryogenic temperatures. The 2195 alloy consists of 4.0 percent copper, 0.4 percent magnesium, 0.4 percent silver, 1.0 percent lithium, and 94.2 percent aluminum.

The same tooling at the Michoud Assembly Facility that was used to manufacture the LWTs is being used with the new alloy without major changes, making the SLWT much more affordable than most other techniques to improve the Shuttle's performance. Unfortunately, the use of the new alloy has increased the production cost from approximately $50 million to $59 million each in FY95 dollars. Total cost to develop and certify the new super light-weight tank (SLWT) was approximately $172.5 million. One of the major obstacles was that there was very little manufacturing experience with the new metal, and no established experience base for the properties of the material or for the welding techniques required during tank manufacturing.

Unlike 2219 aluminum, the new 2195 alloy is very sensitive to contamination during the welding process. Originally, Lockheed Martin used a traveling purge box that

The new waffle-grid design of the LH2 tank for the SLWT is illustrated above, while the front dome of the LO2 tank is shown below. The size of the ET is illustrated by the men in both tanks. (Lockheed Martin Space System Company, Michoud Operations)

End of the line – a LWT just after separation. Note the interesting flow pattern on the tank. Soon the tank would begin tumbling and burn up over the Pacific Ocean. (Lockheed Martin Space Systems Company, Michoud Operations)

was located over a small section to be welded to provide an inert (nitrogen) atmosphere, then moved to the next section – a time consuming process. Subsequently, a static purge chamber was developed that provides an inert atmosphere over the entire length of the piece to be welded. This eliminated all of the moving parts that were jamming, and also provides a more consistent delivery of the purge gas.[164]

The weight of the SLWT is approximately 57,470 pounds – a 7,500-pound reduction from the last LWTs. Approximately 4,500 pounds of the weight reduction is the direct result of the new alloy. An additional 2,500 pounds comes from replacing the LH2 tank T-stiffeners with a waffle-grid design. A new variable output pumping station allows the thickness of the CPR-488 spray-on foam insulation to be varied, resulting in another 500 pound weight savings.

Not surprisingly, the SLWT has had effects on other aspect of the vehicle – specifically, the weight reduction causes the vehicle center-of-gravity to move about 6–8 inches further up into the Z-axis at MECO. This higher c.g. makes controlling the vehicle difficult on abort-to-orbit (press to MECO) cases with SSME #1 failed when the Orbiter Z-c.g. is near the top of the reentry c.g. box. The major problem arises because the thrust structure in the aft-fuselage deforms at slightly different rates for each engine location, limiting the engine gimbal/pitch-change authority under some circumstances. Engineers finally concluded that the easiest solution was to throttle the remaining two engines back to 67 percent power (instead of the normal 91 percent) to minimize this deformation. Although this has eliminated the basic problem, it results in the loss of about 15 fps, resulting in a 6–8 mile lower apogee than before.[166]

Originally, ET #89 was scheduled to be the first SLWT, but funding problems delayed this to ET #96, delivered to KSC on 12 January 1998 – the first SLWT flew on STS-91 in June 1998. In a ceremony on 20 August 1999, Lockheed Martin delivered the 100th ET (the fifth SLWT) to NASA.

The second super lightweight External Tank arrives at KSC and is moved to the VAB after being removed from the barge that brought it from Michoud. The special trailer that transports the ETs is secured at the SRB attach points in front – the ET just rests on the trailer at the back. (NASA photo KSC-98PC-0518)

Solid Rocket Boosters

Two Solid Rocket Boosters (SRB) burn for two minutes with the Space Shuttle Main Engines to provide initial ascent thrust for the vehicle. The SRBs are the largest solid-propellant rocket motors ever flown, and the first to be man-rated. The solid rocket motor (SRM) segments are built by the Thiokol Propulsion division of Cordant Technologies, Brigham City, Utah. The non-motor parts of the SRBs (forward and aft skirts, separation motors, frustrum, parachutes, and the nose cap) were originally manufactured by United Space Boosters Incorporated (USBI), in Huntsville, Alabama, with other portions being made in-house by the MSFC Science and Engineering Directorate. On 1 October 1999 the functions performed by USBI were rolled into the Space Flight Operations Contract (SFOC) under United Space Alliance (USA).

The two SRBs provide the main thrust to lift the Space Shuttle off the pad and up to an altitude of about 150,000 feet. In addition, the two SRBs carry the entire weight of the ET and Orbiter, and transmit the weight load through their structure to the mobile launch platform (MLP). The SRBs provide 71.4 percent of the total vehicle thrust at lift-off and during first stage ascent. Seventy-five seconds after SRB separation, SRB apogee occurs at an altitude of approximately 220,000 feet. SRB impact is approximately 141 miles downrange in the Atlantic Ocean.

The assembled SRB is 149.16 feet long, 12.17 feet (146 inches) in diameter, and weighs approximately 1,250,000 pounds. The original SRBs used for STS-1 through STS-7 produced 2,800,000 lbf in a vacuum. The development of a high-performance motor (HPM) that produces 3,000,000 lbf began in October 1980, was tested beginning in March 1983, and made its first flight on STS-8. This is essentially the same motor (with a slightly different case) used today. Each SRB has a specific impulse of 268.4 seconds. The nozzle expansion ratio for each booster (beginning with STS-8) is 7:79.[167]

Construction

Each SRB is made up of several subassemblies – nose cone, Solid Rocket Motors (SRM), and the 23,932-pound nozzle assembly. Each solid rocket motor case is made of 11 individual cylindrical weldless D6AC steel sections. When assembled they form a tube almost 116 feet long and weigh approximately 150,023 pounds when empty. The eleven sections are the forward dome section, six center cylindrical sections, the aft ET attachment ring section, two stiffener sections, and the aft dome section. The 11 sections are joined by tang-and-clevis joints held together by 177 steel pins around the circumference of each joint.

Four distinct types of cases have been used by the solid

One of the RSRM structural test articles (STA-3) is placed into a test facility at MSFC. From December 1987 to April 1988, STA-3 underwent a series of six tests at the MSFC designed to demonstrate the structural strength of the SRB. (NASA photo MSFC-00348)

rocket motors. Standard-weight cases (STD) were used on STS-1, 2, 3, 4, 5, 8, and 9. Light-weight cases (LTWT), with walls that were 0.003 to 0.005 inches thinner, were used for STS-6 and STS-7, saving approximately 5,000 pounds. However, there were concerns that too much material had been removed, and the program reverted back to the standard-weight cases while additional analysis were conducted. Four interim sets of medium-weight cases (MWC) were constructed with walls 0.001 to 0.002 inches thinner, and used for STS-11/41-B, STS-13/41-C, STS-25/51-G, and STS-26/51F.

Segments used for all other flights prior to STS-26R had walls that were 0.002 to 0.004 inches thinner than standard-weight segments, affording a weight reduction of 4,000 pounds. These were also known as light-weight cases, but were abbreviated LWC instead of LTWT. The current RSRM uses essentially the same case design.

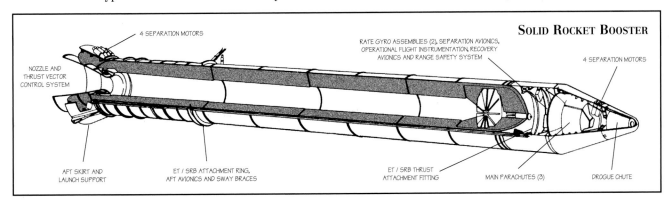

SOLID ROCKET BOOSTER

4 SEPARATION MOTORS

RATE GYRO ASSEMBLIES (2), SEPARATION AVIONICS, OPERATIONAL FLIGHT INSTRUMENTATION, RECOVERY AVIONICS AND RANGE SAFETY SYSTEM

4 SEPARATION MOTORS

NOZZLE AND THRUST VECTOR CONTROL SYSTEM

AFT SKIRT AND LAUNCH SUPPORT

ET / SRB ATTACHMENT RING, AFT AVIONICS AND SWAY BRACES

ET / SRB THRUST ATTACHMENT FITTING

MAIN PARACHUTES (3)

DROGUE CHUTE

SRB DETAILS

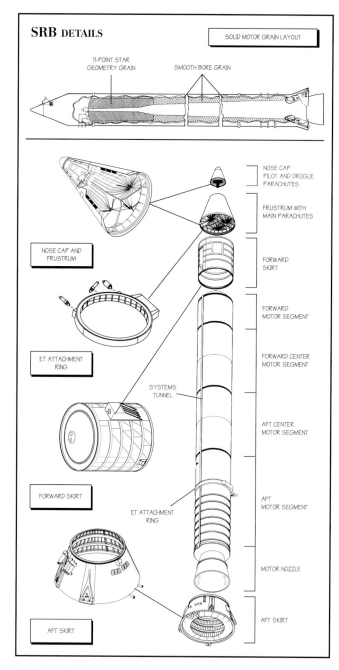

SOLID MOTOR GRAIN LAYOUT

11-POINT STAR GEOMETRY GRAIN

SMOOTH BORE GRAIN

NOSE CAP
PILOT AND DROGUE
PARACHUTES

FRUSTRUM WITH
MAIN PARACHUTES

NOSE CAP AND
FRUSTRUM

FORWARD
SKIRT

ET ATTACHMENT
RING

FORWARD
MOTOR SEGMENT

FORWARD CENTER
MOTOR SEGMENT

SYSTEMS
TUNNEL

AFT CENTER
MOTOR SEGMENT

FORWARD SKIRT

ET ATTACHMENT
RING

AFT
MOTOR SEGMENT

MOTOR NOZZLE

AFT SKIRT

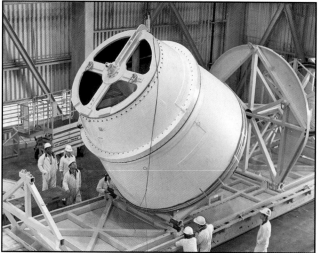

One of the SRB nozzles intended for STS-26R is off-loaded from the rail car it was delivered on. (NASA photo KSC-88P-97)

After the case sections have been machined and fitted, they are partly assembled at the factory into four 'casting' segments. These are the segments into which the propellant (TP-H1148) is poured (or cast). The SRBs are used as matched pairs, and each is made up of four solid rocket motor segments. The pairs are matched by loading each of the four motor segments in pairs from the same batches of propellant ingredients to minimize any thrust imbalance. Each segment is shipped to the launch site on a heavy-duty rail car with a specially built cover.[169]

Joints assembled before the segment is filled with propellant are known as 'factory' joints – joints between the four casting segments are called 'field' joints, and are assembled at the launch site. The case liner material is an asbestos filled carboxyl terminated polybutadiene (CTPB) polymer known as UF-2137.

Due to its part in the *Challenger* accident, the SRM field joint metal parts, internal case insulation and seals were redesigned, and a weather protection system added. In the STS-33/51-L design, the application of actuating pressure to the upstream face of the O-ring was essential for proper joint sealing performance because large sealing gaps were created by pressure-induced deflections, compounded by significantly reduced O-ring sealing performance at low temperature. The major change in the motor case is the new tang capture feature to provide a positive metal-to-metal interference fit around the circumference of the tang and clevis ends of the mating segments. The interference fit limits the deflection between the tang and clevis O-ring sealing surfaces caused by motor pressure and structural loads. The joints are designed so the seals will not leak under 200 percent design structural deflection and rates.

The new design, with the tang capture feature, controls the O-ring sealing gap dimensions. The sealing gap and the O-ring seals are designed so that a positive compression (squeeze) is always on the O-rings. The clevis O-ring groove dimension has been increased so that the O-ring never fills more than 90 percent of the groove. The new field joint design also includes an additional leak-check port to ensure that the primary O-ring is positioned properly. A new O-ring in the capture feature also serves as a thermal barrier in case the sealed insulation is breached. The field joint internal case insulation was modified to be sealed with a pressure-actuated flap called a J-seal, rather than putty as in the earlier configuration. Longer field joint case mating pins, with an improved retainer band, were added to improve the shear strength of the pins. External heaters with integral weather seals were incorporated to maintain the joint and O-ring temperature at a minimum 75 degF. The weather seal also prevents water intrusion into the joint.

The post-*Challenger* solid rocket motors were originally known as Redesigned SRMs (RSRM), but by 1995 had been renamed Reusable SRMs (still RSRM). In May 1999, Thiokol cast the 1,000th shuttle SRM segment – bringing the total to 277 million pounds of propellant cast for the Space Shuttle Program.[170]

Propellant

The 1,106,640 pounds of TP-H1148 contained in each SRB is a composite-type solid propellant formulated of an aluminum powder (16 percent) as a fuel; ammonium perchlorate (69.6 percent) as an oxidizer; iron oxidizer powder (0.4 percent) as a burning rate catalyst; polybutadiene acrylic acid acrylonitrile terpolymer (PBAN – 12.04 percent) as a rubber based binder; and an epoxy curing agent (1.96 percent). The binder and epoxy are also burned as fuel, adding a small amount of thrust, and unfortunately, a significant amount of pollutants. The propellant is an 11-point star-shaped perforation in the forward motor segment and a double-truncated-cone perforation in each of the aft segments and aft closure. This configuration provides high thrust at ignition and then reduces the thrust by approximately thirty-percent 55 seconds after lift-off to prevent overstressing the vehicle during the period of maximum dynamic pressure.[171]

Structure

Each SRB is attached to the ET at the aft end of the forward skirt by a single thrust attachment. The recovery parachute system is attached to the same thrust structure. A non-load bearing sway brace connects the aft portion of the SRB with the bottom of the ET.

While on the launch pad, four 'hold-down' bolts secure each SRB to the MLP. These eight bolts are all that secure the Space Shuttle stack to the MLP. Each bolt is 28 inches long and 3.5 inches in diameter. The top nut of each bolt is frangible, containing two redundant NASA Standard Initiators (NSI) which are ignited at SRB ignition. When the two NSIs are ignited at each hold-down, the hold-down bolt travels downward because of the release of tension in the bolt, NSI gas pressure and gravity. The bolt is stopped by the stud deceleration stand, which contains sand and the frangible nut is captured in a blast container.[172]

During the *Challenger* stand-down, detailed structural analyses were performed on critical structural elements of the SRB. These analyses were primarily focused in areas where anomalies had been noted during post-flight inspection of recovered hardware.[173]

One of the areas was the attach ring where the SRBs connected to the ET. Unanticipated stress was noted in some fasteners where the ring attached to the motor case. This situation was attributed to the high loads encountered during water impact during recovery. To correct the situation, and ensure higher strength margins during ascent, the attach ring was redesigned to completely (360 degrees)

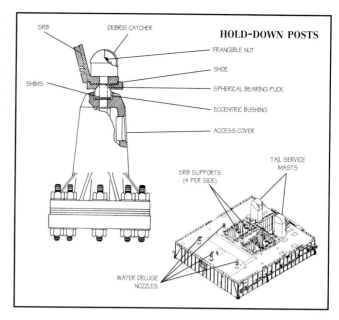

encircle the SRM case. Previously, the attach ring encircled the motor case only 270 degrees, forming a 'C.'[174]

Additionally, special structural tests were performed on the aft skirt, and an anomaly was noted in a critical weld between the hold-down post support and the skin of the skirt. A redesign was implemented to add reinforcement brackets and fittings in the aft ring of the skirt. This was deemed particularly critical in the event of an on-pad abort situation. These modifications added approximately 450 pounds to the weight of each SRB.[175]

The tests did not necessarily end immediately after the Return-to-Flight. As late as FY98, ten tests to measure the buckling characteristics of the SRM cases were conducted. A special test rig was developed that could apply up to 3.4 million pounds of bending forces on an RSRM case to measure its ability to withstand the 'twang' of SSME ignition while the vehicle is still secured to the launch mount (the top of the ET sways as much as three feet at engine ignition). Wind loading (especially southerly winds) can

An SRB comes together in the VAB. At left is the aft motor segment already attached to the aft skirt; the aft center motor segment shows the smooth-bore grain configuration at the rear of the booster; the forward motor segment (left) has a structural dome at the top to contain the pressure. (NASA photos KSC-88PC-203, KSC-88PC-373, and KSC-88P-819)

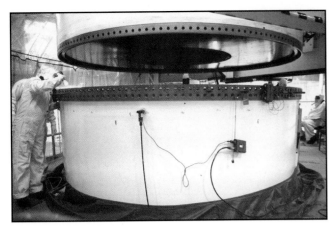

The aft segment and aft center segment show some of the 177 holes for the pins that hold the field-joint segments together. After the pins are inserted, a metal strap is wrapped around them to lock them in place. (NASA photo KSC-00PP-0814)

add to this bending (called buckling capability, or 'twang') before the stack springs back to vertical and the SRBs are ignited. The tests showed that analytical calculations were within five percent, allowing the program to continue with renewed confidence.[176]

Electronics

Each SRB has two integrated electronic assemblies, one forward and one aft. The forward assembly jettisons the nozzle after burnout, releases the nose cap and frustum, detaches the parachutes, and turns on the recovery aids. The aft assembly, mounted in the ET–SRB attach ring, connects with the forward assembly and the Orbiter avionics system for SRB ignition commands and thrust vector control. Each integrated electronics assembly has a multiplexer/demultiplexer (MDM) that sends or receives more than one signal or unit of information on a single communications channel.[177]

Each SRB contains two rate gyro assemblies (RGA), with each RGA containing one pitch and one yaw gyro. These provide an output proportional to angular rates about the pitch and yaw axes during first-stage ascent flight in conjunction with the Orbiter roll rate gyros until SRB separation. At SRB separation, a switchover is made from the SRB RGAs to the Orbiter RGAs. The SRB RGA data are transmitted through the Orbiter flight-aft multiplexer/demultiplexers to the GPCs. Each RGA is designed to be reusable up to 20 times.[178]

Ignition

SRB ignition can occur only after a manual lock pin from each SRB safe and arm device has been removed. The ground crew removes the pin during prelaunch activities, and at T–5 minutes the SRB safe and arm device is remotely rotated to the 'arm' position. The SRB ignition commands can be issued only when all three SSMEs are at or above 90 percent rated thrust, there are no SSME fail indicators, no SRB ignition PIC low-voltage indicators, and there are no 'holds' indicated by the ground-based Launch Processing System.[179]

A PIC single-channel capacitor discharge device controls the firing of each pyrotechnic device. Three signals must be present simultaneously for the PIC to generate the pyro firing output. These signals – ARM, FIRE-1, and FIRE-2 – originate in the Orbiter GPCs, and are transmitted to the main engine controllers. The MECs reformat

them to 28 Vdc signals for the PICs. The arm signal charges the PIC capacitor to 40 Vdc (minimum of 20 Vdc). The fire commands cause the redundant NSIs to fire through a thin barrier seal down a flame tunnel. This ignites a pyro booster charge, which is retained in the safe and arm device behind a perforated plate. The booster charge ignites the propellant in the igniter initiator; and combustion products of this propellant ignite the solid rocket motor initiator, which fires down the length of the SRM, igniting the propellant.

Electrical and Hydraulic Systems

Electrical power distribution in each SRB consists of Orbiter-supplied main dc bus power to each SRB via SRB buses A, B, and C. Orbiter main dc buses A, B, and C supply main dc bus power to corresponding SRB buses A, B, and C. In addition, Orbiter main dc bus C supplies backup power to SRB buses A and B, and Orbiter bus B supplies backup power to SRB bus C. This electrical power distribution arrangement allows all SRB buses to remain powered in the event one Orbiter main bus fails. The nominal voltage is 28 Vdc, with an upper limit of 32 Vdc and a lower limit of 24 Vdc.[180]

Each SRB also contains two self-contained hydraulic power units (HPU), each consisting of an auxiliary power unit, fuel supply, hydraulic pump, hydraulic reservoir, and hydraulic fluid manifold assembly. The APUs are fueled by hydrazine and generate mechanical shaft power to a hydraulic pump that produces hydraulic pressure for the SRB hydraulic system. Each SRB APU/HPU is reusable for 20 missions after refurbishment.[181]

The two separate HPUs and two hydraulic systems are located on the aft end of each SRB between the SRB nozzle and aft skirt. The HPU components are mounted on the aft skirt between the rock and tilt actuators. The two systems operate from T–28 seconds until SRB separation.

The APU controller electronics are located in the SRB aft integrated electronic assemblies on the aft ET attach rings. The APUs and their fuel systems are isolated from each other. Each fuel supply module (tank) contains 22 pounds of hydrazine. The fuel tank is pressurized with gaseous nitrogen at 400 psi, which provides the force to expel (positive expulsion) the fuel from the tank to the fuel distribution line, maintaining a positive fuel supply to the APU throughout its operation.[182]

The APU turbine assembly provides mechanical power to the APU gearbox. The gearbox drives the APU fuel pump, hydraulic pump and lube oil pump. The APU lube oil pump lubricates the gearbox. The turbine exhaust of each APU flows over the exterior of the gas generator, cooling it, and is then directed overboard through an exhaust duct.

Each Hydraulic Power Unit on an SRB is connected to both servoactuators on that SRB. One HPU serves as the primary hydraulic source for the servoactuator, and the other HPU provides as the secondary hydraulics for the servoactuator. Each servoactuator has a switching valve that allows the secondary hydraulics to power the actuator if the primary hydraulic pressure drops below 2,050 psi. The 100 percent APU speed enables one APU/HPU to supply sufficient operating hydraulic pressure to both servoactuators of that SRB. The APU 10 percent speed corresponds to 72,000 rpm, 110 percent to 79,200 rpm, and 112 percent to 80,640 rpm. The hydraulic pump speed is 3,600 rpm and supplies hydraulic pressure of 3,050 ±50 psi. A relief valve provides overpressure protection to the hydraulic system and relieves at 3,750 psi.[183]

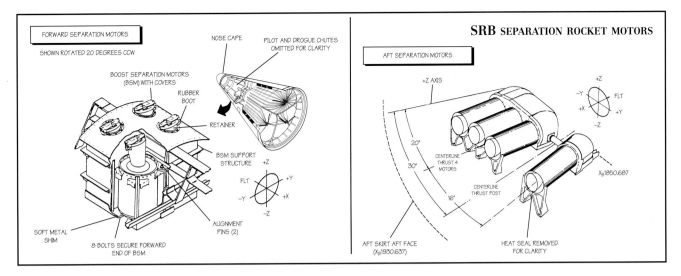

Labels in figure: FORWARD SEPARATION MOTORS / SHOWN ROTATED 20 DEGREES CCW / BOOST SEPARATION MOTORS (BSM) WITH COVERS / RUBBER BOOT / NOSE CAPE / PILOT AND DROGUE CHUTES OMITTED FOR CLARITY / RETAINER / BSM SUPPORT STRUCTURE / SOFT METAL SHIM / 8 BOLTS SECURE FORWARD END OF BSM / ALIGNMENT PINS (2) / AFT SEPARATION MOTORS / +Z AXIS / 20° / 30° / 16° / CENTERLINE THRUST 4 MOTORS / CENTERLINE THRUST POST / AFT SKIRT AFT FACE (X$_B$1930.637) / HEAT SEAL REMOVED FOR CLARITY / X$_B$1850.687

Nozzle

The SRB nozzle is a convergent-divergent, movable design in which an aft pivot-point flexible bearing is the gimbal mechanism. Each nozzle has a carbon cloth phenolic (CCP) liner that erodes and chars during firing. The rayon yarn used as a precursor material in the manufacture of the CCP was taken out of production in September 1997. Prior to this happening, the Space Shuttle Program stockpiled sufficient material to last through the end of 2005. In early 1999 studies were initiated to find a replacement material, and 22 candidates have completed the initial screening process and six have been tested on the MNASA-9 and MNASA-10 test firings. All six were deemed satisfactory. Ultimately the best candidate will be selected in early 2002 and tested in three full-scale static test motors before being used on production RSRM nozzles.[184]

The nozzle of each SRB can be gimbaled up to eight degrees for thrust vector control. Each SRB has two hydraulic gimbal servoactuators: one for rock and one for tilt. The servoactuators provide the force and control to gimbal the nozzle for thrust vector control. Each actuator ram is equipped with multiple transducers that provide position feedback to the thrust vector control system computers. A splashdown load relief assembly is contained within each servoactuator to cushion the nozzle at splashdown, and prevent damage to the nozzle bearing.

SRB Separation and Recovery

SRB separation is initiated when the three solid rocket motor chamber pressure transducers and the head-end chamber pressure of both SRBs is less than or equal to 50 psi. A backup cue is the time elapsed from booster ignition. The residual thrust from each SRB at this point is less than 100,000 lbf. The separation sequence is initiated, commanding the thrust vector control actuators to the null position and putting the main propulsion system into a second-stage configuration (0.8 second from sequence initialization). Orbiter yaw attitude is held for 4 seconds as the SRB thrust drops to less than 60,000 lbf.[185]

The SRBs are released by pyrotechnic separation devices at the forward thrust attachment and the aft sway braces. Separation is complete within 30 milliseconds of the ordnance firing command. The forward attachment point consists of a ball (SRB side) and socket (ET) held together by one bolt. The bolt contains one NSI pressure cartridge at each end. The forward attachment point also carries the range safety system cross-strap wiring connecting each SRB and the ET range safety systems with each other. The aft attachment points consist of three separate struts – upper, diagonal and lower. Each strut contains one bolt with an NSI pressure cartridge at each end. The upper strut also carries the umbilical interface between its SRB and the ET and on to the Orbiter.

Eight solid-fueled separation rockets on each SRB (four in the nose frustum and four in the aft skirt) thrust for 1.02 seconds to provide positive separation of the SRB from the ET. Each separation rocket is 31.1 inches long and 12.8 inches in diameter. The separation rockets in each cluster of four are ignited by firing redundant NSI pressure cartridges into redundant confined detonating fuse manifolds. The OI-28 software featured an interesting addition – to prevent the Orbiter forward and middle window panes against hazing that results from the SRB separation motor exhaust plumes, the forward upward-firing RCS thrusters are briefly fired to deflect the plume.[186]

The recovery sequence begins with the operation of the high-altitude baroswitch, which triggers the functioning of the pyrotechnic nose cap thrusters. This ejects the nose cap, which deploys the pilot parachute. This occurs at 15,700 feet altitude 225 seconds after separation. The 11.5 foot diameter conical ribbon pilot chute provides the force to pull the lanyard activating the zero-second cutter, which cuts the loop securing the drogue retention straps. This allows the pilot chute to pull the drogue pack from the SRB, causing the drogue suspension lines to deploy from their stored position. At full extension of the twelve 95-foot suspension lines, the drogue deployment bag is stripped away from the canopy, and the 54-foot diameter conical ribbon drogue parachute inflates to its initial reefed condition. The drogue disreefs twice after specified time delays, and it reorients/stabilizes the SRB for main chute deployment. The drogue can withstand a load of 270,000 pounds and weighs approximately 1,200 pounds.[187]

After the drogue chute has stabilized the booster in a tail-first attitude, the frustum is separated from the forward skirt by a charge triggered by the low-altitude baroswitch at an altitude of 5,975 feet 248 seconds after separation. It is then pulled away from the SRB by the drogue chute. The main chutes' suspension lines are pulled out from deployment bags that remain in the frustum. At full extension of the lines, which are 204 feet long, the three main chutes are pulled from the deployment bags and inflate to their first reefed condition. The frustum and drogue parachute continue on a separate trajec-

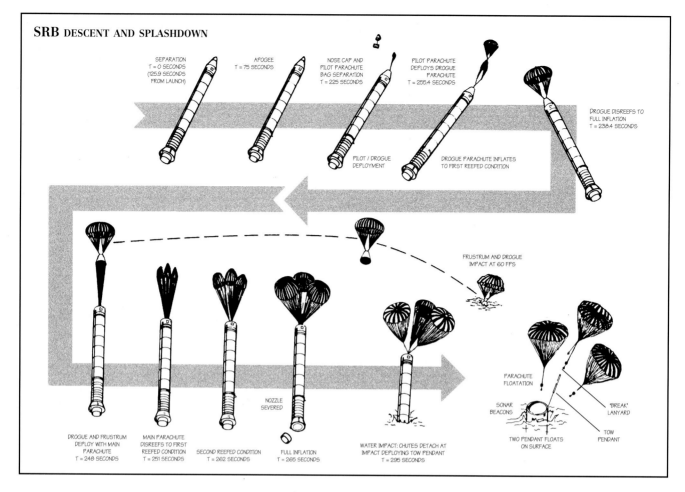

SEPARATION
T = 0 SECONDS
(125.9 SECONDS
FROM LAUNCH)

APOGEE
T = 75 SECONDS

NOSE CAP AND
PILOT PARACHUTE
BAG SEPARATION
T = 225 SECONDS

PILOT PARACHUTE
DEPLOYS DROGUE
PARACHUTE
T = 255.4 SECONDS

DROGUE DISREEFS TO
FULL INFLATION
T = 238.4 SECONDS

PILOT / DROGUE
DEPLOYMENT

DROGUE PARACHUTE INFLATES
TO FIRST REEFED CONDITION

FRUSTRUM AND DROGUE
IMPACT AT 60 FPS

PARACHUTE
FLOATATION

SONAR
BEACONS

'BREAK'
LANYARD

TOW
PENDANT

TWO PENDANT FLOATS
ON SURFACE

DROGUE AND FRUSTRUM
DEPLOY WITH MAIN
PARACHUTE
T = 248 SECONDS

MAIN PARACHUTE
DISREEFS TO FIRST
REEFED CONDITION
T = 251 SECONDS

SECOND REEFED CONDITION
T = 262 SECONDS

NOZZLE
SEVERED

FULL INFLATION
T = 265 SECONDS

WATER IMPACT: CHUTES DETACH AT
IMPACT DEPLOYING TOW PENDANT
T = 295 SECONDS

tory to splashdown. After specified time delays, the main chutes' reefing lines are cut and the chutes inflate to their second reefed and full open configurations. The main chute cluster decelerates the SRB from approximately 230 mph to 50 mph. Each of the three 136 foot diameter, 20 degree conical ribbon parachutes can withstand a load of 180,000 pounds and weighs 2,180 pounds. The nozzle extension is severed by pyrotechnic charge either at apogee or 20 seconds after low baroswitch operation, whichever occurs first.[188]

Water impact occurs 295 seconds after separation at a velocity of 81 fps approximately 141 miles off the eastern coast of Florida in an expected splashdown area about 6.9 by 10.4 miles. Because the parachutes provide for a nozzle-first impact, air is trapped in the empty (burned out) motor casing, causing the booster to float with the forward end approximately 30 feet out of the water.

The main chutes are released from the SRB at impact using the parachute release nut ordnance system. Residual loads in the main chutes deploy the parachute attach fittings with the redundant flotation tethered to each fitting. The drogue and frustum, each main chute with its flotation, and the SRB are buoyant. The SRB recovery aids are the radio beacon and flashing lights, which become operable at frustum separation. The radio transponder in each SRB has a range of 10.35 miles, and the flashing light has a night-time range of 5.75 miles.

The two retrieval ships which perform the SRB recovery are unique vessels that were specifically designed and constructed for this task. Recovery Vessel (R/V) *Freedom Star* and R/V *Liberty Star* are owned by NASA. They were built at Atlantic Marine Shipyard, Fort George Island, near Jacksonville, Florida, in 1980 and 1981. The ships

are 176 feet long, have a 37-foot beam, draw about 12 feet of water, and have a displacement of 1,052 tons. Two 2,900-hp engines turn two 7-foot diameter variable-pitch propellers. The ships also have two water-jet thrusters – installed to protect the endangered manatee population that inhabits regions of the Banana River where the ships are based. The system also allows divers to work near the ship during operations at a greatly reduced risk.[189]

The retrieval ships are on station at the time of splashdown, at about 9–11 miles from the impact area. As soon as the boosters enter the water, the ships accelerate to a speed of 15 knots and quickly close on the boosters. Each ship retrieves one booster. Upon arrival, the team first conducts a visual assessment of the flight hardware. The main parachutes are the first items to be brought on board. Their shroud lines are wound each of three of the four reels on the ship's deck. The drogue parachute, attached to the frustum, is reeled onto the fourth reel until the frustum is approximately 50 feet astern of the ship. The 5,000-pound frustum is then lifted from the water using the ship's power block and deck crane.

With the chutes and frustum recovered, attention turns to the SRB. The dive team prepares for booster recovery. Two small inflatable boats, with eight retrieval divers aboard, are deployed. The job of the first dive team is to install a diver-operated plug (DOP) in the nozzle of the booster. The DOP is launched from the ship and towed to the booster by one of the small boats. An air hose is then deployed from the ship. Once dive preparations are complete, the dive team enters the water for DOP insertion. The DOP is 22 feet long and weighs 1,100 pounds. It is neutrally buoyant in water, meaning it neither floats nor sinks, and is easily guided to the aft skirt at a depth of

The DeepWorker 2000 is a one-man submarine that was tested during STS-101 to perform SRB recovery operations. It is 8.25 feet long, 5.75 feet high, and weighs 3,800 pounds. Inside the sub is Anker Rasmussen. (NASA photo KSC-00PP-0598)

An expended SRB from STS-26R is retrieved by one of the ships (Liberty Star and Freedom Star) operated by Lockheed Space Operations Company for NASA – these ships are now operated by United Space Alliance. (NASA photo KSC-88PC-1049)

about 110 feet by the divers. A quick inspection of the nozzle is conducted. The DOP is then inserted into the booster nozzle. Once the DOP legs are locked in place and the nozzle sealed, an air hose is attached.

The second team double-checks the aft skirt and DOP installation to ensure there are no problems. After the second dive is completed, dewatering operations begin. Air is pumped from the ship through the DOP and into the booster, displacing water within the casing. As the process continues, the booster rises in the water until it becomes top-heavy. It falls horizontally, like a log in the water. Air pumping continues until all water is expelled from the empty casing.

The final step in the ocean retrieval procedure is to connect the ship's tow line. Once the tow connection is made, the divers return to the ship and the trip to Hangar AF on CCAFS begins. The ships enter Port Canaveral, where the booster is changed from the stern tow position to a position alongside the ship, the hip tow position, to allow greater control. The ships then pass through a drawbridge, Canaveral Locks, and transit the Banana River to Hangar AF. The boosters are lifted from the water with straddle-lift cranes, and its components are disassembled and washed with fresh and deionized water to limit salt water corrosion. The motor segments, igniter, and nozzle are shipped back to the manufacturer for refurbishment.[190]

Two new concepts for booster recovery were tested during STS-101. The first was a new enhanced diver operated plug (EDOP) that features a motor-powered locking mechanism that replaces the present manual system to enhance diver safety and reduce workload at depth. It also has been streamlined for easier handling underwater. Since the EDOP is not yet certified, it was removed and taken onboard *Liberty Star* after the test. The second test involved the use of a leased one-person submarine to insert the normal DOP after the EDOP was removed.

The second idea used a small submarine to ease the workload on the divers. The submarine used in the tests was the *DeepWorker 2000*, built by Nuytco Research in North Vancouver, British Columbia. The sub is 8.25 feet long, 5.75 feet high, and weighs 3,800 pounds. Although both tests were reportedly successful, there was no immediate word if either approach will be used in the future.[191]

Beginning in 1998, the retrieval ships are also used to tow the barges that carry ETs from Michoud to KSC. The ships also perform other services, such as towing weath-

One of the recovered SRBs from STS-87 arrives at Hangar AF on Cape Canaveral Air Force Station. Hangar AF was originally used to support Project Mercury. The SRBs are towed by the recovery ships to CCAFS – once at Hangar AF, the SRBs are unloaded onto a hoisting slip and mobile gantry cranes lift them onto tracked dollies where they are safed and undergo their first washing. The boosters are then disassembled, washed, repainted, and sent to Utah. (NASA photos KSC-97PC-1728 and KSC-97PC-1725)

er buoys off the coast. Over the years, both vessels have seen service in side-scan sonar operations, cable-laying, underwater search and salvage, drone aircraft recovery, as platforms for robotic submarine operations and numerous support roles for other government agencies.[192]

Filament-Wound Cases

Early in the program, as the Orbiter continued to gain weight, the payload capacity into polar orbit from the Vandenberg Launch Site was being reduced to a point where the Air Force became worried about being able to lift the large reconnaissance satellites being designed at the time. It was also becoming obvious that the 109 percent certification on the main engines was lagging, so NASA and the Air Force began looking for other methods to increase the payload capacity from Vandenberg.

Morton Thiokol proposed a major change to the Solid Rocket Boosters that would allow an 8,000-pound increase to low-earth orbit. Cases constructed of filament-wound composite weighed 33,000 pounds less than STD steel cases being produced at the time. These cases also eliminated the four 'factory joints' since there were only four FWC cases per boosters instead of eight steel cases. Interestingly, when Thiokol designed the field joints for the new FWC cases, they added a capture feature to the tang and clevis, extremely similar to what would be added to the steel cases after the *Challenger* accident. In fact, the basic design for the FWC joint was used as a starting place for the RSRM joints (the design itself was not suitable since some tolerances and details did not meet the new criteria established after the accident). The filament-wound cases were manufactured by the Hercules Aerospace Company (now part of Alliant Techsystems). Except for the new cases, the motors were largely unchanged from the ones being produced for use at KSC. The same propellant formulation and shape was used, the structural pieces that were used to attach to the ET were the same, as were the forward frustrum and aft skirt.

The development effort went smoothly, culminating in a successful static firing (DM-6) on 25 October 1984. A second successful firing (DM-7) was conducted on 9 May 1985, and the qualification firing (QM-5) was scheduled for early February 1986 to certify the motor for operational use. This firing was first postponed, then cancelled after the *Challenger* accident. In addition to the three test motors, sufficient segments were produced to equip two complete sets of flight motors plus most of a third set.

The first segments using the filament-wound cases had arrived at Vandenberg on 30 May 1985, and all of the components for the first set of boosters were in place in the Solid Rocket Motor Building by the middle of July. There were some concerns over the FWC's ability to withstand the bending forces generated after SSME ignition (while the SRBs are still bolted to the launch mount), but 'twang' tests at Vandenberg using the boosters intended for STS-1V showed that while the characteristics were different than at KSC, they were acceptable. The segments at Vandenberg remained in storage there until mid-1988 when they were finally returned to Utah.

Filament-wound cases were again examined in 1994 after the cancellation of the ASRM project as a way to increase the payload capability of Shuttle to the International Space Station. They were again rejected for a variety of technical and economic reasons.

One set of filament-wound cases was donated to the Space and Rocket Center in Huntsville for display with Pathfinder. Five complete boosters (two flight sets plus the QM-5 motor) and one partial set of motors were left in storage in Utah pending a decision on their fate. Interestingly, it was not until late-2000 that all of the filament-wound cases were finally emptied of propellants and excessed from the inventory.[168]

Propellant for the SRMs is mixed in large batches. It takes numerous batches of propellants to fill one case segment, and the segments are poured in matched pairs to ensure the thrust level is consistent between flight sets. (Thiokol photo 80110-01)

This gives a good indication of how large the segments are. Here workers inspect the insulation that lines each case before the propellant is poured into it. This is the forward segment – the dome at the other end is the top of the motor, and the hole in it is where the igniter will be located. (Thiokol photo 129281-01)

A plug is inserted into the middle of each segment before the propellant is poured into it. This plug creates the tunnel down the center. Here the plug is being 'popped' out of the segment after pouring. (Thiokol photo 80428-05)

An exit cone inside one of the autoclaves that cures the phenolic liner used as insulation. The exit cones are only used once, unlike most other parts of the booster that are refurbished and used over again. (Thiokol photo 136580-02A)

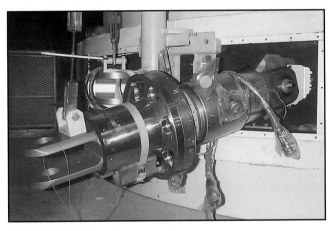

The upper strut that will connect the right SRB with the ET . Black foam surrounds their attachment points within the External Tank Attachment Ring. The upper strut provides the path for electrical connections between the ET and SRBs – some of the cables are hanging loose in the photo. (Kim Keller)

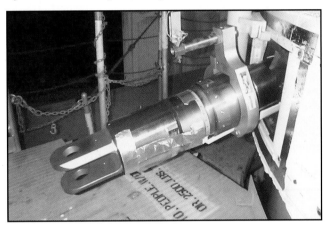

The lower strut that will connect the right SRB with the ET. Note the black line encircling the strut at its mid-point – this is the fracture line where the strut will break when the SRBs are jettisoned. Explosive charges within the strut will force two pistons into each other, setting up a shockwave that will break the strut in half. (Kim Keller)

The moment of ignition of an SRB during a static firing as captured by a video camera at the test site. (Thiokol photo 141078-02)

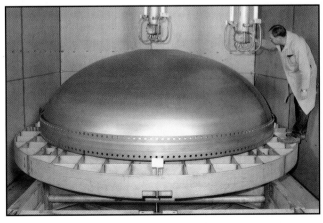

A forward dome being refurbished at Thiokol in Utah. Note the 177 holes that will provide the attachment to the forward segment. (Thiokol photo 142756-01)

Range Safety System

Originally, both SRBs and the ET were fitted with explosive charges capable of destroying the vehicle. The ET system was deleted beginning with ET-80 since it was felt the vehicle was sufficiently far downrange not to present a threat at that point – the SRB systems are retained. This means that STS-78 (ET-79) was the last flight to carry the ET RSS, and STS-79 (ET-82) was the first without it.

The range safety system can be detonated on the command of a Flight Control Officer (FCO – formerly called the Range Safety Officer) if the vehicle crosses the limits established by flight analysis before launch, or if the vehicle is no longer in controlled flight. The determination of 'controllability' is made by the Flight Director in Mission Control. There are FCOs at both Houston and at the Range Control Center at CCAFS. The Launch Commit Criteria that must be met before every mission includes weather constraints to allow both radar and visual tracking of the vehicle to ensure that it does not deviate substantially from its predetermined course.

Destruction of the vehicle is initiated by an encoded 'arm' command sent by multiple transmitting stations to the range safety package, followed by a subsequent encoded 'fire' command. The system is considered completely fail-safe. The 'arm' command also illuminates a light on the flight deck display and control panel in front of the mission commander and pilot stations for their information.

The ET range safety package consisted of linear shaped charges on both the LO2 (8 feet long) and LH2 (20 feet long) tanks on the ET. The SRBs have a linear shaped charge running the length of each booster. The SRB charges split the SRB casing open, effectively terminating thrust. The charges are linked together in such a way that if one RSS package receives a destruct signal, the signal is routed to all packages to ensure complete destruction of the vehicle. On STS-33/51-L, the RSS was commanded to destruct the SRBs approximately 30 seconds after the breakup of *Challenger*, and the ET RSS was recovered intact, so neither system played any role in the accident.[193]

Spacelab

The European-funded and built Spacelab was a non-deployable payload that could be carried in the payload bay. Spacelab was developed to be modular, and could be varied to meet specific mission requirements. The four principal components were the pressurized module, one or more open pallets that exposed materials and equipment to space, a tunnel to gain access to the module, and an instrument pointing system. The first Spacelab was carried on STS-9, and the last on STS-90. There had been 20 Spacelab flights in between (see page 324). Spacelab was retired since the experiments conducted on it may now be performed on the International Space Station.[194]

The pressurized module was available in two segments. The core segment contained supporting systems, such as data processing equipment and utilities for the module and pallets (if pallets were carried), and provided floor-mounted racks and a workbench. The experiment segment provided additional laboratory space, but contained only floor-mounted racks. Each segment was a pressurized cylinder 13.5 feet in outside diameter and 9 feet long. When both segments were assembled with end cones, the maximum outside length was 23 feet. The module was structurally attached to the payload bay by four attach fittings consisting of three sill longeron fitting sets (two pri-

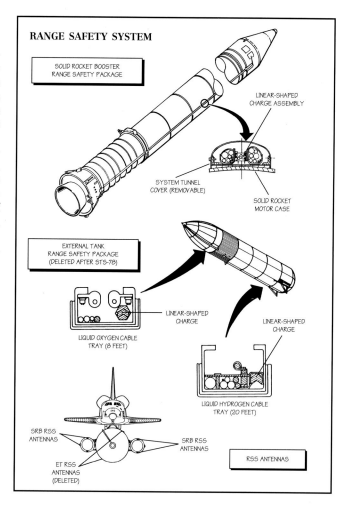

RANGE SAFETY SYSTEM

SOLID ROCKET BOOSTER RANGE SAFETY PACKAGE

LINEAR-SHAPED CHARGE ASSEMBLY

SYSTEM TUNNEL COVER (REMOVABLE)

SOLID ROCKET MOTOR CASE

EXTERNAL TANK RANGE SAFETY PACKAGE (DELETED AFTER STS-78)

LINEAR-SHAPED CHARGE

LINEAR-SHAPED CHARGE

LIQUID OXYGEN CABLE TRAY (8 FEET)

LIQUID HYDROGEN CABLE TRAY (20 FEET)

SRB RSS ANTENNAS

SRB RSS ANTENNAS

ET RSS ANTENNAS (DELETED)

RSS ANTENNAS

mary and one stabilizing) and one keel fitting. The module was covered with multilayer insulation blankets.

The ceiling skin panel of each segment contained a 51.2-inch-diameter opening for mounting a viewport adapter assembly. When the viewport was not used, the openings were closed with cover plates that are bolted in place. The module shell was made from 2219-T851 aluminum, with forward and aft cones bolted to the cylinder segments. The end cones were 30.8-inch-long truncated cones with 161.9-inch outside diameter large ends and 51.2-inch small ends. The modules were designed for a to be used up to 50 times.

Due to center-of-gravity limitations, Spacelab had to be installed in the rear of the payload bay. Therefore, a pressurized tunnel was provided between the crew module airlock and Spacelab. The tunnel was a cylindrical structure with an unobstructed internal diameter of 40 inches. Two tunnel lengths could be used – a 18.88-foot long tunnel and a 8.72-foot short tunnel. A 'joggle' section of the tunnel compensated for the 42.1-inch vertical offset of the mid-deck airlock to the Spacelab centerline. The joggle section was braced to the payload bay sill longeron, as was the forward section of the long tunnel. The joggle section also contained a keel fitting. The tunnel was built by McDonnell Douglas Astronautics in Huntington Beach, California.

The airlock, tunnel adapter, tunnel, and module were at ambient pressure at launch. The tunnel adapter permitted an EVA from the mid-deck airlock without depressurizing the Spacelab module. Nevertheless, mission rules prohibited the Spacelab module being occupied during an EVA.

Spacelab pallets were designed for large instruments, experiments requiring direct exposure to space, or systems needing unobstructed or broad fields of view, such as telescopes, antennas, and sensors (e.g., radiometers and radars). The U-shaped pallets were covered with aluminum honeycomb panels, and a series of hard points were provided for mounting heavy payload equipment. Up to five pallets could be flown on a single mission. Each pallet was held in place by four longeron sill fittings and one keel fitting. Pallet-to-pallet joints were used to connect the pallets to form a single rigid structure called a pallet train. Twelve joints were used to connect two pallets.

The Spacelab pallet portions of essential systems required for supporting experiments (power, experiment control, data handling, communications, etc.) were protected in a pressurized, temperature-controlled igloo housing. The igloo was attached vertically to the forward end frame of the first (forward) pallet. The igloo was approximately 7.9 feet high, 3.6 feet in diameter, was made of aluminum alloy, and was covered with multilayer insulation blankets. Cable ducts and cable support trays could be bolted to the forward and aft frame of each pallet to route electrical cables to and from the experiments and subsystem equipment mounted on the pallet. The ducts and cable trays are made of aluminum alloy sheet metal.

SPACEHAB

In 1990, in response to a need to fly payloads developed by NASA's Commercial Space Centers, McDonnell Douglas built its first Spacehab research module for the Space Shuttle to vastly expand the ability to fly locker sized payloads. The operation was subsequently spun-off as a separate company called SPACEHAB. In 1995, SPACEHAB won a contract to use the Spacehab modules to carry logistics to the Russian *Mir* space station. To support this contract, SPACEHAB created the logistics double module to expand cargo carrying capacity to 10,000 pounds. In 1999, SPACEHAB expanded its assets to include the integrated cargo carrier (ICC) to carry items outside in the payload bay.[195]

The Spacehab module is a pressurized experiment carrier designed to augment mid-deck experiment accommodations. The Spacehab system consists of a module flown in the payload bay that is configured with mid-deck-type lockers, racks, and/or the logistics transportation system (LTS) to accommodate a variety of experiments and equipment. Spacehab offers the Space Shuttle Program three module configurations to accommodate mission-specific requirements.[196]

Configuration 1 uses the Spacehab-SM (single module) with a Spacehab tunnel adapter, transition section, and flex section. The Spacehab module provides the crew with a place to carry out experiments and contains cooling, power, and command and data provisions, in addition to Spacehab housekeeping systems (i.e., power distribution and control, lighting, fire and smoke detection, fire suppression, atmosphere control, status monitoring and control, and thermal control). This configuration can not be carried when an ODS is installed in the payload bay, effectively limiting it to use on *Columbia*.

Configuration 2 allows the Spacehab-SM to be mounted on a different trunnion location to accommodate the ODS. The module is connected to the ODS using a flex section, a long tunnel segment, the regular tunnel segment, and another flex section. All Spacehab module subsystems remain the same as configuration 1, except for a lower air exchange rate with the Orbiter and the addition of two negative pressure relief valves.

Configuration 3 is a Spacehab-DM (double module), consisting of one Spacehab module and one module shell that are jointed by an intermediate adapter. This configuration has the same tunnel configuration and attach points as configuration 2, except for two trunnions that are moved further aft to accommodate the additional module. All Spacehab module subsystems remain the same as configuration 2, except for the addition of a fan and lights in the aft module segment.

ORBITER MAINTENANCE AND REFURBISHMENT

Each of the Orbiters undergo repairs, upgrades, improvements, and other modifications during normal processing flows at KSC. But as the Orbiter fleet grew older, it became obvious that major maintenance and improvements would need to be accomplished outside the normal processing flow at KSC. Therefore, NASA defined a refurbishment schedule for each Orbiter.

The most extensive of these modifications is performed during the Orbiter Maintenance Down Period (OMDP) while the Orbiter is in the Palmdale, California plant. This regular maintenance and upgrade program is scheduled every eight flights for each Orbiter, or approximately every three years. The OMDP provides the most efficient means of comprehensive inspection and major modification for the Orbiter because of the relatively long time available in the OMDP, and because of the scheduling conflicts that can arise in the normal vehicle flow. The nominal OMDP period is approximately 14 months long. Similar to OMDP, and normally accomplished at the same time, are Orbiter Major Modifications (OMM). Most modifications are documented on master change records (MCR) that detail the work to be performed. Each Orbiter is transported to Palmdale aboard an SCA after post-landing processing at KSC. In many instances, technicians at KSC also perform preliminary work for the modifications.

Columbia (OV-102) Micro-Modification

At the completion of STS-4, *Columbia* was modified by installing the equipment (payload signal processor and payload data interleaver) necessary to accommodate the PAM-D being used by the STS-5 payloads, deactivating the ejection seats, installing a specialist seat on the flight deck behind the center console, providing the left mid-deck seat, strengthening the mid-deck floor, and removing parts of the DFI pallet.[197]

Columbia (OV-102) Spacelab Only (SLO)

Columbia was originally scheduled to be extensively modified after STS-9 in preparation for the Orbiter's seventh flight. Further analysis showed that some of the modifications were necessary to support the STS-9 Spacelab mission, and NASA ordered Rockwell to split the modifications into two phases.

The first of these, called OV-102 Spacelab Only (SLO), occurred at KSC after the completion of STS-5, and incorporated 152 MCRs. Contrary to the name of the modification, not all of the changes were in direct support of Spacelab, although that is where the focus was. The changes included adding the remaining payload and mission specialist seats; adding an on-orbit facsimile system; adding a command authentication system to the uplink; strengthening the mid-fuselage structure; adding crew sleep stations; landing gear and brake modifications;

ORBITER EXPERIMENTS

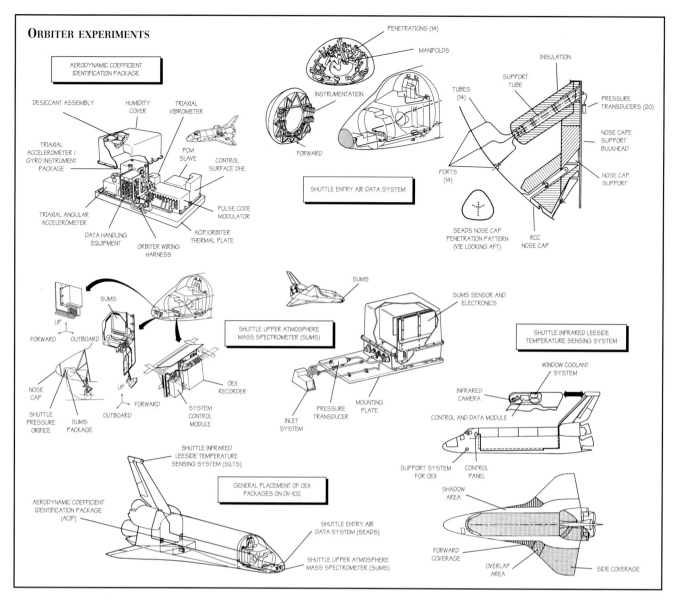

AERODYNAMIC COEFFICIENT IDENTIFICATION PACKAGE

DESICCANT ASSEMBLY
HUMIDITY COVER
TRIAXIAL VIBROMETER
TRIAXIAL ACCELEROMETER / GYRO INSTRUMENT PACKAGE
PCM SLAVE
CONTROL SURFACE DHE
TRIAXIAL ANGULAR ACCELEROMETER
PULSE CODE MODULATOR
DATA HANDLING EQUIPMENT
ORBITER WIRING HARNESS
ACIP/ORBITER THERMAL PLATE

PENETRATIONS (14)
MANIFOLDS
INSTRUMENTATION
FORWARD

SHUTTLE ENTRY AIR DATA SYSTEM

TUBES (14)
SUPPORT TUBE
INSULATION
PRESSURE TRANSDUCERS (20)
NOSE CAPE SUPPORT BULKHEAD
NOSE CAP SUPPORT
PORTS (14)
SEADS NOSE CAP PENETRATION PATTERN (VIE LOOKING AFT)
RCC NOSE CAP

SUMS
UP
FORWARD
OUTBOARD
NOSE CAP
SHUTTLE PRESSURE ORIFICE
SUMS PACKAGE
UP
FORWARD
OUTBOARD
SYSTEM CONTROL MODULE
OEX RECORDER

SHUTTLE UPPER ATMOSPHERE MASS SPECTROMETER (SUMS)

SUMS

SUMS SENSOR AND ELECTRONICS

INLET SYSTEM
PRESSURE TRANSDUCER
MOUNTING PLATE

SHUTTLE INFRARED LEESIDE TEMPERATURE SENSING SYSTEM

WINDOW COOLANT SYSTEM
INFRARED CAMERA
CONTROL AND DATA MODULE
SUPPORT SYSTEM FOR OEX
CONTROL PANEL
SHADOW AREA
FORWARD COVERAGE
OVERLAP AREA
SIDE COVERAGE

SHUTTLE INFRARED LEESIDE TEMPERATURE SENSING SYSTEM (SILTS)

GENERAL PLACEMENT OF OEX PACKAGES ON OV-102

AERODYNAMIC COEFFICIENT IDENTIFICATION PACKAGE (ACIP)
SHUTTLE ENTRY AIR DATA SYSTEM (SEADS)
SHUTTLE UPPER ATMOSPHERE MASS SPECTROMETER (SUMS)

The Shuttle Infrared Leeside Temperature Sensor (SILTS) package was installed on OV-102 (Columbia) in time for STS-32/61-C in January 1986. Noteworthy above is the extensive use of black TPS tiles (compare this to a pre-OEX shot of OV-102). The instrument was removed after STS-52, but the empty pod remains. (NASA photo L-85-12145)

Fourteen collection ports for the Shuttle Entry Air Data System (SEADS) were installed in the carbon-carbon nose cap of Columbia. Behind each port is a system of columbium tubing leading to sensors which sense the local surface pressure across the Orbiter's nose cap. The nose cap was replaced with a standard unit after STS-40. (NASA)

adding structural and electrical provisions for the Spacelab module; removing the remainder of the DFI pallet to make room for Spacelab; and various thermal protection system enhancements.[198]

Columbia (OV-102) Flight 6 Modifications

The remaining modifications were performed at Palmdale in 1984 after the STS-9 mission, including: removing the ejection seats; the installation of the OEX experiment packages (SILTS, SEADS, and SUMS); installing the heads-up displays (HUD); adding provisions for GPS navigation system (although the GPS itself was never installed); replacing the 17-inch disconnect valves with a new design; incorporating the 5.4 loads structural modifications; additional TPS modifications; provisions to carry Manned Maneuvering Units (MMU); and yet more brake system modifications. A total of 231 MCRs were incorporated into OV-102 during this period. *Columbia* returned to KSC in time for STS-32/61-C in 1986. Since Palmdale did not have a mate/demate facility at this time, *Columbia* was moved overland to and from Edwards.[199]

The OEX instrumentation package was installed to measure and record the Orbiter's aerodynamic and thermodynamic characteristics. The instrumentation consists of three experiments developed by NASA Langley and are part of the Orbiter Experiments Program (OEX) managed by NASA's Office of Aeronautics and Shuttle Technology (OAST) in Washington, D.C. The first use of the three experiment packages was on STS-32/61-C in January 1986.[200]

The Shuttle Entry Air Data System (SEADS), developed by a team headed by Paul M. Siemers III, measured the distribution of air pressure around the Orbiter's nose cap during reentry. The SEADS experiment required a new Orbiter nose cap that contained 14 penetration assemblies distributed across the nose surface in a cross-shaped pattern. Each penetration assembly contained a small hole through which local surface air pressure was sensed during ascent and reentry. Each orifice was connected to two pressure transducers – one sensitive to high-level pressures (0–20 psia); the other to low-level pressures (0–1 psia). All components of the penetration assemblies and nose cap internal tubing were made of coated columbium superalloy. HRSI tiles covered the bulkhead behind the nose cap to ensure a leak would not be catastrophic.[201]

The Shuttle Infrared Leeside Temperature Sensing (SILTS) experiment, developed by Langley engineers David A. Throckmorton and E. Vincent Zoby, was designed to obtain high-spatial-resolution infrared images of the leeside (upper) surfaces of the Orbiter's left wing and fuselage during reentry. The information measured aerodynamic heating on the leeside surfaces in flight, something that could not be adequately simulated in ground facilities. The top section of *Columbia*'s vertical stabilizer was replaced with a special pod to house the SILTS sensors. The pod was approximately 20 inches in diameter and was capped at the leading edge with a hemispherical dome that had two infrared-transparent windows. The pod contained an infrared scanning system, an electronics module, and a pressurized nitrogen cooling system. The entire pod, as well as the top ten feet of the vertical, was covered with black HRSI tiles, making *Columbia* easy to identify.[202]

The last of the experiments installed on *Columbia* was the Shuttle Upper Atmospheric Mass Spectrometer (SUMS) developed by Roy J. Duckett and Robert C. Blanchard. SUMS sampled the gases at *Columbia*'s lower surface through a small hole located just aft of the nose cap and forward of the nose wheel doors. The experiment measured and identified the quantities of various gases to allow a determination of atmospheric density. The SUMS instrument was a mass spectrometer originally designed for the Viking Mars mission.[203]

SEADS and SUMS were removed after STS-40 in 1991. The SILTS experiment was originally scheduled to be removed after STS-52 (flight 13) and the tip of the vertical stabilizer fitted with AFRSI insulation. However, it was subsequently decided that it was not necessary to go through the expense of removing the SILTS pod. Instead, the instrumentation was removed at the same time as the rest of the OEX experiments, but the pod remained. The program has since certified (through analysis) the pod for unrestricted use and there are no plans to remove the pod in the foreseeable future.[204]

Columbia (OV-102) Flight 11 EDO Modification

On 10 August, 1991, after the completion of STS-40, *Columbia* was again transported to Palmdale for modifications. The oldest Orbiter in the fleet underwent 62 modifications including the removal of the SEADS and SUMS experiment packages, and replacing the Orbiter nose cap with one without the OEX penetrations. Part of the SILTS instrumentation was removed at KSC while preparing *Columbia* for her trip west, while the remainder of the experiment was removed at Palmdale – except for the vertical stabilizer pod, which was left in place. Other changes included: installing new APUs; the addition of carbon brakes; incorporating the drag chute; improved nose wheel steering; incorporating the Orbiter 6.0 loads structural modifications; deactivating payload bay vent doors #4 and #7; installing AP-101S GPCs; and a refurbishment and enhancement of its thermal protection system. The Orbiter returned to KSC 9 February 1992 to begin processing for STS-50.[205]

Discovery (OV-103) Flight 14 Modification (OMDP-1)

Discovery underwent an extensive checkout and modification period at KSC following her return from STS-42 in February 1992. This is now considered *Discovery*'s OMDP-1. Approximately 78 modifications were completed, the most noticeable being the installation of the drag chute. Other work included a complete structural inspection and a detailed inspection and refurbishment of the thermal protection system. Minor corrosion damage was also repaired.

Atlantis (OV-104) Flight 12 OMDP-1

OV-104 departed KSC on 17 October 1992 and returned on 29 May 1994 to prepare for STS-66. While at Palmdale, the Orbiter was outfitted with improved nose-wheel steering, internal plumbing and electrical connections to accommodate an EDO cryo pallet, and preparations for the installation of the Mir ODS unit at KSC. In addition, payload bay vent doors #4 and #7 were deactivated, the drag chute was installed, and the original APUs were swapped for IAPU models. Structural and electrical provisions were also fitted to enable *Atlantis* to accommodate the Long Duration Orbiter (28-day) pallet, although the pallet itself was never completed. In all, 331 MCRs were incorporated into OV-104, along with 184 maintenance items that had been deferred from KSC. It should be noted that many of the MCRs performed during

OMDPs are of a very minor nature – such as verifying a particular clamp is in place and correctly tightened (MCR 10532). Others are of a larger nature, such as installing the improved APUs (MCR 10063). Still other change the outer mold line of the vehicle, such as installing the drag chute (MCR 12060).[206]

Columbia (OV-102) Flight 17 OMDP-1

OV-102 departed KSC for Palmdale on 8 October 1994. Primarily this down time was a mid-life refurbishment of various Orbiter components and further enhancements to the thermal protection system. Columbia has continued to have a moderate corrosion problem on the wing lead-edge spar, and several modifications were incorporated into this area to try and minimize the problem. Although no decisive cause has ever been established for this corrosion, it is believed that the amount of time the Orbiter spent on Pad 39A before the weather protection system was complete exposed the airframe to excessive salt spray from the Atlantic ocean. A complete inspection of the airframe was also accomplished, and any other corrosion discovered was corrected. In all, 80 MCRs were incorporated into OV-102, along with 143 items that had been deferred from KSC. Other work included 488 structural inspections (469 X-Ray and 19 visual). Columbia returned to KSC on 14 April 1995 to begin processing for STS-73.[207]

Discovery (OV-103) Flight 21 OMDP-2

STS-70 was Discovery's final flight prior to being flown to Palmdale atop N905NA. Discovery departed KSC on 27 September 1995, and returned on 29 June 1996 to begin processing for the second Hubble Space Telescope servicing mission (STS-82).

A total of 96 MCRs were incorporated and 87 deferred maintenance items performed on OV-103 at Palmdale. The most extensive of these was the installation of the definitive Orbiter Docking System (replacing the mid-deck airlock) to support International Space Station operations. This was the first installation of the definitive Orbiter Docking System – Atlantis was still using the interim Mir/ODS. Other work included: thermal protection system repairs and replacements; installation of upgraded hardware for improved payload bay flood lighting; star-tracker shutter replacements; and structural corrosion inspections and repairs.

Endeavour (OV-105) Flight 11 OMDP-1

OV-105 departed KSC on 30 July 1996 and returned on 27 March 1997 to prepare for STS-89. The primary results of the OMDP were the installation of the external airlock and ODS, and a general weight-reduction program to maximize the payload capability to the ISS orbit. A total of 63 modifications were accomplished at Palmdale, and 33 more at KSC (6 before and 20 after the OMDP, plus 7 to the FRCS module at the HMF). The work on ten modifications was 'shared' between KSC and Palmdale.[210]

The airlock installation included significant structural, avionics, electrical, and ECLSS modifications, and is configured so that it may be installed in bay 2 or bay 3 as needed. Palmdale installed the airlock in bay 3 since that was the configuration required to support STS-89, 88, and 96. Provisions were installed to allow a ground cooling hookup (Freon and water) into the payload bay to cool the mini-pressurized logistics module (MPLM).

As part of the weight reduction efforts, the EDO capability was removed from OV-105, although the 'scars' remain to add it back if necessary. This included removing the RCRS and adding LiOH canisters. This leaves OV-102 as the only EDO-capable Orbiter in the fleet.

Almost all of the once-touted AFRSI blankets on the mid-fuselage, aft-fuselage, payload bay doors, and upper wings were replaced with thinner and lighter FRSI blankets. Lightweight crew seats were installed along with the 20-g floor modification. The weight reduction went so far as to replace the LT-80 aluminum foil tape that covered the inside of the landing gear doors with lighter aluminumized Kapton tape. This also improved the thermal response effect on the landing gear and tires.

Some weight was added back however. Doublers were added to several wing spars to eliminate load restrictions and allow heavier payloads (particularly at landing), and two wing glove truss tubes were replaced with units having increased wall thickness. Most of these changes were similar to ones performed on Discovery during OMDP-2.

In addition, the ET range safety decoder was removed since the range safety package is no longer installed on the ET. The OMS/RCS 'sniff tubes' installed after the Challenger accident were removed since the new-design ac motor valves are not susceptible to bellows leakage, eliminating the need for the sniffers.

Atlantis (OV-104) Flight 20 OMDP-2

OV-104 departed KSC on 11 November 1997 and returned on 27 September 1998 to prepare for STS-101. In general this OMDP incorporated the same changes that had been accomplished to Discovery and Endeavour, including the installation of the definitive ISS airlock and ODS, and the removal of the mid-deck airlock. The same weight-saving measures – replacing AFRSI with FRSI, etc. – were also performed. Perhaps the largest difference was that Atlantis also received the first MEDS 'glass cockpit' installation during this OMDP. In addition, while at Palmdale all three TACAN systems were removed and the wiring and controls installed for a three-string GPS system – the actual GPS receivers were to be installed at KSC. However, by the time OV-104 returned to KSC, the decision had been made to defer the installation of the GPS system. Fortunately, the modifications performed during the OMDP made it possible to install either navigation system with only minor changes, and Atlantis carried a single-string GPS and a three-string TACAN on STS-101.[211]

Columbia (OV-102) Flight 26 OMDP-2

OV-102 departed KSC on 24 September 1999, and had been scheduled to return on 3 November 2000. However, inspections of the wiring inside areas of Columbia that had not been accessible at KSC showed more than 4,500 discrepancies, and the OMDP was extended in order to correct these. It was unlikely Columbia would return to KSC before February 2001. During this OMDP, Columbia was scheduled to received MEDS, the remaining lightweight crew seats (some had been installed prior to STS-93 to support the Chandra launch), the 20-g floor, and various thermal protection system upgrades (although Columbia will still retain her unique TPS configuration and markings). Like Atlantis, provisions are being incorporated for a three-string GPS, although no decision has been made as to when such a system may be installed.[212]

After Vandenberg was closed, the lifting frame that had been installed there was moved to Palmdale for use during OMDPs. This is Endeavour being loaded aboard N911NA for her initial flight to KSC after her roll-out. Reports are that when the lifting frame was installed at Palmdale, enough concrete was used to make any further relocation unlikely. (clockwise from top: DVIC photos DF-ST-99-05016, 05014, 05013, 05012, and 05021)

This table summarizes the Orbiter Detail Weight Statement prepared by Boeing for Discovery (OV-103) as of 2 July 1999 (left column). Data from the previous 2 January 1995 statement is included for comparison. The data is generally applicable to Atlantis (OV-104) and Endeavour (OV-105). Columbia is sufficiently different that this statement is not applicable.[213]

WING GROUP		
Outer Panel Interim Section	2,291.4	2,292.2
Outer Panel Torque Box	6,236.4	6,496.6
Outer Panel Leading Edge	188.2	188.2
Outer Panel Trailing Edge	1,306.7	1,018.3
Secondary Structure	2,413.7	2,413.5
Operating Mechanisms and Controls	549.4	549.6
Miscellaneous Provisions and Support	112.8	112.8
Elevon Inboard Surface	1,291.0	1,291.0
Elevon Inboard Support Mechanism	278.2	278.2
Elevon Outboard Surface	921.6	922.6
Elevon Outboard Support Mechanism	339.6	339.6
Total Wings	**15,929.0**	**15,902.6**
TAIL GROUP		
Outer Panel Torque Box	1,406.7	1,406.7
Outer Panel Leading Edge	86.5	88.5
Outer Panel Trailing Edge	287.7	482.7
Secondary Structure	61.2	61.2
Operating Mechanisms and Controls	93.0	93.0
Rudder/Speedbrake	671.9	671.9
Total Vertical Stabilizer	**2,607.0**	**2,804.0**
BODY GROUP		
Forward Fuselage	4,281.5	4,301.5
Crew Module (including hatch and glass)	4,189.2	4,248.3
Crew Module Side Hatch	300.7	†
Crew Module Rear Hatch	113.0	†
Windshields & Windows	1,602.1	†
Airlock (minus hatch)	588.6	†
Airlock Hatch	117.0	†
Mid-Fuselage	10,974.0	11,116.0
Aft-Fuselage (body)	7,250.4	7,305.4
Aft-Fuselage (thrust structure)	3,182.4	3,182.4
Forward RCS Module	343.5	343.5
Body Flap	605.1	471.4
Interior Paint	47.9	†
Exterior Paint	179.0	†
Star Tracker Door and Mechanism	74.0	†
OMS/RCS Pod (right)	1,493.2	1,530.4
OMS/RCS Pod (left)	1,493.2	1,530.4
Payload Bay Doors	4,653.6	4,672.8
Miscellaneous Provisions and Support	1,442.4	†
Total Fuselage	**44,239.0**	**46,938.0**
LANDING GEAR AND ANCILLARY SYSTEMS		
Main Landing Gear	2,053.2	2,328.4
Main Landing Gear Wheels	630.0	†
Main Landing Gear Tires	784.0	†
Main Landing Gear Tire Air	40.0	†
Main Landing Gear Brakes	591.2	†
Main Landing Gear Paint	8.0	†
Main Landing Gear Structure	2,640.4	2,714.4
Main Landing Gear Hydraulics	692.5	712.4
Nose Landing Gear	183.6	187.0
Nose Landing Gear Wheels	75.6	†
Nose Landing Gear Tires	100.0	†
Nose Landing Gear Tire Air	6.0	†
Nose Landing Gear Paint	2.0	†
Nose Landing Gear Structure	600.6	600.6
Nose Landing Gear Hydraulics	147.2	167.2
Drag Chute Installation	–	246.0
External Tank Attachment and Separation	908.3	954.8
Remote Manipulator System	1,320.2	1,320.2
Total Landing Gear and Ancillary Systems	**8,546.0**	**9,231.0**
ASCENT PROPULSION		
SSME Gimbal System	1,763.1	1,758.3
SSME Hydraulic Supply	221.1	221.1
SSME Heat Shields	1,054.5	1,211.5
Helium Pneumatic System	2,091.4	2,101.7
ET Pressurization System	200.9	235.7
LO2 Fill and Drain	2,377.1	2,384.2
LH2 Fill and Drain	2,645.9	2,655.8
SSME (3)	20,884.0	22,278.0
Total Ascent Propulsion	**31,238.0**	**32,883.0**
PROPULSION – OMS/RCS		
Forward RCS	1,514.0	1,474.0
Aft RCS	1,628.0	1,645.0
OMS	3,042.0	3,041.0
Total OMS/RCS	**6,184.0**	**6,160.0**
HYDRAULICS		
Hydraulic Conversion and Distribution	1,865.0	1,855.0
Aero Surface Actuators and Controls	2,785.0	2,785.0
Total Hydraulics	**4,650.0**	**4,640.0**

INDUCED ENVIRONMENTAL PROTECTION		
Wing RCC Panels	1,529.2	†
Wing RCC Installation Hardware	923.8	†
Wing RCC Bulk Insulation	1,180.4	†
Wing Upper Surface HRSI	393.4	†
Wing Upper Surface LRSI	569.6	†
Wing Upper Surface FRSI	238.8	†
Wing Upper Surface AFRSI	31.8	†
Wing Lower Surface HRSI	4,623.2	†
Wing Lower Surface Ablator	15.0	†
Wing Carrier Strips and Panels	260.6	†
Wing Thermal Barriers and Seals	1,201.0	†
Wing Bulk Insulation	380.0	†
Vertical Stabilizer HRSI	237.8	†
Vertical Stabilizer LRSI	559.8	†
Vertical Stabilizer AFRSI	84.0	†
Vertical Stabilizer Aerothermal Seal	243.7	†
Vertical Stabilizer Thermal Barriers	33.5	†
Fuselage RCC Panels	128.1	†
Fuselage RCC Installation Hardware	135.7	†
Fuselage RCC Bulk Insulation	229.7	†
Fuselage HRSI	5,617.9	†
Fuselage LRSI	1,324.7	†
Fuselage FRSI	862.7	†
Fuselage AFRSI	1,362.0	†
Forward RCS Thruster Heat Shields	110.3	†
Fuselage Thermal Barriers and Seals	232.9	†
Forward Body Internal Insulation	299.3	†
Crew Module Internal Insulation	437.0	†
Forward Equipment Compartment Insulation	577.4	†
Payload Bay Insulation	755.7	†
Aft Equipment Compartment Insulation	107.7	†
Landing Gear Compartment Insulation	63.0	†
Hydraulic Line Insulation	221.7	†
OMS/RCS Pod Insulation	449.2	449.6
Miscellaneous Provisions and Support	2,301.4	†
Total Induced Environment Protection	**27,722.0**	**26,703.0**
AVIONICS		
Guidance, Navigation & Control	659.8	918.9
Communications and Tracking	1,896.0	1,488.0
Data Processing System	1,608.6	1,097.6
Inertial Measurement Units	126.0	126.0
Displays and Controls	2,102.5	2,181.3
Miscellaneous	792.9	709.2
Total Avionics	**7,185.8**	**6,521.0**
PRIME POWER		
APUs (including tanks, etc.)	912.0	885.1
Fuel Cells (including tanks, etc.)	2,981.9	3,074.9
Electrical Power Distribution	10,628.0	11,390.0
Total Prime Power	**14,521.9**	**15,350.0**
ENVIRONMENTAL CONTROL		
Cabin and personnel System	2,123.7	2,244.3
Equipment Environment and Heat Transport	3,145.2	3,126.5
Airlock Support	29.1	114.2
Total Environmental Control	**5,298.0**	**5,485.0**
PERSONNEL PROVISIONS		
Galley	164.0	143.0
Smoke Detectors	23.4	46.3
Fire Suppression	18.6	†
Seats	512.0	†
Miscellaneous	1,115.0	†
Total Personnel Provisions	**1,833.0**	**1,644.0**
PAYLOAD SUPPORT	**770.0**	**1,330.0**
Total OV-103 w/ 3 SSME	**170,723.7**	**175,591.6**

† = The 1999 report was not as detailed as the 1995 so individual breakouts were not possible.

ORBITER WEIGHT SUMMARY [209]
CIRCA 2 OCTOBER 2000

Orbiter	Name	Empty Weight w/o SSME	3 SSME	Empty Weight w/ SSME
OV-102	*Columbia*	159,343	22,340	181,683
OV-103	*Discovery*	153,998	22,318	176,316
OV-104	*Atlantis*	154,082	22,323	176,405
OV-105	*Endeavour*	153,593	22,399	175,992

SHUTTLE UPGRADE PROGRAM

The Orbiters were designed for a nominal service life of 100 missions. Even with the anticipated increase to 7–8 flights per year to support the International Space Station, the service life of Space Shuttle may extend until 2030. To ensure its continued viability, a variety of upgrades are being pursued to improve safety, avoid obsolescence problems, and improve supportability.[1]

The Space Shuttle Program Development Office at JSC is responsible for defining and developing the Shuttle Upgrade Program that will keep the Space Shuttle flying safely until at least 2012. A groundrule of the current program is that no unmodified vehicle will fly after 2006, providing a tight schedule for the proposed improvements.

The Shuttle Upgrade Program's primary goal is to improve crew safety and situational awareness, protect people both during flight and on the ground, and to increase the overall reliability of the Space Shuttle system. A variety of upgrades – not only to the vehicle, but also to support systems and processes – have been proposed to the NASA Administrator and the Office of Management and Budget (OMB).

Originally, the upgrades were divided into four phases: Phase I included primarily ongoing modifications initiated during the early 1990s to improve shuttle safety, and upgrades to increase the ability to support the ISS. Phase II upgrades were defined as high-value, low-impact, incremental improvements, primarily to combat obsolescence. Phase III upgrades were major, high-value upgrades that did not affect the outer mold line or basic aerodynamics – for example, a major change to the Orbiter's avionics or communications suite. Phase IV upgrades were also major, high-value changes which might have affected the overall aerodynamic shape – such as liquid fly-back boosters to replace the Solid Rocket Boosters. Only two Phase IV upgrades were seriously discussed – a five-segment version of the existing solid booster (FSB), and a liquid fly-back booster (LFBB).[2]

However, this four-phase approach was rather quickly replaced by a three step program that identified the possible changes as Safety Upgrades, Supportability Upgrades, and an 'Evolved Shuttle' concept that is beyond the scope of current planning.

The candidate upgrades were evaluated and prioritized by a team of experts with in-depth Shuttle experience. These evaluations were augmented by the development of a set of safety and reliability metrics established by a combined NASA–contractor team from each of the human space flight centers (JSC, KSC, and MSFC). To determine the effect of each upgrade on the overall loss-of-vehicle and loss-of-crew statistics, a probabilistic risk assessment was performed for each upgrade candidate. The expert teams then used those metrics together with their own understanding of the history of shuttle hardware problems to develop a prioritized recommendation for safety upgrades to be considered by the Space Shuttle Program.

During FY99 the program funded $95 million for shuttle upgrade activities including almost $18 million that went to new flight hardware – new SRB altitude switch assemblies, display driver units, and mass memory unit projects were all approved to resolve obsolescence issues. An additional $65 million was approved for the continuation of flight and ground upgrade development projects.[3]

Safety Upgrades

The primary objective of safety upgrades is to achieve major reductions in the operational risks inherent in the current system by implementing design or manufacturing changes that eliminate, reduce, or mitigate significant hazards and critical failure modes, and that significantly increase the overall reliability of the system with respect to likelihood of catastrophic failure. The safety upgrades are focused on: (1) major reduction in ascent catastrophic risk (targeting up to 50 percent reduction); (2) significant reductions in orbital and reentry/landing catastrophic risk (targeting up to 30 percent reduction); and (3) major improvement in flight crew situational awareness for managing critical flight situations through cockpit modernization. A total of 16 safety upgrades have been examined, and by the end of 2000, five of these had received formal approval to proceed while the remainder are still being evaluated or are in the approval process.[4]

These changes will lower the probability of a loss-of-vehicle accident during ascent from 1-in-248 to 1-in-483 – a significant improvement over the 1-in-78 that was estimated during the Return-to-Flight of STS-26R assessment in 1988. Incremental improvements between 1988 and 1995, mainly improvements to the SSMEs, had already lowered that number to 1-in-248.[5]

SLWT Friction-Stir Welding Process: Given the sensitivity of the 2195 aluminum-lithium alloy used in the SLWT to welding technique, NASA and Lockheed Martin began investigating other solutions. Friction-stir welding overcomes many of the current problems such as torch control, heat sink, and purge gas delivery. Additionally, this new method changes the welding of aluminum into a machining process that is less labor intensive, less technically complex, and easier to control. It also provides superior strength welds and eliminates the possibility of changing the physical properties of the adjacent material by overheating. In the friction-stir welding process, a welding tool moves along the area to be joined while rotating at a high speed. The action between the tool and the aluminum creates frictional heat, which softens the aluminum but does not melt it. The plasticized material is then, in essence, consolidated to create one piece of metal where there were originally two – it is structurally similar to the original material. The weld is left in a fine-grained, hot worked condition with no entrapped oxides or gas porosity. Friction-stir welding will be used for the large longitudinal welds on both the LO2 and LH2 tanks. This upgrade has been approved for implementation.[6]

Main Landing Gear Tires: NASA has contracted with Michelin to develop a new main landing gear tire that will provide an approximate 20 percent increase in load bearing ability. The design being considered uses a 16-bias-ply design with an inner liner for additional safety. This upgrade results in a tire design that allows higher landing

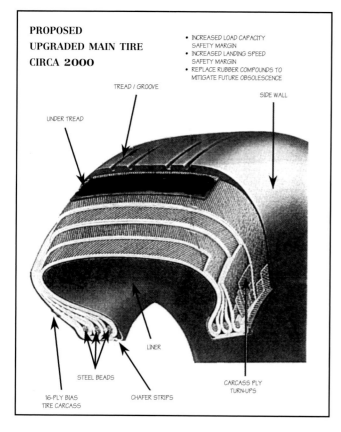

PROPOSED
UPGRADED MAIN TIRE
CIRCA 2000

- INCREASED LOAD CAPACITY SAFETY MARGIN
- INCREASED LANDING SPEED SAFETY MARGIN
- REPLACE RUBBER COMPOUNDS TO MITIGATE FUTURE OBSOLESCENCE

TREAD / GROOVE

SIDE WALL

UNDER TREAD

LINER

STEEL BEADS

16-PLY BIAS
TIRE CARCASS

CHAFER STRIPS

CARCASS PLY
TURN-UPS

speeds, mitigates obsolescence issues, and improves margins for pressure leakage and cold temperature environments. The tire will also require a new wheel design. The program has decided to pursue this approach instead of the extensible nose landing gear (ENLG) because it is less expensive and does not introduce the potential failure modes of the ENLG design (see page 219). This upgrade has been approved for implementation.[7]

Micrometeoroid/Orbital Debris (MMOD) Improvements: The MMOD modifications are a series of Orbiter hardware and software changes intended to minimize the risk to the Orbiter from impacts by small objects while on-orbit. The upgrade consists of adding protection to the radiator cooling tubes, modifying the Freon-21 coolant loop isolation valves, providing flight software to automate the Freon loop isolation, and adding thermal protection blankets in the wing leading edge. This upgrade has been approved for implementation. These changes are discussed further in Chapter X in the payload bay radiator and wing leading edge thermal protection system descriptions.[8]

Orbiter Electric APU: NASA has estimated that the existing APUs are capable of supporting operations through 2014 at current flight rates. After that (or earlier if flight rates increase) the APUs will begin to reach the end of their operational life expectancy. During an exercise to determine the long-term operational costs of the current APUs, NASA estimated that the cost of keeping the current APUs operational until 2030 would be approximately $550 million.[9]

But beyond potential obsolescence, there are other reasons to replace the APUs – they involve high concentrations of mechanical energy and use hazardous liquid hydrazine propellant. And in spite of redundancy against single-point failures, they are spatially vulnerable to common cause failures, such as fire, explosion, and leaks.[10]

To address these concerns, a lithium-ion battery-powered EAPU prototype is under development that eliminates the need to carry hydrazine for the APUs. By eliminating the toxic propellant, and the high-speed gas-driven turbines it powers, safety may be improved by as much as 18 percent. The intent is to keep the physical and functional interface to the hydraulic system transparent, although the batteries will probably cause the overall replacement to weigh slightly more than the existing APUs. As of late 2000 two different concepts were under evaluation – the first uses the existing Orbiter hydraulic pumps with the new EAPU, while the other replaces the existing pumps with a more modern design but keeps the pressure interface the same as the existing system. NASA has spent about $650,000 on the effort to date, and development and implementation costs are estimated at $100 to $150 million. Total costs of developing the system and operating it until 2030 are estimated to be about $350 million.[11]

Cockpit Avionics Upgrade: The introduction of the MEDS 'glass cockpit' allows a variety of upgrades to be considered that present data to the flight crew in a more efficient manner. Among these are a 'smart cockpit' that puts the new displays to better use by indicating the root cause of problems and offering more information to the crew. This will simplify and automate complex procedures that currently require considerable training, are subject to human error, and distract the crew from mission duties. Another effort is to slightly revise the layout of the flight deck to provide a 'symmetric cockpit' for the commander and pilot, simplifying training and providing increased situational awareness.[12]

TPS Lower Surface Tile Study: This study was initiated to examine possible TPS replacements with reduced thermal conductivity, improved dimensional stability, and optimized coating durability and serviceability. Any replacement will need to closely replicate the LI-900 thermal properties and still be lightweight.[13]

Crew Escape Study: Part of the safety upgrades is yet another study of possible crew escape systems to improve the odds of the crew surviving a catastrophic accident. The goals of the study are to develop options for improving crew escape on the existing Orbiter, along with requirements (based upon an operational model for the Orbiter in the post-ISS assembly time frame) for possible crew egress systems. The product of this study will be an assessment of risk reduction versus cost to implement of each viable escape system.[14]

Nozzle/Case Joint J-Leg Insulation Design: This modification of the RSRM nozzle-to-case joint thermal barrier replaces the current polysulfide-filled joint with a design similar to the current field joint (J-leg) and igniter joint insulation configurations. The concept utilizes an insulation J-leg and carbon-fiber rope combination to simplify the present labor-intensive and process-sensitive nozzle installation. The modification will also minimize the potential for gas paths inherent in the existing configuration. This upgrade has been approved for implementation.[15]

RSRM Propellant Geometry Modification: The change in propellant geometry will improve structural margins of safety (from 1.4 to 2.0) in the SRM and reduce risk to personnel through reduced handling and inspection.[16]

Self-contained Atmospheric Protective Ensemble Suit: This project proposes to develop an advanced protective suit (APS) for use in propellant loading operations at KSC and CCAFS. The existing protective suit, called the propellant handler's ensemble (PHE), has been in use since the early 1980s and was derived from the original protective garment called the SCAPE Suit, developed in the 1960s. The term SCAPE is now used to describe a system of components and processes that provide protection to KSC workers during hazardous operations, not just the suit. Other components of the SCAPE system, such as communications improvements and body cooling devices, will also be addressed as part of the project.[17]

Industrial Engineering for Safety: The Industrial Engineering for Safety initiative will apply modern industrial engineering techniques and analysis to study Space Shuttle processing in an attempt to identify, prioritize, and implement high-value upgrade projects that will reduce risk and increase supportability.[18]

SRB Advanced TVC/APU: The SRB thrust vector control (TVC) subsystem consists of two separate hydraulic power units (HPU) that supply power to the TVC servoactuators to position the SRB nozzle during ascent. Each HPU consists of a hydraulic pump driven by a hydrazine-powered APU, similar to the Orbiter unit. MSFC is investigating replacing the APU with either an electric APU, a pressurized-helium-powered APU, or a solid-propellant gas generator powered turbine. An alternate concept is to replace the entire APU/HPU assembly with a pressurized blowdown/helium reccumulation hydraulic system.[19]

SSME Advanced Health Management System Phase I: Phase I includes the modification of 20 advanced Main Engine Controllers to add a high-pressure turbopump synchronous vibration redline capability. This system will be able to monitor vibrations from the SSME in real-time, allowing the operation of the engine to be better controlled – or shut down if significantly out of tolerance – with an overall six percent improvement in ascent safety. This upgrade has been approved for implementation.[20]

SSME Advanced Health Management System Phase II: This follow-on effort consists of a series of prototyping and requirements definition tasks (referred to as Phase IIA) including: (1) development of a brassboard and two protoflight health management computers (HMC), and the flight of a prototype flight HMC; (2) an optical plume anomaly detection system flight experiment; (3) use of a linear engine model as a main combustion chamber analysis tool; and (4) flight system requirements definition. Phase II would culminate in the development and production of a flight-certified HMC (referred to as Phase IIB).[21]

Block III SSME: The Block III SSME will incorporate two major improvements – an extra-large throat main combustion chamber (XLTMCC) and an advanced channel-wall nozzle. The XLTMCC provides another significant reduction (after the LTMCC introduced in the Block II/IIA engines) in the operating environment for the turbopumps and other components, resulting in generally increased engine reliability. This change will lower operating temperatures and pressures, reduce potential hazards in the Orbiter aft-compartment, and eliminate a number of critical welds. In addition, the new design results in lower unit costs and shorter production cycles. The channel-wall nozzle uses a process developed in Russia for the nozzle of the RD-120 engine. Flat stock is roll formed into a conical shape to serve as the nozzle liner. The liner is slotted to provide channels for the nozzle's liquid hydrogen coolant to flow through. A jacket is installed over the liner, welded at the ends, and the entire assembly is then furnace brazed. The channels in the liner take the place of the 1,080 tubes that regeneratively cool the current SSME nozzle. The channel-wall nozzle is a relatively simple design that has fewer parts and welds than the current complex SSME nozzle. (The current SSME nozzle takes 2.5 years to manufacture, costs $7 million, and is currently flown no more than 12 to 15 times because of safety con-

cerns related to hydrogen leaks.) The proposed upgrade would cost an estimated $63 million over four years for development and testing, plus an additional $71 million to build 18 certification and production nozzles.[22]

Abort Improvement Study: This study is focused on developing a prioritized list of feasibility studies that will provide a more complete understanding of the capabilities of the Space Shuttle system. Once these capabilities are better defined, they will be used to generate potentially more useful and reliable contingency abort scenarios.[23]

Supportability Upgrades

Supportability upgrades are primarily targeted to ensure that reliable hardware is available to support the expected mission manifest through at least 2012. This will be accomplished mainly by replacing systems that are becoming obsolete and will not reliably support operations into the second decade of this century. Supportability upgrades are generally in response to an increasing failure rate or decreasing efficiency (age, wearing out), an increasing repair time (component or technology obsolescence), a reduction of serviceable spares through attrition, or a lack of vendor support for repairs or replacement parts.[24]

A secondary category of supportability upgrades are those that provide significant operational improvement with respect to either reliability, mission success, operational processing, or life-cycle cost. This category is lower in priority than those upgrades required to ensure hardware availability to support the manifest. The selection of operational improvement upgrades is based on a strict payback criteria in terms of value of operational improvements or life-cycle cost reductions.

A total of 13 supportability upgrades have been examined, and by the end of 2000, six of these had received formal approval to proceed; the remainder are still being evaluated or are in the approval process.[25]

Robust ET TPS: This upgrade will define a new thermal protection system for the External Tank, primarily in the intertank area. The goals are to find a TPS that is easier to apply, less sensitive to damage, and less costly.[26]

ET Digital Radiography: This upgrade will implement a digital X-ray process to replace the X-ray film currently used to inspect the ET during production.[27]

Orbiter DDU Replacement: Although MEDS is eliminating the need for the dedicated displays on the instrument panel, the existing DDU was retained because it provides power to the rotational hand controller, rudder pedal transducers, and nose-wheel position transducers and amplifiers. However, the DDUs are becoming unsupportable. A new modular unit has been designed and three DDUs will be installed on the flight deck of each Orbiter. This upgrade has been approved for implementation.[28]

Orbiter Modular Memory Unit: The new VME-based solid-state modular memory unit (MMU) will replace the existing mass memory units, the operational (ops) recorders, and the payload recorders with a single unit. The MADS recorders might also be combined into the unit in the future. The new unit eliminates all mechanical moving parts to increase reliability, and also reduces the number of units required and their weight. This upgrade has been approved for implementation.[29]

Long-Life Alkaline Fuel Cell: The fuel cells provide electric power for the Orbiter and water for the crew – 96 fuel cells in 3 stacks convert hydrogen and oxygen into electrical power, water, and heat via an electrolyte. The fuel cells must be removed and overhauled every 2,400 hours (at about $3.5 million per overhaul) and require an average of four repairs (at approximately $100,000 each) per year. With continuing overhauls and repairs, the current inventory of fuel cells could the anticipated manifest beyond 2012. However, supporting the units is becoming increasingly expensive and labor-intensive, and possible mission impacts for fuel cell problems are becoming more significant as the manifest ramps up to support the ISS. Two different upgrades – long-life alkaline fuel cells and proton exchange membrane (PEM) fuel cells – were investigated to replace the current fuel cells. The program subsequently decided to pursue the long-life alkaline fuel cell (LLAFC) concept. It is expected that the replacement design will have an operational life of over 5,000 hours, require less maintenance, and be more reliable. This upgrade has been approved for implementation.[30]

Orbiter Communication System: The Ku-band system electronics are obsolete and have suffered recently from excessively long repair times because there is no vendor support for the equipment and the repair facility has been burdened with higher priority work. There are currently no operational spares and cannibalization is often performed to support the flight schedule. The Ku-band system failed on *Discovery* during STS-92, forcing a manual docking with the ISS. Procurement of additional assets will provide an interim capability to support the manifest while a modern phased array replacement is developed.[31]

Modular Auxiliary Data System: The MADS recorder stores performance and health data of several Orbiter subsystems for post-flight analysis. The MADS equipment is experiencing an increase in failures coupled with a diminishing spare parts inventory. A replacement data recorder will use current solid-state technology and provide supportability through the end of the program. This function may be combined with the new MMU, or will at least use similar equipment.[32]

Pulse Code Modulation Master Unit: The existing PCMMU is obsolete and is experiencing a significant failure rate (relative to flight rate) and long repair time due to limitations in test equipment and parts availability. The upgrade will include new hardware and also provide improvements in payload telemetry capabilities.[33]

Checkout and Launch Control System: This is the third attempt at replacing the existing Launch Processing System (LPS) at KSC with modern technology (the computers used by LPS were designed in the early 1970s – for example, they only have 128 kilobytes of memory). Many of the components of LPS were custom-designed, and are no longer be manufactured. Because of this limited technology, LPS is not capable of supporting many of the planned improvements to the vehicle, and is becoming increasingly expensive to support. The CLCS is addressing this by using modern off-the-shelf processors, networking, and programming techniques. This project has been on-going since the mid-1990s, and was originally intended to be complete by the turn of the century. Delays, primarily with software development, have postponed this by several years.[34]

SRB Altitude Switch Assembly: The new ASA is intended replace a 20-year-old design that has shown a substantial increase in failure rates. This upgrade has been approved for implementation. The design, development, and qualification effort began on 31 March 2000, and the first installation is planned for the second quarter of 2002.[35]

SRB Command Receiver Decoder: The CRD was designed to replace the shuttle range safety system (RSS) integrated receiver/decoder (IRD) and range safety distributor (RSD). The IRD and RSD are beyond their 10-year

design life with ages ranging from 7 to 22 years – and the existing RSS components and their test sets are experiencing an increasing failure rate. The new CRD eliminates 39 obsolete components and also replaces the maintenance intensive test sets currently being used. This upgrade has been approved for implementation. The design has passed development tests and the flight qualification certification effort is underway.[36]

SRB Integrated Electronic Assembly: The IEAs provide power distribution, command/control switching, and data transmission within the SRBs. The current IEAs were designed with 1970s technology, and many component parts and card assemblies are obsolete and no longer available. Possible replacements are being studied.[37]

Automated Booster Assembly Checkout System: The ABACS is a ground-based automated test set that verifies the SRBs are in proper working order and certified for flight. The ABACS has been in use since 1987 and has exceeded its 10-year life expectancy; parts are becoming more difficult to find, and growth capabilities to support more modern SRB subsystems are very limited. A new test set is being studied.[38]

Other Possible Upgrades

There are several other upgrades that have been discussed in detail, although they are not included in the current Upgrades Program, mainly because they are considered too long-term or expensive.

Non-Toxic OMS/RCS: Boeing and NASA are examining the feasibility of eliminating the hypergolic propellants currently used in the OMS/RCS system to reduce potential hazards, but there are many issues remaining to be worked and this upgrade is considered long-term at best. The non-toxic OMS/RCS upgrade would use liquid oxygen and ethanol instead of the current nitrogen tetroxide and monomethyl hydrazine hypergolic propellants. This will involve replacing the two OMS engines, all the RCS thrusters, propellant tanks, valves, and lines. In addition, the FRCS module would be connected to new common propellant tanks that would also be used by the OMS and aft RCS. The FICS modification to the existing OMS/RCS system is currently implementing this capability.

NASA believes that the elimination of toxic and corrosive propellants would reduce hazards on the ground and on-orbit, improve ground operations and turnaround times, decrease corrosion, and result in an annual $24 million savings at KSC per year. The switch to LO2 and ethanol could also improve engine performance, providing better support to the International Space Station. Approximately $4 million has been spent to study the OMS/RCS upgrade, and the total cost of the upgrade is estimated at $90 to $100 million for development, plus $400 million to build eight OMS pods and four FRCS modules. The OMS/RCS upgrade has some disadvantages. Although the modified OMS pod would have fewer parts than the current system, it would be more complex because the LO2 would require additional insulation and thermal controls. Structures and other subsystems in the vicinity of the LO2 tanks and lines might also require thermal protection. Because the nontoxic propellants are not hypergolic, an ignition system would also be required, which might reduce reliability – potentially unacceptable since the OMS engines are the way to initiate a deorbit maneuver.[39]

Integrated Vehicle Health Management/HTD: The first IVHM demonstration (HTD-1) was flown on STS-95 with the second (HTD-2) following on STS-96. These demon-

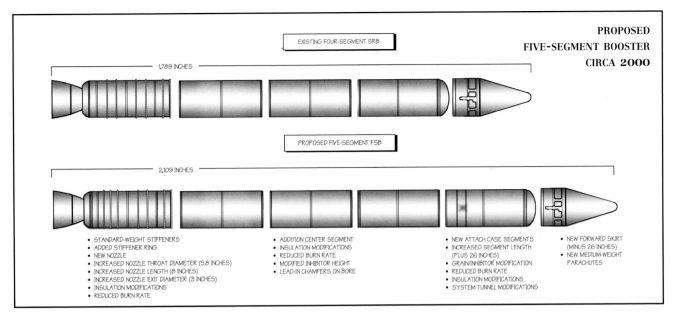

EXISTING FOUR-SEGMENT SRB

1,789 INCHES

PROPOSED FIVE-SEGMENT FSB

2,109 INCHES

- STANDARD-WEIGHT STIFFENERS
- ADDED STIFFENER RING
- NEW NOZZLE
- INCREASED NOZZLE THROAT DIAMETER (5.8 INCHES)
- INCREASED NOZZLE LENGTH (8 INCHES)
- INCREASED NOZZLE EXIT DIAMETER (3 INCHES)
- INSULATION MODIFICATIONS
- REDUCED BURN RATE

- ADDITION CENTER SEGMENT
- INSULATION MODIFICATIONS
- REDUCED BURN RATE
- MODIFIED INHIBITOR HEIGHT
- LEAD-IN CHAMFERS ON BORE

- NEW ATTACH CASE SEGMENTS
- INCREASED SEGMENT LENGTH (PLUS 26 INCHES)
- GRAIN/INHIBITOR MODIFICATION
- REDUCED BURN RATE
- INSULATION MODIFICATIONS
- SYSTEM TUNNEL MODIFICATIONS

- NEW FORWARD SKIRT (MINUS 26 INCHES)
- NEW MEDIUM-WEIGHT PARACHUTES

strated the capability of integrating modern off-the-shelf sensor technologies and operating them in the space environment. The results of these demonstrations will be used to support an analysis of possible IVHM applications to the Shuttle. In addition to the IVHM HTDs, a fiber-optic HTD was flown on STS-95 that captured telemetry data using a fiber optic pathway. The fiber optic data were then compared to telemetry data, and no issues were identified that would preclude the use of fiber optic communications aboard the Orbiter. A nano-micro-electrical mechanical system HTD (NanoMEMs) was flown on STS-93 that used low-weight, low-volume, low-power, and highly-reliable submicron and submillimeter-scale sensors to measure acceleration, temperature, and pressure in the crew module. The data collected by the NanoMEMS sensors matched data collected via normal Orbiter systems. On STS-96, a set of distributed data acquisition health nodes originally developed for the X-33 program was successfully flown in support of IVHM development. On STS-97 a laser dynamic range imager (LDRI) will measure loads during critical ISS assembly tasks, and will likely become an integral part of the space vision system (SVS). On the same flight an HTD will be installed that uses several wireless temperature sensors in the crew module and ODS – transmitters attached to the sensors will send data to the PGSC laptop computer.[40]

FIVE-SEGMENT BOOSTERS

This upgrade, discussed for years and informally proposed by Thiokol Propulsion in early 1999, consists of a variety of modifications to the four-segment SRB intended to improve safety, performance, and reduce overall systems costs. In addition to adding a fifth segment to the SRB, the proposed upgrade would modify the nozzle and insulation, and alter the grain of the solid propellant to provide a more risk-tolerant thrust profile. Estimated total costs are approximately $1,000 million with four years required from authority to proceed until the first flight.[41]

In March 1999 MSFC awarded a $4 million study contract to an USA-led contractor team for an evaluation of the five-segment booster (FSB). The study was to be completed in October 2000, but was later extended to the end of the year. The purpose of the study was to evaluate the overall feasibility of the FSB concept, and to develop a better estimate of the anticipated cost. In addition to USA, the contractor team included Thiokol Propulsion, Boeing North American, and Lockheed Martin at Michoud.[42]

There are several desirable objectives for the FSB. The first is eliminating the need for the RTLS and TAL abort modes by providing sufficient power to perform an AOA even if a main engine fails immediately after lift-off. Alternately, the new design could boost payload performance to high-inclination (ISS) orbits by almost 20,000 pounds (to 60,000 pounds). But there were several constraints put on the study. First, the FSB could not increase the aero, dynamic, thermal, or acoustic loads on the Orbiter or ET. The interfaces to the ET and launch pad could not change, and only minimal changes to other ground support equipment (VAB work stands, etc.) could be tolerated due to funding restrictions.

The final analysis from KSC showed that the RPSF would need modified, both assembly High Bays in the VAB would need work platforms moved, the MLPs would need strengthened, work platforms at each launch pad would need modified, and Hanger AF on CCAFS (where the SRBs are disassembled after being retrieved) would require some changes. It sounds worse than it was.

But adding a segment is not quite as straightforward as it would seem. The major issue is that the ET attachment ring is currently on the forward domed segment. This leads to the need to slightly redesign the forward segment to eliminate the ET attach hardware (thereby making it lighter and less expensive), and also to designing a new second segment that includes the attachment hardware. The approach that was taken also reduces the length of the new forward skirt by 26 inches, and increases the length of the new center segment by 26 inches.

Modifications would also be made to the propellant grain – changing from an 11-point star to a 13-point configuration. The burn rate would be slowed somewhat (from 0.368 to 0.343 inches per second) to limit thrust so that the Orbiter is not overstressed. New parachutes would need designed to handle the increased weight of the boosters. The aft segment that absorbs all of the weight of the stack on the pad would be equipped with the stiffeners originally used on the standard-weight cases early in the program and an additional stiffener would be added. A new nozzle that is 9 inches longer would be designed. This nozzle would have the throat diameter increased from

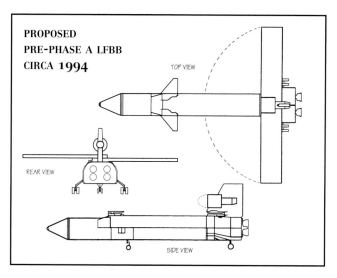

PROPOSED
PRE-PHASE A LFBB
CIRCA 1994

TOP VIEW

REAR VIEW

SIDE VIEW

This is the conceptual liquid fly-back booster that JSC developed during the pre-Phase A studies. The vehicle was 150 feet long and spanned 112.8 feet when the wings were deployed. Each wing folded forward on top of the fuselage – note that the wings deploy at different heights on each side of the fuselage. A single large turbofan engine was mounted in the vertical stabilizer very much like a DC-10 or MD-11 transport aircraft. During ascent the turbofan inlet was covered with a streamline fairing that was jettisoned after the booster slowed down to subsonic speeds. (NASA)

53.68 to 59.62 inches, a 3-inch large exit diameter, and the insulation inside the nozzle would be changed to handle the longer burn time. Interestingly, some items proposed for the FSB – such as the forward thrust bearing – use designs that originated on the abortive ASRM project.

Overall, the FSB would be 320 inches longer than the existing booster, stretching almost all the way to the top of the ET. The new booster would provide 3,799,000 lbf, compared with 3,000,000 lbf for the existing booster. Specific impulse would be lowered from 268.4 to 264.7 seconds.

No decision is expected to be made on whether to proceed with the modified booster in the near future, but the development schedule shows that the first demonstration motor could be fired approximately 12 months after the authority to proceed, with the first qualification motor firing 4 months later. Operational use would begin about 45 months after the project was approved.

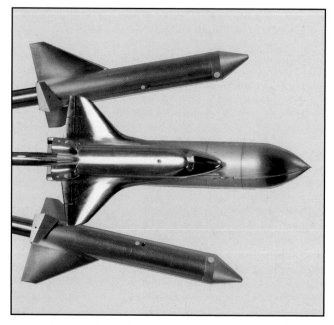

This model of a conceptual LFBB was tested at NASA Langley in early 1998. Note the folding vertical stabilizer on each booster. (NASA photo EL-1998-00050)

LIQUID FLY-BACK BOOSTERS

This proposed upgrade brings the program almost full circle, back to the original Phase A/B concepts of a fully-reusable system using a fly-back booster. In theory, it most closely resembles the Convair Triamese concepts since it uses two booster stages and a parallel-burn Orbiter, although this time the Orbiter keeps the External Tank.

The LFBB would replace the SRBs with two winged liquid-fueled boosters that would autonomously fly back to the launch site using conventional turbofan engines after they have separated from the Orbiter. Both Boeing and Lockheed Martin have been involved in preliminary studies of this concept, and both believe that the LFBBs would improve safety by reducing or eliminating the need for some high-risk abort modes, save money, and increase payload capacity. The companies also predict that the LFBB would enable a three-week turnaround time between missions, and (with three sets of LFBBs) could allow the Space Shuttle to fly 15 times per year.[43]

NASA has spent approximately $12 million to study the LFBB, and both Boeing and Lockheed Martin have also used internal funds to produce some initial design concepts. Both companies estimate that hardware fabrication and testing would take four years and cost about $5,000 million. But this estimate has critics – especially concerning the accuracy of the total costs of the program from development through production and operation. Almost every aspect of the LFBB suggests that the development costs will be high. The LFBB must be extremely reliable ('human rated'); will be a highly complex vehicle that uses both rocket and turbofan propulsion; will have all the systems and subsystems required to fly and land, including wings and landing gear; and will modify the mold line of the shuttle and thus require major testing and analysis of the new configuration. It would not be simple.[44]

Replacements for the SRBs have been studied for almost as long as the Space Shuttle has been in service. See appendix B for details of some proposed liquid rocket boosters (LRB) and the advanced solid rocket motor (ASRM). During the 1993 *Access to Space Study*, the only alternatives to the SRBs that were considered economically viable were liquid fly-back boosters (LFBB). The study concluded that expendable LRBs, LRBs that were recovered from the ocean, improved SRBs, and various hybrid boosters offered possible safety improvements but were much more expensive than the existing SRBs.[45]

In early 1994 a team at JSC began preliminary studies of an LFBB. This team was chartered to develop requirements that would later be released to contractors for further refinement – sort of a 'pre-Phase A' study. The basic objectives of the team were to improve safety, lower costs, and to identify high-risk areas that needed additional study. The LFBBs were primarily intended to improve the payload capacity and abort margins for missions destined for high-inclination ISS orbits (220 miles at 51.6 degrees) since they make up the majority of the manifest.[46]

The JSC team concluded that only two engine choices existed for their concept of an LFBB – an upgraded 2,2022,700 lbf Rocketdyne F-1A and the 1,777,000 lbf RD-170 from Russia. Interestingly, the team only seriously investigated boosters using one or two engines, seemingly ignoring the potential engine-out consequences of such an arrangement. The resulting LFBB was between 16 and 18 feet in diameter with an overall length of 120 to 170 feet. This maximum length was based on several of the earlier LRB studies that had shown severe aerodynamic interference factors with longer boosters – any booster longer than 150 feet (same as the SRB) would change the shock-wave interaction on the rest of the stack during ascent. After some preliminary wind tunnel testing at MSFC, the team decided to concentrate on a 16-foot diameter booster that was 150 feet long powered by two RD-170 engines.[47]

A wide variety of wing planforms were studied – fixed aft wings with canards, swing wings, stand-off wings (the wing was mounted well away from the booster on long struts), oblique-deployable wings, scissor-deployable wings, and deployable wings that folded away from the Orbiter. Because of uncertainties associated with the ascent aerodynamic interference effects of a fixed wing, the team chose a scissor-deployable wing concept, but added that "a wide range of configurations should be considered in Phase A of the LFBB program."[48]

In addition to the in-house efforts at JSC, Space Industries, Inc. performed additional analysis of the LFBB concept in support of the *Access to Space Study* under subcontract from Rockwell International. Both of these efforts, as well as a short KSC-led study of launch site impacts, were completed in 1994 .[49]

The LFBB concept languished for some time waiting to find funding, and also to allow the dust to settle from the decision to combine all of the Space Shuttle contracts into a single prime (United Space Alliance). During this time, MSFC and MSC continued in-house studies, and both Boeing and Lockheed Martin undoubtedly performed some preliminary work using company funds. The $1,000,000 Phase A study contracts were finally issued to Boeing North American in Downey (NAS8-97272) and Lockheed Martin in Michoud (NAS8-97259) on 5 May 1997. Subsequent extensions and options allowed the contractors to continue until the end of 1999. Although the two companies were technically competitors, a great deal of data was shared between them and in several instances they presented joint briefings to the government.[50]

One such joint briefing was to the Space Shuttle Development Conference at NASA Ames on 29 July 1999 – near the end of the study – highlighting the perceived benefits of what had been renamed a Reusable First Stage (RFS). Both contractors agreed that the RFS would be completely reusable and have an autonomous fly-back and landing capability, eliminating the need for water recovery (like the SRBs). All of the booster main engine (BME) candidates used LO2 and RP-1 propellants that were considered safer and easier to handle than LH2. Both contractors had designed their boosters to be tolerate of a single engine out immediately after lift-off, and both boosters eliminated the need for RTLS and TAL aborts by providing an abort-to-orbit capability at all times. Neither contractor felt that any new technology needed to be developed in order to proceed.[51]

Other perceived benefits included eliminating the SSME 'throttle bucket,' thereby substantially reducing the number of SSME failure modes (the ASRMs also claimed this as a benefit, as do the proposed five-segment boosters). The RFS eliminates all of the hazards associated with the SRBs, including handling explosives in the VAB, mid-ocean retrieval, post-retrieval clean-up, and the difficulties of transporting the SRMs to and from Utah. The exhaust plume of the RFS was also thought to be more environmentally friendly than the SRB exhaust.[52]

The Boeing design was a 16.5-foot diameter core that was 152 feet long and used integral aluminum propellant tanks that were derived from the design used on the

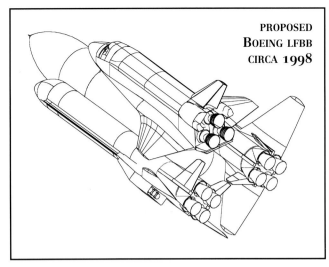

PROPOSED BOEING LFBB CIRCA 1998

An early Boeing LFBB concept. Note the configuration of the vertical stabilizers – later designs used a single centrally-mounted vertical instead of the two small ones. (Boeing)

Delta IV. What was called a delta-wing – but had actually evolved into more of a thick-chord aft-mounted swept wing – had a leading-edge sweep of 35 degrees and spanned 82 feet. The wing was optimized for the subsonic cruise portion of the flight, and small retractable canards (very similar to the stabilizers originally fitted to the B-1A escape capsules) were used for subsonic trim. A single all-moving vertical stabilizer was used. The vehicle empty weight was 204,000 pounds, with a booster lift-off weight of 1,400,000 pounds. The nose gear from the B-1B was used, while the main gear was derived from the Boeing

The initial contractor studies for the LFBBs included two options: 'dual' boosters – two separate winged boosters; and a 'catamaran' booster – a dual fuselage connected by a common wing, remarkably similar to the Martin Marietta SpaceMaster from the Phase A studies. The catamaran concept was dropped relatively early in the LFBB studies, but this 0.4-percent scale stainless steel model was tested at NASA Langley during 1998. (NASA photo EL-1998-00051)

The final Lockheed Martin LFBB in fly-back mode. Almost all of the illustrations released by Lockheed Martin show the use of four Pratt & Whitney RD-180S engines, resulting in eight bell nozzles at the back of each booster. (Lockheed Martin)

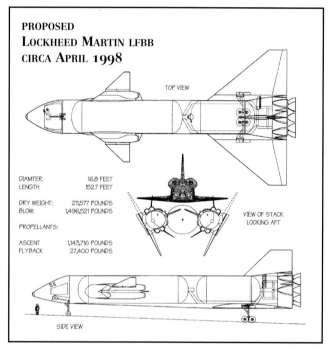

PROPOSED
LOCKHEED MARTIN LFBB
CIRCA **APRIL 1998**

TOP VIEW

DIAMTER: 16.8 FEET
LENGTH   152.7 FEET

DRY WEIGHT: 211,577 POUNDS
BLOW: 1,496,521 POUNDS

PROPELLANTS:

ASCENT   1,143,716 POUNDS
FLYBACK   27,400 POUNDS

VIEW OF STACK
LOOKING AFT

SIDE VIEW

The Lockheed LFBB design in late April 1998. By this time the 'catamaran' design had been abandoned by both contractors. (NASA)

The Lockheed Martin design placed pairs of air-breathing turbofans in semi-flush nacelles on each side of the fuselage, unlike Boeing that elected to use an under-fuselage nacelle that contained all four engines. (Lockheed Martin)

757 design. Four air-breathing engines were housed in a large single nacelle on the lower aft fuselage.[53]

The Lockheed Martin design looked radically different from the Boeing candidate, but was actually very similar in concept. The 151.7-foot long fuselage was 17 feet in diameter based largely on the structure developed for the common core vehicle of the Atlas V. A true delta wing had a 50-degree leading edge sweep, and the trailing edge was slightly swept forward. The vehicle was constructed from 2219 aluminum propellant tanks and a mostly composite structure in order to save weight, but still weighed 204,500 pounds empty with a gross booster lift-off weight of 1,439,000 pounds. All-flying canards protruded from the forward-mounted air-breathing engine nacelles and were used for subsonic trim and also to suppress Orbiter wing loading at max-q during ascent. The single vertical stabilizer was also all-moving.[54]

Both vehicles would be equipped with four booster main engines, either the newly-developed Rocketdyne RS-76 or the Aerojet AJ-800 derived from the Russian NK-33. Alternately, either booster could use the Pratt & Whitney RD-180S or the Aerojet AJ26-58, although this required the installation eight bell nozzles (still only four engines – each having two nozzles but sharing turbomachinery). All of the engines were rated at approximately 860,000 lbf at sea level and 930,000 lbf in a vacuum. The air-breathing engine candidates were the General Electric F118 turbofan engine used on the B-2A and U-2S, or the Pratt & Whitney F100-PW-229A used on the F-15 and F-16. Interestingly, the air-breathing engines (called fly-back engines) were modified to use RP-1 instead of JP-8.[55]

Perhaps not surprisingly, both RFS designs had remarkably similar flight profiles. The maximum dynamic pressure (max-q) would be reached 55 seconds after lift-off at 22,000 feet altitude and Mach 1, a little more than 2.5 miles from the launch site. The SSMEs would remain at 100 percent power through max-q. Booster separation would occur 2.3 minutes after lift-off at 160,000 feet altitude and Mach 5.2, a little more than 36 miles down-range. The Orbiter and ET would continue their ascent to orbit without change. The boosters would continue on a ballistic trajectory to an apogee of 238,000 feet at Mach 5.65, 112 miles down-range. The turn back to the launch site would begin 4.7 minutes after lift-off at a range of 201 miles with the boosters still traveling Mach 4.7 at 119,000 feet. A little over two minutes later the boosters slow to Mach 1 at 57,000 feet, but would be 220 miles from the launch site. The air-breathing engines would be started 9 minutes after lift-off while the boosters are 198 miles from the launch site traveling Mach 0.6 at 30,000 feet. The boosters slow to a steady-state cruise at Mach 0.47 and 10,000 feet, finally landing 56 minutes after launch. A touch-down speed of approximately 220 mph was expected, and the boosters would be sequenced so that there was a separation of about 10 minutes at landing.[56]

Performance estimates for the RFS were excellent, at least according to the contractors. The concept increased payload to high-inclination ISS orbits to 45,000 pounds, limited only by Orbiter landing considerations. The new boosters also allowed efficient dog-leg maneuvers that would permit polar operations from KSC. Both contractors estimated that the use of the RFS would reduce annual operating costs for 8 missions by over $400 million.[57]

The boosters would be equipped with a fairly robust thermal protection system, partially to withstand ascent and reentry heating, but also to prevent the formation of ice on the outside of the LO2 tanks while on the pad. It was felt that ice falling off the boosters during ascent might potentially damage the Orbiter. The boosters would be equipped with range safety systems consisting of linear-shaped charges along the propellant tanks, although these would only be used during ascent. If the need arose to destroy the vehicle during the fly-back portion of flight, the range safety officer would command the vehicle to fly into the ocean (what the X-33 program terms ICFIT – intentional controlled flight into terrain).[58]

The modifications necessary at KSC to support the RFS boosters were fairly substantial. Most options included modifying VAB High Bay #4 into a horizontal processing facility for the boosters, then using High Bay #2 for stacking operations. As more flights used the RFS boosters instead of SRBs, High Bays #1 and #3 would also be modified. The final configuration would be two High Bays dedicated to horizontal processing for the boosters (and also used for processing ETs in the space above the floor), and two High Bays dedicated to stacking. Alternately, a separate horizontal processing facility could be built for the boosters. Perhaps the single most expensive element, however, would be the construction of a new mobile launch platform. The new unit would be built in one of the existing MLP parking areas, and would be substantially similar to the existing MLPs except for the changes necessary to accommodate the RFS boosters (new flame holes, propellant loading systems, etc.). It was estimated that the construction of a new MLP would take five years – longer than the development of the booster itself. However, KSC personnel did not feel that one of the existing MLPs could be removed from service and still maintain a launch rate of 8 missions per year, necessitating the construction of a new MLP. Other changes at KSC included the addition of RP-1 fuel facilities (125,000 gallons) at both launch pads, expanding the amount of LO2 stored at the pads by over 900,000 gallons, changes to the fixed and rotating service structures, modifications to most of the swing-arms to provide clearance for the boosters, and changes to the flame deflectors and sound suppression water system. It would take about 18 months to modify each launch pad. The total cost for modifying KSC facilities was estimated at about $300 million, not counting any new horizontal processing facility.[59]

Perhaps taking a queue from the X-33 program, both Boeing and Lockheed Martin proposed building sub-scale demonstrators of their boosters. The demonstrator would use as much existing hardware as possible to minimize the time and cost associated with developing it. The primary purpose of the demonstrator would be to gain confidence in the proposed RFS concept, but secondary missions included use as a technology testbed and perhaps even as a first stage for small expendable vehicles to orbit payloads.

For instance, the Boeing demonstrator would be based largely on Delta II tanks and would scale out to about 50 percent of the proposed full-size booster. Off-the-shelf avionics, controls, and engines would be used. The landing gear would come from F-4 and F-15 fighter aircraft. The aerodynamic shape would be common with the proposed full-scale booster, but might use a smaller wing in order to maintain the same 110 psf wing loading envisioned for the final vehicle. Four NK-39 boost engines and two air-breathing engines of the same type selected for the full-scale booster would be carried. Boeing estimated that the half-scale demonstrator would weigh 48,000 pounds empty and have a gross lift-off weight of approximately 166,000 pounds. One of the more interesting features of the demonstrator – and Boeing's full-scale vehicle also –

Another major challenge for the LFBB project are the changes required at the launch site. The largest of these is the need to construct a new MLP, and removing a launch pad from service for 18 months for modifications. Once a pad was modified for the liquid boosters it could not be used by a stack equipped with solid boosters. Some of the changes that had been envisioned – extending the crawlerway to High Bay #2 and facility modifications to High Bay #4, have been at least partially completed as part of the Safe Haven project at KSC. But substantial changes would still be required in the VAB to accommodate the new booster Note the incorrect markings on the Orbiter. (Boeing)

The Boeing LFBB design placed the air-breathing engines under the fuselage, and used canards that folded flush with the forward fuselage much like the early B-1A escape capsule stabilizing fins. The landing gear on this vehicle is largely based on existing aircraft – nose gear from the B-1B and main gear from the 757. (Boeing)

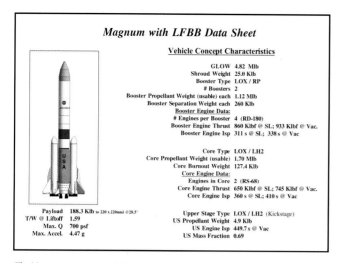

Magnum with LFBB Data Sheet

Vehicle Concept Characteristics

GLOW 4.82 Mlb
Shroud Weight 25.0 Klb
Booster Type LOX / RP
Boosters 2
Booster Propellant Weight (usable) each 1.12 Mlb
Booster Separation Weight each 260 Klb
Booster Engine Data:
Engines per Booster 4 (RD-180)
Booster Engine Thrust 860 Klbf @ SL; 933 Klbf @ Vac.
Booster Engine Isp 311 s @ SL; 338 s @ Vac

Core Type LOX / LH2
Core Propellant Weight (usable) 1.70 Mlb
Core Burnout Weight 127.4 Klb
Core Engine Data:
Engines in Core 2 (RS-68)
Core Engine Thrust 650 Klbf @ SL; 745 Klbf @ Vac.
Core Engine Isp 360 s @ SL; 410 s @ Vac

Upper Stage Type LOX / LH2 (Kickstage)
US Propellant Weight 4.9 Klb
US Engine Isp 449.7 s @ Vac
US Mass Fraction 0.69

Payload 188.3 Klb to 220 x 220nmi @28.5°
T/W @ Liftoff 1.59
Max. Q 700 psf
Max. Accel. 4.47 g

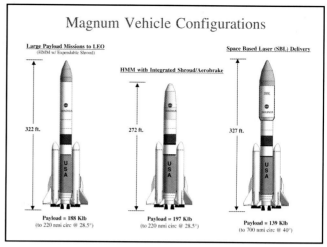

Magnum Vehicle Configurations

Large Payload Missions to LEO
(HMM w/ Expendable Shroud)
322 ft.
Payload = 188 Klb
(to 220 nmi circ @ 28.5°)

HMM with Integrated Shroud/Aerobrake
272 ft.
Payload = 197 Klb
(to 220 nmi circ @ 28.5°)

Space Based Laser (SBL) Delivery
327 ft.
Payload = 139 Klb
(to 700 nmi circ @ 40°)

The Magnum was a new vehicle proposed for the planned Mars Initiative as well as other tasks that required a heavy-lift vehicle. There were several variations to the Magnum, most using the proposed Reusable First Stage booster, but others were equipped with existing SRBs in an attempt to lower development costs. (Lockheed Martin)

was the use of lithium-ion batteries originally developed for the automotive industry for use in electric cars.[60]

The Lockheed sub-scale would use an off-the-self EELV core (Atlas III) including the single RD-180 main engine. Two full-scale air-breathing engines and nacelles would be provided in order to demonstrate the ability to start in-flight. Unlike Boeing, Lockheed planned to equip the demonstrator with a full-scale RFS avionics suite including provisions for the complete ascent profile.[61]

Alternately, at least Boeing proposed building a full-scale demonstrator that initially would not be equipped as a booster – it would contain smaller propellant tanks, less avionics, no thermal protection system, etc. Initially the demonstrator would be used to verify the basic aerodynamic configuration and flight software, along with the new oxygen–alcohol reaction control system and other advanced subsystems. The propellant tanks would be aluminum iso-grid structures very similar to those being used on the Delta IV EELV, The integrated health monitoring system was heavily based on the system installed in the 777 transport, and also on technology developed for the X-33 and X-34 projects. Boeing felt that this vehicle could be incrementally modified to achieve additional performance as confidence was gained, while still minimizing near-year costs. Eventually the airframe could be brought up to the final configuration and used as an operational booster. The

demonstrator would have been launched from Complex 34 at Cape Canaveral and return to the Skid Strip at CCAFS.[62]

As with many of these studies, the two contractors and MSFC quickly expanded the scope to include follow-on vehicles – in this case the future was called Magnum. Two RFS boosters would flank a large core derived from the shuttle ET but including a pair of RS-68 engines underneath it and a large (95 by 25-foot) payload fairing on top. It was expected that this configuration could carry 188,000 pounds to low-earth orbit at a cost of $155 million per flight, and would cost $5,100 million to develop, including the RFS boosters. Alternately, a similar vehicle could be built using the existing shuttle SRBs that could carry 176,000 pounds for a development cost of $2,000 million.[63]

Although the possibility that the RFS/LFBB will ever be approved is very slim, work continues on the concept at a very low level at MSFC and within Boeing and Lockheed Martin. The concept offers several operational advantages for space shuttle, but the $5,000 million development cost is more than NASA can probably justify spending on the 30-year old Orbiter. And the idea comes at an inopportune time since the current pace of space station assembly flights would make it very difficult to take one of the launch pads out of service for a prolong period of time. If shuttle is going to upgrade its boosters, the smart money is on the five-segment solid – and even it is a long shot.

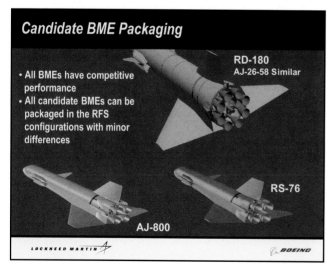

Candidate BME Packaging

• All BMEs have competitive performance
• All candidate BMEs can be packaged in the RFS configurations with minor differences

RD-180
AJ-26-58 Similar

RS-76

AJ-800

LOCKHEED MARTIN   BOEING

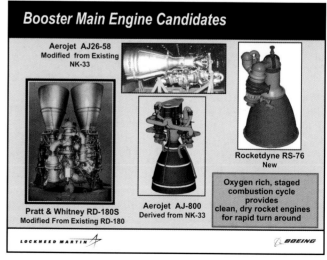

Booster Main Engine Candidates

Aerojet AJ26-58
Modified from Existing NK-33

Rocketdyne RS-76
New

Pratt & Whitney RD-180S
Modified From Existing RD-180

Aerojet AJ-800
Derived from NK-33

Oxygen rich, staged combustion cycle provides clean, dry rocket engines for rapid turn around

LOCKHEED MARTIN   BOEING

The various booster main engine candidates for both the Boeing and Lockheed Martin vehicles, although the booster shown at left is actually the Lockheed Martin design. It is interesting to note that three of the four engines are based on Russian powerplants. (Lockheed Martin)

IMPROVING THE BREED

Almost since the awarding of the original development contracts, NASA and various contractors have been proposing ways to make Space Shuttle 'better.' The definition of better depends on one's perspective – most of the concepts have centered around two issues – increasing payload capabilities, and replacing the SRBs. Although there have been numerous studies conducted along these lines, three separate efforts deserve special attention – Shuttle-C, and the various studies that led up to the decision to build Shuttle-C; the Advanced Solid Rocket Motor (ASRM) developed to replace the RSRMs; and several variations of liquid rocket boosters (LRB) studied in-depth to replace the SRBs. Two newer projects, the five-segment solid booster and the liquid fly-back booster, are covered in Appendix A since they are – remotely – still under consideration.

SIGMA CORPORATION EDIN05

One of the first enhancements to the baseline Space Shuttle vehicle to be studied was the replacement of the SRBs with some sort of liquid rocket booster (LRB). One such study was conducted by the Sigma Corporation in January 1976. This study, designated EDIN05, used a slightly modified Orbiter and an extensively modified External Tank. The ET was larger, both in diameter and length, and had significantly different structural design. Instead of having two SRBs attached to its sides, this ET had a single recoverable liquid booster pack located under its aft end. It should be noted, that unlike later proposals that supplemented the SRBs with liquid engines of some description, the EDIN05 replaced the SRBs.[1]

Two different types of liquid booster packs were investigated by the study. The first used three or four Rocketdyne F-1 engines from the Saturn V first stage, while the other was powered by three new-design high-

---

* There were two variations of this engine, one producing 680,000 lbf (sea level), and another producing 800,000 lbf.
† Apparently, the parallel versus sequential staging debate had not died a quiet death.

pressure engines* powered by LO2 and RP-1 proposed by the System Development Corporation. Two variations† on each booster were considered: one in which the Orbiter's SSMEs were ignited simultaneously with the booster pack, and one where the SSMEs were ignited upon staging. In the F-1 powered configurations the maximum dynamic pressure and longitudinal acceleration were constrained, if necessary, by throttling the SSMEs, or in the event that maximum throttling was not sufficient, by shutting down one F-1 engine. However, there were some concerns about whether the F-1 could be restarted in flight, since this capability had never been tested during the Saturn program. In the new-development high-pressure engine powered configuration these constraints were met by throttling the new engines.

The Orbiter modifications were limited primarily to the structure necessary to accommodate the anticipated increase in payload capacity, and to the electronics necessary to control the new booster. The ET traded the SRB thrust structure design for one optimized to accommodate the new aft-mounted liquid booster pack. Additionally, the ET was also parametrically scaled to increase the available propellant storage capacity. The new booster pack contained the engines and controllers, a parachute recovery system and a retro package to aid in reentry.

One of the major drawbacks to the EDIN05 was that it would have required a major redesign of launch facilities to accommodate it since the fuel systems, hold-down posts, and vehicle interfaces were not compatible with the baseline Space Shuttle vehicle. Although an interesting design, NASA was firmly committed to getting the basic vehicle into operational use, and did not seriously pursue the proposal.

SHUTTLE DERIVED VEHICLES (SDV)

Another possibility was to use as much Space Shuttle hardware as possible to build a derivative vehicle, and a variety of studies were accomplished over a ten year period beginning in 1977. All of these studies had common groundrules – namely, the derivative vehicles had to use as much commercially available and/or Space Shuttle-devel-

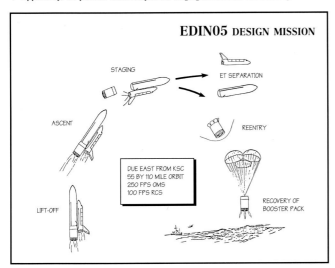

EDIN05 DESIGN MISSION

STAGING

ET SEPARATION

ASCENT

REENTRY

DUE EAST FROM KSC
55 BY 110 MILE ORBIT
250 FPS OMS
100 FPS RCS

LIFT-OFF

RECOVERY OF
BOOSTER PACK

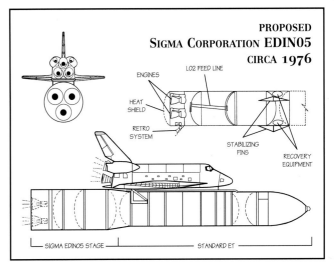

PROPOSED
SIGMA CORPORATION EDIN05
CIRCA 1976

ENGINES

LO2 FEED LINE

HEAT
SHIELD

RETRO
SYSTEM

STABILIZING
FINS

RECOVERY
EQUIPMENT

SIGMA EDIN05 STAGE

STANDARD ET

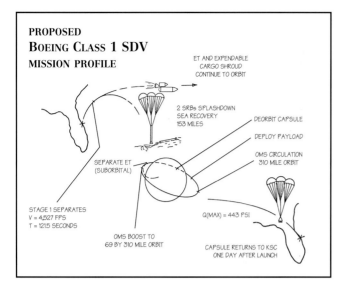

PROPOSED
BOEING CLASS 1 SDV
MISSION PROFILE

oped components as possible in order to keep cost and schedule concerns to a minimum. In addition, modifications to the processing and launch facilities at KSC and Vandenberg had to be kept to a minimum. It should be noted that all of these studies involved unmanned systems. The studies led directly to a vehicle called Shuttle-C that, for a while, seriously looked like it might reach production.

Boeing (NAS9-14710 AND NAS8-32395)

The Boeing Company released two studies on Shuttle Derivative Vehicles (SDV) – one on 28 July 1977 and another in September 1977. The first had been conducted for KSC, while the second was under contract to MSFC. The goals of both were to explore new approaches to lower the cost of lifting a pound into orbit, while at the same time minimizing the impacts on Space Shuttle facilities and the STS program schedule. Although the study for MSFC took a brief look at alternate concepts based on Space Shuttle hardware (including some that looked remarkably like the original two-stage Phase B concepts), in the end all scenarios assumed that both the SDV and Orbiter would co-exist. The studies projected a launch rate of 45 Orbiter and 22 SDV flights per year beginning in 1988. Two distinct vehicles were investigated by Boeing and, somewhat predictably, they received the designations Class 1 and Class 2.[2]

The Class 1 vehicle used the ET and SRBs baselined for use by the normal Space Shuttle program. In place of the Orbiter, one of two different 'expendable cargo shrouds' could be used to house oversize payloads. The first of these was 95 feet long and 23 feet in diameter, with a payload bay length of 68.5 feet. A recoverable capsule containing either three or four SSMEs was connected by a truss structure to the aft end of the cargo shroud. The truss structure was more reminiscent of a Russian design than an American one, and was chosen because it was the simplest and lightest structure that could be designed for the role. The engine capsule was jettisoned at main engine cutoff and parachuted to a waiting recovery vessel for refurbishment and reuse. The other shroud was 125 feet long and 23 feet in diameter with a maximum payload length of 100 feet. The major differences were that this concept had the three main engines attached integral to the aft end of the shroud with clamshell doors that closed to protect the engines during reentry. Also, no payload bay as such was provided, the shroud simply covering the entire payload and being jettisoned once in space.

Boeing envisioned the three-engine derivative of either concept being capable of placing between 130,000 and 200,000 pounds into a 310-mile circular orbit inclined 28.5 degrees from KSC. The four-engine version could boost up to 225,000 pounds. The three-engine Class 1 SDV would have a gross lift-off weight of 2,033,000 pounds, with the four-engine version weighing approximately 2,150,000 pounds at launch. It was expected that the propulsion modules would have a useful life of 300 flights, with a major overhaul performed every 100 flights.

As with the baseline Space Shuttle, the ET would be expended every flight. The SRBs would be production units, with a design life of 20 flights each. The total cost to develop and test the new propulsion module was $930 million, with the first flight article costing $135 million. Production flight articles would have $4.232 million in expendable hardware, with a total cost per flight of $14.536 million. This amounted to a recurring cost of $507 to boost a pound into orbit. This was not much of a savings from the baseline prediction that the Space Shuttle could achieve a recurring cost of only $600 per pound into low-earth orbit.

The proposed Class 2 vehicle used a cargo shroud and propulsion modules similar to the Class 1 vehicle with the exception that the shroud was 27.25 feet in diameter and slightly longer. The major difference was that in place of

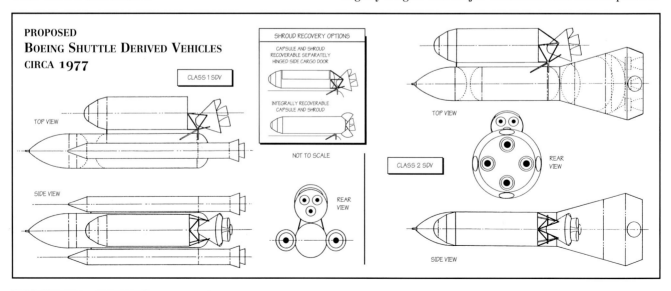

PROPOSED
BOEING SHUTTLE DERIVED VEHICLES
CIRCA **1977**

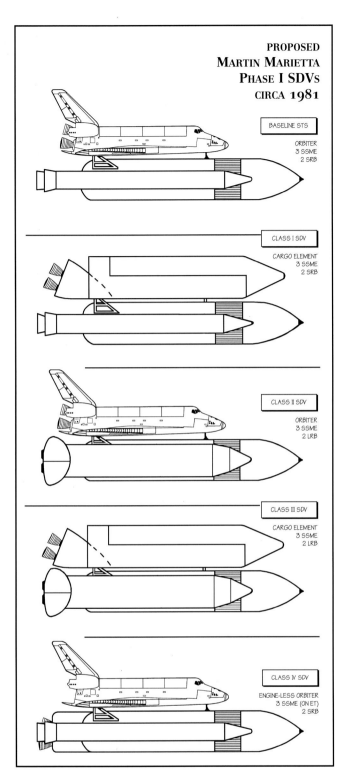

PROPOSED
MARTIN MARIETTA
PHASE I SDVs
CIRCA 1981

BASELINE STS

ORBITER
3 SSME
2 SRB

CLASS I SDV

CARGO ELEMENT
3 SSME
2 SRB

CLASS II SDV

ORBITER
3 SSME
2 LRB

CLASS III SDV

CARGO ELEMENT
3 SSME
2 LRB

CLASS IV SDV

ENGINE-LESS ORBITER
3 SSME (ON ET)
2 SRB

ratio and were designed for 50 flights between major hot-section overhauls. This system had a design gross lift-off weight of 2,502,000 pounds and was to be capable of boosting 200,000 to 300,000 pounds into a 310 mile orbit.

The primary disadvantage of the Class 2 vehicle was that the mobile launch platforms (MLP) would have to be extensively reworked to support the new booster, and that with the retirement of Saturn, there were no longer any RP-1 facilities at Launch Complex 39. In fact, the Boeing study highly recommended the addition of a fourth MLP configured especially for the SDV-2 vehicle. This tact would also be taken by other SDV studies, always without success. The estimated cost to develop the new booster was $2,863 million, with the first article costing $241 million. Production flight articles would use $4.380 million in expendable hardware, with a total cost per flight of $3.437 million. This resulted in a per pound to orbit cost of $269.

The study determined that the Class 1 vehicle could share the majority of the facilities and support equipment being built to support Space Shuttle. In fact, even the time lines and man-loadings used during Shuttle processing could be used to process the SDV. Neither the Class 1 or Class 2 vehicle exceeded the capacity of the crawler-transporter used to carry the stack from the VAB to the launch pad. The study concluded that the Class 1 concept would have fewer program-wide impacts and was probably the better idea, if slightly less efficient. One of the major worries was that having an SDV on the pad instead of a man-rated Orbiter would violate early program rules[*] that stated that a 'rescue' mission could always be launched within 24 hours. As it ended up, this capability was never realized as launch rates never reached their anticipated maximum due to technical limitations, and then actually fell due to budgetary constraints.

Martin Marietta (NAS8-34183)

Another study was undertaken by the Michoud Division of Martin Marietta – already responsible for the design and production of the baseline ET. The study had been initiated by MSFC in late-1980, with the Phase I results presented in July 1981. A small team led by Martin's Frank L. Williams considered four possible Shuttle Derived Vehicles during Phase I:[3]

- Class I: the Orbiter would be replaced with a Cargo Element (CE), and a reusable propulsion/avionics (P/A) package mounted under the Cargo Element
- Class II: the Solid Rocket Boosters were replaced with higher thrust reusable liquid rocket boosters (LRB), but the normal Orbiter was retained
- Class III: a combination of Class I and Class II concepts, the Orbiter is replaced by the Cargo Element and P/A package, and the SRBs are replaced with the LRBs
- Class IV: The propulsion system and appropriate avionics are removed from the Orbiter and installed under the ET, resulting in a 'glider' Orbiter.

The propulsion/avionics package was to be housed in a lifting-body aeroshell that had a hypersonic L/D ratio of 0.83:1. Two options were presented for landing, either tail-first vertically onto a tripod landing gear assembly, or horizontally onto a quadrapod impact attenuation system.

The Class I Shuttle Derived Vehicle simply replaced the Orbiter with a Cargo Element. The resulting vehicle had a much larger payload-to-orbit capacity that the baseline Orbiter simply because the Cargo Element was signifi-

the standard ET and SRBs, a new first stage was proposed. The new stage used a modified ET atop a bell-shaped compartment housing both additional propellant tanks and four engines burning LO2 and RP-1, much the same as the EDIN05 concept. The new high-pressure engines were the same design as investigated by the earlier EDIN05 study. The 680,000-lbf engines burned LO2 and RP-1 in a 2.9:1

---

[*] The rescue concept was based on the assumption that an Orbiter would always be on the pad awaiting launch at any given time, based simply on the flight rate of 45 missions per year. The Orbiter could, with minor difficulties, be made ready for a rescue mission within 24 hours if the need ever arose. This concept quietly went away when it became obvious that the flight rate would never exceed 12–15 per year.

APPENDIX B – STILLBORN

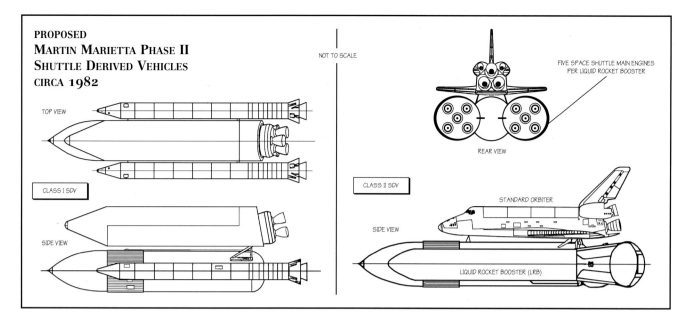

PROPOSED
**Martin Marietta Phase II
Shuttle Derived Vehicles
Circa 1982**

NOT TO SCALE

TOP VIEW

CLASS I SDV

SIDE VIEW

FIVE SPACE SHUTTLE MAIN ENGINES
PER LIQUID ROCKET BOOSTER

REAR VIEW

CLASS II SDV

SIDE VIEW

STANDARD ORBITER

LIQUID ROCKET BOOSTER (LRB)

cantly lighter than the Orbiter. This vehicle required minimal modifications to existing facilities.

Class II configurations replaced the SRBs with two LRBs, each using four SSMEs. These boosters would have given the baseline Orbiter the capability to carry 100,000 pounds into orbit, although it would still have been constrained to the 15 by 60 foot payload bay. There were also concerns about whether the Orbiter structure could reliably support the 100,000 pound payload capacity.

Other proposals called for using fuels besides LH2 in the LRBs, such as methane or propane, or a combination (hybrid) using LH2 and either RP-1 or JP-7. Because of the more delicate nature of the liquid motors, clamshell closures over the aft skirt opening and retro-rockets were baselined in addition to parachutes. A flyback capability for the boosters, using Pratt & Whitney JT9D-7R4D turbofan engines and fixed straight-wings, was also studied briefly. The LRBs would have required extensive modifications to launch facilities.[4]

Class III studies combined the improvements of Class I and Class II, and resulted in a payload of up to 250,000 pounds. This configuration would have been the fastest and easiest way for the United States to recover the heavy lift capability abandoned with the Saturn V.

The Class IV configuration simply removed the SSMEs, and their associated avionics from the Orbiter, and

installed them in a recoverable module under the ET. This freed up some additional room on the existing Orbiters, but offered no real performance improvement. A slight variation on this theme was to leave engines on the Orbiter, and install additional engines under the ET to increase payload capability. Either of these options would have required substantial modifications to the launch facilities.

After full consideration of various mission models, vehicle configurations, technologies, and life cycle cost models, the Phase I effort recommended that the Class I and Class II vehicles be studied further during Phase II. NASA agreed since there was no new technology in either of the other two classes of vehicles.

For Phase II, Martin Marietta further refined their mission models, concentrating on the projected 1985–2000 flight rate. The Phase II report was issued in February 1983, although by this time it was known as Shuttle Derived Cargo Vehicles (SDCV). This report concentrated on the two classes of vehicles approved after Phase I:[5]

- Class I: The Orbiter was replaced with a Payload* Module (similar to the earlier Cargo Element), and a reusable Propulsion/Avionics (P/A) package
- Class II: The Orbiter was retained, but the baseline SRBs were replaced with higher thrust reusable LRBs.

This effort concentrated more on defining the work needed to convert the theoretical studies from Phase I to a production vehicle. As such, it concentrated on performing subsystem trades and design refinements, determining high payoff technology drivers, establishing configuration/technology cost leverages and recommending a finalized SDV configuration. A detailed study of payloads was also performed and found that the majority of the payloads did not require the maximum 65,000 pound lift requirement, leading to the conclusion that volume, not weight, was the limiting factor of the Orbiter.

Various sizes of Payload Modules were examined by the study, which finally settled on one 15 feet in diameter and 90 feet long and a second 25 feet in diameter and 92 feet long. The maximum payload weight was considered to be 150,000 pounds into a due east orbit from KSC. It was

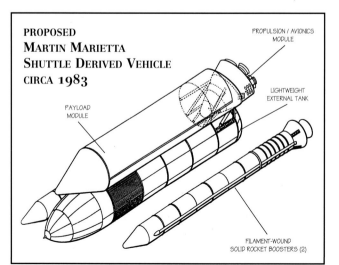

PROPOSED
**Martin Marietta
Shuttle Derived Vehicle
Circa 1983**

PROPULSION / AVIONICS
MODULE

LIGHTWEIGHT
EXTERNAL TANK

PAYLOAD
MODULE

FILAMENT-WOUND
SOLID ROCKET BOOSTERS (2)

---

* NASA tended to not like the term 'cargo' when referring to objects transported by the Space Shuttle. The preferred terminology was 'payload,' hence the change.

expected that the smaller Payload Module could be launched as early as 1985 given an immediate authority to proceed, with the larger version coming on-line in 1987. In both cases the Payload Module was split into two halves, with the upper half being jettisonable to expose the payload. In fact, the shroud was to be jettisoned very early in the flight profile (but still above the atmosphere) to allow the last fraction of the mission to be flown at a reduced weight, slightly increasing the allowable payload to orbit.

The baseline Phase II propulsion/avionics package was essentially a ballistic aeroshell with a hypersonic L/D ratio of 0.19:1. A tail-first landing attitude and tripod landing gear were baselined during Phase I. Phase II landing stability studies resulted in a change to a horizontal landing attitude and a quadrapod impact attenuation gear.

Cross-range requirements for recovery at Edwards AFB were investigated for varying launch inclinations from both KSC and Vandenberg. For the earlier ballistic P/A package, it was determined that any desired cross-range could be made up propulsively with approximately 10,000 pounds of OMS propellants. This was not deemed ideal, and studies of a hypersonic lifting-body configuration was made. The chosen configuration was statically stable during both hypersonic and supersonic flight, minimizing aerodynamic control requirements. This configuration had a hypersonic L/D ratio of 0.8:1, almost identical to the Apollo command module. The lifting-body configuration was six feet longer than the baseline P/A package, resulting in the need to extend the Payload Module by six feet to recover the length. This also allowed some aerodynamic improvements due to a better fineness ratio, but would have resulted in the need to extensively modify some portions of the ground facilities at KSC and Vandenberg.

A slight variation of the Class I theme was presented as a Class IA vehicle. In this concept the separate P/A package was eliminated, and the engines and avionics were installed directly into the Payload Module. Since the payload module was not recoverable, and the engines and avionics constituted a major program expense, a method of retrieving them from orbit had to be found.

The answer was to install the components with quick-disconnect fittings, and to boost the Payload Module to a stable orbit where an Orbiter would eventually recover them. It was envisioned that the Orbiter would be launched with a payload of its own, but after deployment would rendezvous with the expended Payload Module, remove the equipment with the Remote Manipulator System arm, secure the engines and avionics in the payload bay, and return to Earth. This scheme would have required no extravehicular activity by astronauts. If the components needed to be recovered in a hurry, an empty Orbiter could be launched to retrieve them, but it was felt this was not economical.

The LRBs proposed for the Class II vehicle consisted of four or five SSMEs located in the aft end of a structure that was 20 feet in diameter and 104 feet long. This structure contained LO2 and LH2 tanks, parachute recovery system, and all necessary electronics. The four-SSME booster continued to use LO2 and LH2 as propellants, and enabled the Orbiter to lift a total of 100,000 pounds. The thrust-to-weight ratio at lift-off with one engine off in each booster was 1.122:1, which exceeded the 1.05:1 minimum design (safety) requirement.

Recovery of the four-engine SSME booster was extremely similar to recovering the SRBs. The boosters continued to fly a basically ballistic trajectory after they were staged, finally deploying parachutes when the velocity allowed.

Clamshell doors and flotation devices protected the engines from water immersion, and the boosters were towed back to the launch site by the SRB recovery vessels.

During the Class II booster trade studies, a major emphasis was placed on replacing LH2 as a fuel. The two candidate substitutes were methane and propane. Life cycle cost estimates showed that a four engine ballistic booster design using methane or propane fuel were the most cost effective.

The alternate fuel boosters were basically similar to the baseline LRBs with the addition of a set of air-breathing engines. The cruise-back mode would essentially continue the ballistic path established after staging, with a 180-degree turn executed as the booster reentered the atmosphere. After it had descended low enough, the air-breathing engines would ignite for powered flight back to the a recovery area near the launch site.

Some of the concepts studied had small deployable wings that allowed the boosters to glide to a landing at the recovery facility at KSC, while others simply aimed for the recovery area and deployed their parachutes. There were also studies of far more sophisticated flyback boosters, some rivaling the complexity of the earlier 1970 Phase B designs, although all of the SDV concepts were unmanned.

Selected advanced technologies were investigated for the Class I configuration because it was not volume limited, and had better performance capabilities than the Class II vehicle. Technologies investigated included replacing most of the aluminum structure with advanced

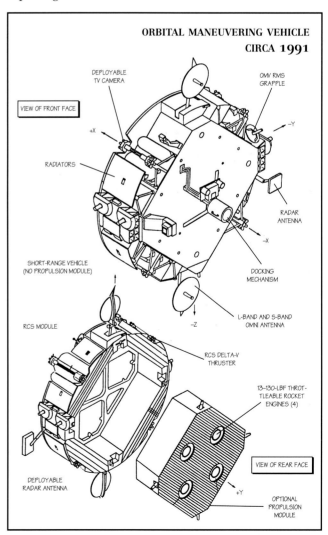

ORBITAL MANEUVERING VEHICLE
CIRCA 1991

DEPLOYABLE TV CAMERA

VIEW OF FRONT FACE

+X

RADIATORS

OMV RMS GRAPPLE

-Y

RADAR ANTENNA

-X

SHORT-RANGE VEHICLE (NO PROPULSION MODULE)

DOCKING MECHANISM

-Z

L-BAND AND S-BAND OMNI ANTENNA

RCS MODULE

RCS DELTA-V THRUSTER

13-130-LBF THROTTLEABLE ROCKET ENGINES (4)

VIEW OF REAR FACE

DEPLOYABLE RADAR ANTENNA

+Y

OPTIONAL PROPULSION MODULE

composites, resulting in a savings of 4,482 pounds for the P/A package and 18,190 pounds for the Payload Module. The use of advanced composites on the Payload Module also eliminated the need for a thermal protection system, eliminating another 6,676 pounds. All electrical signal wiring in both the P/A package and Payload Module could utilize fiber-optic cable instead of conventional coaxial and twisted pair, resulting in a further savings of 2,033 pounds. Advanced computer technologies were also investigated for use as engine controllers.

Another possible improvement was the substitution of uprated SSMEs for the baseline 109 percent engines used by the Orbiter. These SSME-35 engines were rated at 130 percent, primarily by increasing the mixture ratio from 6.0:1 to 6.76:1. This had two primary disadvantages, the first being the need to stretch the ET by four feet to accommodate the needed oxidizer (and also increasing its inert weight by 1,580 pounds), and the second being that the new ET would be totally incompatible with ground facilities at KSC and the production line at Michoud.[6]

However, the incorporation of these advanced technologies had a remarkable effect of performance. Whereas the baseline Class I concept had a payload capacity off 134,536 pounds to a 185 mile 28.5-degree orbit, a vehicle incorporating all of the desired changes could lift 202,081 pounds to the same orbit. This also assumed, incorrectly, that the filament-wound SRB casings being designed for Vandenberg would be available. For this reason, most follow-on studies tended to concentrate on advanced SDV concepts, even though they violated the groundrule regarding usage of existing KSC and VLS facilities.

An artist concept of a late three-engine Shuttle-C concept. The ET and SRBs were stock Space Shuttle items, matched with a new cargo element. (NASA / MSFC)

The NASA 1980 STS mission model for the years 1989–2004 showed a total of 801 flights. The majority of these did not require a man-in-the-loop to perform any nominal-mission tasks. The cost analysis that accompanied the Martin studies contained some interesting observations. Based on the 1980 mission model, increasing the payload beyond 150,000 pounds resulted in a very small reduction in the number of flights, simply because most missions were volume limited, even with the larger payload space available with the Payload Module. This led to a conclusion that increasing the payload capacity to 200,000 pounds was not justifiable on economic grounds.

The external tank/aft cargo carrier (ET/ACC) was studied as a method of increasing the payload volume available on an unmodified Orbiter by providing a large container attached to the aft end of the ET. This would have involved the addition of an 27.9-foot extension to the aft (bottom) end of the ET for use as the ACC. This effectively doubled the available payload volume, and could accommodate payloads up to 25 feet in diameter, compared to 15 feet in the Orbiter. The ACC would add 3,300 pounds to the empty weight of the integrated vehicle. The most favored use of the ACC appeared to be to carry the proposed cryogenically-fueled Orbital Transfer Vehicle (OTV) for use by the Space Station Freedom. The OTV was conceived during 1979 as a method of moving payloads around in space, including transporting them to and from geosynchronous Earth orbit (GEO). By the time the actual development of the vehicle began in 1986, it had been renamed Orbital Maneuvering Vehicle (OMV). In essence, this was a replacement for the Space Tug that had been proposed at the same time as Space Shuttle development was approved.

Martin Marietta completed their Shuttle Derived Vehicle Study with a 10 February 1983 briefing at MSFC. The conclusion was that the best choice for continued development was a vehicle resembling the Class I SDV with a recoverable lifting-body P/A module. This briefing also highlighted several potential missions for the SDV vehicle, including delivering a Very-large Space Telescope then under consideration as an evolution to the Hubble Space Telescope. It was also proposed that the SDV be used as a proof-of-concept vehicle for many of the changes being proposed for the baseline Space Shuttle system, including using it for the first flight of enhanced main engines and improved Solid Rocket Boosters.[7]

SHUTTLE-C

The SDV studies by Boeing and Martin confirmed that building some variation of a shuttle-derived vehicle would be a reasonably economical method of lowering the cost of launching payloads and reducing the risks of manned missions. But the baseline Shuttle program was having problems meeting a demanding launch schedule, and had not lived up to the promises NASA had made during the early studies regarding 'economical' access to space. It was highly unlikely at this point that the Congress, still remembering early promises, would allocate additional funding to develop an SDV capability, especially since NASA was still attempting to increase its fleet of Orbiters. So the SDV studies languished at MSFC, with only a small contingent of NASA engineers continuing to refine the concept and potential uses. This effort was grouped with a variety of others under an 'advanced concepts' category.

The *Challenger* accident in 1986 provided a rationale for NASA to increase the effort expended on SDV studies. Immediately after the accident, NASA was criticized for

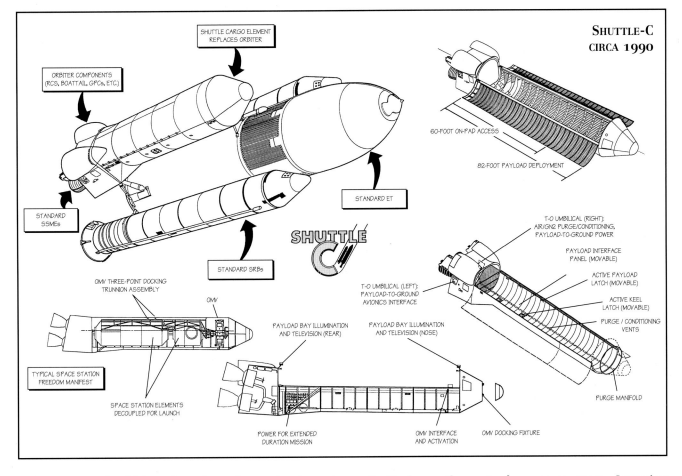

SHUTTLE CARGO ELEMENT
REPLACES ORBITER

ORBITER COMPONENTS
(RCS, BOATTAIL, GPCs, ETC.)

60-FOOT ON-PAD ACCESS

82-FOOT PAYLOAD DEPLOYMENT

STANDARD ET

T-O UMBILICAL (RIGHT):
AIR/GN2 PURGE/CONDITIONING,
PAYLOAD-TO-GROUND POWER

PAYLOAD INTERFACE
PANEL (MOVABLE)

ACTIVE PAYLOAD
LATCH (MOVABLE)

ACTIVE KEEL
LATCH (MOVABLE)

PURGE / CONDITIONING
VENTS

STANDARD
SSMEs

STANDARD SRBs

SHUTTLE C

T-O UMBILICAL (LEFT):
PAYLOAD-TO-GROUND
AVIONICS INTERFACE

OMV THREE-POINT DOCKING
TRUNNION ASSEMBLY

OMV

PAYLOAD BAY ILLUMINATION
AND TELEVISION (REAR)

PAYLOAD BAY ILLUMINATION
AND TELEVISION (NOSE)

PURGE MANIFOLD

TYPICAL SPACE STATION
FREEDOM MANIFEST

SPACE STATION ELEMENTS
DECOUPLED FOR LAUNCH

POWER FOR EXTENDED
DURATION MISSION

OMV INTERFACE
AND ACTIVATION

OMV DOCKING FIXTURE

using a manned vehicle to launch routine payloads, and for placing too much emphasis on a single launch vehicle (Space Shuttle). A shuttle-derived vehicle would help alleviate the criticism regarding using men to launch commercial satellites, although by its very nature it really did not add a second vehicle type. This was primarily because any SDV would make maximum use of Shuttle systems, and any potential problem large enough to ground the Orbiters would most probably ground the SDV also. A possible exception to this was that since the SDV would be unmanned, a greater risk was acceptable since the results of a catastrophic failure were easier to accept. It was also seen as a method of adding to the available lift capability by increasing the launch rates* beyond the limits imposed on Space Shuttle after the *Challenger* accident.

After investigating varied proposals for heavy lift launch vehicles (HLLV) with the Air Force, the concept of a shuttle-derived vehicle was again investigated by NASA. The Air Force had used the *Challenger* accident as a convenient excuse to pull out of the Space Shuttle program, concentrating instead on the development and certification of the unmanned Titan IV, and the conceptual study of an Advanced Launch System (ALS). The primary rationale expressed by the Air Force for not participating in the Shuttle-C studies was that any vehicle based heavily on the man-rated technology used for Space Shuttle could not dra-

matically lower the cost of access to space. Lowering launch costs was one of the capabilities most prized by the Air Force, particularly when the subject of orbiting the Strategic Defense Initiative (SDI) was brought up. However, in an effort to coordinate the activities of the two separate efforts (ALS and Shuttle-C), a joint NASA-DoD-Air Force steering group was set up to monitor the progress of both.

Interestingly, although the Air Force declined to actively participate, most early Shuttle-C manifests showed several potential DoD flights, and many early briefings included details on modifications to SLC-6 at Vandenberg to support polar launches. It was not until the late-1989 decision by the Air Force to terminate all support for SLC-6 that the west coast option disappeared from the studies.[8]

The Air Force's decision left NASA free to proceed and develop whatever conceptual vehicle it felt best met the needs of its commercial customers, and also to optimize the vehicle for launching Space Station elements, a task the Air Force was not willing to fully support. In August 1987, MSFC formalized the Shuttle-C study effort, and assigned additional NASA manpower under Task Team manager Glenn Eudy. This team focused on gathering all available documentation from past SDV studies, and on defining parameters for the upcoming contractor studies.[9]

In November 1987, MSFC awarded study contracts to Martin Marietta, United Technologies (USBI), and Rockwell International for the definition of a Shuttle Derived Vehicle. Each of the study contracts was divided into two phases – the basic contract period (Phase I) covered the first five months of activities, and a Phase II consisting of a negotiated option covering the remaining four months of the contract period. The initial month of the basic contract period was to be a concentrated effort to establish the requirements for the overall vehicle deriva-

---

* After the 1986 *Challenger* accident, NASA set the maximum launch rate from the existing Shuttle facilities at 14 per year. In the months immediately preceding and following the Return-to-Flight of STS-26R it became apparent that this was probably unachievable using the established flows for an Orbiter. After reviewing the flow schedules and determining that little could be done to speed up the process without compromising safety, NASA decided that launching a different, unmanned vehicle (Shuttle-C) could increase the launch rate up to the allowable 14 per year without affecting the primary Orbiter launch flows.

tive, major system elements, and operational concepts. The next two months studied whether an in-line (mounted on top of the ET) or side-mounted (besides the ET) payload element should be pursued. The final two months were devoted to trades and analyses of various vehicle configurations. The Phase II efforts negotiated with each contractor were devoted to further studies of the vehicle defined in the first five months. All of these studies were collectively known as investigations into a Shuttle Cargo Element (SCE), better known as Shuttle-C.[10]

These initial Shuttle-C studies focused on the rapid development of a heavy lift capability making maximum use of existing Space Shuttle systems in order to minimize vehicle development costs and schedule risk, and to ensure payload compatibility with the existing Orbiter payload environment. The new vehicle was targeted at a lift capability of between 100,000 and 150,000 pounds, with elements of the upcoming Space Station Freedom being anticipated as the primary payload. Alternatives investigated included building a 'throw-away' Orbiter fuselage structure (minus wings, crew module and thermal protection system) that would use two SSMEs and a somewhat limited data processing system, along with the normal ET and SRBs. The SRBs would have been recovered per normal procedure, but the rest of the package would have been tumbled back into the atmosphere for destruction at the end of a mission. This was considered an extremely expensive concept, and although it was investigated by all three contractors, it was dropped early in the program.

Other ideas investigated by the contractors included attaching two SSMEs to the bottom of an ET and placing a large payload compartment on top of the tank. This in-line vehicle would have been launched with the aid of two normal SRBs, but would have required extensive modifications to the launch facilities to accommodate its increased height as well as the engines under the ET. Significant structural modifications to the ET itself would also have been required since the thrust vectors would have been significantly different than the baseline Space Shuttle configuration. A somewhat smaller version of this scheme with only one SSME, and a larger version with three SSMEs were also considered, but dropped for similar reasons. A second concept was to mount engines on the payload itself (a true second stage), and to stage the vehicle in the traditional (i.e. Saturn) method. Although this concept required fewer major modifications to the MLP (no central engines), it would still have required very significant changes to the rest of the launch pad and other

processing facilities. Again, significant modifications would have been required to the ET to support the payload, so this concept was also discarded.[11]

Interestingly, the analysis and trade studies resulted in a side-mount configuration similar to the earlier Martin Class I SDV being recommended by all three contractors. At least one of the studies concluded that Shuttle-C could orbit payloads for approximately $2,000 per pound, compared to $3,793 for Delta II and $4,100 for Titan IV.[12]

Based on the results of these studies, a request for proposal (RFP) was issued by MSFC for the continued study of a vehicle with a side-mounted expendable cargo element. The reference payload was 15 feet in diameter by 72 feet in length, and weighed 103,000 pounds. The reference mission was a 253 mile orbit at 28.5 degrees (Space Station Freedom orbit), although an alternate mission required delivering a 114,000 pound payload to a 184 mile orbit. The vehicle was to use a newly designed Cargo Element (CE), two unmodified four-segment RSRMs, a standard ET, a modified Orbiter aft-fuselage (called a 'boattail') with the vertical stabilizer and body flap removed, two SSMEs, two standard OMS pods, and associated avionics from the Space Shuttle and other mature vehicle designs.

The ascent of Shuttle-C flight would be autonomously controlled by the avionics systems onboard the vehicle itself, although it would be closely monitored by ground stations to ensure a safe track was maintained. Once on-orbit, the flight operations would be controlled via an attached Orbital Maneuvering Vehicle (OMV), which would either rendezvous and dock with the Shuttle-C, or be carried up as part of the payload. This would permit control of the Shuttle-C from either the ground or the Space Station. The deorbit thrust for the payload element would be provided by the OMV under control from the ground.[13]

The OMV was an unmanned vehicle under development by TRW. Conceived in 1979 at MSFC as the Teleoperator Retrieval System (TRS), this was a revival of the original Space Tug concept proposed at the beginning of the Space Shuttle Program. Phase A of the TRS program was conducted in-house at MSFC, with Phase B study contracts being awarded to Martin Marietta, LTV, and TRW under the name Orbital Transfer Vehicle (OTV).[14]

The actual development contract was awarded to TRW in October 1986, although the OMV was to become the first budget casualty for Space Station Freedom – this contract also included the development and production of a single OMV with one orbital flight test demonstration. The contract contained options for a second vehicle and nine additional flights. The OMV was designed to allow Freedom to deploy and retrieve spacecraft from low-earth orbit. The vehicle was to be carried into orbit by the Space Shuttle, then left there to perform under the control of Space Station. The Orbiter (or an SDV tanker) would return periodically to refuel the OMV. As originally conceived, the first flight of OMV was to be in 1991, although this was later pushed to 1993 due to budget constraints. The OMV was cancelled during the FY91 budget process when Congress deleted all funding for it.[15]

After evaluating the proposals submitted in response to the RFP, MSFC selected a combined contractor team of Rockwell International, Martin Marietta, Boeing, Teledyne Brown Engineering, Intermetrics, and USBI to continue to study Shuttle-C, and contract NAS8-37144 was awarded in February 1988. Additional engineering personnel from MSFC, KSC, JSC, and NASA Lewis, and their various contractors, were also assigned to the effort. All the primary team members also continued to study the concept individ-

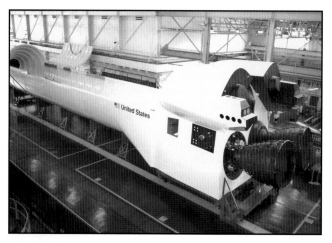

The Engineering Development Model of the Shuttle-C was constructed in 1989. The thrust structure and mid-fuselage framework came from MPTA-098. (NASA / MSFC)

ually, each seeking to gain some advantage for the eventual development and production competition.[16]

During the 18 month study effort that was scheduled to end in August 1989, the Shuttle-C vehicle underwent many changes. When the TRW-developed OMV was cancelled, many groundrules changed, particularly concerning the deorbit and reentry phase of flight. This resulted in the addition of OMS propellants, additional avionics, more complicated software, etc. All of this added significantly to the development process, increased the cost of each flight since additional hardware was being thrown away, and also increased the empty weight of the vehicle, lowering its payload capacity. The Cargo Element itself was continually refined with the final design having two or three SSMEs and a 15 by 81 foot payload bay. The payload bay had 124 mounting locations on each side for trunnion latching and bridge mounting, and 64 locations for keel latching. There were 12 payload latch interface panels, two standard payload interface panels, a payload bay purge manifold, four directable purge spigots, an environmental control system blanket, and two vent doors. The vehicle was to be equipped with electrical power, cooling, fluids, and other utility services. The 81-foot length of the Cargo Element was deemed to be the maximum acceptable without extensive modifications to the processing facilities at KSC.[17]

The Cargo Element was to be of conventional aluminum skin and stringer construction, although advanced composites were also under study for a possible second-generation vehicle. All of the engines and avionics were finally declared expendable, although the recoverable P/A module originally proposed by Boeing and Martin during the SDV studies was maintained for possible use on the second-generation vehicle. Preliminary designs for the follow-on vehicle had three SSMEs and a 24 by 96 foot payload bay. The 96-foot length would have required modifications estimated at $4,000,000 to the facilities at one of the two Space Shuttle launch pads at KSC.[18]

Payload capability for the baseline Shuttle-C to Space Station orbits ranged from 100,000 pounds with two SSMEs, to 170,000 pounds when equipped with three main engines. Various studies were also initiated to investigate increasing this capability by using either the filament-wound SRBs from Vandenberg, or the new ASRMs being developed for Shuttle. The Cargo Element would have had an empty weight of approximately 70,000 pounds with the main engines attached to an production Orbiter thrust structure. Although this is an expensive, robust piece of equipment – built to withstand multiple launches – its use precluded the cost of designing and qualifying a new, 'cheaper' design. Other Orbiter equipment, such as APUs, and the RCS, would have been built minus the 'man-rated' redundancy in order to minimize costs. Shuttle-C was envisioned as using three GPCs, two as a redundant pair, and one running 'backup' flight software. The Shuttle-C concept assumed that the main engines and GPCs would be flight-rated Orbiter components that had reached the end of their economical service lives, and would not be economical to recover. The 24 existing AP-101B GPCs would have been available en masse when the new AP-101S computers came on line in the mid-1990s. It was anticipated that a flight rate of three or four Shuttle-C flights per year could have been sustained in this manner, far below the 10–12 being proposed by NASA.[19]

One of the largest changes in the proposed use of the Shuttle-C came about directly because of a decision made in the baseline Space Shuttle program. This regarded the Centaur upper stage, which was deemed too dangerous to

An artist concept of the two-engine Shuttle-C. Note that the OMS pods do not have OMS engines, and were used only for the RCS function. (Rockwell International)

be carried in the payload bay of the Orbiter after safety analysis conducted in response to the *Challenger* accident. But Shuttle-C was unmanned, and seemed a perfect place to shift the requirements for an advanced high-performance upper stage. Thus, Shuttle-C became Centaur capable, further increasing the complexity and weight of the vehicle. The Centaur G-Prime, the same variant used by Titan IV, was chosen for Shuttle-C.

During early-1989, the study contract was extended one year since it was becoming evident that the Shuttle-C would not be approved by NASA Headquarters for a new-start initiative until FY91 at the earliest. Headquarters' rationale was that the Congress was attempting to cut the budget wherever possible, and NASA was already attempting to secure funding for the final completion of OV-105 and to continue the drastically over-budget Space Station Freedom program. As it turned out, this was the death blow for Shuttle-C.

Because of its high commonality with the existing Space Shuttle, it was thought that Shuttle-C could have been operational within four to five years from authority to proceed (ATP). In 1989 the NASA Office of Technology Assessment (OTA) estimated that the two-engine Shuttle-C could be developed for $985 million, although NASA estimated that $1,800 million would be necessary. The OTA estimated that Shuttle-C would most probably attain a reliability of 97 percent, roughly the same as that estimated for Space Shuttle itself. NASA's revised launch cost estimate was $424 million per launch, resulting in a cost of $4,240 per pound to orbit.

Planners envisioned a launch rate of 14 Orbiters and 10 Shuttle-Cs per year starting in 1995 using the existing facilities at KSC. At this rate, the existing supply of used engines and computers would have been depleted quickly, so Shuttle-C started estimating the costs of using elements that had flown on Shuttle only once or twice.

Considering that SSMEs cost $38 million each at the time, it rapidly became evident that the launch cost of a Shuttle-C would surpass the $500 million mark in the third year of flight if it had to procure new-production SSMEs or GPCs. This made any cost advantage of Shuttle-C minimal compared to Titan IV, negating most of its rationale for existence. Even compared to the baseline Space Shuttle cost estimates, the recurring costs of Shuttle-C were not considered sufficiently lower to amortize the considerable development expense. An increasingly tight NASA budget, and stiff competition from Space Station Freedom for what limited development funds were available, combined to spell the end to Shuttle-C during the FY91 Congressional budget deliberations.[20]

An engineering development model, using the original MPTA-098 thrust structure, was assembled by Essex Corporation, under contract to Boeing at MSFC during 1989–90. This model was designed to be used by engineers at MSFC for the design and integration of subsystems, payload compatibility studies and manufacturing planning. The vehicle was also to be used at KSC to fit-check the various ground processing and launch facilities, but the project was cancelled before this occurred. Planning at that time had the first Shuttle-C launch (designated SCE-1) in early 1995 from KSC. The 1990 decision by Congress to cancel Shuttle-C effectively put an end to the SDV concept, and it is expected that the engineering model will be donated to the Space and Rocket Center, where it will go on display next to the Space Shuttle *Pathfinder* mockup.

It should be noted that Shuttle-C was a real project, not just a 'paper study,' and a great deal of work had been accomplished on it. This not only included development of the flight vehicle, but also flight planning at JSC and processing preparations at KSC. Although no firm launch date had been set, the majority of the program was marching towards a stream of launches beginning in the mid-1990s. However, no actual facility modifications at KSC had been begun by the time the project was cancelled.

SHUTTLE-Z

In August 1989, even before Shuttle-C (designated SHC by JSC) had been totally defined, various groups within NASA and its contractors were investigating ways to improve the concept, most notably for the expected manned Lunar-Mars Initiative. The most obvious improvement that could be made was to substitute the new-development Advanced Solid Rocket Motors (ASRM) for the RSRM that were part of the Shuttle-C baseline. This resulted in an immediate payload increase of 10,000 pounds, but unfortunately most of this was expended on a new payload module measuring 33 by 100 feet. This vehicle was referred to as a Shuttle-C Block-II (SHC-II).[21]

Planning for SHC-II included supplementing the expected 14 Orbiter and 10 Shuttle-C launches with another 6 SHC-II launches using the same two launch pads. It was expected that a new MLP would need to be procured to handle the increased flight rate, but no other significant modifications were anticipated for the launch complex other than reactivating two inactive high bays in the VAB. However, several new payload processing facilities were foreseen to handle the increase in available launch capacity provided by SHC-II.

A truly unique extension to the Shuttle-C concept was the 'low-value cargo vehicle,' also discussed during 1989. This entailed launching an ET without a Payload Element or Orbiter attached to it. Two normal RSRMs would propel a standard Shuttle ET into orbit, assisted by a pair of 'worn-out' SSMEs. Since there was no Payload Element or Orbiter, it was envisioned that the two SSMEs would not use much of the propellants in the ET, making the propellants available to other spacecraft on-orbit. Using the standard ET, it was expected that 163,200 pounds of propellant could be delivered on-orbit. A further 10,000 pounds could be delivered using a stretched ET.

Neither of these concepts evolved very far since NASA was expending most of its energies on getting Shuttle-C approved and ready for flight, a task that eventually did not get very far due to budget limitations. But this did not stop the NASA Space Transportation and Exploration Office from taking the Shuttle-C concept to its logical conclusion. The need for a heavy-lift vehicle to support various 'Exploration Initiatives' such as the Lunar-Mars Initiative led directly to the Shuttle-Z proposal.

This vehicle looked much like a Shuttle-C Block-II on steroids. A launch capability of 300,000 pounds to near-earth orbit was envisioned using a payload shroud with a clear volume of 40 by 60 feet. Additional length (up to 120 feet) was available for the packaging of hardware that was smaller in diameter (such as upper [third] stages attached to the main payload, etc.).[22]

The first stage of Shuttle-Z consisted of two ASRMs, and the vehicle had a gross lift-off weight of 5,249,000 pounds. A maximum acceleration of 2.39-g at 118 seconds into flight was envisioned, and the maximum dynamic pressure (max-q) was listed as 514 pounds per square foot. The second stage utilized four SSMEs running at 104 per-

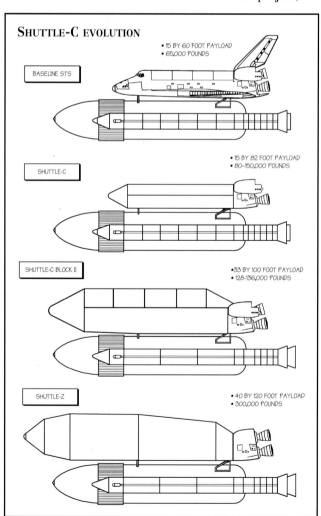

SHUTTLE-C EVOLUTION

BASELINE STS
• 15 BY 60 FOOT PAYLOAD
• 65,000 POUNDS

SHUTTLE-C
• 15 BY 82 FOOT PAYLOAD
• 80-150,000 POUNDS

SHUTTLE-C BLOCK II
• 33 BY 100 FOOT PAYLOAD
• 128-136,000 POUNDS

SHUTTLE-Z
• 40 BY 120 FOOT PAYLOAD
• 300,000 POUNDS

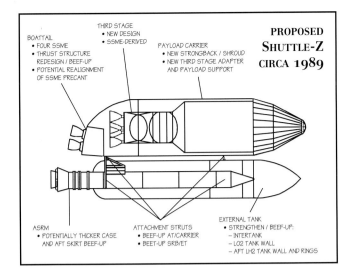

PROPOSED SHUTTLE-Z CIRCA 1989

BOATTAIL
• FOUR SSME
• THRUST STRUCTURE REDESIGN / BEEF-UP
• POTENTIAL REALIGNMENT OF SSME PRECANT

THIRD STAGE
• NEW DESIGN
• SSME-DERIVED

PAYLOAD CARRIER
• NEW STRONGBACK / SHROUD
• NEW THIRD STAGE ADAPTER AND PAYLOAD SUPPORT

ASRM
• POTENTIALLY THICKER CASE AND AFT SKIRT BEEF-UP

ATTACHMENT STRUTS
• BEEF-UP AT/CARRIER
• BEET-UP SRB/ET

EXTERNAL TANK
• STRENGTHEN / BEEF-UP:
 – INTERTANK
 – LO2 TANK WALL
 – AFT LH2 TANK WALL AND RINGS

cent power. A standard ET provided propellant storage, and the SSMEs were housed in the aft end of the payload shroud, mush like that envisioned for Shuttle-C. A total of 1,589,500 pounds of usage propellant was carried. The normal third stage, contained within the payload shroud, consisted of a single advanced SSME and 400,000 pounds of propellant. The payload shroud itself was jettisoned after second stage burnout, but before third stage ignition. This third stage could be swapped for other upper stages depending on the requirements for the particular payload.

Variations on this theme, as well as parametric scaling exercises, continued for about a year in an attempt to find an economical heavy-left vehicle. Concurrent Air Force studies centered around the Advanced Launch System (ALS), some of which looked much like Shuttle-Z (others looked like various ILRV Phase A proposals).

The main drawback of Shuttle-Z was that due to its height (231 feet versus 184 for Space Shuttle), none of the Orbiter processing or launch facilities could handle it without significant modifications. This made it an extremely expensive proposition. Plus, as studies evolved, Shuttle-Z used less and less off-the-shelf Shuttle hardware, making it essentially a new development effort.

This put it in the inevitable position of not being as economical to develop as a Shuttle-derived vehicle, nor as economical to operate as a newly-developed vehicle using all available advanced technology.

ADVANCED SOLID ROCKET MOTOR (ASRM)

Almost from the beginning of the Space Shuttle Program, NASA was aware that the solid rocket motor (SRM) could be improved, both for increased safety and a reduction in direct operating costs. The need for this was significantly reinforced by the findings of the 1986 Rogers' Commission following the *Challenger* accident. Although the Redesigned Solid Rocket Motor (RSRM) rushed into production during the *Challenger* stand-down solved a great many of the complaints voiced after the accident, NASA felt there was still much more that could be accomplished given 15 years advancement in technology.

On 3 September 1986 MSFC awarded five 90-day $500,000 study contracts to Aerojet General, Atlantic Research, Hercules Aerospace, Morton Thiokol, and United Technologies (USBI) for an "alternative or Block II Space Shuttle Solid Rocket Booster." The studies were completed in time for J. R. Thompson, MSFC Director, to brief the results to Congress on 22 January 1987. The NASA Administrator, James C. Fletcher, announced on 2 April that NASA would conduct a Phase B engineering definition study of an advanced SRM, and also initiate studies into possible liquid rocket boosters for the Space Shuttle.[23]

MSFC released an RFP for the ASRM study contracts on 3 June 1987, and set up the LRB task team at the end of that month. Proposals for the studies were received from the same five contractors that had participated in the late-1986 study – on 7 August MSFC awarded 9-month $3.3 million study contracts to all five companies. Independently, NASA conducted a site search for a new government-owned, contractor-operated plant to build the new boosters. On 26 July 1988 MSFC announced the selection of a piece of property owned by the Tennessee Valley Authority known as Yellow Creek (actually in Mississippi) as the production site. Full-scale testing would be conducted at new facilities to be built at the

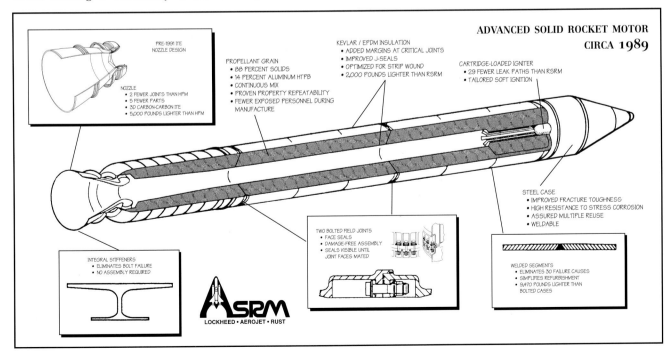

ADVANCED SOLID ROCKET MOTOR CIRCA 1989

PRE-1991 ITE NOZZLE DESIGN

NOZZLE
• 2 FEWER JOINTS THAN HPM
• 5 FEWER PARTS
• 3D CARBON-CARBON ITE
• 5,000 POUNDS LIGHTER THAN HPM

PROPELLANT GRAIN
• 88 PERCENT SOLIDS
• 14 PERCENT ALUMINUM HTPB
• CONTINUOUS MIX
• PROVEN PROPERTY REPEATABILITY
• FEWER EXPOSED PERSONNEL DURING MANUFACTURE

KEVLAR / EPDM INSULATION
• ADDED MARGINS AT CRITICAL JOINTS
• IMPROVED J-SEALS
• OPTIMIZED FOR STRIP WOUND
• 2,000 POUNDS LIGHTER THAN RSRM

CARTRIDGE-LOADED IGNITER
• 29 FEWER LEAK PATHS THAN RSRM
• TAILORED SOFT IGNITION

STEEL CASE
• IMPROVED FRACTURE TOUGHNESS
• HIGH RESISTANCE TO STRESS CORROSION
• ASSURED MULTIPLE REUSE
• WELDABLE

INTEGRAL STIFFENERS
• ELIMINATES BOLT FAILURE
• NO ASSEMBLY REQUIRED

ASRM
LOCKHEED • AEROJET • RUST

TWO BOLTED FIELD JOINTS
• FACE SEALS
• DAMAGE-FREE ASSEMBLY
• SEALS VISIBLE UNTIL JOINT FACES MATED

WELDED SEGMENTS
• ELIMINATES 3D FAILURE CAUSES
• SIMPLIFIES REFURBISHMENT
• 9,470 POUNDS LIGHTER THAN BOLTED CASES

Stennis Space Center (ex-NSTL). The TVA, through the Government Services Administration, transferred 1,297.519 acres to NASA for approximately $5.7 million on 25 September 1990.[24]

Based on the results of the Phase B studies, NASA released an RFP for the development contract on 22 August 1988. The resulting industry competition was rather heated, particularly as some of the major players scrambled to assemble teams to improve the odds of winning. In the end, only two companies submitted bids – teams of Hercules-Atlantic, and Lockheed-Aerojet. On 11 April 1990 NASA announced that the team of Lockheed Missiles and Space Company and Aerojet Space Boosters had been selected to design and manufacture the Advanced Solid Rocket Motor (ASRM). The two team members had a long history together, first teaming for the Navy Polaris submarine launched ballistic missile (SLBM) in the late-1950s, and continuing through the Trident II SLBM that entered service during the 1990s. The competition roughly paralleled the original SRB competition, with basic concepts remarkably similar to those entered in the first round fifteen years earlier again being proposed by roughly the same set of companies. The name of the program, ASRM, was somewhat misleading since the goal of the program was to design and build a totally new Solid Rocket Booster, including most of the non-motor parts. However, the major advances were to the motors themselves, hence the name. In addition to the primary teammates, RUST International was the Yellow Creek facility contractor, Thiokol would manufacture the nozzles at Michoud, the Ladish Company of Cudahy, Indiana, would supply the pre-form case segments, and Babcock & Wilcox would finish the case segments at Mt. Vernon, Indiana.[25]

The total cost for the design, development, and certification of the new motors was anticipated to be in excess of $1,000 million. It was expected to take approximately 6 years to develop and test the new motor, and their first use on a Space Shuttle flight was originally scheduled for 1996. The stated goal of the ASRM project was "… to substantially improve the flight safety, reliability, producibility, and performance of the Space Shuttle's Solid Rocket Motor."

The ASRM was supposed to be a direct replacement for the post-*Challenger* RSRM. Groundrules dictated that the new booster must be compatible with the existing facilities at KSC to the maximum extent possible, since the cost of modifying facilities at KSC could prove prohibitive. The new motors and components had to be reusable over a long period of time, and be easier and safer to handle on the ground than the RSRM components. Perhaps most importantly, the ASRM was being developed in direct response to the need to improve Shuttle performance while assembling the Space Station. NASA maintained that the ASRM was an urgent requirement to keep the cost of Space Station down by limiting the number of assembly flights needed.

The ASRM was to significantly improve Space Shuttle flight safety and reliability by reducing the number of possible catastrophic failure modes. This was achieved in a variety of ways, the primary being a reduction in the number of motor segments from four to three. But more importantly, the existing four RSRM segments were actually made up of 11 individual sections bolted together in 'factory joints.' The ASRM eliminated these entirely, along with 2,267 individual parts. Other RSRM sections eliminated included the ET attachment ring, which was to be built integrally with the motor casings, as were the stiffeners. This produced a stronger case with less tendency to flex, and also reduced the amount of preflight and post-

recovery maintenance. A great deal about the amount of flex desired (some is required to absorb shock otherwise transferred to the vehicle) was learned during the abortive attempt to design lightweight filament-wound composite case versions of the original SRBs for the Vandenberg Launch Site during the early-1980s.[26]

The ASRM also featured an improved igniter and nozzle design which further reduced failure modes. The nozzle was an omni-axis design that eliminated two of the RSRM's five internal joints, and also reduced the current erosion rate by using a carbon-carbon 'integral throat entrance' design. The new nozzle eliminated 72 seals and associated leak paths. But by 16 January 1991 the new ITE nozzle had been cancelled, replaced by an ablative throat design similar to that used on the RSRM. Officially the cancellation was because the ITE was considered too 'high-risk' – in reality, it was just too expensive.[27]

An added benefit to the ASRM was a new propellant grain design, permitting a larger reduction in thrust at max-q and eliminating the need to throttle down the SSMEs during ascent. This was expected to yield a significant increase in the reliability and durability of the SSMEs, as well as eliminating 176 other potential failure modes (e.g., the SSMEs refusing to throttle-up).

The ASRM was to use a different propellant than the TP-H1148 used in the RSRM. The new hydroxy-terminated polybutadiene (HTPB) was composed of 19 percent aluminum powder, 69 percent ammonium perchlorate, 0.17 percent iron oxide powder as a catalyst, and 11.83 percent PBAN rubber binder. As in the RSRM design, the binder was burned as fuel, providing a small amount of additional thrust. The advanced booster carried 1,205,107 pounds of propellant, compared to 1,106,640 pounds for the existing design, and burned about ten seconds longer.[28]

The assembled ASRM booster was four inches larger in diameter (150 versus 146 inches) and exactly the same length (149.16 feet) as the existing SRB. This small increase in diameter was determined not to have any significant impact to the ground facilities at KSC. The ASRM weighed 1,344,597 pounds, compared to 1,256,663 pounds for the redesigned SRB. The new booster generated 3,485,000 lbf and 269.2 seconds specific impulse, compared to 3,000,000 lbf and 268.4 seconds for the RSRM. This should have allowed Shuttle to achieve its full 65,000 pound payload capacity, an increase of 10,000 pounds over the limits set after the 1986 *Challenger* accident. The inert weight of the motor casings of the ASRM were 139,490 pounds, down from 150,023 pounds. The new nozzle weighed only 18,947 pounds, a reduction of almost 5,000 pounds from the current 23,932 pound design.[29]

The direct manufacturing costs would have been reduced through the use of the new highly-automated factory located at Yellow Creek for propellant mixing and pouring. These facilities were constructed expressly for the ASRM project at a cost of over $300 million, with an additional $250 million being spent for state-of-the-art pouring and mixing equipment.

The ASRM propellant would have been mixed and poured with a new technique labeled 'continuous mix.' In theory, continuous mix produced better quality propellant, and the Yellow Creek facility featured an in-line quality control system called 'fourier transform infrared/factor analysis' that constantly sampled propellant as it was produced. This is in contrast to most other large-scale solid motors, which are mixed and poured in multiple batches resulting in a slightly uneven propellant grain under even the best circumstances. The newly-developed model

UK-400 mixer was capable of producing 20,000 pounds of propellant per hour, and a horizontal interrupted screw served as the only blade used by the unit. Oxidizers, pre-mixed fuels, and curing agents were precisely metered by computer, automatically fed, and continuously blended as they traveled the length of the mixer. They emerged as cast-ready propellant, which was then pumped directly to a casting facility and into the motor segments.

This process greatly enhanced manufacturing safety. Since spreading a small quantity over a large surface, the amount of propellant exposed to mechanical action with continuous mix was drastically reduced. Less than 1,000 pounds was exposed to the mix screw for only three minutes at a time, compared to as much as 28,000 pounds churned for three to four hours by batch mix blades. Automated procedures throughout the closed process eliminated most human contact with the exposed propellant. The continuous mix process was expected to achieve a two-thirds reduction in propellant production costs over the industry-standard batch method. Waste propellant would also have been reduced significantly, minimizing disposal cost and the associated environmental concerns.

Aerojet first pioneered the continuous mix process during the 1960s with the construction of a production plant in Sacramento, California, that was used for the Polaris missile program. A second-generation continuous mix plant in Dade County, Florida, successfully served during the 260-inch diameter SRM demonstration project in 1963. Aerojet had proposed using this facility in the early-1970s to build the original Space Shuttle SRBs, but lost the competition to Morton-Thiokol. This original Aerojet concept also proposed casting the entire SRB as a single unit, eliminating all motor joints, and the facility in Dade County was constructed with this in mind. The complete booster would have been transferred to KSC by barge.

During 1988 Aerojet successfully demonstrated third-generation continuous mix technology on ASRM propellant at its UK-150 pilot plant in Sacramento, and later, demonstrated production of an ASRM simulated propellant in a commercial UK-400 unit like that later installed at Yellow Creek. Both tests met all expectations.

The ASRM was scheduled to go through an extensive test and certification program at the Stennis Space Center in Mississippi. This would have largely duplicated the process imposed on the RSRM certification after the *Challenger* accident, but was to be expanded to increase the confidence level in the new booster. Additional analysis of Orbiter and ET structure was also being accomplished to ensure no unforeseen difficulties arose in attempting to integrate the new boosters into the Space Shuttle system. The first ASRM-related test was conducted on 10 April 1991 when a sub-scale nozzle demonstrator was fired at MSFC. On 27 August a 26-foot long materials test motor was fired in the East Test Area at MSFC to evaluate internal insulation materials. Several more tests would follow through the end of 1992.[30]

Things seemed to be going well, at least technically. The ASRM continuous mix process was demonstrated when the pilot plant at Sacramento successfully mixed live propellant during January 1992. But when NASA Administrator Richard H. Truly submitted the agency's FY93 budget request to Congress, it did not contain any funding for ASRM development or production. Congress and the OMB had told NASA to trim its budget, and NASA decided to concentrate on the National Launch System (NLS – also an MSFC project) instead of improving Space Shuttle any more than absolutely necessary. The official rationale was that

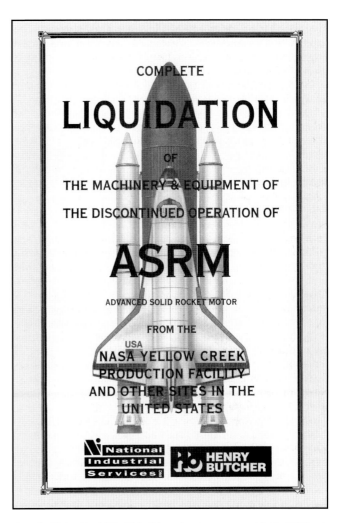

An undignified end for ASRM. All facilities and equipment procured for the ASRM project were auctioned off in early 1995. Most of this equipment was brand new, and had never been used. (National Industrial Services, Inc. and Henry Butcher & Co.)

the RSRM was performing satisfactorily, and NASA's budget could be trimmed by eliminating the ASRM. In its place NASA initiated the development of the super light weight tank (SLWT) by Martin Marietta in Michoud to make up some of the Space Station payload performance that had been expected to be gained from the ASRM.[31]

Nevertheless, construction work at Yellow Creek continued, as did detailed design at Lockheed and Aerojet. The ASRM Preliminary Design Review (PDR) was successfully conducted in February 1992. But the accountants had already determined that work would need to cease by May or June 1992 in order to have sufficient FY92 funds remaining to pay the $465 million program termination costs. Congress was not completely in agreement with the Administration's efforts to kill the project, so NASA elected to continue development and construction in case Congress should reinstate the program. As it ended up, the program did survive for another year, funded at essentially the same level as it had been in FY92.[32]

In March 1993 the estimated cost of the ASRM project rose from $2,200 million to $3,500 million. By this time most of the major facility work at Babcock & Wilcox and Michoud had been completed. The EPA had approved all environmental permits for each of the facilities, and the first kneel-down transporter (KDT) had been delivered to KSC to help move the larger ASRM segments around the processing areas. The facility at Yellow Creek was about 60 percent complete, and the first pieces of major equipment had been delivered. The pilot plant at the Aerojet facility in Sacramento was complete. The first major case segments had been delivered from the Ladish Company to Babcock & Wilcox.[33]

By April 1993 the projected use of an ASRM in support of Space Station assembly flights had slipped to the year 2000 – three years later than originally scheduled. The seemingly-endless Space Station redesign efforts had irritated Congress, and cast a doubt on NASA's ability to manage programs. And now NASA wanted to continue two major development projects for Shuttle – ASRM and SLWT. In planning for the FY94 budget, NASA included just over $280 million for ASRM development work and $34 million for final facility construction. But this time it was Congress that was not sure, although NASA was now indicating that it might take *both* the ASRM and SLWT to make Space Station work – especially if it was placed in a 51.6-degree orbit to cooperate with the Russians. By June it had become clear that ASRM would not survive – Congress had had enough and would not support two major projects. A suggestion to complete the Yellow Creek facility and move existing RSRM production there was defeated, and all that was left was the final political maneuvering within Congress and the Administration.

In October 1993 the ASRM project was officially cancelled – approximately $2,200 million had already been spent on the project. All work on the project was halted, and all facilities and equipment (except at Michoud) procured for the project were auctioned off to the highest bidder. NASA determined that little – if any – of the technology developed for the ASRM could be effectively retrofitted into the RSRM. The major issue was that the designs were just too divergent, and the cost of retrofitting any changes to the existing design and the effort needed to certify it for flight was not worth any small performance increase that might result. The SLWT would need to carry the payload burden of Space Station by itself.[34]

Liquid Rocket Booster

On 2 April 1987 NASA Administrator James C. Fletcher announced that studies would be initiated into possible liquid rocket boosters (LRB) for the Space Shuttle in parallel to beginning development work on the ASRM. The LRB studies would eventually focus on a joint-development for Space Shuttle and the new expendable Advanced Launch System (ALS)* that NASA proposed to supplement Shuttle.[36]

A task team was established within the Advanced Projects Office at MSFC on 30 June 1987 to manage the Phase A study of a liquid rocket booster (LRB). In essence, NASA was reexamining the same options that had been evaluated in 1971–72 – pressure-fed versus pump-fed liquid boosters versus solid rockets (ASRM/RSRM), and whether they should be expendable or recoverable. In the course of 15 years the program had come full-circle, and ironically, the same NASA Administrator was at the helm.†

MSFC awarded Phase A study contracts to General Dynamics and Martin Marietta for the period of 1 October 1987 to 15 July 1988 – a six month extension subsequently brought this to January 1989. In addition, KSC awarded Lockheed Space Operations Company (the SPC contractor) an LRB Integration study contract, while JSC directed the Lockheed Engineering and Management Services Company to support the studies, primarily by providing Orbiter wing loading information and trajectory constraints. The studies were to concentrate on providing replacement boosters that could provide the Shuttle with

the ability to carry 70,500 pounds into a 185-mile due-east orbit from KSC. Only minor modifications were permitted to the Orbiter or ET, and 'reasonable' changes were permitted at KSC (it was recognized early on that the ground systems would require some substantial revisions).[37]

General Dynamics (NAS8-37137)

During the first part of the GD study, 15 different propellant combinations and engine types were evaluated on the basis of safety, performance, and compatibility with existing Shuttle systems. Various recovery techniques were also examined since these greatly drove cost considerations. Three existing pump-fed engines were evaluated – the SSME (LO2/LH2), the Rocketdyne F-1 from the Saturn V (LO2/RP-1) and the Atlas LR87 (NTO/A-50). In addition, eight possible new-development pump-fed engines were examined, all but one using LO2 as an oxidizer – fuels included LH2, CH4, C3H8, and RP-1, and several combinations of the same. The eighth engine used N2O4 as an oxidizer with MMH as a fuel. The new engines were based largely on configurations being developed by Rocketdyne and TRW as part of the Space Transportation Booster Engine (STBE) and Space Transportation Main Engine (STME) studies also being run by MSFC.[38]

Many of the same trade studies conducted 15 years earlier were accomplished again. The pump-fed versus pressure-fed engine debate took on many of the same arguments – was the simplicity of the pressure-fed engine worth the extra weight it imposed on the propellant tanks and structure? Was the efficiency of a pump-fed engine worth developing new turbopumps with their inherent risk and cost? By January 1989 General Dynamics had decided that a pump-fed engine burning LO2/LH2 was the best choice. Major factors in favor of the propellant was its commonality with existing Space Shuttle propellants, minimal environmental impacts, and most importantly, commonality with ALS concepts. General Dynamics felt that the only economic way to develop any LRB for Shuttle was to share the development costs with another program – ALS.

The 70,500-pound payload capacity into a due-east orbit had been chosen by MSFC since it translated into a 65,000 pound capacity into the high-inclination space station orbit at 250 miles (the Orbiter could not have carried 70,500 pounds without significant evaluation of loads data). General Dynamics added the requirement that the LRB-powered Space Shuttle should be able to abort-to-orbit if an LRB engine failed immediately after lift-off. The LRBs were sized to allow derating the SSMEs to 100 percent power during ascent instead of 104 percent, and the max-q throttling requirement was moved to the LRBs instead of the SSME to eliminate the wear-and-tear on those expensive engines (not to mention a great many failure modes).

The whole idea of liquid boosters presented many more options for intact abort scenarios. Unlike the SRBs that can not be shut-down during ascent, the LRBs could be throttled to allow a more rapid RTLS abort, or over-throttled to ensure a TAL or AOA abort could be accomplished at almost any point during ascent. For instance, in the case of multiple SSME failures during ascent, the LRBs could be throttled down to moderate the attach loads and

---

* The National Launch System was renamed after the Air Force objected to yet another 'National' effort that it did not support.

† James C. Fletcher had been Administrator from 27 April 1971 to 1 May 1977. He returned to lead the agency after the *Challenger* accident – from 12 May 1986 to 8 April 1989. He died of lung cancer on 22 December 1991.

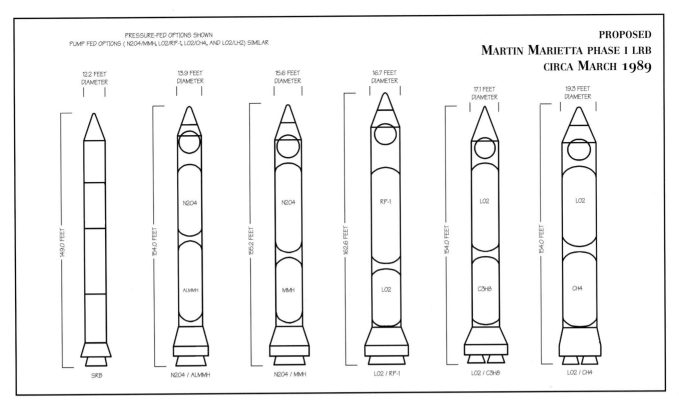

PRESSURE-FED OPTIONS SHOWN
PUMP FED OPTIONS (N2O4/MMH, LO2/RP-1, LO2/CH4, AND LO2/LH2) SIMILAR

could be gimballed to reorient the Orbiter on a glide path back to KSC. Or, in the event of a major LRB failure, both LRBs could be shut down quickly (but not abruptly) to allow the Orbiter to separate from the stack ('fast sep' might actually work under these conditions).

Two different engines were finally chosen by General Dynamics. The first was a 558,000-lbf engine being studied by Rocketdyne under the STBE effort. The engine had a chamber pressure of 2,250 psia and a specific impulse of 411 seconds. The alternate engine was being studied by TRW as part of the STME – 563,900 lbf, 968 psia, and 409.5 seconds. Each was considered to be a relatively simple engine that could be developed at minimal cost and would prove reliable in service.

In both cases, four engines would be mounted in each booster. The basic structure of the booster consisted of a forward LO2 tank and an aft LH2 tank, both of welded construction using 2219 aluminum – the new aluminum-lithium alloys were briefly considered but were rejected due to questions about LO2 compatibility and higher costs. In addition there was a nose cone, intertank adapter, and engine compartment skirt on each booster. The intertank was of simple skin stringer with ring frame construction and contained the thrust fitting (upper attachment) to the ET. The engine skirt contained the engines and the hold-down members to support the vehicle on the pad.

General Dynamics concluded that the LO2/LH2 vehicle had the lowest weight and cost of all the configurations studied. Estimates showed that the initial launch could be made approximately 60 months after authority-to-proceed – slightly less if the decision was made to use a Shuttle-C for first flight since some testing could be deferred. At KSC, modifications would be required to at least one (and probably two) VAB high bays and to one MLP – in addition, it was recommended that an additional MLP be constructed. Modifications were also required to at least one launch pad, plus miscellaneous changes to other facilities. Interestingly, the KSC-directed Lockheed study showed that the required changes were much more extensive, and

concluded that several new processing facilities needed to be constructed to support the LRBs. Simply a matter of different perspectives – GD wanted to keep the apparent cost of the LRBs as low as possible, while Lockheed wanted to provide the best service for the new boosters.

As far as costs went, General Dynamics estimated the total development cost at roughly $2,500 million, including required modifications to facilities and other shuttle components (Orbiter, etc.) – the LRBs themselves would cost approximately $2,000 million to develop. Total recurring

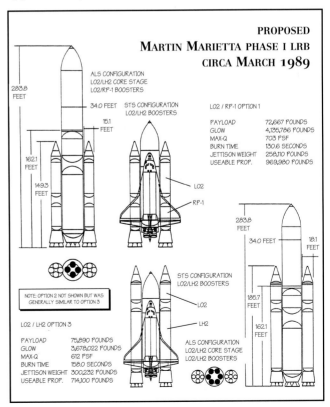

costs were approximately $33 million per LRB, or $66 million per flight. This was based on a production rate sufficient to support 14 flights per year. The total life-cycle costs, including 244 production LRBs, was $11,000 million.

Martin Marietta (NAS8-37136)

Martin Marietta was under the same guidelines as General Dynamics, so it is not surprising that many of the same conclusions were reached. Martin concentrated on two booster designs – one using new-development pump-fed engines and the other using pressure-fed engines. Both boosters were carried through to the completion of conceptual design and all system impacts and program costs were identified for both designs. In addition, the possible application of the pump-fed design to the Advanced Launch System (ALS) program was studied.[39]

Essentially Martin determined that LO2 and RP-1 were the best propellants (as opposed to the LO2/LH2 recommended by General Dynamics), and that both pump-fed and pressure-fed boosters could be developed and flown with no major modifications to the remainder of the Space Shuttle system except ground facilities. There were no 'enabling technology requirements' identified for the pump-fed booster, although the seemingly simpler pressure-fed vehicle required several new technologies, specifically high-strength materials and more efficient pressurization systems. Martin believed that the requirements of the Shuttle and ALS appeared to be compatible and should allow a common engine and booster to be developed, thereby reducing the fiscal requirements on both programs.

The RP-1 fuel has been selected based primarily on a trade study that concluded that it would be significantly

less expensive to develop the turbopumps required for an RP-1 engine than for an LH2 engine. In addition, RP-1 itself was cheaper and hence would reduce the cost of operations, and the fuel was significantly easier to handle at KSC than LH2. The same trade study concluded that the possible hypergolic fuels (ALMMH, MMH) presented significant dangers and development challenges, as did other high-energy chemical fuels (methane, etc.).

The total system costs estimated by Martin roughly matched those from General Dynamics. The pressure-fed booster was estimated at $11,400 million over the course of the program, including development and the production of 244 LRBs. The pump-fed booster was slightly more expensive, coming in at $12,400 million. This worked out to a recurring cost of just more than $36 million per LRB.

In the end, NASA could never justify the idea of investing another $3,000 million into the Space Shuttle, especially since it appeared that the ASRM project would eventually provide a more suitable booster than the existing RSRM. If the ALS project had actually proceeded the way NASA had envisioned, it is possible that Shuttle could have leveraged off of that development to build a new booster at some point in the future, but ALS died a quick death in the budgetary wars. Part of the problem with any of the LRB configurations investigated was that they were all expendable, and that was largely politically unacceptable. Within a few years of the LRB concept fading into the archives at MSFC, another liquid-powered booster concept would arrive on the scene. This time the booster would be recoverable, taking the program almost full-circle to the original Phase B concepts of 1970–71. This liquid fly-back booster (LFBB) is discussed in Appendix A since it is still – remotely – under consideration by the program.

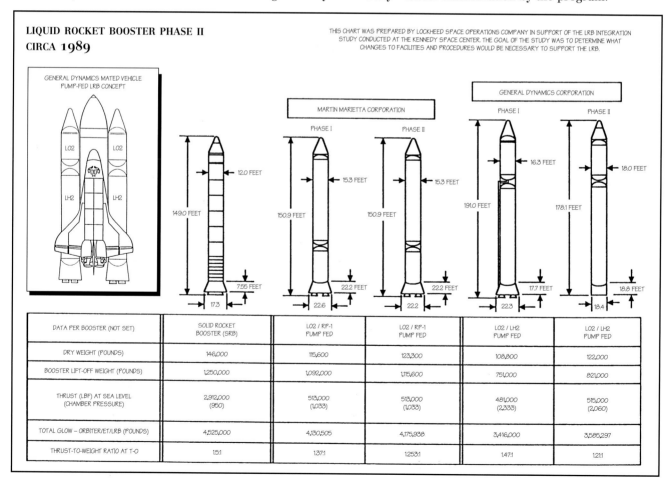

LIQUID ROCKET BOOSTER PHASE II CIRCA 1989

THIS CHART WAS PREPARED BY LOCKHEED SPACE OPERATIONS COMPANY IN SUPPORT OF THE LRB INTEGRATION STUDY CONDUCTED AT THE KENNEDY SPACE CENTER. THE GOAL OF THE STUDY WAS TO DETERMINE WHAT CHANGES TO FACILITIES AND PROCEDURES WOULD BE NECESSARY TO SUPPORT THE LRB.

DATA PER BOOSTER (NOT SET)	SOLID ROCKET BOOSTER (SRB)	LO2 / RP-1 PUMP FED	LO2 / RP-1 PUMP FED	LO2 / LH2 PUMP FED	LO2 / LH2 PUMP FED
DRY WEIGHT (POUNDS)	146,000	115,600	123,300	108,800	122,000
BOOSTER LIFT-OFF WEIGHT (POUNDS)	1,250,000	1,092,000	1,115,600	751,000	821,000
THRUST (LBF) AT SEA LEVEL (CHAMBER PRESSURE)	2,912,000 (950)	513,000 (1,033)	513,000 (1,033)	481,000 (2,333)	515,000 (2,060)
TOTAL GLOW – ORBITER/ET/LRB (POUNDS)	4,525,000	4,130,505	4,175,938	3,416,000	3,585,297
THRUST-TO-WEIGHT RATIO AT T-0	1.5:1	1.37:1	1.253:1	1.47:1	1.21:1

RANCHO DE LA CONCEPCION

Space Launch Complex Six (SLC-6) – 'slick-six' – sits on one of the most picturesque areas of central California. Perched atop a small coastal plain at Point Arguello, about an hour north of Santa Barbara, this launch site has absorbed more government funding than any other space facility in the country, yet for all intents and purposes has never launched anything of consequence.*

Rancho La Espada, or Sudden Ranch, consisted of 14,890 acres of mountainous land; the westerly portion of a 24,992-acre Mexican Land Grant known as Ranch de la Concepcion that extended eastward another 10 miles to the Santa Barbara Channel. Rancho de la Concepcion was granted to Anastacio Carrillo by Juan B. Alvarado, Governor of Alta California on 10 May 1837. A quarter century later, on 13 July 1863, President Abraham Lincoln confirmed the title to the Carrillo family. During the two decades following the Civil War the property changed hands several times and in 1883 the westerly portion (Rancho La Espada) was sold by William Hollister of Santa Barbara to Robert Sudden, a Scottish sea captain.

Rancho de la Concepcion encompassed the heart of the transitional zone between southern and central California. It was a cool, dry, and windy area with sandy and rocky soil suitable mostly for cattle ranching. Separated from inland areas of Central California by the Santa Ynez Mountains, the area was essentially isolated until the com-

---

* The first actual launch from SLC-6 was a small Lockheed Launch Vehicle on 15 August 1995. The vehicle was destroyed by range safety 290 miles downrange after it began tumbling. A second launch on 22 August 1997 was successful, becoming the first vehicle to reach orbit from SLC-6. Another successful launch – this time a larger Athena-2 – occurred on 27 April 1999.

pletion of the Southern Pacific Railroad in 1901. Although fronting on the ocean, there were no natural harbors. In fact, from the sea, the area was noted for inshore currents, dense fog, strong wind, rugged surf pounding against sheer cliffs, and three major headlands – Point Conception, Point Arguello, and Point Pedernales.

The combination of conditions made this a perilous passage. The need for caution was evidenced by numerous disasters, the most tragic of which occurred on 1 October 1854 when 415 people perished as the steamer *Yankee Blade* ran afoul of rocks off Point Pedernales. Seven decades later, on 8 September 1923, in the worst peacetime disaster in the history of the U.S. Navy, seven of 14 destroyers in a follow-the-leader speed run from San Francisco to San Diego ran into the same rocks that had smashed *Yankee Blade*, claiming the lives of 23 sailors.

In view of the maritime disasters, it was not surprising that the first government use of the area was for lighthouses. The United States Lighthouse Service began construction of a facility on top of the 250-foot cliff at Point Conception in 1855 – the same year that lights were set up in San Francisco Bay. Early the following year, a five-wick lantern was installed to illuminate a fresnel cut-glass reflector from France. Supplementing this light was a 5,000-pound fog bell. After a quarter century of use, the Lighthouse Board decided to rebuild the structure because it was dilapidated and it had been determined it was frequently shining out on a cloud layer. Hence, the new lighthouse constructed in 1881 was at the 133-foot level.

With the completion of the Southern Pacific Railroad in 1901, the Lighthouse Service established a light station on the 124-foot level at Point Arguello. During the 1930s a lifeboat station was established, and a lighted sea buoy was anchored off Point Pedernales. The lifeboat facility

A view of SLC-6 probably taken in early 1980. With the exception of the access tower (which has been demolished) the major MOL facilities may still be seen. The MST at extreme right, the single flame trench, the LCC and engineering support building in left-center, and the solid rocket motor building at the extreme left. (DVIC photo DF-SC-83-10697)

---

was discontinued in 1958 and the lighthouse was replaced by a LORAN station during the 1960s.

Military interest in the Sudden Ranch began in 1956 when nearby Camp Cooke was selected by the Air Force as the first operational base for the Atlas intercontinental ballistic missile. The ability offered by Atlas to launch space vehicles southward into polar orbit from the newly-established Vandenberg AFB made Sudden Ranch an ideal extension to the government facility. The remote location separated from the rest of the base offered improved safety, and the Santa Ynez Mountains would provide an effective barrier between the proposed launch sites and the neighboring city of Lompoc. Based on this reasoning, the DoD included a $4.5 million request in its FY65 budget to purchase the property – the House Armed Services committee denied the request.

In the meantime however the Secretary of Defense had cancelled the Dyna-Soar program and directed the Air Force to proceed with the Manned Orbiting Laboratory. This program was expected to use a larger version of the Titan ICBM and needed a large, secure site to launch into polar orbit. The request for funds to purchase Sudden Ranch was resubmitted with the FY66 budget request. This time Congress approved the request.

During December 1965 and January 1966, Air Force and Army Corps of Engineers personnel met with representatives of the Sudden Estate Company to negotiate a price for the property. Following the failure of these negotiations, the Government initiated condemnation proceedings under the power of eminent domain by filing a Declaration of Taking with the Federal District Court in Los Angeles on 28 February 1966. On the following day, a Federal judge issued an order permitting the Air Force to take immediate possession of 4,000 acres, but allowed the Sudden Estate Company to have grazing rights on the remainder of the property until 31 December 1966. In

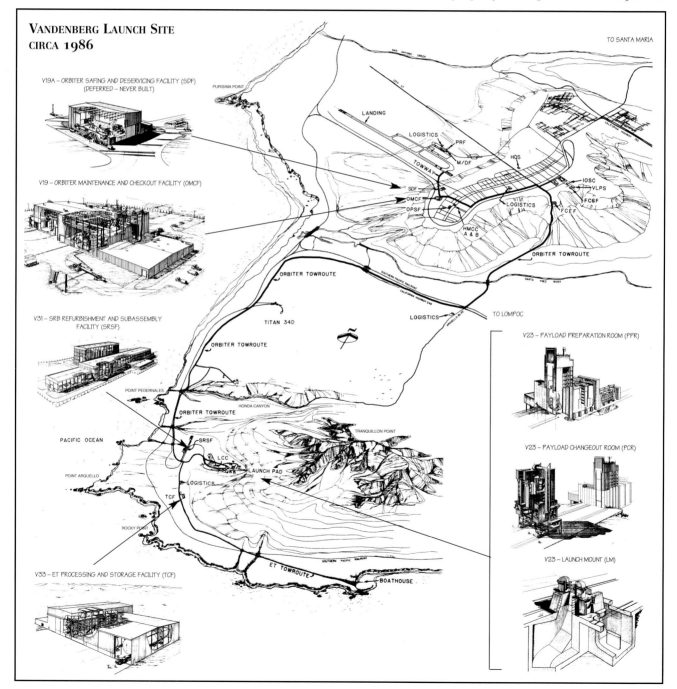

VANDENBERG LAUNCH SITE CIRCA 1986

effect Sudden Ranch became a portion of Vandenberg AFB on 1 March 1966 and expanded the installation to 98,400 acres (154 square miles). The only remaining question – being considered in the condemnation proceedings – was what the total purchase price would ultimately be.

This question was not resolved until December 1968 when, following a 54 day trial and four days of deliberation, a jury ordered the government to pay $9,002,500 as just compensation of the Sudden Ranch property. This was more than double the amount that was authorized for its purchase in FY66.

The original plan for the new property included five plots for launch sites on what would unofficially become known as South Base. Eventually the two northern-most sites would become SLC-3 and SLC-4 for Atlas and Titan, while the next site south was used for small sounding rockets and never boasted a real launch complex. The southern-most site, usually called SLC-7, would remain a vague notion for 35 years. But SLC-6 would become infamous.

MANNED ORBITING LABORATORY

The MOL program was in a hurry. Bids were called for site preparation in January 1966, and on 2 March 1966 a $987,391.33 contract was awarded to Allied-Webb Construction Company of South Gate, California. At this point it was widely expected that the MOL Titan complex would largely replicate – in concept – the integrate-transfer-launch (ITL) facility being built to support Launch Complexes 40 and 41 at Cape Canaveral. The FY66 military budget included $18 million for construction of the Titan IIIM facility at Vandenberg. But further review of projected MOL launch rates (and other Titan III uses at Vandenberg) revealed no particular need for the elaborate ITL capability so in May 1966 a decision was made to construct a single launch pad (with an option for a second) and to erect the vehicle on the pad.

Requests for bids were issued for the construction of SLC-6 and on 17 January 1967 the $17,217,217 contract was issued to S&S Constructors of Lancaster, California (a joint venture of Santa Fe Engineers[*] of Lancaster and Stolte Construction of Los Angeles). This second phase of the construction package (site preparation had been the first) included the launch pad, a 76 by 50-foot flame duct, a 243-foot high umbilical tower, 315-foot high mobile service tower, launch control center, and various other facilities. Construction began on 27 January 1967 and was initially scheduled to be completed on 15 July 1968, although this was later extended to 30 November 1968.

Contracts for the next two phases of construction were awarded, again to S&S, on 30 June 1967. This $2,797,000 award included a solid rocket motor storage and inspection building in phase 3, and a fire station, electrical power plant, and railway service area in phase 4. Construction of these facilities began on 7 July 1967 and were scheduled to be completed on 1 June 1968, a date also extended somewhat. A new hospital being built on North Base also received funds from the MOL program to expand the space medicine facilities.

The FY67 budget included an additional $3.2 million for MOL facilities on North Base including a three-building administration and astronaut housing complex in the industrial area. One of the buildings was for operational training and evaluation, and included a MOL simulator and computer complex and flight surgeon offices. The second building was living quarters for the astronauts and included ten one-bedroom suites, briefing room, squash court, and other recreational facilities. The third building was an administrative building for Air Force contractor personnel working on the program.

But almost as suddenly as it had started, MOL ended. When the MOL program was cancelled in 1969, the 'brick and mortar' construction at Vandenberg was about 90 percent complete. The Air Force decided to complete construction since cancellation clauses in the various contracts amounted to more than remained to be expended to complete the tasks. The facilities were then placed under minimum maintenance (caretaker status) pending possible future use. By the end of 1969, all major construction at both SLC-6 and the Vandenberg industrial area was complete. Approximately $30,000,000 had been spent on construction, exclusive of the land acquisition costs.

ENTER SHUTTLE

It did not take long to find another use. On 14 April 1972 NASA and the Air Force announced that the Kennedy Space Center and Vandenberg AFB had been selected as launch sites for the Space Shuttle. A special task force was established by the Air Force during 1974 to evaluate three possible locations on the vast property at Vandenberg. Cost analysis showed that $100 million could potentially be saved by modifying the mothballed MOL facilities at SLC-6 and reusing the North Base complex. The SLC-6 option was approved in 1975 and demolition in anticipation of construction began in 1979.

The decision to reuse SLC-6 created some interesting

The MOL flame duct being modified for the Space Shuttle. The interior of the duct was not extensively changed, but the outer edge was extended further from the launch mount to reduce the amount of acoustic noise. (DVIC photo DF-SC-83-01702)

---

[*] Interestingly, Santa Fe Engineers would build the Space Shuttle hanger at NASA Dryden in 1976.

challenges – and as events later proved, some expensive problems. SLC-6 was located at almost the extreme south end of the South Base property, nestled in a hilly region at Point Arguello. There were no suitable locations in this area to build a runway, and the cost of a new landing strip would likely have been prohibitive in any case – especially considering that North Base already had an 8,000-foot long runway. So Space Shuttle processing would be divided into pre-launch and post-landing activities on North Base, and launch activities on South Base.

Each physical Space Shuttle facility (or group of facilities) on Vandenberg was assigned a 'station set' designation. For instance, the OMCF was station set V19, while the SDF was V19A. There were also 'functional' station sets, such as the Launch Processing System (V84) and logistics (V88) that were not directly linked to physical locations. The station sets included:

V17	North	Landing Facility (the runway and taxiways)
V18	North	Mate/Demate Facility
V19	North	Orbiter Maintenance and Checkout Facility
V19A	North	Safing and Deservicing Facility
V21	North	Hypergolic Maintenance and Checkout Facility
V23	South	Launch site (SLC-6)
V27	North	Flight Crew Systems
V28	South	Launch Control Center
V31	South	SRB Refurbishment and Subassembly
V32	Pt. H.	SRB Retrieval and Disassembly Facility
V33	South	ET Processing and Storage

The logical location for the administrative offices and flight crew accommodations were the former MOL facilities at the building 8500 complex. Accordingly, building 8500 became the headquarters for the Air Force's 6595th Shuttle Test Group (STG), while building 8505 was slightly refurbished as the crew quarters. The former simulator and computer complex (8510) was converted to house the Launch Processing System that would be used to process the Orbiter in various North Base facilities.

But an array of other facilities would need constructed or modified on North Base. First, the runway was extended from 8,000 to 15,000 feet, and 1,000-foot overruns were added at each end. An Orbiter Maintenance and Checkout Facility (OMCF) was constructed a short distance from the

Enterprise was used to validate the OMCF during her stay at Vandenberg, the only orbiter that was ever inside the building. All of the work platforms had to be isolated in such a manner that an earthquake would not cause them to strike the orbiter – a major driver in the overall design of the OMCF. (Air Force photo via Jacques van Oene)

runway, largely emulating the OPFs at KSC. The primary difference in the OMCF was that it was also a payload integration facility, complete with two underground payload cells that could store, checkout, and prepare horizontal payloads for loading in the Orbiter while it was in the OMCF. Separately, a Safing and Deservicing Facility (SDF) was to be built nearby to perform the hazardous post-landing servicing of the Orbiter, but this facility was deferred from the initial construction effort and was never built. The SRB retrieval facilities would be located at the Navy base at Port Hueneme at Oxnard, California.[1]

Buildings around the airfield were modified to support SSME maintenance and SRB parachute refurbishment. An office building was built near the OMCF, two hypergolic maintenance checkout cells (HMCC – together called V21) were constructed a safe distance away, and a large logistics facility was completed. Numerous buildings, including a

---

* When the first Shuttle personnel arrived at Vandenberg, this building was a fascinating look back in time – it had been finished and completely furnished prior to MOL being shut down, and had remained largely undisturbed for ten years – several bedrooms, a kitchen, briefing room, and squash court.

The V19 Orbiter Maintenance and Checkout Facility (OMCF) was the Air Force's counterpart of the OPFs at KSC. The major difference was that there was a horizontal payload checkout and storage facility co-located with the OMCF. After much debate about whether to install the processing and support equipment in the OMCF (deferring it would have saved money), the Air Force finished the facility in time to support the launch of STS-1V. When the activation of the launch site was cancelled, the work platforms and other equipment were removed from the OMCF, placed on an ocean-going barge, and sent to the Kennedy Space Center where it was used to outfit OPF Bay 3. (Air Force photo via Jacques van Oene)

former Minuteman warehouse and major parts of the old hospital were converted to office space for the thousands of workers – mainly from Martin Marietta, the integration contractor. At least three large modular office complexes were also installed to house additional personnel.

Given that the Orbiter would land at North Base, but be launched at South Base, a method had to be devised to carry it between the facilities. A special 76-wheel transporter was ordered from Fiat in Italy to carry the Orbiter over 17 miles of existing government and public roads. One section of these roads was in particularly hilly terrain and required cutting the hills back to ensure at least two feet of clearance for the Orbiter wing tips. In addition, new road segments bypassed sharp intersections, street signs had hinges installed so they could fold flat, and security guard stations were placed on wheels to provide clearance.

At South Base the existing SLC-6 complex was almost torn down and begun anew, seemingly negating whatever cost savings there should have been. Unlike KSC where the Space Shuttle is stacked inside the VAB then transported to the launch pad, at Vandenberg the entire vehicle would be stacked in-place at the pad. Not completely understanding how delicate the shuttle stacking operation is would later cause some problems with this approach.

New launch pad construction began in late 1979. Given that the launch overpressure from Shuttle was much more severe than that from the Titan IIIM, the MST park position (where it sat during launch) was moved 150 feet further back from the launch mount. Modifications to the MST included shortening the structure 40 feet and replacing the overhead 50-ton crane with a 200-ton crane. Service platforms conforming to the shape of Shuttle components were added. The resulting MST was 28 stories high and weighed over 7,500,000 pounds.

The MOL launch mount was a steel-framed support structure anchored to the center of the pad with openings into a single flame duct. For Shuttle, 87,000 cubic yards of concrete were used to build two additional ducts, each 50 by 70 feet, with walls 9 to 12 feet thick – enough concrete to build a three-foot wide, four-inch thick sidewalk from Los Angeles to San Francisco. The foundation of the launch mount itself was 86 feet thick. After some modifications, the existing MOL flame duct was used for the SSME exhaust, while the two new ducts were for the SRBs. The decision to reuse the MOL duct for the SSMEs would come back to haunt the program.[2]

The MOL launch control center (LCC) was extensively modified and reinforced, and a duplicate of the computers used at KSC were installed to support launch operations (a similar set of computers were installed in 8510 on North Base to support checkout). A 'ready room' constructed during the MOL program was used to house a logistics facility during construction, and also as office space for engineers working on the pad. A nearby solid rocket motor facility constructed for MOL was expanded to serve the same function for Shuttle. The SRM segments were shipped by rail from Thiokol in Utah and stored in the SRSF which had the capacity to store two complete flight sets – 16 segments and related hardware. The motor segments were then transported to the pad individually and stacked by the overhead crane in the MST.

One other major landmark on the pad was also modified. The 'flower box' had been built to protect the Ready Building – made of simple sheet metal – from the launch effects of the Titan IIIM. Essentially this was a large sand-filled concrete structure slightly to the north-west of the launch mount. The Space Shuttle program intended to

Enterprise is towed along a public road between North and South Base. The sides of the hill were graded down to provide adequate clearance for the wing tips. Street signs along the route were hinged in the middle, much like those along the overland route between Palmdale and Edwards AFB. The Orbiter is riding on the special 76-wheel transport that now carries Orbiters between the OPFs and the VAB. (DVIC photo DF-ST-93-03952)

leave the structure intact, mainly because it would cost too much to remove it. However, after the emergency slide-wire escape system was installed (running from the access tower to a set of bunkers to the west), it was found that a fully-loaded slide basket would hit the south corner of the flower box. The solution was to simply cut off about ten feet of the structure to allow the slide basket to pass by without hitting it.

Initially, two major new facilities were to be built on the launch pad. The payload preparation room (PPR), contrary to its name, was a large building with three 100,000-class clean rooms to service and checkout the most sophisticated payloads the DoD could build. The building was arranged with three vertical checkout cells along one side and a central transfer aisle down the other. The payloads arrived horizontally by truck, entered through an airlock, and were hoisted into a checkout cell. When a payload was ready to be installed in an Orbiter, it was taken from the checkout cell, raised through a vertical tunnel in the east

The last of the major facilities added to SLC-6 was the shuttle assembly building (SAB), at right. The PCR is in its park position against the west side of the PPR at the left side of the photo. The LO2 dewar is in the foreground. Most of the trailers and small buildings scattered around the launch site would be removed within a few months of this photo being taken as construction was completed. (DVIC photo DF-SC-86-00118)

Enterprise sits in front of the PCR. The doors to the payload clean-room may be seen on the front of the PCR, which is in its park position on the east side of the PPR. A corner of the LCC may be seen at the right side of the photo. (DVIC DF-ST-86-09420)

end of the building, and transferred to the payload changeout room (PCR).

The PCR was another large building, but this one moved – all 7,500,000 pounds of it. The PCR actually – at least initially – served two functions. After the SRBs and ET had been stacked using the cranes in the MST, the Orbiter was brought to SLC-6 from North Base. The Orbiter would be backed up so that the aft fuselage was only a few feet away from the PCR in its parked position. A strongback (similar to the one used in the VAB at KSC) would be attached to the Orbiter, and using cranes in the PCR, the Orbiter would be rotated to a vertical position snug against the PCR. The entire PCR would then move forward until it was near the SRBs and ET – which were still enclosed on the other three sides by the MST – and the stack would be completed. The PCR's second function was to accept payloads from the PPR via the vertical transfer tunnel, then move 750 feet to the launch mount and transfer the payload to the Orbiter. Doors and seals in the PPR and PCR ensured that a 100,000-class cleanroom environment was maintained during payload transfer.

But by 1980 it had become obvious that stacking the Orbiter in the winds that were almost constant around SLC-6 would probably not work and in 1981 a shuttle assembly building (SAB) was added. This was essentially a large empty shell with a garage-like door at one end and a 125-ton overhead crane. The MST and the SAB moved toward each other on railroad-like tracks from their

parked positions and completely enclosed the launch mount. Together their overhead cranes lifted and stacked the Space Shuttle vehicle. The assembly begins as the SRBs were raised onto the launch mount using cranes in the MST. Next, two cranes, one located in the MST and the other in the SAB, raised the ET and mated it to the SRBs. The crane in the SAB then lifted the Orbiter vertically and completed the stack. Six sliding doors on the west side of the SAB opened to permit the PCR to roll into the building, mate and install payloads. Before launch, the MST and the SAB would roll back to their parked positions approximately 375 feet west and 285 feet east, respectively, from the launch mount. Hydraulic jacks raised the buildings on and off of tiedowns located at the park position and launch mount. Travelling drive systems could move the buildings at speeds up to 40 feet per minute, and at least once the SAB managed to impact the crew cabin access arm when it was moved faster than the lookouts could coordinate with other work on the pad. Movement of the MST and SAB took 50 to 40 minutes including unlocking the tiedowns, travelling, and relocking.

External Tanks arrived at Vandenberg by ocean barge from Michoud across the Gulf of Mexico and through the Panama Canal to a small harbor just south of SLC-6. Part of the 1930s-era Coast Guard lifeboat station, deactivated in 1958 was modified to receive the ETs. An ET processing and storage facility was built about mid-way between the harbor and the pad, capable of storing four tanks while processing a fifth.

The access tower was a steel-framed structure located beside the launch mount that supported various umbilical connections, including the hydrogen vent arm, the gaseous oxygen vent arm, and the crew cabin access arm. The crew cabin access arm was arranged differently than at KSC – at KSC it swings away from the Orbiter, while at SLC-6 it slid backward away from the Orbiter on tracks. The access tower was designed to withstand overpressures of 15,000 pounds per square foot and temperatures over 5,500 degF. The launch mounts were identical to the ones at KSC – in fact they were manufactured at KSC – and contained the same T–0 umbilicals.[3]

Construction of the Vandenberg facilities required approximately 250,000 cubic yards of concrete – the equivalent of a 25 mile, four-lane interstate – 9,000 tons of reinforcing steel, and 15,000 tons of structural steel.

The Problems

By the end of 1984 the vast majority of the major items were pretty much structurally complete, although a great deal of equipment remained to be installed and validated. Most of the construction had been performed or managed by Sverdrup Corporation – six of the major structures at SLC-6 had been designed by Sverdrup and the seventh had been reviewed by them. First launch had originally been scheduled for 1982, but funding problems had delayed construction, and as early as May 1975 it had been rumored that the 1982 date might slip. All reports were that the Air Force would restructure around an initial operational capability (IOC) in 1985. As it turned out, it did not matter much since NASA would have been unable to support Vandenberg operations in 1982 due to development and production delays encountered by Space Shuttle itself.[4]

Although many of the budget problems were eventually overcome, fiscal constraints coupled with ever-increasing costs eventually forced the Air Force to defer or cancel the construction of some facilities, and to defer the outfitting of others. For instance, the V19A deservicing facility was initially postponed, and finally cancelled in order to keep costs down. One of the two hypergolic maintenance cells was deferred, as was outfitting some of the payload cells in the PPR and OMCF. What had begun as a fairly austere launch facility had grown into a fully-capable complex with some systems that rivaled their counterparts at KSC. The late addition of the SAB – at a cost of almost $80 million – did nothing to help the fiscal problems, although it greatly eased concerns about shuttle operations. It should be noted, however, that although the Air Force and Martin Marietta had full confidence in the new launch paradigm being developed at Vandenberg, many engineers at KSC were less convinced. Perhaps just a case of not-invented-here syndrome. Perhaps not.

It should also be explained that although Vandenberg was an Air Force launch site – constructed under Air Force contracts and largely run by Air Force officers – NASA was still heavily involved. Years earlier the Air Force and NASA had reached an agreement where Air Force personnel would train with NASA at KSC to gain experience being test conductors and test directors, as well as in other engineering and management roles. By the time Vandenberg construction was in full-swing, many of these officers had been assigned there and, at times, it was hard to tell who was whom. Many Air Force management and engineering billets were being filled by NASA personnel on temporary loan – conversely, some NASA slots were filled by Air Force officers being trained by the space agency. Martin Marietta at Vandenberg was different than Martin Marietta at KSC, but the two Rockwell organizations were badged out of the same place. The Aerospace Corporation often acted on the Air Force's behalf. You needed a program to tell the players.

By late 1984 the facilities at Vandenberg were considered sufficiently complete to begin the facility verification and validation tests (FVV) using OV-101, which arrived at Vandenberg on 16 November. *Enterprise* was used to validate work platforms in the OMCF, then towed overland to SLC-6 where it was stacked on the pad using the ET intended for STS-3V and a pair of inert steel SRBs. The tests went well, with only minor problems being uncovered.

But analysis were showing other problems with the launch site, in both design and construction. On the design side, the most visible change was the addition of the SAB. The thought of trying to align the Orbiter with the rest of the stack in the constant winds and fog that blew across SLC-6 finally convinced the Air Force of the need for the building. Fortunately, although large and expensive, the SAB was fairly benign from a technical viewpoint and easy to design and construct.

More challenging was how to ensure ice did not build up on the ET. Many times of the year SLC-6 is covered in a very thick morning fog – sometimes the fog stops at the bluffs on the ocean creating a wall of fog several thousand feet high on the ocean side and crystal clear skies over land, but other times the fog simply washes over the cliffs and engulfs the entire coastal plane. Experience at KSC showed that the thermal protection system on the Orbiter was very sensitive to impacts from ice falling off the ET during ascent – the fog at SLC-6 was expected to allow a great deal of ice formation as the cryogenic propellants were loaded. The same gaseous oxygen vent arm installed at KSC was added to the design at Vandenberg, but other measures were also needed. In the end it was decided to install two jet engines in a partially-underground facility just east of the launch mount, and to duct hot air up and around the ET during tanking operations. The installation cost $13 million. There was little doubt the system would work, but the California Air Resources Board (CARB) was not particularly thrilled with the idea. Although technically the Air Force is exempt from local or state regulations, in most instances the federal government attempts to comply. And so it was at Vandenberg. Since the jet engines were stationary, they came under different regulations that they would have if they had been installed in an aircraft. The CARB issued a permit for the engines, but stipulated usage restrictions (hours per year) that essentially limited Vandenberg to four launches annually. This was not perceived to be a problem, and if it had ever become one, the Air Force would just have claimed 'national security' exemptions and proceed as they wished.

Not a constraint to operations, but a concern within the local press and with some employees was the location of the LCC. This building had been reused from MOL, although greatly expanded and reinforced. The concerns centered around its proximity to the launch mount – only 900 feet away. Analysis showed that the LCC could withstand overpressures equivalent to the detonation of 322,842 pounds of TNT at the launch mount – sufficient to withstand anything short of a catastrophic explosion of the vehicle on the pad. Nevertheless, some precautions were taken for the first launch – glass light fixtures were replaced by plastic, and some equipment was bolted down better. Portions of the LPS deemed critical for launch were taken to Wyle Laboratories in Los Angeles and 'shake tested' to verify they could withstand the vibration at launch – expected to move the suspended computer floor over an inch on one second oscillations. The LCC would have been used to support initial launches, but plans were being formulated as early as 1985 to move the critical computer functions to North Base and launch the vehicle from there.

But the largest problem resulted directly from the decision to reuse the MOL launch mount and its flame trench. At KSC as hydrogen passes through the main engines into the flame duct it is dispersed by the prevailing winds since the duct is not tightly enclosed. At SLC-6 the flame trench had been designed for the hypergolic-powered engines of the Titan IIIM, where there was no particular concern about a build up of explosive fumes in the trench. The trench had an opening immediately under the engines, went 50 feet straight down, angled outward towards the side of the pad, then angled upward back to ground level. The majority of the trench was covered top, bottom, and sides with thick concrete walls. Any hydrogen that bled from the main engines would accumulate in the trench and create a large potential explosion hazard. This would become the single most serious design flaw at Vandenberg.

In addition to the design problems, various construction problems received a great deal of attention from the popular press. Several television networks – notably NBC – did exposé shows on these problems, usually seriously distorting the facts in the process.

For instance, it was widely reported that 8,000 welds in critical piping on the launch complex were defective, and had been 'covered up' for over a year. The reality of the situation was somewhat different. In May 1983, a subcontractor to Martin Marietta began welding pipes at SLC-6. These pipes, destined to carry LO2, LH2, helium, nitrogen, and a variety of other gases and fluids, were mostly stainless steel – much of it vacuum jacketed – and required X-ray verification of each weld. Due to other pressing work, Martin did not immediately check the X-rays. By the end of the year it became apparent that the subcontractor had bitten off more than he could chew, and both Martin and the Air Force began to monitor the quality of the work being performed more closely. Finally, on 9 May 1984, the subcontract was terminated by Martin for non-performance, and another subcontractor was hired. The quality of approximately 8,000 welds was suddenly in question. Subsequent evaluation of the X-rays and other documentation by Martin and the Aerospace Corporation confirmed that 5,000 of the welds were good. The remaining 3,000 were X-rayed again, clearing more than 2,300 of them. The remaining welds (less than 700) required rework – less than ten percent of the reported 'problem' – and all were completed by September 1984. The percentages were about average for large construction projects using stainless steel (and especially vacuum-jacketed) pipes.[5]

NBC also reported that the Air Force contract for SLC-6 "pushed the schedule first, the specifications fourth." This was blatantly bad reporting. The particular clause in the contract being quoted was strictly a legal document. In this case 'schedule' referred to a standard form attached to the contract, much like 'Schedule A' attached to your IRS-1040 tax return. It had nothing to do with a calendar, but actually dealt with the business aspects of the contract (prices, etc.). Similarly, 'specifications' in this case referred to another part of the contract, not to the technical statement of work or requirements that needed to be adhered to by Martin Marietta, but rather to contractual conditions and agreements.[6]

Other reports indicated that the entire pad was wired 'out of phase,' forcing the valves to operate incorrectly and causing a catastrophic accident during launch. The charge was ridiculous for two major reasons – first, all of the critical valves were controlled by dc – not ac – and were therefore not sensitive to phasing. Secondly, every component on the launch pad and supporting systems had been well tested prior to first launch. All fluid and gas systems were tested first with inert substances (i.e., the LH2 system was first tested dry, then with liquid nitrogen, then finally with liquid hydrogen) to ensure they operated properly. As with any large construction project, wiring problems – both ac and dc – were found, but all were corrected before critical testing began.

In written testimony to Congress on 11 October 1984, NASA Administrator James Beggs indicated that the "schedules for readying Vandenberg for an October 1985 launch are highly success oriented but achievable." The testimony continued that the second DoD launch was scheduled for September 1986, while the first NASA mission (to repair a LandSat) from Vandenberg was scheduled for January 1987. *Discovery* was to be 'dedicated' to Vandenberg missions, although any Orbiter could be used at either launch site.[7]

But as early as 8 January 1985, NASA and the Air Force had announced that the first flight from Vandenberg would be postponed from 15 October 1985 until no earlier than 29 January 1986. Partially this was in response to continued construction delays at SLC-6, but partially it was because NASA was having a hard time scheduling its own launches from KSC due to various anomalies with the Orbiters. Delaying the Vandenberg launch allowed NASA to use *Discovery* for two additional NASA missions before delivering the Orbiter to Vandenberg – now scheduled for

SLC-6 during the flight vehicle verification tests with OV-101 stacked on the launch mount. The large American flag has yet to be painted on the sides of the shuttle assembly building, although USAF has been. This photo shows the PCR, SAB, and MST all in their park positions where they would be for launch. (U.S. Air Force)

September 1985 instead of May 1985. By this time the first External Tank had already arrived at Vandenberg, as had non-motor pieces of the SRBs. The first filament-wound SRBs would begin arriving on 30 May 1985, and the entire booster set for STS-1V/62-A (also called VLS-1) would be on-site by the end of July.[8]

On 15 February 1985, NASA and the Air Force jointly announced the crew for the STS-1V/62-A mission – Captain Robert L. Crippen (USN), Lieutenant Colonel Guy S. Gardner (USAF), Commander Dale A. Gardner (USN), Major Jerry L. Ross (USAF), and Lieutenant Colonel R. Michael Mullane (USAF) would carry the Teal Ruby satellite into orbit on 29 January 1986. The selection of Crippen was ironic considering he had become an astronaut as part of the MOL program.[9]

On 15 October 1985 – a date that had once been targeted for the first launch – SLC-6 was declared operational – parties were thrown and the workers celebrated. In reality, it probably could have supported a launch, very carefully. Work continued on many systems, and a great deal of analysis and testing laid ahead in an attempt to resolve the hydrogen detonation problem in the flame trench. Various workarounds were identified to support early launches pending a final resolution of the problem.

Planning for the first Vandenberg launch proceeded through late 1985 and early 1986. *Discovery* was being pre-processed at KSC in OPF-2 beginning on 10 September 1985. On 28 October 1985 the first launch was set for 20 March 1986, although five weeks later this had slipped to 15 July. The week prior to the *Challenger* accident, *Discovery* was transferred to VAB High Bay #2 for temporary storage while KSC pushed to clear a backlog of launches in late December and early January. It was expected that *Discovery* would return to OPF-2 for some final work the end of January, then be transported to Vandenberg in preparation for first launch. Before that happened, STS-33/51-L needed to be launch – the ensuing accident would change everything.[10]

The *Challenger* accident presented something of a quandary for the Air Force. Some officials wanted to use it as a convenient excuse to withdraw from the Space Shuttle program and pursue variants of the Atlas and Titan expendable launch vehicles. This was, no doubt, partially the result of the same mentality that had caused NASA to select the 747 as a shuttle carrier aircraft instead of a C-5A – wanting to have complete control over your own destiny. Although the Air Force owned SLC-6, and all of the facilities, computers, and ground support equipment, NASA never let it be forgotten that they 'owned' the Orbiter. This occasionally led to disputes over how the vehicle would be processed at Vandenberg, and probably led some in the Air Force to wonder how the relationship would ever work during operations.

On the other hand, several major payloads had been designed specifically for launch on the Space Shuttle, and could not be launched on any other vehicle. However, everybody's confidence was further shaken when a solid rocket motor failure caused a Titan 34D carrying a KH-9 spy satellite to explode just 15 seconds after lift-off from Vandenberg on 18 April 1986. Now both of the heavy lift launch vehicles were grounded.

Preparations continued at Vandenberg, largely unaffected by the *Challenger* accident. At least for a while. NASA announced that *Columbia* would be the first operational Orbiter to arrive at Vandenberg, mainly to be used for a more extensive facility verification than *Enterprise* provided, and to conduct an FRF in preparation for the

first *Discovery* launch. But as the investigation of the *Challenger* accident dragged on, it was becoming clearer that the days of SLC-6 were numbered.

Other facilities were also under construction. A $7 million expansion of the Easter Island and Hao runways for use as TAL sites was completed by 12 February 1987 except for the installation of the planned navigation aids. Interestingly, the Air Force had procured MSBLS units for the two Pacific TAL sites – something NASA had not done for the Atlantic TAL sites. The MSBLS intended for Easter Island and Hao were eventually transferred to NASA for use at other TAL sites.[11]

The filament-wound SRBs were included in the post-*Challenger* design modifications although a decision was made to halt production of the FWC boosters until two years prior to the first Vandenberg launch. The SRM segments that had been delivered to Vandenberg were eventually returned to Thiokol. Four ETs had been delivered to Vandenberg – serial numbers 23, 27, 33, and 34 – and all were stored for a while, then eventually used to support launches from KSC. Preparations continued for the arrival of *Columbia* for final facility verifications. The intent was to bring the facilities into full readiness, then stand-down pending the decision to return-to-flight.[12]

On 31 July 1986, Secretary of the Air Force Edward C. 'Pete' Aldridge, Jr. announced that the Space Shuttle facilities at Vandenberg would be placed in a 'caretaker' status pending a final decision on when to reactivate the program. By December 1986 plans to bring *Columbia* west had been cancelled, but the Air Force was still undecided about the future of the launch complex.

Testing continued on the potential hydrogen detonation problem in the flame duct and on concerns over the 'twang' of the solidly-built launch mount with the filament-wound SRBs. The latter problem was largely dis-

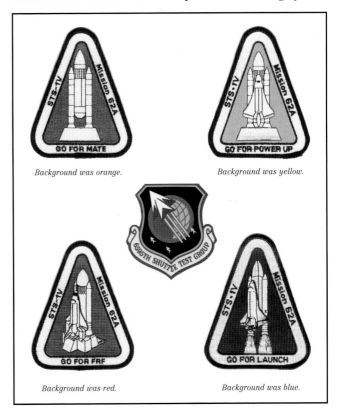

Background was orange.   Background was yellow.

Background was red.   Background was blue.

Some of the patches and insignia from Vandenberg. The 'Go For Stack' decal that was handed out to all employees is shown on page 355. Unfortunately, the program was cancelled before the rest of the patches were distributed. (Courtesy of Liem Bahneman)

missed when a special 'twang test' (pulling two filament-wound boosters with a cable, then releasing them) using the first flight set of filament-wound boosters, showed that the twang was within two percent of predictions – although slightly less than at KSC. The tests were completed on 12 December 1986, but revealed another problem. At KSC the launch mount can give a little under load since the MLP is flexible.* At Vandenberg the launch mount was solidly attached to 68 feet of concrete. While conducting the twang tests, it was found that the loads on the hold-down bolts was greater than predicted. The problem did not appear to be unsolvable, and several alternatives were under consideration at the end of the program.[13]

The hydrogen detonation problem was harder to solve. Testing was conducted on scale models at three commercial laboratories, at least three contractors, and one National laboratory (Sandia). A variety of concepts were evaluated, and on 12 December the Air Force selected an idea know as steam inerting. With this system, 175 degF steam was injected into the flame trench whenever hydrogen was flowing through the main engines. This effectively diluted the hydrogen buildup to non-hazardous levels. It was estimated that the final design of the system would be completed in September 1987, and that installation would take another six months after that. By this time, however, there were no plans to activate SLC-6 so the design was put on the shelf for possible future use.

In late 1986 a proposals were put forth to use the OMCF at Vandenberg to build OV-105. When these were rejected as unnecessarily complicated and expensive – after all, the tooling was still in Palmdale – a proposal was made to use the OMCF for the final checkout of *Endeavour*. This proposal had some merit since Palmdale had never been equipped with an adequate automated checkout system and Vandenberg had a complete and functional Launch Processing System. Eventually it was decided that this was also too expensive since it required maintaining the Vandenberg facilities in a higher state of readiness than was anticipated through the 1988 completion of OV-105. A similar fate awaited proposals to use the OMCF to perform OMDPs of the Orbiter fleet. In the end, the OMCF did contribute – the work stands and most of the equipment installed in the OMCF were removed and transported to KSC were it was installed in the Orbiter Maintenance and Refurbishment Facility (OMRF), creating OPF-3.[14]

Still, as late as December 1986 planning was still underway for the operational use of Vandenberg. Military payloads were identified for launch in the second half of 1992, with a NASA payload in early 1993. The status of the facilities was slightly downgraded† to 'minimum caretaker' on 20 February 1987, signaling the beginning of the final end On 13 May 1988 Aldridge ordered the facilities moth-balled, and on 26 December 1989 the Air Force finally conceded and terminated the Space Shuttle Program at Vandenberg. It was ironic that Aldridge, who had been scheduled as a passenger on STS-1V when he was an assistant SecAF, would be the official to terminate the program. The estimated cost of the Space Shuttle efforts at Vandenberg was a little over $4,000 million.[15]

---

* This is a relative concept. Although the MLP is huge, it does flex under loads.

† Caretaker provides constant maintenance with a minimum crew; mothball only provides major corrosion maintenance and little else.

SLC-6 during 1996. The MST and SAB are in their 'stack' position around the launch mount. The flame duct to the left is the original MOL duct – it was going to be used for the SSMEs. The duct to the right was new, and there was an identical duct on the other side of the pad – these were for the two SRBs. The American Flag on the SAB was paid for by donations from workers at Vandenberg. One of the popular cultural myths concerning Vandenberg is that SLC-6 was built on an ancient Chumash Indian burial ground, and as revenge the Indians had cursed the site and all that came upon it. Given the track record of projects at SLC-6, it is as plausible an explanation as any. In January 1998 the General Services Administration began auctioning off the PCR for scrap – the end of the end. In late 1999 Boeing began modifying SLC-6 to accommodate the new Delta IV expendable launch vehicle. (Dennis R. Jenkins)

ACRONYMS AND ABBREVIATIONS

A

AACB   Aeronautics and Astronautics Coordinating Board (DoD/NASA)
ABPS   Air Breathing Propulsion System
ac   Alternating Current
ACC   Aft Cargo Carrier (Shuttle-C)
ACES   Air-Collection Engine System
ACES   Advanced Crew Escape Suit
AEDC   Arnold Engineering Development Center
AEV   Aerothermoelastic Vehicles (part of the START program)
AFB   Air Force Base
AFFDL   Air Force Flight Dynamics Laboratory (also FDL)
AFFTC   Air Force Flight Test Center
AFHRA   Air Force Historical Research Agency
AFRPL   Air Force Rocket Propulsion Laboratory
AFRSI   Advanced Flexible Reusable Surface Insulation
AFS   Air Force Station
AGL   Above Ground Level
AIAA   American Institute of Aeronautics and Astronautics
ALS   Advanced Launch System
ALT   Approach and Landing Tests
AMLS   Advanced Manned Launch System
AOA   Abort-Once-Around
AOK   Automated Orbiter Kit
APU   Auxiliary Power Unit
ARD   Abort Region Determinator
ARDC   Air Research and Development Command
AS   Air Station (usually Air Force)
ASEE   American Society of Electrical Engineers
ASRM   Abort Solid Rocket Motor
ASSC   Alternate Space Shuttle Concepts
ASSET   Aerothermodynamic/Elastic Structural Systems Environmental Tests
ASV   Aerothermodynamic Structural Vehicles (part of the START program)
ATC   Air Traffic Control
ATCS   Active Thermal Control Subsystem
ATO   Abort-to-Orbit
ATP   Authority to Proceed
AVCO   company name

B

BFS   Backup Flight System
BLOW   Booster Lift-Off Weight
BoB   Bureau of the Budget
BoMi   Bomber-Missile (Bell concept)
BSS   Backup System Software
BTU   British Thermal Unit

C

c.g.   Center of Gravity
CAE   company name
CAPSIM   Computer Application Program Simulation

CalTech   California Institute of Technology
CCSD   Chrysler Corporation Space Division
Cdr.   Commander (of a Shuttle mission)
CDR   Critical Design Review
CDR   Commander (Navy rank)
CE   Cargo Element (Shuttle-C)
CIA   Central Intelligence Agency
CISS   Centaur Integrated Support Structure
CM   Command Module (Apollo)
CNES   Centre National d'Etudes Spatiales (French Space Agency)
CPU   Central Processing Unit
CSDL   Charles Stark Draper Laboratories
CTPB   Carboxyl Terminated Polybutadiene

D

DDT&E   Design, Development, Test and Evaluation
degC   Degrees, Centigrade
degF   Degrees, Fahrenheit
DEU   Display Entry Unit
DFBW   Digital Fly-By-Wire
DFI   Developmental Flight Instrumentation
DFRC   Dryden Flight Research Center (also FRC, HSFS)
DFRF   Dryden Flight Research Facility (now DFRC)
DM   Demonstration Motor (SRB)
DoD   Department of Defense
DPS   Data Processing System
DSS   Decelerator Subsystem (SRB)
Dyna-Soar   Dynamic Soaring (X-20 vehicle)

E

EAPU   Electric Auxiliary Power Unit
ECAL   East Coast Abort Landing
ECLSS   Environmental Control and Life Support System
EES   Ejection Escape Suit
EIU   Engine Interface Unit
ELDO   European Launch Development Organization (now part of ESA)
ELDO   European Launcher Development Organization (now part of ESA)
EMDI   Energy Management Display Indicator (for Dyna-Soar)
EMU   Extravehicular Mobility Unit
EPDC   Electrical Power, Distribution and Control
ERNO   German Space Agency
ERSI   Elastomeric Reusable Surface Insulation
ERSO   European Space Research Organization
ESA   European Space Agency
ET   External Tank
EVA   Extra-Vehicular Activity

F

FAA   Federal Aviation Administration
FBW   Fly-By-Wire

FCO	Flight Control Officer (ex-RSO, at CCAFS)
FD	Flight Director (at the MCC)
FDL	Flight Dynamics Laboratory (USAF)
FDO	Flight Dynamics Officer (at the MCC)
FICS	FRCS Interconnect System
FMOF	First Manned Orbital Flight (of Space Shuttle)
FOD	Foreign Object Damage (or Debris)
fps	feet per second
FRC	Flight Research Center (also DFRC, HSFS)
FRCI	Fibrous Refractory Composite Insulation
FRCS	Forward Reaction Control System
FRF	Flight Readiness Firing
FRL	Fire-Retardant Latex
FRSI	Flexible Reusable Surface Insulation
FSB	Five Segment Booster
FSL	Flight Simulation Laboratory
FSM	Flight Support Motor (SRB)
FVV	Flight Vehicle Verification (VLS)
FWC	Filament Wound Case (SRB)
FY	Fiscal Year

G

GAO	General Accounting Office
GAS	Get-Away Special
GD	General Dynamics
GE	General Electric
Gen.	General (military rank)
GEO	Geosynchronous Earth Orbit
GH2	Gaseous Hydrogen
GHz	Gigahertz
GLOW	Gross Lift-Off Weight
GMT	Greenwich Mean Time (now UTC)
GN	Gaseous Nitrogen
GN&C	Guidance, Navigation and Control
GOR	General Operations Requirement
GPC	General Purpose Computer (IBM)
GRC	Glenn Research Center at Lewis Field, Ohio (formerly LeRC)
GTO	Geosynchronous Transfer Orbit
GVT	Ground Vibration Test

H

HAC	Heading Alignment Circle
HAINS	High Accuracy Inertial Navigation System
HCF	Hard Compacted Fibers
HCF	Hardened Compacted Fiber (tile)
HGVT	Horizontal Ground Vibration Tests
HIRES	Hypersonic In-flight Refueling System
HMCC	Hypergolic Maintenance Checkout Cells (VLS)
HMF	Hypergolic Maintenance Facility (KSC)
HPM	High-Performance Motor (SRB)
HRSI	High -Temperature Reusable Surface Insulation
HSFRS	High-Speed Flight Research Station (later HSFS)
HSFS	High Speed Flight Station (also HSFRS, FRC, DFRC)
HSRA	High-Speed Research Aircraft
HST	Hubble Space Telescope
HUD	Heads-Up Display
HYFAC	Hypersonic Facilities Aircraft
HYTID	Hypersonic Technology Integration Demonstrator

| HYWARDS | Hypersonic Weapon And Research and Development System |

I

IAPU	Improved Auxiliary Power Unit
IBM	International Business Machines Corporation
ICBM	Intercontinental Ballistic Missile
IGV	Incremental Growth Vehicle
ILRV	Integral Launch and Recovery Vehicle
IMAX	type of camera
IOC	Initial Operational Capability
IOP	Input/Output Processor
IRBM	Intermediate Range Ballistic Missile
ISS	International Space Station
ISTB	Integrated Subsystem Test Bed (SSME)
IUS	Interim (later Inertial) Upper Stage

J

JAL	Japan Air Lines
JPL	Jet Propulsion Laboratory
JSC	Johnson Space Center, Texas (formerly MSC)

K

| KSC | Kennedy Space Center, Florida |

L

L/D	Lift-to-Drag (ratio)
LACE	Liquid Air Cycle Engine
LACES	Liquid Air Collection Engine System
LaRC	Langley Research Center, Virginia
LCC	Launch Control Center
LDEF	Long Duration Exposure Facility
LEO	Low-Earth Orbit
LeRC	Lewis Research Center, Ohio (now GRC)
LES	Launch Entry Suit
LFBB	Liquid Fly-Back Booster
LH	Left-Hand
LH2	Liquid Hydrogen (fuel)
LLC	Limited Liability Company
LMSC	Lockheed Missiles and Space Company
LO2	Liquid Oxygen (oxidizer) [obsolete, LOX]
LOC	Launch Operations Center (later KSC)
LPS	Launch Processing System
LRB	Liquid Rocket Booster
LRSI	Low-Temperature Surface Insulation
LSOC	Lockheed Space Operations Company
Lt.Col.	Lieutenant Colonel (military rank)
LTV	Ling-Tempo-Vought (now part of Lockheed Martin)
LWC	Lightweight Case (SRM)
LWC-R	Lightweight Case Redesign (RSRM)
LWT	Lightweight Tank (type of ET)

M

MAC	Mean Aerodynamic Chord
MAF	Michoud Assembly Facility
MajGen.	Major General (military rank)
Mbs	Megabit per Second
MCC	Mission Control Center (Houston)
MCDS	Multifunction CRT Display System

MDM	Multiplexer/Demultiplexer
MEC	Main Engine Controller
MECO	Main Engine Cutoff
MEDS	Multifunction Electronic Display System
MEO	Mid-Earth Orbit
MISS	Man-in-Space-Soonest (Air Force)
MLI	Multi-Layer Insulation
MLP	Mobile Launch Platform
MLS	Microwave Landing System (FAA system – different than MSBLS)
MLU	Memory Load Unit (Skylab)
MMC	Martin Marietta Corporation
MMH	Monomethyl Hydrazine (fuel)
MMU	Manned Maneuvering Unit
MMU	Mass Memory Unit
MNASA	sub-scale test SRB
MOL	Manned Orbiting Laboratory
MOU	Memorandum of Understanding
mph	miles per hour
MPS	Main Propulsion System
MPTA	Main Propulsion Test Article (MPTA-098 and MPTA-ET)
MPTA-098	Main Propulsion Test Article
MPTA-ET	Main Propulsion Test Article – External Tank
MRS	Multipurpose Reusable Spacecraft
MRV	Manned Reconnaissance Vehicle
MSBLS	Microwave Scanning Beam Landing System
MSC	Manned Spacecraft Center, Texas (now JSC)
MSFC	Marshall Space Flight Center, Alabama
msn	manufacturer's serial number
MST	Mobile Service Tower (VLS)
MS	Mission Specialist (on a Shuttle mission)
MTF	Mississippi Test Facility (later NSTL, now SSC)
MTU	Master Timing Unit
MURP	Manned Upper Reusable Payload
MVGVT	Mated Vertical Ground Vibration Tests

N

N2O4	Nitrogen Tetroxide (oxidizer)
NACA	National Advisory Committee on Aeronautics (now NASA)
NAR	North American Rockwell (later RIC)
NASA	National Aeronautics and Space Administration (formerly NACA)
NASDA	National Space Development Agency of Japan
NASM	National Air and Space Museum
NASP	National Aero-Space Plane (X-30 / S-30)
NHFRF	National Flight Research Facility
NLS	National Launch System (later ALS)
NOAA	National Oceanic and Atmospheric Administration
NOPCO	company name
NRO	National Reconnaissance Organization
NSA	National Security Agency
NSDD	National Security Decision Directive
NSI	NASA Standard Initiator
NSTL	National Space Transportation Laboratory (formerly MTF, now SSC)
NSTS	National Space Transportation System (formerly STS, now SSP)
NTSB	National Transportation Safety Board

O

OAST	Office of Aeronautics and Space Technology
OEX	Orbiter Experiments Program
OFT	Orbital Flight Test
OFT	Orbital Flight Tests
OI	Operational Instrumentation
OMB	Office of Management and Budget
OMCF	Orbiter Maintenance and Checkout Facility (VLS)
OMDP	Orbiter Maintenance and Down Period
OMM	Orbiter Major Modification
OMRF	Orbiter Maintenance and Refurbishment Facility (now OPF-3)
OMS	Orbital Maneuvering System
ONR	Office of Naval Research
OPF	Orbiter Processing Facility (KSC)
OSCT	Office of Commercial Space Transportation
OTA	NASA Office of Technology Assessment
OV	Orbiter Vehicle
OV-099	*Challenger*
OV-101	*Enterprise*
OV-102	*Columbia*
OV-103	*Discovery*
OV-104	*Atlantis*
OV-105	*Endeavour*

P

P&W	Pratt & Whitney
PAM	Payload Assist Module (upper stage)
PARD	Piloted Aircraft Research Division
PASS	Primary Avionics System Software
PBAN	Polybutadiene Acrylonitrile
PCM	Pulse Code Modulation
PCMMU	PCM Master Unit
PCR	Payload Changeout Room (VLS)
PDR	Preliminary Design Review
PDU	Power Drive Unit
PIC	Pyrotechnic Initiator Capacitor
PILOT	Piloted Low speed Tests
Plt.	Pilot (of a Shuttle mission)
PM	Phase Modulation
POBATO	Propellants On-Board at Take-Off
PPP	Phased Planning Program
PPR	Payload Preparation Room (VLS)
PRIME	Precision Recovery Including Maneuvering Entry
PRR	Program Readiness Review
PRR	Preliminary Requirements Review
PRSD	Power Reactant Storage and Distribution
PSAC	President's Scientific Advisory Council
psf	pounds per square foot
psi	pounds per square inch
PS	Payload Specialist (on a Shuttle mission)
PVM	Production Verification Motor (SRB)

Q

QM	Qualification Motor (SRB)

R

R&D	Research and Development
RAPT	Reusable Aerospace Passenger Transport

RBSS	Recoverable Booster Space System	SRB	Solid Rocket Booster
RBUS	Rolling Beam Umbilical System (Centaur)	SRM	Solid Rocket Motor (part of SRB)
RCA	Radio Corporation of America	SRTM	Solid Rocket Test Motor (SRB)
RCC	Reinforced Carbon Carbon	SSC	Stennis Space Center, Mississippi
RCS	Reaction Control System	SSD	Space Systems Division
RDT&E	Research, Development, Test and Evaluation	SSME	Space Shuttle Main Engine
RENE	Rocket Engine Nozzle Ejector	SSP	Space Shuttle Program
RFP	Request for Proposals	SSTG	Space Shuttle Task Group
RH	Right-Hand	SSTO	Single-Stage-to-Orbit
RHC	Rotational Hand Controller	SSV	Small Space Vehicle (also used as Space Shuttle Vehicle)
RIC	Rockwell International Corporation (now Boeing)	STA	Shuttle Training Aircraft
RLV	Reusable Launch Vehicle	STA-099	*Challenger* (rebuilt into OV-099)
RMS	Remote Manipulator System	START	Spacecraft Technology and Advanced Reentry Tests
RoBo	Rocket-Bomber (Bell concept)	STDN	Space Tracking and Data Network
ROLS	Recoverable Orbital Launch System	STG	Space Task Group
RP-1	Rocket Propellant	STS	Space Transportation System (also NSTS)
RSI	Reusable Surface Insulation	SUMS	Shuttle Upper Atmosphere Mass Spectrometer (OEX)
RSLV	Reusable Space Launch Vehicle	SWT	Standard-Weight Tank
RSO	Range Safety Officer (at CCAFS)		
RSOC	Rockwell Shuttle Operations Company (now part of USA)	**T**	
RSRM	Redesigned (later Reusable) Solid Rocket Motor	TACAN	Tactical Air Navigation
RSS	Range Safety System; or Rotating Service Structure	TAEM	Terminal Area Energy Management
		TAL	Transoceanic Abort Landing
RTLS	Return-to-Launch-Site Abort	TAOS	Thrust Assisted Orbiter Shuttle
RTV	Room-Temperature Vulcanizing (adhesive)	TCDT	Terminal Countdown Demonstration Test
		TDRS	Tracking and Data Relay Satellite
RUM	Rockwell, USBI, Martin Marietta (original Shuttle contractors)	TDRSS	Tracking and Data Relay Satellite System
RVX	Reentry Vehicle, Experimental	TMc/S	Teledyne-McCormick-Selph
		TNT	TriNitroToluene (explosive)
S		TPS	Thermal Protection System
		TSM	Tail Service Mast
S&E	Science and Engineering Directorate (of MSFC)	TSTO	Two-Stage-to-Orbit
SAB	Scientific Advisory Board	TVC	Thrust Vector Control
SAB	Shuttle Assembly Building (VLS)		
SAIL	Shuttle Avionics Integration Laboratory	**U**	
SAINT	Satellite Inspector (Air Force program)		
SAMOS	Satellite Program	UDMH	Un-dimethylhydrazine (fuel)
SAMSO	Space and Missile Systems Organization	UHF	Ultra-High Frequency
SCA	Shuttle Carrier Aircraft	USA	United Space Alliance
SCAPE	Self-Contained Atmospheric Protective Ensemble	USA	United States Army
		USAF	United States Air Force
SDAM	San Diego Aerospace Museum	USBI	United Space Boosters Incorporated
SDCV	Shuttle Derived Cargo Vehicle	USCG	United States Coast Guard
SDV	Shuttle Derived Vehicle	USN	United States Navy
SEADS	Shuttle Entry Air Data System (OEX)	USTR	United States Trade Representative
SecAF	Secretary of the Air Force	UTC	Universal Coordinated Time
SERV	Single-stage Earth-orbital Reusable Vehicle		
		V	
SFOC	Space Flight Operations Contractor		
SIAT	Space Shuttle Independent Assessment Team	VAB	Vehicle (formerly Vertical) Assembly Building (KSC)
		VAFB	Vandenberg Air Force Base
SILTS	Shuttle Infrared Leeside Temperature Sensing Experiment	VLS	Vandenberg Launch and Landing Site
SLC	Space Launch Complex	**W**	
SLWT	Super Light Weight Tank (ET)		
SMCS	Separation Monitoring and Control System	WADC	Wright Air Development Center
SMTAS	Shuttle Modal Test and Analysis System	WADD	Wright Air Development Division
SOAS	Shuttle Orbiter Arresting System	WSTF	White Sands Test Facility
SOW	Statement of Work		
SPC	Shuttle Processing Contractor	**Y**	
SR	System Requirement		
		Y2K	Year 2000

CHAPTER I – THE BEGINNINGS. PAGES 1–14.

1. Richard P. Hallion, *In the Beginning was the Dream ...*, preface to <u>The Hypersonic Revolution: Case Studies in the History of Hypersonic Technology, Volume 1, From Max Valier to Project PRIME (1924-1967)</u>, Air Force Histories and Museums Program, (Bolling AFB, DC: U.S. Air Force, 1998), pp. xi-xii

2. *Max Valier: A Pioneer of Space Travel*, from the German *Max Valier: ein Vorkampher der Weltraumfahrt, 1895-1930*, translated by NASA as TT F-664, (Washington DC: NASA, 1976), pp. 81-97, 130-135, 248.

3. Walter Hohmann, *The Attainability of Heavenly Bodies*, from the German *Die Erreichbarkeit der Himmelskörper*, translated by NASA as TT F-44, (Washington DC: NASA, 1960); Richard P. Hallion, "The Path to the Space Shuttle: The Evolution of Lifting-Reentry Technology," unpublished manuscript in the files of the AFFTC History Office, April 1983.

4. Eugen Sänger and Irene Bredt, *The Silver Bird Story: A Memoir*, in R. Cargill Hall, editor, <u>Essays of the History of Rocketry and Astronautics: Proceedings of the Third Through Sixth Symposia of the International Academy of Astronautics</u>, volume I, (Washington DC: NASA, 1977), pp. 195-288; Ron Miller, *The Dream Machines: A Pictorial History of the Spaceship in Art, Science, and Literature*, (Malabar, FL: Kreiger Publishing, 1993), pp. 202-203, 238, 260-262.

5. Eugen Sänger, *Rocket Flight Engineering* from the German *Raketenflugtechnik*, translated by NASA as TT F-223, (Washington DC: NASA, 1965); *The Silver Bird Story: A Memoir*, pp. 195-288; Clarence J. Geiger, *Strangled Infant: The Boeing X-20A Dyna-Soar*, Case II of <u>The Hypersonic Revolution: Case Studies in the History of Hypersonic Technology, Volume 1, From Max Valier to Project PRIME (1924-1967)</u>, Air Force Histories and Museums Program, (Bolling AFB, DC: U.S. Air Force, 1998), pp. 185-190; Willey Ley. *Rockets, Missiles, and Space Travel*, second edition, (New York: The Viking Press, 1959), pp 429-434.

6. Eugen Sänger and Irene Bredt, *A Rocket Drive for Long-Range Bombers*, from the German *Über einen Raketenantrieb für Fernbomber* translated by the Naval Technical Information Brach, Bureau of Aeronautics as CGD-32, (Washington DC: U.S. Navy, 1952); *The Silver Bird Story: A Memoir*, pp. 195-288; *Strangled Infant*, pp. 185-190; *Rockets, Missiles, and Space Travel*, pp 429-434.

7. *In the Beginning was the Dream ...*, pp. xxii.

8. Transcript of interview of Dr. Wilhelm Raithel by Dr. John F. Guilmartin, 25 July 1984, in the files of the NASA/JSC History Office, Houston, Texas; various papers and memos in the A-9/A-10 files of the Air Force Museum Research Department, Wright Patterson AFB, Ohio.

9. *Strangled Infant*, pp. 186-186; Walter R. Dornberger, *V-2*, (New York: The Viking Press, 1954), p. 268.

10. *In the Beginning was the Dream ...*, pp. xxiiii.

11. <u>A Collection of Papers from the First Symposium on Space Flight</u>, Hyden Planetarium, October 1951, New York City; <u>Collier's</u> magazine, 21 March 1952, 18 and 25 October 1952, 28 February 1953, 4 and 11 March 1953, 27 June 1953, 30 April 1954; *The Dream Machines*, pp. 334-335.

12. <u>A Collection of Papers from the First Symposium on Space Flight</u>, Hyden Planetarium, October 1951, New York City; <u>Collier's</u> magazine, 21 March 1952, 18 and 25 October 1952, 28 February 1953, 4 and 11 March 1953, 27 June 1953, 30 April 1954.

13. Jay Miller, *The X-Planes: X-1 to X-31*, (Arlington, Texas: Aerofax, Inc., 1988), pp. 9-11; Richard P. Hallion, *On The Frontier*, NASA SP-4303, (Washington DC: NASA, 184), passim.

14. Meeting Minutes of the NACA Committee on Aerodynamics, 4 October 1951, in Group 255, National Archives, Washington DC; *On The Frontier*, pp. 107-108.

15. Ben Guenther, Jay Miller, and Terry Panopalis, *North American X-15/X-15A-2*, Aerofax Datagraph 2, (Arlington, Texas: Aerofax, Inc., 1985), p. 1.

16. Hubert M. Drake and L. Robert Carman, *A Suggestion of Means for Flight Research at Hypersonic Velocities and High Altitudes*, 21 May 1952, (NASA, Flight Research Center, Edwards, California); *On The Frontier*, p. 107.

17. *North American X-15/X-15A-2*, p. 2; *The X-Planes*, p. 114.

18. Meeting Minutes of the NACA Committee on Aerodynamics, 24 June 1953, in Group 255, National Archives, Washington DC; Aircraft Panel, Air Force Scientific Advisory Board, "Some Remarks of the Aircraft Panel on New Technical Developments of the Next Ten Years," (1 October 1954), in the files of the Air Force Museum.

19. John V. Becker, "Review of the Technology Relating to the X-15 Project" (a paper presented at the NACA Conference on the Progress of the X-15 Project, Langley Aeronautical laboratory, Langley Field, Virginia, 25-26 October 1956), pp 1-10.

20. John V. Becker, "The X-15 Program in Retrospect" (a paper presented at the 3rd Eugen Sänger Memorial lecture, Bonn, Germany, 4-5 December 1968), pp. 2-3.

21. Dennis R. Jenkins, *Hypersonics Before the Shuttle: A Concise History of the X-15 Research Airplane*, SP-2000-4518, (Washington DC: NASA, 2000), pp. 10-14.

22. "The X-15 Program in Retrospect," pp. 10-13.

23. Robert S. Houston, *Transiting from Air to Space: The North American X-15*, Case I of <u>The Hypersonic Revolution: Case Studies in the History of Hypersonic Technology, Volume 1, From Max Valier to Project PRIME (1924-1967)</u>, Air Force Histories and Museums Program, (Bolling AFB, DC: U.S. Air Force, 1998), pp. 2-3.

24. Several excellent histories of the X-15 program exist and should be consulted for additional information as required. No specific citations are provided for the X-15 section since the data has been summarized from all the histories. See: Robert S. Houston, *Transiting from Air to Space: The North American X-15*, Case I of <u>The Hypersonic Revolution: Case Studies in the History of Hypersonic Technology, Volume 1, From Max Valier to Project PRIME (1924-1967)</u>, Air Force Histories and Museums Program, (Bolling AFB, DC: U.S. Air Force, 1998); Ben Guenther, Jay Miller, and Terry Panopalis, *North American X-15/X-15A-2*, Aerofax Datagraph 2, (Arlington, Texas: Aerofax, Inc., 1985); Jay Miller, *The X-Planes: X-1 to X-31*, (Arlington, Texas: Aerofax, Inc., 1988); Milton O. Thompson, *At the Edge of Space: The X-15 Flight Program*, (Washington DC: Smithsonian Institution Press, 1992); and Dennis R. Jenkins, *Hypersonics Before the Shuttle: A Concise History of the X-15 Research Airplane*, NASA SP-2000-4518, (Washington DC: NASA, 2000); Richard P. Hallion, *On The Frontier*, NASA SP-4303, (Washington DC: NASA, 184).

25. The Staff at the Ames Aeronautical Laboratory, "Preliminary Investigation of a New Research Airplane for Exploring the Problems of Efficient Hypersonic Flight," (Moffett Field, California: NACA Ames Aeronautical Laboratory, 18 January 1957), pp. 1-2; *The Dream Machines*, p. 384.

26. The Staff at the Ames Aeronautical Laboratory, "Preliminary Investigation of a New Research Airplane for Exploring the Problems of Efficient Hypersonic Flight," (Moffett Field, California: NACA Ames Aeronautical Laboratory, 18 January 1957), p. 2.

27. "Preliminary Investigation," pp. 2-3.

28. "Study of the Feasibility of a Hypersonic Research Airplane," (Moffett Field, California: NACA Ames Aeronautical Laboratory, 3 September 1957), p. 1.

29. "Study of the Feasibility," pp. 6-11.

30. "Study of the Feasibility," pp. 11-17.

31. "Study of the Feasibility," pp. 18-19.

32. "Study of the Feasibility," pp. 19-24.

33. "Preliminary Investigation," p. 32.

CHAPTER II – DYNAMIC SOARING. PAGES 15–32.

1. Alvin Seiff and H. Julian Allen, *Some Aspects of the Design of Hypersonic Boost-Glide Aircraft*, RM-A55E26 (Washington DC: NACA, 15 August 1955), pp. 1-2; H. Julian Allen and Stanford E. Neice, *Problems of Performance and Heating of Hypersonic Vehicle*, RM-A55L15 (Washington DC: NACA, 5 March 1956), pp. 1-2.

2. *Some Aspects of the Design*, pp. 2-21; *Problems of Performance*, pp. 5-6.

3. *Some Aspects of the Design*, pp. 5-21; *Problems of Performance*, pp 2-11.

4. Clarence J. Geiger, *Strangled Infant: The Boeing X-20A Dyna-Soar*, Case II of <u>The Hypersonic Revolution: Case Studies in the History of Hypersonic Technology, Volume 1, From Max Valier to Project PRIME (1924-1967)</u>, Air Force Histories and Museums Program, (Bolling AFB DC: U.S. AIr Force, 1998), pp. 188-189.

5. *Strangled Infant*, pp. 189-190.

6. Letter from Colonel H. Heaton, Headquarters WADC to Brigadier General F. B. Woods, Headquarters ARDC, subject: Rocket Bomber Feasibility Studies, 10 April 1953, in record group 255 of the National Archives, College Park.

7. *Strangled Infant*, pp. 191-192.

8. *Strangled Infant*, p. 194.

9. *Strangled Infant*, p. 191.

10. *Strangled Infant*, pp. 193-194.

11. General Operations Requirement GOR-12, "Piloted Very High Altitude Reconnaissance Weapon System," 12 May 1955, in the files of the ARDC History Office; <u>Advanced Strategic Weapons System Progress Report No. 9</u>, report D143-981-009, (Bell Aircraft Corporation, 15 February 1955).

12. *Strangled Infant*, pp. 194-195.

13. *Strangled Infant*, pp. 195-196.

14. Research and Development Project Card, System 455L, 18 December 1956, in the files of the ARDC History Office.

15. *Strangled Infant*, pp. 199-201.

16. Bell Report D143-945-700, "An Approach to Manned Orbital Flight," presented to the Committee on Advanced Weapons Technology and Environment of the U.S. Air Force Scientific Advisory Board, 29-30 July 1957.

17. "An Approach to Manned Orbital Flight."

18. "An Approach to Manned Orbital Flight."

19. "An Approach to Manned Orbital Flight."

20. "An Approach to Manned Orbital Flight."

21. "An Approach to Manned Orbital Flight."

22. "An Approach to Manned Orbital Flight."

23. "An Approach to Manned Orbital Flight."

24. "An Approach to Manned Orbital Flight."

25. "An Approach to Manned Orbital Flight."

26. "An Approach to Manned Orbital Flight."

27. Convair report FZM-941, "ROBO Feasibility Summary, Volume I," ARDC-TR-57-58, prepared under ARDC contract AF 18(600)-1623, (Fort Worth, Texas: Convair, 1 June 1957). In the files of the San Diego Aerospace Museum.

28. "ROBO Feasibility Summary."

29. "ROBO Feasibility Summary."

30. "ROBO Feasibility Summary."

31. Richard P. Hallion, editor's introduction to Clarence J. Geiger, *Strangled Infant: The Boeing X-20A Dyna-Soar*, Case II of The Hypersonic Revolution: Case Studies in the History of Hypersonic Technology, Volume 1, From Max Valier to Project PRIME (1924-1967), Air Force Histories and Museums Program, (Bolling AFB DC: U.S. Air Force, 1998), pp. II-iii-II-vi; Willy Ley, *Rockets, Missiles, and Men in Space*, (New York: Viking Press, 1968), pp. 449-450; Ron Miller, *The Dream machines: A Pictorial History of the Spaceship in Art, Science, and Literature*, (Malabar, Florida: Kreiger Publishing, 1993), pp. 304-305.

32. Memorandum from John W. Crowley, Associate Director for Research, NACA, to the NACA field centers, 15 November 1957; *Strangled Infant*, pp. 201-202; "This is Dyna-Soar," Boeing News Release S-6826, (Seattle, Washington: The Boeing Company, 1963).

33. *Strangled Infant*, pp. 202-203.

34. *Strangled Infant*, pp. 209-211.

35. *Strangled Infant*, pp. 210-211.

36. *Strangled Infant*, pp. 211-212.

37. *Strangled Infant*, pp. 212-217.

38. System Requirement 201, in the files of the ARDC History Office; *Strangled Infant*, pp. 219-220.

39. *Strangled Infant*, pp. 225-226.

40. *Strangled Infant*, pp. 227-226.

41. *Strangled Infant*, pp. 233-234.

42. *Strangled Infant*, pp. 237-241.

43. *Strangled Infant*, pp. 244-248.

44. *Strangled Infant*, pp. 253-259.

45. Jay Miller, *The X-Planes: X-1 to X-31*, (Arlington, Texas: Aerofax, Inc., 1988), pp. 150-15; *Strangled Infant*, pp. 269-270; "This is Dyna-Soar."

46. "X-20 Dyna-Soar Information Fact Sheet," (Washington DC: U.S. Air Force, January 1963), p. 5; Terry Smith, "Dyna-Soar X-20: A Look at the Hardware and Technology," in Quest: The History of Spaceflight Magazine, Winter 1994, pp. 23-28.

47. Correspondence between Boeing, Weber, and the Air Force Missile Development Center, Holloman AFB, New Mexico, various dates in 1963 and 1964.

48. *The X-Planes*, pp. 150-151.

49. *Strangled Infant*, pp. 261-263.

50. *Strangled Infant*, pp. 269-272.

51. *Strangled Infant*, pp. 276-283 and 291-293.

52. *Strangled Infant*, pp. 290-291.

53. *Strangled Infant*, pp. 294-297.

54. *Strangled Infant*, pp. 300-309.

55. *Strangled Infant*, pp. 310-313.

56. "Dyna-Soar Management Proposal," General Dynamics / Fort Worth Report FZM-871, 26 March 1958.

CHAPTER III – WINGLESS FLIGHT. PAGES 33–48.

1. H. Julian Allen and A. J. Eggers, Jr. "A Study of the Motion and Aerodynamic Heating of Missiles Entering the Earth's Atmosphere at High Supersonic Speeds, NASA RM-A53D28, (Moffett Field, California NASA Ames Aeronautical Laboratory, 25 August 1953); Richard P. Hallion and John L. Vitelli, *The Piloted Lifting Body Demonstrators: Supersonic Predecessors to Hypersonic Lifting Reentry*, Case VII of The Hypersonic Revolution: Case Studies in the History of Hypersonic Technology, Volume 2, From Scramjet to the National Aero-Space Plane (1964-1986), Air Force Histories and Museums Program, (Bolling AFB DC: U.S. Air Force, 1998), p. 864; Richard P. Hallion, *On The Frontier*, NASA SP-4303, (Washington DC: NASA, 1964), p. 147.

2. Maxime A. Faget, *et. al.*, "Preliminary Studies of Manned Satellites – Wingless Configurations: Nonlifting," NACA Conference on High-Speed Aerodynamics: A Compilation of the Papers presented, (Moffett Field, California: NACA, 1958), p. 25.

3. Thomas J. Wong, *et. al.*, "Preliminary Studies of Manned Satellites – Wingless Configurations: Lifting Body," NACA Conference on High-Speed Aerodynamics: A Compilation of the Papers presented, (Moffett Field, California: NACA, 1958), pp. 35-44.

4. Jay Miller, *The X-Planes: X-1 to X-31*, (Arlington, Texas: Aerofax, Inc., 1988), p. 166; John L. Vitelli and Richard P. Hallion, *Project PRIME: Hypersonic Reentry From Space*, Case V of The Hypersonic Revolution: Case Studies in the History of Hypersonic Technology, Volume 1, From Max Valier to Project PRIME (1924-1967), Air Force Histories and Museums Program, (Bolling AFB DC: U.S. Air Force, 1998), p. 529; *Piloted Lifting Body Demonstrators*, p. 866.; Richard P. Hallion, *On The Frontier*, NASA SP-4303, (Washington DC: NASA, 1964), p. 148.

5. Backup Program for Modified 698N Configuration, Martin report ER 12147, (Baltimore, Maryland: Martin Marietta, April 1962), pp. 1-2; *Project PRIME: Hypersonic Reentry*, pp. 535-536.

6. R. Dale Reed with Darlene Lister, *Wingless Flight: The Lifting Body Story*, SP-4220, (Washington DC: NASA, 1997), pp. 14-15.

7. Correspondence between Frederick Raymes and Dennis R. Jenkins 1992-1996; U.S. Patent 203,902, 23 August 1963.

8. SAMOS Satellite Reconnaissance System, Briefing to the Satellite Intelligence Requirements Committee, undated (but in the March-May 1960 timeframe); *Project PRIME: Hypersonic Reentry*, pp. 538-539.

9. *Wingless Flight*, pp. 8-9.

10. Advanced Development Program Development Plan for PRIME: A Project Within START, Program 680A, April 1966, in the files of the Air Force Museum; *The X-Planes*, pp. 163-165; *Project PRIME: Hypersonic Reentry*, pp. 542-555, 572.

11. PRIME: A Project Within START; *The X-Planes*, p. 163; Program START Hypervelocity Test Program White Paper, (Wright-Patterson AFB: U.S. Air Force, 6 April 1964); "History of the START Program and the SV-5 Configuration," AFSC publication Series 67-23-1, (Wright-Patterson AFB: U.S. Air Force, October 1967), in the files of the AFHRA K043.04-20).

12. PRIME: A Project Within START; *The X-Planes*, p. 163; *Project PRIME: Hypersonic Reentry*, pp. iv-vi.

13. PRIME: A Project Within START; *The X-Planes*, p. 163.

14. PRIME: A Project Within START; *Project PRIME: Hypersonic Reentry*, pp. 632-657.

15. PRIME: A Project Within START; PRIME Flight Test No. 1, Flight Analysis, Martin report ER 14461, (Martin, Baltimore, March 1967); *The X-Planes*, p. 163; *Project PRIME: Hypersonic Reentry*, pp. 694-725; see also Joel W. Powell and Ed Hengevold, "ASSET and PRIME: Gliding Re-entry Test Vehicles," Journal of the British Interplanetary Society, XXXVI (1983), pp. 369-376.

16. PRIME: A Project Within START; PRIME Flight Test No. 2, Flight Analysis, Martin report ER 14462, (Martin, Baltimore, June 1967); *The X-Planes*, p. 163.

17. PRIME: A Project Within START; PRIME Flight Test No. 3, Flight Analysis, Martin report ER 14463, (Martin, Baltimore, July 1967); *The X-Planes*, p. 163.

18. *The X-Planes*, p. 163; *Project PRIME: Hypersonic Reentry*, pp. 700-725.

19. *The X-Planes*, p. 163; *Project PRIME: Hypersonic Reentry* pp. 700-725.

20. Interview with Dale Reed published in the DFRC X-Press, 10 March 1967; *Wingless Flight*, passim.

21. Interview with Dale Reed published in the DFRC X-Press, 10 March 1967; *Wingless Flight*, pp. 15-21, 65.

22. *Wingless Flight*, pp. 11-25.

23. *Piloted Lifting Body Demonstrators*, p. 869.

24. *Wingless Flight*, pp. 33-35; *On The Frontier*, N pp. 150-151.

25. *Piloted Lifting Body Demonstrators*, p. 890; *Wingless Flight*, passim.

26. *Wingless Flight*, p. xviii.

27. *Piloted Lifting Body Demonstrators*, pp. 872-873; *Wingless Flight*, passim; *On The Frontier*, pp. 150-152.

28. *Wingless Flight*, pp. 68-69; *On The Frontier*, p. 153.

29. *Wingless Flight*, pp. 66-67.

30. *Piloted Lifting Body Demonstrators*, pp. 873-874; *Wingless Flight*, passim; *On The Frontier*, pp. 153-154.

31. Fred Anderson, *Northrop: An Aeronautical History*, (Northrop Corporation, Los Angeles, July 1976), pp. 232-233; *Piloted Lifting Body Demonstrators*, pp. 874-875; *Wingless Flight*, pp. 72-73; *On The Frontier*, p. 154.

32. *Wingless Flight*, pp. 73-75.

33. *Wingless Flight*, pp. 79-81; *Piloted Lifting Body Demonstrators*, pp. 877-880.

34. *Wingless Flight*, p. 82; *Piloted Lifting Body Demonstrators*, pp. 880-882.

35. "Flight Test Results Pertaining to the Space Shuttlecraft," NASA TM-X-2101, a collection of papers for a symposium at the Flight Research Center, (Edwards, California: NASA, 30 June 1970); *Piloted Lifting Body Demonstrators*, pp. 880-882.

36. *Piloted Lifting Body Demonstrators*, pp. 881-883; *Wingless Flight*, passim; *On The Frontier*, pp. 158-159.

37. *Wingless Flight*, p. 115; *Piloted Lifting Body Demonstrators*, pp. 882-884.

38. *Northrop: An Aeronautical History*, pp. 234-235; *Piloted Lifting Body Demonstrators*, pp. 885-886; *On The Frontier*, pp. 160-167.

39. Fred Anderson, *Northrop: An Aeronautical History*, (Los Angeles: Northrop Corporation, July 1976), pp. 235-236.

40. "Flight Test Results Pertaining to the Space Shuttlecraft;" *Piloted Lifting Body Demonstrators*, pp. 885-886.

41. *Wingless Flight*, p. 102-123; *Piloted Lifting Body Demonstrators*, pp. 885-886.

42. "Flight Test Results Pertaining to the Space Shuttlecraft;" *Piloted Lifting Body Demonstrators*, p. 886; *Wingless Flight*, passim; *On The Frontier*, pp. 161-163.

43. *Piloted Lifting Body Demonstrators*, p. 887; *Wingless Flight* passim.

44. *Wingless Flight*, p. 155; *On The Frontier*, pp. 210-212.

45. *Wingless Flight*, pp. 157-158.

46. *Wingless Flight*, p. 158.

47. *Wingless Flight*, pp. 158-159.

48. *Wingless Flight*, pp. 159-162; *On The Frontier*, pp. 210-212.

49. *Wingless Flight*, pp. 163-164.

50. *Wingless Flight*, pp. 164-165.

51. *Wingless Flight*, pp. 165-166.

52. *Piloted Lifting Body Demonstrators*, pp. 894-898.

53. *Piloted Lifting Body Demonstrators*, pp. 894-898.

54. *Piloted Lifting Body Demonstrators*, pp. 898-900.

55. *Piloted Lifting Body Demonstrators*, p. 900.

56. *Piloted Lifting Body Demonstrators*, pp. 896-900, 923.

57. *The X-Planes*, p. 165; *Piloted Lifting Body Demonstrators* pp. 901-903.

58. *Piloted Lifting Body Demonstrators*, pp. 901-905.

59. *Piloted Lifting Body Demonstrators*, pp. 900-918.

60. *The X-Planes*, pp. 165-166; *Piloted Lifting Body Demonstrators*, p. 918.

61. *Piloted Lifting Body Demonstrators*, p. 919; *On The Frontier*, pp. 163-165.

62. Jon S. Pyle and Lawrence C. Montoya, "Effects of Roughness of Simulated Ablated Material on Low-Speed Performance Characteristics of a Lifting-Body Vehicle," NASA TM S-1810, (Washington DC: NASA, 1969); *Piloted Lifting Body Demonstrators*, p. 919; *On The Frontier*, pp. 163-165.

63. "History of the Air Force Plant Representative, Detachment 23, Martin Marietta Corporation," (Baltimore, Maryland: U.S. Air Force, various dates), in the files of the AFHRA, K243.0707-23; *The X-Planes*, p. 166; *Wingless Flight*, p. 132; *On The Frontier*, pp. 163-165.

64. *The X-Planes*, pp. 166-167; *Piloted Lifting Body Demonstrators*, pp. 919-922.

65. "History of the Air Force Plant Representative;" *Piloted Lifting Body Demonstrators*, p. 923.

66. *The X-Planes*, p. 167; *Piloted Lifting Body Demonstrators*, p. 923.

67. *The X-Planes*, p. 167; *Piloted Lifting Body Demonstrators*, pp. 923-924; *On The Frontier*, pp. 165-167.

68. *Piloted Lifting Body Demonstrators*, pp. 925-928; *On The Frontier*, pp. 165-167.

69. *The X-Planes*, p. 167; *Piloted Lifting Body Demonstrators*, pp. 925-928.

70. *Piloted Lifting Body Demonstrators*, pp. 928-929; *On The Frontier*, pp. 167-168.

71. John A. Manke and Michael V. Love, "X-24B Flight Test Program," 1975 Report to the Aerospace Profession, Society of Experimental Test Pilots, 26 September 1975; *The X-Planes*, pp. 167, 171; *Piloted Lifting Body Demonstrators*, pp. 928-930.

72. *The X-Planes*, pp. 167-168; *Piloted Lifting Body Demonstrators*, p. 930.

73. Richard E. Brackeen and William L. Marcy, X-24B Growth Version Feasibility Study, AFFDL report TR-73-116, (Wright-Patterson AFB: U.S. Air Force, October 1973); *Piloted Lifting Body Demonstrators*, p. 932; *USAF to Begin Hypersonic Testing*, Aviation Week and Space Technology, 17 September 1973, pp. 83-84.

74. L. Robert Jackson and Allan H. Taylor, Structural Design for a Hypersonic Research Airplane, AIAA paper 76-906, presented at the Aircraft Systems and Technology Meeting, Dallas, Texas, 27-29 September 1976 (revised 21 March 1978); J. P. Weidner, W. J. Small, and J. A. Penland, Scramjet Integration on Hypersonic Research Aircraft Concepts, a paper presented at the SAE Propulsion Conference, Palo Alto, California, 26-28 July 1976; Allan H. Taylor and L. Robert Jackson, Heat-Sink Structure Design Concepts for a Hypersonic Research Airplane, AIAA paper 77-392, presented at the AIAA-ASME 18th Structures, Structural Dynamics, and Materials Conference, San Diego, California, 21-23 March 1977.

75. Structural Design; Scramjet Integration; Heat-Sink Structure.

76. *The X-Planes*, p. 168; *Piloted Lifting Body Demonstrators*, pp. 932-935.

77. *Piloted Lifting Body Demonstrators*, pp. 932-935; *On The Frontier*, pp. 170-171.

78. *The X-Planes*, p. 168.

79. X-24C Configuration Development Study, Phase II Review, Prepared under NAS1-14222 by Lockheed Aircraft Corporation, Advanced Development Projects, May 1976; Configuration Development Study of the X-24C Hypersonic Research Airplane, Phase III Review, Prepared under NAS1-14222 by Lockheed Aircraft Corporation, Advanced Development Projects, January 1977.

80. X-24C Configuration Development Study; Configuration Development Study of the X-24C.

81. X-24C Configuration Development Study; Configuration Development Study of the X-24C.

82. X-24C Configuration Development Study; Configuration Development Study of the X-24C.

83. *The X-Planes*, p. 168.

84. The Hypersonic Technology Integration Demonstrator, a briefing prepared by the Air Force Flight Dynamics Laboratory for unknown reasons, undated (probably late-1977 or early-1978).

85. The Hypersonic Technology Integration Demonstrator.

86. The Hypersonic Technology Integration Demonstrator.

87. The Hypersonic Technology Integration Demonstrator.

88. The Hypersonic Technology Integration Demonstrator.

89. The Hypersonic Technology Integration Demonstrator.

90. The Hypersonic Technology Integration Demonstrator.

90. Milton O. Thompson and Joe Weil, Proposal for a Subscale Shuttle, (Edwards, California: NASA, 17 August 1972); *On The Frontier*, pp. 169-170.

92. Proposal for a Subscale Shuttle; *On The Frontier*, pp. 169-170.

93. *Piloted Lifting Body Demonstrators*, p. 938.

CHAPTER IV – CONCEPT EXPLORATION. PAGES 49–76.

1. "ASSET," (preparer unknown, December 1965), in the files of AFHRA as K243-8636-91; Richard P. Hallion, *ASSET: Pioneer of Lifting Reentry*, Case IV of The Hypersonic Revolution: Case Studies in the History of Hypersonic Technology, Volume 1, From Max Valier to Project PRIME (1924-1967), Air Force Histories and Museums Program, (Bolling AFB DC: U.S. Air Force, 1998), pp. 449-450.

2. History of the Aeronautical Systems Division, January Through June 1962, (Wright-Patterson AFB, Ohio: Aeronautical Systems Division, December 1962); *ASSET: Pioneer of Lifting Reentry*, pp. 449-452.

3. "ASSET," pp. 1-2; Advanced Technology Program: Technical Development Plan for Aerothermodynamic/Elastic Structural Systems Environment Tests (ASSET), (Wright-Patterson AFB, Ohio: Air Force Flight Dynamics Laboratory, 1963), pp. 21-22, 43-44; *ASSET: Pioneer of Lifting Reentry*, pp. 449-452.

4. Advanced Technology Program, pp. 21-22, 89; *ASSET: Pioneer of Lifting Reentry*, pp. 451-453; "ASSET," pp. 2-3, 12.

5. Charles J. Cosenza, "ASSET: A Hypersonic Glide Reentry Test Program," a paper presented at the 1964 Annual Fall Meeting of the Ceramic-Metal Systems Division of the American Ceramic Society; "ASSET," pp. 5, 11-12.

6. *ASSET: Pioneer of Lifting Reentry*, pp. 486-495.

7. ASSET: Aerothermoelastic Vehicles (AEV) Results and Conclusions, report 65FD-1197, (Flight Dynamics Laboratory, Wright-Patterson AFB, August 1965); *ASSET: Pioneer of Lifting Reentry*, pp. 483-485.

8. ASSET Final Briefing, report 65FD-850, (Wright-Patterson AFB, Ohio: Air Force Flight Dynamics Laboratory, 5 October 1965); *ASSET: Pioneer of Lifting Reentry*, pp. 500-509.

9. ASSET Final Briefing.

10. Jay Miller, *The X-Planes: X-1 to X-31*, (Arlington, Texas: Aerofax, Inc., 1988), p. 89; Jacob Neufeld, *Ballistic Missiles in the Untied States Air Force: 1945-1960*, (Washington DC: Office of Air Force History, 1990), p. 28.

11. *The X-Planes*, pp. 89-90; *Ballistic Missiles*, pp. 58-59.

12. *The X-Planes* p. 90; *Ballistic Missiles*, p. 48.

13. *The X-Planes*, pp. 90-91.

14. *Ballistic Missiles*, p. 125.

15. *Ballistic Missiles*, pp. 255-258.

16. Richard P. Hallion and James O. Young, *Space Shuttle: Fulfillment of a Dream*, Case VIII of The Hypersonic Revolution: Case Studies in the History of Hypersonic Technology, Volume 1, From Max Valier to Project PRIME (1924-1967), Air Force Histories and Museums Program, (Bolling AFB DC: U.S. Air Force, 1998), p. 949.

17. SR-89774 Reusable Space Launch Vehicle Report, prepared by The Boeing Company, Aero-Space Division, December 1959.

18. Reusable Space Launch Vehicle Systems Study, prepared by the Convair Division of General Dynamics, undated; Reusable Space Launch Vehicle Study, Convair report GD/C-DCJ-65-004, 18 May 1965; "This We Call Experience ..." a promotional brochure produced by Martin Marietta in support of their Phase A SpaceMaster concept. Undated (but probably late 1969, early 1970).

19. *Space Shuttle: Fulfillment of a Dream*, p. 949.

20. Lloyd H. Cornett, Jr., *History of the Air Force Missile Development Center, 1 January – 30 June 1964*, (Holloman AFB, New Mexico: Air Force Missile Development Center, 1964), in the files of the Air Force Historical Research Agency K280.10-61B; *Space Shuttle: Fulfillment of a Dream*, p. 950.

21. *Space Shuttle: Fulfillment of a Dream*, p. 949.

22. Project Spaceplane, report ZP-M-095, (Convair Division of the General Dynamics Corporation, San Diego, California, April 1960).

23. Project Spaceplane.

24. Project Spaceplane.

25. *History of the Air Force Missile Development Center, 1 January – 30 June 1964,*

26. Ron Miller, *The Dream Machines: A Pictorial History of the Spaceship in Art, Science, and Literature*, (Malabar, Florida: Kreiger Publishing, 1993), pp. 447-448.

27. *History of the Air Force Missile Development Center; Space Shuttle: Fulfillment of a Dream*, pp. 949-950.

28. Reusable Space Launch Vehicle Systems Study; *Space Shuttle: Fulfillment of a Dream*, pp. 949-950.

29. General Electric Navigation and Control (GENAC), various progress reports, (New York: General Electric Corporation, 1963), in the files of the AFHRA.

30. *Space Shuttle: Fulfillment of a Dream*, p. 950.

31. *Space Shuttle: Fulfillment of a Dream*, p. 950; *The Dream Machines*, p. 395.

32. Report of the USAF Scientific Advisory Board Aerospace Vehicles/Propulsion Panels on Aeroplane, VTOL, and Strategic Manned Aircraft, (U.S. Air Force, Washington DC, 24 October 1963), pp. 1-3; *The Dream Machines*, p. 395.

33. *Space Shuttle: Fulfillment of a Dream*, p. 950.

34. *History of the Air Force Missile Development Center, 1 January – 30 June 1964.*

35. "USAF Seeks Funds to Speed its Scramjet Hypersonic Flight Program," *Aviation Week and Space Technology*, 12 July 1966, pp. 52-53.

36. *History of the Air Force Missile Development Center, 1 January – 30 June 1964.*

37. *History of the Air Force Missile Development Center, 1 January – 30 June 1964.*

38. *History of the Air Force Missile Development Center, 1 January – 30 June 1964.*

39. *History of the Air Force Missile Development Center, 1 January – 30 June 1964.*

40. *History of the Air Force Missile Development Center, 1 January – 30 June 1964.*

41. *History of the Air Force Missile Development Center, 1 January – 30 June 1964.*

42. *History of the Air Force Missile Development Center, 1 January – 30 June 1964.*

43. *History of the Air Force Missile Development Center, 1 January – 30 June 1964.*

44. C. W. Speith, Astroplane Feasibility Study, report M-62-134, Volume I, (Denver, Colorado: Martin Company Colorado, October 1962).

45. A Study of the Effects of Launch Inclination Angle on the Payload Capabilities of a Reusable Orbital Transport, report GD/C-DCB-65-038, (San Diego, California: General Dynamics/Convair, 14 October 1965), in the files of the AFHRA on microfilm role 40778.

44. Initial Study of a Manned Orbiting Reconnaissance System, report ZP-203, (San Diego, California: General Dynamics/Convair, 18 April 1958).

45. Initial Study of a Manned Orbiting Reconnaissance System.

46. Initial Study of a Manned Orbiting Reconnaissance System.

47. *Space Shuttle: Fulfillment of a Dream*, p. 952.

48. *The Dream Machines*, p. 438.

49. C.W. Speith and W. T. Teegarden, Astrorocket Progress Report, report M-63-1, (Denver, Colorado: Martin Company December 1962).

50. "This We Call Experience ..." a promotional brochure produced by Martin Marietta in support of their Phase A SpaceMaster concept. Undated (but probably late 1969, early 1970).

51. Astrorocket Progress Report.

52. Astrorocket Progress Report.

53. *Space Shuttle: Fulfillment of a Dream*, p. 952; *The Dream Machines*, p. 446.

54. Report of the NASA Special Ad Hoc Panel on Hypersonic Lifting Vehicles with Propulsion, (Houston: NASA, June 1964).

55. A Study of the Conceptual Design of Re-Usable Ten Ton Orbital Carrier Vehicles, report LR-16149, prepared for contract NAS8-2687, (Burbank, California: Lockheed-California, 16 August 1962).

56. Reusable Orbital Transport Second Stage: Detailed Technical Report, report GD/C-DCB-65-018, (San Diego, California: General Dynamics/Convair, April 1965); Reusable Orbital Transport: Definition of the Second Stage of the Baseline Vehicle, report GD/A-DCB-64-078, (San Diego, California: General Dynamics/Convair, 18 September 1964); Baseline Vehicle: First Stage Definition, report LR-18387, (Burbank, California: Lockheed-California, 2 December 1964); Design and Operational Trends of Reusable Aerospace Passenger Transport Systems, report LR-19276, (Burbank, California: Lockheed-California, December 1965).

57. Design and Operational Trends.

58. Design and Operational Trends.

59. Design and Operational Trends.

60. Design and Operational Trends.

61. Reusable Orbital Transport Second Stage: Detailed Technical Report, report GD/C-DCB-65-018, (San Diego, California: General Dynamics/Convair, April 1965).

62. Reusable Orbital Transport Second Stage: Detailed Technical Report.

63. Reusable Aerospace Passenger Transport: Launch mode Comparison Study, report CR-66-35, (Denver, Colorado: Martin Marietta, July 1966).

64. Reusable Aerospace Passenger Transport: Launch mode Comparison Study.

65. Reusable Aerospace Passenger Transport: Launch mode Comparison Study.

66. Reusable Aerospace Passenger Transport: Launch mode Comparison Study

67. Study of a Ten-Passenger Reusable Orbital Carrier, report SID-64-144-3, (El Segundo, California: North American Aviation, Space and Information Systems Division, 19 March 1964).

68. Study of a Ten-Passenger Reusable Orbital Carrier.

69. Mission Requirements of Lifting Systems: Engineering Aspects, report C-100905 (B832), (St. Louis, Missouri: McDonnell Aircraft Company, 18 August 1965).

70. Mission Requirements of Lifting Systems: Engineering Aspects.

71. Mission Requirements of Lifting Systems: Engineering Aspects

72. Report of the Ad Hoc Subpanel on Reusable Launch Vehicle Technology, (Washington DC: Aeronautics and Astronautics Coordinating Board, 14 September 1966).

73. Report of the Ad Hoc Subpanel on Reusable Launch Vehicle Technology.

74. Report of the Ad Hoc Subpanel on Reusable Launch Vehicle Technology

75. Report of the Ad Hoc Subpanel on Reusable Launch Vehicle Technology.

76. Report of the Ad Hoc Subpanel on Reusable Launch Vehicle Technology.

77. Report of the Ad Hoc Subpanel on Reusable Launch Vehicle Technology.

78. *Space Shuttle: Fulfillment of a Dream*, pp. 957-962.

79. Space Transportation System Study, final report (contract F04701-69-C-0382), (Sunnyvale, California: Lockheed Missiles and Space Co., 23 December 1969); J. T. Lloyd, Preliminary Design of Two Hypersonic High L/D Vehicles, report LR-18159, (Burbank, California: Lockheed-California, 10 September 1964); *Space Shuttle: Fulfillment of a Dream*, pp. 957-962; *The Dream Machines*, pp. 445-446.

80. *Space Shuttle: Fulfillment of a Dream*, pp. 957-962.

81. *Space Shuttle: Fulfillment of a Dream*, pp. 957-962.

82. *Space Shuttle: Fulfillment of a Dream*, pp. 957-962.

83. *Space Shuttle: Fulfillment of a Dream*, pp. 957-962.

84. T-18 Vehicle Baseline, Convair report GD/C-DCB68-012-14, (San Diego, California: Convair, August 1965).

85. T-18 Vehicle Baseline.

86. T-18 Vehicle Baseline.

87. *Space Shuttle: Fulfillment of a Dream*, p. 954.

88. *The Dream Machines*, pp. 306, 361, 370, 375, 387-389, 443-444, and 449.

89. *The Dream Machines*, pp. 511-512.

90. Aerospaceplane, ACES/CR Development Planning Study, General Dynamics report GD/A DCB-64-055, (San Diego, California: General Dynamics/Convair, 22 May 1964).

91. Aerospace Plane, Lockheed report LR-15862/LAC-571374, (Burbank, California: Lockheed-California Company, 20 April 1962).

Chapter V – Grand Ambitions. Pages 77–134.

1. Tom. A. Heppenheimer, *The Space Shuttle Decision: NASA's Search for a Reusable Space Vehicle*, NASA SP-4221, (Washington DC: NASA, 1999), p. 86.

2. *The Space Shuttle Decision*, pp. 87-89.

3. George Mueller, *Address by Dr. George E. Mueller Before the British Interplanetary Society*, (London: University College, 10 August 1968).

4. *Address by Dr. George E. Mueller.*

5. *Address by Dr. George E. Mueller*; Helen T. Wells, et. al., Origins of NASA Names, NASA SP-4402, (Washington DC: Scientific and Information Office, 1976).

6. *The Space Shuttle Decision*, pp. 101-103.

7. Milton B. Ames, chairman, Report of the Ad Hoc Subpanel on Reusable Launch Vehicle Technology, (Washington DC: Aeronautics and Astronautics Coordinating Board, 14 September 1966).

8. Max Akridge, *Space Shuttle History*, SHHDC-5013, (Hunstville, Alabama: NASA, 3, 8 January 1970), pp. 25-26.

9. Request for Proposals, MSC-BG721-28-9-96C and MSFC-1-7-21-00020, 30 October 1968. Copies in both the JSC and MSFC History Office files.

10. Ibid.

11. "Four Selected for AF Space Transportation Study," Space Business Daily, 1 August 1969 (Vol. 45, No. 13), page 83; *The Space Shuttle Decision*, pp. 117-118.

12. Charles H. Townes, chairman, *Report of the Task Force on Space*, (Washington DC: 8 January 1969) p. 6 and 21-22, conveniently reprinted in *Exploring the Unknown, Volume I: Organizing for Exploration*, NASA SP-4407, (Washington DC: NASA, 1995), pp. 499-512.

13. "Four Selected for AF Space Transportation Study."

14. *The Space Shuttle Decision* p. 126.

15. Letter, Thomas O. Paine, NASA Administrator, to Robert C. Seamans, Jr., Secretary of the Air Force, 4 April 1969, with attachment: "Terms of References for Joint DoD/NASA Study of Space Transportation Systems;" Grant L. Hansen and George. E. Mueller, Co-Chairmen, *Joint DoD/NASA Study of Space Transportation Systems, Summary Report*, AS-69-0000-02262 (originally classified secret, with top secret appendices), 16 June 1969, p. 3 (in the files of the AFHRA with call number K140.01, July-December 1969, V.7, Pt. 2).

16. George Mueller, *Space Shuttle Contractors Briefing*, (Houston, Texas: NASA, 5 May 1969).

17. *Space Shuttle Contractors Briefing.*

18. *The Space Shuttle Decision* p. 218.

19. Leroy E. Day, manager, Space Shuttle Task Group Report, 5 volumes, (Washington DC: NASA, 12 June 1969), Volume II has been conveniently reprinted in *Exploring the Unknown, Volume IV: Accessing Space*, NASA SP-4407, (Washington DC: NASA, 1999), pp. 206-210.

20. Space Shuttle Task Group Report.

21. *Joint DoD/NASA Study of Space Transportation Systems, Summary Report.*

22. *Joint DoD/NASA Study of Space Transportation Systems, Summary Report.*

23. *Joint DoD/NASA Study of Space Transportation Systems, Summary Report.*

24. *Joint DoD/NASA Study of Space Transportation Systems, Summary Report.*

25. *Joint DoD/NASA Study of Space Transportation Systems, Summary Report.*

26. Leroy E. Day, manager, Summary Report of Recoverable Versus Expendable Booster, Space Shuttle Studies, (Washington DC: NASA, 10 December 1969)

27. Spiro T. Agnew, *The Post-Apollo Space Program: Directions for the Future*, (Space Task Group, Washington DC, September 1969), p. 15; conveniently reprinted in *Exploring the Unknown, Volume I: Organizing for Exploration*, NASA SP-4407, (Washington DC: NASA, 1995), pp. 522-543.

28. *The Post-Apollo Space Program: Directions for the Future,*

29. *The Space Shuttle Decision*, p. 235.

30. *The Space Shuttle Decision*, pp. 235-236.

31. *Space Shuttle History*, pp. 94-96.

32. Statement of Work, Space Shuttle Main Engine Phase A, (Hunstville, Alabama, October 1968).

33. Statement of Work, Space Shuttle Main Engine Phase A.

34. Integral Launch and Reentry Vehicle Systems: Final Report, report MDC-E0049 (multiple volumes), (Saint Louis, Missouri: McDonnell Douglas Astronautics Company, November 1969); Gib G: Logistic Spacecraft Evolving from Gemini, report MDC-H321, (Saint Louis, Missouri: McDonnell Douglas Astronautics Company, 21 August 1969); A Two-Stage Fixed-Wing Space Transportation System, report MDC-E0056, (Saint Louis, Missouri: McDonnell Douglas Astronautics Company, 15 December 1969).

35. Integral Launch and Reentry Vehicle: Second Interim Review, report SP-69-18, (Downey, California: North American Rockwell Space Division, 13 August 1969); Integral Launch and Reentry Vehicle: Second Final Report, report SD-69-573 (multiple volumes), (Downey, California: North American Rockwell Space Division, 22 December 1969).

36. Space Transport and Recovery System, report LMSC-A946632, (Sunnyvale, California: Lockheed, March 1969); Space Shuttle: Integral Launch and Reentry Vehicle, Final Report, report LMSC-A959837, (Sunnyvale, California: Lockheed, 22 December 1969).

37. Space Shuttle Final Technical Report, report GDC-DCB69-046 (multiple volumes), (San Diego, California: General Dynamics/Convair, 23 December 1969).

38. The Convair system was the subject of a very in-depth evaluation by The Aerospace Corporation. See R. T. Blake, *Space Transportation System Fully Recoverable Two-Stage Earth Orbit Shuttle Weight Analysis*, report TOR-0066(5759-02)-2, 15 May 1970, (El Segundo, California: The Aerospace Corporation, 15 May 1970).

39. SpaceMaster: A Two-Stage Fully-Reusable Space Transportation System, report MCR-69-36 (multiple volumes), (Denver, Colorado: Martin Marietta Denver Division, December 1969); Bernard Spencer, Jr., et. al., *Low-Subsonic Longitudinal Aerodynamic Characteristics of a Twin-Body Space-Shuttle Booster Configuration*, report TM-X-2162, (Langley Field, Virginia: NASA, April 1971); "This We Call Experience ..." a promotional brochure produced by Martin Marietta in support of their Phase A SpaceMaster concept, Undated (but probably late 1969, early 1970).

40. *The Post-Apollo Space Program: Directions for the Future*, pp. 522-543.

41. *The Next Decade in Space*, prepared by the Panel on Space Science and Technology, (Washington DC: PSAC, 1970).

42. Richard P. Hallion and James O. Young, *Space Shuttle: Fulfillment of a Dream*, Case VIII of The Hypersonic Revolution: Case Studies in the History of Hypersonic Technology, Volume 1, From Max Valier to Project PRIME (1924-1967), Air Force Histories and Museums Program, (Bolling AFB DC: U.S. Air Force, 1998), pp. 1032-1034.

43. John M. Logsdon, "The Space Shuttle Program: A Policy Failure?," *Science* (Volume 232), p. 1100; *The Space Shuttle Decision*, pp. 270-274.

44. Oskar Morgenstern and Klaus P. Heiss, *Economic Analysis of New Space Transportation* Systems, (Princeton, New Jersey,: Mathematica, 31 May 1971); *Integrated Operations/Payloads/Fleet Analysis Final Report*, report ATR-72(7231)-1, (El Segundo, California: The Aerospace Corporation, 1 August 1971); *Economic Analysis of the Space Shuttle System*, (Washington DC: Mathematica, 31 January 1972), conveniently reprinted in *Exploring the Unknown, Volume IV: Accessing Space*, NASA SP-4407, (Washington DC: NASA, 1999), pp. 239-244; "The Space Shuttle Program: A Policy Failure?," p. 1100; *The Space Shuttle Decision*, pp. 275-280.

45. "The Space Shuttle Program: A Policy Failure?," pp. 1099-1101.

46. R. D. Shaver, et. al, *The Space Shuttle as an Element in the National Space Program*, report RM-6244-1-PR, (Santa Monica, California, Project RAND, October 1970), see p. v for the quote.

47. *The Space Shuttle as an Element in the National Space Program*, pp. 29-33.

48. *The Space Shuttle as an Element in the National Space Program*, pp. 29-33.

49. *The Space Shuttle as an Element in the National Space Program*, pp. 29-33.

50. John M. Logsdon, "The Development of International Space Cooperation," *Exploring the Unknown, Volume II: External Relationships*, NASA SP-4407, (Washington DC: NASA, 1996), pp. 1-6.

51. "The Development of International Space Cooperation," pp. 5-10, 48.

52. "The Development of International Space Cooperation," pp. 7-10, 68-86.

53. "The Development of International Space Cooperation," pp. 80-87.

54. Jacob Neufeld, The Air Force in Space: 1970-1974, (Washington DC: Office of Air Force History, August 1976), p. 12; "The Development of International Space Cooperation," pp. 80-87; the quote from Herman Pollack is found on page 83.

55. "The Development of International Space Cooperation," pp. 80-87.

56. Minutes of the Joint USAF-NASA Shuttle Coordination Board, (Washington DC: STS Committee, 2 October 1970); "The Space Shuttle Program: A Policy Failure?," pp. 1099-1101.

57. I. Rattinger, Summary Report USAF/NASA Manned Space Flight Studies, report TOR-0059(6531-01)-1, (Los Angeles: The Aerospace Corp., 21 August 1970).

58. Summary Report USAF/NASA Manned Space Flight Studies.

59. See most of the individual Phase A and Phase B studies, in the files at KSC, JSC, and MSFC, and most are also at HQ. See also, *Integrated Abort Analyses for a Fully Reusable Space Shuttle*, MSC internal note 71-FM-339, (Houston, Texas: Contingency Analysis Section of the Flight Analysis Branch, 14 September 1971).

60. Summary Report USAF/NASA Manned Space Flight Studies.

61. Summary Report USAF/NASA Manned Space Flight Studies.

62. "The Space Shuttle Program: A Policy Failure?," p. 1101.

63. *Space Shuttle: Fulfillment of a Dream*, p. 1041.

64. "The Space Shuttle Program: A Policy Failure?," p. 1101.

65. *DC-3 Space Shuttle Study*, (Houston, Texas: MSC Engineering and Development Directorate, 27 April 1970); MSC ILRV Space Shuttle, a presentation dated 13 August 1969 in the files of the JSC History Office.

66. 1/10-scale Shuttlecraft Model Drop Test, report TOP-69-42, (Houston, Texas: MSC System Test Division, May 1970)

67. *Space Shuttle: Fulfillment of a Dream*, pp. 1032-1033.

68. Maxine Faget and Milton Silveira, *Fundamental Design Considerations for an Earth-surface to Orbit Shuttle*, (Washington DC: CASI 70A-44618, October 1970).

69. *The Space Shuttle Decision*, p. 210.

70. *The Space Shuttle Decision*, pp. 211-212.

71. *The Space Shuttle Decision*, pp. 211-212.

72. *The Space Shuttle Decision*, pp. 211-212.

73. *The Space Shuttle Decision*, pp. 211-212.

74. Minutes of the Joint USAF-NASA Shuttle Coordination Board, (Washington DC: STS Committee, 12 February 1970)

75. *The Space Shuttle Decision*, p. 226.

76. "NASA Space Shuttle Contracts," a summary of contracts issued, 16 April 1971.

77. *Study Control Document; Space Shuttle System Program Definition (Phase B)*, 15 June 1970, (Washington DC: NASA OMSF, 15 June 1970).

78. Gerald A. Kraft and John B. Whitlow, Jr., "Design Point Study of Auxiliary Airbreathing Engines for a Space Shuttle," NASA TM-X-52810, (Cleveland, Ohio: NASA Lewis Research Center, May 1970).

79. *Study Control Document; Space Shuttle System Program Definition (Phase B)*, 15 June 1970, (Washington DC: NASA OMSF, 15 June 1970).

80. W. F. Wilhelm, "Space Shuttle Orbiter Main Engine Design," a paper (720807) presented in the Society of Automotive Engineers Transactions, Vol. 81, 1982; RFP SSME-70-1, 1 March 1971.

81. *Space Shuttle Main Engine Orientation*, report GP70-271, (West Palm Beach, Florida: Pratt & Whitney, 12 August 1970).

82. Bob Biggs, "Space Shuttle Main Engine: The First Ten Years," a paper written for the American Astronautical Society, National Conference, Los Angeles, California, 2 November 1989; *The Space Shuttle Decision*, p. 240.

83. *The Space Shuttle Decision*, p. 242.

84. For the decision not to use 50 percent power, see the various Phase A and B airframe proposals in the files at KSC, JSC, MSFC, and HQ; see amendments to RFP SSME-70-1 in the files at MSFC for the digital controller requirement.

85. *Study Control Document; Space Shuttle System Program Definition (Phase B)*, 15 June 1970, (Washington DC: NASA OMSF, 1 September 1970).

86. See the various mid-term reports issued by both Phase B airframe contractors for the results of trade studies, for example: Space Shuttle Phase B Systems Study: Mid-Term Review, presented at MSFC, (Saint Louis, Missouri: McDonnell Douglas Corporation, 11 December 1970); Fully Reusable Space Shuttle, report SV-71-28 (Downey, California: North American Rockwell Space Division, 19 July 1971).

87. *The Space Shuttle Decision* p. 233.

88. Proposal to Accomplish Phase B Space Shuttle Program, proposal MDC-E0120, (Saint Louis, Missouri: McDonnell Douglas Corporation, 30 March 1970); Space Shuttle Phase B Systems Study: Mid-Term Review, presented at MSFC, (Saint Louis, Missouri: McDonnell Douglas Corporation, 11 December 1970); Space Shuttle System Phase B Study Final Report, report MDC-E0308, (Saint Louis, Missouri: McDonnell Douglas Corporation, 30 June 1971).

89. Proposal to Accomplish Phase B Space Shuttle Program, proposal SD-70-5, (Downey, California: North American Rockwell Space Division, 27 March 1970); Space Shuttle Program Review, report SD-70-9, (Bonn, Germany: North American Rockwell Space Division, 7 July 1970); Fully Reusable Space Shuttle, report SV-71-28 (Downey, California: North American Rockwell Space Division, 19 July 1971); Space Shuttle Phase B Final Report, (Downey, California: North American Rockwell Space Division, 25 June 1971); Space Shuttle Baseline Descriptions, report SV-71-1, (Downey, California: North American Rockwell Space Division, 25 January 1971); Space Shuttle Booster Configuration Trade Study, report GDC-ACX70-001, (Fort Worth, Texas: General Dynamics/Convair, 15 September 1970); Delta and Stowed VS. Fixed Straight Wing Booster Trade Study, report GDC-76-546-10-001, (Fort Worth, Texas: General Dynamics/Convair, December 1970); Space Shuttle Boost Facilities Utilization and Manufacturing Plan, DRD TM001M, (Fort Worth, Texas: General Dynamics/Convair, 19 February 1971), p. 1-100.

90. "NASA's Internal Organization for the Space Shuttle Project," 10 June 1970, conveniently reprinted in *Exploring the Unknown, Volume IV: Accessing Space*, NASA SP-4407, (Washington DC: NASA, 1999), pp. 250-251.

91. "NASA's Internal Organization for the Space Shuttle Project," pp. 250-251.

92. *Space Shuttle: Fulfillment of a Dream*, p. 1038.

93. *The Space Shuttle Decision*, p. 332.

94. *The Space Shuttle Decision*, p. 221.

95. *The Space Shuttle Decision* p. 265.

96. *Project SERV: A Space Shuttle Feasibility Study*, report AE-PB-69-51 (New Orleans, Louisiana: Chrysler Corp. Space Division, 19 November 1969); *Final Report on Project Single-Stage Earth-Orbital Reusable Vehicle*, report TR-P-71-4, (New Orleans, Louisiana: Chrysler Corp. Space Division, 30 June 1971).

97. *Study of Alternate Space Shuttle Concepts*, report LMSC-A989142, (Sunnyvale, California: Lockheed, 4 June 1971); <u>Alternate Space Shuttle Concepts Study: Interim Review Presentation</u>, report LMSC-A991394, (Sunnyvale, California: Lockheed, 1 September 1971); <u>Alternate Space Shuttle Concepts Study</u>, report LMSC-A995887, (Sunnyvale, California: Lockheed, 3 November 1971)

98. <u>Proposal to Accomplish Phase B Space Shuttle Program</u>, proposal 70-35NAS, (Bethpage, New York: Grumman Aerospace Corporation, 30 March 1970); <u>Alternate Space Shuttle Concepts Mid-Term Report</u>, report B35-43RP-5 (multiple volumes), (Bethpage, New York: Grumman Aerospace Corporation, 31 December 1971). <u>Alternate Space Shuttle Concepts Study Final Report</u>, report B35-43RP-11 (multiple volumes), (Bethpage, New York: Grumman Aerospace Corporation, 6 July 1971); <u>Alternate Space Shuttle Concepts Study: Design requirements and Phase Program Evaluation</u>, report B35-43RP-21, (Bethpage, New York: Grumman Aerospace Corporation, 1 September 1971).

99. *Study Control Document; Space Shuttle System Program Definition (Phase B)*, 15 April 1971, (Houston, Texas: NASA, 15 April 1971).

100. J. Stanley-Dobrzanski, "Integrated Launch Reentry Vehicle Configuration Study" (Hawthorne, California: Northrop Corporation, 1969).

Chapter VI – Lower Expectations. Pages 139–194

1. John Mauer interview with Charles Donlan, Washington DC, 19 October 1983, pp. 19-20, in the files of the NASA History Office.

2. T. A. Heppenheimer, *The Space Shuttle Decision: NASA's Search for a Reusable Space Vehicle*, NASA SP-4221, (Washington DC: NASA, 1999), p. 358.

3. Dennis R. Jenkins, *Lockheed YF-12/SR-71 Blackbird*, Volume 10 in the <u>WarbirdTECH Series</u>, (North Branch, Minnesota: Specialty Press, 1996).

4. John Mauer interview with Charles Donlan, Washington DC, 19 October 1983, pp. 19-20, in the files of the NASA History Office.

5. John M. Logsdon, "The Space Shuttle Program: A Policy Failure?," <u>Science</u>, (Volume 232, 30 May 1986), p. 1101; *The Space Shuttle Decision*, pp. 286-287.

6. "The Space Shuttle Program: A Policy Failure?," p. 1101; *The Space Shuttle Decision*, pp. 288-289.

7. "The Space Shuttle Program: A Policy Failure?," p. 1101; *The Space Shuttle Decision*, pp. 287-288.

8. "The Space Shuttle Program: A Policy Failure?," p. 1101.

9. *The Space Shuttle Decision*, p. 338.

10. <u>Alternate Space Shuttle Concepts Study</u>, (multiple volumes), Executive Summary, report B35-43RP-11, (Bethpage, New York: Grumman Aerospace Corporation, 6 July 1971); *The Space Shuttle Decision*, p. 338.

11. <u>Alternate Space Shuttle Concepts Study</u>.

12. <u>Alternate Space Shuttle Concepts Study</u>.

13. Richard P. Hallion and James O. Young, *Space Shuttle: Fulfillment of a Dream*, Case VIII of <u>The Hypersonic Revolution: Case Studies in the History of Hypersonic Technology, Volume 1, From Max Valier to Project PRIME (1924-1967)</u>, Air Force Histories and Museums Program, (Bolling AFB DC: U.S. Air Force, 1998), pp. 1047-1048.

14. *Space Shuttle: Fulfillment of a Dream*, pp. 1048-1051.

15. Space *Shuttle: Fulfillment of a Dream*, pp. 1048-1051.

16. Letter, Grant L. Hansen, USAF, to Dale D. Meyers, NASA, 21 June 1971; "The Space Shuttle Program: A Policy Failure?," p. 1102.

17. *Space Shuttle: Fulfillment of a Dream*, pp. 1048-1051; *The Space Shuttle Decision*, p. 341.

18. James C. Young and Jimmy M. Underwood, *The Aerodynamic Challenges of the Design and Development of the Space Shuttle Orbiter*, a paper presented at the Space Shuttle Technical Conference (CR-2342) at JSC, 28-30 June 1983. My thanks to Henry Spencer, who was gracious enough to provide a copy of this excellent paper to me.

19. *Space Shuttle: Fulfillment of a Dream*, pp. 1048-1051.

20. <u>Fully Reusable Shuttle</u>, (multiple volumes), report SV71-28, (Downey, California: North American Rockwell, 19 July 1971); <u>Space Shuttle System Phase B Study Final Report</u>, (multiple volumes), report MDC-E0308, (Saint Louis, Missouri: McDonnell Douglas, 30 June 1971).

21. <u>Alternate Space Shuttle Concepts Study</u>, (multiple volumes), Grumman report B35-43RP-11, (Bethpage, New York: Grumman Aerospace Corporation, 6 July 1971).

22. <u>Alternate Space Shuttle Concepts Study</u>, (multiple volumes), Grumman report B35-43RP-11, (Bethpage, New York: Grumman Aerospace Corporation, 6 July 1971).

23. <u>External LH2 Tank Study Final Report</u>, Volume I, Executive Summary, report MDC-E0376, (Saint Louis, Missouri: McDonnell Douglas, 30 June 1971).

24. Letter, James C. Fletcher, NASA, to Donald Rice, OMB, 1 June 1971; *The Space Shuttle Decision*, p. 350.

25. Letter, James C. Fletcher, NASA, to Donald Rice, OMB, 1 June 1971; NASA Press Release, "Statement by James C. Fletcher," 16 June 1971; *The Space Shuttle Decision*, p. 350.

26. *The Space Shuttle Decision*, p. 351.

27. <u>Fully Reusable Shuttle</u>, (multiple volumes), report SV71-28, (Downey, California: North American Rockwell, 19 July 1971); <u>Space Shuttle System Phase B Study Final Report</u>, (multiple volumes), report MDC-E0308, (Saint Louis, Missouri: McDonnell Douglas, 30 June 1971); <u>Alternate Space Shuttle Concepts Study</u>, (multiple volumes), report B35-43RP-11, (Bethpage, New York: Grumman Aerospace Corporation, 6 July 1971); <u>External LH2 Tank Study Final Report</u>, Volume I, Executive Summary, report MDC-E0376, (Saint Louis, Missouri: McDonnell Douglas, 30 June 1971).

28. *The Space Shuttle Decision* p. 353.

29. *The Space Shuttle Decision*, pp. 354-355.

30. <u>SR-89774 Reusable Space Launch Vehicle Report</u>, prepared by The Boeing Company, Aero-Space Division, December 1959; *The Space Shuttle Decision*, pp. 354-355.

31. J.D. Hunley, "Minuteman and Launch Vehicle Technology," a paper that will appear in *To Reach the High Frontier: Case Studies in Launch Vehicle History*, NASA SP-2001-4227, (Washington DC: NASA, 2001). I am deeply indebted to Dill Hunley for allowing me the chance to read a draft version of this paper that is an excellent description of the development of solid rocket technology. No page references are available.

32. "Minuteman and Launch Vehicle Technology."

33. "Minuteman and Launch Vehicle Technology."

34. "Minuteman and Launch Vehicle Technology."

35. "Minuteman and Launch Vehicle Technology."

36. "Minuteman and Launch Vehicle Technology."

37. "Minuteman and Launch Vehicle Technology."

38. "Minuteman and Launch Vehicle Technology;" <u>Proposal for the Solid Rocket Motor for the Space Shuttle System</u>, Volume III, Design, Development, and Verification Proposal, (Dade County, Florida: Aerojet Solid Propulsion Company, 27 August 1973), pp. I-10/25.

39. Richard F. Cottrell, "260 SL-1 Engineering Report," as printed in <u>The Aerojet-General Booster</u>, Volume X, Number 4a, October 1965; "Minuteman and Launch Vehicle Technology;" <u>Proposal for the Solid Rocket Motor for the Space Shuttle System</u>, Volume III, Design, Development, and Verification Proposal, (Dade County, Florida: Aerojet Solid Propulsion Company, 27 August 1973), pp. I-10/25.

40. Claude E. Barfield, "Space Report/NASA Feels Pressure in Deciding on Location For its Space Shuttle Base," <u>CPR National Journal</u>, 24 April 1971, p. 869.

41. "Space Report/NASA Feels Pressure," p. 869.

42. "Space Report/NASA Feels Pressure," pp. 869-870; various Phase A and Phase B contractor reports in the files at KSC, JSC, and MSFC.

43. "Space Report/NASA Feels Pressure," pp. 869-870; various Phase A and Phase B contractor reports in the files at KSC, JSC, and MSFC.

44. House legislative report 91-929, attached to the FY71 NASA budget authorization (HR-17548), 19 March 1970; "Space Report/NASA Feels Pressure," pp. 873-875.

45. "Space Report/NASA Feels Pressure," pp. 873-874.

46. <u>The Space Shuttle at White Sands: A National Range</u>, NASA S69-4052, (Las Cruces, New Mexico: NASA, 1969) – I am very indebted to Glen Swanson at the JSC History Office and Marta-Marie Giles at the JSC Main Library for finding this very obscure document that I first saw hanging on the wall at the Alamogordo Space Museum.

47. Don J. Green, "Space Shuttle Operational Site Selected," NASA press release 72-81, 14 April 1972.

48. "Space Shuttle Operational Site Selected."

49. Letter, John S. Foster, DoD, to James C. Fletcher, NASA, 13 April 1972, in the files of the NASA History Office.

50. G. S. Canetti, study manager, <u>Safety in Earth Orbit Study, Final Report</u>, Volume 1, Technical Summary, report MSC-04477 / SD72-SA-0094-1, (Downey, California: North American Rockwell Space Division, 12 July 1972), pp. intro and 1-3.

51. <u>Safety in Earth Orbit Study, Final Report</u>, pp. 3-4.

52. <u>Safety in Earth Orbit Study, Final Report</u>, pp. 10-14.

53. <u>Safety in Earth Orbit Study, Final Report</u>, pp. 10-14.

54. Safety in Earth Orbit Study, Final Report, pp. 15-16.

55. Safety in Earth Orbit Study, Final Report, pp. 41-42 and 48-50.

56. Safety in Earth Orbit Study, Final Report, pp. 43-45.

57. Safety in Earth Orbit Study, Final Report, pp. 75-82.

58. Safety in Earth Orbit Study, Final Report, p. 103.

59. Safety in Earth Orbit Study, Final Report, p. 103.

60. Safety in Earth Orbit Study, Final Report, p. 103.

61. Safety in Earth Orbit Study, Final Report, pp. 103-104.

62. Safety in Earth Orbit Study, Final Report, pp. 103-104, and 121-124.

63. Safety in Earth Orbit Study, Final Report, pp. 104-105, and 121-124.

64. Safety in Earth Orbit Study, Final Report, pp. 104-106, and 124-128.

65. Safety in Earth Orbit Study, Final Report, pp. 103-129.

66. safety in Earth Orbit Study, Final Report, pp. 141-142.

67. Safety in Earth Orbit Study, Final Report, p. 142.

68. Safety in Earth Orbit Study, Final Report, p. 143.

69. Safety in Earth Orbit Study, Final Report, pp. 143-170.

70. Safety in Earth Orbit Study, Final Report, pp. 143-170.

71. Safety in Earth Orbit Study, Final Report, pp. 143-170.

72. Safety in Earth Orbit Study, Final Report, pp. 143-170.

73. Rolf W. Seiferth, Ablative Heat Shield Design for Space Shuttle, Final Report, report CR-132282, (Denver, Colorado: Martin Marietta Corporation, 17 September 1973), p. 2.

74. Ablative Heat Shield Design, pp. 6-7, 20-21, and 73.

75. Ablative Heat Shield Design, pp. 14-15.

76. Ablative Heat Shield Design, pp. 73-74.

77. Ablative Heat Shield Design, pp. 170-171.

78. Huel H. Chandler, Low-Cost Ablative Heat Shields for Space Shuttles, Final Report, report CR-111800, (Denver, Colorado: Martin Marietta Corporation, 11 January 1971).

79. D. W. Has, Summary Refurbishment Cost Study of the Thermal Protection System of a Space Shuttle Vehicle, report CR-111833, (Saint Louis, Missouri: McDonnell Douglas Astronautics, 6 April 1971).

80. W. M. Pless, Space Shuttle Structural Integrity and Assessment Study, External Thermal Protection System, report CR-134452 / LG73-ER0082, (Marietta, Georgia: Lockheed-Georgia Company, June 1973).

81. Space Shuttle Structural Integrity and Assessment Study.

82. D. Hays, An Assessment of Alternate Thermal Protection Systems for the Space Shuttle Orbiter, Volume I, Executive Summary, report CR-3548, (Downey, California: Industry Team, April 1982).

83. An Assessment of Alternate Thermal Protection Systems, pp. 4-5, and 8-9.

84. An Assessment of Alternate Thermal Protection Systems, pp. 21-24.

85. An Assessment of Alternate Thermal Protection Systems, pp. 25-27.

86. An Assessment of Alternate Thermal Protection Systems, p. 57.

87. "The Space Shuttle Program: A Policy Failure?," p. 1102.

88. The Space Shuttle Decision, pp. 361-369.

89. Letter, James C. Fletcher, NASA, to George Shultz, OMB, 30 September 1971, in the files of the NASA History Office.

90. "The Space Shuttle Program: A Policy Failure?," p. 1101; The Space Shuttle Decision, p. 356.

91. Letter, Alexander Flax, IDA, to Edward David, PSAC, 19 October 1971, in the files of the NASA History Office.

92. The Space Shuttle Decision, p. 372.

93. Letter from Alexander Flax, IDA, to Edward David, PSAC, 19 October 1971, in the files of the NASA History Office.

94. Minutes of the STS Committee, 27 October 1971, in the files of the NASA History Office.

95. Alternate Concepts Study Extension, report LMSC-A995931, (Sunnyvale, California: Lockheed, 15 November 1971); Phase B System Study Extension Final Report, report MDC-E0497, (Saint Louis, Missouri: McDonnell Douglas Astronautics, 15 November 1971); Definition of Mark I/Mark II Orbiters and Ballistic & Flyback Boosters, report B35-43RP-28, (Bethpage, New York: Grumman Aerospace, 15 November 1971); Space Shuttle Phase B Extension Mid-Term Review, report SV71-40, (Downey, California: North American Rockwell, 1 September 1971).

96. Ibid.

97. Ibid.

98. Ibid.

99. Contract Directives to each Phase B contractor dated 16 September 1971, in the files of the NASA MSFC History Office.

100. Andrew J. Dunbar and Stephen P. Waring, Power to Explore: A History of the Marshall Space Flight Center 1960-1990, NASA SP-4313, (Washington DC: NASA, 1999), pp. 284-285.

101. The Space Shuttle Decision p. 360.

102. John Mauer interview with Charles Donlan, Washington DC, 19 October 1983, pp. 23-24, in the files of the NASA History Office.

103. The Space Shuttle Decision p. 373.

104. Oskar Morgenstern and Klaus P. Heiss, Factors for a Decision on a New Reusable Space Transportation System, 28 October 1971, conveniently reprinted in Exploring the Unknown, Volume I: Organizing for Exploration, NASA SP-4407, (Washington DC: NASA, 1995), pp. 549-555.

105. Factors for a Decision, pp. 549-555.

106. Phase B System Study Extension Final Report, report MDC-E0497, (Saint Louis, Missouri: McDonnell Douglas Astronautics, 15 November 1971); Shuttle System Evaluation and Selection, report B35-43RP-30, (Bethpage, New York: Grumman Aerospace, 15 December 1971)

107. "The Space Shuttle Program: A Policy Failure?," p. 1102; The Space Shuttle Decision, pp. 381-387.

108. Letter, James C. Fletcher, NASA, to Caspar Weinberger, OMB, 19 October 1971.

109. Letter from George M. Low, NASA, to Donald Rice, OMB, 22 November 1971, in the files of the NASA History Office, conveniently reprinted in Exploring the Unknown, Volume IV: Accessing Space, NASA SP-4407, (Washington DC: NASA, 1999), pp. 231-238 (quote on p. 234).

110. Letter from George M. Low, pp. 231-238.

111. Letter from George M. Low, pp. 231-238.

112. Letter from George M. Low, pp. 231-238.

113. Memorandum, James C. Fletcher, NASA, to George M. Low, NASA, "Luncheon conversation with Dave Packard," 20 October 1971, in the files of the NASA History Office.

114. Memorandum for the Record, J. Smart, NASA Assistant Administrator for DoD and Interagency Affairs, 6 December 1971; Memorandum, George M. Low, NASA, to James C. Fletcher, NASA. "Discussions with Johnny Foster," 2 December 1971, both in the files of the NASA History Office.

115. Draft of a Letter prepared by J. Smart, would have been from James C. Fletcher, NASA to David Packard, DoD, 1 December 1971.

116. "The Space Shuttle Program: A Policy Failure?," p. 1104.

117. Memorandum, James C. Fletcher, NASA, to J. Rose, Special Assistant to the President, 22 November 1971; "The Space Shuttle Program: A Policy Failure?," p. 1104.

118. Memorandum, James C. Fletcher, NASA, to J. Rose, Special Assistant to the President, 22 November 1971; "The Space Shuttle Program: A Policy Failure?," p. 1104.

119. "The Space Shuttle Program: A Policy Failure?," p. 1104.

120. Memorandum, OMB to President Nixon, 2 December 1971, in the files of the NASA History Office.

121. "The Space Shuttle Program: A Policy Failure?," p. 1103; The Space Shuttle Decision pp. 400-401.

122. Letter from George M. Low, pp. 231-238 (quote on p. 235).

123. Letter from George M. Low, pp. 231-238.

124. Letter from George M. Low, pp. 231-238.

125. Letter from George M. Low, pp. 231-238.

126. Letter from George M. Low, pp. 231-238 (quote on p. 236).

127. Memorandum, George M. Low, NASA, to Dale D. Meyers, NASA-OMSF, 13 December 1971, in the files of the NASA History Office.

128. Letter, James C. Fletcher, NASA, to Caspar W. Weinberger, OMB, 29 December 1971, in the files of the NASA History Office, conveniently reprinted in *Exploring the Unknown, Volume IV: Accessing Space*, NASA SP-4407, (Washington DC: NASA, 1999), pp. 245-249 (quote on p. 245).

129. Letter, James C. Fletcher, NASA, to Caspar W. Weinberger, pp. 245-249.

130. Letter, James C. Fletcher, NASA, to Caspar W. Weinberger, pp. 245-249.

131. Letter, James C. Fletcher, NASA, to Caspar W. Weinberger, pp. 245-249.

132. John Mauer interview with Charles Donlan, Washington DC, 19 October 1983, pp. 19-20, in the files of the NASA History Office; "The Space Shuttle Program: A Policy Failure?," p. 1104; *The Space Shuttle Decision*, pp. 408-411.

133. Memorandum for the Record, George M. Low, NASA, "Meeting with the President on 5 January 1972," 12 January 1972, in the files of the NASA History Office; "The Space Shuttle Program: A Policy Failure?," p. 1104.

134. Richard M. Nixon, "Statement by the President," 5 January 1972; NASA Press Release 72-4, 6 January 1972.

135. *The Space Shuttle Decision*, p. 413.

136. History of the Air Force System Command, 1 July 1974 through 30 June 1975, pp. 330-335, in the files of the AFHRA K243-01.

137. Various mid-term reports and briefing by the airframe contractors, in the technical library files at KSC, JSC, and MSFC.

138. Various mid-term reports and briefing by the airframe contractors, in the technical library files at KSC, JSC, and MSFC.

139. Various mid-term reports and briefing by the airframe contractors, in the technical library files at KSC, JSC, and MSFC; "The Space Shuttle Program: A Policy Failure?," p. 1101.

140. Various mid-term reports and briefing by the airframe contractors, in the technical library files at KSC, JSC, and MSFC.

141. *The Space Shuttle Decision*, p. 417.

142. Quote from letter, Caspar W. Weinberger, OMB, to James C. Fletcher, NASA, 9 February 1972, in the files of the NASA History Office.

143. *The Space Shuttle Decision*, p. 418.

144. *The Space Shuttle Decision*, p. 417.

145. Alternate Space Shuttle Concepts Study; Design Requirements and Phased Program Evaluation, report B35-43RP-21, (Bethpage, New York: Grumman Aerospace Corporation, 1 September 1971); Shuttle Systems Evaluation and Selection; Phase B Extension Final Briefing, (Bethpage, New York: Grumman Aerospace Corporation, 16 February 1972).

146. Alternate Concepts Study Extension, report LMSC-A995931, (Sunnyvale, California: Lockheed, 15 November 1971); Space Shuttle Final Review, report LMSC-D157302, (Sunnyvale, California: Lockheed, 22 February 1972).

147. Space Shuttle Phase B Extension Mid-Term Review, report SV71-40, (Downey, California: North American Rockwell, 1 September 1971); Space Shuttle Phase B Final Briefing, report SV72-14, (Downey, California: North American Rockwell, 122 February 1972).

148. Space Shuttle Design Review: Phase B System Study Extension, (Saint Louis, Missouri: McDonnell Douglas Astronautics, 22 February 1972).

149. Space Shuttle Design Review: Phase B System Study Extension, (Saint Louis, Missouri: McDonnell Douglas Astronautics, 22 February 1972).

150. Memorandum, John Sullivan, OMB, to Donald Rice, OMB, 13 March 1972.

151. NASA News Release 72-61, "Space Shuttle Decisions," 15 March 1972.

152. "The Space Shuttle Program: A Policy Failure?," p. 1102.

153. RFP 9-BC421-67-2-40P, "Space Shuttle Program," (Houston, Texas: NASA 17 March 1972), cover letter; NASA Press Release 72-61, dated 15 March 1972 (but not released until 17 March).

154. RFP "Space Shuttle Program," pp. I-2 – I-7.

155. RFP "Space Shuttle Program," pp. I-2 – I-7.

156. RFP "Space Shuttle Program," pp. III-1 – III-2.

157. RFP "Space Shuttle Program," p. IV-5. These mission profiles had been developed during an on-going series of studies conducted at MSC. See Typical Shuttle Mission Profiles and Attitude Time Lines, MSC Internal Note 71-FM-82, (Houston, Texas: NASA, 1 March 1971) and Space Shuttle Performance Capabilities, MSC Internal Note 71-FM-350, (Houston, Texas: NASA, 16 September 1971) among others for additional information.

158. RFP "Space Shuttle Program," pp. IV-6, IV-9.

159. RFP "Space Shuttle Program," pp. IV-10, IV-12,

160. RFP "Space Shuttle Program," p. IV-18.

161. RFP "Space Shuttle Program," pp. IV-20 – IV-22.

162. RFP "Space Shuttle Program," p. IV-24.

163. Proposal for Space Shuttle Program, (multiple volumes), report 72-74-NAS, (Bethpage, New York: Grumman Aerospace Corporation, 12 May 1972).

164. Proposal to NASA-MSC: Space Shuttle, (multiple volumes), report LMSC-D157364, (Sunnyvale, California: Lockheed Missiles and Space Company, 12 May 1972).

165. Space Shuttle Program, (multiple volumes), report MDC-E0600, (Saint Louis, Missouri: McDonnell Douglas, 12 May 1972).

166. Proposal for the Space Shuttle Program, (multiple volumes), report SD72-SH-50-3, (Downey, California: North American Rockwell Space Division, 12 May 1972).

167. Memorandum for Record, James C. Fletcher, George M. Low, and Richard McCurdy, 18 September 1972, in the files of the NASA History Office, conveniently reprinted in *Exploring the Unknown, Volume IV: Accessing Space*, NASA SP-4407, (Washington DC: NASA, 1999), pp. 262-268.

168. Ibid.

169. Ibid.

170. Ibid.

171. Ibid.

172. Ibid, (quotes on pp. 266-267).

173. NASA News release, 26 July 1972; Major Thomas W. Rutten, History of the Directorate of Space DCS/Research and Development For the Period 1 July 1972 to 31 December 1972, (Los Angeles, California: U.S. Air Force, January 1973), in the files of the AFHRA K140-01.

174. *The Space Shuttle Decision,* pp. 427-428.

175. *The Space Shuttle Decision*, p. 428; Major Thomas W. Rutten, History of the Directorate of Space DCS/Research and Development For the Period 1 January 1972 to 30 June 1972, (Los Angeles, California: U.S. Air Force, July 1972), in the files of the AFHRA K140-01.

176. Andrew J. Dunbar and Stephen P. Waring, *Power to Explore: A History of the Marshall Space Flight Center 1960-1990*, NASA SP-4313, (Washington DC: NASA, 1999), pp. 287-288.

177. *Power to Explore*, pp. 288-289.

178. *Space Shuttle Engine Negotiations*, NASA News release 71-131, 13 July 1971.

179. William O'Donnell, NASA News Release 72-167, "Shuttle Engine Contract," 16 August 1972; *Power to Explore*, pp. 288-289.

180. *Power to Explore*, pp. 288-289.

181. *Power to Explore* pp. 288-289.

182. *Power to Explore* pp. 289-290.

183. *STS Crew Egress and Escape Study: Volume I – Analysis*, Report JSC-22275, (Houston: Shuttle Crew Abort Planned Escape Panel, August 1986), pp. VII-7/8.

184. *Power to Explore*, pp. 289-290.

185. Contractor Proposals for the Solid Rocket Motor Project: Aerojet, Lockheed, Thiokol, and UTC. All dated 27 August 1973, in the files of the NASA History Office.

186. Ibid

187. Proposal for the NASA Solid Rocket Motor for the Space Shuttle Program, (multiple volumes), report AS-73004099, (Dade County, Florida: Aerojet Solid Propulsion Company, 27 August 1973).

188. Ibid.

189. The Comptroller General of the United States, "Decision in the Matter of Protest by Lockheed Propulsion Company, File B-173677," 24 June 1974, in the files of the NASA History Office, conveniently reprinted in *Exploring the Unknown, Volume IV: Accessing Space*, NASA SP-4407, (Washington DC: NASA, 1999), pp. 268-272.

190. "Decision in the Matter of Protest by Lockheed," pp. 268-272.

191. *Power to Explore* pp. 290-291; "Decision in the Matter of Protest by Lockheed," pp. 268-272.

192. Linda Neuman Ezell, *NASA Historical Data Book, Volume III: Programs and Projects 1969-1978* (Washington DC: NASA, 1988), pp. 48-49; Rudy Abramson, "Lockheed Challenges Space Shuttle Award," Los Angeles Times, 10 January 1974; "Decision in the Matter of Protest by Lockheed," pp. 268-272.

193. "Decision in the Matter of Protest by Lockheed," pp. 268-272.

194. "Minuteman and Launch Vehicle Technology."

195. "Minuteman and Launch Vehicle Technology."

196. "Minuteman and Launch Vehicle Technology."

197. *Power to Explore*, pp. 291-292.

198. Memorandum, George M. Low to Dale Meyers, "Selection of External Tank Contractor," 11 October 1972, in the files of the NASA History Office.

199. *Power to Explore*, pp. 292-293.

200. John M. Ryken, *A Study of Air Cushion Landing Systems for Space Shuttle Vehicles*, report CR-111803, contract NAS1-9992, (Buffalo, New York: Bell Aerospace Company, December 1970).

201. *Space Shuttle*, NASA Fact Sheet, (Washington DC: NASA, October 1972). p. 11.

202. *Space Shuttle*, NASA Fact Sheet, (Washington DC: NASA, October 1972). p. 11.

203. *Space Shuttle*, NASA Fact Sheet, (Washington DC: NASA, October 1972). p. 11.

204. *The Aerodynamic Challenges of the Design and Development of the Space Shuttle Orbiter.*

205. *The Aerodynamic Challenges of the Design and Development of the Space Shuttle Orbiter.*

206. A. Miles Whitnah and Ernest R. Hillje, *Space Shuttle Wind Tunnel Test Summary*, NASA Reference Publication 1125 (Washington DC: NASA, 1984), pp. 5-7 and related drawings; *The Aerodynamic Challenges of the Design and Development of the Space Shuttle Orbiter.*

207. *Space Shuttle Wind Tunnel Test Summary*, pp. 5-7 and related drawings.

208. *Space Shuttle Wind Tunnel Test Summary*, pp. 5-7 and related drawings.

209. *Space Shuttle Wind Tunnel Test Summary*, pp. 5-7 and related drawings.

210. *Space Shuttle Wind Tunnel Test Summary*, pp. 5-7 and related drawings; *The Aerodynamic Challenges.*

211. *Space Shuttle Wind Tunnel Test Summary*, pp. 5-7 and related drawings.

212. *The Aerodynamic Challenges.*

213. *Space Shuttle Wind Tunnel Test Summary*, pp. 5-7 and related drawings.

214. *Space Shuttle Wind Tunnel Test Summary*, pp. 5-7 and related drawings.

215. *Space Shuttle Wind Tunnel Test Summary*, pp. 5-7 and related drawings.

216. *Space Shuttle Wind Tunnel Test Summary*, pp. 5-7 and related drawings.

217. *Space Shuttle Wind Tunnel Test Summary*, pp. 5-7 and related drawings.

218. Linda Neuman Ezell, *NASA Historical Data Book, Volume III: Programs and Projects 1969-1978* (Washington DC: NASA, 1988), pp. 122-123.

CHAPTER VII – GETTING READY. PAGES 195–258

1. John W. Paulson, Jr. *Aerodynamic Characteristics of a Large Aircraft to Transport Space Shuttle Orbiter or Other External payloads*, NASA TN-D-7962 (Hampton, Virginia: NASA Langley, August 1975).

2. Ivy Hooks, David Homan, and Paul Romere, *Aerodynamic Challenges of ALT*, a paper presented at the Space Shuttle Technical Conference (CR-2342) at JSC, 28-30 June 1983, p. 295.

3. Letter, George M. Low, NASA, to John L. McLucas, USAF, no subject (but only discussing the shuttle carrier), 24 April 1974; Letter, John L. McLucas, USAF, to George M. Low, NASA, 30 April 1974.

4. *Aerodynamic Challenges of ALT*, p. 297.

5. *Aerodynamic Challenges of ALT*, pp. 297-298.

6. D.J. Homan, D.E. Denison, and K.C. Elchart, *Orbiter/Shuttle Carrier Aircraft Separation: Wind Tunnel, Simulation, and Flight Test Overview and Results,"* NASA TM-55223 (Houston, Texas: NASA, May 1980); *Aerodynamic Challenges of ALT*, p. 298.

7. *Aerodynamic Challenges of ALT*, pp. 298-299.

8. The aircraft was officially registered as N905NA on 17 September 1974. Dennis R. Jenkins, *Boeing 747-100/200/300/SP*, Volume 6 in the AirlinerTech series (North Branch, Minnesota: Specialty Press, 2000), pp. 36-37; John Roach and Anthony B. Eastwood, *Jet Airliner Production List: Volume 1 – Boeing*, (Middlesex, UK: The Aviation Hobby Shop, October 1999), pp. 328 and 333; William A. Rice, "NASA's 747 Shuttle Carrier Aircraft Not New to Heavyweight Ranks," Boeing news release A-0919, undated (probably late 1974).

9. http://www.dfrc.nasa.gov/gallery/747.html, accessed on 1 October 2000.

10. http://www.dfrc.nasa.gov/gallery/747.html, accessed on 1 October 2000.

11. http://www.dfrc.nasa.gov/gallery/747.html, accessed on 1 October 2000.

12. *Aerodynamic Challenges of ALT*, pp. 298-299.

13. "NASA's 747 Shuttle Carrier Aircraft Not New to Heavyweight Ranks;" William A. Rice, Boeing Background Information Sheet A-0776, "747 Shuttle Orbiter Carrier," April 1986.

14. Boeing Background Information Sheet A-0776.

15. Boeing Background Information Sheet A-0776.

16. boeing Background Information Sheet A-0776; NASA Fact Sheet FS-013-95(09)-DFRC, "Shuttle Carrier Aircraft (SCA)," August 1995.

17. Aerodynamic *Challenges of ALT.*

18. *Space Transportation System Background Information* (Washington DC: NASA, 1985), pp. 551-554.

19. Boeing Background Information Sheet A-0776.

20. Teledyne McCormick Selph Fact Sheet "Emergency Crew Escape System for NASA's 747 Shuttle Carrier Aircraft," undated (probably mid-1975); *Space Shuttle Transportation System*, Press Information (Downey, California: Rockwell Space Division, February 1977).

21. *Aerodynamic Challenges of ALT*; email between the author and Larry Buscayart (DFRC Shuttle Support Office) and Alan Brown (DFRC Public Affairs), 13 October 2000.

22. The aircraft was officially registered as N911NA on 28 October 1988; *Boeing 747-100/200/300/SP*, pp. 36-37; *Jet Airliner Production List*, pp. 328 and 333; William P. Rogers, Chairman, *Report of the Presidential Commission on the Space Shuttle Challenger Accident*, Volume 1 (Washington DC: Presidential Commission, 6 August 1986), p. 200.

23. Email between the author and Larry Buscayart (DFRC Shuttle Support Office) and Alan Brown (DFRC Public Affairs), 13 October 2000.

24. NASA Fact Sheet FS-013-95(09)-DFRC.

25. Posting to sci.space.shuttle by Mary Shafer, DFRC, 11 January 2000; email between author and Mary Shafer confirming same topic, 3 October 2000.

26. The two original STAs (msn 146 and 147, respectively, both 1974 builds) appear to have been purchased as new aircraft. The three follow-on Gulfstreams were purchased on the used aircraft market. (N944NA is msn 144, built in 1974; N945NA is msn 118, 1972; and N498NA is msn 222, 1978). http://www.iiscorp.com/projects/control/sta/, accessed on 30 September 2000; Tom A. Heppenheimer, *History of the Space Shuttle – Volume II: Development of the Shuttle, 1972-1981*, a draft manuscript for a proposed second volume of the official NASA shuttle history. Chapter 9 in the 27 April 1999 version.

27. "Shuttle Landing Simulations to Improve with Smart Software," NASA news release 97-229, 14 October 1997; *History of the Space Shuttle – Volume II*, Chapter 3 in the 27 April 1999 version.

28. Conversation between the author and Todd A. Downey, JSC aircraft manager, 2 October 2000; http://www.wstf.nasa.gov/WSSH/default.htm, accessed 30 September 2000.

29. Memorandum for the Record, Gerald J. Mossinghoff, NASA Assistant General Counsel for General Law, subject: Classification of the Space Shuttle as a "Space Vehicle" and not an "Aircraft," 25 September 1975, in the files of the NASA History Office, also conveniently reprinted in *Exploring the Unknown, Volume IV: Accessing Space*, NASA SP-4407, (Washington DC: NASA, 1999), pp. 272-274.

30. "Space Vehicle" and not an "Aircraft," pp. 272-274.

31. C. Thomas Modlin, Jr. and George A. Zupp, Jr., *Shuttle Structural Dynamics Characteristics, the Analysis and Verification*, a paper presented at the Space Shuttle Technical Conference (CR-2342) at JSC, 28-30 June 1983, pp. 325-334.

32. *Shuttle Structural Dynamics Characteristics*, pp. 325-328.

33. *Shuttle Structural Dynamics Characteristics*, pp. 325-334.

34. *Shuttle Structural Dynamics Characteristics*, p. 325.

35. *Shuttle Structural Dynamics Characteristics*, pp. 325-326.

36. Donlad H. Emero, *The Quarter-Scale Space Shuttle Design, Fabrication, and Tests*," Spacecraft & Rockets, Volume 17, Number 4, July-August 1980; B.H. Ujihara, R.J. Guyan, et. al., *Baseline Quarter-Scale Ground Vibration Test*, report STS80-0187 (Downey, California: Rockwell International, July 1980); *Shuttle Structural Dynamics Characteristics*, pp. 325-334.

37. *Shuttle Structural Dynamics Characteristics*, pp. 330-331; Conversation by the author with Dr. Valerie Neal, NASM, and Glen Swanson, JSC History Office.

38. *Shuttle Structural Dynamics Characteristics*, pp. 325-326.

39. *Shuttle Structural Dynamics Characteristics*, pp. 330-331.

40. *Baseline Quarter-Scale Ground Vibration Test; Shuttle Structural Dynamics Characteristics*, pp. 330-331.

41. *Astronautics and Aeronautics*, September 1976, pp. 210-211.

42. *Shuttle Structural Dynamics Characteristics*, pp. 328-329; *Space Shuttle Transportation System*, Press Information (Downey, California: Rockwell Space Division, February 1977).

43. *Shuttle Structural Dynamics Characteristics*, pp. 328-329.

44. *Space Shuttle Transportation System*, Press Information.

45. "OV-101 Structural Integrity Evaluation," Final Report, 26 February 1996, supplied by Julie Kramer, JSC; Charles H. Lowry, "Crew Escape from Space Shuttle Orbiter," SAFE 15th Annual Symposium, December 1972, pp. 169-172; *Space Shuttle Transportation System*, Press Information.

46. *Orbiter/747 Carrier Separation Aerodynamic Data Book: SDM Baseline*, Rockwell report SD75-SH-0033C (Downey, California: Rockwell Space Division, November 1976); *Aerodynamic Challenges of ALT*, p. 297.

47. NASA Press Kit 77-16, "Space Shuttle Orbiter Test Flight Series," 4 February 1977; *Aerodynamic Challenges of ALT*, p. 297; *Space Shuttle Transportation System*, Press Information.

48. *Space Shuttle Transportation System*, Press Information.

49. Personal recollection by the author; *Space Shuttle Transportation System*, Press Information.

50. *Phase I Summary: Space Shuttle Orbiter/747 Shuttle Carrier Aircraft (SCA) Approach and Landing Tests (Orbiter unmanned and systems inactive)*, (Downey, California: Rockwell Space Division, 2 August 1977); *Space Transportation System Background Information* (Washington DC: NASA, 1985), pp. 551-556.

51. *Phase I Summary*; ALT Captive-Inert Test Press Conference, 18 February 1977, with Donald K. Slayton, Fitzhugh Fulton, and Bill Andrews.

52. *Phase I Summary*.

53. *Phase I Summary*.

54. *Phase I Summary*.

55. *Phase I Summary*.

56. Richard P. Hallion and James O. Young, *Space Shuttle: Fulfillment of a Dream*, Case VIII of The Hypersonic Revolution: Case Studies in the History of Hypersonic Technology, Volume 1, From Max Valier to Project PRIME (1924-1967), Air Force Histories and Museums Program, (Bolling AFB DC: U.S. Air Force, 1998), pp. 1149-1150; *AFFTC Evaluation of the Space Shuttle Orbiter and Carrier Aircraft, NASA Approach and Landing Test*, AFFTC-TR-78-14, (Edwards AFB, California: AFFTC, May 1978).

57. *Space Shuttle: Fulfillment of a Dream*, pp. 1153-1154; *AFFTC Evaluation*.

58. *Space Shuttle: Fulfillment of a Dream*, pp. 1154-1155; *AFFTC Evaluation*.

59. *Space Shuttle: Fulfillment of a Dream*, pp. 1154-1155; *AFFTC Evaluation*.

60. *Space Shuttle: Fulfillment of a Dream*, pp. 1155-1156; *AFFTC Evaluation*; Donald E. Fink, "Orbiter Experiences Control Problems," Aviation Week & Space Technology, 31 October 1977, pp. 16-17.

61. Aerodynamic *Challenges of ALT*, pp. 295-302.

62. *Space Transportation System Background Information*, pp. 551-554.

63. C. Thomas Modlin, Jr. and George A. Zupp, Jr., *Shuttle Structural Dynamics Characteristics, the Analysis and Verification*, a paper presented at the Space Shuttle Technical Conference (CR-2342) at JSC, 28-30 June 1983, p. 328.

64. *Shuttle Structural Dynamics Characteristics*, pp. 329-332; Amos Crisp, NASA Fact Sheet 78-47, "Space Shuttle Ground Vibration Tests," March 1978.

65. *Shuttle Structural Dynamics Characteristics*, pp. 329-332; "Space Shuttle Ground Vibration Tests."

66. "Space Shuttle Ground Vibration Tests;" "Shuttle Ground Vibration Tests End," NASA Marshall news release 79-21, 28 February 1979.

67. *Shuttle Structural Dynamics Characteristics*, p. 331.

68. Conversation between the author and Thomas L. Moser, former manager of the Orbiter Structures Group at JSC, 30 September 2000.

69. *Space Transportation System Background Information*, pp. 551-554.

70. NASA Fact Sheet, "Main Propulsion Test Article," undated (probably March 1988); "NASA Asks Museum to Give Shuttle Parts Back," Reuters News Service, 14 February 1999; *Space Transportation System Background Information*.

71. *Space Transportation System Background Information*, p. 560; Conversation by author with Julie Kramer and Chris Hansen at JSC, 26 September 2000.

72. *Space Transportation System Background Information*; Personal recollections by the author.

73. *Space Transportation System Background Information*; Personal recollections by the author.

74. "NASA to Conduct Initial Low-Speed Tests of Shuttle Groundroll Arresting System," Aviation Week & Space Technology, 8 June 1987, p. 20; Space Shuttle Flight and Ground Specification, NSTS-07700, Volume X, Book 3, Requirements for Runways and Navigational Aids, Revision N, 7 September 1995, p. 2-2; "Work Statement/Technical Specification for Space Shuttle Orbiter Arresting System (SOAS), All American Engineering, Inc. 1989.

75. Conversations between the author and Dr. Neal, Glenn Swanson at the JSC History Office, and Julie Kramer, JSC vehicle engineering, September 2000; *Space Transportation System Background Information*.

76. Letter from Ronald D. Dittemore, NASA shuttle program manager, to Dr. Valerie Neal, NASM Curator of OV-101, 4 June 1997; conversations between the author and Dr. Neal, Glenn Swanson at the JSC History Office, and Julie Kramer, JSC vehicle engineering, September 2000; conversations between the author and Carlisle C. Campbell, Jr. at the JSC Orbiter Projects Office, 2October 2000; email between the author and Dr. Neal, 19 October 2000.

77. Letter from D. M. Germany, NASA orbiter manager, to Dr. Valerie Neal, NASM Curator of OV-101, 25 May 1993; conversations between the author and Dr. Neal, Glenn Swanson at the JSC History Office, and Julie Kramer, JSC vehicle engineering, September 2000; email between the author and Dr. Neal, 19 October 2000.

78. "OV-101 Enterprise Assessment Trip Report," 9 February 1996, supplied by Glen Swanson, JSC History Office; "OV-101 Structural Integrity Evaluation," Final Report, 26 February 1996, supplied by Julie Kramer, JSC.

79. Letter from Ronald D. Dittemore, NASA shuttle program manager, to Dr. Valerie Neal, NASM Curator of OV-101, 19 June 1997; conversations between the author and Dr. Neal, Glenn Swanson at the JSC History Office, and Julie Kramer, JSC vehicle engineering, September 2000.

80. Email between the author and Christopher P. Hansen, JSC, various dates in 1999 and 2000.

81. Email between the author and Christopher P. Hansen, JSC, various dates in 1999 and 2000.

82. Email between the author and Christopher P. Hansen, JSC, various dates in 1999 and 2000.

83. Email between the author and Christopher P. Hansen, JSC, various dates in 1999 and 2000.

84. Email between the author and Christopher P. Hansen, JSC, various dates in 1999 and 2000.

85. Email between the author and Christopher P. Hansen, JSC, various dates in 1999 and 2000; Email between the author and Elric N. McHenry, 27 October 2000.

86. Email between the author and Christopher P. Hansen, JSC, various dates in 1999 and 2000.

87. Letter from Ronald D. Dittemore, NASA shuttle program manager, to Dr. Valerie Neal, NASM Curator of OV-101, 1 March 1999; conversations between the author and Dr. Neal, Glenn Swanson at the JSC History Office, and Julie Kramer, JSC vehicle engineering, September 2000; Master Change Request 19358 "FRCS Interconnect System (FICS); email between the author and Kevin C. Templin, FICS project manager at JSC, 2 November 2000."

88. Letter from Ralph R. Roe, NASA Space Shuttle Vehicle Engineering Office, to Dr. Valerie Neal, NASM Curator of OV-101, 1 March 1999; conversations between the author and Dr. Neal, Glenn Swanson at the JSC History Office, and Julie Kramer, JSC vehicle engineering, September 2000; email between the author and Kevin C. Templin, FICS project manager at JSC, 2 November 2000.

89. Conversations between the author and Dr. Valerie Neal, NASM and Julie Kramer, JSC, various dates in September 2000.

90. http://www.nasm.edu, accessed on 27 September 2000; Conversations between the author and Dr. Valerie Neal, NASM, September 2000; "Out Future Takes Off," a promotional brochure produced by the NASM, September 2000.

91. Philip C. Glynn and Thomas L. Moser, "Orbiter Structural Design and Verification," a paper presented at the Space Shuttle Technical Conference (CR-2342) at JSC, 28-30 June 1983, p. 345.

92. Test Data Base (computerized report generator), Holloman AFB; "Orbiter Structural Design and Verification," pp. 345-356; Don J. Green, NASA News Release 73-116, "First Space Shuttle Orbiter Test Article Nearing Completion," 10 September 1973.

93. *Space Transportation System Background Information*, pp. 550-560; *History of the Space Shuttle – Volume II*, chapter 7 in the 27 April 1999 version.

94. *Space Transportation System Background Information*, pp. 550-560; *History of the Space Shuttle – Volume II*, chapter 7 in the 27 April 1999 version.

95. Details on both of these test programs, along with additional details on the other development and test efforts, may be found in Tom A. Heppenheimer, *History of the Space Shuttle – Volume II: Development of the Shuttle, 1972-1981*, when it is published by the NASA History Series in 2001.

96. Bob Biggs, "Space Shuttle Main Engine: The First Ten Years," a paper presented at the American Astronautical Society National Conference & Annual Meeting, 2 November 1989, Los Angeles, California.

97. "Space Shuttle Main Engine: The First Ten Years;" comments to draft manuscript by Bob Biggs, Rocketdyne, 26 December 2000.

98. "Space Shuttle Main Engine: The First Ten Years."

99. *Space Shuttle: Fulfillment of a Dream*, pp. 1157-1158; "Space Shuttle Main Engine: The First Ten Years."

100. "Space Shuttle Main Engine: The First Ten Years;" *Space Shuttle: Fulfillment of a Dream*, p. 1158.

101. "Space Shuttle Main Engine: The First Ten Years."

102. "Space Shuttle Main Engine: The First Ten Years."

103. "Space Shuttle Main Engine: The First Ten Years."

104. "Space Shuttle Main Engine: The First Ten Years;" *Space Shuttle: Fulfillment of a Dream*, p. 1158.

105. "Space Shuttle Main Engine: The First Ten Years."

106. Michael S. Maklkin, teletype RUEANAT0234, "SSME Development Planning," 15 February 1978; comments on manuscript by Bob Biggs, 26 December 2000.

107. "Space Shuttle Main Engine: The First Ten Years;" *Space Transportation System Background Information*, p. 563.

108. "Space Shuttle Main Engine: The First Ten Years;" *Space Transportation System Background Information*, p. 563.

109. *Space Shuttle Solid Rocket Booster*, SA44-80-2, (MSFC, Alabama: NASA, December 1980).

110. "Static test Information," provided by Thiokol Propulsions communications group, 2 November 2000; *Space Shuttle Solid Rocket Booster*.

111. *Space Shuttle Solid Rocket Booster*.

112. *Space Shuttle Solid Rocket Booster*.

113. *Space Shuttle Solid Rocket Booster*.

114. "Booster Improvements," MSFC Fact Sheet 60F1284, December 1984.

115. *History of the Space Shuttle – Volume II*, chapter 5 in the 27 April 1999 version.

116. Mike Starr, "Shuttle Structural Hardware Shipped to Marshall Center," MSFC news release 77-30, 25 February 1977; Curtis Hunt, "First Shuttle Liquid Oxygen Tank Pressure-Tested," MSFC news release 77-58, 7 April 1977; Curtis Hunt, "Liquid Hydrogen Tank Completes Pressure Tests," MSFC news release 77-83, 10 May 1977; Amos Crisp, "External Tank Structural Tests," MSFC new release 77-149, 17 August 1977; Amos Crisp, "External Tank Segment Successfully Tested," MSFC news release 77-212, 11 November 1977; Amos Crisp, "Complete External Tank to Arrive," MSFC news release 78-25, 24 February 1978.

117. Phil J. Baker, "MSFC Conducts Shuttle Tank Tests," MSFC news release 76-142, 29 July 1976.

118. Amos Crisp, "Tanking Test Conducted on Shuttle External Tank," MSFC news release 77-234, 23 December 1977.

119. "Possible External Tank Improvement Concepts," Bethpage, New York: Grumman Corporation, February 1975).

120. "Possible External Tank Improvement Concepts."

121. Harold R. Coldwater, Richard R. Foll, Gayle J. Howell, and Jon A. Dutton, "Space Shuttle External Tank Performance Improvements – The Challenge," a paper presented at the Space Shuttle Technical Conference (CR-2342) at JSC, 28-30 June 1983, p. 357.

122. "Space Shuttle External Tank Performance Improvements," pp. 357-358.

123. "Space Shuttle External Tank Performance Improvements," pp. 357-358.

124. "Space Shuttle External Tank Performance Improvements," pp. 357-358.

125. "Space Shuttle External Tank Performance Improvements," p. 358.

126. "Space Shuttle External Tank Performance Improvements," p. 358.

127. *History of the Space Shuttle – Volume II*, chapter 5 in the 27 April 1999 version.

128. "Space Shuttle External Tank Performance Improvements," p. 358.

129. "Space Shuttle External Tank Performance Improvements," pp. 357-360.

130. *Understanding Computers: Space*, by the editors of Time-Life Books (Alexandria, Virginia: Time-Life, unknown date).

131. *Understanding Computers: Space*.

132. *Understanding Computers: Space*.

133. *Understanding Computers: Space*.

134. *Understanding Computers: Space*.

135. *Understanding Computers: Space*.

136. For an excellent history of the DFBW F-8 program, see James E. Tomayko, *Computers Take Flight: A History of NASA's Pioneering Digital Fly-By-Wire Project*, NASA SP-2000-4224 (Washington DC: NASA, 2000).

137. *Computers Take Flight*, pp. 89-93; John F. Hanaway and Robert W. Moorehead, *Space Shuttle Avionics System*, NASA SP-504 (Washington DC: NASA, 1989), pp. 5-6.

138. *Space Shuttle Avionics System*, p. 3.

139. *Understanding Computers: Space*

140. Don J. Green, NASA News Release 73-126, "JSC Awards New Avionics Contract to IBM," 25 September 1973.

141. *Computers Take Flight*; *Understanding Computers: Space*.

142. *Space Shuttle Avionics System*, p. 3.

143. *Understanding Computers: Space*.

144. *Understanding Computers: Space*.

145. *Understanding Computers: Space*.

146. *Understanding Computers: Space*; *Space Shuttle Avionics System*, pp. 8-10.

147. Michael J. Ryer, *Programming in HAL/S*, (Pasadena, California: Intermetrics/JPL, September 1979), passim; *Understanding Computers*; *Space Shuttle Avionics System*, pp. 8-10.

148. Case Study, "The Space Shuttle Primary Computer System," <u>Communications of the ACM</u>, volume 27, number 9, September 1984, pp. 871-900.

149. William A. Madden and Kyle Y. Rone, "Design, Development, Integration: Space Shuttle Primary Flight Software System," <u>Communications of the ACM</u>, volume 27, number 9, September 1984, p. 918.

150. Edward J. Joyce, "Is Error-Free Software Achievable?," <u>Datamation</u>, 15 February 1989, pp. 53-56; B.G. Kolkhorst and A.J. Macina, "Developing Error-Free Software," <u>IEEE AES Magazine</u>, November 1988, pp. 25-31; A.J. Macina, *Independent Verification and Validation Testing of the Space Shuttle Primary Flight Software System* (Houston, Texas,: IBM, 28 April 1980), passim.

151. "Design, Development, Integration," p. 918.

152. "Is Error-Free Software Achievable?," pp. 53-56; "Developing Error-Free Software," pp. 25-31; *Independent Verification*, passim.

153. Edward S. Chevers, "Shuttle Avionics Software Development Trials, Tribulations, and Successes: The Backup Flight System," a paper presented at the Space Shuttle Technical Conference (CR-2342) at JSC, 28-30 June 1983, pp. 30-37.

154. "Trials, Tribulations, and Successes: The Backup Flight System," pp. 30-37.

155. "Trials, Tribulations, and Successes: The Backup Flight System," pp. 30-37.

156. *Computers Take Flight*, pp. 107-108.

157. "Trials, Tribulations, and Successes,"pp. 30-37.

158. W. David Compton and Charles D. Benson, *Living and Working in Space: A History of Skylab*, SP-4208 (Washington DC: NASA, 1983), pp. 374-375.

159. *Living and Working in Space*, p. 361.

160. *Living and Working in Space*, p. 361.

161. *Living and Working in Space*, pp. 361-362.

162. *Living and Working in Space*, p. 362.

163. *Living and Working in Space*, pp. 362-363.

164. *Living and Working in Space*, p. 363.

165. *Living and Working in Space*, p. 363.

166. *Living and Working in Space*, pp. 363-364.

167. *Living and Working in Space*, pp. 364-365.

168. *Living and Working in Space*, pp. 365-366.

169. *Living and Working in Space*, pp. 366-367.

170. *Living and Working in Space*, p. 367.

171. *Living and Working in Space*, pp. 370-371.

172. Letter, William P. Clements, Jr., Deputy Secretary of Defense, to James C. Fletcher, NASA, 7 August 1974, no subject, but discussing the DoD commitment to shuttle. In the files of the NASA History Office; DoD Shuttle System requirements, Headquarters Space and Missile Systems Organization, USAF, 18 November 1974.

173. NASA Management Instruction 1052-201, NASA/DoD Memorandum of Understanding on Management and Operation of the Space Transportation System, 14 January 1977. The NMI was cancelled in February 1980, but the underlying MoU remained in force. The MoU was signed by John F. Yardley and James C. Fletcher for NASA, and John J. Martin (Assistant Secretary of the Air Force for R&D) and William P. Clements, Jr. (Deputy Secretary of Defense) for the DoD. In the files of the NASA History Office.

174. Draft Memorandum of Agreement, John F. Yardley, NASA, to Brig. Gen. William L. Shields, Jr., Director, Space Systems, Office of the Secretary of the Air Force, 3 August 1976. In the files of the NASA History Office; Memorandum of Agreement Between NASA and DoD, "Basic Principles for NASA/DoD Space Transportation System Launch Reimbursement," 7 March 1977, signed by John F. Yardley and Alan M. Lovelace for NASA, and John J. Martin (Assistant Secretary of the Air Force for R&D), Robert N. Parker (Acting Director, Defense Research and Engineering, DoD) for the DoD. In the files of the NASA History Office.

175. Memorandum for the Record, "Department of Defense Space Programs, Subcommittee on Defense, Senate Appropriations Committee, 15 May 1984," testimony by Dr. Robert S. Cooper, Director of DARPA, and Lt. Gen. James Abrahamson, Jr. Director of the Office of Strategic Defense. In the files of the NASA History Office.

176. *History of the Space Shuttle – Volume II*, chapter 6 in the 27 April 1999 version.

177. *History of the Space Shuttle – Volume II*, chapter 6 in the 27 April 1999 version.

178. *Space Shuttle: Fulfillment of a Dream*, pp. 1159-1160.

179. Paul A. Cooper and Paul F. Holloway, "The Shuttle Tile Story," Astronautics & Aeronautics, XIX, number 1, January 1981, pp. 24-34; *Technology Influence on the Space Shuttle Development*, report 86-125C (Houston, Texas, Eagle Engineering, Inc., 8 June 1986), pp. 6-4/5; *Space Shuttle: Fulfillment of a Dream*, pp. 1159-1160.

180. First quote from "The Shuttle Tile Story," p. 25; Day quote from an interview of LeRoy E. Day by John Mauer, 17 October 1983, pp. 5-6, in the files of the JSC History Office; *Space Shuttle: Fulfillment of a Dream*, pp. 1161-1166.

181. William C. Schneider and Glenn J. Miller, "The Challenging 'Scale of the Bird' (Shuttle Tile Structural Integrity)," a paper presented at the Space Shuttle Technical Conference (CR-2342) at JSC, 28-30 June 1983, pp. 403-413; *Space Shuttle: Fulfillment of a Dream*, pp. 1165-1166.

182. "The Challenging 'Scale of the Bird'," pp. 403-413.

183. "The Challenging 'Scale of the Bird'," pp. 403-413.

184. "The Challenging 'Scale of the Bird'," pp. 410-413.

185. *Space Shuttle: Fulfillment of a Dream*, p. 1166.

186. "The Shuttle Tile Story," pp. 24-27; "The Challenging 'Scale of the Bird'," pp. 409-410; *Space Shuttle: Fulfillment of a Dream*, pp. 1160-1163.

187. "The Challenging 'Scale of the Bird'," pp. 403-413; NASA Dryden Press Release, (no title), 18 January 1980.

188. "The Challenging 'Scale of the Bird'," pp. 403-413.

189. JoAnn Grant, Lockheed news release "Lockheed-California Company Completes Ground Testing of Space Shuttle Vehicle at Palmdale," 4 October 1979.

190. "Lockheed-California Company Completes Ground Testing."

191. Conversation between the author and Thomas L. Moser, former manager of the orbiter structures group at JSC, 30 September 2000.

192. Philip C. Glynn and Thomas L. Moser, "Orbiter Structural Design and Verification," a paper presented at the Space Shuttle Technical Conference (CR-2342) at JSC, 28-30 June 1983, pp. 345-356.

193. "Orbiter Structural Design and Verification," pp. 345-356.

194. "Orbiter Structural Design and Verification," pp. 345-356.

195. "Orbiter Structural Design and Verification," pp. 345-356.

196. "Orbiter Structural Design and Verification," pp. 345-356.

197. Conversations between the author and Julie Kramer, Orbiter Projects Office, JSC, various dates in September and November 2000; conversations between the author and Grant Cates, OV-102 and (acting) OV-103 Flow Manager at KSC, various dates in October and November 2000.

198. Orbiter Production Contract, NAS9-14000; *Space Transportation System Background Information;* Robert Gordon, NASA News Release 79-07, "Space Shuttle Orbiter Procurement Contract Signed," 5 February 1979.

199. *Space Transportation System Background Information.*

200. *Space Transportation System Background Information.*

201. OV-105 Pre-Delivery Review, a presentation at KSC, April 1991.

202. Orbiter Production Contract, NAS9-14000; Fifth Orbiter Contract, NAS917800; *Space Transportation System Background Information.*

203. Janet Dean, OV-105 Fact Sheet, Rockwell International 22 April 1991.

204. NASA DFRC Fact Sheet, NB-52 Carrier Aircraft.

205. Janet Dean, Rockwell Fact Sheet "Endeavour," 18 December 1990; James Hartsfield, NASA JSC press release 91-026, 5 April 1991.

206. "Endeavour;" JSC press release 91-026.

207. Orbiter Production Contract, NAS9-14000.

208. "Automated Orbiter Kit," Rockwell release 3550-A-1, September 1993.

209. Tom. A. Heppenheimer, *The Space Shuttle Decision: NASA's Search for a Reusable Space Vehicle*, NASA SP-4221, (Washington DC: NASA, 1999), pp. 139, and 226-232.

210. *History of the Space Shuttle – Volume II*, chapter 9 in the 27 April 1999 version.

211. Personal Notes of George M. Low, file 150, 9 August 1975, in the files of the NASA History Office.

212. *History of the Space Shuttle – Volume II*, chapter 9 in the 27 April 1999 version.

213. *History of the Space Shuttle – Volume II*, chapter 9 in the 27 April 1999 version.

214. *History of the Space Shuttle – Volume II*, chapter 9 in the 27 April 1999 version.

215. *History of the Space Shuttle – Volume II*, chapter 9 in the 27 April 1999 version.

216. *History of the Space Shuttle – Volume II*, chapter 9 in the 27 April 1999 version.

217. Robert A. Frosch, "Frosch on Centaur," GRC Archives, *Lewis News,* 30 January 1981, p. 1.

218. Two good histories of Centaur – one dealing with the expendable version, and the other with Shuttle/Centaur, may be found in *To Reach the High Frontier: Case Studies in Launch Vehicle History*, NASA SP-2001-4227 (Washington DC: NASA, 2001). See Virginia P. Dawson "Building Centaur: A Technical Decision from the Top," and Mark D. Bowles, "Eclipsed by Tragedy: The Fated Mating of the Shuttle and Centaur."

219. James R. Hansen, *Spaceflight Revolution: NASA Langley Research Center From Sputnik to Apollo*, NASA SP-4308 (Washington, DC: NASA, 1995), p. 283; "Reusable Centaur Study," General Dynamics, Contract NAS 8-30290, 26 September 1973; "Eclipsed by Tragedy."

220. M. Mitchell Waldrop, "Centaur Wars," Science, volume 217, 10 September 1982, p. 1012; "Eclipsed by Tragedy."

221. Letter, Chris C. Kraft Jr., William R. Lucas, Dick G. Smith to Alan M. Lovelace, 19 January 1981, in the files of the JSC History Office; "Shuttle/Centaur Mini-Design Review at San Diego," 2-3 August 1982; "Eclipsed by Tragedy."

222. "NASA/DOD Agreement," in the files of the NASA History Office; "NASA Lewis Research Center/USAF Space Division Agreement for the Management of the Shuttle/Centaur Program 10/25/82," in the files of the NASA History Office; "Shuttle/Centaur Mini-Design Review at San Diego;" "Eclipsed by Tragedy."

223. Harold Hahn, "A New Addition to the Space transportation System" (San Diego, California: General Dynamics, July 1985); Omer F. Spurlock, "Shuttle/Centaur – More Capability for the 1980s" (Cleveland, Ohio: NASA Lewis, undated); "Shuttle/Centaur Mini-Design Review at San Diego;" "Eclipsed by Tragedy."

224. Memorandum, John R. Casani to Galileo Review Board, 6 January 1982, in the files of the NASA History Office.

225. William E. Klein, "Minutes for Shuttle/Centaur-G Safety Certification Process Briefing," 2 June 1986, in the files of the NASA History Office.

226. "Shuttle/Centaur Termination Status," 4 September 1986, in the files of the NASA History Office; Sarah G. Keegan, "NASA Terminates Development of Shuttle/Centaur Upper Stage," 19 June 1986, NASA HQ Press Release, GRC Public Affairs Archives.

227. George M. Low, NASA, and Malcom R. Currie, DoD, "Joint NASA/DoD Position Statement on Space Shuttle Orbiter Procurement," 23 January 1976. In the files of the NASA History Office.

228. Letter, James C. Fletcher, NASA, to James T. Lynn, Director of the OMB, 22 October 1976, with a cover letter from Malcom R. Currie, DoD, indicating "general" concurrence with the contents. In the files of the NASA History Office, also conveniently reprinted in *Exploring the Unknown, Volume IV: Accessing Space*, NASA SP-4407, (Washington DC: NASA, 1999), pp. 286-288.

229. James T. McIntyre, Jr. Acting Director, OMD, to Robert A. Frosch, NASA, 23 December 1977, conveniently reprinted in *Exploring the Unknown, Volume IV: Accessing Space*, NASA SP-4407, (Washington DC: NASA, 1999), pp. 288-289.

230. Contract NAS9-14000, Article B3.

231. *Major Safety Concerns: Space Shuttle Program*, JSC-09990C, 8 November 1976, in the files of the NASA History Office, also conveniently reprinted in *Exploring the Unknown, Volume IV: Accessing Space*, NASA SP-4407, (Washington DC: NASA, 1999), pp. 306-323.

232. *Major Safety Concerns*, pp. 306-323.

233. Eugene E. Covert, Chairman, Ad Hoc Committee for Review of the Space Shuttle Main Engine Development Program, National Research Council, Statement before the Subcommittee on Science, Technology, and Space, Committee on Commerce, Space, and Transportation, U.S. Senate, 22 February 1979, conveniently reprinted in *Exploring the Unknown, Volume IV: Accessing Space*, NASA SP-4407, (Washington DC: NASA, 1999), p. 277; *Major Safety Concerns: Space Shuttle Program*, p. 320.

234. For an example of DoD support, see "Why Shuttle is Needed," a letter from Brigadier General Robert Rosenberg, national Security Council, to the OMB, undated (in November 1979), conveniently reprinted in *Exploring the Unknown, Volume IV: Accessing Space*, NASA SP-4407, (Washington DC: NASA, 1999), pp. 292-293.

235. Background paper, "Meeting on the Space Shuttle," 14 November 1979, conveniently reprinted in *Exploring the Unknown, Volume IV: Accessing Space*, NASA SP-4407, (Washington DC: NASA, 1999), pp. 294-305.

236. "Meeting on the Space Shuttle," pp. 294-305.

237. "Meeting on the Space Shuttle," pp. 294-305; *Major Safety Concerns: Space Shuttle Program*, p. 323.

238. "Meeting on the Space Shuttle," pp. 294-305.

239. "Meeting on the Space Shuttle," pp. 294-305.

240. Robert A. Frosch, NASA Administrator, "Special Announcement: Examination of the Shuttle Program," 18 August 1980.

241. Memorandum, John F. Yardley, NASA, to Director, Public Affairs, NASA, subject: Recommended Orbiter Names, 26 May 1978, in the files of the NASA History Office, also conveniently reprinted in *Exploring the Unknown, Volume IV: Accessing Space*, NASA SP-4407, (Washington DC: NASA, 1999), pp. 274-275.

242. Ibid.

243. Letter, John E. Naugle, NASA Chief Scientist, to Arnold W. Frutkin, NASA Associate Administrator for External Relations, subject: Possible Names for Shuttles, 17 November 1978; Memorandum for Distribution, from Arnold W. Frutkin, subject: Committee Meeting of November 27, 1978, 28 November 1978; Memorandum, Arnold W. Frutkin, NASA Associate Administrator for External Relations to Dr. Alan M. Lovelace, NASA Acting Administrator, subject: Orbiter Names, 11 December 1978. All in the files of the NASA History Office, with many thanks to Steven J. Garber and Colin Fries for finding them for me!

244. *From Ship to Shuttle: NASA Orbiter-Naming Program*, September 1988-May 1989, NASA EP-276 (Washington DC: NASA, October 1991).

245. *From Ship to Shuttle.*

246. *From Ship to Shuttle.*

247. "Endeavour: Naming the New Shuttle," Spaceflight," July 1989, p. 38. Thanks to Colin Burgess in Australia for finding this for me.

248. *From Ship to Shuttle.*

249. The notes on the names themselves come from NASA press releases and from material provided in Memorandum for Distribution, from Arnold W. Frutkin, subject: Committee Meeting of November 27, 1978, 28 November 1978

250. *Space Shuttle 1978: Status Report for the Committee on Science and Technology, U.S. House of Representatives, Ninety-Fifth Congress, Second Session*, Orbiter Project Schedule, (Washington DC: Congress, 1978), pp. 4001-4011.

251. "OV-101 Structural Integrity Evaluation," Final Report, 26 February 1996, supplied by Julie Kramer, Orbiter Projects Office, JSC.

CHAPTER VIII – BROKEN IN MID-STRIDE. PAGES 259–290.

1. *Shuttle Crew Operations Manual*, SFOC-FL0884, Revision B (Houston, Texas: SFOC, 4 October 2000), pp. 1.1-6/7. It should be noted that most technical data on the vehicle is also available in the *National Space Transportation System Reference Manual*, Volume 1, Systems and Facilities, June 1988, or on-line at http://science.ksc.nasa.gov/shuttle/technology/sts-newsref/stsref-toc.html; however, the latter two references (which are actually the same, just reformatted), are currently 12 years out-of-date, so the *Shuttle Crew Operations Manual* takes precedence here unless the subject is not covered.

2. *Shuttle Crew Operations Manual*, p. 1.1-7.

3. *Shuttle Crew Operations Manual*, p. 1.1-2.

4. *Shuttle Crew Operations Manual*, p. 1.1-7.

5. *Shuttle Crew Operations Manual*, p. 1.1-2.

6. *Shuttle Crew Operations Manual*, pp. 1.1-2 and 5.2-2

7. *Shuttle Crew Operations Manual*, pp. 1.1-2 and 9.1-8.

8. *Shuttle Crew Operations Manual*, p. 1.1-2.

9. *Shuttle Flight Data and In-Flight Anomaly List*, JSC-19413 on-line version, "the green book" (Houston, Texas: JSC, November 2000).

10. *Shuttle Crew Operations Manual*, p. 1.1-3.

11. *Shuttle Crew Operations Manual*, pp. 1.1-3/4.

12. *Shuttle Crew Operations Manual*, p. 1.1-4.

13. *Shuttle Crew Operations Manual*, p. 1.1-4.

14. *Shuttle Crew Operations Manual*, p. 1.1-4.

15. *Shuttle Crew Operations Manual*, p. 1.1-4.

16. *Shuttle Crew Operations Manual*, p. 1.1-4.

17. *Shuttle Crew Operations Manual*, p. 1.1-5.

18. *Shuttle Crew Operations Manual*, p. 1.1-5.

19. *Shuttle Crew Operations Manual*, p. 1.1-6.

20. *Shuttle Crew Operations Manual*, p. 1.1-6.

21. *Shuttle Crew Operations Manual*, p. 1.1-6.

22. *Shuttle Crew Operations Manual*, pp. 1.1-6/7.

23. *Shuttle Crew Operations Manual*, p. 1.1-7.

24. *Shuttle Crew Operations Manual*, p. 1.1-5.

25. *Shuttle Crew Operations Manual*, p. 1.1-5.

26. *Shuttle Crew Operations Manual*, p. 1.1-5.

27. "Vandenberg Launch Site, Initial Operating Capability," briefing to the 6595th Shuttle Test Group, October 1985; *National Space Transportation System Reference Manual*, Volume 1, Systems and Facilities, June 1988, pp. 4-5.

28. *Shuttle Crew Operations Manual*, p. 6.1-1.

29. *Shuttle Crew Operations Manual*, p. 6.1-1.

30. *Shuttle Crew Operations Manual*, pp. 6.1-1/6.2-5.

31. *Shuttle Crew Operations Manual*, pp. 6.2-4/5.

32. *Shuttle Crew Operations Manual*, pp. 6.2-4/5.

33. *Shuttle Crew Operations Manual*, p. 6.2-1.

34. *Shuttle Crew Operations Manual*, p. 6.6-1.

35. *Shuttle Crew Operations Manual*, p. 6.5-1.

36. *Shuttle Crew Operations Manual*, pp. 6.4-1/4.

37. Joe H. Jones, "Ascent/Entry Flight Techniques Panel OI-28 Entry ECAL Automation," a briefing prepared by the Mission Operations Directorate, Flight Design and Dynamics Division, JSC, 16 June 2000.

38. "Ascent/Entry Flight Techniques Panel OI-28 Entry ECAL Automation."

39. *Shuttle Crew Operations Manual*, pp. 6.3-1.

40. *Shuttle Crew Operations Manual*, pp. 6.3-1 through 6.6-1; *NSTS Reference Manual*, pp. 13-16.

41. Information for all Space Shuttle missions may be found online at several locations, including : http://spacelink.nasa.gov/NASA.Projects/Human.Exploration.and.Development.of.Space/Human.Space.Flight/Shuttle/Shuttle.Missions/.index.html; http://science.ksc.nasa.gov/shuttle/missions/missions.html; http://www.jps.net/dcfischer/psts.htm. Each is a top-level link listing all missions to-date; *Space Shuttle Mission Chronology: 1981-1996*, NP-1997-12-08-KSC, October 1997; In addition, NASA publishes a press kit for each mission. The definitive source for statistics on each mission is compiled by JSC as the *Shuttle Flight Data and In-Flight Anomaly List*, JSC-19413 on-line version (behind the JSC firewall), "the Green Book" (Houston, Texas: JSC, November 2000). Where necessary, the Green Book has been used to settle disputes between the various public sources. Citations to individual items will be made if they are particularly significant.

42. See note 41.

43. Tom A. Heppenheimer, *History of the Space Shuttle – Volume II: Development of the Shuttle, 1972-1981*, a draft manuscript for an upcoming second volume of the official NASA shuttle history, chapter 10 in the 27 April 1999 version; See note 41.

44. See note 41.

45. See note 41.

46. See note 41.

47. See note 41.

48. See note 41.

49. See note 41.

50. See note 41.

51. See note 41.

52. See note 41.

53. See note 41.

54. See note 41.

55. See note 41.

56. See note 41.

57. See note 41.

58. *Payload Element Database*, prepared by Muniz Engineering for the JSC/DoD Payloads Office under contract F04701-98-D-0102, March 2000; David M. Harland, *The Space Shuttle: Roles, Missions and Accomplishments* (Chichester, England: Praxis Publishing, Ltd. 1998), p. 178; Dwayne A. Day, "Out of the Shadows: The Shuttle's Secret Payloads," <u>Spaceflight</u>, Volume 41, February 1999, pp. 78-83; See note 41.

59. See note 41.

60. See note 41.

61. See note 41.

62. See note 41.

63. See note 41.

64. *Payload Element Database; The Space Shuttle: Roles, Missions and Accomplishments*, p. 180; "Out of the Shadows: The Shuttle's Secret Payloads," pp. 78-83; See note 41.

65. See note 41.

66. See note 41.

67. See note 41.

68. William P. Rogers, Chairman, *Presidential Commission on the Space Shuttle Challenger Accident*, Final Report, 6 June 1986, p. 17; Memorandum, Jesse W. Moore, Coe M/Associate Administrator for Space Flight, "Space Shuttle Mission 51-L Flight Readiness Review Assessment," 17 January 1986.

69. *Presidential Commission Report*, p. 17.

70. *Presidential Commission Report*, p. 17.

71. *Presidential Commission Report*, p. 17.

72. *Presidential Commission Report*, p. 17.

73. *Presidential Commission Report*, pp. 17-19.

74. *Presidential Commission Report*, pp. 17-18.

75. *Presidential Commission Report*, p. 18.

76. *Presidential Commission Report*, p. 18.

77. *Presidential Commission Report*, pp. 66-69.

78. *Presidential Commission Report*, pp. 202-203, 206, and 212-213.

79. *Presidential Commission Report*, p. 40.

80. *Presidential Commission Report*, p. 208.

81. *Presidential Commission Report*, p. 19.

82. *Presidential Commission Report*, pp. 19-20.

83. *Presidential Commission Report*, p. 20.

84. *Presidential Commission Report*, p. 20.

85. *Presidential Commission Report*, p. 20.

86. *Presidential Commission Report*, p. 20.

87. *Presidential Commission Report*, p. 20.

88. *Presidential Commission Report*, pp. 20-21.

89. *Presidential Commission Report*, p. 21.

90. *Presidential Commission Report*, p. 21.

91. *Presidential Commission Report*, p. 21.

92. *Presidential Commission Report*, p. 45.

93. *Presidential Commission Report*, pp. 45-46.

94. *Presidential Commission Report*, passim.

95. *Presidential Commission Report*, p. 148.

96. *Presidential Commission Report*, p. 148.

97. *Presidential Commission Report*, passim.

98. *Presidential Commission Report*, pp. 198-201.

99. *National Space Transportation System Reference*, NASA, June 1988, p. 33.

100. *National Space Transportation System Reference*, NASA, June 1988, pp. 33-33a.

101. *Thirty-Two Months To Discovery*, a brochure published by Morton Thiokol undated (but early 1989); *National Space Transportation System Reference*, NASA, June 1988, pp. 33a-g.

102. *Thirty-Two Months To Discovery*, pp. 33a-g.

103. *Thirty-Two Months To Discovery*, pp. 33a-g.

104. *Thirty-Two Months To Discovery*, pp. 33a-g.

105. *Thirty-Two Months To Discovery*, pp. 33a-g; "Static Test Information," provided by Melodie de Guibert, Thiokol Propulsion, 2 November 2000.

106. *Thirty-Two Months To Discovery*, passim.

107. H. Guyford Stever, "National Research Council Space Shuttle Independent Oversight Panel," various written reports, in the files of the JSC History Office.

108. *Space Shuttle Program 1999 Annual Report* (Houston, Texas: SSP Program Office, 2000), Part 2, pp. 7-8.

109. *STS Crew Egress and Escape Study: Volume I – Analysis*, JSC-22275 (Houston, Texas: Shuttle Crew Abort Planned Escape panel, August 1986).

110. *Shuttle Crew Operations Manual*, passim.

111. *STS Crew Egress and Escape Study: Volume I – Analysis*, Section IV.

112. *NSTS Reference Manual*, pp. 123-136.

113. *NSTS Reference Manual*, pp. 123-136.

114. *NSTS Reference Manual*, pp. 123-136.

115. *NSTS Reference Manual*, pp. 123-136.

116. *NSTS Reference Manual*, pp. 123-136.

117. *NSTS Reference Manual*, pp. 123-136.

118. *NSTS Reference Manual*, pp. 123-136; *Shuttle Crew Operations Manual*, p. 2.14-4.

119. *The Space Shuttle: Roles, Missions and Accomplishments*, passim.

120. *The Space Shuttle: Roles, Missions and Accomplishments*, pp. 111-12.

121. *The Space Shuttle: Roles, Missions and Accomplishments*, pp. 113-14.

122. *The Space Shuttle: Roles, Missions and Accomplishments*, pp. 115-16.

123. *Shuttle Flight Data and In-Flight Anomaly List; The Space Shuttle: Roles, Missions and Accomplishments*, p. 127.

124. *Shuttle Flight Data and In-Flight Anomaly List; The Space Shuttle: Roles, Missions and Accomplishments*, pp. 415-38.

125. *Shuttle Flight Data and In-Flight Anomaly List; The Space Shuttle: Roles, Missions and Accomplishments*, pp. 415-38.

126. *The Space Shuttle: Roles, Missions and Accomplishments*, pp. 124-25.

127. *The Space Shuttle: Roles, Missions and Accomplishments*, p. 128.

128. *The Space Shuttle: Roles, Missions and Accomplishments*, p. 129.

129. *Presidential Commission Report*, passim.

130. U.S. Code, Section 42 – The Public Health and Welfare, Chapter 26 – National Space Program, Section 2465a – Space Shuttle Use Policy. An on-line version may be found at http://www4.law.cornell.edu/uscode/unframed/42/2465a.html

131. *The Space Shuttle: Roles, Missions and Accomplishments*, passim.

132. "United States Space Launch Strategy," National Security Decision Directive 254, 27 December 1986, White House; an on-line version may be found at the X-33 History Project, http://www.hq.nasa.gov/office/pao/History/x-33/facts_1.htm

133. *Shuttle Flight Data and In-Flight Anomaly List; Space Shuttle Mission Chronology: 1999*, NP-2000-12-92-KSC, January 2000.

134. *Presidential Commission Report*, pp. 37-39.

135. *Space Shuttle Flight and Ground Specification*, NSTS-07700, Volume X, Book 3, Requirements for Runways and Navigational Aids, Revision N, 7 September 1995, pp. 1-3/4.

136. See for example, William L. Thomas, *Ditching Investigation of a 1/20th-scale Model of the Space Shuttle Orbiter*, NASA CR-2593 (Bethpage, NY: Grumman Aerospace, October 1975).

137. "Static Test Information," provided by Melodie de Guibert, Thiokol Propulsion, 2 November 2000.

CHAPTER IX – RETURN-TO-FLIGHT. PAGES 291–346.

1. http://science.ksc.nasa.gov/shuttle/missions/sts-26/sts-26-press-kit.txt

2. http://science.ksc.nasa.gov/shuttle/missions/sts-26/sts-26-press-kit.txt

3. http://science.ksc.nasa.gov/shuttle/missions/sts-26/sts-26-press-kit.txt

4. http://science.ksc.nasa.gov/shuttle/missions/sts-26/sts-26-press-kit.txt

5. http://science.ksc.nasa.gov/shuttle/missions/sts-26/sts-26-press-kit.txt

6. http://science.ksc.nasa.gov/shuttle/missions/sts-26/sts-26-press-kit.txt

7. http://science.ksc.nasa.gov/shuttle/missions/sts-26/sts-26-press-kit.txt

8. http://science.ksc.nasa.gov/shuttle/missions/sts-26/sts-26-press-kit.txt

9. http://science.ksc.nasa.gov/shuttle/missions/sts-26/sts-26-press-kit.txt

10. Information for all Space Shuttle missions may be found online at several locations, including : http://spacelink.nasa.gov/NASA.Projects/Human.Exploration .and.Development.of.Space/Human.Space.Flight/Shuttle/Shuttle.Missions/.index .html; http://science.ksc.nasa.gov/shuttle/missions/missions.html; http://www.jps.net/dcfischer/psts.htm. Each is a top-level link listing all missions to-date; *Space Shuttle Mission Chronology: 1981-1996*, NP-1997-12-08-KSC, October 1997; *Space Shuttle Mission Chronology: 1997*, NP-1998-12-92-KSC, February 1998; *Space Shuttle Mission Chronology: 1998*, NP-1999-12-92-KSC, February 1999; *Space Shuttle Mission Chronology: 1999*, NP-2000-12-92-KSC, January 2000; In addition, NASA publishes a press kit for each mission. The definitive source for statistics on each mission is compiled by JSC as the *Shuttle Flight Data and In-Flight Anomaly List*, JSC-19413 on-line version (behind the JSC firewall), "the Green Book" (Houston, Texas: JSC, November 2000). Where necessary, the Green Book has been used to settle disputes between the various public sources. Citations to individual items will be made if they are particularly significant.

11. *Payload Element Database*, prepared by Muniz Engineering for the JSC/DoD Payloads Office under contract F04701-98-D-0102, March 2000; David M. Harland, *The Space Shuttle: Roles, Missions and Accomplishments* (Chichester, England: Praxis Publishing, Ltd. 1998), p. 180; Dwayne A. Day, "Out of the Shadows: The Shuttle's Secret Payloads," Spaceflight, Volume 41, February 1999, pp. 78-83; See note 10.

12. See note 10.

13. See note 10.

14. *Payload Element Database*; *The Space Shuttle: Roles, Missions and Accomplishments*, p. 180; "Out of the Shadows: The Shuttle's Secret Payloads," pp. 78-83; See note 10.

15. See note 10.

16. See note 10.

17. See note 10.

18. *Payload Element Database*; *The Space Shuttle: Roles, Missions and Accomplishments*, p. 181; "Out of the Shadows: The Shuttle's Secret Payloads," pp. 78-83; See note 10.

19. See note 10.

20. See note 10.

21. See note 10.

22. See note 10.

23. See note 10.

24. See note 10.

25. See note 10.

26. See note 10; "Shuttle Rollbacks," a KSC 'Factoid', 22 September 2000.

27. See note 10.

28. *Payload Element Database*; *The Space Shuttle: Roles, Missions and Accomplishments*, p. 180; "Out of the Shadows: The Shuttle's Secret Payloads," pp. 78-83; See note 10.

29. See note 10.

30. See note 10.

31. See note 10.

32. See note 10.

33. See note 10.

34. See note 10.

35. See note 10.
36. See note 10.

37. See note 10.

38. See note 10.

39. See note 10.

40. See note 10.

41. See note 10.

42. See note 10.

43. *Payload Element Database*; *The Space Shuttle: Roles, Missions and Accomplishments*; "Out of the Shadows: The Shuttle's Secret Payloads," p. 180; See note 10.

44. See note 10.

45. See note 10.

46. See note 10.

47. See note 10.

48. See note 10.

49. See note 10.

50. See note 10.

51. See note 10.

52. See note 10.

53. See note 10.

54. See note 10.

55. See note 10.

56. See note 10.

57. See note 10.

58. See note 10.

59. See note 10.

60. See note 10.

61. science.ksc.nasa.gov/shuttle/missions/sts-70/mission-sts-70.html; See note 10.

62. See note 10.

63. See note 10.

64. See note 10.

65. See note 10.

66. See note 10.

67. See note 10.

68. See note 10.

69. See note 10.

70. See note 10.

71. See note 10.

72. See note 10.

73. See note 10.

74. See note 10.

75. See note 10.

76. See note 10.

77. See note 10.

78. See note 10.

79. See note 10.

80. See note 10.

81. See note 10.

82. See note 10.

83. See note 10.

84. See note 10.

85. See note 10.

86. "STS-93 Status Report #11," 23 July 1999, online at http://science.ksc.nasa.gov/shuttle/missions/sts-93/news/sts-93-mcc-01.txt; Joseph C. Anselmo, "Shuttle Scrubs Delay Observatory," Aviation Week and Space Technology, 26 July 1999;

87. "STS-93 Status Report #1."

88. Report to the Associate Administrator, Office of Space Flight, Space Shuttle Independent Assessment Team, 7 March 2000, p. 1-8; Joseph C. Anselmo and Craig Covault, "Chandra Deployed After Liftoff Anomalies," Aviation Week and Space Technology, 2 Aug 19990.

89. Report to the Associate Administrator, Office of Space Flight, p. 1-8.

90. Report to the Associate Administrator, Office of Space Flight, pp. 40-43.

91. Robin Lloyd, "64 Cases of Wiring Problems Found on Shuttle," CNN Interactive, 3 September 1999.

92. "64 Cases of Wiring Problems Found on Shuttle,"

93. "64 Cases of Wiring Problems Found on Shuttle,"

94. "Space Shuttle Independent Assessment Team Report," SpaceDaily Online, 9 March 2000; Report to the Associate Administrator, Office of Space Flight, passim.

95. "Shuttle Launches Delayed Again," SpaceViews Online, 7 October 1999.

96. "Shuttle Launches Delayed Again."

97. "Shuttle Launches Delayed Again."

98. Miles O'Brien, "Shuttle Launch on Indefinite Hold: Y2K Deadline Looms," CNN Interactive, 6 December 1999.

99. Report to the Associate Administrator, Office of Space Flight, passim.

100. Report to the Associate Administrator, Office of Space Flight, pp. 1-5.

101. Report to the Associate Administrator, Office of Space Flight, p. 6.

102. Report to the Associate Administrator, Office of Space Flight, passim.

103. Report to the Associate Administrator, Office of Space Flight, passim.

104. Craig Covault, "KSC Gears for Shuttle Surge," Aviation Week and Space Technology," 28 August 2000, on-line edition.

105. "KSC Gears for Shuttle Surge."

106. "KSC Gears for Shuttle Surge."

107. "NASA Contract Paves the Way for VAB 'Safe Haven' Modifications," KSC News Release 70-99, 24 August 1999; "NASA to Size-Up Shuttle's New "Safe Haven" on Aug. 12," KSC News Release 65-00, 10 August 2000; "Shuttle Atlantis Takes a Detour on the Way to the Launch Pad," KSC News Release 68-00, 14 August 2000.

108. Ibid.; Craig Covault, "KSC Gears for Shuttle Surge," Aviation Week and Space Technology," 28 August 2000; Space Shuttle Program 1999 Annual Report (Houston, TX: Space Shuttle Program Office, 2000), Part 2, p. 42.

109. See note 10.

110. Craig Covault, "Shuttle Radar Goal: Information Superiority," Aviation Week and Space Technology," 31 January 2000; Craig Covault, "Radar Flight Meets mapping Goals," Aviation Week and Space Technology," 28 February 2000; See note 10.

111. Craig Covault, "MEDS Glass Cockpit to Enhance Shuttle Safety," Aviation Week and Space Technology, 6 March 2000; See note 10.

112. See note 10.

113. The Space Shuttle: Roles, Missions and Accomplishments, pp. 128-29.

114. The Space Shuttle: Roles, Missions and Accomplishments, pp. 299-301.

115. The Space Shuttle: Roles, Missions and Accomplishments, pp. 299-301.

116. The Space Shuttle: Roles, Missions and Accomplishments, p. 315.

117. Shuttle Flight Data and In-Flight Anomaly List; Space Shuttle Mission Chronology: 1999, NP-2000-12-92-KSC, January 2000; The Space Shuttle: Roles, Missions and Accomplishments, pp. 415-438.

118. Shuttle Flight Data and In-Flight Anomaly List; Space Shuttle Mission Chronology: 1999, NP-2000-12-92-KSC, January 2000; Shuttle Crew Operations Manual, OI-28 release, SFOC-FL0884, Revision B, (Houston, TX: USA, 4 October 2000), pp 2.25-2; The Space Shuttle: Roles, Missions and Accomplishments, pp. 415-438.

119. Shuttle Flight Data and In-Flight Anomaly List; Space Shuttle Mission Chronology: 1999, NP-2000-12-92-KSC, January 2000; The Space Shuttle: Roles, Missions and Accomplishments, pp. 415-438.

120. Shuttle Flight Data and In-Flight Anomaly List; "Launch Data Summary," prepared by the MSFC Space Shuttle Flight Evaluation Team for each mission, released the day of launch, available on www.spaceflight.nasa.gov and in the files of the KSC Vehicle Engineering Office; Space Shuttle Mission Chronology: 1999, NP-2000-12-92-KSC, January 2000; Processing status reports at http://www-pao.ksc.nasa.gov/kscpao/status/stsstat (various dates); data compiled by Grant Cates, Shuttle Processing Directorate at KSC for the UCF ground processing modeling project.

121. Shuttle Flight Data and In-Flight Anomaly List; "Launch Data Summary," prepared by the MSFC Space Shuttle Flight Evaluation Team for each mission, released the day of launch, available on www.spaceflight.nasa.gov and in the files of the KSC Vehicle Engineering Office; Space Shuttle Mission Chronology: 1999, NP-2000-12-92-KSC, January 2000; ET Delivery Date list supplied by Sharon H. Hansen, Communications Department, Lockheed Martin Space Systems Company, New Orleans, LA, 26 September 2000; "Fact Sheet: Space Shuttle External Tank," 20 September 2000, LMSSC, New Orleans, LA; June Malone, "NASA Completes Purchase of Material for 60 Shuttle External Tanks," MSFC News Release 99-069, 28 April 1999.

122. Shuttle Flight Data and In-Flight Anomaly List; Space Shuttle Mission Chronology: 1999, NP-2000-12-92-KSC, January 2000; "Launch Data Summary," prepared by the MSFC Space Shuttle Flight Evaluation Team for each mission, released the day of launch, available on www.spaceflight.nasa.gov and in the files of the KSC Vehicle Engineering Office; June Malone, "NASA, THiokol Complete $1.7 Billion Shuttle Motor Agreement," MSFC News Release C99-C, 24 August 1999.

123. Shuttle Flight Data and In-Flight Anomaly List; Space Shuttle Mission Chronology: 1999, NP-2000-12-92-KSC, January 2000; "Launch Data Summary," prepared by the MSFC Space Shuttle Flight Evaluation Team for each mission, released the day of launch, available on www.spaceflight.nasa.gov and in the files of the KSC Vehicle Engineering Office.

124. Shuttle Flight Data and In-Flight Anomaly List; Space Shuttle Mission Chronology: 1999, NP-2000-12-92-KSC, January 2000; "Launch Data Summary," prepared by the MSFC Space Shuttle Flight Evaluation Team for each mission, released the day of launch, available on www.spaceflight.nasa.gov and in the files of the KSC Vehicle Engineering Office.

125. http://science.ksc.nasa.gov/shuttle/missions/rollbacks.html

126. Shuttle Flight Data and In-Flight Anomaly List; Space Shuttle Mission Chronology: 1999, NP-2000-12-92-KSC, January 2000; Ferry flight rules posted by the Air Force Space Command, 13 November 2000.

127. Payload Element Database.

128. The Space Shuttle: Roles, Missions and Accomplishments, p. 177.

129. The Space Shuttle: Roles, Missions and Accomplishments, pp. 177-178; http://science.ksc.nasa.gov/shuttle/missions/sts-6/mission-sts-6.html; "Out of the Shadows: The Shuttle's Secret Payloads," p. 80.

130. The Space Shuttle: Roles, Missions and Accomplishments, pp. 177-178; http://science.ksc.nasa.gov/shuttle/missions/51-c/mission-51-c.html.

131. The Space Shuttle: Roles, Missions and Accomplishments, pp. 181-182; http://www.af.mil/news/factsheets/Defense_Satellite_Communicati.html

132. "Out of the Shadows: The Shuttle's Secret Payloads," pp. 83-84; *The Space Shuttle: Roles, Missions and Accomplishments* , pp. 178-179.

133. "Out of the Shadows: The Shuttle's Secret Payloads," pp. 78-79; *The Space Shuttle: Roles, Missions and Accomplishments*, pp. 179-180.

134. *The Space Shuttle: Roles, Missions and Accomplishments*, p. 178.

135. http://www.fas.org/spp/military/program/imint/afp731_jp_960627.htm; *The Space Shuttle: Roles, Missions and Accomplishments*, pp. 180-181.

136. "Out of the Shadows: The Shuttle's Secret Payloads," p. 80; email between the author and Dwayne Day, various dates in October and November 2000.

137. "Out of the Shadows: The Shuttle's Secret Payloads," pp. 80-84; email between the author and Dwayne Day, various dates in October and November 2000.

138. *Payload Element Database*.

139. "Out of the Shadows: The Shuttle's Secret Payloads," pp. 79-80; email between the author and Dwayne Day, various dates in October and November 2000.

140. *The Space Shuttle: Roles, Missions and Accomplishments*, pp. 187-188.

141. "Out of the Shadows: The Shuttle's Secret Payloads," pp. 79-80; *Shuttle Flight Data and In-Flight Anomaly List*; *Space Shuttle Mission Chronology* Email between the author and Dwayne A. Day, October 2000.

142. *Payload Element Database*; "Out of the Shadows: The Shuttle's Secret Payloads," pp. 79-80; *Shuttle Flight Data and In-Flight Anomaly List*; *Space Shuttle Mission Chronology: 1999*, NP-2000-12-92-KSC, January 2000; Email between the author and Dwayne A. Day, October 2000.

143. *Shuttle Flight Data and In-Flight Anomaly List*; "Launch Data Summary," prepared by the MSFC Space Shuttle Flight Evaluation Team for each mission, released the day of launch, available on www.spaceflight.nasa.gov and in the files of the KSC Vehicle Engineering Office.

CHAPTER X – TECHNICAL DESCRIPTION. PAGES 363–440.

1. Personal experience from being there; http://www.unitedspacealliance.com/

2. *Shuttle Crew Operations Manual*, OI-28 release, SFOC-FL0884, Revision B, (Houston, TX: USA, 4 October 2000), pp. 1.1-1/19; It should be noted that most technical data on the vehicle is also available in the *National Space Transportation System Reference Manual*, Volume 1, Systems and Facilities, June 1988, or on-line at http://science.ksc.nasa.gov/shuttle/technology/sts-news-ref/stsref-toc.html; however, the latter two references (which are actually the same, just reformatted), are currently 12 years out-of-date, so the *Shuttle Crew Operations Manual* takes precedence here unless the subject is not covered.

3. *Shuttle Crew Operations Manual*, pp. 1.2-1/3.

4. *Shuttle Crew Operations Manual*, p. 1.2-2.

5. *Shuttle Crew Operations Manual*, p. 1.2-3.

6. *Shuttle Crew Operations Manual*, p. 1.2-3; *OV-101 Structural Integrity Evaluation*, Final Report, (Houston, TX, JSC, 26 February 1996), p. 11.

7. *Shuttle Crew Operations Manual*, p. 1.2-6.

8. *Shuttle Crew Operations Manual*, p. 1.2-6.

9. *Shuttle Crew Operations Manual*, p. 1.2-6.

10. *Shuttle Crew Operations Manual*, p. 1.2-7.

11. *Shuttle Crew Operations Manual*, p. 1.2-7.

12. Orbiter Production Contract, NAS9-14000, Exhibit A, Increment 3, 1 October 1978 amended 1 August 1995, section 1.2.505, p. A-108.

13. *Shuttle Crew Operations Manual*, p. 2.2-6.

14. *Shuttle Crew Operations Manual*, p. 2.10-1.

15. Email between the author and Chris Hansen (the seat designer), JSC-ES, 30 October 2000.

16. *Shuttle Crew Operations Manual*, p. 2.2-1.

17. Daniel M. Barry and John W. Bassick, *NASA Space Shuttle Advanced Crew Escape Suit Development*, a paper prepared for the SAE 25th International Conference on Environmental Systems, San Diego, California, 10-13 July 1995. I am indebted to Mr. Bassick, who is a Vice President of David Clark Company, for providing this paper and other material to me; email between the author and John W. Bassick, David Clark Company, various dates in late November 2000.

18. *NASA Space Shuttle Advanced Crew Escape Suit Development*; email between the author and John W. Bassick, David Clark Company, various dates in late November 2000.

19. *NASA Space Shuttle Advanced Crew Escape Suit Development*.; email between the author and John W. Bassick, David Clark Company, various dates in late November 2000.

20. *Shuttle Crew Operations Manual*, pp. 2.13-30/35.

21. *Shuttle Crew Operations Manual*, p. 2.2-1.

22. *Shuttle Crew Operations Manual*, pp. 2.6-13/14.

23. *Shuttle Crew Operations Manual*, pp. 2.6-13/14.

24. *Shuttle Crew Operations Manual*, pp. 2.7-1/25.

25. *Shuttle Crew Operations Manual*, p. 2.7-22.

26. Craig Covault, "MEDS Glass Cockpit to Enhance Shuttle Safety," Aviation Week & Space Technology, 6 March 2000, pp. 54-56; *Shuttle Crew Operations Manual*, p. 2.18-1.

27. "Space Shuttle Orbiter Mass Properties Status Report," SD72-SH-0120-334, The Boeing Company, Reusable Space Systems Division, 2 October 2000.

28. *Shuttle Crew Operations Manual*, pp. 2.3-1/14.

29. *Shuttle Crew Operations Manual*, pp. 2.5-8/9.

30. *Shuttle Crew Operations Manual*, pp. 1.2-4/5.

31. *Shuttle Crew Operations Manual*, pp. 1.2-5 and 2.27-1.

32. *Shuttle Crew Operations Manual*, p. 1.2-5; *Space Shuttle Program 1999 Annual Report* (Houston, TX: Space Shuttle Program Office, 2000), Part 2, p. 23.

33. *Shuttle Crew Operations Manual*, p. 2.5-4.

34. *Shuttle Crew Operations Manual*, pp. 2.5-10/13.

35. *Shuttle Crew Operations Manual*, p. 2.11-11.

36. *Shuttle Crew Operations Manual*, p. 1.2-5.

37. *Shuttle Crew Operations Manual*, pp. 2.11-3 and 2.11-11.

38. Orbiter Production Contract, NAS9-14000, Exhibit A, Increment 3, 1 October 1978 amended 1 August 1995, Appendix 18 "Orbiter Docking System (ODS) Shuttle/Mir Program Statement of Work," as amended September 1993.

39. Orbiter Production Contract, NAS9-14000, Exhibit A, Increment 3, 1 October 1978 amended 1 August 1995, Appendix 20 "Orbiter Docking System (ODS) External Airlock Systems for the International Space Station Alpha (SSA) Program Statement of Work," June 1994 and Appendix 22 "Orbiter Docking System (ODS) For Internal Space Station Alpha (ISSA) Phase I for Multiple Mir Missions Program Statement of Work," March 1995.

40. *Shuttle Crew Operations Manual*, p. 2.20-1/8; "STS-89 Orbiter Rollout Milestone Review," prepared by the KSC Space Shuttle Vehicle Engineering Office, 2 December 1997.

41. *Shuttle Crew Operations Manual*, p. 1.2-9.

42. *Shuttle Crew Operations Manual*, p. 1.2-10, Orbiter Production Contract, NAS9-14000, Exhibit A, Increment 3, 1 October 1978 amended 1 August 1995, section 1.2.167, p. A-33.

43. *Shuttle Crew Operations Manual*, p. 2.17-9.

44. *Shuttle Crew Operations Manual*, p. 2.17-9.

45. *Shuttle Crew Operations Manual*, p. 2.9-25.

46. "Shuttle Orbiter Radiators," LTV Fact Sheet, 1 February 1990; *Shuttle Crew Operations Manual*, p. 2.9-25; *Space Shuttle Program 1999 Annual Report*, Part 2, p. 25.

47. *Space Shuttle Program 1999 Annual Report*, Part 2, p. 25.

48. *Shuttle Crew Operations Manual*, p. 2.22-1.

49. *Shuttle Crew Operations Manual*, p. 2.22-1.

50. *Shuttle Crew Operations Manual*, p. 2.22-1; *Space Shuttle Program 1999 Annual Report*, Part 2, p. 28.

51. *Shuttle Crew Operations Manual*, p. 2.17-2.

52. *Shuttle Crew Operations Manual*, p. 1.2-10.

53. *Shuttle Crew Operations Manual*, p. 1.2-11.

54. *Shuttle Crew Operations Manual*, p. 1.2-12.

55. *Shuttle Crew Operations Manual*, p. 1.2-12.

56. *Shuttle Crew Operations Manual*, p. 1.2-7; *OV-101 Structural Integrity Evaluation*, p. 51.

57. *Shuttle Crew Operations Manual*, p. 1.2-7; *OV-101 Structural Integrity Evaluation*, p. 45.

58. *Shuttle Crew Operations Manual*, p. 1.2-8; *OV-101 Structural Integrity Evaluation*, p. 45; "STS-89 Orbiter Rollout Milestone Review," prepared by the KSC Space Shuttle Vehicle Engineering Office, 2 December 1997.

59. *OV-101 Structural Integrity Evaluation*, p. 50.

60. *Shuttle Crew Operations Manual*, p. 1.2-8; Orbiter Production Contract, NAS9-14000, Exhibit A, Increment 3, 1 October 1978 amended 1 August 1995, section 1.2.384, p. A-77/78 and 1.2.500, p. A-107.

61. *Shuttle Crew Operations Manual*, p. 1.2-13.

62. *Shuttle Crew Operations Manual*, pp. 2.19-1/3.

63. *Shuttle Crew Operations Manual*, p. 1.2-11.

64. Email between the author and Julie Kramer and Trevor Kott at the Orbiter Projects Office, JSC, 31 October 2000.

65. *Shuttle Crew Operations Manual*, p. 1.2-11.

66. *Shuttle Crew Operations Manual*, p. 2.19-1.

67. *Shuttle Crew Operations Manual*, p. 1.2-2.

68. *Shuttle Crew Operations Manual*, p. 2.23-1.

69. *Shuttle Crew Operations Manual*, p. 2.23-2.

70. *Space Shuttle Program 1999 Annual Report*, Part 1, pp. 23-24, and Part 2, p. 31; "Forward RCS Interconnect System Briefing, 11 September 2000 by Keven C. Templin; Email between the author and Keven C. Templin, JSC FICS program manager, 2 November 2000.

71. *Space Shuttle Program 1999 Annual Report*, Part 1, pp. 23-24, and Part 2, p. 31; Email between the author and Keven C. Templin, JSC FICS program manager, 2 November 2000.

72. *Shuttle Crew Operations Manual*, p. 2.8-1.

73. *Shuttle Crew Operations Manual*, p. 2.8-1; Orbiter Production Contract, NAS9-14000, Exhibit A, Increment 3, 1 October 1978 amended 1 August 1995, Appendix 17 "16 Day Extended Duration Orbiter (EDO) Mod Kits for OV-104."

74. *Shuttle Crew Operations Manual*, p. 2.1-1.

75. *Shuttle Crew Operations Manual*, p. 2.1-2.

76. Orbiter Production Contract, NAS9-14000, Exhibit A, Increment 3, 1 October 1978 amended 1 August 1995, section 1.2.105, p. A-23.

77. *Shuttle Crew Operations Manual*, p. 2.1-12.

78. *Shuttle Crew Operations Manual*, pp. 2.9-1/55.

79. *Shuttle Crew Operations Manual*, p. 2.9-44.

80. *Shuttle Crew Operations Manual*, pp. 2.9-13/14; "STS-89 Orbiter Rollout Milestone Review."

81. *Shuttle Crew Operations Manual*, p. 2.27-1.

82. *Shuttle Crew Operations Manual*, p. 2.27-1.

83. *Shuttle Crew Operations Manual*, p. 2.9-22.

84. *Shuttle Crew Operations Manual*, p. 1.2-14.

85. *Shuttle Crew Operations Manual*, p. 1.2-15; STS-59 Press Kit, April 1994, P. 37; Ames Astrogram (newsletter), 1 April 1994; Charles Redmond, "Improved Shuttle Tile to Fly on STS-59," NASA News Release 94-54, 31 March 1994; "Shuttle Thermal Protection," LTV News Release V84-29, 1 February 1990; "Chin Panel," LTV News Release, 23 March 1988; various material data sheets on TPS products, Lockheed Missiles and Space Company, January 1984.

86. Donald M. Curry, John W. Latchen, and Garland B. Whisenhunt, "Space Shuttle Leading Edge Structural Development," AIAA paper 83-0483, presented at the 21st Aerospace Sciences Meeting, 10-13 January 1983; *Shuttle Crew Operations Manual*, p. 1.2-14.

87. "Space Shuttle Orbiter Mass Properties Status Report;" *Space Shuttle Program 1999 Annual Report*, Part 2, pp. 25-26.

88. *Shuttle Crew Operations Manual*, p. 1.2-15.

88A. "STS-89 Orbiter Rollout Milestone Review."

89. "*Columbia* Modifications and Improvements, KSC Press Release 80-89, 7 August 1989; Orbiter Production Contract, NAS9-14000, Exhibit A, Increment 3, 1 October 1978 amended 1 August 1995, section 1.2.329, p. A-65.

90. *Shuttle Crew Operations Manual*, p. 1.2-14.

91. *Shuttle Crew Operations Manual*, p. 1.2-14.

92. "Space Shuttle Orbiter Mass Properties Status Report."

93. *Shuttle Crew Operations Manual*, pp. 2.13-5/10.

94. "STS-89 Orbiter Rollout Milestone Review;" Orbiter Production Contract, NAS9-14000, Exhibit A, Increment 3, amended 1 August 1995, section 1.2.102, p. A-21.

95. *Shuttle Crew Operations Manual*, p. 2.13-12.

96. *Shuttle Crew Operations Manual*, pp. 2.13-10/11.

97. *Shuttle Crew Operations Manual*, p. 2.4-1.

98. *Shuttle Crew Operations Manual*, p. 2.4-2.

99. *Shuttle Crew Operations Manual*, p. 2.4-9.

100. *Shuttle Crew Operations Manual*, pp. 2.4-13/17.

101. *Shuttle Crew Operations Manual*, p. 2.4-21.

102. *Shuttle Crew Operations Manual*, pp. 2.13-1/3; *Space Shuttle Program 1999 Annual Report*, Part 2, p. 31; emails between the author and Jorge R. Frank, JSC, October 2000.

103. *Shuttle Crew Operations Manual*, pp. 2.13-17/19.

104. *Shuttle Crew Operations Manual*, pp. 2.13-3/4.

105. *Shuttle Crew Operations Manual*, O p. 2.6-1/13; Orbiter Production Contract, NAS9-14000, Exhibit A, Increment 3, 1 October 1978 amended 1 August 1995, section 1.2.368, p. A-72.

106. Orbiter Production Contract, NAS9-14000, Exhibit A, Increment 3, 1 October 1978 amended 1 August 1995, section 1.2.103, p. A-22.

107. *Shuttle Crew Operations Manual*, p. 2.6-1/13.

108. *Shuttle Crew Operations Manual*, p. 2.6-18/21.

109. *Shuttle Crew Operations Manual*, p. 2.6-20/21.

110. *Shuttle Crew Operations Manual*, p. 2.6-15.

111. *Shuttle Crew Operations Manual*, p. 2.21-1.

112. *Shuttle Crew Operations Manual*, p. 2.14-1.

113. *Shuttle Crew Operations Manual*, p. 1.2-2.

114. *Shuttle Crew Operations Manual*, p. 2.14-3.

115. *Shuttle Crew Operations Manual*, pp. 2.14-1/2.

116. *Shuttle Crew Operations Manual*, p. 2.14-3.

117. *Shuttle Crew Operations Manual*, pp. 2.14-9/10; Orbiter Production Contract, NAS9-14000, Exhibit A, Increment 3, 1 October 1978 amended 1 August 1995, section 2.179, p. A-192 and 2.200, p. A-200.

118. *Shuttle Crew Operations Manual*, p. 2.14-3.

119. Jim Cast, "Carbon Brakes to Fly for First Time on STS-36," KSC News Release 50-90, 26 March 1990; Orbiter Production Contract, NAS9-14000, Exhibit A, Increment 3, 1 October 1978 amended 1 August 1995, section 2.178, p. A-192.

120. *Shuttle Crew Operations Manual*, p. 2.14-5.

121. *Shuttle Crew Operations Manual*, p. 2.14-4.

122. *Shuttle Crew Operations Manual*, p. 2.14-13.

123. *Shuttle Crew Operations Manual*, pp. 2.14-4 and 2.14-13.

124. *Shuttle Crew Operations Manual*, pp. 2.16-3, 4.2-1, and E-11; comments on manuscript by Fred H. Jue, Boeing Rocketdyne, 26 December 2000.

125. *Shuttle Crew Operations Manual*, p. 2.16-3.

126. Orbiter Production Contract, NAS9-14000, Exhibit A, Increment 3, 1 October 1978 amended 1 August 1995, section 3.37, P. A-207.

127. *Shuttle Crew Operations Manual*, p. 2.16-5.

128. *Shuttle Crew Operations Manual*, p. 2.16-6; comments on manuscript by Fred H. Jue, Boeing Rocketdyne, 26 December 2000.

129. *Shuttle Crew Operations Manual*, p. 2.16-4.

130. *Shuttle Crew Operations Manual*, p. 2.16-5; comments on manuscript by Fred H. Jue, Boeing Rocketdyne, 26 December 2000.

131. *Shuttle Crew Operations Manual*, p. 2.16-6.

132. *Shuttle Crew Operations Manual*, p. 2.16-7.

133. *Shuttle Crew Operations Manual*, p. 2.16-7; comments on manuscript by Fred H. Jue, Boeing Rocketdyne, 26 December 2000.

134. *Shuttle Crew Operations Manual*, p. 2.16-13.

135. *Shuttle Crew Operations Manual*, pp. 2.16-7/8.

136. *Shuttle Crew Operations Manual*, pp. 2.16-8/9.

137. *Shuttle Crew Operations Manual*, p. 2.16-24/25.

138. *Shuttle Crew Operations Manual*, pp. 2.16-31/32.

139. Byron K. Wood (SSME Development Manager), "Big Throat: Performance Margin for the SSME," Rocketdyne Threshold, Summer 1994, pp. 23-25.

140. "Big Throat: Performance Margin for the SSME," p. 25-27.

141. Fred Jue (SSME Associate Product Manager), "Three of a Kind," Rocketdyne Threshold, Spring 1998, pp. 5-7; Pratt & Whitney Fact Sheet "Space Shuttle Turbopumps," October 2000.

142. "Space Shuttle Main Engine Enhancements," MSFC Fact Sheet FS-2000-07-159-MSFC, August 2000.

143. "Space Shuttle Turbopumps."

144. "Three of a Kind," p. 7, "Space Shuttle Main Engine Enhancements."

145. "Three of a Kind," p. 7, "Space Shuttle Main Engine Enhancements."

146. "Three of a Kind," pp. 7-8.

147. "Three of a Kind," pp. 7-8; Paul Sewell, Large Throat Chamber SSME Tested," Rocketdyne News, 14 November 1990.

148. "Three of a Kind," pp. 7-8.

149. Briefing by Elrich McHenry, Space Shuttle Development Manager, to Joseph Rothenberg, NASA Associate Administrator for Space Flight, early 2000.

150. *Shuttle Crew Operations Manual*, p. 2.16-4; conversation between the author and Fred H. Jue, Boeing Rocketdyne, 2 November 2000.

151. "Three of a Kind," pp. 9-10.

152. "Three of a Kind," pp. 7-8; Briefing by Elrich McHenry, Space Shuttle Development Manager, to Joseph Rothenberg, NASA Associate Administrator for Space Flight, early 2000.

153. Conversation between the author and Fred Jue (SSME Associate Product Manager), Rocketdyne, 2 November 2000.

154. Briefing by McHenry to Rothenberg, early 2000.

155. *Shuttle Crew Operations Manual*, p. 1.3-1.

156. *Shuttle Crew Operations Manual*, p. 1.3-2.

157. *Shuttle Crew Operations Manual*, p. 1.3-2.

158. *Shuttle Crew Operations Manual*, p. 1.3-3.

159. *Shuttle Crew Operations Manual*, p. 1.3-3.

160. *Shuttle Crew Operations Manual*, p. 1.3-3.

161. *Shuttle Crew Operations Manual*, p. 1.3-3.

162. *Shuttle Crew Operations Manual*, p. 2.17-5.

163. "The Lightweight Space Shuttle External Tank," Fact Sheet KSC-26-83, February 1983.

164. June E. Malone, "Super Lightweight External Tank to be Used by Shuttle," MSFC News Release 94-32, 28 March 1994; *Space Shuttle Program 1999 Annual Report*, Part 2, p. 1.

165. *Space Shuttle Program 1999 Annual Report*, Part 2, p. 1.

166. *Shuttle Crew Operations Manual*, pp. E-14/15.

167. *Shuttle Crew Operations Manual*, p. 1.4-2.

168. Andrew J. Dunar and Stephen P. Waring, *Power to Explore: A History of the Marshall Space Flight Center 1960-1990*, NASA SP-4313, (Washington DC: NASA, 1999), pp. 322-323; "Filament Wound Case Segments Arrive at VAFB," Marshall STAR article, 12 June 1985; conversations with Melodie L. de Guibert, Thiokol Propulsion, various dates during October 2000; phone interview with Mr. Rodney Wilks, Thiokol Propulsion, former project manager for the FWC motors, 21 November 2000; N.F. Knight, et. al., *Nonlinear Shell Analysis of the Space Shuttle Solid Rocket Boosters*, NASA TM-101546 (Hampton, VA: NASA, January 1989).

169. *Shuttle Crew Operations Manual*, p. 1.4-2.

170. *Space Shuttle Program 1999 Annual Report*, Part 2, p. 6.

171. Notes from Thiokol on HPM development, supplied by Melodie de Guibert; *Shuttle Crew Operations Manual*, p. 1.4-2.

172. *Shuttle Crew Operations Manual*, p. 1.4-3/4.

173. *Shuttle Crew Operations Manual*, p. 1.4-2.

174. *Shuttle Crew Operations Manual*, p. 1.4-2.

175. *Shuttle Crew Operations Manual*, p. 1.4-2.

176. *Space Shuttle Program 1999 Annual Report*, Part 2, p. 15.

177. *Shuttle Crew Operations Manual*, p. 1.4-2.

178. *Shuttle Crew Operations Manual*, p. 1.4-6.

179. *Shuttle Crew Operations Manual*, p. 1.4-4.

180. *Shuttle Crew Operations Manual*, p. 1.4-5.

181. *Shuttle Crew Operations Manual*, p. 1.4-5.

182. *Shuttle Crew Operations Manual*, p. 1.4-5.

183. *Shuttle Crew Operations Manual*, pp. 1.4-/6.

184. *Space Shuttle Program 1999 Annual Report*, Part 2, p. 11.

185. *Shuttle Crew Operations Manual*, p. 1.4-6.

186. *Shuttle Crew Operations Manual*, p. 1.4-3; *Space Shuttle Program 1999 Annual Report*, Part 2, p. 31.

187. *Shuttle Crew Operations Manual*, p. 1.4-8.

188. "Solid Rocket Booster Retrieval Ships," KSC Fact Sheet 46-81, September 1994.

189. "Solid Rocket Booster Retrieval Ships."

190. *Shuttle Crew Operations Manual*, p. 1.4-3.

191. Bruce Buckingham, "Recovery Team to Test Booster Retrieval by Submarine During STS-101 Mission," KSC News Release 44-00, 22 April 2000.

192. "Solid Rocket Booster Retrieval Ships;" Joel Wells, "Solid Rocket Booster Recovery Ships 'Pull' Double Duty," KSC News Release 81-98, 15 June 1998; George H. Diller, "Weather Buoys Vital for Forecasts But Still Needs Seclusion," KSC News Release 148-99, 13 December 1999.

193. *Shuttle Crew Operations Manual*, pp. 1.3-4/5 and 1.4-3.

194. *Shuttle Crew Operations Manual*, pp. 2.25-1/5.

195. http://www.spacehab.com/

196. *Shuttle Crew Operations Manual*, pp. 2.24-1/4.

197. Orbiter Production Contract, NAS9-14000, Exhibit A, Increment 3, 1 October 1978 amended 1 August 1995, Appendix 7 "Baseline Definition of the OV-102 Micro-Modification."

198. Orbiter Production Contract, NAS9-14000, Exhibit A, Increment 3, 1 October 1978 amended 1 August 1995, Appendix 12 "Baseline Definition of the OV-102 Spacelab Only (SLO) Modification."

199. Orbiter Production Contract, NAS9-14000, Exhibit A, Increment 3, 1 October 1978 amended 1 August 1995, Appendix 13 "OV-102 Palmdale Modification."

200. Maurice Parker, "Langley Experiments Take *Columbia*'s Aerodynamic Pulse," NASA Langley press release 85-99, 13 December 1985.

201. LTV Corporation News Release V85-50, "*Columbia* Gathers Data with LTV-Built One-of-a-Kind Nose Cap," 19 December 1985; Maurice Parker, "Langley Experiments Take *Columbia*'s Aerodynamic Pulse," NASA Langley press release 85-99, 13 December 1985.

202. "Langley Experiments Take *Columbia*'s Aerodynamic Pulse."

203. "Langley Experiments Take *Columbia*'s Aerodynamic Pulse."

204. Orbiter Production Contract, NAS9-14000, Exhibit A, Increment 3, 1 October 1978 amended 1 August 1995, section 1.2.252, p. A-47; Email between the author and Grant Cates, OV-102 Flow Manager at KSC, 7 November 2000.

205. "STS-50 Orbiter Rollout Review," prepared by the JSC Orbiter and GFE Project Office, 19 May 1992; Orbiter Production Contract, NAS9-14000, Exhibit A, Increment 3, 1 October 1978 amended 1 August 1995, Appendix 15 "OV-102 16-Day EDO Palmdale Modification;" Email between the author and Grant Cates, OV-102 Flow Manager at KSC, 7 November 2000.

206. Orbiter Production Contract, NAS9-14000, Exhibit A, Increment 3, 1 October 1978 amended 1 August 1995, Appendix 19 "OV-104 OMDP-1 Structural Inspection and Major Modification," as revised March 1994.

207. "Special Topics for the STS-73 OPF Rollout Review," prepared by the JSC Orbiter Project Office, August 1995; Orbiter Production Contract, NAS9-14000, Exhibit A, Increment 3, 1 October 1978, Appendix 21 "OV-102 OMDP-1 Structural Inspection and Major Modification," as amended December 1994.

208. Orbiter Production Contract, NAS9-14000, Exhibit A, Increment 3, 1 October 1978 amended 1 August 1995, section 1.2-307, p. A-59.

209. "Space Shuttle Orbiter Mass Properties Status Report," Boeing report SD72-SH-0120-33, 2 October 2000.

210. "STS-89 Orbiter Rollout Milestone Review."

211. "STS-101 Orbiter Rollout Milestone Review," prepared by the KSC Space Shuttle Vehicle Engineering Office, 10 February 2000.

212. "OV-102 OMM – Modification Summary," supplied by the KSC Space Shuttle Vehicle Engineering Office, 7 November 2000.

213. "Orbiter Detail Weight Statement (OV-103), Rockwell report SD75-SH-0116-221, 2 January 1995; "Group Mass Properties Summary for OV-103, Flight 27," 2 July 1999.

APPENDIX A – TO CONTINUE SAFELY. PAGES 441–450.

1. *Space Shuttle Program 1999 Annual Report* (Houston, TX: Space Shuttle Program Office, 2000), Part 1, p. 11.

2. Bryan O'Conner, Chairman, *Upgrading the Space Shuttle*, National Research Council (Washington DC: National Academy Press, 1999), electronic version at http://books.nap.edu/books/0309063825/html/r1.html.

3. *Space Shuttle Program 1999 Annual Report*, Part 1, p. 12.

4. *Space Shuttle Program Upgrades Management Plan*, NSTS-37400 (Houston, TX" JSC, 21 July 2000), pp. 3-1 and B-4.

5. *Upgrading the Space Shuttle*; *Space Shuttle Program 1999 Annual Report* , Part 1, p. 11; Briefing by Elrich McHenry, Space Shuttle Development Manager, to Joseph Rothenberg, NASA Associate Administrator for Space Flight, early 2000.

6. Friction Stir Welding is a process invented and patented by TWI, Granta Park, Great Abington, Cambridge, CB1 6AL, UK; *Space Shuttle Program 1999 Annual Report*, Part 2, p. 4; Briefing by Elrich McHenry to Joseph Rothenberg; http://www.frictionstirwelding.com/generalintro.htm; *Space Shuttle Program Upgrades Management Plan*, p. B-5.

7. Briefing by Elrich McHenry to Joseph Rothenberg; Email between the author and Elric N. McHenry, JSC/MA – Shuttle Upgrades Program Manager, 27 October 2000; *Space Shuttle Program Upgrades Management Plan*, p. B-5.

8. *Space Shuttle Program 1999 Annual Report*, Part 1, p. 12; *Space Shuttle Program Upgrades Management Plan*, p. B-5.

9. *Upgrading the Space Shuttle*.

10. *Upgrading the Space Shuttle*.

11. Briefing by Elrich McHenry to Joseph Rothenberg; Email between the author and Elric N. McHenry, JSC/MA – Shuttle Upgrades Program Manager, 27 October 2000; cost estimates from *Upgrading the Space Shuttle*.

12. *Space Shuttle Program Upgrades Management Plan*, p. B-6.

13. *Space Shuttle Program Upgrades Management Plan*, p. B-6.

14. *Space Shuttle Program Upgrades Management Plan*, p. B-6.

15. *Space Shuttle Program Upgrades Management Plan*, pp. B-6/7.

16. Briefing by Elrich McHenry to Joseph Rothenberg; *Space Shuttle Program Upgrades Management Plan*, p. B-7.

17. *Space Shuttle Program Upgrades Management Plan*, p. B-7.

18. *Space Shuttle Program Upgrades Management Plan*, p. B-7.

19. Briefing by Elrich McHenry to Joseph Rothenberg; *Space Shuttle Program Upgrades Management Plan*, p. B-7.

20. *Space Shuttle Program Upgrades Management Plan*, p. B-8.

21. *Space Shuttle Program Upgrades Management Plan*, p. B-8.

22. Briefing by Elrich McHenry to Joseph Rothenberg; *Space Shuttle Program Upgrades Management Plan*, pp. B-8/9; *Upgrading the Space Shuttle*.

23. *Space Shuttle Program Upgrades Management Plan*, p. B-9.

24. *Space Shuttle Program Upgrades Management Plan*, pp. 3-2/3, and B-4.

25. *Space Shuttle Program Upgrades Management Plan*, pp. 3-2/3, and B-4.

26. Briefing by Elrich McHenry to Joseph Rothenberg; *Space Shuttle Program Upgrades Management Plan*, p. B-9.

27. *Space Shuttle Program Upgrades Management Plan*, p. B-9.

28. *Space Shuttle Program Upgrades Management Plan*, pp. B-9/10.

29. *Space Shuttle Program Upgrades Management Plan*, p. B-10.

30. *Space Shuttle Program Upgrades Management Plan*, p. B-10; *Upgrading the Space Shuttle*.

31. *Space Shuttle Program Upgrades Management Plan*, pp. B-10/11.

32. *Space Shuttle Program Upgrades Management Plan*, p. B-11.

33. *Space Shuttle Program Upgrades Management Plan*, p. B-11.

34. Personal experience by the author; *Space Shuttle Program Upgrades Management Plan*, p. B-11.

35. *Space Shuttle Program Upgrades Management Plan*, p. B-12.

36. *Space Shuttle Program Upgrades Management Plan*, p. B-12.

37. *Space Shuttle Program Upgrades Management Plan*, p. B-12.

38. *Space Shuttle Program Upgrades Management Plan*, pp. B-12/13.

39. http://www.boeing.com/defense-space/space/rss_shuttle/upgrades/oms_rcs/; *Upgrading the Space Shuttle*.

40. *Space Shuttle Program 1999 Annual Report*, Part 1, p. 13 and Part 2, p. 26.

41. *Upgrading the Space Shuttle*.

42. "FSB Phase A Study Program Summary," a briefing prepared by USA on 30 October 2000 for presentation to NASA.

43. *Upgrading the Space Shuttle*.

44. *Upgrading the Space Shuttle*.

45. W. Peterson, et. al., *Liquid Flyback Booster Pre-Phase A Study Assessment*, NASA TM-104801, JSC, September 1994.

46. *Liquid Flyback Booster Pre-Phase A Study Assessment*.

47. *Liquid Flyback Booster Pre-Phase A Study Assessment*.

48. *Liquid Flyback Booster Pre-Phase A Study Assessment*.

49. *Liquid Flyback Booster Pre-Phase A Study Assessment*; *Access to Space Study*, Space Industries, Inc. performed under Rockwell purchase order M3D8XXL-453332M, July 1993.

50. "Vehicle Definition Study for a Liquid Fly-Back Booster (LFBB)" Commerce Business Daily announcement, 6 May 1997; as an example of the in-house work, see "LFBB Integrated Avionics," a study by the Flight Data Systems Division at MSC, begun in early 1994 and running through 1995.

51. Tom Hamilton and Tom Healy, "Reusable First Stage – Evolutionary Shuttle Upgrade," a paper presented at the Space Shuttle Development Conference, NASA Ames, 29 July 1999.

52. "Reusable First Stage – Evolutionary Shuttle Upgrade."

53. "Reusable First Stage – Evolutionary Shuttle Upgrade."

54. "Liquid Flyback Booster Operations Concept Document," Lockheed Martin Michoud, Draft dated July 1998 and Preliminary dated September 1998; "Reusable First Stage – Evolutionary Shuttle Upgrade;" LFBB Assessment Study Status, a briefing to MSFC management, April 1998.

55. "Liquid Flyback Booster Operations Concept Document," Lockheed Martin Michoud, Draft dated July 1998 and Preliminary dated September 1998; "Reusable First Stage – Evolutionary Shuttle Upgrade;" "LFBB Assessment Study Status."

56. "Reusable First Stage – Evolutionary Shuttle Upgrade;" "LFBB Assessment Study Status."

57. "Reusable First Stage – Evolutionary Shuttle Upgrade;" "LFBB Assessment Study Status."

58. "Liquid Flyback Booster Operations Concept Document," Lockheed Martin Michoud, Draft dated July 1998 and Preliminary dated September 1998.

59. "Liquid Flyback Booster Operations Concept Document," Lockheed Martin Michoud, Draft dated July 1998 and Preliminary dated September 1998.

60. Ira Victor, "Liquid Fly Back Booster Demonstrator," not dated but probably late September or early October 1998; "Reusable First Stage – Evolutionary Shuttle Upgrade;" "LFBB Assessment Study Status, a briefing to MSFC management, April 1998.; "Reusable First Stage Flight Demonstrator," a joint briefing by Boeing and Lockheed Martin, 30 July 1999.

61. "Reusable First Stage Flight Demonstrator."

62. "Liquid Fly Back Booster Demonstrator;" "Reusable First Stage – Evolutionary Shuttle Upgrade;" "LFBB Assessment Study Status, a briefing to MSFC management, April 1998.

63. "Launch Vehicle Comparisons Vs Lunar / Mars Vehicle Performance Metric," an undated briefing to MSFC but probably mid-1998.

APPENDIX B – STILLBORN. PAGES 451–466.

1. Peter T. Demakes, et. al., *EDIN Design Study: Alternate Space Shuttle Booster Replacement Concepts*, NASA TN-76-108 (also numbered NASA CR-151894), (Houston, TX: Sigma Corporation, January 1976).

2. "Shuttle Derivative Vehicles Study," The Boeing Company, 28 July 1977; "Systems Concepts for STS-Derived Heavy-Lift Launch Vehicles Study," Boeing Aerospace Company, September 1977; "Shuttle Derivative Vehicles Study: Operations, Systems, and Facilities," multiple volumes, DISO-22875-1, The Boeing Company, December 1977.

3. "Shuttle Derived Vehicles (SDV) Technology Requirements Study," Phase I Final Report, multiple volumes, MMC-SDV-DR-6-1, Martin Marietta Corporation, July 1981.

4. "Support on Shuttle Derived Vehicles Technology Requirements Study, ASR-81-121, Rocketdyne Division, September 1981; "Shuttle Derived Vehicles Technology Requirements Study," Monthly Letter Report 2356-M-6, Aerojet Liquid Rocket Company October 1981; "Shuttle Derived Vehicles Technology Requirements Study," Monthly Letter Report 2356-M-7, Aerojet Liquid Rocket Company November 1981.

5. "Shuttle Derived Vehicles (SDV) Technology Requirements Study," Phase II Final Report, multiple volumes, MMC-SDV-DR-6-2, NASA CR-170698, Martin Marietta Corporation, May 1982.

6. "Support on Shuttle Derived Vehicles Technology Requirements Study, ASR-81-121, Rocketdyne Division, September 1981.

7. "Shuttle Derived Cargo Vehicle (SDCV)," Final Review, 10 February 1983. In the files of the KSC Technical Library.

8. For instance, see "Shuttle-C New Start Presentation," a briefing prepared by MSFC for presentation at NASA Headquarters, in November 1987.

9. "A Presentation to the House and Senate Appropriations Committee on Shuttle-C," prepared by MSFC, 15 February 1989.

10. "A Presentation to the House and Senate Appropriations Committee on Shuttle-C," prepared by MSFC, 15 February 1989.

11. "Shuttle-C: Briefing to Hillister Cantus," prepared by MSFC, 27 June 1989.

12. "Shuttle-C: Briefing to Hillister Cantus," prepared by MSFC, 27 June 1989.

13. Terry R. Mitchell, "Shuttle C: A Shuttle Derived Launch Vehicle," a paper in the files of the MSFC History Office.

14. W. Huber, "OMV Background and Status," a paper presented at the "Shuttle-C Users Conference: Today's Launch Vehicle for Tomorrow," at MSFC, 25 May 1989.

15. William E. Galloway, "Orbital Maneuvering Vehicle to Support the Space Station," an unpublished paper in the files of the MSFC History Office; William C. Snoddy, et. al. "Use of the Orbital Maneuvering Vehicle (OMV) for Placement and Retrieval of Spacecraft and Platforms," a paper presented at the Annual AAS Guidance and Control Conference, Keystone, Colorado, February 1986; James R. Turner and William E. Galloway, "Orbital Maneuvering Vehicle (OMV) Servicing Capabilities," a paper presented at the AIAA First Space Logistics Symposium, Huntsville, Alabama, March 1987.

16. Various reports by the contractor team, in the files of the KSC Technical Library and History Office. For instance "Definition of a Space Transportation System Cargo Element," prepared under contract NAS8-37144, 23 February 1989.

17. "Shuttle-C Users Conference: Today's Launch Vehicle for Tomorrow," at MSFC, 25 May 1989.

18. "A Presentation to the House and Senate Appropriations Committee on Shuttle-C," prepared by MSFC, 15 February 1989.

19. "Shuttle-C Status Review – Senior Management Overview," prepared by the contractor team, 24 February 1989; T.J. Lee, "Shuttle-C – Heavy-Lift Vehicle of the 90s," a paper presented at the 2nd European Aerospace Conference on Progress in Space Transportation, Bonn-Bad Godesberg, 22–24 May 1989; "Shuttle-C" MSFC Fact Sheet, 17 April 1989.

20. "Definition of a Space Transportation Systems Cargo Element (Shuttle-C): Operations Support Plan," prepared by Martin Marietta Manned Space Systems, April 1990; "DE Technical Presentation on the Status of Shuttle-C and the Lunar-Mars Initiative," prepared by the Design Engineering Directorate at KSC, 26 February 1990.

21. "DE Technical Presentation."

22. "DE Technical Presentation."

23. "Draft Chronology of Events 1986-1993: Advanced Solid Rocket Motor Program," compiled by Mike Wright, MSFC Historian, 19 November 1993.

24. "Draft Chronology of Events 1986-1993."

25. ASRM Press Kit provided by Aerojet General.

26. N.F. Knight, et. al., *Nonlinear Shell Analyses of the Space Shuttle Solid Rocket Boosters*, NASA TM-101546 (Hampton, Virginia: NASA, January 1989).

27. "Draft Chronology of Events 1986-1993."

28. ASRM Press Kit provided by Aerojet General

29. ASRM Press Kit provided by Aerojet General

30. "Draft Chronology of Events 1986-1993."

31. "Draft Chronology of Events 1986-1993."

32. "Draft Chronology of Events 1986-1993."

33. "Draft Chronology of Events 1986-1993."

34. "Draft Chronology of Events 1986-1993:;" Auction package from National Industry Services and the Henry Butcher Company advertising the equipment to be auctioned at Yellow Creek and Mount Vernon.

35. "Draft Chronology of Events 1986-1993."

36. "Draft Chronology of Events 1986-1993."

37. "Liquid Rocket Booster Study," Final Report, multiple volumes, General Dynamics report GD8-37137-FR3/9, March 1989; "Liquid Rocket Booster (LRB) for the Space Transportation System (STS) Systems Study," Final Report, multiple volumes, Martin Marietta report NAS8-37136-DR4, March 1989.

38. "Liquid Rocket Booster Study," Final Report, multiple volumes, General Dynamics report GD8-37137-FR3/9, March 1989.

39. "Liquid Rocket Booster (LRB) for the Space Transportation System (STS) Systems Study," Martin Marietta.

APPENDIX C – SUDDEN RANCH. PAGES 467–476.

1. The description of facilities comes partially from personal memory of having activated and supported many of them for seven years at VLS, and partially from "GSS Project Familiarization Course: DoD STS Ground Support Systems Integration," a training course prepared by Martin Marietta at Vandenberg, various dates. Other information supplied by various other former Vandenberg employees on both the government and contractor teams.

2. http://www.sverdrup.com/shared/history/spaceshuttle.html

3. http://www.sverdrup.com/shared/history/spaceshuttle.html

4. Meeting Record, "Potential Slip of the Shuttle Initial Operational Capability (IOC) at Vandenberg to FY86," in the files of the NASA History Office; http://www.sverdrup.com/shared/history/spaceshuttle.html

5. Written answers to Congress, provided on 1 September 1984 by Colonel Robert B. Bourne, Commander, 6595th Shuttle Test Group, USAF.

6. Written answers to Congress by Colonel Robert B. Bourne.

7. Letter from Lynn W. Henenger, Acting Director of the NASA Congressional Liaison Division to Lloyd Beasley, Printing Clerk, U.S. Senate, with Mr. Beggs' responses to written questions attached, 11 October 1984.

8. "NASA/USAF Delay First Vandenberg Shuttle Launch," NASA News Release 85-4, 8 January 1985; SRB delivery data supplied by Thiokol Propulsion; ET delivery data supplied by Lockheed Martin Space Systems Company.

9. "Crews for First Vandenberg Mission, DoD Flight Named," NASA News Release 85-25, 15 February 1985.

10. NASA Daily Activities Reports, various dates between 10 September 1985 and 17 January 1986. In the files of the NASA History office.

11. C. R. Gunn, "Vandenberg Shuttle Launch and Landing Site Stand-Down Assessment," 4 June 1986.

12. "Vandenberg Shuttle Launch and Landing Site Stand-Down Assessment."

13. "VLS Status Briefing," July 1986.

14. Richard Kohrs, Space Shuttle Program Manager, "Potential Use of VAFB Shuttle Facilities," 10 December 1986.

15. Vandenberg chronology prepared by the 30SW historian Jeff Geiger at http://www.vafb.af.mil/history/chronology.html

The Death of Dyna-Soar

(to Poe's 'Raven')

Ah, distinctly I remember, it was early last December;
It was felt that very shortly, we would be employed no more;
Every day we feared the morrow; vainly we had sought to borrow;
Funds to budget us tomorrow, for our work on Dyna-Soar …
On the sleek and winged spacecraft we called Dyna-Soar …
Cancelled now, forever more.

From off the duct I pulled the shutter, when, with many a flirt and flutter,
Out there flew a stately raven, of the saintly days of yore;
Not the least obeisance made he; not a minute stopped or stayed he;
But with mien of Lord or Lady, perched beside my office door …
Upon a bust of Eugen Sänger, on the bookcase by the door …
Perched, and sat, and nothing more.

"Prophet !" said I, "Thing of evil !, tell me, agent of the devil,
Whether McNamara axed the program, or just cut us back some more ?
Will he make a presentation, to Congress for appropriation ?
Does he plan continuation, after Fiscal '64 ?
Is the funding in the budget ? Tell me, tell me, I implore … "
Quoth the Raven, "Never more."

Oh, the sleek and winged spacecraft we called Dyna-Soar,
Cancelled now, forever more.

Andy Oberta
Missiles and Rockets
27 January 1964